ALGEBRA

Lines

Slope of the line through $P_1 = (x_1, y_1)$ and $P_2 = (x_2, y_2)$:
$$m = \frac{y_2 - y_1}{x_2 - x_1}$$

Slope-intercept equation of line with slope m and y-intercept b:
$$y = mx + b$$

Point-slope equation of line through $P_1 = (x_1, y_1)$ with slope m:
$$y - y_1 = m(x - x_1)$$

Point-point equation of line through $P_1 = (x_1, y_1)$ and $P_2 = (x_2, y_2)$:
$$y - y_1 = m(x - x_1) \quad \text{where } m = \frac{y_2 - y_1}{x_2 - x_1}$$

Lines of slope m_1 and m_2 are parallel if and only if $m_1 = m_2$.
Lines of slope m_1 and m_2 are perpendicular if and only if $m_1 = -\frac{1}{m_2}$.

Circles

Equation of the circle with center (a, b) and radius r:
$$(x - a)^2 + (y - b)^2 = r^2$$

Distance and Midpoint Formulas

Distance between $P_1 = (x_1, y_1)$ and $P_2 = (x_2, y_2)$:
$$d = \sqrt{(x_2 - x_1)^2 + (y_2 - y_1)^2}$$

Midpoint of $\overline{P_1 P_2}$: $\left(\dfrac{x_1 + x_2}{2}, \dfrac{y_1 + y_2}{2} \right)$

Laws of Exponents

$x^m x^n = x^{m+n}$ $\dfrac{x^m}{x^n} = x^{m-n}$ $(x^m)^n = x^{mn}$

$x^{-n} = \dfrac{1}{x^n}$ $(xy)^n = x^n y^n$ $\left(\dfrac{x}{y}\right)^n = \dfrac{x^n}{y^n}$

$x^{1/n} = \sqrt[n]{x}$ $\sqrt[n]{xy} = \sqrt[n]{x}\,\sqrt[n]{y}$ $\sqrt[n]{\dfrac{x}{y}} = \dfrac{\sqrt[n]{x}}{\sqrt[n]{y}}$

$x^{m/n} = \sqrt[n]{x^m} = \left(\sqrt[n]{x}\right)^m$

Special Factorizations

$x^2 - y^2 = (x + y)(x - y)$
$x^3 + y^3 = (x + y)(x^2 - xy + y^.)$,
$x^3 - y^3 = (x - y)(x^2 + xy + y^2)$

Binomial Theorem

$(x + y)^2 = x^2 + 2xy + y^2$
$(x - y)^2 = x^2 - 2xy + y^2$
$(x + y)^3 = x^3 + 3x^2 y + 3xy^2 + y^3$
$(x - y)^3 = x^3 - 3x^2 y + 3xy^2 - y^3$

$(x + y)^n = x^n + nx^{n-1}y + \dfrac{n(n-1)}{2}x^{n-2}y^2$
$\qquad + \cdots + \binom{n}{k}x^{n-k}y^k + \cdots + nxy^{n-1} + y^n$

where $\dbinom{n}{k} = \dfrac{n(n-1)\cdots(n-k+1)}{1 \cdot 2 \cdot 3 \cdots \cdot k}$

Quadratic Formula

If $ax^2 + bx + c = 0$, then $x = \dfrac{-b \pm \sqrt{b^2 - 4ac}}{2a}$.

Inequalities and Absolute Value

If $a < b$ and $b < c$, then $a < c$.
If $a < b$, then $a + c < b + c$.
If $a < b$ and $c > 0$, then $ca < cb$.
If $a < b$ and $c < 0$, then $ca > cb$.
$|x| = x \quad$ if $x \geq 0$
$|x| = -x \quad$ if $x \leq 0$

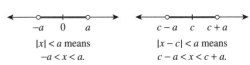

$|x| < a$ means
$-a < x < a$.

$|x - c| < a$ means
$c - a < x < c + a$.

GEOMETRY

Formulas for area A, circumference C, and volume V

Triangle
$A = \frac{1}{2}bh$
$= \frac{1}{2}ab \sin \theta$

Circle
$A = \pi r^2$
$C = 2\pi r$

Sector of Circle
$A = \frac{1}{2}r^2 \theta$
$s = r\theta$
(θ in radians)

Sphere
$V = \frac{4}{3}\pi r^3$
$A = 4\pi r^2$

Cylinder
$V = \pi r^2 h$

Cone
$V = \frac{1}{3}\pi r^2 h$
$A = \pi r \sqrt{r^2 + h^2}$

Cone with arbitrary base
$V = \frac{1}{3}Ah$
where A is the area of the base

Pythagorean Theorem: For a right triangle with hypotenuse of length c and legs of lengths a and b, $c^2 = a^2 + b^2$.

TRIGONOMETRY

Angle Measurement

π radians $= 180°$

$1° = \dfrac{\pi}{180}$ rad $\qquad 1$ rad $= \dfrac{180°}{\pi}$

$s = r\theta \quad (\theta$ in radians$)$

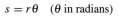

Right Triangle Definitions

$\sin\theta = \dfrac{\text{opp}}{\text{hyp}} \qquad\qquad \cos\theta = \dfrac{\text{adj}}{\text{hyp}}$

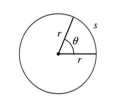

$\tan\theta = \dfrac{\sin\theta}{\cos\theta} = \dfrac{\text{opp}}{\text{adj}} \qquad \cot\theta = \dfrac{\cos\theta}{\sin\theta} = \dfrac{\text{adj}}{\text{opp}}$

$\sec\theta = \dfrac{1}{\cos\theta} = \dfrac{\text{hyp}}{\text{adj}} \qquad \csc\theta = \dfrac{1}{\sin\theta} = \dfrac{\text{hyp}}{\text{opp}}$

Trigonometric Functions

$\sin\theta = \dfrac{y}{r} \qquad\qquad \csc\theta = \dfrac{r}{y}$

$\cos\theta = \dfrac{x}{r} \qquad\qquad \sec\theta = \dfrac{r}{x}$

$\tan\theta = \dfrac{y}{x} \qquad\qquad \cot\theta = \dfrac{x}{y}$

$\displaystyle\lim_{\theta\to 0}\dfrac{\sin\theta}{\theta} = 1 \qquad \lim_{\theta\to 0}\dfrac{1-\cos\theta}{\theta} = 0$

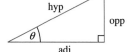

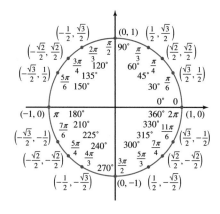

Fundamental Identities

$\sin^2\theta + \cos^2\theta = 1 \qquad\qquad \sin(-\theta) = -\sin\theta$

$1 + \tan^2\theta = \sec^2\theta \qquad\qquad \cos(-\theta) = \cos\theta$

$1 + \cot^2\theta = \csc^2\theta \qquad\qquad \tan(-\theta) = -\tan\theta$

$\sin\left(\dfrac{\pi}{2} - \theta\right) = \cos\theta \qquad \sin(\theta + 2\pi) = \sin\theta$

$\cos\left(\dfrac{\pi}{2} - \theta\right) = \sin\theta \qquad \cos(\theta + 2\pi) = \cos\theta$

$\tan\left(\dfrac{\pi}{2} - \theta\right) = \cot\theta \qquad \tan(\theta + \pi) = \tan\theta$

The Law of Sines

$\dfrac{\sin A}{a} = \dfrac{\sin B}{b} = \dfrac{\sin C}{c}$

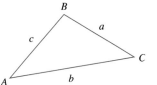

The Law of Cosines

$a^2 = b^2 + c^2 - 2bc\cos A$

Addition and Subtraction Formulas

$\sin(x + y) = \sin x\cos y + \cos x\sin y$

$\sin(x - y) = \sin x\cos y - \cos x\sin y$

$\cos(x + y) = \cos x\cos y - \sin x\sin y$

$\cos(x - y) = \cos x\cos y + \sin x\sin y$

$\tan(x + y) = \dfrac{\tan x + \tan y}{1 - \tan x\tan y}$

$\tan(x - y) = \dfrac{\tan x - \tan y}{1 + \tan x\tan y}$

Double-Angle Formulas

$\sin 2x = 2\sin x\cos x$

$\cos 2x = \cos^2 x - \sin^2 x = 2\cos^2 x - 1 = 1 - 2\sin^2 x$

$\tan 2x = \dfrac{2\tan x}{1 - \tan^2 x}$

$\sin^2 x = \dfrac{1 - \cos 2x}{2} \qquad\qquad \cos^2 x = \dfrac{1 + \cos 2x}{2}$

Graphs of Trigonometric Functions

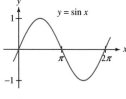

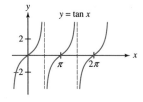

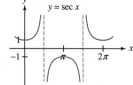

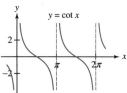

SINGLE VARIABLE
CALCULUS
Early Transcendentals

SINGLE VARIABLE
CALCULUS
Early Transcendentals

JON ROGAWSKI
University of California, Los Angeles

W. H. Freeman and Company
New York

Publisher: *Craig Bleyer*

Executive Editor: *Ruth Baruth*

Senior Acquisitions Editor: *Terri Ward*

Development Editor: *Tony Palermino*

Development Editor: *Bruce Kaplan*

Associate Editor: *Brendan Cady*

Market Development: *Steve Rigolosi*

Executive Marketing Manager: *Robin O'Brien*

Senior Media Editor: *Roland Cheyney*

Assistant Editor: *Laura Capuano*

Photo Editor: *Ted Szczepanski*

Photo Researcher: *Julie Tesser*

Design Manager: *Blake Logan*

Project Editor: *Vivien Weiss*

Illustrations: *Network Graphics*

Illustration Coordinator: *Susan Timmins*

Production Coordinator: *Paul W. Rohloff*

Composition: *Integre Technical Publishing Co.*

Printing and Binding: *RR Donnelley*

Library of Congress Control Number 2006936455
ISBN-13: 978-1-4292-1075-1
ISBN-10: 1-4292-1075-3

ISBN-13: 978-1-4292-1219-9
ISBN-10: 1-4292-1219-5

Printed in the United States of America

Second printing

W. H. Freeman and Company, 41 Madison Avenue, New York, NY 10010
Houndmills, Basingstoke RG21 6XS, England
www.whfreeman.com

To Julie

CONTENTS | SINGLE VARIABLE CALCULUS

Early Transcendentals

As a successful teacher for more than 25 years, Jon Rogawski has listened to and learned much from his own students. These valuable lessons have made an impact on his thinking, his writing, and his shaping of a calculus text.

Jon Rogawski received his undergraduate degree and simultaneously a master's degree in mathematics from Yale University and a Ph.D. in mathematics from Princeton University, where he studied under Robert Langlands. Before joining the Department of Mathematics at UCLA in 1986, where he is currently a full professor, he held teaching positions at Yale University, University of Chicago, the Hebrew University in Jerusalem, and visiting positions at the Institute for Advanced Study, the University of Bonn, and the University of Paris at Jussieu and at Orsay.

Jon's areas of interest are number theory, automorphic forms, and harmonic analysis on semisimple groups. He has published numerous research articles in leading mathematical journals, including a research monograph titled *Automorphic Representations of Unitary Groups in Three Variables* (Princeton University Press). He is the recipient of a Sloan Fellowship and an editor of the *Pacific Journal of Mathematics*.

Jon and his wife, Julie, a physician in family practice, have four children. They run a busy household and, whenever possible, enjoy family vacations in the mountains of California. Jon is a passionate classical music lover and plays the violin and classical guitar.

ABOUT *CALCULUS* by Jon Rogawski

On Teaching Mathematics

As a young instructor, I enjoyed teaching but didn't appreciate how difficult it is to communicate mathematics effectively. Early in my teaching career, I was confronted with a student rebellion when my efforts to explain epsilon-delta proofs were not greeted with the enthusiasm I anticipated. Experiences of this type taught me two basic principles:

1. We should try to teach students as much as possible, but not more so.
2. As math teachers, how we say it is as important as what we say.

When a concept is wrapped in dry mathematical formalism, the majority of students cannot assimilate it. The formal language of mathematics is intimidating to the uninitiated. By presenting the same concept in everyday language, which may be more casual but no less precise, we open the way for students to understand the underlying idea and integrate it into their way of thinking. Students are then in a better position to appreciate the need for formal definitions and proofs and to grasp their logic.

My most valuable resource in writing this text has been my classroom experience of the past 25 years. I have learned how to teach from my students, and I hope my text reflects a true sensitivity to student needs and capabilities. I also hope it helps students experience the joy of understanding and mastering the beautiful ideas of the subject.

On Writing a New Calculus Text

Calculus has a deservedly central role in higher education. It is not only the key to the full range of scientific and engineering disciplines; it is also a crucial component in a student's intellectual development. I hope my text will help to open up this multifaceted world of ideas for the student. Like a great symphony, the ideas in calculus never grow stale. There is always something new to appreciate and delight in, and I enjoy the challenge of finding the best way to communicate these ideas to students.

My text builds on the tradition of several generations of calculus authors. There is no perfect text, given the choices and inevitable compromises that must be made, but many of the existing textbooks reflect years of careful thought, meticulous craft, and accumulated wisdom, all of which cannot and should not be ignored in the writing of a new text.

I have made a sustained effort to communicate the underlying concepts, ideas, and "reasons why things work" in language that is accessible to students. Throughout the text, student difficulties are anticipated and addressed. Problem-solving skills are systematically developed in the examples and problem sets. I have also included a wide range of applications, both innovative and traditional, which provide additional insight into the mathematics and communicate the important message that calculus plays a vital role in the modern world.

My textbook follows a largely traditional organization, with a few exceptions. Two such exceptions are the structure of Chapter 2 on limits and the placement of Taylor polynomials in Chapter 8.

The Limit Concept and the Structure of Chapter 2

Chapter 2 introduces the fundamental concepts of limit and convergence. Sections 2.1 and 2.2 give students a good intuitive grasp of limits. The next two sections discuss limit laws (Section 2.3) and continuity (Section 2.4). The notion of continuity is then used to justify the algebraic approach to limits, developed in Section 2.5. I chose to delay the epsilon-delta definition until Section 2.8, when students have a strong understanding of limits. This placement allows you to treat this topic, if you wish, whenever you would like.

Placement of Taylor Polynomials

Taylor polynomials appear in Chapter 8, before infinite series in Chapter 10. My goal is to present Taylor polynomials as a natural extension of the linear approximation. When I teach infinite series, the primary focus is on convergence, a topic that many students find challenging. By the time we have covered the basic convergence tests and studied the convergence of power series, students are ready to tackle the issues involved in representing a function by its Taylor series. They can then rely on their previous work with Taylor polynomials and the error bound from Chapter 8. However, the section on Taylor polynomials is written so that you can cover this topic together with the materials on infinite series if this order is preferred.

Careful, Precise Development

W. H. Freeman is committed to high quality and precise textbooks and supplements. From this project's inception and throughout its development and production, quality and precision have been given significant priority. We have in place unparalleled procedures to ensure the accuracy of the text.

These are the steps we took to ensure an accurate first edition for you:

- **Exercises and Examples** Rather than waiting until the book was finished before checking it for accuracy (which is often the practice), we have painstakingly checked all the examples, exercises, and their solutions for accuracy in every draft of the manuscript and each phase of production.
- **Exposition** A team of 12 calculus instructors acted as accuracy reviewers, and made four passes through all exposition, confirming the accuracy and precision of the final manuscript.
- **Figures** Tom Banchoff of Brown University verified the appropriateness and accuracy of all figures throughout the production process.
- **Editing** The author worked with an editor with an advanced degree in mathematics to review each line of text, exercise, and figure.
- **Composition** The compositor used the author's original LaTeX files to prevent the introduction of new errors in the production process.
- **Math Clubs** We engaged math clubs at twenty universities to accuracy check all the exercises and solutions.

Together, these procedures far exceed prior industry standards to safeguard the quality and precision of a calculus textbook.

■ SUPPLEMENTS

For Instructors

- Instructor Solutions Manual
  Brian Bradie, Christopher Newport University, and Greg Dresden, Washington and Lee University
  Single Variable: 0-7167-9591-4
  Multivariable: 0-7167-9592-2
  Contains worked-out solutions to all exercises in the text.

- Printed Test Bank
  Calculus: 0-7167-9598-1
  Includes multiple-choice and short-answer test items.

- Test Bank CD-ROM
  Calculus: 0-7167-9895-6
  Available online or on a CD-ROM.

- Instructor's Resource Manual
  Vivien Miller, Mississippi State University; Len Miller, Mississippi State University; and Ted Dobson, Mississippi State University
  Calculus: 0-7167-9589-2
  Offers instructors support material for each section of every chapter. Each section includes suggested class time and content emphasis, selected key points, lecture material, discussion topics and class activities, suggested problems, worksheets, and group projects.

- Instructor's Resource CD-ROM
  Calculus: 1-429-20043-X
  Your one-stop resource. Search and export all resources by key term or chapter. Includes text images, Instructor's Solutions Manual, Instructor's Resource Manual, Lecture PowerPoint slides, Test Bank.

- Instructor's Resource Manual for AP Calculus
  Calculus: 0-7167-9590-6
  In conjunction with the text, this manual provides the opportunity for instructors to prepare their students for the AP Exam.

For Students

- Student Solutions Manual
  Brian Bradie, Christopher Newport University, and Greg Dresden, Washington and Lee University
  Single Variable: 0-7167-9594-9
  Multivariable: 0-7167-9880-8
  Offers worked-out solutions to all odd-numbered exercises in the text.

- Companion website at www.whfreeman.com/rogawski

CalcPortal

One click. One place. For all the tools you need.
CalcPortal is the digital gateway to Rogawski's *Calculus*, designed to enrich your course and improve your students' understanding of calculus. For students, CalcPortal integrates review, diagnostic, and tutorial resources right where they are needed in the eBook. For instructors, CalcPortal provides a powerful but easy-to-use course management system complete with a state-of-the-art algorithmic homework and assessment system.

NEW! Next-Generation eBook

CalcPortal is organized around three main teaching and learning tools: (1) The online eBook is a complete version of Rogawski's *Calculus* and includes a Personalized Study Plan that connects students directly to what they need to learn and to the resources on CalcPortal, including important prerequisite content; (2) the homework and assessment center, with an algorithm problem generator; and (3) the CalcResource center with Just-in-Time algebra and precalculus tutorials.

■ FEATURES

Pedagogical Features in Rogawski's *CALCULUS*

Conceptual Insights
encourage students to develop a conceptual
understanding of calculus by explaining
important ideas clearly but informally.

CONCEPTUAL INSIGHT Are Limits Really Necessary? It is natural to ask whether limits are really necessary. Since the tangent line is so easy to visualize, is there perhaps a better or simpler way to find its equation? History gives one answer. The methods of calculus based on limits have stood the test of time and are used more widely today than ever before.

History aside, there is a more intuitive argument for why limits are necessary. The slope of a line can be computed if the coordinates of *two* points $P = (x_1, y_1)$ and $Q = (x_2, y_2)$ on the line are known:

$$\text{Slope of line} = \frac{y_2 - y_1}{x_2 - x_1}$$

We cannot use this formula to find the slope of the tangent line because we have only one point on the tangent line to work with, namely $P = (a, f(a))$. Limits provide an ingenious way around this obstacle. We choose a point $Q = (a + h, f(a + h))$ on the graph near P and form the secant line. The slope of this secant line is only an approximation to the slope of the tangent line:

$$\text{Slope of secant line} = \frac{f(a + h) - f(a)}{h} \approx \text{slope of tangent line}$$

But this approximation improves as $h \to 0$, and by taking the limit, we convert our approximations into an exact result.

Ch. 3, p. 123

Graphical Insights
enhance students' visual understanding by
making the crucial connection between
graphical properties and the underlying
concept.

GRAPHICAL INSIGHT Keep the graphical interpretation of limits in mind. In Figure 4(A), $f(x)$ approaches L as $x \to c$ because for any $\epsilon > 0$, we can make the gap less than ϵ by taking δ sufficiently small. The function in Figure 4(B) has a jump discontinuity at $x = c$ and the gap cannot be made small, no matter how small δ is taken. Therefore, the limit does not exist.

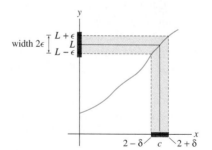

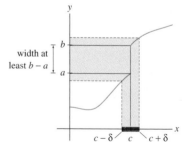

(A) The function is continuous at $x = c$.
By taking δ sufficiently small, we
can make the gap smaller than ϵ.

(B) The function is not continuous at $x = c$.
The gap is always larger than $(b - a)/2$,
no matter how small δ is.

FIGURE 4

Ch. 2, p. 112

■ **EXAMPLE 4** Rotating About the *x*-Axis   Use the Shell Method to compute the volume V of the solid obtained by rotating the area under $y = 9 - x^2$ over $[0, 3]$ about the x-axis.

Solution Since we are rotating about the x-axis rather than the y-axis, the Shell Method gives us an integral with respect to y. Therefore, we solve $y = 9 - x^2$ to obtain $x^2 = 9 - y$ or $x = \sqrt{9 - y}$.

The cylindrical shells are generated by *horizontal* segments. The segment $\overline{AB}$ in Figure 8 generates a cylindrical shell of radius y and height $\sqrt{9 - y}$ (we still use the term "height" although the cylinder is horizontal). Using the substitution $u = 9 - y$, $du = -dy$ in the resulting integral, we obtain

$$V = 2\pi \int_0^9 (\text{radius})(\text{height})\, dy = 2\pi \int_0^9 y\sqrt{9 - y}\, dy = -2\pi \int_9^0 (9 - u)\sqrt{u}\, du$$

$$= 2\pi \int_0^9 (9u^{1/2} - u^{3/2})\, du = 2\pi \left(6u^{3/2} - \frac{2}{5} u^{5/2} \right) \Big|_0^9 = \frac{648}{5}\pi \qquad ■$$

◀·· **REMINDER** *After making the substitution $u = 9 - y$, the limits of integration must be changed. Since $u(0) = 9$ and $u(9) = 0$, we change $\int_0^9$ to $\int_9^0$.*

Ch. 6, p. 404

Reminders are margin notes that link back to important concepts discussed earlier in the text to give students a quick review and make connections with earlier concepts.

| **CAUTION** When using L'Hôpital's Rule, be sure to take the derivative of the numerator and denominator separately. Do not differentiate the quotient function.

$$\lim_{x \to 1} \frac{x^3 - 1}{x - 1} = \lim_{x \to 1} \frac{3x^2}{1} = 3$$

Caution Notes warn students of common pitfalls they can encounter in understanding the material.

Ch. 4, p. 273

Historical Perspectives are brief vignettes that place key discoveries and conceptual advances in their historical context. They give students a glimpse into past accomplishments of great mathematicians and an appreciation for their significance.

HISTORICAL PERSPECTIVE

We take it for granted that the basic laws of physics are best expressed as mathematical relationships. Think of $F = ma$ or the universal law of gravitation. However, the fundamental insight that mathematics could be used to formulate laws of nature (and not just for counting or measuring) developed gradually, beginning with the philosophers of ancient Greece and culminating some 2,000 years later in the discoveries of Galileo and Newton. Archimedes

(287–212 BCE) was one of the first scientists (perhaps the first) to formulate a precise physical law. He considered the following question: If weights of mass m_1 and m_2 are placed at the ends of a weightless lever, where should the fulcrum P be located so that the lever does not tip to one side? Suppose that the distance from P to m_1 and m_2 is L_1 and L_2, respectively (Figure 16). Archimedes's Law states that the lever will balance if

$$\boxed{L_1 m_1 = L_2 m_2}$$

In our terminology, what Archimedes had discovered was the center of mass P of the system of weights (see Exercises 40 and 41).

Ch. 8, p. 498

Assumptions Matter uses short explanations and well-chosen counterexamples to help students appreciate why hypotheses are needed in theorems.

■ **EXAMPLE 10** Assumptions Matter: Choosing the Right Comparison   Determine whether $\displaystyle\int_1^\infty \frac{dx}{\sqrt{x}+e^{3x}}$ converges or diverges.

Solution   Let's try using the inequality

$$\frac{1}{\sqrt{x}+e^{3x}} \le \frac{1}{e^{3x}} \quad (x \ge 0)$$

$$\int_1^\infty \frac{dx}{e^{3x}} = \lim_{R\to\infty} -\frac{1}{3}e^{-3x}\Big|_1^R = \lim_{R\to\infty} \frac{1}{3}(e^{-3} - e^{-3R}) = e^{-3} \quad \text{(converges)}$$

The Comparison Test tells us that our integral converges:

$$\int_1^\infty \frac{dx}{e^{3x}} \quad \text{converges} \quad \Rightarrow \quad \int_1^\infty \frac{dx}{\sqrt{x}+e^{3x}} \quad \text{also converges}$$

Had we not been thinking, we might have tried to use the inequality

$$\frac{1}{\sqrt{x}+e^{3x}} \le \frac{1}{\sqrt{x}} \qquad \boxed{4}$$

The integral $\displaystyle\int_1^\infty \frac{dx}{\sqrt{x}}$ diverges by Theorem 1 (with $p = \frac{1}{2}$), but this tells us nothing about our integral which, by (4), is smaller (Figure 10). ■

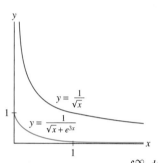

FIGURE 10 The divergence of $\displaystyle\int_1^\infty \frac{dx}{\sqrt{x}}$ says nothing about the divergence or convergence of the smaller integral $\displaystyle\int_1^\infty \frac{dx}{\sqrt{x}+e^{3x}}$.

Ch. 7, p. 472

Section Summaries summarize a section's key points in a concise and useful way and emphasize for students what is most important in the section.

Section Exercise Sets offer a comprehensive set of exercises closely coordinated with the text. These exercises vary in levels of difficulty from routine, to moderate, to more challenging. Also included are problems marked with icons that require the student to give a written response ▱ or require the use of technology GU *CAS*:

> *Preliminary Exercises* begin each exercise set and need little or no computation. They can be used to check understanding of key concepts of a section before problems from the exercise set are attempted.
> *Exercises* offer numerous problems from routine drill problems to moderately challenging problems. These are carefully graded and include many innovative and interesting geometric and real-world applications.
> *Further Insights and Challenges* are more challenging problems that require a deeper level of conceptual understanding and sometimes extend a section's material. Many are excellent for use as small-group projects.

Chapter Review Exercises offer a comprehensive set of exercises closely coordinated with the chapter material to provide additional problems for self-study or assignments.

ACKNOWLEDGMENTS

Jon Rogawski and W. H. Freeman and Company are grateful to the many instructors from across the United States and Canada who have offered comments that assisted in the development and refinement of this book. These contributions included class testing, manuscript reviewing, problems reviewing, and participating in surveys about the book and general course needs.

ALABAMA Tammy Potter, *Gadsden State Community College*; David Dempsey, *Jacksonville State University*; Douglas Bailer, *Northeast Alabama Community College*; Michael Hicks, *Shelton State Community College*; Patricia C. Eiland, *Troy University, Montgomery Campus*; James L. Wang, *The University of Alabama*; Stephen Brick, *University of South Alabama*; Joerg Feldvoss, *University of South Alabama* **ALASKA** Mark A. Fitch, *University of Alaska Anchorage*; Kamal Narang, *University of Alaska Anchorage*; Alexei Rybkin, *University of Alaska Fairbanks*; Martin Getz, *University of Alaska Fairbanks* **ARIZONA** Stefania Tracogna, *Arizona State University*; Bruno Welfert, *Arizona State University*; Daniel Russow, *Arizona Western College*; Garry Carpenter, *Pima Community College, Northwest Campus*; Katie Louchart, *Northern Arizona University*; Donna M. Krawczyk, *The University of Arizona* **ARKANSAS** Deborah Parker, *Arkansas Northeastern College*; J. Michael Hall, *Arkansas State University*; Kevin Cornelius, *Ouachita Baptist University*; Hyungkoo Mark Park, *Southern Arkansas University*; Katherine Pinzon, *University of Arkansas at Fort Smith*; Denise LeGrand, *University of Arkansas at Little Rock*; John Annulis, *University of Arkansas at Monticello*; Erin Haller, *University of Arkansas Fayetteville*; Daniel J. Arrigo, *University of Central Arkansas* **CALIFORNIA** Harvey Greenwald, *California Polytechnic State University, San Luis Obispo*; John M. Alongi, *California Polytechnic State University, San Luis Obispo*; John Hagen, *California Polytechnic State University, San Luis Obispo*; Colleen Margarita Kirk, *California Polytechnic State University, San Luis Obispo*; Lawrence Sze, *California Polytechnic State University, San Luis Obispo*; Raymond Terry, *California Polytechnic State University, San Luis Obispo*; James R. McKinney, *California State Polytechnic University, Pomona*; Charles Lam, *California State University, Bakersfield*; David McKay, *California State University, Long Beach*; Melvin Lax, *California State University, Long Beach*; Wallace A. Etterbeek, *California State University, Sacramento*; Mohamed Allali, *Chapman University*; George Rhys, *College of the Canyons*; Janice Hector, *DeAnza College*; Isabelle Saber, *Glendale Community College*; Peter Stathis, *Glendale Community College*; Kristin Hartford, *Long Beach City College*; Eduardo Arismendi-Pardi, *Orange Coast College*; Mitchell Alves, *Orange Coast College*; Yenkanh Vu, *Orange Coast College*; Yan Tian, *Palomar College*; Donna E. Nordstrom, *Pasadena City College*; Don L. Hancock, *Pepperdine University*; Kevin Iga, *Pepperdine University*; Adolfo J. Rumbos, *Pomona College*; Carlos de la Lama, *San Diego City College*; Matthias Beck, *San Francisco State University*; Arek Goetz, *San Francisco State University*; Nick Bykov, *San Joaquin Delta College*; Eleanor Lang Kendrick, *San Jose City College*; Elizabeth Hodes, *Santa Barbara City College*; William Konya, *Santa Monica College*; John Kennedy, *Santa Monica College*; Peter Lee, *Santa Monica College*; Richard Salome, *Scotts Valley High School*; Norman Feldman, *Sonoma State University*; Elaine McDonald, *Sonoma State University*; Bruno Nachtergaele, *University of California, Davis*; Boumediene Hamzi, *University of California, Davis*; Peter Stevenhagen, *University of California, San Diego*; Jeffrey Stopple, *University of California, Santa Barbara*; Guofang Wei, *University of California, Santa Barbara*; Rick A. Simon, *University of La Verne*; Mohamad A. Alwash, *West Los Angeles College* **COLORADO** Tony Weathers, *Adams State College*; Erica Johnson, *Arapahoe Community College*; Karen Walters, *Arapahoe Community College*; Joshua D. Laison, *Colorado College*; Gerrald G. Greivel, *Colorado School of Mines*; Jim Thomas, *Colorado State University*; Eleanor Storey, *Front Range Community College*; Larry Johnson, *Metropolitan State College of Denver*; Carol Kuper, *Morgan Community*

College; Larry A. Pontaski, *Pueblo Community College*; Terry Reeves, *Red Rocks Community College*; Debra S. Carney, *University of Denver* CONNECTICUT Jeffrey McGowan, *Central Connecticut State University*; Ivan Gotchev, *Central Connecticut State University*; Charles Waiveris, *Central Connecticut State University*; Christopher Hammond, *Connecticut College*; Anthony Y. Aidoo, *Eastern Connecticut State University*; Kim Ward, *Eastern Connecticut State University*; Joan W. Weiss, *Fairfield University*; Theresa M. Sandifer, *Southern Connecticut State University*; Cristian Rios, *Trinity College*; Melanie Stein, *Trinity College* DELAWARE Patrick F. Mwerinde, *University of Delaware* DISTRICT OF COLUMBIA Jeffrey Hakim, *American University*; Joshua M. Lansky, *American University*; James A. Nickerson, *Gallaudet University* FLORIDA Abbas Zadegan, *Florida International University*; Gerardo Aladro, *Florida International University*; Gregory Henderson, *Hillsborough Community College*; Pam Crawford, *Jacksonville University*; Penny Morris, *Polk Community College*; George Schultz, *St. Petersburg College*; Jimmy Chang, *St. Petersburg College*; Carolyn Kistner, *St. Petersburg College*; Aida Kadic-Galeb, *The University of Tampa*; Heath M. Martin, *University of Central Florida*; Constance Schober, *University of Central Florida*; S. Roy Choudhury, *University of Central Florida*; Kurt Overhiser, *Valencia Community College* GEORGIA Thomas T. Morley, *Georgia Institute of Technology*; Ralph Wildy, *Georgia Military College*; Shahram Nazari, *Georgia Perimeter College*; Alice Eiko Pierce, *Georgia Perimeter College Clarkson Campus*; Susan Nelson, *Georgia Perimeter College Clarkson Campus*; Shahram Nazari, *Georgia Perimeter College Dunwoody Campus*; Laurene Fausett, *Georgia Southern University*; Scott N. Kersey, *Georgia Southern University*; Jimmy L. Solomon, *Georgia Southern University*; Allen G. Fuller, *Gordon College*; Marwan Zabdawi, *Gordon College*; Carolyn A. Yackel, *Mercer University*; Shahryar Heydari, *Piedmont College*; Dan Kannan, *The University of Georgia* HAWAII Shuguang Li, *University of Hawaii at Hilo*; Raina B. Ivanova, *University of Hawaii at Hilo* IDAHO Charles Kerr, *Boise State University*; Otis Kenny, *Boise State University*; Alex Feldman, *Boise State University*; Doug Bullock, *Boise State University*; Ed Korntved, *Northwest Nazarene University* ILLINOIS Chris Morin, *Blackburn College*; Alberto L. Delgado, *Bradley University*; John Haverhals, *Bradley University*; Herbert E. Kasube, *Bradley University*; Brenda H. Alberico, *College of DuPage*; Ayse Sahin, *DePaul University*; Marvin Doubet, *Lake Forest College*; Marvin A. Gordon, *Lake Forest Graduate School of Management*; Richard J. Maher, *Loyola University Chicago*; Joseph H. Mayne, *Loyola University Chicago*; Marian Gidea, *Northeastern Illinois University*; Miguel Angel Lerma, *Northwestern University*; Mehmet Dik, *Rockford College*; Tammy Voepel, *Southern Illinois University Edwardsville*; Rahim G. Karimpour, *Southern Illinois University*; Thomas Smith, *University of Chicago* INDIANA Julie A. Killingbeck, *Ball State University*; John P. Boardman, *Franklin College*; Robert N. Talbert, *Franklin College*; Robin Symonds, *Indiana University Kokomo*; Henry L. Wyzinski, *Indiana University Northwest*; Melvin Royer, *Indiana Wesleyan University*; Gail P. Greene, *Indiana Wesleyan University*; David L. Finn, *Rose-Hulman Institute of Technology* IOWA Nasser Dastrange, *Buena Vista University*; Mark A. Mills, *Central College*; Karen Ernst, *Hawkeye Community College*; Richard Mason, *Indian Hills Community College*; Robert S. Keller, *Loras College*; Eric Robert Westlund, *Luther College* KANSAS Timothy W. Flood, *Pittsburg State University*; Sarah Cook, *Washburn University*; Kevin E. Charlwood, *Washburn University* KENTUCKY Alex M. McAllister, *Center College*; Sandy Spears, *Jefferson Community & Technical College*; Leanne Faulkner, *Kentucky Wesleyan College*; Donald O. Clayton, *Madisonville Community College*; Thomas Riedel, *University of Louisville*; Manabendra Das, *University of Louisville*; Lee Larson,

University of Louisville; Jens E. Harlander, *Western Kentucky University*
LOUISIANA   William Forrest, *Baton Rouge Community College*; Paul Wayne Britt, *Louisiana State University*; Galen Turner, *Louisiana Tech University*; Randall Wills, *Southeastern Louisiana University*; Kent Neuerburg, *Southeastern Louisiana University*
MAINE   Andrew Knightly, *The University of Maine*; Sergey Lvin, *The University of Maine*; Joel W. Irish, *University of Southern Maine*; Laurie Woodman, *University of Southern Maine*   MARYLAND   Leonid Stern, *Towson University*; Mark E. Williams, *University of Maryland Eastern Shore*; Austin A. Lobo, *Washington College*
MASSACHUSETTS   Sean McGrath, *Algonquin Regional High School*; Norton Starr, *Amherst College*; Renato Mirollo, *Boston College*; Emma Previato, *Boston University*; Richard H. Stout, *Gordon College*; Matthew P. Leingang, *Harvard University*; Suellen Robinson, *North Shore Community College*; Walter Stone, *North Shore Community College*; Barbara Loud, *Regis College*; Andrew B. Perry, *Springfield College*; Tawanda Gwena, *Tufts University*; Gary Simundza, *Wentworth Institute of Technology*; Mikhail Chkhenkeli, *Western New England College*; David Daniels, *Western New England College*; Alan Gorfin, *Western New England College*; Saeed Ghahramani, *Western New England College*; Julian Fleron, *Westfield State College*; Brigitte Servatius, *Worcester Polytechnic Institute*   MICHIGAN   Mark E. Bollman, *Albion College*; Jim Chesla, *Grand Rapids Community College*; Jeanne Wald, *Michigan State University*; Allan A. Struthers, *Michigan Technological University*; Debra Pharo, *Northwestern Michigan College*; Anna Maria Spagnuolo, *Oakland University*; Diana Faoro, *Romeo Senior High School*; Andrew Strowe, *University of Michigan–Dearborn*; Daniel Stephen Drucker, *Wayne State University*   MINNESOTA   Bruce Bordwell, *Anoka-Ramsey Community College*; Robert Dobrow, *Carleton College*; Jessie K. Lenarz, *Concordia College–Moorhead Minnesota*; Bill Tomhave, *Concordia College–Moorhead Minnesota*; David L. Frank, *University of Minnesota*; Steven I. Sperber, *University of Minnesota*; Jeffrey T. McLean, *University of St. Thomas*; Chehrzad Shakiban, *University of St. Thomas*; Melissa Loe, *University of St. Thomas*   MISSISSIPPI   Vivien G. Miller, *Mississippi State University*; Ted Dobson, *Mississippi State University*; Len Miller, *Mississippi State University*; Tristan Denley, *The University of Mississippi*
MISSOURI   Robert Robertson, *Drury University*; Gregory A. Mitchell, *Metropolitan Community College-Penn Valley*; Charles N. Curtis, *Missouri Southern State University*; Vivek Narayanan, *Moberly Area Community College*; Russell Blyth, *Saint Louis University*; Blake Thornton, *Saint Louis University*; Kevin W. Hopkins, *Southwest Baptist University*   MONTANA   Kelly Cline, *Carroll College*; Richard C. Swanson, *Montana State University*; Nikolaus Vonessen, *The University of Montana*
NEBRASKA   Edward G. Reinke Jr., *Concordia University, Nebraska*; Judith Downey, *University of Nebraska at Omaha*   NEVADA   Rohan Dalpatadu, *University of Nevada, Las Vegas*; Paul Aizley, *University of Nevada, Las Vegas*   NEW HAMPSHIRE   Richard Jardine, *Keene State College*; Michael Cullinane, *Keene State College*; Roberta Kieronski, *University of New Hampshire at Manchester*   NEW JERSEY   Paul S. Rossi, *College of Saint Elizabeth*; Mark Galit, *Essex County College*; Katarzyna Potocka, *Ramapo College of New Jersey*; Nora S. Thornber, *Raritan Valley Community College*; Avraham Soffer, *Rutgers The State University of New Jersey*; Chengwen Wang, *Rutgers The State University of New Jersey*; Stephen J. Greenfield, *Rutgers The State University of New Jersey*; John T. Saccoman, *Seton Hall University*; Lawrence E. Levine, *Stevens Institute of Technology*   NEW MEXICO   Kevin Leith, *Central New Mexico Community College*; David Blankenbaker, *Central New Mexico Community College*; Joseph Lakey, *New Mexico State University*; Jurg Bolli, *University of New Mexico*; Kees Onneweer, *University of New Mexico*   NEW YORK   Robert C. Williams, *Alfred University*; Timmy G. Bremer, *Broome Community College State*

University of New York; Joaquin O. Carbonara, *Buffalo State College*; Robin Sue Sanders, *Buffalo State College*; Daniel Cunningham, *Buffalo State College*; Rose Marie Castner, *Canisius College*; Sharon L. Sullivan, *Catawba College*; Camil Muscalu, *Cornell University*; Maria S. Terrell, *Cornell University*; Margaret Mulligan, *Dominican College of Blauvelt*; Robert Andersen, *Farmingdale State University of New York*; Leonard Nissim, *Fordham University*; Jennifer Roche, *Hobart and William Smith Colleges*; James E. Carpenter, *Iona College*; Peter Shenkin, *John Jay College of Criminal Justice/CUNY*; Gordon Crandall, *LaGuardia Community College/CUNY*; Gilbert Traub, *Maritime College, State University of New York*; Paul E. Seeburger, *Monroe Community College Brighton Campus*; Abraham S. Mantell, *Nassau Community College*; Daniel D. Birmajer, *Nazareth College*; Sybil G. Shaver, *Pace University*; Margaret Kiehl, *Rensselaer Polytechnic Institute*; Carl V. Lutzer, *Rochester Institute of Technology*; Michael A. Radin, *Rochester Institute of Technology*; Hossein Shahmohamad, *Rochester Institute of Technology*; Thomas Rousseau, *Siena College*; Jason Hofstein, *Siena College*; Leon E. Gerber, *St. Johns University*; Christopher Bishop, *Stony Brook University*; James Fulton, *Suffolk County Community College*; John G. Michaels, *SUNY Brockport*; Howard J. Skogman, *SUNY Brockport*; Cristina Bacuta, *SUNY Cortland*; Jean Harper, *SUNY Fredonia*; Kelly Black, *Union College*; Thomas W. Cusick, *University at Buffalo/The State University of New York*; Gino Biondini, *University at Buffalo/The State University of New York*; Robert Koehler, *University at Buffalo/The State University of New York* **NORTH CAROLINA** Jeffrey Clark, *Elon University*; William L. Burgin, *Gaston College*; Manouchehr H. Misaghian, *Johnson C. Smith University*; Legunchim L. Emmanwori, *North Carolina A&T State University*; Drew Pasteur, *North Carolina State University*; Demetrio Labate, *North Carolina State University*; Mohammad Kazemi, *The University of North Carolina at Charlotte*; Richard Carmichael, *Wake Forest University*; Gretchen Wilke Whipple, *Warren Wilson College* **NORTH DAKOTA** Anthony J. Bevelacqua, *The University of North Dakota*; Richard P. Millspaugh, *The University of North Dakota* **OHIO** Christopher Butler, *Case Western Reserve University*; Pamela Pierce, *The College of Wooster*; Tzu-Yi Alan Yang, *Columbus State Community College*; Greg S. Goodhart, *Columbus State Community College*; Kelly C. Stady, *Cuyahoga Community College*; Brian T. Van Pelt, *Cuyahoga Community College*; David Robert Ericson, *Miami University, Middletown Campus*; Frederick S. Gass, *Miami University*; Thomas Stacklin, *Ohio Dominican University*; Vitaly Bergelson, *The Ohio State University*; Darry Andrews, *The Ohio State University*; Robert Knight, *Ohio University*; John R. Pather, *Ohio University, Eastern Campus*; Teresa Contenza, *Otterbein College*; Ali Hajjafar, *The University of Akron*; Jianping Zhu, *The University of Akron*; Ian Clough, *University of Cincinnati Clermont College*; Atif Abueida, *University of Dayton*; Judith McCrory, *The University at Findlay*; Thomas Smotzer, *Youngstown State University*; Angela Spalsbury, *Youngstown State University* **OKLAHOMA** Michael McClendon, *University of Central Oklahoma*; Teri Jo Murphy, *The University of Oklahoma* **OREGON** Lorna TenEyck, *Chemeketa Community College*; Angela Martinek, *Linn-Benton Community College*; Tevian Dray, *Oregon State University* **PENNSYLVANIA** John B. Polhill, *Bloomsburg University of Pennsylvania*; Russell C. Walker, *Carnegie Mellon University*; Jon A. Beal, *Clarion University of Pennsylvania*; Kathleen Kane, *Community College of Allegheny County*; David A. Santos, *Community College of Philadelphia*; David S. Richeson, *Dickinson College*; Christine Marie Cedzo, *Gannon University*; Monica Pierri-Galvao, *Gannon University*; John H. Ellison, *Grove City College*; Gary L. Thompson, *Grove City College*; Dale McIntyre, *Grove City College*; Dennis Benchoff, *Harrisburg Area Community College*; William A. Drumin, *King's College*; Denise Reboli, *King's College*; Chawne Kimber, *Lafeyette College*; David L. Johnson, *Lehigh University*; Zia

Uddin, *Lock Haven University of Pennsylvania*; Donna A. Dietz, *Mansfield University of Pennsylvania*; Samuel Wilcock, *Messiah College*; Neena T. Chopra, *The Pennsylvania State University*; Boris A. Datskovsky, *Temple University*; Dennis M. DeTurck, *University of Pennsylvania*; Jacob Burbea, *University of Pittsburgh*; Mohammed Yahdi, *Ursinus College*; Timothy Feeman, *Villanova University*; Douglas Norton, *Villanova University*; Robert Styer, *Villanova University*   RHODE ISLAND   Thomas F. Banchoff, *Brown University*; Yajni Warnapala-Yehiya, *Roger Williams University*; Carol Gibbons, *Salve Regina University*   SOUTH CAROLINA   Stanley O. Perrine, *Charleston Southern University*; Joan Hoffacker, *Clemson University*; Constance C. Edwards, *Coastal Carolina University*; Thomas L. Fitzkee, *Francis Marion University*; Richard West, *Francis Marion University*; Douglas B. Meade, *University of South Carolina*; George Androulakis, *University of South Carolina*; Art Mark, *University of South Carolina Aiken*   SOUTH DAKOTA   Dan Kemp, *South Dakota State University*   TENNESSEE   Andrew Miller, *Belmont University*; Arthur A. Yanushka, *Christian Brothers University*; Laurie Plunk Dishman, *Cumberland University*; Beth Long, *Pellissippi State Technical Community College*; Judith Fethe, *Pellissippi State Technical Community College*; Andrzej Gutek, *Tennessee Technological University*; Sabine Le Borne, *Tennessee Technological University*; Richard Le Borne, *Tennessee Technological University*; Jim Conant, *The University of Tennessee*; Pavlos Tzermias, *The University of Tennessee*; Jo Ann W. Staples, *Vanderbilt University*   TEXAS   Sally Haas, *Angelina College*; Michael Huff, *Austin Community College*; Scott Wilde, *Baylor University and The University of Texas at Arlington*; Rob Eby, *Blinn College*; Tim Sever, *Houston Community College–Central*; Ernest Lowery, *Houston Community College–Northwest*; Shirley Davis, *South Plains College*; Todd M. Steckler, *South Texas College*; Mary E. Wagner-Krankel, *St. Mary's University*; Elise Z. Price, *Tarrant County College, Southeast Campus*; David Price, *Tarrant County College, Southeast Campus*; Michael Stecher, *Texas A&M University*; Philip B. Yasskin, *Texas A&M University*; Brock Williams, *Texas Tech University*; I. Wayne Lewis, *Texas Tech University*; Robert E. Byerly, *Texas Tech University*; Ellina Grigorieva, *Texas Woman's University*; Abraham Haje, *Tomball College*; Scott Chapman, *Trinity University*; Elias Y. Deeba, *University of Houston Downtown*; Jianping Zhu, *The University of Texas at Arlington*; Tuncay Aktosun, *The University of Texas at Arlington*; John E. Gilbert, *The University of Texas at Austin*; Jorge R. Viramontes-Olivias, *The University of Texas at El Paso*; Melanie Ledwig, *The Victoria College*; Gary L. Walls, *West Texas A&M University*; William Heierman, *Wharton County Junior College*   UTAH   Jason Isaac Preszler, *The University of Utah*   VIRGINIA   Verne E. Leininger, *Bridgewater College*; Brian Bradie, *Christopher Newport University*; Hongwei Chen, *Christopher Newport University*; John J. Avioli, *Christopher Newport University*; James H. Martin, *Christopher Newport University*; Mike Shirazi, *Germanna Community College*; Ramon A. Mata-Toledo, *James Madison University*; Adrian Riskin, *Mary Baldwin College*; Josephine Letts, *Ocean Lakes High School*; Przemyslaw Bogacki, *Old Dominion University*; Deborah Denvir, *Randolph-Macon Woman's College*; Linda Powers, *Virginia Tech*; Gregory Dresden, *Washington and Lee University*; Jacob A. Siehler, *Washington and Lee University*   VERMONT   David Dorman, *Middlebury College*; Rachel Repstad, *Vermont Technical College*   WASHINGTON   Jennifer Laveglia, *Bellevue Community College*; David Whittaker, *Cascadia Community College*; Sharon Saxton, *Cascadia Community College*; Aaron Montgomery, *Central Washington University*; Patrick Averbeck, *Edmonds Community College*; Tana Knudson, *Heritage University*; Kelly Brooks, *Pierce College*; Shana P. Calaway, *Shoreline Community College*; Abel Gage, *Skagit Valley College*; Scott MacDonald, *Tacoma Community College*; Martha A. Gady, *Whitworth College*   WEST VIRGINIA   Ralph

Oberste-Vorth, *Marshall University*; Suda Kunyosying, *Shepard University*; Nicholas Martin, *Shepherd University*; Rajeev Rajaram, *Shepherd University*; Xiaohong Zhang, *West Virginia State University*; Sam B. Nadler, *West Virginia University* WYOMING Claudia Stewart, *Casper College*; Charles Newberg, *Western Wyoming Community College* WISCONSIN Paul Bankston, *Marquette University*; Jane Nichols, *Milwaukee School of Engineering*; Yvonne Yaz, *Milwaukee School of Engineering*; Terry Nyman, *University of Wisconsin–Fox Valley*; Robert L. Wilson, *University of Wisconsin–Madison*; Dietrich A. Uhlenbrock, *University of Wisconsin–Madison*; Paul Milewski, *University of Wisconsin–Madison*; Donald Solomon, *University of Wisconsin–Milwaukee*; Kandasamy Muthuvel, *University of Wisconsin–Oshkosh*; Sheryl Wills, *University of Wisconsin–Platteville*; Kathy A. Tomlinson, *University of Wisconsin–River Falls* CANADA Don St. Jean, *George Brown College*; Len Bos, *University of Calgary*

We would also like to acknowledge the following people for their help in reviewing the Math Marvels and Calculus Chronicles cards:

Gino Biondini, *University at Buffalo/The State University of New York*; Russell Blyth, *Saint Louis University*; Christopher Butler, *Case Western Reserve University*; Timmy G. Bremer, *Broome Community College, The State University of New York*; Robert E. Byerly, *Texas Tech University*; Timothy W. Flood, *Pittsburg State University*; Kevin W. Hopkins, *Southwest Baptist University;* Herbert E. Kasube, *Bradley University*; Scott N. Kersey, *Georgia Southern University*; Jessie K. Lenarz, *Concordia College–Moorehead Minnesota*; Ralph Oberste-Vorth, *Marshall University*; Rachel Repstad, *Vermont Technical College*; Chehrzad Shakiban, *University of St. Thomas*; James W. Thomas, *Colorado State University*; Chengwen Wang, *Rutgers, The State University of New Jersey*; Robert C. Williams, *Alfred University*; Jianping Zhu, *University of Akron*

Finally, we would like to thank the Math Clubs at the following schools for their help in checking the accuracy of the exercises and their solutions:

Arizona State University; California Polytechnic State University, San Luis Obispo; University of California, Berkeley; University of California, Santa Barbara; Florida Atlantic University; University of South Florida; Indiana University; Kansas State University; Louisiana State University; Worcester Polytechnic Institute; University of Missouri–Columbia; Mississippi State University; University of Montana; North Carolina State University; University of South Carolina; Vanderbilt University; Texas State University–San Marcos; University of Vermont; University of Wyoming

It is a pleasant task to thank the many people whose guidance and support were crucial in bringing this book to fruition. First, I thank Tony Palermino, my development editor, for his wisdom and dedication. Tony is responsible for improvements too numerous to detail. Special thanks also go to Professor Thomas Banchoff, who gave generously of his expertise in the development of the graphics throughout the text. I am grateful to the many mathematicians who provided valuable insights, constructive criticism, and innovative exercises. In particular, I thank Professors John Alongi, Chris Bishop, Brian Bradie, Dennis DeTurck, Ted Dobson, Greg Dresden, Mark Fitch, James McKinney, Len Miller, Vivien Miller, Aaron Montgomery, Vivek Narayanan, Emma Previato, Michael Radin, Todd Ruskell, David Wells, and my students Shraddha Chaplot and Sharon Hori. I also thank Rami Zelingher and Dina Lavi of JustAsk!

The production of a first-edition textbook is a major undertaking for author and publisher alike. I thank the staff of W. H. Freeman and Company for unwavering confidence in the project and outstanding editorial support. It is my pleasure to thank Craig Bleyer for signing the project and standing behind it, Terri Ward for skillfully organizing the complex production process, Ruth Baruth and Bruce Kaplan for invaluable knowledge and publishing experience, Steve Rigolosi for expert market development, and Brendan Cady and Laura Capuano for editorial assistance. My thanks are also due to the superb production team: Blake Logan, Bill Page, Paul Rohloff, Ted Szczepanski, Susan Timmins, and Vivien Weiss, as well as Don DeLand and Leslie Galen of Integre Technical Publishing, for their expert composition.

To my wife, Julie, I owe more than I can say. Thank you for everything. To our wonderful children Rivkah, Dvora, Hannah, and Akiva, thank you for putting up with the calculus book, a demanding extra sibling that dwelled in our home for so many years. And to my mother Elise and late father Alexander Rogawski, MD z"l, thank you for love and support from the beginning.

■ TO THE STUDENT

Although I have taught calculus for more than 25 years, I still feel excitement when I enter the classroom on the first day of a new semester, as if a great drama is about to unfold. Does the word "drama" seem out of place in a discussion of mathematics?

There is no doubt that calculus is useful—it is used throughout the sciences and engineering, from weather prediction and space flight to nanotechnology and financial modeling. But what is dramatic about it?

For me, one part of the drama lies in the conceptual and logical development of the subject. Starting with just a few basic concepts such as limits and tangent lines, we gradually build the tools for solving innumerable problems of great practical importance. Along the way, there are high points and moments of suspense—computing a derivative using limits for the first time or learning how the Fundamental Theorem of Calculus unifies differential and integral calculus. We also discover that calculus provides the right language for expressing the universal laws of nature, not just Newton's laws of motion to which it was first applied, but also the laws of electromagnetism and even the strange laws of quantum mechanics.

Another part of the drama is the learning process itself—the personal voyage of discovery. Certainly, one important part of learning calculus is the acquisition of technical skills. You will learn how to compute derivatives and integrals, solve optimization problems, and so on. These are the skills you need to apply calculus in practical situations. But when you study calculus, you also learn the language of science. You gain access to the thoughts of Newton, Euler, Gauss, and Maxwell, the greatest scientific thinkers, all of whom expressed their insights using calculus. The distinguished mathematician I. M. Gelfand put it this way: "The most important thing a student can get from the study of mathematics is the attainment of a higher intellectual level."

This text is designed to develop both skills and conceptual understanding. In fact, the two go hand-in-hand. As you become proficient in problem solving, you will come to appreciate the underlying ideas. And it is equally true that a solid understanding of the concepts will make you a more effective problem solver. You are likely to devote much of your time to studying examples and working exercises. However, the text also contains numerous down-to-earth explanations, sometimes under the heading "Conceptual Insight" or "Graphical Insight." They are designed to show you how and why calculus works. I urge you to take the time to read these explanations and think about them.

A major challenge for me in writing this textbook was to present calculus in a style that students would find comprehensible and interesting. As I wrote, I continually asked myself: Can it be made simpler? Have I assumed something the student may not be aware of? Can I explain the deeper significance of a concept without confusing the student who is learning the subject for the first time?

I'm afraid that no textbook can make learning calculus entirely painless. According to legend, Alexander the Great once asked the mathematician Menaechmus to show him an easy way to learn geometry. Menaechmus replied, "There is no royal road to geometry." Even kings must work hard to learn geometry, and the same is true of calculus—achieving mastery requires time and effort.

However, I hope my efforts have resulted in a textbook that is "student friendly" and that also encourages you to appreciate the big picture—the beautiful and elegant ideas that hold the entire structure of calculus together. Please let me know if you have comments or suggestions for improving the text. I look forward to hearing from you.

Best wishes and good luck!

Jon Rogawski

SINGLE VARIABLE
CALCULUS
Early Transcendentals

1 | PRECALCULUS REVIEW

Functions are one of our most important tools for analyzing phenomena. Biologists have studied the antler weight of male red deer as a function of age (see p. 6).

Calculus builds on the foundation of algebra, analytic geometry, and trigonometry. In this chapter we review some of the basic concepts, facts, and formulas that are used throughout the text. In the last section, we discuss some ways technology can be used to enhance your visual understanding of functions and their properties.

1.1 Real Numbers, Functions, and Graphs

We begin our review with a short discussion of real numbers. This gives us the opportunity to recall some basic properties and standard notation.

A **real number** is a number represented by a decimal or "decimal expansion." There are three types of decimal expansions: finite, repeating, and infinite but nonrepeating. For example,

$$\frac{3}{8} = 0.375, \qquad \frac{1}{7} = 0.142857142857\ldots = 0.\overline{142857}$$

$$\pi = 3.141592653589793\ldots$$

The number $\frac{3}{8}$ is represented by a finite decimal, whereas $\frac{1}{7}$ is represented by a *repeating* or *periodic* decimal. The bar over 142857 indicates that this sequence repeats indefinitely. The decimal expansion of π is infinite but nonrepeating.

The set of all real numbers is denoted by the boldface **R**. When there is no risk of confusion, we refer to a real number simply as a *number*. We also use the standard symbol $\in$ for the phrase "belongs to." Thus,

$$a \in \mathbf{R} \qquad \text{reads} \qquad \text{"}a \text{ belongs to } \mathbf{R}\text{"}$$

Additional properties of real numbers are discussed in Appendix B.

The set of integers is commonly denoted by the letter **Z** (this choice comes from the German word *zahl*, meaning "number"). Thus, $\mathbf{Z} = \{\ldots, -2, -1, 0, 1, 2, \ldots\}$. A **whole number** is a nonnegative integer, i.e., one of the numbers $0, 1, 2, \ldots$.

A real number is called **rational** if it can be represented by a fraction p/q, where p and q are integers with $q \neq 0$. The set of rational numbers is denoted **Q** (for "quotient"). Numbers such as π and $\sqrt{2}$ that are not rational are called **irrational**. We can tell whether or not a number is rational from its decimal expansion: Rational numbers have finite or repeating decimal expansions and irrational numbers have infinite, nonrepeating decimal expansions. Furthermore, the decimal expansion of a number is unique, apart from the following exception: Every finite decimal is equal to an infinite decimal in which the digit 9 repeats. For example,

$$1 = 0.999\ldots, \qquad \frac{3}{8} = 0.375 = 0.374999\ldots, \qquad \frac{47}{20} = 2.35 = 2.34999\ldots$$

FIGURE 1 The set of real numbers represented as a line.

We visualize real numbers as points on a line (Figure 1). For this reason, real numbers are often referred to as **points**. The number 0 is called the **origin** of the real line.

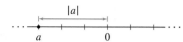

FIGURE 2 The absolute value $|a|$.

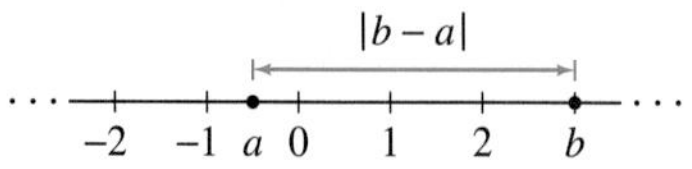

FIGURE 3 The distance between a and b is $|b - a|$.

The **absolute value** of a real number a, denoted $|a|$, is defined by (Figure 2)

$$|a| = \textit{distance from the origin} = \begin{cases} a & \text{if } a \geq 0 \\ -a & \text{if } a < 0 \end{cases}$$

For example, $|1.2| = 1.2$ and $|-8.35| = 8.35$. The absolute value satisfies

$$|a| = |-a|, \qquad |ab| = |a|\,|b|$$

The **distance** between two real numbers a and b is $|b - a|$, which is the length of the line segment joining a and b (Figure 3).

Two real numbers a and b are close to each other if $|b - a|$ is small, and this is the case if their decimal expansions agree to many places. More precisely, *if the decimal expansions of a and b agree to k places (to the right of the decimal point), then the distance $|b - a|$ is at most 10^{-k}*. Thus, the distance between $a = 3.1415$ and $b = 3.1478$ is at most 10^{-2} because a and b agree to two places. In fact, the distance is exactly $|3.1415 - 3.1478| = 0.0063$.

Beware that $|a + b|$ is not equal to $|a| + |b|$ unless a and b have the same sign or at least one of a and b is zero. If they have opposite signs, cancellation occurs in the sum $a + b$ and $|a + b| < |a| + |b|$. For example, $|2 + 5| = |2| + |5|$ but $|-2 + 5| = 3$, which is less than $|-2| + |5| = 7$. In any case, $|a + b|$ is never larger than $|a| + |b|$ and this gives us the simple but important **triangle inequality**:

$$|a + b| \leq |a| + |b| \qquad \boxed{1}$$

We use standard notation for intervals. Given real numbers $a < b$, there are four intervals with endpoints a and b (Figure 4). Each of these intervals has length $b - a$, but they differ according to whether or not one or both endpoints are included.

The **closed interval** $[a, b]$ is the set of all real numbers x such that $a \leq x \leq b$:

$$[a, b] = \{x \in \mathbf{R} : a \leq x \leq b\}$$

We usually write this more simply as $\{x : a \leq x \leq b\}$, it being understood that x belongs to $\mathbf{R}$. The **open** and **half-open intervals** are the sets

$$\underbrace{(a, b) = \{x : a < x < b\}}_{\text{Open interval (endpoints excluded)}}, \qquad \underbrace{[a, b) = \{x : a \leq x < b\}}_{\text{Half-open}}, \qquad \underbrace{(a, b] = \{x : a < x \leq b\}}_{\text{Half-open}}$$

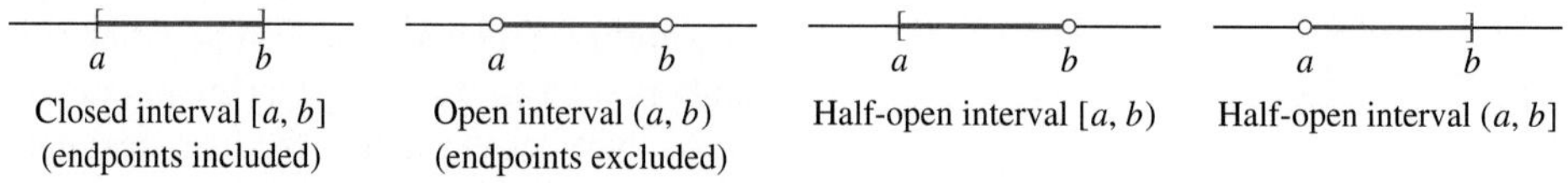

FIGURE 4 The four intervals with endpoints a and b.

Closed interval $[a, b]$ (endpoints included)   Open interval (a, b) (endpoints excluded)   Half-open interval $[a, b)$ Half-open interval $(a, b]$

The infinite interval $(-\infty, \infty)$ is the entire real line $\mathbf{R}$. A half-infinite interval may be open or closed. A closed half-infinite interval contains its finite endpoint:

$$[a, \infty) = \{x : a \leq x < \infty\}, \qquad (-\infty, b] = \{x : -\infty < x \leq b\}$$

FIGURE 5 Closed half-infinite intervals.

$[a, \infty)$ $(-\infty, b]$

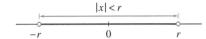

FIGURE 6 The interval
$(-r, r) = \{x : |x| < r\}$.

FIGURE 7 $(a, b) = (c - r, c + r)$, where
$$c = \frac{a + b}{2}, \qquad r = \frac{b - a}{2}$$

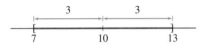

FIGURE 8 The interval $[7, 13]$ is
described by $|x - 10| \leq 3$.

FIGURE 9 The set
$S = \left\{x : \left|\frac{1}{2}x - 3\right| > 4\right\}$.

In Example 2 we use the notation $\cup$ to denote "union": The union $A \cup B$ of sets A and B consists of all elements that belong to either A or B (or both).

Open and closed intervals may be described by inequalities. For example (Figure 6),

$$|x| < r \quad \Leftrightarrow \quad -r < x < r \quad \Leftrightarrow \quad x \in (-r, r) \qquad \boxed{2}$$

More generally, for any c (Figure 7),

$$|x - c| < r \quad \Leftrightarrow \quad c - r < x < c + r \quad \Leftrightarrow \quad x \in (c - r, c + r) \qquad \boxed{3}$$

A similar statement holds for closed intervals with $<$ replaced by $\leq$. We refer to r as the **radius** and c as the **midpoint** or **center**. The intervals (a, b) and $[a, b]$ have midpoint $c = \frac{1}{2}(a + b)$ and radius $r = \frac{1}{2}(b - a)$ (Figure 7).

■ **EXAMPLE 1** Describing Intervals via Inequalities   Describe the intervals $(-4, 4)$ and $[7, 13]$ using inequalities.

Solution We have $(-4, 4) = \{x : |x| < 4\}$. The midpoint of the interval $[7, 13]$ is $c = \frac{1}{2}(7 + 13) = 10$ and its radius is $r = \frac{1}{2}(13 - 7) = 3$ (Figure 8). Therefore

$$[7, 13] = \left\{x \in \mathbf{R} : |x - 10| \leq 3\right\} \qquad ■$$

■ **EXAMPLE 2** Describe the set $S = \left\{x : \left|\frac{1}{2}x - 3\right| > 4\right\}$ in terms of intervals.

Solution It is easier to consider the opposite inequality $\left|\frac{1}{2}x - 3\right| \leq 4$ first. By (2),

$$\left|\frac{1}{2}x - 3\right| \leq 4 \Leftrightarrow -4 \leq \frac{1}{2}x - 3 \leq 4$$

$$-1 \leq \frac{1}{2}x \leq 7 \qquad \text{(add 3)}$$

$$-2 \leq x \leq 14 \qquad \text{(multiply by 2)}$$

Thus, $\left|\frac{1}{2}x - 3\right| \leq 4$ is satisfied when x belongs to $[-2, 14]$. The set S is the *complement*, consisting of all numbers x *not in* $[-2, 14]$. We can describe S as the union of two intervals: $S = (-\infty, -2) \cup (14, \infty)$ (Figure 9). ■

Graphing

Graphing is a basic tool in calculus, as it is in algebra and trigonometry. Recall that rectangular (or Cartesian) coordinates in the plane are defined by choosing two perpendicular axes, the x-axis and the y-axis. To a pair of numbers (a, b) we associate the point P located at the intersection of the line perpendicular to the x-axis at a and the line perpendicular to the y-axis at b [Figure 10(A)]. The numbers a and b are the x- and y-**coordinates** of P. The **origin** is the point with coordinates $(0, 0)$.

The x-coordinate is sometimes called the "abscissa" and the y-coordinate the "ordinate."

The term "Cartesian" refers to the French philosopher and mathematician René Descartes (1596–1650), whose Latin name was Cartesius. He is credited (along with Pierre de Fermat) with the invention of analytic geometry. In his great work La Géométrie, *Descartes used the letters x, y, z for unknowns and a, b, c for constants, a convention that has been followed ever since.*

The axes divide the plane into four quadrants labeled I–IV, determined by the signs of the coordinates [Figure 10(B)]. For example, quadrant III consists of points (x, y) such that $x < 0$ and $y < 0$.

The distance d between two points $P_1 = (x_1, y_1)$ and $P_2 = (x_2, y_2)$ is computed using the Pythagorean Theorem. In Figure 11, we see that $\overline{P_1 P_2}$ is the hypotenuse of a right triangle with sides $a = |x_2 - x_1|$ and $b = |y_2 - y_1|$. Therefore,

$$d^2 = a^2 + b^2 = (x_2 - x_1)^2 + (y_2 - y_1)^2$$

We obtain the distance formula by taking square roots.

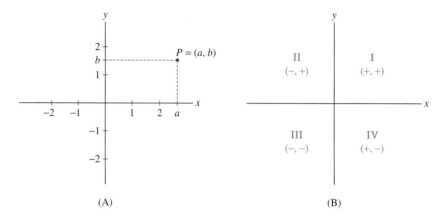

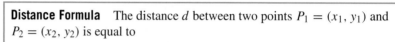

(A)   (B)

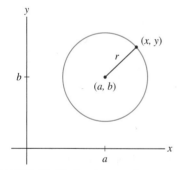

FIGURE 10 Rectangular coordinate system.

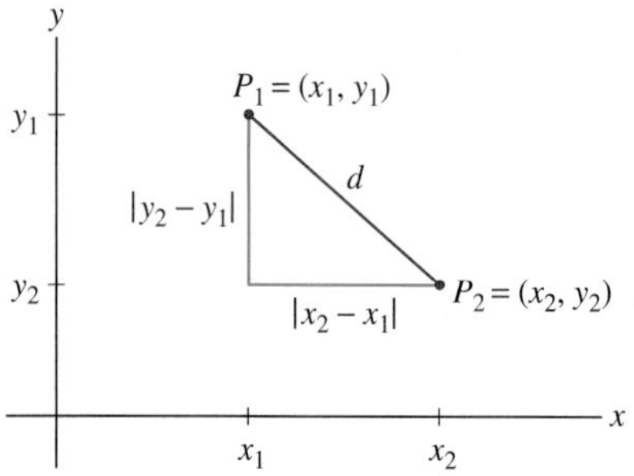

FIGURE 11 Distance d is given by the distance formula.

FIGURE 12 Circle with equation $(x - a)^2 + (y - b)^2 = r^2$.

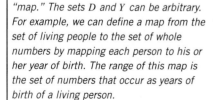

A function $f : D \to Y$ is also called a "map." The sets D and Y can be arbitrary. For example, we can define a map from the set of living people to the set of whole numbers by mapping each person to his or her year of birth. The range of this map is the set of numbers that occur as years of birth of a living person.

Distance Formula   The distance d between two points $P_1 = (x_1, y_1)$ and $P_2 = (x_2, y_2)$ is equal to

$$d = \sqrt{(x_2 - x_1)^2 + (y_2 - y_1)^2}$$

Once we have the distance formula, we can easily derive the equation of a circle of radius r and center (a, b) (Figure 12). A point (x, y) lies on this circle if the distance from (x, y) to (a, b) is r:

$$\sqrt{(x - a)^2 + (y - b)^2} = r$$

Squaring both sides, we obtain the standard equation of the circle:

$$(x - a)^2 + (y - b)^2 = r^2$$

We now review some definitions and notation concerning functions.

DEFINITION Function   A **function** f between two sets D and Y is a rule that assigns, to each element x in D, a unique element $y = f(x)$ in Y. Symbolically, we write

$$f : D \to Y$$

For $x \in D$, $f(x)$ is called the **value** of f at x (Figure 13). We call the set D the **domain** of $f(x)$. The **range** R of f is the subset of Y consisting of all values $f(x)$:

$$R = \{y \in Y : f(x) = y \text{ for some } x \in D\}$$

Informally, we think of f as a "machine" that produces an output y for every input x in the domain D (Figure 14).

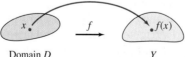

Domain D Y

FIGURE 13 A function is a rule that assigns an element $y = f(x)$ in Y to each $x \in D$.

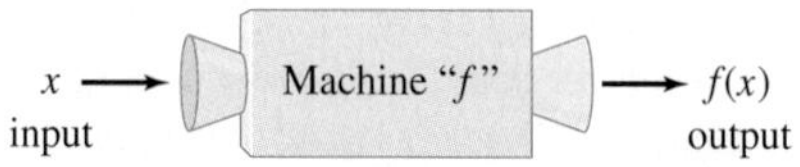

FIGURE 14 Think of f as a "machine" that takes the input x and produces the output $f(x)$.

In multivariable calculus, we deal with functions having a variety of domains and ranges. The domain might be a set of points in three-dimensional space and the range a set of numbers, points, or vectors.

The first part of this text deals with *numerical* functions f, where both the domain and the range are sets of real numbers. We refer to such a function interchangeably as f or $f(x)$. The letter x denotes the **independent variable** that can take on any value in the domain D. We write $y = f(x)$ and refer to y as the **dependent variable** (because its value depends on the choice of x).

When f is defined by a formula, its natural domain is the set of real numbers x for which the formula is meaningful. For example, the function $f(x) = \sqrt{9-x}$ has domain $D = \{x : x \le 9\}$ because $\sqrt{9-x}$ is defined if $9 - x \ge 0$. Here are some other examples of domains and ranges.

$f(x)$	Domain D	Range R
x^2	$\mathbf{R}$	$\{y : y \ge 0\}$
$\cos x$	$\mathbf{R}$	$\{y : -1 \le y \le 1\}$
$\dfrac{1}{x+1}$	$\{x : x \ne -1\}$	$\{y : y \ne 0\}$

The **graph** of a function $y = f(x)$ is obtained by plotting the points $(a, f(a))$ for a in the domain D (Figure 15). If you start at $x = a$ on the x-axis, move up to the graph and then over to the y-axis, you arrive at the value $f(a)$. The value $|f(a)|$ is thus the distance of the graph above or below the x-axis [depending on whether $f(a) \ge 0$ or $f(a) < 0$].

A **zero** or **root** of a function $f(x)$ is a number c such that $f(c) = 0$. The zeros are the values of x where the graph intersects the x-axis.

In Chapter 4, we will use calculus to sketch and analyze graphs. At this stage, to sketch a graph by hand, we make a table of function values, plot the corresponding points (including any zeros), and connect them by a smooth curve.

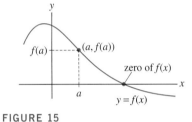

FIGURE 15

■ **EXAMPLE 3** Find the roots and sketch the graph of $f(x) = x^3 - 2x$.

Solution First, we solve $x^3 - 2x = x(x^2 - 2) = 0$. The roots of $f(x)$ are $x = 0$ and $x = \pm\sqrt{2}$. To sketch the graph, we plot the roots and a few values listed in Table 1 and join them by a curve (Figure 16). ■

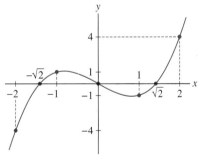

FIGURE 16 Graph of $f(x) = x^3 - 2x$.

TABLE 1

x	$x^3 - 2x$
-2	-4
-1	1
0	0
1	-1
2	4

Functions arising in applications are not always given by formulas. Data collected from observation or experiment define functions that can be displayed either graphically or by a table of values. Figure 17 and Table 2 display data collected by biologist Julian Huxley (1887–1975) in a study of the antler weight W of male red deer as a function of age t. We will see that many of the tools from calculus can be applied to functions constructed from data in this way.

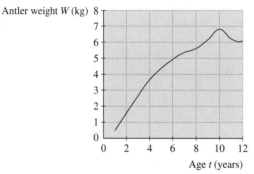

Antler weight W (kg)

t (year)	W (kg)	t (year)	W (kg)
1	0.48	7	5.34
2	1.59	8	5.62
3	2.66	9	6.18
4	3.68	10	6.81
5	4.35	11	6.21
6	4.92	12	6.1

TABLE 2

FIGURE 17 Average antler weight of a male red deer as a function of age.

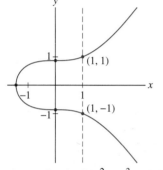

FIGURE 18 Graph of $4y^2 - x^3 = 3$. This graph fails the Vertical Line Test, so it is not the graph of a function.

We can graph any equation relating y and x. For example, to graph the equation $4y^2 - x^3 = 3$, we plot the pairs (x, y) satisfying the equation (Figure 18). This curve is not the graph of a function because there are two values of y for a given x-value such as $x = 1$. In general, a curve is the graph of a function exactly when it passes the **Vertical Line Test:** Every vertical line $x = a$ intersects the curve in at most one point.

We are often interested in whether a function is increasing or decreasing. Roughly speaking, a function $f(x)$ is increasing if its graph goes up as we move to the right and decreasing if its graph goes down [Figures 19(A) and (B)]. More precisely, we define the notion of increase/decrease on an interval:

- **Increasing** on (a, b) if $f(x_1) \le f(x_2)$ for all $x_1, x_2 \in (a, b)$ such that $x_1 \le x_2$
- **Decreasing** on (a, b) if $f(x_1) \ge f(x_2)$ for all $x_1, x_2 \in (a, b)$ such that $x_1 \le x_2$

The function in Figure 19(D) is neither increasing nor decreasing (for all x), but it is decreasing on the interval (a, b). We say that $f(x)$ is **strictly increasing** if $f(x_1) < f(x_2)$ for $x_1 < x_2$. Strictly decreasing functions are defined similarly. The function in Figure 19(C) is increasing but not strictly increasing.

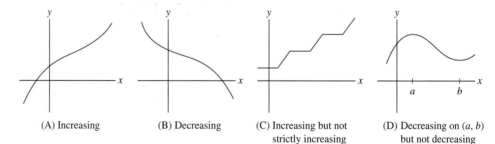

(A) Increasing

(B) Decreasing

(C) Increasing but not strictly increasing

(D) Decreasing on (a, b) but not decreasing everywhere

FIGURE 19

Another important property is **parity**, which refers to whether a function is even or odd:

- $f(x)$ is **even** if $f(-x) = f(x)$
- $f(x)$ is **odd** if $f(-x) = -f(x)$

The graphs of functions with even or odd parity have a special symmetry. The graph of an even function is symmetric with respect to the y-axis (we also say "symmetric *about* the y-axis"). This means that if $P = (a, b)$ lies on the graph, then so does $Q = (-a, b)$

[Figure 20(A)]. The graph of an odd function is symmetric with respect to the origin, which means that if $P = (a, b)$ lies on the graph, then so does $Q = (-a, -b)$ [Figure 20(B)]. A function may be neither even nor odd [Figure 20(C)].

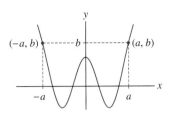
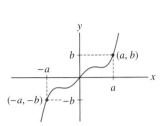
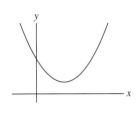

(A) Even function: $f(-x) = f(x)$
Graph is symmetric
about the y-axis.

(B) Odd function: $f(-x) = -f(x)$
Graph is symmetric
about the origin.

(C) Neither even nor odd

FIGURE 20

■ **EXAMPLE 4** Determine whether the function is even, odd, or neither.

(a) $f(x) = x^4$ **(b)** $g(x) = x^{-1}$ **(c)** $h(x) = x^2 + x$

Solution

(a) $f(-x) = (-x)^4 = x^4$. Thus, $f(x) = f(-x)$ and $f(x)$ is even.

(b) $g(-x) = (-x)^{-1} = -x^{-1}$. Thus, $g(-x) = -g(x)$, and $g(x)$ is odd.

(c) $h(-x) = (-x)^2 + (-x) = x^2 - x$. We see that $h(-x)$ is not equal to $h(x) = x^2 + x$ or $-h(x) = -x^2 - x$. Therefore, $h(x)$ is neither even nor odd. ■

■ **EXAMPLE 5** Graph Sketching Using Symmetry Sketch the graph of $f(x) = \dfrac{1}{x^2 + 1}$.

Solution The function $f(x) = \dfrac{1}{x^2 + 1}$ is positive [$f(x) > 0$ for all x] and it is even since $f(x) = f(-x)$. Therefore, the graph of $f(x)$ lies above the x-axis and is symmetric with respect to the y-axis. Furthermore, $f(x)$ is decreasing for $x \geq 0$ (since a larger value of x makes for a larger denominator). We use this information and a short table of values (Table 3) to sketch the graph (Figure 21). Note that the graph approaches the x-axis as we move to the right or left since $f(x)$ is small when $|x|$ is large. ■

Two important ways of modifying a graph are **translation** (or **shifting**) and **scaling**. Translation consists of moving the graph horizontally or vertically:

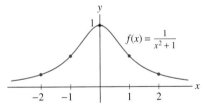

FIGURE 21

TABLE 3

x	$\dfrac{1}{x^2 + 1}$
0	1
± 1	$\dfrac{1}{2}$
± 2	$\dfrac{1}{5}$

DEFINITION Translation (Shifting)

- **Vertical translation** $y = f(x) + c$: This shifts the graph of $f(x)$ *vertically* c units. If $c < 0$, the result is a downward shift.
- **Horizontal translation** $y = f(x + c)$: This shifts the graph of $f(x)$ *horizontally* c units *to the right* if $c < 0$ and c units to the left if $c > 0$.

Figure 22 shows the effect of translating the graph of $f(x) = \dfrac{1}{x^2 + 1}$ vertically and horizontally.

Remember that $f(x) + c$ and $f(x + c)$ are different. The graph of $y = f(x) + c$ is a vertical translation and $y = f(x + c)$ a horizontal translation of the graph of $y = f(x)$.

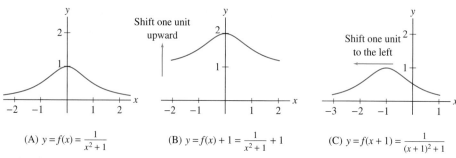

(A) $y = f(x) = \dfrac{1}{x^2 + 1}$ (B) $y = f(x) + 1 = \dfrac{1}{x^2 + 1} + 1$ (C) $y = f(x + 1) = \dfrac{1}{(x+1)^2 + 1}$

FIGURE 22

■ **EXAMPLE 6**   Figure 23(A) is the graph of $f(x) = x^2$, and Figure 23(B) is a horizontal and vertical shift of (A). What is the equation of graph (B)?

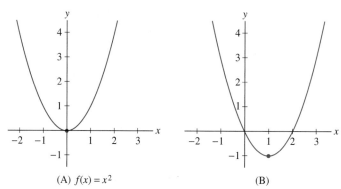

(A) $f(x) = x^2$ (B)

FIGURE 23

Solution   Graph (B) is obtained by shifting graph (A) one unit to the right and one unit down. We can see this by observing that the point $(0, 0)$ on the graph of $f(x)$ is shifted to $(1, -1)$. Therefore, (B) is the graph of $g(x) = (x - 1)^2 - 1$.   ■

Scaling (also called dilation) consists of compressing or expanding the graph in the vertical or horizontal directions:

DEFINITION Scaling

- **Vertical scaling** $y = kf(x)$: If $k > 1$, the graph is expanded vertically by the factor k. If $0 < k < 1$, the graph is compressed vertically. When the scale factor k is negative ($k < 0$), the graph is also reflected across the x-axis (Figure 24).
- **Horizontal scaling** $y = f(kx)$: If $k > 1$, the graph is compressed in the horizontal direction. If $0 < k < 1$, the graph is expanded. If $k < 0$, then the graph is also reflected across the y-axis.

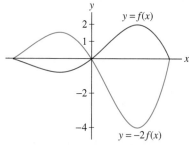

FIGURE 24 Negative vertical scale factor $k = -2$.

We refer to the vertical size of a graph as its **amplitude**. Thus, vertical scaling changes the amplitude by the factor $|k|$.

■ **EXAMPLE 7**   Sketch the graphs of $f(x) = \sin(\pi x)$ and its two dilates $f(3x)$ and $3f(x)$.

Solution   The graph of $f(x) = \sin(\pi x)$ is a sine curve with period 2 [it completes one cycle over every interval of length 2—see Figure 25(A)].

Remember that $kf(x)$ and $f(kx)$ are different. The graph of $y = kf(x)$ is a vertical scaling and $y = f(kx)$ a horizontal scaling of the graph of $y = f(x)$.

- The graph of $y = f(3x) = \sin(3\pi x)$ is a compressed version of $y = f(x)$ [Figure 25(B)]. It completes three cycles over intervals of length 2 instead of just one cycle.
- The graph of $y = 3f(x) = 3\sin(\pi x)$ differs from $y = f(x)$ only in amplitude: It is expanded in the vertical direction by a factor of 3 [Figure 25(C)]. ∎

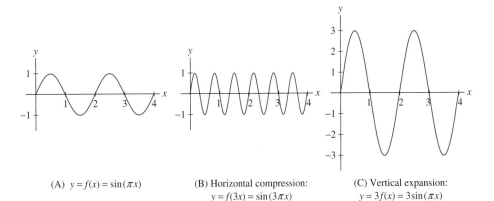

(A) $y = f(x) = \sin(\pi x)$ (B) Horizontal compression: $y = f(3x) = \sin(3\pi x)$ (C) Vertical expansion: $y = 3f(x) = 3\sin(\pi x)$

FIGURE 25 Horizontal and vertical scaling of $f(x) = \sin(\pi x)$.

1.1 SUMMARY

- Absolute value: $|a| = \begin{cases} a & \text{if } a \geq 0 \\ -a & \text{if } a < 0 \end{cases}$

- Triangle inequality: $|a + b| \leq |a| + |b|$
- There are four types of intervals with endpoints a and b:

$$(a, b), \qquad [a, b], \qquad [a, b), \qquad (a, b]$$

- We can express open and closed intervals using inequalities:

$$(a, b) = \{x : |x - c| < r\}, \qquad [a, b] = \{x : |x - c| \leq r\}$$

where $c = \frac{1}{2}(a + b)$ is the midpoint and $r = \frac{1}{2}(b - a)$ is the radius.

- Distance d between (x_1, y_1) and (x_2, y_2): $d = \sqrt{(x_2 - x_1)^2 + (y_2 - y_1)^2}$.

- Equation of circle of radius r with center (a, b): $(x - a)^2 + (y - b)^2 = r^2$.
- A zero or root of a function $f(x)$ is a number c such that $f(c) = 0$.
- Vertical Line Test: A curve in the plane is the graph of a function if and only if each vertical line $x = a$ intersects the curve in at most one point.
- Even function: $f(-x) = f(x)$ (graph is symmetric about the y-axis).
- Odd function: $f(-x) = -f(x)$ (graph is symmetric about the origin).
- Four ways to transform the graph of $f(x)$:

$f(x) + c$	Shifts graph vertically c units
$f(x + c)$	Shifts graph horizontally c units (to the left if $c > 0$)
$kf(x)$	Scales graph vertically by factor k; if $k < 0$, graph is reflected across x-axis
$f(kx)$	Scales graph horizontally by factor k (compresses if $k > 1$); if $k < 0$, graph is reflected across y-axis

1.1 EXERCISES

Preliminary Questions

1. Give an example of numbers a and b such that $a < b$ and $|a| > |b|$.

2. Which numbers satisfy $|a| = a$? Which satisfy $|a| = -a$? What about $|-a| = a$?

3. Give an example of numbers a and b such that $|a + b| < |a| + |b|$.

4. What are the coordinates of the point lying at the intersection of the lines $x = 9$ and $y = -4$?

5. In which quadrant do the following points lie?

(a) $(1, 4)$ **(b)** $(-3, 2)$

(c) $(4, -3)$ **(d)** $(-4, -1)$

6. What is the radius of the circle with equation $(x - 9)^2 + (y - 9)^2 = 9$?

7. The equation $f(x) = 5$ has a solution if (choose one):

(a) 5 belongs to the domain of f.

(b) 5 belongs to the range of f.

8. What kind of symmetry does the graph have if $f(-x) = -f(x)$?

Exercises

1. Use a calculator to find a rational number r such that $|r - \pi^2| < 10^{-4}$.

2. Let $a = -3$ and $b = 2$. Which of the following inequalities are true?

(a) $a < b$ **(b)** $|a| < |b|$ **(c)** $ab > 0$

(d) $3a < 3b$ **(e)** $-4a < -4b$ **(f)** $\dfrac{1}{a} < \dfrac{1}{b}$

In Exercises 3–8, express the interval in terms of an inequality involving absolute value.

3. $[-2, 2]$ **4.** $(-4, 4)$ **5.** $(0, 4)$

6. $[-4, 0]$ **7.** $[1, 5]$ **8.** $(-2, 8)$

In Exercises 9–12, write the inequality in the form $a < x < b$ for some numbers a, b.

9. $|x| < 8$ **10.** $|x - 12| < 8$

11. $|2x + 1| < 5$ **12.** $|3x - 4| < 2$

In Exercises 13–18, express the set of numbers x satisfying the given condition as an interval.

13. $|x| < 4$ **14.** $|x| \le 9$

15. $|x - 4| < 2$ **16.** $|x + 7| < 2$

17. $|4x - 1| \le 8$ **18.** $|3x + 5| < 1$

In Exercises 19–22, describe the set as a union of finite or infinite intervals.

19. $\{x : |x - 4| > 2\}$ **20.** $\{x : |2x + 4| > 3\}$

21. $\{x : |x^2 - 1| > 2\}$ **22.** $\{x : |x^2 + 2x| > 2\}$

23. Match the inequalities (a)–(f) with the corresponding statements (i)–(vi).

(a) $a > 3$ **(b)** $\left|a - 5\right| < \dfrac{1}{3}$

(c) $\left|a - \dfrac{1}{3}\right| < 5$ **(d)** $|a| > 5$

(e) $|a - 4| < 3$ **(f)** $1 < a < 5$

(i) a lies to the right of 3.

(ii) a lies between 1 and 7.

(iii) The distance from a to 5 is less than $\dfrac{1}{3}$.

(iv) The distance from a to 3 is at most 2.

(v) a is less than 5 units from $\dfrac{1}{3}$.

(vi) a lies either to the left of -5 or to the right of 5.

24. Describe the set $\left\{x : \dfrac{x}{x + 1} < 0\right\}$ as an interval.

25. Show that if $a > b$, then $b^{-1} > a^{-1}$, provided that a and b have the same sign. What happens if $a > 0$ and $b < 0$?

26. Which x satisfy $|x - 3| < 2$ and $|x - 5| < 1$?

27. Show that if $|a - 5| < \dfrac{1}{2}$ and $|b - 8| < \dfrac{1}{2}$, then $|(a + b) - 13| < 1$. *Hint:* Use the triangle inequality.

28. Suppose that $|a| \le 2$ and $|b| \le 3$.

(a) What is the largest possible value of $|a + b|$?

(b) What is the largest possible value of $|a + b|$ if a and b have opposite signs?

29. Suppose that $|x - 4| \le 1$.

(a) What is the maximum possible value of $|x + 4|$?

(b) Show that $|x^2 - 16| \le 9$.

30. Prove that $|x| - |y| \le |x - y|$. *Hint:* Apply the triangle inequality to y and $x - y$.

31. Express $r_1 = 0.\overline{27}$ as a fraction. *Hint:* $100r_1 - r_1$ is an integer. Then express $r_2 = 0.2666\ldots$ as a fraction.

32. Represent 1/7 and 4/27 as repeating decimals.

33. The text states the following: *If the decimal expansions of two real numbers a and b agree to k places, then the distance* $|a - b| \leq 10^{-k}$. Show that the converse is not true, that is, for any k we can find real numbers a and b whose decimal expansions *do not agree at all* but $|a - b| \leq 10^{-k}$.

34. Plot each pair of points and compute the distance between them:

(a) $(1, 4)$ and $(3, 2)$
(b) $(2, 1)$ and $(2, 4)$
(c) $(0, 0)$ and $(-2, 3)$
(d) $(-3, -3)$ and $(-2, 3)$

35. Determine the equation of the circle with center $(2, 4)$ and radius 3.

36. Determine the equation of the circle with center $(2, 4)$ passing through $(1, -1)$.

37. Find all points with integer coordinates located at a distance 5 from the origin. Then find all points with integer coordinates located at a distance 5 from $(2, 3)$.

38. Determine the domain and range of the function

$$f : \{r, s, t, u\} \rightarrow \{A, B, C, D, E\}$$

defined by $f(r) = A$, $f(s) = B$, $f(t) = B$, $f(u) = E$.

39. Give an example of a function whose domain D has three elements and range R has two elements. Does a function exist whose domain D has two elements and range has three elements?

In Exercises 40–48, find the domain and range of the function.

40. $f(x) = -x$
41. $g(t) = t^4$

42. $f(x) = x^3$
43. $g(t) = \sqrt{2 - t}$

44. $f(x) = |x|$
45. $h(s) = \dfrac{1}{s}$

46. $f(x) = \dfrac{1}{x^2}$
47. $g(t) = \sqrt{2 + t^2}$

48. $g(t) = \cos \dfrac{1}{t}$

In Exercises 49–52, find the interval on which the function is increasing.

49. $f(x) = |x + 1|$
50. $f(x) = x^3$

51. $f(x) = x^4$
52. $f(x) = \dfrac{1}{x^2 + 1}$

In Exercises 53–58, find the zeros of the function and sketch its graph by plotting points. Use symmetry and increase/decrease information where appropriate.

53. $f(x) = x^2 - 4$
54. $f(x) = 2x^2 - 4$

55. $f(x) = x^3 - 4x$
56. $f(x) = x^3$

57. $f(x) = 2 - x^3$
58. $f(x) = \dfrac{1}{(x - 1)^2 + 1}$

59. Which of the curves in Figure 26 is the graph of a function?

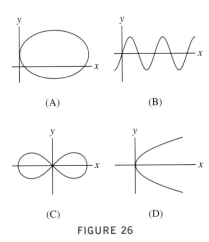

(A)   (B)

(C)   (D)

FIGURE 26

60. State whether the function is increasing, decreasing, or neither.

(a) Surface area of a sphere as a function of its radius
(b) Temperature at a point on the equator as a function of time
(c) Price of an airline ticket as a function of the price of oil
(d) Pressure of the gas in a piston as a function of volume

In Exercises 61–66, let $f(x)$ be the function whose graph is shown in Figure 27.

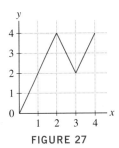

FIGURE 27

61. What are the domain and range of $f(x)$?

62. Sketch the graphs of $f(x + 2)$ and $f(x) + 2$.

63. Sketch the graphs of $f(2x)$, $f\left(\frac{1}{2}x\right)$, and $2f(x)$.

64. Sketch the graphs of $f(-x)$ and $-f(-x)$.

65. Extend the graph of $f(x)$ to $[-4, 4]$ so that it is an even function.

66. Extend the graph of $f(x)$ to $[-4, 4]$ so that it is an odd function.

67. Suppose that $f(x)$ has domain $[4, 8]$ and range $[2, 6]$. What are the domain and range of:

(a) $f(x) + 3$
(b) $f(x + 3)$
(c) $f(3x)$
(d) $3f(x)$

68. Let $f(x) = x^2$. Sketch the graphs of the following functions over $[-2, 2]$:

(a) $f(x + 1)$ (b) $f(x) + 1$

(c) $f(5x)$ (d) $5f(x)$

69. Suppose that the graph of $f(x) = \sin x$ is compressed horizontally by a factor of 2 and then shifted 5 units to the right.

(a) What is the equation for the new graph?

(b) What is the equation if you first shift by 5 and then compress by 2?

(c) 〔GU〕 Verify your answers by plotting your equations.

70. Figure 28 shows the graph of $f(x) = |x| + 1$. Match the functions (a)–(e) with their graphs (i)–(v).

(a) $f(x - 1)$ (b) $-f(x)$ (c) $-f(x) + 2$

(d) $f(x - 1) - 2$ (e) $f(x + 1)$

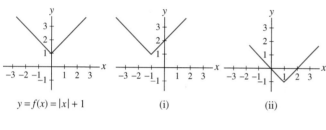

$y = f(x) = |x| + 1$ (i)   (ii)

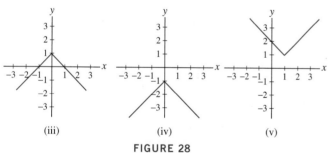

(iii)   (iv)   (v)

FIGURE 28

71. Sketch the graph of $f(2x)$ and $f(\frac{1}{2}x)$, where $f(x) = |x| + 1$ (Figure 28).

72. Find the function $f(x)$ whose graph is obtained by shifting the parabola $y = x^2$ three units to the right and four units down as in Figure 29.

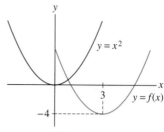

FIGURE 29

73. Define $f(x)$ to be the larger of x and $2 - x$. Sketch the graph of $f(x)$. What are its domain and range? Express $f(x)$ in terms of the absolute value function.

74. For each curve in Figure 30, state whether it is symmetrical with respect to the y-axis, the origin, both, or neither.

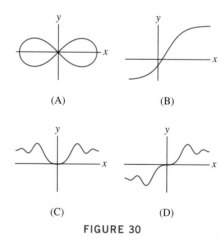

(A)   (B)

(C)   (D)

FIGURE 30

75. Show that the sum of two even functions is even and the sum of two odd functions is odd.

76. Suppose that $f(x)$ and $g(x)$ are both odd. Which of the following functions are even? Which are odd?

(a) $f(x)g(x)$ (b) $f(x)^3$

(c) $f(x) - g(x)$ (d) $\dfrac{f(x)}{g(x)}$

77. Give an example of a curve that is symmetrical with respect to both the y-axis and the origin. Can the graph of a function have both symmetries? *Hint:* Prove algebraically that $f(x) = 0$ is the only such function.

Further Insights and Challenges

78. Prove the triangle inequality by adding the two inequalities

$$-|a| \le a \le |a|, \qquad -|b| \le b \le |b|$$

79. Show that if $r = a/b$ is a fraction in lowest terms, then r has a *finite* decimal expansion if and only if $b = 2^n 5^m$ for some $n, m \ge 0$. *Hint:* Observe that r has a finite decimal expansion when $10^N r$ is an integer for some $N \ge 0$ (and hence b divides 10^N).

80. Let $p = p_1 \ldots p_s$ be an integer with digits $p_1, \ldots, p_s$. Show that

$$\frac{p}{10^s - 1} = 0.\overline{p_1 \ldots p_s}$$

Use this to find the decimal expansion of $r = \dfrac{2}{11}$. Note that

$$r = \frac{2}{11} = \frac{18}{10^2 - 1}$$

81. A function $f(x)$ is symmetrical with respect to the vertical line $x = a$ if $f(a - x) = f(a + x)$.

(a) Draw the graph of a function that is symmetrical with respect to $x = 2$.

(b) Show that if $f(x)$ is symmetrical with respect to $x = a$, then $g(x) = f(x + a)$ is even.

82. Formulate a condition for $f(x)$ to be symmetrical with respect to the point $(a, 0)$ on the x-axis.

1.2 Linear and Quadratic Functions

Linear functions are the simplest of all functions and their graphs (lines) are the simplest of all curves. However, linear functions and lines play an enormously important role in calculus. For this reason, it is important to be thoroughly familiar with the basic properties of linear functions and the different ways of writing the equation of a line.

Let's recall that a **linear function** is a function of the form

$$f(x) = mx + b \quad (m \text{ and } b \text{ constants})$$

The graph of $f(x)$ is a line of slope m, and since $f(0) = b$, the graph intersects the y-axis at the point $(0, b)$ (Figure 1). The number b is called the y-intercept, and the equation $y = mx + b$ for the line is said to be in **slope-intercept form**.

We use the symbols Δx and Δy to denote the *change* (or *increment*) in x and $y = f(x)$ over an interval $[x_1, x_2]$ (Figure 1):

$$\Delta x = x_2 - x_1, \qquad \Delta y = y_2 - y_1 = f(x_2) - f(x_1)$$

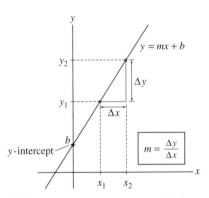

FIGURE 1 The slope m is the ratio "rise over run."

The slope m of a line is the ratio

$$m = \frac{\Delta y}{\Delta x} = \frac{\text{vertical change}}{\text{horizontal change}} = \frac{\text{rise}}{\text{run}}$$

This follows from the formula $y = mx + b$:

$$\frac{\Delta y}{\Delta x} = \frac{y_2 - y_1}{x_2 - x_1} = \frac{(mx_2 + b) - (mx_1 + b)}{x_2 - x_1} = \frac{m(x_2 - x_1)}{x_2 - x_1} = m$$

The slope m measures the *rate of change* of y with respect to x. In fact, by writing

$$\Delta y = m \Delta x$$

we see that a one-unit increase in x (i.e., $\Delta x = 1$) produces an m-unit change Δy in y. For example, if $m = 5$, then y increases by five units per unit increase in x. The rate-of-change interpretation of the slope is fundamental in calculus. We discuss it in greater detail in Section 2.1.

Graphically, the slope m measures the steepness of the line $y = mx + b$. Figure 2(A) shows lines through a point of varying slope m. Note the following properties:

- **Steepness**: The larger the absolute value $|m|$, the steeper the line.
- **Negative slope**: If $m < 0$, the line slants downward from left to right.
- $f(x) = mx + b$ is strictly increasing if $m > 0$ and strictly decreasing if $m < 0$.
- The **horizontal line** $y = b$ has slope $m = 0$ [Figure 2(B)].
- A **vertical line** has equation $x = c$, where c is a constant. Informally speaking, the slope of a vertical line is "infinite," so it is not possible to write the equation of a vertical line in slope-intercept form $y = mx + b$.

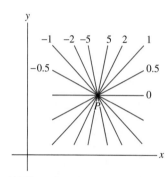
(A) Lines of varying slopes through P

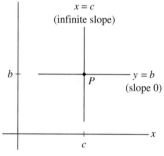
(B) Horizontal and vertical lines through P

FIGURE 2

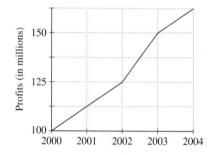

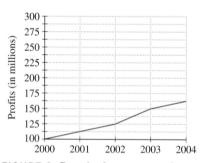

FIGURE 3 Growth of company profits.

CAUTION: Graphs are often plotted using different scales for the x- and y-axes. This is necessary to keep the sizes of graphs within reasonable bounds. However, when the scales are different, lines do not appear with their true slopes.

Scale is especially important in applications because the steepness of a graph depends on the choice of units for the x- and y-axes. We can create very different *subjective* impressions by changing the scale. Figure 3 shows the growth of company profits over a four-year period. The two plots convey the same information, but the upper plot makes the growth look more dramatic.

Next, we recall the relation between the slopes of parallel and perpendicular lines (Figure 4):

- Lines of slopes m_1 and m_2 are **parallel** if and only if $m_1 = m_2$.
- Lines of slopes m_1 and m_2 are **perpendicular** if and only if $m_1 = -1/m_2$ (or $m_1 m_2 = -1$).

FIGURE 4 Parallel and perpendicular lines.

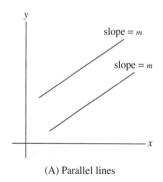
(A) Parallel lines

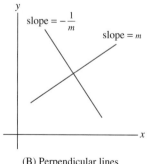
(B) Perpendicular lines

As mentioned above, it is important to be familiar with the standard ways of writing the equation of a line. The general **linear equation** is

$$ax + by = c \qquad \boxed{1}$$

where a and b are not *both* zero. For $b = 0$, this gives the vertical line $ax = c$. When $b \neq 0$, we can rewrite (1) in slope-intercept form. For example, $-6x + 2y = 3$ can be rewritten as $y = 3x + \frac{3}{2}$.

Two other forms we will use frequently are **point-slope** and **point-point** form. Given a point $P = (a, b)$ and a slope m, the equation of the line through P with slope m is $y - b = m(x - a)$. Similarly, the line through two distinct points $P = (a_1, b_1)$ and $Q = (a_2, b_2)$ has slope (Figure 5)

$$m = \frac{b_2 - b_1}{a_2 - a_1}$$

Therefore, we can write its equation as $y - b_1 = m(x - a_1)$.

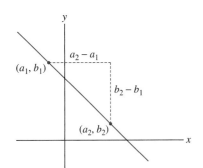

FIGURE 5 Slope of the line between $P = (a_1, b_1)$ and $Q = (a_2, b_2)$ is $m = \dfrac{b_2 - b_1}{a_2 - a_1}$.

Equations for Lines

1. Point-slope form: The line through $P = (a, b)$ with slope m has equation

$$y - b = m(x - a)$$

2. Point-point form: The line through $P = (a_1, b_1)$ and $Q = (a_2, b_2)$ has equation

$$y - b_1 = m(x - a_1) \qquad \text{where } m = \frac{b_2 - b_1}{a_2 - a_1}$$

■ **EXAMPLE 1** Line of Given Slope Through a Given Point Find the equation of the line through $(9, 2)$ with slope $-\frac{2}{3}$.

Solution We can write the equation directly in point-slope form:

$$y - 2 = -\frac{2}{3}(x - 9)$$

In slope-intercept form: $y = -\frac{2}{3}(x - 9) + 2$ or $y = -\frac{2}{3}x + 8$. See Figure 6. ■

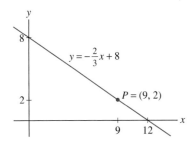

FIGURE 6 Line through $P = (9, 2)$ with slope $m = -\frac{2}{3}$.

■ **EXAMPLE 2** Line Through Two Points Find the equation of the line $\mathcal{L}$ through $(2, 1)$ and $(9, 5)$.

Solution The line $\mathcal{L}$ has slope

$$m = \frac{5 - 1}{9 - 2} = \frac{4}{7}$$

Since $(9, 5)$ lies on $\mathcal{L}$, its equation in point-slope form is $y - 5 = \frac{4}{7}(x - 9)$. ■

CONCEPTUAL INSIGHT We can define the increments Δx and Δy over an interval $[x_1, x_2]$ for any function $f(x)$ (linear or not), but in general, the ratio $\Delta y / \Delta x$ depends on the interval. The characteristic property of a linear function $f(x) = mx + b$ is that

$\Delta y / \Delta x$ has the same value m for every interval (Figure 7). In other words, y has a constant rate of change with respect to x. We can use this property to test if two quantities are related by a linear equation (see Example 3).

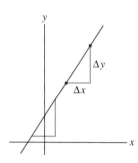

Graph of a linear function. The ratio $\Delta y/\Delta x$ is the same over all intervals.

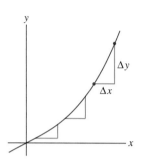

Graph of a nonlinear function. The ratio $\Delta y/\Delta x$ changes, depending on the interval.

FIGURE 7

TABLE 1

Temperature ($^\circ$F)	Pressure (lb/in.^{2})
70	187.42
75	189
85	192.16
100	196.9
110	200.06

Real experimental data are unlikely to reveal perfect linearity, even if the data points do essentially lie on a line. The method of "linear regression" is used to find the linear function that best fits the data.

■ **EXAMPLE 3** Testing for a Linear Relationship   Table 1 gives the pressure reading P of a gas at different temperatures T. Do the data suggest a linear relation between P and T?

Solution We calculate $\dfrac{\Delta P}{\Delta T}$ at successive data points and check whether this ratio is constant:

(T_1, P_1)	(T_2, P_2)	$\dfrac{\Delta P}{\Delta T}$
(70, 187.42)	(75, 189)	$\dfrac{189 - 187.42}{75 - 70} = 0.316$
(75, 189)	(85, 192.16)	$\dfrac{192.16 - 189}{85 - 75} = 0.316$
(85, 192.16)	(100, 196.9)	$\dfrac{196.9 - 192.16}{100 - 85} = 0.316$
(100, 196.9)	(110, 200.06)	$\dfrac{200.06 - 196.9}{110 - 100} = 0.316$

Because $\Delta P / \Delta T$ has the constant value 0.316, the data points lie on a line with slope $m = 0.316$ (this is confirmed in the plot in Figure 8). The line passes through the first data point (70, 187.42), so its equation in point-slope form is

$$P - 187.42 = 0.316(T - 70)$$ ■

A **quadratic function** is a function defined by a quadratic polynomial

$$f(x) = ax^2 + bx + c \quad (a, b, c, \text{ constants with } a \neq 0)$$

The graph of $f(x)$ is a **parabola** (Figure 9). The parabola opens upward if the leading coefficient a is positive and downward if a is negative. The **discriminant** of $f(x)$ is the

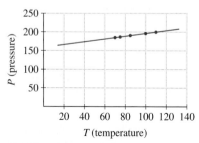

FIGURE 8 Line through pressure-temperature data points.

quantity

$$D = b^2 - 4ac$$

The roots of $f(x)$ are given by the **quadratic formula** (see Exercise 56):

$$\text{Roots of } f(x) = \frac{-b \pm \sqrt{b^2 - 4ac}}{2a} = \frac{-b \pm \sqrt{D}}{2a}$$

The sign of D determines whether or not $f(x)$ has real roots (Figure 9). If $D > 0$, then $f(x)$ has two real roots, and if $D = 0$, it has one real root (a "double root"). If $D < 0$, then $\sqrt{D}$ is imaginary and $f(x)$ has no real roots.

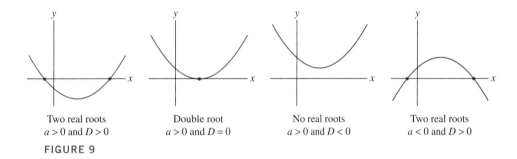

Two real roots
$a > 0$ and $D > 0$

Double root
$a > 0$ and $D = 0$

No real roots
$a > 0$ and $D < 0$

Two real roots
$a < 0$ and $D > 0$

FIGURE 9

When $f(x)$ has two real roots r_1 and r_2, then $f(x)$ factors as

$$f(x) = a(x - r_1)(x - r_2)$$

For example, $f(x) = 2x^2 - 3x + 1$ has discriminant $D = b^2 - 4ac = 9 - 8 = 1 > 0$, and by the quadratic formula, its roots are $(3 \pm 1)/4$ or 1 and $\frac{1}{2}$. Therefore,

$$f(x) = 2x^2 - 3x + 1 = 2(x - 1)\left(x - \frac{1}{2}\right)$$

The technique of **completing the square** consists of writing a quadratic polynomial as a multiple of a square plus a constant:

$$ax^2 + bx + c = a\underbrace{\left(x + \frac{b}{2a}\right)^2}_{\text{Square term}} + \underbrace{\frac{4ac - b^2}{4a}}_{\text{Constant}}$$

$\boxed{2}$

It is not necessary to memorize this formula, but you should know how to carry out the process of completing the square.

Cuneiform texts written on clay tablets show that the method of completing the square was known to ancient Babylonian mathematicians who lived some 4,000 years ago.

■ **EXAMPLE 4** Completing the Square   Complete the square for the quadratic polynomial $4x^2 - 12x + 3$.

Solution First factor out the leading coefficient:

$$4x^2 - 12x + 3 = 4\left(x^2 - 3x + \frac{3}{4}\right)$$

Then complete the square for the term $x^2 - 3x$:

$$x^2 + bx = \left(x + \frac{b}{2}\right)^2 - \frac{b^2}{4}, \qquad x^2 - 3x = \left(x - \frac{3}{2}\right)^2 - \frac{9}{4}$$

Therefore,

$$4x^2 - 12x + 3 = 4\left(\left(x - \frac{3}{2}\right)^2 - \frac{9}{4} + \frac{3}{4}\right) = 4\left(x - \frac{3}{2}\right)^2 - 6 \qquad \blacksquare$$

The method of completing the square can be used to find the minimum or maximum value of a quadratic function.

■ **EXAMPLE 5**   Finding the Minimum of a Quadratic Function   Complete the square and find the minimum value of the quadratic function $f(x) = x^2 - 4x + 9$.

Solution   We have

This term is ≥ 0

$$f(x) = x^2 - 4x + 9 = (x - 2)^2 - 4 + 9 = \overbrace{(x - 2)^2}^{} + 5$$

Thus, $f(x) \geq 5$ for all x and the minimum value of $f(x)$ is $f(2) = 5$ (Figure 10).   ■

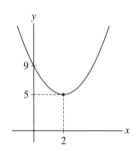

FIGURE 10 Graph of $f(x) = x^2 - 4x + 9$.

1.2 SUMMARY

- A function of the form $f(x) = mx + b$ is called a linear function.
- The general equation of a line is $ax + by = c$. The line $y = c$ is horizontal and $x = c$ is vertical.
- There are three convenient ways of writing the equation of a nonvertical line:

  - Slope-intercept form: $y = mx + b$ (slope m and y-intercept b)
  - Point-slope form: $y - b = m(x - a)$ [slope m, passes through (a, b)]
  - Point-point form: The line through two points $P = (a_1, b_1)$ and $Q = (a_2, b_2)$ has slope $m = \dfrac{b_2 - b_1}{a_2 - a_1}$ and equation $y - b_1 = m(x - a_1)$.

- Two lines of slopes m_1 and m_2 are parallel if and only if $m_1 = m_2$, and they are perpendicular if and only if $m_1 = -1/m_2$.
- The roots of a quadratic function $f(x) = ax^2 + bx + c$ are given by the quadratic formula $x = (-b \pm \sqrt{D})/2a$, where $D = b^2 - 4ac$ is the discriminant. Furthermore, $f(x)$ has distinct real roots if $D > 0$, a double root if $D = 0$, and no real roots if $D < 0$.
- The technique of completing the square consists of writing a quadratic function as a multiple of a square plus a constant.

1.2 EXERCISES

Preliminary Questions

1. What is the slope of the line $y = -4x - 9$?

2. Are the lines $y = 2x + 1$ and $y = -2x - 4$ perpendicular?

3. When is the line $ax + by = c$ parallel to the y-axis? To the x-axis?

4. Suppose $y = 3x + 2$. What is Δy if x increases by 3?

5. What is the minimum of $f(x) = (x + 3)^2 - 4$?

6. What is the result of completing the square for $f(x) = x^2 + 1$?

Exercises

In Exercises 1–4, find the slope, the y-intercept, and the x-intercept of the line with the given equation.

1. $y = 3x + 12$

2. $y = 4 - x$

3. $4x + 9y = 3$

4. $y - 3 = \frac{1}{2}(x - 6)$

In Exercises 5–8, find the slope of the line.

5. $y = 3x + 2$

6. $y = 3(x - 9) + 2$

7. $3x + 4y = 12$

8. $3x + 4y = -8$

In Exercises 9–20, find the equation of the line with the given description.

9. Slope 3, y-intercept 8

10. Slope -2, y-intercept 3

11. Slope 3, passes through $(7, 9)$

12. Slope -5, passes through $(0, 0)$

13. Horizontal, passes through $(0, -2)$

14. Passes through $(-1, 4)$ and $(2, 7)$

15. Parallel to $y = 3x - 4$, passes through $(1, 1)$

16. Passes through $(1, 4)$ and $(12, -3)$

17. Perpendicular to $3x + 5y = 9$, passes through $(2, 3)$

18. Vertical, passes through $(-4, 9)$

19. Horizontal, passes through $(8, 4)$

20. Slope 3, x-intercept 6

21. Find the equation of the perpendicular bisector of the segment joining $(1, 2)$ and $(5, 4)$ (Figure 11). *Hint:* The midpoint Q of the segment joining (a, b) and (c, d) is $\left(\dfrac{a + c}{2}, \dfrac{b + d}{2} \right)$.

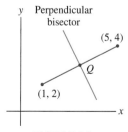

FIGURE 11

22. Intercept-intercept Form   Show that if $a, b \neq 0$, then the line with x-intercept $x = a$ and y-intercept $y = b$ has equation (Figure 12)

$$\frac{x}{a} + \frac{y}{b} = 1$$

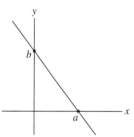

FIGURE 12 Line with equation $\dfrac{x}{a} + \dfrac{y}{b} = 1$.

23. Find the equation of the line with x-intercept $x = 4$ and y-intercept $y = 3$.

24. A line of slope $m = 2$ passes through $(1, 4)$. Find y such that $(3, y)$ lies on the line.

25. Determine whether there exists a constant c such that the line $x + cy = 1$

(a) Has slope 4

(b) Passes through $(3, 1)$

(c) Is horizontal

(d) Is vertical

26. Assume that the number N of concert tickets which can be sold at a price of P dollars per ticket is a linear function $N(P)$ for $10 \leq P \leq 40$. Determine $N(P)$ (called the demand function) if $N(10) = 500$ and $N(40) = 0$. What is the decrease ΔN in the number of tickets sold if the price is increased by $\Delta P = 5$ dollars?

27. Materials expand when heated. Consider a metal rod of length L_0 at temperature T_0. If the temperature is changed by an amount ΔT, then the rod's length changes by $\Delta L = \alpha L_0 \Delta T$, where α is the thermal expansion coefficient. For steel, $\alpha = 1.24 \times 10^{-5}\,°\text{C}^{-1}$.

(a) A steel rod has length $L_0 = 40$ cm at $T_0 = 40°\text{C}$. What is its length at $T = 90°\text{C}$?

(b) Find its length at $T = 50°\text{C}$ if its length at $T_0 = 100°\text{C}$ is 65 in.

(c) Express length L as a function of T if $L_0 = 65$ in. at $T_0 = 100°\text{C}$.

28. Do the points $(0.5, 1)$, $(1, 1.2)$, $(2, 2)$ lie on a line?

29. Find b such that $(2, -1)$, $(3, 2)$, and $(b, 5)$ lie on a line.

30. Find an expression for the velocity v as a linear function of t that matches the following data.

t (s)	0	2	4	6
v (m/s)	39.2	58.6	78	97.4

31. The period T of a pendulum is measured for pendulums of several different lengths L. Based on the following data, does T appear to be a linear function of L?

L (ft)	2	3	4	5
T (s)	1.57	1.92	2.22	2.48

32. Show that $f(x)$ is linear of slope m if and only if

$$f(x + h) - f(x) = mh \quad \text{(for all } x \text{ and } h)$$

33. Find the roots of the quadratic polynomials:

(a) $4x^2 - 3x - 1$
(b) $x^2 - 2x - 1$

In Exercises 34–41, complete the square and find the minimum or maximum value of the quadratic function.

34. $y = x^2 + 2x + 5$

35. $y = x^2 - 6x + 9$

36. $y = -9x^2 + x$

37. $y = x^2 + 6x + 2$

38. $y = 2x^2 - 4x - 7$

39. $y = -4x^2 + 3x + 8$

40. $y = 3x^2 + 12x - 5$

41. $y = 4x - 12x^2$

42. Sketch the graph of $y = x^2 - 6x + 8$ by plotting the roots and the minimum point.

43. Sketch the graph of $y = x^2 + 4x + 6$ by plotting the minimum point, the y-intercept, and one other point.

44. If the alleles A and B of the cystic fibrosis gene occur in a population with frequencies p and $1 - p$ (where p is a fraction between 0 and 1), then the frequency of heterozygous carriers (carriers with both alleles) is $2p(1 - p)$. Which value of p gives the largest frequency of heterozygous carriers?

45. For which values of c does $f(x) = x^2 + cx + 1$ have a double root? No real roots?

46. 🖊 Let $f(x)$ be a quadratic function and c a constant. Which is correct? Explain graphically.

(a) There is a unique value of c such that $y = f(x) - c$ has a double root.

(b) There is a unique value of c such that $y = f(x - c)$ has a double root.

47. Prove that $x + \dfrac{1}{x} \geq 2$ for all $x > 0$. *Hint:* Consider $(x^{1/2} - x^{-1/2})^2$.

48. Let $a, b > 0$. Show that the *geometric mean* $\sqrt{ab}$ is not larger than the *arithmetic mean* $\dfrac{a + b}{2}$. *Hint:* Use a variation of the hint given in Exercise 47.

49. If objects of weights x and w_1 are suspended from the balance in Figure 13(A), the cross-beam is horizontal if $bx = aw_1$. If the lengths a and b are known, we may use this equation to determine an unknown weight x by selecting w_1 so that the cross-beam is horizontal. If a and b are not known precisely, we might proceed as follows. First balance x by w_1 on the left as in (A). Then switch places and balance x by w_2 on the right as in (B). The average $\bar{x} = \frac{1}{2}(w_1 + w_2)$ gives an estimate for x. Show that $\bar{x}$ is greater than or equal to the true weight x.

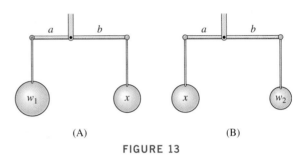

(A)   (B)

FIGURE 13

50. Find numbers x and y with sum 10 and product 24. *Hint:* Find a quadratic polynomial satisfied by x.

51. Find a pair of numbers whose sum and product are both equal to 8.

52. Show that the graph of the parabola $y = x^2$ consists of all points P such that $d_1 = d_2$, where d_1 is the distance from P to $(0, \frac{1}{4})$ and d_2 is the distance from P to the horizontal line $y = -\frac{1}{4}$ (Figure 14).

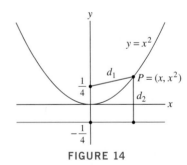

FIGURE 14

Further Insights and Challenges

53. Show that if $f(x)$ and $g(x)$ are linear, then so is $f(x) + g(x)$. Is the same true of $f(x)g(x)$?

54. Show that if $f(x)$ and $g(x)$ are linear functions such that $f(0) = g(0)$ and $f(1) = g(1)$, then $f(x) = g(x)$.

55. Show that the ratio $\Delta y / \Delta x$ for the function $f(x) = x^2$ over the interval $[x_1, x_2]$ is not a constant, but depends on the interval. Determine the exact dependence of $\Delta y / \Delta x$ on x_1 and x_2.

56. Use Eq. (2) to derive the quadratic formula for the roots of $ax^2 + bx + c = 0$.

57. Let $a, c \neq 0$. Show that the roots of $ax^2 + bx + c = 0$ and $cx^2 + bx + a = 0$ are reciprocals of each other.

58. Complete the square to show that the parabolas $y = ax^2 + bx + c$ and $y = ax^2$ have the same shape (show that the first parabola is congruent to the second by a vertical and horizontal translation).

59. Prove **Viète's Formulas**, which state that the quadratic polynomial with given numbers α and β as roots is $x^2 + bx + c$, where $b = -\alpha - \beta$ and $c = \alpha\beta$.

1.3 The Basic Classes of Functions

It would be impossible (and useless) to describe all possible functions $f(x)$. Since the values of a function can be assigned arbitrarily, a function chosen at random would likely be so complicated that we could neither graph it nor describe it in any reasonable way. However, calculus makes no attempt to deal with all possible functions. The techniques of calculus, powerful and general as they are, apply only to functions that are sufficiently "well behaved" (we will see what well-behaved means when we study the derivative in Chapter 3). Fortunately, such functions are adequate for a vast range of applications. In this text, we deal mostly with the following important and familiar classes of well-behaved functions:

polynomials   rational functions   algebraic functions

exponential functions   trigonometric functions

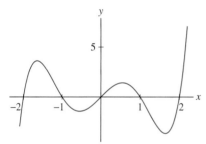

FIGURE 1 The polynomial
$y = x^5 - 5x^3 + 4x$.

- **Polynomials:** For any real number m, the function $f(x) = x^m$ is called the **power function** with exponent m. A polynomial is a sum of multiples of power functions with whole number exponents (Figure 1):

$$f(x) = x^5 - 5x^3 + 4x, \qquad g(t) = 7t^6 + t^3 - 3t - 1$$

Thus, the function $f(x) = x + x^{-1}$ is not a polynomial because it includes a power function x^{-1} with a negative exponent. The general polynomial in the variable x may be written

$$P(x) = a_n x^n + a_{n-1} x^{n-1} + \cdots + a_1 x + a_0$$

- The numbers $a_0, a_1, \ldots, a_n$ are called **coefficients**.
- The **degree** of $P(x)$ is n (assuming that $a_n \neq 0$).
- The coefficient a_n is called the **leading coefficient**.
- The domain of $P(x)$ is **R**.

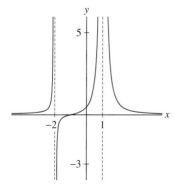

FIGURE 2 The rational function
$f(x) = \dfrac{x + 1}{x^3 - 3x + 2}$.

- **Rational functions:** A rational function is a *quotient* of two polynomials (Figure 2):

$$f(x) = \frac{P(x)}{Q(x)} \quad [P(x) \text{ and } Q(x) \text{ polynomials}]$$

Every polynomial is also a rational function [with $Q(x) = 1$]. The domain of a rational function $\dfrac{P(x)}{Q(x)}$ is the set of numbers x such that $Q(x) \neq 0$. For example,

$$f(x) = \frac{1}{x^2} \qquad \text{domain } \{x : x \neq 0\}$$

$$h(t) = \frac{7t^6 + t^3 - 3t - 1}{t^2 - 1} \qquad \text{domain } \{t : t \neq \pm 1\}$$

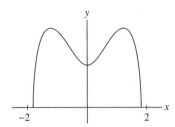

FIGURE 3 The algebraic function $f(x) = \sqrt{1 + 3x^2 - x^4}$.

- **Algebraic functions:** An algebraic function is produced by taking sums, products, and quotients of *roots* of polynomials and rational functions (Figure 3):

$$f(x) = \sqrt{1 + 3x^2 - x^4}, \qquad g(t) = (\sqrt{t} - 2)^{-2}, \qquad h(z) = \frac{z + z^{-5/3}}{5z^3 - \sqrt{z}}$$

A number x belongs to the domain of f if each term in the formula for f is defined and the result does not involve division by zero. For example, $g(t)$ is defined if $t \geq 0$ and $\sqrt{t} \neq 2$, so the domain of $g(t)$ is $D = \{t : t \geq 0 \text{ and } t \neq 4\}$. More generally, algebraic functions are defined by polynomial equations between x and y. In this case, we say that y is **implicitly defined** as a function of x. For example, the equation $y^4 + 2x^2 y + x^4 = 1$ defines y implicitly as a function of x.

- **Exponential functions:** The function $f(x) = b^x$, where $b > 0$, is called the exponential function with base b. Some examples are

$$y = 2^x, \qquad y = 10^x, \qquad y = \left(\frac{1}{3}\right)^x, \qquad y = (\sqrt{5})^x$$

Exponential functions and their *inverses*, the **logarithm functions**, are treated in greater detail in Section 1.6.

- **Trigonometric functions:** Functions built from $\sin x$ and $\cos x$ are called trigonometric functions. These functions are discussed in the next section.

Any function that is not algebraic is called "transcendental." Exponential and trigonometric functions are examples. Other transcendental functions, such as the gamma and Bessel functions, appear in advanced applications to physics, engineering, and statistics. The word "transcendental" was used to describe functions of this type in the 1670s by Gottfried Wilhelm Leibniz (1646–1716).

Constructing New Functions

If f and g are functions, we may construct new functions by forming the sum, difference, product, and quotient functions:

$$(f + g)(x) = f(x) + g(x), \qquad (f - g)(x) = f(x) - g(x)$$

$$(fg)(x) = f(x)\, g(x), \qquad \left(\frac{f}{g}\right)(x) = \frac{f(x)}{g(x)} \quad (\text{where } g(x) \neq 0)$$

For example, if $f(x) = x^2$ and $g(x) = \sin x$, then

$$(f + g)(x) = x^2 + \sin x, \qquad (f - g)(x) = x^2 - \sin x$$

$$(fg)(x) = x^2 \sin x, \qquad \left(\frac{f}{g}\right)(x) = \frac{x^2}{\sin x}$$

We can also multiply functions by constants. A function of the form

$$c_1 f(x) + c_2 g(x) \quad (c_1, c_2 \text{ constants})$$

is called a **linear combination** of $f(x)$ and $g(x)$.

Composition is another important way of constructing new functions. The composition of f and g is the function $f \circ g$ defined by $(f \circ g)(x) = f(g(x))$, defined for values of x in the domain of g such that $g(x)$ lies in the domain of f.

■ **EXAMPLE 1** Compute the composite functions $f \circ g$ and $g \circ f$ and discuss their domains, where

$$f(x) = \sqrt{x}, \qquad g(x) = 1 - x$$

Example 1 shows that the composition of functions is not commutative: The functions $f \circ g$ and $g \circ f$ may be different.

Solution We have

$$(f \circ g)(x) = f(g(x)) = f(1 - x) = \sqrt{1 - x}$$

The square root $\sqrt{1-x}$ is defined if $1 - x \geq 0$ or $x \leq 1$, so the domain of $f \circ g$ is $\{x : x \leq 1\}$. On the other hand,

$$(g \circ f)(x) = g(f(x)) = g(\sqrt{x}) = 1 - \sqrt{x}$$

The domain of $g \circ f$ is $\{x : x \geq 0\}$.   ∎

Elementary Functions

Inverse functions are discussed in Section 1.5.

We have reviewed some of the most basic and familiar functions of mathematics. All of these can be found on any scientific calculator. New functions may be produced using the operations of addition, multiplication, division, as well as composition, extraction of roots, and taking inverses. It is convenient to refer to a function constructed in this way from the basic functions listed above as an **elementary function**. The following functions are elementary:

$$f(x) = \sqrt{2x + \sin x}, \qquad f(x) = 10^{\sqrt{x}}, \qquad f(x) = \frac{1 + x^{-1}}{1 + \cos x}$$

1.3 SUMMARY

- For any real number m, the function $f(x) = x^m$ is the power function with exponent m. A polynomial $P(x)$ is a sum of multiples of power functions x^m, where m is a whole number:

$$P(x) = a_n x^n + a_{n-1} x^{n-1} + \cdots + a_1 x + a_0$$

This polynomial has degree n (if we assume that $a_n \neq 0$) and a_n is called the leading coefficient.
- A rational function is a quotient $P(x)/Q(x)$ of two polynomials.
- An algebraic function is produced by taking sums, products, and nth roots of polynomials and rational functions.
- Exponential function: $f(x) = b^x$, where $b > 0$ (b is called the base).
- The composite function $f \circ g$ is defined by $(f \circ g)(x) = f(g(x))$. The domain of $f \circ g$ is the set of x in the domain of g such that $g(x)$ belongs to the domain of f.

1.3 EXERCISES

Preliminary Questions

1. Give an example of a rational function.

2. Is $|x|$ a polynomial function? What about $|x^2 + 1|$?

3. What is unusual about the domain of $f \circ g$ for $f(x) = x^{1/2}$ and $g(x) = -1 - |x|$?

4. Is $f(x) = \left(\frac{1}{2}\right)^x$ increasing or decreasing?

5. Give an example of a transcendental function.

Exercises

In Exercises 1–12, determine the domain of the function.

1. $f(x) = x^{1/4}$

2. $g(t) = t^{2/3}$

3. $f(x) = x^3 + 3x - 4$

4. $h(z) = z^3 + z^{-3}$

5. $g(t) = \dfrac{1}{t + 2}$

6. $f(x) = \dfrac{1}{x^2 + 4}$

7. $G(u) = \dfrac{1}{u^2 - 4}$

8. $f(x) = \dfrac{\sqrt{x}}{x^2 - 9}$

9. $f(x) = x^{-4} + (x-1)^{-3}$ **10.** $F(s) = \sin\left(\dfrac{s}{s+1}\right)$

11. $g(y) = 10^{\sqrt{y}+y^{-1}}$ **12.** $f(x) = \dfrac{x + x^{-1}}{(x-3)(x+4)}$

In Exercises 13–24, identify each of the following functions as polynomial, rational, algebraic, or transcendental.

13. $f(x) = 4x^3 + 9x^2 - 8$ **14.** $f(x) = x^{-4}$

15. $f(x) = \sqrt{x}$ **16.** $f(x) = \sqrt{1-x^2}$

17. $f(x) = \dfrac{x^2}{x + \sin x}$ **18.** $f(x) = 2^x$

19. $f(x) = \dfrac{2x^3 + 3x}{9 - 7x^2}$ **20.** $f(x) = \dfrac{3x - 9x^{-1/2}}{9 - 7x^2}$

21. $f(x) = \sin(x^2)$ **22.** $f(x) = \dfrac{x}{\sqrt{x}+1}$

23. $f(x) = x^2 + 3x^{-1}$ **24.** $f(x) = \sin(3^x)$

25. Is $f(x) = 2^{x^2}$ a transcendental function?

26. Show that $f(x) = x^2 + 3x^{-1}$ and $g(x) = 3x^3 - 9x + x^{-2}$ are rational functions (show that each is a quotient of polynomials).

In Exercises 27–34, calculate the composite functions $f \circ g$ and $g \circ f$, and determine their domains.

27. $f(x) = \sqrt{x}, \quad g(x) = x + 1$

28. $f(x) = \dfrac{1}{x}, \quad g(x) = x^{-4}$

29. $f(x) = 2^x, \quad g(x) = x^2$

30. $f(x) = |x|, \quad g(\theta) = \sin\theta$

31. $f(\theta) = \cos\theta, \quad g(x) = x^3 + x^2$

32. $f(x) = \dfrac{1}{x^2 + 1}, \quad g(x) = x^{-2}$

33. $f(t) = \dfrac{1}{\sqrt{t}}, \quad g(t) = -t^2$

34. $f(t) = \sqrt{t}, \quad g(t) = 1 - t^3$

35. The population (in millions) of a country as a function of time t (years) is $P(t) = 30 \cdot 2^{kt}$, with $k = 0.1$. Show that the population doubles every 10 years. Show more generally that for any nonzero constants a and k, the function $g(t) = a2^{kt}$ doubles after $1/k$ years.

36. Find all values of c such that the domain of

$$f(x) = \frac{x+1}{x^2 + 2cx + 4}$$

is **R**.

Further Insights and Challenges

In Exercises 37–43, we define the first difference δf of a function $f(x)$ by $\delta f(x) = f(x+1) - f(x)$.

37. Show that if $f(x) = x^2$, then $\delta f(x) = 2x + 1$. Calculate δf for $f(x) = x$ and $f(x) = x^3$.

38. Show that $\delta(10^x) = 9 \cdot 10^x$, and more generally, $\delta(b^x) = c \cdot b^x$ for some constant c.

39. Show that for any two functions f and g, $\delta(f+g) = \delta f + \delta g$ and $\delta(cf) = c\delta(f)$, where c is any constant.

40. First differences can be used to derive formulas for the sum of the kth powers. Suppose we can find a function $P(x)$ such that $\delta P = (x+1)^k$ and $P(0) = 0$. Prove that $P(1) = 1^k$, $P(2) = 1^k + 2^k$, and more generally, for every whole number n,

$$P(n) = 1^k + 2^k + \cdots + n^k \qquad \boxed{1}$$

41. First show that

$$P(x) = \frac{x(x+1)}{2}$$

satisfies $\delta P = (x+1)$. Then apply Exercise 40 to conclude that

$$1 + 2 + 3 + \cdots + n = \frac{n(n+1)}{2}$$

42. Calculate $\delta(x^3)$, $\delta(x^2)$, and $\delta(x)$. Then find a polynomial $P(x)$ of degree 3 such that $\delta P = (x+1)^2$ and $P(0) = 0$. Conclude that

$$P(n) = 1^2 + 2^2 + \cdots + n^2$$

43. This exercise combined with Exercise 40 shows that for all k, there exists a polynomial $P(x)$ satisfying Eq. (1). The solution requires proof by induction and the Binomial Theorem (see Appendix C).

(a) Show that

$$\delta(x^{k+1}) = (k+1)x^k + \cdots$$

where the dots indicate terms involving smaller powers of x.

(b) Show by induction that for all whole numbers k, there exists a polynomial of degree $k + 1$ with leading coefficient $1/(k+1)$:

$$P(x) = \frac{1}{k+1}x^{k+1} + \cdots$$

such that $\delta P = (x+1)^k$ and $P(0) = 0$.

1.4 Trigonometric Functions

We begin our trigonometric review by recalling the two systems of angle measurement: **radians** and **degrees**. They are best described using the relationship between angles and rotation. As is customary, we often use the lowercase Greek letter θ ("theta") to denote angles and rotations.

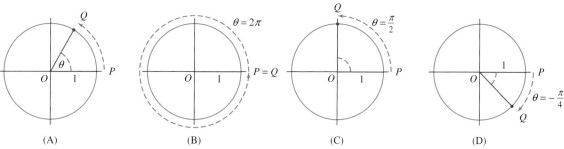

(A)   (B)   (C)   (D)

FIGURE 1 The radian measure θ of a counterclockwise rotation is the length of the arc traversed by P as it rotates into Q.

TABLE 1

Rotation Through	Radian Measure
Two full circles	4π
Full circle	2π
Half circle	π
Quarter circle	$2\pi/4 = \pi/2$
One-sixth circle	$2\pi/6 = \pi/3$

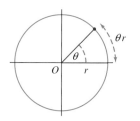

FIGURE 2 On a circle of radius r, the arc traversed by a counterclockwise rotation of θ radians has length θr.

Radians	Degrees
0	$0°$
$\dfrac{\pi}{6}$	$30°$
$\dfrac{\pi}{4}$	$45°$
$\dfrac{\pi}{3}$	$60°$
$\dfrac{\pi}{2}$	$90°$

Figure 1(A) shows a unit circle with radius $\overline{OP}$ rotating counterclockwise into radius $\overline{OQ}$. *The radian measure θ of this rotation is the length of the circular arc traversed by P as it rotates into Q.*

The unit circle has circumference 2π. Therefore, a rotation through a full circle has radian measure $\theta = 2\pi$ [Figure 1(B)]. The radian measure of a rotation through one-quarter of a circle is $\theta = 2\pi/4 = \pi/2$ [Figure 1(C)] and, in general, the rotation through one-nth of a circle has radian measure $2\pi/n$ (Table 1). A negative rotation (with $\theta < 0$) is a rotation in the *clockwise* direction [Figure 1(D)]. On a circle of radius r, *the arc traversed by a counterclockwise rotation of θ radians has length θr* (Figure 2).

Now consider angle $\angle POQ$ in Figure 1(A). The radian measure of angle $\angle POQ$ is defined as the radian measure of a rotation that carries $\overline{OP}$ to $\overline{OQ}$. Notice that every rotation has a unique radian measure but the radian measure of an angle is not unique. For example, the rotations through θ and $\theta + 2\pi$ both carry $\overline{OP}$ to $\overline{OQ}$. Although the rotation $\theta + 2\pi$ takes an extra trip around the circle, θ and $\theta + 2\pi$ represent the same angle. In general, *two angles coincide if the corresponding rotations differ by an integer multiple of 2π.* For example, the angle $\pi/4$ can be represented as $9\pi/4$ or $-15\pi/4$ because

$$\frac{\pi}{4} = \frac{9\pi}{4} - 2\pi = -\frac{15\pi}{4} + 4\pi$$

Every angle has a unique radian measure satisfying $0 \le \theta < 2\pi$. With this choice, the angle subtends an arc of length θr on a circle of radius r.

Degrees are defined by dividing the circle (not necessarily the unit circle) into 360 parts. A degree is $(1/360)$th of a circle. A rotation through θ degrees (denoted $\theta°$) is a rotation through the fraction $\theta/360$ of the complete circle. For example, a rotation through $90°$ is a rotation through the fraction $90/360$, or $\frac{1}{4}$ of a circle.

As with radians, every rotation has a unique degree measure but the degree measure of an angle is not unique. Two angles coincide if their degree measures differ by an integer multiple of 360. Every angle has a unique degree measure θ with $0 \le \theta < 360$. For example, the angles $-45°$ and $675°$ coincide since $675 = -45 + 720$.

To convert between radians and degrees, remember that 2π rad is equal to $360°$. Therefore, 1 rad equals $360/2\pi$ or $180/\pi$ degrees.

Radian measurement is usually the better choice for mathematical purposes, but there are good practical reasons for using degrees. The number 360 has many divisors ($360 = 8 \cdot 9 \cdot 5$) and, consequently, many fractional parts of the circle can be expressed as an integer number of degrees. For example, one-fifth of the circle is $72°$, two-ninths is $80°$, three-eighths is $135°$, etc.

- To convert from radians to degrees, multiply by $180/\pi$.
- To convert from degrees to radians, multiply by $\pi/180$.

■ **EXAMPLE 1** Convert **(a)** $55°$ to radians   and   **(b)** 0.5 rad to degrees.

Solution

(a) $55° \times \dfrac{\pi}{180} \approx 0.9599$ rad   **(b)** 0.5 rad $\times \dfrac{180}{\pi} \approx 28.648°$ ■

Convention: *Unless otherwise stated, we always measure angles in radians.*

The trigonometric functions $\sin\theta$ and $\cos\theta$ are defined in terms of right triangles. Let θ be an acute angle in a right triangle and let us label the sides as in Figure 3. Then

$$\sin\theta = \frac{b}{c} = \frac{\text{opposite}}{\text{hypotenuse}}, \qquad \cos\theta = \frac{a}{c} = \frac{\text{adjacent}}{\text{hypotenuse}}$$

A disadvantage of this definition is that it only makes sense if θ lies between 0 and $\pi/2$ (because an angle in a right triangle cannot exceed $\pi/2$). However, sine and cosine can also be defined in terms of the unit circle and this definition is valid for all angles. Let $P = (x, y)$ be the point on the unit circle corresponding to the angle θ as in Figure 4(A). Then we define

$$\cos\theta = x\text{-coordinate of } P, \qquad \sin\theta = y\text{-coordinate of } P$$

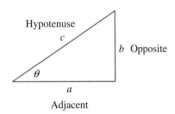

FIGURE 3

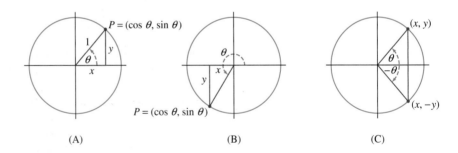

(A)   (B)   (C)

FIGURE 4 The unit circle definition of sine and cosine is valid for all angles θ.

This agrees with the right-triangle definition when $0 < \theta < \pi/2$. Furthermore, we see from Figure 4(C) that $\sin\theta$ is an odd function and $\cos\theta$ is an even function:

$$\sin(-\theta) = -\sin\theta, \qquad \cos(-\theta) = \cos\theta$$

Although we use a calculator to evaluate sine and cosine for general angles, the standard values listed in Figure 5 and Table 2 appear often and should be memorized.

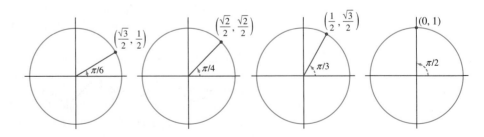

FIGURE 5 Four standard angles: The x- and y-coordinates of the points are $\cos\theta$ and $\sin\theta$.

TABLE 2

θ	0	$\dfrac{\pi}{6}$	$\dfrac{\pi}{4}$	$\dfrac{\pi}{3}$	$\dfrac{\pi}{2}$	$\dfrac{2\pi}{3}$	$\dfrac{3\pi}{4}$	$\dfrac{5\pi}{6}$	π
$\sin\theta$	0	$\dfrac{1}{2}$	$\dfrac{\sqrt{2}}{2}$	$\dfrac{\sqrt{3}}{2}$	1	$\dfrac{\sqrt{3}}{2}$	$\dfrac{\sqrt{2}}{2}$	$\dfrac{1}{2}$	0
$\cos\theta$	1	$\dfrac{\sqrt{3}}{2}$	$\dfrac{\sqrt{2}}{2}$	$\dfrac{1}{2}$	0	$-\dfrac{1}{2}$	$-\dfrac{\sqrt{2}}{2}$	$-\dfrac{\sqrt{3}}{2}$	-1

The functions $\sin\theta$ and $\cos\theta$ are defined for all real numbers θ, and it is not necessary to think of θ as an angle. We often write $\sin x$ and $\cos x$, using x instead of θ. Depending on the application, the angle interpretation may or may not be appropriate.

The graph of $y = \sin\theta$ is the familiar "sine wave" shown in Figure 6. Observe how the graph is generated by the y-coordinate of a point moving around the unit circle. The graph of $y = \cos\theta$ has the same shape, but is shifted to the left $\frac{\pi}{2}$ units (Figure 7). The signs of $\sin\theta$ and $\cos\theta$ vary as the point $P = (\cos\theta, \sin\theta)$ on the unit circle changes quadrant (Figure 7).

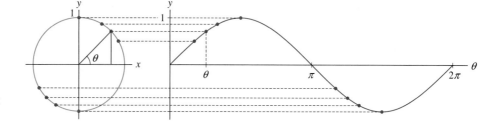

FIGURE 6 A graph of $y = \sin\theta$ is generated as a point moves around the unit circle.

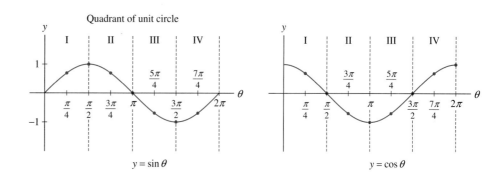

FIGURE 7 Graphs of $y = \sin\theta$ and $y = \cos\theta$ over one *period* of length 2π.

A function $f(x)$ is called **periodic** with period T if $f(x + T) = f(x)$ (for all x) and T is the smallest positive number with this property. The sine and cosine functions are periodic with period $T = 2\pi$ since angles that differ by an integer multiple $2\pi k$ correspond to the same point on the unit circle (Figure 8):

$$\sin x = \sin(x + 2\pi k), \qquad \cos x = \cos(x + 2\pi k)$$

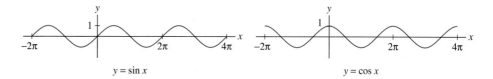

FIGURE 8 Sine and cosine have period 2π.

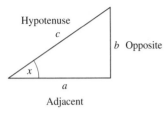

FIGURE 9

Recall that there are four other standard trigonometric functions, each defined in terms of $\sin x$ and $\cos x$ or as ratios of sides in a right triangle (Figure 9):

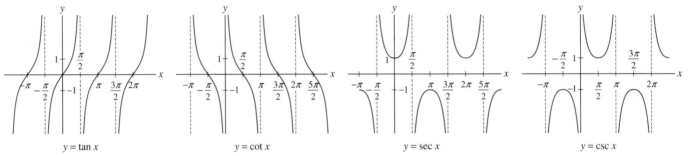

Tangent: $\tan x = \dfrac{\sin x}{\cos x} = \dfrac{b}{a}$,   Cotangent: $\cot x = \dfrac{\cos x}{\sin x} = \dfrac{a}{b}$

Secant: $\sec x = \dfrac{1}{\cos x} = \dfrac{c}{a}$,   Cosecant: $\csc x = \dfrac{1}{\sin x} = \dfrac{c}{b}$

These functions are periodic (Figure 10): $y = \tan x$ and $y = \cot x$ have period π, $y = \sec x$ and $y = \csc x$ have period 2π (see Exercise 51).

| $y = \tan x$ | $y = \cot x$ | $y = \sec x$ | $y = \csc x$ |

FIGURE 10 Graphs of the standard trigonometric functions.

■ **EXAMPLE 2** Computing Values of Trigonometric Functions   Find the values of the six trigonometric functions at $x = 4\pi/3$.

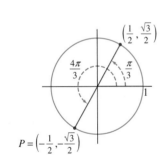

FIGURE 11

Solution The point P on the unit circle corresponding to the angle $x = 4\pi/3$ lies opposite the point with angle $\pi/3$ (Figure 11). Thus, we see that (refer to Table 2)

$$\sin \frac{4\pi}{3} = -\sin \frac{\pi}{3} = -\frac{\sqrt{3}}{2}, \qquad \cos \frac{4\pi}{3} = -\cos \frac{\pi}{3} = -\frac{1}{2}$$

The remaining values are

$$\tan \frac{4\pi}{3} = \frac{\sin 4\pi/3}{\cos 4\pi/3} = \frac{-\sqrt{3}/2}{-1/2} = \sqrt{3}, \qquad \cot \frac{4\pi}{3} = \frac{\cos 4\pi/3}{\sin 4\pi/3} = \frac{\sqrt{3}}{3}$$

$$\sec \frac{4\pi}{3} = \frac{1}{\cos 4\pi/3} = \frac{1}{-1/2} = -2, \qquad \csc \frac{4\pi}{3} = \frac{1}{\sin 4\pi/3} = \frac{-2\sqrt{3}}{3} \qquad ■$$

■ **EXAMPLE 3**   Find the angles x such that $\sec x = 2$.

Solution Since $\sec x = 1/\cos x$, we must solve $\cos x = \frac{1}{2}$. From Figure 12 we see that $x = \frac{\pi}{3}$ and $x = -\frac{\pi}{3}$ are solutions. We may add any integer multiple of 2π, so the general solution is $x = \pm\frac{\pi}{3} + 2\pi k$ for any integer k.   ■

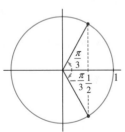

FIGURE 12 $\cos x = 1/2$ for $x = \pm\pi/3$

■ **EXAMPLE 4**   Sketch the graph of $f(x) = 3\cos\left(2\left(x + \dfrac{\pi}{2}\right)\right)$ over $[0, 2\pi]$.

Solution The graph is obtained by scaling and shifting the graph of $y = \cos x$ in three steps (Figure 13):

• Compress horizontally by a factor of 2: $y = \cos 2x$

Note: To shift the graph of $y = \cos 2x$ to the left $\frac{\pi}{2}$ units, we must replace x by $x + \frac{\pi}{2}$ to obtain $\cos\left(2(x + \frac{\pi}{2})\right)$. It is incorrect to take $\cos\left(2x + \frac{\pi}{2}\right)$

- Shift to left $\dfrac{\pi}{2}$ units: $\qquad y = \cos\left(2\left(x + \dfrac{\pi}{2}\right)\right)$

- Expand vertically by factor of 3: $\qquad y = 3\cos\left(2\left(x + \dfrac{\pi}{2}\right)\right)$ ■

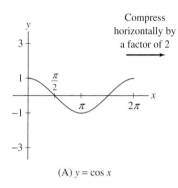

(A) $y = \cos x$

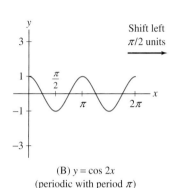

(B) $y = \cos 2x$
(periodic with period π)

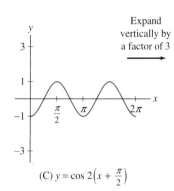

(C) $y = \cos 2\left(x + \frac{\pi}{2}\right)$

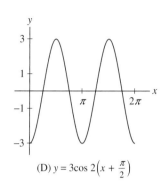

(D) $y = 3\cos 2\left(x + \frac{\pi}{2}\right)$

FIGURE 13

Trigonometric Identities

A key feature of trigonometric functions is that they satisfy a large number of identities. First and foremost, sine and cosine satisfy the following fundamental identity, which is equivalent to the Pythagorean Theorem:

The expression $(\sin\theta)^k$ is usually denoted $\sin^k\theta$. For example, $\sin^2\theta$ is the square of $\sin\theta$. We use similar notation for the other trigonometric functions.

$$\sin^2\theta + \cos^2\theta = 1$$ 　　**1**

Equivalent versions are obtained by dividing Eq. (1) by $\cos^2\theta$ or $\sin^2\theta$:

$$\tan^2\theta + 1 = \sec^2\theta, \qquad 1 + \cot^2\theta = \csc^2\theta$$ 　　**2**

Here is a list of some other commonly used identities.

Basic Trigonometric Identities

Complementary angles: $\quad \sin\left(\dfrac{\pi}{2} - x\right) = \cos x, \quad \cos\left(\dfrac{\pi}{2} - x\right) = \sin x$

Addition formulas: $\quad \sin(x + y) = \sin x \cos y + \cos x \sin y$

$\qquad\qquad\qquad\qquad \cos(x + y) = \cos x \cos y - \sin x \sin y$

Double-angle formulas: $\quad \sin^2 x = \dfrac{1}{2}(1 - \cos 2x), \quad \cos^2 x = \dfrac{1}{2}(1 + \cos 2x)$

$\qquad\qquad\qquad\qquad \cos 2x = \cos^2 x - \sin^2 x, \quad \sin 2x = 2\sin x \cos x$

Shift formulas: $\quad \sin\left(x + \dfrac{\pi}{2}\right) = \cos x, \quad \cos\left(x + \dfrac{\pi}{2}\right) = -\sin x$

FIGURE 14 For complementary angles θ and $\psi = \dfrac{\pi}{2} - \theta$, we have $\sin\theta = \cos\psi$ and $\sin\psi = \cos\theta$.

■ **EXAMPLE 5** Suppose that $\cos\theta = \dfrac{2}{5}$. Calculate $\tan\theta$ if 　(a) $0 \le \theta < \dfrac{\pi}{2}$ 　and
(b) $\pi \le \theta < 2\pi$.

Solution We use the identity $\cos^2\theta + \sin^2\theta = 1$ to determine $\sin\theta$:

$$\sin\theta = \pm\sqrt{1 - \cos^2\theta} = \pm\sqrt{1 - \frac{4}{25}} = \pm\frac{\sqrt{21}}{5}$$

(a) If $0 \le \theta < \pi/2$, then $\sin\theta$ is positive and we take the positive square root:

$$\tan\theta = \frac{\sin\theta}{\cos\theta} = \frac{\sqrt{21}/5}{2/5} = \frac{\sqrt{21}}{2}$$

A good way to visualize this computation is to draw a right triangle with an angle θ such that $\cos\theta = \frac{2}{5}$ as in Figure 15. The opposite side then has length $\sqrt{21} = \sqrt{5^2 - 2^2}$ by the Pythagorean Theorem.

(b) If we assume that $\pi \le \theta < 2\pi$, then $\sin\theta$ is negative and $\tan\theta = -\dfrac{\sqrt{21}}{2}$. ■

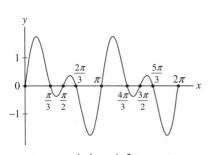

FIGURE 15

■ **EXAMPLE 6** A Trigonometric Equation   Solve the equation $\sin 4x + \sin 2x = 0$ for $0 \le x < 2\pi$.

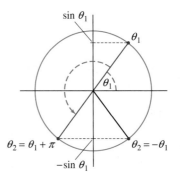

FIGURE 16 $\sin\theta_2 = -\sin\theta_1$ when $\theta_2 = -\theta_1$ or $\theta_2 = \theta_1 + \pi$.

Solution We must find the angles x such that $\sin 4x = -\sin 2x$. First, let's determine when angles θ_1 and θ_2 satisfy $\sin\theta_2 = -\sin\theta_1$. Figure 16 shows that this occurs if $\theta_2 = -\theta_1$ or $\theta_2 = \theta_1 + \pi$. Since the sine function is periodic with period 2π,

$$\sin\theta_2 = -\sin\theta_1 \quad \Leftrightarrow \quad \theta_2 = -\theta_1 + 2\pi k \quad \text{or} \quad \theta_2 = \theta_1 + \pi + 2\pi k$$

where k is an integer. Taking $\theta_2 = 4x$ and $\theta_1 = 2x$, we see that

$$\sin 4x = -\sin 2x \quad \Leftrightarrow \quad 4x = -2x + 2\pi k \quad \text{or} \quad 4x = 2x + \pi + 2\pi k$$

The first equation gives $6x = 2\pi k$ or $x = \frac{\pi}{3}k$ and the second equation gives $2x = \pi + 2\pi k$ or $x = \frac{\pi}{2} + \pi k$. We obtain eight solutions in $[0, 2\pi)$ (Figure 17)

$$x = 0, \quad \frac{\pi}{3}, \quad \frac{2\pi}{3}, \quad \pi, \quad \frac{4\pi}{3}, \quad \frac{5\pi}{3} \quad \text{and} \quad x = \frac{\pi}{2}, \quad \frac{3\pi}{2} \quad ■$$

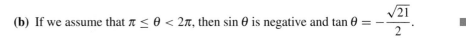

$y = \sin 4x + \sin 2x$

FIGURE 17 Solutions of $\sin 4x + \sin 2x = 0$.

We conclude this section by quoting the **Law of Cosines**, which is a generalization of the Pythagorean Theorem (Figure 18).

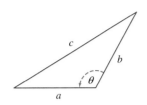

FIGURE 18

THEOREM 1 Law of Cosines If a triangle has sides a, b, and c, and θ is the angle opposite side c, then

$$c^2 = a^2 + b^2 - 2ab\cos\theta$$

If $\theta = 90°$, then $\cos\theta = 0$ and the Law of Cosines reduces to the Pythagorean Theorem.

1.4 SUMMARY

- An angle of θ radians subtends an arc of length θr on a circle of radius r.

- To convert from radians to degrees, multiply by $\dfrac{180}{\pi}$.

- To convert from degrees to radians, multiply by $\dfrac{\pi}{180}$.

- Unless otherwise stated, all angles in this text are given in radians.

- The functions $\cos\theta$ and $\sin\theta$ are defined in terms of right triangles for acute angles and as coordinates of a point on the unit circle for general angles (Figure 19):

$$\sin\theta = \frac{b}{c} = \frac{\text{opposite}}{\text{hypotenuse}}, \qquad \cos\theta = \frac{a}{c} = \frac{\text{adjacent}}{\text{hypotenuse}}$$

- Basic properties of sine and cosine:

  – Periodicity: $\sin(\theta + 2\pi) = \sin\theta$, $\cos(\theta + 2\pi) = \cos\theta$

  – Parity: $\sin(-\theta) = -\sin\theta$, $\cos(-\theta) = \cos\theta$

  – Basic identity: $\sin^2\theta + \cos^2\theta = 1$

- The four additional trigonometric functions:

$$\tan\theta = \frac{\sin\theta}{\cos\theta}, \qquad \cot\theta = \frac{\cos\theta}{\sin\theta}, \qquad \sec\theta = \frac{1}{\cos\theta}, \qquad \csc\theta = \frac{1}{\sin\theta}$$

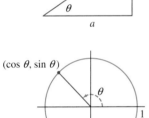

FIGURE 19

1.4 EXERCISES

Preliminary Questions

1. How is it possible for two different rotations to define the same angle?

2. Give two different positive rotations that define the angle $\frac{\pi}{4}$.

3. Give a negative rotation that defines the angle $\frac{\pi}{3}$.

4. The definition of $\cos\theta$ using right triangles applies when (choose the correct answer):

   (a) $0 < \theta < \dfrac{\pi}{2}$ **(b)** $0 < \theta < \pi$ **(c)** $0 < \theta < 2\pi$

5. What is the unit circle definition of $\sin\theta$?

6. How does the periodicity of $\sin\theta$ and $\cos\theta$ follow from the unit circle definition?

Exercises

1. Find the angle between 0 and 2π that is equivalent to $13\pi/4$.

2. Describe the angle $\theta = \frac{\pi}{6}$ by an angle of negative radian measure.

3. Convert from radians to degrees:

(a) 1   **(b)** $\frac{\pi}{3}$ **(c)** $\frac{5}{12}$ **(d)** $-\frac{3\pi}{4}$

4. Convert from degrees to radians:

(a) $1°$ **(b)** $30°$ **(c)** $25°$ **(d)** $120°$

5. Find the lengths of the arcs subtended by the angles θ and ϕ radians in Figure 20.

6. Calculate the values of the six standard trigonometric functions for the angle θ in Figure 21.

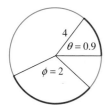

FIGURE 20 Circle of radius 4.

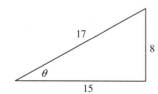

FIGURE 21

7. Fill in the remaining values of $(\cos \theta, \sin \theta)$ for the points in Figure 22.

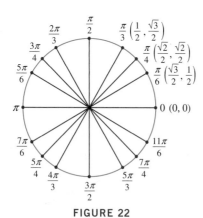

FIGURE 22

8. Find the values of the six standard trigonometric functions at $\theta = 11\pi/6$.

In Exercises 9–14, use Figure 22 to find all angles between 0 and 2π satisfying the given condition.

9. $\cos \theta = \frac{1}{2}$ **10.** $\tan \theta = 1$ **11.** $\tan \theta = -1$

12. $\csc \theta = 2$ **13.** $\sin x = \frac{\sqrt{3}}{2}$ **14.** $\sec t = 2$

15. Fill in the following table of values:

θ	$\dfrac{\pi}{6}$	$\dfrac{\pi}{4}$	$\dfrac{\pi}{3}$	$\dfrac{\pi}{2}$	$\dfrac{2\pi}{3}$	$\dfrac{3\pi}{4}$	$\dfrac{5\pi}{6}$
$\tan \theta$							
$\sec \theta$							

16. Complete the following table of signs for the trigonometric functions:

θ	sin	cos	tan	cot	sec	csc
$0 < \theta < \dfrac{\pi}{2}$	+	+				
$\dfrac{\pi}{2} < \theta < \pi$						
$\pi < \theta < \dfrac{3\pi}{2}$						
$\dfrac{3\pi}{2} < \theta < 2\pi$						

17. Show that if $\tan \theta = c$ and $0 \le \theta < \pi/2$, then $\cos \theta = 1/\sqrt{1+c^2}$. *Hint:* Draw a right triangle whose opposite and adjacent sides have lengths c and 1.

18. Suppose that $\cos \theta = \frac{1}{3}$.

(a) Show that if $0 \le \theta < \pi/2$, then $\sin \theta = 2\sqrt{2}/3$ and $\tan \theta = 2\sqrt{2}$.

(b) Find $\sin \theta$ and $\tan \theta$ if $3\pi/2 \le \theta < 2\pi$.

In Exercises 19–24, assume that $0 \le \theta < \pi/2$.

19. Find $\sin \theta$ and $\tan \theta$ if $\cos \theta = \frac{5}{13}$.

20. Find $\cos \theta$ and $\tan \theta$ if $\sin \theta = \frac{3}{5}$.

21. Find $\sin \theta$, $\sec \theta$, and $\cot \theta$ if $\tan \theta = \frac{2}{7}$.

22. Find $\sin \theta$, $\cos \theta$, and $\sec \theta$ if $\cot \theta = 4$.

23. Find $\cos 2\theta$ if $\sin \theta = \frac{1}{5}$.

24. Find $\sin 2\theta$ and $\cos 2\theta$ if $\tan \theta = \sqrt{2}$.

25. Find $\cos \theta$ and $\tan \theta$ if $\sin \theta = 0.4$ and $\pi/2 \le \theta < \pi$.

26. Find $\cos \theta$ and $\sin \theta$ if $\tan \theta = 4$ and $\pi \le \theta < 3\pi/2$.

27. Find the values of $\sin \theta$, $\cos \theta$, and $\tan \theta$ at the eight points in Figure 23.

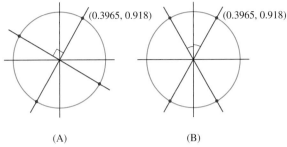

(A) (B)

FIGURE 23

28. Refer to Figure 24(A). Express the functions $\sin\theta$, $\tan\theta$, and $\csc\theta$ in terms of c.

29. Refer to Figure 24(B). Compute $\cos\psi$, $\sin\psi$, $\cot\psi$, and $\csc\psi$.

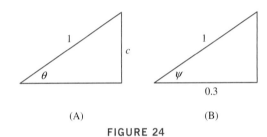

(A) (B)

FIGURE 24

30. Express $\cos\left(\theta + \dfrac{\pi}{2}\right)$ and $\sin\left(\theta + \dfrac{\pi}{2}\right)$ in terms of $\cos\theta$ and $\sin\theta$. *Hint:* Find the relation between the coordinates (a, b) and (c, d) in Figure 25.

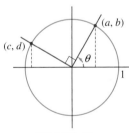

FIGURE 25

In Exercises 31–34, sketch the graph over $[0, 2\pi]$.

31. $2\sin 4\theta$

32. $\cos\left(2\left(\theta - \dfrac{\pi}{2}\right)\right)$

33. $\cos\left(2\theta - \dfrac{\pi}{2}\right)$

34. $\sin\left(2\left(\theta - \dfrac{\pi}{2}\right) + \pi\right) + 2$

35. How many points lie on the intersection of the horizontal line $y = c$ and the graph of $y = \sin x$ for $0 \le x < 2\pi$? *Hint:* The answer depends on c.

36. How many points lie on the intersection of the horizontal line $y = c$ and the graph of $y = \tan x$ for $0 \le x < 2\pi$?

In Exercises 37–40, solve for $0 \le \theta < 2\pi$ (see Example 6).

37. $\sin 2\theta + \sin 3\theta = 0$

38. $\sin\theta = \sin 2\theta$

39. $\cos 4\theta + \cos 2\theta = 0$

40. $\sin\theta = \cos 2\theta$

In Exercises 41–50, derive the identities using the identities listed in this section.

41. $\cos 2\theta = 2\cos^2\theta - 1$

42. $\cos^2\dfrac{\theta}{2} = \dfrac{1 + \cos\theta}{2}$

43. $\sin\dfrac{\theta}{2} = \sqrt{\dfrac{1 - \cos\theta}{2}}$

44. $\sin(\theta + \pi) = -\sin\theta$

45. $\cos(\theta + \pi) = -\cos\theta$

46. $\tan x = \cot\left(\dfrac{\pi}{2} - x\right)$

47. $\tan(\pi - \theta) = -\tan\theta$

48. $\tan 2x = \dfrac{2\tan x}{1 - \tan^2 x}$

49. $\tan x = \dfrac{\sin 2x}{1 + \cos 2x}$

50. $\sin^2 x \cos^2 x = \dfrac{1 - \cos 4x}{8}$

51. Use Exercises 44 and 45 to show that $\tan\theta$ and $\cot\theta$ are periodic with period π.

52. Use the addition formula to compute $\cos\dfrac{\pi}{12}$, noting that $\dfrac{\pi}{12} = \dfrac{\pi}{3} - \dfrac{\pi}{4}$.

53. Use the Law of Cosines to find the distance from P to Q in Figure 26.

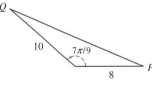

FIGURE 26

Further Insights and Challenges

54. Use the addition formula to prove

$$\cos 3\theta = 4\cos^3\theta - 3\cos\theta$$

55. Use the addition formulas for sine and cosine to prove

$$\tan(a + b) = \dfrac{\tan a + \tan b}{1 - \tan a \tan b}$$

$$\cot(a - b) = \dfrac{\cot a \cot b + 1}{\cot b - \cot a}$$

56. Let θ be the angle between the line $y = mx + b$ and the x-axis [Figure 27(A)]. Prove that $m = \tan \theta$.

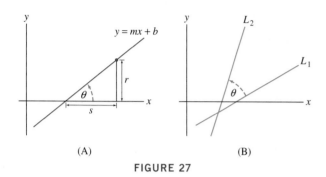

(A)

(B)

FIGURE 27

57. Let L_1 and L_2 be the lines of slope m_1 and m_2 [Figure 27(B)]. Show that the angle θ between L_1 and L_2 satisfies $\cot \theta = \dfrac{m_2 m_1 + 1}{m_2 - m_1}$.

58. Perpendicular Lines   Use Exercise 57 to prove that two lines with nonzero slopes m_1 and m_2 are perpendicular if and only if $m_2 = -1/m_1$.

59. Apply the double-angle formula to prove:

(a) $\cos \dfrac{\pi}{8} = \dfrac{1}{2}\sqrt{2 + \sqrt{2}}$

(b) $\cos \dfrac{\pi}{16} = \dfrac{1}{2}\sqrt{2 + \sqrt{2 + \sqrt{2}}}$

Guess the values of $\cos \dfrac{\pi}{32}$ and of $\cos \dfrac{\pi}{2^n}$ for all n.

1.5 Inverse Functions

Many important functions, such as logarithms or the arcsine, may be defined as **inverses** of other functions. In this section, we review inverse functions and their graphs, and we discuss the inverse trigonometric functions.

The inverse of $f(x)$, denoted $f^{-1}(x)$, is the function that *reverses* the effect of $f(x)$ (Figure 1). For example, the inverse of $f(x) = x^3$ is the cube root function $f^{-1}(x) = x^{1/3}$. Given a table of function values for $f(x)$, we obtain a table for $f^{-1}(x)$ by interchanging the x and y columns:

FIGURE 1 A function and its inverse.

Function			Inverse	
x	$y = x^3$		x	$y = x^{1/3}$
-2	-8		-8	-2
-1	-1	(Interchange columns) $\Longrightarrow$	-1	-1
0	0		0	0
1	1		1	1
2	8		8	2
3	27		27	3

Thus, if we apply both f and f^{-1} to a number x in either order, we get back x. For instance,

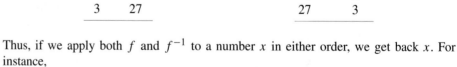

Apply f and then f^{-1}: $2 \xrightarrow{\text{(Apply } x^3)} 8 \xrightarrow{\text{(Apply } x^{1/3})} 2$

Apply f^{-1} and then f: $8 \xrightarrow{\text{(Apply } x^{1/3})} 2 \xrightarrow{\text{(Apply } x^3)} 8$

These two properties are used in the formal definition of the inverse function.

◀·· **REMINDER** *The "domain" is the set of numbers x such that f(x) is defined (the set of allowable inputs), and the "range" is the set of all values f(x) (the set of outputs).*

DEFINITION Inverse   Let $f(x)$ have domain D and range R. The inverse function $f^{-1}(x)$ (if it exists) is the function with domain R such that

$$f^{-1}\big(f(x)\big) = x \quad \text{for } x \in D \qquad \text{and} \qquad f\big(f^{-1}(x)\big) = x \quad \text{for } x \in R$$

If f^{-1} exists, then f is called **invertible**.

■ **EXAMPLE 1** Finding the Inverse Show that $f(x) = 2x - 18$ is invertible. What are the domain and range of $f^{-1}(x)$?

Solution We show that $f(x)$ is invertible by computing f^{-1} in two steps.

Step 1. **Solve the equation $y = f(x)$ for x in terms of y.**

$$y = 2x - 18$$

$$y + 18 = 2x$$

$$x = \frac{1}{2}y + 9$$

This gives us the inverse as a function of the variable y: $f^{-1}(y) = \frac{1}{2}y + 9$.

Step 2. **Interchange variables.**

We usually prefer to write the inverse as a function of x, so we interchange the roles of x and y (Figure 2):

$$f^{-1}(x) = \frac{1}{2}x + 9$$

To check our calculation, let's verify that $f^{-1}(f(x)) = x$ and $f(f^{-1}(x)) = x$:

$$f^{-1}\big(f(x)\big) = f^{-1}(2x - 18) = \frac{1}{2}(2x - 18) + 9 = (x - 9) + 9 = x$$

$$f\big(f^{-1}(x)\big) = f\left(\frac{1}{2}x + 9\right) = 2\left(\frac{1}{2}x + 9\right) - 18 = (x + 18) - 18 = x$$

Since f^{-1} is a linear function, its domain and range are **R**.   ■

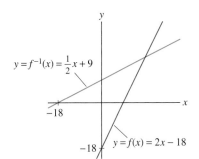

$y = f^{-1}(x) = \frac{1}{2}x + 9$

-18

-18 $y = f(x) = 2x - 18$

FIGURE 2 Graph of $f(x) = 2x - 18$ and its inverse $f^{-1}(x) = \frac{1}{2}x + 9$.

Some functions do not have inverses. For example, when we interchange the columns in a table of values for $f(x) = x^2$ (which should give us a table of values for f^{-1}), the resulting table does not define a function:

Function			Inverse (?)		
x	$y = x^2$		x	y	
-2	4	(Interchange columns) ⟹	4	-2	$f^{-1}(1)$ has two values: 1 and -1.
-1	1		1	-1	
0	0		0	0	
1	1		1	1	
2	4		4	2	

Since 1 occurs twice as an *output* of x^2 in the first table, it also occurs twice as an *input* in the second table, and thus we obtain $f^{-1}(1) = 1$ or $f^{-1}(1) = -1$. However, neither value satisfies the inverse property. For instance, if we set $f^{-1}(1) = 1$, then $f^{-1}(f(-1)) = f^{-1}(1) = 1$, but an inverse would have to satisfy $f^{-1}(f(-1)) = -1$. So when does a function $f(x)$ have an inverse? The answer is: If $f(x)$ is **one-to-one**, which means that $f(x)$ takes on each value at most once (Figure 3).

An equivalent definition of one-to-one: if $a, b \in D$ and $f(a) = f(b)$, then $a = b$.

DEFINITION One-to-One Functions A function $f(x)$ is one-to-one (on its domain D) if for every value c, the equation $f(x) = c$ has at most one solution for $x \in D$.

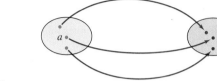
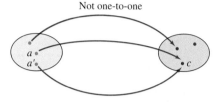

One-to-one   Not one-to-one

FIGURE 3 A one-to-one function takes on each value at most once.

$f(x) = c$ has at most one solution for all c $f(x) = c$ has two solutions: $x = a$ and $x = a'$

Think of a function as a device for "labeling" members of the range by members of the domain. When f is one-to-one, this labeling is unique and f^{-1} maps each number in the range back to its label.

When $f(x)$ is one-to-one on its domain D, the inverse f^{-1} exists and its domain is equal to the range R of f (Figure 4). Indeed, for all $c \in R$, there is a unique $a \in D$ such that $f(a) = c$ and we may define the inverse by setting $f^{-1}(c) = a$. With this definition, $f(f^{-1}(c)) = f(a) = c$. Now consider $a \in D$ and set $c = f(a)$. By definition, $f^{-1}(c) = a$ and thus $f^{-1}(f(a)) = f^{-1}(c) = a$ as required. This proves the following theorem.

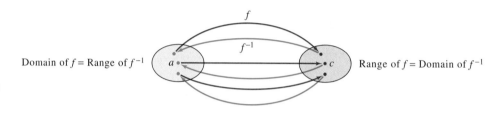

f

f^{-1}

Domain of f = Range of f^{-1} Range of f = Domain of f^{-1}

FIGURE 4 In passing from f to f^{-1}, the domain and range are interchanged.

THEOREM 1 Existence of Inverses   If $f(x)$ is one-to-one on its domain D, then f is invertible. Furthermore,

- Domain of f = range of f^{-1}.
- Range of f = domain of f^{-1}.

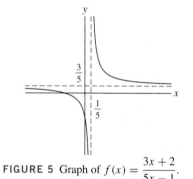

FIGURE 5 Graph of $f(x) = \dfrac{3x + 2}{5x - 1}$.

■ **EXAMPLE 2 Finding the Inverse** Show that $f(x) = \dfrac{3x + 2}{5x - 1}$ is invertible. Determine the domain and range of f and f^{-1}.

Solution The domain of $f(x)$ is $D = \{x : x \neq \frac{1}{5}\}$ (Figure 5). Assume that $x \in D$ and let's solve $y = f(x)$ for x in terms of y:

$$y = \frac{3x + 2}{5x - 1}$$

$$y(5x - 1) = 3x + 2$$

$$5xy - y = 3x + 2$$

$$5xy - 3x = y + 2 \qquad \text{(gather terms involving } x\text{)}$$

$$x(5y - 3) = y + 2 \qquad \text{(factor out } x \text{ in order to solve for } x\text{)} \qquad \boxed{1}$$

$$x = \frac{y + 2}{5y - 3} \qquad\qquad\qquad\qquad\qquad \boxed{2}$$

It is often impossible to solve for x in the equation $y = f(x)$, for example, if $f(x) = xe^x$. In such cases, the inverse may exist but we must make do without an explicit formula for it.

In the last step, we divided by $5y - 3$. This is valid if $y \neq \frac{3}{5}$. However, y cannot equal $\frac{3}{5}$ because Eq. (1) would then give $0 = \frac{3}{5} + 2$, which is false. We conclude that $f(x)$ is invertible on its domain with inverse function $f^{-1}(x) = \dfrac{x + 2}{5x - 3}$. The domain of f^{-1} is

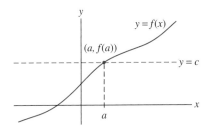

FIGURE 6 The line $y = c$ intersects the graph of $y = f(x)$ at points where $f(a) = c$.

the set $R = \{x : x \neq \frac{3}{5}\}$. Because the domain and range are interchanged under inversion, we conclude that $f(x)$ has range R and f^{-1} has range D. ■

We can tell if a function $f(x)$ is one-to-one by looking at its graph. The horizontal line $y = c$ intersects the graph at the points $(a, f(a))$ such that $f(a) = c$ (Figure 6). If every horizontal line intersects the graph in at most one point, then f is one-to-one.

Horizontal Line Test   A function $f(x)$ is one-to-one if and only if every horizontal line intersects the graph of f in at most one point.

Observe that $f(x) = x^3$ is one-to-one and passes the Horizontal Line Test [Figure 7(A)], whereas $f(x) = x^2$ fails the test [Figure 7(B)].

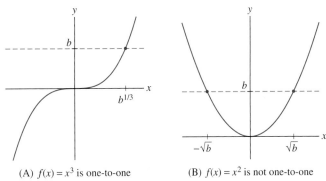

(A) $f(x) = x^3$ is one-to-one   (B) $f(x) = x^2$ is not one-to-one

FIGURE 7 A function is one-to-one if every horizontal line intersects its graph in at most one point.

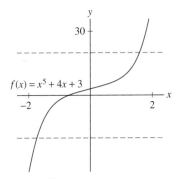

FIGURE 8 The strictly increasing function $f(x) = x^5 + 4x + 3$ satisfies the Horizontal Line Test.

■ **EXAMPLE 3**   Strictly Increasing Functions Are One-to-One   Show that strictly increasing functions are one-to-one. Then show that $f(x) = x^5 + 4x + 3$ is one-to-one.

Solution   If $f(x)$ is not one-to-one, then $f(a) = f(b)$ for some $a < b$. But a strictly increasing function satisfies $f(a) < f(b)$, so this cannot occur.
    Now observe that

- The functions cx^n for n odd and $c > 0$ are strictly increasing.
- A sum of strictly increasing functions is strictly increasing.

Thus, x^5, $4x$, and hence $x^5 + 4x$ are strictly increasing. It follows that the function $f(x) = x^5 + 4x + 3$ is strictly increasing and therefore one-to-one. This is also clear from the graph of $f(x)$ (Figure 8), which, as we observe, satisfies the Horizontal Line Test. ■

Often, it is possible to make a function one-to-one by restricting the domain.

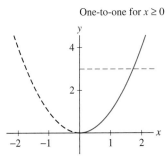

FIGURE 9 $f(x) = x^2$ satisfies the Horizontal Line Test on the domain $\{x : x \geq 0\}$.

■ **EXAMPLE 4**   Restricting the Domain   Find a domain on which $f(x) = x^2$ is one-to-one and determine its inverse on this domain.

Solution   The function $f(x) = x^2$ is one-to-one on the domain $D = \{x : x \geq 0\}$, for if $a^2 = b^2$ and a and b are both nonnegative, then $a = b$ (Figure 9). The inverse of $f(x)$ on D is the positive square root $f^{-1}(x) = \sqrt{x}$. Alternatively, we may restrict $f(x)$ to the domain $\{x : x \leq 0\}$, on which the inverse is $f^{-1}(x) = -\sqrt{x}$. ■

Next we describe the graph of the inverse function. We define the **reflection** of a point (a, b) through the line $y = x$ to be the point (b, a) (Figure 10). If we draw the x- and y-axes to the same scale, then (a, b) and (b, a) are equidistant from the line $y = x$ and the segment joining them is perpendicular to $y = x$.

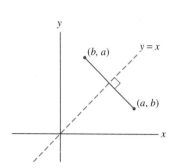

FIGURE 10 Reflection of the point (a, b) through the line $y = x$.

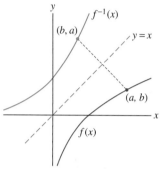

FIGURE 11 Graph of $f^{-1}(x)$ is the reflection of the graph of $f(x)$ through the line $y = x$.

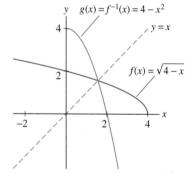

FIGURE 12 Graph of the inverse $g(x)$ of $f(x) = \sqrt{4 - x}$.

The graph of f^{-1} is obtained by reflecting the graph of f through $y = x$. To check this, note that (a, b) lies on the graph of f if $f(a) = b$. In this case, $f^{-1}(b) = a$ and the reflection $(b, f^{-1}(b)) = (b, a)$ lies on the graph of f^{-1} (Figure 11).

■ **EXAMPLE 5** **Sketching the Graph of the Inverse** Sketch the graph of the inverse of $f(x) = \sqrt{4 - x}$.

Solution Let $g(x) = f^{-1}(x)$. Observe that $f(x)$ has domain $\{x : x \le 4\}$ and range $\{x : x \ge 0\}$. We do not need a formula for $g(x)$ to draw its graph. We simply reflect the graph of f through the line $y = x$ as in Figure 12. If desired, however, we can easily solve $y = \sqrt{4 - x}$ to obtain $x = 4 - y^2$ and thus $g(x) = 4 - x^2$ with domain $\{x : x \ge 0\}$. ■

Inverse Trigonometric Functions

We have seen that the inverse $f^{-1}(x)$ exists if $f(x)$ is one-to-one on its domain. The trigonometric functions are not one-to-one. Therefore, we define their inverses by restricting them to domains on which they are one-to-one.

First consider the sine function. Figure 13 shows that $f(\theta) = \sin\theta$ is one-to-one on $[-\pi/2, \pi/2]$. When $\sin\theta$ is restricted to the interval $[-\pi/2, \pi/2]$, its inverse is called the

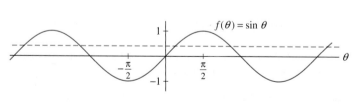
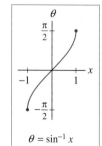

FIGURE 13 $f(\theta) = \sin\theta$ is one-to-one on $[-\pi/2, \pi/2]$ and $\theta = \sin^{-1} x$ is its inverse on $[-\pi/2, \pi/2]$.

Do not confuse the inverse $\sin^{-1} x$ *with the reciprocal*

$$(\sin x)^{-1} = \frac{1}{\sin x}$$

The inverse functions $\sin^{-1} x$, $\cos^{-1} x$, ...
are often denoted $\arcsin x$, $\arccos x$, *etc.*

arcsine function and is denoted $\theta = \sin^{-1} x$. By definition,

$$\boxed{\theta = \sin^{-1} x \text{ is the unique angle in } \left[-\frac{\pi}{2}, \frac{\pi}{2}\right] \text{ such that } \sin \theta = x}$$

A table of values for the arcsine (Table 1) is obtained by reversing the columns in a table of values for $\sin x$.

TABLE 1

x	$\theta = \sin^{-1} x$
-1	$-\frac{\pi}{2}$
$-\frac{\sqrt{3}}{2}$	$-\frac{\pi}{3}$
$-\frac{\sqrt{2}}{2}$	$-\frac{\pi}{4}$
$-\frac{1}{2}$	$-\frac{\pi}{6}$
0	0
$\frac{1}{2}$	$\frac{\pi}{6}$
$\frac{\sqrt{2}}{2}$	$\frac{\pi}{4}$
$\frac{\sqrt{3}}{2}$	$\frac{\pi}{3}$
1	$\frac{\pi}{2}$

Summary of inverse relation between the sine and the arcsine:

$$\sin(\sin^{-1} x) = x \qquad \text{for } -1 \leq x \leq 1$$

$$\sin^{-1}(\sin \theta) = \theta \qquad \text{for } -\frac{\pi}{2} \leq \theta \leq \frac{\pi}{2}$$

■ **EXAMPLE 6** (a) Show that $\sin^{-1}\left(\sin\left(\frac{\pi}{4}\right)\right) = \frac{\pi}{4}$.

(b) Explain why $\sin^{-1}\left(\sin\left(\frac{5\pi}{4}\right)\right) \neq \frac{5\pi}{4}$.

Solution The equation $\sin^{-1}(\sin \theta) = \theta$ is valid only if θ lies in $[-\pi/2, \pi/2]$.

(a) Since $\frac{\pi}{4}$ lies in the required interval, $\sin^{-1}\left(\sin\left(\frac{\pi}{4}\right)\right) = \frac{\pi}{4}$.

(b) Let $\theta = \sin^{-1}\left(\sin\left(\frac{5\pi}{4}\right)\right)$. By definition, θ is the angle in $[-\pi/2, \pi/2]$ such that $\sin \theta = \sin\left(\frac{5\pi}{4}\right)$. By the identity $\sin \theta = \sin(\pi - \theta)$ (Figure 14),

$$\sin\left(\frac{5\pi}{4}\right) = \sin\left(\pi - \frac{5\pi}{4}\right) = \sin\left(-\frac{\pi}{4}\right)$$

The angle $-\frac{\pi}{4}$ lies in the required interval, so $\theta = \sin^{-1}\left(\sin\left(\frac{5\pi}{4}\right)\right) = -\frac{\pi}{4}$. ■

FIGURE 14 $\sin\left(\frac{5\pi}{4}\right) = \sin\left(-\frac{\pi}{4}\right)$.

The cosine function is one-to-one on $[0, \pi]$ rather than $[-\pi/2, \pi/2]$ (Figure 15). When $f(\theta) = \cos \theta$ is restricted to the domain $[0, \pi]$, the inverse is called the **arccosine function** and is denoted $\theta = \cos^{-1} x$. By definition,

$$\boxed{\theta = \cos^{-1} x \text{ is the unique angle in } [0, \pi] \text{ such that } \cos \theta = x}$$

Summary of inverse relation between the cosine and arccosine:

$$\cos(\cos^{-1} x) = x \qquad \text{for } -1 \leq x \leq 1$$

$$\cos^{-1}(\cos \theta) = \theta \qquad \text{for } 0 \leq \theta \leq \pi$$

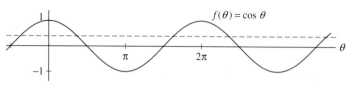

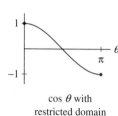

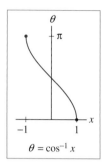

FIGURE 15 $f(\theta) = \cos\theta$ is one-to-one on $[0, \pi]$ and $\theta = \cos^{-1} x$ is its inverse on $[0, \pi]$.

When we study the calculus of inverse trigonometric functions in Section 3.9, we will need to simplify composite expressions such as $\cos(\sin^{-1} x)$ and $\tan(\sec^{-1} x)$. This can be done in two ways: by referring to the appropriate right triangle or by using trigonometric identities.

▪ **EXAMPLE 7** Simplify $\cos(\sin^{-1} x)$ and $\tan(\sin^{-1} x)$.

Solution This problem asks for the values of $\cos\theta$ and $\tan\theta$ at the angle $\theta = \sin^{-1} x$. Consider a right triangle with hypotenuse of length 1 and angle θ such that $\sin\theta = x$ as in Figure 16. By the Pythagorean Theorem, the adjacent side has length $\sqrt{1 - x^2}$. Now we can read off the values from Figure 16:

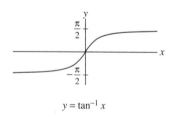

FIGURE 16 Right triangle constructed so that $\sin\theta = x$.

$$\cos(\sin^{-1} x) = \cos\theta = \frac{\text{adjacent}}{\text{hypotenuse}} = \sqrt{1 - x^2}$$

$$\tan(\sin^{-1} x) = \tan\theta = \frac{\text{opposite}}{\text{adjacent}} = \frac{x}{\sqrt{1 - x^2}}$$

$\boxed{3}$

Alternatively, we may argue using trigonometric identities. Because $\sin\theta = x$,

$$\cos(\sin^{-1} x) = \cos\theta = \sqrt{1 - \sin^2\theta} = \sqrt{1 - x^2}$$

We take the positive square root because $\theta = \sin^{-1} x$ lies in $[-\pi/2, \pi/2]$ and $\cos\theta$ is positive in this interval.   ▪

We now address the remaining trigonometric functions. The function $f(\theta) = \tan\theta$ is one-to-one on $(-\pi/2, \pi/2)$, and $f(\theta) = \cot\theta$ is one-to-one on $(0, \pi)$. We define their inverses by restricting to these domains:

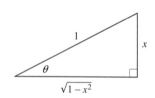

$y = \tan^{-1} x$

$$\theta = \tan^{-1} x \text{ is the unique angle in } \left(-\frac{\pi}{2}, \frac{\pi}{2}\right) \text{ such that } \tan\theta = x$$

$$\theta = \cot^{-1} x \text{ is the unique angle in } (0, \pi) \text{ such that } \cot\theta = x$$

The range of both $\tan\theta$ and $\cot\theta$ is the set of all real numbers $\mathbf{R}$. Therefore, $\theta = \tan^{-1} x$ and $\theta = \cot^{-1} x$ have domain $\mathbf{R}$ (Figure 17).

The function $f(\theta) = \sec\theta$ is not defined at $\theta = \pi/2$, but we see in Figure 18 that it is one-to-one on both $[0, \pi/2)$ and $(\pi/2, \pi]$. Similarly, $f(\theta) = \csc\theta$ is not defined at $\theta = 0$, but it is one-to-one on $[-\pi/2, 0)$ and $(0, \pi/2]$. We define the inverse functions:

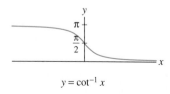

$y = \cot^{-1} x$

FIGURE 17

$$\theta = \sec^{-1} x \text{ is the unique angle in } \left[0, \frac{\pi}{2}\right) \cup \left(\frac{\pi}{2}, \pi\right] \text{ such that } \sec\theta = x$$

$$\theta = \csc^{-1} x \text{ is the unique angle in } \left[-\frac{\pi}{2}, 0\right) \cup \left(0, \frac{\pi}{2}\right] \text{ such that } \csc\theta = x$$

Figure 18 shows that the range of $f(\theta) = \sec\theta$ is the set of real numbers x such that $|x| \geq 1$. The same is true of $f(\theta) = \csc\theta$. It follows that both $\theta = \sec^{-1} x$ and $\theta = \csc^{-1} x$ have domain $\{x : |x| \geq 1\}$.

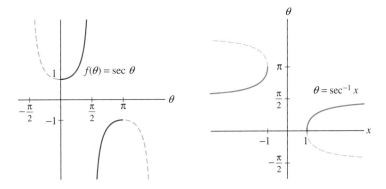

FIGURE 18 $f(\theta) = \sec\theta$ is one-to-one on the interval $[0, \pi]$ (with $\pi/2$ removed).

1.5 SUMMARY

- A function $f(x)$ is *one-to-one* on a domain D if, for all values c, $f(x) = c$ has at most one solution for $x \in D$.
- Let $f(x)$ have domain D and range R. The *inverse* $f^{-1}(x)$ (if it exists) is the function with domain R and range D satisfying $f(f^{-1}(x)) = x$ and $f^{-1}(f(x)) = x$.
- The inverse of $f(x)$ exists if and only if $f(x)$ is one-to-one on its domain.
- To find the inverse of $f(x)$, solve $y = f(x)$ for x in terms of y to obtain $x = g(y)$. The inverse is the function $g(x)$.
- *Horizontal Line Test:* $f(x)$ is one-to-one if and only if every horizontal line intersects the graph of $f(x)$ in at most one point.
- The graph of $f^{-1}(x)$ is obtained by reflecting the graph of $f(x)$ through the line $y = x$.
- The *arcsine* and *arccosine* are defined for $-1 \leq x \leq 1$:

$$\theta = \sin^{-1} x \text{ is the unique angle in } \left[-\frac{\pi}{2}, \frac{\pi}{2}\right] \text{ such that } \sin\theta = x$$

$$\theta = \cos^{-1} x \text{ is the unique angle in } [0, \pi] \text{ such that } \cos\theta = x$$

- $\tan^{-1} x$ and $\cot^{-1} x$ are defined for all x:

$$\theta = \tan^{-1} x \text{ is the unique angle in } \left(-\frac{\pi}{2}, \frac{\pi}{2}\right) \text{ such that } \tan\theta = x$$

$$\theta = \cot^{-1} x \text{ is the unique angle in } (0, \pi) \text{ such that } \cot\theta = x$$

• $\sec^{-1} x$ and $\csc^{-1} x$ are defined for $|x| \geq 1$:

$$\theta = \sec^{-1} x \text{ is the unique angle in } \left[0, \frac{\pi}{2}\right) \cup \left(\frac{\pi}{2}, \pi\right] \text{ such that } \sec\theta = x$$

$$\theta = \csc^{-1} x \text{ is the unique angle in } \left[-\frac{\pi}{2}, 0\right) \cup \left(0, \frac{\pi}{2}\right] \text{ such that } \csc\theta = x$$

1.5 EXERCISES

Preliminary Questions

1. Which of the following satisfy $f^{-1}(x) = f(x)$?

(a) $f(x) = x$

(b) $f(x) = 1 - x$

(c) $f(x) = 1$

(d) $f(x) = \sqrt{x}$

(e) $f(x) = |x|$

(f) $f(x) = x^{-1}$

2. The graph of a function looks like the track of a roller coaster. Is the function one-to-one?

3. Consider the function f that maps teenagers in the United States to their last names. Explain why the inverse of f does not exist.

4. View the following fragment of a train schedule for the New Jersey Transit System as defining a function f from towns to times. Is f one-to-one? What is $f^{-1}(6{:}27)$?

Trenton	6:21
Hamilton Township	6:27
Princeton Junction	6:34
New Brunswick	6:38

5. A homework problem asks for a sketch of the graph of the *inverse* of $f(x) = x + \cos x$. Frank, after trying but failing to find a formula for $f^{-1}(x)$, says it's impossible to graph the inverse. Bianca hands in an accurate sketch without solving for f^{-1}. How did Bianca complete the problem?

6. Which of the following quantities is undefined?

(a) $\sin^{-1}\left(-\frac{1}{2}\right)$

(b) $\cos^{-1}(2)$

(c) $\csc^{-1}\left(\frac{1}{2}\right)$

(d) $\csc^{-1}(2)$

7. Give an example of an angle θ such that $\cos^{-1}(\cos\theta) \neq \theta$. Does this contradict the definition of inverse function?

Exercises

1. Show that $f(x) = 7x - 4$ is invertible by finding its inverse.

2. Is $f(x) = x^2 + 2$ one-to-one? If not, describe a domain on which it is one-to-one.

3. What is the largest interval containing zero on which $f(x) = \sin x$ is one-to-one?

4. Show that $f(x) = \dfrac{x - 2}{x + 3}$ is invertible by finding its inverse.

(a) What is the domain of $f(x)$? The range of $f^{-1}(x)$?

(b) What is the domain of $f^{-1}(x)$? The range of $f(x)$?

5. Verify that $f(x) = x^3 + 3$ and $g(x) = (x - 3)^{1/3}$ are inverses by showing that $f(g(x)) = x$ and $g(f(x)) = x$.

6. Repeat Exercise 5 for $f(t) = \dfrac{t + 1}{t - 1}$ and $g(t) = \dfrac{t + 1}{t - 1}$.

7. The escape velocity from a planet of mass M and radius R is $v(R) = \sqrt{\dfrac{2GM}{R}}$, where G is the universal gravitational constant. Find the inverse of $v(R)$ as a function of R.

In Exercises 8–15, find a domain on which f is one-to-one and a formula for the inverse of f restricted to this domain. Sketch the graphs of f and f^{-1}.

8. $f(x) = 3x - 2$

9. $f(x) = 4 - x$

10. $f(x) = \dfrac{1}{x + 1}$

11. $f(x) = \dfrac{1}{7x - 3}$

12. $f(s) = \dfrac{1}{s^2}$

13. $f(x) = \dfrac{1}{\sqrt{x^2 + 1}}$

14. $f(z) = z^3$

15. $f(x) = \sqrt{x^3 + 9}$

16. For each function shown in Figure 19, sketch the graph of the inverse (restrict the function's domain if necessary).

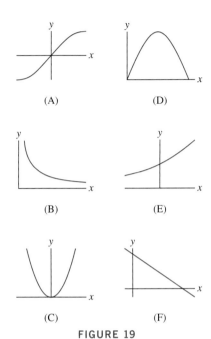

(A)   (D)

(B)   (E)

(C)   (F)

FIGURE 19

17. Which of the graphs in Figure 20 is the graph of a function satisfying $f^{-1} = f$?

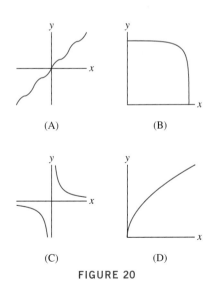

(A)   (B)

(C)   (D)

FIGURE 20

18. Let n be a non-zero integer. Find a domain on which $f(x) = (1 - x^n)^{1/n}$ coincides with its inverse. *Hint:* The answer depends on whether n is even or odd.

19. Let $f(x) = x^7 + x + 1$.

(a) Show that f^{-1} exists (but do not attempt to find it). *Hint:* Show that f is strictly increasing.

(b) What is the domain of f^{-1}?

(c) Find $f^{-1}(3)$.

20. Show that $f(x) = (x^2 + 1)^{-1}$ is one-to-one on $(-\infty, 0]$ and find a formula for f^{-1} for this domain.

21. Let $f(x) = x^2 - 2x$. Determine a domain on which f^{-1} exists and find a formula for f^{-1} on this domain.

22. Show that $f(x) = x + x^{-1}$ is one-to-one on $[1, \infty)$ and find a formula for f^{-1} on this domain. What is the domain of f^{-1}?

In Exercises 23–28, evaluate without using a calculator.

23. $\cos^{-1} 1$ **24.** $\sin^{-1} \frac{1}{2}$

25. $\cot^{-1} 1$ **26.** $\sec^{-1} \dfrac{2}{\sqrt{3}}$

27. $\tan^{-1} \sqrt{3}$ **28.** $\sin^{-1}(-1)$

In Exercises 29–38, compute without using a calculator.

29. $\sin^{-1}\left(\sin \dfrac{\pi}{3}\right)$ **30.** $\sin^{-1}\left(\sin \dfrac{4\pi}{3}\right)$

31. $\cos^{-1}\left(\cos \dfrac{3\pi}{2}\right)$ **32.** $\sin^{-1}\left(\sin\left(-\dfrac{5\pi}{6}\right)\right)$

33. $\tan^{-1}\left(\tan \dfrac{3\pi}{4}\right)$ **34.** $\tan^{-1}(\tan \pi)$

35. $\sec^{-1}(\sec 3\pi)$ **36.** $\sec^{-1}\left(\sec \dfrac{3\pi}{2}\right)$

37. $\csc^{-1}\left(\csc(-\pi)\right)$ **38.** $\cot^{-1}\left(\cot\left(-\dfrac{\pi}{4}\right)\right)$

In Exercises 39–42, simplify by referring to the appropriate triangle or trigonometric identity.

39. $\tan(\cos^{-1} x)$ **40.** $\cos(\tan^{-1} x)$

41. $\cot(\sec^{-1} x)$ **42.** $\cot(\sin^{-1} x)$

In Exercises 43–50, refer to the appropriate triangle or trigonometric identity to compute the given value.

43. $\cos\left(\sin^{-1} \frac{2}{3}\right)$ **44.** $\tan\left(\cos^{-1} \frac{2}{3}\right)$

45. $\tan\left(\sin^{-1} 0.8\right)$ **46.** $\cos\left(\cot^{-1} 1\right)$

47. $\cot\left(\csc^{-1} 2\right)$ **48.** $\tan\left(\sec^{-1}(-2)\right)$

49. $\cot\left(\tan^{-1} 20\right)$ **50.** $\sin\left(\csc^{-1} 20\right)$

Further Insights and Challenges

51. Show that if $f(x)$ is odd and $f^{-1}(x)$ exists, then $f^{-1}(x)$ is odd. Show, on the other hand, that an even function does not have an inverse.

52. A cylindrical tank of radius R and length L lying horizontally as in Figure 21 is filled with oil to height h. Show that the volume $V(h)$ of oil in the tank as a function of height h is

$$V(h) = L\left(R^2 \cos^{-1}\left(1 - \frac{h}{R}\right) - (R-h)\sqrt{2hR - h^2}\right)$$

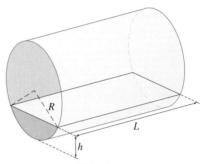

FIGURE 21 Oil in the tank has level h.

1.6 Exponential and Logarithmic Functions

An **exponential function** is a function of the form $f(x) = b^x$, where $b > 0$ and $b \neq 1$. The number b is called the **base**. Some examples are 2^x, $(1.4)^x$, and 10^x. We exclude the case $b = 1$ because $f(x) = 1^x$ is a constant function. Calculators give good decimal approximations to values of exponential functions:

$$2^4 = 16, \quad 2^{-3} = 0.125, \quad (1.4)^3 = 2.744, \quad 10^{4.6} \approx 39,810.717$$

Three properties of exponential functions should be singled out from the start (see Figure 2 for the case $b = 2$):

- *Exponential functions are positive*: $b^x > 0$ for all x.
- $f(x) = b^x$ is *increasing* if $b > 1$ and *decreasing* if $0 < b < 1$.
- The *range* of $f(x) = b^x$ is the set of all positive real numbers.

FIGURE 1 Gordon Moore (1929–).

Moore (who later became chairman of Intel Corporation) predicted that in the decades following 1965, the number of transistors per integrated circuit would grow "exponentially." This prediction has held up for nearly four decades and may well continue for several more years. Moore has said, "Moore's Law is a term that got applied to a curve I plotted in the mid-sixties showing the increase in complexity of integrated circuits versus time. It's been expanded to include a lot more than that, and I'm happy to take credit for all of it."

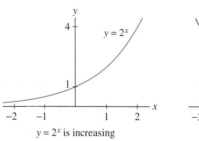

$y = 2^x$ is increasing

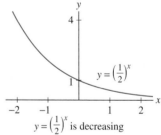

$y = \left(\frac{1}{2}\right)^x$ is decreasing

FIGURE 2

Another important feature of $f(x) = b^x$ (for $b > 1$) is that it increases rapidly. Although "rapid increase" is a subjective term, the following precise statement is true: $f(x) = b^x$ increases more rapidly than every polynomial function (see Section 4.7). For example, Figure 3 shows that $f(x) = 3^x$ eventually overtakes and increases faster than the power functions x^3, x^4, and x^5. Table 1 compares 3^x and x^5.

We now review the laws of exponents. The most important law is

$$\boxed{b^x b^y = b^{x+y}}$$

In other words, *under multiplication, the exponents add*, provided that the bases are the same. This law does not apply to a product such as $3^2 \cdot 5^4$.

TABLE 1		
x	x^5	3^x
1	1	3
5	3,125	243
10	100,000	59,049
15	759,375	14,348,907
25	9,765,625	847,288,609,443

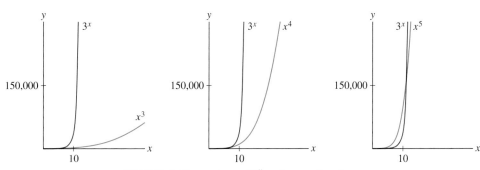

FIGURE 3 Comparison of 3^x and power functions.

THEOREM 1 Laws of Exponents ($b > 0$)

	Rule	Example
Exponent zero	$b^0 = 1$	
Products	$b^x b^y = b^{x+y}$	$2^5 \cdot 2^3 = 2^{5+3} = 2^8$
Quotients	$\dfrac{b^x}{b^y} = b^{x-y}$	$\dfrac{4^7}{4^2} = 4^{7-2} = 4^5$
Negative exponents	$b^{-x} = \dfrac{1}{b^x}$	$3^{-4} = \dfrac{1}{3^4} = \dfrac{1}{81}$
Power to a power	$(b^x)^y = b^{xy}$	$(3^2)^4 = 3^{2(4)} = 3^8$
Roots	$b^{1/n} = \sqrt[n]{b}$	$5^{1/2} = \sqrt{5}$

Be sure you are familiar with the laws of exponents. They are used throughout this text.

■ **EXAMPLE 1** Rewrite as a whole number or fraction (without using a calculator):

(a) $16^{-1/2}$ **(b)** $27^{2/3}$ **(c)** $\dfrac{9^3}{3^7}$ **(d)** $(4^3)^6 \, 4^{-16}$

Solution

(a) $16^{-1/2} = \dfrac{1}{16^{1/2}} = \dfrac{1}{\sqrt{16}} = \dfrac{1}{4}$ **(b)** $27^{2/3} = (27^{1/3})^2 = 3^2 = 9$

(c) $\dfrac{9^3}{3^7} = \dfrac{(3^2)^3}{3^7} = \dfrac{3^6}{3^7} = 3^{-1} = \dfrac{1}{3}$ **(d)** $(4^3)^6 \cdot 4^{-16} = 4^{18} \cdot 4^{-16} = 4^2 = 16$ ■

In the next example, we use the fact that $f(x) = b^x$ is one-to-one. In other words, if $b^x = b^y$, then $x = y$.

■ **EXAMPLE 2** Solve for the unknown:

(a) $2^{3x+1} = 2^5$ **(b)** $b^3 = 5^6$ **(c)** $7^{t+1} = \left(\frac{1}{7}\right)^{2t}$

Solution

(a) If $2^{3x+1} = 2^5$, then $3x + 1 = 5$ and thus $x = \frac{4}{3}$.

(b) Raise both sides of $b^3 = 5^6$ to the $\frac{1}{3}$ power. By the "power to a power" rule, we get

$$b = \left(b^3\right)^{1/3} = \left(5^6\right)^{1/3} = 5^{6/3} = 5^2 = 25$$

(c) Since $\frac{1}{7} = 7^{-1}$, the right-hand side of the equation is $\left(\frac{1}{7}\right)^{2t} = \left(7^{-1}\right)^{2t} = 7^{-2t}$. The equation becomes $7^{t+1} = 7^{-2t}$. Therefore, $t + 1 = -2t$, or $t = -\frac{1}{3}$. ∎

The Number e

Although written references to the number π go back more than 4,000 years, mathematicians first became aware of the special role played by e in the seventeenth century. The notation e was introduced by Leonhard Euler, who discovered many fundamental properties of this important number.

In Chapter 3, we will use calculus to study exponential functions. One of the surprising insights of calculus is that the most convenient or "natural" base for an exponential function is not $b = 10$ or $b = 2$, as one might think at first, but rather a certain irrational number that is denoted by e. The value of e is approximately 2.7. More accurate approximations may be obtained using a calculator or computer algebra system:

$$e \approx 2.718281828459\ldots$$

Similarly, we may use a calculator to evaluate specific values $f(x) = e^x$. For example,

$$e^3 \approx 20.0855, \qquad e^{-1/4} \approx 0.7788$$

How is e defined? There are many different definitions, but they all require the calculus concept of a limit. We shall discuss the definition of e in detail in Section 3.2. For now, we mention two graphical ways of describing e based on Figure 4.

- Using Figure 4(A): Among all exponential functions $y = b^x$, the base $b = e$ is the unique base for which the slope of the tangent line to the graph of $y = e^x$ at the point $(0, 1)$ is equal to 1.
- Using Figure 4(B): The number e is the unique number such that the area of the region under the hyperbola $y = \dfrac{1}{x}$ for $1 \leq x \leq e$ is equal to 1.

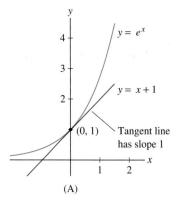
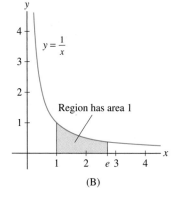

(A)　　　　　　　　(B)

FIGURE 4

We will see how these graphical properties of e arise when we analyze exponential and logarithm functions using calculus. Also, note that the tangent line in Figure 4(A) is the line that touches the graph in one point without crossing the graph. The precise definition of tangent lines also requires calculus and is discussed in Sections 2.1 and 3.1.

Logarithms

Logarithm functions are inverses of exponential functions. More precisely, if $b > 0$ and $b \neq 1$, then the *logarithm to the base b*, denoted $\log_b x$, is the inverse of $f(x) = b^x$. By definition, $y = \log_b x$ if $b^y = x$.

$$b^{\log_b x} = x \qquad \text{and} \qquad \log_b(b^x) = x$$

Thus, $\log_b x$ is the number to which b must be raised in order to get x. For example,

$$\log_2(8) = 3 \quad \text{because} \quad 2^3 = 8$$

$$\log_{10}(1) = 0 \quad \text{because} \quad 10^0 = 1$$

$$\log_3\left(\frac{1}{9}\right) = -2 \quad \text{because} \quad 3^{-2} = \frac{1}{3^2} = \frac{1}{9}$$

The logarithm to the base e, denoted $\ln x$, plays a special role and is called the **natural logarithm**. We use a calculator to evaluate logarithms numerically. For example,

$$\ln 17 \approx 2.83321 \quad \text{because} \quad e^{2.83321} \approx 17$$

Recall that the domain of b^x is **R** and its range is the set of positive real numbers $\{x : x > 0\}$. Since the domain and range are reversed in the inverse function,

- The *domain* of $\log_b x$ is $\{x : x > 0\}$.
- The *range* of $\log_b x$ is the set of all real numbers **R**.

If $b > 1$, then $\log_b x$ is positive for $x > 1$ and negative for $0 < x < 1$. Figure 6 illustrates these facts for the base $b = e$. Note that the logarithm of a negative number does not exist. For example, $\log_{10}(-2)$ does not exist because $10^y > 0$ for all y and thus $10^y = -2$ has no solution.

For each law of exponents, there is a corresponding law for logarithms. The rule $b^{x+y} = b^x b^y$ corresponds to the rule

$$\log_b(xy) = \log_b x + \log_b y$$

In words: *The log of a product is the sum of the logs.* To verify this rule, we write xy as an exponential with base b in two ways:

$$b^{\log_b(xy)} = xy = b^{\log_b x} \cdot b^{\log_b y} = b^{\log_b x + \log_b y}$$

Therefore, the exponents $\log_b(xy)$ and $\log_b x + \log_b y$ are equal as claimed. The remaining logarithm laws are collected in the following table.

FIGURE 5 Renato Solidum, director of the Philippine Institute of Volcanology and Seismology, checks the intensity of the October 8, 2004, Manila earthquake, which registered 6.2 on the Richter scale. The Richter scale is based on the logarithm (to base 10) of the amplitude of seismic waves. Each whole-number increase in Richter magnitude corresponds to a 10-fold increase in amplitude and around 31 times more energy.

In this text, the natural logarithm is denoted $\ln x$. Other common notations are $\log x$ or $\operatorname{Log} x$.

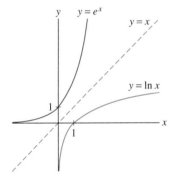

FIGURE 6 $y = \ln x$ is the inverse of $y = e^x$.

Laws of Logarithms

	Law	Example
Log of 1	$\log_b(1) = 0$	
Log of b	$\log_b(b) = 1$	
Products	$\log_b(xy) = \log_b x + \log_b y$	$\log_5(2 \cdot 3) = \log_5 2 + \log_5 3$
Quotients	$\log_b\left(\dfrac{x}{y}\right) = \log_b x - \log_b y$	$\log_2\left(\dfrac{3}{7}\right) = \log_2 3 - \log_2 7$
Reciprocals	$\log_b\left(\dfrac{1}{x}\right) = -\log_b x$	$\log_2 \dfrac{1}{7} = -\log_2 7$
Powers (any n)	$\log_b(x^n) = n \log_b x$	$\log_{10}(8^2) = 2 \cdot \log_{10} 8$

We also note that all logarithm functions are proportional. More precisely, the following **change of base** formula holds (see Exercise 44):

$$\log_b x = \frac{\log_a x}{\log_a b}, \qquad \log_b x = \frac{\ln x}{\ln b} \qquad \boxed{1}$$

■ **EXAMPLE 3** Using the Logarithm Laws Evaluate without using a calculator:

(a) $\log_6 9 + \log_6 4$ **(b)** $\ln\left(\dfrac{1}{\sqrt{e}}\right)$ **(c)** $10\log_b(b^3) - 4\log_b(\sqrt{b})$

Solution

(a) $\log_6 9 + \log_6 4 = \log_6(9 \cdot 4) = \log_6(36) = \log_6(6^2) = 2$

(b) $\ln\left(\dfrac{1}{\sqrt{e}}\right) = \ln(e^{-1/2}) = -\dfrac{1}{2}\ln(e) = -\dfrac{1}{2}$

(c) $10\log_b(b^3) - 4\log_b(\sqrt{b}) = 10(3) - 4\log_b(b^{1/2}) = 30 - 4\left(\dfrac{1}{2}\right) = 28$ ■

■ **EXAMPLE 4** Solving an Exponential Equation The bacteria population in a bottle at time t (in hours) has size $P(t) = 1{,}000e^{0.35t}$. After how many hours will there be 5,000 bacteria?

Solution We must solve $P(t) = 1{,}000e^{0.35t} = 5{,}000$ for t (Figure 7):

$$e^{0.35t} = \frac{5{,}000}{1{,}000} = 5$$

$$\ln(e^{0.35t}) = \ln 5 \qquad \text{(take logarithms of both sides)}$$

$$0.35t = \ln 5 \qquad [\text{because } \ln(e^a) = a]$$

$$t = \frac{\ln 5}{0.35} \approx 4.6 \text{ hours} \qquad ■$$

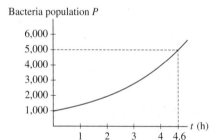

Bacteria population P

FIGURE 7 Bacteria population as a function of time.

FIGURE 8 The St. Louis Arch has the shape of an inverted hyperbolic cosine.

Hyperbolic Functions

The hyperbolic functions are certain special combinations of e^x and e^{-x} that play a role in engineering and physics (see Figure 8 for a real-life example). The hyperbolic sine and cosine, often called "cinch" and "cosh," are defined as follows:

$$\sinh x = \frac{e^x - e^{-x}}{2}, \qquad \cosh x = \frac{e^x + e^{-x}}{2}$$

As the terminology suggests, there are similarities between the hyperbolic and trigonometric functions. Here are some examples:

- **Parity:** The trigonometric functions and their hyperbolic analogs have the same parity. Thus, $\sin x$ and $\sinh x$ are both odd, and $\cos x$ and $\cosh x$ are both even (Figure 9):

$$\sinh(-x) = -\sinh x, \qquad \cosh(-x) = \cosh x$$

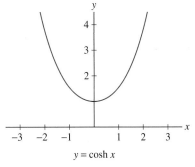

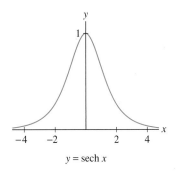

FIGURE 9 $y = \sinh x$ is an odd function, $y = \cosh x$ is an even function.

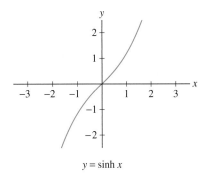

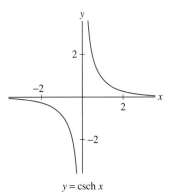

FIGURE 11 The hyperbolic secant and cosecant.

- **Identities:** The basic trigonometric identity $\sin^2 x + \cos^2 x = 1$ has a hyperbolic analog:

$$\boxed{\cosh^2 x - \sinh^2 x = 1} \qquad \boxed{2}$$

The addition formulas satisfied by $\sin \theta$ and $\cos \theta$ also have hyperbolic analogs:

$$\sinh(x + y) = \sinh x \cosh y + \cosh x \sinh y$$
$$\cosh(x + y) = \cosh x \cosh y + \sinh x \sinh y$$

■ **EXAMPLE 5** Verifying the Basic Identity Verify Eq. (2).

Solution It follows from the definitions that

$$\cosh x + \sinh x = e^x, \qquad \cosh x - \sinh x = e^{-x}$$

We obtain Eq. (2) by multiplying these two equations together:

$$\cosh^2 x - \sinh^2 x = (\cosh x + \sinh x)(\cosh x - \sinh x) = e^x \cdot e^{-x} = 1 \qquad ■$$

- **Hyperbola instead of the circle:** The identity $\sin^2 t + \cos^2 t = 1$ tells us that the point $(\cos t, \sin t)$ lies on the unit circle $x^2 + y^2 = 1$. Similarly, the identity $\cosh^2 t - \sinh^2 t = 1$ says that the point $(\cosh t, \sinh t)$ lies on the *hyperbola* $x^2 - y^2 = 1$ (Figure 10).

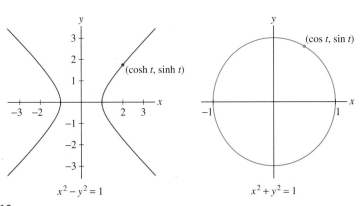

FIGURE 10

- **Other hyperbolic functions:** The hyperbolic tangent, cotangent, secant, and cosecant functions (see Figures 11 and 12) are defined like their trigonometric counterparts:

$$\tanh x = \frac{\sinh x}{\cosh x} = \frac{e^x - e^{-x}}{e^x + e^{-x}}, \quad \operatorname{sech} x = \frac{1}{\cosh x} = \frac{2}{e^x + e^{-x}}$$

$$\coth x = \frac{\cosh x}{\sinh x} = \frac{e^x + e^{-x}}{e^x - e^{-x}}, \quad \operatorname{csch} x = \frac{1}{\sinh x} = \frac{2}{e^x - e^{-x}}$$

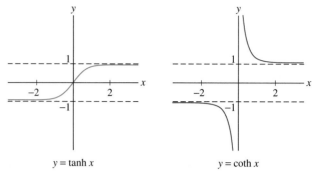

$$y = \tanh x \qquad\qquad y = \coth x$$

FIGURE 12 The hyperbolic tangent and cotangent.

Inverse Hyperbolic Functions

Function	Domain
$y = \sinh^{-1} x$	all x
$y = \cosh^{-1} x$	$x \geq 1$
$y = \tanh^{-1} x$	$|x| < 1$
$y = \coth^{-1} x$	$|x| > 1$
$y = \text{sech}^{-1} x$	$0 < x \leq 1$
$y = \text{csch}^{-1} x$	$x \neq 0$

Einstein's Law (3) reduces to Galileo's Law, $w = u + v$, when u and v are small relative to the velocity of light c. See Exercise 45 for another way of expressing (3).

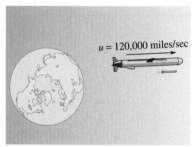

FIGURE 13 What is the missile's velocity relative to the earth?

Inverse Hyperbolic Functions

Each of the hyperbolic functions, except $y = \cosh x$ and $y = \text{sech}\, x$, is one-to-one on its domain and therefore has a well-defined inverse. The functions $y = \cosh x$ and $y = \text{sech}\, x$ are one-to-one on the restricted domain $\{x : x \geq 0\}$. We let $\cosh^{-1} x$ and $\text{sech}^{-1} x$ denote the corresponding inverses.

Einstein's Law of Velocity Addition

The hyperbolic tangent plays a role in the Special Theory of Relativity, developed by Albert Einstein in 1905. One consequence of this theory is that no object can travel faster than the speed of light, $c \approx 186{,}000$ miles/sec. Einstein realized that this contradicts a law stated by Galileo more than 250 years earlier, namely that *velocities add*. Imagine a train traveling at $u = 100$ mph and a man walking down the aisle in the train at $v = 3$ mph. According to Galileo, the man's velocity relative to the ground is $u + v = 100 + 3$ or 103 mph. This agrees with our everyday experience. But now imagine an (unrealistic) rocket traveling away from the earth at $u = 120{,}000$ miles/sec and suppose that the rocket fires a missile with velocity $v = 80{,}000$ miles/sec (relative to the rocket). If Galileo's Law were correct, the velocity of the missile relative to the earth would be $u + v = 200{,}000$ miles/sec, which exceeds Einstein's maximum speed limit of $c \approx 186{,}000$ miles/sec.

However, Einstein's theory replaces Galileo's Law with a new law stating that the *inverse hyperbolic tangents of the velocities add*. More precisely, if u is the rocket's velocity relative to the earth and v is the missile's velocity relative to the rocket, then the velocity of the missile relative to the earth (Figure 13) is w, where

$$\tanh^{-1}\left(\frac{w}{c}\right) = \tanh^{-1}\left(\frac{u}{c}\right) + \tanh^{-1}\left(\frac{v}{c}\right) \qquad \boxed{3}$$

■ **EXAMPLE 6** A space ship travels away from the earth at a velocity of 100,000 mph. An astronaut in the ship fires a gun that emits an electron which travels at 140,000 miles per second in the same direction as the space ship. Use Einstein's Law to find the velocity w of the electron relative to the earth.

Solution The space ship's velocity is $100{,}000/3600 \approx 28$ miles/sec. According to Einstein's Law,

$$\tanh^{-1}\left(\frac{w}{c}\right) = \tanh^{-1}\left(\frac{28}{186{,}000}\right) + \tanh^{-1}\left(\frac{140{,}000}{186{,}000}\right)$$

$$\approx 0.00015 + 0.97913 = 0.97928$$

Therefore, $\dfrac{w}{c} \approx \tanh(0.97928) \approx 0.75275$, and

$$w \approx 0.75275c \approx 0.75275(186{,}000) \approx 140{,}012 \text{ miles/sec}$$

If we had used Galileo's Law, we would have obtained $w = 140{,}028$ miles/sec. ■

■ **EXAMPLE 7** Low Velocities   A plane traveling at 600 mph fires a missile at a velocity of 400 mph. Calculate the missile's velocity w relative to the earth (in mph) using both Einstein's Law and Galileo's Law.

Solution   To apply Einstein's law, we convert all velocities to miles/sec:

$$\tanh^{-1}\left(\frac{w/3600}{186{,}000}\right) = \tanh^{-1}\left(\frac{600/3600}{186{,}000}\right) + \tanh^{-1}\left(\frac{400/3600}{186{,}000}\right)$$

$$\approx 1.493428912784 \cdot 10^{-6}$$

We obtain $w \approx (3600)(186{,}000)\tanh(1.493428912784 \cdot 10^{-6}) \approx 1000 - 6 \cdot 10^{-10}$ mph. This is practically indistinguishable from $w = 600 + 400 = 1000$ mph, obtained using Galileo's Law. ■

1.6 SUMMARY

- $f(x) = b^x$ is the *exponential function* with base b (where $b > 0$ and $b \neq 1$).
- $f(x) = b^x$ is increasing if $b > 1$ and decreasing if $b < 1$.
- The number $e \approx 2.718$.
- For $b > 0$ with $b \neq 1$, the *logarithm function* $\log_b x$ is the inverse of b^x;

$$x = b^y \quad \Leftrightarrow \quad y = \log_b x$$

- The *natural logarithm* is the logarithm to the base e and is denoted $\ln x$.
- Important logarithm laws:

(i) $\log_b(xy) = \log_b x + \log_b y$ **(ii)** $\log_b\left(\dfrac{x}{y}\right) = \log_b x - \log_b y$

(iii) $\log_b(x^n) = n\log_b x$ **(iv)** $\log_b 1 = 0$ and $\log_b b = 1$

- The *hyperbolic sine and cosine:*

$$\sinh x = \frac{e^x - e^{-x}}{2} \quad \text{(odd function)}, \qquad \cosh x = \frac{e^x + e^{-x}}{2} \quad \text{(even function)}$$

The remaining hyperbolic functions:

$$\tanh x = \frac{\sinh x}{\cosh x}, \qquad \coth x = \frac{\cosh x}{\sinh x}, \qquad \operatorname{sech} x = \frac{1}{\cosh x}, \qquad \operatorname{csch} x = \frac{1}{\sinh x}$$

- Basic identity: $\cosh^2 x - \sinh^2 x = 1$.
- The inverse hyperbolic functions:

$$\sinh^{-1} x, \text{ for all } x \qquad \coth^{-1} x, \text{ for } |x| > 1$$

$$\cosh^{-1} x, \text{ for } x \geq 1 \qquad \operatorname{sech}^{-1} x, \text{ for } 0 < x \leq 1$$

$$\tanh^{-1} x, \text{ for } |x| < 1 \qquad \operatorname{csch}^{-1} x, \text{ for } x \neq 0$$

1.6 EXERCISES

Preliminary Questions

1. Which of the following equations is incorrect?

(a) $3^2 \cdot 3^5 = 3^7$ **(b)** $(\sqrt{5})^{4/3} = 5^{2/3}$

(c) $3^2 \cdot 2^3 = 1$ **(d)** $(2^{-2})^{-2} = 16$

2. Compute $\log_{b^2}(b^4)$.

3. When is $\ln x$ negative?

4. What is $\ln(-3)$? Explain.

5. Explain the phrase "the logarithm converts multiplication into addition."

6. What are the domain and range of $\ln x$?

7. Which hyperbolic functions take on only positive values?

8. Which hyperbolic functions are increasing on their domains?

9. Describe three properties of hyperbolic functions that have trigonometric analogs.

Exercises

1. Rewrite as a whole number (without using a calculator):

(a) 7^0 **(b)** $10^2(2^{-2} + 5^{-2})$

(c) $\dfrac{(4^3)^5}{(4^5)^3}$ **(d)** $27^{4/3}$

(e) $8^{-1/3} \cdot 8^{5/3}$ **(f)** $3 \cdot 4^{1/4} - 12 \cdot 2^{-3/2}$

In Exercises 2–10, solve for the unknown variable.

2. $9^{2x} = 9^8$ **3.** $e^{2x} = e^{x+1}$

4. $e^{t^2} = e^{4t-3}$ **5.** $3^x = \left(\frac{1}{3}\right)^{x+1}$

6. $(\sqrt{5})^x = 125$ **7.** $4^{-x} = 2^{x+1}$

8. $b^4 = 10^{12}$ **9.** $k^{3/2} = 27$

10. $(b^2)^{x+1} = b^{-6}$

In Exercises 11–22, calculate directly (without using a calculator).

11. $\log_3 27$ **12.** $\log_5 \frac{1}{25}$

13. $\log_2(2^{5/3})$ **14.** $\log_2(8^{5/3})$

15. $\log_{64} 4$ **16.** $\log_7(49^2)$

17. $\log_8 2 + \log_4 2$ **18.** $\log_{25} 30 + \log_{25} \frac{5}{6}$

19. $\log_4 48 - \log_4 12$ **20.** $\ln(\sqrt{e} \cdot e^{7/5})$

21. $\ln(e^3) + \ln(e^4)$ **22.** $\log_2 \frac{4}{3} + \log_2 24$

23. Write as the natural log of a single expression:

(a) $2\ln 5 + 3\ln 4$ **(b)** $5\ln(x^{1/2}) + \ln(9x)$

24. Solve for x: $\ln(x^2 + 1) - 3\ln x = \ln(2)$.

In Exercises 25–30, solve for the unknown.

25. $7e^{5t} = 100$ **26.** $6e^{-4t} = 2$

27. $2^{x^2 - 2x} = 8$ **28.** $e^{2t+1} = 9e^{1-t}$

29. $\ln(x^4) - \ln(x^2) = 2$ **30.** $\log_3 y + 3\log_3(y^2) = 14$

31. Use a calculator to compute $\sinh x$ and $\cosh x$ for $x = -3, 0, 5$.

32. Compute $\sinh(\ln 5)$ and $\tanh(3\ln 5)$ without using a calculator.

33. For which values of x are $y = \sinh x$ and $y = \cosh x$ increasing and decreasing?

34. Show that $y = \tanh x$ is an odd function.

35. The population of a city (in millions) at time t (years) is $P(t) = 2.4e^{0.06t}$, where $t = 0$ is the year 2000. When will the population double from its size at $t = 0$?

36. According to the Gutenberg–Richter law, the number N of earthquakes worldwide of Richter magnitude M approximately satisfies a relation $\ln N = 16.17 - bM$ for some constant b. Find b, assuming that there are 800 earthquakes of magnitude $M = 5$ per year. How many earthquakes of magnitude $M = 7$ occur per year?

37. The energy E (in joules) radiated as seismic waves from an earthquake of Richter magnitude M is given by the formula $\log_{10} E = 4.8 + 1.5M$.

(a) Express E as a function of M.

(b) Show that when M increases by 1, the energy increases by a factor of approximately 31.

38. ✎ Refer to the graphs to explain why the equation $\sinh x = t$ has a unique solution for every t and $\cosh x = t$ has two solutions for every $t > 1$.

39. Compute $\cosh x$ and $\tanh x$, assuming that $\sinh x = 0.8$.

40. Prove the addition formula for $\cosh x$.

41. Use the addition formulas to prove

$$\sinh(2x) = 2\cosh x \sinh x$$

$$\cosh(2x) = \cosh^2 x + \sinh^2 x$$

42. An (imaginary) train moves along a track at velocity v, and a woman walks down the aisle of the train with velocity u in the direction of the train's motion. Compute the velocity w of the woman relative to the ground using the laws of both Galileo and Einstein in the following cases.

(a) $v = 100$ mph and $u = 5$ mph. Is your calculator accurate enough to detect the difference between the two laws?

(b) $v = 140,000$ mph and $u = 100$ mph.

Further Insights and Challenges

43. Show that $\log_a b \, \log_b a = 1$.

44. Verify the formula $\log_b x = \dfrac{\log_a x}{\log_a b}$ for $a, b > 0$.

45. **(a)** Use the addition formulas for $\sinh x$ and $\cosh x$ to prove

$$\tanh(u + v) = \frac{\tanh u + \tanh v}{1 + \tanh u \tanh v}$$

(b) Use (a) to show that Einstein's Law of Velocity Addition [Eq. (3)] is equivalent to

$$w = \frac{u + v}{1 + \dfrac{uv}{c^2}}$$

46. Prove that every function $f(x)$ is the sum of an even function $f_+(x)$ and an odd function $f_-(x)$. [*Hint:* $f_\pm(x) = \frac{1}{2}(f(x) \pm f(-x))$.] Express $f(x) = 5e^x + 8e^{-x}$ in terms of $\cosh x$ and $\sinh x$.

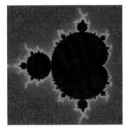

FIGURE 1 Computer-generated image of the Mandelbrot Set, which occurs in the mathematical theory of chaos and fractals.

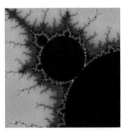

FIGURE 2 Even greater complexity is revealed when we zoom in on a portion of the Mandelbrot Set.

1.7 Technology: Calculators and Computers

Computer technology has vastly extended our ability to calculate and visualize mathematical relationships. In applied settings, computers are indispensable for solving complex systems of equations and analyzing data, as in weather prediction and medical imaging. Mathematicians use computers to study complex structures such as the Mandelbrot set (Figures 1 and 2). We take advantage of this technology to explore the ideas of calculus visually and numerically.

When we plot a function with a graphing calculator, the graph is contained within a **viewing rectangle**, the area determined by the range of x and y values in the plot. We write $[a, b] \times [c, d]$ to denote the rectangle where $a \leq x \leq b$ and $c \leq y \leq d$.

The appearance of the graph depends heavily on the choice of viewing rectangle. Different choices may convey very different impressions, which are sometimes misleading. Compare the three viewing rectangles for the graph of $f(x) = 12 - x - x^2$ in Figure 3. Only (A) successfully displays the shape of the graph as a parabola. In (B), the graph is cut off, and no graph at all appears in (C). Keep in mind that the scales along the axes may change with the viewing rectangle. For example, the unit increment along the y-axis is larger in (B) than (A), and therefore, the graph in (B) is steeper.

There is no single "correct" viewing rectangle. The goal is to select the viewing rectangle that displays the properties you wish to investigate. This usually requires experimentation.

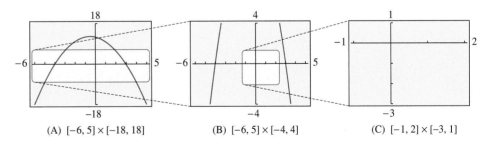

FIGURE 3 Viewing rectangles for the graph of $f(x) = 12 - x - x^2$.

(A) $[-6, 5] \times [-18, 18]$ (B) $[-6, 5] \times [-4, 4]$ (C) $[-1, 2] \times [-3, 1]$

■ **EXAMPLE 1** **How Many Roots and Where?** How many real roots does the function $f(x) = x^9 - 20x + 1$ have? Find their approximate locations.

Solution We experiment with several viewing rectangles (Figure 4). Our first attempt (A) displays a cut off graph, so we try a viewing rectangle that includes a larger range of y values. Plot (B) shows that the roots of $f(x)$ lie somewhere in the interval $[-3, 3]$, but it does not reveal how many real roots there are. Therefore, we try the viewing rectangle in (C). Now we can see clearly that $f(x)$ has three roots. A further zoom in (D) shows

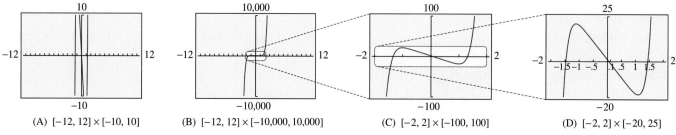

(A) $[-12, 12] \times [-10, 10]$ (B) $[-12, 12] \times [-10{,}000, 10{,}000]$ (C) $[-2, 2] \times [-100, 100]$ (D) $[-2, 2] \times [-20, 25]$

FIGURE 4 Graphs of $f(x) = x^9 - 20x + 1$.

that these roots are located near -1.5, 0.1, and 1.5. Further zooming would provide their locations with greater accuracy.   ∎

❚ **EXAMPLE 2** **Does a Solution Exist?** Does $\cos x = \tan x$ have a solution? Describe the set of all solutions.

Solution The solutions of $\cos x = \tan x$ are the x-coordinates of the points where the graphs of $y = \cos x$ and $y = \tan x$ intersect. Figure 5(A) shows that there are two solutions in the interval $[0, 2\pi]$. By zooming in on the graph as in (B), we see that the first positive root lies between 0.6 and 0.7 and the second positive root lies between 2.4 and 2.5. Further zooming shows that the first root is approximately 0.67 [Figure 5(C)]. Continuing this process, we find that the first two roots are $x \approx 0.666$ and $x \approx 2.475$.

Since $\cos x$ and $\tan x$ are periodic, the picture repeats itself with period 2π. All solutions are obtained by adding multiples of 2π to the two solutions in $[0, 2\pi]$:

$$x \approx 0.666 + 2\pi k \qquad \text{and} \qquad x \approx 2.475 + 2\pi k \quad \text{(for any integer } k)$$ ∎

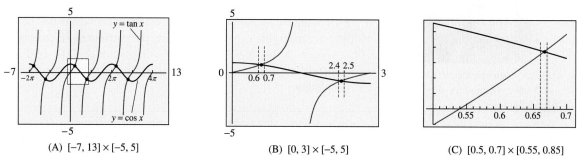

(A) $[-7, 13] \times [-5, 5]$ (B) $[0, 3] \times [-5, 5]$ (C) $[0.5, 0.7] \times [0.55, 0.85]$

FIGURE 5 Graphs of $y = \cos x$ and $y = \tan x$.

❚ **EXAMPLE 3** **Functions with Asymptotes** Plot the function $f(x) = \dfrac{1 - 3x}{x - 2}$ and describe its asymptotic behavior.

Solution First, we plot $f(x)$ in the viewing rectangle $[-10, 20] \times [-5, 5]$ as in Figure 6(A). The vertical line $x = 2$ is called a **vertical asymptote**. Many graphing calculators display this line, but it is *not* part of the graph (and it can usually be eliminated by choosing a smaller range of y-values). We see that $f(x)$ tends to ∞ as x approaches 2 from the left, and to $-\infty$ as x approaches 2 from the right. To display the horizontal asymptotic behavior of $f(x)$, we use the viewing rectangle $[-10, 20] \times [-10, 5]$ [Figure 6(B)]. Here we see that the graph approaches the horizontal line $y = -3$, called a **horizontal asymptote** (which we have added as a dashed horizontal line in the figure).   ∎

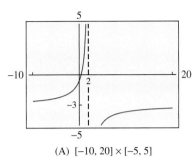

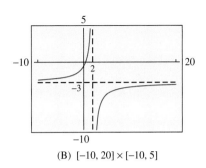

(A) [−10, 20] × [−5, 5]   (B) [−10, 20] × [−10, 5]

FIGURE 6 Graphs of $f(x) = \dfrac{1 - 3x}{x - 2}$.

Calculators give us the freedom to experiment numerically. For instance, we can explore the behavior of a function by constructing a table of values. In the next example, we investigate a function related to exponential functions and compound interest (see Section 7.5).

We will prove in Section 5.8 that $f(n)$ approaches the number e as n tends to infinity. This provides yet another definition of e as the limit *of the function $f(n)$.*

■ **EXAMPLE 4** Investigating the Behavior of a Function How does $f(n) = \left(1 + \dfrac{1}{n}\right)^n$ behave for large whole number values of n? Does $f(n)$ tend to infinity as n gets larger?

Solution First, we make a table of values of $f(n)$ for larger and larger values of n. Table 1 suggests that $f(n)$ does not tend to infinity. Rather, $f(n)$ appears to get closer to some value near 2.718 as n grows larger. This is an example of limiting behavior that we will discuss in Chapter 2. Next, replace n by the variable x and plot the function $f(x) = \left(1 + \dfrac{1}{x}\right)^x$. The graphs in Figure 7 confirm that $f(x)$ approaches a limit of approximately 2.7. ■

TABLE 1

n	$\left(1 + \dfrac{1}{n}\right)^n$
10	2.59374
10^2	2.70481
10^3	2.71692
10^4	2.71815
10^5	2.71827
10^6	2.71828

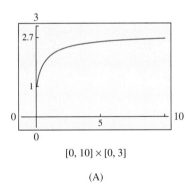

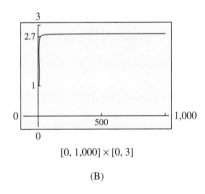

[0, 10] × [0, 3]   [0, 1,000] × [0, 3]

(A)   (B)

FIGURE 7 Graphs of $f(x) = \left(1 + \dfrac{1}{x}\right)^x$.

■ **EXAMPLE 5** Bird Flight: Finding a Minimum Graphically According to one model of bird flight, the power consumed by a pigeon flying at velocity v (in meters per second) is $P(v) = 17v^{-1} + 10^{-3}v^3$ (in joules per second). Use a graph of $P(v)$ to find the velocity that minimizes power consumption.

Solution The velocity that minimizes power consumption corresponds to the lowest point on the graph of $P(v)$. We first plot $P(v)$ in a large viewing rectangle (Figure 8). This figure reveals the general shape of the graph, and we see that $P(v)$ takes on a minimum value for v somewhere between $v = 8$ and $v = 9$. In the viewing rectangle $[8, 9.2] \times [2.6, 2.65]$, we see that the minimum occurs at approximately $v = 8.65$ m/s. ■

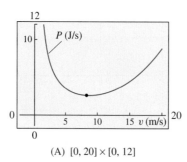

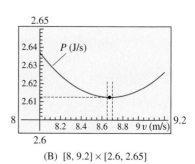

FIGURE 8 Power consumption $P(v)$ as a function of velocity v.

(A) $[0, 20] \times [0, 12]$

(B) $[8, 9.2] \times [2.6, 2.65]$

An important concept in calculus is **local linearity**, the idea that many functions are *nearly linear* over small intervals. Local linearity can be effectively illustrated with a graphing calculator.

■ **EXAMPLE 6** **Illustrating Local Linearity** Illustrate local linearity for the function $f(x) = x^{\sin x}$ at $x = 1$.

Solution First, we plot $f(x) = x^{\sin x}$ in the viewing window of Figure 9(A). The graph moves up and down and appears very wavy. However, as we zoom in, the graph straightens out. Figures (B)–(D) show the result of zooming in on the point $(1, f(1))$. When viewed up close, the graph looks like a straight line. This illustrates local linearity for the graph of $f(x)$ at $x = 1$. ■

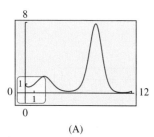

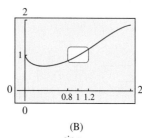

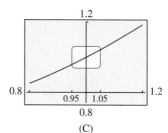

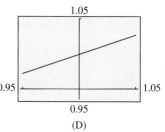

(A)

(B)

(C)

(D)

FIGURE 9 Zooming in on the graph of $f(x) = x^{\sin x}$ near $x = 1$.

1.7 SUMMARY

• The appearance of a graph on a graphing calculator depends on the choice of viewing rectangle. Experiment with different viewing rectangles until you find one that displays the information you want. Keep in mind that the scales along the axes may change as you vary the viewing rectangle.

• The following are some ways in which graphing calculators and computer algebra systems can be used in calculus:

  – Visualizing the behavior of a function
  – Finding solutions graphically or numerically
  – Conducting numerical or graphical experiments
  – Illustrating theoretical ideas (such as local linearity)

1.7 EXERCISES

Preliminary Questions

1. Is there a definite way of choosing the optimal viewing rectangle, or is it best to experiment until you find a viewing rectangle appropriate to the problem at hand?

2. Describe the calculator screen produced when the function $y = 3 + x^2$ is plotted with viewing rectangle:
(a) $[-1, 1] \times [0, 2]$ **(b)** $[0, 1] \times [0, 4]$

3. According to the evidence in Example 4, it appears that $f(n) = (1 + 1/n)^n$ never takes on a value greater than 3 for $n \geq 0$. Does this evidence *prove* that $f(n) \leq 3$ for $n \geq 0$?

4. How can a graphing calculator be used to find the minimum value of a function?

Exercises

The exercises in this section should be done using a graphing calculator or computer algebra system.

1. Plot $f(x) = 2x^4 + 3x^3 - 14x^2 - 9x + 18$ in the appropriate viewing rectangles and determine its roots.

2. How many solutions does $x^3 - 4x + 8 = 0$ have?

3. How many *positive* solutions does $x^3 - 12x + 8 = 0$ have?

4. Does $\cos x + x = 0$ have a solution? Does it have a positive solution?

5. Find all the solutions of $\sin x = \sqrt{x}$ for $x > 0$.

6. How many solutions does $\cos x = x^2$ have?

7. Let $f(x) = (x - 100)^2 + 1{,}000$. What will the display show if you graph $f(x)$ in the viewing rectangle $[-10, 10]$ by $[-10, -10]$? What would be an appropriate viewing rectangle?

8. Plot the graph of $f(x) = \dfrac{8x + 1}{8x - 4}$ in an appropriate viewing rectangle. What are the vertical and horizontal asymptotes (see Example 3)?

9. Plot the graph of $f(x) = \dfrac{x}{4 - x}$ in a viewing rectangle that clearly displays the vertical and horizontal asymptotes.

10. Illustrate local linearity for $f(x) = x^2$ by zooming in on the graph at $x = 0.5$ (see Example 6).

11. Plot $f(x) = \cos(x^2) \sin x$ for $0 \leq x \leq 2\pi$. Then illustrate local linearity at $x = 3.8$ by choosing appropriate viewing rectangles.

12. A bank pays $r = 5\%$ interest compounded monthly. If you deposit P_0 dollars at time $t = 0$, then the value of your account after N months is $P_0 \left(1 + \dfrac{0.05}{12}\right)^N$. Find to the nearest integer N the number of months after which the account value doubles.

In Exercises 13–18, investigate the behavior of the function as n or x grows large by making a table of function values and plotting a graph (see Example 4). Describe the behavior in words.

13. $f(n) = n^{1/n}$

14. $f(n) = \dfrac{4n + 1}{6n - 5}$

15. $f(n) = \left(1 + \dfrac{1}{n}\right)^{n^2}$

16. $f(x) = \left(\dfrac{x + 6}{x - 4}\right)^x$

17. $f(x) = \left(x \tan \dfrac{1}{x}\right)^x$

18. $f(x) = \left(x \tan \dfrac{1}{x}\right)^{x^2}$

19. The graph of $f(\theta) = A \cos \theta + B \sin \theta$ is a sinusoidal wave for any constants A and B. Confirm this for $(A, B) = (1, 1)$, $(1, 2)$, and $(3, 4)$ by plotting $f(\theta)$.

20. Find the maximum value of $f(\theta)$ for the graphs produced in Exercise 19. Can you guess the formula for the maximum value in terms of A and B?

21. Find the intervals on which $f(x) = x(x + 2)(x - 3)$ is positive by plotting a graph.

22. Find the set of solutions to the inequality $(x^2 - 4)(x^2 - 1) < 0$ by plotting a graph.

Further Insights and Challenges

23. $\boxed{\text{C A S}}$ Let $f_1(x) = x$ and define a sequence of functions by $f_{n+1}(x) = \frac{1}{2}(f_n(x) + x/f_n(x))$. For example, $f_2(x) = \frac{1}{2}(x + 1)$. Use a computer algebra system to compute $f_n(x)$ for $n = 3, 4, 5$ and plot $f_n(x)$ together with $\sqrt{x}$ for $x \geq 0$. What do you notice?

24. Set $P_0(x) = 1$ and $P_1(x) = x$. The **Chebyshev polynomials** (useful in approximation theory) are defined successively by the formula $P_{n+1}(x) = 2x P_n(x) - P_{n-1}(x)$.

(a) Show that $P_2(x) = 2x^2 - 1$.

(b) Compute $P_n(x)$ for $3 \leq n \leq 6$ using a computer algebra system or by hand, and plot each $P_n(x)$ over the interval $[-1, 1]$.

(c) Check that your plots confirm two interesting properties of Chebyshev polynomials: (a) $P_n(x)$ has n real roots in $[-1, 1]$ and (b) for $x \in [-1, 1]$, $P_n(x)$ lies between -1 and 1.

CHAPTER REVIEW EXERCISES

1. Express $(4, 10)$ as a set $\{x : |x - a| < c\}$ for suitable a and c.

2. Express as an interval:
(a) $\{x : |x - 5| < 4\}$ **(b)** $\{x : |5x + 3| \leq 2\}$

3. Express $\{x : 2 \leq |x - 1| \leq 6\}$ as a union of two intervals.

4. Give an example of numbers x, y such that $|x| + |y| = x - y$.

5. Describe the pairs of numbers x, y such that $|x + y| = x - y$.

6. Sketch the graph of $y = f(x + 2) - 1$, where $f(x) = x^2$ for $-2 \leq x \leq 2$.

In Exercises 7–10, let $f(x)$ be the function whose graph is shown in Figure 1.

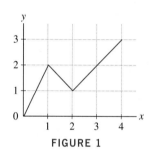

FIGURE 1

7. Sketch the graphs of $y = f(x) + 2$ and $y = f(x + 2)$.

8. Sketch the graphs of $y = \frac{1}{2}f(x)$ and $y = f\left(\frac{1}{2}x\right)$.

9. Continue the graph of $f(x)$ to the interval $[-4, 4]$ as an even function.

10. Continue the graph of $f(x)$ to the interval $[-4, 4]$ as an odd function.

In Exercises 11–14, find the domain and range of the function.

11. $f(x) = \sqrt{x + 1}$ **12.** $f(x) = \dfrac{4}{x^4 + 1}$

13. $f(x) = \dfrac{2}{3 - x}$ **14.** $f(x) = \sqrt{x^2 - x + 5}$

15. Determine whether the function is increasing, decreasing, or neither:
(a) $f(x) = 3^{-x}$ **(b)** $f(x) = \dfrac{1}{x^2 + 1}$
(c) $g(t) = t^2 + t$ **(d)** $g(t) = t^3 + t$

16. Determine whether the function is even, odd, or neither:
(a) $f(x) = x^4 - 3x^2$
(b) $g(x) = \sin(x + 1)$
(c) $f(x) = 2^{-x^2}$

In Exercises 17–22, find the equation of the line.

17. Line passing through $(-1, 4)$ and $(2, 6)$

18. Line passing through $(-1, 4)$ and $(-1, 6)$

19. Line of slope 6 through $(9, 1)$

20. Line of slope $-\dfrac{3}{2}$ through $(4, -12)$

21. Line through $(2, 3)$ parallel to $y = 4 - x$

22. Horizontal line through $(-3, 5)$

23. Does the following table of market data suggest a linear relationship between price and number of homes sold during a one-year period? Explain.

Price (thousands of $)	180	195	220	240
No. of homes sold	127	118	103	91

24. Does the following table of yearly revenue data for a computer manufacturer suggest a linear relation between revenue and time? Explain.

Year	1995	1999	2001	2004
Revenue (billions of $)	13	18	15	11

25. Find the roots of $f(x) = x^4 - 4x^2$ and sketch its graph. On which intervals is $f(x)$ decreasing?

26. Let $h(z) = 2z^2 + 12z + 3$. Complete the square and find the minimum value of $h(z)$.

27. Let $f(x)$ be the square of the distance from the point $(2, 1)$ to a point $(x, 3x + 2)$ on the line $y = 3x + 2$. Show that $f(x)$ is a quadratic function and find its minimum value by completing the square.

28. Prove that $x^2 + 3x + 3 \geq 0$ for all x.

In Exercises 29–34, sketch the graph by hand.

29. $y = t^4$ **30.** $y = t^5$

31. $y = \sin\dfrac{\theta}{2}$ **32.** $y = 10^{-x}$

33. $y = x^{1/3}$ **34.** $y = \dfrac{1}{x^2}$

35. Show that the graph of $y = f\left(\frac{1}{3}x - b\right)$ is obtained by shifting the graph of $y = f\left(\frac{1}{3}x\right)$ to the right $3b$ units. Use this observation to sketch the graph of $y = \left|\frac{1}{3}x - 4\right|$.

36. Let $h(x) = \cos x$ and $g(x) = x^{-1}$. Compute the composite functions $h(g(x))$ and $g(h(x))$, and find their domains.

37. Find functions f and g such that the function

$$f(g(t)) = (12t + 9)^4$$

38. Sketch the points on the unit circle corresponding to the following three angles and find the values of the six standard trigonometric functions at each angle:

(a) $\dfrac{2\pi}{3}$ (b) $\dfrac{7\pi}{4}$ (c) $\dfrac{7\pi}{6}$

39. What is the period of the function $g(\theta) = \sin 2\theta + \sin \dfrac{\theta}{2}$?

40. Assume that $\sin \theta = \dfrac{4}{5}$, where $\dfrac{\pi}{2} < \theta < \pi$. Find:

(a) $\tan \theta$ (b) $\sin 2\theta$ (c) $\csc \dfrac{\theta}{2}$

41. Give an example of values a, b such that

(a) $\cos(a + b) \neq \cos a + \cos b$ (b) $\cos \dfrac{a}{2} \neq \dfrac{\cos a}{2}$

42. Let $f(x) = \cos x$. Sketch the graph of $y = 2f\left(\dfrac{1}{3}x - \dfrac{\pi}{4}\right)$ for $0 \le x \le 6\pi$.

43. Solve $\sin 2x + \cos x = 0$ for $0 \le x < 2\pi$.

44. How does $h(n) = \dfrac{n^2}{2^n}$ behave for large whole number values of n? Does $h(n)$ tend to infinity?

45. $\boxed{\text{GU}}$ Use a graphing calculator to determine whether the equation $\cos x = 5x^2 - 8x^4$ has any solutions.

46. $\boxed{\text{GU}}$ Using a graphing calculator, find the number of real roots and estimate the largest root to two decimal places:

(a) $f(x) = 1.8x^4 - x^5 - x$
(b) $g(x) = 1.7x^4 - x^5 - x$

47. Match each quantity (a)–(d) with (i), (ii), or (iii) if possible, or state that no match exists.

(a) $2^a 3^b$ (b) $\dfrac{2^a}{3^b}$

(c) $(2^a)^b$ (d) $2^{a-b} 3^{b-a}$

(i) 2^{ab} (ii) 6^{a+b} (iii) $\left(\dfrac{2}{3}\right)^{a-b}$

48. Match each quantity (a)–(d) with (i), (ii), or (iii) if possible, or state that no match exists.

(a) $\ln\left(\dfrac{a}{b}\right)$ (b) $\dfrac{\ln a}{\ln b}$

(c) $e^{\ln a - \ln b}$ (d) $(\ln a)(\ln b)$

(i) $\ln a + \ln b$ (ii) $\ln a - \ln b$ (iii) $\dfrac{a}{b}$

49. Find the inverse of $f(x) = \sqrt{x^3 - 8}$ and determine its domain and range.

50. Find the inverse of $f(x) = \dfrac{x - 2}{x - 1}$ and determine its domain and range.

51. Find a domain on which $h(t) = (t - 3)^2$ is one-to-one and determine the inverse on this domain.

52. Show that $g(x) = \dfrac{x}{x + 1}$ is equal to its inverse on the domain $\{x : x \neq -1\}$.

53. Suppose that $g(x)$ is the inverse of $f(x)$. Match the functions (a)–(d) with their inverses (i)–(iv).

(a) $f(x) + 1$ (b) $f(x + 1)$

(c) $4f(x)$ (d) $f(4x)$

(i) $g(x)/4$ (ii) $g(x/4)$

(iii) $g(x - 1)$ (iv) $g(x) - 1$

54. $\boxed{\text{GU}}$ Plot $f(x) = xe^{-x}$ and use the zoom feature to find two solutions of $f(x) = 0.5$.

2 | LIMITS

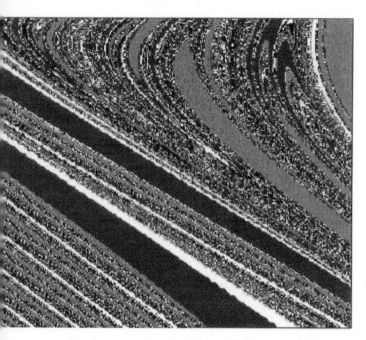

Computer simulation of limiting behavior in a fluid flow with fractal properties.

Calculus is usually divided into two branches, differential and integral, partly for historical reasons. The subject was developed in the seventeenth century to solve two important geometric problems: finding tangent lines to curves (differential calculus) and computing areas under curves (integral calculus). However, calculus is a very broad subject with no clear boundaries. It includes other topics, such as the theory of infinite series, and it has an extraordinarily wide range of applications, especially in the sciences and in engineering. What makes these methods and applications part of calculus is that they all rely ultimately on the concept of a limit. We will see throughout the text how limits allow us to make computations and solve problems that cannot be solved using algebra alone.

This chapter introduces the limit concept and focuses on those aspects of limits that will be used in Chapter 3 to develop differential calculus. The first section, intended as motivation, discusses how limits arise in the study of rates of change and tangent lines.

2.1 Limits, Rates of Change, and Tangent Lines

Rates of change play a role whenever we study the relationship between two changing quantities. Velocity is a familiar example (the rate of change of position with respect to time), but there are many others, such as:

- The infection rate of an epidemic (*newly infected individuals per month*)
- Mileage of an automobile (*miles per gallon*)
- Rate of change of atmospheric temperature with respect to altitude

Roughly speaking, if y and x are related quantities, the rate of change should tell us how much y changes in response to a unit change in x. However, this has a clear meaning only if the rate of change is constant. If a car travels at a constant velocity of 40 mph, then its position changes by 40 miles for each unit change in time (the unit being 1 hour). If the trip lasts only half an hour, its position changes by 20 miles, and in general, the change in position is $40t$ miles, where t is the length of time in hours. In other words,

$$\boxed{\text{Change in position} = \text{velocity} \times \text{time}}$$

This simple formula is no longer valid if the velocity is changing and, in fact, it is not even meaningful. Indeed, if the automobile is accelerating or decelerating, which velocity would we use in the formula? It is possible to compute the change in position when velocity is changing, but this involves both differential and integral calculus. Differential calculus uses the limit concept to define the *instantaneous velocity* of an object, and integral calculus tells us how to compute the change in position in terms of instantaneous velocity. This is just one example of how calculus provides the concepts and tools for modeling real-world phenomena.

In this section, we discuss velocity and other rates of change, emphasizing the graphical interpretation in terms of *tangent lines*. At this stage, we cannot define tangent lines precisely. This will have to wait until Chapter 3. In the meantime, you can think of a tangent line as a line that *skims* a curve at a point as in Figures 1(A) and (B) (but not (C)).

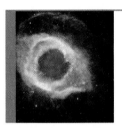

FIGURE 1 The line is tangent in (A) and (B) but not in (C).

(A)   (B)   (C)

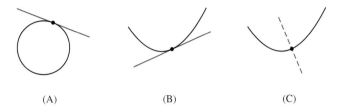

HISTORICAL PERSPECTIVE

Philosophy is written in this grand book—I mean the universe—which stands continually open to our gaze, but it cannot be understood unless one first learns to comprehend the language . . . in which it is written. It is written in the language of mathematics . . .
—GALILEO GALILEI, 1623

The scientific revolution of the sixteenth and seventeenth centuries reached its high point in the work of Isaac Newton (1643–1727), who was the first scientist to show that the physical world, despite its complexity and diversity, is governed by a small number of universal laws, such as the laws of motion and gravitation. One of Newton's great insights was that the universal laws do not describe the world as it is, now or at any particular time, but rather how the world *changes over time* in response to forces.

To express the laws in mathematical form, Newton invented calculus, which he understood to be the mathematics of change.

More than fifty years before the work of Newton, the astronomer Johannes Kepler (1571–1630) discovered his three laws of planetary motion, including the law stating that the path of a planet around the sun is an ellipse. Kepler found these laws through a painstaking analysis of astronomical data, but he could not explain why they were true. Newton's laws explain the motion of any object—planet or pebble—in terms of the forces acting on it. According to Newton, the planets, if left undisturbed, will travel in straight lines. Since their paths are elliptical, some force—in this case, the gravitational force of the sun—must be acting to make them change direction continuously. In a show of unmatched mathematical skill, Newton proved that Kepler's laws follow from Newton's own inverse square law of gravity. Ironically, the proof Newton chose to publish in 1687 in his magnum opus *Principia Mathematica* uses geometrical arguments rather than the calculus Newton himself had invented.

Velocity

The terms "speed" and "velocity," though sometimes used interchangeably, do not have the same meaning. For linear motion, velocity may be positive or negative (indicating the direction of motion). Speed is the absolute value of velocity and is always positive.

Consider an object traveling in a straight line (linear motion). When we speak of its velocity, we usually mean the *instantaneous* velocity, which tells us the speed and direction of the object at a particular moment. As suggested above, however, instantaneous velocity does not have a straightforward definition. It is simpler to define the **average velocity** over a given time interval as the ratio

$$\text{Average velocity} = \frac{\text{change in position}}{\text{length of time interval}}$$

For example, if a car travels 200 miles in 4 hours, then its average velocity during this 4-hour period is $\frac{200}{4} = 50$ mph. At any given moment the car may be going faster or slower than the average.

Unlike average velocity, we cannot define instantaneous velocity as a ratio because we would have to divide by the length of the time interval (which is zero). However, we should be able to estimate instantaneous velocity by computing average velocity over successively smaller time intervals. The guiding principle is the following: *Average velocity over a very small time interval is very close to instantaneous velocity.* To explore this idea further, we introduce some notation.

The Greek letter Δ (delta) is commonly used to denote the *change* in a function or variable. If $s(t)$ is the position of an object (distance from the origin) at time t and $[t_0, t_1]$ is a time interval, we set

$$\Delta s = s(t_1) - s(t_0) = \text{change in position}$$

$$\Delta t = \quad t_1 \quad - \quad t_0 \quad = \text{change in time (length of interval)}$$

For $t_1 \neq t_0$,

$$\boxed{\text{Average velocity over } [t_0, t_1] = \frac{\Delta s}{\Delta t} = \frac{s(t_1) - s(t_0)}{t_1 - t_0}}$$

The change in position Δs is also called the **displacement,** or net change in position.

One motion we will study is the motion of an object falling to earth under the influence of gravity (assuming no air resistance). Galileo discovered that if the object is released at time $t = 0$ from a state of rest (see Figure 2), then the distance traveled after t seconds is given by the formula

$$s(t) = 16t^2 \text{ ft}$$

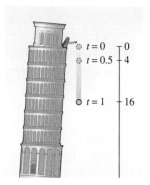

FIGURE 2 Distance traveled by a falling object after t seconds is $s(t) = 16t^2$ feet.

■ **EXAMPLE 1** A stone is released from a state of rest and falls to earth. Estimate the instantaneous velocity at $t = 0.5$ s by calculating average velocity over several small time intervals.

Solution The distance traveled by the stone after t seconds is $s(t) = 16t^2$ by Galileo's formula. We use this to compute the average velocity over the five time intervals listed in Table 1. To find the stone's average velocity over the first interval $[t_0, t_1] = [0.5, 0.6]$, we compute the changes:

$$\Delta s = s(0.6) - s(0.5) = 16(0.6)^2 - 16(0.5)^2 = 5.76 - 4 = 1.76 \text{ ft}$$

$$\Delta t = 0.6 - 0.5 = 0.1 \text{ s}$$

The average velocity over $[0.5, 0.6]$ is the ratio

$$\frac{\Delta s}{\Delta t} = \frac{s(0.6) - s(0.5)}{0.6 - 0.5} = \frac{1.76}{0.1} = 17.6 \text{ ft/s}$$

TABLE 1

Time Interval	Average Velocity
[0.5, 0.6]	17.6
[0.5, 0.55]	16.8
[0.5, 0.51]	16.16
[0.5, 0.505]	16.08
[0.5, 0.5001]	16.0016

Note that there is nothing special about the particular time intervals shown in Table 1. We are looking for a trend and we could have chosen any intervals $[0.5, t]$ for values of t approaching 0.5. We could also have chosen intervals $[t, 0.5]$ for $t < 0.5$.

Table 1 shows the results of similar calculations for several intervals of successively shorter lengths. It is clear that the average velocities get closer to 16 ft/s as the time interval shrinks:

$$17.6, \quad 16.8, \quad 16.16, \quad 16.08, \quad 16.0016$$

We express this by saying that the average velocity *converges* to 16 or that 16 is the *limit* of average velocity as the length of the time interval shrinks to zero. This suggests that the instantaneous velocity at $t = 0.5$ is 16 ft/s.   ■

The conclusion in this example is correct, but we need to base it on more formal and precise reasoning. This will be done in Chapter 3, after we have developed the limit concept.

Graphical Interpretation of Velocity

The idea that average velocity converges to instantaneous velocity as we shorten the time interval has a vivid interpretation in terms of secant lines where, by definition, a **secant line** is a line through two points on a curve.

Consider the graph of position $s(t)$ for an object traveling in a straight line as in Figure 3. The ratio we use to define average velocity over $[t_0, t_1]$ is nothing more than the slope of the secant line through the points $(t_0, s(t_0))$ and $(t_1, s(t_1))$. For $t_1 \neq t_0$,

$$\text{Average velocity} = \text{slope of secant line} = \frac{\Delta s}{\Delta t} = \frac{s(t_1) - s(t_0)}{t_1 - t_0}$$

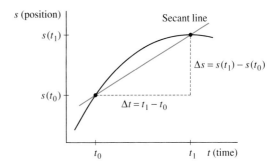

FIGURE 3 Average velocity over $[t_0, t_1]$ is equal to the slope of the secant line.

By interpreting average velocity as a slope, we can visualize what happens as the time interval gets smaller. Figure 4 shows the graph of $s(t) = 16t^2$ together with the secant lines for the time intervals $[0.5, 0.6]$, $[0.5, 0.55]$, and $[0.5, 0.51]$. We see that as the time interval shrinks, the secant lines get closer to—*and seem to rotate into*—the tangent line at $t = 0.5$.

Time Interval	Average Velocity
$[0.5, 0.6]$	17.6
$[0.5, 0.55]$	16.8
$[0.5, 0.51]$	16.16

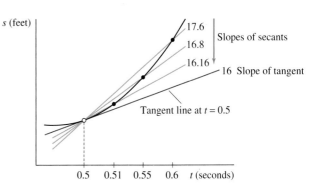

FIGURE 4 The secant line "rotates into" the tangent line as the time interval shrinks. *Note:* The figures are not drawn to scale.

Since the secant lines approach the tangent line, the slopes of the secant lines get closer and closer to the slope of the tangent line. In other words, the statement

> As the time interval shrinks to zero, the average velocity approaches the instantaneous velocity.

has the graphical interpretation

> As the time interval shrinks to zero, the slope of the secant line approaches the slope of the tangent line.

We are thus led to the following interpretation: *Instantaneous velocity is equal to the slope of the tangent line.* This conclusion and its generalization to other rates of change are fundamental to almost all aspects of differential calculus and its applications.

Other Rates of Change

Velocity is only one of many examples of a rate of change (ROC). Our reasoning applies to any quantity y that is a function of a variable x, say, $y = f(x)$. For any interval $[x_0, x_1]$, we set

$$\Delta f = f(x_1) - f(x_0), \qquad \Delta x = x_1 - x_0$$

For $x_1 \neq x_0$, the **average rate of change** of y with respect to x over $[x_0, x_1]$ is the ratio

$$\text{Average ROC} = \frac{\Delta f}{\Delta x} = \underbrace{\frac{f(x_1) - f(x_0)}{x_1 - x_0}}_{\text{Slope of secant line}}$$

The **instantaneous rate of change** is the limit of the average rates of change. We estimate the instantaneous rate of change at $x = x_0$ by computing the average rate of change over smaller and smaller intervals. In the previous example, we considered only right-hand intervals $[x_0, x_1]$. In the next example, we compute the average ROC for intervals lying to both the left and right of x_0.

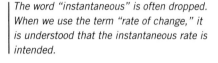

The word "instantaneous" is often dropped. When we use the term "rate of change," it is understood that the instantaneous rate is intended.

Sometimes, we write Δy and $\frac{\Delta y}{\Delta x}$ instead of Δf and $\frac{\Delta f}{\Delta x}$.

An airplane traveling faster than the speed of sound produces a sonic boom. Because temperature decreases linearly with altitudes up to approximately 37,000 ft, pilots of supersonic jets can estimate the speed of sound in terms of altitude.

■ **EXAMPLE 2** **Speed of Sound in Air**   The formula $v = 20\sqrt{T}$ provides a good approximation to the speed of sound v in dry air (in m/s) as a function of air temperature T (in kelvins).

(a) Compute the average ROC of v with respect to T over the interval $[273, 300]$. What are the units of this ROC? Draw the corresponding secant line.

(b) Estimate the instantaneous ROC when $T = 273$ K by computing average rates over small intervals lying to the left and right of $T = 273$.

Solution

(a) The average ROC over the temperature interval $[273, 300]$ is

$$\frac{\Delta v}{\Delta T} = \frac{20\sqrt{300} - 20\sqrt{273}}{300 - 273} \approx \frac{15.95}{27} \approx 0.59 \text{ m/s per K}$$

Since this is the average rate of increase in speed per degree increase in temperature, the appropriate units are *meters per second per kelvin*. This ROC is equal to the slope of the line through the points where $T = 273$ and $T = 300$ on the graph in Figure 5.

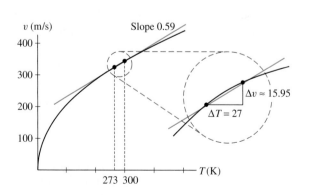

FIGURE 5 The graph of $v = 20\sqrt{T}$.

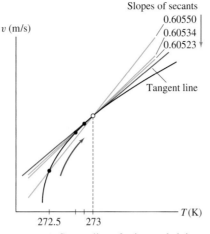

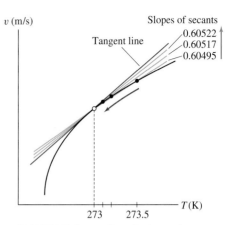

FIGURE 6 Secant lines for intervals lying to the left of $T = 273$.

FIGURE 7 Secant lines for intervals lying to the right of $T = 273$.

TABLE 2 Left-Hand Intervals	
Temperature Interval	**Average ROC**
$[272.5, 273]$	0.60550
$[272.8, 273]$	0.60534
$[272.9, 273]$	0.60528
$[272.99, 273]$	0.60523

TABLE 3 Right-Hand Intervals	
Temperature Interval	**Average ROC**
$[273, 273.5]$	0.60495
$[273, 273.2]$	0.60512
$[273, 273.1]$	0.60517
$[273, 273.01]$	0.60522

(b) To estimate the instantaneous ROC at $T = 273$, we compute the average ROC for several intervals lying to the left (Figure 6) and right (Figure 7) of $T = 273$. The results are shown in Tables 2 and 3. Both tables suggest that the instantaneous rate is approximately 0.605 m/s per K.   ∎

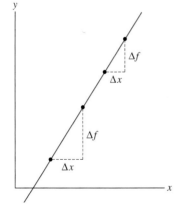

FIGURE 8 For a linear function, the ratio $\Delta f / \Delta x$ is equal to the slope m for every interval.

To conclude this section, we recall an important point discussed in Section 1.2: For any linear function $f(x) = mx + b$, *the average rate of change over every interval is equal to the slope m* (Figure 8). Indeed, for the interval $[x_0, x_1]$ with $x_1 \neq x_0$,

$$\frac{\Delta f}{\Delta x} = \frac{f(x_1) - f(x_0)}{x_1 - x_0} = \frac{(mx_1 + b) - (mx_0 + b)}{x_1 - x_0} = \frac{m(x_1 - x_0)}{x_1 - x_0} = m$$

The instantaneous rate of change at $x = x_0$ is also equal to m because it is the limit of average rates of change. This result makes sense graphically because all secant lines and all tangent lines to the graph of $f(x)$ coincide with the graph itself.

2.1 SUMMARY

- The average rate of change (ROC) of $y = f(x)$ over an interval $[x_0, x_1]$ is

$$\text{Average ROC} = \frac{\Delta f}{\Delta x} = \frac{f(x_1) - f(x_0)}{x_1 - x_0} \qquad (x_1 \neq x_0)$$

Average ROC is interpreted graphically as the slope of the secant line through the points $(x_0, f(x_0))$ and $(x_1, f(x_1))$ on the graph of $f(x)$.
- We estimate the instantaneous rate of change at $x = x_0$ by computing the average ROC over intervals $[x_0, x_1]$ (or $[x_1, x_0]$) for x_1 close to x_0. The instantaneous ROC is the limit of the average ROCs.
- Instantaneous ROC is interpreted graphically as the slope of the tangent line at x_0.
- The velocity of an object traveling in a straight line is the ROC of $s(t)$, where $s(t)$ is the position of the object at time t.
- If $f(x) = mx + b$ is a linear function, then the average rate of change over every interval and the instantaneous rate of change at every point are equal to the slope m.

2.1 EXERCISES

Preliminary Questions

1. Average velocity is defined as a ratio of which two quantities?

2. Average velocity is equal to the slope of a secant line through two points on a graph. Which graph?

3. Can instantaneous velocity be defined as a ratio? If not, how is instantaneous velocity computed?

4. What is the graphical interpretation of instantaneous velocity at a moment $t = t_0$?

5. What is the graphical interpretation of the following statement: The average ROC approaches the instantaneous ROC as the interval $[x_0, x_1]$ shrinks to x_0?

6. The ROC of atmospheric temperature with respect to altitude is equal to the slope of the tangent line to a graph. Which graph? What are possible units for this rate?

Exercises

1. A ball is dropped from a state of rest at time $t = 0$. The distance traveled after t seconds is $s(t) = 16t^2$ ft.

(a) How far does the ball travel during the time interval $[2, 2.5]$?

(b) Compute the average velocity over $[2, 2.5]$.

(c) Compute the average velocity over time intervals $[2, 2.01]$, $[2, 2.005]$, $[2, 2.001]$, $[2, 2.00001]$. Use this to estimate the object's instantaneous velocity at $t = 2$.

2. A wrench is released from a state of rest at time $t = 0$. Estimate the wrench's instantaneous velocity at $t = 1$, assuming that the distance traveled after t seconds is $s(t) = 16t^2$.

3. Let $v = 20\sqrt{T}$ as in Example 2. Estimate the instantaneous ROC of v with respect to T when $T = 300$ K.

4. Compute $\Delta y / \Delta x$ for the interval $[2, 5]$, where $y = 4x - 9$. What is the instantaneous ROC of y with respect to x at $x = 2$?

In Exercises 5–6, a stone is tossed in the air from ground level with an initial velocity of 15 m/s. Its height at time t is $h(t) = 15t - 4.9t^2$ m.

5. Compute the stone's average velocity over the time interval $[0.5, 2.5]$ and indicate the corresponding secant line on a sketch of the graph of $h(t)$.

6. Compute the stone's average velocity over the time intervals $[1, 1.01]$, $[1, 1.001]$, $[1, 1.0001]$ and $[0.99, 1]$, $[0.999, 1]$, $[0.9999, 1]$. Use this to estimate the instantaneous velocity at $t = 1$.

7. With an initial deposit of $100, the balance in a bank account after t years is $f(t) = 100(1.08)^t$ dollars.

(a) What are the units of the ROC of $f(t)$?

(b) Find the average ROC over $[0, 0.5]$ and $[0, 1]$.

(c) Estimate the instantaneous rate of change at $t = 0.5$ by computing the average ROC over intervals to the left and right of $t = 0.5$.

8. The distance traveled by a particle at time t is $s(t) = t^3 + t$. Compute the average velocity over the time interval $[1, 4]$ and estimate the instantaneous velocity at $t = 1$.

In Exercises 9–16, estimate the instantaneous rate of change at the point indicated.

9. $P(x) = 4x^2 - 3$; $x = 2$

10. $f(t) = 3t - 5$; $t = -9$

11. $y(x) = \dfrac{1}{x+2}$; $x = 2$

12. $y(t) = \sqrt{3t + 1}$; $t = 1$

13. $f(x) = e^x$; $x = 0$

14. $f(x) = e^x$; $x = e$

15. $f(x) = \ln x$; $x = 3$

16. $f(x) = \tan^{-1} x$; $x = \frac{\pi}{4}$

17. The atmospheric temperature T (in °F) above a certain point on earth is $T = 59 - 0.00356h$, where h is the altitude in feet (valid for $h \le 37,000$). What are the average and instantaneous rates of change of T with respect to h? Why are they the same? Sketch the graph of T for $h \le 37,000$.

18. The height (in feet) at time t (in seconds) of a small weight oscillating at the end of a spring is $h(t) = 0.5 \cos(8t)$.

(a) Calculate the weight's average velocity over the time intervals [0, 1] and [3, 5].

(b) Estimate its instantaneous velocity at $t = 3$.

19. The number $P(t)$ of *E. coli* cells at time t (hours) in a petri dish is plotted in Figure 9.

(a) Calculate the average ROC of $P(t)$ over the time interval [1, 3] and draw the corresponding secant line.

(b) Estimate the slope m of the line in Figure 9. What does m represent?

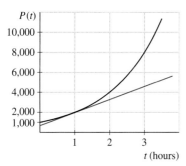

FIGURE 9 Number of *E. coli* cells at time t.

20. Calculate $\sqrt{x + 1} - \sqrt{x}$ for $x = 1, 10, 100, 1,000$. Does $\sqrt{x}$ change more rapidly when x is large or small? Interpret your answer in terms of tangent lines to the graph of $y = \sqrt{x}$.

21. Assume that the period T (in seconds) of a pendulum (the time required for a complete back-and-forth cycle) is $T = \frac{3}{2}\sqrt{L}$, where L is the pendulum's length (in meters).

(a) What are the units for the ROC of T with respect to L? Explain what this rate measures.

(b) Which quantities are represented by the slopes of lines A and B in Figure 10?

(c) Estimate the instantaneous ROC of T with respect to L when $L = 3$ m.

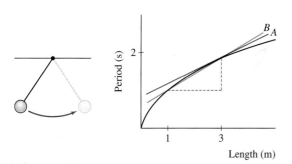

FIGURE 10 The period T is the time required for a pendulum to swing back and forth.

22. The graphs in Figure 11 represent the positions of moving particles as functions of time.

(a) Do the instantaneous velocities at times t_1, t_2, t_3 in (A) form an increasing or decreasing sequence?

(b) Is the particle speeding up or slowing down in (A)?

(c) Is the particle speeding up or slowing down in (B)?

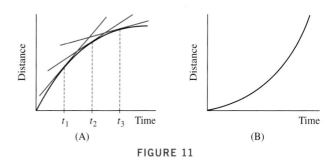

FIGURE 11

23. The graphs in Figure 12 represent the positions s of moving particles as functions of time t. Match each graph with one of the following statements:

(a) Speeding up

(b) Speeding up and then slowing down

(c) Slowing down

(d) Slowing down and then speeding up

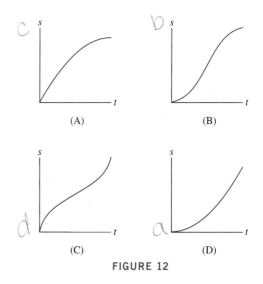

FIGURE 12

24. An epidemiologist finds that the percentage $N(t)$ of susceptible children who were infected on day t during the first three weeks of a measles outbreak is given, to a reasonable approximation, by the formula

$$N(t) = \frac{100t^2}{t^3 + 5t^2 - 100t + 380}$$

A graph of $N(t)$ appears in Figure 13.

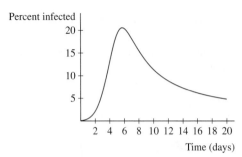

FIGURE 13 Graph of $N(t)$.

(a) Draw the secant line whose slope is the average rate of increase in infected children over the intervals between days 4 and 6 and between days 12 and 14. Then compute these average rates (in units of percent per day).

(b) Estimate the ROC of $N(t)$ on day 12.

25. The fraction of a city's population infected by a flu virus is plotted as a function of time (in weeks) in Figure 14.

(a) Which quantities are represented by the slopes of lines A and B? Estimate these slopes.

(b) Is the flu spreading more rapidly at $t = 1, 2,$ or 3?

(c) Is the flu spreading more rapidly at $t = 4, 5,$ or 6?

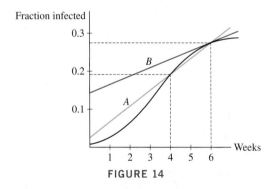

FIGURE 14

26. The fungus *fusarium exosporium* infects a field of flax plants through the roots and causes the plants to wilt. Eventually, the entire field is infected. The percentage $f(t)$ of infected plants as a function of time t (in days) since planting is shown in Figure 15.

(a) What are the units of the rate of change of $f(t)$ with respect to t? What does this rate measure?

(b) Use the graph to rank (from smallest to largest) the average infection rates over the intervals [0, 12], [20, 32], and [40, 52].

(c) Use the following table to compute the average rates of infection over the intervals [30, 40], [40, 50], [30, 50]:

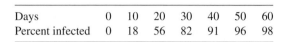

Days	0	10	20	30	40	50	60
Percent infected	0	18	56	82	91	96	98

(d) Draw the tangent line at $t = 40$ and estimate its slope. Choose any two points on the tangent line for the computation.

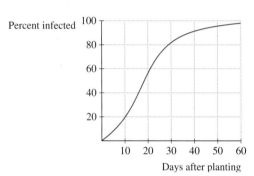

FIGURE 15

27. Let $v = 20\sqrt{T}$ as in Example 2. Is the ROC of v with respect to T greater at low temperatures or high temperatures? Explain in terms of the graph.

28. If an object moving in a straight line (but with changing velocity) covers Δs feet in Δt seconds, then its average velocity is $v_0 = \Delta s / \Delta t$ ft/s. Show that it would cover the same distance if it traveled at constant velocity v_0 over the same time interval of Δt seconds. This is a justification for calling $\Delta s / \Delta t$ the *average velocity*.

29. Sketch the graph of $f(x) = x(1 - x)$ over [0, 1]. Refer to the graph and, without making any computations, find:

(a) The average ROC over [0, 1]

(b) The (instantaneous) ROC at $x = \frac{1}{2}$

(c) The values of x at which the ROC is positive

30. Which graph in Figure 16 has the following property: For all x, the average ROC over $[0, x]$ is greater than the instantaneous ROC at x? Explain.

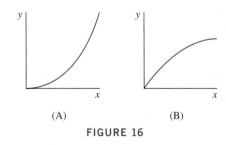

FIGURE 16

Further Insights and Challenges

31. The height of a projectile fired in the air vertically with initial velocity 64 ft/s is $h(t) = 64t - 16t^2$ ft.

(a) Compute $h(1)$. Show that $h(t) - h(1)$ can be factored with $(t - 1)$ as a factor.

(b) Using part (a), show that the average velocity over the interval $[1, t]$ is $-16(t - 3)$.

(c) Use this formula to find the average velocity over several intervals $[1, t]$ with t close to 1. Then estimate the instantaneous velocity at time $t = 1$.

32. Let $Q(t) = t^2$. As in the previous exercise, find a formula for the average ROC of Q over the interval $[1, t]$ and use it to estimate the instantaneous ROC at $t = 1$. Repeat for the interval $[2, t]$ and estimate the ROC at $t = 2$.

33. Show that the average ROC of $f(x) = x^3$ over $[1, x]$ is equal to $x^2 + x + 1$. Use this to estimate the instantaneous ROC of $f(x)$ at $x = 1$.

34. Find a formula for the average ROC of $f(x) = x^3$ over $[2, x]$ and use it to estimate the instantaneous ROC at $x = 2$.

35. Let $T = \frac{3}{2}\sqrt{L}$ as in Exercise 21. The numbers in the second column of Table 4 are increasing and those in the last column are decreasing. Explain why in terms of the graph of T as a function of L. Also, explain graphically why the instantaneous ROC at $L = 3$ lies between 0.4329 and 0.4331.

TABLE 4   Average Rates of Change of T with Respect to L

Interval	Average ROC	Interval	Average ROC
[3, 3.2]	0.42603	[2.8, 3]	0.44048
[3, 3.1]	0.42946	[2.9, 3]	0.43668
[3, 3.001]	0.43298	[2.999, 3]	0.43305
[3, 3.0005]	0.43299	[2.9995, 3]	0.43303

2.2 Limits: A Numerical and Graphical Approach

As we noted in the introduction to this chapter, the limit concept plays a fundamental role in all aspects of calculus. In this section, we define limits and study them using numerical and graphical techniques. We begin with the following question: *How do the values of a function $f(x)$ behave when x approaches a number c, whether or not $f(c)$ is defined?*

To explore this question, we'll experiment with the function

$$f(x) = \frac{\sin x}{x} \quad (x \text{ in radians})$$

The value $f(0)$ is not defined because

The undefined expression $\frac{0}{0}$ is also referred to as an "indeterminate form."

$$f(0) = \frac{\sin 0}{0} = \frac{0}{0} \quad (\text{undefined})$$

Nevertheless, $f(x)$ can be computed for values of x close to 0. When we do this, a clear trend emerges (see the table in the margin and Table 1 on the next page).

To describe the trend, we use the phrases "x approaches 0" or "x tends to 0" to indicate that x takes on values (both positive and negative) that get closer and closer to 0. The notation for this is $x \to 0$. We write:

x	$\dfrac{\sin x}{x}$
1	0.841470985
0.5	0.958851077
0.1	0.998334166
0.05	0.999583385
0.01	0.999983333
0.005	0.999995833
0.001	0.999999833
$x \to 0+$	$f(x) \to 1$

- $x \to 0+$ if x approaches 0 from the right (through positive values).
- $x \to 0-$ if x approaches 0 from the left (through negative values).

Now consider the values listed in Table 1. The table gives the unmistakable impression that $f(x)$ gets closer and closer to 1 as x approaches 0 through positive and negative values.

This conclusion is supported by the graph of $f(x)$ in Figure 1. The point $(0, 1)$ is missing from the graph because $f(x)$ is not defined at $x = 0$, but the graph approaches this missing point as x approaches 0 from the left and right. We say that the *limit* of $f(x)$ as $x \to 0$ is equal to 1, and we write

$$\lim_{x \to 0} f(x) = 1$$

We also say that $f(x)$ *approaches* or *converges to* 1 as $x \to 0$.

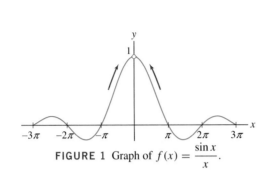

FIGURE 1 Graph of $f(x) = \dfrac{\sin x}{x}$.

	TABLE 1			
x	$\dfrac{\sin x}{x}$	x	$\dfrac{\sin x}{x}$
1	0.841470985	−1	0.841470985
0.5	0.958851077	−0.5	0.958851077
0.1	0.998334166	−0.1	0.998334166
0.05	0.999583385	−0.05	0.999583385
0.01	0.999983333	−0.01	0.999983333
0.005	0.999995833	−0.005	0.999995833
0.001	0.999999833	−0.001	0.999999833
$x \to 0+$	$f(x) \to 1$	$x \to 0-$	$f(x) \to 1$

CONCEPTUAL INSIGHT   The numerical and graphical evidence may have convinced us that $f(x) = \dfrac{\sin x}{x}$ converges to 1 as $x \to 0$. But since $f(0) = 0/0$, could we not have arrived at this conclusion more simply by saying that $0/0 = 1$? The answer is no. It is not correct to say that $0/0$ equals 1 or any other number. In fact, if $0/0$ were equal to 1, we could prove the absurd equality $2 = 1$:

$$2\left(\frac{0}{0}\right) = \frac{2(0)}{0} = \frac{0}{0}$$
$$2 = 1$$

The expression $0/0$ is undefined, and in our example, the limit of $f(x)$ happens to be 1. We will encounter other examples where the formula for a function $f(x)$ produces $0/0$ and the limit is not equal to 1.

Definition of a Limit

We now define limits more formally. Recall that the distance between two numbers a and b is the absolute value $|a - b|$, so we can express the idea that $f(x)$ is close to L by saying that $|f(x) - L|$ is small.

DEFINITION Limit   Assume that $f(x)$ is defined for all x near c (i.e., in some open interval containing c), but not necessarily at c itself. We say that

the limit of $f(x)$ as x approaches c is equal to L

if $|f(x) - L|$ becomes arbitrarily small when x is any number sufficiently close (but not equal) to c. In this case, we write

$$\lim_{x \to c} f(x) = L$$

limit

We also say that $f(x)$ *approaches* or *converges to L* as $x \to c$ (and we write $f(x) \to L$).

If the values of $f(x)$ do not converge to any limit as $x \to c$, we say that $\lim\limits_{x \to c} f(x)$ *does not exist*. It is important to note that the value $f(c)$ itself, which may or may not be defined, plays no role in the limit. All that matters are the values $f(x)$ for x close to c. Furthermore, if $f(x)$ approaches a limit as $x \to c$, then the limiting value L is unique.

Although fundamental in calculus, the limit concept was not fully clarified until the nineteenth century. The French mathematician Augustin-Louis Cauchy (1789–1857, pronounced Koh-shee) gave the following verbal definition: "When the values successively attributed to the same variable approach a fixed value indefinitely, in such a way as to end up differing from it by as little as one could wish, this last value is called the limit of all the others. So, for example, an irrational number is the limit of the various fractions which provide values that approximate it more and more closely." (Translated by J. Grabiner.)

■ **EXAMPLE 1** Use the definition above to verify the following limits:

(a) $\lim_{x \to 7} 5 = 5$

(b) $\lim_{x \to 4} (3x + 1) = 13$

Solution

(a) Let $f(x) = 5$. To show that $\lim_{x \to 7} f(x) = 5$, we must show that $|f(x) - 5|$ becomes arbitrarily small when x is sufficiently close (but not equal) to 7. But $|f(x) - 5| = |5 - 5| = 0$ *for all* x, so what we are required to show is automatic (and it is not necessary to take x close to 7).

(b) Let $f(x) = 3x + 1$. To show that $\lim_{x \to 4} (3x + 1) = 13$, we must show that $|f(x) - 13|$ becomes arbitrarily small when x is sufficiently close (but not equal) to 4. We have

$$|f(x) - 13| = |(3x + 1) - 13| = |3x - 12| = 3|x - 4|$$

Since $|f(x) - 13|$ is a multiple of $|x - 4|$, we can make $|f(x) - 13|$ arbitrarily small by taking x sufficiently close to 4.   ■

Reasoning as in Example 1 but with arbitrary constants, we obtain the following simple but important results:

THEOREM 1   For any constants k and c,   (a) $\lim_{x \to c} k = k$ and   (b) $\lim_{x \to c} x = c$.

Here is a way of stating the limit definition precisely: $\lim_{x \to c} f(x) = L$ *if, for any n, we have* $|f(x) - L| < 10^{-n}$, *provided that* $0 < |x - c| < 10^{-m}$ *(where the choice of m depends on n).*

To deal with more complicated limits in a rigorous manner, it is necessary to make the phrases "arbitrarily small" and "sufficiently close" more precise using inequalities. We do this in Section 2.8.

Graphical and Numerical Investigation

Our goal in the rest of this section is to develop a better intuitive understanding of limits by investigating them graphically and numerically.

Graphical Investigation Use a graphing utility to produce a graph of $f(x)$. The graph should give a visual impression of whether or not a limit exists. It can often be used to estimate the value of the limit.

Numerical Investigation To investigate $\lim_{x \to c} f(x)$,

(i) Make a table of values of $f(x)$ for x close to but less than c.

(ii) Make a second table of values of $f(x)$ for x close to but greater than c.

(iii) If both tables indicate convergence to the same number L, we take L to be an estimate for the limit.

Keep in mind that graphical and numerical investigations provide evidence for a limit but they do not prove that the limit exists or has a given value. This is done using the Limit Laws established in the following sections.

The tables should contain enough values to reveal a clear trend of convergence to a value L. If $f(x)$ approaches a limit, the successive values of $f(x)$ will generally agree to more and more decimal places as x is taken closer to c. If no pattern emerges, then the limit may not exist.

■ **EXAMPLE 2** Investigate graphically and numerically $\lim_{x \to 9} \dfrac{x - 9}{\sqrt{x} - 3}$.

Solution The graph of $f(x) = \dfrac{x-9}{\sqrt{x}-3}$ in Figure 2 has a gap at $x = 9$ because $f(9)$ is not defined:

$$f(9) = \frac{9-9}{\sqrt{9}-3} = \frac{0}{0} \quad \text{(undefined)}$$

However, the graph indicates that $f(x)$ approaches 6 as $x \to 9$. For numerical evidence, we consider a table of values of $f(x)$ for x approaching 9 from both the left and right. The values listed in Table 2 confirm our impression that

$$\lim_{x \to 9} \frac{x-9}{\sqrt{x}-3} = 6 \qquad \blacksquare$$

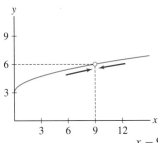

FIGURE 2 Graph of $f(x) = \dfrac{x-9}{\sqrt{x}-3}$.

	TABLE 2		
x	$\dfrac{x-9}{\sqrt{x}-3}$	x	$\dfrac{x-9}{\sqrt{x}-3}$
8.9	**5.98**329	9.1	**6.01**662
8.99	**5.99**833	9.01	**6.001**666
8.999	**5.999**83	9.001	**6.000**167
8.9999	**5.9999**833	9.0001	**6.0000**167

■ **EXAMPLE 3** **Limit Equals Function Value**   Investigate graphically and numerically $\lim\limits_{x \to 4} x^2$.

Solution Let $f(x) = x^2$. Figure 3 and Table 3 both suggest that $\lim\limits_{x \to 4} x^2 = 16$. Of course, in this case $f(4)$ is defined and $f(4) = 16$, so *the limit is equal to the function value*. This pleasant conclusion is valid whenever $f(x)$ is a *continuous* function, a concept treated in Section 2.4. ■

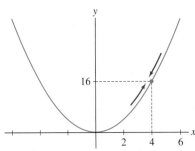

FIGURE 3 Graph of $f(x) = x^2$. The limit is equal to the function value $f(4) = 16$.

	TABLE 3		
x	x^2	x	x^2
3.9	15.21	4.1	16.81
3.99	**15.9**201	4.01	**16.0**801
3.999	**15.99**2001	4.001	**16.00**8001
3.9999	**15.999**20001	4.0001	**16.000**80001

The limit in Example 4 may be taken as the definition of the number e (see Section 3.2). In other words,

$$\lim_{h \to 0} \frac{b^h - 1}{h} = 1 \quad \text{if and only if } b = e$$

■ **EXAMPLE 4** **Defining Property of e**   Verify numerically that $\lim\limits_{h \to 0} \dfrac{e^h - 1}{h} = 1$.

Solution The function $f(h) = \dfrac{e^h - 1}{h}$ is undefined at $h = 0$, but both Figure 4 and Table 4 suggest that $\lim\limits_{h \to 0} \dfrac{e^h - 1}{h} = 1$. ■

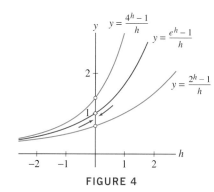

FIGURE 4

TABLE 4			
$h \to 0+$	$\dfrac{e^h - 1}{h}$	$h \to 0-$	$\dfrac{e^h - 1}{h}$
0.02	1.0101	−0.02	0.990
0.005	1.00250	−0.005	0.99750
0.001	1.000500	−0.001	0.999500
0.0001	1.00005000	−0.0001	0.99995000

■ **EXAMPLE 5** A Limit That Does Not Exist Investigate graphically and numerically $\lim\limits_{x \to 0} \sin \dfrac{1}{x}$.

Solution The function $f(x) = \sin \dfrac{1}{x}$ is not defined at $x = 0$, but Figure 5 shows that it oscillates between $+1$ and -1 infinitely often as $x \to 0$. It appears, therefore, that $\lim\limits_{x \to 0} \sin \dfrac{1}{x}$ does not exist. This impression is confirmed by Table 5, which suggests that the values of $f(x)$ bounce around and do not tend toward any limit L as $x \to 0$.   ■

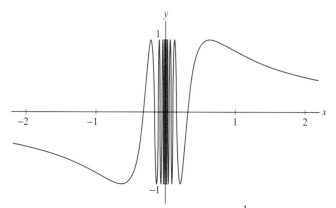

FIGURE 5 Graph of $f(x) = \sin \dfrac{1}{x}$.

TABLE 5	The Function $f(x) = \sin \frac{1}{x}$ Does Not Approach a Limit as $x \to 0$		
x	$\sin \dfrac{1}{x}$	x	$\sin \dfrac{1}{x}$
0.1	0.5440	−0.1	−0.5440
0.05	0.9129	−0.05	−0.9129
0.01	−0.5064	−0.01	0.5064
0.001	0.8269	−0.001	−0.8269
0.0005	0.9300	−0.0005	−0.9300
0.00002	−0.9998	−0.00002	0.9998
0.00001	0.0357	−0.00001	−0.0357

One-Sided Limits

The limits we have discussed so far are *two-sided*. To show that $\lim\limits_{x \to c} f(x) = L$, it is necessary to check that $f(x)$ converges to L as x approaches c through values both larger and smaller than c. In some instances, $f(x)$ may approach L from one side of c without necessarily approaching it from the other side, or $f(x)$ may be defined on only one side of c. For this reason, we define the one-sided limits

$$\lim_{x \to c-} f(x) \quad \text{(left-hand limit)}, \qquad \lim_{x \to c+} f(x) \quad \text{(right-hand limit)}$$

The limit itself exists if both one-sided limits exist and are equal.

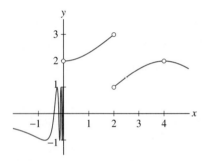

FIGURE 6 Graph of $y = \dfrac{x}{|x|}$.

■ **EXAMPLE 6** Left- and Right-Hand Limits Not Equal   Investigate the one-sided limits of $f(x) = \dfrac{x}{|x|}$ as $x \to 0$. Does $\displaystyle\lim_{x \to 0} f(x)$ exist?

Solution The graph in Figure 6 shows what is going on. For $x > 0$, we have

$$f(x) = \frac{x}{|x|} = \frac{x}{x} = 1$$

Therefore, $f(x) \to 1$ as $x \to 0+$. But for $x < 0$,

$$f(x) = \frac{x}{|x|} = \frac{x}{-x} = -1$$

and thus $f(x) \to -1$ as $x \to 0-$. We see that the one-sided limits exist:

$$\lim_{x \to 0+} f(x) = 1, \qquad \lim_{x \to 0-} f(x) = -1$$

However, these one-sided limits are not equal, so $\displaystyle\lim_{x \to 0} f(x)$ does not exist.   ■

■ **EXAMPLE 7** The function $f(x)$ shown in Figure 7 is not defined at $c = 0, 2, 4$. Do the one- or two-sided limits exist at these points? If so, find their values.

Solution

- $c = 0$: The left-hand limit $\displaystyle\lim_{x \to 0-} f(x)$ does not seem to exist because $f(x)$ appears to oscillate infinitely often to the left of 0. On the other hand, $\displaystyle\lim_{x \to 0+} f(x) = 2$.

- $c = 2$: The one-sided limits exist but are not equal:

$$\lim_{x \to 2-} f(x) = 3 \qquad \text{and} \qquad \lim_{x \to 2+} f(x) = 1$$

  Therefore, $\displaystyle\lim_{x \to 2} f(x)$ does not exist.

- $c = 4$: The one-sided limits exist and both have the value 2. Therefore, the two-sided limit exists and $\displaystyle\lim_{x \to 4} f(x) = 2$.   ■

FIGURE 7

Infinite Limits

Some functions tend to ∞ or $-\infty$ as x approaches a value c. In this case, the limit $\displaystyle\lim_{x \to c} f(x)$ does not exist, but we say that f has an *infinite limit*. More precisely, we write

$$\lim_{x \to c} f(x) = \infty$$

if $f(x)$ increases without bound as $x \to c$. If $f(x)$ tends to $-\infty$ (i.e., $f(x)$ becomes negative and $|f(x)| \to \infty$), then we write

$$\lim_{x \to c} f(x) = -\infty$$

When using this notation, keep in mind that ∞ and $-\infty$ are not numbers. One-sided infinite limits are defined similarly. In the next example, we use the notation $x \to c\pm$ to indicate that the left- and right-hand limits are to be considered separately.

■ **EXAMPLE 8** One- and Two-Sided Infinite Limits   GU   Investigate the one-sided limits graphically:

(a) $\lim\limits_{x \to 2\pm} \dfrac{1}{x-2}$ **(b)** $\lim\limits_{x \to 0\pm} \dfrac{1}{x^2}$ **(c)** $\lim\limits_{x \to 0+} \ln x$

Solution

When $f(x)$ approaches ∞ or $-\infty$ as x approaches c from one or both sides, we say that the line $x = c$ is a vertical asymptote. For example, in Figure 8(A), $x = 2$ is a vertical asymptote. Asymptotic behavior is discussed further in Section 4.5.

(a) The function $f(x) = \dfrac{1}{x-2}$ is negative for $x < 2$ and positive for $x > 2$. As we see in Figure 8(A), $f(x)$ approaches $-\infty$ as x approaches 2 from the left and ∞ as x approaches 2 from the right:

$$\lim_{x \to 2-} \frac{1}{x-2} = -\infty, \qquad \lim_{x \to 2+} \frac{1}{x-2} = \infty$$

(b) The function $f(x) = \dfrac{1}{x^2}$ is positive for all $x \neq 0$ and becomes arbitrarily large as $x \to 0$ from either side [Figure 8(B)]. Therefore, $\lim\limits_{x \to 0} \dfrac{1}{x^2} = \infty$.

(c) The function $f(x) = \ln x$ is negative for $0 < x < 1$ and it tends to $-\infty$ as $x \to 0+$ [Figure 8(C)]. Therefore, $\lim\limits_{x \to 0+} \ln x = -\infty$. ∎

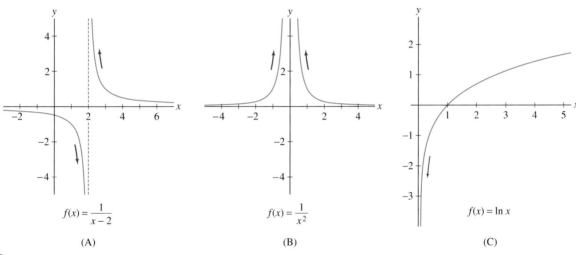

$f(x) = \dfrac{1}{x-2}$

(A)

$f(x) = \dfrac{1}{x^2}$

(B)

$f(x) = \ln x$

(C)

FIGURE 8

CONCEPTUAL INSIGHT You should not think of an infinite limit as a true limit. The notation $\lim\limits_{x \to c} f(x) = \infty$ is a shorthand way of saying that no finite limit exists but $f(x)$ increases beyond all bounds as x approaches c. We must be careful about this because ∞ and $-\infty$ are not numbers, and contradictions may arise if we manipulate them as numbers. For example, if ∞ were a number larger than any finite number, then presumably $\infty + 1 = \infty$. This would give

$$\infty + 1 = \infty$$
$$(\infty + 1) - \infty = \infty - \infty$$
$$1 = 0 \qquad \text{(contradiction!)}$$

For this reason, the Limit Laws discussed in the next section are valid only for finite limits and cannot be applied to infinite limits.

2.2 SUMMARY

- By definition, $\lim_{x \to c} f(x) = L$ if $|f(x) - L|$ becomes arbitrarily small when x is sufficiently close (but not equal) to c. We then say that *the limit of $f(x)$ as x approaches c is L*. Alternate terminology: $f(x)$ *approaches* or *converges to L as x approaches c*.
- If $f(x)$ approaches a limit as $x \to c$, then the limiting value L is unique.
- If $f(x)$ does not approach any limit as $x \to c$, we say that $\lim_{x \to c} f(x)$ does not exist.

- The limit may exist even if $f(c)$ is not defined, as in the example $\lim_{x \to 0} \dfrac{\sin x}{x}$.
- One-sided limits: $\lim_{x \to c-} f(x) = L$ if $f(x)$ converges to L as x approaches c through values less than c and $\lim_{x \to c+} f(x) = L$ if $f(x)$ converges to L as x approaches c through values greater than c.
- The limit exists if and only if both one-sided limits exist and are equal.
- We say that $\lim_{x \to c} f(x) = \infty$ if $f(x)$ increases beyond bound as x approaches c, and $\lim_{x \to c} f(x) = -\infty$ if $f(x)$ becomes arbitrarily large (in absolute value) but negative as x approaches c.

2.2 EXERCISES

Preliminary Questions

1. What is the limit of $f(x) = 1$ as $x \to \pi$?

2. What is the limit of $g(t) = t$ as $t \to \pi$?

3. Can $f(x)$ approach a limit as $x \to c$ if $f(c)$ is undefined? If so, give an example.

4. Is $\lim_{x \to 10} 20$ equal to 10 or 20?

5. What does the following table suggest about $\lim_{x \to 1-} f(x)$ and $\lim_{x \to 1+} f(x)$?

x	0.9	0.99	0.999	1.1	1.01	1.001
$f(x)$	7	25	4317	3.0126	3.0047	3.00011

6. Is it possible to tell if $\lim_{x \to 5} f(x)$ exists by only examining values $f(x)$ for x close to but *greater* than 5? Explain.

7. If you know in advance that $\lim_{x \to 5} f(x)$ exists, can you determine its value just knowing the values of $f(x)$ for all $x > 5$?

8. Which of the following pieces of information is sufficient to determine whether $\lim_{x \to 5} f(x)$ exists? Explain.

(a) The values of $f(x)$ for all x

(b) The values of $f(x)$ for x in $[4.5, 5.5]$

(c) The values of $f(x)$ for all x in $[4.5, 5.5]$ other than $x = 5$

(d) The values of $f(x)$ for all $x \geq 5$

(e) $f(5)$

Exercises

In Exercises 1–6, fill in the tables and guess the value of the limit.

1. $\lim_{x \to 1} f(x)$, where $f(x) = \dfrac{x^3 - 1}{x^2 - 1}$.

x	$f(x)$	x	$f(x)$
1.002		0.998	
1.001		0.999	
1.0005		0.9995	
1.00001		0.99999	

2. $\lim_{t \to 0} h(t)$, where $h(t) = \dfrac{\cos t - 1}{t^2}$. Note that $h(t)$ is even, that is, $h(t) = h(-t)$.

t	± 0.002	± 0.0001	± 0.00005	± 0.00001
$h(t)$				

3. $\lim_{y \to 2} f(y)$, where $f(y) = \dfrac{y^2 - y - 2}{y^2 + y - 6}$.

y	$f(y)$	y	$f(y)$
2.002		1.998	
2.001		1.999	
2.0001		1.9999	

4. $\lim\limits_{\theta \to 0} f(\theta)$, where $f(\theta) = \dfrac{\sin \theta - \theta}{\theta^3}$.

θ	± 0.002	± 0.0001	± 0.00005	± 0.00001
$f(\theta)$				

5. $\lim\limits_{x \to 0} f(x)$, where $f(x) = \dfrac{e^x - x - 1}{x^2}$.

x	± 0.5	± 0.1	± 0.05	± 0.01
$f(x)$				

6. $\lim\limits_{x \to 0+} f(x)$, where $f(x) = x \ln x$.

x	1	0.5	0.1	0.05	0.01	0.005	0.001
$f(x)$							

7. Determine $\lim\limits_{x \to 0.5} f(x)$ for the function $f(x)$ shown in Figure 9.

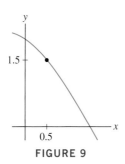

FIGURE 9

8. Do either of the two oscillating functions in Figure 10 appear to approach a limit as $x \to 0$?

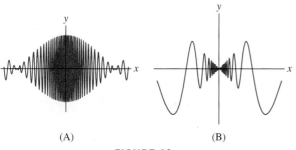

(A)   (B)

FIGURE 10

In Exercises 9–10, evaluate the limit.

9. $\lim\limits_{x \to 21} x$

10. $\lim\limits_{x \to 4.2} \sqrt{3}$

In Exercises 11–20, verify each limit using the limit definition. For example, in Exercise 11, show that $|2x - 6|$ can be made as small as desired by taking x close to 3.

11. $\lim\limits_{x \to 3} 2x = 6$

12. $\lim\limits_{x \to 3} 4 = 4$

13. $\lim\limits_{x \to 2} (4x + 3) = 11$

14. $\lim\limits_{x \to 3} (5x - 7) = 8$

15. $\lim\limits_{x \to 9} (-2x) = -18$

16. $\lim\limits_{x \to -5} (1 - 2x) = 11$

17. $\lim\limits_{x \to 0} x^2 = 0$

18. $\lim\limits_{x \to 0} (2x^2 + 4) = 4$

19. $\lim\limits_{x \to 0} (x^2 + 2x + 3) = 3$

20. $\lim\limits_{x \to 0} (x^3 + 9) = 9$

In Exercises 21–36, estimate the limit numerically or state that the limit does not exist.

21. $\lim\limits_{x \to 1} \dfrac{\sqrt{x} - 1}{x - 1}$

22. $\lim\limits_{x \to -3} \dfrac{2x^2 - 18}{x + 3}$

23. $\lim\limits_{x \to 2} \dfrac{x^2 + x - 6}{x^2 - x - 2}$

24. $\lim\limits_{x \to 3} \dfrac{x^3 - 2x^2 - 9}{x^2 - 2x - 3}$

25. $\lim\limits_{x \to 0} \dfrac{\sin 2x}{x}$

26. $\lim\limits_{x \to 0} \dfrac{\sin 5x}{x}$

27. $\lim\limits_{x \to 0} \dfrac{\sin x}{x^2}$

28. $\lim\limits_{\theta \to 0} \dfrac{\cos \theta - 1}{\theta}$

29. $\lim\limits_{h \to 0} \cos \dfrac{1}{h}$

30. $\lim\limits_{h \to 0} \sin h \cos \dfrac{1}{h}$

31. $\lim\limits_{h \to 0} \dfrac{2^h - 1}{h}$

32. $\lim\limits_{x \to 0} \dfrac{e^x - 2^x}{x}$

33. $\lim\limits_{x \to 1+} \dfrac{\sec^{-1} x}{\sqrt{x - 1}}$

34. $\lim\limits_{x \to 0} |x|^x$

35. $\lim\limits_{x \to 0} \dfrac{\tan^{-1} x}{\sin^{-1} x - x}$

36. $\lim\limits_{x \to 0} \dfrac{\tan^{-1} x - x}{\sin^{-1} x - x}$

37. Determine $\lim\limits_{x \to 2+} f(x)$ and $\lim\limits_{x \to 2-} f(x)$ for the function shown in Figure 11.

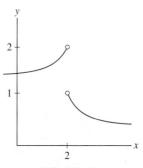

FIGURE 11

38. Determine the one-sided limits at $c = 1, 2, 4, 5$ of the function $g(t)$ shown in Figure 12 and state whether the limit exists at these points.

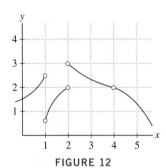

FIGURE 12

39. The greatest integer function is defined by $[x] = n$, where n is the unique integer such that $n \le x < n + 1$. See Figure 13.

(a) For which values of c does $\lim_{x \to c-} [x]$ exist? What about $\lim_{x \to c+} [x]$?

(b) For which values of c does $\lim_{x \to c} [x]$ exist?

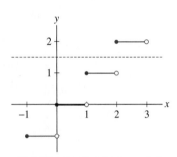

FIGURE 13 Graph of $y = [x]$.

40. Draw a graph of $f(x) = \dfrac{x - 1}{|x - 1|}$ and use it to determine the one-sided limits $\lim_{x \to 1+} f(x)$ and $\lim_{x \to 1-} f(x)$.

In Exercises 41–43, determine the one-sided limits numerically.

41. $\lim_{x \to 0\pm} \dfrac{\sin x}{|x|}$

42. $\lim_{x \to 0\pm} |x|^{1/x}$

43. $\lim_{x \to 0\pm} \dfrac{x - \sin(|x|)}{x^3}$

44. Determine the one- or two-sided infinite limits in Figure 14.

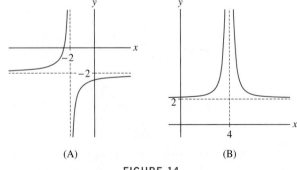

(A)   (B)

FIGURE 14

45. Determine the one-sided limits of $f(x)$ at $c = 2$ and $c = 4$, for the function shown in Figure 15.

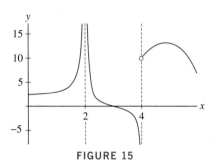

FIGURE 15

46. Determine the infinite one- and two-sided limits in Figure 16.

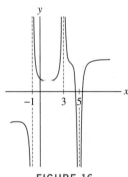

FIGURE 16

In Exercises 47–50, draw the graph of a function with the given limits.

47. $\lim_{x \to 1} f(x) = 2$, $\lim_{x \to 3-} f(x) = 0$, $\lim_{x \to 3+} f(x) = 4$

48. $\lim_{x \to 1} f(x) = \infty$, $\lim_{x \to 3-} f(x) = 0$, $\lim_{x \to 3+} f(x) = -\infty$

49. $\lim_{x \to 2+} f(x) = f(2) = 3$, $\lim_{x \to 2-} f(x) = -1$,

$\lim_{x \to 4} f(x) = 2 \ne f(4)$

50. $\lim_{x \to 1+} f(x) = \infty$, $\lim_{x \to 1-} f(x) = 3$, $\lim_{x \to 4} f(x) = -\infty$

GU *In Exercises 51–56, graph the function and use the graph to estimate the value of the limit.*

51. $\lim_{\theta \to 0} \dfrac{\sin 3\theta}{\sin 2\theta}$

52. $\lim_{x \to 0} \dfrac{2^x - 1}{4^x - 1}$

53. $\lim_{x \to 0} \dfrac{2^x - \cos x}{x}$

54. $\lim_{\theta \to 0} \dfrac{\sin^2 4\theta}{\cos \theta - 1}$

55. $\lim_{\theta \to 0} \dfrac{\cos 3\theta - \cos 4\theta}{\theta^2}$

56. $\lim_{\theta \to 0} \dfrac{\cos 3\theta - \cos 5\theta}{\theta^2}$

Further Insights and Challenges

57. Light waves of frequency λ passing through a slit of width a produce a **Fraunhofer diffraction pattern** of light and dark fringes (Figure 17). The intensity as a function of the angle θ is given by

$$I(\theta) = I_m \left(\frac{\sin(R\sin\theta)}{R\sin\theta} \right)^2$$

where $R = \pi a/\lambda$ and I_m is a constant. Show that the intensity function is not defined at $\theta = 0$. Then check numerically that $I(\theta)$ approaches I_m as $\theta \to 0$ for any two values of R (e.g., choose two integer values).

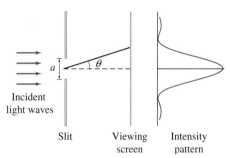

Incident
light waves

Slit   Viewing   Intensity
            screen   pattern

FIGURE 17 Fraunhofer diffraction pattern.

58. Investigate $\displaystyle\lim_{\theta\to 0} \frac{\sin n\theta}{\theta}$ numerically for several values of n and then guess the value in general.

59. Show numerically that $\displaystyle\lim_{x\to 0} \frac{b^x - 1}{x}$ for $b = 3, 5$ appears to equal $\ln 3$, $\ln 5$, where $\ln x$ is the natural logarithm. Then make a conjecture (guess) for the value in general and test your conjecture for two additional values of b.

60. Investigate $\displaystyle\lim_{x\to 1} \frac{x^n - 1}{x^m - 1}$ for (m, n) equal to $(2, 1)$, $(1, 2)$, $(2, 3)$, and $(3, 2)$. Then guess the value of the limit in general and check your guess for at least three additional pairs.

61. Find by experimentation the positive integers k such that $\displaystyle\lim_{x\to 0} \frac{\sin(\sin^2 x)}{x^k}$ exists.

62. [GU] Sketch a graph of $f(x) = \dfrac{2^x - 8}{x - 3}$ with a graphing calculator. Observe that $f(3)$ is not defined.

(a) Zoom in on the graph to estimate $L = \displaystyle\lim_{x\to 3} f(x)$.

(b) Observe that the graph of $f(x)$ is increasing. Explain how this implies that

$$f(2.99999) \le L \le f(3.00001)$$

Use this to determine L to three decimal places.

63. [GU] The function $f(x) = \dfrac{2^{1/x} - 2^{-1/x}}{2^{1/x} + 2^{-1/x}}$ is defined for $x \ne 0$.

(a) Investigate $\displaystyle\lim_{x\to 0+} f(x)$ and $\displaystyle\lim_{x\to 0-} f(x)$ numerically.

(b) Produce a graph of f on a graphing utility and describe its behavior near $x = 0$.

64. Show that $f(x) = \dfrac{\sin x}{x}$ is equal to the slope of a secant line through the origin and the point $(x, \sin x)$ on the graph of $y = \sin x$ (Figure 18). Use this to give a geometric interpretation of $\displaystyle\lim_{x\to 0} f(x)$.

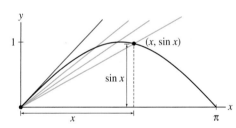

FIGURE 18 Graph of $y = \sin x$.

2.3 Basic Limit Laws

In Section 2.2 we relied on graphical and numerical approaches to investigate limits and estimate their values. In the next four sections we go beyond this intuitive approach and develop tools for computing limits in a precise way. This section discusses the **Basic Limit Laws** used for computing limits of functions constructed as sums, multiples, products, or quotients of other functions.

The proofs of these laws are discussed in Section 2.8 and Appendix D. However, we can understand the underlying idea by arguing informally. For example, to prove the Sum Law, we observe that if $f(x)$ is close to L and $g(x)$ is close to M when $|x - c|$ is sufficiently small, then $f(x) + g(x)$ must be close to $L + M$ when $|x - c|$ is sufficiently small. Similarly, $f(x)g(x)$ must be close to LM, etc.

Handwritten notes in left margin:

Given
- $\lim\limits_{x \to 3} f(x) = -2$
- $\lim\limits_{x \to 3} g(x) = 6$

Find $\lim\limits_{x \to 3} [f(x) + g(x)] = 4$ $-2+6$

$\lim\limits_{x \to 3} (g(x) - f(x)) = 8$ $\frac{6--2}{8}$

$\lim\limits_{x \to 3} 4f(x) = -8$ $(4 \cdot -2)$

$\lim\limits_{x \to 3} (f(x)g(x)) = -12$ $-2 \cdot 6$

THEOREM 1 Basic Limit Laws   Assume that $\lim\limits_{x \to c} f(x)$ and $\lim\limits_{x \to c} g(x)$ exist. Then:

(i) Sum Law:

$$\lim_{x \to c} \big(f(x) + g(x)\big) = \lim_{x \to c} f(x) + \lim_{x \to c} g(x)$$

(ii) Constant Multiple Law: For any number k,

$$\lim_{x \to c} kf(x) = k \lim_{x \to c} f(x)$$

(iii) Product Law:

$$\lim_{x \to c} f(x)g(x) = \left(\lim_{x \to c} f(x)\right)\left(\lim_{x \to c} g(x)\right)$$

(iv) Quotient Law: If $\lim\limits_{x \to c} g(x) \neq 0$, then

$$\lim_{x \to c} \frac{f(x)}{g(x)} = \frac{\lim\limits_{x \to c} f(x)}{\lim\limits_{x \to c} g(x)}$$

Before proceeding to the examples, we make two useful remarks. First, the Sum and Product Laws are valid for any number of functions. For example,

$$\lim_{x \to c} \big(f_1(x) + f_2(x) + f_3(x)\big) = \lim_{x \to c} f_1(x) + \lim_{x \to c} f_2(x) + \lim_{x \to c} f_3(x)$$

Second, the Sum Law has a counterpart for differences:

$$\lim_{x \to c} \big(f(x) - g(x)\big) = \lim_{x \to c} f(x) - \lim_{x \to c} g(x)$$

This is not listed as a separate law because it follows by combining the Sum and Constant Multiple Laws:

$$\lim_{x \to c} \big(f(x) - g(x)\big) = \lim_{x \to c} f(x) + \lim_{x \to c} \big(- g(x)\big) = \lim_{x \to c} f(x) - \lim_{x \to c} g(x)$$

The Limit Laws are applied step by step in the solution to Example 1 to illustrate how they are used. In practice, we take the Limit Laws for granted and skip the intermediate steps.

■ **EXAMPLE 1**   Use the Limit Laws to evaluate the following limits:

(a) $\lim\limits_{x \to 3} x^2$

(b) $\lim\limits_{t \to -1} (4t^3 - t - 5)$

(c) $\lim\limits_{t \to 2} \dfrac{t + 5}{3t}$

Solution

(a) By Theorem 1 of Section 2.2, $\lim\limits_{x \to c} x = c$ for all c. Since x^2 is equal to the product $x^2 = x \cdot x$, we may apply the Product Law

$$\lim_{x \to 3} x^2 = \left(\lim_{x \to 3} x\right)\left(\lim_{x \to 3} x\right) = 3(3) = 9$$

(b) First use the Sum Law (for three functions):

$$\lim_{t \to -1} (4t^3 - t - 5) = \lim_{t \to -1} 4t^3 - \lim_{t \to -1} t - \lim_{t \to -1} 5$$

Then evaluate each limit using the Constant Multiple and Product Laws:

$$\lim_{t \to -1} 4t^3 = 4 \lim_{t \to -1} t^3 = 4\left(\lim_{t \to -1} t\right)\left(\lim_{t \to -1} t\right)\left(\lim_{t \to -1} t\right) = 4(-1)^3 = -4$$

$$\lim_{t \to -1} t = -1$$

$$\lim_{t \to -1} 5 = 5$$

We obtain

$$\lim_{t \to -1} (4t^3 - t - 5) = -4 - (-1) - 5 = -8$$

(c) Use the Quotient, Sum, and Constant Multiple Laws:

$$\lim_{t \to 2} \frac{t + 5}{3t} = \frac{\lim\limits_{t \to 2}(t + 5)}{\lim\limits_{t \to 2} 3t} = \frac{\lim\limits_{t \to 2} t + \lim\limits_{t \to 2} 5}{3 \lim\limits_{t \to 2} t} = \frac{2 + 5}{3 \cdot 2} = \frac{7}{6}$$ ■

The next example reminds us that the Limit Laws apply only when the limits of $f(x)$ and $g(x)$ exist.

■ **EXAMPLE 2** Assumptions Matter   Show that the Product Law cannot be applied to $\lim\limits_{x \to 0} f(x)g(x)$ if $f(x) = x$ and $g(x) = x^{-1}$.

Solution   The product function is $f(x)g(x) = x \cdot x^{-1} = 1$ for $x \neq 0$, so the limit of the product exists:

$$\lim_{x \to 0} f(x)g(x) = \lim_{x \to 0} 1 = 1$$

However, $\lim\limits_{x \to 0} x^{-1}$ does not exist because $g(x) = x^{-1}$ approaches ∞ as $c \to 0+$ and $-\infty$ as $c \to 0-$, so the Product Law cannot be applied and its conclusion does not hold:

$$\left(\lim_{x \to 0} f(x)\right)\left(\lim_{x \to 0} g(x)\right) = \left(\lim_{x \to 0} x\right)\underbrace{\left(\lim_{x \to 0} x^{-1}\right)}_{\text{Does not exist}}$$ ■

2.3 SUMMARY

- The Limit Laws state that if $\lim\limits_{x \to c} f(x)$ and $\lim\limits_{x \to c} g(x)$ both exist, then:

(i) $\lim\limits_{x \to c} \big(f(x) + g(x)\big) = \lim\limits_{x \to c} f(x) + \lim\limits_{x \to c} g(x)$

(ii) $\lim\limits_{x \to c} kf(x) = k \lim\limits_{x \to c} f(x)$

(iii) $\lim\limits_{x \to c} f(x)\, g(x) = \left(\lim\limits_{x \to c} f(x)\right)\left(\lim\limits_{x \to c} g(x)\right)$

(iv) If $\lim\limits_{x \to c} g(x) \neq 0$, then

$$\lim_{x \to c} \frac{f(x)}{g(x)} = \frac{\lim\limits_{x \to c} f(x)}{\lim\limits_{x \to c} g(x)}$$

- If $\lim\limits_{x \to c} f(x)$ or $\lim\limits_{x \to c} g(x)$ does not exist, then the Limit Laws cannot be applied.

2.3 EXERCISES

Preliminary Questions

1. State the Sum Law and Quotient Law.

2. Which of the following is a verbal version of the Product Law?

(a) The product of two functions has a limit.

(b) The limit of the product is the product of the limits.

(c) The product of a limit is a product of functions.

(d) A limit produces a product of functions.

3. Which of the following statements are incorrect (k and c are constants)?

(a) $\lim_{x \to c} k = c$

(b) $\lim_{x \to c} k = k$

(c) $\lim_{x \to c^2} x = c^2$

(d) $\lim_{x \to c} x = x$

4. Which of the following statements are incorrect?

(a) The Product Law does not hold if the limit of one of the functions is zero.

(b) The Quotient Law does not hold if the limit of the denominator is zero.

(c) The Quotient Law does not hold if the limit of the numerator is zero.

Exercises

In Exercises 1–22, evaluate the limits using the Limit Laws and the following two facts, where c and k are constants:

$$\lim_{x \to c} x = c, \qquad \lim_{x \to c} k = k$$

1. $\lim_{x \to 9} x$

2. $\lim_{x \to -3} x$

3. $\lim_{x \to 9} 14$

4. $\lim_{x \to -3} 14$

5. $\lim_{x \to -3} (3x + 4)$

6. $\lim_{y \to -3} 14y$

7. $\lim_{y \to -3} (y + 14)$

8. $\lim_{y \to -3} y(y + 14)$

9. $\lim_{t \to 4} (3t - 14)$

10. $\lim_{x \to -5} (x^3 + 2x)$

11. $\lim_{x \to \frac{1}{2}} (4x + 1)(2x - 1)$

12. $\lim_{x \to -1} (3x^4 - 2x^3 + 4x)$

13. $\lim_{x \to 2} x(x + 1)(x + 2)$

14. $\lim_{x \to 2} (x + 1)(3x^2 - 9)$

15. $\lim_{t \to 9} \dfrac{t}{t + 1}$

16. $\lim_{t \to 4} \dfrac{3t - 14}{t + 1}$

17. $\lim_{x \to 3} \dfrac{1 - x}{1 + x}$

18. $\lim_{x \to -1} \dfrac{x}{x^3 + 4x}$

19. $\lim_{t \to 2} t^{-1}$

20. $\lim_{x \to 5} x^{-2}$

21. $\lim_{x \to 3} (x^2 + 9x^{-3})$

22. $\lim_{z \to 1} \dfrac{z^{-1} + z}{z + 1}$

23. Use the Quotient Law to prove that if $\lim_{x \to c} f(x)$ exists and is nonzero, then

$$\lim_{x \to c} \frac{1}{f(x)} = \frac{1}{\lim_{x \to c} f(x)}$$

24. Assume that $\lim_{x \to 6} f(x) = 4$ and compute:

(a) $\lim_{x \to 6} f(x)^2$

(b) $\lim_{x \to 6} \dfrac{1}{f(x)}$

(c) $\lim_{x \to 6} x f(x)$

In Exercises 25–28, evaluate the limit assuming that $\lim_{x \to -4} f(x) = 3$ and $\lim_{x \to -4} g(x) = 1$.

25. $\lim_{x \to -4} f(x) g(x)$

26. $\lim_{x \to -4} (2f(x) + 3g(x))$

27. $\lim_{x \to -4} \dfrac{g(x)}{x^2}$

28. $\lim_{x \to -4} \dfrac{f(x) + 1}{3g(x) - 9}$

29. Can the Quotient Law be applied to evaluate $\lim_{x \to 0} \dfrac{\sin x}{x}$? Explain.

30. Show that the Product Law cannot be used to evaluate the $\lim_{x \to \pi/2} (x - \pi/2) \tan x$.

31. Give an example where $\lim_{x \to 0} (f(x) + g(x))$ exists but neither $\lim_{x \to 0} f(x)$ nor $\lim_{x \to 0} g(x)$ exists.

32. Assume that the limit $L_a = \lim_{x \to 0} \dfrac{a^x - 1}{x}$ exists and that $\lim_{x \to 0} a^x = 1$ for all $a > 0$. Prove that $L_{ab} = L_a + L_b$ for $a, b > 0$. *Hint:* $(ab)^x - 1 = a^x(b^x - 1) + (a^x - 1)$. Verify numerically that $L_{12} = L_3 + L_4$.

33. Use the Limit Laws and the result $\lim_{x \to c} x = c$ to show that $\lim_{x \to c} x^n = c^n$ for all whole numbers n. If you are familiar with induction, give a formal proof by induction.

34. Extend Exercise 33 to negative integers.

Further Insights and Challenges

35. Show that if both $\lim_{x\to c} f(x)\,g(x)$ and $\lim_{x\to c} g(x)$ exist and $\lim_{x\to c} g(x)$ is nonzero, then $\lim_{x\to c} f(x)$ exists. *Hint:* Write $f(x) = (f(x)\,g(x))/g(x)$ and apply the Quotient Law.

36. Show that if $\lim_{t\to 3} t\,g(t) = 12$, then $\lim_{t\to 3} g(t)$ exists and equals 4.

37. Prove that if $\lim_{t\to 3} \dfrac{h(t)}{t} = 5$, then $\lim_{t\to 3} h(t) = 15$.

38. 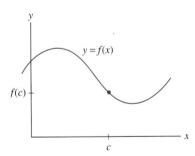 Assuming that $\lim_{x\to 0} \dfrac{f(x)}{x} = 1$, which of the following statements is necessarily true? Why?

(a) $f(0) = 0$ **(b)** $\lim_{x\to 0} f(x) = 0$

39. Prove that if $\lim_{x\to c} f(x) = L \neq 0$ and $\lim_{x\to c} g(x) = 0$, then $\lim_{x\to c} \dfrac{f(x)}{g(x)}$ does not exist.

40. 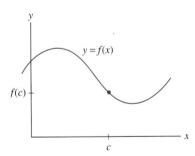 Suppose that $\lim_{h\to 0} g(h) = L$.

(a) Explain why $\lim_{h\to 0} g(ah) = L$ for any constant $a \neq 0$.

(b) If we assume instead that $\lim_{h\to 1} g(h) = L$, is it still necessarily true that $\lim_{h\to 1} g(ah) = L$?

(c) Illustrate the conclusions you reach in (a) and (b) with the function $f(x) = x^2$.

41. 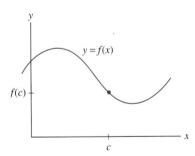 Show that if $\lim_{h\to 0} \dfrac{f(h)}{h} = L$, then $\lim_{h\to 0} \dfrac{f(ah)}{h} = aL$ for all $a \neq 0$. *Hint:* Apply the result of Exercise 40 to $g(h) = \dfrac{f(h)}{h}$.

42. In Section 2.2, we mentioned that the number e is characterized by the property $\lim_{h\to 0} \dfrac{e^h - 1}{h} = 1$. Assuming this, show that for $b > 0$,

$$\lim_{h\to 0} \frac{b^h - 1}{h} = \ln b$$

Hint: Apply Exercise 41 to $f(h) = \dfrac{e^h - 1}{h}$ and $a = \ln b$.

43. 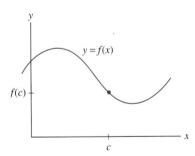 There is a Limit Law for composite functions but it is not stated in the text. Which of the following is the correct statement? Give an intuitive explanation.

(a) $\lim_{x\to c} f(g(x)) = \lim_{x\to c} f(x)$

(b) $\lim_{x\to c} f(g(x)) = \lim_{x\to L} f(x)$, where $L = \lim_{x\to c} g(x)$

(c) $\lim_{x\to c} f(g(x)) = \lim_{x\to L} g(x)$, where $L = \lim_{x\to c} f(x)$

Use the correct version to evaluate

$$\lim_{x\to 2} \sin(g(x)), \quad \text{where} \quad \lim_{x\to 2} g(x) = \frac{\pi}{6}$$

2.4 Limits and Continuity

In everyday speech, the word "continuous" means having no breaks or interruptions. In calculus, continuity is used to describe functions whose graphs have no breaks. If we imagine the graph of a function f as a wavy metal wire, then f is continuous if its graph consists of a single piece of wire as in Figure 1. A break in the wire as in Figure 2 is called a **discontinuity**.

Now observe that the function $g(x)$ in Figure 2 has a discontinuity at $x = c$ and that $\lim_{x\to c} g(x)$ does not exist (the left- and right-hand limits are not equal). By contrast, in Figure 1, $\lim_{x\to c} f(x)$ exists and is equal to the function value $f(c)$. This suggests the following definition of continuity in terms of limits.

FIGURE 1 $f(x)$ is continuous at $x = c$.

FIGURE 2 $g(x)$ has a discontinuity at $x = c$.

DEFINITION Continuity at a Point Assume that $f(x)$ is defined on an open interval containing $x = c$. Then f is **continuous** at $x = c$ if

$$\lim_{x\to c} f(x) = f(c)$$

If the limit does not exist, or if it exists but is not equal to $f(c)$, we say that f has a **discontinuity** (or is **discontinuous**) at $x = c$.

- removable discontinuity
- jump discontinuity

A function $f(x)$ may be continuous at some points and discontinuous at others. If $f(x)$ is continuous at all points in an interval I, then $f(x)$ is said to be continuous on I. Here, if I is an interval $[a, b]$ or $[a, b)$ that includes a as a left-endpoint, we require that $\lim_{x \to a+} f(x) = f(a)$. Similarly, we require that $\lim_{x \to b-} f(x) = f(b)$ if I includes b as a right-endpoint b. If $f(x)$ is continuous at all points in its domain, then $f(x)$ is simply called continuous.

■ **EXAMPLE 1** Show that the following functions are continuous:

(a) $f(x) = k$ (k any constant)   **(b)** $g(x) = x$

Solution

(a) Since $f(x) = k$ for all x,

$$\lim_{x \to c} f(x) = \lim_{x \to c} k = k = f(c)$$

The limit exists and is equal to the function value, so $f(x)$ is continuous at $x = c$ for all c (Figure 3).

(b) Since $g(x) = x$ for all x,

$$\lim_{x \to c} g(x) = \lim_{x \to c} x = c = g(c)$$

Again, the limit exists and is equal to the function value, so $g(x)$ is continuous at c for all c (Figure 4). ■

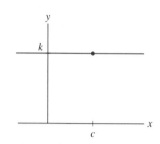

FIGURE 3 Graph of $f(x) = k$.

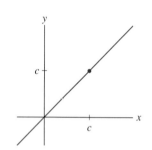

FIGURE 4 Graph of $g(x) = x$.

Examples of Discontinuities

To understand continuity better, let's consider some ways that a function can fail to be continuous. Keep in mind that continuity at a point requires more than just the existence of a limit. For $f(x)$ to be continuous at $x = c$, three conditions must hold:

1. $\lim_{x \to c} f(x)$ exists.   **2.** $f(c)$ exists.   **3.** They are equal.

Otherwise, $f(x)$ is discontinuous at $x = c$.

If the first two conditions hold but the third fails, we say that f has a **removable discontinuity** at $x = c$. The function in Figure 5(A) has a removable discontinuity as $c = 2$ because

$$\underbrace{\lim_{x \to 2} f(x) = 5 \qquad \text{but} \qquad f(2) = 10}_{\text{Limit exists but is not equal to function value}}$$

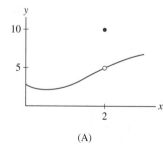

(A)

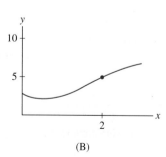

(B)

FIGURE 5 A removable discontinuity: The discontinuity can be removed by redefining $f(2)$.

Removable discontinuities are "mild" in the following sense: We can make f continuous at $x = c$ by redefining $f(c)$. In Figure 5(B), the value $f(2)$ has been redefined as $f(2) = 5$ and this makes f continuous at $x = 2$.

A "worse" type of discontinuity is a **jump discontinuity**, which occurs if the one-sided limits $\lim\limits_{x \to c-} f(x)$ and $\lim\limits_{x \to c+} f(x)$ exist but are not equal. Figure 6 shows two functions with jump discontinuities at $c = 2$. Unlike the removable case, we cannot make $f(x)$ continuous by redefining $f(c)$.

In connection with jump discontinuities, it is convenient to define *one-sided continuity*.

DEFINITION **One-Sided Continuity**   A function $f(x)$ is called:

- **Left-continuous** at $x = c$ if $\lim\limits_{x \to c-} f(x) = f(c)$
- **Right-continuous** at $x = c$ if $\lim\limits_{x \to c+} f(x) = f(c)$

[handwritten: closed circle means continuity.]

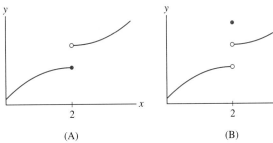

(A)   (B)

FIGURE 6 Jump discontinuities.

[handwritten: Greatest integer func.]

[handwritten: $y = [x]$ $[\![x]\!]$ ↑ input]

[handwritten: output = greatest integer that is ≤ to the input]
[handwritten: $[2.4] = 2$ $[-2.4] = -3$]
[handwritten: $[2.9] = 2$ $[-2.9] = -3$]
[handwritten: $[2] = 2$ $[-2] = -2$]

In Figure 6 above, the function in (A) is left-continuous but the function in (B) is neither left- nor right-continuous. The next example explores one-sided continuity using a piecewise-defined function, that is, a function defined by different formulas on different intervals.

■ **EXAMPLE 2** Piecewise-Defined Function   Discuss the continuity of the function $F(x)$ defined by

[handwritten: step function]

$$F(x) = \begin{cases} x & \text{for } x < 1 \\ 3 & \text{for } 1 \le x \le 3 \\ x & \text{for } x > 3 \end{cases}$$

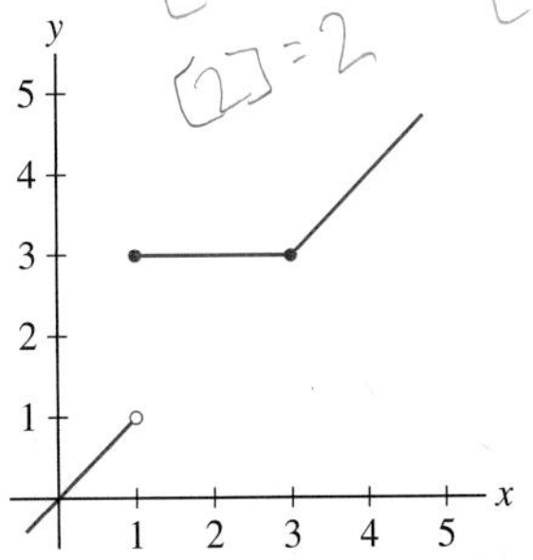

Solution The functions $f(x) = x$ and $g(x) = 3$ are continuous, so $F(x)$ is also continuous, except possibly at the transition points $x = 1$ and $x = 3$ where the formula for $F(x)$ changes (Figure 7). We observe that $F(x)$ has a jump discontinuity at $x = 1$ because the one-sided limits exist but are not equal:

$$\lim_{x \to 1-} F(x) = \lim_{x \to 1-} x = 1, \qquad \lim_{x \to 1+} F(x) = \lim_{x \to 1+} 3 = 3$$

Furthermore, the right-hand limit equals the function value $F(1) = 3$, so $F(x)$ is *right-continuous* at $x = 1$. At $x = 3$,

$$\lim_{x \to 3-} F(x) = \lim_{x \to 3-} 3 = 3, \qquad \lim_{x \to 3+} F(x) = \lim_{x \to 3+} x = 3$$

Both the left- and right-hand limits exist and are equal to $F(3)$, so $F(x)$ is *continuous* at $x = 3$. ■

FIGURE 7 Piecewise-defined function $F(x)$ in Example 2.

We say that $f(x)$ has an **infinite discontinuity** at $x = c$ if one or both of the one-sided limits is infinite (even if $f(x)$ itself is not defined at $x = c$). Figure 8 illustrates three types of infinite discontinuities occurring at $x = 2$. Notice that $x = 2$ does not belong to the domain of the function in cases (A) and (B).

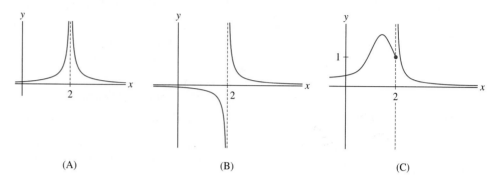

(A)   (B)   (C)

FIGURE 8 Functions with an infinite discontinuity at $x = 2$.

We should mention that some functions have more "severe" types of discontinuity than those discussed above. For example, $f(x) = \sin \dfrac{1}{x}$ oscillates infinitely often between $+1$ and -1 as $x \to 0$ (Figure 9). Neither the left- nor the right-hand limits exist at $x = 0$, so this discontinuity is not a jump discontinuity. See Exercises 90 and 91 for even stranger examples. Although of interest from a theoretical point of view, these discontinuities rarely arise in practice.

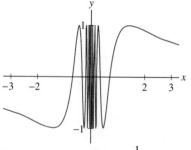

FIGURE 9 Graph of $y = \sin \dfrac{1}{x}$. The discontinuity at $x = 0$ is more "severe." It is neither a jump discontinuity nor a removable or infinite discontinuity.

Building Continuous Functions

Having studied some examples of discontinuities, we focus again on continuous functions. How can we show that a function is continuous? One way is to use the **Laws of Continuity**, which state, roughly speaking, that a function is continuous if it is built out of functions that are known to be continuous.

THEOREM 1 Laws of Continuity   Assume that $f(x)$ and $g(x)$ are continuous at a point $x = c$. Then the following functions are also continuous at $x = c$:

  (i) $f(x) + g(x)$ and $f(x) - g(x)$ **(iii)** $f(x)\, g(x)$

  (ii) $kf(x)$ for any constant k **(iv)** $f(x)/g(x)$ if $g(c) \neq 0$

Proof   These laws follow directly from the corresponding Limit Laws (Theorem 1, Section 2.3). We illustrate by proving the first part of (i) in detail. The remaining laws are proved similarly. By definition, we must show that $\lim\limits_{x \to c} (f(x) + g(x)) = f(c) + g(c)$. Because $f(x)$ and $g(x)$ are both continuous at $x = c$, we have

$$\lim_{x \to c} f(x) = f(c), \qquad \lim_{x \to c} g(x) = g(c)$$

The Sum Law for limits yields the desired result:

$$\lim_{x \to c} (f(x) + g(x)) = \lim_{x \to c} f(x) + \lim_{x \to c} g(x) = f(c) + g(c) \qquad \blacksquare$$

In Section 2.3, we remarked that the Limit Laws for Sums and Products are valid for the sums and products of an arbitrary number of functions. The same is true for continuity, namely, if $f_1(x), \ldots, f_n(x)$ are continuous, then so are the functions

$$f_1(x) + f_2(x) + \cdots + f_n(x), \quad f_1(x) \cdot f_2(x) \cdots f_n(x)$$

We now use the Laws of Continuity to study the continuity of some basic classes of functions.

When a function $f(x)$ is defined and continuous for all values of x, we say that $f(x)$ is continuous on the real line.

THEOREM 2 Continuity of Polynomial and Rational Functions Let $P(x)$ and $Q(x)$ be polynomials. Then:

- $P(x)$ is continuous on the real line.
- $\dfrac{P(x)}{Q(x)}$ is continuous at all values c such that $Q(c) \neq 0$.

*← **REMINDER** A "rational function" is a quotient of two polynomials $\frac{P(x)}{Q(x)}$.*

Continuous at every # in D:
- polynomials
- rational functions
- root functions
- trigonometric functions
- inverse trigonometric func.
- exponential functions
- logarithmic functions
- Absolute value

Proof We use the Laws of Continuity together with the result of Example 1, according to which constant functions and the function $f(x) = x$ are continuous. For any whole number $m \geq 1$, the power function x^m can be written as a product $x^m = x \cdot x \cdots x$ (m factors). Since each factor is continuous, x^m itself is continuous. Similarly, ax^m is continuous for any constant a. Now consider a polynomial

$$P(x) = a_n x^n + a_{n-1}x^{n-1} + \cdots + a_1 x + a_0$$

where $a_0, a_1, \ldots, a_n$ are constants. Each term $a_j x^j$ is continuous, so $P(x)$ is a sum of continuous functions and hence is continuous. Finally, if $Q(x)$ is a polynomial, then by the Continuity Law (iv), $\dfrac{P(x)}{Q(x)}$ is continuous at $x = c$ provided that $Q(c) \neq 0$. ■

This result shows, for example, that $f(x) = 3x^4 - 2x^3 + 8x$ is continuous for all x and that

$$g(x) = \frac{x+3}{x^2-1}$$

is continuous for $x \neq \pm 1$. Furthermore, if n is a whole number, then $f(x) = x^{-n}$ is continuous for $x \neq 0$ since $f(x) = x^{-n} = 1/x^n$ is a rational function.

Most of the basic functions are continuous on their domains. For example, the sine and cosine, exponential and logarithm, and nth root functions are all continuous on their domains, as stated formally in the next theorem. This should not be too surprising because the graphs of these functions have no visible breaks (see Figure 10). However, the formal proofs of continuity are somewhat technical and are thus omitted.

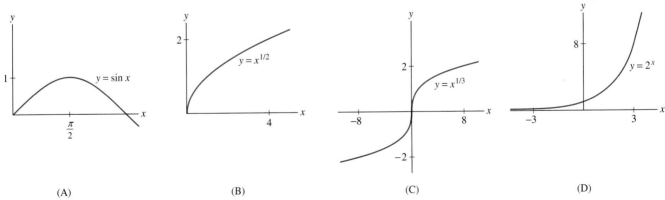

FIGURE 10 As the graphs suggest, these functions are continuous on their domains.

←·· **REMINDER** *The domain of* $y = x^{1/n}$ *is the real line if n is odd and the half-line* $[0, \infty)$ *if n is even.*

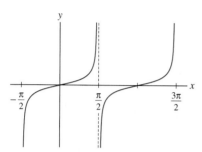

FIGURE 11 Graph of $y = \tan x$.

THEOREM 3 Continuity of Some Basic Functions

- $y = \sin x$ and $y = \cos x$ are continuous on the real line.
- For $b > 0$, $y = b^x$ is continuous on the real line.
- For $b > 0$ and $b \neq 1$, $\log_b x$ is continuous for $x > 0$.
- If n is a natural number, then $y = x^{1/n}$ is continuous on its domain.

Because $\sin x$ and $\cos x$ are continuous, Continuity Law (iv) for Quotients implies that the other standard trigonometric functions are continuous on their domains, that is, at all values of x where their denominators are nonzero:

$$\tan x = \frac{\sin x}{\cos x}, \qquad \cot x = \frac{\cos x}{\sin x}, \qquad \sec x = \frac{1}{\cos x}, \qquad \csc x = \frac{1}{\sin x}$$

They have infinite discontinuities at points where their denominators are zero. For example, $\tan x$ has infinite discontinuities at the points (Figure 11)

$$x = \pm\frac{\pi}{2}, \quad \pm\frac{3\pi}{2}, \quad \pm\frac{5\pi}{2}, \dots.$$

According to the next theorem, the inverse of a continuous function $f(x)$ is continuous. This is to be expected because the graph of the inverse $f^{-1}(x)$ is obtained by reflecting the graph of $f(x)$ through the line $y = x$. If the graph of $f(x)$ "has no breaks," the same ought to be true of the graph of $f^{-1}(x)$. We omit the formal proof here.

THEOREM 4 Continuity of the Inverse Function   If $f(x)$ is a continuous function on an interval I with range R and if the inverse $f^{-1}(x)$ exists, then $f^{-1}(x)$ is continuous on the domain R.

We may use this theorem to conclude that the inverse trigonometric functions $\sin^{-1} x$, $\cos^{-1} x$, and $\tan^{-1} x$, etc. are all continuous on their domains.

Many functions of interest are composite functions, so it is important to know that a composition of continuous functions is again continuous. The following theorem is proved in Appendix D.

THEOREM 5 Continuity of Composite Functions   Let $F(x) = f(g(x))$ be a composite function. If g is continuous at $x = c$ and f is continuous at $x = g(c)$, then $F(x)$ is continuous at $x = c$.

For example, $F(x) = (x^2 + 9)^{1/3}$ is continuous because it is the composite of the continuous functions $f(x) = x^{1/3}$ and $g(x) = x^2 + 9$. The function $F(x) = \cos(x^{-1})$ is the composite of $f(x) = \cos x$ and $g(x) = x^{-1}$, so it is continuous for all $x \neq 0$. Similarly, $e^{\sin x}$ is continuous for all x because it is the composite of two functions that are continuous for all x, namely e^x and $\sin x$.

Substitution: Evaluating Limits Using Continuity

It is easy to evaluate a limit when the function in question is known to be continuous. In this case, by definition, the limit is equal to the function value:

$$\lim_{x \to c} f(x) = f(c)$$

We call this the **Substitution Method** because the limit is evaluated by "plugging in" $x = c$.

■ **EXAMPLE 3** Evaluate   **(a)** $\lim\limits_{y \to \frac{\pi}{3}} \sin y$ and   **(b)** $\lim\limits_{x \to -1} \dfrac{3^x}{\sqrt{x+5}}$.

Solution

(a) Since $f(y) = \sin y$ is continuous, we may evaluate the limit using substitution:

$$\lim_{y \to \frac{\pi}{3}} \sin y = \sin \frac{\pi}{3} = \frac{\sqrt{3}}{2}$$

(b) The function $f(x) = \dfrac{3^x}{\sqrt{x+5}}$ is continuous at $x = -1$ because the numerator and denominator are both continuous at $x = -1$ and the denominator $\sqrt{x+5}$ is nonzero at $x = -1$. Therefore, we may evaluate the limit using substitution:

$$\lim_{x \to -1} \frac{3^x}{\sqrt{x+5}} = \frac{3^{-1}}{\sqrt{-1+5}} = \frac{1}{6}$$ ■

■ **EXAMPLE 4** Assumptions Matter   Can substitution be used to evaluate $\lim\limits_{x \to 2} [x]$, where $[x]$ is the greatest integer function?

Solution   Let $f(x) = [x]$ (Figure 12). Although $f(2) = 2$, we cannot conclude that $\lim\limits_{x \to 2} [x]$ is equal to 2. In fact, $f(x)$ is not continuous at $x = 2$, since the one-sided limits are not equal:

$$\lim_{x \to 2+} [x] = 2 \qquad \text{and} \qquad \lim_{x \to 2-} [x] = 1$$

Therefore, $\lim\limits_{x \to 2} [x]$ does not exist and substitution does not apply.   ■

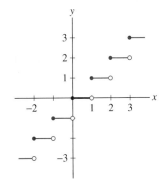

FIGURE 12 Graph of $f(x) = [x]$.

CONCEPTUAL INSIGHT **Real-World Modeling by Continuous Functions**   Continuous functions are often used to represent physical quantities such as velocity, temperature, or voltage. This reflects our everyday experience that change in the physical world tends to occur continuously rather than through abrupt transitions. However, mathematical models are at best approximations to reality and it is important to be aware of their limitations.

For example, in Figure 13, atmospheric temperature is represented as a continuous function of altitude. This is justified for large-scale objects such as the earth's atmosphere because the reading on a thermometer appears to vary continuously as altitude changes. However, temperature is a measure of the average kinetic energy of molecules and thus, at the microsopic level, it would not be meaningful to treat temperature as a quantity that varies continuously from point to point.

Population is another quantity that is often treated as a continuous function of time. The size $P(t)$ of a population at time t is a whole number that changes by ± 1 when an individual is born or dies, so strictly speaking, $P(t)$ is not continuous. If the population is large, the effect of an individual birth or death is small and it is reasonable to treat $P(t)$ as a continuous function of time.

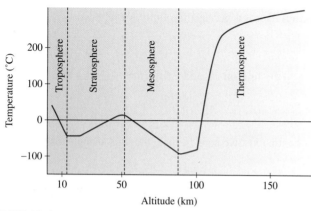

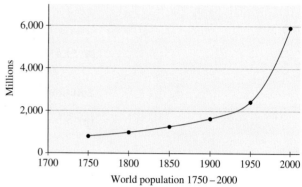

FIGURE 13 Atmospheric temperature and world population are represented by continuous graphs.

2.4 SUMMARY

- By definition, $f(x)$ is *continuous* at $x = c$ if $\lim_{x \to c} f(x) = f(c)$.
- If this limit does not exist, or if it exists but does not equal $f(c)$, then f is *discontinuous* at $x = c$.
- If $f(x)$ is continuous at all points in its domain, f is simply called *continuous*.
- $f(x)$ is *right-continuous* at $x = c$ if $\lim_{x \to c+} f(x) = f(c)$ and is *left-continuous* at $x = c$ if $\lim_{x \to c-} f(x) = f(c)$.
- There are three common types of discontinuities: *removable discontinuity* [$\lim_{x \to c} f(x)$ exists but does not equal $f(c)$], *jump discontinuity* (the one-sided limits both exist but are not equal), and *infinite discontinuity* (the limit is infinite as x approaches c from one or both sides).
- The Laws of Continuity state that sums, products, multiples, and composites of continuous functions are again continuous. The same holds for a quotient $\dfrac{f(x)}{g(x)}$ at points where $g(x) \neq 0$.
- Polynomials, rational functions, trigonometric functions, inverse trigonometric functions, exponential functions, and logarithmic functions are continuous on their domains.
- Substitution Method: If $f(x)$ is known to be continuous at $x = c$, then the value of the limit $\lim_{x \to c} f(x)$ is $f(c)$.

2.4 EXERCISES

Preliminary Questions

1. Which property of $f(x) = x^3$ allows us to conclude that $\lim_{x \to 2} x^3 = 8$?

2. What can be said about $f(3)$ if f is continuous and $\lim_{x \to 3} f(x) = \frac{1}{2}$?

3. Suppose that $f(x) < 0$ if x is positive and $f(x) > 1$ if x is negative. Can f be continuous at $x = 0$?

4. Is it possible to determine $f(7)$ if $f(x) = 3$ for all $x < 7$ and f is right-continuous at $x = 7$?

5. Are the following true or false? If false, state a correct version.

(a) $f(x)$ is continuous at $x = a$ if the left- and right-hand limits of $f(x)$ as $x \to a$ exist and are equal.

(b) $f(x)$ is continuous at $x = a$ if the left- and right-hand limits of $f(x)$ as $x \to a$ exist and equal $f(a)$.

(c) If the left- and right-hand limits of $f(x)$ as $x \to a$ exist, then f has a removable discontinuity at $x = a$.

(d) If $f(x)$ and $g(x)$ are continuous at $x = a$, then $f(x) + g(x)$ is

continuous at $x = a$.

(e) If $f(x)$ and $g(x)$ are continuous at $x = a$, then $f(x)/g(x)$ is continuous at $x = a$.

Exercises

1. Find the points of discontinuity of the function shown in Figure 14 and state whether it is left- or right-continuous (or neither) at these points.

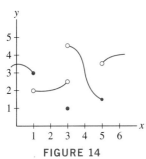

FIGURE 14

In Exercises 2–4, refer to the function $f(x)$ in Figure 15.

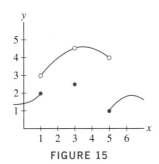

FIGURE 15

2. Find the points of discontinuity of $f(x)$ and state whether $f(x)$ is left- or right-continuous (or neither) at these points.

3. At which point c does $f(x)$ have a removable discontinuity? What value should be assigned to $f(c)$ to make f continuous at $x = c$?

4. Find the point c_1 at which $f(x)$ has a jump discontinuity but is left-continuous. What value should be assigned to $f(c_1)$ to make f right-continuous at $x = c_1$?

5. (a) For the function shown in Figure 16, determine the one-sided limits at the points of discontinuity.
(b) Which of these discontinuities is removable and how should f be redefined to make it continuous at this point?

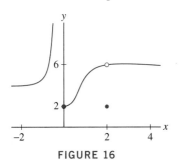

FIGURE 16

6. Let $f(x)$ be the function (Figure 17):

$$f(x) = \begin{cases} x^2 + 3 & \text{for } x < 1 \\ 10 - x & \text{for } 1 \le x \le 2 \\ 6x - x^2 & \text{for } x > 2 \end{cases}$$

Show that $f(x)$ is continuous for $x \ne 1, 2$. Then compute the right- and left-hand limits at $x = 1$ and 2, and determine whether $f(x)$ is left-continuous, right-continuous, or continuous at these points.

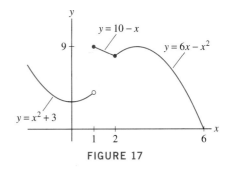

FIGURE 17

In Exercises 7–16, use the Laws of Continuity and Theorems 2–3 to show that the function is continuous.

7. $f(x) = x + \sin x$

8. $f(x) = x \sin x$

9. $f(x) = 3x + 4 \sin x$

10. $f(x) = 3x^3 + 8x^2 - 20x$

11. $f(x) = \dfrac{1}{x^2 + 1}$

12. $f(x) = \dfrac{x^2 - \cos x}{3 + \cos x}$

13. $f(x) = \dfrac{3^x}{1 + 4^x}$

14. $f(x) = 10^x \cos x$

15. $f(x) = e^x \cos 3x$

16. $f(x) = \ln(x^4 + 1)$

In Exercises 17–36, determine the points at which the function is discontinuous and state the type of discontinuity: removable, jump, infinite, or none of these.

17. $f(x) = \dfrac{1}{x}$

18. $f(x) = |x|$

19. $f(x) = \dfrac{x - 2}{|x - 1|}$

20. $f(x) = \dfrac{x - 2}{|x - 2|}$

21. $f(x) = [x]$

22. $f(x) = \left[\dfrac{1}{2}x\right]$

23. $g(t) = \dfrac{1}{t^2 - 1}$

24. $f(x) = \dfrac{x + 1}{4x - 2}$

25. $f(x) = 3x^{3/2} - 9x^3$

26. $g(t) = 3t^{-3/2} - 9t^3$

27. $h(z) = \dfrac{1 - 2z}{z^2 - z - 6}$

28. $h(z) = \dfrac{1 - 2z}{z^2 + 9}$

29. $f(x) = \dfrac{x^2 - 3x + 2}{|x - 2|}$

30. $g(t) = \tan 2t$

31. $f(x) = \csc x^2$

32. $f(x) = \cos \dfrac{1}{x}$

33. $f(x) = \tan(\sin x)$

34. $f(x) = x - |x|$

35. $f(x) = \dfrac{1}{e^x - e^{-x}}$

36. $f(x) = \ln |x - 4|$

In Exercises 37–50, determine the domain of the function and prove that it is continuous on its domain using the Laws of Continuity and the facts quoted in this section.

37. $f(x) = \sqrt{9 - x^2}$

38. $f(x) = \sqrt{x^2 + 9}$

39. $f(x) = \sqrt{x} \sin x$

40. $f(x) = \dfrac{x^2}{x + x^{1/4}}$

41. $f(x) = x^{2/3} 2^x$

42. $f(x) = x^{1/3} + x^{3/4}$

43. $f(x) = x^{-4/3}$

44. $f(x) = \ln(9 - x^2)$

45. $f(x) = \tan^2 x$

46. $f(x) = \cos(2^x)$

47. $f(x) = (x^4 + 1)^{3/2}$

48. $f(x) = e^{-x^2}$

49. $f(x) = \dfrac{\cos(x^2)}{x^2 - 1}$

50. $f(x) = 9^{\tan x}$

51. Suppose that $f(x) = 2$ for $x > 0$ and $f(x) = -4$ for $x < 0$. What is $f(0)$ if f is left-continuous at $x = 0$? What is $f(0)$ if f is right-continuous at $x = 0$?

52. Sawtooth Function Draw the graph of $f(x) = x - [x]$. At which points is f discontinuous? Is it left- or right-continuous at those points?

In Exercises 53–56, draw the graph of a function on $[0, 5]$ with the given properties.

53. $f(x)$ is not continuous at $x = 1$, but $\lim\limits_{x \to 1+} f(x)$ and $\lim\limits_{x \to 1-} f(x)$ exist and are equal.

54. $f(x)$ is left-continuous but not continuous at $x = 2$ and right-continuous but not continuous at $x = 3$.

55. $f(x)$ has a removable discontinuity at $x = 1$, a jump discontinuity at $x = 2$, and

$$\lim\limits_{x \to 3-} f(x) = -\infty, \qquad \lim\limits_{x \to 3+} f(x) = 2$$

56. $f(x)$ is right- but not left-continuous at $x = 1$, left- but not right-continuous at $x = 2$, and neither left- nor right-continuous at $x = 3$.

57. Each of the following statements is *false*. For each statement, sketch the graph of a function that provides a counterexample.

(a) If $\lim\limits_{x \to a} f(x)$ exists, then $f(x)$ is continuous at $x = a$.

(b) If $f(x)$ has a jump discontinuity at $x = a$, then $f(a)$ is equal to either $\lim\limits_{x \to a-} f(x)$ or $\lim\limits_{x \to a+} f(x)$.

(c) If $f(x)$ has a discontinuity at $x = a$, then $\lim\limits_{x \to a-} f(x)$ and $\lim\limits_{x \to a+} f(x)$ exist but are not equal.

(d) The one-sided limits $\lim\limits_{x \to a-} f(x)$ and $\lim\limits_{x \to a+} f(x)$ always exist, even if $\lim\limits_{x \to a} f(x)$ does not exist.

58. According to the Laws of Continuity, if $f(x)$ and $g(x)$ are continuous at $x = c$, then $f(x) + g(x)$ is continuous at $x = c$. Suppose that $f(x)$ and $g(x)$ are discontinuous at $x = c$. Is it true that $f(x) + g(x)$ is discontinuous at $x = c$? If not, give a counterexample.

In Exercises 59–78, evaluate the limit using the substitution method.

59. $\lim\limits_{x \to 5} x^2$

60. $\lim\limits_{x \to 8} x^3$

61. $\lim\limits_{x \to -1} (2x^3 - 4)$

62. $\lim\limits_{x \to 2} (5x - 12x^{-2})$

63. $\lim\limits_{x \to 0} \dfrac{x + 9}{x - 9}$

64. $\lim\limits_{x \to 3} \dfrac{x + 2}{x^2 + 2x}$

65. $\lim\limits_{x \to \pi} \sin\left(\dfrac{x}{2} - \pi\right)$

66. $\lim\limits_{x \to \pi} \sin(x^2)$

67. $\lim\limits_{x \to \frac{\pi}{4}} \tan(3x)$

68. $\lim\limits_{x \to \pi} \dfrac{1}{\cos x}$

69. $\lim\limits_{x \to 4} x^{-5/2}$

70. $\lim\limits_{x \to 2} \sqrt{x^3 + 4x}$

71. $\lim\limits_{x \to -1} (1 - 8x^3)^{3/2}$

72. $\lim\limits_{x \to 2} \left(\dfrac{7x + 2}{4 - x}\right)^{2/3}$

73. $\lim\limits_{x \to 3} 10^{x^2 - 2x}$

74. $\lim\limits_{x \to \frac{\pi}{2}} 3^{\sin x}$

75. $\lim\limits_{x \to 1} e^{x^2 - x}$

76. $\lim\limits_{x \to 5} \ln(ex + e)$

77. $\lim\limits_{x \to 4} \sin^{-1}\left(\dfrac{x}{4}\right)$

78. $\lim\limits_{x \to 0} \tan^{-1}(e^x)$

In Exercises 79–82, sketch the graph of the given function. At each point of discontinuity, state whether f is left- or right-continuous.

79. $f(x) = \begin{cases} x^2 & \text{for } x \le 1 \\ 2 - x & \text{for } x > 1 \end{cases}$

80. $f(x) = \begin{cases} x + 1 & \text{for } x < 1 \\ \dfrac{1}{x} & \text{for } x \ge 1 \end{cases}$

81. $f(x) = \begin{cases} |x - 3| & \text{for } x \le 3 \\ x - 3 & \text{for } x > 3 \end{cases}$

82. $f(x) = \begin{cases} x^3 + 1 & \text{for } -\infty < x \le 0 \\ -x + 1 & \text{for } 0 < x < 2 \\ -x^2 + 10x - 15 & \text{for } x \ge 2 \end{cases}$

In Exercises 83–84, find the value of c that makes the function continuous.

83. $f(x) = \begin{cases} x^2 - c & \text{for } x < 5 \\ 4x + 2c & \text{for } x \geq 5 \end{cases}$

84. $f(x) = \begin{cases} 2x + 9 & \text{for } x \leq 3 \\ -4x + c & \text{for } x > 3 \end{cases}$

85. Find all constants a, b such that the following function has no discontinuities:

$$f(x) = \begin{cases} ax + \cos x & \text{for } x \leq \dfrac{\pi}{4} \\ bx + 2 & \text{for } x > \dfrac{\pi}{4} \end{cases}$$

86. Which of the following quantities would be represented by continuous functions of time and which would have one or more discontinuities?

(a) Velocity of an airplane during a flight

(b) Temperature in a room under ordinary conditions

(c) Value of a bank account with interest paid yearly

(d) The salary of a teacher

(e) The population of the world

87. ✎ In 1993, the amount $T(x)$ of federal income tax owed on an income of x dollars was determined by the formula

$$T(x) = \begin{cases} 0.15x & \text{for } 0 \leq x < 21,450 \\ 3,217.50 + 0.28(x - 21,450) & \text{for } 21,450 \leq x < 51,900 \\ 11,743.50 + 0.31(x - 51,900) & \text{for } x \geq 51,900 \end{cases}$$

Sketch the graph of $T(x)$ and determine if it has any discontinuities. Explain why, if $T(x)$ had a jump discontinuity, it might be advantageous in some situations to earn *less* money.

Further Insights and Challenges

88. ✎ If $f(x)$ has a removable discontinuity at $x = c$, then it is possible to redefine $f(c)$ so that $f(x)$ is continuous at $x = c$. Can $f(c)$ be redefined in more than one way?

89. Give an example of functions $f(x)$ and $g(x)$ such that $f(g(x))$ is continuous but $g(x)$ has at least one discontinuity.

90. Function Continuous at Only One Point Let $f(x) = x$ for x rational, $f(x) = -x$ for x irrational. Show that f is continuous at $x = 0$ and discontinuous at all points $x \neq 0$.

91. Let $f(x) = 1$ if x is rational and $f(x) = -1$ if x is irrational. Show that $f(x)$ is discontinuous at all points, whereas $f(x)^2$ is continuous at all points.

2.5 Evaluating Limits Algebraically

In the previous section, we used substitution to calculate limits. This method applies when the function in question is known to be continuous. For example, $f(x) = x^{-2}$ is continuous at $x = 3$ and therefore

$$\lim_{x \to 3} x^{-2} = 3^{-2} = \frac{1}{9}$$

When we study derivatives in Chapter 3, we will be faced with limits $\lim_{x \to c} f(x)$, where $f(c)$ is not defined. In such cases, substitution cannot be used directly. However, some of these limits can be evaluated using substitution, provided that we first use algebra to rewrite the formula for $f(x)$.

To illustrate, consider the limit

$$\lim_{x \to 4} \frac{x^2 - 16}{x - 4}$$

The function $f(x) = \dfrac{x^2 - 16}{x - 4}$ (Figure 1) is not defined at $x = 4$ because

$$f(4) = \frac{4^2 - 16}{4 - 4} = \frac{0}{0} \quad \text{(undefined)}$$

However, the numerator of $f(x)$ factors and

$$\frac{x^2 - 16}{x - 4} = \frac{(x + 4)(x - 4)}{x - 4} = x + 4 \quad \text{(valid for } x \neq 4)$$

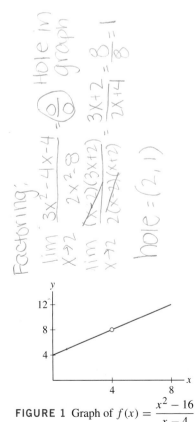

FIGURE 1 Graph of $f(x) = \dfrac{x^2 - 16}{x - 4}$.

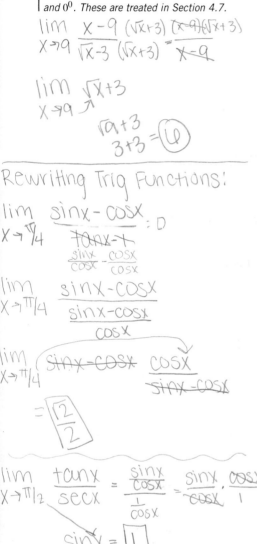

In other words, $f(x)$ coincides with the *continuous* function $x + 4$ for all $x \neq 4$. Since the limit depends only on the values of $f(x)$ for $x \neq 4$, we have

$$\lim_{x \to 4} \frac{x^2 - 16}{x - 4} = \underbrace{\lim_{x \to 4} (x + 4) = 8}_{\text{Evaluate by substitution}}$$

In general, we say that $f(x)$ has an **indeterminate form** at $x = c$ if, when $f(x)$ is evaluated at $x = c$, we obtain an undefined expression of the type

$$\frac{0}{0}, \quad \frac{\infty}{\infty}, \quad \infty \cdot 0, \quad \infty - \infty$$

Other indeterminate forms are 1^∞, ∞^0, and 0^0. These are treated in Section 4.7.

We also say that f is **indeterminate** at $x = c$. Our strategy is to *transform $f(x)$ algebraically if possible into a new expression that is defined and continuous at $x = c$, and then evaluate by substitution ("plugging in")*. As you study the following examples, notice that the critical step in each case is to cancel a common factor from the numerator and denominator at the appropriate moment, thereby removing the indeterminacy.

■ **EXAMPLE 1** Calculate $\lim_{x \to 3} \frac{x^2 - 4x + 3}{x^2 + x - 12}$.

Solution The function is indeterminate at $x = 3$ because

$$\frac{3^2 - 4(3) + 3}{3^2 + 3 - 12} = \frac{0}{0} \quad \text{(indeterminate)}$$

Step 1. **Transform algebraically and cancel.**
Factor the numerator and denominator, and cancel the common factors:

$$\frac{x^2 - 4x + 3}{x^2 + x - 12} = \underbrace{\frac{(x-3)(x-1)}{(x-3)(x+4)}}_{\text{Cancel common factor}} = \underbrace{\frac{x-1}{x+4}}_{\text{Continuous at } x=3} \quad \text{(if } x \neq 3\text{)} \qquad \boxed{1}$$

Step 2. **Substitute (evaluate using continuity).**
Since the expression on the right in Eq. (1) is *continuous* at $x = 3$,

$$\lim_{x \to 3} \frac{x^2 - 4x + 3}{x^2 + x - 12} = \underbrace{\lim_{x \to 3} \frac{(x-1)}{(x+4)} = \frac{2}{7}}_{\text{Evaluate by substitution}}$$ ■

■ **EXAMPLE 2** The Form $\frac{\infty}{\infty}$ Calculate $\lim_{x \to \frac{\pi}{2}} \frac{\tan x}{\sec x}$.

Solution Both the numerator and denominator of $f(x) = \dfrac{\tan x}{\sec x}$ have infinite discontinuities at $x = \frac{\pi}{2}$:

$$\frac{\tan \dfrac{\pi}{2}}{\sec \dfrac{\pi}{2}} = \frac{1/0}{1/0} \quad \text{(indeterminate)}$$

We say that $f(x)$ has the indeterminate form $\frac{\infty}{\infty}$ at $x = \frac{\pi}{2}$.

$\lim_{h\to 1} \frac{(h+3)^2 - 16h}{h-1} = \frac{0}{0}$

$\lim_{h\to 1} \frac{h^2 - 10h + 9}{h-1}$

$\lim_{h\to 1} \frac{(h-9)(h-1)}{h-1}$

$\lim_{h\to 1} h - 9 = -8$

Step 1. **Transform algebraically and cancel.**

$$\frac{\tan x}{\sec x} = \frac{(\sin x)\left(\dfrac{1}{\cos x}\right)}{\dfrac{1}{\cos x}} = \sin x \quad (\text{if } \cos x \neq 0)$$

Step 2. **Substitution (evaluate using continuity).**
Since $\sin x$ is continuous,

$$\lim_{x\to\frac{\pi}{2}} \frac{\tan x}{\sec x} = \lim_{x\to\frac{\pi}{2}} \sin x = \sin\frac{\pi}{2} = 1$$

■

The next example illustrates the algebraic technique of "multiplying by the conjugate," which we use to treat some indeterminate forms involving square roots.

■ **EXAMPLE 3** **Multiplying by the Conjugate** Evaluate:

(a) $\displaystyle\lim_{x\to 4} \frac{\sqrt{x}-2}{x-4}$ **(b)** $\displaystyle\lim_{h\to 5} \frac{h-5}{\sqrt{h+4}-3}$.

Solution

(a) The function $f(x) = \dfrac{\sqrt{x}-2}{x-4}$ is indeterminate at $x = 4$ since

$$f(4) = \frac{\sqrt{4}-2}{2-2} = \frac{0}{0} \quad (\text{indeterminate})$$

Step 1. **Transform algebraically and cancel.**
The conjugate of $\sqrt{x}-2$ is $\sqrt{x}+2$, so we have

In Step 1, observe that $(\sqrt{x}-2)(\sqrt{x}+2) = x-4$.

$$\left(\frac{\sqrt{x}-2}{x-4}\right)\left(\frac{\sqrt{x}+2}{\sqrt{x}+2}\right) = \frac{x-4}{(x-4)(\sqrt{x}+2)} = \frac{1}{\sqrt{x}+2} \quad (\text{if } x \neq 4)$$

Step 2. **Substitute (evaluate using continuity).**
Since $\dfrac{1}{\sqrt{x}+2}$ is continuous at $x = 4$,

$$\lim_{x\to 4} \frac{\sqrt{x}-2}{x-4} = \lim_{x\to 4} \frac{1}{\sqrt{x}+2} = \frac{1}{4}$$

(b) The function $f(h) = \dfrac{h-5}{\sqrt{h+4}-3}$ is indeterminate at $h = 5$ since

$$f(5) = \frac{5-5}{\sqrt{5+4}-3} = \frac{5-5}{3-3} = \frac{0}{0} \quad (\text{indeterminate})$$

Step 1. **Transform algebraically.**
The conjugate of $\sqrt{h+4}-3$ is $\sqrt{h+4}+3$, and

$$\frac{h-5}{\sqrt{h+4}-3} = \left(\frac{h-5}{\sqrt{h+4}-3}\right)\left(\frac{\sqrt{h+4}+3}{\sqrt{h+4}+3}\right) = \frac{(h-5)(\sqrt{h+4}+3)}{(\sqrt{h+4}-3)(\sqrt{h+4}+3)}$$

The denominator is equal to

$$(\sqrt{h+4}-3)(\sqrt{h+4}+3) = (\sqrt{h+4})^2 - 9 = h-5$$

Thus, for $h \neq 5$,

$$f(h) = \frac{h - 5}{\sqrt{h + 4} - 3} = \frac{(h - 5)(\sqrt{h + 4} + 3)}{h - 5} = \sqrt{h + 4} + 3$$

Step 2. Substitute (evaluate using continuity).

$$\lim_{h \to 5} \frac{h - 5}{\sqrt{h + 4} - 3} = \lim_{h \to 5} \left(\sqrt{h + 4} + 3 \right) = \sqrt{9} + 3 = 6 \qquad \blacksquare$$

■ **EXAMPLE 4** The Form $\infty - \infty$ Calculate $\lim\limits_{x \to 1} \left(\dfrac{1}{x - 1} - \dfrac{2}{x^2 - 1} \right)$.

Solution This example gives rise to the indeterminate form $\infty - \infty$ because

$$\frac{1}{1 - 1} - \frac{2}{1^2 - 1} = \frac{1}{0} - \frac{2}{0} \quad \text{(indeterminate)}$$

Step 1. Transform algebraically and cancel.
To rewrite the expression, combine the fractions and simplify (for $x \neq 1$):

$$\frac{1}{x - 1} - \frac{2}{x^2 - 1} = \frac{x + 1}{x^2 - 1} - \frac{2}{x^2 - 1} = \frac{x - 1}{x^2 - 1} = \frac{x - 1}{(x - 1)(x + 1)} = \frac{1}{x + 1}$$

Step 2. Substitute (evaluate using continuity).

$$\lim_{x \to 1} \left(\frac{1}{x - 1} - \frac{2}{x^2 - 1} \right) = \lim_{x \to 1} \frac{1}{x + 1} = \frac{1}{1 + 1} = \frac{1}{2} \qquad \blacksquare$$

■ **EXAMPLE 5** Function Is Infinite but Not Indeterminate Calculate $\lim\limits_{x \to 2} \dfrac{x^2 - x + 5}{x - 2}$.

Solution Let $f(x) = \dfrac{x^2 - x + 5}{x - 2}$. At $x = 2$ we have

$$f(2) = \frac{2^2 - 2 + 5}{2 - 2} = \frac{7}{0} \quad \text{(undefined, but not an indeterminate form)}$$

This is *not* an indeterminate form. In fact, Figure 2 shows that the one-sided limits are infinite:

$$\lim_{x \to 2-} \frac{x^2 - x + 5}{x - 2} = -\infty, \qquad \lim_{x \to 2+} \frac{x^2 - x + 5}{x - 2} = \infty$$

The limit itself does not exist. $\qquad \blacksquare$

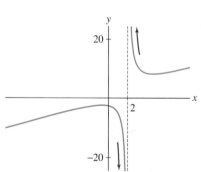

FIGURE 2 Graph of $f(x) = \dfrac{x^2 - x + 5}{x - 2}$.

As preparation for the derivative in Chapter 3, we illustrate how to evaluate a limit that involves a symbolic constant.

■ **EXAMPLE 6** Limit Involving a Symbolic Constant Calculate $\lim\limits_{h \to 0} \dfrac{(h + a)^2 - a^2}{h}$, where a is a constant.

Solution The limit is indeterminate at $h = 0$ because

$$\frac{(0 + a)^2 - a^2}{0} = \frac{a^2 - a^2}{0} = \frac{0}{0} \quad \text{(indeterminate)}$$

Step 1. **Transform algebraically and cancel.**

We expand the numerator and simplify (for $h \neq 0$):

$$\frac{(h+a)^2 - a^2}{h} = \frac{(h^2 + 2ah + a^2) - a^2}{h} = \frac{h^2 + 2ah}{h}$$

$$= \frac{h(h + 2a)}{h} \qquad \text{(factor out } h\text{)}$$

$$= h + 2a \qquad \text{(cancel } h\text{)}$$

Step 2. **Substitute (evaluate using continuity).**

The function $h + 2a$ is continuous (since h is continuous and a is a constant), so

$$\lim_{h \to 0} \frac{(h+a)^2 - a^2}{h} = \lim_{h \to 0} (h + 2a) = 2a \qquad \blacksquare$$

2.5 SUMMARY

- When $f(x)$ is known to be continuous at $x = c$, the limit can be evaluated by substitution: $\lim_{x \to c} f(x) = f(c)$.
- We say that $f(x)$ is *indeterminate* (or has an *indeterminate form*) at $x = c$ if, when we evaluate $f(c)$, we obtain an undefined expression of the type

$$\frac{0}{0}, \quad \frac{\infty}{\infty}, \quad \infty \cdot 0, \quad \infty - \infty$$

- Substitution Method: If $f(x)$ is indeterminate at $x = c$, try using algebra to transform $f(x)$ into a new function that coincides with $f(x)$ near c and is continuous at $x = c$. Then evaluate the limit by applying substitution to the continuous function.

2.5 EXERCISES

Preliminary Questions

1. Which of the following is indeterminate at $x = 1$?

$$\frac{x^2 + 1}{x - 1}, \qquad \frac{x^2 - 1}{x + 2}, \qquad \frac{x^2 - 1}{\sqrt{x + 3} - 2}, \qquad \frac{x^2 + 1}{\sqrt{x + 3} - 2}$$

2. Give counterexamples to show that each of the following statements is false:

(a) If $f(c)$ is indeterminate, then the right- and left-hand limits as $x \to c$ are not equal.

(b) If $\lim_{x \to c} f(x)$ exists, then $f(c)$ is not indeterminate.

(c) If $f(x)$ is undefined at $x = c$, then $f(x)$ has an indeterminate form at $x = c$.

3. Although the method for evaluating limits discussed in this section is sometimes called "simplify and plug in," explain how it actually relies on the property of continuity.

Exercises

In Exercises 1–4, show that the limit leads to an indeterminate form. Then carry out the two-step procedure: Transform the function algebraically and evaluate using continuity.

1. $\lim_{x \to 5} \dfrac{x^2 - 25}{x - 5}$

2. $\lim_{x \to -1} \dfrac{x + 1}{x^2 + 2x + 1}$

3. $\lim_{t \to 7} \dfrac{2t - 14}{5t - 35}$

4. $\lim_{h \to -3} \dfrac{h + 3}{h^2 - 9}$

In Exercises 5–34, evaluate the limit or state that it does not exist.

5. $\lim_{x \to 8} \dfrac{x^2 - 64}{x - 8}$

6. $\lim_{x \to 8} \dfrac{x^2 - 64}{x - 9}$

7. $\lim\limits_{x \to 2} \dfrac{x^2 - 3x + 2}{x - 2}$

8. $\lim\limits_{x \to 2} \dfrac{x^3 - 4x}{x - 2}$

9. $\lim\limits_{x \to 2} \dfrac{x - 2}{x^3 - 4x}$

10. $\lim\limits_{x \to 2} \dfrac{x^3 - 2x}{x - 2}$

11. $\lim\limits_{h \to 0} \dfrac{(1 + h)^3 - 1}{h}$

12. $\lim\limits_{h \to 4} \dfrac{(h + 2)^2 - 9h}{h - 4}$

13. $\lim\limits_{x \to 2} \dfrac{3x^2 - 4x - 4}{2x^2 - 8}$

14. $\lim\limits_{t \to -2} \dfrac{2t + 4}{12 - 3t^2}$

15. $\lim\limits_{y \to 2} \dfrac{(y - 2)^3}{y^3 - 5y + 2}$

16. $\lim\limits_{x \to 16} \dfrac{\sqrt{x} - 4}{x - 16}$

17. $\lim\limits_{h \to 0} \dfrac{\frac{1}{3 + h} - \frac{1}{3}}{h}$

18. $\lim\limits_{h \to 0} \dfrac{\frac{1}{(h + 2)^2} - \frac{1}{4}}{h}$

19. $\lim\limits_{h \to 0} \dfrac{\sqrt{2 + h} - 2}{h}$

20. $\lim\limits_{x \to 10} \dfrac{\sqrt{x - 6} - 2}{x - 10}$

21. $\lim\limits_{x \to 2} \dfrac{x - 2}{\sqrt{x} - \sqrt{4 - x}}$

22. $\lim\limits_{x \to 0} \dfrac{\sqrt{1 + x} - \sqrt{1 - x}}{x}$

23. $\lim\limits_{x \to 2} \dfrac{\sqrt{x^2 - 1} - \sqrt{x + 1}}{x - 3}$

24. $\lim\limits_{x \to 4} \dfrac{\sqrt{5 - x} - 1}{2 - \sqrt{x}}$

25. $\lim\limits_{x \to 4} \left(\dfrac{1}{\sqrt{x} - 2} - \dfrac{4}{x - 4} \right)$

26. $\lim\limits_{x \to 0+} \left(\dfrac{1}{\sqrt{x}} - \dfrac{1}{\sqrt{x^2 + x}} \right)$

27. $\lim\limits_{x \to 0} \dfrac{\cot x}{\csc x}$

28. $\lim\limits_{\theta \to \frac{\pi}{2}} \dfrac{\cot \theta}{\csc \theta}$

29. $\lim\limits_{x \to \frac{\pi}{4}} \dfrac{\sin x - \cos x}{\tan x - 1}$

30. $\lim\limits_{\theta \to \frac{\pi}{2}} \left(\sec \theta - \tan \theta \right)$

31. $\lim\limits_{x \to 0} \dfrac{e^x - e^{2x}}{1 - e^x}$

32. $\lim\limits_{x \to 0} \dfrac{3^{2x} - 1}{3^x - 1}$

33. $\lim\limits_{x \to \frac{\pi}{3}} \dfrac{2\cos^2 x + 3\cos x - 2}{2\cos x - 1}$

34. $\lim\limits_{\theta \to 0} \dfrac{\cos \theta - 1}{\sin \theta}$

35. $\boxed{\text{GU}}$ Use a plot of $f(x) = \dfrac{x - 2}{\sqrt{x} - \sqrt{4 - x}}$ to estimate $\lim\limits_{x \to 2} f(x)$ to two decimal places. Compare with the answer obtained algebraically in Exercise 21.

36. $\boxed{\text{GU}}$ Use a plot of $f(x) = \dfrac{1}{\sqrt{x} - 2} - \dfrac{4}{x - 4}$ to evaluate $\lim\limits_{x \to 4} f(x)$ numerically. Compare with the answer obtained algebraically in Exercise 25.

In Exercises 37–42, use the identity $a^3 - b^3 = (a - b)(a^2 + ab + b^2)$.

37. $\lim\limits_{x \to 1} \dfrac{x^3 - 1}{x - 1}$

38. $\lim\limits_{x \to 2} \dfrac{x^3 - 8}{x^2 - 4}$

39. $\lim\limits_{x \to 1} \dfrac{x^2 - 3x + 2}{x^3 - 1}$

40. $\lim\limits_{x \to 2} \dfrac{x^3 - 8}{x^2 - 6x + 8}$

41. $\lim\limits_{x \to 1} \dfrac{x^4 - 1}{x^3 - 1}$

42. $\lim\limits_{x \to 27} \dfrac{x - 27}{x^{1/3} - 3}$

In Exercises 43–52, evaluate the limits in terms of the constants involved.

43. $\lim\limits_{x \to 0} (2a + x)$

44. $\lim\limits_{h \to -2} (4ah + 7a)$

45. $\lim\limits_{t \to -1} (4t - 2at + 3a)$

46. $\lim\limits_{h \to 0} \dfrac{(3a + h)^2 - 9a^2}{h}$

47. $\lim\limits_{x \to 0} \dfrac{2(x + h)^2 - 2x^2}{h}$

48. $\lim\limits_{x \to a} \dfrac{(x + a)^2 - 4x^2}{x - a}$

49. $\lim\limits_{x \to a} \dfrac{\sqrt{x} - \sqrt{a}}{x - a}$

50. $\lim\limits_{h \to 0} \dfrac{\sqrt{a + 2h} - \sqrt{a}}{h}$

51. $\lim\limits_{x \to 0} \dfrac{(x + a)^3 - a^3}{x}$

52. $\lim\limits_{h \to a} \dfrac{\frac{1}{h} - \frac{1}{a}}{h - a}$

Further Insights and Challenges

In Exercises 53–54, find all values of c such that the limit exists.

53. $\lim\limits_{x \to c} \dfrac{x^2 - 5x - 6}{x - c}$

54. $\lim\limits_{x \to 1} \dfrac{x^2 + 3x + c}{x - 1}$

55. For which sign $\pm$ does the following limit exist?

$$\lim\limits_{x \to 0} \left(\dfrac{1}{x} \pm \dfrac{1}{x(x - 1)} \right)$$

56. For which sign $\pm$ does the following limit exist?

$$\lim\limits_{x \to 0+} \left(\dfrac{1}{x} \pm \dfrac{1}{|x|} \right)$$

2.6 Trigonometric Limits

In our study of the derivative, we will need to evaluate certain limits involving transcendental functions such as sine and cosine. The algebraic techniques of the previous section are often ineffective for such functions and other tools are required. One such tool is the

Squeeze Theorem, which we discuss in this section and use to evaluate the trigonometric limits needed in Section 3.6.

The Squeeze Theorem

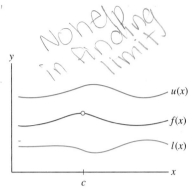

FIGURE 1 $f(x)$ is trapped between $l(x)$ and $u(x)$ (but not squeezed at $x = c$).

Consider a function $f(x)$ that is trapped between two functions $l(x)$ and $u(x)$ on an interval I. In other words,

$$l(x) \le f(x) \le u(x) \quad \text{for all } x \in I$$

In this case, the graph of $f(x)$ lies between the graphs of $l(x)$ and $u(x)$ (Figure 1), with $l(x)$ as the lower and $u(x)$ as the upper function.

The Squeeze Theorem applies when $f(x)$ is not just trapped, but actually **squeezed** at a point $x = c$ by $l(x)$ and $u(x)$ (Figure 2). By this we mean that for all $x \ne c$ in some open interval containing c,

$$l(x) \le f(x) \le u(x) \quad \text{and} \quad \lim_{x \to c} l(x) = \lim_{x \to c} u(x) = L$$

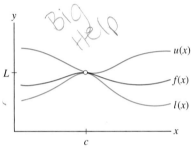

FIGURE 2 $f(x)$ is squeezed by $l(x)$ and $u(x)$ at $x = c$.

We do not require that $f(x)$ be defined at $x = c$, but it is clear graphically that $f(x)$ must approach the limit L (Figure 2). We state this formally in the next theorem. A proof is given in Appendix D.

> **THEOREM 1 Squeeze Theorem** Assume that for $x \ne c$ (in some open interval containing c),
>
> $$l(x) \le f(x) \le u(x) \quad \text{and} \quad \lim_{x \to c} l(x) = \lim_{x \to c} u(x) = L$$
>
> Then $\lim_{x \to c} f(x)$ exists and
>
> $$\lim_{x \to c} f(x) = L$$

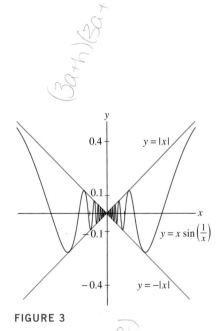

FIGURE 3

■ **EXAMPLE 1** Show that $\lim_{x \to 0} x \sin \dfrac{1}{x} = 0$.

Solution Although $f(x) = x \sin \dfrac{1}{x}$ is a product of two functions, we cannot use the Product Law because $\lim_{x \to 0} \sin \dfrac{1}{x}$ does not exist. Instead, we apply the Squeeze Theorem. The sine function takes on values between 1 and -1 and therefore $\left| \sin \dfrac{1}{x} \right| \le 1$ for all $x \ne 0$. Multiplying by $|x|$, we obtain $\left| x \sin \dfrac{1}{x} \right| \le |x|$ and conclude that (see Figure 3)

$$-|x| \le x \sin \frac{1}{x} \le |x|$$

Since

$$\lim_{x \to 0} |x| = 0 \quad \text{and} \quad \lim_{x \to 0} (-|x|) = -\lim_{x \to 0} |x| = 0$$

we can apply the Squeeze Theorem to conclude that $\lim_{x \to 0} x \sin \dfrac{1}{x} = 0$. ■

The next theorem states two important limits that will be used to develop the calculus of trigonometric functions in Section 3.6.

THEOREM 2 Important Trigonometric Limits

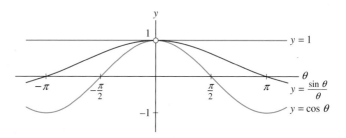

$$\lim_{\theta \to 0} \frac{\sin \theta}{\theta} = 1, \qquad \lim_{\theta \to 0} \frac{1 - \cos \theta}{\theta} = 0$$

Note that both $\dfrac{\sin \theta}{\theta}$ and $\dfrac{\cos \theta - 1}{\theta}$ are indeterminate at $\theta = 0$, so Theorem 2 cannot be proved by substitution.

We use the Squeeze Theorem to evaluate the first limit in Theorem 2. To do so, we must find functions that squeeze $(\sin \theta)/\theta$ at $\theta = 0$. These are provided by the next theorem (see Figure 4).

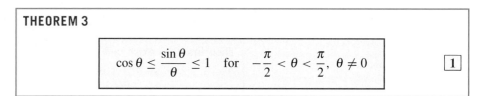

FIGURE 4 Graph illustrating the inequalities of Theorem 3.

THEOREM 3

$$\cos \theta \le \frac{\sin \theta}{\theta} \le 1 \quad \text{for} \quad -\frac{\pi}{2} < \theta < \frac{\pi}{2}, \ \theta \ne 0 \qquad \boxed{1}$$

Proof Assume first that $0 < \theta < \frac{\pi}{2}$. Our proof is based on the following relation between the areas in Figure 5:

$$\text{Area of } \triangle OAB < \text{area of sector } BOA < \text{area of } \triangle OAC \qquad \boxed{2}$$

FIGURE 5

Area of triangle $= \frac{1}{2} \sin \theta$ Area of sector $= \frac{1}{2} \theta$ Area of triangle $= \frac{1}{2} \tan \theta$

Let's compute these three areas. First, $\triangle OAB$ has base 1 and height $\sin \theta$, so its area is $\frac{1}{2} \sin \theta$. Next, recall that a sector of angle θ has area $\frac{1}{2} \theta$. Finally, to compute the area of $\triangle OAC$, we observe that

REMINDER *Let's recall why a sector of angle θ in a circle of radius r has area $\frac{1}{2}r^2\theta$. Since the entire circle is a sector of angle 2π, a sector of angle θ represents a fraction $\frac{\theta}{2\pi}$ of the entire circle. The circle has area πr^2, and hence the sector has area $\left(\frac{\theta}{2\pi}\right)\pi r^2 = \frac{1}{2}r^2\theta$. In the unit circle $(r = 1)$, the sector has area $\frac{1}{2}\theta$.*

$$\tan\theta = \frac{\text{opposite side}}{\text{adjacent side}} = \frac{AC}{OA} = \frac{AC}{1} = AC$$

Thus, $\triangle OAC$ has base 1 and height $\tan\theta$ and its area is $\frac{1}{2}\tan\theta$. We have therefore shown that

$$\underbrace{\frac{1}{2}\sin\theta}_{\text{Area }\triangle OAB} \leq \underbrace{\frac{1}{2}\theta}_{\text{Area of sector}} \leq \underbrace{\frac{1}{2}\frac{\sin\theta}{\cos\theta}}_{\text{Area }\triangle OAC} \qquad \boxed{3}$$

The first inequality gives $\sin\theta \leq \theta$, and since $\theta > 0$, we obtain

$$\frac{\sin\theta}{\theta} \leq 1 \qquad \boxed{4}$$

Next, multiply the second inequality in (3) by $(2\cos\theta)/\theta$ to obtain

$$\cos\theta \leq \frac{\sin\theta}{\theta} \qquad \boxed{5}$$

The combination of (4) and (5) gives us (1) when $0 < \theta < \frac{\pi}{2}$. However, the functions in (1) do not change when θ is replaced by $-\theta$ because both $\cos\theta$ and $\frac{\sin\theta}{\theta}$ are even functions. Indeed, $\cos(-\theta) = \cos\theta$ and

$$\frac{\sin(-\theta)}{-\theta} = \frac{-\sin\theta}{-\theta} = \frac{\sin\theta}{\theta}$$

Therefore, (1) automatically holds for $-\frac{\pi}{2} < \theta < 0$ as well. This completes the proof of Theorem 3. ∎

Proof of Theorem 2   According to Theorem 3,

$$\cos\theta \leq \frac{\sin\theta}{\theta} \leq 1$$

Furthermore,

$$\lim_{\theta\to 0}\cos\theta = \cos 0 = 1 \qquad \text{and} \qquad \lim_{\theta\to 0} 1 = 1$$

Therefore, the Squeeze Theorem gives $\lim_{\theta\to 0}\dfrac{\sin\theta}{\theta} = 1$ as required. For a proof that $\lim_{\theta\to 0}\dfrac{1 - \cos\theta}{\theta} = 0$, see Exercises 42 and 51. ∎

In the next example, we evaluate another trigonometric limit. The key idea is to rewrite the function of h in terms of the new variable $\theta = 4h$.

■ EXAMPLE 2 Evaluating a Limit by Changing Variables   Investigate $\lim_{h\to 0}\dfrac{\sin 4h}{h}$ numerically and then evaluate it exactly.

Solution   The table of values at the left suggests that the limit is equal to 4. To evaluate the limit exactly, we rewrite it in terms of the limit of $(\sin\theta)/\theta$ so that Theorem 2 can be applied. Thus, we set $\theta = 4h$ and write

$$\frac{\sin 4h}{h} = 4\left(\frac{\sin 4h}{4h}\right) = 4\frac{\sin\theta}{\theta}$$

h	$\dfrac{\sin 4h}{h}$
± 14	-0.75680
± 0.5	1.81859
± 0.2	3.58678
± 0.1	$\mathbf{3.89\,418}$
± 0.05	$\mathbf{3.97\,339}$
± 0.01	$\mathbf{3.99\,893}$
± 0.005	$\mathbf{3.999\,73}$

The new variable θ tends to zero as $h \to 0$ since θ is a multiple of h. Therefore, we may change the limit as $h \to 0$ into a limit as $\theta \to 0$ to obtain

$$\lim_{h \to 0} \frac{\sin 4h}{h} = \lim_{\theta \to 0} 4 \frac{\sin \theta}{\theta} = 4 \left(\lim_{\theta \to 0} \frac{\sin \theta}{\theta} \right) = 4(1) = 4 \qquad \boxed{6}$$

■

2.6 SUMMARY

- We say that $f(x)$ is *squeezed* at $x = c$ if there exist functions $l(x)$ and $u(x)$ such that $l(x) \le f(x) \le u(x)$ for all $x \ne c$ in an open interval I containing c, and

$$\lim_{x \to c} l(x) = \lim_{x \to c} u(x) = L$$

The Squeeze Theorem states that in this case, $\lim_{x \to c} f(x) = L$.
- Two important trigonometric limits:

$$\lim_{\theta \to 0} \frac{\sin \theta}{\theta} = 1, \qquad \lim_{\theta \to 0} \frac{1 - \cos \theta}{\theta} = 0$$

2.6 EXERCISES

Preliminary Questions

1. Assume that $-x^4 \le f(x) \le x^2$. What is $\lim_{x \to 0} f(x)$? Is there enough information to evaluate $\lim_{x \to \frac{1}{2}} f(x)$? Explain.

2. State the Squeeze Theorem carefully.

3. Suppose that $f(x)$ is squeezed at $x = c$ on an open interval I by two *constant* functions $u(x)$ and $l(x)$. Is $f(x)$ necessarily constant on I?

4. If you want to evaluate $\lim_{h \to 0} \frac{\sin 5h}{3h}$, it is a good idea to rewrite the limit in terms of the variable (choose one):

(a) $\theta = 5h$ **(b)** $\theta = 3h$ **(c)** $\theta = \frac{5h}{3}$

Exercises

1. In Figure 6, is $f(x)$ squeezed by $u(x)$ and $l(x)$ at $x = 3$? At $x = 2$?

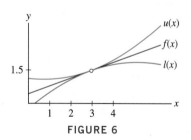

FIGURE 6

2. What is $\lim_{x \to 0} f(x)$ if $f(x)$ satisfies $\cos x \le f(x) \le 1$ for x in $(-1, 1)$?

3. What information about $f(x)$ does the Squeeze Theorem provide if we assume that the graphs of $f(x)$, $u(x)$, and $l(x)$ are related as in

Figure 7 and that $\lim_{x \to 7} u(x) = \lim_{x \to 7} l(x) = 6$? Note that the inequality $f(x) \le u(x)$ is not satisfied for all x. Does this affect the validity of your conclusion?

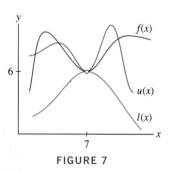

FIGURE 7

4. State whether the inequality provides sufficient information to determine $\lim_{x \to 1} f(x)$ and, if so, find the limit.

(a) $4x - 5 \le f(x) \le x^2$

(b) $2x - 1 \le f(x) \le x^2$

(c) $2x + 1 \le f(x) \le x^2 + 2$

In Exercises 5–8, use the Squeeze Theorem to evaluate the limit.

5. $\lim\limits_{x \to 0} x \cos \dfrac{1}{x}$

6. $\lim\limits_{x \to 0} x^2 \sin \dfrac{1}{x}$

7. $\lim\limits_{x \to 0+} \sqrt{x}\, e^{\cos(\pi/x)}$

8. $\lim\limits_{x \to 1} (x - 1) \sin \dfrac{\pi}{x - 1}$

In Exercises 9–16, evaluate the limit using Theorem 2 as necessary.

9. $\lim\limits_{x \to 0} \dfrac{\sin x \cos x}{x}$

10. $\lim\limits_{t \to 0} \dfrac{\sqrt{t + 2}\, \sin t}{t}$

11. $\lim\limits_{t \to 0} \dfrac{\sin^2 t}{t}$

12. $\lim\limits_{x \to 0} \dfrac{\tan x}{x}$

13. $\lim\limits_{x \to 0} \dfrac{x^2}{\sin^2 x}$

14. $\lim\limits_{t \to \frac{\pi}{2}} \dfrac{1 - \cos t}{t}$

15. $\lim\limits_{t \to \frac{\pi}{4}} \dfrac{\sin t}{t}$

16. $\lim\limits_{t \to 0} \dfrac{\cos t - \cos^2 t}{t}$

17. Let $L = \lim\limits_{x \to 0} \dfrac{\sin 10x}{x}$.

(a) Show, by letting $\theta = 10x$, that $L = \lim\limits_{\theta \to 0} 10 \dfrac{\sin \theta}{\theta}$.

(b) Compute L.

18. Evaluate $\lim\limits_{h \to 0} \dfrac{\sin 2h}{\sin 5h}$. *Hint:* $\dfrac{\sin 2h}{\sin 5h} = \left(\dfrac{2}{5}\right)\left(\dfrac{\sin 2h}{2h}\right)\left(\dfrac{5h}{\sin 5h}\right)$.

In Exercises 19–40, evaluate the limit.

19. $\lim\limits_{h \to 0} \dfrac{\sin 6h}{h}$

20. $\lim\limits_{h \to 0} \dfrac{6 \sin h}{6h}$

21. $\lim\limits_{h \to 0} \dfrac{\sin 6h}{6h}$

22. $\lim\limits_{h \to 0} \dfrac{\sin h}{6h}$

23. $\lim\limits_{x \to 0} \dfrac{\sin 7x}{3x}$

24. $\lim\limits_{x \to \frac{\pi}{4}} \dfrac{x}{\sin 11x}$

25. $\lim\limits_{x \to 0} \dfrac{\tan 4x}{9x}$

26. $\lim\limits_{x \to 0} \dfrac{x}{\csc 25x}$

27. $\lim\limits_{t \to 0} \dfrac{\tan 4t}{t \sec t}$

28. $\lim\limits_{h \to 0} \dfrac{\sin 2h \sin 3h}{h^2}$

29. $\lim\limits_{z \to 0} \dfrac{\sin(z/3)}{\sin z}$

30. $\lim\limits_{\theta \to 0} \dfrac{\sin(-3\theta)}{\sin(4\theta)}$

31. $\lim\limits_{x \to 0} \dfrac{\tan 4x}{\tan 9x}$

32. $\lim\limits_{t \to 0} \dfrac{\csc 8t}{\csc 4t}$

33. $\lim\limits_{x \to 0} \dfrac{\sin 5x \sin 2x}{\sin 3x \sin 5x}$

34. $\lim\limits_{x \to 0} \dfrac{\sin 3x \sin 2x}{x \sin 5x}$

35. $\lim\limits_{h \to 0} \dfrac{1 - \cos 2h}{h}$

36. $\lim\limits_{h \to 0} \dfrac{\sin(2h)(1 - \cos h)}{h}$

37. $\lim\limits_{t \to 0} \dfrac{1 - \cos t}{\sin t}$

38. $\lim\limits_{\theta \to 0} \dfrac{\cos 2\theta - \cos \theta}{\theta}$

39. $\lim\limits_{h \to \frac{\pi}{2}} \dfrac{1 - \cos 3h}{h}$

40. $\lim\limits_{\theta \to 0} \dfrac{1 - \cos 4\theta}{\sin 3\theta}$

41. Calculate (a) $\lim\limits_{x \to 0+} \dfrac{\sin x}{|x|}$ and (b) $\lim\limits_{x \to 0-} \dfrac{\sin x}{|x|}$.

42. Prove the following result stated in Theorem 2:

$$\lim\limits_{\theta \to 0} \dfrac{1 - \cos \theta}{\theta} = 0 \qquad \boxed{7}$$

Hint: $\dfrac{1 - \cos \theta}{\theta} = \dfrac{1}{1 + \cos \theta} \cdot \dfrac{1 - \cos^2 \theta}{\theta}$.

43. Show that $-|\tan x| \le \tan x \cos \dfrac{1}{x} \le |\tan x|$. Then evaluate $\lim\limits_{x \to 0} \tan x \cos \dfrac{1}{x}$ using the Squeeze Theorem.

44. Evaluate $\lim\limits_{x \to 0} \tan x \cos\left(\sin \dfrac{1}{x}\right)$ using the Squeeze Theorem.

45. [GU] Plot the graphs of $u(x) = 1 + |x - \dfrac{\pi}{2}|$ and $l(x) = \sin x$ on the same set of axes. What can you say about $\lim\limits_{x \to \frac{\pi}{2}} f(x)$ if $f(x)$ is squeezed by $l(x)$ and $u(x)$ at $x = \dfrac{\pi}{2}$?

46. Use the Squeeze Theorem to prove that if $\lim\limits_{x \to c} |f(x)| = 0$, then $\lim\limits_{x \to c} f(x) = 0$.

Further Insights and Challenges

47. [GU] Investigate $\lim\limits_{h \to 0} \dfrac{1 - \cos h}{h^2}$ numerically (and graphically if you have a graphing utility). Then prove that the limit is equal to $\dfrac{1}{2}$. Hint: see the hint for Exercise 42.

48. Evaluate $\lim\limits_{h \to 0} \dfrac{\cos 3h - 1}{h^2}$. *Hint:* Use the result of Exercise 47.

49. Evaluate $\lim\limits_{h \to 0} \dfrac{\cos 3h - 1}{\cos 2h - 1}$.

50. Use the result of Exercise 47 to prove that for $m \ne 0$,

$$\lim\limits_{x \to 0} \dfrac{\cos mx - 1}{x^2} = -\dfrac{m^2}{2}$$

51. Using a diagram of the unit circle and the Pythagorean Theorem, show that

$$\sin^2 \theta \le (1 - \cos \theta)^2 + \sin^2 \theta \le \theta^2$$

Conclude that $\sin^2 \theta \leq 2(1 - \cos \theta) \leq \theta^2$ and use this to give an alternate proof of Eq. (7) in Exercise 42. Then give an alternate proof of the result in Exercise 47.

52. (a) Investigate $\lim\limits_{x \to c} \dfrac{\sin x - \sin c}{x - c}$ numerically for the five values $c = 0, \frac{\pi}{6}, \frac{\pi}{4}, \frac{\pi}{3}, \frac{\pi}{2}$.

(b) Can you guess the answer for general c? (We'll see why in Chapter 3.)

(c) Check if your answer to part (b) works for two other values of c.

53. Let $A(n)$ be the area of a regular n-gon inscribed in a unit circle (Figure 8).

(a) Prove that $A(n) = \dfrac{1}{2}n \sin\left(\dfrac{2\pi}{n}\right)$.

(b) Intuitively, why might we expect $A(n)$ to converge to the area of the unit circle as $n \to \infty$?

(c) Use Theorem 2 to evaluate $\lim\limits_{n \to \infty} A(n)$.

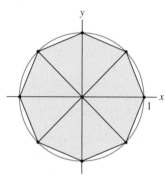

FIGURE 8 Regular n-gon inscribed in a unit circle.

2.7 Intermediate Value Theorem

The **Intermediate Value Theorem (IVT)** is a basic result which states that *a continuous function on an interval cannot skip values*. Consider a plane that takes off and climbs from 0 to 20,000 ft in 20 min. The plane must reach every altitude between 0 and 20,000 during this 20-min interval. Thus, at some moment, the plane's altitude must have been exactly 12,371 ft. Of course, we assume that the plane's motion is continuous, so its altitude cannot jump abruptly, say, from 12,000 to 13,000 ft.

To state this conclusion formally, let $A(t)$ be the plane's altitude at time t. The IVT then asserts that for every altitude M between 0 and 20,000, there is a time t_0 between 0 and 20 such that $A(t_0) = M$. In other words, the graph of $A(t)$ must intersect the horizontal line $y = M$ [Figure 1(A)].

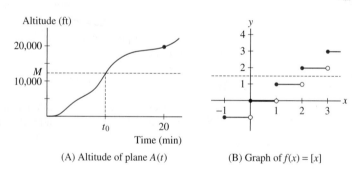

(A) Altitude of plane $A(t)$ (B) Graph of $f(x) = [x]$

FIGURE 1

By contrast, a discontinuous function can skip values. The greatest integer function $f(x) = [x]$ satisfies $[1] = 1$ and $[2] = 2$ [Figure 1(B)] but does not take on the value 1.5 (or any other value between 1 and 2).

| *A proof of the IVT is given in Appendix B.*

> **THEOREM 1 Intermediate Value Theorem** If $f(x)$ is continuous on a closed interval $[a, b]$ and $f(a) \neq f(b)$, then for every value M between $f(a)$ and $f(b)$, there exists at least one value $c \in (a, b)$ such that $f(c) = M$.

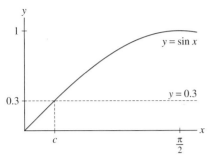

FIGURE 2

A zero or root of a function is a value c such that $f(c) = 0$. Sometimes the word "root" is reserved to refer specifically to the zero of a polynomial.

■ **EXAMPLE 1** Prove that the equation $\sin x = 0.3$ has at least one solution.

Solution We may apply the IVT since $\sin x$ is continuous. We choose an interval where we suspect that a solution exists. The desired value 0.3 lies between the two function values

$$\sin 0 = 0 \quad \text{and} \quad \sin \frac{\pi}{2} = 1$$

so the interval $[0, \frac{\pi}{2}]$ will work (Figure 2). The IVT tells us that $\sin x = 0.3$ has at least one solution in $(0, \frac{\pi}{2})$. Since $\sin x$ is periodic, $\sin x = 0.3$ actually has infinitely many solutions. ■

The IVT can be used to show the existence of zeros of functions. If $f(x)$ is continuous and takes on both positive and negative values, say, $f(a) < 0$ and $f(b) > 0$, then the IVT guarantees that $f(c) = 0$ for some c between a and b.

> **COROLLARY 2 Existence of Zeros** If $f(x)$ is continuous on $[a, b]$ and if $f(a)$ and $f(b)$ are nonzero and have opposite signs, then $f(x)$ has a zero in (a, b).

We can locate zeros of functions to arbitrary accuracy using the **Bisection Method**, as illustrated in the next example.

■ **EXAMPLE 2** The Bisection Method Show that $f(x) = \cos^2 x - 2\sin \frac{x}{4}$ has a zero in $(0, 2)$. Then locate the zero more accurately using the Bisection Method.

Solution Using a calculator, we find that $f(0)$ and $f(2)$ have opposite signs:

$$f(0) = 1 > 0, \qquad f(2) \approx -0.786 < 0$$

Corollary 2 guarantees that $f(x) = 0$ has a solution in $(0, 2)$ (Figure 3).

We can locate a zero more accurately by dividing $[0, 2]$ into two intervals $[0, 1]$ and $[1, 2]$. One of these must contain a zero of $f(x)$. To determine which, we evaluate $f(x)$ at the midpoint $m = 1$. A calculator gives $f(1) \approx -0.203 < 0$, and since $f(0) = 1$, we see that

$$f(x) \text{ takes on opposite signs at the endpoints of } [0, 1]$$

Therefore, $(0, 1)$ must contain a zero. We discard $[1, 2]$ because both $f(1)$ and $f(2)$ are negative. The Bisection Method consists of continuing this process until we narrow down the location of the zero to the desired accuracy. In the following table, the process is carried out three times:

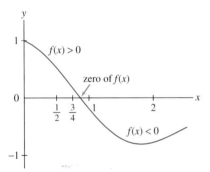

FIGURE 3 Graph of $f(x) = \cos^2 x - 2\sin \frac{x}{4}$.

Computer algebra systems have built-in commands for finding roots of a function or solving an equation numerically. These systems use a variety of methods, including more sophisticated versions of the Bisection Method. Notice that to use the Bisection Method, we must first find an interval containing a root.

Interval	Midpoint of Interval	Function Values	Conclusion
$[0, 1]$	$\frac{1}{2}$	$f\left(\frac{1}{2}\right) \approx 0.521$ $f(1) \approx -0.203$	Zero lies in $\left(\frac{1}{2}, 1\right)$
$\left[\frac{1}{2}, 1\right]$	$\frac{3}{4}$	$f\left(\frac{3}{4}\right) \approx 0.163$ $f(1) \approx -0.203$	Zero lies in $\left(\frac{3}{4}, 1\right)$
$\left[\frac{3}{4}, 1\right]$	$\frac{7}{8}$	$f\left(\frac{7}{8}\right) \approx -0.0231$ $f\left(\frac{3}{4}\right) \approx 0.163$	Zero lies in $\left(\frac{3}{4}, \frac{7}{8}\right)$

We conclude that $f(x)$ has a zero c satisfying $0.75 < c < 0.875$. The Bisection Method can be continued to locate the root with any degree of accuracy. ■

CONCEPTUAL INSIGHT   The IVT seems to state the obvious, namely that a continuous function cannot skip values. Yet its proof (given in Appendix B) is quite subtle because it depends on the *completeness* property of real numbers. To highlight the subtlety, observe that the IVT is *false* for functions defined only on the *rational numbers*. For example, $f(x) = x^2$ does not have the intermediate value property if we restrict its domain to the rational numbers. Indeed, $f(0) = 0$ and $f(2) = 4$ but $f(c) = 2$ *has no solution* for c rational. The solution $c = \sqrt{2}$ is "missing" from the set of rational numbers because it is irrational. From the beginnings of calculus, the IVT was surely regarded as obvious. However, it was not possible to give a genuinely rigorous proof until the completeness property was clarified in the second half of the nineteenth century.

2.7 SUMMARY

- The Intermediate Value Theorem (IVT) says that a continuous function does not *skip* values. More precisely, if $f(x)$ is continuous on $[a, b]$ with $f(a) \neq f(b)$, and if M is a number between $f(a)$ and $f(b)$, then $f(c) = M$ for some $c \in (a, b)$.
- Existence of zeros: If $f(x)$ is continuous on $[a, b]$ and if $f(a)$ and $f(b)$ take opposite signs (one is positive and the other negative), then $f(c) = 0$ for some $c \in (a, b)$.
- Bisection Method: Assume f is continuous. Suppose that $f(a)$ and $f(b)$ have opposite signs, so that f has a zero in (a, b). Then f has a zero in $[a, m]$ or $[m, b]$, where $m = (a + b)/2$ is the midpoint of $[a, b]$. To determine which, compute $f(m)$. A zero lies in (a, m) if $f(a)$ and $f(m)$ have opposite signs and in (m, b) if $f(m)$ and $f(b)$ have opposite signs. Continuing the process, we can locate a zero with arbitrary accuracy.

2.7 EXERCISES

Preliminary Questions

1. Explain why $f(x) = x^2$ takes on the value 0.5 in the interval $[0, 1]$.

2. The temperature in Vancouver was 46° at 6 AM and rose to 68° at noon. What must we assume about temperature to conclude that the temperature was 60° at some moment of time between 6 AM and noon?

3. What is the graphical interpretation of the IVT?

4. Show that the following statement is false by drawing a graph that provides a counterexample: *If $f(x)$ is continuous and has a root in $[a, b]$, then $f(a)$ and $f(b)$ have opposite signs.*

Exercises

1. Use the IVT to show that $f(x) = x^3 + x$ takes on the value 9 for some x in $[1, 2]$.

2. Show that $g(t) = \dfrac{t}{t+1}$ takes on the value 0.499 for some t in $[0, 1]$.

3. Show that $g(t) = t^2 \tan t$ takes on the value $\frac{1}{2}$ for some t in $\left[0, \frac{\pi}{4}\right]$.

4. Show that $f(x) = \dfrac{x^2}{x^7 + 1}$ takes on the value 0.4.

5. Show that $\cos x = x$ has a solution in the interval $[0, 1]$. *Hint:* Show that $f(x) = x - \cos x$ has a zero in $[0, 1]$.

6. Use the IVT to find an interval of length $\frac{1}{2}$ containing a root of $f(x) = x^3 + 2x + 1$.

In Exercises 7–16, use the IVT to prove each of the following statements.

7. $\sqrt{c} + \sqrt{c+1} = 2$ for some number c.

8. For all integers n, $\sin nx = \cos x$ for some $x \in [0, \pi]$.

9. $\sqrt{2}$ exists. *Hint:* Consider $f(x) = x^2$.

10. A positive number c has an nth root for all positive integers n. (This fact is usually taken for granted, but it requires proof.)

11. For all positive integers k, there exists x such that $\cos x = x^k$.

12. $2^x = bx$ has a solution if $b > 2$.

13. $2^x = b$ has a solution for all $b > 0$ (treat $b \geq 1$ first).

14. $\tan x = x$ has infinitely many solutions.

15. The equation $e^x + \ln x = 0$ has a solution in $(0, 1)$.

16. $\tan^{-1} x = \cos^{-1} x$ has a solution.

17. Carry out three steps of the Bisection Method for $f(x) = 2^x - x^3$ as follows:

(a) Show that $f(x)$ has a zero in $[1, 1.5]$.

(b) Show that $f(x)$ has a zero in $[1.25, 1.5]$.

(c) Determine whether $[1.25, 1.375]$ or $[1.375, 1.5]$ contains a zero.

18. Figure 4 shows that $f(x) = x^3 - 8x - 1$ has a root in the interval $[2.75, 3]$. Apply the Bisection Method twice to find an interval of length $\frac{1}{16}$ containing this root.

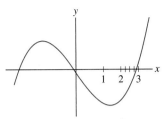

FIGURE 4 Graph of $y = x^3 - 8x - 1$.

19. Find an interval of length $\frac{1}{4}$ in $[0, 1]$ containing a root of $x^5 - 5x + 1 = 0$.

20. Show that $\tan^3 \theta - 8 \tan^2 \theta + 17 \tan \theta - 8 = 0$ has a root in $[0.5, 0.6]$. Apply the Bisection Method twice to find an interval of length 0.025 containing this root.

In Exercises 21–24, draw the graph of a function $f(x)$ on $[0, 4]$ with the given property.

21. Jump discontinuity at $x = 2$ and does not satisfy the conclusion of the IVT.

22. Jump discontinuity at $x = 2$, yet does satisfy the conclusion of the IVT on $[0, 4]$.

23. Infinite one-sided limits at $x = 2$ and does not satisfy the conclusion of the IVT.

24. Infinite one-sided limits at $x = 2$, yet does satisfy the conclusion of the IVT on $[0, 4]$.

25. Corollary 2 is not foolproof. Let $f(x) = x^2 - 1$ and explain why the corollary fails to detect the roots $x = \pm 1$ if $[a, b]$ contains $[-1, 1]$.

Further Insights and Challenges

26. Take any map (e.g., of the United States) and draw a circle on it anywhere. Prove that at any moment in time there exists a pair of diametrically opposite points on that circle corresponding to locations where the temperatures at that moment are equal. *Hint:* Let θ be an angular coordinate along the circle and let $f(\theta)$ be the difference in temperatures at the locations corresponding to θ and $\theta + \pi$.

27. Assume that $f(x)$ is continuous and that $0 \leq f(x) \leq 1$ for $0 \leq x \leq 1$ (see Figure 5). Show that $f(c) = c$ for some c in $[0, 1]$.

28. Use the IVT to show that if $f(x)$ is continuous and one-to-one on an interval $[a, b]$, then $f(x)$ is either an increasing or a decreasing function.

29. **Ham Sandwich Theorem** Figure 6(A) shows a slice of ham. Prove that for any angle θ ($0 \leq \theta \leq \pi$), it is possible to cut the slice of ham in half with a cut of incline θ. *Hint:* The lines of inclination θ are given by the equations $y = (\tan \theta)x + b$, where b varies from $-\infty$ to ∞. Each such line divides the ham slice into two pieces (one of which may be empty). Let $A(b)$ be the amount of ham to the left of the line minus the amount to the right and let A be the total area of the ham. Show that $A(b) = -A$ if b is sufficiently large and $A(b) = A$ if b is sufficiently negative. Then use the IVT. This works if $\theta \neq 0$ or $\frac{\pi}{2}$. If $\theta = 0$, define $A(b)$ as the amount of ham above the line $y = b$ minus the amount below. How can you modify the argument to work when $\theta = \frac{\pi}{2}$ (in which case $\tan \theta = \infty$)?

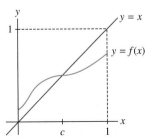

FIGURE 5 A function satisfying $0 \leq f(x) \leq 1$ for $0 \leq x \leq 1$.

30.   Figure 6(B) shows a slice of ham on a piece of bread. Prove that it is possible to slice this open-faced sandwich so that each part has equal amounts of ham and bread. *Hint:* By Exercise 29, for all $0 \le \theta \le \pi$ there is a line $L(\theta)$ of incline θ (which we assume is unique) that divides the ham into two equal pieces. For $\theta \ne 0$ or π, let $B(\theta)$ denote the amount of bread to the left of $L(\theta)$ minus the amount to the right. Notice that $L(\pi) = L(0)$ since $\theta = 0$ and $\theta = \pi$ give the same line, but $B(\pi) = -B(0)$ since left and right get interchanged as the angle moves from 0 to π. Assume that $B(\theta)$ is continuous and apply the IVT. (By a further extension of this argument, one can prove the full "Ham Sandwich Theorem," which states that if you allow the knife to cut at a slant, then it is possible to cut a sandwich consisting of a slice of ham and two slices of bread so that all three layers are divided in half.)

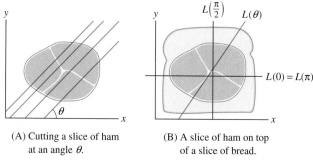

(A) Cutting a slice of ham at an angle θ.

(B) A slice of ham on top of a slice of bread.

FIGURE 6

2.8 The Formal Definition of a Limit

In this section, we reexamine the definition of a limit in order to state it in a rigorous and precise fashion. Why is this necessary? In Section 2.3, we defined limits by saying that $\lim_{x \to c} f(x) = L$ if $|f(x) - L|$ becomes arbitrarily small when x is sufficiently close (but not equal) to c. The problem with this definition lies in the phrases "arbitrarily small" and "sufficiently close." We must find a way to specify just how close is sufficiently close.

The formal limit definition eliminates the vagueness of the informal approach. It is used to develop calculus in a precise and rigorous fashion. Some proofs using the formal limit definition are included in Appendix D, but for a more complete development, we refer to textbooks on the branch of mathematics called "analysis."

The Size of the Gap

Recall that the distance from $f(x)$ to L is $|f(x) - L|$. It is convenient to refer to the quantity $|f(x) - L|$ as the *gap* between the value $f(x)$ and the limit L.

Let's reexamine the basic trigonometric limit

$$\lim_{x \to 0} \frac{\sin x}{x} = 1 \qquad \boxed{1}$$

In this example, $f(x) = \dfrac{\sin x}{x}$ and $L = 1$, so (1) tells us that the gap $|f(x) - 1|$ gets arbitrarily small when x is sufficiently close but not equal to 0.

Suppose we want the gap $|f(x) - 1|$ to be less than 0.2. How close to 0 must x be? Figure 1(B) shows that $f(x)$ lies within 0.2 of $L = 1$ for all values of x in the interval $[-1, 1]$. In other words, the following statement is true:

$$\left| \frac{\sin x}{x} - 1 \right| < 0.2 \qquad \text{if} \qquad 0 < |x| < 1$$

If we insist instead that the gap be smaller than 0.004, we can check by zooming in on the graph as in Figure 1(C) that

$$\left| \frac{\sin x}{x} - 1 \right| < 0.004 \qquad \text{if} \qquad 0 < |x| < 0.15$$

It would seem that this process can be continued: By zooming in on the graph, we can find a small interval around $c = 0$ where the gap $|f(x) - 1|$ is smaller than any prescribed positive number.

To express this in a precise fashion, we follow time-honored tradition and use the Greek letters ϵ (epsilon) and δ (delta) to denote small numbers specifying the size of the

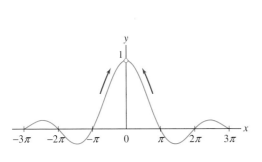

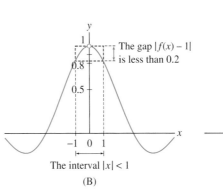

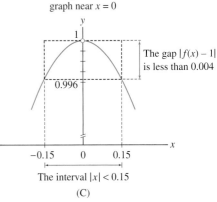

(A)

(B)

(C)

FIGURE 1 Graphs of $y = \frac{\sin x}{x}$. To shrink the gap from 0.2 to 0.004, we require that x lie within 0.15 of 0.

gap and the quantity $|x - c|$, respectively. In our case, $c = 0$ and $|x - c| = |x - 0| = |x|$. The precise meaning of Eq. (1) is that for every choice of $\epsilon > 0$, there exists some δ (depending on ϵ) such that

$$\left| \frac{\sin x}{x} - 1 \right| < \epsilon \qquad \text{if} \qquad 0 < |x| < \delta$$

The number δ tells us how close is sufficiently close for a given ϵ. With this motivation, we are ready to state the formal definition of the limit.

FORMAL DEFINITION OF A LIMIT   Suppose that $f(x)$ is defined for all x in an open interval containing c (but not necessarily at $x = c$). Then

$$\boxed{\lim_{x \to c} f(x) = L}$$

if for all $\epsilon > 0$, there exists $\delta > 0$ such that

$$\boxed{|f(x) - L| < \epsilon \qquad \text{if} \qquad 0 < |x - c| < \delta}$$

The condition $0 < |x - c| < \delta$ in this definition excludes $x = c$. As in our previous informal definition, we formulate it this way so that the limit depends only on values of $f(x)$ near c but not on $f(c)$ itself. As we have seen, in many cases the limit exists even when $f(c)$ is not defined.

The formal definition of a limit is often called the ϵ-δ definition. The tradition of using the symbols ϵ and δ originates in the writings of Augustin-Louis Cauchy on calculus and analysis in the 1820s.

If the symbols ϵ and δ seem to make this definition too abstract, keep in mind that we may take $\epsilon = 10^{-n}$ and $\delta = 10^{-m}$. Thus, $\lim_{x \to c} f(x) = L$ if, for any n, we have $|f(x) - L| < 10^{-n}$, provided that $0 < |x - c| < 10^{-m}$ (where the choice of m depends on n).

■ **EXAMPLE 1** Let $f(x) = 8x + 3$.

(a) Show that $\lim_{x \to 3} f(x) = 27$ using the formal definition of the limit.

(b) Find values of δ that work for $\epsilon = 0.2$ and 0.001.

Solution

(a) We break the proof into two steps.

***Step 1.* Relate the gap to $|x - c|$.**
We must find a relation between two absolute values: $|f(x) - L|$ for $L = 27$ and $|x - c|$ for $c = 3$. Observe that

$$\underbrace{|f(x) - 27|}_{\text{Size of gap}} = |(8x + 3) - 27| = |8x - 24| = 8|x - 3|$$

Thus, the gap is 8 times as large as $|x - 3|$.

Step 2. Choose δ (in terms of ϵ).

We can now see how to make the gap small: If $|x - 3| < \frac{\epsilon}{8}$, then the gap is smaller than $8|x - 3| < 8\left(\dfrac{\epsilon}{8}\right) = \epsilon$. Therefore, for any $\epsilon > 0$, we choose $\delta = \frac{\epsilon}{8}$. The following statement will then hold:

$$\boxed{\quad |f(x) - 27| < \epsilon \quad \text{if} \quad 0 < |x - 3| < \delta, \quad \text{where } \delta = \frac{\epsilon}{8} \quad}$$

Since we have specified δ for all $\epsilon > 0$, we have fulfilled the requirements of the formal definition of the limit, thus proving rigorously that $\lim_{x \to 3}(8x + 3) = 27$.

(b) For the particular choice $\epsilon = 0.2$, we may take $\delta = \dfrac{\epsilon}{8} = \dfrac{0.2}{8} = 0.025$:

$$|f(x) - 27| < 0.2 \quad \text{if} \quad 0 < |x - 3| < 0.025$$

This statement is illustrated in Figure 2. We note that *any positive δ smaller than 0.025 will also work*. For example, the following statement is also true, although it places an unnecessary restriction on x:

$$|f(x) - 27| < 0.2 \quad \text{if} \quad 0 < |x - 3| < 0.019$$

Similarly, to make the gap less than $\epsilon = 0.001$, we may take

$$\delta = \frac{\epsilon}{8} = \frac{0.001}{8} = 0.000125 \qquad \blacksquare$$

The difficulty in applying the limit definition frequently lies in trying to relate $|f(x) - L|$ to $|x - c|$. The next two examples illustrate how this can be done in special cases.

■ **EXAMPLE 2**   A Limit of x^2 Use the limit definition to prove that $\lim_{x \to 2} x^2 = 4$.

Solution Let $f(x) = x^2$.

Step 1. Relate the gap to $|x - c|$.

In this case, we must relate the gap $|f(x) - 4| = |x^2 - 4|$ to the quantity $|x - 2|$ (see Figure 3). This is more difficult than in the previous example because the gap is not a

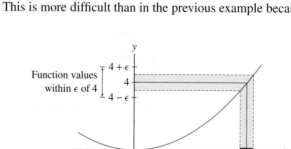

FIGURE 2 To make the gap less than 0.2, we may take $\delta = 0.025$ (not drawn to scale).

FIGURE 3 Graph of $f(x) = x^2$. We may choose δ so that $f(x)$ lies within ϵ of 4 for all x in $[2 - \delta, 2 + \delta]$.

constant multiple of $|x - 2|$. To proceed, consider the factorization

$$|x^2 - 4| = |x + 2|\,|x - 2|$$

Since we are going to require that $|x - 2|$ be small, we may as well assume from the outset that $|x - 2| < 1$, which means that $1 < x < 3$. In this case, $|x + 2|$ is less than 5 and the gap satisfies

$$|x^2 - 4| = |x + 2|\,|x - 2| < 5\,|x - 2| \qquad \text{if} \quad |x - 2| < 1 \qquad \boxed{2}$$

Step 2. Choose δ **(in terms of ϵ).**

We see from (2) that if $|x - 2|$ is smaller than both $\frac{\epsilon}{5}$ and 1, then the gap satisfies

$$|x^2 - 4| < 5|x - 2| < 5\left(\frac{\epsilon}{5}\right) = \epsilon$$

Therefore, the following statement holds for all $\epsilon > 0$:

$$\boxed{\,|x^2 - 4| < \epsilon \qquad \text{if} \qquad 0 < |x - 2| < \delta, \quad \text{where } \delta \text{ is the smaller of } \tfrac{\epsilon}{5} \text{ and } 1\,}$$

We have specified δ for all $\epsilon > 0$, so we have fulfilled the requirements of the formal limit definition, thus proving rigorously that $\displaystyle\lim_{x \to 2} x^2 = 4$.   ■

■ **EXAMPLE 3** Use the limit definition to prove that $\displaystyle\lim_{x \to 3} \frac{1}{x} = \frac{1}{3}$.

Solution

Step 1. Relate the gap to $|x - c|$.

The gap is equal to

$$\left| \frac{1}{x} - \frac{1}{3} \right| = \left| \frac{3 - x}{3x} \right| = |x - 3| \left| \frac{1}{3x} \right|$$

Since we are going to require that $|x - 3|$ be small, we may as well assume from the outset that $|x - 3| < 1$, that is, $2 < x < 4$. Now observe that if $x > 2$, then $3x > 6$ and $\frac{1}{3x} < \frac{1}{6}$, so the following inequality is valid if $|x - 3| < 1$:

◄·· **REMINDER** If $a > b > 0$, then $\frac{1}{a} < \frac{1}{b}$.

Thus, if $3x > 6$, then $\frac{1}{3x} < \frac{1}{6}$.

$$\left| f(x) - \frac{1}{3} \right| = \left| \frac{3 - x}{3x} \right| = \left| \frac{1}{3x} \right| |x - 3| < \frac{1}{6} |x - 3| \qquad \boxed{3}$$

Step 2. Choose δ (in terms of ϵ).

By (3), if $|x - 3| < 1$ and $|x - 3| < 6\epsilon$, then

$$\left| \frac{1}{x} - \frac{1}{3} \right| < \frac{1}{6} |x - 3| < \frac{1}{6}(6\epsilon) = \epsilon$$

Therefore, given any $\epsilon > 0$, we let δ be the smaller of the numbers 6ϵ and 1. Then

$$\boxed{\,\left| \frac{1}{x} - \frac{1}{3} \right| < \epsilon \qquad \text{if} \qquad 0 < |x - 3| < \delta, \quad \text{where } \delta \text{ is the smaller of } 6\epsilon \text{ and } 1\,}$$

Again, we have fulfilled the requirements of the formal limit definition, thus proving rigorously that $\displaystyle\lim_{x \to 3} \frac{1}{x} = \frac{1}{3}$.   ■

GRAPHICAL INSIGHT   Keep the graphical interpretation of limits in mind. In Figure 4(A), $f(x)$ approaches L as $x \to c$ because for any $\epsilon > 0$, we can make the gap less than ϵ by taking δ sufficiently small. The function in Figure 4(B) has a jump discontinuity at $x = c$ and the gap cannot be made small, no matter how small δ is taken. Therefore, the limit does not exist.

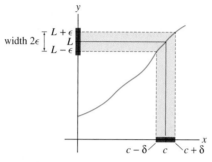

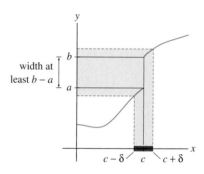

(A) The function is continuous at $x = c$.
By taking δ sufficiently small, we can make the gap smaller than ϵ.

(B) The function is not continuous at $x = c$.
The gap is always larger than $(b - a)/2$, no matter how small δ is.

FIGURE 4

Proving Limit Theorems

In practice, the formal limit definition is rarely used to evaluate limits. Most limits are evaluated using the Limit Laws or other techniques such as the Squeeze Theorem. However, the formal definition allows us to prove these laws in a rigorous fashion and thereby ensure that calculus is based on a solid foundation. We illustrate by proving the Sum Law. Other proofs are given in Appendix D.

Proof of the Sum Law   Assume that

$$\lim_{x \to c} f(x) = L \quad \text{and} \quad \lim_{x \to c} g(x) = M$$

We must prove that $\lim_{x \to c} (f(x) + g(x)) = L + M$.

By the Triangle Inequality [Eq. (1) in Section 1.1], $|a + b| \le |a| + |b|$ for all a and b. Setting $a = f(x) - L$ and $b = g(x) - M$, we obtain

$$|(f(x) + g(x)) - (L + M)| \le |f(x) - L| + |g(x) - M| \qquad \boxed{4}$$

By the limit definition, we can make each term on the right in (4) as small as desired by taking $|x - c|$ small enough. More precisely, given $\epsilon > 0$, we can choose δ such that $|f(x) - L| < \frac{\epsilon}{2}$ and $|g(x) - M| < \frac{\epsilon}{2}$ if $0 < |x - c| < \delta$ (in principle, we might have to choose different δ's for f and g, but we may then use the smaller of the two δ's). Thus, (4) gives

$$|f(x) + g(x) - (L + M)| < \frac{\epsilon}{2} + \frac{\epsilon}{2} = \epsilon \qquad \text{if} \qquad 0 < |x - c| < \delta \qquad \boxed{5}$$

This proves that

$$\lim_{x \to c} \big(f(x) + g(x)\big) = L + M = \lim_{x \to c} f(x) + \lim_{x \to c} g(x) \qquad \blacksquare$$

2.8 SUMMARY

- Informally speaking, the statement $\lim_{x \to c} f(x) = L$ means that the gap $|f(x) - L|$ tends to 0 as x approaches c. To make this precise, we introduce the *formal definition* (called the ϵ-δ definition): $\lim_{x \to c} f(x) = L$ if, for all $\epsilon > 0$, there exists a $\delta > 0$ such that

$$|f(x) - L| < \epsilon \qquad \text{if} \qquad 0 < |x - c| < \delta$$

2.8 EXERCISES

Preliminary Questions

1. Given that $\lim_{x \to 0} \cos x = 1$, which of the following statements is true?

(a) If $|\cos x - 1|$ is very small, then x is close to 0.

(b) There is an $\epsilon > 0$ such that $|x| < 10^{-5}$ if $0 < |\cos x - 1| < \epsilon$.

(c) There is a $\delta > 0$ such that $|\cos x - 1| < 10^{-5}$ if $0 < |x| < \delta$.

(d) There is a $\delta > 0$ such that $|\cos x| < 10^{-5}$ if $0 < |x - 1| < \delta$.

2. Suppose that for a given ϵ and δ, it is known that $|f(x) - 2| < \epsilon$ if $0 < |x - 3| < \delta$. Which of the following statements must also be true?

(a) $|f(x) - 2| < \epsilon$ if $0 < |x - 3| < 2\delta$

(b) $|f(x) - 2| < 2\epsilon$ if $0 < |x - 3| < \delta$

(c) $|f(x) - 2| < \dfrac{\epsilon}{2}$ if $0 < |x - 3| < \dfrac{\delta}{2}$

(d) $|f(x) - 2| < \epsilon$ if $0 < |x - 3| < \dfrac{\delta}{2}$

Exercises

1. Consider $\lim_{x \to 4} f(x)$, where $f(x) = 8x + 3$.

(a) Show that $|f(x) - 35| = 8|x - 4|$.

(b) Show that for any $\epsilon > 0$, $|f(x) - 35| < \epsilon$ if $|x - 4| < \delta$, where $\delta = \frac{\epsilon}{8}$. Explain how this proves rigorously that $\lim_{x \to 4} f(x) = 35$.

2. Consider $\lim_{x \to 2} f(x)$, where $f(x) = 4x - 1$.

(a) Show that $|f(x) - 7| < 4\delta$ if $|x - 2| < \delta$.

(b) Find a δ such that:

$$|f(x) - 7| < 0.01 \qquad \text{if} \qquad |x - 2| < \delta$$

(c) Prove rigorously that $\lim_{x \to 2} f(x) = 7$.

3. Consider $\lim_{x \to 2} x^2 = 4$ (refer to Example 2).

(a) Show that $|x^2 - 4| < 0.05$ if $0 < |x - 2| < 0.01$.

(b) Show that $|x^2 - 4| < 0.0009$ if $0 < |x - 2| < 0.0002$.

(c) Find a value of δ such that $|x^2 - 4|$ is less than 10^{-4} if $0 < |x - 2| < \delta$.

4. With regard to the limit $\lim_{x \to 5} x^2 = 25$,

(a) Show that $|x^2 - 25| < 11|x - 5|$ if $4 < x < 6$. *Hint:* Write $|x^2 - 25| = |x + 5| \cdot |x - 5|$.

(b) Find a δ such that $|x^2 - 25| < 10^{-3}$ if $|x - 5| < \delta$.

(c) Give a rigorous proof of the limit by showing that $|x^2 - 25| < \epsilon$ if $0 < |x - 5| < \delta$, where δ is the smaller of $\frac{\epsilon}{11}$ and 1.

5. Refer to Example 3 to find a value of $\delta > 0$ such that

$$\left| \frac{1}{x} - \frac{1}{3} \right| < 10^{-4} \qquad \text{if} \qquad 0 < |x - 3| < \delta$$

6. [GU] Plot $f(x) = \sqrt{2x - 1}$ together with the horizontal lines $y = 2.9$ and $y = 3.1$. Use this plot to find a value of $\delta > 0$ such that $|\sqrt{2x - 1} - 3| < 0.1$ if $|x - 5| < \delta$.

7. [GU] Plot $f(x) = \tan x$ together with the horizontal lines $y = 0.99$ and $y = 1.01$. Use this plot to find a value of $\delta > 0$ such that $|\tan x - 1| < 0.01$ if $|x - \frac{\pi}{4}| < \delta$.

8. [GU] The number e has the following property: $\lim_{x \to 0} \dfrac{e^x - 1}{x} = 1$. Use a plot of $f(x) = \dfrac{e^x - 1}{x}$ to find a value of $\delta > 0$ such that $|f(x) - 1| < 0.01$ if $|x - 1| < \delta$.

9. [GU] Let $f(x) = \dfrac{4}{x^2 + 1}$. Observe that the limit $L = \lim_{x \to \frac{1}{2}} f(x)$ has the value $L = f(\frac{1}{2}) = \frac{16}{5}$. Let $\epsilon = 0.5$ and, using a plot of $f(x)$, find a value of $\delta > 0$ such that $|f(x) - \frac{16}{5}| < \epsilon$ for $|x - \frac{1}{2}| < \delta$. Repeat for $\epsilon = 0.2$ and 0.1.

10. Consider $\lim_{x \to 2} \dfrac{1}{x}$.

(a) Show that if $|x - 2| < 1$, then

$$\left| \frac{1}{x} - \frac{1}{2} \right| < \frac{1}{2}|x - 2|$$

(b) Let δ be the smaller of 1 and 2ϵ. Prove the following statement:

$$\left| \frac{1}{x} - \frac{1}{2} \right| < \epsilon \quad \text{if} \quad 0 < |x - 2| < \delta$$

(c) Find a $\delta > 0$ such that $\left| \frac{1}{x} - \frac{1}{2} \right| < 0.01$ if $|x - 2| < \delta$.

(d) Prove rigorously that $\lim\limits_{x \to 2} \frac{1}{x} = \frac{1}{2}$.

11. Consider $\lim\limits_{x \to 1} \sqrt{x + 3}$.

(a) Show that $|\sqrt{x+3} - 2| < \frac{1}{2}|x - 1|$ if $|x - 1| < 4$. *Hint:* Multiply the inequality by $|\sqrt{x+3} + 2|$ and observe that $|\sqrt{x+3} + 2| > 2$.

(b) Find $\delta > 0$ such that $|\sqrt{x+3} - 2| < 10^{-4}$ for $|x - 1| < \delta$.

(c) Prove rigorously that the limit is equal to 2.

12. Based on the information conveyed in Figure 5(A), find values of L, ϵ, and $\delta > 0$ such that the following statement holds: $|f(x) - L| < \epsilon$ if $|x| < \delta$.

13. Based on the information conveyed in Figure 5(B), find values of c, L, ϵ, and $\delta > 0$ such that the following statement holds: $|f(x) - L| < \epsilon$ if $|x - c| < \delta$.

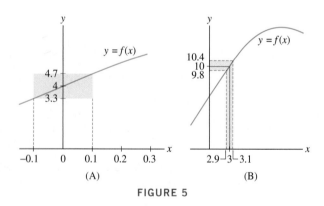

(A)   (B)

FIGURE 5

14. 🖍 A calculator gives the following values for $f(x) = \sin x$:

$$f\left(\frac{\pi}{4} - 0.1\right) \approx 0.633, \quad f\left(\frac{\pi}{4}\right) \approx 0.707, \quad f\left(\frac{\pi}{4} + 0.1\right) \approx 0.774$$

Use these values and the fact that $f(x)$ is increasing on $[0, \frac{\pi}{2}]$ to justify the statement

$$\left| f(x) - f\left(\frac{\pi}{4}\right) \right| < 0.08 \quad \text{if} \quad \left| x - \frac{\pi}{4} \right| < 0.1$$

Then draw a figure like Figure 3 to illustrate this statement.

15. Adapt the argument in Example 1 to prove rigorously that $\lim\limits_{x \to c} (ax + b) = ac + b$, where a, b, c are arbitrary.

16. Adapt the argument in Example 2 to prove rigorously that $\lim\limits_{x \to c} x^2 = c^2$ for all c.

17. Adapt the argument in Example 3 to prove rigorously that $\lim\limits_{x \to c} x^{-1} = \frac{1}{c}$ for all c.

In Exercises 18–23, use the formal definition of the limit to prove the statement rigorously.

18. $\lim\limits_{x \to 4} \sqrt{x} = 2$

19. $\lim\limits_{x \to 1} (3x^2 + x) = 4$

20. $\lim\limits_{x \to 1} x^3 = 1$

21. $\lim\limits_{x \to 2} x^{-2} = \frac{1}{4}$

22. $\lim\limits_{x \to c} |x| = |c|$

23. $\lim\limits_{x \to 0} x \sin \frac{1}{x} = 0$

24. Let $f(x) = \frac{x}{|x|}$. Prove rigorously that $\lim\limits_{x \to 0} f(x)$ does not exist. *Hint:* Show that no number L qualifies as the limit because there always exists some x such that $|x| < \delta$ but $|f(x) - L| \geq \frac{1}{2}$, no matter how small δ is taken.

25. Prove rigorously that $\lim\limits_{x \to 0} \sin \frac{1}{x}$ does not exist.

26. Use the identity

$$\sin x + \sin y = 2 \sin \left(\frac{x + y}{2} \right) \cos \left(\frac{x - y}{2} \right)$$

to verify the relation

$$\sin(a + h) - \sin a = h \frac{\sin(h/2)}{h/2} \cos \left(a + \frac{h}{2} \right) \qquad \boxed{6}$$

Conclude that $|\sin(a + h) - \sin a| < |h|$ for all a and prove rigorously that $\lim\limits_{x \to a} \sin x = \sin a$. You may use the inequality $\left| \frac{\sin x}{x} \right| \leq 1$ for $x \neq 0$.

27. Use Eq. (6) to prove rigorously that

$$\lim\limits_{h \to 0} \frac{\sin(a + h) - \sin a}{h} = \cos a$$

Further Insights and Challenges

28. Uniqueness of the Limit Show that a function converges to at most one limiting value. In other words, use the limit definition to show that if $\lim\limits_{x \to c} f(x) = L_1$ and $\lim\limits_{x \to c} f(x) = L_2$, then $L_1 = L_2$.

In Exercises 29–31, prove the statement using the formal limit definition.

29. The Constant Multiple Law [Theorem 1, part (ii) in Section 2.3, p. 80]

30. The Squeeze Theorem. (Theorem 1 in Section 2.6, p. 99)

31. The Product Law [Theorem 1, part (iii) in Section 2.3, p. 80]. *Hint:* Use the identity

$$f(x)g(x) - LM = (f(x) - L)\,g(x) + L(g(x) - M)$$

32. Let $f(x) = 1$ if x is rational and $f(x) = 0$ if x is irrational. Prove that $\lim_{x \to c} f(x)$ does not exist for any c.

33. Here is a function with strange continuity properties:

$$f(x) = \begin{cases} \dfrac{1}{q} & \text{if } x \text{ is the rational number } \dfrac{p}{q} \text{ in lowest terms} \\ 0 & \text{if } x \text{ is an irrational number} \end{cases}$$

(a) Show that $f(x)$ is discontinuous at c if c is rational. *Hint:* There exist irrational numbers arbitrarily close to c.

(b) Show that $f(x)$ is continuous at c if c is irrational. *Hint:* Let I be the interval $\{x : |x - c| < 1\}$. Show that for any $Q > 0$, I contains at most finitely many fractions $\dfrac{p}{q}$ with $q < Q$. Conclude that there is a δ such that all fractions in $\{x : |x - c| < \delta\}$ have a denominator larger than Q.

CHAPTER REVIEW EXERCISES

1. A particle's position at time t (s) is $s(t) = \sqrt{t^2 + 1}$ m. Compute its average velocity over $[2, 5]$ and estimate its instantaneous velocity at $t = 2$.

2. The price p of natural gas in the United States (in dollars per 1,000 ft^3) on the first day of each month in 2004 is listed in the table below.

J	F	M	A	M	J
9.7	9.84	10	10.52	11.61	13.05

J	A	S	O	N	D
13.45	13.79	13.29	11.68	11.44	11.11

Compute the average rate of change of p (in dollars per 1,000 ft^3 per month) over the quarterly periods January–March, April–June, and July–September.

3. For a whole number n, let $P(n)$ be the number of *partitions* of n, that is, the number of ways of writing n as a sum of one or more whole numbers. For example, $P(4) = 5$ since the number 4 can be partitioned in five different ways: $4, 3 + 1, 2 + 2, 2 + 1 + 1, 1 + 1 + 1 + 1$. Treating $P(n)$ as a continuous function, use Figure 1 to estimate the rate of change of $P(n)$ at $n = 12$.

4. The average velocity v (m/s) of an oxygen molecule in the air at temperature T (°C) is $v = 25.7\sqrt{273.15 + T}$. What is the average

speed at $T = 25°$ (room temperature)? Estimate the rate of change of average velocity with respect to temperature at $T = 25°$. What are the units of this rate?

In Exercises 5–8, estimate the limit numerically to two decimal places or state that the limit does not exist.

5. $\displaystyle\lim_{x \to 0} \frac{1 - \cos^3(x)}{x^2}$

6. $\displaystyle\lim_{x \to 1} x^{1/(x-1)}$

7. $\displaystyle\lim_{x \to 2} \frac{x^x - 4}{x^2 - 4}$

8. $\displaystyle\lim_{x \to 2} \frac{3^x - 9}{5^x - 25}$

In Exercises 9–43, evaluate the limit if possible or state that it does not exist.

9. $\displaystyle\lim_{x \to 4} (3 + x^{1/2})$

10. $\displaystyle\lim_{x \to 1} \frac{5 - x^2}{4x + 7}$

11. $\displaystyle\lim_{x \to -2} \frac{4}{x^3}$

12. $\displaystyle\lim_{x \to -1} \frac{3x^2 + 4x + 1}{x + 1}$

13. $\displaystyle\lim_{x \to 1} \frac{x^3 - x}{x - 1}$

14. $\displaystyle\lim_{x \to 1} \frac{x^3 - 2x}{x - 1}$

15. $\displaystyle\lim_{t \to 9} \frac{\sqrt{t} - 3}{t - 9}$

16. $\displaystyle\lim_{t \to 9} \frac{t - 6}{\sqrt{t} - 3}$

17. $\displaystyle\lim_{x \to 3} \frac{\sqrt{x + 1} - 2}{x - 3}$

18. $\displaystyle\lim_{s \to 0} \frac{1 - \sqrt{s^2 + 1}}{s^2}$

19. $\displaystyle\lim_{h \to 0} \frac{2(a + h)^2 - 2a^2}{h}$

20. $\displaystyle\lim_{t \to -1+} \frac{1}{x + 1}$

21. $\displaystyle\lim_{t \to 3} \frac{1}{t^2 - 9}$

22. $\displaystyle\lim_{x \to b} \frac{x^3 - b^3}{x - b}$

23. $\displaystyle\lim_{a \to b} \frac{a^2 - 3ab + 2b^2}{a - b}$

24. $\displaystyle\lim_{x \to 0} \frac{e^{3x} - e^x}{e^x - 1}$

25. $\displaystyle\lim_{x \to 1} \frac{x^3 - ax^2 + ax - 1}{x - 1}$

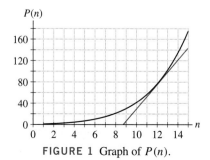

FIGURE 1 Graph of $P(n)$.

26. $\lim\limits_{x \to 0} \left(\dfrac{1}{3x} - \dfrac{1}{x(x+3)} \right)$

27. $\lim\limits_{x \to 1+} \left(\dfrac{1}{\sqrt{x-1}} - \dfrac{1}{\sqrt{x^2-1}} \right)$

28. $\lim\limits_{x \to 1.5} \dfrac{[x]}{x}$

29. $\lim\limits_{x \to 4.3} \dfrac{1}{x - [x]}$

30. $\lim\limits_{t \to 0-} \dfrac{[x]}{x}$

31. $\lim\limits_{t \to 0+} \dfrac{[x]}{x}$

32. $\lim\limits_{\theta \to \frac{\pi}{4}} \sec \theta$

33. $\lim\limits_{\theta \to \frac{\pi}{2}} \theta \sec \theta$

34. $\lim\limits_{\theta \to 0} \dfrac{\cos \theta - 2}{\theta}$

35. $\lim\limits_{\theta \to 0} \dfrac{\sin 5\theta}{\theta}$

36. $\lim\limits_{\theta \to 0} \dfrac{\sin 4x}{\sin 3x}$

37. $\lim\limits_{t \to 0} \dfrac{\sin^2 t}{t^3}$

38. $\lim\limits_{x \to \frac{\pi}{2}} \tan x$

39. $\lim\limits_{t \to 0} \cos \dfrac{1}{t}$

40. $\lim\limits_{t \to 0+} \sqrt{t} \cos \dfrac{1}{t}$

41. $\lim\limits_{x \to 0} \dfrac{\sin 7x}{\sin 3x}$

42. $\lim\limits_{x \to 0} \dfrac{\cos x - 1}{\sin x}$

43. $\lim\limits_{\theta \to 0} \dfrac{\tan \theta - \sin \theta}{\sin^3 \theta}$

44. Find the left- and right-hand limits of the function $f(x)$ in Figure 2 at $x = 0, 2, 4$. State whether $f(x)$ is left- or right-continuous (or both) at these points.

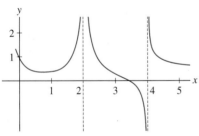

FIGURE 2

45. Sketch the graph of a function $f(x)$ such that

$$\lim\limits_{x \to 2-} f(x) = 1, \qquad \lim\limits_{x \to 2+} f(x) = 3$$

and such that $\lim\limits_{x \to 4} f(x)$ but does not equal $f(4)$.

46. Graph $h(x)$ and describe the type of discontinuity.

$$h(x) = \begin{cases} e^x & \text{for } x \le 0 \\ \ln x & \text{for } x > 0 \end{cases}$$

47. Sketch the graph of a function $g(x)$ such that

$$\lim\limits_{x \to -3-} g(x) = \infty, \qquad \lim\limits_{x \to -3+} g(x) = -\infty, \qquad \lim\limits_{x \to 4} g(x) = \infty$$

48. Find the points of discontinuity of $g(x)$ and determine the type of discontinuity, where

$$g(x) = \begin{cases} \cos\left(\dfrac{\pi x}{2}\right) & \text{for } |x| < 1 \\ |x - 1| & \text{for } |x| \ge 1 \end{cases}$$

49. Show that $f(x) = x e^{\sin x}$ is continuous on its domain. State the domain.

50. Find a constant b such that $h(x)$ is continuous at $x = 2$, where

$$h(x) = \begin{cases} x + 1 & \text{for } |x| < 2 \\ b - x^2 & \text{for } |x| \ge 2 \end{cases}$$

With this choice of b, find all points of discontinuity.

51. Calculate the following limits, assuming that

$$\lim\limits_{x \to 3} f(x) = 6, \qquad \lim\limits_{x \to 3} g(x) = 4$$

(a) $\lim\limits_{x \to 3} (f(x) - 2g(x))$

(b) $\lim\limits_{x \to 3} x^2 f(x)$

(c) $\lim\limits_{x \to 3} \dfrac{f(x)}{g(x) + x}$

(d) $\lim\limits_{x \to 3} (2g(x)^3 - g(x)^2)$

52. Let $f(x)$ and $g(x)$ be functions such that $g(x) \ne 0$ for $x \ne a$, and let

$$A = \lim\limits_{x \to a} f(x), \qquad B = \lim\limits_{x \to a} g(x), \qquad L = \lim\limits_{x \to a} \dfrac{f(x)}{g(x)}$$

Prove that if the limits A, B, and L exist and $L = 1$, then $A = B$. *Hint:* You cannot use the Quotient Law if $B = 0$, so apply the Product Law to L and B instead.

53. In the notation of Exercise 52, give an example in which L exists but neither A nor B exists.

54. Let $f(x)$ be a function defined for all real numbers. Which of the following statements *must be true*, *might be true*, or *are never true*.

(a) $\lim\limits_{x \to 3} f(x) = f(3)$.

(b) If $\lim\limits_{x \to 0} \dfrac{f(x)}{x} = 1$, then $f(0) = 0$.

(c) If $\lim\limits_{x \to -7} f(x) = 8$, then $\lim\limits_{x \to -7} \dfrac{1}{f(x)} = \dfrac{1}{8}$.

(d) If $\lim\limits_{x \to 5+} f(x) = 4$ and $\lim\limits_{x \to 5-} f(x) = 8$, then $\lim\limits_{x \to 5} f(x) = 6$.

(e) If $\lim\limits_{x \to 0} \dfrac{f(x)}{x} = 1$, then $\lim\limits_{x \to 0} f(x) = 0$.

(f) If $\lim\limits_{x \to 5} f(x) = 2$, then $\lim\limits_{x \to 5} f(x)^3 = 8$.

55. Use the Squeeze Theorem to prove that the function $f(x) = x \cos \dfrac{1}{x}$ for $x \ne 0$ and $f(0) = 0$ is continuous at $x = 0$.

56. Let $f(x) = x\left[\frac{1}{x}\right]$, where $[x]$ is the greatest integer function.

(a) Sketch the graph of $f(x)$ on the interval $[\frac{1}{4}, 2]$.

(b) Show that for $x \neq 0$,

$$\frac{1}{x} - 1 < \left[\frac{1}{x}\right] \leq \frac{1}{x}$$

Then use the Squeeze Theorem to prove that

$$\lim_{x \to 0} x\left[\frac{1}{x}\right] = 1$$

57. Let r_1 and r_2 be the roots of the quadratic polynomial $f(x) = ax^2 - 2x + 20$. Observe that $f(x)$ "approaches" the linear function $L(x) = -2x + 20$ as $a \to 0$. Since $r = 10$ is the unique root of $L(x) = 0$, we might expect at least one of the roots of $f(x)$ to approach 10 as $a \to 0$ (Figure 3). Prove that the roots can be labeled so that $\lim\limits_{a \to 0} r_1 = 10$ and $\lim\limits_{a \to 0} r_2 = \infty$.

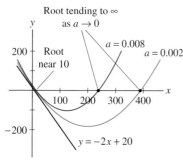

FIGURE 3 Graphs of $f(x) = ax^2 - 2x + 20$.

58. Let $f(x) = \frac{1}{x+2}$.

(a) Show that $\left| f(x) - \frac{1}{4} \right| < \frac{|x-2|}{12}$ if $|x - 2| < 1$. *Hint:* Observe that $|4(x + 2)| > 12$ if $|x - 2| < 1$.

(b) Find $\delta > 0$ such that $\left| f(x) - \frac{1}{4} \right| < 0.01$ for $|x - 2| < \delta$.

(c) Prove rigorously that $\lim\limits_{x \to 2} f(x) = \frac{1}{4}$.

59. GU Plot the function $f(x) = x^{1/3}$. Use the zoom feature to find a $\delta > 0$ such that $|x^{1/3} - 2| < 0.05$ for $|x - 8| < \delta$.

60. Prove rigorously that $\lim\limits_{x \to -1} (4 + 8x) = -4$.

61. Prove rigorously that $\lim\limits_{x \to 3} (x^2 - x) = 6$.

62. Use the IVT to prove that the curves $y = x^2$ and $y = \cos x$ intersect.

63. Use the IVT to prove that $f(x) = x^3 - \frac{x^2+2}{\cos x+2}$ has a root in the interval $[0, 2]$.

64. Use the IVT to show that $e^{-x^2} = x$ has a solution on $(0, 1)$.

65. Use the Bisection Method to locate a root of $x^2 - 7 = 0$ to two decimal places.

66. Give an example of a (discontinuous) function that does not satisfy the conclusion of the IVT on $[-1, 1]$. Then show that the function

$$f(x) = \begin{cases} \sin \dfrac{1}{x} & \text{for } x \neq 0 \\ 0 & \text{for } x = 0 \end{cases}$$

satisfies the conclusion of the IVT on every interval $[-a, a]$, even though f is discontinuous at $x = 0$.

3 | DIFFERENTIATION

Calculus and physics are used to design rides like this one, called Superman the Escape, at Six Flags Magic Mountain in Valencia, California.

| ←·· **REMINDER** A **secant line** is any line through two points on a curve or graph.

FIGURE 1 The secant line has slope $\Delta f / \Delta x$. Our goal is to compute the slope of the tangent line at $(a, f(a))$.

The term "differential calculus" was introduced in its Latin form "calculus differentialis" by Gottfried Wilhelm Leibniz. Newton referred to calculus as the "method of fluxions" (from Latin, meaning "to flow"), but this term was never adopted universally.

Differential calculus is the study of the derivative and its many applications. What is the derivative of a function? This question has three equally important answers: A derivative is a rate of change, it is the slope of a tangent line, and more formally, it is the limit of a difference quotient, as we will explain shortly. In this chapter, we explore these three facets of the derivative and develop the basic rules of differentiation, that is, techniques for computing derivatives. When you master these techniques, you will possess one of the most useful and flexible tools that mathematics has to offer.

3.1 Definition of the Derivative

Let's begin with two questions: How can we define the tangent line to a graph and how can we compute its slope? To answer these questions, we return to the relationship between tangent lines and secant lines first mentioned in Section 2.1.

Recall the formula for the slope of the secant line between two points P and Q on the graph of $y = f(x)$. If $P = (a, f(a))$ and $Q = (x, f(x))$ with $x \neq a$, as in Figure 1(A), then

$$\text{Slope of secant line through } P \text{ and } Q = \frac{\Delta f}{\Delta x} = \frac{f(x) - f(a)}{x - a}$$

where we set

$$\Delta f = f(x) - f(a) \qquad \text{and} \qquad \Delta x = x - a$$

The expression $\dfrac{f(x) - f(a)}{x - a}$ is called the **difference quotient**.

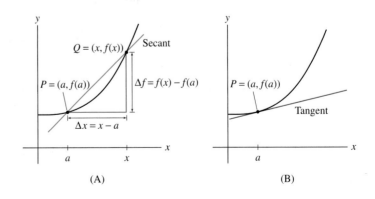

(A) (B)

It is not easy to describe tangent lines accurately in words. Even the Oxford English Dictionary is less than clear when it defines a tangent as "a straight line which touches a curve, i.e. meets it at a point and being produced does not (ordinarily) intersect it at that point." Perhaps it is not surprising that mathematicians found it necessary to use limits to attain the required precision.

Now observe what happens as Q approaches P or, equivalently, as x approaches a. Figure 2 suggests that the secant lines get progressively closer to the tangent line. If we imagine Q moving toward P, then the secant line appears to rotate into the tangent line as in (D). Therefore, we may expect the slopes of the secant lines to approach the slope of the tangent line. Based on this intuition, we define the **derivative** $f'(a)$ (read "f prime of a") as the limit

$$f'(a) = \underbrace{\lim_{x \to a} \frac{f(x) - f(a)}{x - a}}_{\text{Limit of slopes of secant lines}}$$

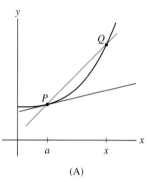
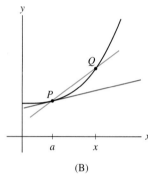
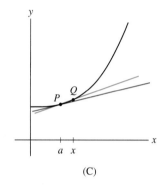
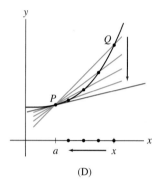

(A)   (B)   (C)   (D)

FIGURE 2 The secant lines approach the tangent line as Q approaches P.

There is another way of writing the difference quotient using the variable h:

$$h = x - a$$

We have $x = a + h$ and for $x \neq a$ (Figure 3)

$$\frac{f(x) - f(a)}{x - a} = \frac{f(a + h) - f(a)}{h}$$

The variable h approaches 0 as $x \to a$, so we can rewrite the derivative as

$$f'(a) = \lim_{h \to 0} \frac{f(a + h) - f(a)}{h}$$

Each way of writing the derivative is useful. The version using h is often more convenient in computations.

FIGURE 3 Secant line and difference quotient in terms of h.

> **DEFINITION The Derivative**   The derivative of a function f at $x = a$ is the limit of the difference quotients (if it exists):
>
> $$f'(a) = \lim_{h \to 0} \frac{f(a + h) - f(a)}{h}$$ **1**
>
> When the limit exists, we say that f **is differentiable** at $x = a$. An equivalent definition of the derivative is
>
> $$f'(a) = \lim_{x \to a} \frac{f(x) - f(a)}{x - a}$$ **2**

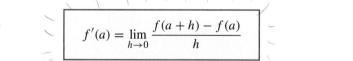

We can now use the derivative to define the tangent line in a precise and unambiguous way as the line of slope $f'(a)$ through $P = (a, f(a))$.

> **DEFINITION Tangent Line**   Assume that $f(x)$ is differentiable at $x = a$. The tangent line to the graph of $y = f(x)$ at $P = (a, f(a))$ is the line through P of slope $f'(a)$. The equation of the tangent line in point-slope form is
>
> $$y - f(a) = f'(a)(x - a)$$ **3**

◄·· **REMINDER** The equation in point-slope form of the line through $P = (a, b)$ of slope m is

$$y - b = m(x - a)$$

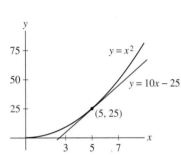

FIGURE 4 Tangent line to $y = x^2$ at $x = 5$.

■ **EXAMPLE 1** **Finding an Equation of a Tangent Line** Find an equation of the tangent line to the graph of $f(x) = x^2$ at $x = 5$.

Solution We need to compute $f'(5)$ and we are free to use either Eq. (1) or (2). Using Eq. (2), we have

$$f'(5) = \lim_{x \to 5} \frac{f(x) - f(5)}{x - 5} = \lim_{x \to 5} \frac{x^2 - 25}{x - 5} = \lim_{x \to 5} \frac{(x - 5)(x + 5)}{x - 5}$$
$$= \lim_{x \to 5} (x + 5) = 10$$

Next, we apply Eq. (3) with $a = 5$. Since $f(5) = 25$, an equation of the tangent line is $y - 25 = 10(x - 5)$, or in slope-intercept form: $y = 10x - 25$ (Figure 4). ■

The process of computing the derivative is called **differentiation**. The next two examples illustrate differentiation using Eq. (1). For clarity, we break up the computations into three steps.

■ **EXAMPLE 2** Compute $f'(3)$, where $f(x) = 4x^2 - 7x$.

Solution We use Eq. (1). The difference quotient at $a = 3$ is

$$\frac{f(a + h) - f(a)}{h} = \frac{f(3 + h) - f(3)}{h} \quad (h \neq 0)$$

Step 1. **Write out the numerator of the difference quotient.**

$$f(3 + h) - f(3) = \left(4(3 + h)^2 - 7(3 + h)\right) - \left(4(3^2) - 7(3)\right)$$
$$= \left(4(9 + 6h + h^2) - (21 + 7h)\right) - 15$$
$$= (15 + 17h + 4h^2) - 15 = 17h + 4h^2 = h(17 + 4h)$$

Step 2. **Divide by h and simplify the difference quotient.**
Simplify the difference quotient by canceling h in the numerator and denominator:

$$\frac{f(3 + h) - f(3)}{h} = \underbrace{\frac{h(17 + 4h)}{h}}_{\text{Cancel } h} = 17 + 4h$$

Step 3. **Compute the derivative by taking the limit.**

$$f'(3) = \lim_{h \to 0} \frac{f(3 + h) - f(3)}{h} = \lim_{h \to 0} (17 + 4h) = 17 \qquad ■$$

point-slope formula
$$y - y_1 = m(x - x_1)$$
$$y - f(a) = f'(a)(x - a)$$

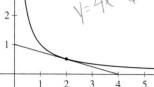

$$y + 2 = 4(x - 1)$$
$$y = 4x - 6$$

FIGURE 5 Graph of $y = \frac{1}{x}$. The equation of the tangent line at $x = 2$ is $y = -\frac{1}{4}x + 1$.

$f(x) = 3x^2 - 2x - 3$ at $x = 1$

$f'(1) = \lim\limits_{x \to 1} \dfrac{f(x) - f(1)}{x - 1}$

$= \lim\limits_{x \to 1} \dfrac{(3x^2 - 2x - 3) - (3 - 2 - 3)}{x - 1}$

$= \lim\limits_{x \to 1} \dfrac{3x^2 - 2x - 1}{x - 1} = \dfrac{(x-1)(3x+1)}{x-1}$

$= \lim\limits_{x \to 1} 3x + 1 = 3(1) + 1 = \boxed{4}$

■ **EXAMPLE 3** Sketch the graph of $f(x) = \dfrac{1}{x}$ and draw the tangent line a

(a) On the basis of the sketch, do you expect $f'(2)$ to be positive or negative?

(b) Find an equation of the tangent line at $x = 2$.

Solution The graph of $f(x)$ and the tangent line at $x = 2$ are sketched in Figure 5.

(a) The tangent line has negative slope, so we expect our calculation of $f'(2)$ to yield a negative number.

(b) We break up the calculation of $f'(2)$ into three steps as before.

Step 1. **Write out the numerator of the difference quotient.**

$$f(2 + h) - f(2) = \frac{1}{2 + h} - \frac{1}{2} = \frac{2}{2(2 + h)} - \frac{2 + h}{2(2 + h)} = -\frac{h}{2(2 + h)}$$

Step 2. **Divide by h and simplify the difference quotient.**

$$\frac{f(2 + h) - f(2)}{h} = \frac{1}{h} \cdot \left(-\frac{h}{2(2 + h)} \right) = -\frac{1}{2(2 + h)}$$

Step 3. **Compute the derivative by taking the limit.**

$$f'(2) = \lim_{h \to 0} \frac{f(2 + h) - f(2)}{h} = \lim_{h \to 0} \frac{-1}{2(2 + h)} = -\frac{1}{4}$$

Since $f(2) = \frac{1}{2}$, the tangent line passes through $(2, \frac{1}{2})$ and has equation

$$y - \frac{1}{2} = -\frac{1}{4}(x - 2)$$

In slope-intercept form: $y = -\frac{1}{4}x + 1$. ■

If $f(x)$ is a linear function, say, $f(x) = mx + b$ (where m and b are constants), then the graph of f is a straight line of slope m. We should expect that $f'(a) = m$ for all a because the tangent line at any point coincides with the line $y = f(x)$ (Figure 6). Let's check this by computing the derivative:

$$f'(a) = \lim_{h \to 0} \frac{f(a + h) - f(a)}{h} = \lim_{h \to 0} \frac{(m(a + h) + b) - (ma + b)}{h}$$

$$= \lim_{h \to 0} \frac{mh}{h} = \lim_{h \to 0} m = m$$

If $m = 0$, then $f(x) = b$ is constant and $f'(a) = 0$ (Figure 7). In summary,

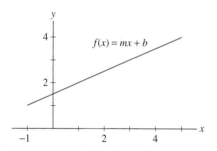

FIGURE 6 The derivative of $f(x) = mx + b$ is $f'(a) = m$.

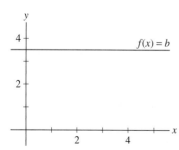

FIGURE 7 The derivative of a constant function $f(x) = b$ is $f'(a) = 0$.

THEOREM 1 Derivative of Linear and Constant Functions

- If $f(x) = mx + b$ is a linear function, then $f'(a) = m$ for all a.
- If $f(x) = b$ is a constant function, then $f'(a) = 0$ for all a.

■ **EXAMPLE 4** Find the derivative of $f(x) = 9x - 5$ at $x = 2$ and $x = 5$.

Solution We have $f'(a) = 9$ for all a. Hence, $f'(2) = f'(5) = 9$. ■

Estimating the Derivative

Approximations to the derivative are useful in situations where we cannot evaluate $f'(a)$ exactly. Since the derivative is the limit of difference quotients, the difference quotient should give a good numerical approximation when h is sufficiently small:

$$f'(a) \approx \frac{f(a+h) - f(a)}{h} \qquad \text{if } h \text{ is small}$$

Graphically, this says that when h is small, the slope of the secant line is nearly equal to the slope of the tangent line (Figure 8).

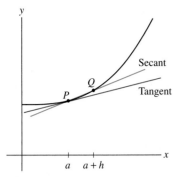

FIGURE 8 When h is small, the secant line has nearly the same slope as the tangent line.

$F(x) = \frac{2}{x}$

$F'(1) = \lim\limits_{x \to 1} \dfrac{f(x) - f(1)}{x-1}$

$= \lim\limits_{x \to 1} \dfrac{\left(\frac{2}{x} - \frac{2}{1}\right)x}{(x-1)x}$

$= \lim\limits_{x \to 1} \dfrac{2 - 2x}{x(x-1)}$

$= \lim\limits_{x \to 1} \dfrac{-2(x-1)}{x(x-1)} = \boxed{-2}$

$y - 2 = -2(x-1)$

$y = -2x + 2$

$y = -2x + 4$

■ **EXAMPLE 5** Estimate the derivative of $f(x) = \sin x$ at $x = \frac{\pi}{6}$.

Solution The difference quotient is

$$\frac{\sin\left(\frac{\pi}{6} + h\right) - \sin\frac{\pi}{6}}{h} = \frac{\sin\left(\frac{\pi}{6} + h\right) - 0.5}{h}$$

We calculate the difference quotient for several small values of h. Table 1 suggests that the limit has a decimal expansion beginning 0.866. In other words, $f'\left(\frac{\pi}{6}\right) \approx 0.866$. ■

TABLE 1 Values of the Difference Quotient for Small h

$h > 0$	$\dfrac{\sin\left(\frac{\pi}{6} + h\right) - 0.5}{h}$	$h < 0$	$\dfrac{\sin\left(\frac{\pi}{6} + h\right) - 0.5}{h}$
0.01	0.863511	−0.01	0.868511
0.001	0.865775	−0.001	0.866275
0.0001	0.866000	−0.0001	0.866050
0.00001	0.8660229	−0.00001	0.8660279

In the previous example, we used a table of values to *suggest* an estimate for $f'\left(\frac{\pi}{6}\right)$. In the next example, we use graphical reasoning to determine the accuracy of this estimate.

■ **EXAMPLE 6** GU **Determining Accuracy Graphically** Let $f(x) = \sin x$. Show that the approximation $f'\left(\frac{\pi}{6}\right) \approx 0.8660$ is accurate to four decimal places by studying the relationship between the secant and tangent lines.

Solution In Figure 9 you will notice that the relationship of the secant line to the tangent line depends on whether h is positive or negative. The slope of the secant line is *smaller* than the slope of the tangent when $h > 0$, but it is *larger* when $h < 0$. This tells us that the difference quotients in the second column of Table 1 are smaller than $f'\left(\frac{\pi}{6}\right)$ and those in the fourth column are greater than $f'\left(\frac{\pi}{6}\right)$. In particular, from the last line in Table 1 we may conclude that

This technique of estimating an unknown quantity by showing that it lies between two known values occurs frequently in calculus.

$$0.866022 \le f'\left(\frac{\pi}{6}\right) \le 0.866028$$

The value of $f'\left(\frac{\pi}{6}\right)$ rounded off to five places is either 0.86602 or 0.86603, so the estimate 0.8660 is accurate to four decimal places. In Section 3.6, we will see that the exact value is $f'\left(\frac{\pi}{6}\right) = \cos\left(\frac{\pi}{6}\right) = \sqrt{3}/2 \approx 0.8660254$. ■

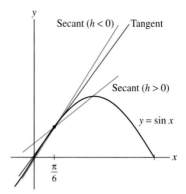

FIGURE 9 The tangent line is squeezed in *between* the secant lines with $h > 0$ and $h < 0$.

CONCEPTUAL INSIGHT Are Limits Really Necessary?   It is natural to ask whether limits are really necessary. Since the tangent line is so easy to visualize, is there perhaps a better or simpler way to find its equation? History gives one answer. The methods of calculus based on limits have stood the test of time and are used more widely today than ever before.

History aside, there is a more intuitive argument for why limits are necessary. The slope of a line can be computed if the coordinates of *two* points $P = (x_1, y_1)$ and $Q = (x_2, y_2)$ on the line are known:

$$\text{Slope of line} = \frac{y_2 - y_1}{x_2 - x_1}$$

We cannot use this formula to find the slope of the tangent line because we have only one point on the tangent line to work with, namely $P = (a, f(a))$. Limits provide an ingenious way around this obstacle. We choose a point $Q = (a + h, f(a + h))$ on the graph near P and form the secant line. The slope of this secant line is only an approximation to the slope of the tangent line:

$$\text{Slope of secant line} = \frac{f(a + h) - f(a)}{h} \approx \text{slope of tangent line}$$

But this approximation improves as $h \to 0$, and by taking the limit, we convert our approximations into an exact result.

3.1 SUMMARY

- The *difference quotient*

$$\frac{f(a + h) - f(a)}{h}$$

is the slope of the secant line through $P = (a, f(a))$ and $Q = (a + h, f(a + h))$ on the graph of $f(x)$.
- The *derivative* $f'(a)$ is defined by the following equivalent limits:

$$f'(a) = \lim_{h \to 0} \frac{f(a + h) - f(a)}{h} = \lim_{x \to a} \frac{f(x) - f(a)}{x - a}$$

If the limit exists, we say that f is *differentiable* at $x = a$.

- By definition, the tangent line at $P = (a, f(a))$ is the line through P with slope $f'(a)$ [assuming that $f'(a)$ exists]. The equation of the tangent line in point-slope form is

$$y - f(a) = f'(a)(x - a)$$

- To calculate $f'(a)$ using the limit definition, it is convenient to follow three steps:

Step 1. Write out the numerator of the difference quotient.
Step 2. Divide by h and simplify the difference quotient.
Step 3. Compute the derivative by taking the limit.

- For small values of h, we have the estimate $f'(a) \approx \dfrac{f(a+h) - f(a)}{h}$.

3.1 EXERCISES

Preliminary Questions

1. Which of the lines in Figure 10 are tangent to the curve?

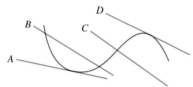

FIGURE 10

2. What are the two ways of writing the difference quotient?

3. For which value of x is

$$\frac{f(x) - f(3)}{x - 3} = \frac{f(7) - f(3)}{4} ?$$

4. What do the following quantities represent in terms of the graph of $f(x) = \sin x$?

(a) $\sin 1.3 - \sin 0.9$

(b) $\dfrac{\sin 1.3 - \sin 0.9}{0.4}$

(c) $f'(1.3)$

5. For which values of a and h is $\dfrac{f(a+h) - f(a)}{h}$ equal to the slope of the secant line between the points $(3, f(3))$ and $(5, f(5))$ on the graph of $f(x)$?

6. To which derivative is the quantity

$$\frac{\tan(\frac{\pi}{4} + 0.00001) - 1}{0.00001}$$

a good approximation?

Exercises

1. Let $f(x) = 3x^2$. Show that

$$f(2 + h) = 3h^2 + 12h + 12$$

Then show that

$$\frac{f(2+h) - f(2)}{h} = 3h + 12$$

and compute $f'(2)$ by taking the limit as $h \to 0$.

2. Let $f(x) = 2x^2 - 3x - 5$. Show that the slope of the secant line through $(2, f(2))$ and $(2 + h, f(2 + h))$ is $2h + 5$. Then use this formula to compute the slope of:

(a) The secant line through $(2, f(2))$ and $(3, f(3))$

(b) The tangent line at $x = 2$ (by taking a limit)

In Exercises 3–6, compute $f'(a)$ in two ways, using Eq. (1) and Eq. (2).

3. $f(x) = x^2 + 9x, \quad a = 0$

4. $f(x) = x^2 + 9x, \quad a = 2$

5. $f(x) = 3x^2 + 4x + 2, \quad a = -1$

6. $f(x) = 9 - 3x^2, \quad a = 0$

In Exercises 7–10, refer to the function whose graph is shown in Figure 11.

7. Calculate the slope of the secant line through the points on the graph where $x = 0$ and $x = 2.5$.

8. Compute $\dfrac{f(2.5) - f(1)}{2.5 - 1}$. What does this quantity represent?

9. Estimate $\dfrac{f(2+h) - f(2)}{h}$ for $h = 0.5$ and $h = -0.5$.

Are these numbers larger or smaller than $f'(2)$? Explain.

y − f(a) = f'(a)(x − a)

10. Estimate $f'(2)$.

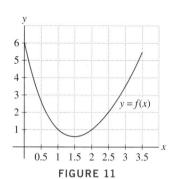

FIGURE 11

In Exercises 11–14, let $f(x)$ be the function whose graph is shown in Figure 12.

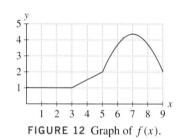

FIGURE 12 Graph of $f(x)$.

11. Determine $f'(a)$ for $a = 1, 2, 4, 7$.

12. Estimate $f'(6)$.

13. Which is larger: $f'(5.5)$ or $f'(6.5)$?

14. Show that $f'(3)$ does not exist.

In Exercises 15–18, use the limit definition to find the derivative of the linear function. (Note: The derivative does not depend on a.)

15. $f(x) = 3x - 2$ **16.** $f(x) = 2$

17. $g(t) = 9 - t$ **18.** $k(z) = 16z + 9$

19. Let $f(x) = \dfrac{1}{x}$. Does $f(-2 + h)$ equal $\dfrac{1}{-2 + h}$ or $\dfrac{1}{-2} + \dfrac{1}{h}$? Compute the difference quotient for $f(x)$ at $a = -2$ with $h = 0.5$.

20. Let $f(x) = \sqrt{x}$. Does $f(5 + h)$ equal $\sqrt{5 + h}$ or $\sqrt{5} + \sqrt{h}$? Compute the difference quotient for $f(x)$ at $a = 5$ with $h = 1$.

21. Let $f(x) = \sqrt{x}$. Show that

$$\frac{f(9 + h) - f(9)}{h} = \frac{1}{\sqrt{9 + h} + 3}$$

Then use this formula to compute $f'(9)$ (by taking the limit).

22. First find the slope and then an equation of the tangent line to the graph of $f(x) = \sqrt{x}$ at $x = 4$.

In Exercises 23–40, compute the derivative at $x = a$ using the limit definition and find an equation of the tangent line.

23. $f(x) = 3x^2 + 2x, \quad a = 2$

24. $f(x) = 3x^2 + 2x, \quad a = -1$

25. $f(x) = x^3, \quad a = 2$

26. $f(x) = x^3, \quad a = 3$

27. $f(x) = x^3 + x, \quad a = 0$

28. $f(t) = 3t^3 + 2t, \quad a = 4$

29. $f(x) = x^{-1}, \quad a = 3$ **30.** $f(x) = x^{-2}, \quad a = 1$

31. $f(x) = 9x - 4, \quad a = -7$ **32.** $f(t) = \sqrt{t + 1}, \quad a = 0$

33. $f(x) = \dfrac{1}{x + 3}, \quad a = -2$ **34.** $f(x) = \dfrac{x + 1}{x - 1}, \quad a = 3$

35. $f(t) = \dfrac{2}{1 - t}, \quad a = -1$ **36.** $f(t) = \dfrac{1}{t + 9}, \quad a = 2$

37. $f(t) = t^{-3}, \quad a = 1$ **38.** $f(t) = t^4, \quad a = 2$

39. $f(x) = \dfrac{1}{\sqrt{x}}, \quad a = 9$

40. $f(x) = (x^2 + 1)^{-1}, \quad a = 0$

41. What is an equation of the tangent line at $x = 3$, assuming that $f(3) = 5$ and $f'(3) = 2$?

42. Suppose that $y = 5x + 2$ is an equation of the tangent line to the graph of $y = f(x)$ at $a = 3$. What is $f(3)$? What is $f'(3)$?

43. Consider the "curve" $y = 2x + 8$. What is the tangent line at the point $(1, 10)$? Describe the tangent line at an arbitrary point.

44. Suppose that $f(x)$ is a function such that $f(2 + h) - f(2) = 3h^2 + 5h$.

(a) What is $f'(2)$?

(b) What is the slope of the secant line through $(2, f(2))$ and $(6, f(6))$?

45. $\boxed{\text{GU}}$ Verify that $P = \left(1, \tfrac{1}{2}\right)$ lies on the graphs of both $f(x) = \dfrac{1}{1 + x^2}$ and $L(x) = \tfrac{1}{2} + m(x - 1)$ for every slope m. Plot $f(x)$ and $L(x)$ on the same axes for several values of m. Experiment until you find a value of m for which $y = L(x)$ appears tangent to the graph of $f(x)$. What is your estimate for $f'(1)$?

46. $\boxed{\text{GU}}$ Plot $f(x) = x^x$ for $0 \le x \le 1.5$.

(a) How many points x_0 are there such that $f'(x_0) = 0$?

(b) Estimate x_0 by plotting several horizontal lines $y = c$ together with the plot of $f(x)$ until you find a value of c for which the line is tangent to the graph.

47. [GU]   Plot $f(x) = x^x$ and the line $y = x + c$ on the same set of axes for several values of c. Experiment until you find a value c_0 such that the line is tangent to the graph. Then:

(a) Use your plot to estimate the value x_0 such that $f'(x_0) = 1$.

(b) Verify that $\dfrac{f(x_0 + h) - f(x_0)}{h}$ is close to 1 for $h = 0.01, 0.001, 0.0001$.

48. Use the following table to calculate a few difference quotients of $f(x)$. Then estimate $f'(1)$ by taking an average of difference quotients at h and $-h$. For example, take the average for $h = 0.03$ and -0.03, etc. Recall that the average of a and b is $\dfrac{a+b}{2}$.

x	1.03	1.02	1.01	1
$f(x)$	0.5148	0.5234	0.5319	0.5403

x	0.97	0.98	0.99	
$f(x)$	0.5653	0.557	0.5486	

49. 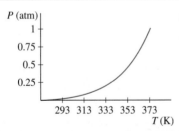   The vapor pressure of water is defined as the atmospheric pressure P at which no net evaporation takes place. The following table and Figure 13 give P (in atmospheres) as a function of temperature T in kelvins.

(a) Which is larger: $P'(300)$ or $P'(350)$? Answer by referring to the graph.

(b) Estimate $P'(T)$ for $T = 303, 313, 323, 333, 343$ using the table and the average of the difference quotients for $h = 10$ and -10:

$$P'(T) \approx \frac{P(T+10) - P(T-10)}{20} \qquad \boxed{4}$$

T (K)	293	303	313	323	333	343	353
P (atm)	0.0278	0.0482	0.0808	0.1311	0.2067	0.3173	0.4754

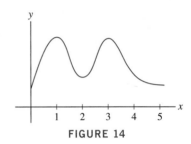

FIGURE 13 Vapor pressure of water as a function temperature T in kelvins.

In Exercises 50–51, traffic speed S along a certain road (in mph) varies as a function of traffic density q (number of cars per mile on the road). Use the following data to answer the questions:

q (density)	100	110	120	130	140
S (speed)	45	42	39.5	37	35

50. Estimate $S'(q)$ when $q = 120$ cars per mile using the average of difference quotients at h and $-h$ as in Exercise 48.

51. 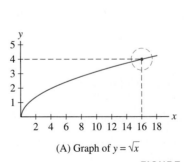 The quantity $V = qS$ is called *traffic volume*. Explain why V is equal to the number of cars passing a particular point per hour. Use the data to compute values of V as a function of q and estimate $V'(q)$ when $q = 120$.

52. For the graph in Figure 14, determine the intervals along the x-axis on which the derivative is positive.

FIGURE 14

In Exercises 53–58, each of the limits represents a derivative $f'(a)$. Find $f(x)$ and a.

53. $\displaystyle\lim_{h \to 0} \frac{(5+h)^3 - 125}{h}$

54. $\displaystyle\lim_{x \to 5} \frac{x^3 - 125}{x - 5}$

55. $\displaystyle\lim_{h \to 0} \frac{\sin(\frac{\pi}{6} + h) - .5}{h}$

56. $\displaystyle\lim_{x \to \frac{1}{4}} \frac{x^{-1} - 4}{x - \frac{1}{4}}$

57. $\displaystyle\lim_{h \to 0} \frac{5^{2+h} - 25}{h}$

58. $\displaystyle\lim_{h \to 0} \frac{5^h - 1}{h}$

59. Sketch the graph of $f(x) = \sin x$ on $[0, \pi]$ and guess the value of $f'\left(\frac{\pi}{2}\right)$. Then calculate the slope of the secant line between $x = \frac{\pi}{2}$ and $x = \frac{\pi}{2} + h$ for at least three small positive and negative values of h. Are these calculations consistent with your guess?

60. Figure 15(A) shows the graph of $f(x) = \sqrt{x}$. The close-up in (B) shows that the graph is nearly a straight line near $x = 16$. Estimate the slope of this line and take it as an estimate for $f'(16)$. Then compute $f'(16)$ and compare with your estimate.

(A) Graph of $y = \sqrt{x}$

(B) Zoom view near (16, 4)

FIGURE 15

61. GU   Let $f(x) = \dfrac{4}{1 + 2^x}$.

(a) Plot $f(x)$ over $[-2, 2]$. Then zoom in near $x = 0$ until the graph appears straight and estimate the slope $f'(0)$.

(b) Use your estimate to find an approximate equation to the tangent line at $x = 0$. Plot this line and the graph on the same set of axes.

62. GU   Repeat Exercise 61 with $f(x) = (2 + x)^x$.

63. GU   Let $f(x) = \cot x$. Estimate $f'(\frac{\pi}{2})$ graphically by zooming in on a plot of $f(x)$ near $x = \frac{\pi}{2}$.

64. ✎   For each graph in Figure 16, determine whether $f'(1)$ is larger or smaller than the slope of the secant line between $x = 1$ and $x = 1 + h$ for $h > 0$. Explain.

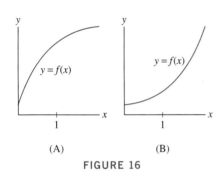

FIGURE 16

65. Apply the method of Example 6 to $f(x) = \sin x$ to determine $f'\left(\frac{\pi}{4}\right)$ accurately to four decimal places.

66. ✎   Apply the method of Example 6 to $f(x) = \cos x$ to determine $f'(\frac{\pi}{5})$ accurately to four decimal places. Use a graph of $f(x)$ to explain how the method works in this case.

67. GU   Sketch the graph of $f(x) = x^{5/2}$ on $[0, 6]$.

(a) Use the sketch to justify the inequalities for $h > 0$:

$$\frac{f(4) - f(4 - h)}{h} \le f'(4) \le \frac{f(4 + h) - f(4)}{h}$$

(b) Use part (a) to compute $f'(4)$ to four decimal places.

(c) Use a graphing utility to plot $f(x)$ and the tangent line at $x = 4$ using your estimate for $f'(4)$.

68. ✎   The graph of $f(x) = 2^x$ is shown in Figure 17.

(a) Referring to the graph, explain why the inequality

$$f'(0) \le \frac{f(h) - f(0)}{h}$$

holds for $h > 0$ and the opposite inequality holds for $h < 0$.

(b) Use part (a) to show that $0.69314 \le f'(0) \le 0.69315$.

(c) Use the same procedure to determine $f'(x)$ to four decimal places for $x = 1, 2, 3, 4$.

(d) Now compute the ratios $f'(x)/f'(0)$ for $x = 1, 2, 3, 4$. Can you guess an (approximate) formula for $f'(x)$ for general x?

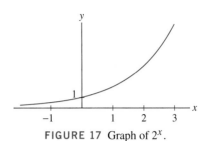

FIGURE 17 Graph of 2^x.

69. The graph in Figure 18 (based on data collected by the biologist Julian Huxley, 1887–1975) gives the average antler weight W of a red deer as a function of age t. Estimate the slope of the tangent line to the graph at $t = 4$. For which values of t is the slope of the tangent line equal to zero? For which values is it negative?

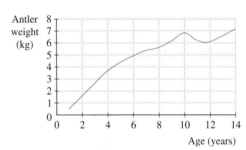

FIGURE 18

In Exercises 70–72, $i(t)$ is the current (in amperes) at time t (seconds) flowing in the circuit shown in Figure 19. According to Kirchhoff's law, $i(t) = Cv'(t) + R^{-1}v(t)$, where $v(t)$ is the voltage (in volts) at time t, C the capacitance (in farads), and R the resistance (in ohms, Ω).

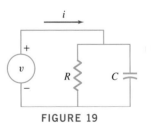

FIGURE 19

70. Calculate the current at $t = 3$ if $v(t) = 0.5t + 4$ V, $C = 0.01$ F, and $R = 100\ \Omega$.

71. Use the following table to estimate $v'(10)$. For a better estimate, take the *average* of the difference quotients for h and $-h$ as described in Exercise 48. Then estimate $i(10)$, assuming $C = 0.03$ and $R = 1,000$.

t	9.8	9.9	10	10.1	10.2
$v(t)$	256.52	257.32	258.11	258.9	259.69

72. Assume that $R = 200 \, \Omega$ but C is unknown. Use the following data to estimate $v'(4)$ as in Exercise 71 and deduce an approximate value for the capacitance C.

t	3.8	3.9	4	4.1	4.2
$v(t)$	388.8	404.2	420	436.2	452.8
$i(t)$	32.34	33.22	34.1	34.98	35.86

Further Insights and Challenges

In Exercises 73–76, we define the symmetric difference quotient (SDQ) at $x = a$ for $h \neq 0$ by

$$\frac{f(a+h) - f(a-h)}{2h} \qquad \boxed{5}$$

73. Explain how SDQ can be interpreted as the slope of a secant line.

74. The SDQ usually gives a better approximation to the derivative than does the ordinary difference quotient. Let $f(x) = 2^x$ and $a = 0$. Compute the SDQ with $h = 0.001$ and the ordinary difference quotients with $h = \pm 0.001$. Compare with the actual value of $f'(0)$ that, to eight decimal places, is 0.69314718.

75. ✎ Show that if $f(x)$ is a quadratic polynomial, then the SDQ at $x = a$ (for any $h \neq 0$) is *equal* to $f'(a)$. Explain the graphical meaning of this result.

76. Let $f(x) = x^{-2}$. Compute $f'(1)$ by taking the limit of the SDQs (with $a = 1$) as $h \to 0$.

77. Which of the two functions in Figure 20 satisfies the inequality

$$\frac{f(a+h) - f(a-h)}{2h} \leq \frac{f(a+h) - f(a)}{h}$$

for $h > 0$? Explain in terms of secant lines.

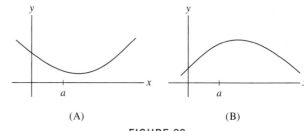

(A)   (B)

FIGURE 20

3.2 The Derivative as a Function

> *"I like to lie down on the sofa for hours at a stretch thinking intently about shapes, relationships and change—rarely about numbers as such. I explore idea after idea in my mind, discarding most. When a concept finally seems promising, I'm ready to try it out on paper. But first I get up and change the baby's diaper."*
>
> ——Charles Fefferman, professor of mathematics at Princeton University. He was appointed full professor at the University of Chicago at age 22, the youngest such appointment ever made in the United States.

Often, the domain of $f'(x)$ is clear from the context. If so, we usually do not mention the domain explicitly.

In the previous section, we computed the derivative $f'(a)$ for specific values of a. It is also useful to view the derivative as a function $f'(x)$ whose value at $x = a$ is $f'(a)$. The function $f'(x)$ is still defined as a limit, but the fixed number a is replaced by the variable x:

$$f'(x) = \lim_{h \to 0} \frac{f(x+h) - f(x)}{h} \qquad \boxed{1}$$

If $y = f(x)$, we also write y' or $y'(x)$ for $f'(x)$.

The domain of $f'(x)$ consists of all values of x in the domain of $f(x)$ for which the limit in Eq. (1) exists. We say that $f(x)$ is **differentiable** on (a, b) if $f'(x)$ exists for all x in (a, b). When $f'(x)$ exists for all x in the interval or intervals on which $f(x)$ is defined, we say simply that $f(x)$ is differentiable.

■ **EXAMPLE 1** Prove that $f(x) = x^3 - 12x$ is differentiable. Compute $f'(x)$ and find an equation of the tangent line at $x = -3$.

Solution We compute $f'(x)$ in three steps as in the previous section.

Step 1. **Write out the numerator of the difference quotient.**

$$f(x+h) - f(x) = \big((x+h)^3 - 12(x+h)\big) - \big(x^3 - 12x\big)$$

$$= (x^3 + 3x^2h + 3xh^2 + h^3 - 12x - 12h) - (x^3 - 12x)$$

$$= 3x^2h + 3xh^2 + h^3 - 12h$$
$$= h(3x^2 + 3xh + h^2 - 12) \quad \text{(factor out } h\text{)}$$

Step 2. **Divide by h and simplify the difference quotient.**

$$\frac{f(x+h) - f(x)}{h} = \frac{h(3x^2 + 3xh + h^2 - 12)}{h} = 3x^2 + 3xh + h^2 - 12 \quad (h \neq 0)$$

Step 3. **Compute the derivative by taking the limit.**

$$f'(x) = \lim_{h \to 0} \frac{f(x+h) - f(x)}{h} = \lim_{h \to 0} (3x^2 + 3xh + h^2 - 12) = 3x^2 - 12$$

In this limit, x is treated as a constant since it does not change as $h \to 0$. We see that the limit exists for all x, so $f(x)$ is differentiable and $f'(x) = 3x^2 - 12$.

Now evaluate:

$$f(-3) = (-3)^3 - 12(-3) = 9, \qquad f'(-3) = 3(-3)^2 - 12 = 15$$

An equation of the tangent line at $x = -3$ is $y - 9 = 15(x + 3)$ (Figure 1).   ■

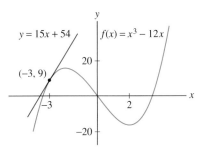

FIGURE 1 Graph of $f(x) = x^3 - 12x$.

■ **EXAMPLE 2** Prove that $y = x^{-2}$ is differentiable and calculate y'.

Solution The domain of $f(x) = x^{-2}$ is $\{x : x \neq 0\}$, so assume that $x \neq 0$. We compute $f'(x)$ directly, without the separate steps of the previous example:

$$y' = \lim_{h \to 0} \frac{f(x+h) - f(x)}{h} = \lim_{h \to 0} \frac{\dfrac{1}{(x+h)^2} - \dfrac{1}{x^2}}{h}$$

$$= \lim_{h \to 0} \frac{\dfrac{x^2 - (x+h)^2}{x^2(x+h)^2}}{h} = \lim_{h \to 0} \frac{1}{h}\left(\frac{x^2 - (x+h)^2}{x^2(x+h)^2}\right)$$

$$= \lim_{h \to 0} \frac{1}{h}\left(\frac{-h(2x+h)}{x^2(x+h)^2}\right) = \lim_{h \to 0} -\frac{2x+h}{x^2(x+h)^2} \quad \text{(cancel } h\text{)}$$

$$= -\frac{2x+0}{x^2(x+0)^2} = -\frac{2x}{x^4} = -\frac{2}{x^3}$$

The limit exists for all $x \neq 0$, so y is differentiable and $y' = -2x^{-3}$.   ■

Leibniz Notation

The "prime" notation y' and $f'(x)$ was introduced by the French mathematician Joseph Louis Lagrange (1736–1813). There is another standard notation for the derivative due to Leibniz (Figure 2):

$$\frac{df}{dx} \quad \text{or} \quad \frac{dy}{dx}$$

In Example 2, we showed that the derivative of $y = x^{-2}$ is $y' = -2x^{-3}$. In Leibniz notation, this result is written

$$\frac{dy}{dx} = -2x^{-3} \quad \text{or} \quad \frac{d}{dx}x^{-2} = -2x^{-3}$$

FIGURE 2 Gottfried Wilhelm von Leibniz (1646–1716), German philosopher and scientist. Newton and Leibniz (pronounced "Libe-nitz") are often regarded as the inventors of calculus (working independently of each other). It is more accurate to credit them with developing calculus into a general and fundamental discipline, since many particular results of calculus had been discovered previously.

To specify the value of the derivative for a fixed value of x, say, $x = 4$, we write

$$\left.\frac{df}{dx}\right|_{x=4} \quad \text{or} \quad \left.\frac{dy}{dx}\right|_{x=4}$$

You should not think of $\dfrac{dy}{dx}$ as the fraction "dy divided by dx." The expressions dy and dx are called **differentials**. They play an important role in more advanced mathematics, but we shall treat them merely as symbols with no independent meaning.

CONCEPTUAL INSIGHT   Leibniz notation is widely used for several reasons. First, it reminds us that the derivative $\dfrac{df}{dx}$, although not itself a ratio, is in fact a *limit* of ratios $\dfrac{\Delta f}{\Delta x}$. Second, the notation specifies the independent variable, which is useful when variables other than x are used. For example, if the independent variable is t, we write $\dfrac{df}{dt}$. Third, we often think of $\dfrac{d}{dx}$ as an "operator" that performs the differentiation operation on functions. In other words, we apply the operator $\dfrac{d}{dx}$ to f to obtain the derivative $\dfrac{df}{dx}$. We will see other advantages of Leibniz notation when we discuss the Chain Rule in Section 3.7.

One of the main goals of this chapter is to develop the basic rules of differentiation. These rules are particularly useful because they allow us to find derivatives without computing limits.

THEOREM 1 The Power Rule   For all exponents n,

$$\frac{d}{dx} x^n = n x^{n-1}$$

The Power Rule is valid for all exponents. Here, we prove it for a whole number n (see Exercise 97 for a negative integer n and p. 193 for arbitrary n).

Proof Assume that n is a whole number and let $f(x) = x^n$. Then

$$f'(a) = \lim_{x \to a} \frac{x^n - a^n}{x - a}$$

To simplify the difference quotient, we need to generalize the following identities:

$$x^2 - a^2 = (x - a)(x + a)$$
$$x^3 - a^3 = (x - a)(x^2 + xa + a^2)$$
$$x^4 - a^4 = (x - a)(x^3 + x^2a + xa^2 + a^3)$$

The generalization is

$$x^n - a^n = (x - a)(x^{n-1} + x^{n-2}a + x^{n-3}a^2 + \cdots + xa^{n-2} + a^{n-1}) \qquad \boxed{2}$$

To verify Eq. (2), observe that the right-hand side is equal to

$$x\left(x^{n-1} + x^{n-2}a + x^{n-3}a^2 + \cdots + xa^{n-2} + a^{n-1}\right)$$
$$- a\left(x^{n-1} + x^{n-2}a + x^{n-3}a^2 + \cdots + xa^{n-2} + a^{n-1}\right)$$

When we carry out the multiplications, all terms cancel except the first and the last, so $x^n - a^n$ remains, as required.

The identity (2) gives us

$$\frac{x^n - a^n}{x - a} = \underbrace{x^{n-1} + x^{n-2}a + x^{n-3}a^2 + \cdots + xa^{n-2} + a^{n-1}}_{n \text{ terms}} \quad (x \neq a) \qquad \boxed{3}$$

Therefore,

$$f'(a) = \lim_{x \to a} \left(x^{n-1} + x^{n-2}a + x^{n-3}a^2 + \cdots + xa^{n-2} + a^{n-1}\right)$$
$$= a^{n-1} + a^{n-2}a + a^{n-3}a^2 + \cdots + aa^{n-2} + a^{n-1} \quad (n \text{ terms})$$
$$= na^{n-1}$$

This proves that $f'(a) = na^{n-1}$, which we may also write as $f'(x) = nx^{n-1}$. ■

You can think of the Power Rule as "bringing down the exponent and subtracting one (from the exponent)":

$$\frac{d}{dx}x^{\text{exponent}} = (\text{exponent})\, x^{\text{exponent}-1}$$

The Power Rule is valid with any variable, not just x. Here are some examples:

$$\frac{d}{dx}x^2 = 2x, \qquad \frac{d}{dz}z^3 = 3z^2, \qquad \frac{d}{dt}t^{20} = 20t^{19}, \qquad \frac{d}{dr}r^5 = 5r^4$$

CAUTION *The Power Rule applies only to the power functions $y = x^n$. It does not apply to exponential functions such as $y = 2^x$. The derivative of $y = 2^x$ **is not** $x2^{x-1}$. We will study the derivatives of exponential functions later in this section.*

We also emphasize that the Power Rule holds for all exponents, whether negative, fractional, or irrational:

$$\frac{d}{dx}x^{-3/5} = -\frac{3}{5}x^{-8/5}, \qquad \frac{d}{dx}x^{\sqrt{2}} = \sqrt{2}\,x^{\sqrt{2}-1}$$

Next, we state the Linearity Rules for derivatives, which are analogous to the linearity laws for limits.

THEOREM 2 Linearity Rules Assume that f and g are differentiable functions.

Sum Rule: The function $f + g$ is differentiable and

$$(f + g)' = f' + g'$$

Constant Multiple Rule: For any constant c, cf is differentiable and

$$(cf)' = cf'$$

Proof By definition,

$$(f+g)'(x) = \lim_{h \to 0} \frac{(f(x+h)+g(x+h)) - (f(x)+g(x))}{h}$$

This difference quotient is equal to a sum:

$$\frac{(f(x+h)+g(x+h)) - (f(x)+g(x))}{h} = \frac{f(x+h)-f(x)}{h} + \frac{g(x+h)-g(x)}{h}$$

Therefore, by the Sum Law for limits,

$$(f+g)'(x) = \lim_{h \to 0} \frac{f(x+h)-f(x)}{h} + \lim_{h \to 0} \frac{g(x+h)-g(x)}{h}$$

$$= f'(x) + g'(x)$$

as claimed. The Constant Multiple Rule is proved similarly.   ■

The Difference Rule is also valid:

$$(f-g)' = f' - g'$$

However, it follows from the Sum and Constant Multiple Rules:

$$(f-g)' = \underbrace{f' + (-g)'}_{\text{Sum Rule}} = \underbrace{f' - g'}_{\text{Constant Multiple Rule}}$$

■ **EXAMPLE 3** Find the points on the graph of $f(t) = t^3 - 12t + 4$ where the tangent line is horizontal.

Solution First, we calculate the derivative:

$$\frac{df}{dt} = \frac{d}{dt}\left(t^3 - 12t + 4\right)$$

$$= \frac{d}{dt}t^3 - \frac{d}{dt}(12t) + \frac{d}{dt}4 \quad \text{(Sum and Difference Rules)}$$

$$= \frac{d}{dt}t^3 - 12\frac{d}{dt}t \quad\quad\quad \text{(Constant Multiple Rule)}$$

$$= 3t^2 - 12 \quad\quad\quad\quad\quad \text{(Power Rule)}$$

Notice that the derivative of the constant 4 is zero.

The tangent line is horizontal at points where the slope $f'(t)$ is zero, so we solve

$$f'(t) = 3t^2 - 12 = 0 \implies t = \pm 2$$

Therefore, the tangent lines are horizontal at $(2, -12)$ and $(-2, 20)$ (Figure 3).   ■

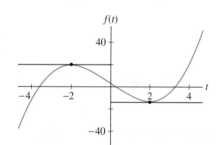

FIGURE 3 Graph of $f(t) = t^3 - 12t + 4$. Tangent lines at $t = \pm 2$ are horizontal.

In the next example, we differentiate term by term using the Power Rule without justifying the intermediate steps, as was done in Example 3.

■ **EXAMPLE 4** Calculate $\dfrac{dg}{dt}\bigg|_{t=1}$, where $g(t) = t^{-3} + 2\sqrt{t} - t^{-4/5}$.

Solution To use the Power Rule, we write $\sqrt{t}$ as $t^{1/2}$:

$$\frac{dg}{dt} = \frac{d}{dt}\left(t^{-3} + 2t^{1/2} - t^{-4/5}\right) = -3t^{-4} + 2\left(\frac{1}{2}\right)t^{-1/2} - \left(-\frac{4}{5}\right)t^{-9/5}$$

$$= -3t^{-4} + t^{-1/2} + \frac{4}{5}t^{-9/5}$$

$$\left.\frac{dg}{dt}\right|_{t=1} = -3 + 1 + \frac{4}{5} = -\frac{6}{5}$$ ∎

The derivative $f'(x)$ gives us important information about the graph of $f(x)$. For example, the sign of $f'(x)$ tells us whether the tangent line has positive or negative slope and the size of $f'(x)$ reveals how steep the slope is.

■ **EXAMPLE 5** Graphical Insight How is the graph of $f(x) = x^3 - 12x$ related to the derivative $f'(x) = 3x^2 - 12$?

Solution We observe that the derivative $f'(x) = 3x^2 - 12 = 3(x^2 - 4)$ is negative for $-2 < x < 2$ and positive for $|x| > 2$ [see Figure 4(B)]. The following table summarizes this sign information [see Figure 4(A)]:

Property of $f'(x)$	Property of the Graph of $f(x)$
$f'(x) < 0$ for $-2 < x < 2$	Tangent line has negative slope for $-2 < x < 2$
$f'(-2) = f'(2) = 0$	Tangent line is horizontal at $x = -2$ and $x = 2$
$f'(x) > 0$ for $x < -2$ and $x > 2$	Tangent line has positive slope for $x < -2$ and $x > 2$

Note also that $f'(x) \to \infty$ as $|x|$ becomes large. This corresponds to the fact that the tangent lines to the graph of $f(x)$ get steeper as $|x|$ grows large. ∎

■ **EXAMPLE 6** Identifying the Derivative The graph of $f(x)$ is shown in Figure 5. Which of (A) or (B) is the graph of $f'(x)$?

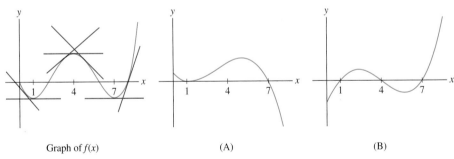

Graph of $f(x)$ (A)   (B)

FIGURE 5

Solution From Figure 5 we glean the following information about the tangent lines to the graph of $f(x)$:

Slope of Tangent Line	Where
Negative	on $(0, 1)$ and $(4, 7)$
Zero	for $x = 1, 4, 7$
Positive	on $(1, 4)$ and $(7, \infty)$

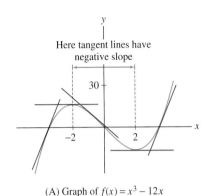

Here tangent lines have negative slope

(A) Graph of $f(x) = x^3 - 12x$

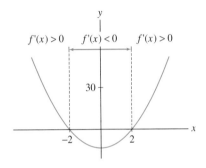

$f'(x) > 0$ $f'(x) < 0$ $f'(x) > 0$

(B) Graph of the derivative $f'(x) = 3x^2 - 12$

FIGURE 4

Therefore, the graph of $f'(x)$ must be negative on $(0, 1)$ and $(4, 7)$ and positive on $(1, 4)$ and $(7, \infty)$. Only (B) has these properties, so (B) is the graph of $f'(x)$. ∎

The Derivative of e^x

The number e and the exponential function e^x were introduced informally in Section 1.6. We now define them more precisely using the derivative. First, let us compute the derivative of $f(x) = b^x$, where $b > 0$ and $b \neq 1$. The difference quotient (for $h \neq 0$) is

$$\frac{f(x+h) - f(x)}{h} = \frac{b^{x+h} - b^x}{h} = \frac{b^x b^h - b^x}{h} = \frac{b^x(b^h - 1)}{h}$$

and the derivative is equal to the limit:

$$f'(x) = \lim_{h \to 0} \frac{f(x+h) - f(x)}{h} = \lim_{h \to 0} \frac{b^x(b^h - 1)}{h} = b^x \lim_{h \to 0} \left(\frac{b^h - 1}{h} \right) \quad \boxed{4}$$

Notice that we took the factor b^x outside the limit. This is legitimate because b^x does not depend on h.

The last limit on the right in Eq. (4) does not depend on x. We assume it exists and denote its value by m_b:

$$m_b = \lim_{h \to 0} \left(\frac{b^h - 1}{h} \right) \quad \boxed{5}$$

What we have shown, then, is that *the derivative of b^x is proportional to b^x*:

$$\boxed{\frac{d}{dx} b^x = m_b\, b^x} \quad \boxed{6}$$

What is m_b? Note, first of all, that m_b has a graphical interpretation as the slope of the tangent line to $y = b^x$ at $x = 0$ (Figure 6) since

$$\left. \frac{d}{dx} b^x \right|_{x=0} = m_b \cdot b^0 = m_b$$

In Section 3.10, we will show that $m_b = \ln b$, the natural logarithm of b. Without assuming this, we may investigate m_b numerically using Eq. (5).

■ **EXAMPLE 7** Investigating m_b Numerically Investigate m_b numerically for $b = 2$, 2.5, 3, and 10.

Solution We create a table of values to estimate m_b numerically:

h	$\dfrac{2^h - 1}{h}$	$\dfrac{(2.5)^h - 1}{h}$	$\dfrac{3^h - 1}{h}$	$\dfrac{10^h - 1}{h}$
0.01	0.69556	0.92050	1.10467	2.32930
0.001	0.69339	0.91671	1.09921	2.30524
0.0001	0.69317	0.91633	1.09867	2.30285
0.00001	0.69315	0.916295	1.09861	2.30261
	$\boxed{m_2 \approx 0.693}$	$\boxed{m_{2.5} \approx 0.916}$	$\boxed{m_3 \approx 1.10}$	$\boxed{m_{10} \approx 2.30}$

■

These computations suggest that the numbers m_b are increasing:

$$m_2 \quad < \quad m_{2.5} \quad < \quad m_3 \quad < \quad m_{10}$$

Let us assume, as is reasonable, that m_b varies *continuously* as a function of b. Then, since $m_2 < 1$ and $m_3 > 1$, the Intermediate Value Theorem says that $m_b = 1$ for some number b between 2 and 3. This number is the number e. As mentioned in Section 1.6, e is irrational. To 40 places,

$$e = 2.7182818284590452353602874713526624977572\ldots$$

For most purposes, the approximation $e \approx 2.718$ is adequate.

Since e is defined by the property $m_e = 1$, Eq. (6) says that $(e^x)' = e^x$. In other words, $y = e^x$ is equal to its own derivative.

Although written reference to the number π goes back more than 4,000 years, mathematicians first became aware of the special role played by e in the seventeenth century. The notation e was introduced around 1730 by Leonhard Euler, who discovered many fundamental properties of this important number.

In many books, e^x is denoted $\exp(x)$. Because of property (7), e^x plays a greater role than the seemingly more natural exponential functions such as 2^x or 10^x. We refer to $f(x) = e^x$ as the exponential function.

> **The Number e**   There is a unique positive real number e ($e \approx 2.718$) with the property
>
> $$\frac{d}{dx}e^x = e^x$$
>
> $\boxed{7}$

GRAPHICAL INSIGHT   All exponential curves $y = b^x$ pass through $(0, 1)$. For $b > 1$, their tangent lines at $(0, 1)$, of slope m_b, become steeper as b increases (Figure 6). By definition, e is the unique base for which the slope m_e of the tangent line at $(0, 1)$ is 1.

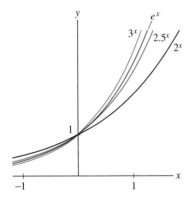

FIGURE 6 The tangent lines to $y = b^x$ at $x = 0$ grow steeper as b increases.

■ **EXAMPLE 8**   Find the equation of the tangent line to the graph of $f(x) = 3e^x - 5x^2$ at $x = 2$.

Solution   We compute the derivative $f'(2)$ and the function value $f(2)$:

$$f'(x) = \frac{d}{dx}(3e^x - 5x^2) = 3\frac{d}{dx}e^x - 5\frac{d}{dx}x^2 = 3e^x - 10x$$

$$f'(2) = 3e^2 - 10(2) \approx 2.17$$

$$f(2) = 3e^2 - 5(2^2) \approx 2.17$$

The equation of the tangent line is $y = f(2) + f'(2)(x - 2)$ (Figure 7). Using the approximate values, we write the equation as

$$y = 2.17 + 2.17(x - 2) \quad \text{or} \quad y = 2.17(x - 1) \qquad ■$$

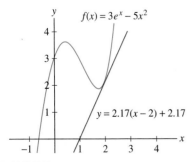

FIGURE 7

CONCEPTUAL INSIGHT   How is b^x defined when x is not a rational number? If n is a whole number, then b^n is simply the product $b \cdot b \cdots b$ (n times), and for any rational number $x = m/n$,

$$b^x = b^{m/n} = \left(b^{1/n}\right)^m = \left(\sqrt[n]{b}\right)^m$$

When x is irrational, this definition does not apply and b^x cannot be defined directly in terms of roots and powers of b. However, it makes sense to view $b^{m/n}$ as an approximation to b^x when m/n is a rational number close to x. For example, $3^{\sqrt{2}}$ should be approximately equal to $3^{1.4142} \approx 4.729$ since 1.4142 is a good rational approximation to $\sqrt{2}$. Formally, then, we may define b^x as a limit over rational numbers m/n approaching x:

$$b^x = \lim_{m/n \to x} b^{m/n}$$

It can be shown that this limit exists and the function $f(x) = b^x$ thus defined is differentiable.

Differentiability, Continuity, and Local Linearity

In the rest of this section, we examine the concept of **differentiability** more closely. We begin by proving that a differentiable function is necessarily continuous. In other words, $f'(c)$ does not exist if f has a discontinuity at $x = c$. Figure 8 shows why in the case of a jump discontinuity: Although the secant lines from the right approach the line L (which is tangent to the right half of the graph), the secant lines from the left approach the vertical (and their slopes tend to ∞).

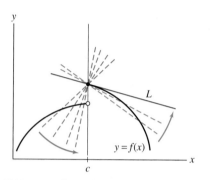

FIGURE 8 Secant lines at a jump discontinuity.

THEOREM 3 Differentiability Implies Continuity   If f is differentiable at $x = c$, then f is continuous at $x = c$.

Proof By definition, if f is differentiable at $x = c$, then the following limit exists:

$$f'(c) = \lim_{x \to c} \frac{f(x) - f(c)}{x - c}$$

Our goal is to prove that f is continuous at $x = c$, which means that $\lim_{x \to c} f(x) = f(c)$. To relate the two limits, consider the equation (valid for $x \neq c$)

$$f(x) - f(c) = (x - c)\frac{f(x) - f(c)}{x - c}$$

Both factors on the right approach a limit as $x \to c$, so we may apply the Limit Laws:

$$\lim_{x \to c}\big(f(x) - f(c)\big) = \lim_{x \to c}\left((x - c)\frac{f(x) - f(c)}{x - c}\right)$$

$$= \left(\lim_{x \to c}(x - c)\right)\left(\lim_{x \to c}\frac{f(x) - f(c)}{x - c}\right)$$

$$= 0 \cdot f'(c) = 0$$

Therefore, $\lim_{x \to c} f(x) = \lim_{x \to c}(f(x) - f(c)) + \lim_{x \to c} f(c) = 0 + f(c) = f(c)$, as desired. ∎

Most of the functions encountered in this text are differentiable, but exceptions exist, as the next example shows.

Theorem 3 states that differentiability implies continuity. Example 9 shows that the converse is false: Continuity does not automatically imply differentiability.

■ **EXAMPLE 9 Continuous But Not Differentiable** Show that $f(x) = |x|$ is continuous but not differentiable at $x = 0$.

Solution The function $f(x)$ is continuous at $x = 0$ since $\lim_{x \to 0} |x| = 0 = f(0)$. On the other hand,

$$f'(0) = \lim_{h \to 0} \frac{f(0 + h) - f(0)}{h} = \lim_{h \to 0} \frac{|0 + h| - |0|}{h} = \lim_{h \to 0} \frac{|h|}{h}$$

This limit does not exist [and hence $f(x)$ is not differentiable at $x = 0$] because

$$\frac{|h|}{h} = \begin{cases} 1 & \text{if } h > 0 \\ -1 & \text{if } h < 0 \end{cases}$$

and the one-sided limits are not equal:

$$\lim_{h \to 0+} \frac{|h|}{h} = 1 \quad \text{and} \quad \lim_{h \to 0-} \frac{|h|}{h} = -1 \qquad \blacksquare$$

GRAPHICAL INSIGHT Differentiability has an important graphical interpretation in terms of local linearity. We say that f is **locally linear** at $x = a$ if the graph looks more and more like a straight line as we zoom in on the point $(a, f(a))$. In this context, the adjective *linear* means "resembling a line," and *local* indicates that we are concerned only with the behavior of the graph near $(a, f(a))$. The graph of a locally linear function may be very wavy or *nonlinear*, as in Figure 9. But as soon as we zoom in on a sufficiently small piece of the graph, it begins to appear straight.

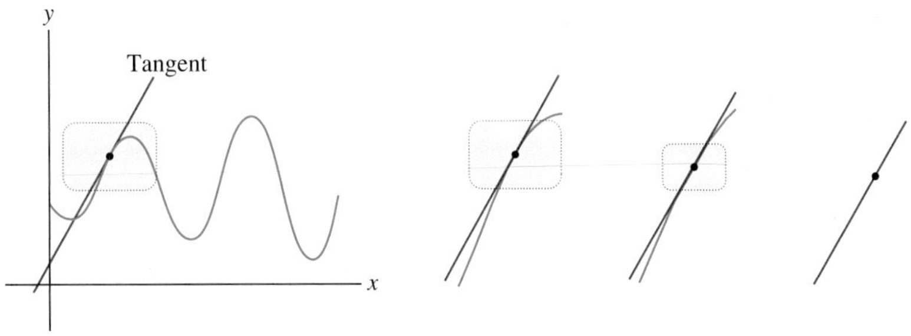

FIGURE 9 As we zoom in on a point, the graph looks more and more like the tangent line.

Not only does the graph look like a line as we zoom in on a point, but as Figure 9 suggests, the "zoom line" is the tangent line. Thus, the relation between differentiability and local linearity can be expressed as follows:

If $f'(a)$ exists, then f is locally linear at $x = a$ in the following sense: As we zoom in on the point $(a, f(a))$, the graph becomes nearly indistinguishable from its tangent line.

Local linearity gives us a graphical way to understand why $f(x) = |x|$ is not differentiable at $x = 0$, as we proved in Example 9. Figure 10 shows that the graph of $f(x) = |x|$ has a corner at $x = 0$. Furthermore, this corner *does not disappear*, no matter how closely we zoom in on the origin. Since the graph does not straighten out under zooming, $f(x)$ is not locally linear at $x = 0$, and we cannot expect $f'(0)$ to exist.

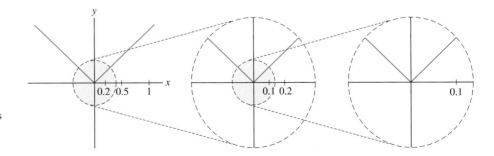

FIGURE 10 The graph of $f(x) = |x|$ is not locally linear at $x = 0$. The corner does not disappear when we zoom in on the origin.

Another way that a continuous function can fail to be differentiable is if the tangent line exists but is vertical, in which case the tangent line has infinite slope.

■ **EXAMPLE 10** Vertical Tangents Show that $f(x) = x^{1/3}$ is not differentiable at $x = 0$ (Figure 11).

Solution The limit defining $f'(0)$ is infinite:

$$\lim_{h \to 0} \frac{f(h) - f(0)}{h} = \lim_{h \to 0} \frac{h^{1/3} - 0}{h} = \lim_{h \to 0} \frac{h^{1/3}}{h} = \lim_{h \to 0} \frac{1}{h^{2/3}} = \infty$$

Note that $h^{-2/3}$ tends to infinity because $h^{2/3}$ tends to zero through positive values as $h \to 0$. Therefore $f'(0)$ does not exist. The tangent line has infinite slope.   ■

As a final remark, we mention that there are more complicated ways in which a continuous function can fail to be differentiable. Figure 12 shows the graph of $f(x) = x \sin \frac{1}{x}$. If we define $f(0) = 0$, then f is continuous but not differentiable at $x = 0$. The secant lines keep oscillating and never settle down to a limiting position (see Exercise 99).

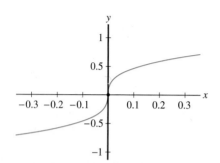

FIGURE 11 The tangent line to $f(x) = x^{1/3}$ at the origin is the (vertical) y-axis.

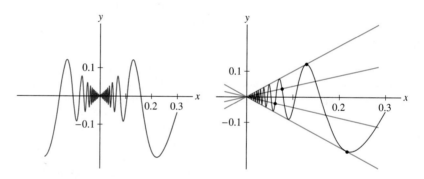

(A) Graph of $f(x) = x \sin \frac{1}{x}$

(B) Secant lines do not settle down to a limiting position.

FIGURE 12

3.2 SUMMARY

- The derivative $f'(x)$ is the function whose value at $x = a$ is the derivative $f'(a)$.
- We have several different notations for the derivative of $y = f(x)$:

$$y', \quad y'(x), \quad f'(x), \quad \frac{dy}{dx}, \quad \frac{df}{dx}$$

The value of the derivative at $x = a$ is written

$$y'(a), \quad f'(a), \quad \frac{dy}{dx}\bigg|_{x=a}, \quad \frac{df}{dx}\bigg|_{x=a}$$

- The Power Rule holds for all exponents n:

$$\frac{d}{dx}x^n = nx^{n-1} \qquad \text{or} \qquad \frac{d}{dx}x^{\text{exponent}} = (\text{exponent})\, x^{\text{exponent}-1}$$

- The Linearity Rules allow us to differentiate term by term:

  Sum Rule: $(f + g)' = f' + g'$, Constant Multiple Rule: $(cf)' = cf'$

- The derivative of b^x is proportional to b^x: $\dfrac{d}{dx} b^x = m_b b^x$, where $m_b = \lim_{h \to 0} \dfrac{b^h - 1}{h}$.

- The number $e \approx 2.718$ is defined by the property $m_e = 1$, so that $\dfrac{d}{dx} e^x = e^x$.

- Differentiability implies continuity: If $f(x)$ is differentiable at $x = a$, then $f(x)$ is continuous at $x = a$. However, there exist continuous functions that are not differentiable.

- If $f'(a)$ exists, then f is locally linear in the following sense: As we zoom in on the point $(a, f(a))$, the graph becomes nearly indistinguishable from its tangent line.

3.2 EXERCISES

Preliminary Questions

1. What is the slope of the tangent line through the point $(2, f(2))$ if f is a function such that $f'(x) = x^3$?

2. Evaluate $(f - g)'(1)$ and $(3f + 2g)'(1)$ assuming that $f'(1) = 3$ and $g'(1) = 5$. Can we evaluate $(fg)'(1)$ using the information given and the rules presented in this section?

3. To which of the following does the Power Rule apply?

(a) $f(x) = x^2$
(b) $f(x) = 2^e$
(c) $f(x) = x^e$
(d) $f(x) = e^x$
(e) $f(x) = x^x$
(f) $f(x) = x^{-4/5}$

4. Which algebraic identity is used to prove the Power Rule for positive integer exponents? Explain how it is used.

5. Does the Power Rule apply to $f(x) = \sqrt[5]{x}$? Explain.

6. In which of the following two cases does the derivative not exist?

(a) Horizontal tangent
(b) Vertical tangent

7. Which property distinguishes $f(x) = e^x$ from all other exponential functions $g(x) = b^x$?

8. What is the slope of the tangent line to $y = e^x$ at $x = 0$? What about $y = 7e^x$?

Exercises

In Exercises 1–8, compute $f'(x)$ using the limit definition.

1. $f(x) = 4x - 3$

2. $f(x) = x^2 + x$

3. $f(x) = 1 - 2x^2$

4. $f(x) = x^3$

5. $f(x) = x^{-1}$

6. $f(x) = \dfrac{x}{x - 1}$

7. $f(x) = \sqrt{x}$

8. $f(x) = x^{-1/2}$

In Exercises 9–16, use the Power Rule to compute the derivative.

9. $\dfrac{d}{dx} x^4 \Big|_{x=-2}$

10. $\dfrac{d}{dx} x^{-4} \Big|_{x=3}$

11. $\dfrac{d}{dt} t^{2/3} \Big|_{t=8}$

12. $\dfrac{d}{dt} t^{-2/3} \Big|_{t=1}$

13. $\dfrac{d}{dx} x^{0.35}$

14. $\dfrac{d}{dx} x^{14/3}$

15. $\dfrac{d}{dt} t^{\sqrt{17}}$

16. $\dfrac{d}{dt} t^{-\pi^2}$

In Exercises 17–20, compute $f'(a)$ and find an equation of the tangent line to the graph at $x = a$.

17. $f(x) = x^5$, $a = 1$

18. $f(x) = x^{-2}$, $a = 3$

19. $f(x) = 3\sqrt{x} + 8x$, $a = 9$

20. $f(x) = \sqrt[3]{x}$, $a = 8$

21. Calculate:

(a) $\dfrac{d}{dx} 9e^x$
(b) $\dfrac{d}{dt}(3t - 4e^t)$
(c) $\dfrac{d}{dt} e^{t+2}$

Hint for (c): Write e^{t+2} as $e^2 e^t$.

22. Find the equation of the tangent line to $y = 4e^x$ at $x = 2$.

In Exercises 23–38, calculate the derivative of the function.

23. $f(x) = x^3 + x^2 - 12$

24. $f(x) = 2x^3 - 3x^2 + 2x$

25. $f(x) = 2x^3 - 10x^{-1}$

26. $f(x) = x^5 - 7x^2 + 10x + 9$

27. $g(z) = 7z^{-3} + z^2 + 5$

28. $h(t) = 6\sqrt{t} + \dfrac{1}{\sqrt{t}}$

29. $f(s) = \sqrt[4]{s} + \sqrt[3]{s}$

30. $W(y) = 6y^4 + 7y^{2/3}$

31. $f(x) = (x + 1)^3$ (*Hint:* Expand)

32. $R(s) = (5s + 1)^2$

33. $P(z) = (3z - 1)(2z + 1)$

34. $q(t) = \sqrt{t}(t + 1)$

35. $g(x) = e^2$

36. $f(x) = 3e^x - x^3$

37. $h(t) = 5e^{t-3}$

38. $f(x) = 9 + 2e^x$

In Exercises 39–46, calculate the derivative indicated.

39. $f'(2), \quad f(x) = \dfrac{3}{x^4}$

40. $y'(16), \quad y = \dfrac{\sqrt{x}+1}{x}$

41. $\left.\dfrac{dT}{dC}\right|_{C=8}, \quad T = 3C^{2/3}$

42. $\left.\dfrac{dP}{dV}\right|_{V=-2}, \quad P = \dfrac{7}{V}$

43. $\left.\dfrac{ds}{dz}\right|_{z=2}, \quad s = 4z - 16z^2$

44. $\left.\dfrac{dR}{dW}\right|_{W=1}, \quad R = W^{\pi}$

45. $\left.\dfrac{dp}{dh}\right|_{h=4}, \quad p = 7e^{h-2}$

46. $\left.\dfrac{dy}{dt}\right|_{t=4}, \quad y = t - e^t$

47. Match the functions in graphs (A)–(D) with their derivatives (I)–(III) in Figure 13. Note that two of the functions have the same derivative. Explain why.

(A)   (B)   (C)   (D)

(I)   (II)   (III)

FIGURE 13

48. Assign the labels $f(x)$, $g(x)$, and $h(x)$ to the graphs in Figure 14 in such a way that $f'(x) = g(x)$ and $g'(x) = h(x)$.

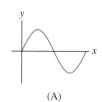 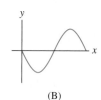 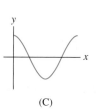

(A)   (B)   (C)

FIGURE 14

49. Sketch the graph of $f'(x)$ for $f(x)$ as in Figure 15, omitting points where $f(x)$ is not differentiable.

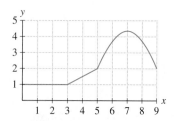

FIGURE 15 Graph of $f(x)$.

50. 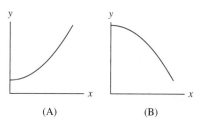 The table below lists values of a function $f(x)$. Does $f'(x)$ appear to increase or decrease as a function of x? Is (A) or (B) in Figure 16 the graph of $f'(x)$? Explain.

x	0	0.5	1	1.5	2	2.5	3	3.5	4
$f(x)$	10	55	98	139	177	210	237	257	268

FIGURE 16 Which is the graph of $f'(x)$?

51. Let R be a variable and r a constant. Compute the derivatives

(a) $\dfrac{d}{dR} R$

(b) $\dfrac{d}{dR} r$

(c) $\dfrac{d}{dR} r^2 R^3$

52. Sketch the graph of $f(x) = x - 3x^2$ and find the values of x for which the tangent line is horizontal.

53. Find the points on the curve $y = x^2 + 3x - 7$ at which the slope of the tangent line is equal to 4.

54. Sketch the graphs of $f(x) = x^2 - 5x + 4$ and $g(x) = -2x + 3$. Find the value of x at which the graphs have parallel tangent lines.

55. Find all values of x where the tangent lines to $y = x^3$ and $y = x^4$ are parallel.

56. Show that there is a unique point on the graph of the function $f(x) = ax^2 + bx + c$ where the tangent line is horizontal (assume $a \neq 0$). Explain graphically.

57. Determine coefficients a and b such that $p(x) = x^2 + ax + b$ satisfies $p(1) = 0$ and $p'(1) = 4$.

58. Find all values of x such that the tangent line to the graph of $y = 4x^2 + 11x + 2$ is steeper than the tangent line to $y = x^3$.

59. Let $f(x) = x^3 - 3x + 1$. Show that $f'(x) \geq -3$ for all x, and that for every $m > -3$, there are precisely two points where $f'(x) = m$. Indicate the position of these points and the corresponding tangent lines for one value of m in a sketch of the graph of $f(x)$.

60. Show that if the tangent lines to the graph of $y = \frac{1}{3}x^3 - x^2$ at $x = a$ and at $x = b$ are parallel, then either $a = b$ or $a + b = 2$.

61. Compute the derivative of $f(x) = x^{-2}$ using the limit definition. *Hint:* Show that

$$\frac{f(x+h) - f(x)}{h} = -\frac{1}{x^2(x+h)^2} \cdot \frac{(x+h)^2 - x^2}{h}$$

62. Compute the derivative of $f(x) = x^{3/2}$ using the limit definition. *Hint:* Show that

$$\frac{f(x+h) - f(x)}{h} = \frac{(x+h)^3 - x^3}{h} \left(\frac{1}{\sqrt{(x+h)^3} + \sqrt{x^3}} \right)$$

63. Find an approximation to m_4 using the limit definition and estimate the slope of the tangent line to $y = 4^x$ at $x = 0$ and $x = 2$.

64. Let $f(x) = xe^x$. Use the limit definition to compute $f'(0)$ and find the equation of the tangent line at $x = 0$.

65. The average speed (in meters per second) of a gas molecule is $v_{\text{avg}} = \sqrt{8RT/(\pi M)}$, where T is the temperature (in kelvin), M is the molar mass (kg/mol) and $R = 8.31$. Calculate dv_{avg}/dT at $T = 300$ K for oxygen, which has a molar mass of 0.032 kg/mol.

66. Biologists have observed that the pulse rate P (in beats per minute) in animals is related to body mass (in kilograms) by the approximate formula $P = 200m^{-1/4}$. This is one of many *allometric scaling laws* prevalent in biology. Is the absolute value $|dP/dm|$ increasing or decreasing as m increases? Find an equation of the tangent line at the points on the graph in Figure 17 that represent goat ($m = 33$) and man ($m = 68$).

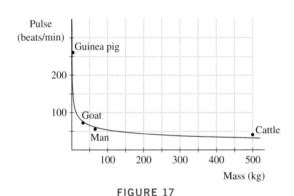

FIGURE 17

67. Some studies suggest that kidney mass K in mammals (in kilograms) is related to body mass m (in kilograms) by the approximate formula $K = 0.007m^{0.85}$. Calculate dK/dm at $m = 68$. Then calculate the derivative with respect to m of the relative kidney-to-mass ratio K/m at $m = 68$.

68. The relation between the *vapor pressure* P (in atmospheres) of water and the temperature T (in kelvin) is given by the Clausius–Clapeyron law:

$$\frac{dP}{dT} = k\frac{P}{T^2}$$

where k is a constant. Use the table below and the approximation

$$\frac{dP}{dT} \approx \frac{P(T+10) - P(T-10)}{20}$$

to estimate dP/dT for $T = 303, 313, 323, 333, 343$. Do your estimates seem to confirm the Clausius–Clapeyron law? What is the approximate value of k? What are the units of k?

T	293	303	313	323	333	343	353
P	0.0278	0.0482	0.0808	0.1311	0.2067	0.3173	0.4754

69. Let L be a tangent line to the hyperbola $xy = 1$ at $x = a$, where $a > 0$. Show that the area of the triangle bounded by L and the coordinate axes does not depend on a.

70. In the notation of Exercise 69, show that the point of tangency is the midpoint of the segment of L lying in the first quadrant.

71. Match the functions (A)–(C) with their derivatives (I)–(III) in Figure 18.

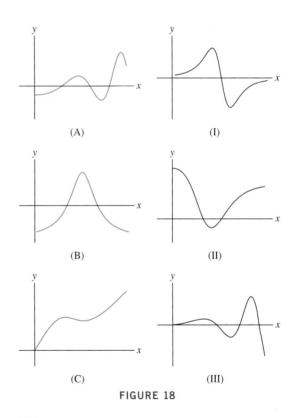

FIGURE 18

72. GU   Plot the derivative $f'(x)$ of $f(x) = 2x^3 - 10x^{-1}$ for $x > 0$ (set the bounds of the viewing box appropriately) and observe that $f'(x) > 0$. What does the positivity of $f'(x)$ tell us about the graph of $f(x)$ itself? Plot $f(x)$ and confirm.

73. Make a rough sketch of the graph of the derivative of the function shown in Figure 19(A).

74. Graph the derivative of the function shown in Figure 19(B), omitting points where the derivative is not defined.

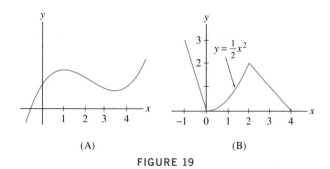

(A)   (B)

FIGURE 19

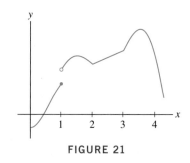

FIGURE 21

75. ✎ Of the two functions f and g in Figure 20, which is the derivative of the other? Justify your answer.

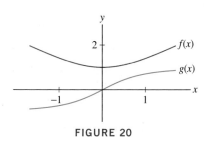

FIGURE 20

76. At which points is the function in Figure 21 discontinuous? At which points is it nondifferentiable?

In Exercises 77–82, find the points c (if any) such that $f'(c)$ does not exist.

77. $f(x) = |x - 1|$

78. $f(x) = [x]$

79. $f(x) = x^{2/3}$

80. $f(x) = x^{3/2}$

81. $f(x) = |x^2 - 1|$

82. $f(x) = |x - 1|^2$

GU   *In Exercises 83–88, zoom in on a plot of $f(x)$ at the point $(a, f(a))$ and state whether or not $f(x)$ appears to be differentiable at $x = a$. If nondifferentiable, state whether the tangent line appears to be vertical or does not exist.*

83. $f(x) = (x - 1)|x|$, $a = 0$

84. $f(x) = (x - 3)^{5/3}$, $a = 3$

85. $f(x) = (x - 3)^{1/3}$, $a = 3$

86. $f(x) = \sin(x^{1/3})$, $a = 0$

87. $f(x) = |\sin x|$, $a = 0$

88. $f(x) = |x - \sin x|$, $a = 0$

89. ✎ Is it true that a nondifferentiable function is not continuous? If not, give a counterexample.

90. Sketch the graph of $y = x\,|x|$ and show that it is differentiable for all x (check differentiability separately for $x < 0$, $x = 0$, and $x > 0$).

Further Insights and Challenges

91. Prove the following theorem of Apollonius of Perga (the Greek mathematician born in 262 BCE who gave the parabola, ellipse, and hyperbola their names): The tangent to the parabola $y = x^2$ at $x = a$ intersects the x-axis at the midpoint between the origin and $(a, 0)$. Draw a diagram.

92. ✎ Apollonius's Theorem in Exercise 91 can be generalized. Show that the tangent to $y = x^3$ at $x = a$ intersects the x-axis at the point $x = \frac{2}{3}a$. Then formulate the general statement for the graph of $y = x^n$ and prove it.

93. CAS   Plot the graph of $f(x) = (4 - x^{2/3})^{3/2}$ (the "astroid"). Let L be a tangent line to a point on the graph in the first quadrant. Use a computer algebra system to compute $f'(x)$ and show that the portion of L in the first quadrant has a constant length 8.

94. Two small arches have the shape of parabolas. The first is given by $f(x) = 1 - x^2$ for $-1 \le x \le 1$ and the second by $g(x) = 4 - (x - 4)^2$ for $2 \le x \le 6$. A board is placed on top of these arches

so it rests on both (Figure 22). What is the slope of the board? *Hint:* Find the tangent line to $y = f(x)$ that intersects $y = g(x)$ in exactly one point.

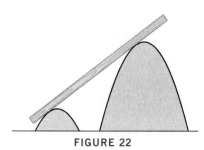

FIGURE 22

95. A vase is formed by rotating $y = x^2$ around the y-axis. If we drop in a marble, it will either touch the bottom point of the vase or be suspended above the bottom by touching the sides (Figure 23). How small must the marble be to touch the bottom?

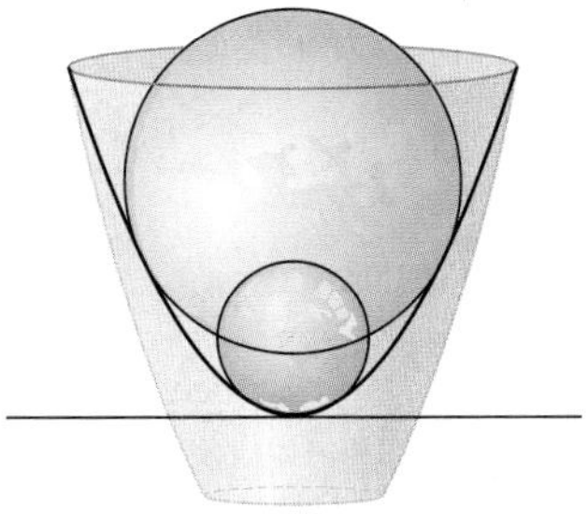

FIGURE 23

96. Let $f(x)$ be a differentiable function and set $g(x) = f(x+c)$, where c is a constant. Use the limit definition to show that $g'(x) = f'(x+c)$. Explain this result graphically, recalling that the graph of $g(x)$ is obtained by shifting the graph of $f(x)$ c units to the left (if $c > 0$) or right (if $c < 0$).

97. Negative Exponents Let n be a whole number. Use the Power Rule for x^n to calculate the derivative of $f(x) = x^{-n}$ by showing that

$$\frac{f(x+h) - f(x)}{h} = \frac{-1}{x^n(x+h)^n} \frac{(x+h)^n - x^n}{h}$$

98. Verify the Power Rule for the exponent $1/n$, where n is a positive integer, using the following trick: Rewrite the difference quotient for $y = x^{1/n}$ at $x = b$ in terms of $u = (b+h)^{1/n}$ and $a = b^{1/n}$.

99. Infinitely Rapid Oscillations Define

$$f(x) = \begin{cases} x \sin \dfrac{1}{x} & \text{if } x \neq 0 \\ 0 & \text{if } x = 0 \end{cases}$$

Show that $f(x)$ is continuous at $x = 0$ but $f'(0)$ does not exist (see Figure 12).

100. Prove that $f(x) = e^x$ is not a polynomial function. *Hint:* Differentiation lowers the degree of a polynomial by 1.

101. Consider the equation $e^x = \lambda x$, where λ is a constant.

(a) For which λ does it have a unique solution? For intuition, draw a graph of $y = e^x$ and the line $y = \lambda x$.

(b) For which λ does it have at least one solution?

3.3 Product and Quotient Rules

One of the main goals of this chapter is to develop efficient techniques for computing derivatives. This section covers the **Product Rule** and **Quotient Rule**. These two rules together with the Chain Rule and implicit differentiation (covered in later sections) make up an extremely effective "differentiation toolkit."

THEOREM 1 Product Rule If f and g are differentiable functions, then fg is differentiable and

$$(fg)'(x) = f(x)\,g'(x) + g(x)\,f'(x)$$

f·g' + f'·g

← *REMINDER The product function fg is defined by $(fg)(x) = f(x)\,g(x)$.*

if h(x) = f(x)g(x), then h'(x) = f(x)g'(x) + f'(x)g(x)

It may be helpful to remember the Product Rule in words: The derivative of a product is equal to *the first function times the derivative of the second function plus the second function times the derivative of the first function*:

$$\text{First} \cdot (\text{second})' + \text{second} \cdot (\text{first})'$$

We prove the Product Rule after presenting three examples.

first·second' + first'·second

■ **EXAMPLE 1** Find the derivative of $h(x) = 3x^2(5x+1)$.

Solution This function is a product

$$h(x) = \overbrace{3x^2}^{\text{First}}\ \overbrace{(5x+1)}^{\text{Second}}$$

By the Product Rule (which we now write in Leibniz notation),

$$h'(x) = \overset{\text{First}}{3x^2} \,\overset{\text{Second}'}{\frac{d}{dx}(5x+1)} \;+\; \overset{\text{Second}}{(5x+1)}\,\overset{\text{First}'}{\frac{d}{dx}(3x^2)}$$

$$= (3x^2)5 + (5x+1)(6x)$$

$$= 15x^2 + (30x^2 + 6x) = 45x^2 + 6x$$

$\blacksquare$

$\blacksquare$ **EXAMPLE 2** Find the derivative of $y = (x^{-1} + 2)(x^{3/2} + 1)$.

Solution Use the Product Rule:

$$y' = \text{first} \cdot (\text{second})' + \text{second} \cdot (\text{first})'$$

$$= (x^{-1}+2)(x^{3/2}+1)' + (x^{3/2}+1)(x^{-1}+2)'$$

$$= (x^{-1}+2)(\tfrac{3}{2}x^{1/2}) + (x^{3/2}+1)(-x^{-2}) \qquad \text{(compute the derivatives)}$$

$$= \tfrac{3}{2}x^{-1/2} + 3x^{1/2} - x^{-1/2} - x^{-2} \qquad\qquad \text{(simplify)}$$

$$= \tfrac{1}{2}x^{-1/2} + 3x^{1/2} - x^{-2}$$

$\blacksquare$

In the previous two examples, we could have avoided using the Product Rule by expanding the function. For example,

$$h(x) = 3x^2(5x+1) = 15x^3 + 3x^2$$

We may use the Product Rule or differentiate directly:

$$h'(x) = \frac{d}{dx}(15x^3 + 3x^2) = 45x^2 + 6x$$

However, in the next example, the function cannot be expanded and we must use the Product Rule (or go back to the limit definition of the derivative).

$\blacksquare$ **EXAMPLE 3** Calculate $\dfrac{d}{dt}t^2 e^t$.

Solution Use the Product Rule:

$$\frac{d}{dt}t^2 e^t = \text{first} \cdot (\text{second})' + \text{second} \cdot (\text{first})'$$

$$= t^2 \frac{d}{dt}e^t + e^t \frac{d}{dt}t^2$$

$$= t^2 e^t + e^t(2t) = (t^2 + 2t)e^t$$

$\blacksquare$

Proof of the Product Rule We prove the Product Rule by writing the difference quotient for $f(x)g(x)$ in a clever way as a sum of two terms. The limit definition of the derivative applied to the product function gives us

$$(fg)'(x) = \lim_{h \to 0} \frac{f(x+h)g(x+h) - f(x)g(x)}{h}$$

The trick is to subtract $f(x+h)g(x)$ and add it back again in the numerator of the difference quotient:

Note how the prime notation is used in the solution to Example 2. We write $(x^{3/2}+1)'$ to denote the derivative of $x^{3/2}+1$, etc.

$h(x) = (x^2 + 2x)e^x$

$= (x^2 + 2x)e^x + (2x+2)e^x$

$= e^x(x^2 + 4x + 2)$

$h(x) = (\sin x)(\sin x)$

$= (\sin x)(\cos x) + (\sin x)(\cos x)$

$= 2\sin x \cos x$

$= \sin 2x$

FIGURE 1 An apparatus of resistance R attached to a battery of voltage V.

of resistance R (Figure 1) is

$$P = \frac{V^2 R}{(R + r)^2}$$

(a) Calculate $\dfrac{dP}{dR}$, assuming that V and r are constants.

(b) Find the value of R at which the tangent to the graph of P versus R is horizontal.

Solution

(a) The key to solving this problem is to keep in mind that P and R are variables, but V and r are constants:

$$\frac{dP}{dR} = \frac{d}{dR}\left(\frac{V^2 R}{(R + r)^2}\right) = V^2 \frac{d}{dR}\left(\frac{R}{(R + r)^2}\right) \qquad (V \text{ is a constant})$$

$$= V^2 \frac{(R + r)^2 \frac{d}{dR} R - R \frac{d}{dR}(R + r)^2}{(R + r)^4} \qquad \text{(Quotient Rule)} \qquad \boxed{4}$$

We have $\dfrac{d}{dR} R = 1$ and, since r is a constant,

$$\frac{d}{dR}(R + r)^2 = \frac{d}{dR}(R^2 + 2rR + r^2)$$

$$= \frac{d}{dR} R^2 + \frac{d}{dR} 2rR + \frac{d}{dR} r^2 \qquad \boxed{5}$$

$$= 2R + 2r + 0 = 2(R + r)$$

In (5), we use

$$\frac{d}{dR} R^2 = 2R$$

and, since r is a constant,

$$\frac{d}{dR} 2rR = 2r \frac{d}{dR} R = 2r$$

$$\frac{d}{dR} r^2 = 0$$

Using this in Eq. (4) and factoring out $(R + r)$, we obtain

$$\frac{dP}{dR} = V^2 \frac{(R + r)^2 - 2R(R + r)}{(R + r)^4} = V^2 \frac{(R + r)\big((R + r) - 2R\big)}{(R + r)^4}$$

$$= V^2 \frac{r - R}{(R + r)^3} \qquad [\text{cancel } (R + r)]$$

(b) The tangent line is horizontal when the derivative is zero. Thus, we set the derivative equal to zero and solve for R:

$$V^2 \frac{r - R}{(R + r)^3} = 0$$

This equation is satisfied only if the numerator is zero, that is, if $R = r$. ∎

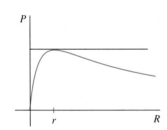

FIGURE 2 Graph of power versus resistance:

$$P = \frac{V^2 R}{(R + r)^2}$$

GRAPHICAL INSIGHT   Figure 2 shows that the point where the tangent line is horizontal is the *maximum point* on the graph. This proves an important result in circuit design: Maximum power is delivered when the resistance of the load (apparatus) is equal to the internal resistance of the battery.

3.3 SUMMARY

- This section deals with two basic rules of differentiation:

  Product Rule: $(fg)' = fg' + gf'$

  Quotient Rule: $\left(\dfrac{f}{g}\right)' = \dfrac{gf' - fg'}{g^2}$

- It is important to remember that the derivative of fg is *not* equal to $f'g'$. Similarly, the derivative of f/g is *not* equal to f'/g'.

3.3 EXERCISES

Preliminary Questions

1. Are the following statements true or false? If false, state the correct version.

(a) The notation fg denotes the function whose value at x is $f(g(x))$.

(b) The notation f/g denotes the function whose value at x is $f(x)/g(x)$.

(c) The derivative of the product is the product of the derivatives.

2. Are the following equations true or false? If false, state the correct version.

(a) $\dfrac{d}{dx}(fg)\Big|_{x=4} = f(4)g'(4) - g(4)f'(4)$

(b) $\dfrac{d}{dx}\dfrac{f}{g}\Big|_{x=4} = \dfrac{f(4)g'(4) + g(4)f'(4)}{(g(4))^2}$

(c) $\dfrac{d}{dx}(fg)\Big|_{x=0} = f(0)g'(0) + g(0)f'(0)$

3. What is the derivative of f/g at $x = 1$ if $f(1) = f'(1) = g(1) = 2$, and $g'(1) = 4$?

4. Suppose that $f(1) = 0$ and $f'(1) = 2$. Find $g(1)$, assuming that $(fg)'(1) = 10$.

Exercises

In Exercises 1–6, use the Product Rule to calculate the derivative.

1. $f(x) = x(x^2 + 1)$

2. $f(x) = \sqrt{x}(1 - x^4)$

3. $f(x) = e^x(x^3 - 1)$

4. $f(x) = x^2(1 + 4e^x)$

5. $\dfrac{dy}{dt}\Big|_{t=3}$, $y = (t^2 + 1)(t + 9)$

6. $\dfrac{dh}{dx}\Big|_{x=4}$, $h(x) = (x^{-1/2} + 2x)(7 - x^{-1})$

In Exercises 7–12, use the Quotient Rule to calculate the derivative.

7. $f(x) = \dfrac{x}{x - 2}$

8. $f(x) = \dfrac{x + 4}{x^2 + x + 1}$

9. $\dfrac{dg}{dt}\Big|_{t=-2}$, $g(t) = \dfrac{t^2 + 1}{t^2 - 1}$

10. $\dfrac{dw}{dz}\Big|_{z=9}$, $w = \dfrac{z^2}{\sqrt{z} + z}$

11. $g(x) = \dfrac{1}{1 + e^x}$

12. $f(x) = \dfrac{e^x}{x^2 + 1}$

In Exercises 13–16, calculate the derivative in two ways: First use the Product or Quotient Rule, then rewrite the function algebraically and apply the Power Rule directly.

13. $f(t) = (2t + 1)(t^2 - 2)$

14. $f(x) = x^2(3 + x^{-1})$

15. $g(x) = \dfrac{x^3 + 2x^2 + 3x^{-1}}{x}$

16. $h(t) = \dfrac{t^2 - 1}{t - 1}$

In Exercises 17–38, calculate the derivative using the appropriate rule or combination of rules.

17. $f(x) = (x^4 - 4)(x^2 + x + 1)$

18. $f(x) = (x^2 + 9)(2 - e^x)$

19. $\dfrac{dy}{dx}\Big|_{x=2}$, $y = \dfrac{1}{x + 4}$

20. $\dfrac{dz}{dx}\Big|_{x=-1}$, $z = \dfrac{x}{x^2 + 1}$

21. $f(x) = (\sqrt{x} + 1)(\sqrt{x} - 1)$

22. $f(x) = \dfrac{3\sqrt{x} - 2}{x}$

23. $\dfrac{dy}{dx}\Big|_{x=2}$, $y = \dfrac{x^4 - 4}{x^2 - 5}$

24. $f(x) = \dfrac{x^4 + e^x}{x + 1}$

25. $\dfrac{dz}{dx}\Big|_{x=1}$, $z = \dfrac{1}{x^3 + 1}$

26. $f(x) = \dfrac{3x^3 - x^2 + 2}{\sqrt{x}}$

27. $h(t) = \dfrac{t}{(t^4 + t^2)(t^7 + 1)}$

28. $f(x) = x^{3/2}(2x^4 - 3x + x^{-1/2})$

29. $f(t) = 3^{1/2} \cdot 5^{1/2}$

30. $h(x) = \pi^2(x - 1)$

31. $f(x) = (x + 3)(x - 1)(x - 5)$

32. $f(x) = x(x^2 + 1)(x + 4)$

33. $f(x) = \dfrac{e^x}{(e^x + 1)(x + 1)}$

34. $g(x) = \dfrac{e^{x+1} + e^x}{e + 1}$

35. $g(z) = \left(\dfrac{z^2 - 4}{z - 1}\right)\left(\dfrac{z^2 - 1}{z + 2}\right)$ *Hint: Simplify first.*

36. $\dfrac{d}{dx}\left((ax + b)(abx^2 + 1)\right)$ (*a*, *b* constants)

37. $\dfrac{d}{dt}\left(\dfrac{xt - 4}{t^2 - x}\right)$ (*x* constant)

38. $\dfrac{d}{dx}\left(\dfrac{ax + b}{cx + d}\right)$ (*a*, *b*, *c*, *d* constants)

39. ⬚GU⬚   Plot the derivative of $f(x) = \dfrac{x}{x^2 + 1}$ over $[-4, 4]$. Use the graph to determine the intervals on which $f'(x) > 0$ and $f'(x) < 0$. Then plot $f(x)$ and describe how the sign of $f'(x)$ is reflected in the graph of $f(x)$.

40. ⬚GU⬚   Plot $f(x) = \dfrac{x}{x^2 - 1}$ (in a suitably bounded viewing box). Use the plot to determine whether $f'(x)$ is positive or negative on its domain $\{x : x \neq \pm 1\}$. Then compute $f'(x)$ and confirm your conclusion algebraically.

41. Let $P = \dfrac{V^2 R}{(R + r)^2}$ as in Example 7. Calculate dP/dr, assuming that r is variable and R is constant.

42. Find the value $a > 0$ such that the tangent line to $f(x) = x^2 e^{-x}$ passes through the origin (Figure 3).

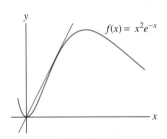

FIGURE 3

43. According to Ohm's Law, current I (in amps), voltage V (in volts), and resistance R (in ohms) in a circuit are related by $I = V/R$.

(a) Calculate $\left.\dfrac{dI}{dR}\right|_{R=6}$, assuming that V has the constant value $V = 24$.

(b) Calculate $\left.\dfrac{dV}{dR}\right|_{R=6}$, assuming that I has the constant value $I = 4$.

44. Find an equation of the tangent line to the graph $y = \dfrac{x}{x + x^{-1}}$ at $x = 2$.

45. The curve $y = \dfrac{1}{x^2 + 1}$ is called the *witch of Agnesi* (Figure 4) after the Italian mathematician Maria Agnesi (1718–1799) who wrote one of the first books on calculus. This strange name is the result of a mistranslation of the Italian word *la versiera*, meaning "that which turns." Find equations of the tangent lines at $x = \pm 1$.

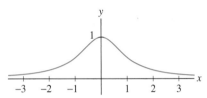

FIGURE 4 Graph of the witch of Agnesi.

46. Let $f(x) = g(x) = x$. Show that $(f/g)' \neq f'/g'$.

47. Use the Product Rule to show that $(f^2)' = 2ff'$.

48. Show that $(f^3)' = 3f^2 f'$.

In Exercises 49–52, use the following function values:

$f(4)$	$f'(4)$	$g(4)$	$g'(4)$
2	−3	5	−1

49. Find the derivative of fg and f/g at $x = 4$.

50. Calculate $F'(4)$, where $F(x) = xf(x)$.

51. Calculate $G'(4)$, where $G(x) = xg(x)f(x)$.

52. Calculate $H'(4)$, where $H(x) = \dfrac{x}{g(x)f(x)}$.

53. Calculate $F'(0)$, where

$$F(x) = \dfrac{x^9 + x^8 + 4x^5 - 7x}{x^4 - 3x^2 + 2x + 1}$$

Hint: Do not calculate $F'(x)$. Instead, write $F(x) = f(x)/g(x)$ and express $F'(0)$ directly in terms of $f(0)$, $f'(0)$, $g(0)$, $g'(0)$.

54. Proceed as in Exercise 53 to calculate $F'(0)$, where

$$F(x) = \left(1 + x + x^{4/3} + x^{5/3}\right)\dfrac{3x^5 + 5x^4 + 5x + 1}{8x^9 - 7x^4 + 1}$$

Further Insights and Challenges

55. Let f, g, h be differentiable functions. Show that $(fgh)'(x)$ is equal to

$$f(x)g(x)h'(x) + f(x)g'(x)h(x) + f'(x)g(x)h(x)$$

Hint: Write fgh as $f(gh)$.

56. Prove the Quotient Rule using the limit definition of the derivative.

57. Derivative of the Reciprocal   Use the limit definition to show that if $f(x)$ is differentiable and $f(x) \neq 0$, then $1/f(x)$ is differentiable and

$$\frac{d}{dx}\left(\frac{1}{f(x)}\right) = -\frac{f'(x)}{f^2(x)} \qquad \boxed{6}$$

Hint: Show that the difference quotient for $1/f(x)$ is equal to

$$\frac{f(x) - f(x+h)}{h f(x) f(x+h)}$$

58. Derive the Quotient Rule using Eq. (6) and the Product Rule.

59. Show that Eq. (6) is a special case of the Quotient Rule.

60. Carry out the details of Agnesi's proof of the Quotient Rule from her book on calculus, published in 1748: Assume that f, g, and $h = f/g$ are differentiable. Compute the derivative of $hg = f$ using the Product Rule and solve for h'.

61. The Power Rule Revisited   If you are familiar with *proof by induction*, use induction to prove the Power Rule for all whole numbers n. Show that the Power Rule holds for $n = 1$, then write x^n as $x \cdot x^{n-1}$ and use the Product Rule.

In Exercises 62–63, let $f(x)$ be a polynomial. A basic fact of algebra states that c is a root of $f(x)$ if and only if $f(x) = (x - c)g(x)$

for some polynomial $g(x)$. We say that c is a multiple root if $f(x) = (x - c)^2 h(x)$, where $h(x)$ is a polynomial.

62. Show that c is a multiple root of $f(x)$ if and only if c is a root of both $f(x)$ and $f'(x)$.

63. Use Exercise 62 to determine whether $c = -1$ is a multiple root of the polynomials
(a) $x^5 + 2x^4 - 4x^3 - 8x^2 - x + 2$
(b) $x^4 + x^3 - 5x^2 - 3x + 2$

64. 🖉   Figure 5 is the graph of a polynomial with roots at A, B, and C. Which of these is a multiple root? Explain your reasoning using the result of Exercise 62.

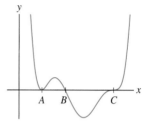

FIGURE 5

65. In Section 3.1, we showed that $\dfrac{d}{dx}b^x = m_b\,b^x$, where

$$m_b = \lim_{h\to 0} \frac{b^h - 1}{h}$$

Prove in two ways that the numbers m_b satisfy the relation

$$m_{ab} = m_a + m_b$$

3.4 Rates of Change

Recall the notation for the average rate of change (ROC) of a function $y = f(x)$ over an interval $[x_0, x_1]$:

$$\Delta y = \text{change in } y = f(x_1) - f(x_0)$$

$$\Delta x = \text{change in } x = x_1 - x_0$$

$$\boxed{\text{Average ROC} = \frac{\Delta y}{\Delta x} = \frac{f(x_1) - f(x_0)}{x_1 - x_0}}$$

We usually omit the word "instantaneous" and refer to the derivative simply as the rate of change (ROC). This is shorter and also more accurate when applied to general ROCs, since the term instantaneous would seem to refer to only ROCs with respect to time.

In our prior discussion in Section 2.1, limits and derivatives had not yet been introduced. Now that we have them at our disposal, we are ready to define the **instantaneous** ROC of y with respect to x at $x = x_0$:

$$\boxed{\text{Instantaneous ROC} = f'(x_0) = \lim_{\Delta x \to 0} \frac{\Delta y}{\Delta x} = \lim_{x_1 \to x_0} \frac{f(x_1) - f(x_0)}{x_1 - x_0}}$$

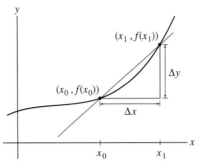

FIGURE 1 The average ROC over $[x_0, x_1]$ is the slope of the secant line.

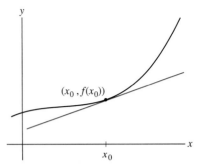

FIGURE 2 The instantaneous ROC at x_0 is the slope of the tangent line.

It is important to keep the geometric interpretation of rate of change in mind. The average ROC is the slope of the secant line (Figure 1) and the instantaneous ROC is the slope of the tangent line (Figure 2).

Whenever we measure the rate of change of a quantity y, it is always with respect to some other variable x (the independent variable). The rate dy/dx is measured in units of y per unit of x. For example, the ROC of temperature with respect to time has units such as degrees per minute and the ROC of temperature with respect to altitude has units such as degrees per kilometer.

Leibniz notation is particularly convenient when dealing with rates of change because it specifies the independent variable. The notation dy/dx indicates that we are considering the ROC of y with respect to the quantity x.

■ **EXAMPLE 1** Table 1 contains data on the temperature T (in degrees Celsius) on the surface of Mars at Martian time t, collected by the NASA Pathfinder space probe. A Martian hour is approximately 61.6 minutes.

(a) Use the data in Table 1 to calculate the average ROC of temperature over the time interval from 6:11 AM to 9:05 AM.

(b) Use Figure 3 to estimate the instantaneous ROC at $t = 12:28$ PM.

Solution

(a) The time interval from 6:11 to 9:05 has a length of 174 min or $\Delta t = 2.9$ hours. According to Table 1, the change in temperature from 6:11 to 9:05 is

$$\Delta T = -44.3 - (-71.6) = 27.3°C$$

TABLE 1   Data from Mars Pathfinder Mission, July 1997

Time	Temperature
5:42	−74.7
6:11	−71.6
6:40	−67.2
7:09	−63.7
7:38	−59.5
8:07	−53
8:36	−47.7
9:05	−44.3
9:34	−42

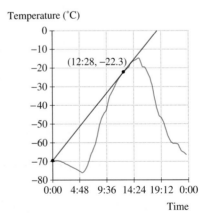

FIGURE 3 Temperature variation on the surface of Mars on July 6, 1997.

The average rate of temperature change is the ratio

$$\frac{\Delta T}{\Delta t} = \frac{27.3}{2.9} \approx 9.4°C/h$$

(b) The instantaneous ROC is the slope of the tangent line through the point (12:28, −22.3) shown in Figure 3. To compute its slope, we must choose a second point on the tangent line. A convenient choice is (00:00, −70). The time interval between these two points is 12 hours and 28 min, or approximately 12.47 hours. Therefore,

$$ROC = \text{slope of tangent line} \approx \frac{-22.3 - (-70)}{12.47} \approx 3.8°C/hr \qquad ■$$

■ **EXAMPLE 2** Let $A = \pi r^2$ be the area of a circle of radius r.

(a) Calculate the ROC of area with respect to radius.

(b) Compute dA/dr for $r = 2$ and $r = 5$, and explain why dA/dr is larger at $r = 5$.

Solution

(a) The ROC of area with respect to radius is the derivative

$$\frac{dA}{dr} = \frac{d}{dr}(\pi r^2) = 2\pi r$$

Note that the derivative of the area is equal to the circumference.

(b) At $r = 2$ and $r = 5$, we have

$$\frac{dA}{dr}\bigg|_{r=2} = 2\pi(2) \approx 12.57 \qquad \text{and} \qquad \frac{dA}{dr}\bigg|_{r=5} = 2\pi(5) \approx 31.42$$

Why is the derivative larger at $r = 5$? The derivative dA/dr measures how quickly the area changes when r increases. When the radius is increased from r to $r + \Delta r$, the area of the circle increases by a band of thickness Δr. We see in Figure 4 that the area of the band is greater when $r = 5$ than when $r = 2$. Thus, area increases more rapidly as a function of radius when r is larger. ■

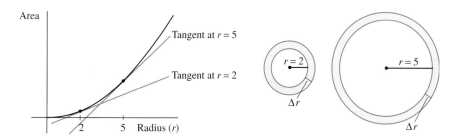

FIGURE 4 The blue bands represent the change in area when r is increased by Δr.

The Effect of a One-Unit Change

For small values of h, the difference quotient is close to the derivative itself:

$$f'(x_0) \approx \frac{f(x_0 + h) - f(x_0)}{h} \qquad \boxed{1}$$

This approximation generally improves as h gets smaller, but in some applications, the approximation is already useful with $h = 1$. Setting $h = 1$ in Eq. (1) gives

$$f'(x_0) \approx f(x_0 + 1) - f(x_0)$$

$\boxed{2}$

In other words, $f'(x_0)$ *is approximately equal to the change in f caused by a one-unit change in x when $x = x_0$.*

■ **EXAMPLE 3**   **Stopping Distance**   The stopping distance of an automobile after the brakes are applied (in feet) is given approximately by the function $F(s) = 1.1s + 0.05s^2$ for speeds s between 30 and 75 mph.

(a) Calculate the stopping distance and the ROC of stopping distance (with respect to speed) when $s = 30$ mph. Then estimate the change in stopping distance if the speed is increased by 1 mph.

(b) Do the same for $s = 60$ mph.

Solution   The ROC of stopping distance with respect to speed is the derivative

$$F'(s) = \frac{d}{ds}(1.1s + 0.05s^2) = 1.1 + 0.1s \text{ ft/mph}$$

In this example, we use the derivative $F'(s)$ to estimate the change in stopping distance for a one-unit increase in speed. Compare the estimates with the following exact values:

Actual Change (ft)	$F'(s)$
$F(31) - F(30) = 4.15$	4.1
$F(61) - F(60) = 7.15$	7.1

(a) The stopping distance at $s = 30$ mph is $F(30) = 1.1(30) + 0.05(30)^2 = 78$ ft. Since $F'(30) = 1.1 + 0.1(30) = 4.1$ ft/mph, Eq. (2) gives us the approximation:

$$\underbrace{F(31) - F(30)}_{\text{Change in stopping distance}} \approx F'(30) = 4.1 \text{ ft}$$

Thus, when traveling at 30 mph, stopping distance increases by roughly 4 ft if the speed is increased by 1 mph.

(b) The stopping distance at $s = 60$ mph is $F(60) = 1.1(60) + 0.05(60)^2 = 246$ ft. We have $F'(60) = 1.1 + 0.1(60) = 7.1$ ft/mph. Thus, when traveling at 60 mph, stopping distance increases by roughly 7 ft if the speed is increased by 1 mph.   ■

Marginal Cost in Economics

Many products can be manufactured more cheaply if produced in large quantities. The number of units manufactured is called the **production level**. To study the relation between costs and production level, economists define a rate of change called **marginal cost**. Let $C(x)$ denote the dollar cost (including labor and parts) of producing x units of a particular product. By definition, *the marginal cost at production level x_0 is the cost of producing one additional unit*:

Although $C(x)$ is meaningful only when x is whole number, we usually treat $C(x)$ as a differentiable function of x so that the techniques of calculus can be applied.

$$\text{Marginal cost} = C(x_0 + 1) - C(x_0)$$

In this setting, Eq. (2) usually gives a good approximation, so we take $C'(x_0)$ as an estimate of the marginal cost.

■ **EXAMPLE 4**   **Cost of an Air Flight**   Company data suggest that the total dollar cost of a 1,200-mile flight is approximately $C(x) = 0.0005x^3 - 0.38x^2 + 120x$, where x is the number of passengers (Figure 5).

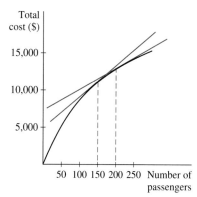

FIGURE 5 Cost of an air flight. The slopes of the tangent lines are decreasing, so marginal cost is decreasing.

(a) Estimate the marginal cost of an additional passenger if the flight already has 150 passengers.

(b) Compare your estimate with the actual cost.

(c) Is it more expensive to add a passenger when $x = 150$ or 200?

Solution The derivative is $C'(x) = 0.0015x^2 - 0.76x + 120$.

(a) We estimate the marginal cost at $x = 150$ by the derivative

$$C'(150) = 0.0015(150)^2 - 0.76(150) + 120 = 39.75$$

Thus, it costs approximately \$39.75 to add one additional passenger.

(b) The actual cost of adding one additional passenger is

$$C(151) - C(150) \approx 11{,}177.10 - 11{,}137.50 = 39.60$$

Our estimate \$39.75 is very close.

(c) We estimate the marginal cost at $x = 200$ by the derivative

$$C'(200) = 0.0015(200)^2 - 0.76(200) + 120 = 28$$

Thus, it costs approximately \$28 to add one additional passenger when $x = 200$. It is more expensive to add a passenger when $x = 150$. ∎

Linear Motion

In his "Lectures on Physics," Nobel laureate Richard Feynman (1918–1988) uses the following dialogue to make a point about instantaneous velocity:

Policeman: "My friend, you were going 75 miles an hour."

Driver: "That's impossible, sir, I was traveling for only seven minutes."

Recall that *linear motion* refers to the motion of an object along a straight line. This includes the horizontal motion of a vehicle along a straight highway or the vertical motion of a falling object. Let $s(t)$ denote the position or distance from the origin at time t. Velocity is the ROC of position with respect to time:

$$v(t) = \text{velocity} = \frac{ds}{dt}$$

The *sign* of $v(t)$ indicates the direction of motion. For example, if $s(t)$ is the height above ground, then $v(t) > 0$ indicates that the object is rising. **Speed** is defined as the absolute value of velocity $|v(t)|$.

The graph of $s(t)$ conveys a great deal of information about velocity and how it changes. Consider Figure 6, showing the position of a car as a function of time. Keep in mind that the height of the graph represents the car's distance from the point of origin.

Here are some facts we can glean from the graph:

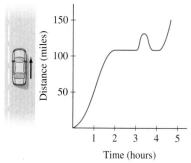

FIGURE 6 Graph of distance versus time.

- **Speeding up or slowing down?** The tangent lines get steeper in the interval [0, 1], indicating that *the car was speeding up during the first hour*. They get flatter in the interval [1, 2], indicating that the car slowed down in the second hour.
- **Standing still** The graph is horizontal over [2, 3] (perhaps the driver stopped at a restaurant for an hour).
- **Returning to the same spot** The graph rises and falls in the interval [3, 4], indicating that the driver returned to the restaurant (perhaps she left her wallet there).
- **Average velocity** The graph rises more over [0, 2] than over [3, 5], indicating that the average velocity was greater over the first two hours than the last two hours.

■ **EXAMPLE 5** A large truck enters the off-ramp of a freeway at $t = 0$. Its position after t seconds is $s(t) = 84t - t^3$ ft for $0 \leq t \leq 5$.

(a) How fast is the truck going at the moment it enters the off-ramp?

(b) Is the truck speeding up or slowing down?

Solution The truck's velocity at time t is $v(t) = \dfrac{d}{dt}(84t - t^3) = 84 - 3t^2$.

(a) We have $v(0) = 84$, so the truck enters the off-ramp with a velocity of 84 ft/s.

(b) The function $v(t) = 84 - 3t^2$ is decreasing (Figure 7), so the truck is slowing down. ∎

■

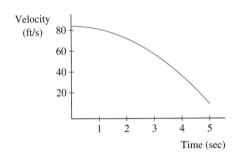

FIGURE 7 Graph of velocity $v(t) = 84 - 3t^2$.

Motion under the Influence of Gravity

Galileo discovered that the height $s(t)$ and velocity $v(t)$ of an object tossed vertically in the air are given as functions of time by the formulas

Galileo's formulas are valid when air resistance is negligible. We assume this to be the case in all examples here.

$$s(t) = s_0 + v_0 t - \frac{1}{2}gt^2, \qquad v(t) = \frac{ds}{dt} = v_0 - gt \qquad \boxed{3}$$

The constants s_0 and v_0 are the *initial values*:

- $s_0 = s(0)$ is the position at time $t = 0$.
- $v_0 = v(0)$ is the velocity at $t = 0$.
- g is the acceleration due to gravity on the surface of the earth, with the value

$$g \approx 32 \text{ ft/s}^2 \qquad \text{or} \qquad 9.8 \text{ m/s}^2$$

A simple observation allows us to find the object's maximum height. Since velocity is positive as the object rises and negative as it falls back to earth, the object reaches its maximum height at the moment of transition, when it is no longer rising and has not yet begun to fall. At that moment, its velocity is zero. In other words, *the maximum height is attained when $v(t) = 0$*. Since $v(t) = s'(t)$, the object reaches its maximum height when the tangent line to the graph of $s(t)$ is horizontal (Figure 8).

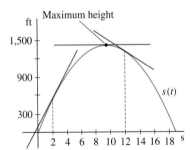

FIGURE 8 Maximum height occurs when $s'(t) = v(t) = 0$.

■ **EXAMPLE 6** Finding the Maximum Height A slingshot launches a stone vertically with an initial velocity of 300 ft/s from an initial height of 6 ft.

(a) Find the stone's velocity at $t = 2$ and at $t = 12$. Explain the change in sign.

(b) What is the stone's maximum height and when does it reach that height?

Solution Apply Eq. (3) with $s_0 = 6$, $v_0 = 300$, and $g = 32$:

$$s(t) = 6 + 300t - 16t^2, \qquad v(t) = 300 - 32t$$

(a) Therefore,

$$v(2) = 300 - 32(2) = 236 \text{ ft/s}, \qquad v(12) = 300 - 32(12) = -84 \text{ ft/s}$$

At $t = 2$, the stone is rising and its velocity is positive (Figure 8). At $t = 12$, the stone is already on the way down and its velocity is negative.

(b) Since the maximum height is attained when the velocity is zero, we set the velocity equal to zero and solve for t:

$$v(t) = 300 - 32t = 0 \qquad \text{gives} \qquad t = \frac{300}{32} = 9.375$$

Therefore, the stone reaches its maximum height at $t = 9.375$ s. The maximum height is

$$s(9.375) = 6 + 300(9.375) - 16(9.375)^2 \approx 1{,}412 \text{ ft} \qquad \blacksquare$$

In the previous example, we specified the initial values of position and velocity. In the next example, the goal is to determine initial velocity.

> *How important are units? In September 1999, the $125 million Mars Climate Orbiter spacecraft burned up in the Martian atmosphere before completing its scientific mission. According to Arthur Stephenson, NASA chairman of the Mars Climate Orbiter Mission Failure Investigation Board, 1999, "The 'root cause' of the loss of the spacecraft was the failed translation of English units into metric units in a segment of ground-based, navigation-related mission software."*

■ **EXAMPLE 7** Finding Initial Conditions   A bullet is fired vertically from an initial height $s_0 = 0$. What initial velocity v_0 is required for the bullet to reach a maximum height of 2 km?

Solution We need to derive a formula for the maximum height as a function of initial velocity v_0. Since $s_0 = 0$, the bullet's height is $s(t) = v_0 t - \frac{1}{2} g t^2$. Its maximum height is attained when the velocity is zero:

$$v(t) = v_0 - gt = 0 \qquad \text{or} \qquad t = \frac{v_0}{g}$$

We obtain the maximum height by computing $s(t)$ at $t = v_0/g$:

$$s\left(\frac{v_0}{g}\right) = v_0\left(\frac{v_0}{g}\right) - \frac{1}{2}g\left(\frac{v_0}{g}\right)^2 = \frac{v_0^2}{g} - \frac{1}{2}\frac{v_0^2}{g} = \frac{v_0^2}{2g}$$

Now we can solve for v_0 using the value $g = 9.8 \text{ m/s}^2$ (note that 2 km = 2,000 m):

> *In Eq. (4), distance must be in meters rather than kilometers since the units of g are m/s^2.*

$$\text{Maximum height} = \frac{v_0^2}{2g} = \frac{v_0^2}{2(9.8)} = 2{,}000 \text{ m} \qquad \boxed{4}$$

This yields $v_0 = \sqrt{(2)(9.8)2{,}000} \approx 198 \text{ m/s}.$ ■

HISTORICAL PERSPECTIVE

Around 1600 Galileo Galilei (1564–1642) discovered the laws of motion for falling objects on the earth's surface. This was a major event in the history of science because it gave us, for the first time, a mathematical description of motion as a function of time. Galileo's discoveries paved the way for Newton's general laws of motion. How did Galileo arrive at his formulas? The motion of a falling object is too rapid to measure directly. Today, we can photograph an object in motion and examine the trajectory frame by frame, but Galileo had to be more resourceful. He experimented with balls rolling down an incline. For a sufficiently flat incline, the motion is slow enough to measure accurately, and Galileo found that the velocity of the rolling ball is proportional to time. He then reasoned that motion in free-fall is just a faster version of motion down an incline, and deduced the formula

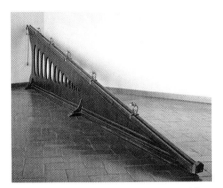

FIGURE 9 Apparatus of the type used by Galileo to study the motion of falling objects.

$v(t) = -gt$ for falling objects (assuming that the initial velocity v_0 is 0). This formula has a striking consequence.

Prior to Galileo, it had been incorrectly assumed that heavy objects fall more quickly than lighter ones. Galileo realized that this was not true, as long as air resistance is negligible. The formula $v(t) = -gt$ shows clearly that the velocity depends on time but not on the weight of the object. Interestingly, 300 years later, another great physicist, Albert Einstein, was deeply puzzled by Galileo's discovery that objects fall at the same rate regardless of their weight. He called this the Principle of Equivalence and sought to understand why it was true. In 1916, after a decade of intensive work, Einstein developed the General Theory of Relativity, which finally gave a full explanation of the Principle of Equivalence in terms of the geometry of space and time.

3.4 SUMMARY

- The instantaneous rate of change (ROC) of $y = f(x)$ with respect to x at $x = x_0$ is defined as the derivative

$$f'(x_0) = \lim_{\Delta x \to 0} \frac{\Delta y}{\Delta x} = \lim_{x_1 \to x_0} \frac{f(x_1) - f(x_0)}{x_1 - x_0}$$

- The rate dy/dx is measured in *units of y per unit of unit of x*.
- For linear motion, velocity $v(t)$ is the ROC of position $s(t)$ with respect to time, that is, $v(t) = s'(t)$.
- In some applications, $f'(x_0)$ provides a good estimate of the change in f due to a one-unit increase in x when $x = x_0$:

$$f'(x_0) \approx f(x_0 + 1) - f(x_0)$$

- Marginal cost is the cost of producing one additional unit. If $C(x)$ is the cost of producing x units, then the marginal cost at production level x_0 is $C(x_0 + 1) - C(x_0)$. The derivative $C'(x_0)$ is often a good estimate for marginal cost.
- Galileo's formulas for the height and velocity at time t of an object rising or falling under the influence of gravity near the earth's surface ($s_0 = $ initial position, $v_0 = $ initial velocity) are

$$s(t) = s_0 + v_0 t - \frac{1}{2}gt^2, \qquad v(t) = v_0 - gt$$

3.4 EXERCISES

Preliminary Questions

1. What units might be used to measure the ROC of:

(a) Pressure (in atmospheres) in a water tank with respect to depth?

(b) The reaction rate of a chemical reaction (the ROC of concentration with respect to time), where concentration is measured in moles per liter?

2. Suppose that $f(2) = 4$ and the average ROC of f between 2 and 5 is 3. What is $f(5)$?

3. Two trains travel from New Orleans to Memphis in 4 hours. The first train travels at a constant velocity of 90 mph, but the velocity of the second train varies. What was the second train's average velocity during the trip?

4. Estimate $f(26)$, assuming that $f(25) = 43$ and $f'(25) = 0.75$.

5. The population $P(t)$ of Freedonia in 1933 was $P(1933) = 5$ million.

(a) What is the meaning of the derivative $P'(1933)$?

(b) Estimate $P(1934)$ if $P'(1933) = 0.2$. What if $P'(1933) = 0$?

Exercises

1. Find the ROC of the area of a square with respect to the length of its side s when $s = 3$ and $s = 5$.

2. Find the ROC of the volume of a cube with respect to the length of its side s when $s = 3$ and $s = 5$.

3. Find the ROC of $y = x^{-1}$ with respect to x for $x = 1, 10$.

4. At what rate is the cube root $\sqrt[3]{x}$ changing with respect to x when $x = 1, 8, 27$?

In Exercises 5–8, calculate the ROC.

5. $\dfrac{dV}{dr}$, where V is the volume of a cylinder whose height is equal to its radius (the volume of a cylinder of height h and radius r is $\pi r^2 h$)

6. ROC of the volume V of a cube with respect to its surface area A

7. ROC of the volume V of a sphere with respect to its radius (the volume of a sphere is $V = \frac{4}{3}\pi r^3$)

8. $\dfrac{dA}{dD}$, where A is the surface area of a sphere of diameter D (the surface area of a sphere of radius r is $4\pi r^2$)

In Exercises 9–10, refer to Figure 10, which shows the graph of distance (in kilometers) versus time (in hours) for a car trip.

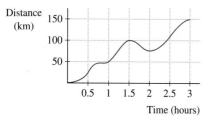

FIGURE 10 Graph of distance versus time for a car trip.

9. (a) Estimate the average velocity over [0.5, 1].

(b) Is average velocity greater over [1, 2] or [2, 3]?

(c) At what time is velocity at a maximum?

10. Match the description with the interval (a)–(d).

(i) Velocity increasing   **(ii)** Velocity decreasing

(iii) Velocity negative

(iv) Average velocity of 50 mph

(a) [0, 0.5]   **(b)** [0, 1]

(c) [1.5, 2]   **(d)** [2.5, 3]

11. Figure 11 displays the voltage across a capacitor as a function of time while the capacitor is being charged. Estimate the ROC of voltage at $t = 20$ s. Indicate the values in your calculation and include proper units. Does voltage change more quickly or more slowly as time goes on? Explain in terms of tangent lines.

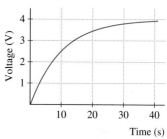

FIGURE 11

12. Use Figure 12 to estimate dT/dh at $h = 30$ and 70, where T is atmospheric temperature (in degrees Celsius) and h is altitude (in kilometers). Where is dT/dh equal to zero?

13. A stone is tossed vertically upward with an initial velocity of 25 ft/s from the top of a 30-ft building.

(a) What is the height of the stone after 0.25 s?

(b) Find the velocity of the stone after 1 s.

(c) When does the stone hit the ground?

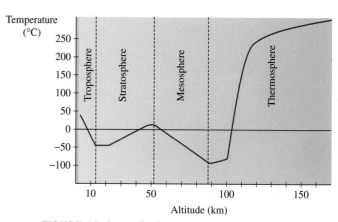

FIGURE 12 Atmospheric temperature versus altitude.

14. The height (in feet) of a skydiver at time t (in seconds) after opening his parachute is $h(t) = 2,000 - 15t$ ft. Find the skydiver's velocity after the parachute opens.

15. The temperature of an object (in degrees Fahrenheit) as a function of time (in minutes) is $T(t) = \frac{3}{4}t^2 - 30t + 340$ for $0 \le t \le 20$. At what rate does the object cool after 10 min (give correct units)?

16. The velocity (in centimeters per second) of a blood molecule flowing through a capillary of radius 0.008 cm is given by the formula $v = 6.4 \times 10^{-8} - 0.001r^2$, where r is the distance from the molecule to the center of the capillary. Find the ROC of velocity as a function of distance when $r = 0.004$ cm.

17. The earth exerts a gravitational force of $F(r) = \dfrac{2.99 \times 10^{16}}{r^2}$ (in Newtons) on an object with a mass of 75 kg, where r is the distance (in meters) from the center of the earth. Find the ROC of force with respect to distance at the surface of the earth, assuming the radius of the earth is 6.77×10^6 m.

18. The escape velocity at a distance r meters from the center of the earth is $v_{esc} = (2.82 \times 10^7) r^{-1/2}$ m/s. Calculate the rate at which v_{esc} changes with respect to distance at the surface of the earth.

19. The power delivered by a battery to an apparatus of resistance R (in ohms) is $P = \dfrac{2.25 R}{(R + 0.5)^2}$ W. Find the rate of change of power with respect to resistance for $R = 3$ and $R = 5$ Ω.

20. The position of a particle moving in a straight line during a 5-s trip is $s(t) = t^2 - t + 10$ cm.
(a) What is the average velocity for the entire trip?
(b) Is there a time at which the instantaneous velocity is equal to this average velocity? If so, find it.

21. By Faraday's Law, if a conducting wire of length ℓ meters moves at velocity v m/s perpendicular to a magnetic field of strength B (in teslas), a voltage of size $V = -B\ell v$ is induced in the wire. Assume that $B = 2$ and $\ell = 0.5$.
(a) Find the rate of change dV/dv.
(b) Find the rate of change of V with respect to time t if $v = 4t + 9$.

22. The height (in feet) of a helicopter at time t (in minutes) is $s(t) = -3t^3 + 400t$ for $0 \le t \le 10$.
(a) Plot the graphs of height $s(t)$ and velocity $v(t)$.
(b) Find the velocity at $t = 6$ and $t = 7$.
(c) Find the maximum height of the helicopter.

23. The population $P(t)$ of a city (in millions) is given by the formula $P(t) = 0.00005t^2 + 0.01t + 1$, where t denotes the number of years since 1990.
(a) How large is the population in 1996 and how fast is it growing?
(b) When does the population grow at a rate of 12,000 people per year?

24. According to Ohm's Law, the voltage V, current I, and resistance R in a circuit are related by the equation $V = IR$, where the units are volts, amperes, and ohms. Assume that voltage is constant with $V = 12$ V. Calculate (specifying the units):
(a) The average ROC of I with respect to R for the interval from $R = 8$ to $R = 8.1$
(b) The ROC of I with respect to R when $R = 8$
(c) The ROC of R with respect to I when $I = 1.5$

25. Ethan finds that with h hours of tutoring, he is able to answer correctly $S(h)$ percent of the problems on a math exam. What is the meaning of the derivative $S'(h)$? Which would you expect to be larger: $S'(3)$ or $S'(30)$? Explain.

26. Suppose $\theta(t)$ measures the angle between a clock's minute and hour hands. What is $\theta'(t)$ at 3 o'clock?

27. Table 2 gives the total U.S. population during each month of 1999 as determined by the U.S. Department of Commerce.
(a) Estimate $P'(t)$ for each of the months January–November.
(b) Plot these data points for $P'(t)$ and connect the points by a smooth curve.
(c) Write a newspaper headline describing the information contained in this plot.

TABLE 2   Total U.S. Population in 1999

t	$P(t)$ in Thousands
January	271,841
February	271,987
March	272,142
April	272,317
May	272,508
June	272,718
July	272,945
August	273,197
September	273,439
October	273,672
November	273,891
December	274,076

28. The tangent lines to the graph of $f(x) = x^2$ grow steeper as x increases. At what rate do the slopes of the tangent lines increase?

29. According to a formula widely used by doctors to determine drug dosages, a person's body surface area (BSA) (in meters squared) is given by the formula BSA $= \sqrt{hw}/60$, where h is the height in centimeters and w the weight in kilograms. Calculate the ROC of BSA with respect to weight for a person of constant height $h = 180$. What is this ROC for $w = 70$ and $w = 80$? Express your result in the correct units. Does BSA increase more rapidly with respect to weight at lower or higher body weights?

30. A slingshot is used to shoot a pebble in the air vertically from ground level with an initial velocity 200 m/s. Find the pebble's maximum velocity and height.

31. What is the velocity of an object dropped from a height of 300 m when it hits the ground?

32. It takes a stone 3 s to hit the ground when dropped from the top of a building. How high is the building and what is the stone's velocity upon impact?

33. A ball is tossed up vertically from ground level and returns to earth 4 s later. What was the initial velocity of the stone and how high did it go?

34. An object is tossed up vertically from ground level and hits the ground T s later. Show that its maximum height was reached after $T/2$ s.

35. A man on the tenth floor of a building sees a bucket (dropped by a window washer) pass his window and notes that it hits the ground 1.5 s later. Assuming a floor is 16 ft high (and neglecting air friction), from which floor was the bucket dropped?

36. Which of the following statements is true for an object falling under the influence of gravity near the surface of the earth? Explain.

(a) The object covers equal distance in equal time intervals.

(b) Velocity increases by equal amounts in equal time intervals.

(c) The derivative of velocity increases with time.

37. Show that for an object rising and falling according to Galileo's formula in Eq. (3), the average velocity over any time interval $[t_1, t_2]$ is equal to the average of the instantaneous velocities at t_1 and t_2.

38. A weight oscillates up and down at the end of a spring. Figure 13 shows the height y of the weight through one cycle of the oscillation. Make a rough sketch of the graph of the velocity as a function of time.

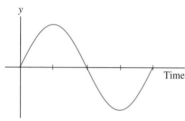

FIGURE 13

In Exercises 39–46, use Eq. (2) to estimate the unit change.

39. Estimate $\sqrt{2} - \sqrt{1}$ and $\sqrt{101} - \sqrt{100}$. Compare your estimates with the actual values.

40. Suppose that $f(x)$ is a function with $f'(x) = 2^{-x}$. Estimate $f(7) - f(6)$. Then estimate $f(5)$, assuming that $f(4) = 7$.

41. Let $F(s) = 1.1s + 0.03s^2$ be the stopping distance as in Example 3. Calculate $F(65)$ and estimate the increase in stopping distance if speed is increased from 65 to 66 mph. Compare your estimate with the actual increase.

42. According to Kleiber's Law, the metabolic rate P (in kilocalories per day) and body mass m (in kilograms) of an animal are related by a *three-quarter power law* $P = 73.3m^{3/4}$. Estimate the increase in metabolic rate when body mass increases from 60 to 61 kg.

43. The dollar cost of producing x bagels is $C(x) = 300 + 0.25x - 0.5(x/1,000)^3$. Determine the cost of producing 2,000 bagels and estimate the cost of the 2001st bagel. Compare your estimate with the actual cost of the 2001st bagel.

44. Suppose the dollar cost of producing x video cameras is $C(x) = 500x - 0.003x^2 + 10^{-8}x^3$.

(a) Estimate the marginal cost at production level $x = 5,000$ and compare it with the actual cost $C(5,001) - C(5,000)$.

(b) Compare the marginal cost at $x = 5,000$ with the average cost per camera, defined as $C(x)/x$.

45. The demand for a commodity generally decreases as the price is raised. Suppose that the demand for oil (per capita per year) is $D(p) = 900/p$ barrels, where p is the price per barrel in dollars. Find the demand when $p = \$40$. Estimate the decrease in demand if p rises to $\$41$ and the increase if p is decreased to $\$39$.

46. The reproduction rate of the fruit fly *Drosophila melanogaster*, grown in bottles in a laboratory, decreases as the bottle becomes more crowded. A researcher has found that when a bottle contains p flies, the number of offspring per female per day is

$$f(p) = (34 - 0.612p)p^{-0.658}$$

(a) Calculate $f(15)$ and $f'(15)$.

(b) Estimate the decrease in daily offspring per female when p is increased from 15 to 16. Is this estimate larger or smaller than the actual value $f(16) - f(15)$?

(c) ⃞GU Plot $f(p)$ for $5 \le p \le 25$ and verify that $f(p)$ is a decreasing function of p. Do you expect $f'(p)$ to be positive or negative? Plot $f'(p)$ and confirm your expectation.

47. Let $A = s^2$. Show that the estimate of $A(s + 1) - A(s)$ provided by Eq. (2) has error exactly equal to 1. Explain this result using Figure 14.

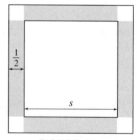

FIGURE 14

48. According to Steven's Law in psychology, the perceived magnitude of a stimulus (how strong a person feels the stimulus to be) is proportional to a power of the actual intensity I of the stimulus. Although not an exact law, experiments show that the *perceived brightness* B of a light satisfies $B = kI^{2/3}$, where I is the light intensity, whereas the *perceived heaviness* H of a weight W satisfies $H = kW^{3/2}$ (k is a constant that is different in the two cases). Compute dB/dI and dH/dW and state whether they are increasing or decreasing functions. Use this to justify the statements:

(a) A one-unit increase in light intensity is felt more strongly when I is small than when I is large.

(b) Adding another pound to a load W is felt more strongly when W is large than when W is small.

49. Let $M(t)$ be the mass (in kilograms) of a plant as a function of time (in years). Recent studies by Niklas and Enquist have suggested that for a remarkably wide range of plants (from algae and grass to palm trees), the growth rate during the life span of the organism satisfies a *three-quarter power law*, that is, $dM/dt = CM^{3/4}$ for some constant C.

(a) If a tree has a growth rate of 6 kg/year when $M = 100$ kg, what is its growth rate when $M = 125$ kg?

(b) If $M = 0.5$ kg, how much more mass must the plant acquire to double its growth rate?

Further Insights and Challenges

50. As an epidemic spreads through a population, the percentage p of infected individuals at time t (in days) satisfies the equation (called a *differential equation*)

$$\frac{dp}{dt} = 4p - 0.06p^2 \quad 0 \le p \le 100$$

(a) How fast is the epidemic spreading when $p = 10\%$ and when $p = 70\%$?

(b) For which p is the epidemic neither spreading nor diminishing?

(c) Plot dp/dt as a function of p.

(d) What is the maximum possible rate of increase and for which p does this occur?

51. The size of a certain animal population $P(t)$ at time t (in months) satisfies $\dfrac{dP}{dt} = 0.2(300 - P)$.

(a) Is P growing or shrinking when $P = 250$? when $P = 350$?

(b) Sketch the graph of dP/dt as a function of P for $0 \le P \le 300$.

(c) Which of the graphs in Figure 15 is the graph of $P(t)$ if $P(0) = 200$?

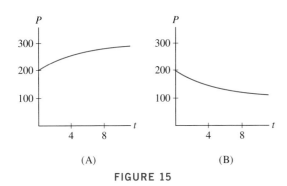

(A)   (B)

FIGURE 15

52. ⌂⌐⌐ Studies of internet usage show that website popularity is described quite well by Zipf's Law, according to which the nth most popular website receives roughly the fraction $1/n$ of all visits. Suppose that on a particular day, the nth most popular site had approximately $V(n) = 10^6/n$ visitors (for $n \le 15,000$).

(a) Verify that the top 50 websites received nearly 45% of the visits. *Hint:* Let $T(N)$ denote the sum of $V(n)$ for $1 \le n \le N$. Use a computer algebra system to compute $T(45)$ and $T(15,000)$.

(b) Verify, by numerical experimentation, that when Eq. (2) is used to estimate $V(n + 1) - V(n)$, the error in the estimate decreases as n grows larger. Find (again, by experimentation) an N such that the error is at most 10 for $n \ge N$.

(c) Using Eq. (2), show that for $n \ge 100$, the nth website received at most 100 more visitors than the $(n + 1)$st web site.

In Exercises 53–54, the average cost per unit at production level x is defined as $C_{\text{avg}}(x) = C(x)/x$, where $C(x)$ is the cost function. Average cost is a measure of the efficiency of the production process.

53. Show that $C_{\text{avg}}(x)$ is equal to the slope of the line through the origin and the point $(x, C(x))$ on the graph of $C(x)$. Using this interpretation, determine whether average cost or marginal cost is greater at points A, B, C, D in Figure 16.

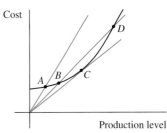

FIGURE 16 Graph of $C(x)$.

54. The cost in dollars of producing alarm clocks is

$$C(x) = 50x^3 - 750x^2 + 3,740x + 3,750$$

where x is in units of 1,000.

(a) Calculate the average cost at $x = 4, 6, 8$, and 10.

(b) Use the graphical interpretation of average cost to find the production level x_0 at which average cost is lowest. What is the relation between average cost and marginal cost at x_0 (see Figure 17)?

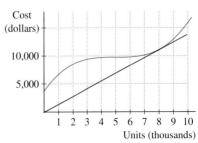

FIGURE 17 Cost function $C(x) = 50x^3 - 750x^2 + 3,740x + 3,750$.

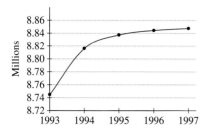

FIGURE 1 Population $P(t)$ of Sweden (in millions). The rate of increase declined in the period 1993–1997.

3.5 Higher Derivatives

Higher derivatives are the functions obtained by repeatedly differentiating a function $y = f(x)$. If f' is differentiable, the **second derivative**, denoted f'' or y'', is the derivative of f':

$$f''(x) = \frac{d}{dx}\left(f'(x)\right)$$

The second derivative is the rate of change of $f'(x)$. We refer to $f(x)$ as the zeroth derivative and $f'(x)$ as the first derivative.

To clarify the distinction between the first and second derivatives, consider the population $P(t)$ of Sweden (Figure 1). Government statistics show that $P(t)$ has increased every year for more than 100 years. Therefore, the first derivative $P'(t)$ is positive. However, Table 1 shows that the rate of yearly increase declined dramatically in the years 1993–1997. So although $P'(t)$ was still positive in these years, $P'(t)$ decreased and therefore the second derivative $P''(t)$ was negative in the period 1993–1997.

TABLE 1   Population of Sweden

Year	1993	1994	1995	1996	1997
Population	8,745,109	8,816,381	8,837,496	8,844,499	8,847,625
Yearly increase		71,272	21,115	7,003	3,126

Differentiation can be continued beyond the second derivative, provided that the derivatives exist. The third derivative is the derivative of $f''(x)$ and is denoted $f'''(x)$ or $f^{(3)}(x)$. More generally, the nth-order derivative is the derivative of the $(n-1)$st derivative and is denoted $f^{(n)}(x)$. In Leibniz notation, the higher derivatives are denoted

$$\frac{df}{dx},\quad \frac{d^2 f}{dx^2},\quad \frac{d^3 f}{dx^3},\quad \frac{d^4 f}{dx^4},\dots$$

- dy/dx has units of y per unit of x.
- d^2y/dx^2 has units of dy/dx per unit of x, or units of y per unit of x-squared.

■ **EXAMPLE 1** Find $f''(x)$ and $f'''(x)$ for $f(x) = x^4 + 2x - 9x^{-2}$ and evaluate $f'''(-1)$.

Solution We must calculate the derivatives, beginning with $f'(x)$:

$$f'(x) = \frac{d}{dx}\left(x^4 + 2x - 9x^{-2}\right) = 4x^3 + 2 + 18x^{-3}$$

$$f''(x) = \frac{d}{dx}\left(4x^3 + 2 + 18x^{-3}\right) = 12x^2 - 54x^{-4}$$

$$f'''(x) = \frac{d}{dx}\left(12x^2 - 54x^{-4}\right) = 24x + 216x^{-5}$$

At $x = -1$, $f'''(-1) = -24 - 216 = -240$.   ■

If $f(x)$ is a polynomial of degree k, then $f^{(k)}(x)$ is a constant and, therefore, the higher derivatives $f^{(n)}(x)$ for $n > k$ are all zero. Indeed, each time we differentiate a polynomial, the degree decreases by 1. Table 2 shows the higher derivatives of $f(x) = x^5$, and we note that the nth derivative is zero for $n > 5$. By contrast, if k is not a whole number, the higher derivatives of $f(x) = x^k$ are all nonzero, as illustrated in the next example.

TABLE 2   Derivatives of x^5

$f(x)$	$f'(x)$	$f''(x)$	$f'''(x)$	$f^{(4)}(x)$	$f^{(5)}(x)$	$f^{(6)}(x)$
x^5	$5x^4$	$20x^3$	$60x^2$	$120x$	120	0

■ **EXAMPLE 2** Calculate the first four derivatives of $y = x^{-1}$. Then find the pattern and determine a general formula for the nth derivative $y^{(n)}$.

Solution By the Power Rule,

$$y'(x) = -x^{-2}, \quad y'' = 2x^{-3}, \quad y''' = -2(3)x^{-4}, \quad y^{(4)} = 2(3)(4)x^{-5}$$

We see that $y^{(n)}(x)$ is equal to $\pm n!\, x^{-n-1}$. Now observe that the sign alternates. Since the odd-order derivatives occur with a minus sign, the sign of $y^{(n)}(x)$ is $(-1)^n$. In general, therefore, $y^{(n)}(x) = (-1)^n n!\, x^{-n-1}$.   ■

■ **EXAMPLE 3** Calculate $f'''(x)$ for $f(x) = xe^x$.

Solution We calculate the derivatives using the Product Rule, beginning with $f'(x)$:

$$f'(x) = \frac{d}{dx}(xe^x) = xe^x + e^x = (x+1)e^x$$

$$f''(x) = \frac{d}{dx}\big((x+1)e^x\big) = (x+1)e^x + e^x = (x+2)e^x$$

$$f'''(x) = \frac{d}{dx}\big((x+2)e^x\big) = (x+2)e^x + e^x = (x+3)e^x$$ ■

Acceleration, defined as the ROC of velocity, is a familiar second derivative. If an object in linear motion has position $s(t)$ at time t, then it has velocity $v(t) = s'(t)$ and acceleration $a(t) = v'(t) = s''(t)$. The unit of acceleration is the unit of velocity per unit of time or "distance per time squared." Thus, if velocity is measured in units of m/s, then acceleration has units of m/s per second or m/s^2.

■ **EXAMPLE 4** Acceleration due to Gravity Calculate the acceleration $a(t)$ of a ball tossed vertically in the air from ground level with an initial velocity of 40 ft/s. How does $a(t)$ describe the change in the ball's velocity as it rises and falls?

Solution By Galileo's formula, the height of the ball at time t is $s(t) = s_0 + v_0 t - 16t^2$ [Figure 2(A)]. In our case, $s_0 = 0$ and $v_0 = 40$, so $s(t) = 40t - 16t^2$ ft. Therefore, $v(t) = s'(t) = 40 - 32t$ ft/s and the ball's acceleration is

$$a(t) = s''(t) = \frac{d}{dt}(40 - 32t) = -32 \text{ ft/s}^2$$

We see that the acceleration is constant with value $a = -32$ ft/s^2. Although the ball rises and falls, its velocity decreases from 40 to -40 ft/s at a constant rate [Figure 2(B)]. Note that a is the slope of the graph of velocity.   ■

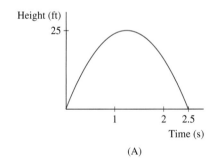

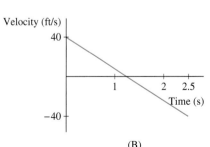

Height (ft)

(A)

Velocity (ft/s)

(B)

FIGURE 2 Height and velocity of the ball.

GRAPHICAL INSIGHT   Can we visualize the rate represented by a second derivative? Since the second derivative is the rate at which $f'(x)$ is changing, $f''(x)$ is large when the slopes of the tangent lines change rapidly, as in Figure 3(A). Similarly, f'' is small if the slopes of the tangent lines change slowly—in this case, the curve is relatively flat, as in Figure 3(B). If f is a linear function, the tangent line does not change at all and $f''(x) = 0$.

(A) Large second derivative:
    Tangent lines turn rapidly.

(B) Smaller second derivative:
    Tangent lines turn slowly.

(C) Second derivative is zero:
    Tangent line does not change.

FIGURE 3

With this understanding, we can see how the graphs of f, f', and f'' are related. In Figure 4, the slopes of the tangent lines to the graph of f are *increasing* on the interval $[a, b]$. This is reflected in the graphs of both $f'(x)$ and $f''(x)$. Namely, $f'(x)$ is increasing on $[a, b]$ and $f''(x)$ is positive on $[a, b]$.

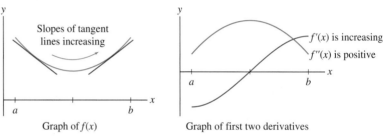

Graph of $f(x)$

Graph of first two derivatives

FIGURE 4

3.5 SUMMARY

- The higher derivatives f', f'', f''', ... are defined by successive differentiation:

$$f''(x) = \frac{d}{dx} f'(x), \qquad f'''(x) = \frac{d}{dx} f''(x), \ldots .$$

The nth derivative is also denoted $f^{(n)}(x)$.
- The second derivative plays an important role: It is the rate at which f' varies. Graphically, f'' measures how fast the tangent lines change direction and is thus a measure of how curved the graph is.
- If $s(t)$ is the position of an object at time t, then the first derivative $s'(t)$ is velocity, and the second derivative $s''(t)$ is acceleration.

3.5 EXERCISES

Preliminary Questions

1. An economist who announces that "America's economic growth is slowing" is making a statement about the gross national product (GNP) as a function of time. Is the second derivative of the GNP positive? What about the first derivative?

2. On September 4, 2003, the *Wall Street Journal* printed the headline "Stocks Go Higher, Though the Pace of Their Gains Slows." Rephrase as a statement about the first and second time derivatives of stock prices and sketch a possible graph.

3. Is the following statement true or false? The third derivative of position with respect to time is zero for an object falling to earth under the influence of gravity. Explain.

4. Which type of polynomial satisfies $f''(x) = 0$ for all x?

5. What is the millionth derivative of $f(x) = e^x$?

Exercises

In Exercises 1–14, calculate the second and third derivatives.

1. $y = 14x^2$

2. $y = 7 - 2x$

3. $y = x^4 - 25x^2 + 2x$

4. $y = 4t^3 - 9t^2 + 7$

5. $y = \dfrac{4}{3}\pi r^3$

6. $y = \sqrt{x}$

7. $y = 20t^{4/5} - 6t^{2/3}$

8. $y = x^{-9/5}$

9. $y = z - \dfrac{1}{z}$

10. $y = t^2(t^2 + t)$

11. $y = (x^2 + x)(x^3 + 1)$

12. $y = \dfrac{1}{1+x}$

13. $y = 2x + e^x$

14. $y = x^4 e^x$

In Exercises 15–28, calculate the derivative indicated.

15. $f^{(4)}(1), \quad f(x) = x^4$

16. $g'''(1), \quad g(t) = 4t^{-3}$

17. $\dfrac{d^2 y}{dt^2}\Big|_{t=1}, \quad y = 4t^{-3} + 3t^2$

18. $\dfrac{d^4 f}{dt^4}\Big|_{t=1}, \quad f(t) = 6t^9 - 2t^5$

19. $h'''(9), \quad h(x) = \sqrt{x}$

20. $g'''(9), \quad g(x) = x^{-1/2}$

21. $\dfrac{d^4 x}{dt^4}\Big|_{t=16}, \quad x = t^{-3/4}$

22. $f'''(4), \quad f(t) = 2t^2 - t$

23. $g''(1), \quad g(x) = \dfrac{x}{x+1}$

24. $f''(1), \quad f(t) = \dfrac{1}{t^3 + 1}$

25. $h''(1), \quad h(x) = \dfrac{1}{\sqrt{x}+1}$

26. $F''(2), \quad F(x) = \dfrac{x^2}{x-3}$

27. $f''(0), \quad f(x) = \dfrac{x}{e^x + 1}$

28. $f'''(0), \quad f(x) = 4e^x - x^3$

29. Calculate $y^{(k)}(0)$ for $0 \le k \le 5$, where $y = x^4 + ax^3 + bx^2 + cx + d$ (with a, b, c, d the constants).

30. Which of the following functions satisfy $f^{(k)}(x) = 0$ for all $k \ge 6$?

(a) $f(x) = 7x^4 + 4 + x^{-1}$

(b) $f(x) = x^3 - 2$

(c) $f(x) = \sqrt{x}$

(d) $f(x) = 1 - x^6$

(e) $f(x) = x^{9/5}$

(f) $f(x) = 2x^2 + 3x^5$

31. Use the result in Example 2 to find $\dfrac{d^6}{dx^6} x^{-1}$.

32. Calculate the first five derivatives of $f(x) = \sqrt{x}$.

(a) Show that $f^{(n)}(x)$ is a multiple of $x^{-n+1/2}$.

(b) Show that $f^{(n)}(x)$ alternates in sign as $(-1)^{n-1}$ for $n \ge 1$.

(c) Find a formula for $f^{(n)}(x)$ for $n \ge 2$. *Hint:* Verify that the coefficient of $x^{-n+1/2}$ is $\pm 1 \cdot 3 \cdot 5 \cdots \dfrac{2n-3}{2^n}$.

In Exercises 33–36, find a general formula for $f^{(n)}(x)$.

33. $f(x) = (x+1)^{-1}$

34. $f(x) = x^{-2}$

35. $f(x) = x^{-1/2}$

36. $f(x) = x^{-3/2}$

37. (a) Find the acceleration at time $t = 5$ min of a helicopter whose height (in feet) is $h(t) = -3t^3 + 400t$.

(b) $\boxed{\text{GU}}$ Plot the acceleration $h''(t)$ for $0 \le t \le 6$. How does this graph show that the helicopter is slowing down during this time interval?

38. Find an equation of the tangent to the graph of $y = f'(x)$ at $x = 3$, where $f(x) = x^4$.

39. Figure 5 shows f, f', and f''. Determine which is which.

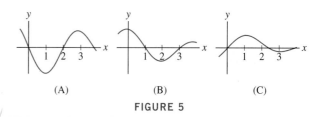

(A) (B) (C)

FIGURE 5

40. The second derivative f'' is shown in Figure 6. Determine which graph, (A) or (B), is f and which is f'.

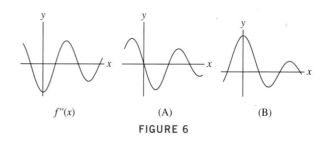

$f''(x)$ (A) (B)

FIGURE 6

41. Figure 7 shows the graph of the position of an object as a function of time. Determine the intervals on which the acceleration is positive.

42. Find the second derivative of the volume of a cube with respect to the length of a side.

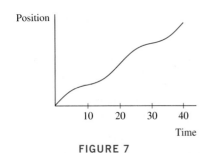

FIGURE 7

43. Find a polynomial $f(x)$ satisfying the equation $xf''(x) + f(x) = x^2$.

44. Find a value of n such that $y = x^n e^x$ satisfies the equation $xy' = (x - 3)y$.

45. 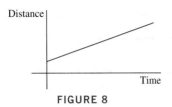 Which of the following descriptions could *not* apply to Figure 8? Explain.

(a) Graph of acceleration when velocity is constant

(b) Graph of velocity when acceleration is constant

(c) Graph of position when acceleration is zero

FIGURE 8

46. A servomotor controls the vertical movement of a drill bit that will drill a pattern of holes in sheet metal. The maximum vertical speed of the drill bit is 4 in./s, and while drilling the hole, it must move no more than 2.6 in./s to avoid warping the metal. During a cycle, the bit begins and ends at rest, quickly approaches the sheet metal, and quickly returns to its initial position after the hole is drilled. Sketch possible graphs of the drill bit's vertical velocity and acceleration. Label the point where the bit enters the sheet metal.

In Exercises 47–48, refer to the following. In their 1997 study, Boardman and Lave related the traffic speed S on a two-lane road to traffic density Q (number of cars per mile of road) by the formula $S = 2{,}882Q^{-1} - 0.052Q + 31.73$ for $60 \le Q \le 400$ (Figure 9).

47. Calculate dS/dQ and d^2S/dQ^2.

48. (a) 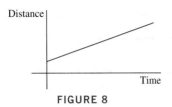 Explain intuitively why we should expect that $dS/dQ < 0$.

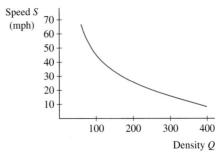

FIGURE 9 Speed as a function of traffic density.

(b) Show that $d^2S/dQ^2 > 0$. Then use the fact that $dS/dQ < 0$ and $d^2S/dQ^2 > 0$ to justify the following statement: *A one-unit increase in traffic density slows down traffic more when Q is small than when Q is large.*

(c) [GU] Plot dS/dQ. Which property of this graph shows that $d^2S/dQ^2 > 0$?

49. 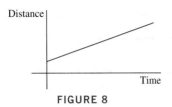 According to one model that attempts to account for air resistance, the distance $s(t)$ (in feet) traveled by a falling raindrop satisfies

$$\frac{d^2s}{dt^2} = g - \frac{0.0005}{D}\left(\frac{ds}{dt}\right)^2$$

where D is the raindrop diameter and $g = 32$ ft/s^2. Terminal velocity v_{term} is defined as the velocity at which the drop has zero acceleration (one can show that velocity approaches v_{term} as time proceeds).

(a) Show that $v_{\text{term}} = \sqrt{2000gD}$.

(b) Find v_{term} for drops of diameter 0.003 and 0.0003 ft.

(c) In this model, do raindrops accelerate more rapidly at higher or lower velocities?

50. [CAS] Use a computer algebra system to compute $f^{(k)}(x)$ for $k = 1, 2, 3$ for the functions:

(a) $f(x) = (1 + x^3)^{5/3}$

(b) $f(x) = \dfrac{1 - x^4}{1 - 5x - 6x^2}$

51. [CAS] Let $f(x) = \dfrac{x + 2}{x - 1}$. Use a computer algebra system to compute the $f^{(k)}(x)$ for $1 \le k \le 4$. Can you find a general formula for $f^{(k)}(x)$?

Further Insights and Challenges

52. Find the 100th derivative of

$$p(x) = (x + x^5 + x^7)^{10}(1 + x^2)^{11}(x^3 + x^5 + x^7)$$

53. What is the $p^{(99)}(x)$ for $p(x)$ as in Exercise 52?

54. Use the Product Rule twice to find a formula for $(fg)''$ in terms of the first and second derivative of f and g.

55. Use the Product Rule to find a formula for $(fg)'''$ and compare your result with the expansion of $(a + b)^3$. Then try to guess the general formula for $(fg)^{(n)}$.

56. Compute

$$\Delta f(x) = \lim_{h \to 0} \frac{f(x+h) + f(x-h) - 2f(x)}{h^2}$$

for the following functions:

(a) $f(x) = x$ **(b)** $f(x) = x^2$ **(c)** $f(x) = x^3$

Based on these examples, can you formulate a conjecture about what Δf is?

57. Calculate the first four derivatives of $f(x) = x^2 e^x$. Then guess the formula for $f^{(n)}(x)$ (use induction to prove it if you are familiar with this method of proof).

3.6 Trigonometric Functions

We can use the rules developed so far to differentiate functions involving powers of x, but we cannot yet handle the trigonometric functions. What is missing are the formulas for the derivatives of $\sin x$ and $\cos x$. Fortunately, these functions have simple derivatives—each is the derivative of the other up to a sign.

Recall our convention concerning angles: *All angles are measured in radians, unless otherwise specified.*

CAUTION Theorem 1 applies only when radians are used. The derivatives of sine and cosine with respect to degrees involves an extra, unwieldy factor of $\pi/180$ (see Example 6 in Section 3.7).

THEOREM 1 Derivative of Sine and Cosine The functions $y = \sin x$ and $y = \cos x$ are differentiable and

$$\frac{d}{dx} \sin x = \cos x \quad \text{and} \quad \frac{d}{dx} \cos x = -\sin x$$

Proof We must go back to the definition of the derivative:

$$\frac{d}{dx} \sin x = \lim_{h \to 0} \frac{\sin(x+h) - \sin x}{h} \qquad \boxed{1}$$

Unlike the derivative computations in Sections 3.1 and 3.2, it is not possible to cancel h by rewriting the difference quotient. Instead, we use the addition formula

$$\sin(x+h) = \sin x \cos h + \cos x \sin h$$

to rewrite the numerator of the difference quotient as a sum of two terms:

$$\sin(x+h) - \sin x = \sin x \cos h + \cos x \sin h - \sin x \qquad \text{(addition formula)}$$
$$= (\sin x \cos h - \sin x) + \cos x \sin h \qquad \text{(rearrange)}$$
$$= \sin x(\cos h - 1) + \cos x \sin h \qquad \text{(sum of two terms)}$$

This gives us

$$\frac{\sin(x+h) - \sin x}{h} = \frac{\sin x \,(\cos h - 1)}{h} + \frac{\cos x \,\sin h}{h}$$

Now apply the Sum Law for Limits:

$$\frac{d \sin x}{dx} = \lim_{h \to 0} \frac{\sin x \,(\cos h - 1)}{h} + \lim_{h \to 0} \frac{\cos x \,\sin h}{h}$$

$$= (\sin x) \underbrace{\lim_{h \to 0} \frac{\cos h - 1}{h}}_{\text{This equals 0}} + (\cos x) \underbrace{\lim_{h \to 0} \frac{\sin h}{h}}_{\text{This equals 1}} \qquad \boxed{2}$$

Handwritten annotations:

$\frac{d}{dx}(\tan x) = \frac{d}{dx} = \left(\frac{\sin x}{\cos x}\right)$

$= \frac{\cos x \cos x - \sin x(-\sin x)}{\cos^2 x}$

$= \frac{\cos^2 x + \sin^2 x}{\cos^2 x}$

$= \frac{1}{\cos^2 x} = \boxed{\sec^2 x}$

$\frac{d}{dx}(\sec) = \frac{d}{dx}\left(\frac{1}{\cos x}\right)$

$= \frac{(\cos x)(0) - (1)(\sin x)}{\cos^2 x}$

$= \frac{-\sin x}{\cos^2 x} = \frac{\sin x}{\cos x} \cdot \frac{1}{\cos x} = \boxed{\tan x \sec x}$

$\boxed{\sec x \tan x}$

$\frac{d}{dx}(\cot) = \boxed{-\csc^2 x} \qquad \frac{d}{dx}(\csc) = \boxed{-\csc x \cot x}$

Note that $\sin x$ and $\cos x$ do not depend on h and may be taken outside the limits in Eq. (2). We are justified in using the Sum Law because the final two limits exist. Indeed, by Theorem 2 in Section 2.6,

$$\lim_{h \to 0} \frac{\cos h - 1}{h} = 0 \quad \text{and} \quad \lim_{h \to 0} \frac{\sin h}{h} = 1$$

Therefore, Eq. (2) reduces to the formula $\dfrac{d}{dx} \sin x = \cos x$ as desired. The formula $\dfrac{d}{dx} \cos x = -\sin x$ is proved similarly (see Exercise 52). ▪

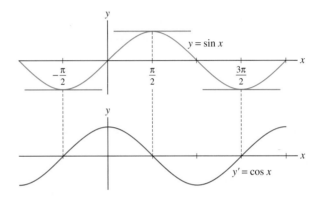

FIGURE 1 Compare the graphs of $y = \sin x$ and its derivative $y' = \cos x$.

GRAPHICAL INSIGHT   The formula $(\sin x)' = \cos x$ makes sense when we compare the graphs of sine and cosine (Figure 1). The tangent lines to the graph of $y = \sin x$ have positive slope on the interval $(-\frac{\pi}{2}, \frac{\pi}{2})$, and on this interval, the derivative $y' = \cos x$ is positive. Similarly, the tangent lines have negative slope on the interval $(\frac{\pi}{2}, \frac{3\pi}{2})$, where $y' = \cos x$ is negative. The tangent lines are horizontal at $x = -\frac{\pi}{2}, \frac{\pi}{2}, \frac{3\pi}{2}$, where $\cos x = 0$.

▪ **EXAMPLE 1**   Calculate $f'(x)$ and $f''(x)$, where $f(x) = x \cos x$.

Solution By the Product Rule,

$$f'(x) = x \frac{d}{dx} \cos x + \cos x \frac{d}{dx} x = x(-\sin x) + \cos x = \cos x - x \sin x$$

$$f''(x) = \frac{d}{dx} (\cos x - x \sin x) = \frac{d}{dx} \cos x - \frac{d}{dx} x \sin x$$

$$= -\sin x - \left(x(\sin x)' + \sin x \right) = -2 \sin x - x \cos x \qquad ▪$$

◀┈ *REMINDER The standard trigonometric functions are defined in Section 1.4.*

The derivatives of the other **standard trigonometric functions** may be computed using the Quotient Rule. We derive the formula for $(\tan x)'$ in Example 2 and leave the remaining formulas for the exercises (Exercises 39–41).

THEOREM 2 Derivatives of Standard Trigonometric Functions

$$\frac{d}{dx} \tan x = \sec^2 x, \qquad \frac{d}{dx} \sec x = \sec x \tan x$$

$$\frac{d}{dx} \cot x = -\csc^2 x \qquad \frac{d}{dx} \csc x = -\csc x \cot x$$

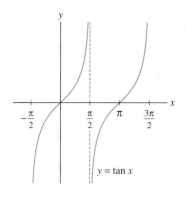

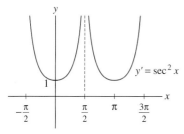

FIGURE 2 Graphs of $y = \tan x$ and its derivative $y' = \sec^2 x$.

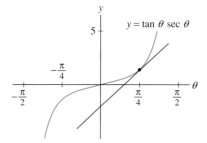

FIGURE 3 Tangent line to $y = \tan \theta \sec \theta$ at $\theta = \frac{\pi}{4}$.

■ **EXAMPLE 2** Verify the formula $\dfrac{d}{dx} \tan x = \sec^2 x$ (Figure 2).

Solution We use the Quotient Rule and the identity $\cos^2 x + \sin^2 x = 1$:

$$\frac{d}{dx} \tan x = \left(\frac{\sin x}{\cos x} \right)' = \frac{\cos x \cdot (\sin x)' - \sin x \cdot (\cos x)'}{\cos^2 x}$$

$$= \frac{\cos x \cos x - \sin x (-\sin x)}{\cos^2 x}$$

$$= \frac{\cos^2 x + \sin^2 x}{\cos^2 x} = \frac{1}{\cos^2 x} = \sec^2 x \qquad ■$$

■ **EXAMPLE 3** Find an equation of the tangent line to the graph of $y = \tan \theta \sec \theta$ at $\theta = \frac{\pi}{4}$.

Solution By the Product Rule,

$$y' = \tan \theta \, (\sec \theta)' + \sec \theta \, (\tan \theta)' = \tan \theta \, (\sec \theta \tan \theta) + \sec \theta \sec^2 \theta$$

$$= \tan^2 \theta \sec \theta + \sec^3 \theta$$

Now use the values $\sec \frac{\pi}{4} = \sqrt{2}$ and $\tan \frac{\pi}{4} = 1$ to compute

$$y \left(\frac{\pi}{4} \right) = \tan \left(\frac{\pi}{4} \right) \sec \left(\frac{\pi}{4} \right) = \sqrt{2}$$

$$y' \left(\frac{\pi}{4} \right) = \tan^2 \left(\frac{\pi}{4} \right) \sec \left(\frac{\pi}{4} \right) + \sec^3 \left(\frac{\pi}{4} \right) = \sqrt{2} + 2\sqrt{2} = 3\sqrt{2}$$

An equation of the tangent line (Figure 3) is $y - \sqrt{2} = 3\sqrt{2} \left(\theta - \frac{\pi}{4} \right)$. ■

■ **EXAMPLE 4** Calculate $f'(0)$, where $f(x) = e^x \cos x$.

Solution Use the Product Rule:

$$f'(x) = e^x \cdot (\cos x)' + \cos x \cdot (e^x)' = -e^x \sin x + \cos x \cdot e^x = e^x (\cos x - \sin x)$$

Then $f'(0) = e^0 (1 - 0) = 1$. ■

3.6 SUMMARY

- The basic trigonometric derivatives are

$$\boxed{\frac{d}{dx} \sin x = \cos x, \qquad \frac{d}{dx} \cos x = -\sin x}$$

- These two formulas are used to derive the additional formulas:

$$\frac{d}{dx} \tan x = \sec^2 x, \qquad \frac{d}{dx} \sec x = \sec x \tan x,$$

$$\frac{d}{dx} \cot x = -\csc^2 x, \qquad \frac{d}{dx} \csc x = -\csc x \cot x.$$

- Remember: *These formulas are valid only when the angle x is measured in radians.*

3.6 EXERCISES

Preliminary Questions

1. Determine the sign $\pm$ that yields the correct formula for the following:

(a) $\dfrac{d}{dx}(\sin x + \cos x) = \pm \sin x \pm \cos x$

(b) $\dfrac{d}{dx}\tan x = \pm \sec^2 x$

(c) $\dfrac{d}{dx}\sec x = \pm \sec x \tan x$

(d) $\dfrac{d}{dx}\cot x = \pm \csc^2 x$

2. Which of the following functions can be differentiated using the rules we have covered so far?

(a) $y = 3\cos x \cot x$ **(b)** $y = \cos(x^2)$ **(c)** $y = e^x \sin x$

3. Compute $\dfrac{d}{dx}(\sin^2 x + \cos^2 x)$ without using the derivative formulas for $\sin x$ and $\cos x$.

4. How is the addition formula used in deriving the formula $(\sin x)' = \cos x$?

Exercises

In Exercises 1–4, find an equation of the tangent line at the point indicated.

1. $y = \sin x, \quad x = \dfrac{\pi}{4}$

2. $y = \cos x, \quad x = \dfrac{\pi}{3}$

3. $y = \tan x, \quad x = \dfrac{\pi}{4}$

4. $y = \sec x, \quad x = \dfrac{\pi}{6}$

In Exercises 5–26, use the Product and Quotient Rules as necessary to find the derivative of each function.

5. $f(x) = \sin x \cos x$

6. $f(x) = x^2 \cos x$

7. $f(x) = \sin^2 x$

8. $f(x) = 9\sec x + 12\cot x$

9. $f(x) = x^3 \sin x$

10. $f(x) = \dfrac{\sin x}{x}$

11. $f(\theta) = \tan \theta \sec \theta$

12. $g(\theta) = \dfrac{e^\theta}{\cos \theta}$

13. $h(\theta) = e^{-\theta}\cos^2 \theta$

14. $k(\theta) = \theta^2 \sin^2 \theta$

15. $f(x) = (x - x^2)\cot x$

16. $f(z) = z \tan z$

17. $f(x) = \dfrac{\sec x}{x^2}$

18. $f(x) = \dfrac{x}{\tan x}$

19. $g(t) = \sin t - \dfrac{2}{\cos t}$

20. $R(y) = \dfrac{\cos y - 1}{\sin y}$

21. $f(x) = \dfrac{x}{\sin x + 2}$

22. $f(x) = (x^2 - x - 4)\csc x$

23. $f(x) = \dfrac{1 + \sin x}{1 - \sin x}$

24. $f(x) = \dfrac{1 + \tan x}{1 - \tan x}$

25. $g(x) = \dfrac{\sec x}{x}$

26. $h(x) = \dfrac{\sin x}{4 + \cos x}$

In Exercises 27–30, calculate the second derivative.

27. $f(x) = 3\sin x + 4\cos x$

28. $f(x) = \tan x$

29. $g(\theta) = \theta \sin \theta$

30. $h(\theta) = \csc \theta$

In Exercises 31–38, find an equation of the tangent line at the point specified.

31. $y = x^2 + \sin x, \quad x = 0$

32. $y = \theta \tan \theta, \quad \theta = \dfrac{\pi}{4}$

33. $y = 2\sin x + 3\cos x, \quad x = \dfrac{\pi}{3}$

34. $y = \theta^2 + \sin \theta, \quad \theta = 0$

35. $y = \csc x - \cot x, \quad x = \dfrac{\pi}{4}$

36. $y = \dfrac{\cos \theta}{1 + \sin \theta}, \quad \theta = \dfrac{\pi}{3}$

37. $y = e^x \cos x, \quad x = 0$

38. $y = e^\theta \sec \theta, \quad \theta = \dfrac{\pi}{4}$

In Exercises 39–41, verify the formula using $(\sin x)' = \cos x$ and $(\cos x)' = -\sin x$.

39. $\dfrac{d}{dx}\cot x = -\csc^2 x$

40. $\dfrac{d}{dx}\sec x = \sec x \tan x$

41. $\dfrac{d}{dx}\csc x = -\csc x \cot x$

42. Find the values of x between 0 and 2π where the tangent line to the graph of $y = \sin x \cos x$ is horizontal.

43. Calculate the first five derivatives of $f(x) = \cos x$. Then determine $f^{(8)}$ and $f^{(37)}$.

44. Find $y^{(157)}$, where $y = \sin x$.

45. Calculate $f''(x)$ and $f'''(x)$, where $f(x) = \tan x$.

46. GU Let $g(t) = t - \sin t$.

(a) Plot the graph of g with a graphing utility for $0 \le t \le 4\pi$.

(b) Show that the slope of the tangent line is always positive and verify this on your graph.

(c) For which values of t in the given range is the tangent line horizontal?

47. CAS Let $f(x) = \dfrac{\sin x}{x}$ for $x \ne 0$ and $f(0) = 1$.

(a) Plot $f(x)$ on $[-3\pi, 3\pi]$.

(b) Show that $f'(c) = 0$ if $c = \tan c$. Use the numerical root finder on a computer algebra system to find a good approximation to the smallest *positive* value c_0 such that $f'(c_0) = 0$.

(c) Verify that the horizontal line $y = f(c_0)$ is tangent to the graph of $y = f(x)$ at $x = c_0$ by plotting them on the same set of axes.

48. Show that no tangent line to the graph of $f(x) = \tan x$ has zero slope. What is the least slope of a tangent line? Justify your response by sketching the graph of $(\tan x)'$.

49. The height at time t (s) of a weight, oscillating up and down at the end of a spring, is $s(t) = 300 + 40 \sin t$ cm. Find the velocity and acceleration at $t = \frac{\pi}{3}$ s.

50. The horizontal range R of a projectile launched from ground level at an angle θ and initial velocity v_0 m/s is $R = (v_0^2/9.8) \sin \theta \cos \theta$. Calculate $dR/d\theta$. If $\theta = 7\pi/24$, will the range increase or decrease if the angle is increased slightly? Base your answer on the sign of the derivative.

51. If you stand 1 m from a wall and mark off points on the wall at equal increments δ of angular elevation (Figure 4), then these points grow increasingly far apart. Explain how this illustrates the fact that the derivative of $\tan \theta$ is increasing.

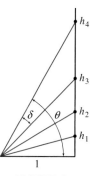

FIGURE 4

Further Insights and Challenges

52. Use the limit definition of the derivative and the addition law for the cosine to prove that $(\cos x)' = -\sin x$.

53. Show that a nonzero polynomial function $y = f(x)$ *cannot* satisfy the equation $y'' = -y$. Use this to prove that neither $\sin x$ nor $\cos x$ is a polynomial.

54. Verify the following identity and use it to give another proof of the formula $\sin' x = \cos x$:

$$\sin(x + h) - \sin x = 2 \cos \left(x + \tfrac{1}{2}h \right) \sin \left(\tfrac{1}{2}h \right)$$

Hint: Use the addition formula to prove that $\sin(a + b) - \sin(a - b) = 2 \cos a \sin b$.

55. Let $f(x) = x \sin x$ and $g(x) = x \cos x$.

(a) Show that $f'(x) = g(x) + \sin x$ and $g'(x) = -f(x) + \cos x$.

(b) Verify that $f''(x) = -f(x) + 2 \cos x$ and $g''(x) = -g(x) - 2 \sin x$.

(c) By further experimentation, try to find formulas for all higher derivatives of f and g. *Hint:* The kth derivative depends on whether $k = 4n, 4n + 1, 4n + 2,$ or $4n + 3$.

56. Show that if $\pi/2 < \theta < \pi$, then the distance along the x-axis between θ and the point where the tangent line intersects the x-axis is equal to $|\tan \theta|$ (Figure 5).

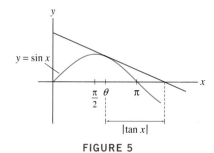

FIGURE 5

3.7 The Chain Rule

The **Chain Rule** is used to differentiate composite functions such as $y = \cos(x^2)$ and $y = \sqrt{x^3 + 1}$, which cannot be differentiated using any of the rules covered thus far. Recall that a *composite function* is obtained by "plugging" one function into another. More formally, the composite of f and g, denoted $f \circ g$, is defined by

$$(f \circ g)(x) = f\big(g(x)\big)$$

For convenience, we refer to g as the *inside* function and f as the *outside* function. It is convenient to use u rather than x as the variable for the outside function f. We then regard the composite as $f(u)$, where $u = g(x)$. For example, $y = \sqrt{x^3 + 1}$ is the function $y = \sqrt{u}$, where $u = x^3 + 1$.

A proof of the Chain Rule is given at the end of this section.

THEOREM 1 Chain Rule   If f and g are differentiable, then $(f \circ g)(x) = f(g(x))$ is differentiable and

$$\big(f(g(x))\big)' = f'\big(g(x)\big)\, g'(x)$$

$$\left(f\big(g\,(h(x))\big)\right)' = f'\big(g(h(x))\big) \cdot g'(h(x)) \cdot h'(x)$$

Setting $u = g(x)$, we may also write the Chain Rule as

$$\frac{d}{dx} f(u) = f'(u)\, \frac{du}{dx}$$

Here is a verbal form of the Chain Rule:

$$\big(f(g(x))\big)' = \text{outside}'(\text{inside}) \cdot \text{inside}'$$

■ **EXAMPLE 1**   Calculate the derivative of $y = \sqrt{x^3 + 1}$.

Solution   As noted above, $y = \sqrt{x^3 + 1}$ is a composite $f(g(x))$ with

$$f(u) = \sqrt{u},$$
$$u = g(x) = x^3 + 1$$

Since $f'(u) = \frac{1}{2}u^{-1/2}$ and $g'(x) = 3x^2$, we have

$$\text{Outside}'(\text{inside}) = f'(g(x)) = \frac{1}{2}u^{-1/2} = \frac{1}{2}(x^3 + 1)^{-1/2}$$

$$\text{Inside}' = g'(x) = 3x^2$$

By the Chain Rule,

$$\frac{d}{dx}\sqrt{x^3 + 1} = \text{outside}'(\text{inside}) \cdot \text{inside}' = \underbrace{\frac{1}{2}(x^3 + 1)^{-1/2}}_{f'(g(x))}\ \underbrace{(3x^2)}_{g'(x)} = \frac{3x^2}{2\sqrt{x^3 + 1}} \qquad ■$$

■ **EXAMPLE 2**   Calculate $\dfrac{dy}{dx}$ for   (a) $y = \cos(x^2)$ and   (b) $y = \tan\left(\dfrac{x}{x + 1}\right)$.

Solution

(a) We have $\cos(x^2) = f(g(x))$, where

$$\begin{array}{llll} \text{Outside:} & f(u) = \cos u, & \text{inside:} & u = g(x) = x^2 \\ \text{Outside}': & f'(u) = -\sin u, & \text{inside}': & g'(x) = 2x \end{array}$$

By the Chain Rule,

$$\frac{d}{dx}\cos(x^2) = f'(g(x))\, g'(x) = -\sin(x^2)\, 2x = -2x\sin(x^2)$$

(b) The outside function for $y = \tan\left(\dfrac{x}{x+1}\right)$ is $f(u) = \tan u$. Since $f'(u) = \sec^2 u$, the Chain Rule gives us

$$\frac{d}{dx}\tan\left(\frac{x}{x+1}\right) = \sec^2\left(\frac{x}{x+1}\right)\underbrace{\frac{d}{dx}\left(\frac{x}{x+1}\right)}_{\substack{\text{Derivative of}\\\text{inside function}}}$$

Now, by the Quotient Rule,

$$\frac{d}{dx}\left(\frac{x}{x+1}\right) = \frac{(x+1)\dfrac{d}{dx}x - x\dfrac{d}{dx}(x+1)}{(x+1)^2} = \frac{1}{(x+1)^2}$$

We obtain

$$\frac{d}{dx}\tan\left(\frac{x}{x+1}\right) = \sec^2\left(\frac{x}{x+1}\right)\ \frac{1}{(x+1)^2} = \frac{\sec^2\left(\dfrac{x}{x+1}\right)}{(x+1)^2} \qquad \blacksquare$$

It is instructive to write the Chain Rule in Leibniz notation. Suppose that $u = g(x)$ and

$$y = f(u) = f(g(x))$$

Then, by the Chain Rule,

$$\frac{dy}{dx} = f'(u)\, g'(x) = \frac{df}{du}\frac{du}{dx}$$

or

$$\boxed{\ \frac{dy}{dx} = \frac{dy}{du}\frac{du}{dx}\ }$$

Christiaan Huygens (1629–1695), one of the greatest scientists of his age, was Leibniz's teacher in mathematics and physics. He greatly admired Isaac Newton but did not accept Newton's theory of gravitation, which he referred to as the "improbable principle of attraction" because it did not explain how two masses separated by a distance could influence each other.

CONCEPTUAL INSIGHT   In Leibniz notation, it appears as if we are multiplying fractions and the Chain Rule is simply a matter of "cancelling the du." Since the symbolic expressions dy/du and du/dx are not fractions, this does not make sense literally, but it does suggest that derivatives behave *as if they were fractions* (this is reasonable because a derivative is a *limit* of fractions, namely of the difference quotients). Leibniz's form also emphasizes a key aspect of the Chain Rule: *rates of change multiply*. To illustrate this point, suppose that your salary increases three times as fast as the gross domestic product (GDP) and that the GDP increases by 2% per year. Then your salary will increase at the rate of 3×2 or 6% per year. In terms of derivatives,

$$\frac{d(\text{salary})}{dt} = \frac{d(\text{salary})}{d(\text{GDP})} \times \frac{d(\text{GDP})}{dt}$$

$$6\% \text{ per year} = 3 \times 2\% \text{ per year}$$

❚ **EXAMPLE 3** Imagine a sphere whose radius r increases at a rate of 3 cm/s. At what rate is the volume V of the sphere increasing when $r = 10$ cm?

Solution Since we are asked to determine the rate at which V is increasing, we must find dV/dt. The rate dr/dt is given as 3 cm/s, so we use the Chain Rule to express dV/dt in terms of dr/dt:

$$\underbrace{\frac{dV}{dt}}_{\substack{\text{ROC of volume} \\ \text{with respect to time}}} = \underbrace{\frac{dV}{dr}}_{\substack{\text{ROC of volume} \\ \text{with respect to radius}}} \times \underbrace{\frac{dr}{dt}}_{\substack{\text{ROC of radius} \\ \text{with respect to time}}}$$

To compute dV/dr, we use the formula for the volume of a sphere, $V = \frac{4}{3}\pi r^3$:

$$\frac{dV}{dr} = \frac{d}{dr}\left(\frac{4}{3}\pi r^3\right) = 4\pi r^2$$

Since $dr/dt = 3$, we obtain

$$\frac{dV}{dt} = \frac{dV}{dr}\frac{dr}{dt} = 4\pi r^2(3) = 12\pi r^2$$

When $r = 10$,

$$\left.\frac{dV}{dt}\right|_{r=10} = (12\pi)10^2 = 1{,}200\pi \approx 3{,}770 \text{ cm}^3/\text{s}$$ **■**

We now discuss some important special cases of the Chain Rule that arise frequently.

COROLLARY 2 General Power and Exponential Rules   If $u = g(x)$ is a differentiable function, then

- $\dfrac{d}{dx}g(x)^n = n(g(x))^{n-1}g'(x)$ (for any number n)

- $\dfrac{d}{dx}e^{g(x)} = g'(x)e^{g(x)}$

Proof To prove the General Power Rule, observe that $g(x)^n = f(g(x))$, where $f(u) = u^n$. Since $f'(u) = nu^{n-1}$, the Chain Rule yields

$$\frac{d}{dx}g(x)^n = f'(g(x))g'(x) = n(g(x))^{n-1}g'(x)$$

On the other hand, $e^{g(x)} = h(g(x))$, where $h(u) = e^u$. In this case, $h'(u) = h(u) = e^u$ and the Chain Rule yields

$$\frac{d}{dx}e^{g(x)} = h'(g(x))g'(x) = e^{g(x)}g'(x) = g'(x)e^{g(x)}$$ **■**

❚ **EXAMPLE 4** **Using the General Power Rule** Find the derivatives:
(a) $y = (x^3 + 9x + 2)^{-1/3}$ and   **(b)** $y = \sec^4 t$.

Solution We use the General Power Rule.

(a)
$$\frac{d}{dx}(x^3 + 9x + 2)^{-1/3} = -\frac{1}{3}(x^3 + 9x + 2)^{-4/3}\frac{d}{dx}(x^3 + 9x + 2)$$

$$= -\frac{1}{3}(x^3 + 9x + 2)^{-4/3}(3x^2 + 9)$$

$$= -(x^2 + 3)(x^3 + 9x + 2)^{-4/3}$$

(b)
$$\frac{d}{dt}\sec^4 t = 4\sec^3 t\,\frac{d}{dt}\sec t = 4\sec^3 t\,(\sec t\,\tan t) = 4\sec^4 t\,\tan t \qquad\blacksquare$$

■ **EXAMPLE 5** Using the General Exponential Rule Differentiate **(a)** $f(x) = e^{9x}$ and **(b)** $f(x) = e^{\cos x}$.

Solution

(a) $\dfrac{d}{dx}\left(e^{9x}\right) = 9e^{9x}$ **(b)** $\dfrac{d}{dx}\left(e^{\cos x}\right) = -(\sin x)e^{\cos x}$ ■

The next special case of the Chain Rule deals with functions of the form $f(kx + b)$, where k and b are constants. Recall that the graph of $f(kx + b)$ is related to the graph of $f(x)$ by scaling and shifting (Section 1.1). To compute the derivative, we view $f(kx + b)$ as a composite function with inside function $u = kx + b$ and apply the Chain Rule:

$$\frac{d}{dx}f(kx + b) = f'(kx + b)\frac{d}{dx}(kx + b) = kf'(kx + b)$$

COROLLARY 3 Shifting and Scaling Rule   If $f(x)$ is differentiable, then for any constants k and b,

$$\frac{d}{dx}f(kx + b) = k\,f'(kx + b)$$

In particular, $\dfrac{d}{dx}e^{kx+b} = ke^{kx+b}$.

For example,

$$\frac{d}{dx}\sin\left(2x + \frac{\pi}{4}\right) = 2\cos\left(2x + \frac{\pi}{4}\right)$$

$$\frac{d}{dx}(9x - 2)^5 = (9)(5)(9x - 2)^4 = 45(9x - 2)^4$$

$$\frac{d}{dt}e^{7-4t} = -4e^{7-4t}$$

According to our convention, $\sin x$ denotes the sine of x radians, and with this convention, the formula $(\sin x)' = \cos x$ holds. In the next example, we derive a formula for the derivative of the sine function when x is measured in degrees.

■ **EXAMPLE 6** Trigonometric Derivatives in Degrees Calculate the derivative of the sine function as a function of degrees rather than radians.

Solution It is convenient to underline a trigonometric function to indicate that the variable represents degrees rather than radians. For example,

$$\underline{\sin} x = \text{the sine of } x \text{ degrees}$$

A similar calculation shows that the factor $\pi/180$ appears in the formulas for the derivatives of the other standard trigonometric functions with respect to degrees. For example,

$$\frac{d}{dx}\underline{\tan}\, x = \left(\frac{\pi}{180}\right)\underline{\sec^2}\, x$$

The functions $\sin x$ and $\underline{\sin}\, x$ are different, but they are related. In fact, for any number x, we have

$$\underline{\sin}\, x = \sin\left(\frac{\pi}{180}\, x\right)$$

since x degrees corresponds to $\frac{\pi}{180}x$ radians. By Corollary 3,

$$\frac{d}{dx}\underline{\sin}\, x = \frac{d}{dx}\sin\left(\frac{\pi}{180}\, x\right) = \left(\frac{\pi}{180}\right)\cos\left(\frac{\pi}{180}x\right) = \left(\frac{\pi}{180}\right)\underline{\cos}\, x \qquad \blacksquare$$

When the inside function is itself a composite function, we apply the Chain Rule more than once, as in the next example.

■ **EXAMPLE 7** Using the Chain Rule Twice Find the derivative of $y = \sqrt{x + \sqrt{x^2 + 1}}$.

Solution First apply the Chain Rule with inside function $u = x + \sqrt{x^2 + 1}$:

$$\frac{d}{dx}\sqrt{x + \sqrt{x^2 + 1}} = \frac{1}{2}(x + \sqrt{x^2 + 1})^{-1/2}\frac{d}{dx}\left(x + \sqrt{x^2 + 1}\right)$$

Then use the Chain Rule again to compute the remaining derivative:

$$\frac{d}{dx}\sqrt{x + \sqrt{x^2 + 1}} = \frac{1}{2}(x + \sqrt{x^2 + 1})^{-1/2}\left(1 + \frac{1}{2}(x^2 + 1)^{-1/2}\frac{d}{dx}(x^2 + 1)\right)$$

$$= \frac{1}{2}(x + \sqrt{x^2 + 1})^{-1/2}\left(1 + \frac{1}{2}(x^2 + 1)^{-1/2}(2x)\right)$$

$$= \frac{1}{2}(x + \sqrt{x^2 + 1})^{-1/2}\left(1 + x(x^2 + 1)^{-1/2}\right) \qquad \blacksquare$$

Proof of the Chain Rule The difference quotient for the composite $f \circ g$ is

$$\frac{f(g(x + h)) - f(g(x))}{h} \qquad (h \neq 0)$$

Our goal is to show that $(f \circ g)'$ is the product of $f'(g(x))$ and $g'(x)$, so it makes sense to write the difference quotient as a product:

$$\frac{f(g(x + h)) - f(g(x))}{h} = \frac{f(g(x + h)) - f(g(x))}{g(x + h) - g(x)} \times \frac{g(x + h) - g(x)}{h} \qquad \boxed{1}$$

This is legitimate only if the denominator $g(x + h) - g(x)$ is nonzero. Therefore, to continue our proof, we make the extra assumption that $g(x + h) - g(x) \neq 0$ for all h near but not equal to 0. This assumption is not necessary, but without it, the argument is more technical (see Exercise 103).

Under our assumption, we may use Eq. (1) to write $(f \circ g)'(x)$ as a product of two limits:

$$(f \circ g)'(x) = \underbrace{\lim_{h \to 0}\frac{f(g(x + h)) - f(g(x))}{g(x + h) - g(x)}}_{\text{Show that this equals } f'(g(x)).} \times \underbrace{\lim_{h \to 0}\frac{g(x + h) - g(x)}{h}}_{\text{This is } g'(x).} \qquad \boxed{2}$$

The second limit on the right is $g'(x)$. The Chain Rule will follow if we show that the first limit equals $f'(g(x))$. To verify this, set

$$k = g(x + h) - g(x)$$

Then $g(x + h) = g(x) + k$ and

$$\frac{f(g(x+h)) - f(g(x))}{g(x+h) - g(x)} = \frac{f(g(x) + k) - f(g(x))}{k}$$

Since $g(x)$ is differentiable, it is also continuous. Therefore, $g(x + h)$ tends to $g(x)$ and $k = g(x + h) - g(x)$ tends to zero as $h \to 0$. Thus, we may rewrite the limit in terms of k to obtain the desired result:

$$\lim_{h \to 0} \frac{f(g(x+h)) - f(g(x))}{g(x+h) - g(x)} = \lim_{k \to 0} \frac{f(g(x) + k) - f(g(x))}{k} = f'(g(x)) \qquad \blacksquare$$

3.7 SUMMARY

- The Chain Rule expresses $(f \circ g)'$ in terms of f' and g':

$$(f(g(x)))' = f'(g(x))\, g'(x)$$

- In Leibniz notation: $\dfrac{dy}{dx} = \dfrac{dy}{du} \dfrac{du}{dx}$, where $y = f(u)$ and $u = g(x)$

- General Power Rule: $\dfrac{d}{dx} g(x)^n = n(g(x))^{n-1} g'(x)$.

- General Exponential Rule: $\dfrac{d}{dx} e^{g(x)} = g'(x) e^{g(x)}$.

- Shifting and Scaling Rule: $\dfrac{d}{dx} f(kx + b) = k f'(kx + b)$. In particular,

$$\frac{d}{dx} e^{kx+b} = k e^{kx+b}$$

3.7 EXERCISES

Preliminary Questions

1. Identify the outside and inside functions for each of these composite functions.

(a) $y = \sqrt{4x + 9x^2}$

(b) $y = \tan(x^2 + 1)$

(c) $y = \sec^5 x$

(d) $y = (1 + e^x)^4$

2. Which of the following can be differentiated easily *without* using the Chain Rule?

$$y = \tan(7x^2 + 2), \qquad y = \frac{x}{x+1},$$

$$y = \sqrt{x} \cdot \sec x, \qquad y = \sqrt{x \cos x},$$

$$y = xe^x, \qquad y = e^{\sin x}$$

3. Which is the derivative of $f(5x)$?

(a) $5 f'(x)$ **(b)** $5 f'(5x)$ **(c)** $f'(5x)$

4. How many times must the Chain Rule be used to differentiate each function?

(a) $y = \cos(x^2 + 1)$

(b) $y = \cos((x^2 + 1)^4)$

(c) $y = \sqrt{\cos((x^2 + 1)^4)}$

(d) $y = \tan(e^{5-6x})$

5. Suppose that $f'(4) = g(4) = g'(4) = 1$. Do we have enough information to compute $F'(4)$, where $F(x) = f(g(x))$? If not, what is missing?

Exercises

In Exercises 1–4, fill in a table of the following type:

$f(g(x))$	$f'(u)$	$f'(g(x))$	$g'(x)$	$(f \circ g)'$

1. $f(u) = u^{3/2}$, $g(x) = x^4 + 1$

2. $f(u) = u^3$, $g(x) = 3x + 5$

3. $f(u) = \tan u$, $g(x) = x^4$

4. $f(u) = u^4 + u$, $g(x) = \cos x$

In Exercises 5–6, write the function as a composite $f(g(x))$ and compute the derivative using the Chain Rule.

5. $y = (x + \sin x)^4$

6. $y = \cos(x^3)$

7. Calculate $\dfrac{d}{dx} \cos u$ for the following choices of $u(x)$:

(a) $u = 9 - x^2$ (b) $u = x^{-1}$ (c) $u = \tan x$

8. Calculate $\dfrac{d}{dx} f(x^2 + 1)$ for the following choices of $f(u)$:

(a) $f(u) = \sin u$ (b) $f(u) = 3u^{3/2}$

(c) $f(u) = u^2 - u$

In Exercises 9–14, use the General Power Rule or the Shifting and Scaling Rule to find the derivative.

9. $y = (x^2 + 9)^4$

10. $y = \sin^5 x$

11. $y = \sqrt{11x + 4}$

12. $y = (7x - 9)^5$

13. $y = e^{10 - x^2}$

14. $y = (x^4 - x + 2)^{-3/2}$

In Exercises 15–18, find the derivative of $f \circ g$.

15. $f(u) = \sin u$, $g(x) = 2x + 1$

16. $f(u) = 2u + 1$, $g(x) = \sin x$

17. $f(u) = e^u$, $g(x) = x + x^{-1}$

18. $f(u) = \dfrac{u}{u - 1}$, $g(x) = \csc x$

In Exercises 19–20, find the derivatives of $f(g(x))$ and $g(f(x))$.

19. $f(u) = \cos u$, $g(x) = x^2 + 1$

20. $f(u) = u^3$, $g(x) = \dfrac{1}{x + 1}$

In Exercises 21–34, use the Chain Rule to find the derivative.

21. $y = \sin(x^2)$

22. $y = \sin^2 x$

23. $y = \cot(4t^2 + 9)$

24. $y = \sqrt{1 - t^2}$

25. $y = (t^2 + 3t + 1)^{-5/2}$

26. $y = (x^4 - x^3 - 1)^{2/3}$

27. $y = \left(\dfrac{x + 1}{x - 1}\right)^4$

28. $y = \sec \dfrac{1}{x}$

29. $y = \cos^3(e^{4\theta})$

30. $y = \tan(\theta + \cos \theta)$

31. $y = \dfrac{1}{\sqrt{\cos(x^2) + 1}}$

32. $y = (\sqrt{x + 1} - 1)^{3/2}$

33. $y = e^{1/x}$

34. $y = e^{x^2}$

In Exercises 35–72, find the derivative using the appropriate rule or combination of rules.

35. $y = \tan 5x$

36. $y = \sin(x^2 + 4x)$

37. $y = x \cos(1 - 3x)$

38. $y = \sin(x^2) \cos(x^2)$

39. $y = (4t + 9)^{1/2}$

40. $y = \sin(\cos \theta)$

41. $y = (x^3 + \cos x)^{-4}$

42. $y = \sin(\cos(\sin x))$

43. $y = \sqrt{\sin x \cos x}$

44. $y = x^2 \tan 2x$

45. $y = (z + 1)^4 (2z - 1)^3$

46. $y = 3 + 2s\sqrt{s}$

47. $y = (x + x^{-1})\sqrt{x + 1}$

48. $y = \cos^2(8x)$

49. $y = (\cos 6x + \sin x^2)^{1/2}$

50. $y = \dfrac{(x + 1)^{1/2}}{x + 2}$

51. $y = \tan^3 x + \tan(x^3)$

52. $y = \sqrt{4 - 3\cos x}$

53. $y = \sqrt{\dfrac{z + 1}{z - 1}}$

54. $y = (\cos^3 x + 3\cos x + 7)^9$

55. $y = \dfrac{\cos(1 + x)}{1 + \cos x}$

56. $y = \sec(\sqrt{t^2 - 9})$

57. $y = \cot^7(x^5)$

58. $y = \dfrac{\cos(x^2)}{1 + x^2}$

59. $y = (1 + (x^2 + 2)^5)^3$

60. $y = \left(1 + \cot^5(x^4 + 1)\right)^9$

61. $y = 4e^{-x} + 7e^{-2x}$

62. $y = e^{\tan \theta}$

63. $y = (2e^{3x} + 3e^{-2x})^4$

64. $y = \dfrac{1}{1 - e^{-3t}}$

65. $y = \cos(te^{-2t})$

66. $y = \tan(e^{5 - 6x})$

67. $y = e^{(x^2 + 2x + 3)^2}$

68. $y = e^{e^x}$

69. $y = \sqrt{1 + \sqrt{1 + \sqrt{x}}}$

70. $y = \sqrt{\sqrt{x + 1} + 1}$

71. $y = \sqrt{kx + b}$; k and b any constants

72. $y = \dfrac{1}{\sqrt{kt^4 + b}}$; k, b constants, not both zero

73. Compute $\dfrac{df}{dx}$ if $\dfrac{df}{du} = 2$ and $\dfrac{du}{dx} = 6$.

74. The average molecular velocity v of a gas in a certain container is given by $v = 29\sqrt{T}$ m/s, where T is the temperature in kelvins. The temperature is related to the pressure (in atmospheres) by $T = 200P$. Find $\dfrac{dv}{dP}\bigg|_{P=1.5}$.

75. With notation as in Example 6, calculate

(a) $\dfrac{d}{d\theta} \sin \theta \bigg|_{\theta=60°}$

(b) $\dfrac{d}{d\theta} \left(\theta + \tan \theta \right) \bigg|_{\theta=45°}$

76. Assume that $f(0) = 2$ and $f'(0) = 3$. Find the derivatives of $(f(x))^3$ and $f(7x)$ at $x = 0$.

77. Compute the derivative of $h(\sin x)$ at $x = \frac{\pi}{6}$, assuming that $h'(0.5) = 10$.

78. Let $F(x) = f(g(x))$, where the graphs of f and g are shown in Figure 1. Estimate $g'(2)$ and $f'(g(2))$ from the graph and compute $F'(2)$.

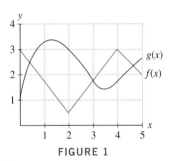

FIGURE 1

In Exercises 79–82, use the table of values to calculate the derivative of the function at the given point.

x	1	4	6
$f(x)$	4	0	6
$f'(x)$	5	7	4
$g(x)$	4	1	6
$g'(x)$	5	$\frac{1}{2}$	3

79. $f(g(x))$, $x = 6$

80. $e^{f(x)}$, $x = 4$

81. $g(\sqrt{x})$, $x = 16$

82. $f(2x + g(x))$, $x = 1$

In Exercises 83–86, compute the indicated higher derivatives.

83. $\dfrac{d^2}{dx^2} \sin(x^2)$

84. $\dfrac{d^2}{dx^2} (x^2 + 9)^5$

85. $\dfrac{d^3}{dx^3} (3x + 9)^{11}$

86. $\dfrac{d^3}{dx^3} \sin(2x)$

87. ⌂⌂⌂ Use a computer algebra system to compute $f^{(k)}(x)$ for $k = 1, 2, 3$ for the following functions:

(a) $f(x) = \cot(x^2)$

(b) $f(x) = \sqrt{x^3 + 1}$

88. Use the Chain Rule to express the second derivative of $f \circ g$ in terms of the first and second derivatives of f and g.

89. Compute the second derivative of $\sin(g(x))$ at $x = 2$, assuming that $g(2) = \frac{\pi}{4}$, $g'(2) = 5$, and $g''(2) = 3$.

90. An expanding sphere has radius $r = 0.4t$ cm at time t (in seconds). Let V be the sphere's volume. Find dV/dt when (a) $r = 3$ and (b) $t = 3$.

91. The power P in a circuit is $P = Ri^2$, where R is resistance and i the current. Find dP/dt at $t = 2$ if $R = 1,000 \ \Omega$ and i varies according to $i = \sin(4\pi t)$ (time in seconds).

92. The price (in dollars) of a computer component is $P = 2C - 18C^{-1}$, where C is the manufacturer's cost to produce it. Assume that cost at time t (in years) is $C = 9 + 3t^{-1}$ and determine the ROC of price with respect to time at $t = 3$.

93. The force F (in Newtons) between two charged objects is $F = 100/r^2$, where r is the distance (in meters) between them. Find dF/dt at $t = 10$ if the distance at time t (in seconds) is $r = 1 + 0.4t^2$.

94. According to the U.S. standard atmospheric model, developed by the National Oceanic and Atmospheric Administration for use in aircraft and rocket design, atmospheric temperature T (in degrees Celsius), pressure P (kPa = 1,000 Pascals), and altitude h (meters) are related by the formulas (valid in the troposphere $h \le 11,000$):

$$T = 15.04 - 0.000649h, \qquad P = 101.29 + \left(\frac{T + 273.1}{288.08} \right)^{5.256}$$

Calculate dP/dh. Then estimate the change in P (in Pascals, Pa) per additional meter of altitude when $h = 3,000$.

95. Conservation of Energy The position at time t (in seconds) of a weight of mass m oscillating at the end of a spring is $x(t) = L\sin(2\pi\omega t)$. Here, L is the maximum length of the spring, and ω the frequency (number of oscillations per second). Let v and a be the velocity and acceleration of the weight.

(a) By Hooke's Law, the spring exerts a force of magnitude $F = -kx$ on the weight, where k is the *spring constant*. Use Newton's Second Law, $F = ma$, to show that $2\pi\omega = \sqrt{k/m}$.

(b) The weight has kinetic energy $K = \frac{1}{2}mv^2$ and potential energy $U = \frac{1}{2}kx^2$. Prove that the total energy $E = K + U$ is conserved, that is, $dE/dt = 0$.

Further Insights and Challenges

96. Show that if f, g, and h are differentiable, then

$$[f(g(h(x)))]' = f'(g(h(x)))g'(h(x))h'(x)$$

97. Recall that $f(x)$ is *even* if $f(-x) = f(x)$ and *odd* if $f(-x) = -f(x)$.

(a) Sketch a graph of any even function and explain graphically why its derivative is an odd function.

(b) Show that differentiation reverses parity: If f is even, then f' is odd, and if f is odd, then f' is even. *Hint:* Differentiate $f(-x)$.

(c) Suppose that f' is even. Is f necessarily odd? *Hint:* Check if this is true for linear functions.

98. Power Rule for Fractional Exponents Let $f(u) = u^q$ and $g(x) = x^{p/q}$. Show that $f(g(x)) = x^p$ (recall the laws of exponents). Apply the Chain Rule and the Power Rule for integer exponents to show that $f'(g(x))g'(x) = px^{p-1}$. Then derive the Power Rule for $x^{p/q}$.

In Exercises 99–101, use the following fact (proved in Chapter 4): If a differentiable function f satisfies $f'(x) = 0$ for all x, then f is a constant function.

99. Differential Equation of Sine and Cosine Suppose that $f(x)$ satisfies the following equation (called a **differential equation**):

$$f''(x) = -f(x) \qquad \boxed{3}$$

(a) Show that $f(x)^2 + f'(x)^2 = f(0)^2 + f'(0)^2$. *Hint:* Show that the function on the left has zero derivative.

(b) Verify that $\sin x$ and $\cos x$ satisfy Eq. (3) and deduce that $\sin^2 x + \cos^2 x = 1$.

100. Suppose that both f and g satisfy Eq. (3) and have the same initial values, that is, $f(0) = g(0)$ and $f'(0) = g'(0)$. Show that $f(x) = g(x)$ for all x. *Hint:* Show that $f - g$ satisfies Eq. (3) and apply Exercise 99(a).

101. Use the result of Exercise 100 to show the following: $f(x) = \sin x$ is the unique solution of Eq. (3) such that $f(0) = 0$ and $f'(0) = 1$; and $g(x) = \cos x$ is the unique solution such that $g(0) = 1$ and $g'(0) = 0$. This result provides a means of defining and developing all the properties of the trigonometric functions without reference to triangles or the unit circle.

102. A Discontinuous Derivative Use the limit definition to show that $g'(0)$ exists but $g'(0) \neq \lim_{x \to 0} g'(x)$, where

$$g(x) = \begin{cases} x^2 \sin \dfrac{1}{x} & \text{if } x \neq 0 \\ 0 & \text{if } x = 0 \end{cases}$$

103. Chain Rule This exercise proves the Chain Rule without the special assumption made in the text. For any number b, define a new function

$$F(u) = \frac{f(u) - f(b)}{u - b} \qquad \text{for all } u \neq b$$

Observe that $F(u)$ is equal to the slope of the secant line through $(b, f(b))$ and $(u, f(u))$.

(a) Show that if we define $F(b) = f'(b)$, then $F(u)$ is continuous at $u = b$.

(b) Take $b = g(a)$. Show that if $x \neq a$, then for all u,

$$\frac{f(u) - f(g(a))}{x - a} = F(u)\frac{u - g(a)}{x - a} \qquad \boxed{4}$$

Note that both sides are zero if $u = g(a)$.

(c) Substitute $u = g(x)$ in Eq. (4) to obtain

$$\frac{f(g(x)) - f(g(a))}{x - a} = F(g(x))\frac{g(x) - g(a)}{x - a}$$

Derive the Chain Rule by computing the limit of both sides as $x \to a$.

3.8 Implicit Differentiation

To differentiate using the methods covered thus far, we must have a formula for y in terms of x, for instance, $y = x^3 + 1$. But suppose that y is determined instead by an equation such as

$$y^4 + xy = x^3 - x + 2 \qquad \boxed{1}$$

In this case, we say that y is defined *implicitly*. How can we find the slope of the tangent line at a point on the graph (Figure 1)? It may be inconvenient or even impossible to solve for y explicitly as a function of x, but we can find dy/dx using the method of **implicit differentiation**.

To illustrate, consider the equation of the unit circle:

$$x^2 + y^2 = 1$$

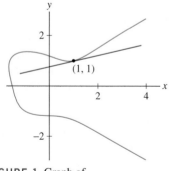

FIGURE 1 Graph of
$$y^4 + xy = x^3 - x + 2$$

To compute $\dfrac{dy}{dx}$, first take the derivative of both sides of the equation and evaluate:

$$\frac{d}{dx}\left(x^2 + y^2\right) = \frac{d}{dx}(1)$$

$$\frac{d}{dx}\left(x^2\right) + \frac{d}{dx}\left(y^2\right) = 0$$

$$2x + \frac{d}{dx}\left(y^2\right) = 0 \qquad \boxed{2}$$

What should we do with the term $\dfrac{d}{dx}(y^2)$? We use the Chain Rule. If we think of y as a function $y = f(x)$, then $y^2 = f(x)^2$, and by the Chain Rule,

$$\frac{d}{dx}y^2 = \frac{d}{dx}f(x)^2 = 2f(x)\frac{df}{dx} = 2y\frac{dy}{dx}$$

Equation (2) becomes $2x + 2y\dfrac{dy}{dx} = 0$, and we may solve for $\dfrac{dy}{dx}$ if $y \neq 0$:

$$\boxed{\dfrac{dy}{dx} = -\dfrac{x}{y}} \qquad \boxed{3}$$

■ **EXAMPLE 1** Use Eq. (3) to find the slope of the tangent line at the point $P = \left(\frac{3}{5}, \frac{4}{5}\right)$ on the unit circle (Figure 2).

Solution Set $x = \frac{3}{5}$ and $y = \frac{4}{5}$ in Eq. (3):

$$\left.\frac{dy}{dx}\right|_P = -\frac{x}{y} = -\frac{\frac{3}{5}}{\frac{4}{5}} = -\frac{3}{4} \qquad ■$$

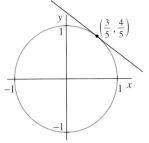

FIGURE 2 Tangent line to unit circle at $\left(\frac{3}{5}, \frac{4}{5}\right)$ has slope $-\frac{3}{4}$.

In the previous example, we could have computed dy/dx directly without using implicit differentitation. The upper semicircle is the graph of $y = \sqrt{1 - x^2}$ and

$$\frac{dy}{dx} = \frac{d}{dx}\sqrt{1 - x^2} = \frac{1}{2}\left(1 - x^2\right)^{-1/2}\frac{d}{dx}\left(1 - x^2\right)$$

$$= \frac{1}{2}\left(1 - x^2\right)^{-1/2}(-2x) = -\frac{x}{\sqrt{1 - x^2}}$$

This is a formula for the derivative in terms of x alone, whereas Eq. (3) expresses the derivative in terms of both x and y, as is typical when we use implicit differentiation. The two formulas agree because $y = \sqrt{1 - x^2}$. Although it was possible to solve for y in terms of x in this particular case, it is often difficult or impossible to do so. In such cases, implicit differentiation is the only available method for computing the derivative.

Before presenting additional examples, let's pause to examine again how the factor dy/dx arises when you differentiate an expression involving y with respect to x. It would not appear if you were differentiating with respect to y. Thus,

$$\frac{d}{dy}\sin y = \cos y \qquad \text{but} \qquad \frac{d}{dx}\sin y = (\cos y)\frac{dy}{dx}$$

Notice what happens if we insist on applying the Chain Rule to $\dfrac{d}{dy}\sin y$. The extra factor appears but it is equal to 1:

$$\frac{d}{dy}\sin y = (\cos y)\frac{dy}{dy} = \cos y$$

Similarly, we may apply the Product Rule to xy to obtain (where y' denotes dy/dx):

$$\frac{d}{dx}(xy) = x\frac{dy}{dx} + y\frac{dx}{dx} = x\frac{dy}{dx} + y = xy' + y$$

or we may apply the Quotient Rule to t/y (where now y' denotes dy/dt):

$$\frac{d}{dt}\left(\frac{t}{y}\right) = \frac{y\left(\dfrac{dt}{dt}\right) - t\left(\dfrac{dy}{dt}\right)}{y^2} = \frac{y - ty'}{y^2}$$

■ **EXAMPLE 2** Find an equation of the tangent line at the point $P = (1, 1)$ on the curve (Figure 1)

$$y^4 + xy = x^3 - x + 2$$

Solution We break up the derivative calculation into two steps.

Step 1. **Differentiate both sides of the equation with respect to x.**

$$\frac{d}{dx}y^4 + \frac{d}{dx}(xy) = \frac{d}{dx}\left(x^3 - x + 2\right)$$

$$4y^3\, y' + \left(xy' + y\right) = 3x^2 - 1 \qquad \boxed{4}$$

Step 2. **Solve for y'.**

Group the terms in Eq. (4) involving y' on the left-hand side and place the remaining terms on the right-hand side:

$$4y^3 y' + xy' = 3x^2 - 1 - y$$

Then factor out y':

$$\left(4y^3 + x\right)y' = 3x^2 - 1 - y$$

and divide:

$$y' = \frac{3x^2 - 1 - y}{4y^3 + x} \qquad \boxed{5}$$

To find the derivative at $P = (1, 1)$, we apply Eq. (5) with $x = 1$ and $y = 1$:

$$\left.\frac{dy}{dx}\right|_{(1,1)} = \frac{3 \cdot 1^2 - 1 - 1}{4 \cdot 1^3 + 1} = \frac{1}{5}$$

The equation of the tangent line may be written $y - 1 = \dfrac{1}{5}(x - 1)$ or $y = \dfrac{1}{5}x + \dfrac{4}{5}$. ■

CONCEPTUAL INSIGHT   The graph of an equation need not define a function because there may be more than one y-value for a given value of x. However, the graph is generally made up of several pieces called **branches**, each of which does define a function. For example, the branches of the unit circle $x^2 + y^2 = 1$ are the graphs of the functions $y = \sqrt{1 - x^2}$ and $y = -\sqrt{1 - x^2}$. Similarly, the graph in Figure 3 has upper and lower branches. We have not discussed the precise conditions under which these branches are differentiable—a more advanced topic that lies beyond the scope of this text. But in the examples we present, it will be clear that y' exists either for all x or for all but a certain number of exceptional values of x that are identifiable from the graph.

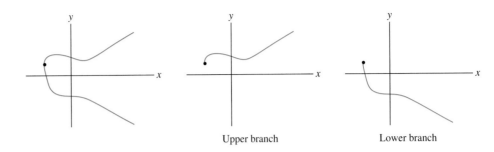

FIGURE 3 Each branch of the graph of $y^4 + xy = x^3 - x + 2$ defines a function of x.

Upper branch

Lower branch

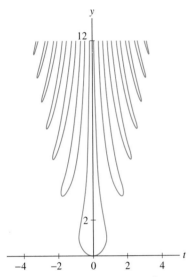

FIGURE 4 Graph of $\cos(ty) = \dfrac{t^2}{y}$.

■ **EXAMPLE 3** Find $\dfrac{dy}{dt}$, where $\cos(ty) = \dfrac{t^2}{y}$ (Figure 4).

Solution The two steps used in the previous example also apply here.

Step 1. **Differentiate both sides of the equation with respect to t.**
We obtain

$$\frac{d}{dt}\cos(ty) = \frac{d}{dt}\left(\frac{t^2}{y}\right)$$

$\boxed{6}$

We calculate each derivative separately, writing y' for dy/dt:

$$\frac{d}{dt}(\cos(ty)) = -\big(\sin(ty)\big)\frac{d}{dt}(ty) = -\sin(ty)\big(ty' + y\big) \quad \text{(Chain Rule)}$$

$$\frac{d}{dt}\left(\frac{t^2}{y}\right) = \frac{y\dfrac{d}{dt}t^2 - t^2\dfrac{dy}{dt}}{y^2} = \frac{2ty - t^2 y'}{y^2} \quad \text{(Quotient Rule)}$$

Equation (6) becomes

$$-\big(\sin(ty)\big)\big(ty' + y\big) = \frac{2ty - t^2 y'}{y^2}$$

$\boxed{7}$

Step 2. **Solve for y'.**

$$-t\big(\sin(ty)\big)y' - y\sin(ty) = \frac{2t}{y} - \frac{t^2}{y^2}y' \quad \text{[expand Eq. (7)]}$$

$$-t\sin(ty)\,y' + \frac{t^2}{y^2}\,y' = \frac{2t}{y} + y\sin(ty) \quad \text{(place all y' terms on left)}$$

$$y'\left(-t\sin(ty) + \frac{t^2}{y^2}\right) = \frac{2t}{y} + y\sin(ty) \quad \text{(factor out y')}$$

Now we can solve for y':

$$y' = \frac{\dfrac{2t}{y} + y\sin(ty)}{-t\sin(ty) + \dfrac{t^2}{y^2}} = \frac{2ty + y^3\sin(ty)}{t^2 - ty^2\sin(ty)}$$

■

■ **EXAMPLE 4**   Find the slope of the tangent line to the curve $e^{xy} = x + y$ at the point $P = (-1, 1.28)$ (the y-coordinate is approximate).

Solution   We use implicit differentiation (Figure 5):

$$\frac{d}{dx}e^{xy} = \frac{d}{dx}(x+y)$$

$$(xy' + y)e^{xy} = 1 + y'$$

$$(xe^{xy} - 1)y' = 1 - ye^{xy}$$

$$y' = \frac{1 - ye^{xy}}{xe^{xy} - 1}$$

The slope of the tangent line at $P = (-1, 1.28)$ is (approximately)

$$\left.\frac{dy}{dx}\right|_{(-1,1.28)} \approx \frac{1 - (1.28)e^{-1.28}}{e^{-1.28} - 1} \approx -0.89$$ ■

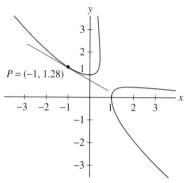

$P = (-1, 1.28)$

FIGURE 5 Graph of $e^{xy} = x + y$.

3.8 SUMMARY

- Implicit differentiation is used to compute dy/dx when x and y are related by an equation.

Step 1.   Take the derivative of both sides of the equation with respect to x.

Step 2.   Solve for y' by collecting the terms involving y' on one side and the remaining terms on the other side of the equation.

- Remember to include the factor dy/dx when differentiating expressions involving y with respect to x. For instance,

$$\frac{d}{dx}\sin y = (\cos y)\frac{dy}{dx}$$

3.8 EXERCISES

Preliminary Questions

1. Which differentiation rule is used to show $\dfrac{d}{dx}\sin y = \cos y\,\dfrac{dy}{dx}$?

2. One of (a)–(c) is incorrect. Find the mistake and correct it.

(a) $\dfrac{d}{dy}\sin(y^2) = 2y\cos(y^2)$

(b) $\dfrac{d}{dx}\sin(x^2) = 2x\cos(x^2)$

(c) $\dfrac{d}{dx}\sin(y^2) = 2y\cos(y^2)$

3. On an exam, Jason was asked to differentiate the equation

$$x^2 + 2xy + y^3 = 7$$

What are the errors in Jason's answer?

$$2x + 2xy' + 3y^2 = 0$$

4. Which of (a) or (b) is equal to $\dfrac{d}{dx}(x\sin t)$?

(a) $(x\cos t)\dfrac{dt}{dx}$

(b) $(x\cos t)\dfrac{dt}{dx} + \sin t$

Exercises

1. Show that if you differentiate both sides of $x^2 + 2y^3 = 6$, the result is $2x + 6y^2y' = 0$. Then solve for y' and calculate dy/dx at the point $(2, 1)$.

2. Show that if you differentiate both sides of the equation $xy + 3x + y = 1$, the result is $(x + 1)y' + y + 3 = 0$. Then solve for y' and calculate dy/dx at the point $(1, -1)$.

$$4x^3 - 6x^2 + 2x = 0 \qquad 2x(2x+1)(x-1) = 0$$
$$2x(2x^2 - 3x + 1) = 0$$
$$x = 0, \tfrac{1}{2}, 1$$

In Exercises 3–8, differentiate the expression with respect to x.

3. $x^2 y^3$

4. $\dfrac{x^3}{y}$

5. $(x^2 + y^2)^{3/2}$

6. $\tan(xt)$

7. $z + z^2$

8. $e^{y/x}$

In Exercises 9–26, calculate the derivative of y (or other variable) with respect to x.

9. $3y^3 + x^2 = 5$

10. $y^4 - y = x^3 + x$

11. $x^2 y + 2xy^2 = x + y$

12. $\sin(xt) = t$

13. $x^2 y + y^4 - 3x = 8$

14. $x^3 R^5 = 1$

15. $x^4 + z^4 = 1$

16. $\dfrac{y}{x} + \dfrac{x}{y} = 2y$

17. $y^{-3/2} + x^{3/2} = 1$

18. $x^{1/2} + y^{2/3} = x + y$

19. $\sqrt{x + s} = \dfrac{1}{x} + \dfrac{1}{s}$

20. $\sqrt{x + y} + \sqrt{y} = 2x$

21. $y + \dfrac{x}{y} = 1$

22. $\dfrac{x}{y} + \dfrac{y^2}{x + 1} = 0$

23. $\sin(x + y) = x + \cos y$

24. $\tan(x^2 y) = x + y$

25. $xe^y = 2xy + y^3$

26. $e^{xy} = x^2 + y^2$

In Exercises 27–28, find dy/dx at the given point.

27. $(x + 2)^2 - 6(2y + 3)^2 = 3$, $(1, -1)$

28. $\sin^2(3y) = x + y$, $\left(\dfrac{2 - \pi}{4}, \dfrac{\pi}{4}\right)$

In Exercises 29–34, find an equation of the tangent line at the given point.

29. $xy - 2y = 1$, $(3, 1)$

30. $x^2 y^3 + 2y = 3x$, $(2, 1)$

31. $x^{2/3} + y^{2/3} = 2$, $(1, 1)$

32. $ye^x + xe^y = 4$, $(4, 0)$

33. $x^{1/2} + y^{-1/2} = 2xy$, $(1, 1)$

34. $ye^{x^2 - 4} + y^2 = 6$, $(2, 2)$

35. Find the points on the graph of $y^2 = x^3 - 3x + 1$ (Figure 6) where the tangent line is horizontal.
(a) First show that $2yy' = 3x^2 - 3$, where $y' = dy/dx$.
(b) Do not solve for y'. Rather, set $y' = 0$ and solve for x. This gives two possible values of x where the slope may be zero.
(c) Show that the positive value of x does not correspond to a point on the graph.
(d) The negative value corresponds to the two points on the graph where the tangent line is horizontal. Find the coordinates of these two points.

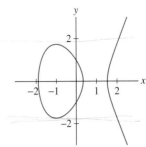

FIGURE 6 Graph of $y^2 = x^3 - 3x + 1$.

36. Find all points on the graph of $3x^2 + 4y^2 + 3xy = 24$ where the tangent line is horizontal (Figure 7).
(a) By differentiating the equation of the curve implicitly and setting $y' = 0$, show that if the tangent line is horizontal at (x, y), then $y = -2x$.
(b) Solve for x by substituting $y = -2x$ in the equation of the curve.

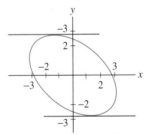

FIGURE 7 Graph of $3x^2 + 4y^2 + 3xy = 24$.

37. Show that no point on the graph of $x^2 - 3xy + y^2 = 1$ has a horizontal tangent line.

38. Figure 1 shows the graph of $y^4 + xy = x^3 - x + 2$. Find dy/dx at the two points on the graph with x-coordinate 0 and find an equation of the tangent line at $(1, 1)$.

39. If the derivative dx/dy exists at a point and $dx/dy = 0$, then the tangent line is vertical. Calculate dx/dy for the equation $y^4 + 1 = y^2 + x^2$ and find the points on the graph where the tangent line is vertical.

40. Differentiate the equation $xy = 1$ with respect to the variable t and derive the relation $\dfrac{dy}{dt} = -\dfrac{y}{x}\dfrac{dx}{dt}$.

41. Differentiate the equation $x^3 + 3xy^2 = 1$ with respect to the variable t and express dy/dt in terms of dx/dt, as in Exercise 40.

In Exercises 42–43, differentiate the equation with respect to t to calculate dy/dt in terms of dx/dt.

42. $x^2 - y^2 = 1$

43. $y^4 + 2xy + x^2 = 0$

44. 📖   The volume V and pressure P of gas in a piston (which vary in time t) satisfy $PV^{3/2} = C$, where C is a constant. Prove that

$$\frac{dP/dt}{dV/dt} = -\frac{3}{2}\frac{P}{V}$$

The ratio of the derivatives is negative. Could you have predicted this from the relation $PV^{3/2} = C$?

45. The **folium of Descartes** is the curve with equation $x^3 + y^3 = 3xy$ (Figure 8). It was first discussed in 1638 by the French philosopher-mathematician René Descartes. The name "folium" means leaf. Descartes's scientific colleague Gilles de Roberval called it the "jasmine flower." Both men believed incorrectly that the leaf shape in the first quadrant was repeated in each quadrant, giving the appearance of petals of a flower. Find an equation of the tangent line to this curve at the point $\left(\frac{2}{3}, \frac{4}{3}\right)$.

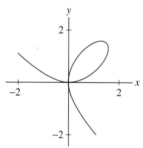

FIGURE 8 Folium of Descartes: $x^3 + y^3 = 3xy$.

46. Find all points on the folium $x^3 + y^3 = 3xy$ at which the tangent line is horizontal.

47. GU   Plot the equation $x^3 + y^3 = 3xy + b$ for several values of b.

(a) Describe how the graph changes as $b \to 0$.

(b) Compute dy/dx (in terms of b) at the point on the graph where $y = 0$. How does this value change as $b \to \infty$? Do your plots confirm this conclusion?

48. Show that the tangent lines at $x = 1 \pm \sqrt{2}$ to the so-called *conchoid* (Figure 9) with equation $(x-1)^2(x^2 + y^2) = 2x^2$ are vertical.

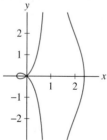

FIGURE 9 Conchoid: $(x-1)^2(x^2 + y^2) = 2x^2$.

49. The equation $xy = x^3 - 5x^2 + 2x - 1$ defines a *trident curve* (Figure 10), named by Isaac Newton in his treatise on curves published in 1710. Find the points where the tangent to the trident is horizontal as follows.

(a) Show that $xy' + y = 3x^2 - 10x + 2$.

(b) Set $y' = 0$ in (a), replace y by $x^{-1}(x^3 - 5x^2 + 2x - 1)$, and solve the resulting equation for x.

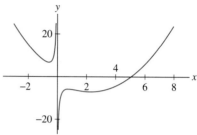

FIGURE 10 Trident curve: $xy = x^3 - 5x^2 + 2x - 1$.

50. Find an equation of the tangent line at the four points on the curve $(x^2 + y^2 - 4x)^2 = 2(x^2 + y^2)$, where $x = 1$. This curve (Figure 11) is an example of a *limaçon of Pascal*, named after the father of the French philosopher Blaise Pascal, who first described it in 1650.

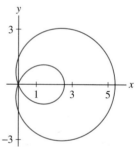

FIGURE 11 Limaçon: $(x^2 + y^2 - 4x)^2 = 2(x^2 + y^2)$.

51. Find equations of the tangent lines at the points where $x = 1$ on the so-called folium (Figure 12):

$$(x^2 + y^2)^2 = \frac{25}{4}xy^2$$

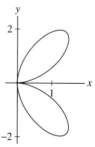

FIGURE 12

52. CAS   Use a computer algebra system to plot $(x^2 + y^2)^2 = 12(x^2 - y^2) + 2$ for $-4 \le x, y \le 4$. How many horizontal tangent lines does the curve appear to have? Find the points where these occur.

53. $\boxed{CAS}$ Use a computer algebra system to plot $y^2 = x^3 - 4x$ for $-4 \le x, y \le 4$. Show that if $dx/dy = 0$, then $y = 0$. Conclude that the tangent line is vertical at the points where the curve intersects the x-axis. Does your plot confirm this conclusion?

In Exercises 54–57, use implicit differentiation to calculate higher derivatives.

54. Consider the equation $y^3 - \frac{3}{2}x^2 = 1$.

(a) Show that $y' = x/y^2$ and differentiate again to show that

$$y'' = \frac{y^2 - 2xyy'}{y^4}$$

(b) Express y'' in terms of x and y using part (a).

55. Use the method of the previous exercise to show that $y'' = -y^{-3}$ if $x^2 + y^2 = 1$.

56. Calculate y'' at the point $(1, 1)$ on the curve $xy^2 + y - 2 = 0$ by the following steps:

(a) Find y' by implicit differentiation and calculate y' at the point $(1, 1)$.

(b) Differentiate the expression for y' found in (a) and then compute y'' at $(1, 1)$ by substituting $x = 1$, $y = 1$, and the value of y' found in (a).

57. Use the method of the previous exercise to compute y'' at the point $\left(\frac{2}{3}, \frac{4}{3}\right)$ on the folium of Descartes with the equation $x^3 + y^3 = 3xy$.

Further Insights and Challenges

58. Show that if P lies on the intersection of the two curves $x^2 - y^2 = c$ and $xy = d$ (c, d constants), then the tangents to the curves at P are perpendicular.

59. The *lemniscate curve* $(x^2 + y^2)^2 = 4(x^2 - y^2)$ was discovered by Jacob Bernoulli in 1694, who noted that it is "shaped like a figure 8, or a knot, or the bow of a ribbon." Find the coordinates of the four points at which the tangent line is horizontal (Figure 13).

60. Divide the curve in Figure 14 into five branches, each of which is the graph of a function. Sketch the branches.

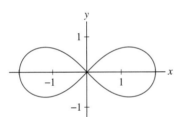

FIGURE 13 Lemniscate curve: $(x^2 + y^2)^2 = 4(x^2 - y^2)$.

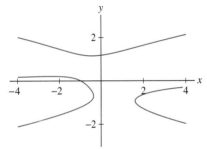

FIGURE 14 Graph of $y^5 - y = x^2y + x + 1$.

3.9 Derivatives of Inverse Functions

In this section, we begin with an important formula for the derivative of the inverse $f^{-1}(x)$ of a function $f(x)$. We then apply this formula to determine the derivatives of the inverse trigonometric functions. In the next section, we will use the formula to compute the derivatives of logarithmic functions.

> **THEOREM 1 Derivative of the Inverse** Assume that $f(x)$ is differentiable and one-to-one with inverse $g(x) = f^{-1}(x)$. If b belongs to the domain of $g(x)$ and $f'(g(b)) \neq 0$, then $g'(b)$ exists and
>
> $$g'(b) = \frac{1}{f'(g(b))} \qquad \boxed{1}$$

CONCEPTUAL INSIGHT   A proof of Theorem 1 is given in Appendix D, but we can explain the result graphically as follows. Recall that the graph of the inverse $g(x)$ is obtained by reflecting the graph of $f(x)$ through the line $y = x$. Now, consider a line L of slope m and let L' be its reflection through $y = x$ as in Figure 1(A). Then the slope of L' is $1/m$. Indeed, if (a, b) and (c, d) are any two points on L, then (b, a) and (d, c) lie on L' and

> **Derivative of the Inverse:** If $g(x) = f^{-1}(x)$ is the inverse of $f(x)$, then
> $$g'(x) = \frac{1}{f'(g(x))}$$

$$\underbrace{\text{Slope of } L = \frac{b-d}{a-c}, \qquad \text{Slope of } L' = \frac{a-c}{b-d}}_{\text{Reciprocal slopes}}$$

Figure 1(B) tells the rest of the story. Let $g(x) = f^{-1}(x)$. The reflection of the tangent line to $y = f(x)$ at $x = a$ is the tangent line to $y = g(x)$ at $x = b$ [where $b = f(a)$ and $a = g(b)$]. These tangent lines have reciprocal slopes, and thus $g'(b) = 1/f'(a) = 1/f'(g(b))$ as claimed in Theorem 1.

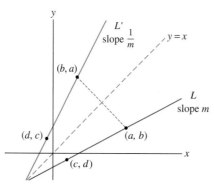

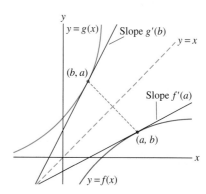

FIGURE 1 Graphical illustration of the formula $g'(b) = 1/f'(g(b))$.

(A) If L has slope m, then its reflection L' has slope $1/m$.

(B) The tangent line to the inverse $y = g(x)$ is the reflection of the tangent line to $y = f(x)$

■ **EXAMPLE 1**   Calculate $g'(x) = f^{-1}(x)$, where $g(x)$ is the inverse of $f(x) = x^2 + 4$ on the domain $\{x : x \geq 0\}$.

Solution   We solve $y = x^2 + 4$ to obtain $x = \sqrt{y - 4}$ and thus $g(x) = \sqrt{x - 4}$. Since $f'(x) = 2x$, we have $f'(g(x)) = 2g(x)$ and by the formula for the derivative of the inverse:

$$g'(x) = \frac{1}{f'(g(x))} = \frac{1}{2g(x)} = \frac{1}{2\sqrt{x-4}} = \frac{1}{2}(x-4)^{-1/2}$$

We obtain the same result by differentiating $g(x) = \sqrt{x - 4}$ directly.   ■

■ **EXAMPLE 2**   **Calculating $g'(x)$ Without Solving for $g(x)$**   Calculate $g'(1)$, where $g(x)$ is the inverse of $f(x) = x + e^x$.

Solution   We have $f'(x) = 1 + e^x$, and thus

$$g'(1) = \frac{1}{f'(g(1))} = \frac{1}{f'(c)} = \frac{1}{1 + e^c} \qquad \text{where } c = g(1)$$

We do not have to solve for $g(x)$ (which cannot be done explicitly in this case) if we can compute $c = g(1)$ directly. By definition of the inverse, $f(c) = 1$ and we see by

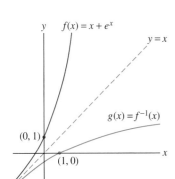

FIGURE 2 Graph of $f(x) = x + e^x$ and its inverse $g(x)$.

◄⋯ **REMINDER** *Recall from Example 7 in Section 1.5 that to simplify the composite* $\cos(\sin^{-1} x)$, *we refer to the right triangle in Figure 3 to obtain*

$$\cos(\sin^{-1} x) = \cos\theta = \frac{\text{adjacent}}{\text{hypotenuse}}$$

$$= \sqrt{1 - x^2}$$

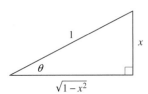

FIGURE 3 Right triangle constructed so that $\sin\theta = x$.

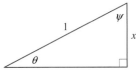

FIGURE 4 The angles $\theta = \sin^{-1} x$ and $\psi = \cos^{-1} x$ are complementary and thus sum to $\pi/2$.

inspection that $f(0) = 0 + e^0 = 1$ (Figure 2). Therefore, $c = 0$ and the formula above yields $g'(1) = 1/(1 + e^0) = \frac{1}{2}$. ■

Derivatives of Inverse Trigonometric Functions

We now apply Theorem 1 to the inverse trigonometric functions. An interesting feature of these functions is that their derivatives are not trigonometric. Rather, they involve quadratic expressions and their square roots.

THEOREM 2 Derivatives of Arcsine and Arccosine

$$\frac{d}{dx}\sin^{-1} x = \frac{1}{\sqrt{1 - x^2}}, \qquad \frac{d}{dx}\cos^{-1} x = -\frac{1}{\sqrt{1 - x^2}} \qquad \boxed{2}$$

Proof We apply Eq. (1) to $f(x) = \sin x$ and $g(x) = \sin^{-1} x$. Then $f'(x) = \cos x$, and by the equation in the margin,

$$\frac{d}{dx}\sin^{-1} x = \frac{1}{f'(g(x))} = \frac{1}{\cos(\sin^{-1} x)} = \frac{1}{\sqrt{1 - x^2}}$$

The derivative of $\cos^{-1} x$ is computed in a similar fashion (see Exercise 38 or the next example). ■

■ **EXAMPLE 3 Complementary Angles** The derivatives of $\sin^{-1} x$ and $\cos^{-1} x$ are equal up to a minus sign. Explain this by proving that

$$\sin^{-1} x + \cos^{-1} x = \frac{\pi}{2}$$

Solution In Figure 4, we have $\theta = \sin^{-1} x$ and $\psi = \cos^{-1} x$. These angles are complementary, so $\theta + \psi = \pi/2$ as claimed. Therefore,

$$\frac{d}{dx}\cos^{-1} x = \frac{d}{dx}\left(\frac{\pi}{2} - \sin^{-1} x\right) = -\frac{d}{dx}\sin^{-1} x \qquad ■$$

■ **EXAMPLE 4** Calculate $f'\left(\frac{1}{2}\right)$, where $f(x) = \arcsin(x^2)$.

Solution Recall that $\arcsin x$ is another notation for $\sin^{-1} x$. By the Chain Rule,

$$\frac{d}{dx}\arcsin(x^2) = \frac{d}{dx}\sin^{-1}(x^2) = \frac{1}{\sqrt{1 - x^4}}\frac{d}{dx}x^2 = \frac{2x}{\sqrt{1 - x^4}}$$

$$f'\left(\frac{1}{2}\right) = \frac{2\left(\frac{1}{2}\right)}{\sqrt{1 - \left(\frac{1}{2}\right)^4}} = \frac{1}{\sqrt{\frac{15}{16}}} = \frac{4}{\sqrt{15}} \qquad ■$$

The next theorem states the formulas for the derivatives of the remaining inverse trigonometric functions.

The proofs of the formulas in Theorem 3 are similar to the proof of Theorem 2. See Exercises 39 and 41.

THEOREM 3 Derivatives of Inverse Trigonometric Functions

$$\frac{d}{dx}\tan^{-1}x = \frac{1}{x^2+1}, \qquad \frac{d}{dx}\cot^{-1}x = -\frac{1}{x^2+1}$$

$$\frac{d}{dx}\sec^{-1}x = \frac{1}{|x|\sqrt{x^2-1}}, \qquad \frac{d}{dx}\csc^{-1}x = -\frac{1}{|x|\sqrt{x^2-1}}$$

■ **EXAMPLE 5** Calculate **(a)** $\dfrac{d}{dx}\tan^{-1}(3x+1)$ and **(b)** $\dfrac{d}{dx}\csc^{-1}(e^x+1)\Big|_{x=0}$.

Solution

(a) Apply the Chain Rule using the formula $\dfrac{d}{du}\tan^{-1}u = \dfrac{1}{u^2+1}$:

$$\frac{d}{dx}\tan^{-1}(3x+1) = \frac{1}{(3x+1)^2+1}\frac{d}{dx}(3x+1) = \frac{3}{9x^2+6x+2}$$

(b) Apply the Chain Rule using the formula $\dfrac{d}{du}\csc^{-1}u = -\dfrac{1}{|u|\sqrt{u^2-1}}$:

$$\frac{d}{dx}\csc^{-1}(e^x+1) = -\frac{1}{|e^x+1|\sqrt{(e^x+1)^2-1}}\frac{d}{dx}(e^x+1)$$

$$= -\frac{e^x}{(e^x+1)\sqrt{e^{2x}+2e^x}}$$

We have replaced $|e^x+1|$ by e^x+1 since this quantity is positive. Now we have

$$\frac{d}{dx}\csc^{-1}(e^x+1)\Big|_{x=0} = -\frac{e^0}{(e^0+1)\sqrt{e^0+2e^0}} = -\frac{1}{2\sqrt{3}} \qquad ■$$

3.9 SUMMARY

• Derivative of the inverse: If $f(x)$ is differentiable and one-to-one with inverse $g(x)$, then for x such that $f'(g(x)) \neq 0$,

$$g'(x) = \frac{1}{f'(g(x))}$$

• Derivative formulas:

$$\frac{d}{dx}\sin^{-1}x = \frac{1}{\sqrt{1-x^2}},$$

$$\frac{d}{dx}\cos^{-1}x = \frac{-1}{\sqrt{1-x^2}},$$

$$\frac{d}{dx}\tan^{-1}x = \frac{1}{x^2+1}, \qquad \frac{d}{dx}\cot^{-1}x = -\frac{1}{x^2+1}$$

$$\frac{d}{dx}\sec^{-1}x = \frac{1}{|x|\sqrt{x^2-1}}, \qquad \frac{d}{dx}\csc^{-1}x = -\frac{1}{|x|\sqrt{x^2-1}}$$

$$\frac{d}{dx}\left(\sin^{-1}v\right) = \frac{v'}{\sqrt{1-v^2}} \qquad \frac{d}{dx}\left(\tan^{-1}v\right) = \frac{v'}{1+v^2} \qquad \begin{array}{l} \cos = -\sin \\ \cot = -\tan \end{array}$$

3.9 EXERCISES

Preliminary Questions

1. What is the slope of the line obtained by reflecting the line $y = x/2$ through the line $y = x$?

2. Suppose that $P = (2, 4)$ lies on the graph of $f(x)$ and that the slope of the tangent line through P is $m = 3$. Assuming that $f^{-1}(x)$ exists, what is the slope of the tangent line to the graph of $f^{-1}(x)$ at the point $Q = (4, 2)$?

3. Which inverse trigonometric function $g(x)$ has the derivative $g'(x) = \dfrac{1}{x^2 + 1}$?

4. What is the geometric interpretation of the identity
$$\sin^{-1} x + \cos^{-1} x = \frac{\pi}{2}?$$

Exercises

1. Find the inverse $g(x)$ of $f(x) = \sqrt{x^2 + 9}$ with domain $x \geq 0$ and calculate $g'(x)$ in two ways: using Theorem 1 and by direct calculation.

2. Let $g(x)$ be the inverse of $f(x) = x^3 + 1$. Find a formula for $g(x)$ and calculate $g'(x)$ in two ways: using Theorem 1 and then by direct calculation.

In Exercises 3–8, use Theorem 1 to calculate $g'(x)$, where $g(x)$ is the inverse of $f(x)$.

3. $f(x) = 7x + 6$

4. $f(x) = \sqrt{3 - x}$

5. $f(x) = x^{-5}$

6. $f(x) = 4x^3 - 1$

7. $f(x) = \dfrac{x}{x + 1}$

8. $f(x) = 2 + x^{-1}$

9. Let $g(x)$ be the inverse of $f(x) = x^3 + 2x + 4$. Calculate $g(7)$ [without finding a formula for $g(x)$] and then calculate $g'(7)$.

10. Find $g'(-\frac{1}{2})$, where $g(x)$ is the inverse of $f(x) = \dfrac{x^3}{x^2 + 1}$.

In Exercises 11–16, calculate $g(b)$ and $g'(b)$, where g is the inverse of f (in the given domain, if indicated).

11. $f(x) = x + \cos x, \quad b = 1$

12. $f(x) = 4x^3 - 2x, \quad b = -2$

13. $f(x) = \sqrt{x^2 + 6x}$ for $x \geq 0, \quad b = 4$

14. $f(x) = \sqrt{x^2 + 6x}$ for $x \leq -6, \quad b = 4$

15. $f(x) = \dfrac{1}{x + 1}, \quad b = \dfrac{1}{4}$

16. $f(x) = e^x, \quad b = e$

17. Let $f(x) = x^n$ and $g(x) = x^{1/n}$. Compute $g'(x)$ using Theorem 1 and check your answer using the Power Rule.

In Exercises 18–21, compute the derivative at the point indicated without using a calculator.

18. $y = \sin^{-1} x, \quad x = \frac{3}{5}$

19. $y = \tan^{-1} x, \quad x = \frac{1}{2}$

20. $y = \sec^{-1} x, \quad x = 4$

21. $y = \arccos(4x), \quad x = \frac{1}{5}$

In Exercises 22–37, find the derivative.

22. $y = \sin^{-1}(7x)$

23. $y = \arctan\left(\dfrac{x}{3}\right)$

24. $y = \cos^{-1}(x^2)$

25. $y = \sec^{-1}(t + 1)$

26. $y = x \tan^{-1} x$

27. $y = \dfrac{\cos^{-1} x}{\sin^{-1} x}$

28. $y = \arcsin(e^x)$

29. $y = \csc^{-1}(x^{-1})$

30. $y = \tan^{-1}\left(\dfrac{z}{1 - z^2}\right)$

31. $y = \tan^{-1}\left(\dfrac{1 + t}{1 - t}\right)$

32. $y = \cos^{-1} t^{-1} - \sec^{-1} t$

33. $y = e^{\cos^{-1} x}$

34. $y = \cos^{-1}(x + \sin^{-1} x)$

35. $y = \ln(\sin^{-1} t)$

36. $y = \sqrt{1 - t^2} + \sin^{-1} t$

37. $y = x^{\sin^{-1} x}$

38. Use Figure 5 to prove that $(\cos^{-1} x)' = -\dfrac{1}{\sqrt{1 - x^2}}$.

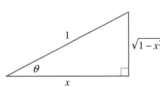

FIGURE 5 Right triangle with $\theta = \cos^{-1} x$.

39. Show that $(\tan^{-1} x)' = \cos^2(\tan^{-1} x)$ and then use Figure 6 to prove that $(\tan^{-1} x)' = (x^2 + 1)^{-1}$.

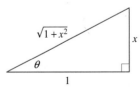

FIGURE 6 Right triangle with $\theta = \tan^{-1} x$.

40. Let $\theta = \sec^{-1} x$. Show that $\tan\theta = \sqrt{x^2 - 1}$ if $x \geq 1$ and $\tan\theta = -\sqrt{x^2 - 1}$ if $x \leq -1$. *Hint:* $\tan\theta \geq 0$ on $(0, \pi/2)$ and $\tan\theta \leq 0$ on $(\pi/2, \pi)$.

41. Use Exercise 40 to verify the formula

$$(\sec^{-1} x)' = \frac{1}{|x|\sqrt{x^2 - 1}}$$

Further Insights and Challenges

42. Let g be the inverse of a function f satisfying $f'(x) = f(x)$. Show that $g'(x) = x^{-1}$. We will apply this in the next section to show that the inverse of $f(x) = e^x$ (the natural logarithm) has the derivative $f'(x) = x^{-1}$.

3.10 Derivatives of General Exponential and Logarithmic Functions

In Section 3.2, we proved the formula for any base $b > 0$:

$$\frac{d}{dx} b^x = m_b \, b^x, \qquad \text{where} \quad m_b = \lim_{h \to 0} \frac{b^h - 1}{h}$$

This established the important result that the derivative of an exponential function $f(x) = b^x$ is proportional to b^x, but we were not able to identify the factor of proportionality m_b (other than to say that e is the unique number for which $m_e = 1$). We now use the Chain Rule to prove that $m_b = \ln b$. Recall that $\ln x$ is the natural logarithm, that is, $\ln x = \log_e x$.

THEOREM 1 Derivative of $f(x) = b^x$

$$\frac{d}{dx} b^x = (\ln b)b^x \qquad \text{for } b > 0$$ $\boxed{1}$

Proof Since $b = e^{\ln b}$, we have $b^x = e^{\ln b^x} = e^{(\ln b)x}$. Now use the Chain Rule:

$$\frac{d}{dx} b^x = \frac{d}{dx} e^{(\ln b)x} = (\ln b)e^{(\ln b)x} = (\ln b)b^x \qquad \blacksquare$$

For example, $(10^x)' = (\ln 10)10^x$.

■ **EXAMPLE 1** Find the derivative of **(a)** $f(x) = 4^{3x}$ and **(b)** $f(x) = 5^{x^2}$.

Solution

(a) The function $f(x) = 4^{3x}$ is a composite of 4^u and $u = 3x$:

$$\frac{d}{dx} 4^{3x} = \frac{d}{du} 4^u \frac{du}{dx} = (\ln 4)4^u (3x)' = (3\ln 4)4^{3x}$$

(b) The function $f(x) = 5^{x^2}$ is a composite of 5^u and $u = x^2$:

$$\frac{d}{dx} 5^{x^2} = (\ln 5)5^{x^2}(x^2)' = (2\ln 5)\, x \, 5^{x^2} \qquad \blacksquare$$

Next, we calculate the derivative of $y = \ln x$. Recall that if $g(x)$ is the inverse of $f(x)$, then $g'(x) = 1/f'(g(x))$. Let $f(x) = e^x$ and $g(x) = \ln x$. Then $f'(x) = f(x)$ and

$$\frac{d}{dx} \ln x = g'(x) = \frac{1}{f'(g(x))} = \frac{1}{f(g(x))} = \frac{1}{x}$$

The two most important calculus facts about exponentials and logs are

$$\frac{d}{dx}e^x = e^x, \qquad \frac{d}{dx}\ln x = \frac{1}{x}$$

THEOREM 2 Derivative of the Natural Logarithm

$$\frac{d}{dx}\ln x = \frac{1}{x} \qquad \text{for } x > 0 \qquad \boxed{2}$$

Handwritten annotations:

$d(5^{\sin x}) = 5^{\sin x}\ln 5 \cdot \cos x$

$\frac{d}{dx}(\ln u) = \frac{1}{u} \cdot u' = \boxed{\frac{u'}{u}}$

$\frac{d}{dx}(\log_a u) = \frac{d}{dx}\left(\frac{\ln u}{\ln a}\right) = \boxed{\frac{u'}{\ln a \cdot u}}$

EXAMPLE 2 Find the derivative of **(a)** $y = x\ln x$ and **(b)** $y = (\ln x)^2$.

Solution

(a) Use the Product Rule:

$$\frac{d}{dx}(x\ln x) = x \cdot (\ln x)' + (x)' \cdot \ln x = x \cdot \frac{1}{x} + \ln x = 1 + \ln x$$

(b) By the General Power Rule,

$$\frac{d}{dx}(\ln x)^2 = 2\ln x \cdot \frac{d}{dx}\ln x = \frac{2\ln x}{x}$$

In Section 3.2, we proved the Power Rule for whole number exponents. We now prove it for all real numbers n and $x > 0$ by writing x^n as an exponential:

$$x^n = (e^{\ln x})^n = e^{n\ln x}$$

$$\frac{d}{dx}x^n = \frac{d}{dx}e^{n\ln x} = \left(\frac{d}{dx}n\ln x\right)e^{n\ln x}$$

$$= \left(\frac{n}{x}\right)x^n = nx^{n-1}$$

Handwritten annotation:

$\frac{d}{dx}(\ln \cos x) = \frac{-\sin x}{\cos x} = \boxed{-\tan x}$

The Chain Rule gives us a useful formula for the derivative of a composite function of the form $\ln(f(x))$. Taking $u = f(x)$, we have

$$\frac{d}{dx}\ln(f(x)) = \frac{d}{du}\ln(u)\frac{du}{dx} = \frac{1}{u} \cdot u' = \frac{1}{f(x)}f'(x)$$

$$\boxed{\frac{d}{dx}\ln(f(x)) = \frac{f'(x)}{f(x)}} \qquad \boxed{3}$$

EXAMPLE 3 Find the derivative of **(a)** $y = \ln(x^3 + 1)$ and **(b)** $y = \ln(\sqrt{\sin x})$.

Solution Use Eq. (3):

(a) $\dfrac{d}{dx}\ln(x^3 + 1) = \dfrac{(x^3 + 1)'}{x^3 + 1} = \dfrac{3x^2}{x^3 + 1}$

(b) The algebra is simpler if we write $\ln(\sqrt{\sin x}) = \ln((\sin x)^{1/2}) = \frac{1}{2}\ln(\sin x)$:

$$\frac{d}{dx}\ln(\sqrt{\sin x}) = \frac{1}{2}\frac{d}{dx}\ln(\sin x) = \frac{1}{2}\frac{(\sin x)'}{\sin x} = \frac{1}{2}\frac{\cos x}{\sin x} = \frac{1}{2}\cot x$$

According to Eq. (1) in Section 1.6, we have the "change of base" formulas:

$$\log_b x = \frac{\log_a x}{\log_a b}, \qquad \log_b x = \frac{\ln x}{\ln b}$$

It follows, as in Example 4, that for any base $b > 0$, $b \neq 1$:

$$\frac{d}{dx}\log_b x = \frac{1}{(\ln b)x}$$

Logarithmic differentiation saves work when the function is a product of several factors or a quotient, as in Example 5.

EXAMPLE 4 **Derivative of the Logarithm to Another Base** Use Eq. (1) from Section 1.6 to calculate $\dfrac{d}{dx}\log_{10} x$.

Solution By Eq. (1) from Section 1.6, $\log_{10} x = \dfrac{\ln x}{\ln 10}$, so

$$\frac{d}{dx}\log_{10} x = \frac{d}{dx}\frac{\ln x}{\ln 10} = \frac{1}{\ln 10}\frac{d}{dx}\ln x = \frac{1}{(\ln 10)x}$$

The next example illustrates the technique of **logarithmic differentiation**.

EXAMPLE 5 **Logarithmic Differentiation** Find the derivative of

$$f(x) = \frac{(x + 1)^2(2x^2 - 3)}{\sqrt{x^2 + 1}}$$

$\frac{d}{dx}(\ln x^2) = \frac{2x}{x^2} = \boxed{\frac{2}{x}}$

$\downarrow = \uparrow^{x^2}$

$\frac{d}{dx} 2\ln x = 2 \cdot \frac{1}{x} = \boxed{\frac{2}{x}}$

$f(x) = \left(\ln \frac{x+1}{\sqrt{x-2}}\right)$ find $f'(x)$

$= \ln(x+1) - \ln(\sqrt{x-2})$

$= \ln(x+1) - \frac{1}{2}\ln(x-2)$

$f'(x) = \frac{1}{x+1} - \frac{1}{2}\left(\frac{1}{(x-2)}\right)$

$\boxed{\frac{1}{x+1} , \frac{1}{2x-4}}$

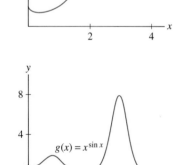

Solution In logarithmic differentiation, we differentiate $\ln(f(x))$ rather than $f(x)$ itself. We have

$$\ln(f(x)) = \ln\left((x+1)^2\right) + \ln\left(2x^2 - 3\right) - \ln\left(\sqrt{x^2+1}\right)$$

Now use Eq. (3):

$$\frac{f'(x)}{f(x)} = \frac{d}{dx}\ln(f(x)) = \frac{d}{dx}\ln\left((x+1)^2\right) + \frac{d}{dx}\ln\left(2x^2-3\right) - \frac{d}{dx}\ln\left(\sqrt{x^2+1}\right)$$

$$= 2\frac{d}{dx}\ln(x+1) + \frac{d}{dx}\ln\left(2x^2-3\right) - \frac{1}{2}\frac{d}{dx}\ln\left(x^2+1\right)$$

$$\frac{f'(x)}{f(x)} = 2\frac{1}{x+1} + \frac{4x}{2x^2-3} - \frac{1}{2}\frac{2x}{x^2+1}$$

To obtain $f'(x)$, multiply through by $f(x)$:

$$f'(x) = \left(\frac{(x+1)^2(2x^2-3)}{\sqrt{x^2+1}}\right)\left(\frac{2}{x+1} + \frac{4x}{2x^2-3} - \frac{x}{x^2+1}\right)$$

∎

■ **EXAMPLE 6** Find the derivative (for $x > 0$) of **(a)** $f(x) = x^x$ and **(b)** $g(x) = x^{\sin x}$.

Solution The two problems are similar. We illustrate two different methods (Figure 1).

(a) Method 1: Use the equation $x = e^{\ln x}$ to rewrite $f(x)$ as an exponential and apply the Chain Rule:

$$f(x) = x^x = (e^{\ln x})^x = e^{x\ln x}$$

$$f'(x) = (x\ln x)'e^{x\ln x} = (1 + \ln x)e^{x\ln x} = (1+\ln x)x^x$$

(b) Method 2: Use logarithmic differentiation, that is, differentiate $\ln(g(x))$ rather than $g(x)$. Since $\ln(g(x)) = \ln(x^{\sin x}) = (\sin x)\ln x$, Eq. (3) yields

$$\frac{g'(x)}{g(x)} = \frac{d}{dx}\ln(g(x)) = \frac{d}{dx}(\sin x)(\ln x) = \frac{\sin x}{x} + (\cos x)\ln x$$

$$g'(x) = \left(\frac{\sin x}{x} + (\cos x)\ln x\right)g(x) = \left(\frac{\sin x}{x} + (\cos x)\ln x\right)x^{\sin x}$$

∎

Derivatives of Hyperbolic Functions

Recall from Section 1.6 that the hyperbolic functions are special combinations of e^x and e^{-x} that arise in engineering and physics applications. We can now compute their derivatives easily using our rules of differentiation. First, we treat the hyperbolic sine and cosine:

$$\sinh x = \frac{e^x - e^{-x}}{2}, \qquad \cosh x = \frac{e^x + e^{-x}}{2}$$

The formulas for their derivatives are similar to those for the corresponding trigonometric functions, differing at most by a sign:

$$\boxed{\frac{d}{dx}\sinh x = \cosh x, \qquad \frac{d}{dx}\cosh x = \sinh x}$$

FIGURE 1 Graphs of $f(x) = x^x$ and $g(x) = x^{\sin x}$.

*See Section 1.6 (p. 48) to review
hyperbolic functions. The basic identity is*

$$\cosh^2 x - \sinh^2 x = 1$$

For example,

$$\frac{d}{dx}\left(\frac{e^x - e^{-x}}{2}\right) = \left(\frac{e^x - e^{-x}}{2}\right)' = \frac{e^x + e^{-x}}{2} = \cosh x$$

The hyperbolic tangent, cotangent, secant, and cosecant functions are defined like their trigonometric counterparts:

$$\tanh x = \frac{\sinh x}{\cosh x} = \frac{e^x - e^{-x}}{e^x + e^{-x}}, \quad \operatorname{sech} x = \frac{1}{\cosh x} = \frac{2}{e^x + e^{-x}}$$

$$\coth x = \frac{\cosh x}{\sinh x} = \frac{e^x + e^{-x}}{e^x - e^{-x}}, \quad \operatorname{csch} x = \frac{1}{\sinh x} = \frac{2}{e^x - e^{-x}}$$

We can also calculate the derivatives of these additional hyperbolic functions directly. They too differ from their trigonometric counterparts by a sign at most.

Derivatives of Hyperbolic and Trigonometric Functions

$$\frac{d}{dx}\tanh x = \operatorname{sech}^2 x, \qquad \frac{d}{dx}\tan x = \sec^2 x$$

$$\frac{d}{dx}\coth x = -\operatorname{csch}^2 x, \qquad \frac{d}{dx}\cot x = -\csc^2 x$$

$$\frac{d}{dx}\operatorname{sech} x = -\operatorname{sech} x \tanh x, \qquad \frac{d}{dx}\sec x = \sec x \tan x$$

$$\frac{d}{dx}\operatorname{csch} x = -\operatorname{csch} x \coth x, \qquad \frac{d}{dx}\csc x = -\csc x \cot x$$

■ **EXAMPLE 7** Verify the formula $\dfrac{d}{dx}\coth x = -\operatorname{csch}^2 x$.

Solution By the Quotient Rule and the identity $\cosh^2 x - \sinh^2 x = 1$,

$$\frac{d}{dx}\coth x = \left(\frac{\cosh x}{\sinh x}\right)' = \frac{(\sinh x)(\cosh x)' - (\cosh x)(\sinh x)'}{\sinh^2 x}$$

$$= \frac{\sinh^2 x - \cosh^2 x}{\sinh^2 x} = \frac{-1}{\sinh^2 x} = -\operatorname{csch}^2 x \qquad ■$$

■ **EXAMPLE 8** Calculate **(a)** $\dfrac{d}{dx}\cosh(3x^2 + 1)$ and **(b)** $\dfrac{d}{dx}\sinh x \tanh x$.

Solution

(a) By the Chain Rule, $\dfrac{d}{dx}\cosh(3x^2 + 1) = 6x \sinh(3x^2 + 1)$

(b) By the Product Rule,

$$\frac{d}{dx}(\sinh x \tanh x) = \sinh x \operatorname{sech}^2 x + \tanh x \cosh x = \operatorname{sech} x \tanh x + \sinh x \qquad ■$$

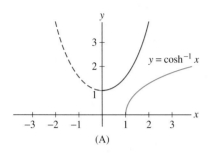

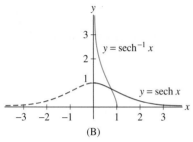

FIGURE 2

Inverse Hyperbolic Functions

Recall that a function $f(x)$ with domain D has an inverse if it is one-to-one on D. Each of the hyperbolic functions except $\cosh x$ and $\operatorname{sech} x$ is one-to-one on its domain and therefore has a well-defined inverse. The functions $\cosh x$ and $\operatorname{sech} x$ are one-to-one on the restricted domain $\{x : x \geq 0\}$. We let $\cosh^{-1} x$ and $\operatorname{sech}^{-1} x$ denote the corresponding inverses (Figure 2). In reading the following table, keep in mind that the domain of the inverse is equal to the range of the function.

Inverse Hyperbolic Functions and Their Derivatives

Function	Domain	Derivative
$y = \sinh^{-1} x$	all x	$\dfrac{d}{dx} \sinh^{-1} x = \dfrac{1}{\sqrt{x^2 + 1}}$
$y = \cosh^{-1} x$	$x \geq 1$	$\dfrac{d}{dx} \cosh^{-1} x = \dfrac{1}{\sqrt{x^2 - 1}}$
$y = \tanh^{-1} x$	$\lvert x \rvert < 1$	$\dfrac{d}{dx} \tanh^{-1} x = \dfrac{1}{1 - x^2}$
$y = \coth^{-1} x$	$\lvert x \rvert > 1$	$\dfrac{d}{dx} \coth^{-1} x = \dfrac{1}{1 - x^2}$
$y = \operatorname{sech}^{-1} x$	$0 < x \leq 1$	$\dfrac{d}{dx} \operatorname{sech}^{-1} x = -\dfrac{1}{x\sqrt{1 - x^2}}$
$y = \operatorname{csch}^{-1} x$	$x \neq 0$	$\dfrac{d}{dx} \operatorname{csch}^{-1} x = -\dfrac{1}{\lvert x \rvert \sqrt{x^2 + 1}}$

■ **EXAMPLE 9** (a) Verify the formula $\dfrac{d}{dx} \tanh^{-1} x = \dfrac{1}{1 - x^2}$.

(b) The functions $y = \tanh^{-1} x$ and $y = \coth^{-1} x$ appear to have the same derivative. Does this imply that they differ by a constant?

Solution

(a) We apply the formula for the derivative of an inverse [Eq. (1) in Section 3.9]. Since $(\tanh x)' = \operatorname{sech}^2 x$,

$$\frac{d}{dx} \tanh^{-1} x = \frac{1}{\operatorname{sech}^2(\tanh^{-1} x)}$$

Now let $t = \tanh^{-1} x$. Then

$$\cosh^2 t - \sinh^2 t = 1 \qquad \text{(basic identity)}$$

$$1 - \tanh^2 t = \operatorname{sech}^2 t \qquad \text{(divide by } \cosh^2 t\text{)}$$

$$1 - x^2 = \operatorname{sech}^2(\tanh^{-1} x) \quad \text{(because } x = \tanh t\text{)}$$

This gives the desired result:

$$\frac{d}{dx} \tanh^{-1} x = \frac{1}{\operatorname{sech}^2(\tanh^{-1} x)} = \frac{1}{1 - x^2}$$

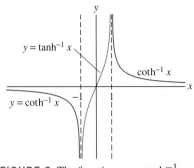

FIGURE 3 The functions $y = \tanh^{-1} x$ and $y = \coth^{-1} x$ have disjoint domains.

(b) A similar computation shows that $\dfrac{d}{dx} \coth^{-1} x = \dfrac{1}{1 - x^2}$. Thus, the derivatives of $y = \tanh^{-1} x$ and $y = \coth^{-1} x$ are given by the same formula. Functions defined on a common interval with the same derivative differ by a constant. However, the functions $y = \tanh^{-1} x$ and $y = \coth^{-1} x$ have disjoint domains and therefore do not differ by a constant (Figure 3). ∎

3.10 SUMMARY

- Derivative formulas:

$$\frac{d}{dx} e^x = e^x, \qquad \frac{d}{dx} \ln x = \frac{1}{x}, \qquad \frac{d}{dx} b^x = (\ln b) b^x, \qquad \frac{d}{dx} \log_b x = \frac{1}{(\ln b)\, x}$$

- Derivative formulas:

$$\frac{d}{dx} \sinh x = \cosh x, \qquad\qquad \frac{d}{dx} \cosh x = \sinh x$$

$$\frac{d}{dx} \tanh x = \operatorname{sech}^2 x, \qquad\qquad \frac{d}{dx} \coth x = -\operatorname{csch}^2 x$$

$$\frac{d}{dx} \operatorname{sech} x = -\operatorname{sech} x \tanh x, \qquad\qquad \frac{d}{dx} \operatorname{csch} x = -\operatorname{csch} x \coth x$$

$$\frac{d}{dx} \sinh^{-1} x = \frac{1}{\sqrt{x^2 + 1}}, \qquad\qquad \frac{d}{dx} \cosh^{-1} x = \frac{1}{\sqrt{x^2 - 1}} \quad (x > 1)$$

$$\frac{d}{dx} \tanh^{-1} x = \frac{1}{1 - x^2} \quad (|x| < 1), \qquad\qquad \frac{d}{dx} \coth^{-1} x = \frac{1}{1 - x^2} \quad (|x| > 1)$$

$$\frac{d}{dx} \operatorname{sech}^{-1} x = \frac{-1}{x\sqrt{1 - x^2}} \quad (0 < x < 1), \qquad\qquad \frac{d}{dx} \operatorname{csch}^{-1} x = -\frac{1}{|x|\sqrt{x^2 + 1}} \quad (x \neq 0)$$

3.10 EXERCISES

Preliminary Questions

1. What is the slope of the tangent line to $y = 4^x$ at $x = 0$?

2. For which $b > 0$ does the tangent line to $y = b^x$ at $x = 0$ have slope 2?

3. What is the rate of change of $y = \ln x$ at $x = 10$?

4. For which base b is $(\log_b x)' = \dfrac{1}{3x}$?

5. What are $y^{(100)}$ and $y^{(101)}$ for $y = \cosh x$?

Exercises

In Exercises 1–18, find the derivative.

1. $y = x \ln x$

2. $y = t \ln t - t$

3. $y = (\ln x)^2$

4. $y = \ln(x^2)$

5. $y = \ln(x^3 + 3x + 1)$

6. $y = \ln(2^s)$

7. $y = \ln(\sin t + 1)$

8. $y = x^2 \ln x$

9. $y = \dfrac{\ln x}{x}$

10. $y = e^{(\ln x)^2}$

11. $y = \ln(\ln x)$

12. $y = (\ln(\ln x))^3$

13. $y = \ln((\ln x)^3)$

14. $y = \ln((x + 1)(2x + 9))$

15. $y = \ln(\tan x)$

16. $y = \ln\left(\dfrac{x + 1}{x^3 + 1}\right)$

17. $y = 5^x$

18. $y = 5^{x^2 - x}$

In Exercises 19–22, compute the derivative using Eq. (1).

19. $f'(x), \quad f(x) = \log_2 x$

20. $\dfrac{d}{dx} \log_{10}(x^3 + x^2)$

21. $f'(3)$, $f(x) = \log_5 x$

22. $\dfrac{d}{dt} \log_3(\sin t)$

In Exercises 23–34, find an equation of the tangent line at the point indicated.

23. $f(x) = 4^x$, $x = 3$

24. $f(x) = (\sqrt{2})^x$, $x = \sqrt{2}$

25. $s(t) = 3^{7t}$, $t = 2$

26. $f(x) = \pi^{3x+9}$, $x = 1$

27. $f(x) = 5^{x^2-2x+9}$, $x = 1$

28. $s(t) = \ln t$, $t = 5$

29. $s(t) = \ln(8 - 4t)$, $t = 1$

30. $f(x) = \ln(x^2)$, $x = 4$

31. $f(x) = 4\ln(9x + 2)$, $x = 2$

32. $f(x) = \ln(\sin x)$, $x = \dfrac{\pi}{4}$

33. $f(x) = \log_5 x$, $x = 2$

34. $f(x) = \log_2(x + x^{-1})$, $x = 1$

In Exercises 35–40, find the derivative using the methods of Example 6.

35. $f(x) = x^{2x}$

36. $f(x) = x^{\cos x}$

37. $f(x) = x^{e^x}$

38. $f(x) = x^{x^2}$

39. $f(x) = x^{2^x}$

40. $f(x) = e^{x^x}$

In Exercises 41–48, evaluate the derivative using logarithmic differentiation as in Example 5.

41. $y = (x + 2)(x + 4)$

42. $y = (x + 1)(x + 2)(x + 4)$

43. $y = \dfrac{x(x + 1)^3}{(3x - 1)^2}$

44. $y = \dfrac{x(x^2 + 1)}{\sqrt{x + 1}}$

45. $y = (2x + 1)(4x^2)\sqrt{x - 9}$

46. $y = \sqrt{\dfrac{x(x + 2)}{(2x + 1)(2x + 2)}}$

47. $y = (x^2 + 1)(x^2 + 2)(x^2 + 3)^2$

48. $y = \dfrac{x \cos x}{(x + 1) \sin x}$

49. ✎   Use the formula $(\ln f(x))' = f'(x)/f(x)$ to show that $\ln x$ and $\ln(2x)$ have the same derivative. Is there a simpler explanation of this result?

In Exercises 50–75, calculate the derivative.

50. $y = \sinh(3x)$

51. $y = \sinh(x^2)$

52. $y = \cosh(1 - 4t)$

53. $y = \tanh(t^2 + 1)$

54. $y = \sqrt{\cosh x + 1}$

55. $y = \sinh x \tanh x$

56. $y = \dfrac{\sinh t}{1 + \cosh t}$

57. $y = \ln(\cosh x)$

58. $y = \tanh(2x)$

59. $y = \sinh(\ln x)$

60. $y = e^{\tanh x}$

61. $y = \tanh(e^x)$

62. $y = \sinh(xe^x)$

63. $y = \sinh(\cosh x)$

64. $y = \operatorname{sech}(\sqrt{x})$

65. $y = \coth(x^2)$

66. $y = \ln(\coth x)$

67. $y = \operatorname{sech} x \coth x$

68. $y = \tanh(3x^2 - 9)$

69. $y = x^{\sinh x}$

70. $y = \tanh^{-1} 3x$

71. $y = \sinh^{-1}(x^2)$

72. $y = (\operatorname{csch}^{-1} 3x)^4$

73. $y = e^{\cosh^{-1} x}$

74. $y = \sinh^{-1}(\sqrt{x^2 + 1})$

75. $y = \ln(\tanh^{-1} x)$

In Exercises 76–78, prove the formula.

76. $\dfrac{d}{dx}(\coth x) = -\operatorname{csch}^2 x$

77. $\dfrac{d}{dt} \sinh^{-1} t = \dfrac{1}{\sqrt{t^2 + 1}}$

78. $\dfrac{d}{dt} \cosh^{-1} t = \dfrac{1}{\sqrt{t^2 - 1}}$ for $t > 1$

79. The energy E (in joules) radiated as seismic waves from an earthquake of Richter magnitude M is given by the formula $\log_{10} E = 4.8 + 1.5M$.

(a) Express E as a function of M.

(b) Show that when M increases by 1, the energy increases by a factor of approximately 31.

(c) Calculate $\dfrac{dE}{dM}$.

80. The Palermo Technical Impact Hazard Scale P is used to quantify the risk associated with the impact of an asteroid colliding with the earth:

$$P = \log_{10}\left(\dfrac{p_i E^{0.8}}{0.03T}\right)$$

where p_i is the probability of impact occurring in T years and E is the energy of impact (in megatons of TNT). The risk is greater than a random event of similar magnitude if $P > 0$ (asteroids observed in 2006 have $P \le -2$, indicating low risk).

(a) Calculate $\dfrac{dP}{dT}$, assuming that $p_i = 2 \times 10^{-5}$ and $E = 2$ megatons.

(b) Use the derivative to estimate the approximate change in P if T increases to 26 years.

Further Insights and Challenges

81. (a) Show that if f and g are differentiable, then

$$\frac{d}{dx}\ln(f(x)g(x)) = \frac{f'(x)}{f(x)} + \frac{g'(x)}{g(x)}\qquad \boxed{4}$$

(b) Give a new proof of the Product Rule by observing that the left-hand side of Eq. (4) is equal to $\dfrac{(f(x)g(x))'}{f(x)g(x)}$.

82. Use the formula $\log_b x = \dfrac{\log_a x}{\log_a b}$ for $a, b > 0$ to verify the formula

$$\frac{d}{dx}\log_b x = \frac{1}{(\ln b)x}$$

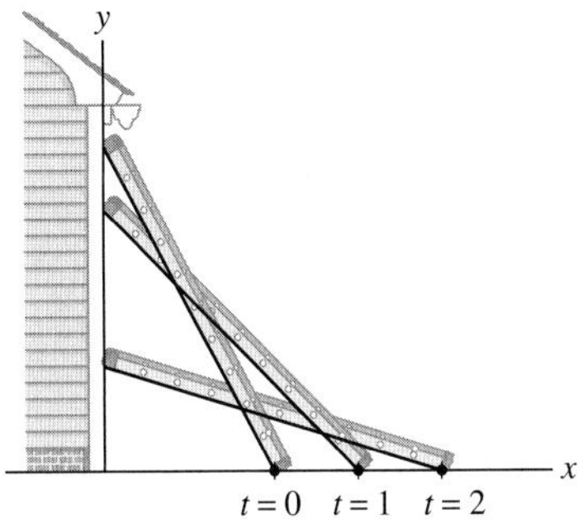

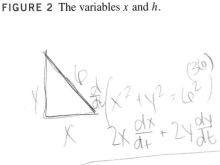

FIGURE 1 Positions of a ladder at times $t = 0, 1, 2$.

$t = 0 \quad t = 1 \quad t = 2$

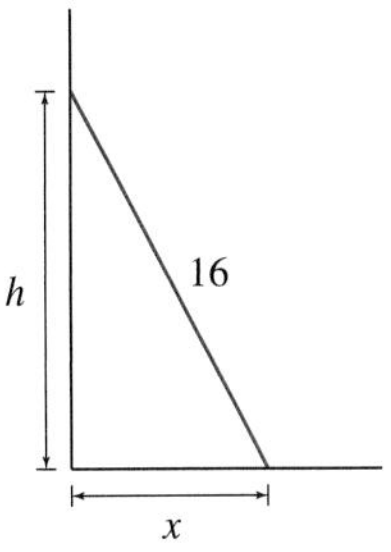

FIGURE 2 The variables x and h.

3.11 Related Rates

In *related rate* problems, the goal is to calculate an unknown rate of change in terms of other rates of change that are known. One typical problem involves a ladder leaning against a wall. The question is: *How fast does the top of the ladder move if the bottom of the ladder is pulled away from the wall at constant speed?* What is interesting and perhaps surprising is that the top and bottom travel at different speeds. Figure 1 shows this clearly: the bottom travels the same distance over each time interval, but the top travels farther during the second time interval than the first. In other words, the top is speeding up while the bottom moves at a constant speed. In the next example, we use calculus to find the speed of the ladder's top.

■ **EXAMPLE 1** Ladder Problem A 16-ft ladder leans against a wall. The bottom of the ladder is 5 ft from the wall at time $t = 0$ and slides away from the wall at a rate of 3 ft/s. Find the velocity of the top of the ladder at time $t = 1$.

Solution The first step in any related rate problem is to choose variables for the relevant quantities. Since we are considering how the top and bottom of the ladder change position, we use variables (Figure 2):

- $x = x(t)$ distance from the bottom of the ladder to the wall
- $h = h(t)$ height of the ladder's top

Both x and h are functions of time. The velocity of the bottom is $dx/dt = 3$ ft/s. Since the velocity of the top is dh/dt and the initial distance from the bottom to the wall is $x(0) = 5$, we can restate the problem as

$$\text{Compute } \frac{dh}{dt} \text{ at } t = 1 \qquad \text{given that} \qquad \frac{dx}{dt} = 3 \text{ ft/s and } x(0) = 5 \text{ ft}$$

To solve this problem, we need an equation relating x and h (Figure 2). This is provided by the Pythagorean Theorem:

$$x^2 + h^2 = 16^2$$

To calculate dh/dt, we differentiate both sides of this equation *with respect to t*:

$$\frac{d}{dt}x^2 + \frac{d}{dt}h^2 = \frac{d}{dt}16^2$$

$$2x\frac{dx}{dt} + 2h\frac{dh}{dt} = 0$$

t	x	h	dh/dt
0	5	15.20	−0.99
1	8	13.86	−1.73
2	11	11.62	−2.84
3	14	7.75	−5.42

This table of values confirms that the top of the ladder is speeding up.

This yields $\dfrac{dh}{dt} = -\dfrac{x}{h}\dfrac{dx}{dt}$, and since $\dfrac{dx}{dt} = 3$ ft/s, the velocity of the top is

$$\boxed{\dfrac{dh}{dt} = -3\dfrac{x}{h}\text{ ft/s}}\qquad \boxed{1}$$

To apply this formula, we must find x and h at time $t = 1$. Since the bottom slides away at 3 ft/s and $x(0) = 5$, we have $x(1) = 8$ and $h(1) = \sqrt{16^2 - 8^2} \approx 13.86$. Therefore,

$$\left.\dfrac{dh}{dt}\right|_{t=1} = -3\dfrac{x(1)}{h(1)} \approx -3\dfrac{8}{13.86} \approx -1.7\text{ ft/s} \qquad ■$$

In the next examples, we divide the solution into three steps that can be used when working the exercises.

■ **EXAMPLE 2** Filling a Rectangular Tank Water pours into a fish tank at a rate of 3 ft^3/min. How fast is the water level rising if the base of the tank is a rectangle of dimensions 2×3 ft?

Solution To solve a related rate problem, it is useful to draw a diagram if possible. Figure 3 illustrates our problem.

If possible, choose variables that are either related to or traditionally associated with the quantity represented. For example, we may use V for volume, θ for an angle, h or y for height, and r for radius.

Step 1. Assign variables and restate the problem.
First, we must recognize that the rate at which water pours into the tank is the derivative of water volume with respect to time. Therefore, let V be the volume and h the height of the water at time t. Then

$$\dfrac{dV}{dt} = \text{rate at which water is added to the tank}$$

$$\dfrac{dh}{dt} = \text{rate at which the water level is rising}$$

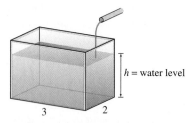

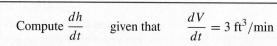

Now we can restate our problem in terms of derivatives:

$$\boxed{\text{Compute }\dfrac{dh}{dt} \quad \text{given that} \quad \dfrac{dV}{dt} = 3\text{ ft}^3/\text{min}}$$

Step 2. Find an equation that relates the variables and differentiate.
We need a relation between V and h. We have $V = 6h$ since the tank's base has area 6 ft^2. Therefore,

$$\dfrac{dV}{dt} = 6\dfrac{dh}{dt}$$

Step 3. Use the data to find the unknown derivative.
Since $dV/dt = 3$,

$$\dfrac{dh}{dt} = \dfrac{1}{6}\dfrac{dV}{dt} = \dfrac{1}{6}(3) = 0.5\text{ ft/min}$$

Note that dh/dt has units of ft/min since h and t are in feet and minutes, respectively.
                                                                                            ■

FIGURE 3 $V = $ water volume at time t.

$\dfrac{dV}{dt} = 2\,cm^3/min$ $\dfrac{dh}{dt} = ?$

The set-up in the next example is similar but more complicated because the water tank has the shape of a circular cone. We use similar triangles to derive a relation between the volume and height of the water. We also need the formula $V = \frac{1}{3}\pi h r^2$ for the volume of a circular cone of height h and radius r.

■ **EXAMPLE 3** 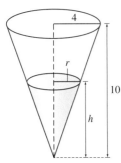 Filling a Conical Tank   Water pours into a conical tank of height 10 ft and radius 4 ft at a rate of 10 ft^3/min.

(a) How fast is the water level rising when it is 5 ft high?

(b) As time passes, what happens to the rate at which the water level rises? Explain.

Solution

(a) *Step 1.* **Assign variables and restate the problem.**

As in the previous example, let V and h be the volume and height of the water in the tank at time t. The problem in part (a) is

$$\boxed{\text{Compute } \frac{dh}{dt} \text{ at } h = 5 \quad \text{given that} \quad \frac{dV}{dt} = 10 \text{ ft}^3/\text{min}}$$

Step 2. **Find an equation that relates the variables and differentiate.**

The volume is $V = \frac{1}{3}\pi h r^2$, where r is the radius of the cone at height h, but *we cannot use this relation unless we eliminate the variable r*. Using similar triangles in Figure 4, we see that

$$\frac{r}{h} = \frac{4}{10} \quad \text{or} \quad r = 0.4\,h$$

Therefore,

$$V = \frac{1}{3}\pi h (0.4\,h)^2 = \frac{1}{3}\pi(0.16)h^3 \qquad \boxed{2}$$

$$\frac{dV}{dt} = (0.16)\pi h^2 \frac{dh}{dt} \qquad \boxed{3}$$

Step 3. **Use the data to find the unknown derivative.**

Using the given data $dV/dt = 10$ in Eq. (3), we have

$$(0.16)\pi h^2 \frac{dh}{dt} = 10$$

$$\frac{dh}{dt} = \frac{10}{(0.16)\pi h^2} \approx \frac{20}{h^2} \qquad \boxed{4}$$

When $h = 5$, the level is rising at $dh/dt \approx 20/5^2 = 0.8$ ft^3/min.

(b) Eq. (4) shows that dh/dt is inversely proportional to h^2. As h increases, the water level rises more slowly. This is reasonable if you consider that a thin slice of the cone of width Δh has more volume when h is large, so more water is needed to raise the level when h is large (Figure 5). ■

■ **EXAMPLE 4** Tracking a Rocket   A spy tracks a rocket through a telescope to determine its velocity. The rocket is traveling vertically from a launching pad located 10 km away, as in Figure 6. At a certain moment, the spy's instruments show that the angle between the telescope and the ground is equal to $\frac{\pi}{3}$ and is changing at a rate of 0.5 rad/min. What is the rocket's velocity at that moment?

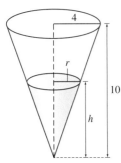

FIGURE 4 By similar triangles,

$$\frac{h}{r} = \frac{10}{4}$$

CAUTION *A common mistake is substituting $h = 5$ in Eq. (2). Do not set $h = 5$ until the end of the problem, after the derivatives have been computed. This applies to all related rate problems.*

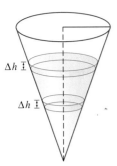

FIGURE 5 When h is larger, it takes more water to raise the level by an amount Δh.

Solution

Step 1. Assign variables and restate the problem.

Let θ be the angle between the telescope and the ground, and let y be the height of the rocket at time t. Then our goal is to compute the rocket's velocity dy/dt when $\theta = \frac{\pi}{3}$. We restate the problem as follows:

> Compute $\dfrac{dy}{dt}\bigg|_{\theta=\frac{\pi}{3}}$ given that $\dfrac{d\theta}{dt} = 0.5$ rad/min when $\theta = \dfrac{\pi}{3}$

Step 2. Find an equation that relates the variables and differentiate.

We need a relation between θ and y. As we see in Figure 6,

$$\tan \theta = \frac{y}{10}$$

Now differentiate with respect to time:

$$\sec^2 \theta \frac{d\theta}{dt} = \frac{1}{10} \frac{dy}{dt}$$

$$\frac{dy}{dt} = \frac{10}{\cos^2 \theta} \frac{d\theta}{dt}$$

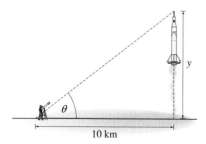

FIGURE 6 Tracking a rocket through a telescope.

$$\boxed{5}$$

Step 3. Use the given data to find the unknown derivative.

At the given moment, $\theta = \frac{\pi}{3}$ and $d\theta/dt = 0.5$, so Eq. (5) yields

$$\frac{dy}{dt} = \frac{10}{\cos^2(\pi/3)}(0.5) = \frac{10}{(0.5)^2}(0.5) = 20 \text{ km/min.}$$

The rocket's velocity at this moment is 20 km/min or 1,200 km/hour. ∎

For the next example, let's recall the following fact: If an object is located at $x = A$ at time $t = 0$ and moves along the x-axis with constant velocity v, then its position at time t is $x = A + vt$. For example, if $v = 3$ mph, then the object's position is $A + 3t$ miles (see Figure 7).

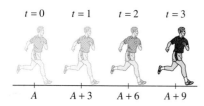

FIGURE 7 Moving along the x-axis (starting at A) with constant velocity 3.

■ **EXAMPLE 5** One train travels west toward Denver at 120 mph, while a second train travels north away from Denver at 90 mph (Figure 8). At time $t = 0$, the first train is 10 miles east and the second train 20 miles north of the Denver station. Calculate the rate at which the distance between the trains is changing:

(a) At time $t = 0$ **(b)** 10 min later

Solution

Step 1. Assign variables and restate the problem.

$$x = \text{distance in miles from the westbound train to the station}$$

$$y = \text{distance in miles from the northbound train to the station}$$

$$\ell = \text{distance in miles between the two trains}$$

The velocities of the trains are dx/dt and dy/dt, so our problem can be restated as follows:

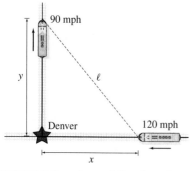

FIGURE 8 Location and direction of trains.

(handwritten left margin notes:)

1. Read problem carefully
2. Draw a diagram if possible
3. Assign variables to all quantities that change as time goes by.
4. Express the given info and required rate in terms of derivatives
5. Write an equation that relates the various quantities of the problem. If necessary, eliminate one of the variables by substitution.
6. Use the chain rule to differentiate both sides of the equation w/ respect to

---

$$\text{Compute } \frac{d\ell}{dt} \text{ given that } \frac{dx}{dt} = -120 \text{ mph} \quad \text{and} \quad \frac{dy}{dt} = 90 \text{ mph}$$

Note that dx/dt is negative since the train is moving west (to the left).

Step 2. Find an equation that relates the variables and differentiate.

The variables are related by the Pythagorean Theorem

$$\ell^2 = x^2 + y^2$$

(handwritten:) $2(10)(-120) + 2(20)(90) = 2(\sqrt{500})\frac{dz}{dt}$
$-2400 + 3600 = 2\sqrt{500}$
$12\sqrt{5} = \frac{dz}{dt}$

Differentiate with respect to t:

$$2\ell\frac{d\ell}{dt} = 2x\frac{dx}{dt} + 2y\frac{dy}{dt}$$

$$\frac{d\ell}{dt} = \frac{x\frac{dx}{dt} + y\frac{dy}{dt}}{\ell} \qquad \boxed{6}$$

Step 3. Use the given data to find the unknown derivative.

Substituting $dx/dt = -120$ and $dy/dt = 90$ in Eq. (6), we obtain

$$\frac{d\ell}{dt} = \frac{-120x + 90y}{\ell} = \frac{-120x + 90y}{\sqrt{x^2 + y^2}} \qquad \boxed{7}$$

We can now solve the two parts of the problem.

(a) At $t = 0$, we are given $x = 10$ and $y = 20$. By (7), the distance between the trains is changing at a rate of

$$\frac{d\ell}{dt}\bigg|_{t=0} = \frac{-120(10) + 90(20)}{\sqrt{10^2 + 20^2}} \approx \frac{600}{22.36} \approx 27 \text{ mph}$$

(b) By the remark preceding this example, the coordinates of the trains as a function of time are

$$x = 10 - 120t, \qquad y = 20 + 90t$$

After 10 min, or $t = \frac{1}{6}$ hour,

$$x = 10 - 120\left(\frac{1}{6}\right) = -10 \text{ miles} \quad \text{and} \quad y = 20 + 90\left(\frac{1}{6}\right) = 35 \text{ miles}$$

Therefore, by (7), the distance between the trains is changing at a rate of

$$\frac{d\ell}{dt}\bigg|_{t=\frac{1}{6}} = \frac{-120(-10) + 90(35)}{\sqrt{(-10)^2 + 35^2}} \approx \frac{4{,}350}{36.4} \approx 120 \text{ mph} \qquad \blacksquare$$

3.11 SUMMARY

- Related rate problems present us with situations in which one or more variables are related by an equation. We are asked to compute the ROC of one of the variables in terms of the ROCs of the other variables.
- Draw a diagram if possible. It may also be useful to break the solution into three steps:

Step 1. Assign variables and restate the problem.

Step 2. Find an equation that relates the variables and differentiate.

This gives us an equation relating the known and unknown derivatives. Remember not to substitute values for the variables until after you have computed all derivatives.

Step 3. Use the given data to find the unknown derivative.

• The two facts from geometry that arise most often in related rate problems are the Pythagorean Theorem and the Theorem of Similar Triangles (ratios of corresponding sides are equal).

3.11 EXERCISES

Preliminary Questions

1. Assign variables and restate the following problem in terms of known and unknown derivatives (but do not solve it): How fast is the volume of a cube increasing if its side increases at a rate of 0.5 cm/s?

2. What is the relation between dV/dt and dr/dt if

$$V = \left(\frac{4}{3}\right)\pi r^3$$

In Questions 3–4, suppose that water pours into a cylindrical glass of radius 4 cm. The variables V and h denote the volume and water level at time t, respectively.

3. Restate in terms of the derivatives dV/dt and dh/dt: How fast is the water level rising if water pours in at a rate of 2 cm^3/min?

4. Repeat the same for this problem: At what rate is water pouring in if the water level rises at a rate of 1 cm/min?

Exercises

In Exercises 1–2, consider a rectangular bathtub whose base is 18 ft^2.

1. How fast is the water level rising if water is filling the tub at a rate of 0.7 ft^3/min?

2. At what rate is water pouring into the tub if the water level rises at a rate of 0.8 ft/min?

3. The radius of a circular oil slick expands at a rate of 2 m/min.

(a) How fast is the area of the oil slick increasing when the radius is 25 m?

(b) If the radius is 0 at time $t = 0$, how fast is the area increasing after 3 min?

4. At what rate is the diagonal of a cube increasing if its edges are increasing at a rate of 2 cm/s?

In Exercises 5–8, assume that the radius r of a sphere is expanding at a rate of 14 in./min. The volume of a sphere is $V = \frac{4}{3}\pi r^3$ and its surface area is $4\pi r^2$.

5. Determine the rate at which the volume is changing with respect to time when $r = 8$ in.

6. Determine the rate at which the volume is changing with respect to time at $t = 2$ min, assuming that $r = 0$ at $t = 0$.

7. Determine the rate at which the surface area is changing when the radius is $r = 8$ in.

8. Determine the rate at which the surface area is changing with respect to time at $t = 2$ min, assuming that $r = 3$ at $t = 0$.

9. A road perpendicular to a highway leads to a farmhouse located 1 mile away (Figure 9). An automobile travels past the farmhouse at a speed of 60 mph. How fast is the distance between the automobile and the farmhouse increasing when the automobile is 3 miles past the intersection of the highway and the road?

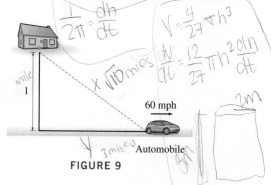

FIGURE 9

10. A conical tank has height 3 m and radius 2 m at the top. Water flows in at a rate of 2 m^3/min. How fast is the water level rising when it is 2 m?

11. Follow the same set-up as Exercise 10, but assume that the water level is rising at a rate of 0.3 m/min when it is 2 m. At what rate is water flowing in?

12. Sonya and Isaac are in motorboats located at the center of a lake. At time $t = 0$, Sonya begins traveling south at a speed of 32 mph. At the same time, Isaac takes off, heading east at a speed of 27 mph.

(a) How far have Sonya and Isaac each traveled after 12 min?

(b) At what rate is the distance between them increasing at $t = 12$ min?

13. Answer (a) and (b) in Exercise 12 assuming that Sonya begins moving 1 minute after Isaac takes off.

14. A 6-ft man walks away from a 15-ft lamppost at a speed of 3 ft/s (Figure 10). Find the rate at which his shadow is increasing in length.

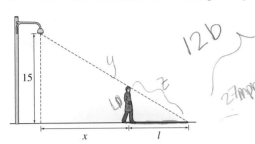

FIGURE 10

15. At a given moment, a plane passes directly above a radar station at an altitude of 6 miles.

(a) If the plane's speed is 500 mph, how fast is the distance between the plane and the station changing half an hour later?

(b) How fast is the distance between the plane and the station changing when the plane passes directly above the station?

16. In the setting of Exercise 15, suppose that the line through the radar station and the plane makes an angle θ with the horizontal. How fast is θ changing 10 min after the plane passes over the radar station?

17. A hot air balloon rising vertically is tracked by an observer located 2 miles from the lift-off point. At a certain moment, the angle between the observer's line-of-sight and the horizontal is $\frac{\pi}{5}$, and it is changing at a rate of 0.2 rad/min. How fast is the balloon rising at this moment?

18. As a man walks away from a 12-ft lamppost, the tip of his shadow moves twice as fast as he does. What is the man's height?

In Exercises 19–23, refer to a 16-ft ladder sliding down a wall, as in Figures 1 and 2. The variable h is the height of the ladder's top at time t, and x is the distance from the wall to the ladder's bottom.

19. Assume the bottom slides away from the wall at a rate of 3 ft/s. Find the velocity of the top of the ladder at $t = 2$ if the bottom is 5 ft from the wall at $t = 0$.

20. Suppose that the top is sliding down the wall at a rate of 4 ft/s. Calculate dx/dt when $h = 12$.

21. Suppose that $h(0) = 12$ and the top slides down the wall at a rate of 4 ft/s. Calculate x and dx/dt at $t = 2$ s.

22. What is the relation between h and x at the moment when the top and bottom of the ladder move at the same speed?

23. Show that the velocity dh/dt approaches infinity as the ladder slides down to the ground (assuming dx/dt is constant). This suggests that our mathematical description is unrealistic, at least for small values of h. What would, in fact, happen as the top of the ladder approaches the ground?

24. The radius r of a right circular cone of fixed height $h = 20$ cm is increasing at a rate of 2 cm/s. How fast is the volume increasing when $r = 10$?

25. Suppose that both the radius r and height h of a circular cone change at a rate of 2 cm/s. How fast is the volume of the cone increasing when $r = 10$ and $h = 20$?

26. A particle moves counterclockwise around the ellipse $9x^2 + 16y^2 = 25$ (Figure 11).

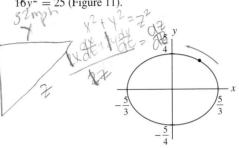

FIGURE 11

(a) In which of the four quadrants is the derivative dx/dt positive? Explain your answer.

(b) Find a relation between dx/dt and dy/dt.

(c) At what rate is the x-coordinate changing when the particle passes the point $(1, 1)$ if its y-coordinate is increasing at a rate of 6 ft/s?

(d) What is dy/dt when the particle is at the top and bottom of the ellipse?

27. A searchlight rotates at a rate of 3 revolutions per minute. The beam hits a wall located 10 miles away and produces a dot of light that moves horizontally along the wall. How fast is this dot moving when the angle θ between the beam and the line through the searchlight perpendicular to the wall is $\frac{\pi}{6}$? Note that $d\theta/dt = 3(2\pi) = 6\pi$.

28. A rocket travels vertically at a speed of 800 mph. The rocket is tracked through a telescope by an observer located 10 miles from the launching pad. Find the rate at which the angle between the telescope and the ground is increasing 3 min after lift-off.

29. A plane traveling at an altitude of 20,000 ft passes directly overhead at time $t = 0$. One minute later you observe that the angle between the vertical and your line of sight to the plane is 1.14 rad and that this angle is changing at a rate of 0.38 rad/min. Calculate the velocity of the airplane.

30. Calculate the rate (in cm^2/s) at which area is swept out by the second hand of a circular clock as a function of the clock's radius.

31. A jogger runs around a circular track of radius 60 ft. Let (x, y) be her coordinates, where the origin is at the center of the track. When the jogger's coordinates are $(36, 48)$, her x-coordinate is changing at a rate of 14 ft/s. Find dy/dt.

32. A car travels down a highway at 55 mph. An observer is standing 500 ft from the highway.

(a) How fast is the distance between the observer and the car increasing at the moment the car passes in front of the observer? Can you justify your answer without relying on any calculations?

(b) How fast is the distance between the observer and the car increasing 1 min later?

In Exercises 33–34, assume that the pressure P (in kilopascals) and volume V (in cubic centimeters) of an expanding gas are related by $PV^b = C$, where b and C are constants (this holds in adiabatic expansion, without heat gain or loss).

33. Find dP/dt if $b = 1.2$, $P = 8$ kPa, $V = 100$ cm^2, and $dV/dt = 20$ cm^3/min.

34. Find b if $P = 25$ kPa, $dP/dt = 12$ kPa/min, $V = 100$ cm^2, and $dV/dt = 20$ cm^3/min.

35. A point moves along the parabola $y = x^2 + 1$. Let $\ell(t)$ be the distance between the point and the origin. Calculate $\ell'(t)$, assuming that the x-coordinate of the point is increasing at a rate of 9 ft/s.

36. The base x of the right triangle in Figure 12 increases at a rate of 5 cm/s, while the height remains constant at $h = 20$. How fast is the angle θ changing when $x = 20$?

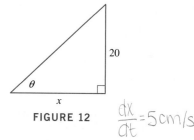

FIGURE 12 $\dfrac{dx}{dt} = 5 \text{cm/s}$

37. A water tank in the shape of a right circular cone of radius 300 cm and height 500 cm leaks water from the vertex at a rate of 10 cm^3/min. Find the rate at which the water level is decreasing when it is 200 cm.

38. Two parallel paths 50 ft apart run through the woods. Shirley jogs east on one path at 6 mph, while Jamail walks west on the other at 4 mph. If they pass each other at time $t = 0$, how far apart are they 3 s later, and how fast is the distance between them changing at that moment?

Further Insights and Challenges

39. Henry is pulling on a rope that passes through a pulley on a 10-ft pole and is attached to a wagon (Figure 13). Assume that the rope is attached to a loop on the wagon 2 ft off the ground. Let x be the distance between the loop and the pole.

(a) Find a formula for the speed of the wagon in terms of x and the rate at which Henry pulls the rope.

(b) Find the speed of the wagon when it is 12 ft from the pole, assuming that Henry pulls the rope at a rate of 1.5 ft/s.

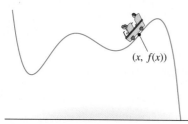

FIGURE 13

40. A roller coaster has the shape of the graph in Figure 14. Show that when the roller coaster passes the point $(x, f(x))$, the vertical velocity of the roller coaster is equal to $f'(x)$ times its horizontal velocity.

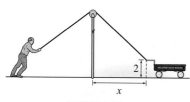

FIGURE 14 Graph of $f(x)$ as a roller coaster track.

41. Using a telescope, you track a rocket that was launched 2 miles away, recording the angle θ between the telescope and the ground at half-second intervals. Estimate the velocity of the rocket if $\theta(10) = 0.205$ and $\theta(10.5) = 0.225$.

42. Two trains leave a station at $t = 0$ and travel with constant velocity v along straight tracks that make an angle θ.

(a) Show that the trains are separating from each other at a rate $v\sqrt{2 - 2\cos\theta}$.

(b) What does this formula give for $\theta = \pi$?

43. A baseball player runs from home plate toward first base at 20 ft/s. How fast is the player's distance from second base changing when the player is halfway to first base? See Figure 15.

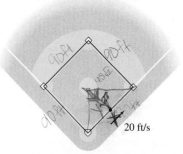

FIGURE 15 Baseball diamond.

44. As the wheel of radius r cm in Figure 16 rotates, the rod of length L attached at the point P drives a piston back and forth in a straight line. Let x be the distance from the origin to the point Q at the end of the rod as in the figure.

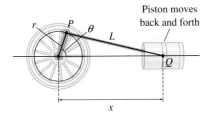

FIGURE 16

(a) Use the Pythagorean Theorem to show that

$$L^2 = (x - r\cos\theta)^2 + r^2\sin^2\theta \qquad \boxed{8}$$

(b) Differentiate Eq. (8) with respect to t to prove that

$$2(x - r\cos\theta)\left(\frac{dx}{dt} + r\sin\theta\frac{d\theta}{dt}\right) + 2r^2\sin\theta\cos\theta\frac{d\theta}{dt} = 0$$

(c) Calculate the speed of the piston when $\theta = \frac{\pi}{2}$, assuming that $r = 10$ cm, $L = 30$ cm, and the wheel rotates at 4 revolutions per minute.

45. A spectator seated 300 m away from the center of a circular track of radius 100 m watches an athlete run laps at a speed of 5 m/s. How fast is the distance between the spectator and athlete changing when the runner is approaching the spectator and the distance between them is 250 m? *Hint:* The diagram for this problem is similar to Figure 16, with $r = 100$ and $x = 300$.

46. A cylindrical tank of radius R and length L lying horizontally as in Figure 17 is filled with oil to height h.

(a) Show that the volume $V(h)$ of oil in the tank as a function of height h is

$$V(h) = L\left(R^2\cos^{-1}\left(1 - \frac{h}{R}\right) - (R - h)\sqrt{2hR - h^2}\right)$$

(b) Show that $\dfrac{dV}{dh} = 2L\sqrt{h(2R - h)}$.

(c) Suppose that $R = 4$ ft and $L = 30$ ft, and that the tank is filled at a constant rate of 10 ft^3/min. How fast is the height h increasing when $h = 5$?

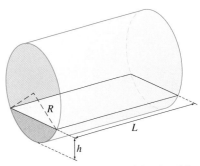

FIGURE 17 Oil in the tank has level h.

CHAPTER REVIEW EXERCISES

In Exercises 1–4, refer to the function $f(x)$ whose graph is shown in Figure 1.

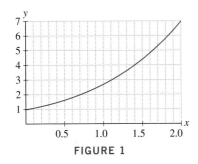

FIGURE 1

1. Compute the average ROC of $f(x)$ over $[0, 2]$. What is the graphical interpretation of this average ROC?

2. For which value of h is $\dfrac{f(0.7 + h) - f(0.7)}{h}$ equal to the slope of the secant line between the points where $x = 0.7$ and $x = 1.1$.

3. Estimate $\dfrac{f(0.7 + h) - f(0.7)}{h}$ for $h = 0.3$. Is this number larger or smaller than $f'(0.7)$?

4. Estimate $f'(0.7)$ and $f'(1.1)$.

In Exercises 5–8, compute $f'(a)$ using the limit definition and find an equation of the tangent line to the graph of $f(x)$ at $x = a$.

5. $f(x) = x^2 - x,\quad a = 1$

6. $f(x) = 5 - 3x,\quad a = 2$

7. $f(x) = x^{-1},\quad a = 4$

8. $f(x) = x^3,\quad a = -2$

In Exercises 9–12, compute dy/dx using the limit definition.

9. $y = 4 - x^2$

10. $y = \sqrt{2x + 1}$

11. $y = \dfrac{1}{2 - x}$

12. $y = \dfrac{1}{(x - 1)^2}$

In Exercises 13–16, express the limit as a derivative.

13. $\lim\limits_{h \to 0} \dfrac{\sqrt{1 + h} - 1}{h}$

14. $\lim\limits_{x \to -1} \dfrac{x^3 + 1}{x + 1}$

15. $\lim\limits_{t \to \pi} \dfrac{\sin t \cos t}{t - \pi}$

16. $\lim\limits_{\theta \to \pi} \dfrac{\cos\theta - \sin\theta + 1}{\theta - \pi}$

17. Find $f(4)$ and $f'(4)$ if the tangent line to the graph of $f(x)$ at $x = 4$ has equation $y = 3x - 14$.

18. Each graph in Figure 2 shows the graph of a function $f(x)$ and its derivative $f'(x)$. Determine which is the function and which is the derivative.

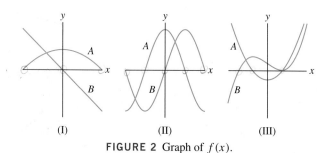

(I)   (II)   (III)

FIGURE 2 Graph of $f(x)$.

19. Is (A), (B), or (C) the graph of the derivative of the function $f(x)$ shown in Figure 3?

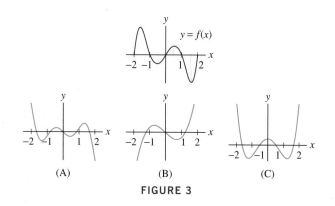

(A)   (B)   (C)

FIGURE 3

20. Let $N(t)$ be the percentage of a state population infected with a flu virus on week t of an epidemic. What percentage is likely to be infected in week 4 if $N(3) = 8$ and $N'(3) = 1.2$?

21. A girl's height $h(t)$ (in centimeters) is measured at time t (years) for $0 \leq t \leq 14$:

$$52, 75.1, 87.5, 96.7, 104.5, 111.8, 118.7, 125.2,$$

$$131.5, 137.5, 143.3, 149.2, 155.3, 160.8, 164.7$$

(a) What is the girl's average growth rate over the 14-year period?

(b) Is the average growth rate larger over the first half or the second half of this period?

(c) Estimate $h'(t)$ (in centimeters per year) for $t = 3, 8$.

22. A planet's period P (time in years to complete one revolution around the sun) is approximately $0.0011A^{3/2}$, where A is the average distance (in millions of miles) from the planet to the sun.

(a) Calculate P and dP/dA for Mars using the value $A = 142$.

(b) Estimate the increase in P if A is increased to 143.

In Exercises 23–24, use the following table of values for the number $A(t)$ of automobiles (in millions) manufactured in the United States in year t.

t	1970	1971	1972	1973	1974	1975	1976
$A(t)$	6.55	8.58	8.83	9.67	7.32	6.72	8.50

23. What is the interpretation of $A'(t)$? Estimate $A'(1971)$. Does $A'(1974)$ appear to be positive or negative?

24. Given the data, which of (A)–(C) in Figure 4 could be the graph of the derivative $A'(t)$? Explain.

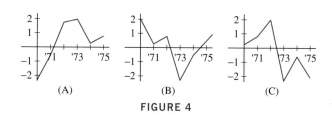

(A)   (B)   (C)

FIGURE 4

25. Which of the following is equal to $\dfrac{d}{dx}2^x$?

(a) 2^x

(b) $(\ln 2)2^x$

(c) $x2^{x-1}$

(d) $\dfrac{1}{\ln 2}2^x$

26. Describe the graphical interpretation of the relation $g'(x) = 1/f'(g(x))$, where $f(x)$ and $g(x)$ are inverses of each other.

27. Find $g'(8)$, where $g(x)$ is the inverse of a differentiable function $f(x)$ such that $f(-1) = 8$ and $f'(-1) = 12$.

In Exercises 28–53, compute the derivative.

28. $y = 3x^5 - 7x^2 + 4$

29. $y = 4x^{-3/2}$

30. $y = t^{-7.3}$

31. $y = 4x^2 - x^{-2}$

32. $y = \dfrac{x+1}{x^2+1}$

33. $y = \dfrac{3t-2}{4t-9}$

34. $y = (x^4 - 9x)^6$

35. $y = (3t^2 + 20t^{-3})^6$

36. $y = (2 + 9x^2)^{3/2}$

37. $y = (x+1)^3(x+4)^4$

38. $y = \dfrac{z}{\sqrt{1-z}}$

39. $y = \left(1 + \dfrac{1}{x}\right)^3$

40. $y = \dfrac{x^4 + \sqrt{x}}{x^2}$

41. $y = \dfrac{1}{(1-x)\sqrt{2-x}}$

42. $y = \tan(t^{-3})$

43. $y = 4\cos(2 - 3x)$

44. $y = \sin(2x)\cos^2 x$

45. $y = \sin(\sqrt{x^2 + 1})$

46. $y = \tan^3 \theta$

47. $y = \sin\left(\dfrac{4}{\theta}\right)$

48. $y = \dfrac{t}{1 + \sec t}$

49. $y = z \csc(9z + 1)$

50. $y = \dfrac{8}{1 + \cot \theta}$

51. $y = \tan(\cos x)$

52. $y = \sqrt{x + \sqrt{x + \sqrt{x}}}$

53. $y = \cos(\cos(\cos(\theta)))$

In Exercises 54–75, find the derivative.

54. $f(x) = 9e^{-4x}$

55. $f(x) = \ln(4x^2 + 1)$

56. $f(x) = \dfrac{e^{-x}}{x}$

57. $f(x) = \ln(x + e^x)$

58. $G(s) = (\ln(s))^2$

59. $G(s) = \ln(s^2)$

60. $g(t) = e^{4t - t^2}$

61. $g(t) = t^2 e^{1/t}$

62. $f(\theta) = \ln(\sin \theta)$

63. $f(\theta) = \sin(\ln \theta)$

64. $f(x) = e^{x + \ln x}$

65. $f(x) = e^{\sin^2 x}$

66. $h(y) = 2^{1-y}$

67. $h(y) = \dfrac{1 + e^y}{1 - e^y}$

68. $G(s) = \cos^{-1}(s^{-1})$

69. $G(s) = \tan^{-1}(\sqrt{s})$

70. $f(x) = \ln(\csc^{-1} x)$

71. $f(x) = e^{\sec^{-1} x}$

72. $g(t) = \sinh(t^2)$

73. $h(y) = y \tanh(4y)$

74. $g(x) = \tanh^{-1}(e^x)$

75. $g(t) = \sqrt{t^2 - 1} \sinh^{-1} t$

76. Suppose that $f(g(x)) = e^{x^2}$, where $g(1) = 2$ and $g'(1) = 4$. Find $f'(2)$.

In Exercises 77–78, let $f(x) = xe^{-x}$.

77. Show that $f(x)$ has an inverse on $[1, \infty)$. Let $g(x)$ be this inverse. Find the domain and range of $g(x)$ and compute $g'(2e^{-2})$.

78. Show that $f(x) = c$ has two solutions if $0 < c < 1$.

In Exercises 79–84, use the table of values to calculate the derivative of the given function at $x = 2$.

x	$f(x)$	$g(x)$	$f'(x)$	$g'(x)$
2	5	4	−3	9
4	3	2	−2	3

79. $S(x) = 3f(x) - 2g(x)$

80. $H(x) = f(x)g(x)$

81. $R(x) = \dfrac{f(x)}{g(x)}$

82. $G(x) = f(g(x))$

83. $F(x) = f(g(2x))$

84. $K(x) = f(x^2)$

In Exercises 85–88, let $f(x) = x^3 - 3x^2 + x + 4$.

85. Find the points on the graph of $f(x)$ where the tangent line has slope 10.

86. For which values of x are the tangent lines to the graph of $f(x)$ horizontal?

87. Find all values of b such that $y = 25x + b$ is tangent to the graph of $f(x)$.

88. Find all values of k such that $f(x)$ has only one tangent line of slope k.

89. (a) Show that there is a unique value of a such that $f(x) = x^3 - 2x^2 + x + 1$ has the same slope at both a and $a + 1$.
(b) $\boxed{\text{GU}}$ Plot $f(x)$ together with the tangent lines at $x = a$ and $x = a + 1$ and confirm your answer to part (a).

90. ✏ The following table gives the percentage of voters supporting each of two candidates in the days before an election. Compute the average ROC of A's percentage over the intervals from day 20 to 15, day 15 to 10, and day 10 to 5. If this trend continues over the last 5 days before the election, will A win?

Days before Election	20	15	10	5
Candidate A	44.8%	46.8%	48.3%	49.3%
Candidate B	55.2%	53.2%	51.7%	50.7%

In Exercises 91–96, calculate y''.

91. $y = 12x^3 - 5x^2 + 3x$

92. $y = x^{-2/5}$

93. $y = \sqrt{2x + 3}$

94. $y = \dfrac{4x}{x + 1}$

95. $y = \tan(x^2)$

96. $y = \sin^2(x + 9)$

97. In Figure 5, label the graphs f, f', and f''.

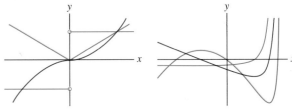

FIGURE 5

98. Let q be a differentiable function of p, say, $q = f(p)$. If p changes by an amount Δp, then the percentage change in p is $(100\Delta p)/p$, and the corresponding percentage change in q is $(100\Delta q)/q$, where $\Delta q = f(p + \Delta p) - f(p)$. The *percentage rate of change* of q with respect to p is the limit of the ratio of percentage changes:

$$\lim_{\Delta p \to 0} \frac{(100\Delta q)/q}{(100\Delta p)/p}$$

Show that this limit is equal to $\left(\dfrac{p}{q}\right)\dfrac{dq}{dp}$.

99. The number q of frozen chocolate cakes that a commercial bakery can sell per week depends on the price p. In this setting, the percentage ROC of q with respect to p is called the *price elasticity of demand* $E(p)$. Assume that $q = 50p(10 - p)$ for $5 < p < 10$.

(a) Show that $E(p) = \dfrac{2p - 10}{p - 10}$.

(b) Compute the value of $E(8)$ and explain how it justifies the following statement: If the price is set at $8, then a 1% increase in price reduces demand by 3%.

100. The daily demand for flights between Chicago and St. Louis at the price p is $q = 350 - 0.7p$ (in hundreds of passengers). Calculate the price elasticity of demand when $p = \$250$ and estimate the number of additional passengers if the ticket price is lowered by 1%.

In Exercises 101–108, compute $\dfrac{dy}{dx}$.

101. $x^3 - y^3 = 4$

102. $4x^2 - 9y^2 = 36$

103. $y = xy^2 + 2x^2$

104. $\dfrac{y}{x} = x + y$

105. $x + y = \sqrt{3x^2 + 2y^2}$

106. $y = \sin(x + y)$

107. $\tan(x + y) = xy$

108. $(\sin x)(\cos y) = x^2 - y$

109. Find the points on the graph of $x^3 - y^3 = 3xy - 3$ where the tangent line is horizontal.

110. Find the points on the graph of $x^{2/3} + y^{2/3} = 1$ where the tangent line has slope 1.

111. For which values of α is $f(x) = |x|^\alpha$ differentiable at $x = 0$?

112. Let $f(x) = x^2 \sin(x^{-1})$ for $x \neq 0$ and $f(0) = 0$. Show that $f'(x)$ exists for all x (including $x = 0$) but that $f'(x)$ is not continuous at $x = 0$ (Figure 6).

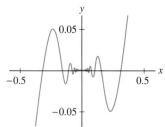

FIGURE 6 Graph of $f(x) = x^2 \sin(x^{-2})$

113. Water pours into the tank in Figure 7 at a rate of 20 m^3/min. How fast is the water level rising when the water level is $h = 4$ m?

114. The minute hand of a clock is 4 in. long and the hour hand is 3 in. long. How fast is the distance between the tips of the hands changing at 3 o'clock?

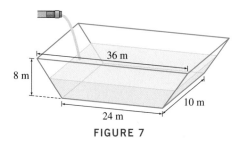

FIGURE 7

115. A light moving at 3 ft/s approaches a 6-ft man standing 12 ft from a wall (Figure 8). The light is 3 ft above the ground. How fast is the tip P of the man's shadow moving when the light is 24 ft from the wall?

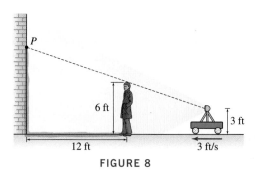

FIGURE 8

116. A bead slides down the curve $xy = 10$. Find the bead's horizontal velocity if its height at time t seconds is $y = 80 - 16t^2$ cm.

117. (a) Side x of the triangle in Figure 9 is increasing at 2 cm/s and side y is increasing at 3 cm/s. Assume that θ decreases in such a way that the area of the triangle has the constant value 4 cm^2. How fast is θ decreasing when $x = 4$, $y = 4$?

(b) How fast is the distance between P and Q changing when $x = 2$, $y = 3$?

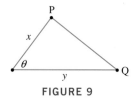

FIGURE 9

In Exercises 118–123, use logarithmic differentiation to find the derivative.

118. $y = \dfrac{(x + 1)^3}{(4x - 2)^2}$

119. $y = \dfrac{(x + 1)(x + 2)^2}{(x + 3)(x + 4)}$

120. $y = e^{(x-1)^2} e^{(x-3)^2}$

121. $y = \dfrac{e^x \sin^{-1} x}{\ln x}$

122. $y = \dfrac{e^{3x}(x - 2)^2}{(x + 1)^2}$

123. $y = x^{\sqrt{x}}(x^{\ln x})$

The honeycomb structure is designed to minimize the amount of wax required (see Exercise 72 in Section 4.2). As far as we know, bees found this structure without using calculus.

4 | APPLICATIONS OF THE DERIVATIVE

The previous chapter developed the basic tools of differentiation. In this chapter, we put these tools to work. We use the first and second derivatives to analyze functions and their graphs, and we develop techniques for solving optimization problems (problems that involve finding the minimum or maximum value of a function). Newton's Method in Section 4.8 employs the derivative to find numerical approximations to solutions of equations. In Section 4.9, we introduce antidifferentiation, the inverse operation to differentiation, to prepare for the study of integration in Chapter 5.

4.1 Linear Approximation and Applications

In some situations we are interested in determining the "effect of a small change." For example:

- A basketball player releases the ball at a certain angle θ. How does a small change in θ affect the distance of the shot? (Exercises 64–65.)
- The price of a movie ticket is $9.00. How will revenues at the box office be affected if the ticket price is increased to $9.50?
- The cube root of 27 is 3. How much larger is the cube root of 27.2? (Exercise 25.)

In each case, we have a function $f(x)$ and we're interested in the change

$$\Delta f = f(a + \Delta x) - f(a), \qquad \text{where } \Delta x \text{ is small}$$

The **Linear Approximation** uses the derivative to estimate Δf without computing it exactly. By definition, the derivative is the limit

$$f'(a) = \lim_{\Delta x \to 0} \frac{f(a + \Delta x) - f(a)}{\Delta x} = \lim_{\Delta x \to 0} \frac{\Delta f}{\Delta x}$$

So when Δx is small, we have $\Delta f / \Delta x \approx f'(a)$, and thus,

$$\Delta f \approx f'(a)\Delta x$$

◄·· **REMINDER** The notation $\approx$ means "approximately equal to." The accuracy of the linear approximation is discussed at the end of this section.

Linear Approximation of Δf If f is differentiable at $x = a$ and Δx is small, then

$$\boxed{\Delta f \approx f'(a)\Delta x}$$

where $\Delta f = f(a + \Delta x) - f(a)$.

The linear approximation states

$$\Delta f \approx f'(a)\,\Delta x$$

where $\Delta f = f(a + \Delta x) - f(a)$.

GRAPHICAL INSIGHT   The Linear Approximation is often called the **tangent line approximation** because of the following interpretation. The quantity Δf is the vertical change from $x = a$ to $x = a + \Delta x$ in the graph of $f(x)$ (Figure 1). Recall that for a nonvertical straight line, the vertical change is equal to the slope times the horizontal change. Since the tangent line has slope $f'(a)$, the vertical change in the tangent line is $f'(a)\Delta x$. What the Linear Approximation does, therefore, is use the vertical change in the tangent line as an approximation to the vertical change in the graph of $f(x)$. When Δx is small, the two quantities are nearly equal.

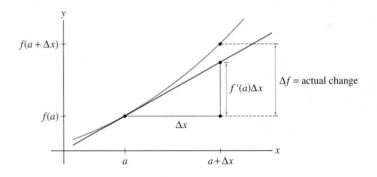

FIGURE 1 Graphical meaning of the Linear Approximation $\Delta f \approx f'(a)\Delta x$.

When we use a quantity E as an estimate for another quantity A, the "error" is defined as the absolute value $|A - E|$.

Keep in mind the different roles played by Δf and $f'(a)\Delta x$. The quantity of interest is Δf and we estimate it by $f'(a)\,\Delta x$. Most importantly, the Linear Approximation tells us that there is a nearly linear relationship between Δf and Δx when Δx is small.

■ **EXAMPLE 1**   Use the Linear Approximation to estimate $\dfrac{1}{10.2} - \dfrac{1}{10}$. How accurate is your estimate?

Solution   We apply the Linear Approximation to $f(x) = \dfrac{1}{x}$ with $a = 10$ and $\Delta x = 0.2$:

$$\Delta f = f(10.2) - f(10) = \frac{1}{10.2} - \frac{1}{10} \approx f'(10)\Delta x$$

Since $f'(x) = -x^{-2}$, $f'(10) = -0.01$ and we obtain the estimate

$$\frac{1}{10.2} - \frac{1}{10} \approx f'(10)\Delta x = -(0.01)(0.2) = -0.002$$

How accurate is this estimate? A calculator gives the value

$$\frac{1}{10.2} - \frac{1}{10} \approx -0.00196$$

The error in the linear approximation is the quantity

$$Error = \left| \Delta f - f'(a)\Delta x \right|$$

Therefore, the error is less than 10^{-4}:

$$Error \approx \left| -0.00196 - (-0.002) \right| = 0.00004 < 10^{-4}$$ ■

Recall the formula for the distance traveled when the velocity is constant:

$$\text{Distance traveled} = \text{velocity} \times \text{time}$$

The Linear Approximation tells us that this equation is *approximately* true over small time intervals even if the velocity is changing. If $s(t)$ is the position of an object at time

t and Δs is the change in position over a small time interval $[t_0, t_0 + \Delta t]$, then

$$\Delta s \approx \underbrace{s'(t_0)}_{\text{Velocity at } t_0} \times \underbrace{\Delta t}_{\text{Length of time}}$$

■ **EXAMPLE 2** Position and Velocity   The position of an object in linear motion at time t (in seconds) is $s(t) = t^3 - 20t + 8$ m. Estimate the distance traveled over the time interval $[3, 3.025]$.

Solution We have $s'(t) = 3t^2 - 20$ and hence $s'(3) = 7$ m/s. The time interval $[3, 3.025]$ has length $\Delta t = 0.025$, so the Linear Approximation gives

$$\Delta s \approx s'(3)\Delta t = 7(0.025) = 0.175 \text{ m}$$

■

When engineers need to monitor the change in position of an object with great accuracy, they may use a cable position transducer (Figure 2). This device detects and records the movement of a metal cable attached to the object. The accuracy is affected by changes in the ambient temperature because heat causes the cable to stretch. The Linear Approximation can be used to estimate these effects.

■ **EXAMPLE 3** Thermal Expansion   A thin metal cable has length $L = 6$ in. when the ambient temperature is $T = 70°$F. Estimate the change in length when T rises to $75°$F, assuming that

$$\frac{dL}{dT} = kL \qquad \boxed{1}$$

where $k = 9.6 \times 10^{-6}°\text{F}^{-1}$ (k is called the coefficient of thermal expansion).

Solution How does the Linear Approximation apply here? We want to estimate the change in length ΔL when T increases from $70°$ to $75°$, that is, when $\Delta T = 5°$. We use Eq. (1) to find dL/dT when $L = 6$ in.:

$$\frac{dL}{dT} = kL = (9.6 \times 10^{-6})(6) \approx 5.8 \times 10^{-5} \text{ in./}°\text{F}$$

According to the Linear Approximation, ΔL is estimated by the derivative times ΔT:

$$\Delta L \approx \frac{dL}{dT}\Delta T \approx (5.8 \times 10^{-5})(5) = 2.9 \times 10^{-4} \text{ in.}$$

■

FIGURE 2 Cable position transducers (manufactured by Space Age Control, Inc.). In one application, a transducer was used to compare the changes in throttle position on a Formula 1 race car with the shifting actions of the driver.

In the next example, we set $h = \Delta x$ and write the Linear Approximation in the equivalent form:

$$\boxed{f(a + h) - f(a) \approx f'(a)h} \qquad \boxed{2}$$

■ **EXAMPLE 4** Weight Loss in an Airplane   Newton's Law of Gravitation can be used to show that if an object weighs w pounds on the surface of the earth, then its weight at distance x from the center of the earth is

$$W(x) = \frac{wR^2}{x^2} \qquad \text{(for } x \geq R)$$

where $R = 3{,}960$ miles is the radius of the earth. Estimate the weight lost by a 200-lb football player flying in a jet at an altitude of 7 miles.

TABLE 1   Actual versus Approximate Weight Loss for a 200-lb Person

Altitude (miles)	ΔW (pounds)	Linear Approx.
0	0	0
1	−0.10097	−0.101
2	−0.20187	−0.202
3	−0.30269	−0.303
4	−0.40343	−0.404
5	−0.50410	−0.505
6	−0.60469	−0.606
7	−0.70520	−0.707

Solution We use Eq. (2) with $a = 3{,}960$. In moving from the earth's surface ($x = 3{,}960$) to altitude h ($x = 3{,}960 + h$), an object's weight changes by the amount

$$\Delta W = W(3{,}960 + h) - W(3{,}960) \approx W'(3{,}960)h$$

We have

$$W'(x) = \frac{d}{dx}\left(\frac{wR^2}{x^2}\right) = -\frac{2wR^2}{x^3}$$

$$W'(3{,}960) = -\frac{2wR^2}{3{,}960^3} = -\frac{2w(3{,}960)^2}{3{,}960^3} \approx -0.000505w$$

The player's weight is $w = 200$, so $W'(3{,}960) \approx -(0.000505)(200) = -0.101$. By Eq. (2), $\Delta W \approx -0.101h$, or more simply, $\Delta W \approx -0.1h$. In other words, *the player loses approximately one-tenth of a pound per mile of altitude gained*. At an altitude of $h = 7$ miles, the weight loss is $\Delta W \approx -0.7$ lb. Table 1 compares ΔW with the approximation $-0.101h$. ∎

Suppose we measure the *diameter* of a circle to be D and use this result to compute the *area* of the circle. If our measurement of D is inexact, the area computation will also be inexact. What is the effect of the measurement error on the resulting area computation? This can be estimated using the Linear Approximation, as in the next example.

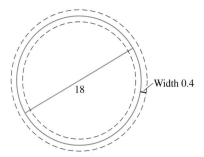

FIGURE 3 The border of the actual pizza lies between the dashed circles.

■ EXAMPLE 5 Effect of an Inexact Measurement   The Bonzo Pizza Company claims that its pizzas are circular with diameter 18 in.

(a) What is the area of the pizza?

(b) Estimate the amount of pizza lost or gained if Bonzo's chefs err in the diameter by at most 0.4 in. (Figure 3).

Solution First, we need a formula for the area A of a circle in terms of its diameter D. Since the radius is $r = D/2$, the area is

$$A(D) = \pi r^2 = \pi\left(\frac{D}{2}\right)^2 = \frac{\pi}{4}D^2$$

(a) The area of an 18-in. pizza is $A(18) = (\frac{\pi}{4})(18)^2 \approx 254.5$ in.^{2}.

(b) If the chefs make a mistake and the actual diameter is equal to $18 + \Delta D$, then the loss or gain in pizza area is $\Delta A = A(18 + \Delta D) - A(18)$. Observe that $A'(D) = \frac{\pi}{2}D$ and $A'(18) = 9\pi \approx 28.3$ in., so the Linear Approximation yields

$$\Delta A = A(18 + \Delta D) - A(18) \approx (28.3)\,\Delta D$$

Because ΔD is at most ± 0.4 in., the gain or loss in pizza is no more than around

$$\Delta A \approx \pm(28.3)(0.4) \approx \pm 11.3 \text{ in.}^2$$

This is a loss or gain of approximately $4\frac{1}{2}\%$. ∎

In this example, we interpret ΔA as the possible error in the computation of $A(D)$. This should not be confused with the error in the Linear Approximation. This latter error refers to the accuracy in using $A'(D)\,\Delta D$ to approximate ΔA.

Some textbooks express the Linear Approximation using the "differentials" dx and dy, where dy denotes the vertical change in the tangent line corresponding to a change dx in x. Thus, $dy = f'(x)dx$ since the tangent line has slope $f'(x)$. In this notation, the Linear Approximation is the statement $\Delta y \approx dy$ or $\Delta y \approx f'(x)dx$, where Δy is the change in $f(x)$ for a given change dx in x.

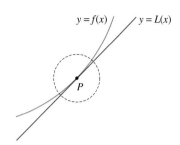

FIGURE 4 The tangent line is a good approximation in a small neighborhood of $P = (a, f(a))$.

Linearization

We can approximate $f(x)$ itself rather than the change Δf by rewriting the Linear Approximation in terms of the variable $x = a + \Delta x$. Then

$$f(a + \Delta x) - f(a) \approx f'(a)\,\Delta x$$

$$f(x) - f(a) \approx f'(a)(x - a) \qquad \text{(since } \Delta x = x - a\text{)}$$

$$f(x) \approx f(a) + f'(a)(x - a)$$

The function on the right, denoted $L(x)$, is called the **linearization** of $f(x)$ at $x = a$:

$$L(x) = f'(a)(x - a) + f(a)$$

We refer to $x = a$ as the *center* of the linearization. Notice that $y = L(x)$ is the equation of the tangent line to the graph of $f(x)$ at $x = a$ (Figure 4).

Approximating $f(x)$ by Its Linearization   Assume that f is differentiable at $x = a$. If x is close to a, then

$$f(x) \approx L(x) = f'(a)(x - a) + f(a)$$

■ **EXAMPLE 6** Compute the linearization of $f(x) = \sqrt{x}\,e^{x-1}$ at $a = 1$.

Solution We compute the derivative of $f(x)$ using the Product Rule:

$$f'(x) = x^{1/2}e^{x-1} + \frac{1}{2}x^{-1/2}e^{x-1} = \left(x^{1/2} + \frac{1}{2}x^{-1/2}\right)e^{x-1}$$

Therefore,

$$f(1) = \sqrt{1}e^0 = 1, \qquad f'(1) = \left(1 + \frac{1}{2}\right)e^0 = \frac{3}{2}$$

and

$$L(x) = f(1) + f'(1)(x - 1) = 1 + \frac{3}{2}(x - 1) = \frac{3}{2}x - \frac{1}{2} \qquad ■$$

The linearization can be used to approximate function values. Consider the linearization $L(x)$ of $f(x) = \sqrt{x}$ at $a = 1$. Since $f(1)$ and $f'(x) = \frac{1}{2}x^{-1/2}$, we have $f'(1) = \frac{1}{2}$ and

$$L(x) = \frac{1}{2}(x - 1) + 1 = \frac{1}{2}x + \frac{1}{2}$$

The following table compares three approximations with the value of $\sqrt{x}$ obtained from a calculator. As expected, the error is large for $x = 9$ since 9 is not close to the center $a = 1$ (Figure 5).

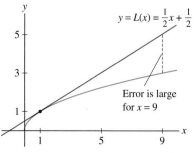

FIGURE 5 Graph of linearization of $f(x) = \sqrt{x}$ at $a = 1$.

x	$\sqrt{x}$	**Linearization at $a = 1$**	**Calculator**	**Error**
1.1	$\sqrt{1.1}$	$L(1.1) = \frac{1}{2}(1.1) + \frac{1}{2} = 1.05$	1.0488	0.0012
0.98	$\sqrt{0.98}$	$L(0.98) = \frac{1}{2}(0.98) + \frac{1}{2} = 0.99$	0.98995	$5 \cdot 10^{-5}$
9	$\sqrt{9}$	$L(9) = \frac{1}{2}(9) + \frac{1}{2} = 5$	3	2

In the next example, we compute the **percentage error**, which is often more important than the error itself. By definition,

$$\text{Percentage error} = \left| \frac{\text{error}}{\text{actual value}} \right| \times 100\%$$

■ **EXAMPLE 7** Use the linearization to estimate $\tan(\frac{\pi}{4} + 0.02)$ and compute the percentage error.

Solution We find the linearization of $f(x) = \tan x$ at $a = \frac{\pi}{4}$:

$$f\left(\frac{\pi}{4}\right) = \tan\left(\frac{\pi}{4}\right) = 1, \qquad f'\left(\frac{\pi}{4}\right) = \sec^2\left(\frac{\pi}{4}\right) = \left(\sqrt{2}\right)^2 = 2$$

$$L(x) = f\left(\frac{\pi}{4}\right) + f'\left(\frac{\pi}{4}\right)\left(x - \frac{\pi}{4}\right) = 1 + 2\left(x - \frac{\pi}{4}\right)$$

The linearization yields the estimate

$$\tan\left(\frac{\pi}{4} + 0.02\right) \approx L\left(\frac{\pi}{4} + 0.02\right) = 1 + 2(0.02) = 1.04$$

A calculator gives $\tan(\frac{\pi}{4} + 0.02) \approx 1.0408$, so

$$\text{Percentage error} \approx \left| \frac{1.0408 - 1.04}{1.0408} \right| \times 100 \approx 0.08\%$$ ■

The Size of the Error

The examples in this section may have convinced you that the Linear Approximation gives a good approximation to $\Delta f = f(a + h) - f(a)$ when h is small, but if we want to rely on the Linear Approximation, we need to know more about the size of the error:

$$E = \text{error} = \left| \Delta f - f'(a)h \right|$$

Graphically, the error in the Linear Approximation is the vertical distance between the graph and the tangent line (Figure 6). Now compare the two graphs in Figure 7. When the graph is relatively flat as in (A), the Linear Approximation is quite accurate, especially within the dashed circle. By contrast, when the graph bends sharply as in (B), the Linear Approximation is much less accurate.

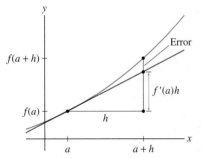

FIGURE 6 Graphical interpretation of the error in the Linear Approximation.

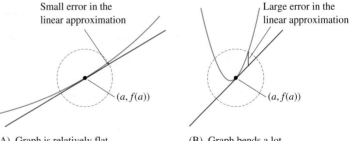

(A) Graph is relatively flat.   (B) Graph bends a lot.

FIGURE 7 The accuracy of the linear approximation depends on how much the curve bends.

The bending or curvature of a graph is determined by how quickly the tangent lines change direction, which is related to $f''(x)$. When $|f''(x)|$ is smaller, the graph is flatter and the Linear Approximation is more accurate over a larger interval around $x = a$.

In Section 8.4, we will prove that if $f''(x)$ is continuous (as is usually the case in practice), then the error satisfies the following **Error Bound**:

$$E \leq \frac{1}{2}K\,h^2$$

3

where K is the maximum value of $|f''(x)|$ on the interval from a to $a + h$. This is a quantitative relation between the error E and the size of the second derivative. It also shows that the error E is of *order two* in h, meaning that E is no larger than a constant times h^2. Thus, if h is small, say, $h = 10^{-n}$, then E is substantially smaller because it has order of magnitude $h^2 = 10^{-2n}$.

The Error Bound also tells us that the accuracy improves as we take smaller and smaller intervals around $x = a$. This is a precise version of the "local linearity" property of differentiable functions discussed in Section 3.2.

4.1 SUMMARY

- Let $\Delta f = f(a + \Delta x) - f(a)$. The *Linear Approximation* is the estimate

$$\Delta f \approx f'(a)\Delta x$$

for small Δx. It can also be written $\Delta y \approx \dfrac{dy}{dx}\Delta x$.

- $L(x) = f'(a)(x - a) + f(a)$ is the *linearization* of f at $x = a$. The Linear Approximation can be rewritten as the estimate $f(x) \approx L(x)$ for small $|x - a|$.

- The error in the Linear Approximation is the quantity

$$\text{Error} = \left| \Delta f - f'(a)\Delta x \right|$$

In many cases, the percentage error is more relevant than the error itself:

$$\text{Percentage error} = \left| \frac{\text{error}}{\text{actual value}} \right| \times 100\%$$

4.1 EXERCISES

Preliminary Questions

1. Estimate $g(1.2) - g(1)$ if $g'(1) = 4$.

2. Estimate $f(2.1)$, assuming that $f(2) = 1$ and $f'(2) = 3$.

3. The velocity of a train at a given instant is 110 ft/s. How far does the train travel during the next half-second (use the Linear Approximation)?

4. Discuss how the Linear Approximation makes the following statement more precise: The sensitivity of the output to a small change in input depends on the derivative.

5. Suppose that the linearization of $f(x)$ at $a = 2$ is $L(x) = 2x + 4$. What are $f(2)$ and $f'(2)$?

Exercises

In Exercises 1–6, use the Linear Approximation to estimate $\Delta f = f(3.02) - f(3)$ for the given function.

1. $f(x) = x^2$

2. $f(x) = x^4$

3. $f(x) = x^{-1}$

4. $f(x) = \dfrac{1}{x + 1}$

5. $f(x) = e^{2x}$

6. $f(x) = x \ln x$

7. Let $f(x) = \ln x$. Use the Linear Approximation to estimate $\Delta f = \ln(e^2 + 0.1) - \ln(e^2)$.

In Exercises 8–15, estimate Δf using the Linear Approximation and use a calculator to compute both the error and the percentage error.

8. $f(x) = x - 2x^2$, $a = 5$, $\Delta x = -0.4$

9. $f(x) = \sqrt{1 + x}$, $a = 8$, $\Delta x = 1$

10. $f(x) = \dfrac{1}{1 + x^2}$, $a = 3$, $\Delta x = 0.5$

11. $f(x) = \sin x$, $a = 0$, $\Delta x = 0.02$

12. $f(x) = \tan x$, $a = \frac{\pi}{4}$, $\Delta x = 0.013$

13. $f(x) = \cos x$, $a = \frac{\pi}{4}$, $\Delta x = 0.03$

14. $f(x) = \ln x$, $a = 1$, $\Delta x = 0.02$

15. $f(x) = x^{1/3} e^{x-1}$, $a = 1$, $\Delta x = 0.1$

In Exercises 16–24, estimate the quantity using the Linear Approximation and find the error using a calculator.

16. $\sqrt{26} - \sqrt{25}$

17. $16.5^{1/4} - 16^{1/4}$

18. $\dfrac{1}{\sqrt{101}} - \dfrac{1}{10}$

19. $\dfrac{1}{\sqrt{98}} - \dfrac{1}{10}$

20. $\tan^{-1}(1.05) - \tan^{-1} 1$

21. $\sin(0.023)$ *Hint: Estimate $\sin(0.023) - \sin(0)$.*

22. $(15.8)^{1/4}$

23. $\ln(0.97)$

24. $e^{0.02}$

25. The cube root of 27 is 3. How much larger is the cube root of 27.2? Estimate using the Linear Approximation.

26. Which is larger: $\sqrt{2.1} - \sqrt{2}$ or $\sqrt{9.1} - \sqrt{9}$? Explain using the Linear Approximation.

27. Estimate $\sin 61° - \sin 60°$ using the Linear Approximation. *Note:* You must express $\Delta\theta$ in radians.

28. A thin silver wire has length $L = 18$ cm when the temperature is $T = 30°C$. Estimate the length when $T = 25°C$ if the coefficient of thermal expansion is $k = 1.9 \times 10^{-5}°C^{-1}$ (see Example 3).

29. The atmospheric pressure P (in kilopascals) at altitudes h (in kilometers) for $11 \le h \le 25$ is approximately $P(h) = 128e^{-0.157h}$.

(a) Use the Linear Approximation to estimate the change in pressure at $h = 20$ when $\Delta h = 0.5$.

(b) Compute the actual change and compute the percentage error in the Linear Approximation.

30. The resistance R of a copper wire at temperature $T = 20°C$ is $R = 15\ \Omega$. Estimate the resistance at $T = 22°C$, assuming that $dR/dT\big|_{T=20} = 0.06\ \Omega/°C$.

31. The side s of a square carpet is measured at 6 ft. Estimate the maximum error in the area A of the carpet if s is accurate to within half an inch.

32. A spherical balloon has a radius of 6 in. Estimate the change in volume and surface area if the radius increases by 0.3 in.

33. A stone tossed vertically in the air with initial velocity v ft/s reaches a maximum height of $h = v^2/64$ ft.

(a) Estimate Δh if v is increased from 25 to 26 ft/s.

(b) Estimate Δh if v is increased from 30 to 31 ft/s.

(c) In general, does a 1 ft/s increase in initial velocity cause a greater change in maximum height at low or high initial velocities? Explain.

34. If the price of a bus pass from Albuquerque to Los Alamos is set at x dollars, a bus company takes in a monthly revenue of $R(x) = 1.5x - 0.01x^2$ (in thousands of dollars).

(a) Estimate the change in revenue if the price rises from \$50 to \$53.

(b) Suppose that $x = 80$. How will revenue be affected by a small increase in price? Explain using the Linear Approximation.

35. The *stopping distance* for an automobile (after applying the brakes) is approximately $F(s) = 1.1s + 0.054s^2$ ft, where s is the speed in mph. Use the Linear Approximation to estimate the change in stopping distance per additional mph when $s = 35$ and when $s = 55$.

36. Juan measures the circumference C of a spherical ball at 40 cm and computes the ball's volume V. Estimate the maximum possible error in V if the error in C is at most 2 cm. Recall that $C = 2\pi r$ and $V = \frac{4}{3}\pi r^3$, where r is the ball's radius.

37. Estimate the weight loss per mile of altitude gained for a 130-lb pilot. At which altitude would she weigh 129.5 lb? See Example 4.

38. How much would a 160-lb astronaut weigh in a satellite orbiting the earth at an altitude of 2,000 miles? Estimate the astronaut's weight loss per additional mile of altitude beyond 2,000.

39. The volume of a certain gas (in liters) is related to pressure P (in atmospheres) by the formula $PV = 24$. Suppose that $V = 5$ with a possible error of ± 0.5 L.

(a) Compute P and estimate the possible error.

(b) Estimate the maximum allowable error in V if P must have an error of at most 0.5 atm.

40. The dosage D of diphenhydramine for a dog of body mass w kg is $D = kw^{2/3}$ mg, where k is a constant. A cocker spaniel has mass $w = 10$ kg according to a veterinarian's scale. Estimate the maximum allowable error in w if the percentage error in the dosage D must be less than 5%.

In Exercises 41–50, find the linearization at $x = a$.

41. $y = \cos x \sin x$, $a = 0$

42. $y = \cos x \sin x$, $a = \dfrac{\pi}{4}$

43. $y = (1 + x)^{-1/2}$, $a = 0$

44. $y = (1 + x)^{-1/2}$, $a = 3$

45. $y = (1 + x^2)^{-1/2}$, $a = 0$

46. $y = \dfrac{\sin x}{x}$, $a = \dfrac{\pi}{2}$

47. $y = \sin^{-1} x$, $a = \frac{1}{2}$

48. $y = e^{\sqrt{x}}$, $a = 1$

49. $y = e^x \ln x$, $a = 1$ **50.** $y = \tan^{-1} x$, $a = 1$

51. GU Estimate $\sqrt{16.2}$ using the linearization $L(x)$ of $f(x) = \sqrt{x}$ at $a = 16$. Plot $f(x)$ and $L(x)$ on the same set of axes and determine if the estimate is too large or too small.

52. GU Estimate $1/\sqrt{15}$ using a suitable linearization of $f(x) = 1/\sqrt{x}$. Plot $f(x)$ and $L(x)$ on the same set of axes and determine if the estimate is too large or too small. Use a calculator to compute the percentage error.

In Exercises 53–61, approximate using linearization and use a calculator to compute the percentage error.

53. $\sqrt{17}$ **54.** $\dfrac{1}{\sqrt{17}}$ **55.** $(17)^{1/4}$

56. $(27.03)^{1/3}$ **57.** $(27.001)^{1/3}$ **58.** $(1.2)^{5/3}$

59. $\ln 1.07$ **60.** $e^{-0.012}$ **61.** $\cos^{-1}(0.52)$

62. GU Plot $f(x) = \tan x$ and its linearization $L(x)$ at $a = \dfrac{\pi}{4}$ on the same set of axes.

(a) Does the linearization overestimate or underestimate $f(x)$?

(b) Show, by graphing $y = f(x) - L(x)$ and $y = 0.1$ on the same set of axes, that the error $|f(x) - L(x)|$ is at most 0.1 for $0.55 \le x \le 0.95$.

(c) Find an interval of x-values on which the error is at most 0.05.

63. GU Compute the linearization $L(x)$ of $f(x) = x^2 - x^{3/2}$ at $a = 2$. Then plot $f(x) - L(x)$ and find an interval around $a = 1$ such that $|f(x) - L(x)| \le 0.1$.

In Exercises 64–65, use the following fact derived from Newton's Laws: An object released at an angle θ with initial velocity v ft/s travels a total distance $s = \frac{1}{32} v^2 \sin 2\theta$ ft (Figure 8).

64. A player located 18.1 ft from a basket launches a successful jump shot from a height of 10 ft (level with the rim of the basket), at an angle $\theta = 34°$ and initial velocity of $v = 25$ ft/s.

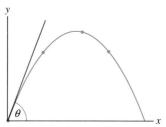

FIGURE 8 Trajectory of an object released at an angle θ.

(a) Show that the distance s of the shot changes by approximately $0.255 \Delta\theta$ ft if the angle changes by an amount $\Delta\theta$. Remember to convert the angles to radians in the Linear Approximation.

(b) Is it likely that the shot would have been successful if the angle were off by $2°$?

65. Estimate the change in the distance s of the shot if the angle changes from $50°$ to $51°$ for $v = 25$ ft/s and $v = 30$ ft/s. Is the shot more sensitive to the angle when the velocity is large or small? Explain.

66. Compute the linearization of $f(x) = 3x - 4$ at $a = 0$ and $a = 2$. Prove more generally that a linear function coincides with its linearization at $x = a$ for all a.

67. According to (3), the error in the Linear Approximation is of "order two" in h. Show that the Linear Approximation to $f(x) = \sqrt{x}$ at $x = 9$ yields the estimate $\sqrt{9+h} - 3 \approx \frac{1}{6}h$. Then compute the error E for $h = 10^{-n}$, $1 \le n \le 4$, and verify numerically that $E \le 0.006h^2$.

68. Show that the Linear Approximation to $f(x) = \tan x$ at $x = \frac{\pi}{4}$ yields the estimate $\tan(\frac{\pi}{4} + h) - 1 \approx 2h$. Compute the error E for $h = 10^{-n}$, $1 \le n \le 4$, and verify that E satisfies the Error Bound (3) with $K = 6.2$.

Further Insights and Challenges

69. Show that for any real number k, $(1 + x)^k \approx 1 + kx$ for small x. Estimate $(1.02)^{0.7}$ and $(1.02)^{-0.3}$.

70. GU **(a)** Show that $f(x) = \sin x$ and $g(x) = \tan x$ have the same linearization at $a = 0$.

(b) Which function is approximated more accurately? Explain using a graph over $[0, \frac{\pi}{6}]$.

(c) Calculate the error in these linearizations at $x = \frac{\pi}{6}$. Does the answer confirm your conclusion in (b)?

71. Let $\Delta f = f(5 + h) - f(5)$, where $f(x) = x^2$, and let $E = |\Delta f - f'(5)h|$ be the error in the Linear Approximation. Verify directly that E satisfies (3) with $K = 2$ (thus E is of order two in h).

72. Let $f(x) = x^{-1}$ and let $E = |\Delta f - f'(1)h|$ be the error in the Linear Approximation at $a = 1$. Show directly that $E = h^2/(1 + h)$.

Then prove that $E \le 2h^2$ if $-\frac{1}{2} \le h \le \frac{1}{2}$. *Hint:* In this case, $\frac{1}{2} \le 1 + h \le \frac{3}{2}$.

73. Consider the curve $y^4 + x^2 + xy = 13$.

(a) Calculate dy/dx at $P = (3, 1)$ and find the linearization $y = L(x)$ at P.

(b) GU Plot the curve and $y = L(x)$ on the same set of axes.

(c) Although y is only defined implicitly as a function of x, we can use $y = L(x)$ to approximate y for x near 3. Use this idea to find an approximate solution of $y^4 + 3.1^2 + 3.1y = 13$. Based on your plot in (b), is your approximation too large or too small?

(d) CAS Solve $y^4 + 3.1^2 + 3.1y = 13$ numerically using a computer algebra system and use the result to find the percentage error in your approximation.

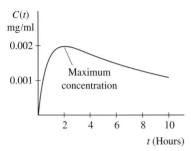

FIGURE 1 $C(t) =$ drug concentration in bloodstream.

We often drop the word "absolute" and speak simply of the min or max on an interval I. When no interval is mentioned, it is understood that we refer to the extreme values on the entire domain of the function.

4.2 Extreme Values

An important problem in many applications is to find the minimum or maximum value of a function $f(x)$. For example, a physician must determine the maximum drug concentration in a patient's bloodstream when a drug is administered. This amounts to finding the highest point on the graph of $C(t)$, the concentration at time t (Figure 1).

We refer to the maximum and minimum values (max and min for short) as **extreme values** or **extrema** (singular: extremum) and to the process of finding them as **optimization**. Sometimes, we are interested in finding the min or max for x in a particular interval I, rather than on the entire domain of $f(x)$.

> **DEFINITION Extreme Values on an Interval**   Let $f(x)$ be a function on an interval I and let $a \in I$. We say that $f(a)$ is the
>
> • **Absolute minimum** of $f(x)$ on I if $f(a) \le f(x)$ for all $x \in I$.
> • **Absolute maximum** of $f(x)$ on I if $f(a) \ge f(x)$ for all $x \in I$.

Does every function have a minimum or maximum value? Clearly not, as we see by taking $f(x) = x$. Indeed, $f(x) = x$ increases without bound as $x \to \infty$ and decreases without bound as $x \to -\infty$. In fact, extreme values do not always exist even if we restrict ourselves to an interval I. Figure 2 illustrates what can go wrong if I is open or f has a discontinuity:

• **Discontinuity**: (A) shows a discontinuous function with no maximum value. The values of $g(x)$ get arbitrarily close to 3 from below, but 3 is not the maximum value because $g(x)$ never actually takes on the value 3.
• **Open interval:** In (B), $f(x)$ is defined on the *open* interval (a, b) and $f(x)$ has no min or max because it tends to ∞ on the right and $-\infty$ on the left.

Fortunately, the following theorem guarantees that extreme values exist when $f(x)$ is continuous and I is closed [Figure 2(C)].

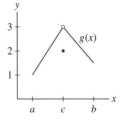

(A) $g(x)$ is discontinuous and has no max on $[a, b]$.

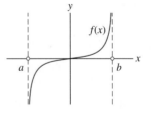

(B) $f(x)$ has no min or max on the open interval (a, b).

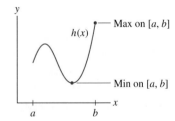

(C) $h(x)$ is continuous and $[a, b]$ is closed. Therefore, $h(x)$ has a min and max on $[a, b]$.

FIGURE 2

◄··· REMINDER *A closed, bounded interval is an interval $I = [a, b]$ (endpoints included) where a and b are both finite. Often, we drop the word "bounded" and refer to a closed, bounded interval more simply as a closed interval. An open interval (a, b) (endpoints not included) may have one or two infinite endpoints.*

> **THEOREM 1 Existence of Extrema on a Closed Interval**   If $f(x)$ is a continuous function on a closed (bounded) interval $I = [a, b]$, then $f(x)$ takes on a minimum and a maximum value on I.

CONCEPTUAL INSIGHT   The idea in Theorem 1 is that a continuous function on a closed interval must take on a min and a max because it cannot skip values or go off to $\pm\infty$, as in Figures 2(A) and (B). However, a rigorous proof relies on the *completeness property* of the real numbers (see Appendix D for a proof).

Local Extrema and Critical Points

We now focus on the problem of finding extreme values. A key concept is that of a local minimum or maximum.

DEFINITION Local Extrema   We say that $f(x)$ has a

- **Local minimum** at $x = c$ if $f(c)$ is the minimum value of f on some open interval (in the domain of f) containing c.
- **Local maximum** at $x = c$ if $f(c)$ is the maximum value of $f(x)$ on some open interval (in the domain of f) containing c.

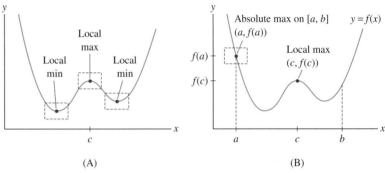

(A)   (B)

FIGURE 3

A local min or max occurs at $x = c$ if $(c, f(c))$ is the highest or lowest point on the graph within some small box [Figure 3(A)]. To be a local max, $f(c)$ must be greater than or equal to all other *nearby* values, but it does not have to be the absolute maximum value (and similarly for a local min). Figure 3(B) illustrates the difference between local and absolute extrema. Note that $f(a)$ is the absolute max on $[a, b]$ but it is not a local max because $f(x)$ takes on larger values to the left of $x = a$.

How do we find the local extrema of a function? The crucial observation is that *the tangent line at a local min or max is horizontal* [Figure 4(A)]. In other words, if $f(c)$

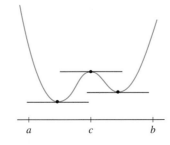
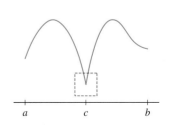

(A) Tangent line is horizontal   (B) This local minimum occurs at a point
    at the local extrema.   where the function is not differentiable.

FIGURE 4

is a local min or max, then $f'(c) = 0$. However, this assumes $f(x)$ is differentiable. Otherwise, the tangent line may not exist at a local min or max, as in Figure 4(B). To take both possibilities into account, we define the notion of a critical point.

DEFINITION Critical Points   A number c in the domain of f is called a **critical point** if *either* $f'(c) = 0$ or $f'(c)$ does not exist.

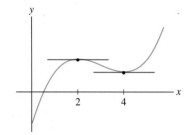

FIGURE 5 Graph of
$f(x) = x^3 - 9x^2 + 24x - 10$.

■ **EXAMPLE 1**   Find the critical points of $f(x) = x^3 - 9x^2 + 24x - 10$ (Figure 5).

Solution The function $f(x)$ is differentiable everywhere, so the critical points are the solutions of $f'(x) = 0$:

$$f'(x) = 3x^2 - 18x + 24 = 3(x^2 - 6x + 8) = 3(x - 2)(x - 4) = 0$$

The critical points are the roots $c = 2$ and $c = 4$.   ■

■ **EXAMPLE 2**   A Nondifferentiable Function   Find the critical points of $f(x) = |x|$.

Solution As we see in Figure 6, $f'(x) = -1$ for $x < 0$ and $f'(x) = 1$ for $x > 0$. Therefore, $f'(x) = 0$ has no solutions with $x \neq 0$. However, $f'(0)$ does not exist and therefore $c = 0$ is a critical point.   ■

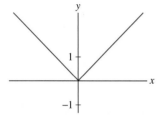

FIGURE 6 Graph of $f(x) = |x|$.

We now state a basic result that we will use repeatedly to find extreme values.

THEOREM 2 Fermat's Theorem on Local Extrema   If $f(c)$ is a local min or max, then c is a critical point of f.

Proof Suppose that $f(c)$ is a local minimum (a local maximum is treated similarly). If $f'(c)$ does not exist, then c is a critical point and there is nothing more to prove. So assume that $f'(c)$ exists. We must then prove that $f'(c) = 0$.

Because $f(c)$ is a local minimum, we have $f(c + h) \geq f(c)$ for all sufficiently small $h \neq 0$. Equivalently, $f(c + h) - f(c) \geq 0$. Now divide this inequality by h:

$$\frac{f(c + h) - f(c)}{h} \geq 0 \qquad \text{if } h > 0 \qquad \boxed{1}$$

$$\frac{f(c + h) - f(c)}{h} \leq 0 \qquad \text{if } h < 0 \qquad \boxed{2}$$

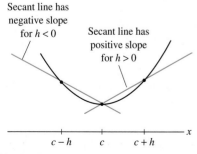

Secant line has negative slope for $h < 0$

Secant line has positive slope for $h > 0$

FIGURE 7

Figure 7 shows the graphical interpretation of these inequalities. Taking the one-sided limits of both sides of (1) and (2), we obtain

$$f'(c) = \lim_{h \to 0+} \frac{f(c + h) - f(c)}{h} \geq \lim_{h \to 0+} 0 = 0$$

$$f'(c) = \lim_{h \to 0-} \frac{f(c + h) - f(c)}{h} \leq \lim_{h \to 0-} 0 = 0$$

This shows that $f'(c) \geq 0$ and $f'(c) \leq 0$. The only possibility is $f'(c) = 0$ as claimed.   ■

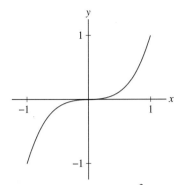

FIGURE 8 Graph of $f(x) = x^3$. The tangent line at $(0, 0)$ is horizontal.

In this section, we restrict our attention to closed intervals because in this case extreme values are guaranteed to exist (Theorem 1). Optimization on open intervals is discussed in Section 4.6.

CONCEPTUAL INSIGHT   Fermat's Theorem *does not claim* that all critical points yield local extrema. "False positives" may exist, by which we mean that $f'(c) = 0$ but $f(c)$ is not a local extremum. In this case, f may have a point of inflection at $x = c$, where the tangent line is horizontal but crosses the graph. For example, if $f(x) = x^3$, then $f'(x) = 3x^2$ and $f'(0) = 0$, but $f(0)$ is neither a local min nor a local max (Figure 8).

Optimizing on a Closed Interval

Finally, we have all the tools needed for optimizing a continuous function on a closed interval. Theorem 1 tells us that the extreme values exist, and the following theorem tells us where to find them, namely among the critical points or endpoints of the interval.

THEOREM 3 Extreme Values on a Closed Interval   Assume that $f(x)$ is continuous on $[a, b]$ and let $f(c)$ be the minimum or maximum value on $[a, b]$. Then c is either a critical point or one of the endpoints a or b.

Proof Assume first that $f(c)$ is a max on $[a, b]$. Either c is an endpoint a or b, or c belongs to the open interval (a, b). In the latter case, $f(c)$ is also a local maximum because it is the maximum value of $f(x)$ on (a, b). By Fermat's Theorem, c is a critical point. The case of a minimum is similar.   ∎

■ **EXAMPLE 3**   Find the extreme values of $f(x) = 2x^3 - 15x^2 + 24x + 7$ on $[0, 6]$.

Solution The extreme values occur at critical points or endpoints by Theorem 3, so we can break up the problem neatly into two steps.

Step 1. Find the critical points.
    The function $f(x)$ is differentiable, so we find the critical points by solving

$$f'(x) = 6x^2 - 30x + 24 = 6(x - 1)(x - 4) = 0$$

The critical points are $c = 1$ and 4.

Step 2. Calculate $f(x)$ at the critical points and endpoints and compare.

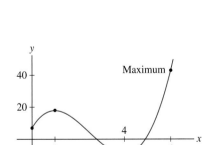

FIGURE 9 Graph of
$f(x) = 2x^3 - 15x^2 + 24x + 7$.

x-value	Value of f	
1 (critical point)	$f(1) = 18$	
4 (critical point)	$f(4) = -9$	min
0 (endpoint)	$f(0) = 7$	
6 (endpoint)	$f(6) = 43$	max

The maximum of $f(x)$ on $[0, 6]$ is the largest of the values in this table, namely $f(6) = 43$ (Figure 9). Similarly, the minimum is $f(4) = -9$.   ■

■ **EXAMPLE 4**   Function with a Cusp Find the maximum of $f(x) = 1 - (x - 1)^{2/3}$ on $[-1, 2]$.

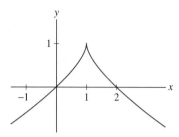

FIGURE 10 Graph of
$f(x) = 1 - (x-1)^{2/3}$.

Solution We have $f(x) = 1 - (x-1)^{2/3}$.

Step 1. Find the critical points.

The derivative is

$$f'(x) = -\frac{2}{3}(x-1)^{-1/3} = -\frac{2}{3(x-1)^{1/3}}$$

The equation $f'(x) = 0$ has no solutions since the numerator of $f'(x)$ is never zero. However, $f'(x)$ does not exist at $x = 1$, so $c = 1$ is a critical point (Figure 10).

Step 2. Calculate $f(x)$ at the critical points and endpoints and compare.

x-value	Value of f	
1 (critical point)	$f(1) = 1$	max
−1 (endpoint)	$f(-1) \approx -0.59$	min
2 (endpoint)	$f(2) = 0$	

■

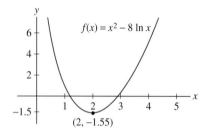

FIGURE 11

■ **EXAMPLE 5** Find the extreme values of $f(x) = x^2 - 8\ln x$ on $[1, 4]$ (Figure 11).

Solution First, solve for the critical points:

$$f'(x) = 2x - \frac{8}{x} = 0 \quad \Rightarrow \quad 2x = \frac{8}{x} \quad \Rightarrow \quad x = \pm 2$$

There is only one critical point $c = 2$ in the interval $[1, 4]$. Next, we compare $f(x)$ at the critical points and endpoints:

x-value	Value of f	
2 (critical point)	$f(2) \approx 1.55$	
1 (endpoint)	$f(1) = 1$	min
4 (endpoint)	$f(4) \approx 4.9$	max

We see that the minimum value on $[1, 4]$ is $f(2) \approx 1.55$ and the maximum value is $f(4) \approx 4.9$.

■

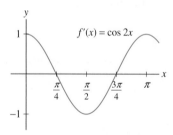

FIGURE 12 Two solutions of $\cos 2x = 0$ lie in the interval $[0, \pi]$, namely $\frac{\pi}{4}$ and $\frac{3\pi}{4}$.

■ **EXAMPLE 6** Trigonometric Example   Find the min and max of $f(x) = \sin x \cos x$ on $[0, \pi]$.

Solution First, we rewrite $f(x)$ using a trigonometric identity as $f(x) = \frac{1}{2}\sin 2x$ and solve $f'(x) = \cos 2x = 0$. This gives $2x = \frac{\pi}{2}$ or $\frac{3\pi}{2}$, so the critical points in the interval $[0, \pi]$ are $c = \frac{\pi}{4}$ and $\frac{3\pi}{4}$ (Figure 12). Next, we evaluate $f(x)$ at the critical points and endpoints to find the min and max:

x-value	Value of f	
$\frac{\pi}{4}$ (critical point)	$f\left(\frac{\pi}{4}\right) = \frac{1}{2}$	max
$\frac{3\pi}{4}$ (critical point)	$f\left(\frac{3\pi}{4}\right) = -\frac{1}{2}$	min
0 (endpoint)	$f(0) = 0$	
π (endpoint)	$f(\pi) = 0$	

■

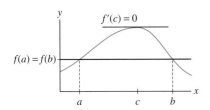

FIGURE 13 Rolle's Theorem: If $f(a) = f(b)$, then $f'(c) = 0$ for some c between a and b.

Rolle's Theorem

As an application of our optimization methods, we prove Rolle's Theorem, which states that if $f(x)$ is differentiable and $f(a) = f(b)$, then the tangent line is horizontal at some point c between a and b (Figure 13).

> **THEOREM 4 Rolle's Theorem**   Assume that $f(x)$ is continuous on $[a, b]$ and differentiable on (a, b). If $f(a) = f(b)$, then there exists a number c between a and b such that $f'(c) = 0$.

Proof Since $f(x)$ is continuous and $[a, b]$ is closed, $f(x)$ has a min and a max in $[a, b]$. Where do they occur? There are two cases. If an extreme value occurs at a point c in the *open* interval (a, b), then $f(c)$ is a local extreme value and $f'(c) = 0$ by Fermat's Theorem (Theorem 2). Otherwise, both the min and the max occur at the endpoints. However, $f(a) = f(b)$, so the maximum and minimum values coincide and $f(x)$ must be a constant function. Therefore, $f'(x) = 0$ for all $x \in (a, b)$ and Rolle's Theorem is satisfied for all $c \in (a, b)$. ∎

Pierre de Fermat
(1601–1665)

René Descartes
(1596–1650)

**HISTORICAL
PERSPECTIVE**

We can hardly expect a more general method . . . . This method never fails and could be extended to a number of beautiful problems; with its aid we have found the centers of gravity of figures bounded by straight lines or curves, as well as those of solids, and a number of other results which we may treat elsewhere if we have the time to do so.

—From Fermat's *On Maxima and Minima and on Tangents*

Sometime in the 1630s, in the decade before Isaac Newton was born, the French mathematician Pierre de Fermat (1601–1665) invented a general method for finding extreme values. In essence, Fermat said that if you want to find extrema, you must set the derivative equal to zero and solve for the critical points, just as we have done in this section. He also described a general method for finding tangent lines that is not essentially different from our method of derivatives. For this reason, Fermat is often regarded as one of the inventors of calculus, along with Newton and Leibniz.

At around the same time, René Descartes (1596–1650) developed a different but much less effective approach to finding tangent lines. Descartes, after whom Cartesian coordinates are named, was the leading philosopher and scientist of his time in Europe and is regarded today as the father of modern philosophy. He was a profound thinker and brilliant writer, but he was also known for his sarcastic wit and disdainful attitude, believing that he had little

to learn from other people. Fermat had angered Descartes because he had once dismissed Descartes's work on optics as "groping about in the shadows." Perhaps in retaliation, Descartes attacked Fermat, claiming that his method of finding tangents was wrong. After some third-party refereeing, Descartes was forced to admit that it was correct after all. He wrote back to Fermat:

. . . Seeing the last method that you use for finding tangents to curved lines, I can reply to it in no other way than to say that it is very good and that, if you had explained it in this manner at the outset, I would have not contradicted it at all.

Unfortunately, Descartes, who was the more famous and influential of the two men, continued to try to damage Fermat's reputation. Today, Fermat is recognized as one of the greatest mathematicians of his age who made far-reaching contributions in several areas of mathematics.

■ **EXAMPLE 7** Illustrating Rolle's Theorem   Verify Rolle's Theorem for

$$f(x) = x^4 - x^2 \qquad \text{on } [-2, 2]$$

Solution First, we check that $f(2) = f(-2)$:

$$f(2) = 2^4 - 2^2 = 12, \qquad f(-2) = (-2)^4 - (-2)^2 = 12$$

Second, we note that $f(x)$ is differentiable everywhere. Therefore, the hypotheses of Rolle's Theorem are satisfied and $f'(c) = 0$ must have a solution in $(-2, 2)$. To verify this, we solve $f'(x) = 4x^3 - 2x = 2x(2x^2 - 1) = 0$. The solutions $c = 0$ and $c = \pm 1/\sqrt{2} \approx \pm 0.707$ all lie in $(-2, 2)$, so Rolle's Theorem is satisfied with three values of c.   ■

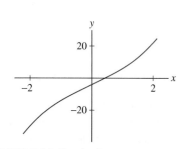

FIGURE 14 Graph of $f(x) = x^3 + 9x - 4$. This function has one real root.

■ **EXAMPLE 8** Using Rolle's Theorem   Show that $f(x) = x^3 + 9x - 4$ has at most one real root.

Solution We use proof by contradiction. If $f(x)$ had two real roots a and b, then $f(a) = f(b) = 0$ and Rolle's Theorem would imply that $f'(c) = 0$ for some $c \in (a, b)$. However, $f'(x) = 3x^2 + 9 \geq 9$, so $f'(c) = 0$ has no solutions. We conclude that $f(x)$ cannot have more than one real root (Figure 14).   ■

4.2 SUMMARY

- The *extreme values* of $f(x)$ on an interval I are the minimum and maximum values of $f(x)$ for $x \in I$ (also called *absolute* extrema on I).
- Basic Theorem: If $f(x)$ is continuous on a closed interval $[a, b]$, then $f(x)$ has both a min and max on $[a, b]$.
- $f(c)$ is a *local minimum* if $f(x) \geq f(c)$ for all x in some open interval around c. Local maxima are defined similarly.
- $x = c$ is a *critical point* of $f(x)$ if either $f'(c) = 0$ or $f'(c)$ does not exist.
- Fermat's Theorem: If $f(c)$ is a local min or max, then c is a critical point.
- To find the extreme values of a continuous function $f(x)$ on a closed interval $[a, b]$:

Step 1. Find the critical points of $f(x)$ in $[a, b]$.

Step 2. Calculate $f(x)$ at the critical points in $[a, b]$ and at the endpoints.

The min and max on $[a, b]$ are the smallest and largest among the values computed in Step 2.

- Rolle's Theorem: If $f(x)$ is continuous on $[a, b]$ and differentiable on (a, b), and if $f(a) = f(b)$, then there exists c between a and b such that $f'(c) = 0$.

4.2 EXERCISES

Preliminary Questions

1. If $f(x)$ is continuous on $(0, 1)$, then (choose the correct statement):

(a) $f(x)$ has a minimum value on $(0, 1)$.

(b) $f(x)$ has no minimum value on $(0, 1)$.

(c) $f(x)$ might not have a minimum value on $(0, 1)$.

2. If $f(x)$ is not continuous on $[0, 1]$, then (choose the correct state-

ment):

(a) $f(x)$ has no extreme values on $[0, 1]$.

(b) $f(x)$ might not have any extreme values on $[0, 1]$.

3. What is the definition of a critical point?

4. True or false: If $f(x)$ is differentiable and $f'(x) = 0$ has no solutions, then $f(x)$ has no local minima or maxima.

5. Fermat's Theorem *does not* claim that if $f'(c) = 0$, then $f(c)$ is a local extreme value (this is false). What *does* Fermat's Theorem assert?

6. If $f(x)$ is continuous but has no critical points in $[0, 1]$, then (choose the correct statement):

(a) $f(x)$ has no min or max on $[0, 1]$.

(b) Either $f(0)$ or $f(1)$ is the minimum value on $[0, 1]$.

Exercises

1. The following questions refer to Figure 15.

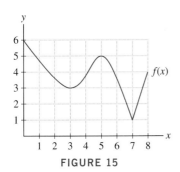

FIGURE 15

(a) How many critical points does $f(x)$ have?

(b) What is the maximum value of $f(x)$ on $[0, 8]$?

(c) What are the local maximum values of $f(x)$?

(d) Find a closed interval on which both the minimum and maximum values of $f(x)$ occur at critical points.

(e) Find an interval on which the minimum value occurs at an endpoint.

2. State whether $f(x) = x^{-1}$ (Figure 16) has a minimum or maximum value on the following intervals:

(a) $(0, 2)$ **(b)** $(1, 2)$ **(c)** $[1, 2]$

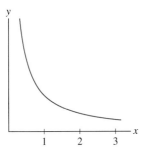

FIGURE 16 Graph of $f(x) = x^{-1}$.

In Exercises 3–14, find all critical points of the function.

3. $f(x) = x^2 - 2x + 4$ **4.** $f(x) = 7x - 2$

5. $f(x) = x^3 - \dfrac{9}{2}x^2 - 54x + 2$ **6.** $f(t) = 8t^3 - t^2$

7. $f(x) = \dfrac{x}{x^2 + 1}$ **8.** $f(t) = 4t - \sqrt{t^2 + 1}$

9. $f(x) = x^{1/3}$ **10.** $f(x) = \dfrac{x^2}{x + 1}$

11. $f(x) = x \ln x$ **12.** $f(x) = xe^{2x}$

13. $f(x) = \sin^{-1} x - 2x$ **14.** $f(x) = \sec^{-1} x - \ln x$

15. Let $f(x) = x^2 - 4x + 1$.

(a) Find the critical point c of $f(x)$ and compute $f(c)$.

(b) Compute the value of $f(x)$ at the endpoints of the interval $[0, 4]$.

(c) Determine the min and max of $f(x)$ on $[0, 4]$.

(d) Find the extreme values of $f(x)$ on $[0, 1]$.

16. Find the extreme values of $2x^3 - 9x^2 + 12x$ on $[0, 3]$ and $[0, 2]$.

17. Find the minimum value of $y = \tan^{-1}(x^2 - x)$.

18. Find the critical points of $f(x) = \sin x + \cos x$ and determine the extreme values on $[0, \frac{\pi}{2}]$.

19. Compute the critical points of $h(t) = (t^2 - 1)^{1/3}$. Check that your answer is consistent with Figure 17. Then find the extreme values of $h(t)$ on $[0, 1]$ and $[0, 2]$.

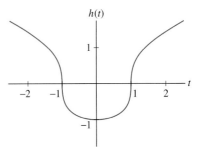

FIGURE 17 Graph of $h(t) = (t^2 - 1)^{1/3}$.

20. [GU] Plot $f(x) = 2\sqrt{x} - x$ on $[0, 4]$ and determine the maximum value graphically. Then verify your answer using calculus.

21. [GU] Plot $f(x) = \ln x - 5 \sin x$ on $[0, 2\pi]$ (choose an appropriate viewing rectangle) and approximate both the critical points and extreme values.

22. [CAS] Approximate the critical points of $g(x) = x \arccos x$ and estimate the maximum value of $g(x)$.

In Exercises 23–56, find the maximum and minimum values of the function on the given interval.

23. $y = 2x^2 - 4x + 2$, $[0, 3]$

24. $y = -x^2 + 10x + 43$, $[3, 8]$

25. $y = x^2 - 6x - 1$, $[-2, 2]$

26. $y = x^2 - 6x - 1$, $[-2, 0]$

27. $y = -4x^2 + 3x + 4$, $[-1, 1]$

28. $y = x^3 - 3x + 1$, $[0, 2]$

29. $y = x^3 - 6x + 1$, $[-1, 1]$

30. $y = x^3 - 12x^2 + 21x$, $[0, 2]$

31. $y = x^3 + 3x^2 - 9x + 2$, $[-1, 1]$

32. $y = x^3 + 3x^2 - 9x + 2$, $[0, 2]$

33. $y = x^3 + 3x^2 - 9x + 2$, $[-4, 4]$

34. $y = x^5 - x$, $[0, 2]$

35. $y = x^5 - 3x^2$, $[-1, 5]$

36. $y = 2s^3 - 3s^2 + 3$, $[-4, 4]$

37. $y = \dfrac{x^2 + 1}{x - 4}$, $[5, 6]$

38. $y = \dfrac{1 - x}{x^2 + 3x}$, $[1, 4]$

39. $y = x - \dfrac{4x}{x + 1}$, $[0, 3]$

40. $y = 2\sqrt{x^2 + 1} - x$, $[0, 2]$

41. $y = (2 + x)\sqrt{2 + (2 - x)^2}$, $[0, 2]$

42. $y = \sqrt{1 + x^2} - 2x$, $[3, 6]$

43. $y = \sqrt{x + x^2} - 2\sqrt{x}$, $[0, 4]$

44. $y = (t - t^2)^{1/3}$, $[-1, 2]$

45. $y = \sin x \cos x$, $[0, \frac{\pi}{2}]$

46. $y = x - \sin x$, $[0, 2\pi]$

47. $y = \sqrt{2}\,\theta - \sec \theta$, $[0, \frac{\pi}{3}]$

48. $y = \cos \theta + \sin \theta$, $[0, 2\pi]$

49. $y = \theta - 2\sin \theta$, $[0, 2\pi]$

50. $y = \sin^3 \theta - \cos^2 \theta$, $[0, 2\pi]$

51. $y = \tan x - 2x$, $[0, 1]$

52. $y = xe^{-x}$, $[0, 2]$

53. $y = \dfrac{\ln x}{x}$, $[1, 3]$

54. $y = 3e^x - e^{2x}$, $[-\frac{1}{2}, 1]$

55. $y = 5\tan^{-1} x - x$, $[1, 5]$

56. $y = x^3 - 24\ln x$, $[\frac{1}{2}, 3]$

57. Let $f(\theta) = 2\sin 2\theta + \sin 4\theta$.

(a) Show that θ is a critical point if $\cos 4\theta = -\cos 2\theta$.

(b) Show, using a unit circle, that $\cos \theta_1 = -\cos \theta_2$ if and only if $\theta_1 = \pi \pm \theta_2 + 2\pi k$ for an integer k.

(c) Show that $\cos 4\theta = -\cos 2\theta$ if and only if $\theta = \pi/2 + \pi k$ or $\theta = \pi/6 + (\pi/3)k$.

(d) Find the six critical points of $f(\theta)$ on $[0, 2\pi]$ and find the extreme values of $f(\theta)$ on this interval.

(e) $\boxed{\text{GU}}$ Check your results against a graph of $f(\theta)$.

58. $\boxed{\text{GU}}$ Find the critical points of $f(x) = 2\cos 3x + 3\cos 2x$. Check your answer against a graph of $f(x)$.

In Exercises 59–62, find the critical points and the extreme values on $[0, 3]$. *In Exercises 61 and 62, refer to Figure 18.*

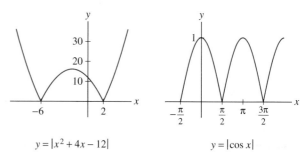

$y = |x^2 + 4x - 12|$ $y = |\cos x|$

FIGURE 18

59. $y = |x - 2|$

60. $y = |3x - 9|$

61. $y = |x^2 + 4x - 12|$

62. $y = |\cos x|$

63. Let $f(x) = 3x - x^3$. Check that $f(-2) = f(1)$. What may we conclude from Rolle's Theorem? Verify this conclusion.

In Exercises 64–67, verify Rolle's Theorem for the given interval.

64. $f(x) = x + x^{-1}$, $[\frac{1}{2}, 2]$

65. $f(x) = \sin x$, $[\frac{\pi}{4}, \frac{3\pi}{4}]$

66. $f(x) = \dfrac{x^2}{8x - 15}$, $[3, 5]$

67. $f(x) = \sin^2 x - \cos^2 x$, $[\frac{\pi}{4}, \frac{3\pi}{4}]$

68. Use Rolle's Theorem to prove that $f(x) = x^5 + 2x^3 + 4x - 12$ has at most one real root.

69. Use Rolle's Theorem to prove that $f(x) = \dfrac{x^3}{6} + \dfrac{x^2}{2} + x + 1$ has at most one real root.

70. The concentration $C(t)$ (in mg/cm^3) of a drug in a patient's bloodstream after t hours is

$$C(t) = \frac{0.016t}{t^2 + 4t + 4}$$

Find the maximum concentration and the time at which it occurs.

71. Maximum of a Resonance Curve   The size of the response of a circuit or other oscillatory system to an input signal of frequency ω ("omega") is described in suitable units by the function

$$\phi(\omega) = \frac{1}{\sqrt{(\omega_0^2 - \omega^2)^2 + 4D^2\omega^2}}$$

Both ω_0 (the natural frequency of the system) and D (the damping factor) are positive constants. The graph of ϕ is called a resonance curve, and the positive frequency $\omega_r > 0$ where ϕ takes its maximum value, if such a positive frequency exists, is the **resonant frequency**. Show that a resonant frequency exists if and only if $0 < D < \omega_0/\sqrt{2}$ (Figure 19). Then show that if this condition is satisfied, $\omega_r = \sqrt{\omega_0^2 - 2D^2}$.

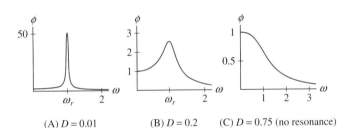

(A) $D = 0.01$ (B) $D = 0.2$ (C) $D = 0.75$ (no resonance)

FIGURE 19 Resonance curves with $\omega_0 = 1$.

72. Bees build honeycomb structures out of cells with a hexagonal base and three rhombus-shaped faces on top as in Figure 20. Using geometry, we can show that the surface area of this cell is

$$A(\theta) = 6hs + \frac{3}{2}s^2(\sqrt{3}\,\csc\theta - \cot\theta)$$

where h, s, and θ are as indicated in the figure. It is a remarkable fact that bees "know" which angle θ minimizes the surface area (and therefore requires the least amount of wax).

FIGURE 20 A cell in a honeycomb constructed by bees.

(a) Show that this angle is approximately 54.7° by finding the critical point of $A(\theta)$ for $0 < \theta < \pi/2$ (assume h and s are constant).

(b) GU   Confirm, by graphing $f(\theta) = \sqrt{3}\,\csc\theta - \cot\theta$, that the critical point indeed minimizes the surface area.

73. Migrating fish tend to swim at a velocity v that minimizes the total expenditure of energy E. According to one model, E is proportional to $f(v) = \dfrac{v^3}{v - v_r}$, where v_r is the velocity of the river water.

(a) Find the critical points of $f(v)$.

(b) GU   Choose a value of v_r (say, $v_r = 10$) and plot $f(v)$. Confirm that $f(v)$ has a minimum value at the critical point.

74. Find the maximum of $y = x^a - x^b$ on $[0, 1]$ where $0 < a < b$. In particular, find the maximum of $y = x^5 - x^{10}$ on $[0, 1]$.

In Exercises 75–77, plot the function using a graphing utility and find its critical points and extreme values on $[-5, 5]$.

75. GU $y = \dfrac{1}{1 + |x - 1|}$

76. GU $y = \dfrac{1}{1 + |x - 1|} + \dfrac{1}{1 + |x - 4|}$

77. GU $y = \dfrac{x}{|x^2 - 1| + |x^2 - 4|}$

78. (a) Use implicit differentiation to find the critical points on the curve $27x^2 = (x^2 + y^2)^3$.

(b) GU   Plot the curve and the horizontal tangent lines on the same set of axes.

79. Sketch the graph of a continuous function on $(0, 4)$ with a minimum value but no maximum value.

80. Sketch the graph of a continuous function on $(0, 4)$ having a local minimum but no absolute minimum.

81. Sketch the graph of a function on $[0, 4]$ having

(a) Two local maxima and one local minimum.

(b) An absolute minimum that occurs at an endpoint, and an absolute maximum that occurs at a critical point.

82. Sketch the graph of a function $f(x)$ on $[0, 4]$ with a discontinuity such that $f(x)$ has an absolute minimum but no absolute maximum.

Further Insights and Challenges

83. Show that the extreme values of $f(x) = a\sin x + b\cos x$ are $\pm\sqrt{a^2 + b^2}$.

84. Show, by considering its minimum, that $f(x) = x^2 - 2x + 3$ takes on only positive values. More generally, find the conditions on r and s under which the quadratic function $f(x) = x^2 + rx + s$ takes

on only positive values. Give examples of r and s for which f takes on both positive and negative values.

85. Show that if the quadratic polynomial $f(x) = x^2 + rx + s$ takes on both positive and negative values, then its minimum value occurs at the midpoint between the two roots.

86. Generalize Exercise 85: Show that if the horizontal line $y = c$ intersects the graph of $f(x) = x^2 + rx + s$ at two points $(x_1, f(x_1))$ and $(x_2, f(x_2))$, then $f(x)$ takes its minimum value at the midpoint $M = \dfrac{x_1 + x_2}{2}$ (Figure 21).

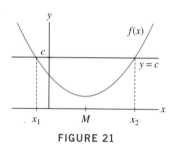

FIGURE 21

87. The graphs in Figure 22 show that a cubic polynomial may have a local min and max, or it may have neither. Find conditions on the coefficients a and b of $f(x) = \frac{1}{3}x^3 + \frac{1}{2}ax^2 + bx + c$ that ensure f has neither a local min nor max. *Hint:* Apply Exercise 84 to $f'(x)$.

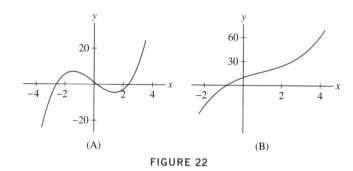

FIGURE 22

88. Find the minimum and maximum values of $f(x) = x^p(1 - x)^q$ on $[0, 1]$, where p and q are positive numbers.

89. Prove that if f is continuous and $f(a)$ and $f(b)$ are local minima where $a < b$, then there exists a value c between a and b such that $f(c)$ is a local maximum. (*Hint:* Apply Theorem 1 to the interval $[a, b]$.) Show that continuity is a necessary hypothesis by sketching the graph of a function (necessarily discontinuous) with two local minima but no local maximum.

4.3 The Mean Value Theorem and Monotonicity

We know intuitively that the derivative determines whether a function is increasing or decreasing. Since $f'(x)$ is the rate of change, $f(x)$ is increasing if the rate is positive and decreasing if the rate is negative. In this section, we prove this statement rigorously using an important theoretical result called the Mean Value Theorem (MVT). We then develop a method for "testing" critical points, that is, for determining whether they correspond to local minima or maxima.

Consider the secant line through points $(a, f(a))$ and $(b, f(b))$ on a graph as in Figure 1. According to the MVT, *there exists at least one tangent line that is parallel to the secant line in the interval* (a, b). Because two lines are parallel if they have the same slope, what the MVT claims is that there exists a point c between a and b such that

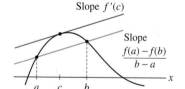

FIGURE 1 By the MVT, there exists at least one tangent line parallel to the secant line.

$$\underbrace{f'(c)}_{\text{Slope of tangent line}} = \underbrace{\frac{f(b) - f(a)}{b - a}}_{\text{Slope of secant line}}$$

THEOREM 1 The Mean Value Theorem   Assume that f is continuous on the closed interval $[a, b]$ and differentiable on (a, b). Then there exists at least one value c in (a, b) such that

$$f'(c) = \frac{f(b) - f(a)}{b - a}$$

Note that Rolle's Theorem is the special case of the MVT where $f(a) = f(b)$. In this case, $f'(c) = 0$.

FIGURE 2 Move the secant line in a parallel fashion until it becomes tangent to the curve.

GRAPHICAL INSIGHT   To see that the MVT is plausible, imagine what happens when the secant line is moved parallel to itself. Eventually, it becomes a tangent line as in Figure 2. We present a formal proof at the end of this section.

■ **EXAMPLE 1**   Illustrate the MVT with $f(x) = \sqrt{x}$ and the points $a = 1$ and $b = 9$.

Solution   The slope of the secant line is

$$\frac{f(b) - f(a)}{b - a} = \frac{\sqrt{9} - \sqrt{1}}{9 - 1} = \frac{3 - 1}{9 - 1} = \frac{1}{4}$$

The MVT states that $f'(c) = \frac{1}{4}$ for at least one value c in $(1, 9)$. Because $f'(x) = \frac{1}{2}x^{-1/2}$, we can find c by solving

$$f'(c) = \frac{1}{2\sqrt{c}} = \frac{1}{4}$$

This gives $2\sqrt{c} = 4$ or $c = 4$. Notice that $c = 4$ lies in the interval $(1, 9)$ as required.   ■

As a first application, we prove that a function with zero derivative is constant.

COROLLARY   Assume that $f(x)$ is differentiable and $f'(x) = 0$ for all $x \in (a, b)$. Then $f(x)$ is constant on (a, b). In other words, $f(x) = C$ for some constant C.

Proof   If a_1 and b_1 are any two distinct points in (a, b), then, by the MVT, there exists c between a_1 and b_1 such that

$$f'(c) = \frac{f(b_1) - f(a_1)}{b_1 - a_1} = 0$$

The last equality holds because $f'(c) = 0$. Therefore, $f(b_1) = f(a_1)$ for all a_1 and b_1 in (a, b). This shows that $f(x)$ is constant on (a, b).   ■

Increasing/Decreasing Behavior of Functions

We now prove that the sign of the derivative determines whether $f(x)$ is increasing or decreasing. Recall that $f(x)$ is called:

- **Increasing on (a, b)**   if $f(x_1) < f(x_2)$ for all $x_1, x_2 \in (a, b)$ such that $x_1 < x_2$
- **Decreasing on (a, b)**   if $f(x_1) > f(x_2)$ for all $x_1, x_2 \in (a, b)$ such that $x_1 < x_2$

We say that $f(x)$ is **monotonic** on (a, b) if it is either increasing or decreasing on (a, b).

We say that f is "nondecreasing" if

$$f(x_1) \le f(x_2) \quad \text{for} \quad x_1 \le x_2$$

The term "nonincreasing" is defined similarly. In Theorem 2, if we assume that $f'(x) \ge 0$ (instead of > 0), then $f(x)$ is nondecreasing on (a, b). Similarly, if $f'(x) \le 0$, then $f(x)$ is nonincreasing on (a, b).

THEOREM 2 The Sign of the Derivative   Let f be a differentiable function on the open interval (a, b).

- If $f'(x) > 0$ for $x \in (a, b)$, then f is increasing on (a, b).
- If $f'(x) < 0$ for $x \in (a, b)$, then f is decreasing on (a, b).

Proof   Suppose first that $f'(x) > 0$ for all $x \in (a, b)$. For any two points $x_1 < x_2$ in (a, b), the MVT tells us that there exists c between x_1 and x_2 such that

$$f'(c) = \frac{f(x_2) - f(x_1)}{x_2 - x_1} > 0$$

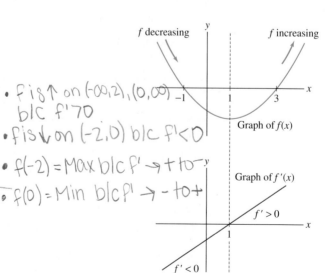

Increasing function
Tangent lines have positive slope.

Decreasing function
Tangent lines have negative slope.

FIGURE 3

Here, by our hypothesis, $f'(c) > 0$. Since $(x_2 - x_1)$ is positive, we may conclude that $f(x_2) - f(x_1) > 0$ or $f(x_2) > f(x_1)$ as required. The case $f'(x) < 0$ is similar.   ▪

GRAPHICAL INSIGHT   Theorem 2 confirms our graphical intuition: $f(x)$ is increasing if the tangent lines to the graph have positive slope and decreasing if they have negative slope (Figure 3).

▪ **EXAMPLE 2**   Describe the graph of $f(x) = \ln x$. Is $f'(x)$ increasing or decreasing?

Solution   The derivative $f'(x) = x^{-1}$ is positive on the domain $\{x : x > 0\}$, and thus $f(x) = \ln x$ is increasing. However, $f'(x) = x^{-1}$ is decreasing, so the graph of $f(x)$ grows flatter as $x \to \infty$ (Figure 4).   ▪

▪ **EXAMPLE 3**   Find the intervals on which $f(x) = x^2 - 2x - 3$ is monotonic.

Solution   The derivative $f'(x) = 2x - 2 = 2(x - 1)$ is positive for $x > 1$ and negative for $x < 1$. By Theorem 2, f is decreasing on the interval $(-\infty, 1)$ and increasing on the interval $(1, \infty)$.

Note how the graph of $f'(x)$ reflects the increasing/decreasing behavior of $f(x)$ (Figure 5). The graph of $f'(x)$ lies *below* the x-axis on the interval $(-\infty, 1)$ where f is decreasing, and it lies *above* the x-axis on the interval $(1, \infty)$ where f is increasing.   ▪

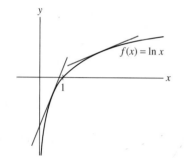

$f(x) = \ln x$

FIGURE 4 The tangent lines to $y = \ln x$ get flatter as $x \to \infty$.

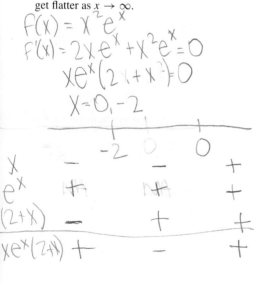

$f(x) = x^2 e^x$

$f'(x) = 2xe^x + x^2 e^x = 0$

$xe^x(2x + x^2) = 0$

$x = 0, -2$

	-2	0	
x	$-$	$-$	$+$
e^x	$+$	$+$	$+$
$(2+x)$	$-$	$+$	$+$
$xe^x(2+x)$	$+$	$-$	$+$

• f is ↑ on $(-\infty, 2), (0, \infty)$ b/c f' > 0
• f is ↓ on $(-2, 0)$ b/c f' < 0
• f(-2) = Max b/c f' → + to −
• f(0) = Min b/c f' → − to +

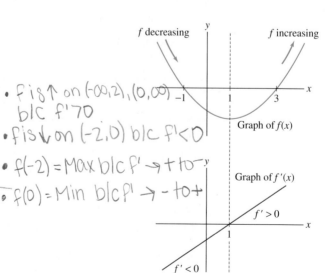

f decreasing f increasing

Graph of $f(x)$

Graph of $f'(x)$

$f' > 0$

$f' < 0$

FIGURE 5 Graph of $f(x) = x^2 - 2x - 3$ and its derivative $f'(x) = 2x - 2$.

Testing Critical Points

There is a useful test for determining whether a critical point of a function $f(x)$ is a min or max (or neither) based on the *sign change* of $f'(x)$.

To explain the term "sign change," suppose that a function $F(x)$ satisfies $F(c) = 0$. We say that $F(x)$ *changes from positive to negative* at $x = c$ if $F(x) > 0$ to the left of c and $F(x) < 0$ to the right of c for x within a small open interval around c (Figure 6). A sign change from negative to positive is defined similarly. Another possibility is that $F(x)$ does not change sign (as in Figure 6 at $x = 5$).

Now suppose that $f'(c) = 0$ and that $f'(x)$ changes sign at $x = c$, say, from $+$ to $-$. Then $f(x)$ is increasing to the left of c and decreasing to the right; therefore, $f(c)$ is a local maximum [Figure 7(A)]. Similarly, if $f'(x)$ changes sign from $-$ to $+$, then $f(c)$ is a local minimum. If $f'(x)$ does not change sign at $x = c$ [Figure 7(B)], then $f(x)$ is either nondecreasing or nonincreasing near $x = c$, and $f(c)$ is neither a local min nor a local max [unless $f(x)$ is constant to the right or left of $x = c$].

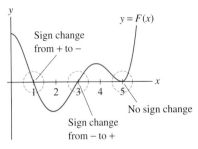

Sign change
from + to −

Sign change
from − to +

No sign change

$y = F(x)$

FIGURE 6

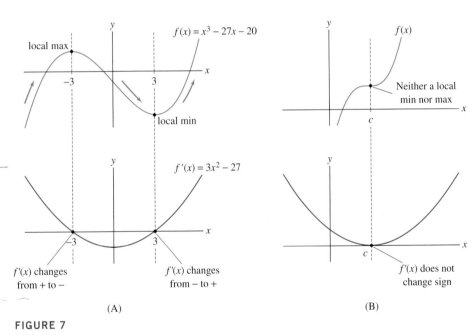

local max

$f(x) = x^3 - 27x - 20$

-3 3

local min

y

$f'(x) = 3x^2 - 27$

-3 3

$f'(x)$ changes
from + to −

$f'(x)$ changes
from − to +

(A)

$f(x)$

Neither a local
min nor max

c

$f'(x)$ does not
change sign

(B)

FIGURE 7

THEOREM 3 First Derivative Test for Critical Points   Assume that $f(x)$ is differentiable and let c be a critical point of $f(x)$. Then:

- $f'(x)$ changes from $+$ to $-$ at $c \Rightarrow f(c)$ is a local maximum.
- $f'(x)$ changes from $-$ to $+$ at $c \Rightarrow f(c)$ is a local minimum.

The sign of $f'(x)$ may change at a critical point, but it cannot change on the interval between two consecutive critical points [one can prove this is true even if $f'(x)$ is not assumed to be continuous]. Therefore, to determine the sign of $f'(x)$ on an interval between consecutive critical points, we need only evaluate $f'(x)$ at an arbitrarily chosen *test point* x_0 inside the interval. The sign of $f'(x_0)$ is the sign of $f'(x)$ on the entire interval.

■ **EXAMPLE 4** Analyzing Critical Points   Analyze the critical points of

$$f(x) = x^3 - 27x - 20$$

Solution The derivative is $f'(x) = 3x^2 - 27 = 3(x^2 - 9)$ and the critical points are $c = \pm 3$ [Figure 7(A)]. These critical points divide the real line into three intervals

$$(-\infty, -3), \qquad (-3, 3), \qquad (3, \infty)$$

We determine the sign of $f'(x)$ on each of these intervals by evaluating $f'(x)$ at a test point inside each interval. For example, in $(-\infty, -3)$ we choose the test point -4. Because $f'(-4) = 21 > 0$, $f'(x)$ is positive on the entire interval $(-3, \infty)$. Similarly,

$$f'(-4) = 21 > 0 \quad \Rightarrow \quad f'(x) > 0 \quad \text{for all } x \in (-\infty, -3)$$

$$f'(0) \;\; = -27 < 0 \quad \Rightarrow \quad f'(x) < 0 \quad \text{for all } x \in (-3, 3)$$

$$f'(4) \;\; = 21 > 0 \quad \Rightarrow \quad f'(x) > 0 \quad \text{for all } x \in (3, \infty)$$

We chose the test points -4, 0, and 4 arbitrarily. For example, to find the sign of $f'(x)$ on $(-\infty, -3)$, we could just as well have computed $f'(-5)$ or any other value of f' in the interval in $(-\infty, -3)$.

This information is displayed in the following sign diagram:

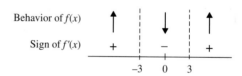

Now use the First Derivative Test:

- $c = -3$: $f'(x)$ changes from $+$ to $-$ $\Rightarrow$ $f(-3)$ is a local max.
- $c = 3$: $f'(x)$ changes from $-$ to $+$ $\Rightarrow$ $f(3)$ is a local min.   ■

■ **EXAMPLE 5** Finding Intervals of Increase and Decrease   Find the intervals of increase/decrease for $f(x) = \cos^2 x + \sin x$ on $[0, \pi]$.

Solution The increasing/decreasing behavior of $f(x)$ can change only at critical points, so first we find the critical points by solving

$$f'(x) = -2\cos x \sin x + \cos x = -(\cos x)(2\sin x - 1) = 0$$

This equation is satisfied if

$$\cos x = 0 \qquad \text{or} \qquad 2\sin x - 1 = 0$$

On $[0, \pi]$ we have

$$\cos x = 0 \quad \text{for } x = \frac{\pi}{2}$$

and

$$\sin x = \frac{1}{2} \quad \text{for } x = \frac{\pi}{6}, \frac{5\pi}{6}$$

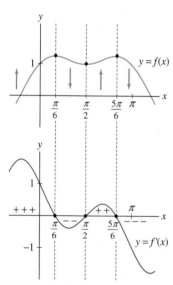

FIGURE 8 Graph of $f(x) = \cos^2 x + \sin x$ and its derivative.

In the following table, we calculate test values to determine the sign of $f'(x)$ on the intervals between the critical points:

[Handwritten margin notes, top:]
$f(x) = x^3$
$f'(x) = 3x^2 = 0$
$x = 0$ | $(-\infty, 0)$ $(0, \infty)$
x | -1 | 1
sign of f' | $+$ | $+$
f is ↑ on $(-\infty, \infty)$ b/c $f' > 0$
f is ↑ NONE
Max: None
Min: None

[Handwritten graph annotation near Figure 9:]
$f'(1) = 0$

Interval	Test Value	Sign of $f'(x)$	Behavior of $f(x)$
$\left(0, \frac{\pi}{6}\right)$	$f'(0.5) \approx 0.04$	$+$	↑
$\left(\frac{\pi}{6}, \frac{\pi}{2}\right)$	$f'(1) \approx -0.37$	$-$	↓
$\left(\frac{\pi}{2}, \frac{5\pi}{6}\right)$	$f'(2) \approx 0.34$	$+$	↑
$\left(\frac{5\pi}{6}, \pi\right)$	$f'(3) \approx -0.71$	$-$	↓

The behavior of $f(x)$ and $f'(x)$ is reflected in the graphs in Figure 8. ■

The next example illustrates the case where $f'(c) = 0$ but $f'(x)$ does not change sign at $x = c$. In this case, $f(c)$ is neither a local min nor a local max.

■ **EXAMPLE 6** A Critical Point Without a Sign Transition Analyze the critical points of $f(x) = \frac{1}{3}x^3 - x^2 + x$.

Solution The derivative is $f'(x) = x^2 - 2x + 1 = (x - 1)^2$, so $c = 1$ is the only critical point. However, $(x - 1)^2 \geq 0$, so $f'(x)$ does not change sign at $c = 1$ and $f(1)$ is neither a local min nor a local max (Figure 9). ■

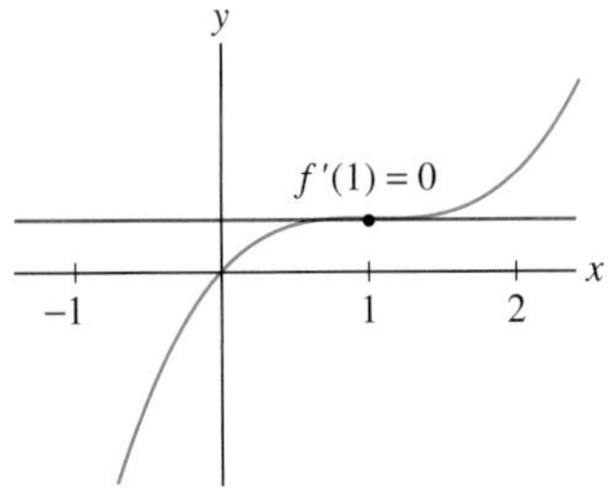

FIGURE 9 Graph of $f(x) = \frac{1}{3}x^3 - x^2 + x$. The function does not have an extreme value at the critical point.

Proof of the MVT

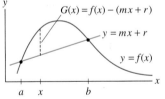

Let $m = \dfrac{f(b) - f(a)}{b - a}$ be the slope of the secant line joining $(a, f(a))$ and $(b, f(b))$. We may write the equation of the secant line in the form $y = mx + r$ for some r (Figure 10). The precise value of r is not important, but you can check that $r = f(a) - ma$. To show that $f'(c) = m$ for some c in (a, b), consider the function

$$G(x) = f(x) - (mx + r) \qquad \boxed{1}$$

As indicated in Figure 10, $G(x)$ is the vertical distance between the graph and the secant line at x (it is negative at points where the graph of f lies below the secant line). Furthermore, this distance is zero at the endpoints and therefore $G(a) = G(b) = 0$. By Rolle's Theorem (Section 4.2), there exists a point c in (a, b) such that $G'(c) = 0$. By Eq. (1), $G'(x) = f'(x) - m$, so we obtain

$$G'(c) = f'(c) - m = 0$$

Thus, $f'(c) = m$ as desired. ■

FIGURE 10 $G(x)$ is the vertical distance between the graph and the secant line.

4.3 SUMMARY

- The *sign* of $f'(x)$ determines whether $f(x)$ is increasing or decreasing:

$$f'(x) > 0 \text{ for } x \in (a, b) \quad \Rightarrow \quad f \text{ is increasing on } (a, b)$$
$$f'(x) < 0 \text{ for } x \in (a, b) \quad \Rightarrow \quad f \text{ is decreasing on } (a, b)$$

- The sign of $f'(x)$ can change only at the critical points, so $f(x)$ is *monotonic* (increasing or decreasing) on the intervals between the critical points.

- To find the sign of $f'(x)$ on the interval between two critical points, calculate the sign of $f'(x_0)$ at any test point x_0 in that interval.
- *First Derivative Test:* Assume that $f(x)$ is differentiable. If c is a critical point, then:

Sign Change of f' at c	Type of Critical Point
From $+$ to $-$	Local maximum
From $-$ to $+$	Local minimum

If f' does not change sign at $x = c$, then $f(c)$ is neither a local min nor max (unless f is constant to the left or right of c).

4.3 EXERCISES

Preliminary Questions

1. Which value of m makes the following statement correct? If $f(2) = 3$ and $f(4) = 9$, where $f(x)$ is differentiable, then the graph of f has a tangent line of slope m.

2. Which of the following conclusions does *not* follow from the MVT (assume that f is differentiable)?

(a) If f has a secant line of slope 0, then $f'(c) = 0$ for some value of c.

(b) If $f(5) < f(9)$, then $f'(c) > 0$ for some $c \in (5, 9)$.

(c) If $f'(c) = 0$ for some value of c, then there is a secant line whose slope is 0.

(d) If $f'(x) > 0$ for all x, then every secant line has positive slope.

3. Can a function that takes on only negative values have a positive derivative? Sketch an example or explain why no such functions exist.

4. (a) Use the graph of $f'(x)$ in Figure 11 to determine whether $f(c)$ is a local minimum or maximum.

(b) Can you conclude from Figure 11 that $f(x)$ is a decreasing function?

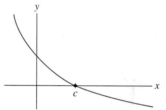

FIGURE 11 Graph of derivative $f'(x)$.

Exercises

In Exercises 1–10, find a point c satisfying the conclusion of the MVT for the given function and interval.

1. $y = x^{-1}$, $[1, 4]$

2. $y = \sqrt{x}$, $[4, 9]$

3. $y = (x - 1)(x - 3)$, $[1, 3]$

4. $y = \cos x - \sin x$, $[0, 2\pi]$

5. $y = \dfrac{x}{x + 1}$, $[3, 6]$

6. $y = x^3$, $[-3, -1]$

7. $y = x \ln x$, $[1, 2]$

8. $y = e^{-2x}$, $[0, 3]$

9. $y = \cosh x$, $[-1, 1]$

10. $y = e^x - x$, $[-1, 1]$

11. [GU] Let $f(x) = x^5 + x^2$. Check that the secant line between $x = 0$ and $x = 1$ has slope 2. By the MVT, $f'(c) = 2$ for some

(handwritten notes) 12, F'(x) is + on (0,1), (3,5) f(x) is − on (1,3),(5,6)

13. f(x) ↑ on [0,2) (4,6] f(x) ↓ on (2,4)

$c \in (0, 1)$. Estimate c graphically as follows. Plot $f(x)$ and the secant line on the same axes. Then plot the lines $y = 2x + b$ for different values of b until you find a value of b for which it is tangent to $y = f(x)$. Zoom in on the point of tangency to find its x-coordinate.

12. Determine the intervals on which $f'(x)$ is positive and negative, assuming that Figure 12 is the graph of $f(x)$.

13. Determine the intervals on which $f(x)$ is increasing or decreasing, assuming that Figure 12 is the graph of the derivative $f'(x)$.

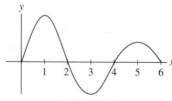

FIGURE 12

14. [GU] Plot the derivative $f'(x)$ of $f(x) = 3x^5 - 5x^3$ and describe the sign changes of $f'(x)$. Use this to determine the local extreme values of $f(x)$. Then graph $f(x)$ to confirm your conclusions.

In Exercises 15–18, sketch the graph of a function $f(x)$ whose derivative $f'(x)$ has the given description.

15. $f'(x) > 0$ for $x > 3$ and $f'(x) < 0$ for $x < 3$

16. $f'(x) > 0$ for $x < 1$ and $f'(x) < 0$ for $x > 1$

17. $f'(x)$ is negative on $(1, 3)$ and positive everywhere else.

18. $f'(x)$ makes the sign transitions $+, -, +, -$.

In Exercises 19–22, use the First Derivative Test to determine whether the function attains a local minimum or local maximum (or neither) at the given critical point.

19. $y = 7 + 4x - x^2$, $c = 2$

20. $y = x^3 - 27x + 2$, $c = -3$

21. $y = \dfrac{x^2}{x + 1}$, $c = 0$

22. $y = \sin x \cos x$, $c = \dfrac{\pi}{4}$

23. Assuming that Figure 12 is the graph of the derivative $f'(x)$, state whether $f(2)$ and $f(4)$ are local minima or maxima.

24. Figure 13 shows the graph of the derivative $f'(x)$ of a function $f(x)$. Find the critical points of $f(x)$ and determine whether they are local minima, maxima, or neither.

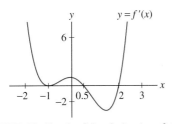

FIGURE 13 Graph of the *derivative* of $f(x)$.

In Exercises 25–52, find the critical points and the intervals on which the function is increasing or decreasing, and apply the First Derivative Test to each critical point.

25. $y = -x^2 + 7x - 17$

26. $y = 5x^2 + 6x - 4$

27. $y = x^3 - 6x^2$

28. $y = x(x + 1)^3$

29. $y = 3x^4 + 8x^3 - 6x^2 - 24x$

30. $y = x^2 + (10 - x)^2$

31. $y = \frac{1}{3}x^3 + \frac{3}{2}x^2 + 2x + 4$

32. $y = x^5 + x^3 + 1$

33. $y = x^4 + x^3$

34. $y = x^2 - x^4$

35. $y = \dfrac{1}{x^2 + 1}$

36. $y = \dfrac{2x + 1}{x^2 + 1}$

37. $y = x + x^{-1}$ $(x > 0)$

38. $y = x^4 - 4x^{3/2}$ $(x > 0)$

39. $y = x^{5/2} - x^2$ $(x > 0)$

40. $y = x^{-2} - 4x^{-1}$ $(x > 0)$

41. $y = \sin \theta \cos \theta$, $[0, 2\pi]$

42. $y = \cos \theta + \sin \theta$, $[0, 2\pi]$

43. $y = \theta + \cos \theta$, $[0, 2\pi]$

44. $y = \theta - 2\cos \theta$, $[0, 2\pi]$

45. $y = x + e^{-x}$

46. $y = \dfrac{e^x}{x}$ for $x > 0$

47. $y = e^{-x} \cos x$, $[-\frac{\pi}{2}, \frac{\pi}{2}]$

48. $y = x^2 e^x$

49. $y = \tan^{-1} x - \frac{1}{2}x$

50. $y = (x^3 - 2x)e^x$

51. $y = x - \ln x$ for $x > 0$

52. $y = \dfrac{\ln x}{x}$ for $x > 0$

53. Show that $f(x) = x^2 + bx + c$ is decreasing on $(-\infty, -\frac{b}{2})$ and increasing on $(-\frac{b}{2}, \infty)$.

54. Show that $f(x) = x^3 - 2x^2 + 2x$ is an increasing function. *Hint:* Find the minimum value of $f'(x)$.

55. Find conditions on a and b that ensure that $f(x) = x^3 + ax + b$ is increasing on $(-\infty, \infty)$.

56. [GU]   Suppose that $h(x) = \dfrac{x(x^2 - 1)}{x^2 + 1}$ is the derivative of a function $f(x)$. Plot $h(x)$ and use the plot to describe the local extrema and the increasing/decreasing behavior of $f(x)$. Sketch a plausible graph for $f(x)$ itself.

57. Sam made two statements that Deborah found dubious.
(a) "Although the average velocity for my trip was 70 mph, at no point in time did my speedometer read 70 mph."
(b) "Although a policeman clocked me going 70 mph, my speedometer never read 65 mph."
In each case, which theorem did Deborah apply to prove Sam's statement false: the Intermediate Value Theorem or the Mean Value Theorem? Explain.

58. Determine where $f(x) = (1{,}000 - x)^2 + x^2$ is increasing. Use this to decide which is larger: $1{,}000^2$ or $998^2 + 2^2$.

59. Show that $f(x) = 1 - |x|$ satisfies the conclusion of the MVT on $[a, b]$ if both a and b are positive or negative, but not if $a < 0$ and $b > 0$.

60. Which values of c satisfy the conclusion of the MVT on the interval $[a, b]$ if $f(x)$ is a linear function?

61. Show that if f is a quadratic polynomial, then the midpoint $c = \dfrac{a + b}{2}$ satisfies the conclusion of the MVT on $[a, b]$ for any a and b.

62. Suppose that $f(0) = 4$ and $f'(x) \le 2$ for $x > 0$. Apply the MVT to the interval $[0, 3]$ to prove that $f(3) \le 10$. Prove more generally that $f(x) \le 4 + 2x$ for all $x > 0$.

63. Suppose that $f(2) = -2$ and $f'(x) \ge 5$. Show that $f(4) \ge 8$.

64. Find the minimum value of $f(x) = x^x$ for $x > 0$.

Further Insights and Challenges

65. Show that the cubic function $f(x) = x^3 + ax^2 + bx + c$ is increasing on $(-\infty, \infty)$ if $b > a^2/3$.

66. Prove that if $f(0) = g(0)$ and $f'(x) \leq g'(x)$ for $x \geq 0$, then $f(x) \leq g(x)$ for all $x \geq 0$. *Hint:* Show that $f(x) - g(x)$ is nonincreasing.

67. Use Exercise 66 to prove that $\sin x \leq x$ for $x \geq 0$.

68. Use Exercises 66 and 67 to establish the following assertions for all $x \geq 0$ (each assertion follows from the previous one):

(a) $\cos x \geq 1 - \frac{1}{2}x^2$

(b) $\sin x \geq x - \frac{1}{6}x^3$

(c) $\cos x \leq 1 - \frac{1}{2}x^2 + \frac{1}{24}x^4$

(d) Can you guess the next inequality in the series?

69. Let $f(x) = e^{-x}$. Use the method of Exercise 68 to prove the following inequalities for $x \geq 0$:

(a) $e^{-x} \geq 1 - x$

(b) $e^{-x} \leq 1 - x + \frac{1}{2}x^2$

(c) $e^{-x} \geq 1 - x + \frac{1}{2}x^2 - \frac{1}{6}x^3$

Can you guess what the next inequality in the series is?

70. Assume that f'' exists and $f''(x) = 0$ for all x. Prove that $f(x) = mx + b$, where $m = f'(0)$ and $b = f(0)$.

71. Define $f(x) = x^3 \sin(\frac{1}{x})$ for $x \neq 0$ and $f(0) = 0$.

(a) Show that $f'(x)$ is continuous at $x = 0$ and $x = 0$ is a critical point of f.

(b) GU Examine the graphs of $f(x)$ and $f'(x)$. Can the First Derivative Test be applied?

(c) Show that $f(0)$ is neither a local min nor max.

4.4 The Shape of a Graph

In the previous section, we studied the increasing/decreasing behavior of a function, as determined by the sign of the derivative. Another important property of a function is its concavity, which refers to the way its graph bends. Informally, a curve is called *concave up* if it bends up and *concave down* if it bends down (Figure 1).

Concave down   Concave up

FIGURE 1

To analyze concavity in a precise fashion, let's examine how concavity is related to tangent lines and derivatives. Observe in Figure 2 that when $f(x)$ is concave up, $f'(x)$ is increasing (the slopes of the tangent lines get larger as we move to the right). Similarly, when $f(x)$ is concave down, $f'(x)$ is decreasing. Thus we make the following definition.

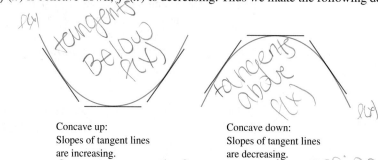

Concave up:
Slopes of tangent lines
are increasing.

Concave down:
Slopes of tangent lines
are decreasing.

FIGURE 2

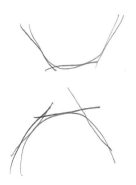

> **DEFINITION Concavity**   Let $f(x)$ be a differentiable function on an open interval (a, b). Then
>
> - f is **concave up** on (a, b) if $f'(x)$ is increasing on (a, b).
> - f is **concave down** on (a, b) if $f'(x)$ is decreasing on (a, b).

■ **EXAMPLE 1** Concavity and Stock Prices Two stocks, A and B, went up in value and both currently sell for \$75 (Figure 3). However, one is clearly a better investment than the other. Explain in terms of concavity.

fis concave up when,
f' is increasing
f'' is positive

fis concave down,
f' is decreasing
f'' is negative

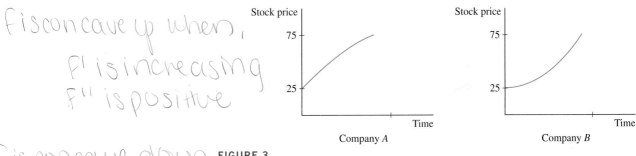

FIGURE 3

Solution The growth rate (first derivative) of Stock A is declining as time goes on (since the graph is concave down) and the growth rate of Stock B is increasing (since the graph is concave up). If these trends continue, then Stock B is the better investment.   ■

The concavity of a function is determined by the *sign* of its second derivative. Indeed, if $f''(x) > 0$, then $f'(x)$ is increasing and hence $f(x)$ is concave up. Similarly, if $f''(x) < 0$, then $f'(x)$ is decreasing and $f(x)$ is concave down.

> **THEOREM 1 Test for Concavity**   Suppose that $f''(x)$ exists for all $x \in (a, b)$.
>
> - If $f''(x) > 0$ for all $x \in (a, b)$, then f is concave up on (a, b).
> - If $f''(x) < 0$ for all $x \in (a, b)$, then f is concave down on (a, b).

Of special interest are the points where the concavity of the graph changes. We say that $P = (c, f(c))$ is a **point of inflection** of $f(x)$ if the concavity changes from up to down or vice versa at $x = c$. In other words, $f'(x)$ is increasing on one side of $x = c$ and decreasing on the other. In Figure 4, the curve on the right is made up of two arcs—one is concave down and one is concave up (the word "arc" refers to a piece of a curve). The point P where the two arcs are joined is a point of inflection. We will denote points of inflection in graphs by a solid square ■.

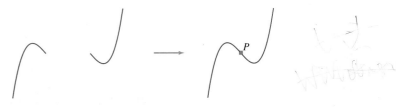

FIGURE 4 P is a point of inflection.

Concave down   Concave up

Concave down   Concave up   Concave down

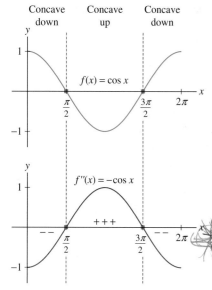

FIGURE 5

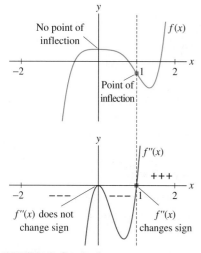

No point of inflection

Point of inflection

$f''(x)$ does not change sign

$f''(x)$ changes sign

FIGURE 6 Graph of $f(x) = 3x^5 - 5x^4 + 1$ and its second derivative.

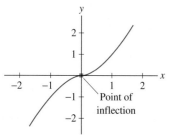

Point of inflection

FIGURE 7 The concavity of $f(x) = x^{5/3}$ changes at $x = 0$ even though $f''(0)$ does not exist.

According to Theorem 1, the concavity of f is determined by the sign of f''. Thus, a point of inflection is a point where $f''(x)$ changes sign.

> **THEOREM 2 Test for Inflection Points**   Assume that $f''(x)$ exists for all $x \in (a, b)$ and let $c \in (a, b)$. If $f''(c) = 0$ and $f''(x)$ changes sign at $x = c$, then $f(x)$ has a point of inflection at $x = c$.

■ **EXAMPLE 2**   Find the points of inflection of $f(x) = \cos x$ on $[0, 2\pi]$.

Solution   We have $f'(x) = -\sin x$, $f''(x) = -\cos x$, and $f''(x) = 0$ for $x = \frac{\pi}{2}, \frac{3\pi}{2}$. Figure 5 shows that $f''(x)$ changes sign at both $x = \frac{\pi}{2}$ and $\frac{3\pi}{2}$, so $f(x)$ has a point of inflection at both points.   ■

■ **EXAMPLE 3**   Finding Points of Inflection and Intervals of Concavity   Find the points of inflection of $f(x) = 3x^5 - 5x^4 + 1$ and determine the intervals where $f(x)$ is concave up and down.

Solution   The first derivative is $f'(x) = 15x^4 - 20x^3$ and

$$f''(x) = 60x^3 - 60x^2 = 60x^2(x - 1)$$

The zeros $x = 0, 1$ of $f''(x)$ divide the x-axis into three intervals: $(-\infty, 0)$, $(0, 1)$, and $(1, \infty)$. We determine the sign of $f''(x)$ and the concavity of f by computing "test values" within each interval (Figure 6):

Interval	Test Value	Sign of $f''(x)$	Behavior of $f(x)$
$(-\infty, 0)$	$f''(-1) = -120$	$-$	Concave down
$(0, 1)$	$f''\left(\frac{1}{2}\right) = -\frac{15}{2}$	$-$	Concave down
$(1, \infty)$	$f''(2) = 240$	$+$	Concave up

We can read off the points of inflection from this table:

- $c = 0$: no point of inflection, because $f''(x)$ does not change sign at 0.
- $c = 1$: point of inflection, because $f''(x)$ changes sign at 1.   ■

Usually, we find the inflection points by solving $f''(x) = 0$ and checking whether $f''(x)$ changes sign. However, an inflection point may also occur at a point c where $f''(c)$ does not exist.

■ **EXAMPLE 4**   A Case Where the Second Derivative Does Not Exist   Find the points of inflection of $f(x) = x^{5/3}$.

Solution   In this case, $f'(x) = \frac{5}{3}x^{2/3}$ and $f''(x) = \frac{10}{9}x^{-1/3}$. Although $f''(0)$ does not exist, $f''(x)$ does change sign at $x = 0$:

$$f''(x) = \frac{10}{9x^{1/3}} = \begin{cases} > 0 & \text{for } x > 0 \\ < 0 & \text{for } x < 0 \end{cases}$$

Therefore, the concavity of $f(x)$ changes at $x = 0$, and $(0, 0)$ is a point of inflection (Figure 7).   ■

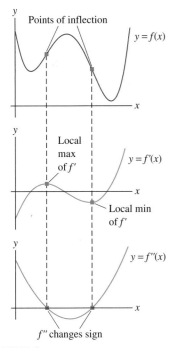

FIGURE 8

Handwritten note: Inflection is when f'' changes sign.

GRAPHICAL INSIGHT   Points of inflection are easy to spot on the graph of the first derivative $f'(x)$. If $f''(c) = 0$ and $f''(x)$ changes sign at $x = c$, then the increasing/decreasing behavior of $f'(x)$ changes at $x = c$. Thus, *inflection points of f occur where $f'(x)$ has a local min or max* (Figure 8).

Second Derivative Test for Critical Points

There is a simple test for critical points based on concavity. Suppose that $f'(c) = 0$. As we see in Figure 9, if $f(x)$ is concave down, then $f(x)$ is a local max, and if $f(x)$ is concave up, then $f(x)$ is a local min. Because concavity is determined by the sign of f'', we obtain the Second Derivative Test. (See Exercise 60 for a detailed proof.)

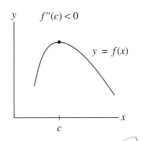
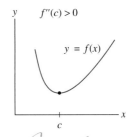

Concave down—local max   Concave up—local min

FIGURE 9 Concavity determines whether the critical point is a local minimum or maximum.

> **THEOREM 3 Second Derivative Test**   Assume that $f(x)$ is differentiable and let c be a critical point. If $f''(c)$ exists, then
>
> - $f''(c) > 0$ $\Rightarrow$ $f(c)$ is a local minimum *concave up*
> - $f''(c) < 0$ $\Rightarrow$ $f(c)$ is a local maximum *concave down*
> - $f''(c) = 0$ $\Rightarrow$ inconclusive: $f(c)$ may be a local min, max, or neither

■ **EXAMPLE 5** Analyze the critical points of $f(x) = (2x - x^2)e^x$.

Solution First we find the critical points by solving

$$f'(x) = (2x - x^2)e^x + e^x(2 - 2x) = (2 - x^2)e^x = 0$$

The critical points are $c = \pm\sqrt{2}$ (Figure 10). Next, we determine the sign of the second derivative at the critical points:

$$f''(x) = (2 - x^2)e^x + e^x(-2x) = (2 - 2x - x^2)e^x$$

We have

$$f''(-\sqrt{2}) = (2 - 2(-\sqrt{2}) - (-\sqrt{2})^2)e^{-\sqrt{2}} = 2\sqrt{2}e^{-\sqrt{2}} \quad > 0 \quad \text{(local min)}$$

$$f''(\sqrt{2}) = (2 - 2\sqrt{2} - (\sqrt{2})^2)e^{\sqrt{2}} \quad = -2\sqrt{2}e^{\sqrt{2}} \quad < 0 \quad \text{(local max)}$$

By the Second Derivative Test, $f(x)$ has a local min at $c = -\sqrt{2}$ and a local max at $c = \sqrt{2}$ (Figure 10). ■

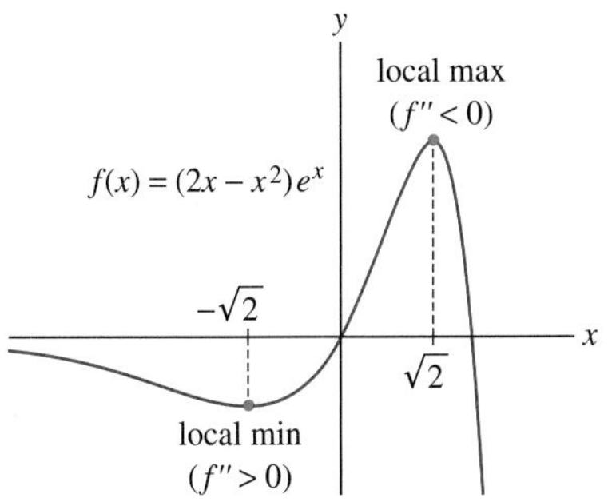

Figure 10 (left margin):

local max $(f'' < 0)$

$f(x) = (2x - x^2)e^x$

$-\sqrt{2}$

$\sqrt{2}$

local min $(f'' > 0)$

FIGURE 10 At $c = \sqrt{2}$, $f''(c) < 0$ and f has a local maximum by the Second Derivative Test.

■ **EXAMPLE 6** **Second Derivative Test Inconclusive** Analyze the critical points of $f(x) = x^5 - 5x^4$.

Solution The first two derivatives of $f(x) = x^5 - 5x^4$ are

$$f'(x) = 5x^4 - 20x^3 \;= 5x^3(x - 4)$$
$$f''(x) = 20x^3 - 60x^2$$

The critical points are $c = 0, 4$ and the Second Derivative Test yields

$$f''(0) = 0 \qquad \Rightarrow \qquad \text{Second Derivative Test fails}$$
$$f''(4) = 320 > 0 \quad \Rightarrow \quad f(4) \text{ is a local min}$$

The Second Derivative Test fails at $c = 0$, so we fall back on the First Derivative Test. Choosing test points to the left and right of $c = 0$, we find

$$f'(-1) = 5 + 20 = 25 > 0 \qquad \Rightarrow \qquad f'(x) \text{ is positive on } (-\infty, 0)$$
$$f'(1) = 5 - 20 = -15 < 0 \quad \Rightarrow \quad f'(x) \text{ is negative on } (0, 4)$$

Since $f'(x)$ changes from $+$ to $-$ at $c = 0$, $f(0)$ is a local max (Figure 11). ∎

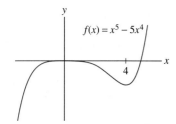

FIGURE 11 Graph of $f(x) = x^5 - 5x^4$.

4.4 SUMMARY

- A differentiable function $f(x)$ is *concave up* on (a, b) if $f'(x)$ is increasing and *concave down* if $f'(x)$ is decreasing on (a, b).
- The signs of the first two derivatives provide the following information:

First Derivative	Second Derivative
$f' > 0 \Rightarrow f$ is increasing	$f'' > 0 \Rightarrow f$ is concave up
$f' < 0 \Rightarrow f$ is decreasing	$f'' < 0 \Rightarrow f$ is concave down

- A *point of inflection* is a point where the concavity changes, from concave up to concave down or vice versa.
- If $f''(c) = 0$ and $f''(x)$ changes sign at c, then c is a point of inflection.
- Second Derivative Test: If $f'(c) = 0$ and $f''(c)$ exists, then
  - $f(c)$ is a local maximum if $f''(c) < 0$.
  - $f(c)$ is a local minimum if $f''(c) > 0$.
  - The test fails if $f''(c) = 0$.

4.4 EXERCISES

Preliminary Questions

1. Choose the correct answer: If f is concave up, then f' is:
(a) increasing   **(b)** decreasing

2. If x_0 is a critical point and f is concave down, then $f(x_0)$ is a local:
(a) min   **(b)** max   **(c)** undetermined

In Questions 3–8, state whether true or false and explain. Assume that $f''(x)$ exists for all x.

3. If $f'(c) = 0$ and $f''(c) < 0$, then $f(c)$ is a local minimum.

4. A function that is concave down on $(-\infty, \infty)$ can have no minimum value.

5. If $f''(c) = 0$, then f must have a point of inflection at $x = c$.

6. If f has a point of inflection at $x = c$, then $f''(c) = 0$.

7. If f is concave up and f' changes sign at $x = c$, then f' changes sign from negative to positive at $x = c$.

8. If $f(c)$ is a local maximum, then $f''(c)$ must be negative.

9. Suppose that $f''(c) = 0$ and $f''(x)$ changes sign from $+$ to $-$ at $x = c$. Which of the following statements are correct?

(a) $f(x)$ has a local maximum at $x = a$.

(b) $f'(x)$ has a local minimum at $x = a$.

(c) $f'(x)$ has a local maximum at $x = a$.

(d) $f(x)$ has a point of inflection at $x = a$.

Exercises

1. Match the graphs in Figure 12 with the description:

(a) $f''(x) < 0$ for all x.

(b) $f''(x)$ goes from $+$ to $-$.

(c) $f''(x) > 0$ for all x.

(d) $f''(x)$ goes from $-$ to $+$.

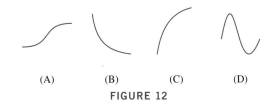

(A)   (B)   (C)   (D)

FIGURE 12

2. Match each statement with a graph in Figure 13 that represents company profits as a function of time.

(a) The outlook is great: The growth rate keeps increasing.

(b) We're losing money, but not as quickly as before.

(c) We're losing money, and it's getting worse as time goes on.

(d) We're doing well, but our growth rate is leveling off.

(e) Business had been cooling off, but now it's picking up.

(f) Business had been picking up, but now it's cooling off.

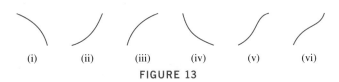

(i)   (ii)   (iii)   (iv)   (v)   (vi)

FIGURE 13

3. Sketch the graph of an increasing function such that $f''(x)$ changes from $+$ to $-$ at $x = 2$ and from $-$ to $+$ at $x = 4$. Do the same for a decreasing function.

4. If Figure 14 is the graph of a function $f(x)$, where do the points of inflection of $f(x)$ occur, and on which interval is $f(x)$ concave down?

5. If Figure 14 is the graph of the *derivative* $f'(x)$, where do the points of inflection of $f(x)$ occur, and on which interval is $f(x)$ concave down?

6. If Figure 14 is the graph of the *second derivative* $f''(x)$, where do the points of inflection of $f(x)$ occur, and on which interval is $f(x)$ concave down?

In Exercises 7–18, determine the intervals on which the function is concave up or down and find the points of inflection.

7. $y = x^2 + 7x + 10$

8. $y = t^3 - 3t^2 + 1$

9. $y = x - 2\cos x$

10. $y = 4x^5 - 5x^4$

11. $y = x(x - 8\sqrt{x})$

12. $y = (x - 2)(1 - x^3)$

13. $y = \dfrac{1}{x^2 + 3}$

14. $y = x^{7/5}$

15. $y = xe^{-x}$

16. $y = \dfrac{\ln x}{x^2}$ for $x > 0$

17. $y = (x^2 - 3)e^x$

18. $y = x - \ln x$ for $x > 0$

19. Sketch the graph of $f(x) = x^4$ and state whether f has any points of inflection. Verify your conclusion by showing that $f''(x)$ does not change sign.

20. Through her website, Leticia has been selling bean bag chairs with monthly sales as recorded below. In a report to investors, she states, "Sales reached a point of inflection when I started using pay-per-click advertising." In which month did that occur? Explain.

Month	1	2	3	4	5	6	7	8
Sales	2	20	30	35	38	44	60	90

21. The growth of a sunflower during its first 100 days is modeled well by the *logistic curve* $y = h(t)$ shown in Figure 15. Estimate

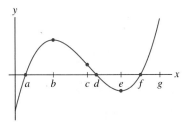

FIGURE 14

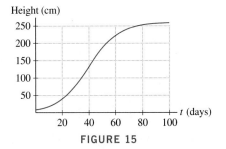

FIGURE 15

the growth rate at the point of inflection and explain its significance. Then make a rough sketch of the first and second derivatives of $h(t)$.

22. Figure 16 shows the graph of the *derivative* $f'(x)$ on [0, 1.2]. Locate the points of inflection of $f(x)$ and the points where the local minima and maxima occur. Determine the intervals on which $f(x)$ has the following properties:

(a) Increasing   **(b)** Decreasing

(c) Concave up   **(d)** Concave down

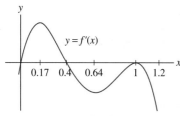

FIGURE 16

In Exercises 23–36, find the critical points of $f(x)$ and use the Second Derivative Test (if possible) to determine whether each corresponds to a local minimum or maximum.

23. $f(x) = x^3 - 12x^2 + 45x$

24. $f(x) = x^4 - 8x^2 + 1$

25. $f(x) = 3x^4 - 8x^3 + 6x^2$

26. $f(x) = x^5 - x^2$

27. $f(x) = x^5 - x^3$

28. $f(x) = \sin^2 x + \cos x$, $[0, \pi]$

29. $f(x) = \dfrac{1}{\cos x + 2}$

30. $f(x) = \dfrac{1}{x^2 - x + 2}$

31. $f(x) = 3x^{3/2} - x^{1/2}$

32. $f(x) = x^{7/4} - x$

33. $f(x) = xe^{-x^2}$

34. $f(x) = e^x + e^{-2x}$

35. $f(x) = x^3 \ln x$ for $x > 0$

36. $f(x) = \ln x + \ln(4 - x^2)$ for $0 < x < 2$

In Exercises 37–50, find the intervals on which f is concave up or down, the points of inflection, and the critical points, and determine whether each critical point corresponds to a local minimum or maximum (or neither).

37. $f(x) = x^3 - 2x^2 + x$

38. $f(x) = x^2(x - 4)$

39. $f(t) = t^2 - t^3$

40. $f(x) = 2x^4 - 3x^2 + 2$

41. $f(x) = x^2 - x^{1/2}$

42. $f(x) = \dfrac{x}{x^2 + 2}$

43. $f(t) = \dfrac{1}{t^2 + 1}$

44. $f(x) = \dfrac{1}{x^4 + 1}$

45. $f(\theta) = \theta + \sin \theta$ for $0 \le \theta \le 2\pi$

46. $f(t) = \sin^2 t$ for $0 \le t \le \pi$

47. $f(x) = x - \sin x$ for $0 \le x \le 2\pi$

48. $f(x) = \tan x$ for $-\dfrac{\pi}{2} < x < \dfrac{\pi}{2}$

49. $f(x) = e^{-x} \cos x$ for $-\dfrac{\pi}{2} \le x \le \dfrac{3\pi}{2}$

50. $f(x) = (2x^2 - 1)e^{-x^2}$

51. An infectious flu spreads slowly at the beginning of an epidemic. The infection process accelerates until a majority of the susceptible individuals are infected, at which point the process slows down.

(a) If $R(t)$ is the number of individuals infected at time t, describe the concavity of the graph of R near the beginning and end of the epidemic.

(b) Write a one-sentence news bulletin describing the status of the epidemic on the day that $R(t)$ has a point of inflection.

52. Water is pumped into a sphere at a constant rate (Figure 17). Let $h(t)$ be the water level at time t. Sketch the graph of $h(t)$ (approximately, but with the correct concavity). Where does the point of inflection occur?

53. Water is pumped into a sphere at a variable rate in such a way that the water level rises at a constant rate c (Figure 17). Let $V(t)$ be the volume of water at time t. Sketch the graph of $V(t)$ (approximately, but with the correct concavity). Where does the point of inflection occur?

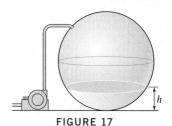

FIGURE 17

54. GU (Continuation of Exercise 53) When the water level in a sphere of radius R is h, the volume of water is $V = \pi(Rh^2 - \frac{1}{3}h^3)$. Assume the level rises at a constant rate $c = 1$ (i.e., $h = t$).

(a) Find the inflection point of $V(t)$. Does this agree with your conclusion in Exercise 53?

(b) GU Plot $V(t)$ for $R = 1$.

55. Let $f(x) = \tan^{-1} x$. Compute $f'(x)$ and $f''(x)$, and determine the increasing/decreasing and concavity behavior of $f(x)$.

In Exercises 56–58, sketch the graph of a function $f(x)$ satisfying all of the given conditions.

56. $f'(x) > 0$ and $f''(x) < 0$ for all x.

57. **(i)** $f'(x) > 0$ for all x, and

(ii) $f''(x) < 0$ for $x < 0$ and $f''(x) > 0$ for $x > 0$.

58. (i) $f'(x) < 0$ for $x < 0$ and $f'(x) > 0$ for $x > 0$, and
(ii) $f''(x) < 0$ for $|x| > 2$, and $f''(x) > 0$ for $|x| < 2$.

59. [icon] Use graphical reasoning to determine if the following statements are true or false. If false, modify the statement to make it correct.

(a) If $f(x)$ is increasing, then $f^{-1}(x)$ is decreasing.

(b) If $f(x)$ is decreasing, then $f^{-1}(x)$ is decreasing.

(c) If $f(x)$ is concave up, then $f^{-1}(x)$ is concave up.

(d) If $f(x)$ is concave down, then $f^{-1}(x)$ is concave up.

Further Insights and Challenges

In Exercises 60–62, assume that $f(x)$ is differentiable.

60. Proof of the Second Derivative Test Let c be a critical point in (a, b) such that $f''(c) > 0$ (the case $f''(c) < 0$ is similar).

(a) Show that $f''(c) = \lim\limits_{h \to 0} \dfrac{f'(c+h)}{h}$.

(b) Use part (a) to show that there exists an open interval (a, b) containing c such that $f'(x) < 0$ if $a < x < c$ and $f'(x) > 0$ if $c < x < b$. Conclude that $f(c)$ is a local minimum.

61. [icon] Assume that $f''(x)$ exists and $f''(x) > 0$. Prove that the graph of $f(x)$ "sits above" its tangent lines as follows.

(a) For any c, set $G(x) = f(x) - f'(c)(x - c) - f(c)$. It is sufficient to prove that $G(x) \ge 0$ for all c. Explain why with a sketch.

(b) Show that $G(c) = G'(c) = 0$ and $G''(x) > 0$ for all x. Use this to conclude that $G'(x) < 0$ for $x < c$ and $G'(x) > 0$ for $x > c$. Then deduce, using the MVT, that $G(x) > G(c)$ for $x \ne c$.

62. [icon] Assume that $f''(x)$ exists and let c be a point of inflection of $f(x)$.

(a) Use the method of Exercise 61 to prove that the tangent line at $x = c$ *crosses the graph* (Figure 18). *Hint:* Show that $G(x)$ changes sign at $x = c$.

FIGURE 18 Tangent line crosses graph at point of inflection.

(b) [GU] Verify this conclusion for $f(x) = \dfrac{x}{3x^2 + 1}$ by graphing $f(x)$ and the tangent line at each inflection point on the same set of axes.

63. Let $C(x)$ be the cost of producing x units of a certain good. Assume that the graph of $C(x)$ is concave up.

(a) Show that the average cost $A(x) = C(x)/x$ is minimized at that production level x_0 for which average cost equals marginal cost.

(b) Show that the line through $(0, 0)$ and $(x_0, C(x_0))$ is tangent to the graph of $C(x)$.

64. Let $f(x)$ be a polynomial of degree n.

(a) Show that if n is odd and $n > 1$, then $f(x)$ has at least one point of inflection.

(b) Show by giving an example that $f(x)$ need not have a point of inflection if n is even.

65. Critical Points and Inflection Points If $f'(c) = 0$ and $f(c)$ is neither a local min or max, must $x = c$ be a point of inflection? This is true of most "reasonable" examples (including the examples in this text), but it is not true in general. Let

$$f(x) = \begin{cases} x^2 \sin \frac{1}{x} & \text{for } x \ne 0 \\ 0 & \text{for } x = 0 \end{cases}$$

(a) Use the limit definition of the derivative to show that $f'(0)$ exists and $f'(0) = 0$.

(b) Show that $f(0)$ is neither a local min nor max.

(c) Show that $f'(x)$ changes sign infinitely often near $x = 0$ and conclude that $f(x)$ does not have a point of inflection at $x = 0$.

4.5 Graph Sketching and Asymptotes

In this section, our goal is to sketch graphs using the information provided by the first two derivatives f' and f''. We will see that a useful sketch can be produced without plotting a large number of points. Although nowadays almost all graphs are produced by computer (including, of course, the graphs in this textbook), sketching graphs by hand is useful as a way of solidifying your understanding of the basic concepts in this chapter.

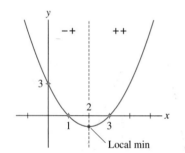

	f'' + Concave up	− Concave down
+ Increasing	++	+−
− Decreasing	−+	−−

FIGURE 1 The four basic shapes.

FIGURE 2 The graph of $f(x)$ with transition points and sign combinations of f' and f''.

FIGURE 3 Graph of $f(x) = x^2 - 4x + 3$.

Most graphs are made up of smaller *arcs* that have one of the four basic shapes, corresponding to the four possible sign combinations of f' and f'' (Figure 1). Since f' and f'' can each have sign + or −, the sign combinations are

$$++ \qquad +- \qquad -+ \qquad --$$

In this notation, the first sign refers to f' and the second sign to f''. For instance, $-+$ indicates that $f'(x) < 0$ and $f''(x) > 0$.

In graph sketching, we pay particular attention to the **transition points**, where the basic shape changes due to a sign change in either f' (local min or max) or f'' (point of inflection). In this section, local extrema are indicated by solid dots and points of inflection by solid squares (Figure 2).

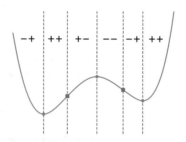

❚ **EXAMPLE 1** Quadratic Polynomial Sketch the graph of $f(x) = x^2 - 4x + 3$ (Figure 3).

$$2(x-2)<0$$

Solution We have $f'(x) = 2x - 4 = 2(x - 2)$, so

$$f'(x) < 0 \quad \text{for} \quad x < 2, \qquad f'(x) > 0 \quad \text{for} \quad x > 2$$

Thus, f decreases to the left and increases to the right of $x = 2$, with a local min at $x = 2$. The second derivative $f''(x) = 2$ is positive, so $f(x)$ is concave up. To sketch the graph, we plot the local minimum $(2, -1)$, the y-intercept, and the roots $x = 1, 3$ (Figure 3). ❚

❚ **EXAMPLE 2** A Cubic Polynomial Sketch the graph of

$$f(x) = \frac{1}{3}x^3 - \frac{1}{2}x^2 - 2x + 3$$

Solution

Step 1. **Determine the signs of f' and f''.**

First, set the derivative equal to zero to find the critical points:

$$f'(x) = x^2 - x - 2 = (x + 1)(x - 2) = 0$$

The critical points $c = -1, 2$ divide the x-axis into intervals $(-\infty, -1)$, $(-1, 2)$, and $(2, \infty)$. As in previous sections, we determine the sign of f' by computing "test values" within the intervals:

Interval	Test Value	Sign of f'
$(-\infty, -1)$	$f'(-2) = 4$	+
$(-1, 2)$	$f'(0) = -2$	−
$(2, \infty)$	$f'(3) = 4$	+

In this example, we sketch the graph of $f(x)$ *where*

$$f(x) = \frac{1}{3}x^3 - \frac{1}{2}x^2 - 2x + 3$$

$$f'(x) = (x+1)(x-2)$$

$$f''(x) = 2x - 1$$

Next, set the second derivative equal to zero and solve: $f''(x) = 2x - 1 = 0$ has the solution $c = \frac{1}{2}$ and the sign of f'' is as follows:

Interval	Test Value	Sign of f''
$(-\infty, \frac{1}{2})$	$f''(0) = -1$	$-$
$(\frac{1}{2}, \infty)$	$f''(1) = 1$	$+$

***Step 2.* Note transition points and sign combinations.**

This step merges the information about f' and f'' in a sign diagram (Figure 4). There are three transition points:

- $c = -1$: local max since f' changes from $+$ to $-$ at $c = -1$.
- $c = \frac{1}{2}$: point of inflection since f'' changes sign at $c = \frac{1}{2}$.
- $c = 2$: local min since f' changes from $-$ to $+$ at $c = 2$.

We plot the transition points $c = -1, \frac{1}{2}, 2$ together with the y-intercept as in Figure 5. It is necessary to compute y-values at these points:

$$f(-1) = \frac{25}{6}, \qquad f\left(\frac{1}{2}\right) = \frac{23}{12}, \qquad f(0) = 3, \qquad f(2) = -\frac{1}{3}$$

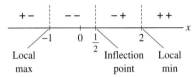

FIGURE 4 Sign combinations of f' and f''.

***Step 3.* Draw arcs of appropriate shape.**

To create the sketch, it remains only to connect the transition points by arcs of the appropriate concavity as in Figure 5.

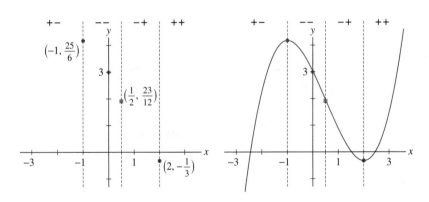

FIGURE 5 Graph of $f(x) = \frac{1}{3}x^3 - \frac{1}{2}x^2 - 2x + 3$.

Notice that the graph of our cubic is built out of four arcs, each with the appropriate increase/decrease and concavity behavior (Figure 6). ∎

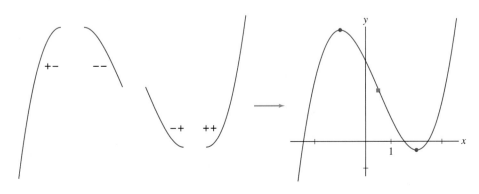

FIGURE 6 The graph of the cubic is composed of four arcs.

The graph of a polynomial $f(x)$ has the following general characteristics. It "wiggles" up and down a finite number of times and then tends to positive or negative infinity (Figure 7). To analyze the behavior of $f(x)$ as $|x|$ tends to infinity (this is called the **asymptotic behavior**), consider first the simplest polynomials x^n:

$$\lim_{x \to \infty} x^n = \infty \quad \text{and} \quad \lim_{x \to -\infty} x^n = \begin{cases} \infty & \text{if } n \text{ is even} \\ -\infty & \text{if } n \text{ is odd} \end{cases}$$

Now consider a general polynomial

$$f(x) = a_n x^n + a_{n-1} x^{n-1} + \cdots + a_1 x + a_0$$

where the leading coefficient a_n is nonzero. We may write $f(x)$ as a product:

$$f(x) = x^n \cdot \left(a_n + \frac{a_{n-1}}{x} + \cdots + \frac{a_1}{x^{n-1}} + \frac{a_0}{x^n} \right) \qquad \boxed{1}$$

As $|x|$ grows large, the second factor approaches a_n and thus $f(x)$ tends to infinity in the same way as its leading term $a_n x^n$ (Figure 7). The lower-order terms have no effect on the asymptotic behavior. Therefore, if $a_n > 0$,

$$\lim_{x \to \infty} f(x) = \infty \quad \text{and} \quad \lim_{x \to -\infty} f(x) = \begin{cases} \infty & \text{if } n \text{ is even} \\ -\infty & \text{if } n \text{ is odd} \end{cases}$$

If $a_n < 0$, these limits have opposite signs.

Strictly speaking, we have not yet defined limits as $x \to \pm\infty$. This is done more formally below. However,

$$\lim_{x \to \infty} f(x) = \infty$$

means that $f(x)$ gets arbitrarily large as x increases beyond all bounds. The limit $\lim_{x \to \infty} f(x) = -\infty$ has a similar meaning.

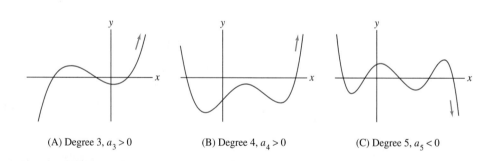

(A) Degree 3, $a_3 > 0$ (B) Degree 4, $a_4 > 0$ (C) Degree 5, $a_5 < 0$

FIGURE 7 Graphs of polynomials.

■ **EXAMPLE 3** A Polynomial of Degree 4   Sketch the graph of

$$f(x) = 3x^4 - 8x^3 + 6x^2 + 1.$$

Solution

Step 1. Determine the signs of f' and f''.
First, set the derivative equal to zero:

$$f'(x) = 12x^3 - 24x^2 + 12x = 12x(x - 1)^2 = 0$$

The critical points are $c = 0, 1$. Next, set the second derivative equal to zero:

$$f''(x) = 36x^2 - 48x + 12 = 12(x - 1)(3x - 1) = 0$$

The zeros of $f''(x)$ are $x = \frac{1}{3}$ and 1. The signs of f' and f'' are recorded in the following tables:

In this example, we sketch the graph of
f(x) where

$$f(x) = 3x^4 - 8x^3 + 6x^2 + 1$$

$$f'(x) = 12x(x-1)^2$$

$$f''(x) = 12(x-1)(3x-1)$$

Interval	Test Value	Sign of f'
$(-\infty, 0)$	$f'(-1) = -48$	$-$
$(0, 1)$	$f'(\frac{1}{2}) = \frac{3}{2}$	$+$
$(1, \infty)$	$f'(2) = 24$	$+$

Interval	Test Value	Sign of f''
$\left(-\infty, \frac{1}{3}\right)$	$f''(0) = 12$	$+$
$(\frac{1}{3}, 1)$	$f''(\frac{1}{2}) = -3$	$-$
$(1, \infty)$	$f''(2) = 60$	$+$

Step 2. Note transition points and sign combinations.

There are three transition points $c = 0, \frac{1}{3}, 1$, which divide the x-axis into four intervals (Figure 8). The type of sign change determines the nature of the transition point:

- $c = 0$: local min since f' changes from $-$ to $+$ at $c = 0$.
- $c = \frac{1}{3}$: point of inflection since f'' changes sign at $c = \frac{1}{3}$.
- $c = 1$: neither a local min nor max since f' does not change sign, but it is a point of inflection since $f''(x)$ changes sign at $c = 1$.

We plot the transition points $c = 0, \frac{1}{3}, 1$ in Figure 9 using function values $f(0) = 1$, $f(\frac{1}{3}) = \frac{38}{27}$, and $f(1) = 2$.

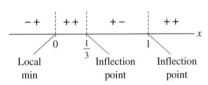

FIGURE 8

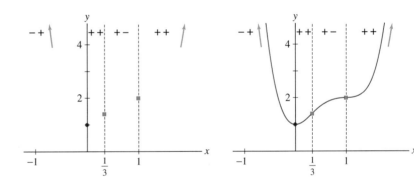

FIGURE 9 $f(x) = 3x^4 - 8x^3 + 6x^2 + 1$.

Step 3. Draw arcs of appropriate shape and asymptotic behavior.

Before connecting the transition points by arcs, we note the asymptotic behavior of $f(x)$. Since $f(x)$ is a polynomial of degree 4 whose leading term $3x^4$ has a positive coefficient, $f(x)$ tends to ∞ as $x \to \infty$ and as $x \to -\infty$. This asymptotic behavior is noted by the arrows in Figure 9. ∎

■ **EXAMPLE 4** A Trigonometric Function   Sketch the graph of $f(x) = \cos x + \frac{1}{2}x$ over the interval $[0, \pi]$.

Solution First, we solve on $[0, \pi]$:

$$f'(x) = -\sin x + \frac{1}{2} = 0 \quad \Rightarrow \quad x = \frac{\pi}{6}, \frac{5\pi}{6}$$

$$f''(x) = -\cos x = 0 \quad \Rightarrow \quad x = \frac{\pi}{2}$$

and determine the sign combinations:

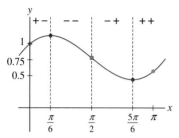

FIGURE 10 $f(x) = \cos x + \frac{1}{2}x$.

Interval	Test Value	Sign of f'
$(0, \frac{\pi}{6})$	$f'(\frac{\pi}{12}) \approx 0.24$	$+$
$(\frac{\pi}{6}, \frac{5\pi}{6})$	$f'(\frac{\pi}{2}) = -\frac{1}{2}$	$-$
$(\frac{5\pi}{6}, \pi)$	$f'(\frac{11\pi}{12}) \approx 0.24$	$+$

Interval	Test Value	Sign of f''
$(0, \frac{\pi}{2})$	$f''(\frac{\pi}{4}) = -\frac{\sqrt{2}}{2}$	$-$
$(\frac{\pi}{2}, \pi)$	$f''(\frac{3\pi}{4}) = \frac{\sqrt{2}}{2}$	$+$

We record the sign changes and transition points in Figure 10, and sketch the graph using the values:

$$f(0) = 1, \quad f\left(\frac{\pi}{6}\right) \approx 1.13, \quad f\left(\frac{\pi}{2}\right) \approx 0.79, \quad f\left(\frac{5\pi}{6}\right) \approx 0.44, \quad f(\pi) \approx 0.57 \quad \blacksquare$$

■ **EXAMPLE 5** Sketching a Function Involving e^x Sketch the graph of $f(x) = xe^x$ on the interval $[-4, 2]$.

Solution As usual, the first step is to solve for the critical points:

$$f'(x) = \frac{d}{dx} xe^x = xe^x + e^x = (x + 1)e^x = 0$$

Since $e^x > 0$ for all x, the unique critical point is $x = -1$ and

$$f'(x) = \begin{cases} < 0 & \text{for} \quad x < -1 \\ > 0 & \text{for} \quad x > -1 \end{cases}$$

Thus, $f'(x)$ changes sign from $-$ to $+$ at $x = -1$ and $f(-1)$ is a local minimum. For the second derivative, we have

$$f''(x) = (x + 1) \cdot (e^x)' + e^x \cdot (x + 1)' = (x + 1)e^x + e^x = (x + 2)e^x$$

$$f''(x) = \begin{cases} < 0 & \text{for} \quad x < -2 \\ > 0 & \text{for} \quad x > -2 \end{cases}$$

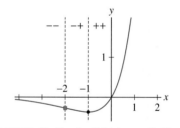

FIGURE 11 Graph of $f(x) = xe^x$. The sign combinations $--$, $-+$, $++$ indicate the signs of f' and f''.

Thus, $x = -2$ is a point of inflection and the graph changes from concave down to concave up at $x = -2$. Figure 11 shows the graph with its local minimum and point of inflection.   ■

Asymptotic Behavior

Asymptotic behavior refers to the behavior of a function $f(x)$ as either x or $f(x)$ approaches $\pm\infty$. A horizontal line $y = L$ is called a **horizontal asymptote** if one or both of the following limits exist:

$$\lim_{x \to \infty} f(x) = L \qquad \text{or} \qquad \lim_{x \to -\infty} f(x) = L$$

By definition, in this first limit, $|f(x) - L|$ gets arbitrarily small as x increases without bound (that is, for any $\epsilon > 0$, there exists a number M such that $|f(x) - L| < \epsilon$ for all $x > M$). In the second limit, $|f(x) - L|$ gets arbitrarily small as x decreases without bound. In Figure 12, $\lim_{x \to \pm\infty} f(x) = 1$ and $y = 1$ is a horizontal asymptote.

We are also interested in vertical asymptotic behavior, where the function value $f(x)$ approaches $\pm\infty$. A vertical line $x = L$ is called a **vertical asymptote** if $f(x)$ has an

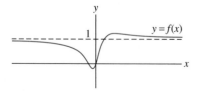

FIGURE 12 The line $y = 1$ is a horizontal asymptote.

infinite limit as x approaches L from the left or right (or both) (Figure 13). That is, if

$$\lim_{x \to L+} f(x) = \pm\infty \qquad \text{and/or} \qquad \lim_{x \to L-} f(x) = \pm\infty$$

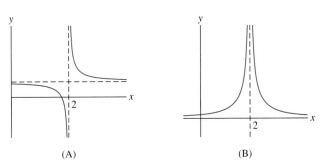

FIGURE 13 The line $x = 2$ is a vertical asymptote for both graphs.

(A)   (B)

For example, we have

$$\lim_{x \to -\infty} e^x = 0 \qquad \text{and} \qquad \lim_{x \to 0+} \ln x = -\infty$$

It follows that the x-axis, $y = 0$, is a horizontal asymptote of $f(x) = e^x$ and the y-axis, $x = 0$, is a vertical asymptote of $f(x) = \ln x$ (Figure 14).

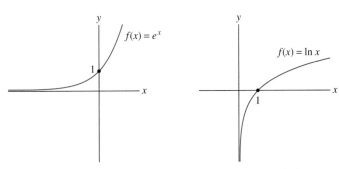

(A) The x-axis is a horizontal asymptote.   (B) The y-axis is a vertical asymptote.

FIGURE 14

■ **EXAMPLE 6** Calculate $\displaystyle\lim_{x \to \pm\infty} \frac{20x^2 - 3x}{3x^5 - 4x^2 + 5}$.

Solution It would be nice if we could apply the Quotient Law for limits, but this law is valid only if the denominator has a finite, nonzero limit. In our case, both the numerator and the denominator approach infinity:

$$\lim_{x \to \infty} (20x^2 - 3x) = \infty \qquad \text{and} \qquad \lim_{x \to \infty} (3x^5 - 4x^2 + 5) = \infty$$

The way around this difficulty is to divide the numerator and denominator by x^5 (the highest power of x in the denominator):

$$\frac{20x^2 - 3x}{3x^5 - 4x^2 + 5} = \frac{x^{-5}(20x^2 - 3x)}{x^{-5}(3x^5 - 4x^2 + 5)} = \frac{20x^{-3} - 3x^{-4}}{3 - 4x^{-3} + 5x^{-5}}$$

After rewriting the function in this way, we can use the Quotient Law:

$$\lim_{x \to \pm\infty} \frac{20x^2 - 3x}{3x^5 - 4x^2 + 5} = \frac{\lim_{x \to \pm\infty} (20x^{-3} - 3x^{-4})}{\lim_{x \to \pm\infty} (3 - 4x^{-3} + 5x^{-5})} = \frac{0}{3} = 0$$ ■

The method of the previous example applies to any rational function:

$$f(x) = \frac{a_n x^n + a_{n-1} x^{n-1} + \cdots + a_0}{b_m x^m + b_{m-1} x^{m-1} + \cdots + b_0}$$

where $a_n \neq 0$ and $b_m \neq 0$. Divide the numerator and denominator by x^m:

$$f(x) = \frac{a_n x^{n-m} + a_{n-1} x^{n-1-m} + \cdots + a_0 x^{-m}}{b_m + b_{m-1} x^{-1} + \cdots + b_0 x^{-m}}$$

Now, in the denominator, each term after the first tends to zero, and thus

$$\lim_{x \to \infty} (b_m + b_{m-1} x^{-1} + \cdots + b_0 x^{-m}) = b_m$$

The limit of the numerator depends only on the leading term $a_n x^{n-m}$, so

$$\lim_{x \to \pm\infty} f(x) = \frac{\lim_{x \to \pm\infty} (a_n x^{n-m} + a_{n-1} x^{n-1-m} + \cdots + a_0 x^{-m})}{b_m}$$

$$= \frac{\lim_{x \to \pm\infty} a_n x^{n-m}}{b_m} = \frac{a_n}{b_m} \lim_{x \to \pm\infty} x^{n-m}$$

THEOREM 1 Asymptotic Behavior of a Rational Function   The asymptotic behavior of a rational function depends only on the leading terms of its numerator and denominator. If $a_n, b_m \neq 0$, then

$$\lim_{x \to \pm\infty} \left(\frac{a_n x^n + a_{n-1} x^{n-1} + \cdots + a_0}{b_m x^m + b_{m-1} x^{m-1} + \cdots + b_0} \right) = \frac{a_n}{b_m} \lim_{x \to \pm\infty} x^{n-m}$$

The following examples illustrate this theorem:

- $n = m$: $\displaystyle \lim_{x \to \infty} \frac{3x^4 - 7x + 9}{7x^4 - 4} = \frac{3}{7} \lim_{x \to \infty} x^0 = \frac{3}{7}$

- $n < m$: $\displaystyle \lim_{x \to \infty} \frac{3x^3 - 7x + 9}{7x^4 - 4} = \frac{3}{7} \lim_{x \to \infty} x^{-1} = 0$

- $n > m$, $n - m$ odd: $\displaystyle \lim_{x \to -\infty} \frac{3x^8 - 7x + 9}{7x^3 - 4} = \frac{3}{7} \lim_{x \to -\infty} x^5 = -\infty$

- $n > m$, $n - m$ even: $\displaystyle \lim_{x \to -\infty} \frac{3x^7 - 7x + 9}{7x^3 - 4} = \frac{3}{7} \lim_{x \to -\infty} x^4 = \infty$

In the next example, we apply our method to quotients involving noninteger exponents and algebraic functions.

■ **EXAMPLE 7** Calculate the limits   **(a)** $\displaystyle \lim_{x \to \infty} \frac{3x^{5/2} + 7x^{-1/2}}{x - x^{1/2}}$ **(b)** $\displaystyle \lim_{x \to \infty} \frac{x^2}{\sqrt{x^3 + 1}}$

Solution

(a) As before, we divide the numerator and denominator by the largest power of x occurring in the denominator (multiply the top and bottom by x^{-1}):

$$\frac{3x^{5/2} + 7x^{-1/2}}{x - x^{1/2}} = \left(\frac{x^{-1}}{x^{-1}}\right)\frac{3x^{5/2} + 7x^{-1/2}}{x - x^{1/2}} = \frac{3x^{3/2} + 7x^{-3/2}}{1 - x^{-1/2}}$$

We obtain

$$\lim_{x \to \infty} \frac{3x^{3/2} + 7x^{-3/2}}{1 - x^{-1/2}} = \frac{\lim\limits_{x \to \infty}(3x^{3/2} + 7x^{-3/2})}{\lim\limits_{x \to \infty}(1 - x^{-1/2})} = \frac{\infty}{1} = \infty$$

(b) The key is to observe that the denominator of $\dfrac{x^2}{\sqrt{x^3 + 1}}$ "behaves" like $x^{3/2}$:

$$\sqrt{x^3 + 1} = \sqrt{x^3(1 + x^{-3})} = x^{3/2}\sqrt{1 + x^{-3}}$$

This suggests that we handle the quotient by dividing the numerator and denominator by $x^{3/2}$:

$$\frac{x^2}{\sqrt{x^3 + 1}} = \left(\frac{x^{-3/2}}{x^{-3/2}}\right)\frac{x^2}{x^{3/2}\sqrt{1 + x^{-3}}} = \frac{x^{1/2}}{\sqrt{1 + x^{-3}}}$$

We may now apply the Rule for Quotients:

$$\lim_{x \to \infty}\frac{x^2}{\sqrt{x^3 + 1}} = \lim_{x \to \infty}\frac{x^{1/2}}{\sqrt{1 + x^{-3}}} = \frac{\infty}{1} = \infty \qquad ■$$

The next two examples carry out graph sketching for functions with horizontal and vertical asymptotes.

■ **EXAMPLE 8** Sketch the graph of $f(x) = \dfrac{3x + 2}{2x - 4}$.

Solution The function $f(x)$ is not defined for all x. This plays a role in our analysis so we add a Step 0 to our procedure:

Step 0. **Determine the domain of f.**
   Since $f(x)$ is not defined for $x = 2$, the domain of f consists of the two intervals $(-\infty, 2)$ and $(2, \infty)$. We must analyze the behavior of f on these intervals separately.

Step 1. **Determine the signs of f' and f''.**
   A calculation shows that

$$f'(x) = -\frac{4}{(x - 2)^2}, \qquad f''(x) = \frac{8}{(x - 2)^3}$$

The derivative is not defined at $x = 2$, but we do not call $x = 2$ a critical point because 2 is not in the domain of f. The denominator of $f'(x)$ is positive, so $f'(x) < 0$ for $x \neq 2$. Therefore, $f(x)$ is decreasing for all $x \neq 2$ and $f(x)$ has no critical points.
   On the other hand, $f''(x) > 0$ for $x > 2$ and $f''(x) < 0$ for $x < 2$. Although $f''(x)$ changes sign at $x = 2$, we do not call $x = 2$ a point of inflection because 2 is not in the domain of f.

Step 2. **Note transition points and sign combinations.**
   There are no transition points in the domain of f.

$(-\infty, 2)$	$f'(x) < 0$ and $f''(x) < 0$
$(2, \infty)$	$f'(x) < 0$ and $f''(x) > 0$

***Step 3.* Draw arcs of appropriate shape and asymptotic behavior.**
The following limits show that $y = \frac{3}{2}$ is a horizontal asymptote:

$$\lim_{x \to \pm\infty} \frac{3x + 2}{2x - 4} = \lim_{x \to \pm\infty} \frac{3 + 2x^{-1}}{2 - 4x^{-1}} = \frac{3}{2}$$

The line $x = 2$ is a vertical asymptote because $f(x)$ has infinite one-sided limits as x approaches 2:

$$\lim_{x \to 2-} \frac{3x + 2}{2x - 4} = -\infty, \qquad \lim_{x \to 2+} \frac{3x + 2}{2x - 4} = \infty$$

To verify these limits, observe that as x approaches 2, the numerator $3x + 2$ is positive, whereas the denominator $2x - 4$ is small negative if $x < 2$ and small positive if $x > 2$.
    To create the sketch, first draw the asymptotes as in Figure 15(A). What does the graph look like to the right of $x = 2$? It is decreasing and concave up since $f' < 0$ and $f'' > 0$, and it approaches the asymptotes. The only possibility is the right-hand curve in Figure 15(B). To the left of $x = 2$, the graph is decreasing, concave down, and approaches the asymptotes. The x-intercept is $x = -\frac{3}{2}$ because $f(-\frac{3}{2}) = 0$ and the y-intercept is $y = f(0) = -\frac{1}{2}$.   ▮

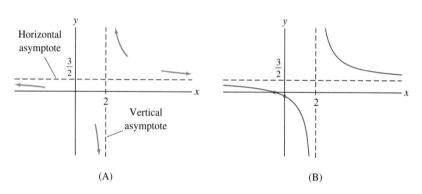

FIGURE 15 Graph of $y = \dfrac{3x + 2}{2x - 4}$.

(A)   (B)

▮ **EXAMPLE 9** Sketch the graph of $f(x) = \dfrac{1}{x^2 - 1}$.

Solution The function $f(x)$ is defined for $x \neq \pm 1$. By calculation,

$$f'(x) = -\frac{2x}{(x^2 - 1)^2}, \qquad f''(x) = \frac{6x^2 + 2}{(x^2 - 1)^3}$$

For $x \neq \pm 1$, the denominator of $f'(x)$ is positive. Therefore, $f'(x)$ and x have opposite signs:

• $f'(x) > 0$ for $x < 0$ and $f'(x) < 0$ for $x > 0$

Because $f'(x)$ changes sign from $+$ to $-$ at $x = 0$, $f(0) = -1$ is a local max. The sign of $f''(x)$ is equal to the sign of $x^2 - 1$ because $6x^2 + 2$ is positive:

• $f''(x) > 0$ for $x < -1$ or $x > 1$ and $f''(x) < 0$ for $-1 < x < 1$

Figure 16 summarizes the sign information.

In this example, we sketch the graph of $f(x)$ *where*

$$f(x) = \frac{1}{x^2 - 1}$$

$$f'(x) = -\frac{2x}{(x^2 - 1)^2}$$

$$f''(x) = \frac{6x^2 + 2}{(x^2 - 1)^3}$$

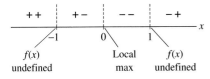

FIGURE 16

The x-axis $y = 0$ is a horizontal asymptote because

$$\lim_{x \to \infty} \frac{1}{x^2 - 1} = 0 \quad \text{and} \quad \lim_{x \to -\infty} \frac{1}{x^2 - 1} = 0$$

The function $f(x)$ has infinite one-sided limits as $x \to \pm 1$. We observe that $f(x) < 0$ if $-1 < x < 1$ and that $f(x) > 0$ if $|x| > 1$. Thus, $f(x)$ approaches $-\infty$ as $x \to \pm 1$ from within the interval $(-1, 1)$, and it approaches ∞ as $x \to \pm 1$ from outside $(-1, 1)$ (Figure 17). We obtain the sketch in Figure 18.

Vertical Asymptote	Left-Hand Limit	Right-Hand Limit
$x = -1$	$\displaystyle\lim_{x \to -1-} \frac{1}{x^2 - 1} = \infty$	$\displaystyle\lim_{x \to -1+} \frac{1}{x^2 - 1} = -\infty$
$x = 1$	$\displaystyle\lim_{x \to 1-} \frac{1}{x^2 - 1} = -\infty$	$\displaystyle\lim_{x \to 1+} \frac{1}{x^2 - 1} = \infty$

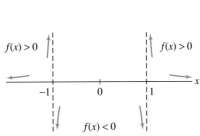

FIGURE 17 Behavior at vertical asymptotes.

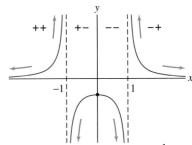

FIGURE 18 Graph of $y = \dfrac{1}{x^2 - 1}$.

4.5 SUMMARY

- Most graphs are made up of arcs that have one of the four basic shapes (Figure 19):

Sign Combination		Curve Type
$++$	$f' > 0, f'' > 0$	Increasing and concave up
$+-$	$f' > 0, f'' < 0$	Increasing and concave down
$-+$	$f' < 0, f'' > 0$	Decreasing and concave up
$--$	$f' < 0, f'' < 0$	Decreasing and concave down

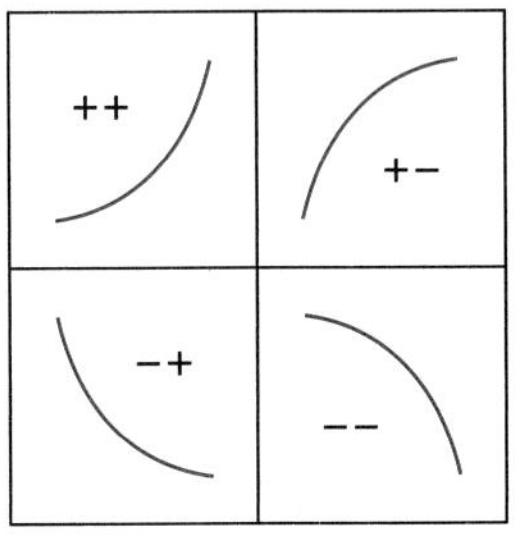

FIGURE 19 The four basic shapes.

- A *transition point* is a point in the domain of f at which either f' changes sign (local min or max) or f'' changes sign (point of inflection).
- It is convenient to break up the curve sketching process into steps:

Step 0. Determine the domain of f.

Step 1. Determine the signs of f' and f''.

Step 2. Note transition points and sign combinations.

Step 3. Draw arcs of appropriate shape and asymptotic behavior.

- A horizontal line $y = L$ is a *horizontal asymptote* if

$$\lim_{x \to \infty} f(x) = L \qquad \text{and/or} \qquad \lim_{x \to -\infty} f(x) = L$$

- A vertical line $x = L$ is a *vertical asymptote* if

$$\lim_{x \to L+} f(x) = \pm\infty \qquad \text{and/or} \qquad \lim_{x \to L-} f(x) = \pm\infty$$

- For a rational function $f(x) = \dfrac{a_n x^n + a_{n-1}x^{n-1} + \cdots + a_0}{b_m x^m + b_{m-1}x^{m-1} + \cdots + b_0}$ with $a_n, b_m \neq 0$,

$$\lim_{x \to \pm\infty} f(x) = \frac{a_n}{b_m} \lim_{x \to \pm\infty} x^{n-m}$$

4.5 EXERCISES

Preliminary Questions

1. Sketch an arc where f' and f have the sign combination ++. Do the same for −+.

2. If the sign combination of f' and f'' changes from ++ to +− at $x = c$, then (choose the correct answer):
(a) $f(c)$ is a local min
(b) $f(c)$ is a local max
(c) c is a point of inflection

3. What are the following limits?
(a) $\lim\limits_{x \to \infty} x^3$
(b) $\lim\limits_{x \to -\infty} x^3$
(c) $\lim\limits_{x \to -\infty} x^4$

4. What is the sign of a_2 if $f(x) = a_2 x^2 + a_1 x + a_0$ satisfies $\lim\limits_{x \to -\infty} f(x) = -\infty$?

5. What is the sign of the leading coefficient a_7 if $f(x)$ is a polynomial of degree 7 such that $\lim\limits_{x \to -\infty} f(x) = \infty$?

6. The second derivative of the function $f(x) = (x - 4)^{-1}$ is $f''(x) = 2(x - 4)^{-3}$. Although $f''(x)$ changes sign at $x = 4$, $f(x)$ does not have a point of inflection at $x = 4$. Why not?

Exercises

1. Determine the sign combinations of f' and f'' for each interval A–G in Figure 20.

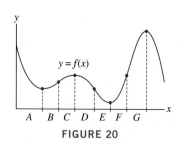

FIGURE 20

2. State the sign change at each transition point A–G in Figure 21. Example: $f'(x)$ goes from + to − at A.

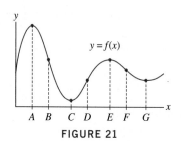

FIGURE 21

In Exercises 3–6, draw the graph of a function for which f' and f'' take on the given sign combinations.

3. ++, +−, −−

4. +−, −−, −+

5. −+, −−, −+

6. −+, ++, +−

7. Sketch the graph of $y = x^2 - 2x + 3$.

8. Sketch the graph of $y = 3 + 5x - 2x^2$.

9. Sketch the graph of the cubic $f(x) = x^3 - 3x^2 + 2$. For extra accuracy, plot the zeros of $f(x)$, which are $x = 1$ and $x = 1 \pm \sqrt{3}$ or $x \approx -0.73, 2.73$.

10. Show that the cubic $x^3 - 3x^2 + 6x$ has a point of inflection but no local extreme values. Sketch the graph.

11. Extend the sketch of the graph of $f(x) = \cos x + \frac{1}{2}x$ over $[0, \pi]$ in Example 4 to the interval $[0, 5\pi]$.

12. Sketch the graphs of $y = x^{2/3}$ and $y = x^{4/3}$.

In Exercises 13–36, sketch the graph of the function. Indicate the transition points (local extrema and points of inflection).

13. $y = x^3 + \frac{3}{2}x^2$

14. $y = x^3 - 3x + 5$

15. $y = x^2 - 4x^3$

16. $y = \frac{1}{3}x^3 + x^2 + 3x$

17. $y = 4 - 2x^2 + \frac{1}{6}x^4$

18. $y = 7x^4 - 6x^2 + 1$

19. $y = x^5 + 5x$

20. $y = x^5 - 5x^3$

21. $y = x^4 - 3x^3 + 4x$

22. $y = x^2(x-4)^2$

23. $y = 6x^7 - 7x^6$

24. $y = x^6 - 3x^4$

25. $y = x - \sqrt{x}$

26. $y = \sqrt{x+2}$

27. $y = \sqrt{x} + \sqrt{9-x}$

28. $y = x(1-x)^{1/3}$

29. $y = (x^2 - x)^{1/3}$

30. $y = (x^3 - 4x)^{1/3}$

31. $y = 4x - \ln x$

32. $y = \ln(x^2 + 1)$

33. $y = xe^{-x^2}$

34. $y = x(4-x) - 3\ln x$

35. $y = (2x^2 - 1)e^{-x^2}$

36. $y = x - 2\ln(x^2 + 1)$

37. Sketch the graph of $f(x) = 18(x-3)(x-1)^{2/3}$ using the following formulas:

$$f'(x) = \frac{30(x - \frac{9}{5})}{(x-1)^{1/3}}, \qquad f''(x) = \frac{20(x - \frac{3}{5})}{(x-1)^{4/3}}.$$

CAS *In Exercises 38-41, sketch the graph of the function. Indicate all transition points. If necessary, use a graphing utility or computer algebra system to locate the transition points numerically.*

38. $y = x^2 - 10\ln(x^2 + 1)$

39. $y = e^x - 4x - 8$

40. $y = (x^2 - 1)\sin^{-1}x$ $(-1 \le x \le 1)$

41. $y = \tan^{-1}(x^2 - x)$

In Exercises 42–47, sketch the graph over the given interval. Indicate the transition points.

42. $y = x + \sin x$, $[0, 2\pi]$

43. $y = \sin x + \cos x$, $[0, 2\pi]$

44. $y = 2\sin x - \cos^2 x$, $[0, 2\pi]$

45. $y = \sin x + \frac{1}{2}x$, $[0, 2\pi]$

46. $y = \sin x + \sqrt{3}\cos x$, $[0, \pi]$

47. $y = \sin x - \frac{1}{2}\sin 2x$, $[0, \pi]$

Hint for Exercise 47: We find numerically that there is one point of inflection in the interval, occurring at $x \approx 1.3182$.

48. Are all sign transitions possible? Explain with a sketch why the transitions $++ \rightarrow -+$ and $-- \rightarrow +-$ do not occur if the function is differentiable. (See Exercise 97 for a proof.)

49. Suppose that f is twice differentiable satisfying (i) $f(0) = 1$, (ii) $f'(x) > 0$ for all $x \ne 0$, and (iii) $f''(x) < 0$ for $x < 0$ and $f''(x) > 0$ for $x > 0$. Let $g(x) = f(x^2)$.

(a) Sketch a possible graph of $f(x)$.

(b) Prove that $g(x)$ has no points of inflection and a unique local extreme value at $x = 0$. Sketch a possible graph of $g(x)$.

50. Which of the graphs in Figure 22 *cannot* be the graph of a polynomial? Explain.

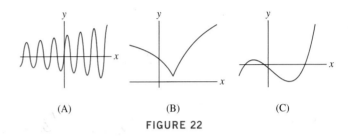

(A)   (B)   (C)

FIGURE 22

In Exercises 51–60, calculate the following limits (divide the numerator and denominator by the highest power of x appearing in the denominator).

51. $\lim\limits_{x \to \infty} \dfrac{x}{x+9}$

52. $\lim\limits_{x \to \infty} \dfrac{3x^2 + 20x}{4x^2 + 9}$

53. $\lim\limits_{x \to \infty} \dfrac{3x^2 + 20x}{2x^4 + 3x^3 - 29}$

54. $\lim\limits_{x \to \infty} \dfrac{4}{x+5}$

55. $\lim\limits_{x \to \infty} \dfrac{7x - 9}{4x + 3}$

56. $\lim\limits_{x \to \infty} \dfrac{7x^2 - 9}{4x + 3}$

57. $\lim\limits_{x \to -\infty} \dfrac{7x^2 - 9}{4x + 3}$

58. $\lim\limits_{x \to \infty} \dfrac{5x - 9}{4x^3 + 1}$

59. $\lim\limits_{x \to -\infty} \dfrac{x^2 - 1}{x + 4}$

60. $\lim\limits_{x \to \infty} \dfrac{2x^5 + 3x^4 - 31x}{8x^4 - 31x^2 + 12}$

In Exercises 61–70, calculate the limit.

61. $\lim\limits_{x \to \infty} \dfrac{\sqrt{x^2 + 1}}{x + 1}$

62. $\lim\limits_{x \to -\infty} \dfrac{\sqrt{x^4 + 1}}{x^3 + 1}$

63. $\lim\limits_{x \to \infty} \dfrac{x + 1}{\sqrt[3]{x^2 + 1}}$

64. $\lim\limits_{x \to \infty} \dfrac{x^{4/3} + 4x^{1/3}}{x^{5/4} - 2x}$

65. $\lim\limits_{x \to -\infty} \dfrac{x}{(x^6 + 1)^{1/3}}$

66. $\lim\limits_{x \to -\infty} \dfrac{4x - 9}{(3x^4 - 2)^{1/3}}$

67. $\lim\limits_{x \to 0+} \dfrac{\ln x}{x^2}$

68. $\lim\limits_{x \to -\infty} (x^2 + 4x - 8)e^{-x}$

69. $\lim\limits_{x \to \infty} \dfrac{e^{-x^2/2}}{x^2 - 5}$

70. $\lim\limits_{x \to \infty} (x^3 + 4x^2)e^x$

71. Which curve in Figure 23 is the graph of $f(x) = \dfrac{2x^4 - 1}{1 + x^4}$?
Explain on the basis of horizontal asymptotes.

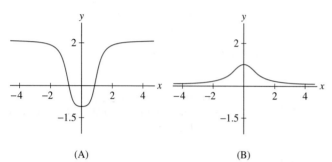

FIGURE 23

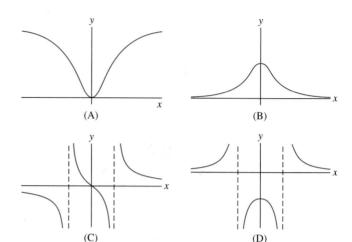

FIGURE 25

72. Match the graphs in Figure 24 with the two functions $y = \dfrac{3x}{x^2 - 1}$
and $y = \dfrac{3x^2}{x^2 - 1}$. Explain.

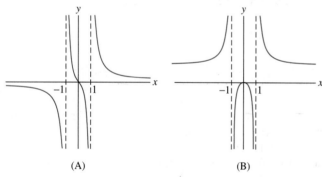

FIGURE 24

73. Match the functions with their graphs in Figure 25.

(a) $y = \dfrac{1}{x^2 - 1}$ **(b)** $y = \dfrac{x^2}{x^2 + 1}$

(c) $y = \dfrac{1}{x^2 + 1}$ **(d)** $y = \dfrac{x}{x^2 - 1}$

74. Sketch the graph of $f(x) = \dfrac{2x - 1}{x + 1}$.

75. Sketch the graph of $f(x) = \dfrac{x}{x^2 + 1}$ using the formulas

$$f'(x) = \frac{1 - x^2}{(1 + x^2)^2}, \qquad f''(x) = \frac{2x(x^2 - 3)}{(x^2 + 1)^3}$$

In Exercises 76–91, sketch the graph of the function. Indicate the asymptotes, local extrema, and points of inflection.

76. $y = \dfrac{1}{2x - 1}$ **77.** $y = \dfrac{x}{2x - 1}$

78. $y = \dfrac{x + 1}{2x - 1}$ **79.** $y = \dfrac{x + 3}{x - 2}$

80. $y = x + \dfrac{1}{x}$ **81.** $y = \dfrac{1}{x} + \dfrac{1}{x - 1}$

82. $y = \dfrac{1}{x} - \dfrac{1}{x - 1}$ **83.** $y = 1 - \dfrac{3}{x} + \dfrac{4}{x^3}$

84. $y = \dfrac{1}{x^2 - 6x + 8}$ **85.** $y = \dfrac{1}{x^2} + \dfrac{1}{(x - 2)^2}$

86. $y = \dfrac{1}{x^2} - \dfrac{1}{(x - 2)^2}$ **87.** $y = \dfrac{4}{x^2 - 9}$

88. $y = \dfrac{1}{(x^2 + 1)^2}$ **89.** $y = \dfrac{x^2}{(x^2 - 1)(x^2 + 1)}$

90. $y = \dfrac{1}{\sqrt{x^2 + 1}}$ **91.** $y = \dfrac{x}{\sqrt{x^2 + 1}}$

Further Insights and Challenges

*In Exercises 92–96, we explore functions whose graphs approach a nonhorizontal line as $x \to \infty$. A line $y = ax + b$ is called a **slant asymptote** if*

$$\lim_{x \to \infty} (f(x) - (ax + b)) = 0$$

or

$$\lim_{x \to -\infty} (f(x) - (ax + b)) = 0$$

92. Let $f(x) = \dfrac{x^2}{x-1}$. Verify the following:

(a) f is concave down on $(-\infty, 1)$ and concave up on $(1, \infty)$ as in Figure 26.

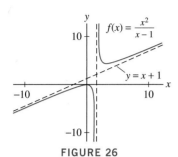

FIGURE 26

(b) $f(0)$ is a local max and $f(2)$ a local min.

(c) $\displaystyle\lim_{x \to 1-} f(x) = -\infty$ and $\displaystyle\lim_{x \to 1+} f(x) = \infty$.

(d) $y = x + 1$ is a slant asymptote of $f(x)$ as $x \to \pm\infty$.

(e) The slant asymptote lies above the graph of $f(x)$ for $x < 1$ and below the graph for $x > 1$.

93. If $f(x) = P(x)/Q(x)$, where P and Q are polynomials of degrees $m + 1$ and m, then by long division, we can write

$$f(x) = (ax + b) + P_1(x)/Q(x)$$

where P_1 is a polynomial of degree $< m$. Show that $y = ax + b$ is the slant asymptote of $f(x)$. Use this procedure to find the slant asymptotes of the functions:

(a) $y = \dfrac{x^2}{x+2}$ **(b)** $y = \dfrac{x^3 + x}{x^2 + x + 1}$

94. Sketch the graph of $f(x) = \dfrac{x^2}{x+1}$. Proceed as in the previous exercise to find the slant asymptote.

95. Show that $y = 3x$ is a slant asymptote for $f(x) = 3x + x^{-2}$. Determine whether $f(x)$ approaches the slant asymptote from above or below and make a sketch of the graph.

96. Sketch the graph of $f(x) = \dfrac{1 - x^2}{2 - x}$.

97. Assume that $f'(x)$ and $f''(x)$ exist for all x and let c be a critical point of $f(x)$. Show that $f(x)$ cannot make a transition from $++$ to $-+$ at $x = c$. *Hint:* Apply the MVT to $f'(x)$.

98. Assume that $f''(x)$ exists and $f''(x) > 0$ for all x. Show that $f(x)$ cannot be negative for all x. *Hint:* Show $f'(b) \neq 0$ for some b and use the result of Exercise 61 in Section 4.4.

4.6 Applied Optimization

Optimization problems arise in a wide range of disciplines, including physics, engineering, economics, and biology. For example, ornithologists have found that migrating birds solve an optimization problem: They fly at a velocity v that maximizes the distance they can travel without stopping, given the energy that can be stored as body fat (Figure 1).

In many optimization problems, a key step is to determine the function whose minimum or maximum we need. Once we find the function, we can apply the optimization techniques developed in this chapter. We first consider problems that lead to the optimization of a continuous function $f(x)$ on a closed interval $[a, b]$. Let's recall the steps for finding extrema developed in Section 4.2:

(i) Find the critical points of $f(x)$ in $[a, b]$.

(ii) Evaluate $f(x)$ at the critical points and the endpoints a and b.

(iii) The largest and smallest of these values are the extreme values of $f(x)$ in $[a, b]$.

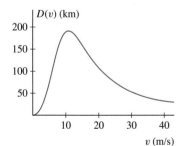

FIGURE 1 Scientists use principles of physiology and aerodynamics to obtain a plausible formula for bird migration distance $D(v)$ as a function of velocity v. The optimal velocity corresponds to the maximum point on the graph (see Exercise 64).

■ **EXAMPLE 1** A piece of wire of length L is bent into the shape of a rectangle (Figure 2). Which dimensions produce the rectangle of maximum area?

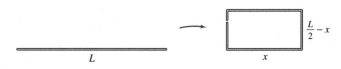

FIGURE 2

Solution If the rectangle has sides of length x and y, then its area is $A = xy$. Since A depends on two variables x and y, we cannot find the maximum until we eliminate one of the variables. This can be done because the variables are related: The perimeter of the rectangle is L and $2x + 2y = L$. This gives us $y = L/2 - x$, and we can rewrite the area as a function of x alone:

$$A(x) = x\left(\frac{L}{2} - x\right) = \left(\frac{L}{2}\right)x - x^2$$

An equation relating two or more variables in an optimization problem is called a "constraint equation." In Example 1, the constraint equation is

$$2x + 2y = L$$

Over which interval does the optimization take place? Since the sides of the rectangle cannot have negative length, we must require both $x \geq 0$ and $L/2 - x \geq 0$. Thus, $0 \leq x \leq L/2$ and our problem reduces to finding the maximum of $A(x)$ on the closed interval $[0, L/2]$.

Solving $A'(x) = L/2 - 2x = 0$, we find that $x = L/4$ is the critical point. It remains to calculate the values of $A(x)$ at the endpoints and critical points:

Endpoints: $\quad A(0) \quad = 0$

$$A\left(\frac{L}{2}\right) = \frac{L}{2}\left(\frac{L}{2} - \frac{L}{2}\right) = 0$$

Critical point: $\quad A\left(\frac{L}{4}\right) = \left(\frac{L}{4}\right)\left(\frac{L}{2} - \frac{L}{4}\right) = \frac{L^2}{16}$

The largest area occurs for $x = L/4$. The corresponding value of y is

$$y = \frac{L}{2} - x = \frac{L}{2} - \frac{L}{4} = \frac{L}{4}$$

This shows that the rectangle of maximum area is the square of sides $x = y = L/4$. ■

━━━

┃ **EXAMPLE 2** Minimizing Travel Time Cowboy Clint wants to build a dirt road from his ranch to the highway so that he can drive to the city in the shortest amount of time (Figure 3). The perpendicular distance from the ranch to the highway is 4 miles, and the city is located 9 miles down the highway. Where should Clint join the dirt road to the highway if the speed limit is 20 mph on the dirt road and 55 mph on the highway?

Solution This problem is more complicated than the previous one, so we'll analyze it in three steps. You can follow these steps to solve other optimization problems.

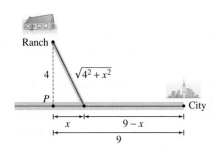

FIGURE 3

Step 1. **Choose variables.**
We need to decide where the dirt road should join the highway. So let P be the point on the highway nearest the ranch and let x be the distance from P to the point where the dirt road joins the highway.

Step 2. **Find the function and the interval.**
Since we want to minimize the travel time, we need to compute the travel time $T(x)$ of the trip as a function of x. By the Pythagorean Theorem, the length of the dirt road is $\sqrt{4^2 + x^2}$. How long will it take to traverse the road at 20 mph? Remember that at constant velocity v, the distance traveled is $d = vt$. Therefore, the *time* required to travel a distance d is $t = d/v$. Applying this with $v = 20$ mph, we see that it will take

$$\frac{\sqrt{4^2 + x^2}}{20} \quad \text{hours to traverse the dirt road}$$

[Handwritten margin notes:]

1. Write primary equation; (what is it that is being optimized?). It may have many variables!

2. Write a secondary equation: this is what is used to reduce the # of variables in our primary equation. You may have more than 1 of these.

3.

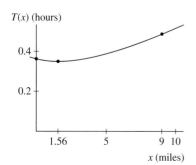

FIGURE 4 Graph of time of trip as function of x.

The strip of highway has length $9 - x$. At a speed of 55 mph, it will take

$$\frac{9 - x}{55} \text{ hours to traverse the strip of highway}$$

The total number of hours for the trip is

$$T(x) = \frac{\sqrt{16 + x^2}}{20} + \frac{9 - x}{55}$$

Over which interval does the optimization take place? Since the dirt road joins the highway somewhere between P and the city, we have $0 \le x \le 9$. So our problem is to find the minimum of $T(x)$ on $[0, 9]$ (Figure 4).

Step 3. **Optimize the function.**

We solve $T'(x) = 0$ to find the critical points:

$$T'(x) = \frac{x}{20\sqrt{16 + x^2}} - \frac{1}{55} = 0$$

$$55x = 20\sqrt{16 + x^2}$$

$$11x = 4\sqrt{16 + x^2}$$

$$121x^2 = 16(16 + x^2)$$

Thus, $105x^2 = 16^2$ or $x = 16/\sqrt{105} \approx 1.56$ miles. To find the minimum value of $T(x)$, we compute $T(x)$ at the critical point and endpoints of $[0, 9]$:

$$T(0) \approx 0.36 \text{ h}, \qquad T(1.56) \approx 0.35 \text{ h}, \qquad T(9) \approx 0.49 \text{ h}$$

We conclude that the travel time is minimized if the dirt road joins the highway at a distance $x \approx 1.56$ miles from P. ■

Calculus is widely used to solve optimization problems that arise in business decision making. The next example is adapted from a 1957 analysis of corn production in Kansas carried out by the American economist Earl Heady.

■ **EXAMPLE 3** Optimization in Agriculture Experiments show that if fertilizer made from N pounds of nitrogen and P pounds of phosphate is used on an acre of Kansas farmland, then the yield of corn is $B = 8 + 0.3\sqrt{NP}$ bushels per acre. Suppose that nitrogen costs 25 cents/lb and phosphate costs 20 cents/lb. A farmer intends to spend $30 per acre on fertilizer. Which combination of nitrogen and phosphate will produce the highest yield of corn?

Solution The formula for the yield B depends on two variables N and P, so we must eliminate one of the variables as before. The farmer will spend $30 per acre, so we have the constraint equation

$$0.25N + 0.2P = 30 \qquad \text{or} \qquad P = 150 - 1.25N$$

We substitute in the formula for B to obtain a function of N alone:

$$B(N) = 8 + 0.3\sqrt{N(150 - 1.25N)}$$

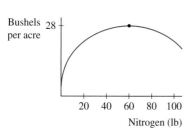

FIGURE 5 Yield of corn per acre as a function of N, with a budget of $30 per acre.

Both N and P must be nonnegative. Since $P = 150 - 1.25N \ge 0$, we require that $0 \le N \le 120$. In other words, $B(N)$ must be optimized over $[0, 120]$ (Figure 5).

To find the critical points of $B(N)$, set the derivative equal to zero:

$$\frac{dB}{dN} = \frac{0.3(150 - 2.5N)}{2\sqrt{N(150 - 1.25N)}} = 0$$

This quotient is zero if the numerator is zero:

$$150 - 2.5N = 0 \qquad \text{or} \qquad N = 60$$

The derivative is undefined at $N = 0$ and $N = 120$, so these are also critical points (which happen to be endpoints). We evaluate $B(N)$ at the critical points:

$$B(0) = 8, \qquad B(60) = 8 + 0.3\sqrt{60(150 - 1.25 \cdot 60)} \approx 28, \qquad B(120) = 8$$

The maximum occurs for $N = 60$. By the constraint equation, $P = 150 - 1.25(60) = 75$, so the combination $N = 60$, $P = 75$ is optimal. ■

Whenever possible, check that your answer is plausible. The values $N = 60$ and $P = 75$ are plausible for the following reason: Phosphate is cheaper so it makes sense to use more phosphate than nitrogen—but we cannot reduce the nitrogen to zero because the yield depends on the product NP.

Open Versus Closed Intervals

Some problems require us to optimize a continuous function on an open interval. In this case, there is no guarantee that a min or max exists. Remember that Theorem 1 in Section 4.2 states that a continuous function $f(x)$ has extreme values on a closed, bounded interval, but it says nothing about open intervals. Often, we can show that a min or max exists by examining $f(x)$ near the endpoints of the open interval. If $f(x)$ tends to infinity as x approaches the endpoints (see Figure 7), then $f(x)$ must take on a minimum at a critical point.

■ **EXAMPLE 4** Design a cylindrical can of volume 10 ft^3 so that it uses the least amount of metal (Figure 6). In other words, minimize the surface area of the can (including its top and bottom).

FIGURE 6 Cylinders with the same volume but different surface areas.

Solution

Step 1. **Choose variables.**
   We must specify the can's radius and height. Therefore, let r be the radius and h the height. Let A be the surface area of the can.

Step 2. **Find the function and the interval.**
   We compute A as a function of r and h:

$$A = \underbrace{\pi r^2}_{\text{Top}} + \underbrace{\pi r^2}_{\text{Bottom}} + \underbrace{2\pi r h}_{\text{Side}} = 2\pi r^2 + 2\pi r h$$

The can's volume is $V = \pi r^2 h$. Since we require that $V = 10$ ft^3, we obtain the constraint equation $\pi r^2 h = 10$. Thus $h = (10/\pi)r^{-2}$ and

$$A(r) = 2\pi r^2 + 2\pi r \left(\frac{10}{\pi r^2} \right) = 2\pi r^2 + \frac{20}{r}$$

The radius r can take on any positive value, so the interval of optimization is $(0, \infty)$.

Step 3. Optimize the function.

We observe that $A(r)$ tends to infinity as r approaches the endpoints of $(0, \infty)$ (Figure 7).

- $A(r) \to \infty$ as $r \to \infty$ (because of the r^2 term)
- $A(r) \to \infty$ as $r \to 0$ (because of the $1/r$ term)

We conclude that $A(r)$ takes on a minimum value at a critical point in the open interval $(0, \infty)$. Solving for the critical point, we obtain

$$\frac{dA}{dr} = 4\pi r - \frac{20}{r^2} = 0 \qquad \text{or} \qquad r^3 = \frac{5}{\pi}$$

Therefore, the value of r that minimizes the surface area is

$$r = \left(\frac{5}{\pi} \right)^{1/3} \approx 1.17 \text{ ft}$$

We also need to calculate the height:

$$h = \frac{10}{\pi} r^{-2} = 2 \left(\frac{5}{\pi} \right) r^{-2} = 2 \left(\frac{5}{\pi} \right) \left(\frac{5}{\pi} \right)^{-2/3} = 2 \left(\frac{5}{\pi} \right)^{1/3} \approx 2.34 \text{ ft}$$

Notice that the optimal dimensions satisfy $h = 2r$. In other words, the optimal can is as tall as it is wide. ∎

■ **EXAMPLE 5** Optimization Problem with No Solution   Is it possible to design a cylinder of volume 10 ft^3 with the largest possible surface area?

Solution   The answer is no. In the previous example, we showed that a cylinder of volume 10 and radius r has surface area

$$A(r) = 2\pi r^2 + \frac{20}{r}$$

This function has no maximum value because it tends to infinity as $r \to 0$ or $r \to \infty$ (Figure 7). In physical terms, this means that a cylinder of fixed volume has a large surface area if it is either very fat and short (r large) or very tall and skinny (r small). ∎

The **Principle of Least Distance** states that a light beam reflected in a mirror travels along the shortest path. In other words, if the beam originates at point A in Figure 8 and

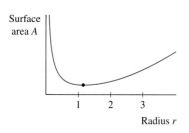

FIGURE 7 Surface area increases as r tends to zero or ∞. The minimum value exists.

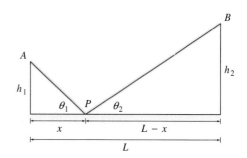

FIGURE 8 Reflection of a light beam in a mirror.

The Principle of Least Distance was known to the Greek mathematician Heron of Alexandria (c. 100 AD). See Exercise 50 for Snell's Law, a more general optical law based on the Principle of Least Time.

ends at point B, then the point of reflection P has the following minimality property:

$$AP + PB = \text{as small as possible}$$

The next example uses this principle to show that the angle of incidence θ_1 is equal to the angle of reflection θ_2.

■ **EXAMPLE 6** Angle of Incidence Equals Angle of Reflection Use the Principle of Least Distance to show that the angle of incidence θ_1 is equal to the angle of reflection θ_2.

Solution Let h_1 and h_2 be the distances from A and B to the mirror as in Figure 8, and let x be the distance from P to the left end of the mirror (Figure 8). By the Pythagorean Theorem, the length of the path from A to B is

$$f(x) = AP + PB = \sqrt{x^2 + h_1^2} + \sqrt{(L - x)^2 + h_2^2}$$

The light beam strikes the mirror at the point P for which $f(x)$ has a minimum value. The graph of $f(x)$ shows that the minimum occurs at a critical point (see Figure 9). Thus, we set the derivative equal to zero:

$$f'(x) = \frac{x}{\sqrt{x^2 + h_1^2}} - \frac{L - x}{\sqrt{(L - x)^2 + h_2^2}} = 0$$

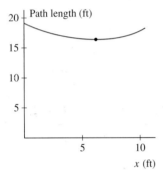

FIGURE 9 Graph of path length for $h_1 = 10$, $h_2 = 20$, $L = 40$.

If we wanted to find the critical point (giving the actual point on the mirror where the light beam strikes), we would have to solve this equation for x. However, our goal is not to find x, but rather to show that $\theta_1 = \theta_2$. Therefore, we rewrite this equation as

$$\underbrace{\frac{x}{\sqrt{x^2 + h_1^2}}}_{\cos\theta_1} = \underbrace{\frac{L - x}{\sqrt{(L - x)^2 + h_2^2}}}_{\cos\theta_2}$$

Referring to Figure 8, we see that this equation says $\cos\theta_1 = \cos\theta_2$, and since θ_1 and θ_2 lie between 0 and $\frac{\pi}{2}$, we have $\theta_1 = \theta_2$ as required.   ■

CONCEPTUAL INSIGHT The examples in this section were selected because they lead to optimization problems where the min or max occurs at a critical point. In problems of this type, the critical point often represents the best compromise between "competing factors." In Example 3, we maximized corn yield by finding the compromise between phosphate (which is cheaper) and nitrogen (without which the yield is low). In Example 4, we minimized the surface area of a metal can by finding the best compromise between height and width. In daily life, however, we often encounter endpoint rather than critical point solutions. For example, to minimize your time in the 40-yard dash, you should run as fast as you can—the solution is not a critical point but rather an endpoint (your maximum speed).

4.6 SUMMARY

• There are usually three main steps in solving an applied optimization problem:

Step 1. Choose variables.

Determine which quantities are relevant, often by drawing a diagram, and assign appropriate variables.

Step 2. Find the function and the interval.

Restate as an optimization problem for a function f over some interval. If the function depends on more than one variable, use a *constraint equation* to write f as a function of just one variable.

Step 3. Optimize the function.

- If the interval is open, f does not necessarily take on a minimum or maximum value. But if it does, these must occur at critical points within the interval. To determine if a min or max exists, analyze the behavior of f as x approaches the endpoints of the interval.

4.6 EXERCISES

Preliminary Questions

1. The problem is to find the right triangle of perimeter 10 whose area is as large as possible. What is the constraint equation relating the base b and height h of the triangle?

2. What are the relevant variables if the problem is to find a right circular cone of surface area 20 and maximum volume?

3. Does a continuous function on an open interval always have a maximum value?

4. Describe a way of showing that a continuous function on an open interval (a, b) has a minimum value.

Exercises

1. Find the dimensions of the rectangle of maximum area that can be formed from a 50-in. piece of wire.
(a) What is the constraint equation relating the lengths x and y of the sides?
(b) Find a formula for the area in terms of x alone.
(c) Does this problem require optimization over an open interval or a closed interval?
(d) Solve the optimization problem.

2. A 100-in. piece of wire is divided into two pieces and each piece is bent into a square. How should this be done in order to minimize the sum of the areas of the two squares?
(a) Express the sum of the areas of the squares in terms of the lengths x and y of the two pieces.
(b) What is the constraint equation relating x and y?
(c) Does this problem require optimization over an open or closed interval?
(d) Solve the optimization problem.

3. Find the positive number x such that the sum of x and its reciprocal is as small as possible. Does this problem require optimization over an open interval or a closed interval?

4. The legs of a right triangle have lengths a and b satisfying $a + b = 10$. Which values of a and b maximize the area of the triangle?

5. Find positive numbers x, y such that $xy = 16$ and $x + y$ is as small as possible.

6. A 20-in. piece of wire is bent into an L-shape. Where should the bend be made to minimize the distance between the two ends?

7. Let S be the set of all rectangles with area 100.
(a) What are the dimensions of the rectangle in S with the least perimeter?
(b) Is there a rectangle in S with the greatest perimeter? Explain.

8. A box has a square base of side x and height y.
(a) Find the dimensions x, y for which the volume is 12 and the surface area is as small as possible.
(b) Find the dimensions for which the surface area is 20 and the volume is as large as possible.

9. Suppose that 600 ft of fencing are used to enclose a corral in the shape of a rectangle with a semicircle whose diameter is a side of the rectangle as in Figure 10. Find the dimensions of the corral with maximum area.

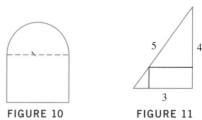

FIGURE 10   FIGURE 11

10. Find the rectangle of maximum area that can be inscribed in a right triangle with legs of length 3 and 4 if the sides of the rectangle are parallel to the legs of the triangle, as in Figure 11.

11. A landscape architect wishes to enclose a rectangular garden on one side by a brick wall costing $30/ft and on the other three sides by a metal fence costing $10/ft. If the area of the garden is 1,000 ft², find the dimensions of the garden that minimize the cost.

12. Find the point on the line $y = x$ closest to the point $(1, 0)$.

13. Find the point P on the parabola $y = x^2$ closest to the point $(3, 0)$ (Figure 12).

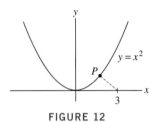

FIGURE 12

14. *CAS*   Find a good numerical approximation to the coordinates of the point on the graph of $y = \ln x - x$ closest to the origin (Figure 13).

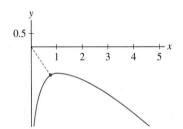

FIGURE 13 Graph of $y = \ln x - x$.

15. A box is constructed out of two different types of metal. The metal for the top and bottom, which are both square, costs $1/ft² and the metal for the sides costs $2/ft². Find the dimensions that minimize cost if the box has a volume of 20 ft³.

16. Find the dimensions of the rectangle of maximum area that can be inscribed in a circle of radius r (Figure 14).

FIGURE 14

17. Problem of Tartaglia (1500–1557)   Among all positive numbers a, b whose sum is 8, find those for which the product of the two numbers and their difference is largest. *Hint:* Let $x = a - b$ and express abx in terms of x alone.

18. Find the angle θ that maximizes the area of the isosceles triangle whose legs have length ℓ (Figure 15).

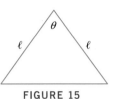

FIGURE 15

19. The volume of a right circular cone is $\frac{\pi}{3}r^2 h$ and its surface area is $S = \pi r\sqrt{r^2 + h^2}$. Find the dimensions of the cone with surface area 1 and maximal volume (Figure 16).

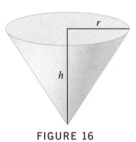

FIGURE 16

20. Rice production requires both labor and capital investment in equipment and land. Suppose that if x dollars per acre are invested in labor and y dollars per acre are invested in equipment and land, then the yield P of rice per acre is given by the formula $P = 100\sqrt{x} + 150\sqrt{y}$. If a farmer invests $40/acre, how should he divide the $40 between labor and capital investment in order to maximize the amount of rice produced?

21. Kepler's Wine Barrel Problem   The following problem was stated and solved in the work *Nova stereometria doliorum vinariorum* (New Solid Geometry of a Wine Barrel), published in 1615 by the astronomer Johannes Kepler (1571–1630). What are the dimensions of the cylinder of largest volume that can be inscribed in the sphere of radius R? *Hint:* Show that the volume of an inscribed cylinder is $2\pi x(R^2 - x^2)$, where x is one-half the height of the cylinder.

22. Find the dimensions x and y of the rectangle inscribed in a circle of radius r that maximizes the quantity xy^2.

23. Find the angle θ that maximizes the area of the trapezoid with a base of length 4 and sides of length 2, as in Figure 17.

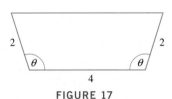

FIGURE 17

24. Consider a rectangular industrial warehouse consisting of three separate spaces of equal size as in Figure 18. Assume that the wall materials cost $200 per linear ft and the company allocates $2,400,000 for the project.

(a) Which dimensions maximize the total area of the warehouse?

(b) What is the area of each compartment in this case?

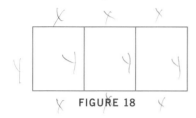

FIGURE 18

25. Suppose, in the previous exercise, that the warehouse consists of n separate spaces of equal size. Find a formula in terms of n for the maximum possible area of the warehouse.

26. The amount of light reaching a point at a distance r from a light source A of intensity I_A is I_A/r^2. Suppose that a second light source B of intensity $I_B = 4I_A$ is located 10 ft from A. Find the point on the segment joining A and B where the total amount of light is at a minimum.

27. Find the area of the largest rectangle that can be inscribed in the region bounded by the graph of $y = \dfrac{4 - x}{2 + x}$ and the coordinate axes (Figure 19).

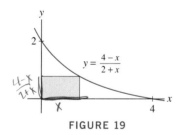

FIGURE 19

28. According to postal regulations, a carton is classified as "oversized" if the sum of its height and girth (the perimeter of its base) exceeds 108 in. Find the dimensions of a carton with square base that is not oversized and has maximum volume.

29. Find the maximum area of a triangle formed in the first quadrant by the x-axis, y-axis, and a tangent line to the graph of $y = (x + 1)^{-2}$.

30. What is the area of the largest rectangle that can be circumscribed around a rectangle of sides L and H? *Hint:* Express the area of the circumscribed rectangle in terms of the angle θ (Figure 20).

31. Optimal Price   Let r be the monthly rent per unit in an apartment building with 100 units. A survey reveals that all units can be rented when $r = \$900$ and that one unit becomes vacant with each

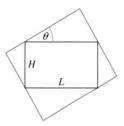

FIGURE 20

$10 increase in rent. Suppose that the average monthly maintenance per occupied unit is $100/month.

(a) Show that the number of units rented is $n = 190 - r/10$ for $900 \le r \le 1,900$.

(b) Find a formula for the net cash intake (revenue minus maintenance) and determine the rent r that maximizes intake.

32. An 8-billion-bushel corn crop brings a price of $2.40/bushel. A commodity broker uses the following rule of thumb: If the crop is reduced by x percent, then the price increases by $10x$ cents. Which crop size results in maximum revenue and what is the price per bushel? *Hint:* Revenue is equal to price times crop size.

33. Given n numbers $x_1, \ldots, x_n$, find the value of x minimizing the sum of the squares:

$$(x - x_1)^2 + (x - x_2)^2 + \cdots + (x - x_n)^2$$

First solve the problem for $n = 2, 3$ and then solve it for arbitrary n.

34. Let $P = (a, b)$ be a point in the first quadrant.

(a) Find the slope of the line through P such that the triangle bounded by this line and the axes in the first quadrant has minimal area.

(b) Show that P is the midpoint of the hypotenuse of this triangle.

35. A truck gets 10 mpg (miles per gallon) traveling along an interstate highway at 50 mph, and this is reduced by 0.15 mpg for each mile per hour increase above 50 mph.

(a) If the truck driver is paid $30/hour and diesel fuel costs $P = \$3/\text{gal}$, which speed v between 50 and 70 mph will minimize the cost of a trip along the highway? Notice that the actual cost depends on the length of the trip but the optimal speed does not.

(b) $\boxed{\text{GU}}$ Plot cost as a function of v (choose the length arbitrarily) and verify your answer to part (a).

(c) $\boxed{\text{GU}}$ Do you expect the optimal speed v to increase or decrease if fuel costs go down to $P = \$2/\text{gal}$? Plot the graphs of cost as a function of v for $P = 2$ and $P = 3$ on the same axis and verify your conclusion.

36. Figure 21 shows a rectangular plot of size 100×200 feet. Pipe is to be laid from A to a point P on side BC and from there to C. The cost of laying pipe through the lot is $30/ft (since it must be underground) and the cost along the side of the plot is $15/ft.

(a) Let $f(x)$ be the total cost, where x is the distance from P to B. Determine $f(x)$, but note that f is discontinuous at $x = 0$ (when $x = 0$, the cost of the entire pipe is $15/ft).

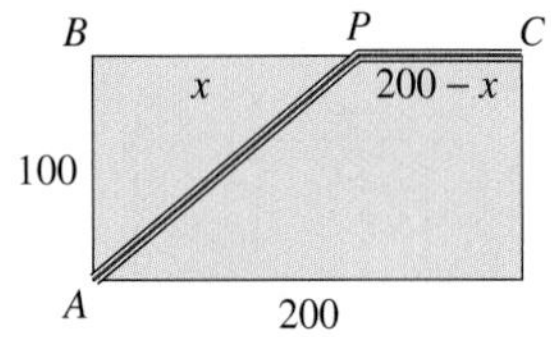

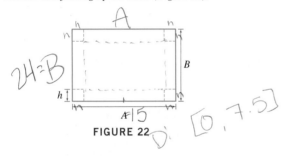

FIGURE 21

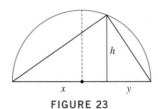

FIGURE 23

(b) What is the most economical way to lay the pipe? What if the cost along the sides is $24/ft?

37. Find the dimensions of a cylinder of volume 1 m^3 of minimal cost if the top and bottom are made of material that costs twice as much as the material for the side.

38. In Example 6 in this section, find the x-coordinate of the point P where the light beam strikes the mirror if $h_1 = 10$, $h_2 = 5$, and $L = 20$.

In Exercises 39–41, a box (with no top) is to be constructed from a piece of cardboard of sides A and B by cutting out squares of length h from the corners and folding up the sides (Figure 22).

FIGURE 22

39. Find the value of h that maximizes the volume of the box if $A = 15$ and $B = 24$. What are the dimensions of the resulting box?

40. Which value of h maximizes the volume if $A = B$?

41. Suppose that a box of height $h = 3$ in. is constructed using 144 in.^{2} of cardboard (i.e., $AB = 144$). Which values A and B maximize the volume?

42. The monthly output P of a light bulb factory is given by the formula $P = 350LK$, where L is the amount invested in labor and K the amount invested in equipment (in thousands of dollars). If the company needs to produce 10,000 units per month, how should the investment be divided among labor and equipment to minimize the cost of production? The cost of production is $L + K$.

43. Use calculus to show that among all right triangles with hypotenuse of length 1, the isosceles triangle has maximum area. Can you see more directly why this must be true by reasoning from Figure 23?

44. Janice can swim 3 mph and run 8 mph. She is standing at one bank of a river that is 300 ft wide and wants to reach a point located 200 ft downstream on the other side as quickly as possible. She will swim

diagonally across the river and then jog along the river bank. Find the best route for Janice to take.

45. Optimal Delivery Schedule A gas station sells Q gallons of gasoline per year, which is delivered N times per year in equal shipments of Q/N gallons. The cost of each delivery is d dollars and the yearly storage costs are sQT, where T is the length of time (a fraction of a year) between shipments and s is a constant. Show that costs are minimized for $N = \sqrt{sQ/d}$. (*Hint:* Express T in terms of N.) Find the optimal number of deliveries if $Q = 2$ million gal, $d = \$8,000$, and $s = 30$ cents/gal-yr. Your answer should be a whole number, so compare costs for the two integer values of N nearest the optimal value.

46. (a) Find the radius and height of a cylindrical can of total surface area A whose volume is as large as possible.

(b) Can you design a cylinder with total surface area A and minimal total volume?

47. Find the area of the largest isosceles triangle that can be inscribed in a circle of radius r.

48. A billboard of height b is mounted on the side of a building with its bottom edge at a distance h from the street. At what distance x should an observer stand from the wall to maximize the angle of observation θ (Figure 24)?

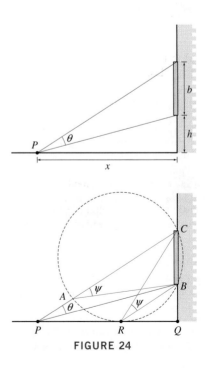

FIGURE 24

(a) Find x using calculus.

(b) Solve the problem again using geometry (without any calculation!). There is a unique circle passing through points B and C which is tangent to the street. Let R be the point of tangency. Show that θ is maximized at the point R. *Hint:* The two angles labeled ψ are, in fact, equal because they subtend equal arcs on the circle. Let A be the intersection of the circle with PC and show that $\psi = \theta + \angle PBA > \theta$.

(c) Prove that the two answers in (a) and (b) agree.

49. Use the result of Exercise 48 to show that θ is maximized at the value of x for which the angles $\angle QPB$ and $\angle QCP$ are equal.

50. Snell's Law   Figure 25 represents the surface of a swimming pool. A light beam travels from point A located above the pool to point B located underneath the water. Let v_1 be the velocity of light in air and let v_2 be the velocity of light in water (it is a fact that $v_1 > v_2$). Prove Snell's Law of Refraction according to which the path from A to B that takes the *least time* satisfies the relation

$$\frac{\sin \theta_1}{v_1} = \frac{\sin \theta_2}{v_2}$$

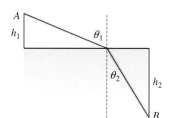

FIGURE 25

51. Snell's Law, derived in Exercise 50, explains why it is impossible to see above the water if the angle of vision θ_2 is such that $\sin \theta_2 > v_2/v_1$.

(a) Show that if this inequality holds, then there is no angle θ_1 for which Snell's Law holds.

(b) What will you see if you look through the water with an angle of vision θ_2 such that $\sin \theta_2 > v_2/v_1$?

52. A poster of area 6 ft^2 has blank margins of width 6 in. on the top and bottom and 4 in. on the sides. Find the dimensions that maximize the printed area.

53. Vascular Branching   A small blood vessel of radius r branches off at an angle θ from a larger vessel of radius R to supply blood along a path from A to B. According to Poiseuille's Law, the total resistance to blood flow is proportional to

$$T = \left(\frac{a - b \cot \theta}{R^4} + \frac{b \csc \theta}{r^4} \right)$$

where a and b are as in Figure 26. Show that the total resistance is minimized when $\cos \theta = (r/R)^4$.

54. Find the minimum length ℓ of a beam that can clear a fence of height h and touch a wall located b ft behind the fence (Figure 27).

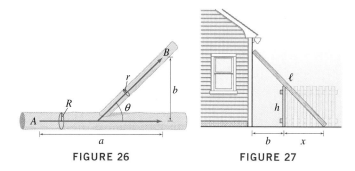

FIGURE 26   **FIGURE 27**

55. Let (a, b) be a fixed point in the first quadrant and let $S(d)$ be the sum of the distances from $(d, 0)$ to the points $(0, 0)$, (a, b), and $(a, -b)$.

(a) Find the value of d for which $S(d)$ is minimal. The answer depends on whether $b < \sqrt{3}a$ or $b \ge \sqrt{3}a$. *Hint:* Show that $d = 0$ when $b \ge \sqrt{3}a$.

(b) $\boxed{\text{GU}}$ Let $a = 1$. Plot $S(d)$ for $b = 0.5$, $\sqrt{3}$, 3 and describe the position of the minimum.

56. The minimum force required to drive a wedge of angle α into a block (Figure 28) is proportional to

$$F(\alpha) = \frac{1}{\sin \alpha + f \cos \alpha}$$

where f is a positive constant. Find the angle α for which the least force is required, assuming $f = 0.4$.

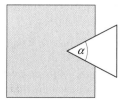

FIGURE 28

57. In the setting of Exercise 56, show that for any f the minimal force required is proportional to $1/\sqrt{1 + f^2}$.

58. The problem is to put a "roof" of side s on an attic room of height h and width b. Find the smallest length s for which this is possible. See Figure 29.

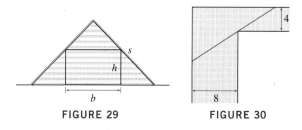

FIGURE 29   **FIGURE 30**

59. Find the maximum length of a pole that can be carried horizontally around a corner joining corridors of widths 8 ft and 4 ft (Figure 30).

60. Redo Exercise 59 for corridors of arbitrary widths a and b.

61. Find the isosceles triangle of smallest area that circumscribes a circle of radius 1 (from Thomas Simpson's *The Doctrine and Application of Fluxions*, a calculus text that appeared in 1750). See Figure 31.

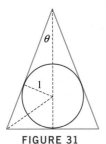

FIGURE 31

62. A basketball player stands d feet from the basket. Let h and α be as in Figure 32. Using physics, one can show that if the player releases the ball at an angle θ, then the initial velocity required to make the ball go through the basket satisfies

$$v^2 = \frac{16d}{\cos^2 \theta (\tan \theta - \tan \alpha)}$$

(a) Explain why this formula is only meaningful for $\alpha < \theta < \frac{\pi}{2}$. Why does v approach infinity at the endpoints of this interval?

(b) [GU] Take $\alpha = \frac{\pi}{6}$ and plot v^2 as a function of θ for $\frac{\pi}{6} < \theta < \frac{\pi}{2}$. Verify that the minimum occurs at $\theta = \frac{\pi}{3}$.

(c) Set $F(\theta) = \cos^2 \theta (\tan \theta - \tan \alpha)$. Explain why v is minimized for θ such that $F(\theta)$ is maximized.

(d) Verify that $F'(\theta) = \cos(\alpha - 2\theta) \sec \alpha$ (you will need to use the addition formula for cosine) and show that the maximum value of $F(\theta)$ on $[\alpha, \frac{\pi}{2}]$ occurs at $\theta_0 = \frac{\alpha}{2} + \frac{\pi}{4}$.

(e) For a given α, the optimal angle for shooting the basket is θ_0 because it minimizes v^2 and therefore minimizes the energy required to make the shot (energy is proportional to v^2). Show that the velocity v_{opt} at the optimal angle θ_0 satisfies

$$v_{\text{opt}}^2 = \frac{32d \cos \alpha}{1 - \sin \alpha} = \frac{32 d^2}{-h + \sqrt{d^2 + h^2}}$$

(f) [GU] Show with a graph that for fixed d (say, $d = 15$ ft, the distance of a free throw), v_{opt}^2 is an increasing function of h. Use this to explain why taller players have an advantage and why it can help to jump while shooting.

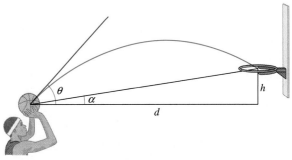

FIGURE 32

63. Tom drives with his friend Ali along a highway represented by the graph of a differentiable function $y = f(x)$ as in Figure 33. During the trip, Ali views a billboard represented by the segment $\overline{BC}$ along the y-axis. Let Q be the y-intercept of the tangent line to $y = f(x)$. Show that θ is maximized at the value of x for which the angles $\angle QPB$ and $\angle QCP$ are equal. This is a generalization of Exercise 49 [which corresponds to the case $f(x) = 0$]. *Hints:*

(a) Compute $d\theta/dx$ and check that it equals

$$(b - c) \cdot \frac{(x^2 + (xf'(x))^2) - (b - (f(x) - xf'(x)))(c - (f(x) - xf'(x)))}{(x^2 + (b - f(x))^2)(x^2 + (c - f(x))^2)}$$

(b) Show that the y-coordinate of Q is $f(x) - xf'(x)$.

(c) Show that the condition $\dfrac{d\theta}{dx} = 0$ is equivalent to

$$PQ^2 = BQ \cdot CQ$$

(d) Use (c) to conclude that triangles $\triangle QPB$ and $\triangle QCP$ are similar.

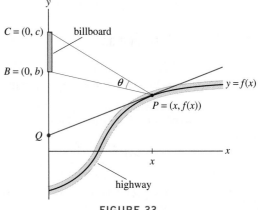

FIGURE 33

Further Insights and Challenges

64. Bird Migration Power P is the rate at which energy E is consumed per unit time. Ornithologists have found that the power consumed by a certain pigeon flying at velocity v m/s is described well by the function $P(v) = 17v^{-1} + 10^{-3}v^3$ J/s. Assume that the pigeon can store 5×10^4 J of usable energy as body fat.

(a) Find the velocity v_{pmin} that *minimizes* power consumption.

(b) Show that a pigeon flying at velocity v and using all of its stored energy can fly a total distance of $D(v) = (5 \times 10^4)v/P(v)$.

(c) Migrating birds are smart enough to fly at the velocity that maximizes distance traveled rather than minimizes power consumption.

Show that the velocity v_{dmax} which maximizes total distance $D(v)$ satisfies $P'(v) = P(v)/v$. Show that v_{dmax} is obtained graphically as the velocity coordinate of the point where a line through the origin is tangent to the graph of $P(v)$ (Figure 34).

(d) Find v_{dmax} and the maximum total distance that the bird can fly.

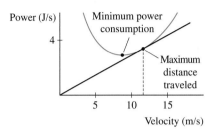

FIGURE 34

Seismic Prospecting *Exercises 65–67 are concerned with determining the thickness d of a layer of soil that lies on top of a rock formation. Geologists send two sound pulses from point A to point D separated by a distance s. The first pulse travels directly from A to D along the surface of the earth. The second pulse travels down to the rock formation, then along its surface, and then back up to D (path ABCD), as in Figure 35. The pulse travels with velocity v_1 in the soil and v_2 in the rock.*

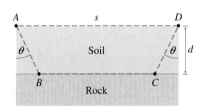

FIGURE 35

65. (a) Show that the time required for the first pulse to travel from A to D is $t_1 = s/v_1$.

(b) Show that the time required for the second pulse is

$$t_2 = \frac{2d}{v_1}\sec\theta + \frac{s - 2d\tan\theta}{v_2}$$

provided that

$$\tan\theta \le \frac{s}{2d} \qquad \boxed{1}$$

(*Note:* If this inequality is not satisfied, then point B does not lie to the left of C.)

(c) Show that t_2 is minimized when $\sin\theta = v_1/v_2$.

66. In this exercise, assume that $v_2/v_1 \ge \sqrt{1 + 4(d/s)^2}$.

(a) Show that inequality (1) holds if $\sin\theta = v_1/v_2$.

(b) Show that the minimal time for the second pulse is

$$t_2 = \frac{2d}{v_1}(1 - k^2)^{1/2} + \frac{s}{v_2}$$

where $k = v_1/v_2$.

(c) Conclude that $\dfrac{t_2}{t_1} = \dfrac{2d(1 - k^2)^{1/2}}{s} + k$.

67. Continue with the assumption of the previous exercise.

(a) Find the thickness of the soil layer, assuming that $v_1 = 0.7v_2$, $t_2/t_1 = 1.3$, and $s = 1,500$ ft.

(b) The times t_1 and t_2 are measured experimentally. The equation in Exercise 66(c) shows that t_2/t_1 is a linear function of $1/s$. What might you conclude if experiments were formed for several values of s and the points $(1/s, t_2/t_1)$ did *not* lie on a straight line?

68. Three towns A, B, and C are to be joined by an underground fiber cable as illustrated in Figure 36(A). Assume that C is located directly below the midpoint of $\overline{AB}$. Find the junction point P that minimizes the total amount of cable used.

(a) First show that P must lie directly above C. Show that if the junction is placed at a point Q as in Figure 36(B), then we can reduce the cable length by moving Q horizontally over to the point P lying above C. You may want to use the result of Example 6.

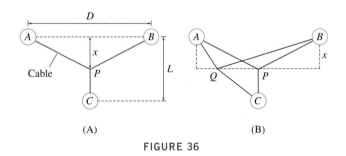

FIGURE 36

(b) With x as in Figure 36(A), let $f(x)$ be the total length of cable used. Show that $f(x)$ has a unique critical point c. Compute c and show that $0 \le c \le L$ if and only if $D \le 2\sqrt{3}\,L$.

(c) Find the minimum of $f(x)$ on $[0, L]$ in two cases: $D = 2$, $L = 4$ and $D = 8$, $L = 2$.

69. A jewelry designer plans to incorporate a component made of gold in the shape of a frustum of a cone of height 1 cm and fixed lower radius r (Figure 37). The upper radius x can take on any value between

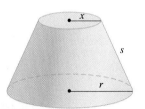

FIGURE 37 Frustrum of height 1 cm.

0 and r. Note that $x = 0$ and $x = r$ correspond to a cone and cylinder, respectively. As a function of x, the surface area (not including the top and bottom) is $S(x) = \pi s(r + x)$, where s is the *slant height* as indicated in the figure. Which value of x yields the least expensive design [the minimum value of $S(x)$ for $0 \le x \le r$]?

(a) Show that $S(x) = \pi(r + x)\sqrt{1 + (r - x)^2}$.

(b) Show that if $r < \sqrt{2}$, then $S(x)$ is an increasing function. Conclude that the cone ($x = 0$) has minimal area in this case.

(c) Assume that $r > \sqrt{2}$. Show that $S(x)$ has two critical points $x_1 < x_2$ in $(0, r)$, and that $S(x_1)$ is a local maximum, and $S(x_2)$ is a local minimum.

(d) Conclude that the minimum occurs at $x = 0$ or x_2.

(e) Find the minimum in the cases $r = 1.5$ and $r = 2$.

(f) Challenge: Let $c = \sqrt{(5 + 3\sqrt{3})/4} \approx 1.597$. Prove that the minimum occurs at $x = 0$ (cone) if $\sqrt{2} < r < c$, but the minimum occurs at $x = x_2$ if $r > c$.

4.7 L'Hôpital's Rule

L'Hôpital's Rule is named for the French mathematician Guillaume François Antoine Marquis de L'Hôpital (1661–1704) who wrote the first textbook on calculus in 1696. The name L'Hôpital is pronounced "Lo-pee-tal."

L'Hôpital's Rule is used to evaluate limits of certain types of quotients:

$$\lim_{x \to a} \frac{f(x)}{g(x)}$$

Roughly speaking, L'Hôpital's Rule states that when $f(x)/g(x)$ has an indeterminate form at $x = a$ of type $0/0$ or ∞/∞, then *we may replace $f(x)/g(x)$ by the quotient of the derivatives $f'(x)/g'(x)$.*

THEOREM 1 L'Hôpital's Rule   Assume that $f(x)$ and $g(x)$ are differentiable on an open interval containing a and that

$$f(a) = g(a) = 0$$

Also assume that $g'(x) \ne 0$ for x near but not equal to a. Then

$$\lim_{x \to a} \frac{f(x)}{g(x)} = \lim_{x \to a} \frac{f'(x)}{g'(x)}$$

provided that the limit on the right exists. This conclusion also holds if $f(x)$ and $g(x)$ are differentiable for x near (but not equal to) a and

$$\lim_{x \to a} f(x) = \pm\infty \quad \text{and} \quad \lim_{x \to a} g(x) = \pm\infty$$

Furthermore, these limits may be replaced by one-sided limits.

■ **EXAMPLE 1** Use L'Hôpital's Rule to evaluate $\lim\limits_{x \to 1} \dfrac{x^3 - 1}{x - 1}$.

Solution Both $f(x) = x^3 - 1$ and $g(x) = x - 1$ are differentiable and $f(1) = g(1) = 0$, so the quotient is indeterminate at $x = 1$:

$$\frac{x^3 - 1}{x - 1}\bigg|_{x=1} = \frac{1^3 - 1}{1 - 1} = \frac{0}{0} \quad \text{(indeterminate)}$$

Furthermore, $g'(x) = 1$, and thus $g'(x)$ is nonzero. Therefore, L'Hôpital's Rule applies. We may replace the numerator and denominator by their derivatives to obtain

CAUTION When using L'Hôpital's Rule, be sure to take the derivative of the numerator and denominator separately. Do not differentiate the quotient *function.*

$$\lim_{x \to 1} \frac{x^3 - 1}{x - 1} = \lim_{x \to 1} \frac{3x^2}{1} = 3 \qquad ■$$

■ **EXAMPLE 2** Evaluate $\lim\limits_{x \to \pi/2} \dfrac{\cos^2 x}{\sin x - 1}$.

Solution L'Hôpital's Rule applies because $\cos^2 x$ and $\sin x - 1$ are differentiable but

$$\frac{\cos^2 x}{\sin x - 1}\bigg|_{x=\pi/2} = \frac{\cos^2(\frac{\pi}{2})}{\sin \frac{\pi}{2} - 1} = \frac{0^2}{1 - 1} = \frac{0}{0} \quad \text{(indeterminate)}$$

We replace the numerator and denominator by their derivatives:

$$\lim_{x \to \pi/2} \frac{\cos^2 x}{\sin x - 1} = \underbrace{\lim_{x \to \pi/2} \frac{-2\cos x \, \sin x}{\cos x}}_{\text{Simplify}} = \lim_{x \to \pi/2} (-2\sin x) = -2 \qquad ∎$$

■ **EXAMPLE 3** Indeterminate Form $\frac{\infty}{\infty}$ Evaluate $\lim\limits_{x \to 0+} x \ln x$.

Solution This is a one-sided limit because $f(x) = x \ln x$ is not defined for $x \leq 0$. L'Hôpital's Rule is valid for one-sided limits, but $f(x)$ is not a quotient, so we rewrite it as $f(x) = (\ln x)/x^{-1}$. Since $\ln x$ approaches $-\infty$ and x^{-1} approaches ∞ as $x \to 0+$, we have the indeterminate form $-\infty/\infty$ and L'Hôpital's Rule applies:

$$\lim_{x \to 0+} x \ln x = \lim_{x \to 0+} \frac{\ln x}{x^{-1}} = \lim_{x \to 0+} \left(\frac{x^{-1}}{-x^{-2}}\right) = \lim_{x \to 0+} (-x) = 0 \qquad ∎$$

■ **EXAMPLE 4** Using L'Hôpital's Rule Twice Evaluate $\lim\limits_{x \to 0} \dfrac{e^x - x - 1}{\cos x - 1}$.

Solution L'Hôpital's Rule applies because

$$\frac{e^x - x - 1}{\cos x - 1}\bigg|_{x=0} = \frac{e^0 - 0 - 1}{\cos 0 - 1} = \frac{1 - 0 - 1}{1 - 1} = \frac{0}{0} \quad \text{(indeterminate)}$$

A first application of L'Hôpital's Rule gives

$$\lim_{x \to 0} \frac{e^x - x - 1}{\cos x - 1} = \lim_{x \to 0} \left(\frac{e^x - 1}{-\sin x}\right) = \lim_{x \to 0} \frac{1 - e^x}{\sin x}$$

This limit is again indeterminate of type $0/0$, so we apply L'Hôpital's Rule again:

$$\lim_{x \to 0} \frac{1 - e^x}{\sin x} = \lim_{x \to 0} \frac{-e^x}{\cos x} = \frac{-e^0}{\cos 0} = -1 \qquad ∎$$

■ **EXAMPLE 5** Assumptions Matter Can L'Hôpital's Rule be applied to $\lim\limits_{x \to 1} \dfrac{x^2 + 1}{2x + 1}$?

Solution The answer is no. The function does *not* have an indeterminate form because

$$\frac{x^2 + 1}{2x + 1}\bigg|_{x=1} = \frac{1^2 + 1}{2 \cdot 1 + 1} = \frac{2}{3}$$

However, the limit can be evaluated directly by substitution:

$$\lim_{x \to 1} \frac{x^2 + 1}{2x + 1} = \frac{2}{3}$$

An incorrect application of L'Hôpital's Rule would lead to the following limit with a different value:

$$\lim_{x \to 1} \frac{2x}{2} = 1 \quad \text{(not equal to original limit)} \qquad ∎$$

■ **EXAMPLE 6** The Form $\infty - \infty$ Evaluate $\lim\limits_{x\to 0}\left(\dfrac{1}{\sin x}-\dfrac{1}{x}\right)$.

Solution L'Hôpital's Rule does not apply directly because the indeterminate form is neither $0/0$ nor ∞/∞:

$$\left(\frac{1}{\sin x}-\frac{1}{x}\right)\bigg|_{x=0}=\frac{1}{0}-\frac{1}{0}\quad\text{(indeterminate)}$$

However,

$$\frac{1}{\sin x}-\frac{1}{x}=\frac{x-\sin x}{x\sin x}$$

This quotient has the indeterminate form $0/0$ at $x=0$, and L'Hôpital's Rule now yields

$$\lim_{x\to 0}\left(\frac{1}{\sin x}-\frac{1}{x}\right)=\lim_{x\to 0}\frac{x-\sin x}{x\sin x}=\lim_{x\to 0}\frac{1-\cos x}{x\cos x+\sin x}$$

The limit on the right is still indeterminate at $x=0$, but a second application of L'Hôpital's Rule finishes the job (Figure 1):

$$\lim_{x\to 0}\frac{1-\cos x}{x\cos x+\sin x}=\lim_{x\to 0}\frac{\sin x}{-x\sin x+2\cos x}=\frac{0}{2}=0\qquad\blacksquare$$

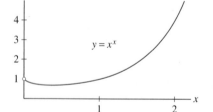

FIGURE 1 The graph confirms that $y=\dfrac{1}{\sin x}-\dfrac{1}{x}$ approaches 0 as $x\to 0$.

Limits of functions of the form $f(x)^{g(x)}$ can lead to indeterminate forms of the type 0^0, 1^∞, or ∞^0. In such cases, we may take the logarithm and then apply L'Hôpital's Rule.

■ **EXAMPLE 7** The Form 0^0 Evaluate $\lim\limits_{x\to 0+}x^x$.

Solution First, compute the limit of the logarithm. If $y=x^x$, then $\ln y=x\ln x$ and

$$\lim_{x\to 0+}\ln(x^x)=\lim_{x\to 0+}x\ln x=\lim_{x\to 0+}\frac{\ln x}{x^{-1}}\qquad\text{(rewrite as a quotient)}$$

$$=\lim_{x\to 0+}\frac{x^{-1}}{-x^{-2}}=\lim_{x\to 0+}(-x)=0\quad\text{(apply L'Hôpital's Rule)}$$

By the continuity of e^x, we may exponentiate to obtain the desired limit (see Figure 2):

$$\lim_{x\to 0+}x^x=\lim_{x\to 0+}e^{\ln(x^x)}=e^{\lim_{x\to 0+}\ln(x^x)}=e^0=1\qquad\blacksquare$$

FIGURE 2 The function $y=x^x$ approaches 1 as $x\to 0$.

Comparing Growth of Functions

Sometimes, we are interested in determining which of two functions, $f(x)$ and $g(x)$, grows faster. For example, there are two standard computer algorithms for sorting data (alphabetizing, ordering according to rank, etc.): **Quick Sort** and **Bubble Sort.** The average time required to sort a list of size n has order of magnitude $n\ln n$ for Quick Sort and n^2 for Bubble Sort. Which algorithm is faster when the size n is large? Although n is a whole number, this problem amounts to comparing the growth of $f(x)=x\ln x$ and $g(x)=x^2$ as $x\to\infty$.

We say that $f(x)$ grows *faster* than $g(x)$ if

$$\lim_{x\to\infty}\frac{f(x)}{g(x)}=\infty\qquad\text{or equivalently,}\qquad\lim_{x\to\infty}\frac{g(x)}{f(x)}=0$$

In this case, we write $g \ll f$. For example, $x \ll x^2$ because

$$\lim_{x\to\infty} \frac{x^2}{x} = \lim_{x\to\infty} x = \infty$$

To compare the growth of functions, we need a version of L'Hôpital's Rule that applies to limits as $x \to \infty$.

THEOREM 2 L'Hôpital's Rule for Limits as $x \to \infty$ Assume that $f(x)$ and $g(x)$ are differentiable in an interval (b, ∞) and that $g'(x) \neq 0$ for $x > b$. If $\lim_{x\to\infty} f(x)$ and $\lim_{x\to\infty} g(x)$ exist and either both are zero or both are infinite, then

$$\lim_{x\to\infty} \frac{f(x)}{g(x)} = \lim_{x\to\infty} \frac{f'(x)}{g'(x)}$$

provided that the limit on the right exists. A similar result holds for limits as $x \to -\infty$.

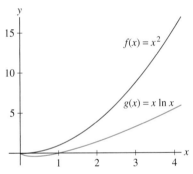

FIGURE 3

■ **EXAMPLE 8** Does $f(x) = x^2$ or $g(x) = x \ln x$ grow faster as $x \to \infty$?

Solution Both $f(x)$ and $g(x)$ approach infinity as $x \to \infty$, so L'Hôpital's Rule applies to the quotient:

$$\lim_{x\to\infty} \frac{f(x)}{g(x)} = \lim_{x\to\infty} \frac{x^2}{x \ln x} = \underbrace{\lim_{x\to\infty} \frac{x}{\ln x} = \lim_{x\to\infty} \frac{1}{x^{-1}}}_{\text{L'Hôpital's Rule}} = \lim_{x\to\infty} x = \infty$$

We conclude that $x \ln x \ll x^2$ (Figure 3). ■

■ **EXAMPLE 9** Jonathan is interested in comparing two computer algorithms whose average run times are approximately $(\ln n)^2$ and $\sqrt{n}$. Which algorithm takes less time for large values of n?

Solution We replace n by the continuous variable x and apply L'Hôpital's Rule:

$$\lim_{x\to\infty} \frac{\sqrt{x}}{(\ln x)^2} = \lim_{x\to\infty} \frac{\frac{1}{2}x^{-1/2}}{2x^{-1}\ln x} \qquad \text{(L'Hôpital's Rule)}$$

$$= \lim_{x\to\infty} \frac{x^{1/2}}{4\ln x} \qquad \text{(simplify)}$$

$$= \lim_{x\to\infty} \frac{\frac{1}{2}x^{-1/2}}{4x^{-1}} \qquad \text{(L'Hôpital's Rule again)}$$

$$= \lim_{x\to\infty} \frac{1}{8}x^{1/2} = \infty \qquad \text{(simplify)}$$

This shows that $(\ln x)^2 \ll \sqrt{x}$. We conclude that the algorithm whose average time is proportional to $(\ln n)^2$ takes less time for large n. ■

In Section 1.6, we asserted that exponential functions increase more rapidly than polynomial functions. We now prove this by showing that $x^n \ll e^x$ for every exponent n (Figure 4).

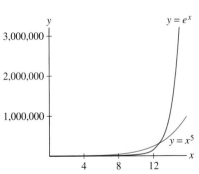

FIGURE 4 Graph illustrating $x^5 \ll e^x$.

> **THEOREM 3 Growth of e^x**
>
> $$x^n \ll e^x \qquad \text{for every exponent } n$$

Proof For $n = 1$, L'Hôpital's Rule gives

$$\lim_{x \to \infty} \frac{e^x}{x} = \lim_{x \to \infty} \frac{e^x}{1} = \infty$$

To treat higher powers, we observe that L'Hôpital's Rule reduces the limit with x^n to the limit for x^{n-1}. For example,

$$\lim_{x \to \infty} \frac{e^x}{x^2} = \lim_{x \to \infty} \frac{e^x}{2x} = \frac{1}{2} \lim_{x \to \infty} \frac{e^x}{x} = \infty$$

Similarly, having proved that $e^x/x^2 \to \infty$, we conclude

$$\lim_{x \to \infty} \frac{e^x}{x^3} = \lim_{x \to \infty} \frac{e^x}{3x^2} = \frac{1}{3} \lim_{x \to \infty} \frac{e^x}{x^2} = \infty$$

Proceeding in this way, we may prove the result for all n. A more formal proof would use the principle of induction. ∎

Proof of L'Hôpital's Rule

We prove L'Hôpital's Rule here only in the first case of Theorem 1, namely, in the case that $f(a) = g(a) = 0$. We also assume that f' and g' are continuous at $x = a$ and that $g'(a) \neq 0$. Then $g(x) \neq g(a)$ for x near but not equal to a. In this case, L'Hôpital's Rule states

$$\lim_{x \to a} \frac{f(x)}{g(x)} = \lim_{x \to a} \frac{f'(x)}{g'(x)} = \frac{f'(a)}{g'(a)}$$

To prove this, we observe that if $f(a) = g(a) = 0$, then for x near but not equal to a,

$$\frac{f(x)}{g(x)} = \frac{f(x) - f(a)}{g(x) - g(a)} = \frac{\dfrac{f(x) - f(a)}{x - a}}{\dfrac{g(x) - g(a)}{x - a}}$$

By the Quotient Law for Limits and the definition of the derivative, we obtain

$$\lim_{x \to a} \frac{f(x)}{g(x)} = \frac{\displaystyle\lim_{x \to a} \frac{f(x) - f(a)}{x - a}}{\displaystyle\lim_{x \to a} \frac{g(x) - g(a)}{x - a}} = \frac{f'(a)}{g'(a)} \qquad ∎$$

4.7 SUMMARY

- *L'Hôpital's Rule:* Assume that f and g are differentiable near a and that

$$f(a) = g(a) = 0$$

Assume also that $g'(x) \neq 0$ for x near (but not equal to) a. Then

$$\lim_{x \to a} \frac{f(x)}{g(x)} = \lim_{x \to a} \frac{f'(x)}{g'(x)}$$

provided that the limit on the right exists.
- L'Hôpital's Rule remains valid if $\lim_{x \to a} f(x)$ and $\lim_{x \to a} g(x)$ are both infinite. It also applies to limits as $x \to \infty$.
- Limits involving the indeterminate forms 0^0, 1^∞, or ∞^0 can often be evaluated by first taking the logarithm and then applying L'Hôpital's Rule.
- In comparing the growth rates of functions, we write $g \ll f$ if

$$\lim_{x \to \infty} \frac{f(x)}{g(x)} = \infty$$

4.7 EXERCISES

Preliminary Questions

1. Which of the following two limits can be evaluated using L'Hôpital's Rule?

$$\lim_{x \to 4} \frac{3x - 12}{x^2 - 16}, \qquad \lim_{x \to 4} \frac{12x - 3}{x - 4}$$

2. What is wrong with evaluating

$$\lim_{x \to 0} \frac{x^2 - 2x}{3x - 2}$$

using L'Hôpital's Rule?

Exercises

In Exercises 1–10, show that L'Hôpital's Rule is applicable and use it to evaluate the limit.

1. $\lim_{x \to 1} \dfrac{2x^2 + x - 3}{x - 1}$

2. $\lim_{x \to 2} \dfrac{2x^2 + x - 10}{2x^5 - 40x + 16}$

3. $\lim_{x \to -1} \dfrac{6x^3 + 13x^2 + 9x + 2}{6x^3 - x^2 - 5x + 2}$

4. $\lim_{x \to 0} \dfrac{\sin 4x}{x^2 + 3x}$

5. $\lim_{x \to 9} \dfrac{\sqrt{x} - 3}{2x^2 - 17x - 9}$

6. $\lim_{x \to 0} \dfrac{x^3}{\sin x - x}$

7. $\lim_{x \to 0} \dfrac{x^2}{1 - \cos x}$

8. $\lim_{x \to 0} \dfrac{\cos 2x - 1}{\sin 5x}$

9. $\lim_{x \to 0} \dfrac{\cos x - \cos^2 x}{\sin x}$

10. $\lim_{x \to 3} \dfrac{\sqrt{x + 1} - 2}{x^3 - 7x - 6}$

In Exercises 11–16, show that L'Hôpital's Rule is applicable to the limit as $x \to \pm\infty$ and evaluate.

11. $\lim_{x \to \infty} \dfrac{3x - 1}{7 - 12x}$

12. $\lim_{x \to \infty} \dfrac{\ln x}{\sqrt{x}}$

13. $\lim_{x \to \infty} \dfrac{x}{e^x}$

14. $\lim_{x \to -\infty} \dfrac{\ln(x^4 + 1)}{x}$

15. $\lim_{x \to \infty} \dfrac{x^2}{e^x}$

16. $\lim_{x \to -\infty} x \sin \dfrac{1}{x}$

In Exercises 17–48, apply L'Hôpital's Rule to evaluate the limit. In some cases, it may be necessary to apply it more than once.

17. $\lim_{x \to 0} \dfrac{\sin 4x}{\sin 3x}$

18. $\lim_{x \to \pi/2} \dfrac{\tan 3x}{\tan 5x}$

19. $\lim_{x \to 0} \dfrac{\tan x}{x}$

20. $\lim_{x \to 1} \dfrac{\sqrt{8 + x} - 3x^{1/4}}{x^2 + 3x - 4}$

21. $\lim_{x \to 0} \left(\cot x - \dfrac{1}{x} \right)$

22. $\lim_{x \to 2} \dfrac{e^{x^2} - e^4}{x - 2}$

23. $\lim_{x \to -\infty} \dfrac{3x - 2}{1 - 5x}$

24. $\lim_{x \to \infty} \dfrac{x^2 + 4x}{9x^3 + 4}$

25. $\lim_{x \to -\infty} \dfrac{7x^2 + 4x}{9 - 3x^2}$

26. $\lim_{x \to 1} \tan \left(\dfrac{\pi x}{2} \right) \ln x$

27. $\lim_{x \to \infty} \dfrac{3x^3 + 4x^2}{4x^3 - 7}$

28. $\lim_{x \to 4} \left[\dfrac{1}{\sqrt{x} - 2} - \dfrac{4}{x - 4} \right]$

29. $\lim_{x \to 1} \dfrac{x(\ln x - 1) + 1}{(x - 1) \ln x}$

30. $\lim_{x \to \pi/2} \left(x - \dfrac{\pi}{2} \right) \tan x$

31. $\lim_{x \to 0} \dfrac{\cos(x + \frac{\pi}{2})}{\sin x}$

32. $\lim_{x \to 0} \dfrac{e^x - 1}{\sin x}$

33. $\lim_{x \to 0} \dfrac{\sin x - x \cos x}{x - \sin x}$

34. $\lim_{x \to \infty} \left(\dfrac{1}{x^2} - \csc^2 x \right)$

35. $\lim\limits_{x \to 1} \dfrac{e^x - e}{\ln x}$

36. $\lim\limits_{x \to 0} \dfrac{x^2}{1 - \cos x}$

37. $\lim\limits_{x \to \pi/2} \dfrac{\cos x}{\sin(2x)}$

38. $\lim\limits_{x \to 0} \dfrac{e^{2x} - 1}{x}$

39. $\lim\limits_{x \to \infty} e^{-x}(x^3 - x^2 + 9)$

40. $\lim\limits_{x \to 0} \dfrac{a^x - 1}{x}$ $(a > 0)$

41. $\lim\limits_{x \to 1} \dfrac{x^{1/3} - 1}{x^2 + 3x - 4}$

42. $\lim\limits_{x \to \pi/2} (\sec x - \tan x)$

43. $\lim\limits_{x \to \infty} x^{1/x}$

44. $\lim\limits_{x \to 1} (1 + \ln x)^{1/(x-1)}$

45. $\lim\limits_{t \to 0} (\sin t)(\ln t)$

46. $\lim\limits_{x \to 0} x^{\sin x}$

47. $\lim\limits_{x \to 0} x^{\sin x}$

48. $\lim\limits_{x \to \infty} \left(\dfrac{x}{x+1} \right)^x$

In Exercises 49–52, evaluate the limit using L'Hôpital's Rule if necessary.

49. $\lim\limits_{x \to 0} \dfrac{\sin^{-1} x}{x}$

50. $\lim\limits_{x \to 0} \dfrac{\tan^{-1} x}{\sin^{-1} x}$

51. $\lim\limits_{x \to 1} \dfrac{\tan^{-1} x - \frac{\pi}{4}}{\tan \frac{\pi}{4} x - 1}$

52. $\lim\limits_{x \to 0} \ln x \tan^{-1} x$

53. Evaluate $\lim\limits_{x \to \pi/2} \dfrac{\cos mx}{\cos nx}$, where m and n are nonzero whole numbers.

54. Evaluate $\lim\limits_{x \to 1} \dfrac{x^n - 1}{x^m - 1}$ for any numbers $n, m \neq 0$.

55. $\boxed{\text{GU}}$ Can L'Hôpital's Rule be applied to $\lim\limits_{x \to 0+} x^{\sin(1/x)}$? Does a graphical or numerical investigation suggest that the limit exists?

56. Sketch the graph of $y = \dfrac{\ln x}{x}$ by determining the increasing/decreasing behavior, concavity, and asymptotic behavior as in Section 4.5.

57. Let $f(x) = x^{1/x}$ in the domain $\{x : x > 0\}$.

(a) Calculate $\lim\limits_{x \to 0+} f(x)$ and $\lim\limits_{x \to \infty} f(x)$.

(b) Find the maximum value of $f(x)$ and determine the intervals on which $f(x)$ is increasing or decreasing.

(c) Use (a) and (b) to prove that $x^{1/x} = c$ has a unique solution if $0 < c \leq 1$ or $c = e^{1/e}$, two solutions if $1 < c < e^{1/e}$, and no solutions if $c > e^{1/e}$.

(d) $\boxed{\text{GU}}$ Plot the graph of $f(x)$. Explain how the graph confirms the conclusions in (c).

58. Determine if $f \ll g$ or $g \ll f$ (or neither) for the functions $f(x) = \log_{10} x$ and $g(x) = \ln x$.

59. Show that $(\ln x)^2 \ll \sqrt{x}$ and $(\ln x)^4 \ll x^{1/10}$.

60. Just as exponential functions are distinguished by their rapid rate of increase, the logarithm functions grow particularly slowly. Show that $\ln x \ll x^a$ for all $a > 0$.

61. Show that $(\ln x)^N \ll x^a$ for all N and all $a > 0$.

62. Determine whether $\sqrt{x} \ll e^{\sqrt{\ln x}}$ or $e^{\sqrt{\ln x}} \ll \sqrt{x}$. *Hint:* Do not use L'Hôpital's Rule. Instead, make a substitution $u = \ln x$.

63. Show that $\lim\limits_{x \to \infty} x^n e^{-x} = 0$ for all whole numbers $n > 0$.

64. Assumptions Matter   Let $f(x) = x(2 + \sin x)$ and $g(x) = x^2 + 1$.

(a) Show directly that $\lim\limits_{x \to \infty} f(x)/g(x) = 0$.

(b) Show that $\lim\limits_{x \to \infty} f(x) = \lim\limits_{x \to \infty} g(x) = \infty$, but $\lim\limits_{x \to \infty} f'(x)/g'(x)$ does not exist.

Do (a) and (b) contradict L'Hôpital's Rule? Explain.

65. Show that $\lim\limits_{t \to \infty} t^k e^{-t^2} = 0$ for all k. *Hint:* Compare with $\lim\limits_{t \to \infty} t^k e^{-t} = 0$.

In Exercises 66–68, let

$$f(x) = \begin{cases} e^{-1/x^2} & \text{for } x \neq 0 \\ 0 & \text{for } x = 0 \end{cases}$$

These exercises show that $f(x)$ has an unusual property: All of its higher derivatives at $x = 0$ exist and are equal to zero.

66. Show that $\lim\limits_{x \to 0} \dfrac{f(x)}{x^k} = 0$ for all k. *Hint:* Let $t = x^{-1}$ and apply the result of Exercise 65.

67. Show that $f'(0)$ exists and is equal to zero. Also, verify that $f''(0)$ exists and is equal to zero.

68. Show that for $k \geq 1$ and $x \neq 0$,

$$f^{(k)}(x) = \frac{P(x)e^{-1/x^2}}{x^r}$$

for some polynomial $P(x)$ and some exponent $r \geq 1$. Use the result of Exercise 66 to show that $f^{(k)}(0)$ exists and is equal to zero for all $k \geq 1$.

Further Insights and Challenges

69. Show that L'Hôpital's Rule applies to $\lim\limits_{x \to \infty} \dfrac{x + \cos x}{x - \cos x}$, but that it is of no help. Then evaluate the limit directly.

70. Use L'Hôpital's Rule to evaluate the following limit, assuming that

f is differentiable and f' is continuous:

$$\lim\limits_{x \to a} \frac{f(x) - f(a)}{x - a}$$

71. Resonance A spring oscillates with a natural frequency $\lambda/2\pi$. If we drive the spring with a sinusoidal force $C\sin(\omega t)$, where $\omega \neq \lambda$, then the spring oscillates according to

$$y(t) = \frac{C}{\lambda^2 - \omega^2}\left(\lambda\sin(\omega t) - \omega\sin(\lambda t)\right)$$

(a) Use L'Hôpital's Rule to determine $y(t)$ in the limit as $\omega \to \lambda$.

(b) Define $y_0(t) = \lim\limits_{\omega \to \lambda} y(t)$. Show that $y_0(t)$ ceases to be periodic and that its amplitude $|y_0(t)|$ tends to infinity as $t \to \infty$ (the system is said to be in resonance; eventually, the spring is stretched beyond its limits).

(c) $\boxed{CAS}$ Plot $y(t)$ for $\lambda = 1$ and $\omega = 0.5$, 0.8, 0.9, 0.99, and 0.999. How do the graphs change? Do the graphs confirm your conclusion in (b)?

72. We expended a lot of effort to evaluate $\lim\limits_{x \to 0} \dfrac{\sin x}{x}$ in Chapter 2. Show that we could have evaluated it easily using L'Hôpital's Rule. Then explain why this method would involve *circular reasoning*.

73. Suppose that f and g are polynomials such that $f(a) = g(a) = 0$. In this case, it is a fact from algebra that $f(x) = (x - a)f_1(x)$ and $g(x) = (x - a)g_1(x)$ for some polynomials f_1 and g_1. Use this to verify L'Hôpital's Rule directly for f and g.

74. In this exercise, we verify the following limit formula for e:

$$e = \lim_{x \to 0} (1 + x)^{1/x} \qquad \boxed{1}$$

(a) Use L'Hôpital's Rule to prove that $\lim\limits_{x \to 0} \dfrac{\ln(1 + x)}{x} = 1$.

(b) Use (a) to prove that $\lim\limits_{x \to 0} \ln\left((1 + x)^{1/x}\right) = 1$.

(c) Use (b) to verify Eq. (1).

(d) Find a value of x such that $|(1 + x)^{1/x} - e| \leq 0.001$.

75. Patience Required Use L'Hôpital's Rule to evaluate and check your answers numerically:

(a) $\lim\limits_{x \to 0+} \left(\dfrac{\sin x}{x}\right)^{1/x^2}$ **(b)** $\lim\limits_{x \to 0} \left(\dfrac{1}{\sin^2 x} - \dfrac{1}{x^2}\right)$

4.8 Newton's Method

◄·· *REMINDER A "zero" or "root" of a function $f(x)$ is a solution to the equation $f(x) = 0$.*

Newton's Method is a procedure for finding numerical approximations to zeros of functions. Numerical approximations are important because it is often impossible to find the zeros exactly. For example, the polynomial $f(x) = x^5 - x - 1$ has one real root c (see Figure 1), but we can prove, using an advanced branch of mathematics called *Galois Theory*, that there is no algebraic formula for this root. Newton's Method shows that $c \approx 1.1673$, and with enough computation, we can compute c to any desired degree of accuracy.

In Newton's Method, we begin by choosing a number x_0, which we believe is close to a root of $f(x)$. This starting value x_0 is called the **initial guess**. Newton's Method then produces a sequence $x_0, x_1, x_2, \ldots$ of successive approximations that, in favorable situations, converge to a root.

Figure 2 illustrates the procedure. Given an initial guess x_0, we draw the tangent line to the graph at $(x_0, f(x_0))$. The approximation x_1 is defined as the x-coordinate of the

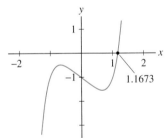

FIGURE 1 Graph of $y = x^5 - x - 1$. The value 1.1673 is a good numerical approximation to the root.

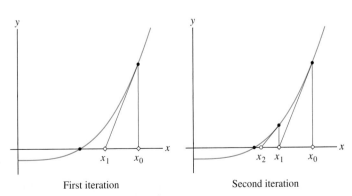

First iteration   Second iteration

FIGURE 2 The sequence produced by iteration converges to a root.

point where the tangent line intersects the x-axis. To produce the second approximation x_2 (also called the second iterate), we apply this procedure to x_1.

Let's derive a formula for x_1. The equation of the tangent line at $(x_0, f(x_0))$ is

$$y = f(x_0) + f'(x_0)(x - x_0)$$

The tangent line crosses the x-axis at x_1, where

$$y = f(x_0) + f'(x_0)(x_1 - x_0) = 0$$

If $f'(x_0) \neq 0$, we may solve for x_1 to obtain $x_1 - x_0 = -f(x_0)/f'(x_0)$, or

$$x_1 = x_0 - \frac{f(x_0)}{f'(x_0)}$$

The second iterate x_2 is obtained by applying this formula to x_1 instead of x_0:

$$x_2 = x_1 - \frac{f(x_1)}{f'(x_1)}$$

and so on. Notice in Figure 2 that x_1 is closer to the root than x_0 and x_2 is closer still. This is typical: The successive approximations usually converge to the actual root. However, there are cases where Newton's Method fails (see Figure 4).

Newton's Method is an example of an iterative procedure. To "iterate" means to repeat, and in Newton's Method we use (1) repeatedly to produce the sequence of approximations.

Newton's Method   To find a numerical approximation to a root of $f(x) = 0$:

Step 1. Choose initial guess x_0 (close to the desired root if possible).
Step 2. Generate successive approximations $x_1, x_2, \dots$ where

$$x_{n+1} = x_n - \frac{f(x_n)}{f'(x_n)} \qquad \boxed{1}$$

■ **EXAMPLE 1** Approximating $\sqrt{5}$ Calculate the first three approximations x_1, x_2, x_3 to a root of $f(x) = x^2 - 5$ using the initial guess $x_0 = 2$.

Solution We have $f'(x) = 2x$. Therefore:

$$x_1 = x_0 - \frac{f(x_0)}{f'(x_0)} = x_0 - \frac{x_0^2 - 5}{2x_0}$$

We compute the successive approximations as follows:

$$x_1 = x_0 - \frac{f(x_0)}{f'(x_0)} = 2 - \frac{2^2 - 5}{2 \cdot 2} \qquad\qquad = 2.25$$

$$x_2 = x_1 - \frac{f(x_1)}{f'(x_1)} = 2.25 - \frac{2.25^2 - 5}{2 \cdot 2.25} \qquad \approx 2.23611$$

$$x_3 = x_2 - \frac{f(x_2)}{f'(x_2)} = 2.23611 - \frac{2.23611^2 - 5}{2 \cdot 2.23611} \qquad \approx \mathbf{2.23606797789}$$

This sequence provides successive approximations to the root

$$\sqrt{5} = \mathbf{2.23606797}7499789696\dots$$

Observe that x_3 is accurate to within an error of less than 10^{-9}. This is impressive accuracy for just three iterations of Newton's Method.   ■

How Many Iterations Are Required?

How many iterations of Newton's Method are required to approximate a root to within a given accuracy? There is no definitive answer, but in many cases, the number of correct decimal places roughly doubles at each step. In practice, it is usually safe to assume that if x_n and x_{n+1} agree to m decimal places, then the approximation x_n is correct to these m places.

■ **EXAMPLE 2** GU Let c be the smallest positive solution of $\sin 3x = \cos x$.

(a) Use a computer-generated graph to choose an initial guess x_0 for c.

(b) Use Newton's Method to approximate c to within an error of at most 10^{-6}.

Solution

(a) A solution of $\sin 3x = \cos x$ is a zero of the function $f(x) = \sin 3x - \cos x$. Figure 3 shows that the smallest zero is approximately halfway between 0 and $\frac{\pi}{4}$. Because $\frac{\pi}{4} \approx$ 0.785, a good initial guess is $x_0 = 0.4$.

(b) Since $f'(x) = 3 \cos 3x + \sin x$, Eq. (1) yields the formula

$$x_{n+1} = x_n - \frac{\sin 3x_n - \cos x_n}{3 \cos 3x_n + \sin x_n}$$

With $x_0 = 0.4$ as the initial guess, the first four iterates are

$$x_1 \approx \mathbf{0.3925}647447$$

$$x_2 \approx \mathbf{0.3926990}382$$

$$x_3 \approx \mathbf{0.3926990}817$$

$$x_4 \approx \mathbf{0.3926990}816$$

Stopping here, we can be fairly confident that 0.392699 approximates the smallest positive root c with an error of less than 10^{-6}. As a further test, we check that $\sin(3x_4) \approx \cos(x_4)$:

$$\sin(3x_4) \approx \sin(3(0.392699)) \approx 0.92387944$$

$$\cos(x_4) \approx \cos(0.392699) \approx 0.92387956$$ ■

Which Root Does Newton's Method Compute?

Sometimes, Newton's Method computes no root at all. In Figure 4, the iterates diverge to infinity. However, Newton's Method usually converges quickly and if a particular choice of x_0 does not lead to a root, the best strategy is to try a different initial guess, consulting a graph if possible. If $f(x) = 0$ has more than one root, different initial guesses x_0 may lead to different roots.

■ **EXAMPLE 3** Figure 5 shows that $f(x) = x^4 - 6x^2 + x + 5$ has four real roots.

(a) Show that with $x_0 = 0$, Newton's Method converges to the root near -2.

(b) Show that with $x_0 = -1$, Newton's Method converges to the root near -1.

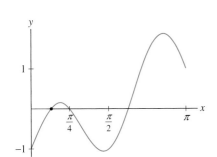

FIGURE 3 Graph of $f(x) = \sin 3x - \cos x$.

There is no single "correct" initial guess. In Example 2, we chose $x_0 = 0.4$ but another possible choice is $x_0 = 0$, leading to the sequence

$$x_1 \approx 0.3333333333$$

$$x_2 \approx 0.3864547725$$

$$x_3 \approx 0.3926082513$$

$$x_4 \approx 0.3926990816$$

You can check, however, that $x_0 = 1$ yields a sequence converging to $\frac{\pi}{4}$, which is the second positive solution of $\sin 3x = \cos x$.

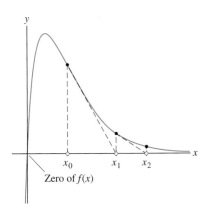

FIGURE 4 Function has only one zero but the sequence of Newton iterates goes off to infinity.

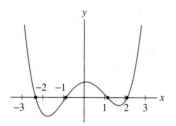

FIGURE 5 Graph of
$f(x) = x^4 - 6x^2 + x + 5$.

Solution We have $f'(x) = 4x^3 - 12x + 1$ and

$$x_{n+1} = x_n - \frac{x_n^4 - 6x_n^2 + x_n + 5}{4x_n^3 - 12x_n + 1} = \frac{3x_n^4 - 6x_n^2 - 5}{4x_n^3 - 12x_n + 1}$$

(a) On the basis of Table 1, we can be confident that when $x_0 = 0$, Newton's Method converges to a root near -2.3. Notice in Figure 5 that this is not the closest root to x_0.

(b) Table 2 suggests that with $x_0 = -1$, Newton's Method converges to the root near -0.9. ∎

TABLE 1	
x_0	0
x_1	-5
x_2	-3.9179954
x_3	-3.1669480
x_4	-2.6871270
x_5	-2.4363303
x_6	-2.3572979
x_7	-2.3495000

TABLE 2	
x_0	-1
x_1	-0.8888888888
x_2	-0.8882866140
x_3	-0.88828656234358
x_4	-0.888286562343575

4.8 SUMMARY

- Newton's Method: To find a sequence of numerical approximations to a solution of $f(x) = 0$, begin with an initial guess x_0. Then construct the sequence $x_0, x_1, x_2, \dots$ using the formula

$$x_{n+1} = x_n - \frac{f(x_n)}{f'(x_n)}$$

In favorable cases, the sequence converges to a solution. The initial guess x_0 should be chosen as close as possible to a root, possibly by referring to a graph.
- In most cases, the sequence $x_0, x_1, x_2, \dots$ converges rapidly to a solution. If x_n and x_{n+1} agree to m decimal places, you may be reasonably confident that x_n agrees with a true solution to m decimal places.

4.8 EXERCISES

Preliminary Questions

1. How many iterations of Newton's Method are required to compute a root if $f(x)$ is a linear function?

2. What happens in Newton's Method if your initial guess happens to be a zero of f?

3. What happens in Newton's Method if your initial guess happens to be a local min or max of f?

4. Is the following a reasonable description of Newton's Method: "A root of the equation of the tangent line to $f(x)$ is used as an approximation to a root of $f(x)$ itself"? Explain.

Exercises

In Exercises 1–4, use Newton's Method with the given function and initial value x_0 to calculate x_1, x_2, x_3.

1. $f(x) = x^2 - 2$, $x_0 = 1$

2. $f(x) = x^2 - 7$, $x_0 = 2.5$

3. $f(x) = x^3 - 5$, $x_0 = 1.6$

4. $f(x) = \cos x - x$, $x_0 = 0.8$

5. Use Figure 6 to choose an initial guess x_0 to the unique real root of $x^3 + 2x + 5 = 0$. Then compute the first three iterates of Newton's Method.

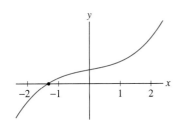

FIGURE 6 Graph of $y = x^3 + 2x + 5$.

6. Use Newton's Method to find a solution to $\sin x = \cos 2x$ in the interval $[0, \frac{\pi}{2}]$ to three decimal places. Then guess the exact solution and compare with your approximation.

7. Use Newton's Method to find the two solutions of $e^x = 5x$ to three decimal places (Figure 7).

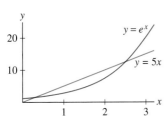

FIGURE 7 Graphs of e^x and $5x$.

8. Use Newton's Method to approximate the positive solution to the equation $\ln(x + 4) = x$ to three decimal places.

In Exercises 9–12, use Newton's Method to approximate the root to three decimal places and compare with the value obtained from a calculator.

9. $\sqrt{10}$

10. $7^{1/4}$

11. $5^{1/3}$

12. $2^{-1/2}$

13. Use Newton's Method to approximate the largest positive root of $f(x) = x^4 - 6x^2 + x + 5$ to within an error of at most 10^{-4}. Refer to Figure 5.

14. GU Sketch the graph of $f(x) = x^3 - 4x + 1$ and use Newton's Method to approximate the largest positive root to within an error of at most 10^{-3}.

15. GU Use a graphing calculator to choose an initial guess for the unique positive root of $x^4 + x^2 - 2x - 1 = 0$. Calculate the first three iterates of Newton's Method.

16. The first positive solution of $\sin x = 0$ is $x = \pi$. Use Newton's Method to calculate π to four decimal places.

17. GU Estimate the smallest positive solution of $\dfrac{\sin \theta}{\theta} = 0.9$ to three decimal places. Use a graphing calculator to choose the initial guess.

18. In 1535, the mathematician Antonio Fior challenged his rival Niccolo Tartaglia to solve this problem: A tree stands 12 *braccia* high; it is broken into two parts at such a point that the height of the part left standing is the cube root of the length of the part cut away. What is the height of the part left standing? Show that this is equivalent to solving $x^3 + x = 12$ and find the height to three decimal places. Tartaglia, who had discovered the secrets of cubic equations, was able to determine the exact answer:

$$x = \left(\sqrt[3]{\sqrt{2{,}919} + 54} - \sqrt[3]{\sqrt{2{,}919} - 54} \right) \bigg/ \sqrt[3]{9}$$

19. Let x_1, x_2 be the estimates to a root obtained by applying Newton's Method with $x_0 = 1$ to the function graphed in Figure 8. Estimate the numerical values of x_1 and x_2, and draw the tangent lines used to obtain them.

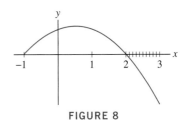

FIGURE 8

20. Find the coordinates to two decimal places of the point P in Figure 9 where the tangent line to $y = \cos x$ passes through the origin.

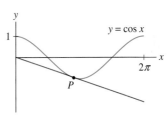

FIGURE 9

21. Find the x-coordinate to two decimal places of the first point in the region $x > 0$ where $y = x$ intersects $y = \tan x$ (draw a graph).

Newton's Method is often used to determine interest rates in financial calculations. In Exercises 22–24, r denotes a yearly interest rate expressed as a decimal (rather than as a percent).

22. If P dollars are deposited every month in an account earning interest at the yearly rate r, then the value S of the account after N years is

$$S = P \left(\frac{b^{12N+1} - b}{b - 1} \right) \qquad \text{where } b = 1 + \frac{r}{12}$$

You have decided to deposit $P = 100$ dollars per month.

(a) What is the value after 5 years if the yearly interest rate is $r = 0.07$ (i.e., 7%)?

(b) Show that to save \$10,000 after 5 years, you must earn interest at a rate r determined by the equation $b^{61} - 101b + 100 = 0$. Use Newton's Method to solve for b (note that $b = 1$ is a root, but you want the root satisfying $b > 1$). Then find r.

23. If you borrow L dollars for N years at a yearly interest rate r, your monthly payment of P dollars is calculated using the equation

$$L = P\left(\frac{1 - b^{-12N}}{b - 1}\right) \qquad \text{where } b = 1 + \frac{r}{12}$$

(a) What is the monthly payment if $L = \$5,000$, $N = 3$, and $r = 0.08$ (8%)?

(b) You are offered a loan of $L = \$5,000$ to be paid back over 3 years with monthly payments of $P = \$200$. Use Newton's Method to compute b and find the implied interest rate r of this loan. *Hint:* Show that $(L/P)b^{12N+1} - (1 + L/P)b^{12N} + 1 = 0$.

24. If you deposit P dollars in a retirement fund every year for N years with the intention of then withdrawing Q dollars per year for M years, you must earn interest at a rate r satisfying $P(b^N - 1) = Q(1 - b^{-M})$, where $b = 1 + r$. Assume that \$2,000 is deposited each year for 30 years and the goal is to withdraw \$8,000 per year for 20 years. Use Newton's Method to compute b and then find r.

25. Kepler's Problem   Although planetary motion is beautifully described by Kepler's three laws (see Section 13.6), there is no simple formula for the position of a planet P along its elliptical orbit as a function of time. Kepler developed a method for locating P at time t by drawing the auxiliary dashed circle in Figure 10 and introducing the angle θ (note that P determines θ, which is the central angle of the point B on the circle). Let $a = OA$ and $e = OS/OA$ (the eccentricity of the orbit).

(a) Show that sector BSA has area $(a^2/2)(\theta - e \sin \theta)$.

(b) It follows from Kepler's Second Law that the area of sector BSA is proportional to the time t elapsed since the planet passed point A. More precisely, since the circle has area πa^2, BSA has area $(\pi a^2)(t/T)$, where T is the period of the orbit. Deduce that

$$\frac{2\pi t}{T} = \theta - e \sin \theta$$

(c) The eccentricity of Mercury's orbit is approximately $e = 0.2$. Use Newton's Method to find θ after a quarter of Mercury's year has elapsed ($t = T/4$). Convert θ to degrees. Has Mercury covered more than a quarter of its orbit at $t = T/4$?

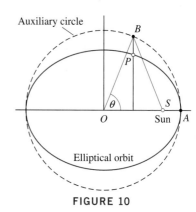

FIGURE 10

26. What happens when you apply Newton's Method to find a zero of $f(x) = x^{1/3}$? Note that $x = 0$ is the only zero.

27. What happens when you apply Newton's Method to the equation $x^3 - 20x = 0$ with the unlucky initial guess $x_0 = 2$?

Further Insights and Challenges

28. Let c be a positive number and let $f(x) = x^{-1} - c$.
(a) Show that $x - (f(x)/f'(x)) = 2x - cx^2$. Thus, Newton's Method provides a way of computing reciprocals without performing division.
(b) Calculate the first three iterates of Newton's Method with $c = 10.324$ and the two initial guesses $x_0 = 0.1$ and $x_0 = 0.5$.
(c) Explain graphically why $x_0 = 0.5$ does not yield a sequence of approximations to the reciprocal $1/10.324$.

29. The roots of $f(x) = \frac{1}{3}x^3 - 4x + 1$ to three decimal places are -3.583, 0.251, and 3.332 (Figure 11). Determine the root to which

Newton's Method converges for the initial choices $x_0 = 1.85$, 1.7, and 1.55. The answer shows that a small change in x_0 can have a significant effect on the outcome of Newton's Method.

In Exercises 30–31, consider a metal rod of length L inches that is fastened at both ends. If you cut the rod and weld on an additional m inches of rod, leaving the ends fixed, the rod will bow up into a circular arc of radius R (unknown), as indicated in Figure 12.

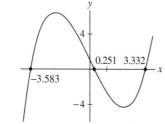

FIGURE 11 Graph of $f(x) = \frac{1}{3}x^3 - 4x + 1$.

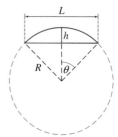

FIGURE 12 The bold circular arc has length $L + m$.

30. Let h be the maximum vertical displacement of the rod.

(a) Show that $L = 2R \sin \theta$ and conclude that

$$h = \frac{L(1 - \cos \theta)}{2 \sin \theta}$$

(b) Show $L + m = 2R\theta$ and then prove

$$\frac{\sin \theta}{\theta} = \frac{L}{L + m} \qquad \boxed{2}$$

31. Let $L = 3$ and $m = 1$. Apply Newton's Method to Eq. (2) to estimate θ and use this to estimate h.

32. Quadratic Convergence to Square Roots   Let $f(x) = x^2 - c$ and let $e_n = x_n - \sqrt{c}$ be the error in x_n.

(a) Show that $x_{n+1} = \frac{1}{2}(x_n + c/x_n)$ and that $e_{n+1} = e_n^2/2x_n$.

(b) Show that if $x_0 > \sqrt{c}$, then $x_n > \sqrt{c}$ for all n. Explain this graphically.

(c) Assuming that $x_0 > \sqrt{c}$, show that $e_{n+1} \le e_n^2/2\sqrt{c}$.

*In Exercises 33–35, a flexible chain of length L is suspended between two poles of equal height separated by a distance $2M$ (Figure 13). By Newton's laws, the chain describes a curve (called a **catenary**) with equation $y = a \cosh\left(\dfrac{x}{a}\right) + C$. The constant C is arbitrary and a is the number such that $L = 2a \sinh\left(\dfrac{M}{a}\right)$. The sag s is the vertical distance from the highest to the lowest point on the chain.*

33. Suppose that $L = 120$ and $M = 50$.

(a) Use Newton's Method to find a value of a (to two decimal places) satisfying $L = 2a \sinh(M/a)$.

(b) Show that the sag is given by $s = a \cosh(M/a) - a$. Compute s.

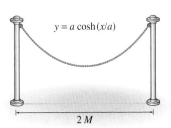

FIGURE 13 Chain hanging between two poles describes the curve $y = a \cosh(x/a)$.

34. Let M be a fixed constant. Show that the sag is given by $s = a \cosh\left(\dfrac{M}{a}\right) - a$.

(a) Calculate $\dfrac{ds}{da}$.

(b) Calculate $\dfrac{da}{dL}$ by implicit differentiation using the relation $L = 2a \sinh\left(\dfrac{M}{a}\right)$.

(c) Use (a) and (b) and the Chain Rule to show that

$$\frac{ds}{dL} = \frac{ds}{da}\frac{da}{dL} = \frac{\cosh(M/a) - (M/a)\sinh(M/a) - 1}{2\sinh(M/a) - (2M/a)\cosh(M/a)} \qquad \boxed{3}$$

35. Suppose that $L = 160$ and $M = 50$.

(a) Use Newton's Method to find a value of a (to two decimal places) satisfying $L = 2a \sinh(M/a)$.

(b) Use Eq. (3) and the Linear Approximation to estimate the increase in cable sag s if L is increased from $L = 160$ to $L = 161$ and from $L = 160$ to $L = 165$.

(c) $\mathcal{CAS}$ Compute $s(161) - s(160)$ and $s(165) - s(160)$ directly and compare with your estimates in (a).

4.9 Antiderivatives

In addition to finding derivatives, there is an important "inverse" problem, namely *given the derivative, find the function itself*. In physics, for example, we may want to compute the position $s(t)$ of an object given its velocity $v(t)$. Since $s'(t) = v(t)$, this amounts to finding a function whose derivative is $v(t)$. A function $F(x)$ whose derivative is $f(x)$ is called an antiderivative of $f(x)$.

> **DEFINITION Antiderivatives**   A function $F(x)$ is an **antiderivative** of $f(x)$ on (a, b) if $F'(x) = f(x)$ for all $x \in (a, b)$.

For example, $F(x) = \frac{1}{3}x^3$ is an antiderivative of $f(x) = x^2$ because

$$F'(x) = \frac{d}{dx}\left(\frac{1}{3}x^3\right) = x^2 = f(x)$$

Similarly, $F(x) = -\cos x$ is an antiderivative of $f(x) = \sin x$ because

$$F'(x) = \frac{d}{dx}(-\cos x) = \sin x = f(x)$$

One of the first characteristics to note about antiderivatives is that they are not unique. We are free to add a constant C: If $F'(x) = f(x)$, then $(F(x) + C)' = f(x)$ because the derivative of the constant is zero. Thus, each of the following is an antiderivative of x^2:

$$\frac{1}{3}x^3, \qquad \frac{1}{3}x^3 + 5, \qquad \frac{1}{3}x^3 - 4$$

Are there any antiderivatives of $f(x)$ other than those obtained by adding a constant to a given antiderivative $F(x)$? According to the following theorem, the answer is no if $f(x)$ is defined on an interval (a, b).

THEOREM 1   The General Antiderivative   Let $F(x)$ be an antiderivative of $f(x)$ on (a, b). Then every other antiderivative on (a, b) is of the form $F(x) + C$ for some constant C.

Proof   Let $F(x)$ and $G(x)$ be antiderivatives of $f(x)$ and set $H(x) = G(x) - F(x)$. Then $H'(x) = G'(x) - F'(x) = f(x) - f(x) = 0$. By the Corollary in Section 4.3, $H(x)$ must be a constant, say, $H(x) = C$, and this gives $G(x) = F(x) + C$.   ∎

GRAPHICAL INSIGHT   The graph of $F(x) + C$ is obtained by shifting the graph of $F(x)$ vertically by C units. Since vertical shifting moves the tangent lines without changing their slopes, it makes sense that all of the functions $F(x) + C$ have the same derivative (Figure 1). Conversely, Theorem 1 tells us that if two graphs have parallel tangent lines, then one graph is obtained from the other by a vertical shift.

We often describe the *general* antiderivative of a function in terms of an arbitrary constant C, as in the following example.

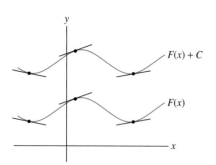

FIGURE 1   The tangent lines to the graphs of $y = F(x)$ and $y = F(x) + C$ are parallel.

■ **EXAMPLE 1**   Find two antiderivatives of $f(x) = \cos x$. Then determine the general antiderivative.

Solution   The functions $F(x) = \sin x$ and $G(x) = \sin x + 2$ are both antiderivatives of $f(x)$. The general antiderivative is $F(x) = \sin x + C$, where C is any constant.   ■

The process of finding an antiderivative is called **integration**. We will see why in Chapter 5, when we discuss the connection between antiderivatives and areas under curves given by the Fundamental Theorem of Calculus. Anticipating this result, we now begin using the integral sign $\int$, the standard notation for antiderivatives.

The terms "antiderivative" and "indefinite integral" are used interchangeably. In some textbooks, an antiderivative is called a "primitive function."

NOTATION   Indefinite Integral   The notation

$$\int f(x)\,dx = F(x) + C \quad \text{means that} \quad F'(x) = f(x)$$

We say that $F(x) + C$ is the general antiderivative or **indefinite integral** of $f(x)$.

The function $f(x)$ appearing in the integral sign is called the **integrand**. The symbol "dx," called a *differential*, is part of the integral notation. It indicates the independent variable, but we do not assign it any other meaning. The constant C is called the *constant of integration*.

Some indefinite integrals can be evaluated by reversing the familiar derivative formulas. For example, we obtain the indefinite integral of x^n by reversing the Power Rule for derivatives.

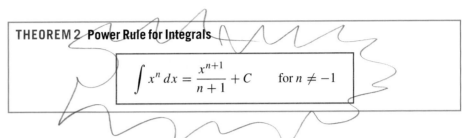

> **THEOREM 2 Power Rule for Integrals**
>
> $$\int x^n \, dx = \frac{x^{n+1}}{n+1} + C \qquad \text{for } n \neq -1$$

Proof We verify the Power Rule for Integrals by taking the derivative

$$\frac{d}{dx}\left(\frac{x^{n+1}}{n+1} + C\right) = \frac{1}{n+1}\left((n+1)x^n\right) = x^n \qquad\blacksquare$$

In words, the Power Rule for Integrals says that to integrate a power of x, "add one to the exponent and then divide by the new exponent." Here are some examples:

$$\int x^5 \, dx = \frac{1}{6}x^6 + C, \qquad \int x^{-9} \, dx = -\frac{1}{8}x^{-8} + C, \qquad \int x^{3/5} \, dx = \frac{5}{8}x^{8/5} + C$$

The Power Rule is not valid for $n = -1$. In fact, for $n = -1$, we obtain the meaningless result

$$\int x^{-1} \, dx = \frac{x^{n+1}}{n+1} + C = \frac{x^0}{0} + C \qquad \text{(meaningless)}$$

Recall, however, that the derivative of the natural logarithm is $\dfrac{d}{dx} \ln x = x^{-1}$. This shows that $\ln x$ is an antiderivative of $y = x^{-1}$, and thus, for $n = -1$, instead of the Power Rule we have

$$\int \frac{dx}{x} = \ln x + C$$

This formula is valid for $x > 0$, where $f(x) = \ln x$ is defined. We would like to have an antiderivative of $y = x^{-1}$ on its full domain, namely on $\{x : x \neq 0\}$. To achieve this end, we extend $f(x)$ to an even function by setting $f(x) = \ln|x|$ (Figure 2). Then $f(x) = f(-x)$, and by the Chain Rule, $f'(x) = -f'(-x)$. For $x < 0$, we obtain

$$\frac{d}{dx} \ln|x| = f'(x) = -f'(-x) = -\frac{1}{-x} = \frac{1}{x}$$

This proves that $f'(x) = x^{-1}$ for all $x \neq 0$.

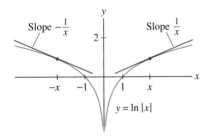

FIGURE 2

THEOREM 3 Antiderivative of $y = \dfrac{1}{x}$ The function $f(x) = \ln|x|$ is an antiderivative of $y = \dfrac{1}{x}$ in the domain $\{x : x \neq 0\}$, that is,

$$\int \frac{dx}{x} = \ln|x| + C$$

[1]

The indefinite integral obeys the usual linearity rules that allow us to integrate "term by term." This follows directly from the linearity rules for the derivative (see Exercise 81.)

THEOREM 4 Linearity of the Indefinite Integral

- **Sum Rule:** $\displaystyle \int (f(x) + g(x))\,dx = \int f(x)\,dx + \int g(x)\,dx$

- **Multiples Rule:** $\displaystyle \int cf(x)\,dx = c \int f(x)\,dx$

┃ EXAMPLE 2 Evaluate $\int (3x^4 - 5x^{2/3} + x^{-3})\,dx$.

Solution We integrate term by term and use the Power Rule:

$$\int (3x^4 - 5x^{2/3} + x^{-3})\,dx = \int 3x^4\,dx - \int 5x^{2/3}\,dx + \int x^{-3}\,dx \quad \text{(Sum Rule)}$$

$$= 3\int x^4\,dx - 5\int x^{2/3}\,dx + \int x^{-3}\,dx \quad \text{(Multiples Rule)}$$

$$= 3\frac{x^5}{5} - 5\frac{x^{5/3}}{5/3} + \frac{x^{-2}}{-2} + C \quad \text{(Power Rule)}$$

$$= \frac{3}{5}x^5 - 3x^{5/3} - \frac{1}{2}x^{-2} + C$$

When we break up an indefinite integral into a sum of several integrals as in Example 2, it is not necessary to include separate constants of integration for each integral.

To check the answer, we verify that the derivative is equal to the integrand:

$$\frac{d}{dx}\left(\frac{3}{5}x^5 - 3x^{5/3} - \frac{1}{2}x^{-2} + C\right) = 3x^4 - 5x^{2/3} + x^{-3}$$

■

┃ EXAMPLE 3 Evaluate $\displaystyle \int \left(\frac{5}{x} - 3x^{-10}\right)\,dx$.

Solution Apply Eq. (1) and the Power Rule:

$$\int \left(\frac{5}{x} - 3x^{-10}\right)\,dx = 5\int \frac{dx}{x} - 3\int x^{-10}\,dx$$

$$= 5\ln|x| - 3\left(\frac{x^{-9}}{-9}\right) + C = 5\ln|x| + \frac{1}{3}x^{-9} + C$$

■

The differentiation formulas for the trigonometric functions give us the following integration formulas. Each formula can be checked by differentiation.

Basic Trigonometric Integrals

$$\int \sin x \, dx = -\cos x + C \qquad \int \cos x \, dx = \sin x + C$$

$$\int \sec^2 x \, dx = \tan x + C \qquad \int \csc^2 x \, dx = -\cot x + C$$

$$\int \sec x \tan x \, dx = \sec x + C \qquad \int \csc x \cot x \, dx = -\csc x + C$$

Similarly, for any constants b and k with $k \neq 0$, the formulas

$$\frac{d}{dx} \sin(kx + b) = k \cos(kx + b), \qquad \frac{d}{dx} \cos(kx + b) = -k \sin(kx + b)$$

translate to the following indefinite integral formulas:

$$\int \cos(kx + b) \, dx = \frac{1}{k} \sin(kx + b) + C$$

$$\int \sin(kx + b) \, dx = -\frac{1}{k} \cos(kx + b) + C$$

■ **EXAMPLE 4** Evaluate $\int \left(\sin(2t - 9) + 20 \cos 3t \right) dt$.

Solution

$$\int \left(\sin(2t - 9) + 20 \cos 3t \right) dt = \int \sin(2t - 9) \, dt + 20 \int \cos 3t \, dt$$

$$= -\frac{1}{2} \cos(2t - 9) + \frac{20}{3} \sin 3t + C \qquad ■$$

Integrals Involving e^x

The formula $(e^x)' = e^x$ says that e^x is its own derivative. But this means that e^x is also *its own antiderivative*. In other words,

$$\int e^x \, dx = e^x + C$$

More generally, for any constants b and $k \neq 0$,

$$\int e^{kx+b} \, dx = \frac{1}{k} e^{kx+b} + C$$

■ **EXAMPLE 5** Evaluate **(a)** $\int (3e^x - 4) \, dx$ and **(b)** $\int 12e^{7-3x} \, dx$.

Solution

(a) $\displaystyle\int (3e^x - 4)\,dx = 3\int e^x\,dx - \int 4\,dx = 3e^x - 4x + C$

(b) $\displaystyle\int 12e^{7-3x}\,dx = 12\int e^{7-3x}\,dx = 12\left(\frac{1}{-3}e^{7-3x}\right) = -4e^{7-3x} + C$ ∎

Initial Conditions

We can think of an antiderivative as a solution to the **differential equation**

$$\frac{dy}{dx} = f(x) \qquad \boxed{2}$$

In this equation, the unknown is the function $y = y(x)$ and a solution is a function $y = F(x)$ whose derivative is $f(x)$, that is, $F(x)$ is an antiderivative of $f(x)$. The anti-derivative is not unique, but we can specify a particular solution by imposing an **initial condition**, that is, by requiring that the solution satisfy a condition $y(x_0) = y_0$ for some fixed x_0 and y_0.

An initial condition is like the y-intercept of a line, which specifies that line among all lines with the same slope. Similarly, the graphs of the antiderivatives of $f(x)$ are all parallel (Figure 1), and the initial condition specifies one of them.

■ **EXAMPLE 6** Solve $\dfrac{dy}{dx} = 4x^7$ subject to the initial condition $y(0) = 4$.

Solution Since y is an antiderivative of $4x^7$, the general solution is

$$y = \int 4x^7\,dx = \frac{1}{2}x^8 + C$$

We choose C so that the initial condition $y(0) = 4$ is satisfied. Since $y(0) = 0 + C = 4$, we have $C = 4$ and the particular solution is $y = \frac{1}{2}x^8 + 4$. ∎

■ **EXAMPLE 7** Solve $\dfrac{dy}{dt} = \sin(\pi t)$ with initial condition $y(2) = 2$.

Solution The general solution is

$$y(t) = \int \sin(\pi t)\,dt = -\frac{1}{\pi}\cos(\pi t) + C$$

We solve for C by evaluating at $t = 2$:

$$y(2) = -\frac{1}{\pi}\cos(2\pi) + C = 2 \quad \Rightarrow \quad C = 2 + \frac{1}{\pi}$$

Therefore, the solution is $y(t) = -\dfrac{1}{\pi}\cos(\pi t) + 2 + \dfrac{1}{\pi}$. ∎

■ **EXAMPLE 8** At time $t = 0$, a car traveling with velocity 96 ft/s begins to slow down with constant deceleration $a = -12$ ft/s^2. Find the velocity $v(t)$ at time t and the distance traveled before the car comes to a halt.

Solution The derivative of velocity is acceleration, so $v'(t) = a = -12$ and (writing C_1 for the constant of integration)

$$v(t) = \int a\,dt = \int (-12)\,dt = -12t + C_1$$

The initial condition is $v(0) = C_1 = 96$, so $v(t) = -12t + 96$. Next we compute the car's position $s(t)$. Since $s'(t) = v(t)$,

$$s(t) = \int v(t)\,dt = \int (-12t + 96)\,dt = -6t^2 + 96t + C_2$$

We use the initial condition $s(0) = 0$ since we want to compute the distance traveled from time $t = 0$. This gives $C_2 = 0$ and $s(t) = -6t^2 + 96t$.

Finally, to compute how far the car traveled, we observe that the car came to a halt when its velocity was zero, so we solve:

$$v(t) = -12t + 96 = 0 \quad \Rightarrow \quad t = \frac{96}{12} = 8 \text{ s}$$

Therefore, the car came to a halt after 8 s and traveled $s(8) = -6(8^2) + 96(8) = 384$ ft.
∎

■ **EXAMPLE 9** Solve $y' = 10e^{-2x}$ with initial condition $y(0) = 12$.

Solution The general solution is

$$y = \int 10e^{-2x}\,dx = 10 \int e^{-2x}\,dx = 10\left(\frac{e^{-2x}}{-2}\right) + C = -5e^{-2x} + C$$

To determine C, we use the initial condition:

$$y(0) = -5e^0 + C = -5 + C = 12$$

Therefore, $C = 17$ and $y = -5e^{-2x} + 17$.
∎

4.9 SUMMARY

- $F(x)$ is called an *antiderivative* of $f(x)$ if $F'(x) = f(x)$.
- Any two antiderivatives of $f(x)$ on an interval (a, b) differ by a constant.
- The general antiderivative is denoted by the indefinite integral

$$\int f(x)\,dx = F(x) + C$$

- Integration formulas:

$$\int x^n\,dx = \frac{x^{n+1}}{n+1} + C \qquad (n \neq -1)$$

$$\int \sin(kx + b)\,dx = -\frac{1}{k}\cos(kx + b) + C \qquad (k \neq 0)$$

$$\int \cos(kx + b)\,dx = \frac{1}{k}\sin(kx + b) + C \qquad (k \neq 0)$$

$$\int e^{kx+b}\,dx = \frac{1}{k}e^{kx+b} + C \qquad (k \neq 0)$$

$$\int \frac{dx}{x} = \ln|x| + C$$

- To solve a differential equation $\dfrac{dy}{dx} = f(x)$ with initial condition $y(x_0) = y_0$, first find the general antiderivative $y = F(x) + C$. Then determine C using the initial condition $F(x_0) + C = y_0$.

4.9 EXERCISES

Preliminary Questions

1. Does $f(x) = x^{-1}$ have an antiderivative for $x < 0$? If so, describe one.

2. Find an antiderivative of the function $f(x) = 0$.

3. What is the difference, if any, between finding the general anti-derivative of a function $f(x)$ and evaluating $\int f(x)\,dx$?

4. Jacques happens to know that $f(x)$ and $g(x)$ have the same deriva-tive, and he would like to know if $f(x) = g(x)$. Does Jacques have sufficient information to answer his question?

5. Write any two antiderivatives of $\cos x$. Which initial conditions do they satisfy at $x = 0$?

6. Suppose that $F'(x) = f(x)$ and $G'(x) = g(x)$. Are the following statements true or false? Explain.

(a) If $f = g$, then $F = G$.

(b) If F and G differ by a constant, then $f = g$.

(c) If f and g differ by a constant, then $F = G$.

7. Determine if $y = x^2$ is a solution to the differential equation with initial condition

$$\frac{dy}{dx} = 2x, \qquad y(0) = 1$$

Exercises

In Exercises 1–8, find the general antiderivative of $f(x)$ and check your answer by differentiating.

1. $f(x) = 12x$

2. $f(x) = x^2$

3. $f(x) = x^2 + 3x + 2$

4. $f(x) = x^2 + 1$

5. $f(x) = 8x^{-4}$

6. $f(x) = \cos x + 3\sin x$

7. $f(x) = 5e^x + x^2$

8. $f(x) = 3x^3 + \frac{4}{x}$

In Exercises 9–12, match the function with its antiderivative (a)–(d).

(a) $F(x) = \cos(1 - x)$

(b) $F(x) = -\cos x$

(c) $F(x) = -\frac{1}{2}\cos(x^2)$

(d) $F(x) = \sin x - x\cos x$

9. $f(x) = \sin x$

10. $f(x) = x\sin(x^2)$

11. $f(x) = \sin(1 - x)$

12. $f(x) = x\sin x$

In Exercises 13–42, evaluate the indefinite integral.

13. $\displaystyle\int (x + 1)\,dx$

14. $\displaystyle\int (9 - 5x)\,dx$

15. $\displaystyle\int (t^5 + 3t + 2)\,dt$

16. $\displaystyle\int 8s^{-4}\,ds$

17. $\displaystyle\int t^{-9/5}\,dt$

18. $\displaystyle\int (5x^3 - x^{-2} - x^{3/5})\,dx$

19. $\displaystyle\int 2\,dx$

20. $\displaystyle\int \frac{1}{\sqrt{x}}\,dx$

21. $\displaystyle\int (5t - 9)\,dt$

22. $\displaystyle\int (x^3 + 4x^{-2})\,dx$

23. $\displaystyle\int x^{-2}\,dx$

24. $\displaystyle\int \sqrt{x}\,dx$

25. $\displaystyle\int (x + 3)^{-2}\,dx$

26. $\displaystyle\int (4t - 9)^{-3}\,dt$

27. $\displaystyle\int \frac{3}{z^5}\,dz$

28. $\displaystyle\int \frac{3}{x^{3/2}}\,dx$

29. $\displaystyle\int \sqrt{x}(x - 1)\,dx$

30. $\displaystyle\int (x + x^{-1})(3x^2 - 5x)\,dx$

31. $\displaystyle\int \frac{t - 7}{\sqrt{t}}\,dt$

32. $\displaystyle\int \frac{x^2 + 2x - 3}{x^4}\,dx$

33. $\displaystyle\int (4\sin x - 3\cos x)\,dx$

34. $\displaystyle\int \sin 9x\,dx$

35. $\displaystyle\int \cos(6t + 4)\,dt$

36. $\displaystyle\int (4\theta + \cos 8\theta)\,d\theta$

37. $\displaystyle\int \cos(3 - 4t)\,dt$

38. $\displaystyle\int 18\sin(3z + 8)\,dz$

39. $\displaystyle\int (\cos x - e^x)\,dx$

40. $\displaystyle\int \left(\frac{8}{x} + 3e^x\right)\,dx$

41. $\displaystyle\int 25e^{5x}\,dx$

42. $\displaystyle\int (2x + e^{14 - 2x})\,dx$

43. In Figure 3, which of (A) or (B) is the graph of an antiderivative of $f(x)$?

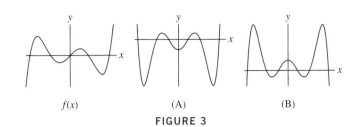

f(x)   (A)   (B)

FIGURE 3

44. In Figure 4, which of (A), (B), (C) is not the graph of an antiderivative of $f(x)$? Explain.

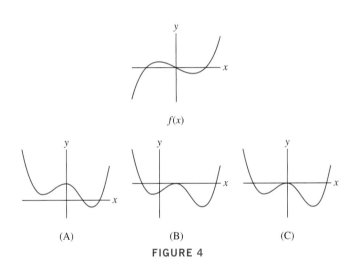

f(x)

(A)   (B)   (C)

FIGURE 4

45. Use the formulas for the derivatives of $f(x) = \tan x$ and $f(x) = \sec x$ to evaluate the integrals.

(a) $\displaystyle \int \sec^2(3x)\, dx$

(b) $\displaystyle \int \sec(x + 3)\tan(x + 3)\, dx$

46. Use the formulas for the derivatives of $f(x) = \cot x$ and $f(x) = \csc x$ to evaluate the integrals.

(a) $\displaystyle \int \csc^2 x\, dx$

(b) $\displaystyle \int \csc x \cot x\, dx$

In Exercises 47–62, solve the differential equation with initial condition.

47. $\dfrac{dy}{dx} = \cos 2x, \quad y(0) = 3$

48. $\dfrac{dy}{dx} = x^3, \quad y(0) = 2$

49. $\dfrac{dy}{dx} = x, \quad y(0) = 5$

50. $\dfrac{dy}{dt} = 0, \quad y(3) = 5$

51. $\dfrac{dy}{dt} = 5 - 2t^2, \quad y(1) = 2$

52. $\dfrac{dy}{dx} = 8x^3 + 3x^2 - 3, \quad y(1) = 1$

53. $\dfrac{dy}{dt} = 4t + 9, \quad y(0) = 1$

54. $\dfrac{dy}{dt} = \sqrt{t}, \quad y(1) = 1$

55. $\dfrac{dy}{dx} = \sin x, \quad y\left(\dfrac{\pi}{2}\right) = 1$

56. $\dfrac{dy}{dz} = \sin 2z, \quad y\left(\dfrac{\pi}{4}\right) = 4$

57. $\dfrac{dy}{dx} = \cos 5x, \quad y(\pi) = 3$

58. $\dfrac{dy}{dx} = \sec^2 3x, \quad y\left(\dfrac{\pi}{4}\right) = 2$

59. $\dfrac{dy}{dx} = e^x, \quad y(0) = 4$

60. $\dfrac{dy}{dx} = e^x, \quad y(2) = 4$

61. $\dfrac{dy}{dx} = e^{5x}, \quad y(0) = -3$

62. $\dfrac{dy}{dx} = e^{-x}, \quad y(1) = 1$

In Exercises 63–68, first find f' and then find f.

63. $f''(x) = x, \quad f'(0) = 1, \quad f(0) = 0$

64. $f''(x) = x^3 - 2x + 1, \quad f'(0) = 1, \quad f(0) = 0$

65. $f''(x) = x^3 - 2x + 1, \quad f'(1) = 0, \quad f(1) = 4$

66. $f''(t) = t^{-3/2}, \quad f'(4) = 1, \quad f(4) = 4$

67. $f''(\theta) = \cos \theta, \quad f'\left(\dfrac{\pi}{2}\right) = 1, \quad f\left(\dfrac{\pi}{2}\right) = 6$

68. $f''(t) = t - \cos t, \quad f'(0) = 2, \quad f(0) = -2$

69. Show that $f(x) = \tan^2 x$ and $g(x) = \sec^2 x$ have the same derivative. What can you conclude about the relation between f and g? Verify this conclusion directly.

70. Show, by computing derivatives, that $\sin^2 x = -\frac{1}{2}\cos 2x + C$ for some constant C. Find C by setting $x = 0$.

71. A particle located at the origin at $t = 0$ begins moving along the x-axis with velocity $v(t) = \frac{1}{2}t^2 - t$ ft/s. Let $s(t)$ be its position at time t. State the differential equation with initial condition satisfied by $s(t)$ and find $s(t)$.

72. Repeat Exercise 71, but replace the initial condition $s(0) = 0$ with $s(2) = 3$.

73. A particle moves along the x-axis with velocity $v(t) = 25t - t^2$ ft/s. Let $s(t)$ be the position at time t.

(a) Find $s(t)$, assuming that the particle is located at $x = 5$ at time $t = 0$.

(b) Find $s(t)$, assuming that the particle is located at $x = 5$ at time $t = 2$.

74. A particle located at the origin at $t = 0$ moves in a straight line with *acceleration* $a(t) = 4 - \frac{1}{2}t$ ft/s^2. Let $v(t)$ be the velocity and $s(t)$ the position at time t.

(a) State and solve the differential equation for $v(t)$ assuming that the particle is at rest at $t = 0$.

(b) Find $s(t)$.

75. A car traveling 84 ft/s begins to decelerate at a constant rate of 14 ft/s^2. After how many seconds does the car come to a stop and how far will the car have traveled before stopping?

76. Beginning at rest, an object moves in a straight line with constant acceleration a, covering 100 ft in 5 s. Find a.

77. A 900-kg rocket is released from a spacecraft. As the rocket burns fuel, its mass decreases and its velocity increases. Let $v(m)$ be the velocity (in meters per second) as a function of mass m. Find the velocity when $m = 500$ if $dv/dm = -50m^{-1/2}$. Assume that $v(900) = 0$.

78. As water flows through a tube of radius $R = 10$ cm, the velocity of an individual water particle depends on its distance r from the center

of the tube according to the formula $\dfrac{dv}{dr} = -0.06r$. Determine $v(r)$, assuming that particles at the walls of the tube have zero velocity.

79. Find constants c_1 and c_2 such that $F(x) = c_1 x \sin x + c_2 \cos x$ is an antiderivative of $f(x) = x \cos x$.

80. Find the general antiderivative of $(2x + 9)^{10}$.

81. Verify the linearity properties of the indefinite integral stated in Theorem 4.

Further Insights and Challenges

82. Suppose that $F'(x) = f(x)$ and $G'(x) = g(x)$. Is it true that $F(x)G(x)$ is an antiderivative of $f(x)g(x)$? Confirm or provide a counterexample.

83. Suppose that $F'(x) = f(x)$.
(a) Show that $\frac{1}{2} F(2x)$ is an antiderivative of $f(2x)$.
(b) Find the general antiderivative of $f(kx)$ for any constant k.

84. Find an antiderivative for $f(x) = |x|$.

85. Let $F(x) = \dfrac{x^{n+1} - 1}{n + 1}$.

(a) Show that if $n \neq -1$, then $F(x)$ is an antiderivative of $y = x^n$.
(b) Use L'Hôpital's Rule to prove that

$$\lim_{n \to -1} F(x) = \ln x$$

In this limit, x is fixed and we treat n as a variable tending to -1. This shows that although the Power Rule breaks down for $n = -1$, we may view the antiderivative of $y = x^{-1}$ as a limit of antiderivatives of x^n as $n \to -1$.

CHAPTER REVIEW EXERCISES

In Exercises 1–6, estimate using the Linear Approximation or linearization and use a calculator to compute the error.

1. $8.1^{1/3} - 2$

2. $\dfrac{1}{\sqrt{4.1}} - \dfrac{1}{2}$

3. $625^{1/4} - 624^{1/4}$

4. $\sqrt{101}$

5. $\dfrac{1}{1.02}$

6. $\sqrt[5]{33}$

In Exercises 7–12, find the linearization at the point indicated.

7. $y = \sqrt{x}, \quad a = 25$

8. $v(t) = 32t - 4t^2, \quad a = 2$

9. $A(r) = \frac{4}{3}\pi r^3, \quad a = 3$

10. $V(h) = 4h(2 - h)(4 - 2h), \quad a = 1$

11. $P(x) = e^{-x^2/2}, \quad a = 1$

12. $f(x) = \ln(x + e), \quad a = e$

In Exercises 13–17, use the Linear Approximation.

13. The position of an object in linear motion at time t is $s(t) = 0.4t^2 + (t + 1)^{-1}$. Estimate the distance traveled over the time interval $[4, 4.2]$.

14. A bond that pays \$10,000 in 6 years is offered for sale at a price P. The percentage yield Y of the bond is

$$Y = 100 \left(\left(\frac{10{,}000}{P} \right)^{1/6} - 1 \right)$$

Verify that if $P = \$7{,}500$, then $Y = 4.91\%$. Estimate the drop in yield if the price rises to \$7,700.

15. A store sells 80 MP3 players per week when the players are priced at $P = \$75$. Estimate the number N sold if P is raised to \$80, assuming that $dN/dP = -4$. Estimate N if the price is lowered to \$69.

16. The circumference of a sphere is measured at $C = 100$ cm. Estimate the maximum percentage error in V if the error in C is at most 3 cm.

17. Show that $\sqrt{a^2 + b} \approx a + \frac{b}{2a}$ if b is small. Use this to estimate $\sqrt{26}$ and find the error using a calculator.

18. Use the Intermediate Value Theorem to prove that $\sin x - \cos x = 3x$ has a solution and use Rolle's Theorem to show that this solution is unique.

19. Show that $f(x) = x + \dfrac{x}{x^2 + 1}$ has precisely one real root.

20. Verify the MVT for $f(x) = \ln x$ on $[1, 4]$.

21. Suppose that $f(1) = 5$ and $f'(x) \geq 2$ for $x \geq 1$. Use the MVT to show that $f(8) - f(1) \geq 19$.

22. Use the MVT to prove that if $f'(x) \le 2$ for $x > 0$ and $f(0) = 4$, then $f(x) \le 2x + 4$ for all $x \ge 0$.

In Exercises 23–26, find the local extrema and determine whether they are minima, maxima, or neither.

23. $f(x) = x^3 - 4x^2 + 4x$

24. $s(t) = t^4 - 8t^2$

25. $f(x) = x^2(x + 2)^3$

26. $f(x) = x^{2/3}(1 - x)$

In Exercises 27–34, find the extreme values on the interval.

27. $f(x) = x(10 - x)$, $\quad [-1, 3]$

28. $f(x) = 6x^4 - 4x^6$, $\quad [-2, 2]$

29. $g(\theta) = \sin^2 \theta - \cos \theta$, $\quad [0, 2\pi]$

30. $R(t) = \dfrac{t}{t^2 + t + 1}$, $\quad [0, 3]$

31. $f(x) = x^{2/3} - 2x^{1/3}$, $\quad [-1, 3]$

32. $f(x) = x - \tan x$, $\quad [-1, 1]$

33. $f(x) = x - 12 \ln x$, $\quad [5, 40]$

34. $f(x) = e^x - 20x - 1$, $\quad [0, 5]$

35. Find the critical points and extreme values of $f(x) = |x - 1| + |2x - 6|$ in $[0, 8]$.

36. A function $f(x)$ has derivative $f'(x) = \dfrac{1}{x^4 + 1}$. Where on the interval $[1, 4]$ does $f(x)$ take on its maximum value?

In Exercises 37–42, find the points of inflection.

37. $y = x^3 - 4x^2 + 4x$

38. $y = x - 2 \cos x$

39. $y = \dfrac{x^2}{x^2 + 4}$

40. $y = \dfrac{x}{(x^2 - 4)^{1/3}}$

41. $f(x) = (x^2 - x)e^{-x}$

42. $f(x) = x(\ln x)^2$

43. Match the description of $f(x)$ with the graph of its derivative $f'(x)$ in Figure 1.

(a) $f(x)$ is increasing and concave up.

(b) $f(x)$ is decreasing and concave up.

(c) $f(x)$ is increasing and concave down.

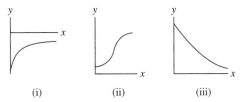

(i)   (ii)   (iii)

FIGURE 1 Graphs of the derivative.

44. Draw a curve $y = f(x)$ for which f' and f'' have signs as indicated in Figure 2.

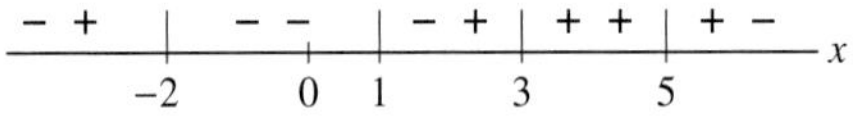

$$\frac{-\ +\ \ |\ \ -\ -\ \ |\ -\ +\ |\ +\ +\ |\ +\ -}{\qquad -2 \quad\ \ 0\ \ 1 \qquad 3 \qquad 5}\ x$$

FIGURE 2

In Exercises 45–60, evaluate the limit.

45. $\displaystyle\lim_{x \to \infty} (9x^4 - 12x^3)$

46. $\displaystyle\lim_{x \to -\infty} (7x^2 - 9x^5)$

47. $\displaystyle\lim_{x \to \infty} \frac{x^3 + 2x}{4x^2 - 9}$

48. $\displaystyle\lim_{x \to -\infty} \frac{x^3 + 2x}{4x^2 - 9}$

49. $\displaystyle\lim_{x \to \infty} \frac{x^3 + 2x}{4x^3 - 9}$

50. $\displaystyle\lim_{x \to -\infty} \frac{x^3 + 2x}{4x^5 - 40x^3}$

51. $\displaystyle\lim_{x \to \infty} \frac{x^2 - 9x}{\sqrt{x^2 + 4x}}$

52. $\displaystyle\lim_{x \to -\infty} \frac{12x + 1}{\sqrt{4x^2 + 4x}}$

53. $\displaystyle\lim_{x \to \infty} \frac{x^{1/2}}{\sqrt{4x - 9}}$

54. $\displaystyle\lim_{x \to \infty} \frac{x^{3/2}}{(16x^6 - 9x^4)^{1/4}}$

55. $\displaystyle\lim_{x \to 3+} \frac{1 - 2x}{x - 3}$

56. $\displaystyle\lim_{x \to 2-} \frac{1}{x^2 - 4}$

57. $\displaystyle\lim_{x \to 2+} \frac{x - 5}{x - 2}$

58. $\displaystyle\lim_{x \to -3+} \frac{1}{(x + 3)^3}$

59. $\displaystyle\lim_{x \to \infty} x^2 e^{-x}$

60. $\displaystyle\lim_{x \to 1} \frac{x^2 - 1}{e^{1 - x^2} - 1}$

In Exercises 61–70, sketch the graph, noting the transition points and asymptotic behavior.

61. $y = 12x - 3x^2$

62. $y = 8x^2 - x^4$

63. $y = x^3 - 2x^2 + 3$

64. $y = 4x - x^{3/2}$

65. $y = \dfrac{x}{x^3 + 1}$

66. $y = \dfrac{x}{(x^2 - 4)^{2/3}}$

67. $y = \dfrac{1}{|x + 2| + 1}$

68. $y = \sqrt{2 - x^3}$

69. $y = 2 \sin x - \cos x$ on $[0, 2\pi]$

70. $y = 2x - \tan x$ on $[0, 2\pi]$

71. Find the maximum volume of a right-circular cone placed upside-down in a right-circular cone of radius R and height H (Figure 3). The volume of a cone of radius r and height h is $\frac{4}{3}\pi r^2 h$.

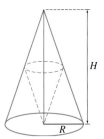

FIGURE 3

72. On a certain farm, the corn yield is

$$Y = -0.118x^2 + 8.5x + 12.9 \quad \text{(bushels per acre)}$$

where x is the number of corn plants per acre (in thousands). Assume that corn seed costs \$1.25 (per thousand seeds) and that corn can be sold for \$1.50/bushel.

(a) Find the value x_0 that maximizes yield Y. Then compute the profit (revenue minus the cost of seeds) at planting level x_0.

(b) Compute the profit $P(x)$ as a function of x and find the value x_1 that maximizes profit.

(c) Compare the profit at levels x_0 and x_1. Does a maximum yield lead to maximum profit?

73. Show that the maximum area of a parallelogram ADEF inscribed in a triangle $\triangle ABC$, as in Figure 4, is equal to one-half the area of $\triangle ABC$.

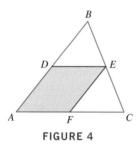

FIGURE 4

74. The area of the shaded region in Figure 5 is $\frac{1}{2}r^2(\theta - \sin\theta)$. What is the maximum possible area of this region if the length of the circular arc is 1?

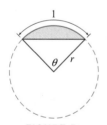

FIGURE 5

75. Let $f(x)$ be a function whose graph does not pass through the x-axis and let $Q = (a, 0)$. Let $P = (x_0, f(x_0))$ be the point on the graph closest to Q (Figure 6). Prove that $\overline{PQ}$ is perpendicular to the

tangent line to the graph of x_0. *Hint:* Let $q(x)$ be the distance from $(x, f(x))$ to $(a, 0)$ and observe that x_0 is a critical point of $q(x)$.

76. Take a circular piece of paper of radius R, remove a sector of angle θ (Figure 7), and fold the remaining piece into a cone-shaped cup. Which angle θ produces the cup of largest volume?

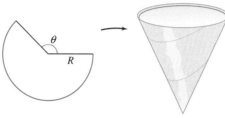

FIGURE 7

77. Use Newton's Method to estimate $\sqrt[3]{25}$ to four decimal places.

78. Use Newton's Method to find a root of $f(x) = x^2 - x - 1$ to four decimal places.

In Exercises 79–92, calculate the indefinite integral.

79. $\displaystyle\int \left(4x^3 - 2x^2\right) dx$

80. $\displaystyle\int x^{9/4}\, dx$

81. $\displaystyle\int \sin(\theta - 8)\, d\theta$

82. $\displaystyle\int \cos(5 - 7\theta)\, d\theta$

83. $\displaystyle\int \left(4t^{-3} - 12t^{-4}\right) dt$

84. $\displaystyle\int \left(9t^{-2/3} + 4t^{7/3}\right) dt$

85. $\displaystyle\int \sec^2 x\, dx$

86. $\displaystyle\int \tan 3\theta \sec 3\theta\, d\theta$

87. $\displaystyle\int (y + 2)^4\, dy$

88. $\displaystyle\int \frac{3x^3 - 9}{x^2}\, dx$

89. $\displaystyle\int (e^x - x)\, dx$

90. $\displaystyle\int e^{-4x}\, dx$

91. $\displaystyle\int 4x^{-1}\, dx$

92. $\displaystyle\int \sin(4x - 9)\, dx$

In Exercises 93–98, solve the differential equation with initial condition.

93. $\dfrac{dy}{dx} = 4x^3, \quad y(1) = 4$

94. $\dfrac{dy}{dt} = 3t^2 + \cos t, \quad y(0) = 12$

95. $\dfrac{dy}{dx} = x^{-1/2}, \quad y(1) = 1$

96. $\dfrac{dy}{dx} = \sec^2 x, \quad y(\frac{\pi}{4}) = 2$

97. $\dfrac{dy}{dx} = e^{-x}, \quad y(0) = 3$

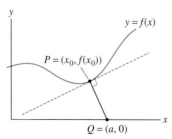

FIGURE 6

98. $\dfrac{dy}{dx} = e^{4x}$, $y(1) = 1$

99. Find $f(t)$, assuming that $f''(t) = 1 - 2t$, $f(0) = 2$, and $f'(0) = -1$.

100. The driver of an automobile applies the brakes at time $t = 0$ and comes to a halt after traveling 500 ft. Find the automobile's velocity at $t = 0$ assuming that the rate of deceleration was a constant -10 ft/s^2.

101. Find the local extrema of $f(x) = \dfrac{e^{2x} + 1}{e^{x+1}}$.

102. Find the points of inflection of $f(x) = \ln(x^2 + 1)$ and determine whether the concavity changes from up to down or vice versa.

In Exercises 103–106, find the local extrema and points of inflection, and sketch the graph. Use L'Hôpital's Rule to determine the limits as $x \to 0+$ or $x \to \pm\infty$ if necessary.

103. $y = x \ln x$, $x > 0$

104. $y = e^{x-x^2}$

105. $y = x(\ln x)^2$, $x > 0$

106. $y = \tan^{-1}\left(\dfrac{x^2}{4}\right)$

107. 🖎   Explain why L'Hôpital's Rule gives no information about $\displaystyle\lim_{x \to \infty} \dfrac{2x - \sin x}{3x + \cos 2x}$. Evaluate the limit by another method.

108. Let $f(x)$ be a differentiable function with inverse $g(x)$ such that $f(0) = 0$ and $f'(0) \neq 0$. Prove that

$$\lim_{x \to 0} \frac{f(x)}{g(x)} = f'(0)^2$$

In Exercises 109–120, verify that L'Hôpital's Rule applies and evaluate the limit.

109. $\displaystyle\lim_{x \to 3} \dfrac{4x - 12}{x^2 - 5x + 6}$

110. $\displaystyle\lim_{x \to -2} \dfrac{x^3 + 2x^2 - x - 2}{x^4 + 2x^3 - 4x - 8}$

111. $\displaystyle\lim_{x \to 0+} x^{1/2} \ln x$

112. $\displaystyle\lim_{t \to \infty} \dfrac{\ln(e^t + 1)}{t}$

113. $\displaystyle\lim_{\theta \to 0} \dfrac{2 \sin \theta - \sin 2\theta}{\sin \theta - \theta \cos \theta}$

114. $\displaystyle\lim_{x \to 0} \dfrac{\sqrt{4 + x} - 2\sqrt[8]{1 + x}}{x^2}$

115. $\displaystyle\lim_{t \to \infty} \dfrac{\ln(t + 2)}{\log_2 t}$

116. $\displaystyle\lim_{x \to 0} \left(\dfrac{e^x}{e^x - 1} - \dfrac{1}{x}\right)$

117. $\displaystyle\lim_{y \to 0} \dfrac{\sin^{-1} y - y}{y^3}$

118. $\displaystyle\lim_{x \to 1} \dfrac{\sqrt{1 - x^2}}{\cos^{-1} x}$

119. $\displaystyle\lim_{x \to 0} \dfrac{\sinh(x^2)}{\cosh x - 1}$

120. $\displaystyle\lim_{x \to 0} \dfrac{\tanh x - \sinh x}{\sin x - x}$

121. Let $f(x) = e^{-Ax^2/2}$, where $A > 0$. Given any n numbers $a_1, a_2, \ldots, a_n$, set

$$\Phi(x) = f(x - a_1)f(x - a_2) \cdots f(x - a_n)$$

(a) Assume $n = 2$ and prove that $\Phi(x)$ attains its maximum value at the average $x = \frac{1}{2}(a_1 + a_2)$. *Hint:* Show that $\dfrac{d}{dx} \ln(f(x)) = -Ax$ and calculate $\Phi'(x)$ using logarithmic differentiation.

(b) Show that for any n, $\Phi(x)$ attains its maximum value at $x = \frac{1}{n}(a_1 + a_2 + \cdots + a_n)$. This fact is related to the role of $f(x)$ (whose graph is a bell-shaped curve) in statistics.

5 | THE INTEGRAL

The starting point in integral calculus is the problem of finding the area under a curve. You may wonder why calculus is concerned with two seemingly unrelated topics such as areas and tangent lines. One reason is that both are computed using limits. However, another deeper connection between areas and tangent lines is revealed by the Fundamental Theorem of Calculus, discussed in Sections 5.3 and 5.4. This theorem expresses the fundamental "inverse" relationship between integration and differentiation that plays a key role in almost all applications of calculus, both theoretical and practical.

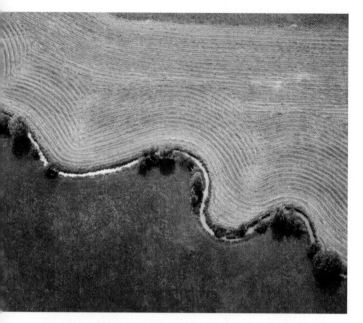

Integration solves an ancient mathematical problem—finding the area of an irregular region.

5.1 Approximating and Computing Area

We begin by discussing a basic example in which the area under a curve plays a role. If an object moves in a straight line with *constant velocity* v (assumed to be positive), then the distance traveled over a time interval $[t_1, t_2]$ is equal to $v(t_2 - t_1)$. This is the well-known formula

$$\boxed{\text{Distance traveled} = \text{velocity} \times \text{time}} \qquad \boxed{1}$$

We may interpret this formula in terms of area by observing that the graph of velocity is a horizontal line at height v, and $v(t_2 - t_1)$ is the area of the rectangular region under the graph lying over the interval $[t_1, t_2]$ (Figure 1). Thus,

$$\boxed{\text{Distance traveled during } [t_1, t_2] = \text{area under the graph of velocity over } [t_1, t_2]}$$

$$\boxed{2}$$

Although Eqs. (1) and (2) are equivalent, the advantage of Eq. (2) is that it remains true, even if the velocity is not constant (Figure 2). This is discussed in Section 5.5, but we can understand the main idea by considering the graph of velocity in Figure 3. In this case, the velocity changes over time but is constant on intervals (of course, this graph is not realistic because an object's velocity cannot change so abruptly). The distance traveled over each time interval is equal to the area of the rectangle above that interval, and the total distance traveled is the sum of the areas of the rectangles:

$$\text{Distance traveled over } [0, 8] = \underbrace{10 + 15 + 30 + 10}_{\text{Sum of areas of rectangles}} = 65 \text{ ft}$$

This reasoning shows more generally that distance traveled is equal to the area under the graph whenever the velocity is constant on intervals. This argument does not work when the velocity changes continuously as in Figure 2. Our strategy in this case is to *approximate* the area under the graph by sums of areas of rectangles and then pass to a limit.

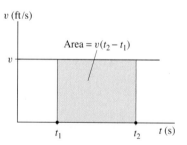

FIGURE 1 Distance traveled equals the area of the rectangle.

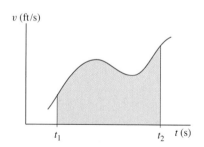

FIGURE 2 Distance traveled is the area under the graph of velocity.

Approximating Area by Rectangles

For the rest of this section, we assume that $f(x)$ is continuous and *positive*, so that the graph of $f(x)$ lies above the x-axis. Our goal is to compute the area "under the graph,"

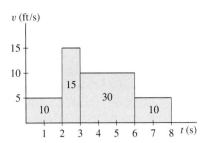

FIGURE 3 Distance traveled is the sum of the areas of the rectangles.

that is, the area between the graph and the x-axis. As a first step, we approximate the area using rectangles.

First, choose a whole number N and divide $[a, b]$ into N subintervals of equal width, as in Figure 4. Each subinterval has width $\Delta x = \dfrac{b-a}{N}$ since $[a, b]$ has width $b - a$. The right endpoints of the subintervals are

$$a + \Delta x, \ a + 2\Delta x, \ \ldots, \ a + (N-1)\Delta x, \ a + N\Delta x$$

Notice that the last right endpoint is b because $a + N\Delta x = a + N\left(\dfrac{b-a}{N}\right) = b$. Next, above each subinterval, construct the rectangle whose height is the value of $f(x)$ at the *right endpoint* of the subinterval.

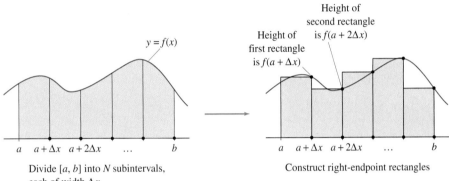

Divide $[a, b]$ into N subintervals, each of width Δx

Construct right-endpoint rectangles

FIGURE 4 First subdivide the interval, then construct the rectangles.

Each rectangle has width Δx. The height of the first rectangle is $f(a + \Delta x)$ and its area is $f(a + \Delta x)\Delta x$. Similarly, the second rectangle has height $f(a + 2\Delta x)$ and area $f(a + 2\Delta x)\,\Delta x$, etc. The sum of the areas of the rectangles is

$$f(a + \Delta x)\Delta x + f(a + 2\Delta x)\Delta x + \cdots + f(a + N\Delta x)\Delta x$$

This sum is called the **Nth right-endpoint approximation** and is denoted R_N. Factoring out Δx, we may write R_N as

$$R_N = \Delta x \left(f(a + \Delta x) + f(a + 2\Delta x) + \cdots + f(a + N\Delta x) \right)$$

In words, R_N is equal to Δx times the sum of the function values at the right endpoints of the subintervals.

To summarize,

$a = $ *left endpoint of interval* $[a, b]$

$b = $ *right endpoint of interval* $[a, b]$

$N = $ *number of subintervals in* $[a, b]$

$\Delta x = \dfrac{b-a}{N}$

■ **EXAMPLE 1** Computing Right-Endpoint Approximations   Calculate R_4 and R_6 for $f(x) = x^2$ on the interval $[1, 3]$.

Solution

Step 1. **Determine Δx and the endpoints.**

To calculate R_4, we divide $[1, 3]$ into four subintervals of width $\Delta x = \frac{3-1}{4} = \frac{1}{2}$. The right endpoints are the numbers $a + j\Delta x = 1 + j(\frac{1}{2})$ for $j = 1, 2, 3, 4$. They are spaced at intervals of $\frac{1}{2}$ beginning at $\frac{3}{2}$, so as we see in Figure 5, the right endpoints are $\frac{3}{2}, \frac{4}{2}, \frac{5}{2}, \frac{6}{2}$.

Step 2. **Sum the function values.**

R_4 is Δx times the sum of the values of $f(x) = x^2$ at the right endpoints:

$$R_4 = \frac{1}{2}\left(f\left(\frac{3}{2}\right) + f\left(\frac{4}{2}\right) + f\left(\frac{5}{2}\right) + f\left(\frac{6}{2}\right)\right)$$

$$= \frac{1}{2}\left(\frac{9}{4} + \frac{16}{4} + \frac{25}{4} + \frac{36}{4}\right) = \frac{43}{4} = 10.75$$

The calculation of R_6 is similar: $\Delta x = \frac{3-1}{6} = \frac{1}{3}$, and the right endpoints are spaced at intervals of $\frac{1}{3}$ beginning at $\frac{4}{3}$ and ending at 3, as in Figure 5. Thus,

$$R_6 = \frac{1}{3}\left(f\left(\frac{4}{3}\right) + f\left(\frac{5}{3}\right) + f\left(\frac{6}{3}\right) + f\left(\frac{7}{3}\right) + f\left(\frac{8}{3}\right) + f\left(\frac{9}{3}\right)\right)$$

$$= \frac{1}{3}\left(\frac{16}{9} + \frac{25}{9} + \frac{36}{9} + \frac{49}{9} + \frac{64}{9} + \frac{81}{9}\right) = \frac{271}{27} \approx 10.037 \qquad \blacksquare$$

FIGURE 5 Approximations to the area under the graph of $f(x) = x^2$ over $[1, 3]$.

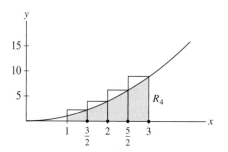
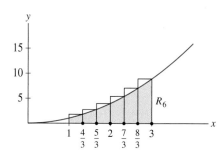

Summation Notation

Since approximations to area involve sums of function values, we introduce **summation notation**, which is a standard way of writing sums in compact form. The sum of numbers $a_m, \ldots, a_n$ $(m \le n)$ is denoted

$$\sum_{j=m}^{n} a_j = a_m + a_{m+1} + \cdots + a_n$$

The capital Greek letter $\sum$ (sigma) stands for "sum" and the notation $\displaystyle\sum_{j=m}^{n}$ tells us to start the summation at $j = m$ and end it at $j = n$. For example,

$$\sum_{j=1}^{5} j^2 = 1^2 + 2^2 + 3^2 + 4^2 + 5^2 = 55$$

In this summation, the jth term is given by the formula $a_j = j^2$. We call j^2 the **general term**. The letter j plays the role of index and is sometimes called a **dummy variable** because any other letter can be used instead. For example,

$$\sum_{j=4}^{6}(j^3 - 2j) = \sum_{k=4}^{6}\left(k^3 - 2k\right) = \left(4^3 - 2(4)\right) + \left(5^3 - 2(5)\right) + \left(6^3 - 2(6)\right) = 375$$

The following rules are used to manipulate summations. They are consequences of the usual commutative, associative, and distributive laws of addition.

Linearity of Summations

- $$\sum_{j=m}^{n} (a_j + b_j) = \sum_{j=m}^{n} a_j + \sum_{j=m}^{n} b_j$$

- $$\sum_{j=m}^{n} C a_j = C \sum_{j=m}^{n} a_j \qquad (C \text{ any constant})$$

- $$\sum_{j=1}^{n} k = nk \qquad (k \text{ any constant and } n \geq 1)$$

For example,

$$\sum_{i=3}^{5} (i^2 + i) = \overbrace{(3^2 + 3)}^{i=3} + \overbrace{(4^2 + 4)}^{i=4} + \overbrace{(5^2 + 5)}^{i=5}$$

is equal to

$$\sum_{i=3}^{5} i^2 + \sum_{i=3}^{5} i = \left(3^2 + 4^2 + 5^2\right) + (3 + 4 + 5)$$

We often use linearity to write a single summation as a sum of several summations. For example,

$$\sum_{k=0}^{100} (7k^2 - 4k + 9) = \sum_{k=0}^{100} 7k^2 + \sum_{k=0}^{100} (-4k) + \sum_{k=0}^{100} 9$$

$$= 7 \sum_{k=0}^{100} k^2 - 4 \sum_{k=0}^{100} k + 9 \sum_{k=0}^{100} 1$$

It is convenient to write R_N in summation notation. By definition,

$$R_N = \Delta x \left[f(a + \Delta x) + f(a + 2\Delta x) + \cdots + f(a + N\Delta x) \right]$$

The general term is $f(a + j\Delta x)$ and the summation extends from $j = 1$ to $j = N$, so

$$R_N = \Delta x \sum_{j=1}^{N} f(a + j\Delta x)$$

We now discuss two other approximations to area: the left-endpoint and the midpoint approximations. To calculate the **left-endpoint approximation**, we divide $[a, b]$ into N subintervals as before, but the heights of the rectangles are the values of $f(x)$ at the left endpoints [Figure 6(A)]. The left endpoints are

$$a, \ a + \Delta x, \ a + 2\Delta x, \ldots, \ a + (N - 1)\Delta x$$

and the sum of the areas of the left-endpoint rectangles is

$$L_N = \Delta x \left(f(a) + f(a + \Delta x) + f(a + 2\Delta x) + \cdots + f(a + (N - 1)\Delta x) \right)$$

Note that both R_N and L_N have general term $f(a + j\Delta x)$, but the sum for L_N runs from $j = 0$ to $j = N - 1$ rather than from $j = 1$ to $j = N$. Thus,

$$L_N = \Delta x \sum_{j=0}^{N-1} f(a + j\Delta x)$$

In the **midpoint approximation** M_N, the heights of the rectangles are the values of $f(x)$ at the midpoints of the subintervals rather than at the endpoints [Figure 6(B)]. The midpoints are

$$a + \frac{1}{2}\Delta x, \ a + \frac{3}{2}\Delta x, \ \dots, \ a + \left(N - \frac{1}{2}\right)\Delta x$$

and the sum of the areas of the midpoint rectangles is

$$M_N = \Delta x \left(f\left(a + \frac{1}{2}\Delta x\right) + f\left(a + \frac{3}{2}\Delta x\right) + \cdots + f\left(a + \left(N - \frac{1}{2}\right)\Delta x\right) \right)$$

In summation notation,

$$M_N = \Delta x \sum_{j=1}^{N} f\left(a + \left(j - \frac{1}{2}\right)\Delta x\right)$$

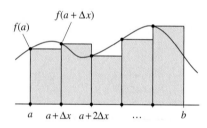

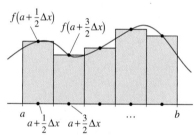

FIGURE 6 Left-endpoint and midpoint approximations.

(A) Left-endpoint rectangles

(B) Midpoint rectangles

■ **EXAMPLE 2** Calculate R_6, L_6, and M_6 for $f(x) = x^{-1}$ on $[2, 4]$.

Solution In this case, $a = 2$, $b = 4$, and $\Delta x = \frac{4-2}{6} = \frac{1}{3}$. To compute R_6, let's evaluate the general term in the sum representing R_6:

$$f(a + j\Delta x) = f\left(2 + j\left(\frac{1}{3}\right)\right) = \frac{1}{2 + \frac{j}{3}}$$

Therefore,

$$R_6 = \frac{1}{3} \sum_{j=1}^{6} f\left(2 + \left(\frac{1}{3}\right)j\right) = \frac{1}{3} \sum_{j=1}^{6} \frac{1}{2 + \frac{j}{3}}$$

$$= \frac{1}{3}\left(\frac{1}{2 + \frac{1}{3}} + \frac{1}{2 + \frac{2}{3}} + \frac{1}{2 + \frac{3}{3}} + \frac{1}{2 + \frac{4}{3}} + \frac{1}{2 + \frac{5}{3}} + \frac{1}{2 + \frac{6}{3}} \right)$$

$$= \frac{1}{3}\left(\frac{3}{7} + \frac{3}{8} + \frac{3}{9} + \frac{3}{10} + \frac{3}{11} + \frac{3}{12} \right) \approx 0.653$$

The left-endpoint approximation sum is nearly the same, but the sum begins with $j = 0$ and ends at $j = 5$:

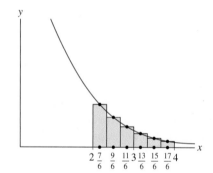

FIGURE 7 M_6 for $f(x) = x^{-1}$ on $[2, 4]$.

$$L_6 = \frac{1}{3} \sum_{j=0}^{5} \frac{1}{2 + \frac{j}{3}} = \frac{1}{3}\left(\frac{3}{6} + \frac{3}{7} + \frac{3}{8} + \frac{3}{9} + \frac{3}{10} + \frac{3}{11}\right) \approx 0.737$$

To compute M_6, we note that the general term is

$$f\left(a + \left(j - \frac{1}{2}\right)\Delta x\right) = f\left(2 + \left(j - \frac{1}{2}\right)\frac{1}{3}\right) = \frac{1}{2 + \frac{j}{3} - \frac{1}{6}}$$

Summing up from $j = 1$ to 6, we obtain (Figure 7)

$$M_6 = \frac{1}{3} \sum_{j=1}^{6} f\left(2 + \left(j - \frac{1}{2}\right)\frac{1}{3}\right) = \frac{1}{3}\sum_{j=1}^{6} \frac{1}{2 + \frac{j}{3} - \frac{1}{6}}$$

$$= \frac{1}{3}\left(\frac{1}{2 + \frac{1}{6}} + \frac{1}{2 + \frac{3}{6}} + \frac{1}{2 + \frac{5}{6}} + \frac{1}{2 + \frac{7}{6}} + \frac{1}{2 + \frac{9}{6}} + \frac{1}{2 + \frac{11}{6}}\right)$$

$$= \frac{1}{3}\left(\frac{6}{13} + \frac{6}{15} + \frac{6}{17} + \frac{6}{19} + \frac{6}{21} + \frac{6}{23}\right) \approx 0.692$$

■

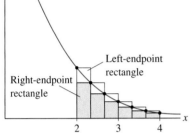

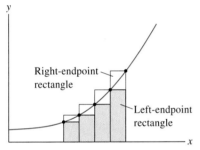

FIGURE 8 L_6 and R_6 for $f(x) = x^{-1}$ on $[2, 4]$.

FIGURE 9 When $f(x)$ is increasing, the left-endpoint rectangles lie below the graph and right-endpoint rectangles lie above it.

GRAPHICAL INSIGHT Monotonic Functions Observe in Figure 8 that the left-endpoint rectangles for $f(x) = x^{-1}$ extend above the graph and the right-endpoint rectangles lie below it. Therefore, the exact area A lies between R_6 and L_6 and so, according to the previous example, $0.65 \le A \le 0.74$. More generally, *when $f(x)$ is monotonic (increasing or decreasing), the exact area lies between R_N and L_N* (Figures 8 and 9):

- $f(x)$ increasing $\Rightarrow$ $L_N \le$ area under graph $\le R_N$
- $f(x)$ decreasing $\Rightarrow$ $R_N \le$ area under graph $\le L_N$

Computing Area as the Limit of Approximations

What can be said about the accuracy of our approximations to the area under a graph? Figure 10 shows the error for several right-endpoint approximations as the yellow region above the graph. We see that the error gets smaller as the number of rectangles increases. Furthermore, it appears that *we can achieve arbitrarily high accuracy by taking the number N of rectangles large enough*. If so, it makes sense to consider the limit as $N \to \infty$, which should give us the exact area under the curve. The following theorem guarantees that the limit exists (see Exercise 91 for a proof for monotonic functions).

FIGURE 10 The error decreases as we use more rectangles.

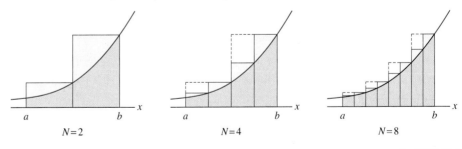

$N=2$ $N=4$ $N=8$

> **THEOREM 1** If $f(x)$ is continuous on $[a, b]$, then the endpoint and midpoint approximations approach one and the same limit as $N \to \infty$. In other words, there is a value L such that
>
> $$\lim_{N\to\infty} R_N = \lim_{N\to\infty} L_N = \lim_{N\to\infty} M_N = L$$

In Theorem 1, it is not assumed that $f(x) \geq 0$. However, if $f(x)$ takes on negative values, the limit L no longer represents an area under a graph. We can interpret it as a "signed area," a concept discussed in the next section.

Theorem 1 gives us a precise definition of the area under a graph as a limit L. Recall that the tangent line is another basic concept which is defined as a limit. When we studied the derivative, we found that the slope of the tangent line is the limit of slopes of secant lines.

In the next two examples, we illustrate Theorem 1 by finding the area under the graphs of some simple polynomials. These calculations rely on the formulas for the **power sums.** The following are the first three power sum formulas.

DEFINITION Power Sums The kth power sum is the sum of the kth powers of the first N integers.

$$\sum_{j=1}^{N} j = 1 + 2 + \cdots + N = \frac{N(N+1)}{2} = \frac{N^2}{2} + \frac{N}{2} \qquad \boxed{3}$$

$$\sum_{j=1}^{N} j^2 = 1^2 + 2^2 + \cdots + N^2 = \frac{N(N+1)(2N+1)}{6} = \frac{N^3}{3} + \frac{N^2}{2} + \frac{N}{6} \qquad \boxed{4}$$

$$\sum_{j=1}^{N} j^3 = 1^3 + 2^3 + \cdots + N^3 = \frac{N^2(N+1)^2}{4} = \frac{N^4}{4} + \frac{N^3}{2} + \frac{N^2}{4} \qquad \boxed{5}$$

For example, by Eq. (4),

$$\sum_{j=1}^{6} j^2 = 1^2 + 2^2 + 3^2 + 4^2 + 5^2 + 6^2 = \underbrace{\frac{6^3}{3} + \frac{6^2}{2} + \frac{6}{6}}_{\frac{N^3}{3} + \frac{N^2}{2} + \frac{N}{6} \text{ for } N=6} = 91$$

As a first illustration, we compute the area of a right triangle "the hard way."

■ **EXAMPLE 3** Calculating Area as a Limit Calculate the area under the graph of $f(x) = x$ over $[0, 4]$ in three ways:

(a) $\lim_{N \to \infty} R_N$ **(b)** $\lim_{N \to \infty} L_N$ **(c)** Using geometry

◀·· REMINDER

$$R_N = \Delta x \sum_{j=1}^{N} f(a + j\Delta x)$$

$$L_N = \Delta x \sum_{j=0}^{N-1} f(a + j\Delta x)$$

Solution The interval is $[a, b] = [0, 4]$, so $\Delta x = \dfrac{b-a}{N} = \dfrac{4}{N}$, and since $f(x) = x$,

$$f(a + j\Delta x) = f\left(0 + \frac{4j}{N}\right) = \frac{4j}{N}$$

Therefore,

$$R_N = \Delta x \sum_{j=1}^{N} f(a + j\Delta x) = \frac{4}{N} \sum_{j=1}^{N} \frac{4j}{N} = \frac{16}{N^2} \sum_{j=1}^{N} j$$

In the last equality, we factor out $\dfrac{4}{N}$ from the sum. This is valid because $\dfrac{4}{N}$ is a constant that does not depend on j. Now use formula (3) for $\displaystyle\sum_{j=1}^{N} j$ to obtain

$$R_N = \frac{16}{N^2} \sum_{j=1}^{N} j = \frac{16}{N^2} \underbrace{\left(\frac{N^2}{2} + \frac{N}{2}\right)}_{\text{Formula for power sum}} = 8 + \frac{8}{N} \qquad \boxed{6}$$

The second term $\dfrac{8}{N}$ tends to zero as N approaches ∞, so the limit is equal to

$$\text{Area} = \lim_{N \to \infty} R_N = \lim_{N \to \infty} \left(8 + \frac{8}{N}\right) = 8$$

The left-endpoint approximation is nearly the same as Eq. (6), but the sum begins at $j = 0$ and ends at $j = N - 1$ (we leave the algebra of the last step to you):

$$L_N = \frac{16}{N^2} \sum_{j=0}^{N-1} j = \frac{16}{N^2} \sum_{j=1}^{N-1} j = \frac{16}{N^2}\left(\frac{(N-1)^2}{2} + \frac{(N-1)}{2}\right) = 8 - \frac{8}{N}$$

Notice that in the second step, we replaced the sum beginning at $j = 0$ with a sum beginning at $j = 1$. This is valid because the general term j is zero for $j = 0$ and may be dropped. Again, we find that $\lim_{N \to \infty} L_N = \lim_{N \to \infty} (8 - 8/N) = 8$.

Finally, we note that the region under the graph is a right triangle with base 4 and height 4 (Figure 11), so its area is $(\frac{1}{2})(4)(4) = 8$. All three methods of computing the area yield the same value of 8. ∎

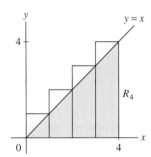 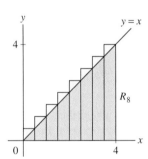 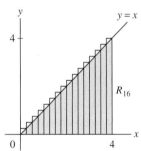 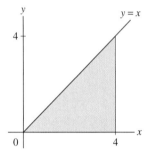

FIGURE 11 The right-endpoint approximations approach the area of the triangle.

In the next example, we compute the area under a curved graph. Unlike the previous example, it is not possible to compute the area directly using geometry.

■ **EXAMPLE 4** Let A be the area under the graph of $f(x) = 2x^2 - x + 3$ over $[2, 4]$ (Figure 12). Find a formula for R_N and compute A as the limit $\lim_{N \to \infty} R_N$.

Solution For clarity, we break up the solution into steps.

Step 1. **Express R_N in terms of power sums.**

The interval is $[2, 4]$, so $\Delta x = \dfrac{4 - 2}{N} = \dfrac{2}{N}$ and

$$R_N = \Delta x \sum_{j=1}^{N} f(a + j\Delta x) = \frac{2}{N} \sum_{j=1}^{N} f\left(2 + \frac{2j}{N}\right)$$

Let's compute the general term in this sum. Since $f(x) = 2x^2 - x + 3$,

$$f\left(2 + \frac{2j}{N}\right) = 2\left(2 + \frac{2j}{N}\right)^2 - \left(2 + \frac{2j}{N}\right) + 3$$

$$= 2\left(4 + \frac{8j}{N} + \frac{4j^2}{N^2}\right) - \left(2 + \frac{2j}{N}\right) + 3 = \frac{8}{N^2}j^2 + \frac{14}{N}j + 9$$

FIGURE 12 Area under the graph of $f(x) = 2x^2 - x + 3$ over $[2, 4]$.

Now we can express R_N in terms of power sums:

$$R_N = \frac{2}{N} \sum_{j=1}^{N} \left(\frac{8}{N^2} j^2 + \frac{14}{N} j + 9 \right) = \frac{2}{N} \sum_{j=1}^{N} \frac{8}{N^2} j^2 + \frac{2}{N} \sum_{j=1}^{N} \frac{14}{N} j + \frac{2}{N} \sum_{j=1}^{N} 9$$

$$= \frac{16}{N^3} \sum_{j=1}^{N} j^2 + \frac{28}{N^2} \sum_{j=1}^{N} j + \frac{18}{N} \sum_{j=1}^{N} 1 \qquad \boxed{7}$$

Step 2. **Use the formulas for the power sums.**

We use formulas (3) and (4) for the power sums in Eq. (7) to obtain

$$R_N = \frac{16}{N^3} \left(\frac{N^3}{3} + \frac{N^2}{2} + \frac{N}{6} \right) + \frac{28}{N^2} \left(\frac{N^2}{2} + \frac{N}{2} \right) + \frac{18}{N} (N)$$

$$= \left(\frac{16}{3} + \frac{8}{N} + \frac{8}{3N^2} \right) + \left(14 + \frac{14}{N} \right) + 18$$

$$= \frac{112}{3} + \frac{22}{N} + \frac{8}{3N^2}$$

Step 3. **Calculate the limit.**

$$\text{Area} = \lim_{N \to \infty} R_N = \lim_{N \to \infty} \left(\frac{112}{3} + \frac{22}{N} + \frac{8}{3N^2} \right) = \frac{112}{3} \qquad \blacksquare$$

We can calculate the area under the graph of a polynomial using the formulas for power sums as in the previous example. For more general functions, the area can be expressed as a limit, but the limit may be more complicated. Consider $f(x) = \sin x$ on the interval $[\frac{\pi}{4}, \frac{3\pi}{4}]$ (Figure 13). In this case,

$$\Delta x = \frac{3\pi/4 - \pi/4}{N} = \frac{\pi}{2N}$$

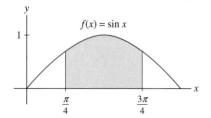

FIGURE 13 This area is more difficult to compute as a limit of endpoint approximations.

In the calculations above, we used the formulas for the kth power sums for $k = 1, 2, 3$. You may be curious whether or not similar formulas exist for all powers k. The problem of finding such formulas was studied in the seventeenth century and eventually solved around 1690 by the great Swiss mathematician Jacob Bernoulli (1654–1705). Of this discovery, he wrote

With the help of [these formulas] it took me less than half of a quarter of an hour to find that the 10th

powers of the first 1000 numbers being added together will yield the sum

914099242414242434242419242425 00

Bernoulli's formula has the general form

$$\sum_{j=1}^{n} j^k = \frac{1}{k+1} n^{k+1} + \frac{1}{2} n^k + \frac{k}{12} n^{k-1} + \cdots$$

The dots indicate terms involving smaller powers of n whose coefficients are expressed in terms of the so-called Bernoulli numbers. For example,

$$\sum_{j=1}^{n} j^4 = \frac{1}{5} n^5 + \frac{1}{2} n^4 + \frac{1}{3} n^3 - \frac{1}{30} n$$

These formulas are available on most computer algebra systems.

and the area A under the graph is equal to

$$A = \lim_{N\to\infty} R_N = \lim_{N\to\infty} \Delta x \sum_{j=1}^{N} f(a + j\Delta x) = \lim_{N\to\infty} \frac{\pi}{2N} \sum_{j=1}^{N} \sin\left(\frac{\pi}{4} + \frac{\pi j}{2N}\right)$$

Although this limit can be evaluated with some work, we will see in Section 5.3 that it is much easier to use the Fundamental Theorem of Calculus, which reduces area computations to the problem of finding antiderivatives.

5.1 SUMMARY

- Approximations to the area under the graph of $f(x)$ over $[a, b]$ $\left(\Delta x = \dfrac{b-a}{N}\right)$:

$$R_N = \Delta x \sum_{j=1}^{N} f(a + j\Delta x) = \Delta x\big(f(a + \Delta x) + f(a + 2\Delta x) + \cdots + f(a + N\Delta x)\big)$$

$$L_N = \Delta x \sum_{j=0}^{N-1} f(a + j\Delta x) = \Delta x\big(f(a) + f(a + \Delta x) + \cdots + f(a + (N-1)\Delta x)\big)$$

$$M_N = \Delta x \sum_{j=1}^{N} f\left(a + \left(j - \frac{1}{2}\right)\Delta x\right)$$

$$= \Delta x \left(f\left(a + \frac{1}{2}\Delta x\right) + \cdots + f\left(a + \left(N - \frac{1}{2}\right)\Delta x\right)\right)$$

- If $f(x)$ is continuous on $[a, b]$, then the endpoint and midpoint approximations approach one and the same limit L:

$$\lim_{N\to\infty} R_N = \lim_{N\to\infty} L_N = \lim_{N\to\infty} M_N = L$$

- If $f(x) \geq 0$ on $[a, b]$, we take L as the definition of the area under the graph of $y = f(x)$ over $[a, b]$.

5.1 EXERCISES

Preliminary Questions

1. Suppose that $[2, 5]$ is divided into six subintervals. What are the right and left endpoints of the subintervals?

2. If $f(x) = x^{-2}$ on $[3, 7]$, which is larger: R_2 or L_2?

3. Which of the following pairs of sums are *not* equal?

(a) $\displaystyle\sum_{i=1}^{4} i, \quad \sum_{\ell=1}^{4} \ell$

(b) $\displaystyle\sum_{j=1}^{4} j^2, \quad \sum_{k=2}^{5} k^2$

(c) $\displaystyle\sum_{j=1}^{4} j, \quad \sum_{i=2}^{5} (i-1)$

(d) $\displaystyle\sum_{i=1}^{4} i(i+1), \quad \sum_{j=2}^{5} (j-1)j$

4. Explain why $\displaystyle\sum_{j=1}^{100} j$ is equal to $\displaystyle\sum_{j=0}^{100} j$ but $\displaystyle\sum_{j=1}^{100} 1$ is not equal to $\displaystyle\sum_{j=0}^{100} 1$.

5. We divide the interval $[1, 5]$ into 16 subintervals.

(a) What are the left endpoints of the first and last subintervals?

(b) What are the right endpoints of the first two subintervals?

6. Are the following statements true or false?

(a) The right-endpoint rectangles lie below the graph if $f(x)$ is increasing.

(b) If $f(x)$ is monotonic, then the area under the graph lies between R_N and L_N.

(c) If $f(x)$ is constant, then the right-endpoint rectangles all have the same height.

Exercises

1. An athlete runs with velocity 4 mph for half an hour, 6 mph for the next hour, and 5 mph for another half-hour. Compute the total distance traveled and indicate on a graph how this quantity can be interpreted as an area.

2. Figure 14 shows the velocity of an object over a 3-min interval. Determine the distance traveled over the intervals [0, 3] and [1, 2.5] (remember to convert from miles per hour to miles per minute).

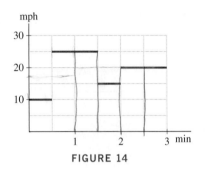

FIGURE 14

3. A rainstorm hit Portland, Maine, in October 1996, resulting in record rainfall. The rainfall rate $R(t)$ on October 21 is recorded, in inches per hour, in the following table, where t is the number of hours since midnight. Compute the total rainfall during this 24-hour period and indicate on a graph how this quantity can be interpreted as an area.

t	0–2	2–4	4–9	9–12	12–20	20–24
$R(t)$	0.2	0.1	0.4	1.0	0.6	0.25

4. The velocity of an object is $v(t) = 32t$ ft/s. Use Eq. (2) and geometry to find the distance traveled over the time intervals [0, 2] and [2, 5].

5. Compute R_6, L_6, and M_3 to estimate the distance traveled over [0, 3] if the velocity at half-second intervals is as follows:

t (s)	0	0.5	1	1.5	2	2.5	3
v (ft/s)	0	12	18	25	20	14	20

6. Use the following table of values to estimate the area under the graph of $f(x)$ over [0, 1] by computing the average of R_5 and L_5.

x	0	0.2	0.4	0.6	0.8	1
$f(x)$	50	48	46	44	42	40

7. Consider $f(x) = 2x + 3$ on [0, 3].

(a) Compute R_6 and L_6 over [0, 3].

(b) Find the error in these approximations by computing the area exactly using geometry.

8. Let $f(x) = x^2 + x - 2$.

(a) Calculate R_3 and L_3 over [2, 5].

(b) Sketch the graph of f and the rectangles that make up each approximation. Is the area under the graph larger or smaller than R_3? Than L_3?

9. Estimate R_6, L_6, and M_6 over [0, 1.5] for the function in Figure 15.

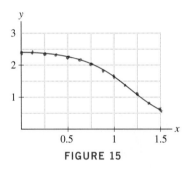

FIGURE 15

10. Estimate R_2, M_3, and L_6 for the graph in Figure 16.

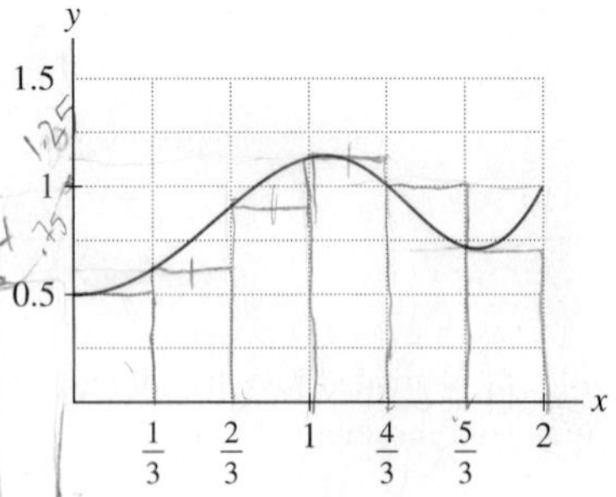

FIGURE 16

11. Let $f(x) = \sqrt{x^2 + 1}$ and $\Delta x = \frac{1}{3}$. Sketch the graph of $f(x)$ and draw the rectangles whose area is represented by the sum $\sum_{i=1}^{6} f(1 + i\Delta x)\Delta x$.

12. Calculate the area of the shaded rectangles in Figure 17. Which approximation do these rectangles represent?

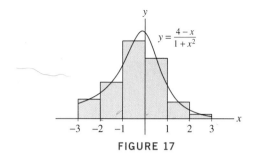

$$y = \frac{4-x}{1+x^2}$$

$-3 \quad -2 \quad -1 \qquad 1 \quad 2 \quad 3 \qquad x$

FIGURE 17

In Exercises 13–24, calculate the approximation for the given function and interval.

13. R_8, $f(x) = 7 - x$, $[3, 5]$

14. M_4, $f(x) = 7 - x$, $[3, 5]$

15. M_4, $f(x) = x^2$, $[0, 1]$

16. M_6, $f(x) = \sqrt{x}$, $[2, 5]$

17. R_6, $f(x) = 2x^2 - x + 2$, $[1, 4]$

18. L_6, $f(x) = 2x^2 - x + 2$, $[1, 4]$

19. L_5, $f(x) = x^{-1}$, $[1, 2]$

20. M_4, $f(x) = x^{-2}$, $[1, 3]$

21. L_4, $f(x) = \cos x$, $[\frac{\pi}{4}, \frac{\pi}{2}]$

22. R_6, $f(x) = e^x$, $[0, 2]$

23. M_6, $f(x) = \ln x$, $[1, 2]$

24. L_5, $f(x) = x^2 + 3|x|$, $[-3, 2]$

In Exercises 25–28, use the Graphical Insight on page 303 to obtain bounds on the area.

25. Let A be the area under the graph of $f(x) = \sqrt{x}$ over $[0, 1]$. Prove that $0.51 \le A \le 0.77$ by computing R_4 and L_4. Explain your reasoning.

26. Use R_6 and L_6 to show that the area A under $y = x^{-2}$ over $[10, 12]$ satisfies $0.0161 \le A \le 0.0172$.

27. Use R_4 and L_4 to show that the area A under the graph of $y = \sin x$ over $[0, \pi/2]$ satisfies $0.79 \le A \le 1.19$.

28. Show that the area A under the graph of $f(x) = x^{-1}$ over $[1, 8]$ satisfies

$$\frac{1}{2} + \frac{1}{3} + \frac{1}{4} + \frac{1}{5} + \frac{1}{6} + \frac{1}{7} + \frac{1}{8} \le A \le 1 + \frac{1}{2} + \frac{1}{3} + \frac{1}{4} + \frac{1}{5} + \frac{1}{6} + \frac{1}{7}$$

29. $\mathcal{LAS}$ Show that the area A in Exercise 25 satisfies $L_N \le A \le R_N$ for all N. Then use a computer algebra system to calculate L_N and R_N for $N = 100$ and 150. Which of these calculations allows you to conclude that $A \approx 0.66$ to two decimal places?

30. $\mathcal{LAS}$ Show that the area A in Exercise 26 satisfies $R_N \le A \le L_N$ for all N. Use a computer algebra system to calculate L_N and R_N for N sufficiently large to determine A to within an error of at most 10^{-4}.

31. Calculate the following sums:

(a) $\sum\limits_{i=1}^{5} 3$ (b) $\sum\limits_{i=0}^{5} 3$ (c) $\sum\limits_{k=2}^{4} k^3$

(d) $\sum\limits_{j=3}^{4} \sin\left(j\frac{\pi}{2}\right)$ (e) $\sum\limits_{k=2}^{4} \frac{1}{k-1}$ (f) $\sum\limits_{j=0}^{3} 3^j$

32. Let $b_1 = 3$, $b_2 = 1$, $b_3 = 17$, and $b_4 = -17$. Calculate the sums.

(a) $\sum\limits_{i=2}^{4} b_i$ (b) $\sum\limits_{j=1}^{2} (b_j + 2b_j)$ (c) $\sum\limits_{k=1}^{3} \frac{b_k}{b_{k+1}}$

33. Calculate $\sum\limits_{j=101}^{200} j$ by writing it as a difference of two sums and using formula (3).

In Exercises 34–39, write the sum in summation notation.

34. $4^7 + 5^7 + 6^7 + 7^7 + 8^7$

35. $(2^2 + 2) + (3^2 + 3) + (4^2 + 4) + (5^2 + 5)$

36. $(2^2 + 2) + (2^3 + 2) + (2^4 + 2) + (2^5 + 2)$

37. $\sqrt{1 + 1^3} + \sqrt{2 + 2^3} + \cdots + \sqrt{n + n^3}$

38. $\frac{1}{2 \cdot 3} + \frac{2}{3 \cdot 4} + \cdots + \frac{n}{(n+1)(n+2)}$

39. $e^\pi + e^{\pi/2} + e^{\pi/3} + \cdots + e^{\pi/n}$

In Exercises 40–47, use linearity and formulas (3)–(5) to rewrite and evaluate the sums.

40. $\sum\limits_{j=1}^{15} 12j^3$ **41.** $\sum\limits_{k=1}^{20} (2k + 1)$

42. $\sum\limits_{k=51}^{150} (2k + 1)$ **43.** $\sum\limits_{k=100}^{200} k^3$

44. $\sum\limits_{\ell=1}^{10} (\ell^3 - 2\ell^2)$ **45.** $\sum\limits_{j=2}^{30} \left(6j + \frac{4j^2}{3}\right)$

46. $\sum\limits_{j=0}^{50} j(j-1)$ **47.** $\sum\limits_{s=1}^{30} (3s^2 - 4s - 1)$

In Exercises 48–51, calculate the sum, assuming that $a_1 = -1$, $\sum\limits_{i=1}^{10} a_i = 10$, and $\sum\limits_{i=1}^{10} b_i = 7$.

48. $\sum\limits_{i=1}^{10} 2a_i$ **49.** $\sum\limits_{i=1}^{10} (a_i - b_i)$

50. $\displaystyle\sum_{\ell=1}^{10}(3a_\ell + 4b_\ell)$

51. $\displaystyle\sum_{i=2}^{10}a_i$

In Exercises 52–55, use formulas (3)–(5) to evaluate the limit.

52. $\displaystyle\lim_{N\to\infty}\sum_{i=1}^{N}\frac{i}{N^2}$

53. $\displaystyle\lim_{N\to\infty}\sum_{j=1}^{N}\frac{j^3}{N^4}$

54. $\displaystyle\lim_{N\to\infty}\sum_{i=1}^{N}\frac{i^2-i+1}{N^3}$

55. $\displaystyle\lim_{N\to\infty}\sum_{i=1}^{N}\left(\frac{i^3}{N^4}-\frac{20}{N}\right)$

In Exercises 56–59, calculate the limit for the given function and interval. Verify your answer by using geometry.

56. $\displaystyle\lim_{N\to\infty}R_N,\quad f(x)=5x,\quad [0,3]$

57. $\displaystyle\lim_{N\to\infty}L_N,\quad f(x)=5x,\quad [1,3]$

58. $\displaystyle\lim_{N\to\infty}L_N,\quad f(x)=6-2x,\quad [0,2]$

59. $\displaystyle\lim_{N\to\infty}M_N,\quad f(x)=x,\quad [0,1]$

In Exercises 60–69, find a formula for R_N for the given function and interval. Then compute the area under the graph as a limit.

60. $f(x)=x^2,\quad [0,1]$

61. $f(x)=x^3,\quad [0,1]$

62. $f(x)=x^3+2x^2,\quad [0,3]$

63. $f(x)=1-x^3,\quad [0,1]$

64. $f(x)=3x^2-x+4,\quad [0,1]$

65. $f(x)=3x^2-x+4,\quad [1,5]$

66. $f(x)=2x+7,\quad [3,6]$

67. $f(x)=x^2,\quad [2,4]$

68. $f(x)=2x+1,\quad [a,b]\quad (a,b\text{ constants with }a<b)$

69. $f(x)=x^2,\quad [a,b]\quad (a,b\text{ constants with }a<b)$

70. Let A be the area under the graph of $y=e^x$ for $0\le x\le 1$ [Figure 18(A)]. In this exercise, we evaluate A using the formula for a geometric sum (valid for $r\ne 1$):

$$1+r+r^2+\cdots+r^{N-1}=\sum_{j=1}^{N-1}r^j=\frac{r^N-1}{r-1}\qquad\boxed{8}$$

(a) Show that the left-endpoint approximation to A is

$$L_N=\frac{1}{N}\sum_{j=0}^{N-1}e^{j/N}$$

(b) Apply Eq. (8) with $r=e^{1/N}$ to prove that

$$A=(e-1)\lim_{N\to\infty}\frac{1}{N(e^{1/N}-1)}$$

(c) Evaluate the limit in Figure 18(B) and calculate A. *Hint:* Show that L'Hôpital's Rule may be used after writing

$$\frac{1}{N(e^{1/N}-1)}=\frac{N^{-1}}{e^{1/N}-1}$$

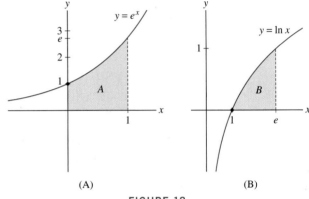

(A)   (B)

FIGURE 18

71. Use the result of Exercise 70 to show that the area B under the graph of $f(x)=\ln x$ over $[1,e]$ is equal to 1. *Hint:* Relate B to the area A computed in Exercise 70.

In Exercises 72–75, describe the area represented by the limits.

72. $\displaystyle\lim_{N\to\infty}\frac{1}{N}\sum_{j=1}^{N}\left(\frac{j}{N}\right)^4$

73. $\displaystyle\lim_{N\to\infty}\frac{3}{N}\sum_{j=1}^{N}\left(2+\frac{3j}{N}\right)^4$

74. $\displaystyle\lim_{N\to\infty}\frac{5}{N}\sum_{j=0}^{N-1}e^{-2+5j/N}$

75. $\displaystyle\lim_{N\to\infty}\frac{\pi}{2N}\sum_{j=1}^{N}\sin\left(\frac{\pi}{3}+\frac{j\pi}{2N}\right)$

76. Evaluate $\displaystyle\lim_{N\to\infty}\frac{1}{N}\sum_{j=1}^{N}\sqrt{1-\left(\frac{j}{N}\right)^2}$ by interpreting it as the area of part of a familiar geometric figure.

In Exercises 77–82, use the approximation indicated (in summation notation) to express the area under the graph as a limit but do not evaluate.

77. $R_N,\quad f(x)=\sin x$ over $[0,\pi]$

78. $R_N,\quad f(x)=x^{-1}$ over $[1,7]$

79. $M_N,\quad f(x)=\tan x$ over $[\frac{1}{2},1]$

80. $M_N,\quad f(x)=x^{-2}$ over $[3,5]$

81. L_N, $f(x) = \cos x$ over $[\frac{\pi}{8}, \pi]$

82. L_N, $f(x) = \cos x$ over $[\frac{\pi}{8}, \frac{\pi}{4}]$

In Exercises 83–85, let $f(x) = x^2$ and let R_N, L_N, and M_N be the approximations for the interval $[0, 1]$.

83. Show that $R_N = \frac{1}{3} + \frac{1}{2N} + \frac{1}{6N^2}$. Interpret the quantity $\frac{1}{2N} + \frac{1}{6N^2}$ as the area of a region.

84. Show that

$$L_N = \frac{1}{3} - \frac{1}{2N} + \frac{1}{6N^2}, \qquad M_N = \frac{1}{3} - \frac{1}{12N^2}$$

Then rank the three approximations R_N, L_N, and M_N in order of increasing accuracy (use the formula for R_N in Exercise 83).

85. For each of R_N, L_N, and M_N, find the smallest integer N for which the error is less than 0.001.

Further Insights and Challenges

86. Although the accuracy of R_N generally improves as N increases, this need not be true for small values of N. Draw the graph of a positive continuous function $f(x)$ on an interval such that R_1 is closer than R_2 to the exact area under the graph. Can such a function be monotonic?

87. Draw the graph of a positive continuous function on an interval such that R_2 and L_2 are both smaller than the exact area under the graph. Can such a function be monotonic?

88. Explain the following statement graphically: *The endpoint approximations are less accurate when $f'(x)$ is large.*

89. Assume that $f(x)$ is monotonic. Prove that M_N lies between R_N and L_N and that M_N is closer to the actual area under the graph than both R_N and L_N. *Hint:* Argue from Figure 19; the part of the error in R_N due to the ith rectangle is the sum of the areas $A + B + D$, and for M_N it is $|B - E|$.

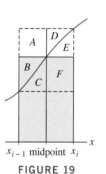

FIGURE 19

90. Prove that for any function $f(x)$ on $[a, b]$,

$$R_N - L_N = \frac{b - a}{N}(f(b) - f(a)) \qquad \boxed{9}$$

91. In this exercise, we prove that the limits $\lim_{N\to\infty} R_N$ and $\lim_{N\to\infty} L_N$ exist and are equal if $f(x)$ is positive and increasing [the case of $f(x)$ decreasing is similar]. We use the concept of a least upper bound discussed in Appendix B.

(a) Explain with a graph why $L_N \le R_M$ for all $N, M \ge 1$.

(b) By part (a), the sequence $\{L_N\}$ is bounded by R_M for any M, so it has a least upper bound L. By definition, L is the smallest number such that $L_N \le L$ for all N. Show that $L \le R_M$ for all M.

(c) According to part (b), $L_N \le L \le R_N$ for all N. Use Eq. (9) to show that $\lim_{N\to\infty} L_N = L$ and $\lim_{N\to\infty} R_N = L$.

92. Assume that $f(x)$ is positive and monotonic, and let A be the area under its graph over $[a, b]$. Use Eq. (9) to show that

$$|R_N - A| \le \frac{b - a}{N}|f(b) - f(a)| \qquad \boxed{10}$$

In Exercises 93–94, use Eq. (10) to find a value of N such that $|R_N - A| < 10^{-4}$ for the given function and interval.

93. $f(x) = \sqrt{x}$, $[1, 4]$

94. $f(x) = \sqrt{9 - x^2}$, $[0, 3]$

5.2 The Definite Integral

In the previous section we saw that if $f(x)$ is continuous on an interval $[a, b]$, then the endpoint and midpoint approximations approach a common limit L as $N \to \infty$:

$$L = \lim_{N\to\infty} R_N = \lim_{N\to\infty} L_N = \lim_{N\to\infty} M_N \qquad \boxed{1}$$

Furthermore, when $f(x) \ge 0$, we interpret L as the area under the graph of $f(x)$ over $[a, b]$. In a moment, we will state formally that the limit L is the *definite integral* of $f(x)$ over $[a, b]$. Before doing so, we introduce **Riemann sums**, which are more general approximations to area.

Georg Friedrich Riemann (1826–1866)

One of the greatest mathematicians of the nineteenth century and perhaps second only to his teacher C. F. Gauss, Riemann transformed the fields of geometry, analysis, and number theory. In 1916, Albert Einstein based his General Theory of Relativity *on Riemann's geometry. The "Riemann hypothesis" dealing with prime numbers is one of the great unsolved problems in present-day mathematics. The Clay Foundation has offered a $1 million prize for its solution (http://www.claymath.org/millennium).*

In the endpoint and midpoint approximations, we approximate the area under the graph by rectangles of equal width. The heights of the rectangles are the values of $f(x)$ at the endpoints or midpoints of the subintervals. In Riemann sum approximations, we relax these requirements; the rectangles need not have equal width and the height of a rectangle may be *any* value of $f(x)$ within the subinterval. More formally, to define a Riemann sum, we choose a partition and a set of intermediate points. A **partition** P of length N is any choice of points in $[a, b]$ that divides the interval into N subintervals:

$$\text{Partition } P: \quad a = x_0 < x_1 < x_2 \quad \cdots \quad < x_N = b$$

$$\text{Subintervals:} \quad [x_0, x_1], \quad [x_1, x_2], \quad \ldots \quad [x_{N-1}, x_N]$$

We denote the length of the ith interval by

$$\Delta x_i = x_i - x_{i-1}$$

The maximum of the lengths Δx_i is called the **norm** of the partition P and is denoted $\|P\|$. A **set of intermediate points** is a set of points $C = \{c_1, \ldots, c_N\}$, where c_i belongs to $[x_{i-1}, x_i]$ [Figure 1(A)].

Now, over each subinterval $[x_{i-1}, x_i]$ we construct the rectangle of height $f(c_i)$ and base Δx_i, as in Figure 1(B). This rectangle has area $f(c_i)\Delta x_i$. If $f(c_i) < 0$, the rectangle extends below the x-axis and the "area" $f(c_i)\Delta x_i$ is negative. The Riemann sum is the sum of these positive or negative areas:

$$R(f, P, C) = \sum_{i=1}^{N} f(c_i)\Delta x_i = f(c_1)\Delta x_1 + f(c_2)\Delta x_2 + \cdots + f(c_N)\Delta x_N$$

2

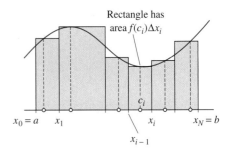

(A) Partition of $[a, b]$ into subintervals

(B) Construct rectangle above each subinterval of height $f(c_i)$

FIGURE 1 The Riemann sum $R(f, P, C)$ is the sum of the areas of the rectangles.

The endpoint and midpoint approximations are examples of Riemann sums. They correspond to the partition of $[a, b]$ into N subintervals of equal length $\Delta x = \dfrac{b - a}{N}$ and the choice of right endpoints, left endpoints, or midpoints as intermediate points. In the next example, we consider a Riemann sum with intervals of unequal length.

■ **EXAMPLE 1** Let $f(x) = 8 + 12\sin x - 4x$. Calculate $R(f, P, C)$ for the partition P of $[0, 4]$:

$$P: \quad x_0 = 0 < x_1 = 1 < x_2 = 1.8 < x_3 = 2.9 < x_4 = 4$$

and intermediate points $C = \{0.4, 1.2, 2, 3.5\}$. What is the norm $\|P\|$?

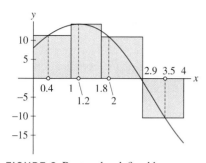

FIGURE 2 Rectangles defined by a Riemann sum for
$f(x) = 8 + 12\sin x - 4x$ on $[0, 4]$.

Solution The widths of the subintervals in the partition (Figure 2) are

$$\Delta x_1 = x_1 - x_0 = 1 - 0 = 1, \qquad \Delta x_2 = x_2 - x_1 = 1.8 - 1 = 0.8$$

$$\Delta x_3 = x_3 - x_2 = 2.9 - 1.8 = 1.1, \qquad \Delta x_4 = x_4 - x_3 = 4 - 2.9 = 1.1$$

The norm of the partition is $\|P\| = 1.1$ since the two longest subintervals have width 1.1. Using a calculator, we obtain:

$$R(f, P, C) = f(0.4)\Delta x_1 + f(1.2)\Delta x_2 + f(2)\Delta x_3 + f(3.5)\Delta x_4$$

$$\approx 11.07(1) + 14.38(0.8) + 10.91(1.1) - 10.2(1.1) \approx 23.35 \qquad ■$$

Having introduced Riemann sums, we are ready to make the following definition: A function is **integrable** over $[a, b]$ if *all* of the Riemann sums (not just the endpoint and midpoint approximations) approach one and the same limit L as the norm of the partition tends to zero. More formally, we write

$$L = \lim_{\|P\| \to 0} R(f, P, C) = \lim_{\|P\| \to 0} \sum_{i=1}^{N} f(c_i)\,\Delta x_i \qquad \boxed{3}$$

if $|R(f, P, C) - L|$ gets arbitrarily small as the norm $\|P\|$ tends to zero, no matter how we choose the partition and intermediate points. Note that as $\|P\| \to 0$, the number N of intervals tends to ∞. The limit L is called the **definite integral** of $f(x)$ over $[a, b]$.

The notation $\int f(x)\,dx$, introduced by Leibniz in 1686, is appropriate because it reminds us of the Riemann sums used to define the definite integral. The symbol $\int$ is an elongated S standing for summation, and dx corresponds to the lengths Δx_i along the x-axis.

DEFINITION Definite Integral   The definite integral of $f(x)$ over $[a, b]$ is the limit of Riemann sums and is denoted by the integral sign:

$$\int_a^b f(x)\,dx = \lim_{\|P\| \to 0} R(f, P, C) = \lim_{\|P\| \to 0} \sum_{i=1}^{N} f(c_i)\,\Delta x_i$$

When this limit exists, we say that $f(x)$ is integrable over $[a, b]$.

Remarks

- We often refer to the definite integral more simply as the *integral* of f over $[a, b]$.
- The function $f(x)$ inside the integral sign is called the **integrand**.
- The numbers a and b representing the interval $[a, b]$ are called the **limits of integration**.
- Any variable may be used as a variable of integration (this is a "dummy" variable). Thus, the following three integrals all denote the same quantity:

$$\int_a^b f(x)\,dx, \qquad \int_a^b f(t)\,dt, \qquad \int_a^b f(u)\,du$$

- Recall that the *indefinite integral* $\int f(x)\,dx$ denotes the general antiderivative of $f(x)$ (Section 4.9). The connection between definite and indefinite integrals is explained in Section 5.3.

Most functions that arise in practice are integrable. In particular, every continuous function is integrable, as stated in the following theorem. A proof is given in Appendix D.

> **THEOREM 1**   If $f(x)$ is continuous on $[a, b]$, then $f(x)$ is integrable over $[a, b]$.

CONCEPTUAL INSIGHT   General Riemann sums are rarely used for computations. In practice, we compute definite integrals using either particular approximations such as the midpoint approximation or the Fundamental Theorem of Calculus discussed in the next section. If so, why bother introducing Riemann sums? The answer is that Riemann sums play a theoretical rather than a computational role. They are useful in proofs and for dealing rigorously with certain discontinuous functions. In later sections, Riemann sums are used to show that volumes and other quantities can be expressed as definite integrals.

Graphical Interpretation of the Definite Integral

When $f(x)$ is continuous and positive, we interpret the definite integral as the area under the graph. If $f(x)$ is not positive, then the definite integral is not equal to an area in the usual sense, but we may interpret it as the **signed area** between the graph and the x-axis. Signed area is defined as follows:

> Signed area of a region = (area above x-axis) − (area below x-axis)

Thus, signed area treats regions under the x-axis as "negative area" (Figure 3).

To see why the definite integral gives us signed area, suppose first that $f(x)$ is negative on $[a, b]$ as in Figure 4(A) and consider a Riemann sum:

$$R(f, C, P) = f(c_1)\Delta x_1 + f(c_2)\Delta x_2 + \cdots + f(c_N)\Delta x_N$$

Since $f(c_i) < 0$, each term is equal to the negative of the area of a rectangle:

$$f(c_i)\Delta x_i = -\big(\text{area of rectangle of height } |f(c_i)|\big)$$

Therefore, the Riemann sums $R(f, C, P)$ converge to the negative of the area between the graph and the x-axis. If $f(x)$ takes on both positive and negative values as in Figure 4(B), then $f(c_i)\Delta x_i$ is positive or negative, depending on whether the corresponding rectangle lies above or below the x-axis. In this case, the Riemann sums converge to the signed area. In summary,

FIGURE 3 The signed area is the area above the x-axis minus the area below the x-axis.

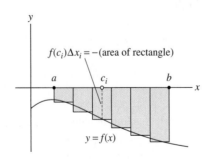

(A) Approximation to the definite integral of a negative function

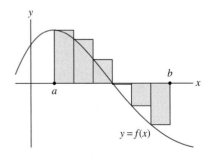

(B) Approximation to the definite integral of a function taking positive and negative values

FIGURE 4

$$\int_a^b f(x)\,dx = \text{signed area of region between the graph and } x\text{-axis over } [a, b]$$

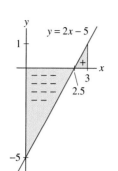

FIGURE 5 Signed area under the graph of $f(x) = 2x - 5$.

■ **EXAMPLE 2** Calculating Signed Area Calculate $\int_0^3 (2x - 5)\,dx$ using the interpretation as signed area.

Solution Figure 5 shows that the region between the graph and the x-axis consists of two triangles. The first triangle has area $(\frac{1}{2})(2.5)(5) = \frac{25}{4}$, but it lies below the x-axis, so its signed area is $-\frac{25}{4}$. The second triangle lies above the x-axis and has area $(\frac{1}{2})(0.5)(1) = \frac{1}{4}$. Therefore,

$$\int_0^3 (2x - 5)\,dx = -\frac{25}{4} + \frac{1}{4} = -6$$ ■

■ **EXAMPLE 3** Integral of an Absolute Value Calculate $\int_0^4 |3 - x|\,dx$.

Solution The graph of $y = |3 - x|$ in Figure 6(B) is derived from the graph of $y = 3 - x$ by reflecting it through the x-axis for $x \geq 3$. We see that the integral is equal to the sum of the areas of two triangles:

$$\int_0^4 |3 - x|\,dx = \frac{1}{2}(3)(3) + \frac{1}{2}(1)(1) = 5$$ ■

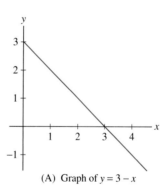   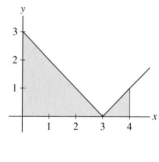

(A) Graph of $y = 3 - x$ (B) Graph of $y = |3 - x|$

FIGURE 6

In the next example, we compute the integral of e^x as a limit of left-endpoint approximations (Figure 7). In the next section, we will derive this same result with less effort using the Fundamental Theorem of Calculus.

■ **EXAMPLE 4** The Integral of $f(x) = e^x$ Prove the formula

$$\boxed{\int_0^b e^x\,dx = e^b - 1}$$ **4**

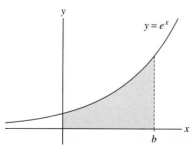

FIGURE 7 The area of the shaded region is equal to $e^b - 1$.

Solution We compute the integral as a limit of left-endpoint approximations L_N, using the formula for a geometric sum to evaluate L_N and then L'Hôpital's Rule.

Step 1. **Evaluate L_N as a geometric sum.**

We integrate over $[0, b]$, so $\Delta x = b/N$ and the left-endpoint approximation is

$$L_N = \frac{b}{N}\left(f(0) + f\left(\frac{b}{N}\right) + + f\left(\frac{2b}{N}\right) + \cdots + f\left(\frac{(N-1)b}{N}\right)\right)$$

$$= \frac{b}{N}\left(e^0 + e^{b/N} + e^{2b/N} + \cdots + e^{(N-1)b/N}\right)$$

The following formula for the geometric sum *is valid for all $r \neq 1$*:

$$1 + r + r^2 + \cdots + r^{N-1} = \frac{r^N - 1}{r - 1} \quad \boxed{5}$$

Now let $r = e^{b/N}$. Then $r^j = e^{jb/N}$ and $r^N = e^b$. We may write L_N in terms of r and evaluate using Eq. (5) in the margin:

$$L_N = \frac{b}{N}\left(1 + r + r^2 + \cdots + r^{N-1}\right) = \frac{b}{N}\left(\frac{r^N - 1}{r - 1}\right) = \frac{b}{N}\left(\frac{e^b - 1}{e^{b/N} - 1}\right)$$

The integral is equal to the limit

$$\int_0^b e^x\, dx = \lim_{N\to\infty} L_N = \lim_{N\to\infty} \frac{b}{N}\left(\frac{e^b - 1}{e^{b/N} - 1}\right) = b(e^b - 1)\lim_{N\to\infty}\frac{1}{N(e^{b/N} - 1)}$$

Step 2. **Use L'Hôpital's Rule to evaluate the limit.**

We observe that

$$\frac{1}{N(e^{b/N} - 1)} = \frac{N^{-1}}{e^{b/N} - 1}$$

The fraction on the right yields the indeterminate form $0/0$ as $N \to \infty$ since $e^{b/N} - 1$ approaches $e^0 - 1 = 0$. Therefore, we may apply L'Hôpital's Rule (with respect to the variable N):

$$\lim_{N\to\infty}\frac{1}{N(e^{b/N} - 1)} = \overbrace{\lim_{N\to\infty}\frac{N^{-1}}{e^{b/N} - 1} = \lim_{N\to\infty}\left(\frac{-N^{-2}}{-bN^{-2}e^{b/N}}\right)}^{\text{L'Hôpital's Rule}}$$

$$= \frac{1}{b}\lim_{N\to\infty}e^{-b/N} = \frac{1}{b}e^0 = \frac{1}{b}$$

Thus, we obtain the desired result:

$$\int_0^b e^x\, dx = b(e^b - 1)\lim_{N\to\infty}\frac{1}{N(e^{b/N} - 1)} = b(e^b - 1)\frac{1}{b} = e^b - 1 \qquad \blacksquare$$

Properties of the Definite Integral

In the rest of this section, we discuss some basic properties of definite integrals. First, we note that the integral of a constant function $f(x) = C$ over $[a, b]$ is the signed area $C(b - a)$ of a rectangle as shown in Figure 8.

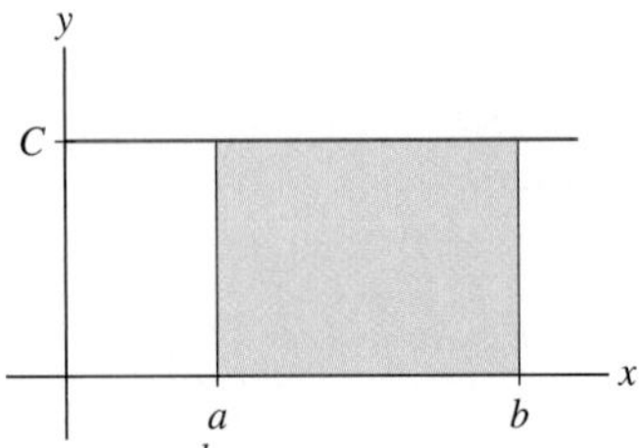

FIGURE 8 $\displaystyle\int_a^b C\, dx = C(b - a).$

> **THEOREM 2 Integral of a Constant**   For any constant C, $\displaystyle\int_a^b C\, dx = C(b - a)$.

Next, we state the linearity properties of the definite integral.

> **THEOREM 3 Linearity of the Definite Integral** If $f(x)$ and $g(x)$ are integrable over $[a, b]$, then
>
> $$\int_a^b \left(f(x) + g(x) \right) dx = \int_a^b f(x)\, dx + \int_a^b g(x)\, dx$$
>
> $$\int_a^b C f(x)\, dx = C \int_a^b f(x)\, dx \quad \text{for any constant } C$$

Proof These properties follow from the corresponding linearity properties of sums and limits. To prove additivity, we observe that Riemann sums are additive:

$$R(f + g, P, C) = \sum_{i=1}^N \left(f(c_i) + g(c_i) \right) \Delta x_i = \sum_{i=1}^N f(c_i)\, \Delta x_i + \sum_{i=1}^N g(c_i)\, \Delta x_i$$

$$= R(f, P, C) + R(g, P, C)$$

By the additivity of limits,

$$\int_a^b (f(x) + g(x))\, dx = \lim_{\|P\| \to 0} R(f + g, P, C)$$

$$= \lim_{\|P\| \to 0} R(f, P, C) + \lim_{\|P\| \to 0} R(g, P, C)$$

$$= \int_a^b f(x)\, dx + \int_a^b g(x)\, dx$$

The second property is proved similarly. ∎

In Example 5, the function $f(x) = x^2$ is continuous, hence integrable, so all Riemann sum approximations approach the same limit. We would obtain the same answer using L_N or M_N. The formula (6) is valid for all b (see Exercise 87).

■ **EXAMPLE 5** Use the limit of right-endpoint approximations to prove that for $b > 0$

$$\int_0^b x^2\, dx = \frac{b^3}{3} \qquad \boxed{6}$$

Then use the Linearity Rules to calculate $\int_0^3 (2x^2 - 5)\, dx$.

Solution In the right-endpoint approximation for $f(x) = x^2$ on $[0, b]$, we have $\Delta x = b/N$. Using the formula for the power sum recalled in the margin, we obtain

←·· *REMINDER*

$$\sum_{j=1}^N j^2 = \frac{N^3}{3} + \frac{N^2}{2} + \frac{N}{6}$$

$$R_N = \Delta x \sum_{j=1}^N f(0 + j\, \Delta x) = \frac{b}{N} \sum_{j=1}^N \left(\frac{jb}{N} \right)^2 = \frac{b^3}{N^3} \sum_{j=1}^N j^2$$

$$= \frac{b^3}{N^3} \left(\frac{N^3}{3} + \frac{N^2}{2} + \frac{N}{6} \right) = \frac{b^3}{3} + \frac{b^3}{2N} + \frac{b^3}{6N^2}$$

We prove Eq. (6) by taking the limit:

$$\int_0^b x^2\, dx = \lim_{N \to \infty} \left(\frac{b^3}{3} + \frac{b^3}{2N} + \frac{b^3}{6N^2} \right) = \frac{b^3}{3}$$

Next, by linearity and Theorem 2,

$$\int_0^3 (2x^2 - 5)\, dx = \int_0^3 2x^2\, dx + \int_0^3 (-5)\, dx = 2\int_0^3 x^2\, dx - \int_0^3 5\, dx$$

$$= 2\left(\frac{3^3}{3}\right) - 5(3 - 0) = 3 \qquad \blacksquare$$

So far we have used the symbol $\int_a^b f(x)\, dx$ with the understanding that $a < b$. It is convenient to define the definite integral for arbitrary a and b.

DEFINITION Reversing the Limits of Integration   For $a < b$, we set

$$\int_b^a f(x)\, dx = -\int_a^b f(x)\, dx$$

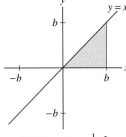

(A) Signed area $\frac{1}{2}b^2$

For example, using the formula $\int_0^b x^2\, dx = b^3/3$ from Example 5 with $b = 5$, we have

$$\int_5^0 x^2\, dx = -\int_0^5 x^2\, dx = -\frac{5^3}{3} = -\frac{125}{3}$$

When $a = b$, the interval $[a, b] = [a, a]$ has length zero and we define the definite integral to be zero:

$$\int_a^a f(x)\, dx = 0$$

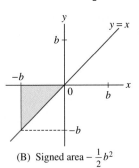

(B) Signed area $-\frac{1}{2}b^2$

FIGURE 9

■ **EXAMPLE 6**   Prove for all b (positive or negative)

$$\int_0^b x\, dx = \frac{1}{2}b^2 \qquad \boxed{7}$$

Solution   If $b > 0$, $\int_0^b x\, dx$ is the area of the triangle of height b and base b in Figure 9(A). The area is $\frac{1}{2}b^2$. For $b < 0$, $\int_b^0 x\, dx$ is the signed area of the triangle in Figure 9(B), which equals $-\frac{1}{2}b^2$. Therefore,

$$\int_0^b x\, dx = -\int_b^0 x\, dx = -\left(-\frac{1}{2}b^2\right) = \frac{1}{2}b^2 \qquad \blacksquare$$

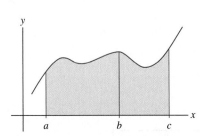

FIGURE 10 The area over $[a, c]$ is the *sum* of the areas over $[a, b]$ and $[b, c]$.

Definite integrals satisfy an important additivity property. If $f(x)$ is integrable and $a \leq b \leq c$ as in Figure 10, then the integral from a to c is equal to the integral from a to b *plus* the integral from b to c. We state this formally in the next theorem (proof omitted).

The property stated in Theorem 4 remains true as stated for all a, b, c even if the condition $a \leq b \leq c$ is not satisfied (Exercise 88).

THEOREM 4 Additivity for Adjacent Intervals For $a \leq b \leq c$,

$$\int_a^c f(x)\,dx = \int_a^b f(x)\,dx + \int_b^c f(x)\,dx$$

■ **EXAMPLE 7** Calculate $\int_4^7 x^2\,dx$ using the formula $\int_0^b x^2 = \dfrac{b^3}{3}$ from Example 5.

Solution By additivity for adjacent intervals, we have

$$\underbrace{\int_0^4 x^2\,dx}_{\left(\frac{1}{3}\right)4^3} + \int_4^7 x^2\,dx = \underbrace{\int_0^7 x^2\,dx}_{\left(\frac{1}{3}\right)7^3}$$

Therefore,

$$\int_4^7 x^2\,dx = \int_0^7 x^2\,dx - \int_0^4 x^2\,dx = \left(\frac{1}{3}\right)7^3 - \left(\frac{1}{3}\right)4^3 = 93 \qquad ■$$

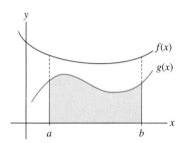

FIGURE 11 The integral of $f(x)$ is larger than the integral of $g(x)$.

Sometimes, we wish to compare the integrals of two functions. If $f(x)$ and $g(x)$ are integrable and $g(x) \leq f(x)$, then the larger function $f(x)$ has a larger integral (Figure 11).

THEOREM 5 Comparison Theorem If $g(x) \leq f(x)$ on an interval $[a, b]$, then

$$\int_a^b g(x)\,dx \leq \int_a^b f(x)\,dx$$

Proof For any partition P of $[a, b]$ with endpoints $x_1, \ldots, x_N$ and choice of intermediate points $\{c_i\}$, we have $g(c_i)\,\Delta x_i \leq f(c_i)\,\Delta x_i$ for all i. Therefore, the Riemann sums satisfy

$$\sum_{i=1}^N g(c_i)\,\Delta x_i \leq \sum_{i=1}^N f(c_i)\,\Delta x_i$$

Taking the limit as the norm $\|P\|$ tends to zero, we obtain

$$\int_a^b g(x)\,dx = \lim_{\|P\| \to 0} \sum_{i=1}^N g(c_i)\,\Delta x_i \leq \lim_{\|P\| \to 0} \sum_{i=1}^N f(c_i)\,\Delta x_i = \int_a^b f(x)\,dx \qquad ■$$

Here is a simple but useful application of the Comparison Theorem: Suppose that $f(x)$ satisfies $m \leq f(x) \leq M$ for x in $[a, b]$. The numbers m and M are called lower and upper bounds for $f(x)$ on $[a, b]$. In this case, the integral lies between the areas of two rectangles (Figure 12):

$$\int_a^b m\,dx \leq \int_a^b f(x)\,dx \leq \int_a^b M\,dx$$

This gives us

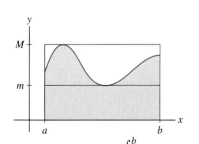

FIGURE 12 The integral $\int_a^b f(x)\,dx$ lies between the areas of the rectangles of heights m and M.

$$\boxed{m(b-a) \leq \int_a^b f(x)\,dx \leq M(b-a)} \qquad \boxed{8}$$

■ **EXAMPLE 8** Prove the inequalities:

(a) $\displaystyle\int_1^5 \frac{1}{x^2}\,dx \le \int_1^5 \frac{1}{x}\,dx$

(b) $\displaystyle\frac{3}{4} \le \int_{1/2}^2 \frac{1}{x}\,dx \le 3.$

Solution

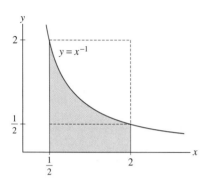

y = x⁻¹

FIGURE 13

(a) If $x \ge 1$, then $x^2 \ge x$, and hence by taking inverses, we obtain $x^{-2} \le x^{-1}$. By the Comparison Theorem,

$$\int_1^5 \frac{1}{x^2}\,dx \le \int_1^5 \frac{1}{x}\,dx$$

(b) Since $f(x) = x^{-1}$ is decreasing, its maximum on $\left[\frac{1}{2}, 2\right]$ is $M = f\left(\frac{1}{2}\right) = 2$ and its minimum is $m = f(2) = \frac{1}{2}$ (Figure 13). By Eq. (8),

$$\underbrace{\frac{1}{2}\left(2 - \frac{1}{2}\right)}_{m(b-a)} = \frac{3}{4} \le \int_{1/2}^2 \frac{1}{x}\,dx \le \underbrace{2\left(2 - \frac{1}{2}\right)}_{M(b-a)} = 3 \qquad ■$$

5.2 SUMMARY

- A Riemann sum $R(f, P, C)$ for the interval $[a, b]$ is defined by choosing a *partition*

$$P: \qquad a = x_0 < x_1 < x_2 < \cdots < x_N = b$$

and *intermediate points* $C = \{c_i\}$, where $c_i \in [x_{i-1}, x_i]$. Let $\Delta x_i = x_i - x_{i-1}$. Then

$$R(f, P, C) = \sum_{i=1}^N f(c_i)\,\Delta x_i$$

- The maximum of the widths Δx_i is called the norm $\|P\|$ of the partition.
- The *definite integral* is the limit of the Riemann sums (if it exists):

$$\int_a^b f(x)\,dx = \lim_{\|P\| \to 0} R(f, P, C)$$

We say that $f(x)$ is *integrable* over $[a, b]$ if the limit exists.

- Theorem: If $f(x)$ is continuous on $[a, b]$, then $f(x)$ is integrable over $[a, b]$.

- The definite integral $\displaystyle\int_a^b f(x)\,dx$ is equal to the signed area of the region between the graph of $f(x)$ and the x-axis.

- Properties of definite integrals [for $f(x)$ and $g(x)$ integrable]:

$$\int_a^b \big(f(x) + g(x)\big)\,dx = \int_a^b f(x)\,dx + \int_a^b g(x)\,dx$$

$$\int_a^b Cf(x)\,dx = C\int_a^b f(x)\,dx \quad \text{for any constant } C$$

$$\int_a^b f(x)\,dx + \int_b^c f(x)\,dx = \int_a^c f(x)\,dx \quad \text{for all } a, b, c$$

$$\int_a^b f(x)\,dx = -\int_b^a f(x)\,dx$$

$$\int_a^a f(x)\,dx = 0$$

- Formulas:

$$\int_a^b C\,dx = C(b-a) \quad (C \text{ any constant})$$

$$\int_0^b x\,dx = \frac{1}{2}b^2$$

$$\int_0^b x^2\,dx = \frac{1}{3}b^3$$

$$\int_0^b e^x\,dx = e^b - 1$$

- Comparison Theorem: If $f(x) \le g(x)$ on $[a, b]$, then

$$\int_a^b f(x)\,dx \le \int_a^b g(x)\,dx$$

If $m \le f(x) \le M$ on $[a, b]$, then

$$m(b-a) \le \int_a^b f(x)\,dx \le M(b-a)$$

5.2 EXERCISES

Preliminary Questions

1. What is $\int_a^b dx$ [here the function is $f(x) = 1$]?

2. Are the following statements true or false [assume that $f(x)$ is continuous]?

(a) $\int_a^b f(x)\,dx$ is the area between the graph and the x-axis over $[a, b]$.

(b) $\int_a^b f(x)\,dx$ is the area between the graph and the x-axis over $[a, b]$ if $f(x) \ge 0$.

(c) If $f(x) \le 0$, then $-\int_a^b f(x)\,dx$ is the area between the graph of $f(x)$ and the x-axis over $[a, b]$.

3. Explain graphically why $\int_0^\pi \cos x\,dx = 0$.

4. Is $\int_{-5}^{-1} 8\,dx$ negative?

5. What is the largest possible value of $\int_0^6 f(x)\,dx$ if $f(x) \le \frac{1}{3}$?

Exercises

In Exercises 1–10, draw a graph of the signed area represented by the integral and compute it using geometry.

1. $\int_{-3}^3 2x\,dx$

2. $\int_{-2}^3 (2x+4)\,dx$

3. $\int_{-2}^1 (3x+4)\,dx$

4. $\int_{-2}^1 4\,dx$

5. $\int_6^8 (7-x)\,dx$

6. $\int_{\pi/2}^{3\pi/2} \sin x\,dx$

7. $\int_0^5 \sqrt{25-x^2}\,dx$

8. $\int_{-2}^3 |x|\,dx$

9. $\int_{-2}^2 (2-|x|)\,dx$

10. $\int_{-1}^1 (2x-|x|)\,dx$

11. Calculate $\int_0^6 (4-x)\,dx$ in two ways:

(a) As the limit $\lim\limits_{N\to\infty} R_N$

(b) By sketching the relevant signed area and using geometry

12. Calculate $\int_2^5 (2x+1)\,dx$ in two ways: As the limit $\lim\limits_{N\to\infty} R_N$ and using geometry.

13. Evaluate the integrals for $f(x)$ shown in Figure 14.

(a) $\int_0^2 f(x)\,dx$ **(b)** $\int_0^6 f(x)\,dx$

(c) $\int_1^4 f(x)\,dx$ **(d)** $\int_1^6 |f(x)|\,dx$

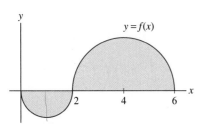

FIGURE 14 The two parts of the graph are semicircles.

In Exercises 14–15, refer to Figure 15.

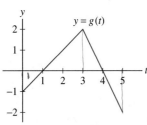

FIGURE 15

14. Evaluate $\int_0^3 g(t)\,dt$ and $\int_3^5 g(t)\,dt$.

15. Find a, b, and c such that $\int_0^a g(t)\,dt$ and $\int_b^c g(t)\,dt$ are as large as possible.

16. Describe the partition P and the set of intermediate points C for the Riemann sum shown in Figure 16. Compute the value of the Riemann sum.

In Exercises 17–22, sketch the signed area represented by the integral. Indicate the regions of positive and negative area.

17. $\int_0^2 (x-x^2)\,dx$ **18.** $\int_0^3 (2x-x^2)\,dx$

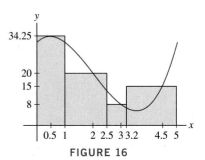

FIGURE 16

19. $\int_\pi^{2\pi} \sin x\,dx$ **20.** $\int_0^{3\pi} \sin x\,dx$

21. $\int_{1/2}^2 \ln x\,dx$ **22.** $\int_{-1}^1 \tan^{-1} x\,dx$

In Exercises 23–26, determine the sign of the integral without calculating it. Draw a graph if necessary.

23. $\int_{-2}^1 x^4\,dx$ **24.** $\int_{-2}^1 x^3\,dx$

25. GU $\int_0^{2\pi} x \sin x\,dx$ **26.** GU $\int_0^{2\pi} \dfrac{\sin x}{x}\,dx$

In Exercises 27–30, calculate the Riemann sum $R(f, P, C)$ for the given function, partition, and choice of intermediate points. Also, sketch the graph of f and the rectangles corresponding to $R(f, P, C)$.

27. $f(x) = x$, $P = \{1, 1.2, 1.5, 2\}$, $C = \{1.1, 1.4, 1.9\}$

28. $f(x) = x^2 + x$, $P = \{2, 3, 4.5, 5\}$, $C = \{2, 3.5, 5\}$

29. $f(x) = x + 1$, $P = \{-2, -1.6, -1.2, -0.8, -0.4, 0\}$, $C = \{-1.7, -1.3, -0.9, -0.5, 0\}$

30. $f(x) = \sin x$, $P = \{0, \frac{\pi}{6}, \frac{\pi}{3}, \frac{\pi}{2}\}$, $C = \{0.4, 0.7, 1.2\}$

In Exercises 31–40, use the basic properties of the integral and the formulas in the summary to calculate the integrals.

31. $\int_0^4 x^2\,dx$ **32.** $\int_1^4 x^2\,dx$

33. $\int_0^3 (3t+4)\,dt$ **34.** $\int_{-2}^3 (3x+4)\,dx$

35. $\int_0^1 (u^2 - 2u)\,du$ **36.** $\int_0^3 (6y^2 + 7y + 1)\,dy$

37. $\int_{-a}^1 (x^2 + x)\,dx$ **38.** $\int_a^{a^2} x^2\,dx$

39. $\int_0^4 e^x\,dx$ **40.** $\int_2^0 (x^2 - e^x)\,dx$

41. Prove by computing the limit of right-endpoint approximations:

$$\int_0^b x^3\,dx = \frac{b^4}{4}$$

$\boxed{9}$

In Exercises 42–49, use the formulas in the summary and Eq. (9) to evaluate the integral.

42. $\int_0^3 x^2 \, dx$

43. $\int_0^2 (x^2 + 2x) \, dx$

44. $\int_0^3 x^3 \, dx$

45. $\int_0^2 (x - x^3) \, dx$

46. $\int_0^1 (2x^3 - x + 4) \, dx$

47. $\int_{-3}^0 (2x - 5) \, dx$

48. $\int_1^3 x^3 \, dx$

49. $\int_1^2 (x - x^3) \, dx$

In Exercises 50–54, calculate the integral, assuming that

$$\int_0^5 f(x) \, dx = 5, \qquad \int_0^5 g(x) \, dx = 12$$

50. $\int_0^5 (f(x) + g(x)) \, dx$

51. $\int_0^5 (f(x) + 4g(x)) \, dx$

52. $\int_5^0 g(x) \, dx$

53. $\int_0^5 (3f(x) - 5g(x)) \, dx$

54. Is it possible to calculate $\int_0^5 g(x) f(x) \, dx$ from the information given?

In Exercises 55–58, calculate the integral, assuming that

$$\int_0^1 f(x) \, dx = 1, \qquad \int_0^2 f(x) \, dx = 4, \qquad \int_1^4 f(x) \, dx = 7$$

55. $\int_0^4 f(x) \, dx$

56. $\int_1^2 f(x) \, dx$

57. $\int_4^1 f(x) \, dx$

58. $\int_2^4 f(x) \, dx$

In Exercises 59–62, express each integral as a single integral.

59. $\int_0^3 f(x) \, dx + \int_3^7 f(x) \, dx$

60. $\int_2^9 f(x) \, dx - \int_4^9 f(x) \, dx$

61. $\int_2^9 f(x) \, dx - \int_2^5 f(x) \, dx$

62. $\int_7^3 f(x) \, dx + \int_3^9 f(x) \, dx$

In Exercises 63–66, calculate the integral, assuming that f is an integrable function such that $\int_1^b f(x) \, dx = 1 - b^{-1}$ for all $b > 0$.

63. $\int_1^3 f(x) \, dx$

64. $\int_2^4 f(x) \, dx$

65. $\int_1^4 (4f(x) - 2) \, dx$

66. $\int_{1/2}^1 f(x) \, dx$

67. Use the result of Example 4 and Theorem 4 to prove that for $b > a > 0$,

$$\int_a^b e^x \, dx = e^b - e^a$$

68. Use the result of Exercise 67 to evaluate $\int_2^4 (x - e^x) \, dx$.

69. Explain the difference in graphical interpretation between $\int_a^b f(x) \, dx$ and $\int_a^b |f(x)| \, dx$.

70. Let $I = \int_0^{2\pi} \sin^2 x \, dx$ and $J = \int_0^{2\pi} \cos^2 x \, dx$. Use the following trick to prove that $I = J = \pi$: First show with a graph that $I = J$ and then prove $I + J = \int_0^{2\pi} dx$.

In Exercises 71–74, calculate the integral.

71. $\int_0^6 |3 - x| \, dx$

72. $\int_1^3 |2x - 4| \, dx$

73. $\int_{-1}^1 |x^3| \, dx$

74. $\int_0^2 |x^2 - 1| \, dx$

75. Use the Comparison Theorem to show that

$$\int_0^1 x^5 \, dx \le \int_0^1 x^4 \, dx, \qquad \int_1^2 x^4 \, dx \le \int_1^2 x^5 \, dx$$

76. Prove that $\dfrac{1}{3} \le \displaystyle\int_4^6 \dfrac{1}{x} \, dx \le \dfrac{1}{2}$.

77. Prove that $0.0198 \le \displaystyle\int_{0.2}^{0.3} \sin x \, dx \le 0.0296$. *Hint:* Show that $0.198 \le \sin x \le 0.296$ for x in $[0.2, 0.3]$.

78. Prove that $0.277 \le \displaystyle\int_{\pi/8}^{\pi/4} \cos x \, dx \le 0.363$.

79. GU   Prove that

$$\int_{\pi/4}^{\pi/2} \frac{\sin x}{x} \, dx \le \frac{\sqrt{2}}{2}$$

Hint: Graph $y = \dfrac{\sin x}{x}$ and observe that it is decreasing on $[\frac{\pi}{4}, \frac{\pi}{2}]$.

80. Find upper and lower bounds for $\displaystyle\int_0^1 \dfrac{dx}{\sqrt{x^3 + 4}}$.

81. Suppose that $f(x) \le g(x)$ on $[a, b]$. By the Comparison Theorem, $\int_a^b f(x) \, dx \le \int_a^b g(x) \, dx$. Is it also true that $f'(x) \le g'(x)$ for $x \in [a, b]$? If not, give a counterexample.

82. State whether true or false. If false, sketch the graph of a counterexample.

(a) If $f(x) > 0$, then $\displaystyle\int_a^b f(x)\,dx > 0$.

(b) If $\displaystyle\int_a^b f(x)\,dx > 0$, then $f(x) > 0$.

Further Insights and Challenges

83. Explain graphically: $\displaystyle\int_{-a}^a f(x)\,dx = 0$ if $f(x)$ is an odd function.

84. Compute $\displaystyle\int_{-1}^1 \sin(\sin(x))(\sin^2(x) + 1)\,dx$.

85. Let k and b be positive. Show, by comparing the right-endpoint approximations, that

$$\int_0^b x^k\,dx = b^{k+1}\int_0^1 x^k\,dx$$

86. Verify by interpreting the integral as an area:

$$\int_0^b \sqrt{1 - x^2}\,dx = \frac{1}{2}b\sqrt{1 - b^2} + \frac{1}{2}\theta$$

Here, $0 \le b \le 1$ and θ is the angle between 0 and $\frac{\pi}{2}$ such that $\sin\theta = b$.

87. Show that Eq. (6) holds for $b \le 0$.

88. Theorem 4 remains true without the assumption $a \le b \le c$. Verify this for the cases $b < a < c$ and $c < a < b$.

5.3 The Fundamental Theorem of Calculus, Part I

The FTC was first stated clearly by Isaac Newton in 1666, although other mathematicians, including Newton's teacher Isaac Barrow, had come close to discovering it.

The Fundamental Theorem of Calculus (FTC) establishes a relationship between the two main operations of calculus: differentiation and integration. The theorem has two parts. Although they are closely related, we discuss them in separate sections to emphasize their different roles. To motivate the first part of the FTC, recall a formula proved in Example 5 of Section 5.2:

$$\int_0^b x^2\,dx = \frac{1}{3}b^3$$

Notice that $F(x) = \frac{1}{3}x^3$ is an *antiderivative* of x^2, and since $F(0) = 0$, this formula can be rewritten as

$$\int_0^b x^2\,dx = F(b) - F(0)$$

According to the FTC, Part I, this is no coincidence—such a relationship between the definite integral of $f(x)$ and its antiderivative holds in general.

THEOREM 1   The Fundamental Theorem of Calculus, Part I   Assume that $f(x)$ is continuous on $[a, b]$ and let $F(x)$ be an antiderivative of $f(x)$ on $[a, b]$. Then

$$\int_a^b f(x)\,dx = F(b) - F(a)$$

This result allows us to compute definite integrals without calculating any limits, provided that we can find an antiderivative F. It also reveals an unexpected connection between two thus far unrelated concepts: antiderivatives and areas under curves. Furthermore, the FTC shows that the notation $\displaystyle\int f(x)\,dx$ for the indefinite integral is appropriate because the antiderivative is used to evaluate the definite integral.

Notation: We denote $F(b) - F(a)$ by $F(x)\big|_a^b$. In this notation, the FTC reads

$$\int_a^b f(x)\,dx = F(x)\Big|_a^b$$

Proof Consider an arbitrary partition P of $[a, b]$:

$$x_0 = a < x_1 < x_2 < \cdots < x_N = b$$

The first key step is to write the change $F(b) - F(a)$ as the sum of the changes of $F(x)$ over the subintervals $[x_{i-1}, x_i]$ (Figure 1):

$$F(b) - F(a) = F(x_N) - F(x_0)$$

$$= \big(F(x_1) - F(x_0)\big) + \big(F(x_2) - F(x_1)\big) + \cdots + \big(F(x_N) - F(x_{N-1})\big)$$

In this equality, $F(x_1)$ is cancelled by $-F(x_1)$ in the next term, $F(x_2)$ is cancelled by $-F(x_2)$, etc. In summation notation,

$$F(b) - F(a) = \sum_{i=1}^{N} \big(F(x_i) - F(x_{i-1})\big) \qquad \boxed{1}$$

In this proof, the MVT shows that for any partition P, there is a choice of intermediate points $C = \{c_i^*\}$ such that

$$F(b) - F(a) = R(f, P, C)$$

Once we know this, FTC I follows by taking a limit.

The second key step is to apply the Mean Value Theorem (MVT): In each subinterval $[x_{i-1}, x_i]$, there exists an intermediate point c_i^* such that

$$\frac{F(x_i) - F(x_{i-1})}{x_i - x_{i-1}} = F'(c_i^*) = f(c_i^*)$$

Therefore, $F(x_i) - F(x_{i-1}) = f(c_i^*)(x_i - x_{i-1})$, and Eq. (1) can be written

$$F(b) - F(a) = \sum_{i=1}^{N} f(c_i^*)(x_i - x_{i-1}) = \sum_{i=1}^{N} f(c_i^*)\,\Delta x_i$$

This is the Riemann sum approximation to the integral $\int_a^b f(x)\,dx$ with $C = \{c_i^*\}$, so

$$F(b) - F(a) = R(f, P, C)$$

Since $f(x)$ is integrable, $R(f, P, C)$ approaches $\int_a^b f(x)\,dx$ as the norm $\|P\|$ tends to zero. On the other hand, $R(f, P, C)$ is actually *equal* to $F(b) - F(a)$ by our choice of

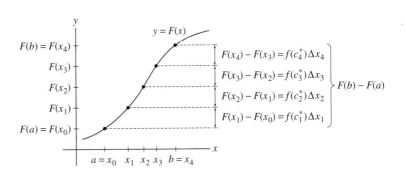

FIGURE 1 The total change $F(b) - F(a)$ is the sum of terms in a Riemann sum for f.

intermediate points. This proves the desired result:

$$F(b) - F(a) = \lim_{\|P\| \to 0} R(f, P, C) = \int_a^b f(x)\, dx \qquad \blacksquare$$

Area computations are much easier with the help of the FTC—but only in cases where an antiderivative can be found.

◀·· REMINDER *The Power Rule for Integrals (valid for n ≠ −1) states that*

$$\int x^n\, dx = \frac{x^{n+1}}{n+1} + C$$

■ EXAMPLE 1 Calculate the area under the graph:

(a) $f(x) = x^3$ over $[2, 4]$ **(b)** $g(x) = x^{-3/4} + 3x^{5/3}$ over $[1, 3]$

Solution

(a) Since $F(x) = \frac{1}{4}x^4$ is an antiderivative of $f(x) = x^3$, the FTC I gives us

$$\int_2^4 x^3\, dx = F(4) - F(2) = \frac{1}{4}x^4 \Big|_2^4 = \frac{1}{4}4^4 - \frac{1}{4}2^4 = 60$$

(b) Since $G(x) = 4x^{1/4} + \frac{9}{8}x^{8/3}$ is an antiderivative of $g(x) = x^{-3/4} + 3x^{5/3}$,

$$\int_1^3 (x^{-3/4} + 3x^{5/3})\, dx = G(x) \Big|_1^3 = \left(4x^{1/4} + \frac{9}{8}x^{8/3} \right) \Big|_1^3$$

$$= \left(4 \cdot 3^{1/4} + \frac{9}{8} \cdot 3^{8/3} \right) - \left(4 \cdot 1^{1/4} + \frac{9}{8} \cdot 1^{8/3} \right)$$

$$\approx 26.325 - 5.125 = 21.2$$

This is the area of the region shown in Figure 2.   ■

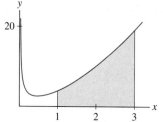

FIGURE 2 The area under the graph of $y = x^{-3/4} + 3x^{5/3}$ over $[1, 3]$ is approximately 21.2.

■ EXAMPLE 2 Sketch the region under $y = \sec^2 x$ for $-\frac{\pi}{4} \le x \le \frac{\pi}{4}$ and find its area.

Solution The region is shown in Figure 3. Recall that $(\tan x)' = \sec^2 x$. Therefore,

$$\int_{-\pi/4}^{\pi/4} \sec^2 x\, dx = \tan x \Big|_{-\pi/4}^{\pi/4} = \tan\left(\frac{\pi}{4} \right) - \tan\left(-\frac{\pi}{4} \right)$$

$$= 1 - (-1) = 2 \qquad \blacksquare$$

FIGURE 3 Graph of $y = \sec^2 x$.

In the previous section, we saw that the definite integral is equal to the *signed* area between the graph and the x-axis. Needless to say, the FTC "knows" this: When you evaluate an integral using the FTC, you obtain the signed area. This is the principal reason we consider signed area.

■ EXAMPLE 3 Evaluate   **(a)** $\int_0^\pi \sin x\, dx$ and   **(b)** $\int_0^{2\pi} \sin x\, dx$.

Solution

(a) Since $(-\cos x)' = \sin x$, the area of one "hump" (Figure 4) is

$$\int_0^\pi \sin x\, dx = -\cos x \Big|_0^\pi = -\cos \pi - (-\cos 0) = -(-1) - (-1) = 2$$

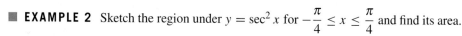

FIGURE 4 Area of one hump is 2. Signed area over $[0, 2\pi]$ is zero.

(b) We expect the signed area over $[0, 2\pi]$ to be zero since the second hump lies below the x-axis, and indeed,

$$\int_0^{2\pi} \sin x \, dx = -\cos x \Big|_0^{2\pi} = (-\cos(2\pi) - (-\cos 0)) = -1 - (-1) = 0 \quad \blacksquare$$

■ **EXAMPLE 4** Integral of the Exponential Function Evaluate $\int_{-2}^1 e^x \, dx$.

Solution The region is shown in Figure 5. Since $F(x) = e^x$ is an antiderivative of $f(x) = e^x$, we have

$$\int_{-2}^1 e^x \, dx = F(1) - F(-2) = e - e^{-2} \approx 2.58 \quad \blacksquare$$

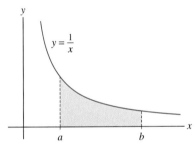

FIGURE 5

Recall, from Section 4.9, that $F(x) = \ln |x|$ is an antiderivative of $f(x) = x^{-1}$ in the domain $\{x : x \neq 0\}$. Therefore, the FTC yields the following formula, valid if both a and b are positive or both are negative:

$$\boxed{\int_a^b \frac{dx}{x} = \ln |b| - \ln |a| = \ln \frac{b}{a}} \qquad \boxed{2}$$

In other words, the area under the hyperbola $y = x^{-1}$ over $[a, b]$ is equal to $\ln \dfrac{b}{a}$ (Figure 6).

■ **EXAMPLE 5** The Logarithm as an Antiderivative Evaluate

(a) $\displaystyle\int_2^8 \frac{dx}{x}$ and   **(b)** $\displaystyle\int_{-4}^{-2} \frac{dx}{x}$.

Solution By Eq. (2),

$$\int_2^8 \frac{dx}{x} = \ln \frac{8}{2} = \ln 4 \approx 1.39$$

$$\int_{-4}^{-2} \frac{dx}{x} = \ln \left(\frac{-2}{-4} \right) = \ln \frac{1}{2} \approx -0.69$$

The area represented by these integrals is shown in Figures 7(A) and (B). ■

FIGURE 6 The area under $y = x^{-1}$ over $[a, b]$ is $\ln \dfrac{b}{a}$.

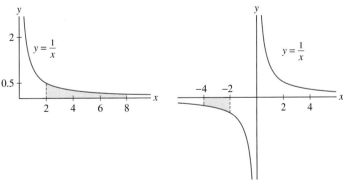

(A) Shaded region has area ln 4.   (B) Shaded region has signed area $\ln \dfrac{1}{2}$.

FIGURE 7

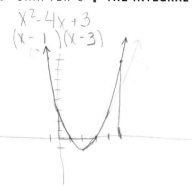

$$X^2 - 4x + 3$$
$$(X - 1)(X - 3)$$

$$\int_0^1 (x^2 - 4x + 3)dx - \int_1^3 (x^2 - 4x + 3)dx +$$

$$\int_3^5 (x^2 - 4x + 3)dx$$

CONCEPTUAL INSIGHT **Which Antiderivative?** We saw in Chapter 4 that the antiderivative of a function is unique only to within an additive constant. Does it matter which antiderivative is used in the FTC? The answer is no. If $F(x)$ and $G(x)$ are both antiderivatives of $f(x)$, then $F(x) = G(x) + C$ for some constant C. Therefore,

$$F(b) - F(a) = \underbrace{(G(b) + C) - (G(a) + C)}_{\text{The constant cancels}} = G(b) - G(a)$$

and

$$\int_a^b f(x)\,dx = F(b) - F(a) = G(b) - G(a)$$

For example, $F(x) = \frac{1}{3}x^3$ and $G(x) = \frac{1}{3}x^3 - 9$ are both antiderivatives of x^2. Either one may be used to evaluate the definite integral:

$$\int_1^3 x^2\,dx = F(x)\Big|_1^3 = \frac{1}{3}3^3 - \frac{1}{3}1^3 = \frac{26}{3}$$

$$\int_1^3 x^2\,dx = G(x)\Big|_1^3 = \left(\frac{1}{3}3^3 - 9\right) - \left(\frac{1}{3}1^3 - 9\right) = \frac{26}{3}$$

5.3 SUMMARY

• The Fundamental Theorem of Calculus, Part I, states that

$$\int_a^b f(x)\,dx = F(b) - F(a)$$

where $F(x)$ is an antiderivative of $f(x)$. The FTC I is used to evaluate definite integrals in cases where we can find an antiderivative of the integrand.

• Basic antiderivative formulas for evaluating definite integrals:

$$\int x^n\,dx = \frac{x^{n+1}}{n+1} + C \qquad \text{for } n \neq -1$$

$$\int \frac{dx}{x} = \ln|x| + C$$

$$\int e^x\,dx = e^x + C$$

$$\int \sin x\,dx = -\cos x + C, \qquad \int \cos x\,dx = \sin x + C$$

$$\int \sec^2 x\,dx = \tan x + C, \qquad \int \csc^2 x\,dx = -\cot x + C$$

$$\int \sec x \tan x\,dx = \sec x + C, \qquad \int \csc x \cot x\,dx = -\csc x + C$$

5.3 EXERCISES

Preliminary Questions

1. Assume that $f(x) \geq 0$. What is the area under the graph of $f(x)$ over $[0, 2]$ if $f(x)$ has an antiderivative $F(x)$ such that $F(0) = 3$ and $F(2) = 7$?

2. Suppose that $F(x)$ is an antiderivative of $f(x)$. What is the graphical interpretation of $F(4) - F(1)$ if $f(x)$ takes on both positive and negative values?

3. Evaluate $\int_0^7 f(x)\,dx$ and $\int_2^7 f(x)\,dx$, assuming that $f(x)$ has an antiderivative $F(x)$ with values from the following table:

x	0	2	7
$F(x)$	3	7	9

4. Are the following statements true or false? Explain.

(a) The FTC I is only valid for positive functions.

(b) To use the FTC I, you have to choose the right antiderivative.

(c) If you cannot find an antiderivative of $f(x)$, then the definite integral does not exist.

5. What is the value of $\int_2^9 f'(x)\,dx$ if $f(x)$ is differentiable and $f(2) = f(9) = 4$?

Exercises

In Exercises 1–4, sketch the region under the graph of the function and find its area using the FTC I.

1. $f(x) = x^2$, $[0, 1]$

2. $f(x) = 2x - x^2$, $[0, 2]$

3. $f(x) = \sin x$, $[0, \pi/2]$

4. $f(x) = \cos x$, $[0, \pi/2]$

In Exercises 5–40, evaluate the integral using the FTC I.

5. $\int_3^6 x\,dx$

6. $\int_0^9 2\,dx$

7. $\int_{-3}^2 u^2\,du$

8. $\int_0^1 (x - x^2)\,dx$

9. $\int_3^5 e^x\,dx$

10. $\int_1^4 \left(x + \frac{1}{x}\right)dx$

11. $\int_{-2}^0 (3x - 2e^x)\,dx$

12. $\int_{-12}^{-4} \frac{dx}{x}\,dx$

13. $\int_1^3 (t^3 - t^2)\,dt$

14. $\int_0^1 (4 - 5u^4)\,du$

15. $\int_{-3}^4 (x^2 + 2)\,dx$

16. $\int_0^4 (3x^5 + x^2 - 2x)\,dx$

17. $\int_{-2}^2 (10x^9 + 3x^5)\,dx$

18. $\int_{-1}^1 (5u^4 - 6u^2)\,du$

19. $\int_3^1 (4t^{3/2} + t^{7/2})\,dt$

20. $\int_1^2 (x^2 - x^{-2})\,dx$

21. $\int_1^4 \frac{1}{t^2}\,dt$

22. $\int_0^4 \sqrt{y}\,dy$

23. $\int_1^{27} x^{1/3}\,dx$

24. $\int_1^4 x^{-4}\,dx$

25. $\int_1^9 t^{-1/2}\,dt$

26. $\int_4^9 \frac{8}{x^3}\,dx$

27. $\int_{0.2}^{10} \frac{dx}{3x}$

28. $\int_0^1 (9e^x)\,dx$

29. $\int_{-2}^{-1} \frac{1}{x^3}\,dx$

30. $\int_2^4 \pi^2\,dx$

31. $\int_1^{27} \frac{t+1}{\sqrt{t}}\,dt$

32. $\int_0^{\pi/2} \cos\theta\,d\theta$

33. $\int_{-\pi/2}^{\pi/2} \cos x\,dx$

34. $\int_0^{2\pi} \cos t\,dt$

35. $\int_{\pi/4}^{3\pi/4} \sin\theta\,d\theta$

36. $\int_{2\pi}^{4\pi} \sin x\,dx$

37. $\int_0^{\pi/4} \sec^2 t\,dt$

38. $\int_0^{\pi/4} \sec\theta\tan\theta\,d\theta$

39. $\int_{\pi/6}^{\pi/3} \csc x \cot x\,dx$

40. $\int_{\pi/6}^{\pi/2} \csc^2 y\,dy$

In Exercises 41–46, write the integral as a sum of integrals without absolute values and evaluate.

41. $\int_{-2}^1 |x|\,dx$

42. $\int_0^5 |3 - x|\,dx$

43. $\displaystyle\int_{-2}^{3} |x^3|\, dx$

44. $\displaystyle\int_{0}^{3} |x^2 - 1|\, dx$

45. $\displaystyle\int_{0}^{\pi} |\cos x|\, dx$

46. $\displaystyle\int_{0}^{5} |x^2 - 4x + 3|\, dx$

In Exercises 47–52, evaluate the integral in terms of the constants.

47. $\displaystyle\int_{1}^{b} x^3\, dx$

48. $\displaystyle\int_{b}^{a} x^4\, dx$

49. $\displaystyle\int_{1}^{b} x^5\, dx$

50. $\displaystyle\int_{-x}^{x} (t^3 + t)\, dt$

51. $\displaystyle\int_{a}^{5a} \frac{dx}{x}$

52. $\displaystyle\int_{b}^{b^2} \frac{dx}{x}$

53. Use the FTC I to show that $\displaystyle\int_{-1}^{1} x^n\, dx = 0$ if n is an odd whole number. Explain graphically.

54. What is the area (a positive number) between the x-axis and the graph of $f(x)$ on $[1, 3]$ if $f(x)$ is a *negative* function whose antiderivative F has the values $F(1) = 7$ and $F(3) = 4$?

55. Show that the area of a parabolic arch (the shaded region in Figure 8) is equal to four-thirds the area of the triangle shown.

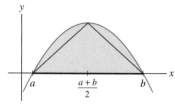

FIGURE 8 Graph of $y = (x - a)(b - x)$.

56. Does $\displaystyle\int_{0}^{1} x^n\, dx$ get larger or smaller as n increases? Explain graphically.

57. Calculate $\displaystyle\int_{-2}^{3} f(x)\, dx$, where

$$f(x) = \begin{cases} 12 - x^2 & \text{for } x \leq 2 \\ x^3 & \text{for } x > 2 \end{cases}$$

58. ⌨ **CAS** Plot the function $f(x) = \sin 3x - x$. Find the positive root of $f(x)$ to three places and use it to find the area under the graph of $f(x)$ in the first quadrant.

Further Insights and Challenges

59. In this exercise, we generalize the result of Exercise 55 by proving the famous result of Archimedes: For $r < s$, the area of the shaded region in Figure 9 is equal to four-thirds the area of triangle $\triangle ACE$, where C is the point on the parabola at which the tangent line is parallel to secant line $\overline{AE}$.

(a) Show that C has x-coordinate $(r + s)/2$.

(b) Show that $ABDE$ has area $(s - r)^3/4$ by viewing it as a parallelogram of height $s - r$ and base of length $\overline{CF}$.

(c) Show that $\triangle ACE$ has area $(s - r)^3/8$ by observing that it has the same base and height as the parallelogram.

(d) Compute the shaded area as the area under the graph minus the area of a trapezoid and prove Archimedes's result.

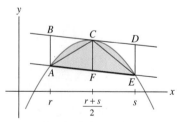

FIGURE 9 Graph of $f(x) = (x - a)(b - x)$.

60. (a) Apply the Comparison Theorem (Theorem 5 in Section 5.2) to the inequality $\sin x \leq x$ (valid for $x \geq 0$) to prove

$$1 - \frac{x^2}{2} \leq \cos x \leq 1$$

(b) Apply it again to prove

$$x - \frac{x^3}{6} \leq \sin x \leq x \quad (\text{for } x \geq 0)$$

(c) Verify these inequalities for $x = 0.3$.

61. Use the method of Exercise 60 to prove that

$$1 - \frac{x^2}{2} \leq \cos x \leq 1 - \frac{x^2}{2} + \frac{x^4}{24}$$

$$x - \frac{x^3}{6} \leq \sin x \leq x - \frac{x^3}{6} + \frac{x^5}{120} \quad (\text{for } x \geq 0)$$

Verify these inequalities for $x = 0.1$. Why have we specified $x \geq 0$ for $\sin x$ but not $\cos x$?

62. Calculate the next pair of inequalities for $\sin x$ and $\cos x$ by integrating the results of Exercise 61. Can you guess the general pattern?

63. Assume that $|f'(x)| \leq K$ for $x \in [a, b]$. Use FTC I to prove that $|f(x) - f(a)| \leq K|x - a|$ for $x \in [a, b]$.

64. (a) Prove that $|\sin a - \sin b| \leq |a - b|$ for all a, b (use Exercise 63).

(b) Let $f(x) = \sin(x + a) - \sin x$. Use part (a) to show that the graph of $f(x)$ lies between the horizontal lines $y = \pm a$.

(c) ⊞ **GU** Produce a graph of $f(x)$ and verify part (b) for $a = 0.5$ and $a = 0.2$.

5.4 The Fundamental Theorem of Calculus, Part II

Part I of the Fundamental Theorem shows that we can compute definite integrals using antiderivatives. Part II turns this relationship around: It tells us that we can use the definite integral to *construct* antiderivatives.

To state Part II, we introduce the **area function** (or **cumulative area function**) associated to a function $f(x)$ and lower limit a (Figure 1):

$$A(x) = \int_a^x f(t)\,dt = \text{signed area from } a \text{ to } x$$

In essence, we turn the definite integral into a function by treating the upper limit x as a variable rather than a constant. Notice that $A(a) = 0$ because $A(a) = \int_a^a f(t)\,dt = 0$.

Although $A(x)$ is defined by an integral, in many cases we can find an explicit formula for it.

■ **EXAMPLE 1**   Find a formula for the area function $A(x) = \int_3^x t^2\,dt$.

Solution   The function $F(t) = \frac{1}{3}t^3$ is an antiderivative for $f(t) = t^2$. By FTC I (Figure 2):

$$A(x) = \int_3^x t^2\,dt = F(x) - F(3) = \frac{1}{3}x^3 - \frac{1}{3}\cdot 3^3 = \frac{1}{3}x^3 - 9 \qquad ■$$

Notice in the previous example that *the derivative of the area function is $f(x)$ itself:*

$$A'(x) = \frac{d}{dx}\left(\frac{1}{3}x^3 - 9\right) = x^2$$

FTC II states that this relation always holds: The derivative of the area function is equal to the original function.

THEOREM 1 Fundamental Theorem of Calculus, Part II   Let $f(x)$ be a continuous function on $[a,b]$. Then $A(x) = \int_a^x f(t)\,dt$ is an antiderivative of $f(x)$, that is, $A'(x) = f(x)$, or equivalently,

$$\frac{d}{dx}\int_a^x f(t)\,dt = f(x)$$

Furthermore, $A(x)$ satisfies the initial condition $A(a) = 0$.

Proof   To simplify the proof, we assume that $f(x)$ is increasing (see Exercise 50 for the general case). We compute $A'(x)$ using the limit definition of the derivative. First, let's analyze the numerator of the difference quotient. By the additivity property of the definite integral,

$$A(x+h) - A(x) = \int_a^{x+h} f(t)\,dt - \int_a^x f(t)\,dt = \int_x^{x+h} f(t)\,dt$$

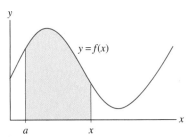

In the definition of $A(x)$, we use t as the variable of integration to avoid confusion with x, which is the upper limit of integration. In fact, t is a dummy variable and may be replaced by any other variable.

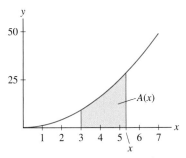

FIGURE 1 $A(x)$ is the area under the graph from a to x.

FIGURE 2   The area under $y = x^2$ from 3 to x is $A(x) = \frac{1}{3}x^3 - 9$.

In this proof,

$$A(x) = \int_a^x f(t)\, dt$$

$$A(x + h) - A(x) = \int_x^{x+h} f(t)\, dt$$

$$A'(x) = \lim_{h \to 0} \frac{A(x + h) - A(x)}{h}$$

if c_i exists in $[x_{i-1}, x_i]$, then:

$$F'(c_i) = \frac{F(x_i) - F(x_{i-1})}{x_i - x_{i-1}}$$

$$f(c_i) = \frac{F(x_i) - F(x_{i-1})}{x_i - x_{i-1}}$$

$$f(c_i)(x_i - x_{i-1}) = F(x_i) - F(x_{i-1})$$

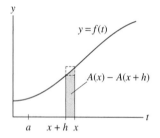

FIGURE 5 Here, $h < 0$ and the area of the thin sliver is $A(x) - A(x + h)$.

$$F(b) - F(a) = \lim_{n \to \infty} \sum_{i=1}^{n} f(c_i)\,\Delta x = \int_a^b f(x)\, dx$$

Assume first that $h > 0$. Then this numerator is equal to the signed area of the thin sliver between the graph and the x-axis from x to $x + h$ (Figure 3).

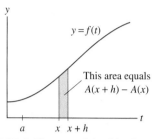

FIGURE 3 The area of the thin sliver equals $A(x + h) - A(x)$.

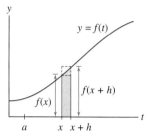

FIGURE 4 The shaded sliver lies between the rectangles of heights $f(x)$ and $f(x + h)$.

This thin sliver lies between the two rectangles of heights $f(x)$ and $f(x + h)$ in Figure 4. Since the rectangles have base h,

$$\underbrace{hf(x)}_{\text{Area of smaller rectangle}} \leq \underbrace{A(x + h) - A(x)}_{\text{Area of sliver}} \leq \underbrace{hf(x + h)}_{\text{Area of larger rectangle}}$$

Now divide by h to obtain the difference quotient:

$$f(x) \leq \frac{A(x + h) - A(x)}{h} \leq f(x + h)$$

and apply the Squeeze Theorem. The continuity of $f(x)$ implies that $\lim_{h \to 0} f(x + h) = f(x)$ and $\lim_{h \to 0} f(x) = f(x)$. We conclude that

$$\lim_{h \to 0+} \frac{A(x + h) - A(x)}{h} = f(x) \qquad \boxed{1}$$

If $h < 0$, then $x + h$ lies to the left of x. In this case, the sliver in Figure 5 has signed area $A(x) - A(x + h)$, and a comparison with the rectangles yields

$$-hf(x + h) \leq A(x) - A(x + h) \leq -hf(x)$$

Dividing by $-h$ (which is positive), we obtain

$$f(x + h) \leq \frac{A(x + h) - A(x)}{h} \leq f(x)$$

and again, the Squeeze Theorem gives us

$$\lim_{h \to 0-} \frac{A(x + h) - A(x)}{h} = f(x) \qquad \boxed{2}$$

Equations (1) and (2) show that $A'(x)$ exists and $A'(x) = f(x)$. ■

CONCEPTUAL INSIGHT   Although most elementary functions are defined by formulas or geometry (the trigonometric functions), many new and important functions are defined by integrals. We can graph such functions and compute their values numerically using a computer algebra system. Figure 7 shows a computer-generated graph of the antiderivative of $f(x) = \sin(x^2)$ obtained as an area function (Figure 6).

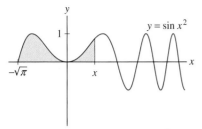

FIGURE 6 Graph of $f(x) = \sin(x^2)$.

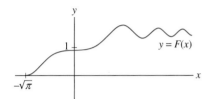

FIGURE 7 Computer-generated graph of
$F(x) = \int_{-\sqrt{\pi}}^{x} \sin(t^2)\, dt$.

■ **EXAMPLE 2** Expressing an Antiderivative as an Integral   There is no elementary formula for an antiderivative of $f(x) = \sin(x^2)$. Express the antiderivative $F(x)$ satisfying $F(-\sqrt{\pi}) = 0$ as an integral.

Solution According to FTC II, the area function with lower limit $a = -\sqrt{\pi}$ is an antiderivative satisfying $F(-\sqrt{\pi}) = 0$:

$$F(x) = \int_{-\sqrt{\pi}}^{x} \sin(t^2)\, dt$$

■ **EXAMPLE 3** Differentiating an Integral   Find the derivative of

$$A(x) = \int_{2}^{x} \sqrt{1 + t^3}\, dt$$

and calculate $A'(2)$, $A'(3)$, and $A(2)$.

Solution By FTC II, $A'(x) = \sqrt{1 + x^3}$. In particular,

$$A'(2) = \sqrt{1 + 2^3} = 3 \qquad \text{and} \qquad A'(3) = \sqrt{1 + 3^3} = \sqrt{28}$$

On the other hand, $A(2) = \int_{2}^{2} \sqrt{1 + t^3}\, dt = 0$.

CONCEPTUAL INSIGHT   The FTC shows that integration and differentiation are *inverse operations*. By FTC II, if you start with a continuous function $f(x)$ and form the integral $\int_{a}^{x} f(x)\, dx$, then you get back the original function $f(x)$ by differentiating:

$$f(x) \xrightarrow{\text{Integrate}} \int_{a}^{x} f(t)\, dt \xrightarrow{\text{Differentiate}} \frac{d}{dx} \int_{a}^{x} f(t)\, dt = f(x)$$

On the other hand, by FTC I, if you first differentiate $f(x)$ and then integrate, you also recover $f(x)$ [but only up to a constant $f(a)$]:

$$f(x) \xrightarrow{\text{Differentiate}} f'(x) \xrightarrow{\text{Integrate}} \int_{a}^{x} f'(t)\, dt = f(x) - f(a)$$

When the upper limit of the integral is a *function* of x rather than x itself, we use FTC II together with the Chain Rule to differentiate the integral.

■ **EXAMPLE 4** Combining the FTC and the Chain Rule   Find the derivative of

$$G(x) = \int_{-2}^{x^2} \sin t\, dt.$$

Solution FTC II does not apply directly because the upper limit is x^2 rather than x. It is necessary to recognize that $G(x)$ is a *composite function*. To see this clearly, define the usual area function $A(x) = \int_{-2}^{x} \sin t\, dt$. Then

$$G(x) = A(x^2) = \int_{-2}^{x^2} \sin t\, dt$$

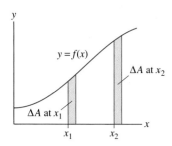

FIGURE 8 The change in area ΔA for a given Δx is larger when $f(x)$ is larger.

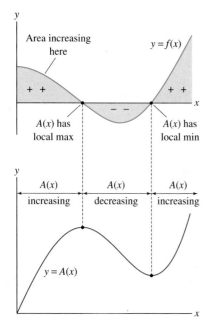

FIGURE 9 The sign of $f(x)$ determines the increasing/decreasing behavior of $A(x)$.

FTC II tells us that $A'(x) = \sin x$, so by the Chain Rule,

$$G'(x) = A'(x^2) \cdot (x^2)' = \sin(x^2) \cdot (2x) = 2x\sin(x^2)$$

Alternatively, we may set $u = x^2$ and use the Chain Rule as follows:

$$\frac{dG}{dx} = \frac{d}{dx}\int_{-2}^{x^2} \sin t\, dt = \left(\frac{d}{du}\int_{-2}^{u} \sin t\, dt\right)\frac{du}{dx} = (\sin u)2x = 2x\sin(x^2) \quad \blacksquare$$

GRAPHICAL INSIGHT   FTC II claims that $A'(x) = f(x)$, which means $f(x)$ is the rate at which $A(x)$ increases or decreases. Figures 8 and 9 provide "evidence" for this claim. First, Figure 8 shows that for a given change Δx, the increase in area is larger at x_2 [where $f(x)$ is larger] than at x_1. Thus, the size of $f(x)$ indeed determines how quickly $A(x)$ changes.

Second, Figure 9 shows that the sign of $f(x)$ determines the increasing/decreasing behavior of $A(x)$. Indeed, $A(x)$ is increasing when $f(x) > 0$ because positive area is added as x moves to the right. When $f(x)$ turns negative, $A(x)$ begins to decrease because we start adding negative area. In particular, $A(x)$ has a local maximum at points where $f(x)$ changes sign from $+$ to $-$, in agreement with the First Derivative Test. Similarly, when $f(x)$ changes sign from $-$ to $+$, $A(x)$ reaches a local minimum. Thus, $f(x)$ "behaves" like the derivative of $A(x)$, as claimed by FTC II.

5.4 SUMMARY

- The *area function* with lower limit a is $A(x) = \displaystyle\int_a^x f(t)\, dt$ and satisfies $A(a) = 0$.

- The FTC II states that $A'(x) = f(x)$, or equivalently, $\dfrac{d}{dx}\displaystyle\int_a^x f(t)\, dt = f(x)$.

- The FTC II shows that every continuous function has an antiderivative, namely the associated area function (with any lower limit).

- To differentiate a function of the form $G(x) = \displaystyle\int_a^{g(x)} f(t)\, dt$, write it as a composite $G(x) = A(g(x))$, where $A(x) = \displaystyle\int_a^x f(t)\, dt$. Then use the Chain Rule:

$$G'(x) = A'(g(x))g'(x) = f(g(x))g'(x)$$

5.4 EXERCISES

Preliminary Questions

1. What is $A(-2)$, where $A(x) = \displaystyle\int_{-2}^{x} f(t)\, dt$?

2. Let $G(x) = \displaystyle\int_{4}^{x} \sqrt{t^3 + 1}\, dt$.

(a) Is the FTC needed to calculate $G(4)$?

(b) Is the FTC needed to calculate $G'(4)$?

3. Which of the following defines an antiderivative $F(x)$ of $f(x) = x^2$ satisfying $F(2) = 0$?

(a) $\displaystyle\int_{2}^{x} 2t\, dt$ **(b)** $\displaystyle\int_{0}^{2} t^2\, dt$ **(c)** $\displaystyle\int_{2}^{x} t^2\, dt$

4. True or false? Some continuous functions do not have antiderivatives. Explain.

5. Let $G(x) = \displaystyle\int_{4}^{x^3} \sin t\, dt$. Which of the following statements are correct?

(a) $G(x)$ is the composite function $\sin(x^3)$.

(b) $G(x)$ is the composite function $A(x^3)$, where

$$A(x) = \int_4^x \sin(t)\, dt$$

(c) $G(x)$ is too complicated to differentiate.
(d) The Product Rule is used to differentiate $G(x)$.

(e) The Chain Rule is used to differentiate $G(x)$.

(f) $G'(x) = 3x^2 \sin(x^3)$.

6. Trick question: Find the derivative of $\int_1^3 t^3\, dt$ at $x = 2$.

Exercises

1. Write the area function of $f(x) = 2x + 4$ with lower limit $a = -2$ as an integral and find a formula for it.

2. Find a formula for the area function of $f(x) = 2x + 4$ with lower limit $a = 0$.

3. Let $G(x) = \int_1^x (t^2 - 2)\, dt$.
(a) What is $G(1)$?
(b) Use FTC II to find $G'(1)$ and $G'(2)$.
(c) Find a formula for $G(x)$ and use it to verify your answers to (a) and (b).

4. Find $F(0)$, $F'(0)$, and $F'(3)$, where $F(x) = \int_0^x \sqrt{t^2 + t}\, dt$.

5. Find $G(1)$, $G'(0)$, and $G'(\pi/4)$, where $G(x) = \int_1^x \tan t\, dt$.

6. Find $H(-2)$ and $H'(-2)$, where $H(x) = \int_{-2}^x \dfrac{du}{u^2 + 1}$.

In Exercises 7–14, find formulas for the functions represented by the integrals.

7. $\displaystyle\int_2^x u^3\, du$

8. $\displaystyle\int_0^x \sin u\, du$

9. $\displaystyle\int_1^{x^2} t\, dt$

10. $\displaystyle\int_2^x (t^2 - t)\, dt$

11. $\displaystyle\int_x^5 e^t\, dt$

12. $\displaystyle\int_{\pi/4}^x \cos u\, du$

13. $\displaystyle\int_{-\pi/4}^x \sec^2 \theta\, d\theta$

14. $\displaystyle\int_2^{\sqrt{x}} \dfrac{dt}{t}$

In Exercises 15–18, express the antiderivative $F(x)$ of $f(x)$ satisfying the given initial condition as an integral.

15. $f(x) = \sqrt{x^4 + 1}$, $F(3) = 0$

16. $f(x) = \dfrac{x + 1}{x^2 + 9}$, $F(7) = 0$

17. $f(x) = \sec x$, $F(0) = 0$

18. $f(x) = e^{-x^2}$, $F(4) = 0$

In Exercises 19–22, calculate the derivative.

19. $\dfrac{d}{dx} \displaystyle\int_0^x (t^3 - t)\, dt$

20. $\dfrac{d}{dx} \displaystyle\int_1^x \sin(t^2)\, dt$

21. $\dfrac{d}{dt} \displaystyle\int_{100}^t \cos 5x\, dx$

22. $\dfrac{d}{ds} \displaystyle\int_{-2}^s \tan\left(\dfrac{1}{1 + u^2}\right) du$

23. Sketch the graph of $A(x) = \int_0^x f(t)\, dt$ for each of the functions shown in Figure 10.

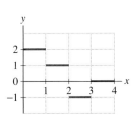

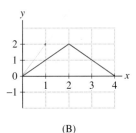

$\qquad$ (A) $\qquad\qquad\qquad$ (B)

FIGURE 10

24. Let $A(x) = \int_0^x f(t)\, dt$ for $f(x)$ shown in Figure 11. Calculate $A(2)$, $A(3)$, $A'(2)$, and $A'(3)$. Then find a formula for $A(x)$ (actually two formulas, one for $0 \le x \le 2$ and one for $2 \le x \le 4$) and sketch the graph of $A(x)$.

$\int_2^x x\, dt = \frac{1}{2} t^2 \Big|_2^x = \frac{1}{2}x^2 - 2$

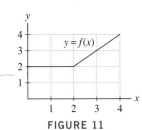

FIGURE 11

25. Make a rough sketch of the graph of the area function of $g(x)$ shown in Figure 12.

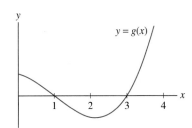

FIGURE 12

26. Show that $\int_0^x |t|\,dt$ is equal to $\frac{1}{2}x|x|$. *Hint:* Consider $x \geq 0$ and $x \leq 0$ separately.

27. Find $G'(x)$, where $G(x) = \int_3^{x^3} \tan t\,dt$.

28. Find $G'(1)$, where $G(x) = \int_0^{x^2} \sqrt{t^3 + 3}\,dt$.

In Exercises 29–36, calculate the derivative.

29. $\dfrac{d}{dx} \int_0^{x^2} \sin^2 t\,dt$

30. $\dfrac{d}{dx} \int_1^{1/x} \sin(t^2)\,dt$

31. $\dfrac{d}{ds} \int_{-6}^{\cos s} (u^4 - 3u)\,du$

32. $\dfrac{d}{dx} \int_x^0 \sin^2 t\,dt$

33. $\dfrac{d}{dx} \int_{x^3}^0 \sin^2 t\,dt$

34. $\dfrac{d}{dx} \int_{x^2}^{x^4} \sqrt{t}\,dt$

Hint for Exercise 34: $F(x) = A(x^4) - A(x^2)$, where

$$A(x) = \int_0^x \sqrt{t}\,dt$$

35. $\dfrac{d}{dx} \int_{\sqrt{x}}^{x^2} \tan t\,dt$

36. $\dfrac{d}{du} \int_{-u}^{3u+9} \sqrt{x^2 + 1}\,dx$

In Exercises 37–38, let $A(x) = \int_0^x f(t)\,dt$ and $B(x) = \int_2^x f(t)\,dt$, with $f(x)$ as in Figure 13.

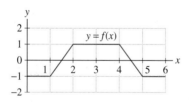

FIGURE 13

37. Find the min and max of $A(x)$ on $[0, 6]$.

38. Find formulas for $A(x)$ and $B(x)$ valid on $[2, 4]$.

39. Let $A(x) = \int_0^x f(t)\,dt$, with $f(x)$ as in Figure 14.
(a) Does $A(x)$ have a local maximum at P?
(b) Where does $A(x)$ have a local minimum?
(c) Where does $A(x)$ have a local maximum?
(d) True or false? $A(x) < 0$ for all x in the interval shown.

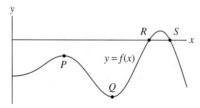

FIGURE 14 Graph of $f(x)$.

40. Find the smallest positive critical point of

$$F(x) = \int_0^x \cos(t^{3/2})\,dt$$

and determine whether it is a local min or max.

In Exercises 41–42, let $A(x) = \int_a^x f(t)\,dt$, where $f(x)$ is continuous.

41. 　Area Functions and Concavity Explain why the following statements are true. Assume $f(x)$ is differentiable.
(a) If c is an inflection point of $A(x)$, then $f'(c) = 0$.
(b) $A(x)$ is concave up if $f(x)$ is increasing.
(c) $A(x)$ is concave down if $f(x)$ is decreasing.

42. Match the property of $A(x)$ with the corresponding property of the graph of $f(x)$. Assume $f(x)$ is differentiable.

Area function $A(x)$
(a) $A(x)$ is decreasing.
(b) $A(x)$ has a local maximum.
(c) $A(x)$ is concave up.
(d) $A(x)$ goes from concave up to concave down.

Graph of $f(x)$
(i) Lies below the x-axis.
(ii) Crosses the x-axis from positive to negative.
(iii) Has a local maximum.
(iv) $f(x)$ is increasing.

43. Let $A(x) = \int_0^x f(t)\,dt$, with $f(x)$ as in Figure 15. Determine:
(a) The intervals on which $A(x)$ is increasing and decreasing
(b) The values x where $A(x)$ has a local min or max
(c) The inflection points of $A(x)$
(d) The intervals where $A(x)$ is concave up or concave down

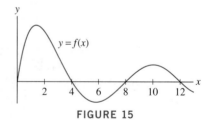

FIGURE 15

44. Let $f(x) = x^2 - 5x - 6$ and $F(x) = \int_0^x f(t)\,dt$.
(a) Find the critical points of $F(x)$ and determine whether they are local minima or maxima.
(b) Find the points of inflection of $F(x)$ and determine whether the concavity changes from up to down or vice versa.

(c) GU Plot $f(x)$ and $F(x)$ on the same set of axes and confirm your answers to (a) and (b).

45. Sketch the graph of an increasing function $f(x)$ such that both $f'(x)$ and $A(x) = \int_0^x f(t)\, dt$ are decreasing.

46. 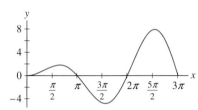 Figure 16 shows the graph of $f(x) = x \sin x$. Let $F(x) = \int_0^x t \sin t\, dt$.

FIGURE 16 Graph of $y = x \sin x$.

(a) Locate the local maxima and absolute maximum of $F(x)$ on $[0, 3\pi]$.

(b) Justify graphically that $F(x)$ has precisely one zero in the interval $[\pi, 2\pi]$.

(c) How many zeros does $F(x)$ have in $[0, 3\pi]$?

(d) Find the inflection points of $F(x)$ on $[0, 3\pi]$ and, for each one, state whether the concavity changes from up to down or vice versa.

47. GU Find the smallest positive inflection point of
$$F(x) = \int_0^x \cos(t^{3/2})\, dt$$

Use a graph of $y = \cos(x^{3/2})$ to determine whether the concavity changes from up to down or vice versa at this point of inflection.

48. Determine $f(x)$, assuming that $\int_0^x f(t)\, dt$ is equal to $x^2 + x$.

49. Determine the function $g(x)$ and all values of c such that
$$\int_c^x g(t)\, dt = x^2 + x - 6$$

Further Insights and Challenges

50. Proof of FTC II The proof in the text assumes that $f(x)$ is increasing. To prove it for all continuous functions, let $m(h)$ and $M(h)$ denote the *minimum* and *maximum* of $f(x)$ on $[x, x+h]$ (Figure 17). The continuity of $f(x)$ implies that $\lim_{h \to 0} m(h) = \lim_{h \to 0} M(h) = f(x)$. Show that for $h > 0$,
$$hm(h) \le A(x+h) - A(x) \le hM(h)$$
For $h < 0$, the inequalities are reversed. Prove that $A'(x) = f(x)$.

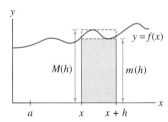

FIGURE 17 Graphical interpretation of $A(x+h) - A(x)$.

51. Proof of FTC I FTC I asserts that $\int_a^b f(t)\, dt = F(b) - F(a)$ if $F'(x) = f(x)$. Assume FTC II and give a new proof of FTC I as follows. Set $A(x) = \int_a^x f(t)\, dt$.

(a) Show that $F(x) = A(x) + C$ for some constant.

(b) Show that $F(b) - F(a) = A(b) - A(a) = \int_a^b f(t)\, dt$.

52. Can Every Antiderivative Be Expressed as an Integral? The area function $\int_a^x f(t)\, dt$ is an antiderivative of $f(x)$ for every value of a. However, not all antiderivatives are obtained in this way. The general antiderivative of $f(x) = x$ is $F(x) = \frac{1}{2}x^2 + C$. Show that $F(x)$ is an area function if $C \le 0$ but not if $C > 0$.

53. Find the values $a \le b$ such that $\int_a^b (x^2 - 9)\, dx$ has minimal value.

5.5 Net or Total Change as the Integral of a Rate

So far we have focused on the area interpretation of the integral. In this section, we study the integral as a tool for computing net or total change.

Consider the following problem: Water flows into an empty bucket at a rate of $r(t)$ gallons per second. How much water is in the bucket after 4 s? This would be easy to answer if the rate of water flow were *constant*, say, 0.3 gal/s. We would have

$$\text{Total amount in bucket} = \text{flow rate} \times \text{length of time} = (0.3)4 = 1.2 \text{ gal}$$

Suppose, however, that the flow rate $r(t)$ varies as in Figure 1. Then *the quantity of water is equal to the area under the graph of $r(t)$*. To prove this, let $s(t)$ be the amount of water

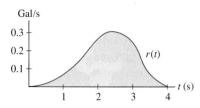

FIGURE 1 The amount of water in the bucket is equal to the area under the graph of the flow rate $r(t)$.

In Theorem 1, the variable t does not have to be a time variable.

in the bucket at time t. Then $s'(t) = r(t)$ because $s'(t)$ is the rate at which the quantity of water is changing, and $s(0) = 0$ because the bucket is initially empty. By FTC I,

$$\underbrace{\int_0^4 s'(t)\,dt}_{\substack{\text{Area under the graph} \\ \text{of the flow rate}}} = s(4) - s(0) = \underbrace{s(4)}_{\substack{\text{Water in bucket} \\ \text{at } t = 4}}$$

More generally, we call the quantity $s(t_2) - s(t_1)$ the **net change** in $s(t)$ over the interval $[t_1, t_2]$. The FTC I yields the following result.

THEOREM 1 Net Change as the Integral of a Rate   The net change in $s(t)$ over an interval $[t_1, t_2]$ is given by the integral

$$\underbrace{\int_{t_1}^{t_2} s'(t)\,dt}_{\text{Integral of the rate of change}} = \underbrace{s(t_2) - s(t_1)}_{\text{Total change over } [t_1, t_2]}$$

■ **EXAMPLE 1**   At 7 AM, water begins leaking from a tank at a rate of $2 + 0.25t$ gal/hour (t is the number of hours after 7 AM). How much water is lost between 9 and 11 AM?

Solution Let $s(t)$ be the quantity of water in the tank at time t. Then $s(t)$ is decreasing and thus $s'(t) = -(2 + 0.25t)$. Since 9 AM corresponds to $t = 2$, the net change in $s(t)$ between 9 and 11 AM is

$$s(4) - s(2) = \int_2^4 s'(t)\,dt = -\int_2^4 (2 + 0.25t)\,dt$$

$$= -\left.\left(2t + \frac{1}{8}t^2\right)\right|_2^4 = -(10 - 4.5) = -5.5 \text{ gal}$$

Thus, the quantity of water lost between 9 and 11 AM is 5.5 gal.   ■

In the next example, an integral is estimated using numerical data. We compute both the left- and right-endpoint approximations and take their average. This average, which is usually more accurate than either endpoint approximation alone, is called the **Trapezoidal Approximation;** we will study it in Chapter 7.

■ **EXAMPLE 2** Traffic Flow   The number of cars per hour passing an observation point along a highway is called the rate of traffic flow $q(t)$ (in cars per hour).

(a) Which quantity is represented by the integral $\int_{t_1}^{t_2} q(t)\,dt$?

(b) The flow rate is recorded at 15-min intervals between 7:00 and 9:00 AM. Estimate the number of cars using the highway during this 2-hour period by taking the average of the left- and right-endpoint approximations.

t	7:00	7:15	7:30	7:45	8:00	8:15	8:30	8:45	9:00
$q(t)$	1,044	1,297	1,478	1,844	1,451	1,378	1,155	802	542

Solution

(a) The integral represents the total number of cars that passed the observation point during the time interval $[t_1, t_2]$.

(b) The data values are spaced at intervals of $\Delta t = 0.25$ hour. Thus,

$$L_N = 0.25\Big(1{,}044 + 1{,}297 + 1{,}478 + 1{,}844 + 1{,}451 + 1{,}378 + 1{,}155 + 802\Big)$$

$$\approx 2{,}612$$

$$R_N = 0.25\Big(1{,}297 + 1{,}478 + 1{,}844 + 1{,}451 + 1{,}378 + 1{,}155 + 802 + 542\Big)$$

$$\approx 2{,}487$$

In Example 2, L_N is the sum of the values of $q(t)$ at the left endpoints

7:00, 7:15, ..., 8:45

and R_N is the sum of the values of $q(t)$ at the right endpoints

7:15, ..., 8:45, 9:00

We estimate the number of cars that passed the observation point between 7 and 9 AM by taking the average of R_N and L_N:

$$\frac{1}{2}(R_N + L_N) = \frac{1}{2}(2{,}612 + 2{,}487) \approx 2{,}550 \text{ cars passed between 7 and 9 AM} \quad \blacksquare$$

The Integral of Velocity

At the beginning of this chapter, we claimed that the area under the graph of velocity is equal to the distance traveled. We can now verify this claim, with the understanding that if the velocity is not positive, then the integral computes the *net change in position* or *displacement*. Indeed, if $s(t)$ is the position at time t, then $v(t) = s'(t)$ is the velocity and

$$\int_{t_1}^{t_2} v(t)\,dt = \int_{t_1}^{t_2} s'(t)\,dt = \underbrace{s(t_2) - s(t_1)}_{\text{Displacement or net change in position}}$$

To calculate the actual distance traveled rather than displacement, we must integrate the absolute value of velocity, which is the *speed* $|v(t)|$.

THEOREM 2 The Integral of Velocity   If an object travels in a straight line with velocity $v(t)$, then

$$\text{Displacement during } [t_1, t_2] = \int_{t_1}^{t_2} v(t)\,dt$$

$$\text{Total distance traveled during } [t_1, t_2] = \int_{t_1}^{t_2} |v(t)|\,dt$$

■ **EXAMPLE 3**   The velocity of a particle is $v(t) = t^3 - 10t^2 + 24t$ ft/s (Figure 2). Compute the

(a) Displacement over [0, 4], [4, 6], and [0, 6]

(b) Total distance traveled over [0, 6]

Indicate the particle's trajectory with a motion diagram.

Solution   We first compute an indefinite integral:

$$\int v(t)\,dt = \int (t^3 - 10t^2 + 24t)\,dt = \frac{1}{4}t^4 - \frac{10}{3}t^3 + 12t^2 + C$$

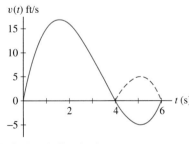

$v(t)$ ft/s

FIGURE 2 Graph of $v(t) = t^3 - 10t^2 + 24t$. Over [4, 6], the broken curve is the graph of $|v(t)|$.

(a) The displacements over the given intervals are

$$[0, 4]: \quad \int_0^4 v(t)\, dt = \left(\frac{1}{4}t^4 - \frac{10}{3}t^3 + 12t^2 \right)\Bigg|_0^4 = 42\frac{2}{3} \text{ ft}$$

$$[4, 6]: \quad \int_4^6 v(t)\, dt = \left(\frac{1}{4}t^4 - \frac{10}{3}t^3 + 12t^2 \right)\Bigg|_4^6 = -6\frac{2}{3} \text{ ft}$$

$$[0, 6]: \quad \int_0^6 v(t)\, dt = \left(\frac{1}{4}t^4 - \frac{10}{3}t^3 + 12t^2 \right)\Bigg|_0^6 = 36 \text{ ft}$$

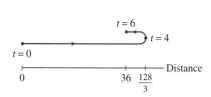

FIGURE 3 Path of the particle along a straight line.

Figure 3 is a motion diagram indicating the particle's trajectory. It travels $42\frac{2}{3}$ ft during the first 4 s and then backtracks $6\frac{2}{3}$ ft over the next 2 s.

(b) To integrate $|v(t)|$, we refer to Figure 2, which shows that $v(t)$ is negative on $[4, 6]$. Although the displacement over $[0, 6]$ is 36 ft, the total distance traveled is

$$\int_0^6 |v(t)|\, dt = \int_0^4 v(t)\, dt - \int_4^6 v(t)\, dt = 42\frac{2}{3} - \left(-6\frac{2}{3} \right) = 49\frac{1}{3} \text{ ft} \qquad \blacksquare$$

Total Versus Marginal Cost

Economists and business managers often study how the cost of manufacturing a product varies with the number of units produced. The cost function $C(x)$ is defined as the cost of producing x units of a commodity and the **marginal cost** is the derivative $C'(x)$. We then have

$$\text{Cost of increasing production from } a \text{ to } b = \int_a^b C'(x)\, dx$$

■ **EXAMPLE 4**   The marginal cost of producing x computer chips (in units of 1,000) is $C'(x) = 150x^2 - 3{,}000x + 17{,}500$ (dollars per thousand chips).

(a) Find the cost of increasing production from 10,000 to 15,000 chips.

(b) Determine the total cost of producing 15,000 chips, assuming that $C(0) = 35{,}000$ (in other words, it costs \$35,000 to set up the manufacturing run).

Solution

(a) The cost of increasing production from 10,000 ($x = 10$) to 15,000 ($x = 15$) is

$$C(15) - C(10) = \int_{10}^{15} (150x^2 - 3{,}000x + 17{,}500)\, dx$$

$$= (50x^3 - 1{,}500x^2 + 17{,}500x)\Bigg|_{10}^{15}$$

$$= 93{,}750 - 75{,}000 = \$18{,}750$$

(b) We have

$$C(15) - C(0) = \int_0^{15} (150x^2 - 3{,}000x + 17{,}500)\, dx$$

$$= (50x^3 - 1{,}500x^2 + 17{,}500x)\Bigg|_0^{15}$$

$$= \$93{,}750$$

This is the increase in cost from 0 to 15,000 chips. Since setup costs are $C(0) = \$35,000$, the total cost of producing 15,000 chips is

$$C(15) = C(0) + 93,750 = 35,000 + 93,750 = \$128,750$$ ■

5.5 SUMMARY

- Many applications are based on the following principle: *The integral of the rate of change is the net change:*

$$\underbrace{s(t_2) - s(t_1)}_{\text{Net change over } [t_1, t_2]} = \int_{t_1}^{t_2} s'(t)\, dt$$

- For an object traveling in a straight line at velocity $v(t)$,

$$\text{Displacement during } [t_1, t_2] = \int_{t_1}^{t_2} v(t)\, dt$$

$$\text{Total distance traveled during } [t_1, t_2] = \int_{t_1}^{t_2} |v(t)|\, dt$$

- If $C(x)$ is the cost of producing x units of a commodity, then $C'(x)$ is the marginal cost and

$$\text{Cost of increasing production from } a \text{ to } b = \int_a^b C'(x)\, dx$$

5.5 EXERCISES

Preliminary Questions

1. An airplane makes the 350-mile trip from Los Angeles to San Francisco in 1 hour. Assuming that the plane's velocity at time t is $v(t)$ mph, what is the value of the integral $\int_0^1 v(t)\, dt$?

2. A hot metal object is submerged in cold water. The rate at which the object cools (in degrees per minute) is a function $f(t)$ of time. Which quantity is represented by the integral $\int_0^T f(t)\, dt$?

3. Which of the following quantities would be naturally represented

as derivatives and which as integrals?

(a) Velocity of a train

(b) Rainfall during a 6-month period

(c) Mileage per gallon of an automobile

(d) Increase in the population of Los Angeles from 1970 to 1990

4. Two airplanes take off at $t = 0$ from the same place and in the same direction. Their velocities are $v_1(t)$ and $v_2(t)$, respectively. What is the physical interpretation of the area between the graphs of $v_1(t)$ and $v_2(t)$ over an interval $[0, T]$?

Exercises

1. Water flows into an empty reservoir at a rate of $3,000 + 5t$ gal/hour. What is the quantity of water in the reservoir after 5 hours?

2. Find the displacement of a particle moving in a straight line with velocity $v(t) = 4t - 3$ ft/s over the time interval $[2, 5]$.

3. A population of insects increases at a rate of $200 + 10t + 0.25t^2$ insects per day. Find the insect population after 3 days, assuming that there are 35 insects at $t = 0$.

4. A survey shows that a mayoral candidate is gaining votes at a rate of $2,000t + 1,000$ votes per day, where t is the number of days since she announced her candidacy. How many supporters will the candidate have after 60 days, assuming that she had no supporters at $t = 0$?

5. A factory produces bicycles at a rate of $95 + 0.1t^2 - t$ bicycles per week (t in weeks). How many bicycles were produced from day 8 to 21?

6. Find the displacement over the time interval $[1, 6]$ of a helicopter whose (vertical) velocity at time t is $v(t) = 0.02t^2 + t$ ft/s.

7. A cat falls from a tree (with zero initial velocity) at time $t = 0$. How far does the cat fall between $t = 0.5$ and $t = 1$ s? Use Galileo's formula $v(t) = -32t$ ft/s.

8. A projectile is released with initial (vertical) velocity 100 m/s. Use the formula $v(t) = 100 - 9.8t$ for velocity to determine the distance traveled during the first 15 s.

In Exercises 9–12, assume that a particle moves in a straight line with given velocity. Find the total displacement and total distance traveled over the time interval, and draw a motion diagram like Figure 3 (with distance and time labels).

9. $12 - 4t$ ft/s, $[0, 5]$ **10.** $32 - 2t^2$ ft/s, $[0, 6]$

11. $t^{-2} - 1$ m/s, $[0.5, 2]$ **12.** $\cos t$ m/s, $[0, 4\pi]$

13. The rate (in liters per minute) at which water drains from a tank is recorded at half-minute intervals. Use the average of the left- and right-endpoint approximations to estimate the total amount of water drained during the first 3 min.

t (min)	0	0.5	1	1.5	2	2.5	3
l/min	50	48	46	44	42	40	38

14. The velocity of a car is recorded at half-second intervals (in feet per second). Use the average of the left- and right-endpoint approximations to estimate the total distance traveled during the first 4 s.

t	0	0.5	1	1.5	2	2.5	3	3.5	4
$v(t)$	0	12	20	29	38	44	32	35	30

15. Let $a(t)$ be the acceleration of an object in linear motion at time t. Explain why $\int_{t_1}^{t_2} a(t)\,dt$ is the net change in velocity over $[t_1, t_2]$. Find the net change in velocity over $[1, 6]$ if $a(t) = 24t - 3t^2$ ft/s^2.

16. Show that if acceleration a is constant, then the change in velocity is proportional to the length of the time interval.

17. The traffic flow rate past a certain point on a highway is $q(t) = 3{,}000 + 2{,}000t - 300t^2$, where t is in hours and $t = 0$ is 8 AM. How many cars pass by during the time interval from 8 to 10 AM?

18. Suppose that the marginal cost of producing x video recorders is $0.001x^2 - 0.6x + 350$ dollars. What is the cost of producing 300 units if the setup cost is \$20,000 (see Example 4)? If production is set at 300 units, what is the cost of producing 20 additional units?

19. Carbon Tax   To encourage manufacturers to reduce pollution, a carbon tax on each ton of CO_2 released into the atmosphere has been proposed. To model the effects of such a tax, policymakers study the *marginal cost of abatement* $B(x)$, defined as the cost of increasing CO_2

reduction from x to $x + 1$ tons (in units of ten thousand tons—Figure 4). Which quantity is represented by $\int_0^3 B(t)\,dt$?

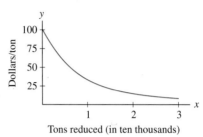

FIGURE 4   Marginal cost of abatement $B(x)$.

20. Power is the rate of energy consumption per unit time. A megawatt of power is 10^6 W or 3.6×10^9 J/hour. Figure 5 shows the power supplied by the California power grid over a typical 1-day period. Which quantity is represented by the area under the graph?

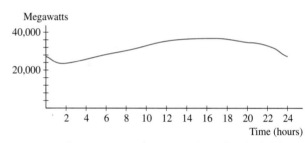

FIGURE 5   Power consumption over 1-day period in California.

21. [✎]   Figure 6 shows the migration rate $M(t)$ of Ireland during the period 1988–1998. This is the rate at which people (in thousands per year) move in or out of the country.

(a) What does $\int_{1988}^{1991} M(t)\,dt$ represent?

(b) Did migration over the 11-year period 1988–1998 result in a net influx or outflow of people from Ireland? Base your answer on a rough estimate of the positive and negative areas involved.

(c) During which year could the Irish prime minister announce, "We are still losing population but we've hit an inflection point—the trend is now improving."

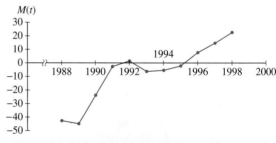

FIGURE 6   Irish migration rate (in thousands per year).

22. Figure 7 shows the graph of $Q(t)$, the rate of retail truck sales in the United States (in thousands sold per year).

(a) What does the area under the graph over the interval [1995, 1997] represent?

(b) Express the total number of trucks sold in the period 1994–1997 as an integral (but do not compute it).

(c) Use the following data to compute the average of the right- and left-endpoint approximations as an estimate for the total number of trucks sold during the 2-year period 1995–1996.

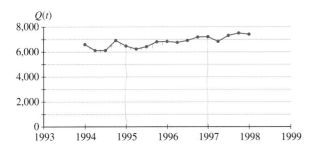

Year (qtr.)	$Q(t)$ ($)	Year (qtr.)	$Q(t)$ ($)
1995(1)	6,484	1996(1)	7,216
1995(2)	6,255	1996(2)	6,850
1995(3)	6,424	1996(3)	7,322
1995(4)	6,818	1996(4)	7,537

FIGURE 7 Quarterly retail sales rate of trucks in the United States (in thousands per year).

23. Heat Capacity   The heat capacity $C(T)$ of a substance is the amount of energy (in joules) required to raise the temperature of 1 g by 1°C at temperature T.

(a) Explain why the energy required to raise the temperature from T_1 to T_2 is the area under the graph of $C(T)$ over $[T_1, T_2]$.

(b) How much energy is required to raise the temperature from 50 to 100°C if $C(T) = 6 + 0.2\sqrt{T}$?

In Exercises 24 and 25, consider the following. Paleobiologists have studied the extinction of marine animal families during the phanerozoic period, which began 544 million years ago. A recent study suggests that the extinction rate $r(t)$ may be modeled by the function

$r(t) = 3,130/(t + 262)$ *for $0 \leq t \leq 544$. Here, t is time elapsed (in millions of years) since the beginning of the phanerozoic period. Thus, $t = 544$ refers to the present time, $t = 540$ is 4 million years ago, etc.*

24. Use R_N or L_N with $N = 10$ (or their average) to estimate the total number of families that became extinct in the periods $100 \leq t \leq 150$ and $350 \leq t \leq 400$.

25. $\mathcal{CAS}$ Estimate the total number of extinct families from $t = 0$ to the present, using M_N with $N = 544$.

26. 📖   Cardiac output is the rate R of volume of blood pumped by the heart per unit time (in liters per minute). Doctors measure R by injecting A mg of dye into a vein leading into the heart at $t = 0$ and recording the concentration $c(t)$ of dye (in milligrams per liter) pumped out at short regular time intervals (Figure 8).

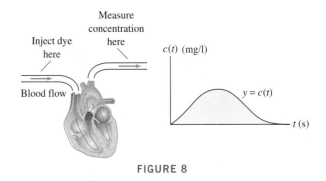

FIGURE 8

(a) The quantity of dye pumped out in a small time interval $[t, t + \Delta t]$ is approximately $Rc(t)\Delta t$. Explain why.

(b) Show that $A = R \int_0^T c(t)\, dt$, where T is large enough that all of the dye is pumped through the heart but not so large that the dye returns by recirculation.

(c) Use the following data to estimate R, assuming that $A = 5$ mg:

t (s)	0	1	2	3	4	5
$c(t)$	0	0.4	2.8	6.5	9.8	8.9
t (s)	6	7	8	9	10	
$c(t)$	6.1	4	2.3	1.1	0	

Further Insights and Challenges

27. A particle located at the origin at $t = 0$ moves along the x-axis with velocity $v(t) = (t + 1)^{-2}$. Show that the particle will never pass the point $x = 1$.

28. A particle located at the origin at $t = 0$ moves along the x-axis with velocity $v(t) = (t + 1)^{-1/2}$. Will the particle be at the point $x = 1$ at any time t? If so, find t.

5.6 Substitution Method

In calculus, the term "integration" is used in two ways. It denotes the process of finding signed area under a graph (computing a definite integral) and the process of finding an antiderivative (evaluating an indefinite integral).

Integration (antidifferentiation) is generally more difficult than differentiation. There are no sure-fire methods for finding an antiderivative, and, in some cases, there is no elementary formula for the antiderivative. However, there are several general techniques that are widely applicable. One such technique is the **Substitution Method**, which uses the Chain Rule "in reverse."

Consider the integral $\int 2x \cos(x^2)\, dx$. We can evaluate it if we remember the Chain Rule calculation

$$\frac{d}{dx} \sin(x^2) = 2x \cos(x^2)$$

This says that $\sin(x^2)$ is an antiderivative of $2x \cos(x^2)$. Therefore,

$$\int \underbrace{2x}_{\substack{\text{Derivative of}\\\text{inside function}}} \cos \underbrace{(x^2)}_{\substack{\text{Inside}\\\text{function}}} dx = \sin(x^2) + C$$

A similar Chain Rule calculation shows that

$$\int \underbrace{(1 + 3x^2)}_{\substack{\text{Derivative of}\\\text{inside function}}} \cos \underbrace{(x + x^3)}_{\substack{\text{Inside}\\\text{function}}} dx = \sin(x + x^3) + C$$

REMINDER *A "composite function" is a function of the form $f(g(x))$. For convenience, we call $g(x)$ the "inside function" and $f(u)$ the "outside function."*

In both cases, the integrand is the product of a composite function *and* the derivative of the inside function. The Chain Rule does not help if the derivative of the inside function is missing. For instance, we cannot use the Chain Rule to compute $\int \cos(x^2)\, dx$.

In general, this method works when the integrand is of the form $f(u(x))u'(x)$. If $F(u)$ is an antiderivative of $f(u)$, then by the Chain Rule,

$$\frac{d}{dx} F(u(x)) = F'(u(x))u'(x) = f(u(x))u'(x)$$

This translates into the following integration formula:

THEOREM 1 The Substitution Method   If $F'(x) = f(x)$, then

$$\int f(u(x))u'(x)\, dx = F(u(x)) + C$$

Substitution Using Differentials

Before proceeding to examples, we discuss the procedure for carrying out substitution using differentials. A **differential** is a symbol such as dx or du that occurs in Leibniz's notations du/dx and $\int f(x)\, dx$. We treat differentials as symbols that do not represent actual quantities, but we manipulate them algebraically as if they were related by an equation in which we may cancel dx:

$$du = \frac{du}{dx}\, dx$$

Equivalently,

$$du = u'(x)\,dx \qquad \boxed{1}$$

For example,

If $u = x^2$, then $du = 2x\,dx$ [since $u'(x) = 2x$]

If $u = \cos(x^3)$, then $du = -3x^2\sin(x^3)\,dx$ [since $u'(x) = -3x^2\sin(x^3)$]

Now, when the integrand has the form $f(u(x))\,u'(x)$, we may use Eq. (1) to rewrite the entire integral (including the term dx) in terms of u and its differential du:

The symbolic calculus of substitution using differentials was invented by Leibniz and is considered one of his greatest achievements. It reduces the otherwise complicated process of transforming integrals to a convenient set of rules.

$$\int \underbrace{f(u(x))}_{f(u)}\ \underbrace{u'(x)\,dx}_{du} = \int f(u)\,du$$

This is often referred to as the **Change of Variables Formula**. It transforms an integral in the variable x into a (hopefully simpler) integral in the new variable u.

In substitution, the key step is to choose the appropriate inside function u.

■ **EXAMPLE 1** Evaluate $\displaystyle\int 3x^2\sin(x^3)\,dx$ and check the answer.

Solution The integrand contains the composite function $\sin(x^3)$, so we set $u = x^3$. Since the differential $du = 3x^2 dx$ also appears in the integrand, we can carry out the substitution:

$$\int 3x^2\sin(x^3)\,dx = \int \underbrace{\sin(x^3)}_{\sin u}\ \underbrace{3x^2\,dx}_{du} = \int \sin u\,du$$

Next, we evaluate the integral in the u variable and replace u by x^3 in the answer:

$$\int 3x^2\sin(x^3)\,dx = \int \sin u\,du = -\cos u + C = -\cos(x^3) + C$$

We can check the answer by differentiating:

$$\frac{d}{dx}(-\cos(x^3)) = \sin(x^3)\frac{d}{dx}x^3 = 3x^2\sin(x^3) \qquad ■$$

■ **EXAMPLE 2** Evaluate $\displaystyle\int 2x(x^2 + 9)^5\,dx$.

Solution For clarity, we break up the solution into three steps.

Step 1. **Choose the function u and compute du.**
　In this case, we let $u = x^2 + 9$. This is a good choice because both the composite $u^5 = (x^2 + 9)^5$ and the differential $du = 2x\,dx$ appear in the integral.

Step 2. **Rewrite the integral in terms of u and du, and evaluate.**

$$\int 2x(x^2 + 9)^5\,dx = \int \overbrace{(x^2 + 9)^5}^{u^5}\ \overbrace{2x\,dx}^{du} = \int u^5\,du = \frac{1}{6}u^6 + C$$

Step 3. **Express the final answer in terms of x.**

Rewrite the previous answer in terms of x by substituting $u = x^2 + 9$:

$$\int 2x(x^2 + 9)^5 \, dx = \frac{1}{6}u^6 + C = \frac{1}{6}(x^2 + 9)^6 + C$$

∎

■ **EXAMPLE 3** Integral of $\tan \theta$ Evaluate $\displaystyle\int \tan \theta \, d\theta$.

Solution Use the substitution $u = \cos x$, $du = -\sin x \, dx$:

$$\int \tan x \, dx = \int \frac{\sin x}{\cos x} \, dx = -\int \frac{du}{u} = -\ln|u| + C = -\ln|\cos x| + C$$

Now recall that $-\ln u = \ln \frac{1}{u}$. Thus, $-\ln|\cos x| = \ln \frac{1}{|\cos x|}$, and we obtain

$$\int \tan x \, dx = \ln \left| \frac{1}{\cos x} \right| + C = \ln|\sec x| + C$$

∎

Applying Substitution:

(1) Choose u and compute du.

(2) Rewrite the integral in terms of u and du, and evaluate.

(3) Express the final answer in terms of x.

■ **EXAMPLE 4** Multiplying du by a Constant Evaluate $\displaystyle\int \frac{x^2 + 2x}{(x^3 + 3x^2 + 9)^4} \, dx$.

Solution The composite function $(x^3 + 3x^2 + 9)^{-4}$ in the integrand suggests the substitution $u = x^3 + 3x^2 + 9$. However, the differential

$$du = (3x^2 + 6x) \, dx = 3(x^2 + 2x) \, dx \qquad \boxed{2}$$

does *not* appear in the integrand. What does appear is $(x^2 + 2x) \, dx$. To remedy the situation, we multiply Eq. (2) by $\frac{1}{3}$ to obtain

$$\frac{1}{3} \, du = (x^2 + 2x) \, dx$$

Now we can carry out the substitution:

$$\int \frac{x^2 + 2x}{(x^3 + 3x^2 + 9)^4} \, dx = \int \overbrace{(x^3 + 3x^2 + 9)^{-4}}^{u^{-4}} \overbrace{(x^2 + 2x) \, dx}^{\frac{1}{3} \, du}$$

$$= \frac{1}{3} \int u^{-4} \, du = -\frac{1}{9}u^{-3} + C$$

$$= -\frac{1}{9}(x^3 + 3x^2 + 9)^{-3} + C$$

∎

■ **EXAMPLE 5** Evaluate $\displaystyle\int \sin(7\theta + 5) \, d\theta$.

Solution Let $u = 7\theta + 5$. Then $du = 7 \, d\theta$ and $\frac{1}{7}du = d\theta$. We obtain

$$\int \sin(7\theta + 5) \, d\theta = \int (\sin u) \left(\frac{1}{7} \, du \right) = \frac{1}{7} \int \sin u \, du$$

$$= -\frac{1}{7} \cos u + C = -\frac{1}{7} \cos(7\theta + 5) + C$$

∎

■ **EXAMPLE 6** Evaluate $\int e^{-9x}\, dx$.

Solution We use the substitution $u = -9x$, $du = -9\, dx$:

$$\int e^{-9x}\, dx = \int e^u \left(-\frac{1}{9}\, du \right) = -\frac{1}{9} \int e^u\, du = -\frac{1}{9} e^u + C = -\frac{1}{9} e^{-9x} + C \qquad ■$$

■ **EXAMPLE 7** Additional Step Necessary Evaluate $\int x\sqrt{5x+1}\, dx$.

Solution We recognize that $\sqrt{5x+1}$ is a composite function, so it makes sense to try $u = 5x + 1$. Then $\frac{1}{5}\, du = dx$ and $\sqrt{5x+1} = u^{1/2}$, and

$$\sqrt{5x+1}\, dx = \frac{1}{5} u^{1/2}\, du$$

Unfortunately, the integrand is not $\sqrt{5x+1}$ but $x\sqrt{5x+1}$. To take care of the extra factor of x, we solve $u = 5x + 1$ to obtain $x = \frac{1}{5}(u - 1)$. Then

$$x\sqrt{5x+1}\, dx = \left(\frac{1}{5}(u-1) \right) \frac{1}{5} u^{1/2}\, du = \frac{1}{25}(u-1) u^{1/2}\, du$$

and we evaluate as follows:

$$\int x\sqrt{5x+1}\, dx = \frac{1}{25} \int (u-1) u^{1/2}\, du = \frac{1}{25} \int (u^{3/2} - u^{1/2})\, du$$

$$= \frac{1}{25} \left(\frac{2}{5} u^{5/2} - \frac{2}{3} u^{3/2} \right) + C$$

$$= \frac{2}{125}(5x+1)^{5/2} - \frac{2}{75}(5x+1)^{3/2} + C \qquad ■$$

Change of Variables Formula for Definite Integrals

There are two ways of using substitution to compute definite integrals. One way is to carry out the Substitution Method fully to find an antiderivative in terms of the x-variable. However, it is often more efficient to evaluate the definite integral directly in terms of the u-variable. This can be done provided that the limits of integration are changed as indicated in the next theorem.

Change of Variables Formula for Definite Integrals If $u(x)$ is differentiable on $[a, b]$ and $f(x)$ is integrable on the range of $u(x)$, then

$$\int_a^b f(u(x)) u'(x)\, dx = \int_{u(a)}^{u(b)} f(u)\, du$$

The new limits of integration with respect to the u-variable are $u(a)$ and $u(b)$. Think of it this way: as x varies from a to b, the variable $u = u(x)$ varies from $u(a)$ to $u(b)$.

Proof If $F(x)$ is an antiderivative of $f(x)$, then $F(u(x))$ is an antiderivative of $f(u(x))u'(x)$. By FTC I,

$$\int_a^b f(u(x)) u'(x)\, dx = F(u(b)) - F(u(a))$$

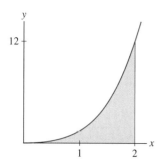

FIGURE 1 Region represented by
$\int_0^2 x^2\sqrt{x^3+1}\,dx$.

FIGURE 2 Region represented by
$\frac{1}{3}\int_1^9 \sqrt{u}\,du$.

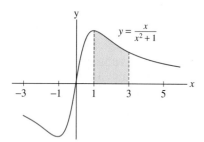

FIGURE 3 Area under the graph of
$y = \dfrac{x}{x^2+1}$ over $[1, 3]$.

The FTC applied to the u variable also yields

$$\int_{u(a)}^{u(b)} f(u)\,du = F(u(b)) - F(u(a))$$

■ **EXAMPLE 8** Evaluate $\int_0^2 x^2\sqrt{x^3+1}\,dx$.

Solution Use the substitution $u = x^3 + 1$, $du = 3x^2\,dx$ to obtain

$$x^2\sqrt{x^3+1}\,dx = \frac{1}{3}\sqrt{u}\,du$$

According to the Change of Variables Formula, the new limits of integration are $u(0) = 0^3 + 1 = 1$ and $u(2) = 2^3 + 1 = 9$. We obtain

$$\int_0^2 x^2\sqrt{x^3+1}\,dx = \frac{1}{3}\int_1^9 \sqrt{u}\,du = \frac{2}{9}u^{3/2}\Big|_1^9 = \frac{52}{9}$$

This substitution shows that the regions in Figures 1 and 2 have equal area.   ■

■ **EXAMPLE 9** Evaluate $\int_0^{\pi/4} \tan^3 x \sec^2 x\,dx$.

Solution To carry out the substitution, we must remember that $\sec^2 x$ is the derivative of $\tan x$. This suggests setting $u = \tan x$, so that $u^3 = \tan^3 x$ and $du = \sec^2 x\,dx$. The new limits of integration are $u(0) = \tan 0 = 0$ and $u(\frac{\pi}{4}) = \tan(\frac{\pi}{4}) = 1$. Thus,

$$\int_0^{\pi/4} \tan^3 x \sec^2 x\,dx = \int_0^1 u^3\,du = \frac{u^4}{4}\Big|_0^1 = \frac{1}{4}$$ ■

■ **EXAMPLE 10** Calculate the area under the graph of $y = \dfrac{x}{x^2+1}$ over $[1, 3]$.

Solution The area in question (the shaded region in Figure 3) is equal to $\int_1^3 \dfrac{x}{x^2+1}\,dx$. We use the substitution

$$u = x^2 + 1, \qquad \frac{1}{2}du = x\,dx, \qquad \frac{1}{2}\frac{du}{u} = \frac{x\,dx}{x^2+1}$$

The new limits of integration are $u(1) = 1^2 + 1 = 2$ and $u(3) = 3^2 + 1 = 10$. By the Change of Variables Formula,

$$\int_1^3 \frac{x}{x^2+1}\,dx = \frac{1}{2}\int_2^{10} \frac{du}{u} = \frac{1}{2}\ln|u|\Big|_2^{10} = \frac{1}{2}\ln 10 - \frac{1}{2}\ln 2 \approx 0.805$$ ■

5.6 SUMMARY

• Try the Substitution Method when the integrand has the form $f(u(x))\,u'(x)$. If F is an antiderivative of f, then:

$$\int f(u(x))\,u'(x)\,dx = F(u(x)) + C$$

- The differential of $u(x)$ is the symbol $du = u'(x)\,dx$. The Substitution Method is expressed by the Change of Variables Formula:

$$\int f(u(x))\,u'(x)\,dx = \int f(u)\,du$$

- Definite integrals may be evaluated using the Change of Variables Formula, provided that the limits of integration are changed appropriately:

$$\int_a^b f(u(x))\,u'(x)\,dx = \int_{u(a)}^{u(b)} f(u)\,du$$

5.6 EXERCISES

Preliminary Questions

1. Which of the following integrals is a candidate for the Substitution Method?

(a) $\displaystyle\int 5x^4 \sin(x^5)\,dx$

(b) $\displaystyle\int \sin^5 x\,\cos x\,dx$

(c) $\displaystyle\int x^5 \sin x\,dx$

2. Write each of the following functions in the form $cg(u(x))u'(x)$, where c is a constant.

(a) $x(x^2+9)^4$

(b) $x^2 \sin(x^3)$

(c) $\sin x\,\cos^2 x$

3. Which of the following is equal to $\displaystyle\int_0^2 x^2(x^3+1)\,dx$ for a suitable substitution?

(a) $\dfrac{1}{3}\displaystyle\int_0^2 u\,du$

(b) $\displaystyle\int_0^9 u\,du$

(c) $\dfrac{1}{3}\displaystyle\int_1^9 u\,du$

Exercises

In Exercises 1–6, calculate du for the given function.

1. $u = 1 - x^2$

2. $u = \sin x$

3. $u = x^3 - 2$

4. $u = 2x^4 + 8x$

5. $u = \cos(x^2)$

6. $u = \tan x$

In Exercises 7–28, write the integral in terms of u and du. Then evaluate.

7. $\displaystyle\int (x-7)^3\,dx, \quad u = x - 7$

8. $\displaystyle\int 2x\sqrt{x^2+1}\,dx, \quad u = x^2 + 1$

9. $\displaystyle\int (x+1)^{-2}\,dx, \quad u = x + 1$

10. $\displaystyle\int x(x+1)^9\,dx, \quad u = x + 1$

11. $\displaystyle\int \sin(2x-4)\,dx, \quad u = 2x - 4$

12. $\displaystyle\int \frac{x^3}{(x^4+1)^4}\,dx, \quad u = x^4 + 1$

13. $\displaystyle\int \frac{x+1}{(x^2+2x)^3}\,dx, \quad u = x^2 + 2x$

14. $\displaystyle\int \frac{x}{(8x+5)^3}\,dx, \quad u = 8x + 5$

15. $\displaystyle\int \sqrt{4x-1}\,dx, \quad u = 4x - 1$

16. $\displaystyle\int x\sqrt{4x-1}\,dx, \quad u = 4x - 1$

17. $\displaystyle\int x^2\sqrt{4x-1}\,dx, \quad u = 4x - 1$

18. $\displaystyle\int x\cos(x^2)\,dx, \quad u = x^2$

19. $\displaystyle\int \sin^2 x\,\cos x\,dx, \quad u = \sin x$

20. $\displaystyle\int \sec^2 x\,\tan x\,dx, \quad u = \tan x$

21. $\displaystyle\int \tan 2x\,dx, \quad u = \cos 2x$

22. $\displaystyle\int \cot x\,dx, \quad u = \sin x$

23. $\displaystyle\int xe^{-x^2}\,dx, \quad u = -x^2$

24. $\displaystyle\int (\sec^2\theta)\,e^{\tan\theta}\,d\theta, \quad u = \tan\theta$

25. $\displaystyle\int \frac{e^t\,dt}{e^{2t}+2e^t+1}, \quad u = e^t$

26. $\displaystyle\int \frac{(\ln x)^2\, dx}{x}, \quad u = \ln x$

27. $\displaystyle\int \frac{dx}{x(\ln x)^2}, \quad u = \ln x$

28. $\displaystyle\int \frac{(\tan^{-1} x)^2\, dx}{x^2 + 1}, \quad u = \tan^{-1} x$

In Exercises 29–32, show that each of the following integrals is equal to a multiple of $\sin(u(x)) + C$ for an appropriate choice of $u(x)$.

29. $\displaystyle\int x^3 \cos(x^4)\, dx$

30. $\displaystyle\int x^2 \cos(x^3 + 1)\, dx$

31. $\displaystyle\int x^{1/2} \cos(x^{3/2})\, dx$

32. $\displaystyle\int \cos x \cos(\sin x)\, dx$

In Exercises 33–70, evaluate the indefinite integral.

33. $\displaystyle\int (4x + 3)^4\, dx$

34. $\displaystyle\int x^2(x^3 + 1)^3\, dx$

35. $\displaystyle\int \frac{1}{\sqrt{x - 7}}\, dx$

36. $\displaystyle\int \sin(x - 7)\, dx$

37. $\displaystyle\int x\sqrt{x^2 - 4}\, dx$

38. $\displaystyle\int (2x + 1)(x^2 + x)^3\, dx$

39. $\displaystyle\int \frac{dx}{(x + 9)^2}$

40. $\displaystyle\int \frac{x}{\sqrt{x^2 + 9}}\, dx$

41. $\displaystyle\int \frac{2x^2 + x}{(4x^3 + 3x^2)^2}\, dx$

42. $\displaystyle\int (3x^2 + 1)(x^3 + x)^2\, dx$

43. $\displaystyle\int \frac{5x^4 + 2x}{(x^5 + x^2)^3}\, dx$

44. $\displaystyle\int x^2(x^3 + 1)^4\, dx$

45. $\displaystyle\int (3x + 9)^{10}\, dx$

46. $\displaystyle\int x(3x + 9)^{10}\, dx$

47. $\displaystyle\int x(x + 1)^{1/4}\, dx$

48. $\displaystyle\int x^2(x + 1)^7\, dx$

49. $\displaystyle\int x^3(x^2 - 1)^{3/2}\, dx$

50. $\displaystyle\int x^2 \sin(x^3)\, dx$

51. $\displaystyle\int \sin^5 x \cos x\, dx$

52. $\displaystyle\int x^2 \sin(x^3 + 1)\, dx$

53. $\displaystyle\int \tan 3x\, dx$

54. $\displaystyle\int \frac{\tan(\ln x)}{x}\, dx$

55. $\displaystyle\int \sec^2(4x + 9)\, dx$

56. $\displaystyle\int \sec^2 x \tan^4 x\, dx$

57. $\displaystyle\int \frac{\cos 2x}{(1 + \sin 2x)^2}\, dx$

58. $\displaystyle\int \sin 4x\sqrt{\cos 4x + 1}\, dx$

59. $\displaystyle\int \cos x(3 \sin x - 1)\, dx$

60. $\displaystyle\int \frac{\cos \sqrt{x}}{\sqrt{x}}\, dx$

61. $\displaystyle\int \sec^2 x(4 \tan^3 x - 3 \tan^2 x)\, dx$

62. $\displaystyle\int e^{14x - 7}\, dx$

63. $\displaystyle\int (x + 1)e^{x^2 + 2x}\, dx$

64. $\displaystyle\int \frac{dx}{(x + 1)^4}$

65. $\displaystyle\int \frac{e^x\, dx}{(e^x + 1)^4}$

66. $\displaystyle\int \frac{\sec^2(\sqrt{x})\, dx}{\sqrt{x}}$

67. $\displaystyle\int \frac{(\ln x)^4\, dx}{x}$

68. $\displaystyle\int \frac{dx}{x\sqrt{\ln x}}$

69. $\displaystyle\int \frac{dx}{x \ln x}$

70. $\displaystyle\int (\cot x) \ln(\sin x)\, dx$

71. Evaluate $\displaystyle\int x^5\sqrt{x^3 + 1}\, dx$ using $u = x^3 + 1$. *Hint:* $x^5\, dx = x^3 \cdot x^2\, dx$ and $x^3 = u - 1$.

72. Evaluate $\displaystyle\int (x^3 + 1)^{1/4} x^5\, dx$.

73. Can They Both Be Right?   Hannah uses the substitution $u = \tan x$ and Akiva uses $u = \sec x$ to evaluate $\displaystyle\int \tan x \sec^2 x\, dx$. Show that they obtain different answers and explain the apparent contradiction.

74. Evaluate $\displaystyle\int \sin x \cos x\, dx$ using substitution in two different ways: first using $u = \sin x$ and then $u = \cos x$. Reconcile the two different answers.

75. Some Choices Are Better Than Others   Evaluate

$$\int \sin x \cos^2 x\, dx$$

twice. First use $u = \sin x$ to show that

$$\int \sin x \cos^2 x\, dx = \int u\sqrt{1 - u^2}\, du$$

and evaluate the integral on the right by a further substitution. Then show that $u = \cos x$ is a better choice.

76. What are the new limits of integration if we apply the substitution $u = 3x + \pi$ to the integral $\displaystyle\int_0^\pi \sin(3x + \pi)\, dx$?

77. Which of the following is the result of applying the substitution $u = 4x - 9$ to the integral $\displaystyle\int_2^8 (4x - 9)^{20}\, dx$?

(a) $\displaystyle\int_2^8 u^{20}\, du$

(b) $\displaystyle\frac{1}{4}\int_2^8 u^{20}\, du$

(c) $\displaystyle 4\int_{-1}^{23} u^{20}\, du$

(d) $\displaystyle\frac{1}{4}\int_{-1}^{23} u^{20}\, du$

In Exercises 78–91, use the Change of Variables Formula to evaluate the definite integral.

78. $\displaystyle\int_1^3 (x + 2)^3\, dx$

79. $\displaystyle\int_1^6 \sqrt{x + 3}\, dx$

80. $\int_0^1 \dfrac{x}{(x^2+1)^3}\,dx$

81. $\int_{-1}^2 \sqrt{5x+6}\,dx$

82. $\int_0^4 x\sqrt{x^2+9}\,dx$

83. $\int_0^2 \dfrac{x+3}{(x^2+6x+1)^3}\,dx$

84. $\int_1^2 (x+1)(x^2+2x)^3\,dx$

85. $\int_{10}^{17} (x-9)^{-2/3}\,dx$

86. $\int_0^{\pi/4} \tan\theta\,d\theta$

87. $\int_0^1 \theta\tan(\theta^2)\,d\theta$

88. $\int_0^{\pi/2} \cos 3x\,dx$

89. $\int_0^{\pi/2} \cos\left(3x+\dfrac{\pi}{2}\right)dx$

90. $\int_0^{\pi/2} \cos^3 x \sin x\,dx$

91. $\int_0^{\pi/4} \tan^2 x \sec^2 x\,dx$

92. Evaluate $\int \dfrac{dx}{(2+\sqrt{x})^3}$ using $u=2+\sqrt{x}$.

93. Evaluate $\int_0^2 r\sqrt{5-\sqrt{4-r^2}}\,dr$.

In Exercises 94–95, use substitution to evaluate the integral in terms of $f(x)$.

94. $\int f(x)^3 f'(x)\,dx$

95. $\int \dfrac{f'(x)}{f(x)^2}\,dx$

96. Show that $\int_0^{\pi/6} f(\sin\theta)\,d\theta = \int_0^{1/2} f(u)\dfrac{1}{\sqrt{1-u^2}}\,du$.

97. Evaluate $\int_0^{\pi/2} \sin^n x \cos x\,dx$, where n is an integer, $n \neq -1$.

Further Insights and Challenges

98. Use the substitution $u=1+x^{1/n}$ to show that

$$\int \sqrt{1+x^{1/n}}\,dx = n\int u^{1/2}(u-1)^{n-1}\,du$$

Evaluate for $n=2,3$.

99. Evaluate $I = \int_0^{\pi/2} \dfrac{d\theta}{1+\tan^{6,000}\theta}$. *Hint:* Use substitution to

show that I is equal to $J = \int_0^{\pi/2} \dfrac{d\theta}{1+\cot^{6,000}\theta}$ and then check that

$I+J = \int_0^{\pi/2} d\theta$.

100. Show that $\int_{-a}^a f(x)\,dx = 0$ if f is an odd function.

101. (a) Use the substitution $u = x/a$ to prove that the hyperbola $y = x^{-1}$ (Figure 4) has the following special property: If $a, b > 0$, then $\int_a^b \dfrac{1}{x}\,dx = \int_1^{b/a} \dfrac{1}{x}\,dx$.

(b) Show that the areas under the hyperbola over the intervals $[1, 2]$, $[2, 4]$, $[4, 8]$, ... are all equal.

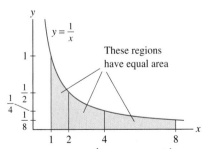

FIGURE 4 The area under $y = \frac{1}{x}$ over $[2^n, 2^{n+1}]$ is the same for all $n = 0, 1, 2 \ldots$.

102. Show that the two regions in Figure 5 have the same area. Then use the identity $\cos^2 u = \frac{1}{2}(1+\cos 2u)$ to compute the second area.

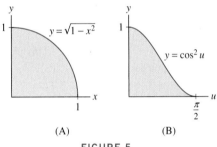

(A)   (B)

FIGURE 5

103. Area of a Circle   The number π is defined as one-half the *circumference* of the unit circle. Prove that the area of a circle of radius r is $A = \pi r^2$. The case $r = 1$ follows from Exercise 102. Prove it for all $r > 0$ by showing that

$$\int_0^r \sqrt{r^2-x^2}\,dx = r^2 \int_0^1 \sqrt{1-x^2}\,dx$$

104. Area of an Ellipse   Prove the formula $A = \pi ab$ for the area of the ellipse with equation

$$\frac{x^2}{a^2} + \frac{y^2}{b^2} = 1$$

Hint: Show that $A = 2b\int_{-a}^a \sqrt{1-(x/a)^2}\,dx$, change variables, and use the formula for the area of a circle (Figure 6).

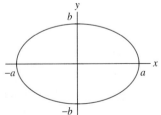

FIGURE 6 Graph of $\dfrac{x^2}{a^2} + \dfrac{y^2}{b^2} = 1$.

5.7 Further Transcendental Functions

In Section 5.3, we used the FTC to derive the formula

$$\int_a^b \frac{dx}{x} = \ln \frac{b}{a}$$

Setting $a = 1$ and $b = x$, we obtain the following formula of $\ln x$ as a definite integral:

$$\ln x = \int_1^x \frac{dt}{t} \qquad \text{for } x > 0 \qquad \boxed{1}$$

In other words, $\ln x$ is equal to the area under the hyperbola over $[1, x]$ as shown in Figure 1. In some treatments of logarithms, Eq. (1) is taken as the definition of the logarithm. All the properties of $\ln x$ may be derived from this integral expression (see Exercise 77).

In a similar fashion, the inverse trigonometric functions can be expressed as integrals. For example, we showed in Section 3.9 that

$$\frac{d}{dx} \sin^{-1} x = \frac{1}{\sqrt{1 - x^2}}$$

This yields the formula

$$\int \frac{dx}{\sqrt{1 - x^2}} = \sin^{-1} x + C$$

Since $\sin^{-1} 0 = 0$, we may express the arcsine as a definite integral:

$$\sin^{-1} x = \int_0^x \frac{dt}{\sqrt{1 - t^2}} \qquad \text{for } -1 < x < 1$$

Thus, $\sin^{-1} x$ is equal to the area under the graph of $y = (1 - x^2)^{-1/2}$ over $[0, x]$ as shown in Figure 2. The following integral formulas follow from the derivative formulas derived in Section 3.9.

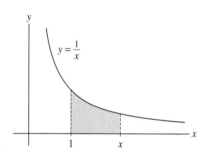

FIGURE 1 Shaded region has area $\ln x$.

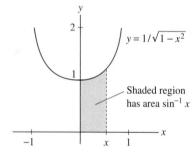

FIGURE 2

THEOREM 1 Inverse Trigonometric Functions

$$\frac{d}{dx} \sin^{-1} x = \frac{1}{\sqrt{1 - x^2}}, \qquad \int \frac{dx}{\sqrt{1 - x^2}} = \sin^{-1} x + C \qquad \boxed{2}$$

$$\frac{d}{dx} \tan^{-1} x = \frac{1}{x^2 + 1}, \qquad \int \frac{dx}{x^2 + 1} = \tan^{-1} x + C \qquad \boxed{3}$$

$$\frac{d}{dx} \sec^{-1} x = \frac{1}{|x|\sqrt{x^2 - 1}}, \qquad \int \frac{dx}{|x|\sqrt{x^2 - 1}} = \sec^{-1} x + C \qquad \boxed{4}$$

In this list, we omit the integral formulas corresponding to the derivatives of $\cos^{-1} x$, $\cot^{-1}$, and $\csc^{-1} x$. The resulting integrals differ only by a minus sign from those already on the list. For example,

$$\frac{d}{dx} \cos^{-1} x = -\frac{1}{\sqrt{1 - x^2}}, \qquad \int \frac{dx}{\sqrt{1 - x^2}} = -\cos^{-1} x + C$$

Thus, we may evaluate integrals of the type $\int \dfrac{dx}{\sqrt{1-x^2}}$ using $\sin^{-1} x$ or $\cos^{-1} x$.

■ **EXAMPLE 1** Evaluate **(a)** $\displaystyle\int_0^1 \dfrac{dx}{x^2+1}$ and **(b)** $\displaystyle\int_0^2 \dfrac{dx}{\sqrt{9-x^2}}$.

Solution

(a) Use Eq. (3):

$$\int_0^1 \frac{dx}{x^2+1} = \tan^{-1} x \Big|_0^1 = \tan^{-1} 1 - \tan^{-1} 0 = \frac{\pi}{4} - 0 = \frac{\pi}{4}$$

This integral represents the shaded area in Figure 3.

(b) Since the integral involves $\sqrt{9-x^2}$ instead of $\sqrt{1-x^2}$, Eq. (2) cannot be applied directly. To take care of the 9, we use the substitution

$$x = 3u, \qquad dx = 3\,du \qquad\qquad \boxed{5}$$

$$\sqrt{9-x^2} = \sqrt{9-(3u)^2} = \sqrt{9(1-u^2)} = 3\sqrt{1-u^2}$$

Since $u = \frac{1}{3}x$, the new limits of integration are $u(0) = 0$ and $u(2) = \frac{2}{3}$:

$$\int_0^2 \frac{dx}{\sqrt{9-x^2}} = \int_0^{2/3} \frac{3\,du}{3\sqrt{1-u^2}} = \sin^{-1}\frac{2}{3} - \sin^{-1} 0 \approx 0.73 - 0 = 0.73 \qquad ■$$

■ **EXAMPLE 2** Using Substitution Evaluate $\displaystyle\int_{1/\sqrt{2}}^1 \dfrac{dx}{x\sqrt{4x^2-1}}$.

Solution To apply Eq. (4), we rewrite the integral using the substitution

$$u = 2x, \qquad du = 2\,dx$$

The new limits are $u\!\left(\frac{1}{\sqrt{2}}\right) = 2\!\left(\frac{1}{\sqrt{2}}\right) = \sqrt{2}$ and $u(1) = 2$. Thus, we obtain

$$\int_{1/\sqrt{2}}^1 \frac{dx}{x\sqrt{4x^2-1}} = \int_{\sqrt{2}}^2 \frac{\frac{1}{2}\,du}{(u/2)\sqrt{u^2-1}}$$

$$= \int_{\sqrt{2}}^2 \frac{du}{u\sqrt{u^2-1}}$$

$$= \sec^{-1} 2 - \sec^{-1}\sqrt{2} = \frac{\pi}{3} - \frac{\pi}{4} = \frac{\pi}{12} \qquad ■$$

Integrals Involving $f(x) = b^x$

We have seen that the exponential function to the base e is particularly convenient because $f(x) = e^x$ is both its own derivative and its own antiderivative. Thus,

$$\frac{d}{dx}e^x = e^x, \qquad \int e^x\,dx = e^x + C$$

We now consider an arbitrary base b, where $b > 0$ and $b \neq 1$. Recall that to differentiate $f(x) = b^x$, we write $b^x = (e^{\ln b})^x = e^{(\ln b)x}$ and use the Chain Rule:

$$\frac{d}{dx}b^x = \frac{d}{dx}e^{(\ln b)x} = (\ln b)e^{(\ln b)x} = (\ln b)b^x$$

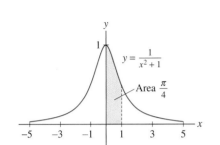

FIGURE 3 The shaded region has an area equal to $\tan^{-1} 1 = \frac{\pi}{4}$.

In substitution, we usually define u as a function of x. Sometimes, it is more convenient to define x as a function of u. We do this in Eq. (5), where we set $x = 3u$.

Therefore, $\dfrac{d}{dx}\left(\dfrac{b^x}{\ln b}\right) = b^x$ and we obtain the integral formula

$$\boxed{\int b^x \, dx = \frac{b^x}{\ln b} + C}$$

6

■ **EXAMPLE 3** Evaluate   **(a)** $\displaystyle\int_2^5 10^x \, dx$ and   **(b)** $\displaystyle\int_0^{\pi/2} (\cos \theta) 10^{\sin \theta} \, d\theta$.

Solution We apply Eq. (6) with $b = 10$.

(a) $\displaystyle\int_2^5 10^x \, dx = \frac{10^x}{\ln 10}\Big|_2^5 = \frac{10^5 - 10^2}{\ln 10} \approx 43{,}386.0$

(b) Use the substitution $u = \sin \theta$, $du = \cos \theta \, d\theta$. The new limits of integration become $u(0) = 0$ and $u(\pi/2) = 1$:

$$\int_0^{\pi/2} (\cos \theta) 10^{\sin \theta} \, d\theta = \int_0^1 10^u \, du = \frac{10^u}{\ln 10}\Big|_0^1 = \frac{10^1 - 10^0}{\ln 10} \approx 3.9 \qquad ■$$

5.7 SUMMARY

- Formula for the logarithm as a definite integral:

$$\ln x = \int_1^x \frac{dt}{t}$$

- Integral formulas for inverse trigonometric functions:

$$\int \frac{dx}{\sqrt{1 - x^2}} = \sin^{-1} x + C$$

$$\int \frac{dx}{x^2 + 1} = \tan^{-1} x + C$$

$$\int \frac{dx}{|x|\sqrt{x^2 - 1}} = \sec^{-1} x + C$$

- Integrals of exponential functions ($b > 0$, $b \neq 1$):

$$\int e^x \, dx = e^x + C, \qquad \int b^x \, dx = \frac{b^x}{\ln b} + C$$

5.7 EXERCISES

Preliminary Questions

1. What is the general antiderivative of the function?

(a) $f(x) = 2^x$ **(b)** $f(x) = x^{-1}$

(c) $f(x) = (1 - x^2)^{-1/2}$

2. Find a value of b such that $\displaystyle\int_1^b \frac{dx}{x}$ is equal to

(a) $\ln 3$ **(b)** 3

3. For which value of b is $\displaystyle\int_0^b \frac{dx}{1 + x^2} = \frac{\pi}{3}$?

4. Which of the following integrals should be evaluated using substitution?

(a) $\displaystyle\int \frac{9 \, dx}{1 + x^2}$ **(b)** $\displaystyle\int \frac{dx}{1 + 9x^2}$

5. If we set $x = 3u$, then $\sqrt{9 - x^2} = 3\sqrt{1 - u^2}$. Which relation between x and u yields the equality $\sqrt{16 + x^2} = 4\sqrt{1 + u^2}$?

Exercises

In Exercises 1–10, evaluate the definite integral.

1. $\int_1^2 \frac{1}{x}\,dx$

2. $\int_4^{12} \frac{1}{x}\,dx$

3. $\int_1^e \frac{1}{x}\,dx$

4. $\int_2^4 \frac{dt}{3t+4}$

5. $\int_{-e^2}^{-e} \frac{1}{t}\,dt$

6. $\int_e^{e^2} \frac{1}{t\ln t}\,dt$

7. $\int_0^{1/2} \frac{dx}{\sqrt{1-x^2}}$

8. $\int_{\tan 1}^{\tan 8} \frac{dx}{x^2+1}$

9. $\int_{-2}^{-2/\sqrt{3}} \frac{dx}{|x|\sqrt{x^2-1}}$

10. $\int_{-1/2}^{\sqrt{3}/2} \frac{dx}{\sqrt{1-x^2}}$

11. Use the substitution $u = x/3$ to prove

$$\int \frac{dx}{9+x^2} = \frac{1}{3}\tan^{-1}\frac{x}{3} + C$$

12. Use the substitution $u = 2x$ to evaluate $\int \frac{dx}{4x^2+1}$.

In Exercises 13–32, calculate the indefinite integral.

13. $\int_0^2 \frac{dx}{x^2+4}$

14. $\int_{1/\sqrt{3}}^{1/\sqrt{2}} \frac{dx}{x\sqrt{x^2-4}}$

15. $\int \frac{dt}{\sqrt{16-t^2}}$

16. $\int \frac{dt}{\sqrt{1-16t^2}}$

17. $\int \frac{dt}{\sqrt{25-4t^2}}$

18. $\int \frac{dx}{x\sqrt{1-4x^2}}$

19. $\int \frac{dx}{\sqrt{1-4x^2}}$

20. $\int \frac{dx}{4+x^2}$

21. $\int \frac{(x+1)dx}{\sqrt{1-x^2}}$

22. $\int \frac{dx}{x\sqrt{1-x^4}}$

23. $\int \frac{e^x\,dx}{1+e^{2x}}$

24. $\int \frac{\ln(\cos^{-1}x)\,dx}{(\cos^{-1}x)\sqrt{1-x^2}}$

25. $\int \frac{\tan^{-1}x\,dx}{1+x^2}$

26. $\int \frac{dx}{(\tan^{-1}x)(1+x^2)}$

27. $\int_0^1 3^x\,dx$

28. $\int_0^1 3^{-x}\,dx$

29. $\int_0^{\log_4(3)} 4^x\,dx$

30. $\int_{-2}^2 x10^{x^2}\,dx$

31. $\int 9^x \sin(9^x)\,dx$

32. $\int \frac{dx}{\sqrt{5^{2x}-1}}$

In Exercises 33–70, evaluate the integral using the methods covered in the text so far.

33. $\int (e^x + 2)\,dx$

34. $\int e^{4x}\,dx$

35. $\int 7^{-x}\,dx$

36. $\int ye^{y^2}\,dy$

37. $\int (e^{4x}+1)\,dx$

38. $\int \frac{4x\,dx}{x^2+1}$

39. $\int e^{-9t}\,dt$

40. $\int (e^x + e^{-x})\,dx$

41. $\int \frac{dx}{\sqrt{1-16x^2}}$

42. $\int \frac{dx}{\sqrt{9-16x^2}}$

43. $\int e^t\sqrt{e^t+1}\,dt$

44. $\int (e^{-x}-4x)\,dx$

45. $\int (7-e^{10x})\,dx$

46. $\int \frac{e^{2x}-e^{4x}}{e^x}\,dx$

47. $\int \frac{dx}{x\sqrt{25x^2-1}}$

48. $\int \frac{x\,dx}{\sqrt{4x^2+9}}$

49. $\int xe^{-4x^2}\,dx$

50. $\int e^x \cos(e^x)\,dx$

51. $\int \frac{e^x}{\sqrt{e^x+1}}\,dx$

52. $\int e^x(e^{2x}+1)^3\,dx$

53. $\int \frac{dx}{2x+4}$

54. $\int \frac{t\,dt}{t^2+4}$

55. $\int \frac{x^2\,dx}{x^3+2}$

56. $\int \frac{(3x-1)\,dx}{9-2x+3x^2}$

57. $\int \tan(4x+1)\,dx$

58. $\int \cot x\,dx$

59. $\int \frac{\cos x}{2\sin x+3}\,dx$

60. $\int \frac{\ln x}{x}\,dx$

61. $\int \frac{4\ln x+5}{x}\,dx$

62. $\int \frac{(\ln x)^2}{x}\,dx$

63. $\int \frac{dx}{x\ln x}$

64. $\int \frac{dx}{(4x-1)\ln(8x-2)}$

65. $\int \frac{\ln(\ln x)}{x\ln x}\,dx$

66. $\int \cot x\ln(\sin x)\,dx$

67. $\int 3^x\,dx$

68. $\int x3^{x^2}\,dx$

69. $\int \cos x\,3^{\sin x}\,dx$

70. $\int \left(\frac{1}{2}\right)^{3x+2}\,dx$

71. Use Figure 4 on the following page to prove the formula

$$\int_0^x \sqrt{1-t^2}\,dt = \frac{1}{2}x\sqrt{1-x^2} + \frac{1}{2}\sin^{-1}x$$

Hint: The area represented by the integral is the sum of a triangle and a sector.

72. Show that $G(t) = \sqrt{1 - t^2} + t \sin^{-1} t$ is an antiderivative of $\sin^{-1} t$.

73. Verify by differentiation:

$$\int_0^T t e^{rt} dt = \frac{e^{rT}(rT - 1) + 1}{r^2}$$

Then use L'Hôpital's Rule to show that the limit of the right-hand side as $r \to 0$ is equal to the value of the integral for $r = 0$.

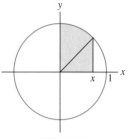

FIGURE 4

Further Insights and Challenges

74. Recall the following property of integrals: If $f(t) \geq g(t)$ for all $t \geq 0$, then for all $x \geq 0$,

$$\int_0^x f(t) dt \geq \int_0^x g(t) dt \qquad \boxed{7}$$

The inequality $e^t \geq 1$ holds for $t \geq 0$ because $e > 1$. Use (7) to prove that $e^x \geq 1 + x$ for $x \geq 0$. Then prove, by successive integration, the following inequalities (for $x \geq 0$):

$$e^x \geq 1 + x + \frac{1}{2}x^2, \qquad e^x \geq 1 + x + \frac{1}{2}x^2 + \frac{1}{6}x^3$$

75. Generalize Exercise 74; that is, use induction (if you are familiar with this method of proof) to prove that for all $n \geq 0$,

$$e^x \geq 1 + x + \frac{1}{2}x^2 + \frac{1}{6}x^3 + \cdots + \frac{1}{n!}x^n \quad (x \geq 0)$$

76. Use Exercise 74 to show that $\dfrac{e^x}{x^2} \geq \dfrac{x}{6}$ and conclude that $\lim\limits_{x \to \infty} \dfrac{e^x}{x^2} = \infty$. Then use Exercise 75 to prove more generally that $\lim\limits_{x \to \infty} \dfrac{e^x}{x^n} = \infty$ for all n.

77. Defining ln x as an Integral   Define a function $\varphi(x)$ in the domain $x > 0$:

$$\varphi(x) = \int_1^x \frac{1}{t} dt$$

This exercise proceeds as if we didn't know that $\varphi(x) = \ln x$ and shows directly that $\varphi(x)$ has all the basic properties of the logarithm. Prove the following statements:

(a) $\displaystyle\int_1^b \frac{1}{t} dt = \int_a^{ab} \frac{1}{t} dt$ for all $a, b > 0$. *Hint:* Use the substitution $u = t/a$.

(b) $\varphi(ab) = \varphi(a) + \varphi(b)$. *Hint:* Break up the integral from 1 to ab into two integrals and use (a).

(c) $\varphi(1) = 0$ and $\varphi(a^{-1}) = -\varphi(a)$ for $a > 0$.

(d) $\varphi(a^n) = n\varphi(a)$ for all $a > 0$ and integers n.

(e) $\varphi(a^{1/n}) = \frac{1}{n}\varphi(a)$ for all $a > 0$ and integers $n \neq 0$.

(f) $\varphi(a^r) = r\varphi(a)$ for all $a > 0$ and rational number r.

(g) There exists x such that $\varphi(x) > 1$. *Hint:* Show that $\varphi(a) > 0$ for every $a > 1$. Then take $x = a^m$ for $m > 1/\varphi(a)$.

(h) Show that $\varphi(t)$ is increasing and use the Intermediate Value Theorem to show that there exists a unique number e such that $\varphi(e) = 1$.

(i) $\varphi(e^r) = r$ for any rational number r.

78. Show that if $f(x)$ is increasing and satisfies $f(xy) = f(x) + f(y)$, then its inverse $g(x)$ satisfies $g(x + y) = g(x)g(y)$.

79. This is a continuation of the previous two exercises. Let $g(x)$ be the inverse of $\varphi(x)$. Show that

(a) $g(x)g(y) = g(x + y)$.

(b) $g(r) = e^r$ for any rational number.

(c) $g'(x) = g(x)$.

Exercises 77–79 provide a mathematically elegant approach to the exponential and logarithm functions, which avoids the problem of defining e^x for irrational x and of proving that e^x is differentiable.

80. The formula $\displaystyle\int x^n dx = \frac{x^{n+1}}{n+1} + C$ is valid for $n \neq -1$. Use L'Hôpital's Rule to show that the exceptional case $n = -1$ is a limit of the general case in the following sense: For fixed $x > 0$,

$$\lim_{n \to -1} \int_1^x t^n dt = \int_1^x t^{-1} dt$$

Note that the integral on the left is equal to $\dfrac{x^{n+1} - 1}{n + 1}$.

81. $\boxed{\text{CAS}}$ The integral on the left in Exercise 80 is equal to $f_n(x) = \dfrac{x^{n+1} - 1}{n + 1}$. Investigate the limit graphically by plotting $f_n(x)$ for $n = 0, -0.3, -0.6,$ and -0.9 together with $\ln x$ on a single plot.

82. Use the substitution $u = \tan x$ to evaluate $\displaystyle\int \frac{dx}{1 + \sin^2 x}$. *Hint:* Show that

$$\frac{dx}{1 + \sin^2 x} = \frac{du}{1 + 2u^2}$$

5.8 Exponential Growth and Decay

In this section, we explore some of the applications of the exponential function to biology, physics, and economics. Consider a quantity $P(t)$ that depends exponentially on time:

$$P(t) = P_0 e^{kt}$$

The constant k has units of "inverse time"; if t is measured in days, then k has units of $(\text{days})^{-1}$.

where P_0 and k are constants. If $k > 0$, then $P(t)$ *grows exponentially* with *growth constant k*, and if $k < 0$, then $P(t)$ *decays exponentially* with *decay constant k*. The coefficient P_0 is the initial size since $P(0) = P_0 e^{k \cdot 0} = P_0$. Exponential growth or decay may also be expressed by the formula $P(t) = P_0 b^t$ with $b = e^k$, since $P_0 b^t = P_0 (e^k)^t = P_0 e^{kt}$.

Population is a typical example of a quantity that grows exponentially, at least under suitable conditions. To understand why, consider a cell colony with initial population $P_0 = 100$ and assume that each cell divides into two cells after 1 hour. Then with each passing hour, the population $P(t)$ doubles:

$$
\begin{aligned}
P(0) &= 100 &&\text{(initial population)} \\
P(1) = 2(100) &= 200 &&\text{(population doubles)} \\
P(2) = 2(200) &= 400 &&\text{(population doubles again)}
\end{aligned}
$$

After t hours, $P(t) = (100)2^t$.

FIGURE 1 *E. coli* bacteria.

Exponential growth cannot continue over long periods of time. A colony starting with one E. coli cell would grow to 5×10^{89} cells after 3 weeks, much more than the estimated number of atoms in the observable universe. In actual cell growth, the exponential phase is followed by a period in which growth slows and may decline.

■ **EXAMPLE 1** When the *Escherichia coli* bacteria (found in the human intestine; see Figure 1) is grown in the laboratory, it increases exponentially with a growth constant of $k = 0.41$ $(\text{hours})^{-1}$. Assume that 1,000 bacteria are present at time $t = 0$.

(a) Find the formula for the number of bacteria $P(t)$ at time t.

(b) How large is the population after 5 hours?

(c) When will the population reach 10,000?

Solution

(a) The initial size is $P_0 = 1,000$ and $k = 0.41$, so $P(t) = 1,000e^{0.41t}$ (t in hours).

(b) After 5 hours, $P(5) = 1,000e^{0.41 \cdot 5} = 1,000e^{2.05} \approx 7767.9$. Because the number of bacteria must be a whole number, we round off the answer to 7,768.

(c) The problem asks for the time t such that $P(t) = 10,000$, so we must solve the equation $1,000e^{0.41t} = 10,000$, or

$$e^{0.41t} = \frac{10,000}{1,000} = 10$$

Taking the logarithm of both sides, we obtain $\ln\left(e^{0.41t}\right) = \ln 10$, or

$$0.41t = \ln 10 \Rightarrow t = \frac{\ln 10}{0.41} \approx 5.62$$

Therefore, $P(t)$ reaches 10,000 after approximately 5 hours and 37 min (Figure 2). ■

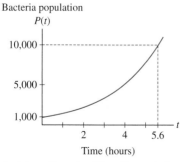

Bacteria population
$P(t)$

10,000

5,000

1,000

Time (hours)

FIGURE 2 Growth of *E. coli* population.

The important role played by exponential functions is best understood in terms of the differential equation $y' = ky$. The function $y = P_0 e^{kt}$ satisfies this differential equation,

A differential equation is an equation relating a function $y = f(x)$ to its derivative y' (or higher derivatives y', y'', y''', ...).

as we can check directly:

$$y' = \frac{d}{dt}\left(P_0 e^{kt}\right) = k P_0 e^{kt} = ky$$

Theorem 1 goes further and asserts that the exponential functions $y = P_0 e^{kt}$ are the *only* functions that satisfy this differential equation.

Theorem 1 is an example of a "uniqueness theorem," stating that the solution to a differential equation with given initial conditions is unique.

THEOREM 1   If $y(t)$ is a differentiable function satisfying the differential equation

$$\boxed{y' = ky}$$

then $y(t) = P_0 e^{kt}$, where P_0 is the initial value $P_0 = y(0)$.

Proof Compute the derivative of ye^{-kt}. If $y' = ky$, then

$$\frac{d}{dt}\left(ye^{-kt}\right) = y'e^{-kt} - ke^{-kt}y = e^{-kt}(y' - ky) = 0 \quad \text{(since } y' = ky\text{)}$$

Because the derivative is zero, $y(t)e^{-kt} = P_0$ for some constant P_0 and $y(t) = P_0 e^{kt}$ as claimed. ■

CONCEPTUAL INSIGHT   Theorem 1 tells us that a process obeys an exponential law precisely when *its rate of change is proportional to the amount present at time t*. This formulation helps us understand why certain quantities grow exponentially. A population grows exponentially because its growth rate is proportional to the size of the population (each organism contributes to the growth through reproduction). However, this is true only if the organisms do not interact. If they do interact, say, by competing for food or mates, then the growth rate may not be proportional to population size and we cannot expect exponential growth. Similarly, the experimental fact that radioactive substances decay exponentially is consistent with the hypothesis that atoms decay randomly and independently. Under this hypothesis, a fixed fraction of the atoms decay per unit time and the total number decaying per unit time is proportional to the total number of atoms present, leading to the observed exponential decay. If exponential decay were not observed, we might suspect that the decay is influenced by some interaction between the atoms.

■ **EXAMPLE 2** Find all solutions of $y' = 3y$. Which particular solution satisfies $y(0) = 9$?

Solution The solutions to $y' = 3y$ are the functions $y(t) = Ce^{3t}$, where C is the initial value $C = y(0)$. The particular solution satisfying $y(0) = 9$ is $y(t) = 9e^{3t}$. ■

■ **EXAMPLE 3** Modeling Penicillin Pharmacologists have shown that the rate at which penicillin leaves a person's bloodstream is proportional to the amount of penicillin present.

(a) Express this statement as a differential equation.

(b) Find the decay constant if 50 mg of penicillin remain in the bloodstream 7 hours after an initial injection of 450 mg.

(c) Under the hypothesis of (b), at what time was 200 mg of penicillin present?

Solution

(a) Let's denote the quantity of penicillin present in the bloodstream at time t by $A(t)$. Then the rate at which penicillin leaves the bloodstream is $A'(t)$. If this rate is proportional to $A(t)$, then

$$A'(t) = -kA(t) \qquad \boxed{1}$$

for some constant k. Note that $k > 0$ because $A(t)$ is decreasing.

(b) Since $A(t)$ satisfies Eq. (1) and $A(0) = 450$, $A(t) = 450e^{-kt}$. We use the condition $A(7) = 50$ to solve for k:

$$A(7) = 450e^{-7k} = 50 \quad \Rightarrow \quad e^{-7k} = \frac{1}{9}$$

Thus, $-7k = \ln\frac{1}{9}$ and $k = -\ln(\frac{1}{9})/7 \approx 0.31$.

(c) To find the time at which 200 mg was present, we solve

$$A(t) = 450e^{-0.31t} = 200$$

$$e^{-0.31t} = \frac{4}{9}$$

$$t = -\frac{1}{0.31}\ln\left(\frac{4}{9}\right) \approx 2.62$$

This shows that 200 mg was present at time $t = 2.62$ hours (Figure 3). ■

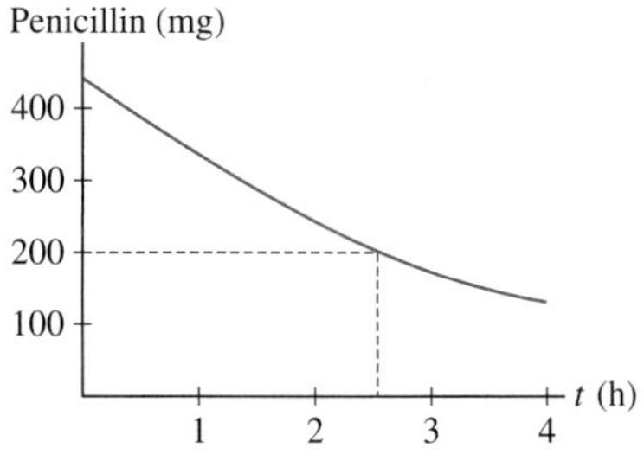

Penicillin (mg)

FIGURE 3

Quantities that grow exponentially possess an important property: There is a doubling time T such that $P(t)$ doubles in size over every time interval of length T. To prove this, let $P(t) = P_0e^{kt}$ and solve for T in the equation $P(t + T) = 2P(t)$:

$$P_0e^{k(t+T)} = 2P_0e^{kt}$$

$$e^{kt}e^{kT} = 2e^{kt}$$

$$e^{kT} = 2$$

The constant k has units of time^{-1}, so the doubling time $T = (\ln 2)/k$ has units of time, as we should expect.

We obtain $kT = \ln 2$ or $T = \ln 2/k$.

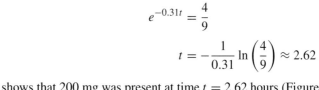

Doubling Time   If $P(t) = P_0e^{kt}$ with $k > 0$, then the doubling time of P is:

$$\text{Doubling time} = \frac{\ln 2}{k}$$

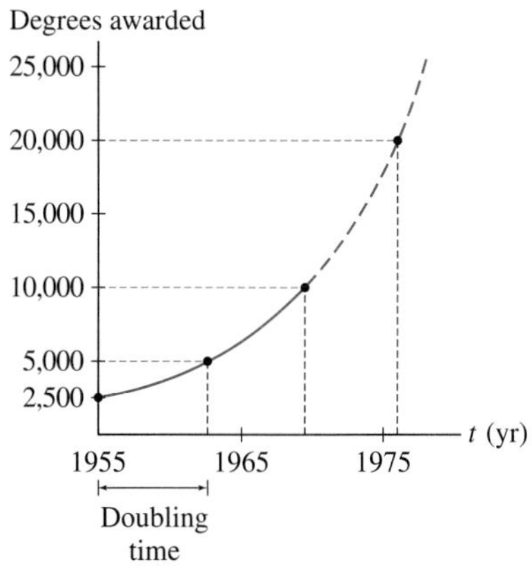

Degrees awarded

FIGURE 4 Doubling (from 2,500 to 5,000 to 10,000, etc.) occurs at equal time intervals.

■ **EXAMPLE 4** Computing Doubling Time from k Some studies have suggested that from 1955 to 1970, the number of bachelor's degrees in physics awarded per year by U.S. universities grew exponentially, with growth constant $k = 0.1$ (approximately 2,500 degrees awarded in 1955). If this was true,

(a) What was the doubling time?

(b) How long would it take for the number of degrees awarded per year to increase 8-fold, assuming that exponential growth continued?

Solution

(a) The doubling time is $(\ln 2)/0.1 \approx 0.693/0.1 = 6.93$ (Figure 4). Since degrees are usually awarded once a year, we round off the doubling time to 7 years.

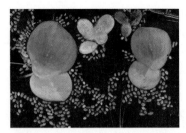

FIGURE 5 The tiny plants are *Wolffia*, with plant bodies smaller than the head of a pin.

(b) The number doubles after 7 years, quadruples after 14 years, and increases 8-fold after 21 years. ■

■ **EXAMPLE 5** Calculating k from the Doubling Time   One of the world's smallest flowering plants, *Wolffia globosa* (Figure 5), has a doubling time of approximately 30 hours. Find the growth constant k and determine the initial population if the population grew to 1,000 after 48 hours.

Solution By the formula for the doubling time, $30 = \dfrac{\ln 2}{k}$. Therefore, $k = \dfrac{\ln 2}{30} \approx 0.023$. The plant population after t hours is $P(t) = P_0 e^{0.023t}$. If $P(48) = 1,000$, then

$$P_0 e^{(0.023)48} = 1,000 \quad \Rightarrow \quad P_0 = 1,000 e^{-(0.023)48} \approx 332$$ ■

Scientists have observed that if a quantity R_0 of a radioactive isotope is present at time $t = 0$, then the quantity present at time t is $R(t) = R_0 e^{-kt}$, where $k > 0$. In this situation of exponential decay, we have a **half-life** rather than a doubling time. The half-life is the time it takes for the quantity to decrease by a factor of $\frac{1}{2}$. The calculation of doubling time above shows similarly that

$$\boxed{\text{Half-life} = \dfrac{\ln 2}{k}}$$

Fraction present

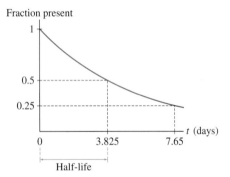

FIGURE 6 Fraction of Radon-222 present at time t.

■ **EXAMPLE 6** The isotope Radon-222 has a half-life of 3.825 days. Find the decay constant and determine how long will it take for 80% of the isotope to decay.

Solution By the equation for half-life, $k = \dfrac{\ln 2}{3.825} \approx 0.181$. Therefore, the quantity of Radon-222 at time t is $R(t) = R_0 e^{-0.181t}$, where R_0 is the amount present at $t = 0$ (Figure 6). To determine when 20% remains, we solve for t in the equation $P(t) = 0.2R_0$:

$$R_0 e^{-0.181t} = 0.2 R_0$$
$$e^{-0.181t} = 0.2$$
$$-0.181t = \ln(0.2) \quad \Rightarrow \quad t = \frac{\ln(0.2)}{-0.181} \approx 8.9 \text{ days}$$

Therefore, the quantity of Radon-222 decreases by 80% after 8.9 days. ■

Carbon Dating

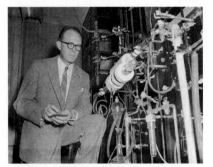

FIGURE 7 American chemist Willard Libby (1908–1980) developed the technique of carbon dating in 1946 to determine the age of fossils and was awarded the Nobel Prize in Chemistry for this work in 1960. Since then the technique has been considerably refined.

Carbon dating relies on the fact that all living organisms contain carbon. The carbon, which enters the food chain through the carbon dioxide absorbed by plants from the atmosphere, is made up mostly of nonradioactive C^{12} and a much smaller amount of radioactive C^{14} that decays into nitrogen. The ratio R of C^{14} to C^{12} in the atmosphere is approximately $R = 10^{-12}$. The ratio of C^{14} to C^{12} in a living organism is also R because the organism's carbon originates in the atmosphere.

When the organism dies, its carbon is no longer replenished. The C^{14} begins to decay exponentially while the C^{12} remains unchanged. Therefore, the ratio of C^{14} to C^{12} in the organism decreases exponentially as a function of the time t since the organism died. By measuring this ratio, we can determine when the death occurred. The decay constant for

FIGURE 8 Detail of bison and other animals from a replica of the Lascaux cave mural.

Decay of C^{14}

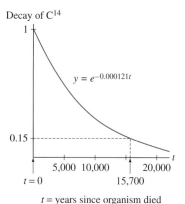

$$y = e^{-0.000121t}$$

$t = $ years since organism died

FIGURE 9 If only 15% of the C^{14} remains, the object is approximately 16,000 years old.

Convention: Time t is measured in years and interest rates are given as yearly rates, either as a decimal or as a percentage. Thus, r = 0.05 corresponds to an interest rate of 5% per year.

After t years, the principal increases by the factor

$$\left(1 + \frac{r}{M}\right)^{Mt}$$

This factor is called the "multiplier."

C^{14} is $k = 0.000121$ years^{-1}, so we obtain

$$\text{Ratio of } C^{14} \text{ to } C^{12} \text{ after } t \text{ years} = Re^{-0.000121t}$$

■ **EXAMPLE 7** Lascaux Cave Paintings   A remarkable gallery of prehistoric paintings of animals was discovered in the Lascaux cave in Dordogne, France, in 1940 (Figure 8). Scientists determined that a charcoal sample taken from the cave walls had a C^{14} to C^{12} ratio equal to 15% of that found in the atmosphere. What did they estimate the age of the paintings to be?

Solution Let t be the age of the paintings. Then the C^{14} to C^{12} ratio decreased by a factor of $e^{-0.000121t}$. If the ratio in the sample decreased by a factor of 0.15, then

$$e^{-0.000121t} = 0.15$$
$$-0.000121t = \ln(0.15) \approx -1.9$$

Thus $t = 1.9/0.000121 \approx 15,700$. The scientists concluded that the cave paintings are approximately 16,000 years old (Figure 9).   ■

Compound Interest and Present Value

Exponential functions play an important role in financial calculations. For example, they are used to compute compound interest and present value.

The initial sum of money P_0 placed in an account or investment is called the **principal**. When a principal of P_0 dollars is deposited into a bank account, the amount or *balance* in the account at time t depends on two factors: the **interest rate** r and how often interest is **compounded**. If interest is paid out once a year at the end of the year, we say that the interest is *compounded annually*. If the principal and interest are retained in the account, then the balance grows exponentially:

$$\text{Balance after 1 year} = \overbrace{P_0}^{\text{Principal}} + \overbrace{rP_0}^{\text{Interest}} = P_0(1 + r)$$

$$\text{Balance after 2 years} = P_0(1 + r) + rP_0(1 + r) = P_0(1 + r)^2$$

$$\text{Balance after } t \text{ years} = P_0(1 + r)^{t-1} + rP_0(1 + r)^{t-1} = P_0(1 + r)^t$$

When interest is paid out quarterly, the interest earned after 3 months (the first quarter) is $\frac{r}{4}P_0$ dollars. The principal increases by the factor $\left(1 + \frac{r}{4}\right)$ after each 3-month period, and

$$\text{Balance after 1 year} = P_0\left(1 + \frac{r}{4}\right)\left(1 + \frac{r}{4}\right)\left(1 + \frac{r}{4}\right)\left(1 + \frac{r}{4}\right) = P_0\left(1 + \frac{r}{4}\right)^4$$

$$\text{Balance after } t \text{ years} = P_0\left(1 + \frac{r}{4}\right)^{4t}$$

In general, if interest is compounded M times a year, the principal increases by the factor $\left(1 + \frac{r}{M}\right)^M$ after each year, and after t years, $P(t) = P_0\left(1 + \frac{r}{M}\right)^{Mt}$.

TABLE 1   Compound Interest with Principal $P_0 = \$100$ and $r = 0.09$

	Principal after 1 Year
Annual	$100(1 + 0.09) = \$109$
Quarterly	$100\left(1 + \frac{0.09}{4}\right)^4 \approx \$109.31$
Monthly	$100\left(1 + \frac{0.09}{12}\right)^{12} \approx \$109.38$
Weekly	$100\left(1 + \frac{0.09}{52}\right)^{52} \approx \$109.41$
Daily	$100\left(1 + \frac{0.09}{365}\right)^{365} \approx \$109.42$

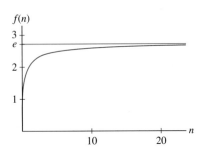

FIGURE 10 The function $f(n) = \left(1 + \dfrac{1}{n}\right)^n$ approaches e as $n \to \infty$.

Compound Interest   If P_0 dollars are deposited into an account earning interest at an annual rate r, compounded M times yearly, then the value of the account after t years is

$$P(t) = P_0\left(1 + \frac{r}{M}\right)^{Mt}$$

Table 1 shows that there is a benefit to more frequent compounding. What happens in the limit as M tends to infinity? The yearly multiplier approaches the limit

$$\lim_{M\to\infty}\left(1 + \frac{r}{M}\right)^M \qquad \boxed{2}$$

Theorem 2 relates this limit to the exponential function (a proof is given at the end of this section). Figure 10 illustrates the first limit in Theorem 2 graphically.

THEOREM 2 Limit Formula for e and e^x

$$e = \lim_{n\to\infty}\left(1 + \frac{1}{n}\right)^n \qquad \text{and} \qquad e^x = \lim_{n\to\infty}\left(1 + \frac{x}{n}\right)^n \qquad \text{for all } x$$

To evaluate (2), we apply Theorem 2 with $x = r$ and $n = M$:

$$\lim_{M\to\infty}\left(1 + \frac{r}{M}\right)^M = e^r$$

Thus, as we compound more and more frequently, the yearly multiplier approaches e^r and the multiplier for t years approaches $(e^r)^t = e^{rt}$. We use this result to define **continuously compounded interest**.

Continuously Compounded Interest   If P_0 dollars are deposited into an account earning interest at an annual rate r, compounded continuously, then the value of the account after t years is

$$P(t) = P_0 e^{rt}$$

■ **EXAMPLE 8**   A principal $P_0 = \$10,000$ is deposited into an account paying 6% interest. Find the balance after 3 years if interest is compounded quarterly and if interest is compounded continuously.

Solution   After 3 years, the balance is

$$\text{Quarterly compounding:} \qquad 10,000\left(1 + \frac{0.06}{4}\right)^{4(3)} \approx \$11,956.18$$

$$\text{Continuous compounding:} \qquad 10,000e^{(0.06)3} \approx \$11,972.17 \qquad ■$$

Present Value

The financial world has many different interest rates (federal funds rate, prime rate, etc.). We simplify the discussion by assuming that there is just one rate.

The concept of *present value* (PV) is used in business and finance to compare the values of payments made at different times. Assume that there is an interest rate r at which any investor can lend or borrow money and that interest is compounded continuously. Then the PV of P dollars to be received t years in the future is defined to be Pe^{-rt}:

> The PV of P dollars received at time t is Pe^{-rt}

What is the reasoning behind this definition? When you invest at the interest rate r for t years, your principal increases by the factor e^{rt}, so if you invest Pe^{-rt} dollars, your principal grows to $Pe^{-rt}e^{rt} = P$ dollars at time t. Thus, the present value Pe^{-rt} is the amount you would have to invest *today* in order to have P dollars at time t.

■ **EXAMPLE 9** Is it better to receive \$2,000 today or \$2,200 in 2 years? Consider $r = 0.07$ and $r = 0.03$.

Solution We compare \$2,000 today with the present value of \$2,200 received in 2 years. If $r = 0.07$, the PV is $2,200e^{-(0.07)2} \approx \$1,912.59$. This PV is less than \$2,000, so it is better to receive \$2,000 today.

If $r = 0.03$, the PV is $2,200e^{-(0.03)2} \approx \$2,071.88$. In this case, a payment of \$2,200 in 2 years is preferable to a \$2,000 payment today. ■

■ **EXAMPLE 10** Deciding Whether to Invest Chief Operating Officer Marjorie Bean must decide whether to invest \$400,000 to upgrade her company's computer system. The upgrade will save \$150,000 a year for each of the next 3 years. Is this a good investment if $r = 7\%$? What if $r = 4\%$?

Solution Marjorie must compare the PV of the money saved (\$150,000 annually for 3 years) with the \$400,000 cost of the upgrade. For simplicity, assume that the \$150,000 in savings is received as a lump sum at the end of each year.

If $r = 0.07$, the PV of the savings over 3 years is

$$150{,}000e^{-(0.07)} + 150{,}000e^{-(0.07)2} + 150{,}000e^{-(0.07)3} \approx \$391{,}850$$

Since this is *less* than the initial cost, the upgrade is not a worthwhile investment.

If $r = 0.04$, then the present value of the savings is

$$150{,}000e^{-(0.04)} + 150{,}000e^{-(0.04)2} + 150{,}000e^{-(0.04)3} \approx \$415{,}624$$

In this "interest-rate environment," the upgrade is a good investment because it saves the company \$15,624. ■

An **income stream** is a sequence of periodic payments that continue over an interval of T years. Consider an investment that produces income at a rate of \$800/year for 5 years. A total of \$4,000 is paid out over 5 years but the PV of the income stream is less. For instance, if $r = 0.06$ and payments are made at the end of the year, then the PV is

$$800e^{-0.06} + 800e^{-(0.06)2} + 800e^{-(0.06)3} + 800e^{-(0.06)4} + 800e^{-(0.06)5} \approx \$3{,}353.12$$

It is more convenient mathematically to assume that payments are made *continuously* at a rate of $R(t)$ dollars/year. We may then calculate PV as an integral. Divide the time interval $[0, T]$ into N subintervals of length $\Delta t = T/N$. If Δt is small, the amount paid out between time t and $t + \Delta t$ is approximately equal to

$$\underbrace{R(t)}_{\text{Rate}} \times \underbrace{\Delta t}_{\text{Time interval}} = R(t)\,\Delta t$$

The PV of this payment is approximately $e^{-rt}R(t)\Delta t$. If $t_0 = 0, t_1, \dots, t_N$ are the end-points of the subintervals, we obtain the approximation

$$\text{PV of income stream} \approx \sum_{i=1}^{N} e^{-rt_i} R(t_i)\,\Delta t$$

This is a Riemann sum whose value approaches $\int_0^T R(t)e^{-rt}\,dt$ as Δt tends to zero.

PV of an Income Stream The present value at interest rate r of an income stream paying out $R(t)$ dollars/year continuously for T years is

$$PV = \int_0^T R(t)e^{-rt}\,dt \qquad \boxed{3}$$

■ **EXAMPLE 11** An investment pays out \$800/year continuously for 5 years. Find the PV of the investment for $r = 0.06$ and $r = 0.04$.

Solution If $r = 0.06$, the PV is equal to

$$\int_0^5 800 e^{-0.06t}\,dt = -800\frac{e^{-0.06t}}{0.06}\Big|_0^5 \approx -9{,}877.58 - (-13{,}333.33) = \$3{,}455.76$$

If $r = 0.04$, the PV is equal to

$$\int_0^5 800 e^{-0.04t}\,dt = -800\frac{e^{-0.04t}}{0.04}\Big|_0^5 \approx -16{,}374.62 - (-20{,}000) = \$3{,}625.38 \quad ■$$

Proof of Theorem 2 To verify the first limit, $e = \lim_{n\to\infty}(1 + 1/n)^n$, we use the formula $\ln b = \int_1^b t^{-1}\,dt$ with $b = 1 + 1/n$:

$$\ln\left(1 + \frac{1}{n}\right) = \int_1^{1+1/n} \frac{dt}{t}$$

Figure 11 shows that the area represented by this integral lies between the areas of two rectangles of heights $n/(n+1)$ and 1, both of base $1/n$. These rectangles have areas $1/(n+1)$ and $1/n$, so

$$\frac{1}{n+1} \le \ln\left(1 + \frac{1}{n}\right) = \int_1^{1+1/n}\frac{dt}{t} \le \frac{1}{n}$$

Since e^x is increasing, these inequalities remain true when we apply e^x to each term:

$$e^{1/(n+1)} \le 1 + \frac{1}{n} \le e^{1/n} \qquad \boxed{4}$$

$$e^{n/(n+1)} \le \left(1 + \frac{1}{n}\right)^n \le e \quad \text{(raise both sides to the nth power)} \qquad \boxed{5}$$

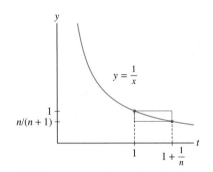

FIGURE 11

Now apply the Squeeze Theorem. Since e^x is continuous,

$$\lim_{n\to\infty} e^{n/(n+1)} = e^{\lim_{n\to\infty}(n/(n+1))} = e^1 = e$$

We conclude that $\lim_{n\to\infty}(1+\frac{1}{n})^n = e$ as desired. See Exercise 58 for a proof of the more general formula $e^x = \lim_{n\to\infty}(1+\frac{x}{n})^n$. ∎

5.8 SUMMARY

- $P(t)$ *grows exponentially* with growth constant $k > 0$ if $P(t) = P_0 e^{kt}$. The initial amount is P_0.
- The solutions of the differential equation $y' = ky$ are the exponential functions $y = Ce^{kt}$, where C is a constant.
- A quantity $P(t)$ grows exponentially if it grows at a rate proportional to its size, that is, if $P'(t) = kP(t)$.
- If $P(t)$ grows exponentially, its *doubling time* is $\dfrac{\ln 2}{k}$.
- $P(t)$ is said to *decay exponentially* with decay constant $k > 0$ if $P(t) = P_0 e^{-kt}$. If $P(t)$ decays exponentially, then $P'(t) = -kP(t)$. The *half-life* of $P(t)$ is $(\ln 2)/k$.
- If a principal P_0 is deposited in an account bearing interest at the rate r, compounded M times per year, the value of the account after t years is $P(t) = P_0(1 + r/M)^{Mt}$.
- When interest is compounded continuously, $P(t) = P_0 e^{rt}$.
- The *present value* (PV) of P dollars, to be paid t years in the future, is Pe^{-rt}.
- The present value of an income stream paying out $R(t)$ dollars/year continuously for T years is

$$PV = \int_0^T R(t)e^{-rt}\,dt$$

5.8 EXERCISES

Preliminary Questions

1. Two quantities increase exponentially with growth constants $k = 1.2$ and $k = 3.4$, respectively. Which quantity doubles more rapidly?

2. If you are given both the doubling time and the growth constant of a quantity that increases exponentially, can you determine the initial amount?

3. A cell population grows exponentially beginning with one cell. Does it take less time for the population to increase from one to two cells than from 10 million to 20 million cells?

4. Referring to his popular book *A Brief History of Time*, the renowned physicist Stephen Hawking said, "Someone told me that each equation I included in the book would halve its sales." If this is so, write a differential equation satisfied by the sales function $S(n)$, where n is the number of equations in the book.

5. Carbon dating is based on the assumption that the ratio R of C^{14} to C^{12} in the atmosphere has been constant over the past 50,000 years.

If R were actually smaller in the past than it is today, would the age estimates produced by carbon dating be too ancient or too recent?

6. Which is preferable: an interest rate of 12% compounded quarterly, or an interest rate of 11% compounded continuously?

7. Find the yearly multiplier if $r = 9\%$ and interest is compounded (a) continuously and (b) quarterly.

8. The PV of N dollars received at time T is (choose the correct answer):
(a) The value at time T of N dollars invested today
(b) The amount you would have to invest today in order to receive N dollars at time T

9. A year from now, \$1 will be received. Will its PV increase or decrease if the interest rate goes up?

10. Xavier expects to receive a check for \$1,000 1 year from today. Explain, using the concept of PV, whether he will be happy or sad to learn that the interest rate has just increased from 6% to 7%.

Exercises

1. A certain bacteria population P obeys the exponential growth law $P(t) = 2{,}000e^{1.3t}$ (t in hours).

(a) How many bacteria are present initially?

(b) At what time will there be 10,000 bacteria?

2. A quantity P obeys the exponential growth law $P(t) = e^{5t}$ (t in years).

(a) At what time t is $P = 10$?

(b) At what time t is $P = 20$?

(c) What is the doubling time for P?

3. A certain RNA molecule replicates every 3 minutes. Find the differential equation for the number $N(t)$ of molecules present at time t (in minutes). Starting with one molecule, how many will be present after 10 min?

4. A quantity P obeys the exponential growth law $P(t) = Ce^{kt}$ (t in years). Find the formula for $P(t)$, assuming that the doubling time is 7 years and $P(0) = 100$.

5. The decay constant of Cobalt-60 is 0.13 years^{-1}. What is its half-life?

6. Find the decay constant of Radium-226, given that its half-life is 1,622 years.

7. Find all solutions to the differential equation $y' = -5y$. Which solution satisfies the initial condition $y(0) = 3.4$?

8. Find the solution to $y' = \sqrt{2}y$ satisfying $y(0) = 20$.

9. Find the solution to $y' = 3y$ satisfying $y(2) = 4$.

10. Find the function $y = f(t)$ that satisfies the differential equation $y' = -0.7y$ and initial condition $y(0) = 10$.

11. The population of a city is $P(t) = 2 \cdot e^{0.06t}$ (in millions), where t is measured in years.

(a) Calculate the doubling time of the population.

(b) How long does it take for the population to triple in size?

(c) How long does it take for the population to quadruple in size?

12. The population of Washington state increased from 4.86 million in 1990 to 5.89 million in 2000. Assuming exponential growth,

(a) What will the population be in 2010?

(b) What is the doubling time?

13. Assuming that population growth is approximately exponential, which of the two sets of data is most likely to represent the population (in millions) of a city over a 5-year period?

Year	2000	2001	2002	2003	2004
Data I	3.14	3.36	3.60	3.85	4.11
Data II	3.14	3.24	3.54	4.04	4.74

14. Light Intensity The intensity of light passing through an absorbing medium decreases exponentially with the distance traveled. Suppose the decay constant for a certain plastic block is $k = 2$ when the distance is measured in feet. How thick must the block be to reduce the intensity by a factor of one-third?

15. The **Beer–Lambert Law** is used in spectroscopy to determine the molar absorptivity α or the concentration c of a compound dissolved in a solution at low concentrations (Figure 12). The law states that the intensity I of light as it passes through the solution satisfies $\ln(I/I_0) = \alpha cx$, where I_0 is the initial intensity and x is the distance traveled by the light. Show that I satisfies a differential equation $dI/dx = -kx$ for some constant k.

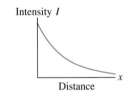

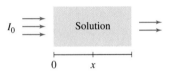

FIGURE 12 Light of intensity passing through a solution.

16. An insect population triples in size after 5 months. Assuming exponential growth, when will it quadruple in size?

17. A 10-kg quantity of a radioactive isotope decays to 3 kg after 17 years. Find the decay constant of the isotope.

18. Measurements showed that a sample of sheepskin parchment discovered by archaeologists had a C^{14} to C^{12} ratio equal to 40% of that found in the atmosphere. Approximately how old is the parchment?

19. Chauvet Caves In 1994, rock climbers in southern France stumbled on a cave containing prehistoric cave paintings. A C^{14}-analysis carried out by French archeologist Helene Valladas showed that the paintings are between 29,700 and 32,400 years old, much older than any previously known human art. Given that the C^{14} to C^{12} ratio of the atmosphere is $R = 10^{-12}$, what range of C^{14} to C^{12} ratios did Valladas find in the charcoal specimens?

20. A paleontologist has discovered the remains of animals that appear to have died at the onset of the Holocene ice age. She applies carbon dating to test her theory that the Holocene age started between 10,000 and 12,000 years ago. What range of C^{14} to C^{12} ratio would she expect to find in the animal remains?

21. Atmospheric Pressure The atmospheric pressure $P(h)$ (in pounds per square inch) at a height h (in miles) above sea level on earth satisfies a differential equation $P' = -kP$ for some positive constant k.

(a) Measurements with a barometer show that $P(0) = 14.7$ and $P(10) = 2.13$. What is the decay constant k?

(b) Determine the atmospheric pressure 15 miles above sea level.

22. Inversion of Sugar When cane sugar is dissolved in water, it converts to invert sugar over a period of several hours. The percentage $f(t)$ of unconverted cane sugar at time t decreases exponentially. Suppose that $f' = -0.2f$. What percentage of cane sugar remains after 5 hours? After 10 hours?

23. A quantity P increases exponentially with doubling time 6 hours. After how many hours has P increased by 50%?

24. Two bacteria colonies are cultivated in a laboratory. The first colony has a doubling time of 2 hours and the second a doubling time of 3 hours. Initially, the first colony contains 1,000 bacteria and the second colony 3,000 bacteria. At what time t will sizes of the colonies be equal?

25. Moore's Law In 1965, Gordon Moore predicted that the number N of transistors on a microchip would increase exponentially.

(a) Does the table of data below confirm Moore's prediction for the period from 1971 to 2000? If so, estimate the growth constant k.

(b) *CAS* Plot the data in the table.

(c) Let $N(t)$ be the number of transistors t years after 1971. Find an approximate formula $N(t) \approx Ce^{kt}$, where t is the number of years after 1971.

(d) Estimate the doubling time in Moore's Law for the period from 1971 to 2000.

(e) If Moore's Law continues to hold until the end of the decade, how many transistors will a chip contain in 2010?

(f) Can Moore have expected his prediction to hold indefinitely?

Transistors	Year	No. Transistors
4004	1971	2,250
8008	1972	2,500
8080	1974	5,000
8086	1978	29,000
286	1982	120,000
386 processor	1985	275,000
486 DX processor	1989	1,180,000
Pentium processor	1993	3,100,000
Pentium II processor	1997	7,500,000
Pentium III processor	1999	24,000,000
Pentium 4 processor	2000	42,000,000

26. Assume that in a certain country, the rate at which jobs are created is proportional to the number of people who already have jobs. If there are 15 million jobs at $t = 0$ and 15.1 million jobs 3 months later, how many jobs will there be after two years?

*In Exercises 27–28, we consider the **Gompertz differential equation**:*

$$\frac{dy}{dt} = ky \ln\left(\frac{y}{M}\right)$$

(where M and k are constants), introduced in 1825 by the English mathematician Benjamin Gompertz and still used today to model aging and mortality.

27. Show that $y = Me^{ae^{kt}}$ is a solution for any constant a.

28. To model mortality in a population of 200 laboratory rats, a scientist assumes that the number $P(t)$ of rats alive at time t (in months) satisfies the Gompertz equation with $M = 204$ and $k = 0.15$ months^{-1} (Figure 13). Find $P(t)$ [note that $P(0) = 200$] and determine the population after 20 months.

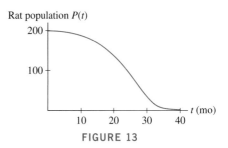

FIGURE 13

29. A certain quantity increases quadratically: $P(t) = P_0 t^2$.

(a) Starting at time $t_0 = 1$, how long will it take for P to double in size? How long will it take starting at $t_0 = 2$ or 3?

(b) In general, starting at time t_0, how long will it take for P to double in size?

30. Verify that the half-life of a quantity that decays exponentially with decay constant k is equal to $\ln 2 / k$.

31. Compute the balance after 10 years if $2,000 is deposited in an account paying 9% interest and interest is compounded (a) quarterly, (b) monthly, and (c) continuously.

32. Suppose $500 is deposited into an account paying interest at a rate of 7%, continuously compounded. Find a formula for the value of the account at time t. What is the value of the account after 3 years?

33. A bank pays interest at a rate of 5%. What is the yearly multiplier if interest is compounded

(a) yearly?   **(b)** three times a year?

(c) continuously?

34. How long will it take for $4,000 to double in value if it is deposited in an account bearing 7% interest, continuously compounded?

35. Show that if interest is compounded continuously at a rate r, then an account doubles after $(\ln 2)/r$ years.

36. How much must be invested today in order to receive $20,000 after 5 years if interest is compounded continuously at the rate $r = 9\%$?

37. An investment increases in value at a continuously compounded rate of 9%. How large must the initial investment be in order to build up a value of $50,000 over a seven-year period?

38. Compute the PV of $5,000 received in 3 years if the interest rate is (a) 6% and (b) 11%. What is the PV in these two cases if the sum is instead received in 5 years?

39. Is it better to receive $1,000 today or $1,300 in 4 years? Consider $r = 0.08$ and $r = 0.03$.

40. Find the interest rate r if the PV of $8,000 to be received in 1 year is $7,300.

41. If a company invests $2 million to upgrade its factory, it will earn additional profits of $500,000/year for 5 years. Is the investment worthwhile, assuming an interest rate of 6% (assume that the savings are received as a lump sum at the end of each year)?

42. A new computer system costing $25,000 will reduce labor costs by $7,000/year for 5 years.
(a) Is it a good investment if $r = 8\%$?
(b) How much money will the company actually save?

43. After winning $25 million in the state lottery, Jessica learns that she will receive five yearly payments of $5 million beginning immediately.
(a) What is the PV of Jessica's prize if $r = 6\%$?
(b) How much more would the prize be worth if the entire amount were paid today?

44. An investment group purchased an office building in 1998 for $17 million and sold it 5 years later for $26 million. Calculate the annual (continuously compounded) rate of return on this investment.

45. Use Eq. (3) to compute the PV of an income stream paying out $R(t) = \$5,000$/year continuously for 10 years and $r = 0.05$.

46. Compute the PV of an income stream if income is paid out continuously at a rate $R(t) = \$5,000e^{0.1t}$/year for 5 years and $r = 0.05$.

47. Find the PV of an investment that produces income continuously at a rate of $800/year for 5 years, assuming an interest rate of $r = 0.08$.

48. The rate of yearly income generated by a commercial property is $50,000/year at $t = 0$ and increases at a continuously compounded rate of 5%. Find the PV of the income generated in the first four years if $r = 8\%$.

49. Show that the PV of an investment that pays out R dollars/year continuously for T years is $R(1 - e^{-rT})/r$, where r is the interest rate.

50. [icon]   Explain this statement: If T is very large, then the PV of the income stream described in Exercise 49 is approximately R/r.

51. Suppose that $r = 0.06$. Use the result of Exercise 50 to estimate the payout rate R needed to produce an income stream whose PV is $20,000, assuming that the stream continues for a large number of years.

52. Verify by differentiation

$$\int te^{-rt}\,dt = -\frac{e^{-rt}(1 + rt)}{r^2} + C \qquad \boxed{6}$$

Use Eq. (6) to compute the PV of an investment that pays out income continuously at a rate $R(t) = (5,000 + 1,000t)$ dollars/year for 5 years and $r = 0.05$.

53. Use Eq. (6) to compute the PV of an investment that pays out income continuously at a rate $R(t) = (5,000 + 1,000t)e^{0.02t}$ dollars/year for 10 years and $r = 0.08$.

54. [icon]   **Banker's Rule of 70** Bankers have a rule of thumb that if you receive R percent interest, continuously compounded, then your money doubles after approximately $70/R$ years. For example, at $R = 5\%$, your money doubles after $70/5$ or 14 years. Use the concept of doubling time to justify the Banker's Rule. (*Note:* Sometimes, the approximation $72/R$ is used. It is less accurate but easier to apply because 72 is divisible by more numbers than 70.)

Further Insights and Challenges

55. [icon]   **Isotopes for Dating** Which of the following isotopes would be most suitable for dating extremely old rocks: Carbon-14 (half-life 5,570 years), Lead-210 (half-life 22.26 years), and Potassium-49 (half-life 1.3 billion years)? Explain why.

56. Let $P = P(t)$ be a quantity that obeys an exponential growth law with growth constant k. Show that P increases m-fold after an interval of $(\ln m)/k$ years.

57. [icon]   **Average Time of Decay** Physicists use the radioactive decay law $R = R_0 e^{-kt}$ to compute the average or *mean time* M until an atom decays. Let $F(t) = R/R_0 = e^{-kt}$ be the fraction of atoms that have survived to time t without decaying.
(a) Find the inverse function $t(F)$.
(b) The error in the following approximation tends to zero as $N \to \infty$:

$$M = \text{mean time to decay} \approx \frac{1}{N}\sum_{j=1}^{N} t\left(\frac{j}{N}\right)$$

Argue that $M = \int_0^1 t(F)\,dF$.

(c) Verify the formula $\int \ln x\,dx = x \ln x - x$ by differentiation and use it to show that for $c > 0$,

$$\int_c^1 t(F)\,dF = \frac{1}{k} + \frac{1}{k}(c \ln c - c)$$

(d) Verify numerically that $\lim_{c \to 0}(c - c \ln c) = 0$.

(e) The integral defining M is "improper" because $t(0)$ is infinite. Show that $M = 1/k$ by computing the limit

$$M = \lim_{c \to 0} \int_c^1 t(F)\,dF$$

$$\left(1 + \frac{1}{n}\right)^n \le e \le \left(1 + \frac{1}{n}\right)^{n+1}$$

(f) What is the mean time to decay for Radon (with a half-life of 3.825 days)?

58. The text proves that $e = \lim_{n \to \infty} (1 + \frac{1}{n})^n$. Use a change of variables to show that for any x,

$$\lim_{n \to \infty} \left(1 + \frac{x}{n}\right)^n = \lim_{n \to \infty} \left(1 + \frac{1}{n}\right)^{nx}$$

Use this to conclude that $e^x = \lim_{n \to \infty} (1 + \frac{x}{n})^n$.

59. Use Eq. (4) to prove that for $n > 0$,

60. A bank pays interest at the rate r, compounded M times yearly. The **effective interest rate** r_e is the rate at which interest, if compounded annually, would have to be paid to produce the same yearly return.

(a) Find r_e if $r = 9\%$ compounded monthly.

(b) Show that $r_e = (1 + r/M)^M - 1$ and that $r_e = e^r - 1$ if interest is compounded continuously.

(c) Find r_e if $r = 11\%$ compounded continuously.

(d) Find the rate r, compounded weekly, that would yield an effective rate of 20%.

CHAPTER REVIEW EXERCISES

In Exercises 1–4, refer to the function $f(x)$ whose graph is shown in Figure 1.

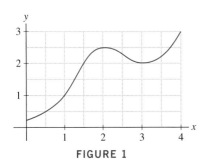

FIGURE 1

1. Estimate L_4 and M_4 on $[0, 4]$.

2. Estimate R_4, L_4, and M_4 on $[1, 3]$.

3. Find an interval $[a, b]$ on which R_4 is larger than $\int_a^b f(x)\,dx$. Do the same for L_4.

4. Justify $\dfrac{7}{4} \le \displaystyle\int_1^2 f(x)\,dx \le \dfrac{9}{4}$.

In Exercises 5–8, let $f(x) = x^2 + 4x$.

5. Calculate R_6, M_6, and L_6 for $f(x)$ on the interval $[1, 4]$. Sketch the graph of $f(x)$ and the corresponding rectangles for each approximation.

6. Find a formula for R_N for $f(x)$ on $[1, 4]$ and compute $\int_1^4 f(x)\,dx$ by taking the limit.

7. Find a formula for L_N for $f(x)$ on $[0, 2]$ and compute $\int_0^2 f(x)\,dx$ by taking the limit.

8. Use FTC I to evaluate $A(x) = \displaystyle\int_{-2}^x f(t)\,dt$.

9. Calculate R_6, M_6, and L_6 for $f(x) = (x^2 + 1)^{-1}$ on the interval $[0, 1]$.

10. Let R_N be the Nth right-endpoint approximation for $f(x) = x^3$ on $[0, 4]$ (Figure 2).

(a) Prove that $R_N = \dfrac{64(N + 1)^2}{N^2}$.

(b) Prove that the area of the region below the right-endpoint rectangles and above the graph is equal to

$$\frac{64(2N + 1)}{N^2}$$

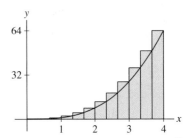

FIGURE 2 Approximation R_N for $f(x) = x^3$ on $[0, 4]$.

11. Which approximation to the area is represented by the shaded rectangles in Figure 3? Compute R_5 and L_5.

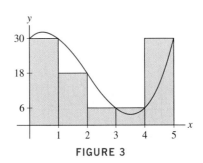

FIGURE 3

12. Calculate any two Riemann sums for $f(x) = x^2$ on the interval $[2, 5]$, but choose partitions with at least five subintervals of unequal widths and intermediate points that are neither endpoints nor midpoints.

In Exercises 13–34, evaluate the integral.

13. $\int (6x^3 - 9x^2 + 4x)\,dx$

14. $\int_0^1 (4x^3 - 2x^5)\,dx$

15. $\int (2x^3 - 1)^2\,dx$

16. $\int_1^4 (x^{5/2} - 2x^{-1/2})\,dx$

17. $\int \dfrac{x^4 + 1}{x^2}\,dx$

18. $\int_1^4 r^{-2}\,dr$

19. $\int_{-1}^4 |x^2 - 9|\,dx$

20. $\int_1^3 [t]\,dt$

21. $\int \csc^2 \theta\,d\theta$

22. $\int_0^{\pi/4} \sec t \tan t\,dt$

23. $\int \sec^2(9t - 4)\,dt$

24. $\int_0^{\pi/3} \sin 4\theta\,d\theta$

25. $\int (9t - 4)^{11}\,dt$

26. $\int_6^2 \sqrt{4y + 1}\,dy$

27. $\int \sin^2(3\theta) \cos(3\theta)\,d\theta$

28. $\int_0^{\pi/2} \sec^2(\cos \theta) \sin \theta\,d\theta$

29. $\int \dfrac{(2x^3 + 3x)\,dx}{(3x^4 + 9x^2)^5}$

30. $\int_{-4}^{-2} \dfrac{12x\,dx}{(x^2 + 2)^3}$

31. $\int \sin \theta \sqrt{4 - \cos \theta}\,d\theta$

32. $\int_0^{\pi/3} \dfrac{\sin \theta}{\cos^{2/3} \theta}\,d\theta$

33. $\int y\sqrt{2y + 3}\,dy$

34. $\int_1^8 t^2\sqrt{t + 8}\,dt$

35. Combine to write as a single integral

$$\int_0^8 f(x)\,dx + \int_{-2}^0 f(x)\,dx + \int_8^6 f(x)\,dx$$

36. Let $A(x) = \int_0^x f(x)\,dx$, where $f(x)$ is the function shown in Figure 4. Indicate on the graph of f where the local minima, maxima, and points of inflection of $A(x)$ occur and identify the intervals where $A(x)$ is increasing, decreasing, concave up, or concave down.

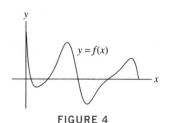

FIGURE 4

37. Find inflection points of $A(x) = \displaystyle\int_3^x \dfrac{t\,dt}{t^2 + 1}$.

38. A particle starts at the origin at time $t = 0$ and moves with velocity $v(t)$ as shown in Figure 5.

(a) How many times does the particle return to the origin in the first 12 s?

(b) Where is the particle located at time $t = 12$?

(c) At which time t is the particle's distance to the origin at a maximum?

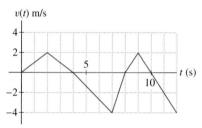

FIGURE 5

39. On a typical day, a city consumes water at the rate of $r(t) = 100 + 72t - 3t^2$ (in thousands of gallons per hour), where t is the number of hours past midnight. What is the daily water consumption? How much water is consumed between 6 PM and midnight?

40. The learning curve for producing bicycles in a certain factory is $L(x) = 12x^{-1/5}$ (in hours per bicycle), which means that it takes a bike mechanic $L(n)$ hours to assemble the nth bicycle. If 24 bicycles are produced, how long does it take to produce the second batch of 12?

41. Cost engineers at NASA have the task of projecting the cost P of major space projects. It has been found that the cost C of developing a projection increases with P at the rate $dC/dP \approx 21P^{-0.65}$, where C is in thousands of dollars and P in millions of dollars. What is the cost of developing a projection for a project whose cost turns out to be $P = \$35$ million?

42. The cost of jet fuel increased dramatically in 2005. Figure 6 displays Department of Transportation estimates for the rate of percentage price increase $R(t)$ (in units of percentage per year) during the first 6 months of the year. Express the total percentage price increase I during the first 6 months as an integral and calculate I. When determining the limits of integration, keep in mind that t is in years since $R(t)$ is a yearly rate.

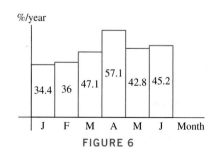

FIGURE 6

43. 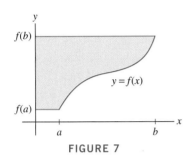 Let $f(x)$ be a positive increasing continuous function on $[a, b]$, where $0 \le a < b$ as in Figure 7. Show that the shaded region has area

$$I = bf(b) - af(a) - \int_a^b f(x)\,dx \qquad \boxed{1}$$

FIGURE 7

44.   How can we interpret the quantity I in Eq. (1) if $a < b \le 0$? Explain with a graph.

In Exercises 45–49, express the limit as an integral (or multiple of an integral) and evaluate.

45. $\displaystyle \lim_{N \to \infty} \frac{2}{N} \sum_{j=1}^{N} \sin\left(\frac{2j}{N}\right)$

46. $\displaystyle \lim_{N \to \infty} \frac{4}{N} \sum_{k=1}^{N} \left(3 + \frac{4k}{N}\right)$

47. $\displaystyle \lim_{N \to \infty} \frac{\pi}{N} \sum_{j=0}^{N-1} \sin\left(\frac{\pi}{2} + \frac{\pi j}{N}\right)$

48. $\displaystyle \lim_{N \to \infty} \frac{4}{N} \sum_{k=1}^{N} \frac{1}{(3 + \frac{4k}{N})^2}$

49. $\displaystyle \lim_{N \to \infty} \frac{1^k + 2^k + \cdots N^k}{N^{k+1}} \qquad (k > 0)$

50. Evaluate $\displaystyle \int_{-\pi/4}^{\pi/4} \frac{x^9\,dx}{\cos^2 x}$, using the properties of odd functions.

51. Evaluate $\displaystyle \int_0^1 f(x)\,dx$, assuming that $f(x)$ is an even continuous function such that

$$\int_1^2 f(x)\,dx = 5, \qquad \int_{-2}^1 f(x)\,dx = 8$$

52. [GU] Plot the graph of $f(x) = \sin mx \sin nx$ on $[0, \pi]$ for the pairs $(m, n) = (2, 4), (3, 5)$ and in each case guess the value of $I = \int_0^\pi f(x)\,dx$. Experiment with a few more values (including two cases with $m = n$) and formulate a conjecture for when I is zero.

53. Show that

$$\int x\, f(x)\,dx = xF(x) - G(x)$$

where $F'(x) = f(x)$ and $G'(x) = F(x)$. Use this to evaluate $\int x \cos x\,dx$.

54. Prove

$$2 \le \int_1^2 2^x\,dx \le 4 \qquad \text{and} \qquad \frac{1}{9} \le \int_1^2 3^{-x}\,dx \le \frac{1}{3}$$

55. [GU] Plot the graph of $f(x) = x^{-2}\sin x$ and show that $0.2 \le \int_1^2 f(x)\,dx \le 0.9$.

56. Find upper and lower bounds for $\int_0^1 f(x)\,dx$, where $f(x)$ has the graph shown in Figure 8.

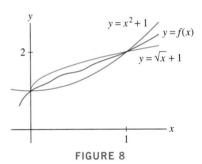

FIGURE 8

In Exercises 57–62, find the derivative.

57. $A'(x)$, where $A(x) = \int_3^x \sin(t^3)\,dt$

58. $A'(\pi)$, where $A(x) = \int_2^x \frac{\cos t}{1+t}\,dt$

59. $\displaystyle \frac{d}{dy} \int_{-2}^y 3^x\,dx$

60. $G'(x)$, where $G(x) = \int_{-2}^{\sin x} t^3\,dt$

61. $G'(2)$, where $G(x) = \int_0^{x^3} \sqrt{t+1}\,dt$

62. $H'(1)$, where $H(x) = \int_{4x^2}^9 \frac{1}{t}\,dt$

63.   Explain with a graph: If $f(x)$ is increasing and concave up on $[a, b]$, then L_N is more accurate than R_N. Which is more accurate if $f(x)$ is increasing and concave down?

64.   Explain with a graph: If $f(x)$ is linear on $[a, b]$, then the $\int_a^b f(x)\,dx = \frac{1}{2}(R_N + L_N)$ for all N.

In Exercises 65–70, use the given substitution to evaluate the integral.

65. $\displaystyle\int \frac{(\ln x)^2 dx}{x}, \quad u = \ln x$

66. $\displaystyle\int \frac{dx}{4x^2 + 9}, \quad u = \frac{2x}{3}$

67. $\displaystyle\int \frac{dx}{\sqrt{e^{2x} - 1}}, \quad u = e^x$

68. $\displaystyle\int \frac{\cos^{-1} t \, dt}{\sqrt{1 - t^2}}, \quad u = \cos^{-1} t$

69. $\displaystyle\int \frac{dt}{t(1 + (\ln t)^2)}, \quad u = \ln t$

70. $\displaystyle\int \sec^2(2\theta) \tan(2\theta) \, d\theta, \quad u = \tan(2\theta)$

In Exercises 71–92, calculate the integral.

71. $\displaystyle\int e^{9-2x} \, dx$

72. $\displaystyle\int x^2 e^{x^3} \, dx$

73. $\displaystyle\int e^{-2x} \sin(e^{-2x}) \, dx$

74. $\displaystyle\int \frac{\cos(\ln x) \, dx}{x}$

75. $\displaystyle\int_1^e \frac{\ln x \, dx}{x}$

76. $\displaystyle\int_0^{\ln 3} e^{x - e^x} \, dx$

77. $\displaystyle\int_{1/3}^{2/3} \frac{dx}{\sqrt{1 - x^2}}$

78. $\displaystyle\int_4^{12} \frac{dx}{x\sqrt{x^2 - 1}}$

79. $\displaystyle\int_0^{\pi/3} \tan\theta \, d\theta$

80. $\displaystyle\int_{\pi/6}^{2\pi/3} \cot\left(\frac{1}{2}\theta\right) d\theta$

81. $\displaystyle\int_0^1 \cos 2t \, dt$

82. $\displaystyle\int_0^2 \frac{dt}{4t + 12}$

83. $\displaystyle\int_0^3 \frac{x \, dx}{x^2 + 9}$

84. $\displaystyle\int_0^3 \frac{dx}{x^2 + 9}$

85. $\displaystyle\int \frac{x \, dx}{\sqrt{1 - x^4}}$

86. $\displaystyle\int e^x 10^x \, dx$

87. $\displaystyle\int \frac{\sin^{-1} x \, dx}{\sqrt{1 - x^2}}$

88. $\displaystyle\int \tan 5x \, dx$

89. $\displaystyle\int \sin x \cos^3 x \, dx$

90. $\displaystyle\int_0^1 \frac{dx}{25 + x^2}$

91. $\displaystyle\int_0^4 \frac{dx}{2x^2 + 1}$

92. $\displaystyle\int_5^8 \frac{dx}{x\sqrt{x^2 - 16}}$

93. In this exercise, we prove that for all $x > 0$,

$$x - \frac{x^2}{2} \leq \ln(1 + x) \leq x \qquad \boxed{2}$$

(a) Show that $\ln(1 + x) = \displaystyle\int_0^x \frac{dt}{1 + t}$ for $x > 0$.

(b) Verify that $1 - t \leq \dfrac{1}{1 + t} \leq 1$ for all $t > 0$.

(c) Use (b) to prove Eq. (2).

(d) Verify Eq. (2) for $x = 0.5, 0.1$, and 0.01.

94. Let

$$F(x) = x\sqrt{x^2 - 1} - 2\int_1^x \sqrt{t^2 - 1} \, dt$$

Prove that $F(x)$ and $\cosh^{-1} x$ differ by a constant by showing that they have the same derivative. Then prove they are equal by evaluating both at $x = 1$.

95. Let

$$F(x) = \int_2^x \frac{dt}{\ln t} \qquad \text{and} \qquad G(x) = \frac{x}{\ln x}$$

Verify that L'Hôpital's Rule may be applied to the limit $L = \displaystyle\lim_{x \to \infty} \frac{F(x)}{G(x)}$ and evaluate L.

96. The isotope Thorium-234 has a half-life of 24.5 days.

(a) Find the differential equation satisfied by the amount $y(t)$ of Thorium-234 in a sample at time t.

(b) At $t = 0$, a sample contains 2 kg of Thorium-234. How much remains after 1 year?

97. The Oldest Snack Food   In Bat Cave, New Mexico, archaeologists found ancient human remains, including cobs of popping corn, that had a C^{14} to C^{12} ratio equal to around 48% of that found in living matter. Estimate the age of the corn cobs.

98. The C^{14} to C^{12} ratio of a sample is proportional to the disintegration rate (number of beta particles emitted per minute) that is measured directly with a Geiger counter. The disintegration rate of carbon in a living organism is 15.3 beta particles/min per gram. Find the age of a sample that emits 9.5 beta particles/min per gram.

99. An investment pays out $5,000 at the end of the year for 3 years. Compute the PV, assuming an interest rate of 8%.

100. Use Eq. (3) of Section 5.8 to show that the PV of an investment which pays out income continuously at a constant rate of R dollars/year for T years is $\text{PV} = R\dfrac{1 - e^{-rT}}{r}$, where r is the interest rate. Use L'Hôpital's Rule to prove that the PV approaches RT as $r \to 0$.

101. In a first-order chemical reaction, the quantity $y(t)$ of reactant at time t satisfies $y' = -ky$, where $k > 0$. The dependence of k on temperature T (in kelvins) is given by the **Arrhenius equation** $k = Ae^{-E_a/(RT)}$, where E_a is the activation energy (J-mol^{-1}), $R = 8.314$ J-mol^{-1}-K^{-1}, and A is a constant. Assume that $A = 72 \times 10^{12}$ hour^{-1} and $E_a = 1.1 \times 10^5$. Calculate $\dfrac{dk}{dT}$ for $T = 500$ and use the Linear Approximation to estimate the change in k if T is raised from 500 to 510 K.

102. Find the interest rate if the PV of $50,000 to be received in 3 years is 43,000.

103. An equipment upgrade costing $1 million will save a company $320,000 per year for 4 years. Is this a good investment if the interest rate is $r = 5\%$? What is the largest interest rate that would make the investment worthwhile? Assume that the savings are received as a lump sum at the end of each year.

104. Calculate the limit:

(a) $\lim\limits_{n \to \infty} \left(1 + \dfrac{4}{n}\right)^n$

(b) $\lim\limits_{n \to \infty} \left(1 + \dfrac{1}{n}\right)^{4n}$

(c) $\lim\limits_{n \to \infty} \left(1 + \dfrac{4}{n}\right)^{3n}$

6 | APPLICATIONS OF THE INTEGRAL

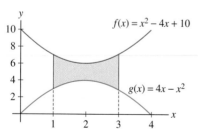

The CAT scan is based on tomography, a mathematical technique for combining a large series of X-rays of the body taken at different angles into a single cross-sectional image.

The integral, like the derivative, has a wide variety of applications. In the previous chapter, we used the integral to compute areas under curves and net change. In this chapter, we discuss some of the other quantities that are represented by integrals, including volume, average value, work, total mass, population, and fluid flow.

6.1 Area Between Two Curves

In Section 5.2, we learned that the definite integral $\int_a^b f(x)\,dx$ represents the signed area between the graph of $f(x)$ and the x-axis over the interval $[a, b]$. Sometimes, we are interested in computing the area between two graphs (Figure 1). If $f(x) \geq g(x) \geq 0$ for $x \in [a, b]$, then the graph of $f(x)$ lies above the graph of $g(x)$, and the region between the two graphs is obtained by removing the region under $y = g(x)$ from the region under $y = f(x)$. Therefore,

$$\text{Area between the graphs} = \int_a^b f(x)\,dx - \int_a^b g(x)\,dx$$

$$= \int_a^b (f(x) - g(x))\,dx \qquad \boxed{1}$$

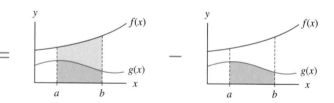

FIGURE 1 The area between the graphs is a difference of two areas.

Region between the graphs

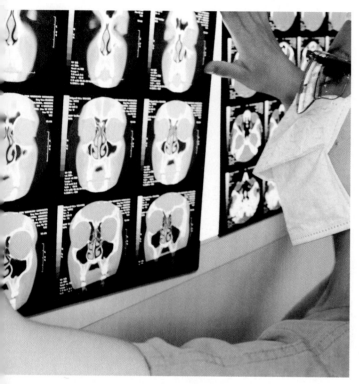

FIGURE 2 Region between the graphs of $f(x) = x^2 - 4x + 10$ and $g(x) = 4x - x^2$ over $[1, 3]$.

■ **EXAMPLE 1** Calculate the area of the region between the graphs of

$$f(x) = x^2 - 4x + 10 \qquad \text{and} \qquad g(x) = 4x - x^2$$

over $[1, 3]$.

Solution To calculate the area between the graphs, we must first determine which graph lies on top. Figure 2 shows that $f(x) \geq g(x)$. We can verify this without appealing to the graph by completing the square:

$$f(x) - g(x) = (x^2 - 4x + 10) - (4x - x^2) = 2x^2 - 8x + 10 = 2(x - 2)^2 + 2 > 0$$

By Eq. (1), the area between the graphs is

$$\int_1^3 \big(f(x) - g(x)\big)\,dx = \int_1^3 \big((x^2 - 4x + 10) - (4x - x^2)\big)\,dx$$

$$= \int_1^3 (2x^2 - 8x + 10)\,dx = \left(\frac{2}{3}x^3 - 4x^2 + 10x\right)\bigg|_1^3 = 12 - \frac{20}{3} = \frac{16}{3} \quad\blacksquare$$

Before continuing with more examples, we use Riemann sums to explain why Eq. (1) remains valid if $f(x) \geq g(x)$ but $f(x)$ and $g(x)$ are not assumed to be positive:

$$\int_a^b \big(f(x) - g(x)\big)\,dx = \lim_{\|P\| \to 0} R(f - g, P, C) = \lim_{N \to \infty} \sum_{i=1}^N \big(f(c_i) - g(c_i)\big)\Delta x_i$$

Recall that P denotes a partition of $[a, b]$:

$$\text{Partition } P: \quad a = x_0 < x_1 < x_2 < \cdots < x_N = b$$

$C = \{c_1, \ldots, c_N\}$ is a choice of intermediate points where $c_i \in [x_{i-1}, x_i]$, and $\Delta x_i = x_i - x_{i-1}$. The ith term in the Riemann sum is equal to the area of a thin vertical rectangle of height $(f(c_i) - g(c_i))$ and width Δx_i (Figure 3):

$$\big(f(c_i) - g(c_i)\big)\Delta x_i = \text{height} \times \text{width}$$

Therefore, $R(f - g, P, C)$ is an approximation to the area between the graphs using thin vertical rectangles. As the norm $\|P\|$ (the maximum width of the rectangles) approaches zero, the Riemann sum converges to the area between the graphs and we obtain Eq. (1).

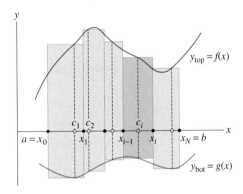

FIGURE 3 The ith rectangle has width $\Delta x_i = x_i - x_{i-1}$ and height $f(c_i) - g(c_i)$.

To help identify the functions, we sometimes denote the upper graph by $y_{\text{top}} = f(x)$ and the lower graph by $y_{\text{bot}} = g(x)$:

Keep in mind that $(y_{\text{top}} - y_{\text{bot}})$ is the height of a thin vertical slice of the region.

$$\boxed{\text{Area between the graphs} = \int_a^b (y_{\text{top}} - y_{\text{bot}})\,dx = \int_a^b \big(f(x) - g(x)\big)\,dx}$$

■ **EXAMPLE 2** Find the area between the graphs of $f(x) = x^2 - 5x - 7$ and $g(x) = x - 12$ over $[-2, 5]$.

Solution To calculate the area, we must first determine which graph lies on top.

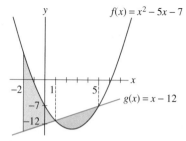

FIGURE 4

In Example 2, we found the intersection points of $y = f(x)$ and $y = g(x)$ algebraically. For more complicated functions, it may be necessary to use a computer algebra system.

Step 1. Sketch the region (especially, find any points of intersection).

We know that $y = f(x)$ is a parabola with y-intercept -7 and $y = g(x)$ is a line with y-intercept -12. To determine where the graphs intersect, we solve $f(x) = g(x)$:

$$x^2 - 5x - 7 = x - 12 \quad \text{or} \quad x^2 - 6x + 5 = (x - 1)(x - 5) = 0$$

Thus, the points of intersection are $x = 1, 5$ (Figure 4).

Step 2. Set up the integrals and evaluate.

Figure 4 shows that

$$f(x) \geq g(x) \text{ on } [-2, 1] \quad \text{and} \quad g(x) \geq f(x) \text{ on } [1, 5]$$

Therefore, we write the area as a sum of integrals over the two intervals:

$$\int_{-2}^{5} (y_{\text{top}} - y_{\text{bot}}) \, dx$$

$$= \int_{-2}^{1} \left(f(x) - g(x) \right) dx + \int_{1}^{5} \left(g(x) - f(x) \right) dx$$

$$= \int_{-2}^{1} \left((x^2 - 5x - 7) - (x - 12) \right) dx + \int_{1}^{5} \left((x - 12) - (x^2 - 5x - 7) \right) dx$$

$$= \int_{-2}^{1} (x^2 - 6x + 5) \, dx + \int_{1}^{5} (-x^2 + 6x - 5) \, dx$$

$$= \left(\frac{1}{3}x^3 - 3x^2 + 5x \right) \Big|_{-2}^{1} + \left(-\frac{1}{3}x^3 + 3x^2 - 5x \right) \Big|_{1}^{5}$$

$$= \left(\frac{7}{3} - \frac{(-74)}{3} \right) + \left(\frac{25}{3} - \frac{(-7)}{3} \right) = \frac{113}{3} \qquad \blacksquare$$

■ **EXAMPLE 3** Calculating Area by Dividing the Region Find the area of the region bounded by the graphs of $y = \dfrac{8}{x^2}$, $y = 8x$, and $y = x$.

Solution

Step 1. Sketch the region (especially, find any points of intersection).

The curve $y = \dfrac{8}{x^2}$ cuts off a region in the sector between the two lines $y = 8x$ and $y = x$ (Figure 5). To find the intersection of $y = \dfrac{8}{x^2}$ and $y = 8x$, we solve

$$\frac{8}{x^2} = 8x \Rightarrow x^3 = 1 \Rightarrow x = 1$$

FIGURE 5 Area bounded by $y = \dfrac{8}{x^2}$, $y = 8x$, and $y = x$ as a sum of two areas.

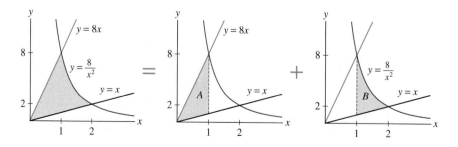

To find the intersection of $y = \dfrac{8}{x^2}$ and $y = x$, we solve

$$\frac{8}{x^2} = x \Rightarrow x^3 = 8 \Rightarrow x = 2$$

Step 2. **Set up the integrals and evaluate.**

Figure 5 shows that $y_{\text{bot}} = x$, but y_{top} changes at $x = 1$ from $y_{\text{top}} = 8x$ to $y_{\text{top}} = \dfrac{8}{x^2}$. Therefore, we break up the regions into two parts, A and B, and compute their areas separately:

$$\text{Area of } A = \int_0^1 \left(y_{\text{top}} - y_{\text{bot}} \right) dx = \int_0^1 (8x - x)\, dx = \int_0^1 7x\, dx = \frac{7}{2} x^2 \Big|_0^1 = \frac{7}{2}$$

$$\text{Area of } B = \int_1^2 \left(y_{\text{top}} - y_{\text{bot}} \right) dx = \int_1^2 \left(\frac{8}{x^2} - x \right) dx = \left(-\frac{8}{x} - \frac{1}{2} x^2 \right) \Big|_1^2 = \frac{5}{2}$$

The total area bounded by the curves is the sum $\frac{7}{2} + \frac{5}{2} = 6$. ■

Integration Along the y-Axis

Suppose we are given x as a function of y, say, $x = g(y)$. What is the meaning of the integral $\displaystyle\int_c^d g(y)\, dy$? This integral may be interpreted as *signed area*, where regions to the *right* of the y-axis have positive area and regions to the *left* have negative area:

$$\int_c^d g(y)\, dy = \text{signed area between graph and } y\text{-axis for } c \le y \le d$$

Figure 6(A) shows the graph of $g(y) = y^2 - 1$. The region to the left of the y-axis has negative signed area. The integral is equal to the signed area:

$$\underbrace{\int_{-2}^2 (y^2 - 1)\, dy}_{\substack{\text{Area to the right of } y\text{-axis minus} \\ \text{area to the left of } y\text{-axis}}} = \left(\frac{1}{3} y^3 - y \right) \Big|_{-2}^2 = \frac{4}{3}$$

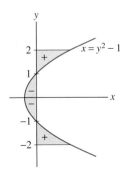

(A) Region between $x = y^2 - 1$
    and the y-axis

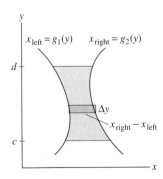

(B) Region between $x = g_2(y)$
    and $x = g_1(y)$

FIGURE 6

More generally, if $g_2(y) \geq g_1(y)$ as in Figure 6(B), then the graph of $x = g_2(y)$ lies to the right of the graph of $x = g_1(y)$. As a reminder, we write $x_{\text{right}} = g_2(y)$ and $x_{\text{left}} = g_1(y)$. The area between the two graphs for $c \leq y \leq d$ is equal to

$$\text{Area between the graphs} = \int_c^d \big(g_2(y) - g_1(y)\big)\, dy = \int_c^d \big(x_{\text{right}} - x_{\text{left}}\big)\, dy \qquad \boxed{2}$$

In this case, the Riemann sums approximate the area by thin horizontal rectangles of width $x_{\text{right}} - x_{\text{left}}$ and height Δy.

■ **EXAMPLE 4** Calculate the area between the graphs of $g_1(y) = y^3 - 4y$ and $g_2(y) = y^3 + y^2 + 8$ for $-2 \leq y \leq 2$.

Solution We confirm that $g_2(y) \geq g_1(y)$ as shown in Figure 7:

$$g_2(y) - g_1(y) = (y^3 + y^2 + 8) - (y^3 - 4y) = y^2 + 4y + 8 = (y + 2)^2 + 4 > 0$$

Therefore, $x_{\text{right}} = g_2(y)$ and $x_{\text{left}} = g_1(y)$, and the area is

$$\int_{-2}^{2} (x_{\text{right}} - x_{\text{left}})\, dy = \int_{-2}^{2} (y^2 + 4y + 8)\, dy = \left(\frac{1}{3}y^3 + 2y^2 + 8y \right)\Bigg|_{-2}^{2}$$

$$= \frac{80}{3} - \frac{-32}{3} = \frac{112}{3} \qquad ■$$

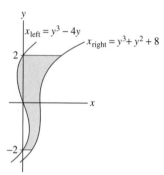

FIGURE 7 Region between $g_1(x) = y^3 - 4y$ and $g_2(x) = y^3 + y^2 + 8$ for $-2 \leq y \leq 2$.

It would be more difficult to calculate the area of the region in Figure 7 as an integral with respect to x because the curves are not graphs of functions of x.

6.1 SUMMARY

- If $f(x) \geq g(x)$ on $[a, b]$, then the area between the graphs of f and g over $[a, b]$ is

$$\text{Area between graphs} = \int_a^b \big(f(x) - g(x)\big)\, dx = \int_a^b \big(y_{\text{top}} - y_{\text{bot}}\big)\, dx$$

- To calculate the area between two graphs $y = f(x)$ and $y = g(x)$, sketch the region to find y_{top}. If necessary, find points of intersection by solving $f(x) = g(x)$.
- The integral along the y-axis, $\displaystyle\int_c^d g(y)\, dy$, is equal to the signed area between the graph and the y-axis for $c \leq y \leq d$, where area to the right of the y-axis is positive and area to the left is negative.
- If $g_2(y) \geq g_1(y)$, then the graph of $x = g_2(y)$ lies to the right of the graph of $x = g_1(y)$ and the area between the graphs for $c \leq y \leq d$ is

$$\text{Area between graphs} = \int_c^d \big(g_2(y) - g_1(y)\big)\, dy = \int_c^d \big(x_{\text{right}} - x_{\text{left}}\big)\, dy$$

6.1 EXERCISES

Preliminary Questions

1. What is the area interpretation of $\displaystyle\int_a^b \big(f(x) - g(x)\big)\, dx$ if $f(x) \geq g(x)$?

2. Is $\displaystyle\int_a^b \big(f(x) - g(x)\big)\, dx$ still equal to the area between the graphs of f and g if $f(x) \geq 0$ but $g(x) \leq 0$?

3. Suppose that $f(x) \ge g(x)$ on [0, 3] and $g(x) \ge f(x)$ on [3, 5]. Express the area between the graphs over [0, 5] as a sum of integrals.

4. Suppose that the graph of $x = f(y)$ lies to the left of the *y*-axis. Is $\int_a^b f(y)\,dy$ positive or negative?

Exercises

1. Find the area of the region between $y = 3x^2 + 12$ and $y = 4x + 4$ over [−3, 3] (Figure 8).

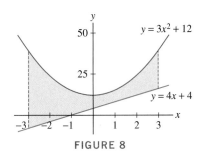

FIGURE 8

2. Compute the area of the region in Figure 9(A), which lies between $y = 2 - x^2$ and $y = -2$ over [−2, 2].

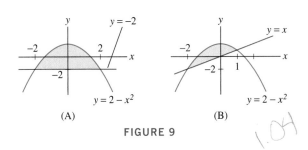

(A)   (B)

FIGURE 9

3. Let $f(x) = x$ and $g(x) = 2 - x^2$ [Figure 9(B)].
(a) Find the points of intersection of the graphs.
(b) Find the area enclosed by the graphs of f and g.

4. Let $f(x) = 8x - 10$ and $g(x) = x^2 - 4x + 10$.
(a) Find the points of intersection of the graphs.
(b) Compute the area of the region *below* the graph of f and *above* the graph of g.

In Exercises 5–7, find the area between $y = \sin x$ and $y = \cos x$ over the interval. Sketch the curves if necessary.

5. $\left[0, \dfrac{\pi}{4}\right]$ **6.** $\left[\dfrac{\pi}{4}, \dfrac{\pi}{2}\right]$ **7.** [0, π]

In Exercises 8–10, let $f(x) = 20 + x - x^2$ and $g(x) = x^2 - 5x$.

8. Find the area between the graphs of f and g over [1, 3].

9. Find the area of the region enclosed by the two graphs.

10. Compute the area of the region between the two graphs over [4, 8] as a sum of two integrals.

11. Find the area between $y = e^x$ and $y = e^{2x}$ over [0, 1].

12. Find the area of the region bounded by $y = e^x$ and $y = 12 - e^x$ and the *y*-axis.

13. Sketch the region bounded by $y = \dfrac{1}{\sqrt{1 - x^2}}$ and $y = -\dfrac{1}{\sqrt{1 - x^2}}$ for $-\frac{1}{2} \le x \le \frac{1}{2}$ and find its area.

14. Sketch the region bounded by $y = \sec^2 x$ and $y = 2$ and find its area.

In Exercises 15–18, find the area of the shaded region in the figure.

15.

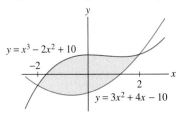

FIGURE 10

16.

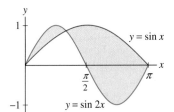

FIGURE 11

17.

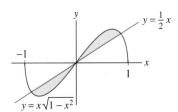

FIGURE 12

18.

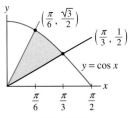

FIGURE 13

19. Find the area of the region enclosed by the curves $y = x^3 - 6x$ and $y = 8 - 3x^2$.

20. Find the area of the region enclosed by the *semicubical parabola* $y^2 = x^3$ and the line $x = 1$.

In Exercises 21–22, find the area between the graphs of $x = \sin y$ and $x = 1 - \cos y$ over the given interval (Figure 14).

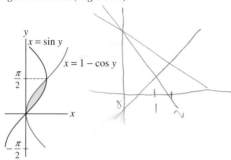

FIGURE 14

21. $0 \le y \le \dfrac{\pi}{2}$

22. $-\dfrac{\pi}{2} \le y \le \dfrac{\pi}{2}$

23. Find the area of the region lying to the right of $x = y^2 + 4y - 22$ and the left of $x = 3y + 8$.

24. Find the area of the region lying to the right of $x = y^2 - 5$ and the left of $x = 3 - y^2$.

25. Calculate the area enclosed by $x = 9 - y^2$ and $x = 5$ in two ways: as an integral along the y-axis and as an integral along the x-axis.

26. Figure 15 shows the graphs of $x = y^3 - 26y + 10$ and $x = 40 - 6y^2 - y^3$. Match the equations with the curve and compute the area of the shaded region.

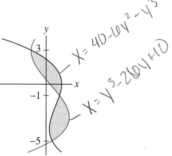

FIGURE 15

In Exercises 27–28, find the area of the region using the method (integration along either the x- or y-axis) that requires you to evaluate just one integral.

27. Region between $y^2 = x + 5$ and $y^2 = 3 - x$

28. Region between $y = x$ and $x + y = 8$ over $[2, 3]$

In Exercises 29–45, sketch the region enclosed by the curves and compute its area as an integral along the x- or y-axis.

29. $y = 4 - x^2$, $y = x^2 - 4$

30. $y = x^2 - 6$, $y = 6 - x^3$, y-axis

31. $x = \sin y$, $x = \dfrac{2}{\pi} y$

32. $x + y = 4$, $x - y = 0$, $y + 3x = 4$

33. $y = 3x^{-3}$, $y = 4 - x$, $y = \dfrac{x}{3}$

34. $y = 2 - \sqrt{x}$, $y = \sqrt{x}$, $x = 0$

35. $y = x\sqrt{x - 2}$, $y = -x\sqrt{x - 2}$, $x = 4$

36. $y = |x|$, $y = x^2 - 6$

37. $x = |y|$, $x = 6 - y^2$

38. $x = |y|$, $x = 1 - |y|$

39. $x = 12 - y$, $x = y$, $x = 2y$

40. $x = y^3 - 18y$, $y + 2x = 0$

41. $x = 2y$, $x + 1 = (y - 1)^2$

42. $x + y = 1$, $x^{1/2} + y^{1/2} = 1$

43. $y = 6$, $y = x^{-2} + x^2$ (in the region $x > 0$)

44. $y = \cos x$, $y = \cos(2x)$, $x = 0$, $x = \dfrac{2\pi}{3}$

45. $y = \sin x$, $y = \csc^2 x$, $x = \dfrac{\pi}{4}$, $x = \dfrac{3\pi}{4}$

46. *CAS* Plot $y = \dfrac{x}{\sqrt{x^2 + 1}}$ and $y = (x - 1)^2$ on the same set of axes. Use a computer algebra system to find the points of intersection numerically and compute the area between the curves.

47. Sketch a region whose area is represented by

$$\int_{-\sqrt{2}/2}^{\sqrt{2}/2} \left(\sqrt{1 - x^2} - |x| \right) dx$$

and evaluate using geometry.

48. Two athletes run in the same direction along a straight track with velocities $v_1(t)$ and $v_2(t)$ (in ft/s). Assume that

$$\int_0^5 (v_1(t) - v_2(t))\, dt = 2, \qquad \int_0^{20} (v_1(t) - v_2(t))\, dt = 5$$

$$\int_{30}^{35} (v_1(t) - v_2(t))\, dt = -2$$

(a) Give a verbal interpretation of the integral $\displaystyle\int_{t_1}^{t_2} (v_1(t) - v_2(t))\, dt$.

(b) Is enough information given to determine the distance between the two runners at time $t = 5$ s?

(c) If the runners begin at the same time and place, how far ahead is runner 1 at time $t = 20$ s?

(d) Suppose that runner 1 is 8 ft ahead at $t = 30$ s. How far ahead or behind is she at $t = 35$ s?

49. Express the area (not signed) of the shaded region in Figure 16 as a sum of three integrals involving the functions f and g.

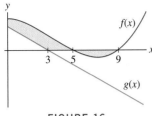

FIGURE 16

50. Find the area enclosed by the curves $y = c - x^2$ and $y = x^2 - c$ as a function of c. Find the value of c for which this area is equal to 1.

51. Set up (but do not evaluate) an integral that expresses the area between the circles $x^2 + y^2 = 2$ and $x^2 + (y - 1)^2 = 1$.

52. Set up (but do not evaluate) an integral that expresses the area between the graphs of $y = (1 + x^2)^{-1}$ and $y = x^2$.

53. ⌐﹁5 Find a numerical approximation to the area above $y = 1 - (x/\pi)$ and below $y = \sin x$ (find the points of intersection numerically).

54. ⌐﹁5 Find a numerical approximation to the area above $y = |x|$ and below $y = \cos x$.

55. ⌐﹁5 Use a computer algebra system to find a numerical approximation to the number c (besides zero) in $[0, \frac{\pi}{2}]$, where the curves $y = \sin x$ and $y = \tan^2 x$ intersect. Then find the area enclosed by the graphs over $[0, c]$.

56. The back of Jon's guitar (Figure 17) has a length 19 in. He measured the widths at 1-in. intervals, beginning and ending $\frac{1}{2}$ in. from the ends, obtaining the results

6, 9, 10.25, 10.75, 10.75, 10.25, 9.75, 9.5, 10, 11.25,

12.75, 13.75, 14.25, 14.5, 14.5, 14, 13.25, 11.25, 9

Use the midpoint rule to estimate the area of the back.

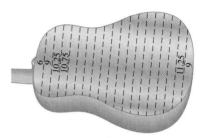

FIGURE 17 Back of guitar.

Further Insights and Challenges

57. Find the line $y = mx$ that divides the area under the curve $y = x(1 - x)$ over $[0, 1]$ into two regions of equal area.

58. ⌐﹁5 Let c be the number such that the area under $y = \sin x$ over $[0, \pi]$ is divided in half by the line $y = cx$ (Figure 18). Find an

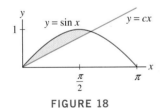

FIGURE 18

equation for c and solve this equation *numerically* using a computer algebra system.

59. ✎ Explain geometrically (without calculation) why the following holds for any $n > 0$:

$$\int_0^1 x^n \, dx + \int_0^1 x^{1/n} \, dx = 1$$

60. Let $f(x)$ be a strictly increasing function with inverse $g(x)$. Explain the equality geometrically:

$$\int_0^a f(x) \, dx + \int_{f(0)}^{f(a)} g(x) \, dx = af(a)$$

6.2 Setting Up Integrals: Volume, Density, Average Value

In this section, we use the integral to compute quantities such as volume, total mass, and fluid flow. The common thread in these diverse applications is that we approximate the relevant quantity by a Riemann sum and then pass to the limit to obtain an exact value.

Volume

The term "solid" or "solid body" refers to a solid three-dimensional object.

We begin by showing how integration can be used to compute the **volume** of a solid body. Before proceeding, let's recall that the volume of a *right cylinder* (Figure 1) is Ah, where A is the area of the base and h is the height, measured perpendicular to the base. Here we use the term "cylinder" in a general sense; the base does not have to be circular.

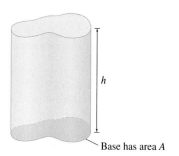

FIGURE 1 The volume of a right cylinder of height h is Ah.

Now let V be the volume of a solid body that extends from height $y = a$ to $y = b$ along the y-axis as in Figure 2. The intersection of the solid with the horizontal plane at height y is called the **horizontal cross section** at height y. Let $A(y)$ be its area.

To compute V, we divide the solid into N horizontal slices of thickness $\Delta y = \dfrac{b - a}{N}$. The ith slice extends from y_{i-1} to y_i, where $y_i = a + i \, \Delta y$. Let V_i be the volume of the slice. If N is very large, then Δy is very small and the slices are very thin. In this case, the ith slice is nearly a right cylinder of base $A(y_{i-1})$ and height Δy, and therefore $V_i \approx A(y_{i-1})\Delta y$. Summing up, we obtain

$$V = \sum_{i=1}^{N} V_i \approx \sum_{i=1}^{N} A(y_{i-1})\Delta y$$

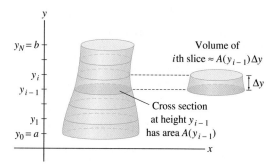

FIGURE 2 Divide the solid into thin horizontal slices. Each slice is nearly a right cylinder whose volume can be approximated.

The sum on the right is a left-endpoint approximation to the integral $\displaystyle\int_a^b A(y) \, dy$. We assume that $A(y)$ is a continuous function. Then the approximation improves in accuracy and converges to the integral as $N \to \infty$. We conclude that *the volume of the solid is equal to the integral of its cross-sectional area.*

> **Volume as the Integral of Cross-Sectional Area**   Suppose that a solid body extends from height $y = a$ to $y = b$. Let $A(y)$ be the area of the horizontal cross section at height y. Then
>
> $$\text{Volume of the solid body} = \int_a^b A(y) \, dy \qquad \boxed{1}$$

■ **EXAMPLE 1**   Volume of a Pyramid: Horizontal Cross Sections   Calculate the volume V of a pyramid of height 12 ft whose base is a square of side 4 ft.

Solution   To use Eq. (1), we need a formula for $A(y)$.

Step 1. **Find a formula for $A(y)$.**
We see in Figure 3 that the horizontal cross section at height y is a square. Let $\ell(y)$ be the length of its sides. We apply the law of similar triangles to $\triangle ABC$ and the triangle of height $12 - y$ whose base of length $\frac{1}{2}\ell(y)$ lies on the cross section:

$$\frac{\text{Base}}{\text{Height}} = \frac{2}{12} = \frac{\frac{1}{2}\ell(y)}{12 - y} \quad \Rightarrow \quad 2(12 - y) = 6\ell(y)$$

We find that $\ell(y) = \frac{1}{3}(12 - y)$ and therefore $A(y) = \ell(y)^2 = \frac{1}{9}(12 - y)^2$.

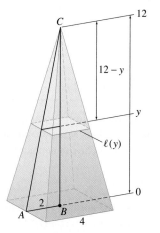

FIGURE 3 A horizontal cross section of the pyramid is a square.

Step 2. **Compute V as the integral of $A(y)$.**

$$V = \int_0^{12} A(y)\,dy = \int_0^{12} \frac{1}{9}(12 - y)^2\,dy = -\frac{1}{27}(12 - y)^3\Big|_0^{12} = 64 \text{ ft}^3$$

This is the same result we would obtain using the formula $V = \frac{1}{3}Ah$ for the volume of a pyramid of base A and height h, since $\frac{1}{3}Ah = \frac{1}{3}(4^2)(12) = 64$. ∎

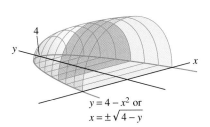

■ **EXAMPLE 2** The base of a solid is the region between the x-axis and the inverted parabola $y = 4 - x^2$. The vertical cross sections of the solid perpendicular to the y-axis are semicircles (Figure 4). Compute the volume of the solid.

Solution First, we find a formula for the cross section at height y. Note that $y = 4 - x^2$ can be written $x = \pm\sqrt{4 - y}$. We see in Figure 4 that the cross section at height y is a semicircle of radius $r = \sqrt{4 - y}$. The area of the semicircle is $A(y) = \frac{1}{2}\pi r^2 = \frac{\pi}{2}(4 - y)$, and the volume of the solid is

$$V = \int_0^4 A(y)\,dy = \frac{\pi}{2}\int_0^4 (4 - y)\,dy = \frac{\pi}{2}\left(4y - \frac{1}{2}y^2\right)\Big|_0^4 = 4\pi$$ ∎

FIGURE 4 A solid whose base is the region between $y = 4 - x^2$ and the x-axis.

The volume of a solid body may be computed using vertical rather than horizontal cross sections. We then obtain an integral with respect to x rather than y.

■ **EXAMPLE 3** Volume of a Sphere: Vertical Cross Sections Compute the volume of a sphere of radius R as an integral of cross-sectional area.

Solution As we see in Figure 5, the vertical cross section of the sphere at x is a circle whose radius r satisfies $x^2 + r^2 = R^2$ or $r = \sqrt{R^2 - x^2}$. The area of the cross section is $A(x) = \pi r^2 = \pi(R^2 - x^2)$. Therefore, the volume of the sphere is

$$\int_{-R}^R \pi(R^2 - x^2)\,dx = \pi\left(R^2 x - \frac{x^3}{3}\right)\Big|_{-R}^R = 2\left(\pi R^3 - \pi\frac{R^3}{3}\right) = \frac{4}{3}\pi R^3$$ ∎

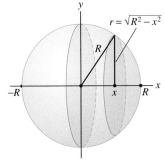

FIGURE 5 Vertical cross section is a circle of radius $\sqrt{R^2 - x^2}$.

Density

Next, we show how an integral may be used to compute the total mass of an object given its mass density. Consider a rod of length ℓ. The rod's **linear mass density** ρ is defined as the mass per unit length. If the density ρ is constant, then

$$\text{Total mass} = \text{linear mass density} \times \text{length} = \rho \cdot \ell \qquad \boxed{2}$$

For example, if $\ell = 10$ cm and $\rho = 9$ g/cm, then the total mass is $\rho\ell = 9 \cdot 10 = 90$ g.

Integration is needed to compute total mass when the density is not constant. Consider a rod extending along the x-axis from $x = a$ to $x = b$ whose density $\rho(x)$ depends on x, as in Figure 6. To compute the total mass M, we decompose the rod into N small segments of length $\Delta x = \dfrac{b - a}{N}$. Then $M = \displaystyle\sum_{i=1}^N M_i$, where M_i is the mass of the ith segment. Although Eq. (2) cannot be used when $\rho(x)$ is not constant, we can argue that if Δx is small, then $\rho(x)$ is nearly constant along the ith segment. Therefore, if the ith segment extends from x_{i-1} to x_i and if c_i is any sample point in $[x_{i-1}, x_i]$, then $\Delta x = x_i - x_{i-1}$ and $M_i \approx \rho(c_i)\Delta x$. We obtain

$$\text{Total mass } M = \sum_{i=1}^{N} M_i \approx \sum_{i=1}^{N} \rho(c_i)\Delta x$$

As $N \to \infty$, the accuracy of the approximation improves. However, the sum on the right is a Riemann sum whose value approaches $\int_a^b \rho(x)\,dx$. We conclude that *the total mass of a rod is equal to the integral of its linear mass density*:

$$\text{Total mass } M = \int_a^b \rho(x)\,dx$$

Do you see the similarity in the way we use thin slices to compute volume and small pieces to compute total mass?

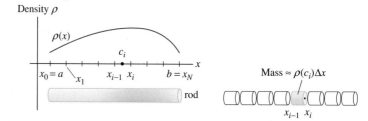

FIGURE 6 The total mass of the rod is equal to the area under the graph of mass density ρ.

▮ **EXAMPLE 4** Total Mass Find the total mass of a 2-m rod of linear density $\rho(x) = 1 + x(2 - x)$ kg/m, where x is the distance from one end of the rod.

Solution The total mass is

$$\int_0^2 \rho(x)\,dx = \int_0^2 \left(1 + x(2 - x)\right) dx = \left(x + x^2 - \frac{1}{3}x^3\right)\Big|_0^2 = \frac{10}{3} \text{ kg} \qquad ▮$$

In some situations, density is a function of distance to the origin. For example, in the study of urban populations, it might be assumed that the population density $\rho(r)$ (in people per square mile) depends only on the distance r from the center of a city. Such a density function is called a **radial density function.**

More general density functions depend on two variables $\rho(x, y)$. In this case, total mass or population is computed using double integration, a topic in multivariable calculus.

We now derive a formula for the total population P within a radius R of the city center, assuming that the population density $\rho(r)$ is a radial density function. To compute P, it makes sense to divide the circle of radius R into N thin rings of equal width $\Delta r = R/N$ as in Figure 7.

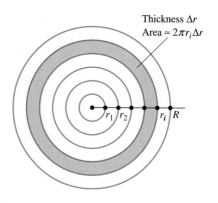

FIGURE 7 Dividing the circle of radius R into N thin rings of thickness $\Delta r = \dfrac{R}{N}$.

Let P_i be the population within the ith ring, so that $P = \displaystyle\sum_{i=1}^{N} P_i$. If the outer radius of the ith ring is r_i, then circumference is $2\pi r_i$, and if Δr is small, the area of this ring is *approximately* $2\pi r_i \Delta r$ (outer circumference times thickness). Furthermore, the population density within the thin ring is nearly constant with value $\rho(r_i)$. With these approximations,

$$P_i \approx \underbrace{2\pi r_i \Delta r}_{\text{Area}} \times \underbrace{\rho(r_i)}_{\substack{\text{Population} \\ \text{density}}} = 2\pi r_i \rho(r_i) \Delta r$$

and

$$P = \sum_{i=1}^{N} P_i \approx 2\pi \sum_{i=1}^{N} r_i \rho(r_i) \Delta r$$

This sum is a right-endpoint approximation to the integral $2\pi \int_0^R r\rho(r)\,dr$. As N tends to ∞, the approximation improves in accuracy and the sum converges to the integral. We conclude that for a population with a radial density function $\rho(r)$,

Remember that for a radial density function, the total population is obtained by integrating $2\pi r\rho(r)$ rather than $\rho(r)$.

$$\boxed{\text{Population } P \text{ within a radius } R = 2\pi \int_0^R r\rho(r)\,dr} \qquad \boxed{3}$$

■ **EXAMPLE 5** Computing Total Population from Population Density   The density function for the population in a certain city is $\rho(r) = 13.2(1 + r^2)^{-1/2}$, where r is the distance from the center in miles and ρ has units of thousands per square mile. How many people live within a 20-mile radius of the city center?

Solution   The population (in thousands) within a 20-mile radius is

$$2\pi \int_0^{20} r\left(13.2(1 + r^2)^{-1/2}\right) dr = 2\pi(13.2) \int_0^{20} \frac{r}{(1 + r^2)^{1/2}} \, dr$$

To evaluate the integral, use the substitution $u = 1 + r^2$, $du = 2r\,dr$. The limits of integration become $u(0) = 1$ and $u(20) = 401$, and we obtain

$$13.2\pi \int_1^{401} u^{-1/2}\,du = 26.4\pi\, u^{1/2}\Big|_1^{401} \approx 1{,}577.9 \text{ thousand}$$

In other words, the population is nearly 1.6 million people.   ■

Flow Rate

When liquid flows through a tube, the **flow rate** Q is the *volume per unit time* of fluid passing through the tube. The flow rate depends on the velocity of the fluid particles. If all particles of the liquid travel with the same velocity v (say, in units of centimeters per minute), then the flow rate through a tube of radius R is

$$\underbrace{\text{Flow rate } Q}_{\text{Volume per unit time}} = \text{cross-sectional area} \times \text{velocity} = \pi R^2 v \text{ cm}^3/\text{min}$$

How do we obtain this formula? Fix an observation point P in the tube and ask the following question: Which liquid particles flow past P in a 1-min interval? The particles passing P during this minute are located at most v centimeters to the left of P since each particle travels v centimeters per minute (assuming the liquid flows from left to right).

Therefore, the column of liquid flowing past P in a 1-min interval is a cylinder of radius R and length v, which has volume $\pi R^2 v$ (Figure 8).

FIGURE 8 The column of fluid flowing past P in one unit of time is a cylinder of volume $\pi R^2 v$.

In reality, the particles of liquid do not all travel at the same velocity because of friction. However, for a slowly moving liquid, the flow is **laminar**, by which we mean that the velocity $v(r)$ depends only on the distance r from the center of the tube. The particles traveling along the center of the tube travel most quickly and the velocity tapers off to zero near the walls of the tube (Figure 9).

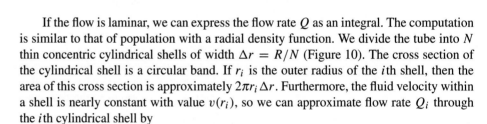

FIGURE 9 Laminar flow: Velocity of liquid increases toward the center of the tube.

If the flow is laminar, we can express the flow rate Q as an integral. The computation is similar to that of population with a radial density function. We divide the tube into N thin concentric cylindrical shells of width $\Delta r = R/N$ (Figure 10). The cross section of the cylindrical shell is a circular band. If r_i is the outer radius of the ith shell, then the area of this cross section is approximately $2\pi r_i \Delta r$. Furthermore, the fluid velocity within a shell is nearly constant with value $v(r_i)$, so we can approximate flow rate Q_i through the ith cylindrical shell by

$$Q_i \approx \text{cross-sectional area} \times \text{velocity} \approx 2\pi r_i \Delta r \, v(r_i)$$

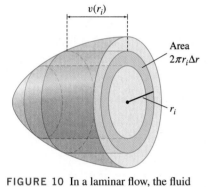

FIGURE 10 In a laminar flow, the fluid particles in a thin cylindrical shell all travel at nearly the same velocity.

We obtain

$$Q \approx \sum_{i=1}^{N} Q_i = 2\pi \sum_{i=1}^{N} r_i \, v(r_i) \Delta r$$

The sum on the right is a right-endpoint approximation to the integral $2\pi \int_0^R r v(r) \, dr$. Once again, we let N tend to ∞ to obtain the formula

$$\boxed{\text{Flow rate } Q = 2\pi \int_0^R r v(r) \, dr} \qquad \boxed{4}$$

The French physician Jean Poiseuille (1799–1869) discovered the law of laminar flow that cardiologists use to study blood flow in humans. Poiseuille's Law highlights the danger of cholesterol buildup in blood vessels: The flow rate through a blood vessel of radius R is proportional to R^4, so if R is reduced by one-half, the flow is reduced by a factor of 16.

■ **EXAMPLE 6** Poiseuille's Law of Laminar Flow   According to Poiseuille's Law, the velocity of blood flowing in a blood vessel of radius R cm is $v(r) = k(R^2 - r^2)$, where r is the distance from the center of the vessel (in centimeters) and k is a constant. Calculate the flow rate Q as function of R, assuming that $k = 0.5$ (cm-s)$^{-1}$.

Solution By Eq. (4),

$$Q = 2\pi \int_0^R (0.5) r (R^2 - r^2) \, dr = \pi \left(R^2 \frac{r^2}{2} - \frac{r^4}{4} \right) \Bigg|_0^R = \frac{\pi}{4} R^4 \text{ cm}^3/\text{s}$$

This shows that the flow rate is proportional to R^4 (this is true for any value of k). ■

Average Value

As a final example, we discuss the *average value* of a function. Recall that the average of N numbers $a_1, a_2, \ldots, a_N$ is the sum divided by N:

$$\frac{a_1 + a_2 + \cdots + a_N}{N} = \frac{1}{N} \sum_{j=1}^{N} a_j$$

For example, the average of 18, 25, 22, and 31, is $\frac{1}{4}(18 + 25 + 22 + 31) = 24$.

We cannot define the average value of a function $f(x)$ on an interval $[a, b]$ as a sum because there are infinitely many values of x to consider. However, the right-endpoint approximation may be interpreted as an average value (Figure 11):

$$R_N = \frac{b-a}{N} \big(f(x_1) + f(x_2) + \cdots + f(x_N) \big)$$

where $x_i = a + i\left(\dfrac{b-a}{N}\right)$. Dividing by $(b-a)$, we obtain the average of the function values $f(x_i)$:

$$\frac{1}{b-a} R_N = \underbrace{\frac{f(x_1) + f(x_2) + \cdots + f(x_N)}{N}}_{\text{Average of the function values}}$$

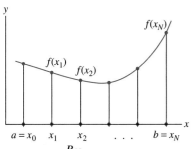

FIGURE 11 $\dfrac{R_N}{b-a}$ is the average of the values of $f(x)$ at the right endpoints.

If N is large, it is reasonable to think of this quantity as an *approximation* to the average of $f(x)$ on $[a, b]$. Therefore, we define the average value itself as the limit:

$$\text{Average value} = \lim_{N \to \infty} \frac{1}{b-a} R_N(f) = \frac{1}{b-a} \int_a^b f(x)\, dx$$

The average value is also called the **mean value**.

DEFINITION Average Value   The **average value** of an integrable function $f(x)$ on $[a, b]$ is the quantity

$$\text{Average value} = \frac{1}{b-a} \int_a^b f(x)\, dx$$

■ **EXAMPLE 7**   Find the average value of $f(x) = \sin x$ on **(a)** $[0, \pi]$ and **(b)** $[0, 2\pi]$.

Solution

(a) The average value of $\sin x$ on $[0, \pi]$ is

$$\frac{1}{\pi} \int_0^\pi \sin x\, dx = -\frac{1}{\pi} \cos x \Big|_0^\pi = \frac{1}{\pi}\big(-(-1) - (-1)\big) = \frac{2}{\pi} \approx 0.637$$

This answer is reasonable because $\sin x$ varies from 0 to 1 on the interval $[0, \pi]$ and the average 0.637 lies somewhere between the two extremes (Figure 12).

(b) The average value over $[0, 2\pi]$ is

$$\frac{1}{2\pi} \int_0^{2\pi} \sin x\, dx = -\frac{1}{2\pi} \cos x \Big|_0^{2\pi} = -\frac{1}{2\pi}\big(1 - (1)\big) = 0$$

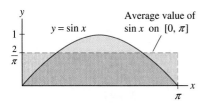

FIGURE 12 The area under the graph is equal to the area of the rectangle whose height is the average value.

This answer is also reasonable: The positive and negative values of $\sin x$ on $[0, 2\pi]$ cancel each other out, yielding an average of zero. ∎

GRAPHICAL INSIGHT   The average value M of a function $f(x)$ on $[a, b]$ is the average height of its graph over $[a, b]$. Furthermore, the signed area of the rectangle of height M over $[a, b]$, $M(b-a)$, is equal to $\int_a^b f(x)\,dx$ (Figure 15).

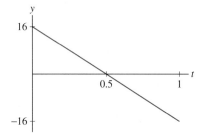

FIGURE 13 The frog's velocity is $h'(t) = 16 - 32t$.

■ **EXAMPLE 8**   A frog hops straight up from the ground with an initial velocity of $v_0 = 16$ ft/s. Use Galileo's formula for the height $h(t) = v_0 t - 16t^2$ to find the frog's average speed.

Solution   The first step is to determine the time interval of the jump. Since $v_0 = 16$, the frog's height at time t is $h(t) = 16t - 16t^2 = 16t(1 - t)$. We have $h(0) = h(1) = 0$. Therefore, the jump begins at $t = 0$ s and ends at $t = 1$ s.

The frog's velocity is $h'(t) = 16 - 32t$ (Figure 13). The frog's average speed is the average of $|h'(t)| = |16 - 32t|$, which we evaluate as the sum of two integrals since $h'(t)$ is positive on $[0, \frac{1}{2}]$ and negative on $[\frac{1}{2}, 1]$ (Figure 14):

$$\frac{1}{1-0} \int_0^1 |16 - 32t|\,dt = \int_0^{0.5} (16 - 32t)\,dt + \int_{0.5}^1 (32t - 16)\,dt$$

$$= (16t - 16t^2)\big|_0^{0.5} + (16t^2 - 16t)\big|_{0.5}^1$$

$$= 4 + 4 = 8 \text{ ft/s}$$ ∎

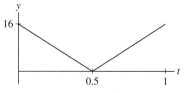

FIGURE 14 The frog's speed is $|h'(t)| = |16 - 32t|$.

Our next result, the Mean Value Theorem (MVT) for Integrals, highlights an important difference between the average of a list of numbers and the average value of a continuous function. If the average score on an exam is 84, then 84 lies between the highest and lowest scores, but it is possible that no student received a score of 84. By contrast, the MVT for Integrals asserts that a continuous function always takes on its average value at some point in the interval (Figure 15).

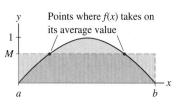

FIGURE 15 The function $f(x)$ takes on its average value M at the points where the upper edge of the rectangle intersects the graph.

THEOREM 1   Mean Value Theorem for Integrals   If $f(x)$ is continuous on $[a, b]$, then there exists a value $c \in [a, b]$ such that

$$f(c) = \frac{1}{b-a} \int_a^b f(x)\,dx$$

Proof   Because $f(x)$ is continuous and $[a, b]$ is closed, there exist points $c_{\min}$ and $c_{\max}$ in $[a, b]$ such that $f(c_{\min})$ and $f(c_{\max})$ are the minimum and maximum values of $f(x)$ on $[a, b]$. Thus, $f(c_{\min}) \le f(x) \le f(c_{\max})$ for all $x \in [a, b]$ and therefore,

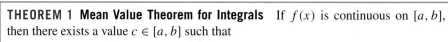

$$\int_a^b f(c_{\min})\,dx \le \int_a^b f(x)\,dx \le \int_a^b f(c_{\max})\,dx$$

$$f(c_{\min})(b-a) \le \int_a^b f(x)\,dx \le f(c_{\max})(b-a)$$

Notice how the proof of Theorem 1 uses important parts of the theory we have developed so far: the existence of extreme values on a closed interval, the IVT, and the basic property that if $f(x) \leq g(x)$, then

$$\int_a^b f(x)\,dx \leq \int_a^b g(x)\,dx$$

Now divide by $(b - a)$:

$$f(c_{\min}) \leq \underbrace{\frac{1}{b-a}\int_a^b f(x)\,dx}_{\text{Average value } M} \leq f(c_{\max}) \qquad \boxed{5}$$

The expression in the middle is the average value M of $f(x)$ on $[a, b]$. We see that M lies between the min and max of $f(x)$ on $[a, b]$. Because $f(x)$ is continuous, the Intermediate Value Theorem (IVT) guarantees that $f(c) = M$ for some $c \in [a, b]$ as claimed. ∎

6.2 SUMMARY

- The volume V of a solid body is equal to the integral of the area of the horizontal (or vertical) cross sections $A(y)$:

$$V = \int_a^b A(y)\,dy$$

- *Linear mass density* $\rho(x)$ is defined as mass per unit length. If a rod (or other object) with density ρ extends from $x = a$ to $x = b$, then its total mass is $M = \int_a^b \rho(x)\,dx$.

- If the density function $\rho(r)$ depends only on the distance r from the origin (*radial density function*), then the total amount (of population, mass, etc.) within a radius R of the center is equal to $2\pi\int_0^R r\rho(r)\,dr$.

- The volume of fluid passing through a tube of radius R per unit time is called the *flow rate* Q. The flow is *laminar* if the velocity $v(r)$ of a fluid particle depends only on its distance r from the center of the tube. For a laminar flow, $Q = 2\pi\int_0^R rv(r)\,dr$.

- The *average* (or *mean*) value on $[a, b]$: $M = \dfrac{1}{b-a}\int_a^b f(x)\,dx$.

- The MVT for Integrals: If $f(x)$ is continuous on $[a, b]$ with average value M, then $f(c) = M$ for some $c \in [a, b]$.

6.2 EXERCISES

Preliminary Questions

1. What is the average value of $f(x)$ on $[1, 4]$ if the area between the graph of $f(x)$ and the x-axis is equal to 9?

2. Find the volume of a solid extending from $y = 2$ to $y = 5$ if the cross section at y has area $A(y) = 5$ for all y.

3. Describe the horizontal cross sections of an ice cream cone and the vertical cross sections of a football (when it is held horizontally).

4. What is the formula for the total population within a circle of radius R around a city center if the population has a radial function?

5. What is the definition of flow rate?

6. Which assumption about fluid velocity did we use to compute the flow rate as an integral?

Exercises

1. Let V be the volume of a pyramid of height 20 whose base is a square of side 8.

(a) Use similar triangles as in Example 1 to find the area of the horizontal cross section at a height y.

(b) Calculate V by integrating the cross-sectional area.

2. Let V be the volume of a right circular cone of height 10 whose base is a circle of radius 4 (Figure 16).

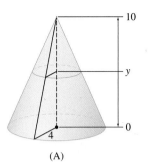
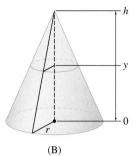

FIGURE 16 Right circular cones.

(a) Use similar triangles to find the area of a horizontal cross section at a height y.

(b) Calculate V by integrating the cross-sectional area.

3. Use the method of Exercise 2 to find the formula for the volume of a right circular cone of height h whose base is a circle of radius r (Figure 16).

4. Calculate the volume of the ramp in Figure 17 in three ways by integrating the area of the cross sections:

(a) Perpendicular to the x-axis (rectangles)

(b) Perpendicular to the y-axis (triangles)

(c) Perpendicular to the z-axis (rectangles)

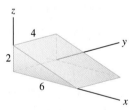

FIGURE 17 Ramp of length 6, width 4, and height 2.

5. Find the volume of liquid needed to fill a sphere of radius R to height h (Figure 18).

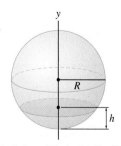

FIGURE 18 Sphere filled with liquid to height h.

6. Find the volume of the wedge in Figure 19(A) by integrating the area of vertical cross sections.

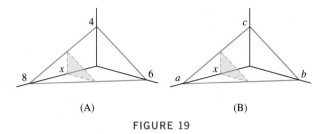

FIGURE 19

7. Derive a formula for the volume of the wedge in Figure 19(B) in terms of the constants a, b, and c.

8. Let B be the solid whose base is the unit circle $x^2 + y^2 = 1$ and whose vertical cross sections perpendicular to the x-axis are equilateral triangles. Show that the vertical cross sections have area $A(x) = \sqrt{3}(1 - x^2)$ and compute the volume of B.

In Exercises 9–14, find the volume of the solid with given base and cross sections.

9. The base is the unit circle $x^2 + y^2 = 1$ and the cross sections perpendicular to the x-axis are triangles whose height and base are equal.

10. The base is the triangle enclosed by $x + y = 1$, the x-axis, and the y-axis. The cross sections perpendicular to the y-axis are semicircles.

11. The base is the semicircle $y = \sqrt{9 - x^2}$, where $-3 \le x \le 3$. The cross sections perpendicular to the x-axis are squares.

12. The base is a square, one of whose sides is the interval $[0, \ell]$ along the x-axis. The cross sections perpendicular to the x-axis are rectangles of height $f(x) = x^2$.

13. The base is the region enclosed by $y = x^2$ and $y = 3$. The cross sections perpendicular to the y-axis are squares.

14. The base is the region enclosed by $y = x^2$ and $y = 3$. The cross sections perpendicular to the y-axis are rectangles of height y^3.

15. Find the volume of the solid whose base is the region $|x| + |y| \le 1$ and whose vertical cross sections perpendicular to the y-axis are semi-circles (with diameter along the base).

16. Show that the volume of a pyramid of height h whose base is an equilateral triangle of side s is equal to $\dfrac{\sqrt{3}}{12}hs^2$.

17. Find the volume V of a *regular* tetrahedron whose face is an equilateral triangle of side s (Figure 20).

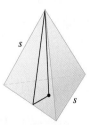

FIGURE 20 Regular tetrahedron.

18. The area of an ellipse is πab, where a and b are the lengths of the semimajor and semiminor axes (Figure 21). Compute the volume of a cone of height 12 whose base is an ellipse with semimajor axis $a = 6$ and semiminor axis $b = 4$.

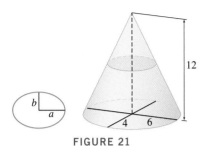

FIGURE 21

19. A frustum of a pyramid is a pyramid with its top cut off [Figure 22(A)]. Let V be the volume of a frustum of height h whose base is a square of side a and top is a square of side b with $a > b \geq 0$.

(a) Show that if the frustum were continued to a full pyramid, it would have height $\dfrac{ha}{a - b}$ [Figure 22(B)].

(b) Show that the cross section at height x is a square of side $(1/h)(a(h - x) + bx)$.

(c) Show that $V = \frac{1}{3}h(a^2 + ab + b^2)$. A papyrus dating to the year 1850 BCE indicates that Egyptian mathematicians had discovered this formula almost 4,000 years ago.

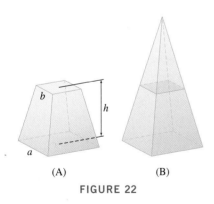

(A)   (B)

FIGURE 22

20. A plane inclined at an angle of 45° passes through a diameter of the base of a cylinder of radius r. Find the volume of the region within the cylinder and below the plane (Figure 23).

FIGURE 23

21. Figure 24 shows the solid S obtained by intersecting two cylinders of radius r whose axes are perpendicular.

(a) The horizontal cross section of each cylinder at distance y from the central axis is a rectangular strip. Find the strip's width.

(b) Find the area of the horizontal cross section of S at distance y.

(c) Find the volume of S as a function of r.

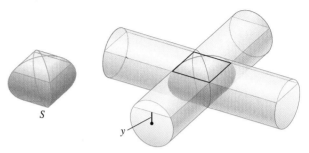

FIGURE 24 Intersection of two cylinders intersecting at right angles.

22. Let S be the solid obtained by intersecting two cylinders of radius r whose axes intersect at an angle θ. Find the volume of S as a function of r and θ.

23. Calculate the volume of a cylinder inclined at an angle $\theta = 30°$ whose height is 10 and whose base is a circle of radius 4 (Figure 25).

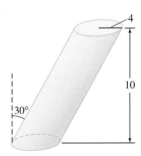

FIGURE 25 Cylinder inclined at an angle $\theta = 30°$.

24. Find the total mass of a 1-m rod whose linear density function is $\rho(x) = 10(x + 1)^{-2}$ kg/m for $0 \leq x \leq 1$.

25. Find the total mass of a 2-m rod whose linear density function is $\rho(x) = 1 + 0.5\sin(\pi x)$ kg/m for $0 \leq x \leq 2$.

26. A mineral deposit along a strip of length 6 cm has density $s(x) = 0.01x(6 - x)$ g/cm for $0 \leq x \leq 6$. Calculate the total mass of the deposit.

27. Calculate the population within a 10-mile radius of the city center if the radial population density is $\rho(r) = 4(1 + r^2)^{1/3}$ (in thousands per square mile).

28. Odzala National Park in the Congo has a high density of gorillas. Suppose that the radial population density is $\rho(r) = 52(1 + r^2)^{-2}$ gorillas per square kilometer, where r is the distance from a large grassy clearing with a source of food and water. Calculate the number of gorillas within a 5-km radius of the clearing.

29. Table 1 lists the population density (in people per squared kilometer) as a function of distance r (in kilometers) from the center of a rural town. Estimate the total population within a 2-km radius of the center by taking the average of the left- and right-endpoint approximations.

TABLE 1 Population Density

r	$\rho(r)$	r	$\rho(r)$
0.0	125.0	1.2	37.6
0.2	102.3	1.4	30.8
0.4	83.8	1.6	25.2
0.6	68.6	1.8	20.7
0.8	56.2	2.0	16.9
1.0	46.0		

30. Find the total mass of a circular plate of radius 20 cm whose mass density is the radial function $\rho(r) = 0.03 + 0.01\cos(\pi r^2)$ g/cm^2.

31. The density of deer in a forest is the radial function $\rho(r) = 150(r^2 + 2)^{-2}$ deer per km^2, where r is the distance (in kilometers) to a small meadow. Calculate the number of deer in the region $2 \le r \le 5$ km.

32. Show that a circular plate of radius 2 cm with radial mass density $\rho(r) = \dfrac{4}{r}$ g/cm has finite total mass, even though the density becomes infinite at the origin.

33. Find the flow rate through a tube of radius 4 cm, assuming that the velocity of fluid particles at a distance r cm from the center is $v(r) = 16 - r^2$ cm/s.

34. Let $v(r)$ be the velocity of blood in an arterial capillary of radius $R = 4 \times 10^{-5}$ m. Use Poiseuille's Law (Example 6) with $k = 10^6$ (m·s)$^{-1}$ to determine the velocity at the center of the capillary and the flow rate (use correct units).

35. A solid rod of radius 1 cm is placed in a pipe of radius 3 cm so that their axes are aligned. Water flows through the pipe and around the rod. Find the flow rate if the velocity of the water is given by the radial function $v(r) = 0.5(r - 1)(3 - r)$ cm/s.

36. To estimate the volume V of Lake Nogebow, the Minnesota Bureau of Fisheries created the depth contour map in Figure 26 and determined the area of the cross section of the lake at the depths recorded in the table below. Estimate V by taking the average of the right- and left-endpoint approximations to the integral of cross-sectional area.

Depth (ft)	0	5	10	15	20
Area (million ft^2)	2.1	1.5	1.1	0.835	0.217

In Exercises 37–46, calculate the average over the given interval.

37. $f(x) = x^3$, $[0, 1]$

38. $f(x) = x^3$, $[-1, 1]$

39. $f(x) = \cos x$, $[0, \frac{\pi}{2}]$

40. $f(x) = \sec^2 x$, $[0, \frac{\pi}{4}]$

41. $f(s) = s^{-2}$, $[2, 5]$

42. $f(x) = \dfrac{\sin(\pi/x)}{x^2}$, $[1, 2]$

43. $f(x) = 2x^3 - 3x^2$, $[-1, 3]$

44. $f(x) = x^n$, $[0, 1]$

45. $f(x) = \dfrac{1}{x^2 + 1}$, $[-1, 1]$

46. $f(x) = e^{-nx}$, $[-1, 1]$

47. Let M be the average value of $f(x) = x^3$ on $[0, A]$, where $A > 0$. Which theorem guarantees that $f(c) = M$ has a solution c in $[0, A]$? Find c.

48. *CAS* Let $f(x) = 2\sin x - x$. Use a computer algebra system to plot $f(x)$ and estimate:

(a) The positive root α of $f(x)$.

(b) The average value M of $f(x)$ on $[0, \alpha]$.

(c) A value $c \in [0, \alpha]$ such that $f(c) = M$.

49. Which of $f(x) = x\sin^2 x$ and $g(x) = x^2\sin^2 x$ has a larger average value over $[0, 1]$? Over $[1, 2]$?

50. Show that the average value of $f(x) = \dfrac{\sin x}{x}$ over $[\frac{\pi}{2}, \pi]$ is less than 0.41. Sketch the graph if necessary.

51. ✎ Sketch the graph of a function $f(x)$ such that $f(x) \ge 0$ on $[0, 1]$ and $f(x) \le 0$ on $[1, 2]$, whose average on $[0, 2]$ is negative.

52. Find the average of $f(x) = ax + b$ over the interval $[-M, M]$, where a, b, and M are arbitrary constants.

53. The temperature $T(t)$ at time t (in hours) in an art museum varies according to $T(t) = 70 + 5\cos\left(\dfrac{\pi}{12}t\right)$. Find the average over the time periods $[0, 24]$ and $[2, 6]$.

54. A ball is thrown in the air vertically from ground level with initial velocity 64 ft/s. Find the average height of the ball over the time interval extending from the time of the ball's release to its return to ground level. Recall that the height at time t is $h(t) = 64t - 16t^2$.

55. What is the average area of the circles whose radii vary from 0 to 1?

56. An object with zero initial velocity accelerates at a constant rate of 10 m/s^2. Find its average velocity during the first 15 s.

57. The acceleration of a particle is $a(t) = t - t^3$ m/s^2 for $0 \le t \le 1$. Compute the average acceleration and average velocity over the time interval $[0, 1]$, assuming that the particle's initial velocity is zero.

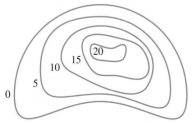

FIGURE 26 Depth contour map of Lake Nogebow.

58. Let M be the average value of $f(x) = x^4$ on $[0, 3]$. Find a value of c in $[0, 3]$ such that $f(c) = M$.

59. Let $f(x) = \sqrt{x}$. Find a value of c in $[4, 9]$ such that $f(c)$ is equal

to the average of f on $[4, 9]$.

60. Give an example of a function (necessarily discontinuous) that does not satisfy the conclusion of the MVT for Integrals.

Further Insights and Challenges

61. An object is tossed in the air vertically from ground level with initial velocity v_0 ft/s at time $t = 0$. Find the average speed of the object over the time interval $[0, T]$, where T is the time the object returns to earth.

62. Review the MVT stated in Section 4.3 (Theorem 1, p. 230) and show how it can be used, together with the Fundamental Theorem of Calculus, to prove the MVT for integrals.

6.3 Volumes of Revolution

We use the terms "revolve" and "rotate" interchangeably.

A **solid of revolution** is a solid obtained by rotating a region in the plane about an axis. The sphere and right circular cone are familiar examples of such solids. Each of these is "swept out" as a plane region revolves around an axis (Figure 1).

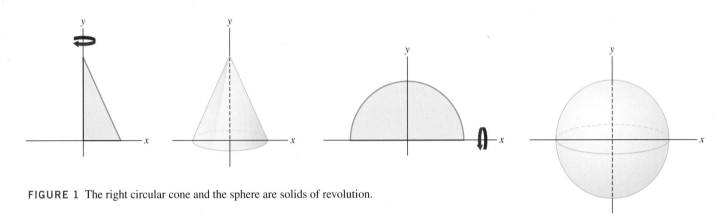

FIGURE 1 The right circular cone and the sphere are solids of revolution.

This method for computing the volume is often referred to as the "disk method" because the vertical slices of the solid are circular disks.

In general, if $f(x) \geq 0$ for $a \leq x \leq b$, then the region under the graph lies above the x-axis. Rotating this region around the x-axis produces a solid with a special feature: All vertical cross sections are circles (Figure 2). In fact, the vertical cross section at location x is a circle of radius $R = f(x)$ and has area

$$\text{Area of the vertical cross section} = \pi R^2 = \pi f(x)^2$$

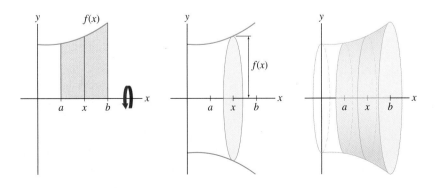

FIGURE 2 The cross section of a solid of revolution is a circle of radius $f(x)$ and area $\pi f(x)^2$.

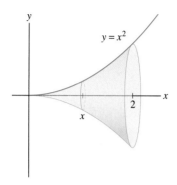

As we saw in Section 6.2, the total volume V is equal to the integral of cross-sectional area. Therefore, $V = \int_a^b \pi f(x)^2\,dx$.

Volume of a Solid of Revolution: Disk Method If $f(x)$ is continuous and $f(x) \geq 0$ on $[a, b]$, then the volume V obtained by rotating the region under the graph about the x-axis is [with $R = f(x)$]:

$$V = \pi \int_a^b R^2\,dx = \pi \int_a^b f(x)^2\,dx \qquad \boxed{1}$$

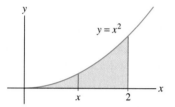

FIGURE 3 Region under $y = x^2$ rotated about the x-axis.

■ **EXAMPLE 1** Calculate the volume V of the solid obtained by rotating the region under $y = x^2$ about the x-axis for $0 \leq x \leq 2$.

Solution In this case, $f(x) = x^2$, and by Eq. (1), the volume is (Figure 3)

$$V = \pi \int_0^2 R^2\,dx = \pi \int_0^2 (x^2)^2\,dx = \pi \int_0^2 x^4\,dx = \pi \frac{x^5}{5}\bigg|_0^2 = \pi \frac{2^5}{5} = \frac{32}{5}\pi \qquad ■$$

We now consider some variations on the formula for a volume of revolution. First, suppose that the region *between* two curves $y = f(x)$ and $y = g(x)$, where $f(x) \geq g(x) \geq 0$, is rotated about the x-axis [Figure 5(C)]. Then the vertical cross section of the solid at x is generated as the segment $\overline{AB}$ in Figure 5 revolves around the x-axis. This cross section is a **washer** of outer radius $R = f(x)$ and inner radius $r = g(x)$ [Figure 5(B)]. The area of this washer is $\pi R^2 - \pi r^2$ or $\pi(f(x)^2 - g(x)^2)$, and we obtain the volume of the solid as the integral of the cross-sectional area:

$$V = \pi \int_a^b \left(R^2 - r^2\right)dx = \pi \int_a^b \left(f(x)^2 - g(x)^2\right)dx \qquad \boxed{2}$$

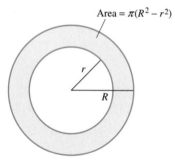

FIGURE 4 The region between two concentric circles is called an "annulus," or more informally, a "washer."

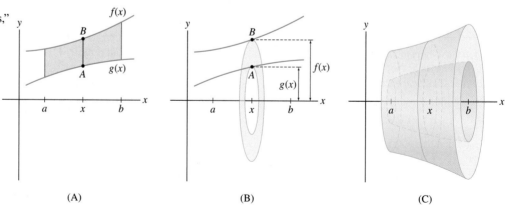

(A)   (B)   (C)

FIGURE 5 The vertical cross section is the washer generated when $\overline{AB}$ is rotated about the x-axis.

■ **EXAMPLE 2** Rotating the Area Between Two Curves Find the volume V of the solid obtained by rotating the region between $y = x^2 + 4$ and $y = 2$ about the x-axis for $1 \leq x \leq 3$.

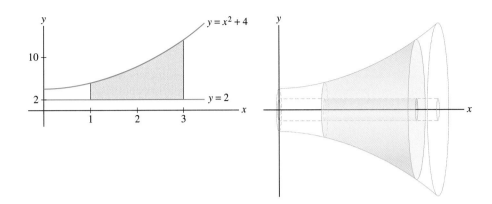

FIGURE 6 The area between $y = x^2 + 4$ and $y = 2$ over $[1, 3]$ rotated about the x-axis.

Solution The graph of $y = x^2 + 4$ lies above the graph of $y = 2$ (Figure 6). Therefore, the outer radius is $R = x^2 + 4$ and the inner radius is $r = 2$. By Eq. (2),

$$V = \pi \int_1^3 \left(R^2 - r^2\right) dx = \pi \int_1^3 \left((x^2 + 4)^2 - 2^2\right) dx$$

$$= \pi \int_1^3 \left(x^4 + 8x^2 + 12\right) dx = \pi \left(\frac{1}{5}x^5 + \frac{8}{3}x^3 + 12x\right)\Bigg|_1^3 = \frac{2,126}{15}\pi \quad \blacksquare$$

Equation (2) can be modified to compute volumes of revolution about horizontal lines parallel to the x-axis.

■ **EXAMPLE 3** Revolving About a Horizontal Axis Find the volume V of the "wedding band" in Figure 7(C), obtained by rotating the region between the graphs of $f(x) = x^2 + 2$ and $g(x) = 4 - x^2$ about the horizontal line $y = -3$.

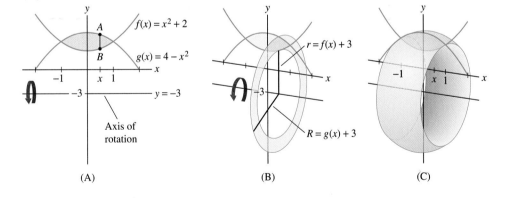

FIGURE 7 Segment $\overline{AB}$ generates a washer when it is rotated about the axis $y = -3$.

(A)   (B)   (C)

Solution We first determine the points of intersection of the two graphs by solving

$$x^2 + 2 = 4 - x^2 \quad \text{or} \quad x^2 = 1$$

The graphs intersect at $x = \pm 1$. Figure 7(A) shows that the graph of $g(x) = 4 - x^2$ lies above the graph of $f(x) = x^2 + 2$ on $[-1, 1]$. *If we revolved the region between the graphs about the x-axis,* the volume would be given by Eq. (2):

$$V \text{ (about } x\text{-axis)} = \pi \int_{-1}^{1} \left(g(x)^2 - f(x)^2\right) dx = \pi \int_{-1}^{1} \left((4 - x^2)^2 - (x^2 + 2)^2\right) dx$$

When you set up the integral for a volume of revolution, visualize the cross sections. These cross sections are washers (or disks) whose inner and outer radii depend on the axis of rotation.

The formula is similar when we rotate about a horizontal line other than the x-axis, but we must use the appropriate outer and inner radii. When we rotate about $y = -3$, $\overline{AB}$ generates a washer whose outer and inner radii are both three units longer [Figure 7(B)]:

- Outer radius extends from $y = -3$ to $y = g(x)$, so $R = g(x) + 3 = 7 - x^2$.
- Inner radius extends from $y = -3$ to $y = f(x)$, so $r = f(x) + 3 = x^2 + 5$.

The volume of revolution is obtained by integrating the area of this washer:

$$V \text{ (about } y = -3) = \pi \int_{-1}^{1} (R^2 - r^2) \, dx = \pi \int_{-1}^{1} \left((g(x) + 3)^2 - (f(x) + 3)^2 \right) dx$$

$$= \pi \int_{-1}^{1} \left((7 - x^2)^2 - (x^2 + 5)^2 \right) dx$$

$$= \pi \int_{-1}^{1} \left((49 - 14x^2 + x^4) - (25 + 10x^2 + x^4) \right) dx$$

$$= \pi \int_{-1}^{1} (24 - 24x^2) \, dx = \pi (24x - 8x^3) \Big|_{-1}^{1} = 32\pi \qquad \blacksquare$$

■ **EXAMPLE 4** Find the volume of the solid obtained by rotating the region between the graph of $f(x) = 9 - x^2$ and the line $y = 12$ for $0 \le x \le 3$ about **(a)** the line $y = 12$ and **(b)** the line $y = 15$.

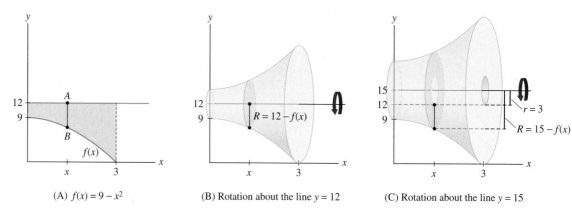

(A) $f(x) = 9 - x^2$ (B) Rotation about the line $y = 12$ (C) Rotation about the line $y = 15$

FIGURE 8 Segment $\overline{AB}$ generates a disk when rotated about $y = 12$, but it generates a washer when rotated about $y = 15$.

Solution To set up the integrals, we must first visualize whether the cross section is a disk or washer.

In Figure 8, the length of $\overline{AB}$ is $12 - f(x)$ rather than $f(x) - 12$ because the line $y = 12$ lies above the graph of $f(x)$.

(a) Figure 8(B) shows that when $\overline{AB}$ is rotated about $y = 12$, it generates a disk of radius

$$R = \text{length of } \overline{AB} = 12 - f(x) = 12 - (9 - x^2) = 3 + x^2$$

The volume of the solid of revolution about $y = 12$ is

$$\pi \int_{0}^{3} R^2 \, dx = \pi \int_{0}^{3} (3 + x^2)^2 \, dx = \pi \int_{0}^{3} (9 + 6x^2 + x^4) \, dx$$

$$= \pi \left(9x + 2x^3 + \frac{1}{5}x^5 \right) \Big|_{0}^{3} = \frac{648}{5} \pi$$

(b) Figure 8(C) shows that $\overline{AB}$ generates a washer when rotated about $y = 15$. The outer radius R of this washer is the distance from B to the line $y = 15$:

$$R = 15 - f(x) = 15 - (9 - x^2) = 6 + x^2$$

The inner radius is $r = 3$, so the volume of revolution about $y = 15$ is

$$\pi \int_0^3 (R^2 - r^2)\,dx = \pi \int_0^3 \left((6 + x^2)^2 - 3^2\right)dx = \pi \int_0^3 \left(36 + 12x^2 + x^4 - 9\right)dx$$

$$= \pi \left(27x + 4x^3 + \frac{1}{5}x^5\right)\Big|_0^3 = \frac{1{,}188}{5}\pi \qquad \blacksquare$$

To use the disk and washer methods for solids of revolution about vertical axes, the graph must be described as a function of y rather than x.

■ **EXAMPLE 5** Revolving About a Vertical Axis   Find the volume of the solid obtained by rotating the region under the graph of $f(x) = 9 - x^2$ for $0 \le x \le 3$ about the vertical axis $x = -2$.

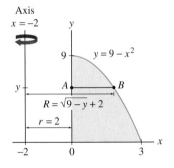

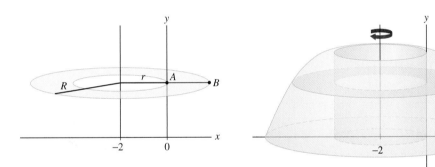

FIGURE 9

Solution In Figure 9, the horizontal segment $\overline{AB}$ at height y sweeps out a horizontal washer when it is rotated about the vertical line $x = -2$. To find the area of this washer as a function of y, we first solve for x in $y = 9 - x^2$ to obtain $x^2 = 9 - y$ or $x = \sqrt{9 - y}$. Thus

$$\text{Outer radius } R = \sqrt{9 - y} + 2, \qquad \text{inner radius } r = 2$$

Since the region extends from $y = 0$ to $y = 9$ along the y-axis, the volume is obtained by integrating from $y = 0$ to $y = 9$:

$$\pi \int_0^9 (R^2 - r^2)\,dy = \pi \int_0^9 \left((\sqrt{9 - y} + 2)^2 - 2^2\right)dy$$

$$= \pi \int_0^9 \left(9 - y + 4\sqrt{9 - y}\right)dy$$

$$= \pi \left(9y - \frac{1}{2}y^2 - \frac{8}{3}(9 - y)^{3/2}\right)\Big|_0^9 = \frac{225}{2}\pi \qquad \blacksquare$$

6.3 SUMMARY

- *Disk method:* When we rotate the region under the graph of $f(x)$ about the x-axis for $a \le x \le b$, we obtain a solid whose vertical cross section is a circle of radius $R = f(x)$ and area $\pi R^2 = \pi f(x)^2$. The volume V of the solid is

$$V = \pi \int_a^b R^2 \, dx = \pi \int_a^b f(x)^2 \, dx$$

- *Washer method:* Assume that $f(x) \ge g(x) \ge 0$ for $a \le x \le b$. When we rotate the region between the graphs of $f(x)$ and $g(x)$ about the x-axis, we obtain a solid whose vertical cross section is a washer of outer radius $R = f(x)$ and inner radius $r = g(x)$. The volume V of the solid is

$$V = \pi \int_a^b (R^2 - r^2) \, dx = \pi \int_a^b \left(f(x)^2 - g(x)^2 \right) dx$$

- These formulas apply when we rotate about a horizontal line $y = c$, but the radii R and r must be modified appropriately.

6.3 EXERCISES

Preliminary Questions

1. Which of the following is a solid of revolution?
(a) Sphere   **(b)** Pyramid   **(c)** Cylinder   **(d)** Cube

2. True or false? When a solid is formed by rotating the region under a graph about the x-axis, the cross sections perpendicular to the x-axis are circular disks.

3. True or false? When a solid is formed by rotating the region between two graphs about the x-axis, the cross sections perpendicular to the x-axis are circular disks.

4. Which of the following integrals expresses the volume of the solid obtained by rotating the area between $y = f(x)$ and $y = g(x)$ over $[a, b]$ around the x-axis [assume $f(x) \ge g(x) \ge 0$]?

(a) $\pi \int_a^b \left(f(x) - g(x) \right)^2 dx$

(b) $\pi \int_a^b \left(f(x)^2 - g(x)^2 \right) dx$

Exercises

In Exercises 1–4, (a) sketch the solid obtained by revolving the region under the graph of $f(x)$ about the x-axis over the given interval, (b) describe the cross section perpendicular to the x-axis located at x, and (c) calculate the volume of the solid.

1. $f(x) = x + 1$, $[0, 3]$ **2.** $f(x) = x^2$, $[1, 3]$

3. $f(x) = \sqrt{x + 1}$, $[1, 4]$ **4.** $f(x) = x^{-1}$, $[1, 2]$

In Exercises 5–12, find the volume of the solid obtained by rotating the region under the graph of the function about the x-axis over the given interval.

5. $f(x) = x^2 - 3x$, $[0, 3]$ **6.** $f(x) = \dfrac{1}{x^2}$, $[1, 4]$

7. $f(x) = x^{5/3}$, $[1, 8]$ **8.** $f(x) = 4 - x^2$, $[0, 2]$

9. $f(x) = \dfrac{2}{x + 1}$, $[1, 3]$

10. $f(x) = \sqrt{x^4 + 1}$, $[1, 3]$

11. $f(x) = e^x$, $[0, 1]$

12. $f(x) = \sqrt{\cos x \sin x}$, $[0, \pi/2]$

In Exercises 13–18, (a) sketch the region enclosed by the curves, (b) describe the cross section perpendicular to the x-axis located at x, and (c) find the volume of the solid obtained by rotating the region about the x-axis.

13. $y = x^2 + 2$, $y = 10 - x^2$

14. $y = x^2$, $y = 2x + 3$

15. $y = 16 - x$, $y = 3x + 12$, $x = -1$

16. $y = \dfrac{1}{x}$, $y = \dfrac{5}{2} - x$

17. $y = \sec x$, $y = 0$, $x = -\dfrac{\pi}{4}$, $x = \dfrac{\pi}{4}$

18. $y = \sec x$, $y = \csc x$, $y = 0$, $x = 0$,   and $x = \dfrac{\pi}{2}$.

In Exercises 19–22, find the volume of the solid obtained by rotating the region enclosed by the graphs about the y-axis over the given interval.

19. $x = \sqrt{y}$, $x = 0$; $1 \le y \le 4$

20. $x = \sqrt{\sin y}$, $x = 0$; $0 \le y \le \pi$

21. $x = y^2$, $x = \sqrt{y}$; $0 \le y \le 1$

22. $x = 4 - y$, $x = 16 - y^2$; $-3 \le y \le 4$

In Exercises 23–28, find the volume of the solid obtained by rotating region A in Figure 10 about the given axis.

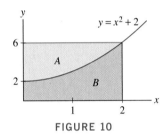

FIGURE 10

23. x-axis

24. $y = -2$

25. $y = 2$

26. y-axis

27. $x = -3$

28. $x = 2$

In Exercises 29–34, find the volume of the solid obtained by rotating region B in Figure 10 about the given axis.

29. x-axis

30. $y = -2$

31. $y = 6$

32. y-axis

Hint for Exercise 32: Express the volume as a sum of two integrals along the y-axis, or use Exercise 26.

33. $x = 2$

34. $x = -3$

In Exercises 35–48, find the volume of the solid obtained by rotating the region enclosed by the graphs about the given axis.

35. $y = x^2$, $y = 12 - x$, $x = 0$,   about $y = -2$

36. $y = x^2$, $y = 12 - x$, $x = 0$,   about $y = 15$

37. $y = 16 - x$, $y = 3x + 12$, $x = 0$,   about y-axis

38. $y = 16 - x$, $y = 3x + 12$, $x = 0$,   about $x = 2$

39. $y = \dfrac{9}{x^2}$, $y = 10 - x^2$,   about x-axis

40. $y = \dfrac{9}{x^2}$, $y = 10 - x^2$,   about $y = 12$

41. $y = \dfrac{1}{x}$, $y = \dfrac{5}{2} - x$,   about y-axis

42. $x = 2$, $x = 3$, $y = 16 - x^4$, $y = 0$,   about y-axis

43. $y = x^3$, $y = x^{1/3}$,   about y-axis

44. $y = x^3$, $y = x^{1/3}$,   about $x = -2$

45. $y = e^{-x}$, $y = 1 - e^{-x}$, $x = 0$,   about $y = 4$

46. $y = \cosh x$, $x = \pm 2$,   about x-axis

47. $y^2 = 4x$, $y = x$, $y = 0$,   about x-axis

48. $y^2 = 4x$, $y = x$,   about $y = 8$

49. ⎣GU⎦   Sketch the hypocycloid $x^{2/3} + y^{2/3} = 1$ and find the volume of the solid obtained by revolving it about the x-axis.

50. The solid generated by rotating the region between the branches of the hyperbola $y^2 - x^2 = 1$ about the x-axis is called a **hyperboloid** (Figure 11). Find the volume of the hyperboloid for $-a \le x \le a$.

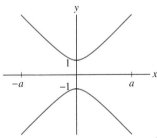

FIGURE 11 The hyperbola with equation $y^2 - x^2 = 1$.

51. A "bead" is formed by removing a cylinder of radius r from the center of a sphere of radius R (Figure 12). Find the volume of the bead with $r = 1$ and $R = 2$.

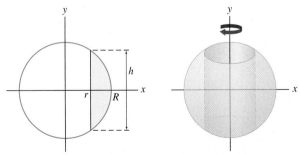

FIGURE 12 A bead is a sphere with a cylinder removed.

Further Insights and Challenges

52. Find the volume V of the bead (Figure 12) in terms of r and R. Then show that $V = \frac{\pi}{6} h^3$, where h is the height of the bead. This formula has a surprising consequence: Since V can be expressed in terms of h alone, it follows that two beads of height 2 in., one formed from a sphere the size of an orange and the other the size of the earth would have the same volume! Can you explain intuitively how this is possible?

53. The solid generated by rotating the region inside the ellipse with equation $\left(\frac{x}{a}\right)^2 + \left(\frac{y}{b}\right)^2 = 1$ around the x-axis is called an **ellipsoid**. Show that the ellipsoid has volume $\frac{4}{3}\pi a b^2$. What is the volume if the ellipse is rotated around the y-axis?

54. A doughnut-shaped solid is called a **torus** (Figure 13). Use the washer method to calculate the volume of the torus obtained by rotating the region inside the circle with equation $(x - a)^2 + y^2 = b^2$ around the y-axis (assume that $a > b$). *Hint:* Evaluate the integral by interpreting it as the area of a circle.

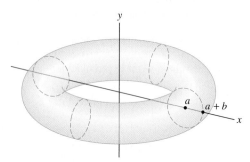

FIGURE 13 Torus obtained by rotating a circle about the y-axis.

55. The curve $y = f(x)$ in Figure 14, called a **tractrix**, has the following property: the tangent line at each point (x, y) on the curve has slope

$$\frac{dy}{dx} = \frac{-y}{\sqrt{1 - y^2}}.$$

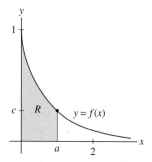

FIGURE 14 The tractrix.

Let R be the shaded region under the graph of $0 \leq x \leq a$ in Figure 14. Compute the volume V of the solid obtained by revolving R around the x-axis in terms of the constant $c = f(a)$. *Hint:* Use the disk method and the substitution $u = f(x)$ to show that

$$V = \pi \int_c^1 u\sqrt{1 - u^2}\, du$$

56. Verify the formula

$$\int_{x_1}^{x_2} (x - x_1)(x - x_2)\, dx = \frac{1}{6}(x_1 - x_2)^3 \qquad \boxed{3}$$

Then prove that the solid obtained by rotating the shaded region in Figure 15 about the x-axis has volume $V = \frac{\pi}{6} B H^2$, with B and H as in the figure. *Hint:* Let x_1 and x_2 be the roots of $f(x) = ax + b - (mx + c)^2$, where $x_1 < x_2$. Show that

$$V = \pi \int_{x_1}^{x_2} f(x)\, dx$$

and use Eq. (3).

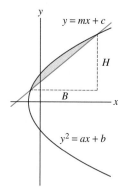

FIGURE 15 The line $y = mx + c$ intersects the parabola $y^2 = ax + b$ at two points above the x-axis.

57. Let R be the region in the unit circle lying above the cut with the line $y = mx + b$ (Figure 16). Assume the points where the line intersects the circle lie above the x-axis. Use the method of Exercise 56 to show that the solid obtained by rotating R about the x-axis has volume $V = \frac{\pi}{6} h d^2$, with h and d as in the figure.

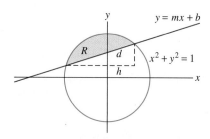

FIGURE 16

6.4 The Method of Cylindrical Shells

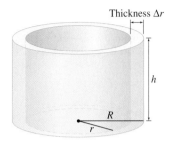

FIGURE 1 The volume of the cylindrical shell is *approximately* $2\pi R h \, \Delta r$, where $\Delta r = R - r$.

Observe that the outer circumference of the cylindrical shell is $2\pi R$. So Eq. (2) says that the volume is approximately equal to

$(2\pi R)(h)(\Delta r)$

$= (circumference) \times (height) \times (thickness)$

In the previous two sections, we computed volumes by integrating cross-sectional area. The Shell Method is based on a different idea and is more convenient in some cases.

In the **Shell Method**, we use cylindrical shells like the one in Figure 1 to approximate a volume of revolution. To this end, let us first derive an approximation to the volume of a cylindrical shell of height h, outer radius R, and inner radius r. Since the shell is obtained by removing a cylinder of radius r from the wider cylinder of radius R, its volume is equal to

$$\pi R^2 h - \pi r^2 h = \pi h (R^2 - r^2) = \pi h (R + r)(R - r) = \pi h (R + r) \Delta r \qquad \boxed{1}$$

where $\Delta r = R - r$ is the shell's thickness. If the shell is very thin, then R and r are nearly equal and we may replace $(R + r)$ by $2R$ in Eq. (1) to obtain the approximation

$$\text{Volume of shell} \approx 2\pi R h \, \Delta r \qquad \boxed{2}$$

Now consider a solid obtained by rotating the region under $y = f(x)$ from $x = a$ to $x = b$ around the y-axis as in Figure 2. The idea is to divide the solid into thin concentric shells. More precisely, we divide $[a, b]$ into N subintervals of length $\Delta x = \dfrac{b - a}{N}$ with endpoints $x_0, x_1, \ldots, x_N$. When we rotate the thin strip of area above $[x_{i-1}, x_i]$ about the y-axis, we obtain a thin shell whose volume we denote by V_i. The total volume V of the solid is equal to $V = \displaystyle\sum_{i=1}^{N} V_i$.

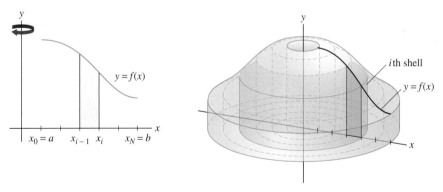

FIGURE 2 The shaded strip, when rotated about the y-axis, generates a "thin shell."

The top rim of the ith thin shell in Figure 2 is curved. However, when Δx is small, we may approximate this thin shell by the cylindrical shell (with a flat rim) of height $f(x_i)$ and use (2) to obtain

$$V_i \approx (\text{circumference})(\text{height})(\text{thickness}) = 2\pi x_i f(x_i) \, \Delta x$$

Therefore,

$$V = \sum_{i=1}^{N} V_i \approx \sum_{i=1}^{N} 2\pi x_i f(x_i) \, \Delta x$$

The sum on the right is the volume of the cylindrical approximation to V illustrated in Figure 3. We complete the argument in the usual way. As $N \to \infty$, the accuracy of the approximation improves, and the sum on the right is a right-endpoint approximation that converges to $V = \displaystyle\int_a^b 2\pi x f(x) \, dx$.

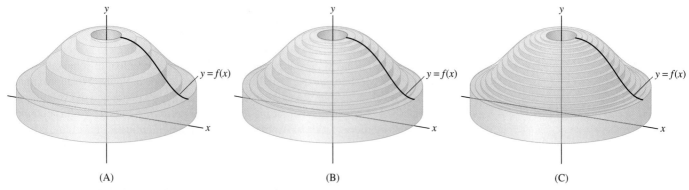

(A)   (B)   (C)

FIGURE 3 The volume is approximated by the sum of volumes of thin cylinders.

In the Shell Method, we integrate with respect to x even though we are rotating about the y-axis.

Volume of a Solid of Revolution: The Shell Method   The volume V of the solid obtained by rotating the region under the graph of $y = f(x)$ over the interval $[a, b]$ about the y-axis is equal to

$$V = 2\pi \int_a^b x f(x)\, dx \qquad \boxed{3}$$

It would be hard to use the disk method in Example 1. Since the axis of revolution is the y-axis, we would have to integrate with respect to y. This would require finding the inverse function $g(y) = f^{-1}(y)$.

■ **EXAMPLE 1**   Find the volume V of the solid obtained by rotating the area under the graph of $f(x) = 1 - 2x + 3x^2 - 2x^3$ over [0, 1] about the y-axis.

Solution   The solid is shown in Figure 4. By Eq. (3),

$$V = 2\pi \int_0^1 x f(x)\, dx = 2\pi \int_0^1 x(1 - 2x + 3x^2 - 2x^3)\, dx$$

$$= 2\pi \left(\frac{1}{2}x^2 - \frac{2}{3}x^3 + \frac{3}{4}x^4 - \frac{2}{5}x^5 \right) \bigg|_0^1 = \frac{11}{30}\pi \qquad ■$$

FIGURE 4 The graph of $f(x) = 1 - 2x + 3x^2 - 2x^3$ rotated about the y-axis.

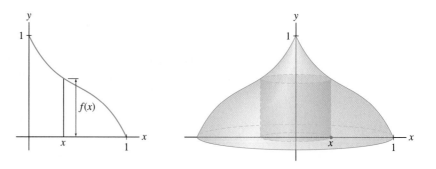

GRAPHICAL INSIGHT   When we rotate the vertical segment at location x about the y-axis, we obtain a cylinder of radius x and height $f(x)$. Thus, we have the interpretation

$$V = 2\pi \int_a^b x f(x)\, dx = 2\pi \int_a^b (\text{radius})(\text{height of shell})\, dx \qquad \boxed{4}$$

For some solids, it is necessary to modify Eq. (4) in various ways. As a first example, let us rotate the region between the graphs of functions $f(x)$ and $g(x)$ over $[a, b]$ around the y-axis, where $f(x) \geq g(x)$. The vertical segment at location x generates a cylinder of radius x and height $f(x) - g(x)$ (Figure 5), so the volume V is

$$V = 2\pi \int_a^b \left(\text{radius}\right)\left(\text{height of shell}\right) dx = 2\pi \int_a^b x\left(f(x) - g(x)\right) dx \qquad \boxed{5}$$

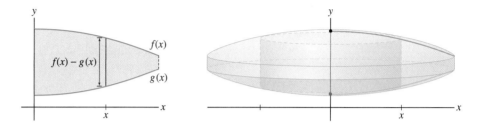

FIGURE 5 The vertical segment at location x generates a cylinder of radius x and height $f(x) - g(x)$.

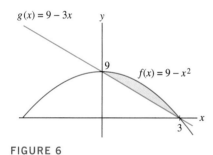

FIGURE 6

■ **EXAMPLE 2** Rotating the Area Between Two Curves   Use the Shell Method to calculate the volume V of the solid obtained by rotating the area enclosed by the graphs of $f(x) = 9 - x^2$ and $g(x) = 9 - 3x$ about the y-axis.

Solution We solve $9 - x^2 = 9 - 3x$ to find the points of intersection $x = 0, 3$. Figure 6 shows that $9 - x^2 \geq 9 - 3x$ for x in $[0, 3]$, so $f(x) = 9 - x^2$ is the upper curve and $g(x) = 9 - 3x$ is the lower curve. We have

$$\text{Height} = f(x) - g(x) = (9 - x^2) - (9 - 3x) = 3x - x^2$$

$$V = 2\pi \int_0^3 (\text{radius})(\text{height}) \, dx = 2\pi \int_0^3 x\left(f(x) - g(x)\right) dx = 2\pi \int_0^3 (3x^2 - x^3) \, dx$$

$$= 2\pi \left(x^3 - \frac{1}{4}x^4 \right) \Big|_0^3 = \frac{27}{2}\pi \qquad ■$$

■ **EXAMPLE 3** Rotating Around a Vertical Axis   Use the Shell Method to calculate the volume V of the solid obtained by rotating the region under the graph of $f(x) = x^{-1/2}$ over $[1, 4]$ about $x = -3$.

The reasoning in Example 3 shows that if we rotate the region under $y = f(x)$ over $[a, b]$ about the vertical line $x = c$ with $c \leq a$, then the volume is

$$V = 2\pi \int_a^b (x - c) f(x) \, dx$$

Solution If we revolved around the y-axis, the volume would be

$$V \text{ (around } y\text{-axis)} = 2\pi \int_1^4 (\text{radius})(\text{height}) \, dx = 2\pi \int_1^4 x \cdot x^{-1/2} \, dx$$

The formula is similar when we rotate around $x = -3$, but we must use the correct radius (distance to the axis of rotation). The height is still $f(x) = x^{-1/2}$, but the radius is now three units longer, that is, $x + 3$ (Figure 7), so

$$V \text{ (around } x = -3) = 2\pi \int_1^4 (\text{radius})(\text{height}) \, dx$$

$$= 2\pi \int_1^4 (x + 3)x^{-1/2} \, dx = 2\pi \left(\frac{2}{3}x^{3/2} + 6x^{1/2} \right) \Big|_1^4 = \frac{64\pi}{3} \qquad ■$$

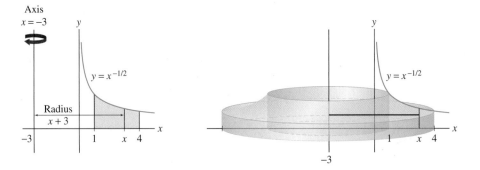

FIGURE 7 Region under the graph of $y = x^{-1/2}$ over [1, 4] rotated about the axis $x = -3$.

The method of cylindrical shells can be applied to rotations about horizontal axes, but in this case, the graph must be described in the form $x = g(y)$.

■ **EXAMPLE 4** Rotating About the x-Axis Use the Shell Method to compute the volume V of the solid obtained by rotating the area under $y = 9 - x^2$ over [0, 3] about the x-axis.

Solution Since we are rotating about the x-axis rather than the y-axis, the Shell Method gives us an integral with respect to y. Therefore, we solve $y = 9 - x^2$ to obtain $x^2 = 9 - y$ or $x = \sqrt{9 - y}$.

The cylindrical shells are generated by *horizontal* segments. The segment $\overline{AB}$ in Figure 8 generates a cylindrical shell of radius y and height $\sqrt{9 - y}$ (we still use the term "height" although the cylinder is horizontal). Using the substitution $u = 9 - y$, $du = -dy$ in the resulting integral, we obtain

⬅·· *REMINDER After making the substitution $u = 9 - y$, the limits of integration must be changed. Since $u(0) = 9$ and $u(9) = 0$, we change $\int_0^9$ to $\int_9^0$.*

$$V = 2\pi \int_0^9 (\text{radius})(\text{height}) \, dy = 2\pi \int_0^9 y \sqrt{9 - y} \, dy = -2\pi \int_9^0 (9 - u) \sqrt{u} \, du$$

$$= 2\pi \int_0^9 (9u^{1/2} - u^{3/2}) \, du = 2\pi \left(6u^{3/2} - \frac{2}{5} u^{5/2} \right) \Big|_0^9 = \frac{648}{5} \pi \qquad ■$$

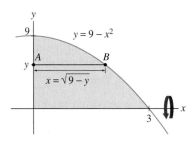

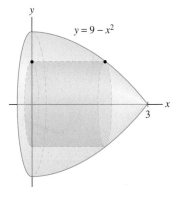

FIGURE 8 Shell generated by a horizontal segment in the region under the graph of $y = 9 - x^2$.

6.4 SUMMARY

• If $f(x) \geq 0$, then the volume V of the solid obtained by rotating the region underneath the graph of $y = f(x)$ over $[a, b]$ about the y-axis is

$$V = 2\pi \int_a^b (\text{radius})(\text{height})\, dx = \boxed{2\pi \int_a^b x f(x)\, dx}$$

- If we revolve the region about the vertical axis $x = c$ rather than the y-axis, then the radius of the shell (distance to the axis of rotation) is no longer x. For example, if $c < a$, the radius is $(x - c)$ and

$$V = 2\pi \int_a^b (\text{radius})(\text{height})\, dx = \boxed{2\pi \int_a^b (x - c) f(x)\, dx}$$

- We can use the Shell Method to compute volumes of revolution about the x-axis. It is necessary to express the curve in the form $x = g(y)$:

$$V = 2\pi \int_c^d (\text{radius})(\text{height})\, dy = \boxed{2\pi \int_c^d y\, g(y)\, dy}$$

6.4 EXERCISES

Preliminary Questions

1. Consider the region $\mathcal{R}$ under the graph of the constant function $f(x) = h$ over the interval $[0, r]$. What are the height and radius of the cylinder generated when $\mathcal{R}$ is rotated about:

(a) the x-axis   **(b)** the y-axis

2. Let V be the volume of a solid of revolution about the y-axis.

(a) Does the Shell Method for computing V lead to an integral with respect to x or y?

(b) Does the Disk or Washer Method for computing V lead to an integral with respect to x or y?

Exercises

In Exercises 1–10, sketch the solid obtained by rotating the region underneath the graph of the function over the given interval about the y-axis and find its volume.

1. $f(x) = x^3$, $[0, 1]$

2. $f(x) = \sqrt{x}$, $[0, 4]$

3. $f(x) = 3x + 2$, $[2, 4]$

4. $f(x) = 1 + x^2$, $[1, 3]$

5. $f(x) = 4 - x^2$, $[0, 2]$

6. $f(x) = \sqrt{x^2 + 9}$, $[0, 3]$

7. $f(x) = \sin(x^2)$, $[0, \sqrt{\pi}]$

8. $f(x) = x^{-1}$, $[1, 3]$

9. $f(x) = x + 1 - 2x^2$, $[0, 1]$

10. $f(x) = \dfrac{x}{\sqrt{1 + x^3}}$, $[1, 4]$

In Exercises 11–14, use the Shell Method to compute the volume of the solids obtained by rotating the region enclosed by the graphs of the functions about the y-axis.

11. $y = x^2$, $y = 8 - x^2$, $x = 0$

12. $y = 8 - x^3$, $y = 8 - 4x$

13. $y = \sqrt{x}$, $y = x^2$

14. $y = 1 - |x - 1|$, $y = 0$

GU   *In Exercises 15–16, use the Shell Method to compute the volume of rotation of the region enclosed by the curves about the y-axis. Use a computer algebra system or graphing utility to find the points of intersection numerically.*

15. $y = \frac{1}{2}x^2$, $y = \sin(x^2)$

16. $y = e^{-x^2/2}$, $y = x$, $x = 0$

In Exercises 17–22, sketch the solid obtained by rotating the region underneath the graph of the function over the interval about the given axis and calculate its volume using the Shell Method.

17. $f(x) = x^3$, $[0, 1]$, $x = 2$

18. $f(x) = x^3$, $[0, 1]$, $x = -2$

19. $f(x) = x^{-4}$, $[-3, -1]$, $x = 4$

20. $f(x) = \dfrac{1}{\sqrt{x^2 + 1}}$, $[0, 2]$, $x = 0$

21. $f(x) = a - bx$, $[0, a/b]$, $x = -1$, $a, b > 0$

22. $f(x) = 1 - x^2$, $[-1, 1]$, $x = c$ (with $c > 1$)

In Exercises 23–28, use the Shell Method to calculate the volume of rotation about the x-axis for the region underneath the graph.

23. $y = x, \quad 0 \le x \le 1$

24. $y = 4 - x^2, \quad 0 \le x \le 2$

25. $y = x^{1/3} - 2, \quad 8 \le x \le 27$

26. $y = x^{-1}, \quad 1 \le x \le 4$. Sketch the region and express the volume as a sum of two integrals.

27. $y = x^{-2}, \quad 2 \le x \le 4$

28. $y = \sqrt{x}, \quad 1 \le x \le 4$

29. Use both the Shell and Disk Methods to calculate the volume of the solid obtained by rotating the region under the graph of $f(x) = 8 - x^3$ for $0 \le x \le 2$ about:

(a) the x-axis   **(b)** the y-axis

30. Sketch the solid of rotation about the y-axis for the region under the graph of the constant function $f(x) = c$ (where $c > 0$) for $0 \le x \le r$.

(a) Find the volume without using integration.

(b) Use the Shell Method to compute the volume.

31. Assume that the graph in Figure 9(A) can be described by both $y = f(x)$ and $x = h(y)$. Let V be the volume of the solid obtained by rotating the region under the curve about the y-axis.

(a) Describe the figures generated by rotating segments $\overline{AB}$ and $\overline{CB}$ about the y-axis.

(b) Set up integrals that compute V by the Shell and Disk Methods.

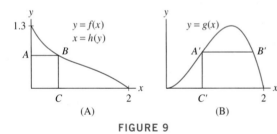

(A)   (B)

FIGURE 9

32. 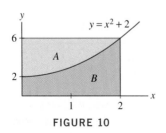 Let W be the volume of the solid obtained by rotating the region under the curve in Figure 9(B) about the y-axis.

(a) Describe the figures generated by rotating segments $\overline{A'B'}$ and $\overline{A'C'}$ about the y-axis.

(b) Set up an integral that computes W by the Shell Method.

(c) Explain the difficulty in computing W by the Washer Method.

In Exercises 33–38, use the Shell Method to find the volume of the solid obtained by rotating region A in Figure 10 about the given axis.

33. y-axis   **34.** $x = -3$

35. $x = 2$ **36.** x-axis

37. $y = -2$ **38.** $y = 6$

In Exercises 39–44, use the Shell Method to find the volumes of the solids obtained by rotating region B in Figure 10 about the given axis.

39. y-axis   **40.** $x = -3$

41. $x = 2$ **42.** x-axis

43. $y = -2$ **44.** $y = 8$

45. Use the Shell Method to compute the volume of a sphere of radius r.

46. Use the Shell Method to calculate the volume V of the "bead" formed by removing a cylinder of radius r from the center of a sphere of radius R (compare with Exercise 51 in Section 6.3).

47. Use the Shell Method to compute the volume of the torus obtained by rotating the interior of the circle $(x - a)^2 + y^2 = r^2$ about the y-axis, where $a > r$. *Hint:* Evaluate the integral by interpreting part of it as the area of a circle.

48. Use the Shell or Disk Method (whichever is easier) to compute the volume of the solid obtained by rotating the region in Figure 11 about:

(a) the x-axis   **(b)** the y-axis

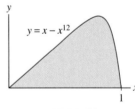

FIGURE 11

49. Use the most convenient method to compute the volume of the solid obtained by rotating the region in Figure 12 about the axis:

(a) $x = 4$ **(b)** $y = -2$

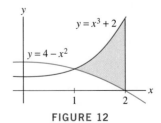

FIGURE 10

FIGURE 12

Further Insights and Challenges

50. 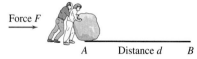 The surface area of a sphere of radius r is $4\pi r^2$. Use this to derive the formula for the volume V of a sphere of radius R in a new way.

(a) Show that the volume of a thin spherical shell of inner radius r and thickness Δx is approximately $4\pi r^2 \Delta x$.

(b) Approximate V by decomposing the sphere of radius R into N thin spherical shells of thickness $\Delta x = R/N$.

(c) Show that the approximation is a Riemann sum which converges to an integral. Evaluate the integral.

51. Let R be the region bounded by the ellipse $\left(\dfrac{x}{a}\right)^2 + \left(\dfrac{y}{b}\right)^2 = 1$ (Figure 13). Show that the solid obtained by rotating R about the y-axis (called an **ellipsoid**) has volume $\frac{4}{3}\pi a^2 b$.

52. The bell-shaped curve in Figure 14 is the graph of a certain function $y = f(x)$ with the following property: The tangent line at a point (x, y) on the graph has slope $dy/dx = -xy$. Let R be the shaded region under the graph for $0 \le x \le a$ in Figure 14. Use the Shell Method and the substitution $u = f(x)$ to show that the solid obtained by revolving R around the y-axis has volume $V = 2\pi(1 - c)$, where $c = f(a)$. Observe that as $c \to 0$, the region R becomes infinite but the volume V approaches 2π.

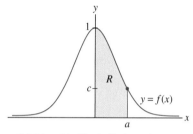

FIGURE 14 The bell-shaped curve.

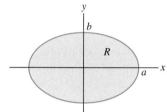

FIGURE 13 The ellipse $\left(\dfrac{x}{a}\right)^2 + \left(\dfrac{y}{b}\right)^2 = 1$.

6.5 Work and Energy

> "For those who want some proof that physicists are human, the proof is in the idiocy of all the different units which they use for measuring energy."
>
> —Richard Feynman,
> *The Character of Physical Law*

All physical tasks, such as boiling water or lighting a fire, require an expenditure of energy. When a force is applied to an object to move it, the energy expended is called **work**. If a *constant* force F is applied through a distance d, then the work W is defined as "force times distance" (Figure 1)

$$\boxed{W = F \cdot d} \qquad \boxed{1}$$

In the metric system, the unit of force is the *newton* (abbreviated N), defined as 1 kg-m/s^2. Energy and work are both measured in units of the *joule* (J), equal to 1 N-m. In the British system, the unit of force is the pound, and both energy and work are measured in foot-pounds (ft-lb). Another unit of energy is the *calorie*. One ft-lb is approximately 0.738 J or 3.088 calories.

To become familiar with the units, let's calculate the work W required to lift a 2-kg stone 3 m above the ground. Gravity pulls down on the stone of mass m with a force equal to $-mg$, where $g = 9.8$ m/s^2. Therefore, lifting the stone requires an upward vertical force $F = mg$, and the work expended is

$$W = \underbrace{(mg)h}_{F \cdot d} = (2\,\text{kg})(9.8\,\text{m/s}^2)(3\,\text{m}) = 58.8\,\text{J}$$

While the kilogram is a unit of mass, the pound is a unit of force rather than mass, so the factor g does not appear when computing work against gravity in the British system. The

Force F

FIGURE 1 The work expended to move the object from A to B is $W = F \cdot d$.

A Distance d B

work required to lift a 2-lb stone 3 ft above ground is

$$W = \underbrace{(2 \text{ lb})(3 \text{ ft})}_{F \cdot d} = 6 \text{ ft-lb}$$

We use integration to calculate work when the force is not constant. Suppose that the force $F(x)$ varies as the object moves from a to b along the x-axis. Then Eq. (1) does not apply directly, but we may break up the task into a large number of smaller tasks where Eq. (1) gives a good approximation. Divide $[a, b]$ into N subintervals of length $\Delta x = \dfrac{b - a}{N}$, with endpoints: $a = x_0, x_1, x_2, \ldots, x_{N-1}, x_N = b$. Let W_i be the work required to move the object from x_{i-1} to x_i (Figure 2). If Δx is small, then the force $F(x)$ is nearly constant on the interval $[x_{i-1}, x_i]$ with value $F(x_i)$, so $W_i \approx F(x_i) \, \Delta x$. Summing the contributions, we obtain

$$W = \sum_{i=1}^{N} W_i \approx \underbrace{\sum_{i=1}^{N} F(x_i) \, \Delta x}_{\text{Right-endpoint approximation}}$$

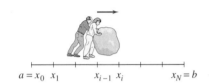

$a = x_0 \; x_1 \qquad x_{i-1} \; x_i \qquad x_N = b$

FIGURE 2 The work to move an object from x_{i-1} to x_i is approximately $F(x_i) \, \Delta x$.

The sum on the right is a right-endpoint approximation converging to $\displaystyle\int_a^b F(x) \, dx$. This leads to the following definition.

DEFINITION Work   The work performed in moving an object along the x-axis from a to b by applying a force of magnitude $F(x)$ is

$$W = \int_a^b F(x) \, dx \qquad \boxed{2}$$

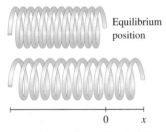

Equilibrium position

$0 \qquad x$

FIGURE 3 According to Hooke's Law, a spring stretched x units past equilibrium exerts a restoring force $-kx$ in the opposite direction.

One typical calculation involves finding the work required to stretch a spring. Assume that the end of the spring has position $x = 0$ at equilibrium (when no force is acting) and that it can be stretched x units in either direction (Figure 3). **Hooke's Law** states that the spring exerts a restoring force $-kx$ in the opposite direction, where k is the **spring constant**, measured in units of kilograms per second squared.

■ **EXAMPLE 1** Hooke's Law   Assuming a spring constant of $k = 400 \text{ kg/s}^2$, find the work (in joules) required to **(a)** stretch the spring 10 cm beyond equilibrium and **(b)** compress the spring 2 additional cm when it is already compressed 3 cm.

Solution By Hooke's Law, the spring exerts a restoring force of $-400x$ N when it is stretched x units. Therefore, we must apply a force $F(x) = 400x$ N to stretch the spring further. To compute the work in joules, we must convert from centimeters to meters since 1 J is equal to a Newton-meter. Therefore,

(a) The work required to stretch the spring 10 cm (0.1 m) beyond equilibrium is

$$W = \int_0^{0.1} 400x \, dx = 200x^2 \Big|_0^{0.1} = 2 \text{ J}$$

(b) If the spring is at position $x = -0.03$ m, then the work W required to compress it further to $x = -0.05$ m is

$$W = \int_{-0.03}^{-0.05} 400x \, dx = 200x^2 \Big|_{-0.03}^{-0.05} = 0.5 - 0.18 = 0.32 \text{ J}$$

Hooke's Law is named after the English scientist, inventor, and architect Robert Hooke (1635–1703) who made important discoveries in physics, astronomy, chemistry, and biology. He was a pioneer in the use of the microscope to study organisms. Unfortunately, Hooke was involved in several bitter disputes with other scientists, most notably with his contemporary Isaac Newton. Newton was furious when Hooke criticized his work on optics. Later, Hooke told Newton that he believed Kepler's Laws would follow from an inverse square law of gravitation. Newton refused to acknowledge Hooke's contributions in his masterwork Principia. *It was in a letter to Hooke that Newton made his famous remark "If I have seen further it is by standing on the shoulders of giants."*

Note that we integrate from right to left (the lower limit −0.03 is *larger* than the upper limit −0.05) because we're compressing the spring to the left.   ■

In the next two examples, we compute work in a different way. In these examples, the formula $W = \int_a^b F(x)\,dx$ cannot be used because we are not moving a single object through a fixed distance. Rather, each thin layer of the object is moved through a different distance. We compute total work by "summing" (i.e., *integrating*) the work performed on each thin layer.

■ **EXAMPLE 2** Building a Cement Column   Compute the work (against gravity) required to build a cement column of height 5 m and square base of side 2 m. Assume that cement has density 1,500 kg/m³.

Solution   Think of the column as a stack of N thin layers of thickness $\Delta y = 5/N$. The work consists of lifting up these layers and placing them on the stack (Figure 4), but the work performed on a given layer depends on how high we lift it. To find the work required to lift a thin layer of thickness Δy to height y, we compute the volume, mass, and gravitational force on the layer:

$$\text{Volume of layer} = \text{area} \times \text{thickness} = 4\,\Delta y \text{ m}^3$$
$$\text{Mass of layer} = \text{density} \times \text{volume} = 1{,}500 \cdot 4\,\Delta y \text{ kg}$$
$$\text{Force on layer} = g \times \text{mass} = 9.8 \cdot 1{,}500 \cdot 4\,\Delta y = 58{,}800\,\Delta y \text{ N}$$

The work required to raise this layer to height y is approximately equal to force times the distance y, that is, $58{,}800y\,\Delta y$. We set $W(y) = 58{,}800y$ and write

> Work performed lifting layer to height $y \approx W(y)\,\Delta y$

This is only an approximation because the layer has a nonzero thickness and the cement particles at the top have been lifted a little bit higher than those at the bottom.

The ith layer is lifted to height $y_i = i\,\Delta y$, and the total work performed is

$$W \approx \sum_{i=1}^{N} W(y_i)\,\Delta y$$

The sum on the right is a right-endpoint approximation to $\int_0^5 W(y)\,dy$. Letting N tend to ∞, we obtain

$$W = \int_0^5 W(y)\,dy = \int_0^5 58{,}800y\,dy = 58{,}800\frac{y^2}{2}\Big|_0^5 = 735{,}000 \text{ J}$$ ■

■ **EXAMPLE 3** Pumping Water out of a Tank   A spherical tank of radius R meters with a small hole at the top is filled with water. How much work (against gravity) is done pumping the water out through the hole? The density of water is 1,000 kg/m³.

Solution   This is similar to the previous example. We divide the sphere into N thin layers of thickness Δy (Figure 5) and show that the work performed on the layer at height y is approximately $W(y)\Delta y$ for some function $W(y)$.

On the earth's surface, work against gravity is equal to the force mg times the vertical distance through which the object is moved. No work against gravity is done when an object is moved sideways.

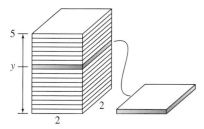

FIGURE 4 Total work is the sum of the work performed on each layer of the column.

Step 1. **Approximate the work performed on a single layer.**

We place the origin at the center of the sphere so that y ranges from $-R$ to R (Figure 5). The ith layer at height $y = y_i$ is nearly cylindrical of thickness $\Delta y = \dfrac{2R}{N}$ and some radius which we call x. By the Pythagorean Theorem, $x^2 + y^2 = R^2$. Therefore, $x^2 = R^2 - y^2$ and the volume of the ith layer is approximately

$$\text{Volume of } i\text{th layer} \approx \pi x^2 \, \Delta y = \pi(R^2 - y^2) \, \Delta y \text{ m}^3$$

Furthermore,

$$\text{Mass of } i\text{th layer} = \text{density} \times \text{volume} \approx 1{,}000\pi(R^2 - y^2) \, \Delta y \text{ kg}$$

$$\text{Force on } i\text{th layer} = g \times \text{mass} \qquad \approx (9.8)1{,}000\pi(R^2 - y^2) \, \Delta y \text{ N}$$

◄·· *REMINDER In the MKS system of units, the force due to gravity on an object of mass m kg is 9.8m N.*

We must raise up this thin layer a vertical distance of approximately $R - y$ (no work against gravity is required to move an object sideways), so the work performed on the layer is approximately

$$\overbrace{9{,}800\pi(R^2 - y^2)\,\Delta y}^{\text{Force against gravity}} \cdot \overbrace{(R - y)}^{\text{Vertical distance moved}} = \overbrace{9{,}800\pi(R^3 - R^2 y - R y^2 + y^3)\,\Delta y}^{\text{Call this } W(y)}$$

With $W(y)$ as indicated and $y = y_i$,

$$\boxed{\text{Work performed on } i\text{th layer} \approx W(y_i)\,\Delta y}$$

Step 2. **Integrate the work performed on the layers.**

We complete the argument as in Example 2. The total work W is the sum of the work performed on the N layers. Thus $W \approx \displaystyle\sum_{i=1}^{N} W(y_i)\,\Delta y$ and in the limit as $N \to \infty$, we obtain

The limits of integration in Eq. (3) are $-R$ to R because the y-coordinate along the sphere varies from $-R$ to R.

$$W = \int_{-R}^{R} W(y)\,dy = 9{,}800\pi \int_{-R}^{R} (R^3 - R^2 y - R y^2 + y^3)\,dy \qquad \boxed{3}$$

$$= 9{,}800\pi \left(R^3 y - \frac{1}{2} R^2 y^2 - \frac{1}{3} R y^3 + \frac{1}{4} y^4 \right) \Bigg|_{-R}^{R} = \frac{39{,}200\pi}{3} R^4 \text{ J} \qquad ■$$

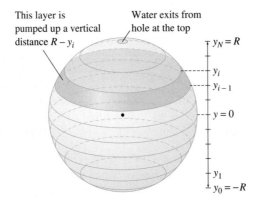

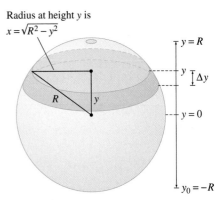

FIGURE 5 The radius of a thin layer at height y is $x = \sqrt{R^2 - y^2}$.

6.5 SUMMARY

- The work W performed when a force F is applied to move an object along a straight line:

$$\text{Constant force:} \quad W = F \cdot d, \qquad \text{Variable force:} \quad W = \int_a^b F(x)\, dx$$

- In some cases, the work is computed by decomposing an object into N thin layers of thickness $\Delta y = \dfrac{b - a}{N}$ (where the object extends from $y = a$ to $y = b$). We approximate the work W_i performed on the ith layer as $W_i \approx W(y_i)\, \Delta y$ for some function $W(y)$. The total work is equal to $W = \int_a^b W(y)\, dy$.

6.5 EXERCISES

Preliminary Questions

1. Why is integration needed to compute the work performed in stretching a spring?

2. Why is integration needed to compute the work performed in pumping water out of a tank but not to compute the work performed in lifting up the tank?

3. Which of the following represents the work required to stretch a spring (with spring constant k) a distance x beyond its equilibrium position: kx, $-kx$, $\frac{1}{2}mk^2$, $\frac{1}{2}kx^2$, or $\frac{1}{2}mx^2$?

Exercises

1. How much work is done raising a 4-kg mass to a height of 16 m above ground?

2. How much work is done raising a 4-lb mass to a height of 16 ft above ground?

In Exercises 3–6, compute the work (in joules) required to stretch or compress a spring as indicated, assuming that the spring constant is $k = 150$ kg/s^2.

3. Stretching from equilibrium to 12 cm past equilibrium

4. Compressing from equilibrium to 4 cm past equilibrium

5. Stretching from 5 to 15 cm past equilibrium

6. Compressing the spring 4 more cm when it is already compressed 5 cm

7. If 5 J of work are needed to stretch a spring 10 cm beyond equilibrium, how much work is required to stretch it 15 cm beyond equilibrium?

8. If 5 J of work are needed to stretch a spring 10 cm beyond equilibrium, how much work is required to compress it 5 cm beyond equilibrium?

9. If 10 ft-lb of work are needed to stretch a spring 1 ft beyond equilibrium, how far will the spring stretch if a 10-lb weight is attached to its end?

10. Show that the work required to stretch a spring from position a to position b is $\frac{1}{2}k(b^2 - a^2)$, where k is the spring constant. How do you interpret the negative work obtained when $|b| < |a|$?

In Exercises 11–14, calculate the work against gravity required to build the structure out of brick using the method of Examples 2 and 3. Assume that brick has density 80 lb/ft^3.

11. A tower of height 20 ft and square base of side 10 ft

12. A cylindrical tower of height 20 ft and radius 10 ft

13. A 20-ft-high tower in the shape of a right circular cone with base of radius 4 ft

14. A structure in the shape of a hemisphere of radius 4 ft

15. Built around 2600 BCE, the Great Pyramid of Giza in Egypt is 485 ft high (due to erosion, its current height is slightly less) and has a square base of side 755.5 ft (Figure 6). Find the work needed to build the pyramid if the density of the stone is estimated at 125 lb/ft^3.

FIGURE 6 The Great Pyramid in Giza, Egypt.

In Exercises 16–20, calculate the work (in joules) required to pump all of the water out of the tank. Assume that the tank is full, distances are measured in meters, and the density of water is $1,000$ kg/m^3.

16. The box in Figure 7; water exits from a small hole at the top.

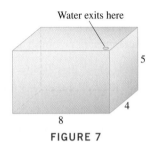

Water exits here

5

4

8

FIGURE 7

17. The hemisphere in Figure 8; water exits from the spout as shown.

2   10

FIGURE 8

18. The conical tank in Figure 9; water exits through the spout as shown.

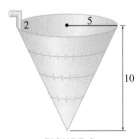

2   5

10

FIGURE 9

19. The horizontal cylinder in Figure 10; water exits from a small hole at the top. *Hint:* Evaluate the integral by interpreting part of it as the area of a circle.

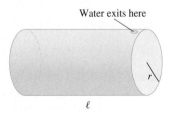

Water exits here

r

ℓ

FIGURE 10

20. The trough in Figure 11; water exits by pouring over the sides.

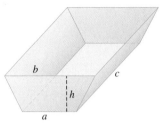

b   c

h

a

FIGURE 11

21. Find the work W required to empty the tank in Figure 7 if it is half full of water.

22. 🖊 Assume the tank in Figure 7 is full of water and let W be the work required to pump out half of the water. Do you expect W to equal the work computed in Exercise 21? Explain and then compute W.

23. Find the work required to empty the tank in Figure 9 if it is half full of water.

24. Assume the tank in Figure 9 is full of water and find the work required to pump out half of the water.

25. Assume that the tank in Figure 9 is full.

(a) Calculate the work $F(y)$ required to pump out water until the water level has reached level y.

(b) *CAS* Plot $F(y)$.

(c) 🖊 What is the significance of $F'(y)$ as a rate of change?

(d) *CAS* If your goal is to pump out all of the water, at which water level y_0 will half of the work be done?

26. How much work is done lifting a 25-ft chain over the side of a building (Figure 12)? Assume that the chain has a density of 4 lb/ft. *Hint:* Break up the chain into N segments, estimate the work performed on a segment, and compute the limit as $N \to \infty$ as an integral.

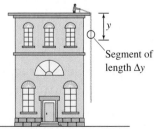

y

Segment of length Δy

FIGURE 12 The small segment of the chain of length Δy located y feet from the top is lifted through a vertical distance y.

27. How much work is done lifting a 3-m chain over the side of a building if the chain has mass density 4 kg/m?

28. An 8-ft chain weighs 16 lb. Find the work required to lift the chain over the side of a building.

29. A 20-ft chain with mass density 3 lb/ft is initially coiled on the ground. How much work is performed in lifting the chain so that it is fully extended (and one end touches the ground)?

30. How much work is done lifting a 20-ft chain with mass density 3 lb/ft (initially coiled on the ground) so that its top end is 30 ft above the ground?

31. A 1,000-lb wrecking ball hangs from a 30-ft cable of density 10 lb/ft attached to a crane. Calculate the work done if the crane lifts the ball from ground level to 30 ft in the air by drawing in the cable.

In Exercises 32–34, use Newton's Universal Law of Gravity, according to which the gravitational force between two objects of mass m and M separated by a distance r has magnitude GMm/r^2, where $G = 6.67 \times 10^{-11}$ m³kg⁻¹s⁻¹. Although the Universal Law refers to point masses, Newton proved that it also holds for uniform spherical objects, where r is the distance between their centers.

32. Two spheres of mass M and m are separated by a distance r_1. Show that the work required to increase the separation to a distance r_2 is equal to $W = GMm(r_1^{-1} - r_2^{-1})$.

33. Use the result of Exercise 32 to calculate the work required to place a 2,000-kg satellite in an orbit 1,200 km above the surface of the earth. Assume that the earth is a sphere of mass $M_e = 5.98 \times 10^{24}$ kg and radius $r_e = 6.37 \times 10^6$ m. Treat the satellite as a point mass.

34. Use the result of Exercise 32 to compute the work required to move a 1,500-kg satellite from an orbit 1,000 to 1,500 km above the surface of the earth.

35. Assume that the pressure P and volume V of the gas in a 30-in. cylinder of radius 3 in. with a movable piston are related by $PV^{1.4} = k$, where k is a constant (Figure 13). When the cylinder is full, the gas pressure is 200 lb/in.²

(a) Calculate k.

(b) Calculate the force on the piston as a function of the length x of the column of gas (the force is PA, where A is the piston's area).

(c) Calculate the work required to compress the gas column from 30 to 20 in.

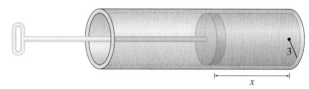

FIGURE 13 Gas in a cylinder with a piston.

Further Insights and Challenges

36. A 20-ft chain with linear mass density

$$\rho(x) = 0.02x(20 - x) \text{ lb/ft}$$

lies on the ground.

(a) How much work is done lifting the chain so that it is fully extended (and one end touches the ground)?

(b) How much work is done lifting the chain so that its top end has a height of 30 ft?

37. Work-Kinetic Energy Theorem The **kinetic energy** of an object of mass m moving with velocity v is KE $= \frac{1}{2}mv^2$.

(a) Suppose that the object moves from x_1 to x_2 during the time interval $[t_1, t_2]$ due to a net force $F(x)$ acting along the interval $[x_1, x_2]$. Let $x(t)$ be the position of the object at time t. Use the Change of Variables formula to show that the work performed is equal to

$$W = \int_{x_1}^{x_2} F(x)\, dx = \int_{t_1}^{t_2} F(x(t))v(t)\, dt$$

(b) By Newton's Second Law, $F(x(t)) = ma(t)$, where $a(t)$ is the acceleration at time t. Show that

$$\frac{d}{dt}\left(\frac{1}{2}mv(t)^2\right) = F(x(t))v(t)$$

(c) Use the FTC to show that the change in kinetic energy during the time interval $[t_1, t_2]$ is equal to

$$\int_{t_1}^{t_2} F(x(t))v(t)\, dt.$$

(d) Prove the Work-Kinetic Energy Theorem: The change in KE is equal to the work W performed.

38. A model train of mass 0.5 kg is placed at one end of a straight 3-m electric track. Assume that a force $F(x) = 3x - x^2$ N acts on the train at distance x along the track. Use the Work-Kinetic Energy Theorem (Exercise 37) to determine the velocity of the train when it reaches the end of the track.

39. With what initial velocity v_0 must we fire a rocket so it attains a maximum height r above the earth? *Hint:* Use the results of Exercises 32 and 37. As the rocket reaches its maximum height, its KE decreases from $\frac{1}{2}mv_0^2$ to zero.

40. With what initial velocity must we fire a rocket so it attains a maximum height of $r = 20$ km above the surface of the earth?

41. Calculate **escape velocity**, the minimum initial velocity of an object to ensure that it will continue traveling into space and never fall back to earth (assuming that no force is applied after takeoff). *Hint:* Take the limit as $r \to \infty$ in Exercise 39.

CHAPTER REVIEW EXERCISES

In Exercises 1–6, find the area of the region bounded by the graphs of the functions.

1. $y = \sin x$, $\quad y = \cos x$, $\quad 0 \le x \le \dfrac{5\pi}{4}$

2. $f(x) = x^3 - 2x^2 + x$, $\quad g(x) = x^2 - x$

3. $f(x) = x^2 + 2x$, $\quad g(x) = x^2 - 1$, $\quad h(x) = x^2 + x - 2$

4. $f(x) = \sin x$, $\quad g(x) = \sin 2x$, $\quad \dfrac{\pi}{3} \le x \le \pi$

5. $y = e^x$, $\quad y = 1 - x$, $\quad x = 1$

6. $y = \cosh 1 - \cosh x$, $\quad y = \cosh x - \cosh 1$

In Exercises 7–10, sketch the region bounded by the graphs of the functions and find its area.

7. $f(x) = x^3 - x^2 - x + 1$, $\quad g(x) = \sqrt{1 - x^2}$, $\quad 0 \le x \le 1$
Hint: Use geometry to evaluate the integral.

8. $x = \dfrac{1}{2}y$, $\quad x = y\sqrt{1 - y^2}$, $\quad 0 \le y \le 1$

9. $y = 4 - x^2$, $\quad y = 3x$, $\quad y = 4$

10. $x = y^3 - 2y^2 + y$, $\quad x = y^2 - y$

11. $\boxed{\text{GU}}$ Use a graphing utility to locate the points of intersection of $y = e^{-x}$ and $y = 1 - x^2$ and find the area between the two curves (approximately).

12. Figure 1 shows a solid whose horizontal cross section at height y is a circle of radius $(1 + y)^{-2}$ for $0 \le y \le H$. Find the volume of the solid.

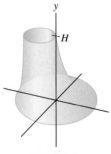

FIGURE 1

13. Find the total weight of a 3-ft metal rod of linear density

$$\rho(x) = 1 + 2x + \frac{2}{9}x^3 \text{ lb/ft.}$$

14. Find the flow rate (in the correct units) through a pipe of diameter 6 cm if the velocity of fluid particles at a distance r from the center of the pipe is $v(r) = (3 - r)$ cm/s.

In Exercises 15–20, find the average value of the function over the interval.

15. $f(x) = x^3 - 2x + 2$, $\quad [-1, 2]$

16. $f(x) = \sqrt{9 - x^2}$, $\quad [0, 3]$ *Hint: Use geometry to evaluate the integral.*

17. $f(x) = |x|$, $\quad [-4, 4]$ **18.** $f(x) = x[x]$, $\quad [0, 3]$

19. $f(x) = x \cosh(x^2)$, $\quad [0, 1]$ **20.** $f(x) = \dfrac{e^x}{1 + e^{2x}}$, $\quad \left[0, \dfrac{1}{2}\right]$

21. The average value of $g(t)$ on $[2, 5]$ is 9. Find $\displaystyle\int_2^5 g(t)\, dt$.

22. For all $x \ge 0$, the average value of $R(x)$ over $[0, x]$ is equal to x. Find $R(x)$.

23. Use the Shell Method to find the volume of the solid obtained by revolving the region between $y = x^2$ and $y = mx$ about the x-axis (Figure 2).

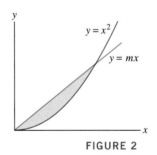

FIGURE 2

24. Use the washer method to find the volume of the solid obtained by revolving the region between $y = x^2$ and $y = mx$ about the y-axis (Figure 2).

25. Let R be the intersection of the circles of radius 1 centered at $(1, 0)$ and $(0, 1)$. Express as an integral (but do not evaluate): **(a)** the area of R and **(b)** the volume of revolution of R about the x-axis.

26. Use the Shell Method to set up an integral (but do not evaluate) expressing the volume of the solid obtained by rotating the region under $y = \cos x$ over $[0, \pi/2]$ about the line $x = \pi$.

In Exercises 27–35, find the volume of the solid obtained by rotating the region enclosed by the curves about the given axis.

27. $y = 2x$, $\quad y = 0$, $\quad x = 8$; $\quad x$-axis

28. $y = 2x$, $\quad y = 0$, $\quad x = 8$; $\quad$ axis $x = -3$

29. $y = x^2 - 1$, $\quad y = 2x - 1$, $\quad$ axis $x = -2$

30. $y = x^2 - 1$, $\quad y = 2x - 1$, $\quad$ axis $y = 4$

31. $y^2 = x^3$, $\quad y = x$, $\quad x = 8$; $\quad$ axis $x = -1$

32. $y^2 = x^{-1}$, $\quad x = 1$, $\quad x = 3$; $\quad$ axis $y = -3$

33. $y = -x^2 + 4x - 3$, $\quad y = 0$; $\quad$ axis $y = -1$

34. $x = 4y - y^3$, $\quad y = 0$, $\quad y = 2$; $\quad$ y-axis

35. $y^2 = x^{-1}$, $\quad x = 1$, $\quad x = 3$; $\quad$ axis $x = -3$

In Exercises 36–38, the regions refer to the graph of the hyperbola $y^2 - x^2 = 1$ in Figure 3. Calculate the volume of revolution about both the x- and y-axes.

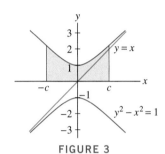

FIGURE 3

36. The shaded region between the upper branch of the hyperbola and the x-axis for $-c \le x \le c$.

37. The region between the upper branch of the hyperbola and the line $y = x$ for $0 \le x \le c$.

38. The region between the upper branch of the hyperbola and $y = 2$.

39. Let $a > 0$. Show that when the region between $y = a\sqrt{x - ax^2}$ and the x-axis is rotated about the x-axis, the resulting volume is independent of the constant a.

40. A spring whose equilibrium length is 15 cm exerts a force of 50 N when it is stretched to 20 cm. Find the work required to stretch the spring from 22 to 24 cm.

In Exercises 41–42, water is pumped into a spherical tank of radius 5 ft from a source located 2 ft below a hole at the bottom (Figure 4). The density of water is 64.2 lb/ft³.

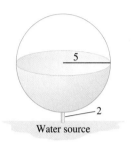

Water source

FIGURE 4

41. Calculate the work required to fill the tank.

42. Calculate the work $F(h)$ required to fill the tank to height h ft from the bottom of the sphere.

43. A container weighing 50 lb is filled with 20 ft³ of water. The container is raised vertically at a constant speed of 2 ft/s for 1 min, during which time it leaks water at a rate of $\frac{1}{3}$ ft³/s. Calculate the total work performed in raising the container. The density of water is 64.2 lb/ft³.

44. Let W be the work (against the sun's gravitational force) required to transport an 80-kg person from Earth to Mars when the two planets are aligned with the sun at their minimal distance of 55.7×10^6 km. Use Newton's Universal Law of Gravity (see Exercises 32–34 in Section 6.5) to express W as an integral and evaluate it. The sun has mass $M_s = 1.99 \times 10^{30}$ kg, and the distance from the sun to the earth is 149.6×10^6 km.

7 | TECHNIQUES OF INTEGRATION

In this chapter, we develop some basic techniques of integration. The two most important general techniques are integration by parts and substitution (which was covered in Chapter 5). We also have specialized techniques for treating particular classes of functions such as trigonometric or rational functions. However, there is no surefire method for finding antiderivatives; in fact, it is not always possible to express an antiderivative in elementary terms. Therefore, we begin with a discussion of numerical integration. Every definite integral can be approximated numerically to any desired degree of accuracy.

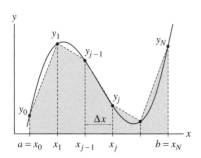

Computer simulation of the 1996 Kamchatka tsunami by the National Oceanic and Atmospheric Administration, using models of wave motion based on advanced calculus.

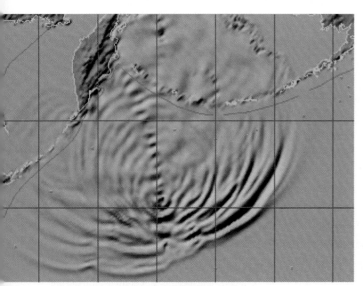

FIGURE 1 There is no explicit formula for areas under the bell-shaped curve.

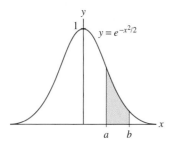

FIGURE 2 T_N approximates the area under the graph by trapezoids.

7.1 Numerical Integration

Numerical integration is the process of approximating a definite integral using well-chosen sums of function values. We use numerical integration when we cannot find a formula for an antiderivative and thus cannot apply the FTC. For instance, the Gaussian function $f(x) = e^{-x^2/2}$, whose graph is a bell-shaped curve (Figure 1), does not have an elementary antiderivative. Numerical methods are needed to approximate $\int_a^b e^{-x^2/2}\,dx$.

Let N be a whole number. The **Trapezoidal Rule** T_N consists of approximating $\int_a^b f(x)\,dx$ by the average of the left- and right-endpoint approximations:

$$T_N = \frac{1}{2}(R_N + L_N)$$

This approximation was used informally in previous chapters, where we observed that the average was likely to give a better approximation than either of R_N or L_N alone. To derive a formula for T_N, let's recall the formulas for the endpoint approximations. We divide $[a, b]$ into N subintervals of length $\Delta x = \dfrac{b-a}{N}$ (Figure 2) with endpoints

$$x_0 = a, \qquad x_1 = a + \Delta x, \qquad x_2 = a + 2\Delta x, \qquad \ldots, \qquad x_N = b$$

For convenience, we denote the values of $f(x)$ at the endpoints by y_j:

$$y_j = f(x_j) = f(a + j\Delta x) \quad (j = 0, 1, 2, \ldots, N)$$

In particular, $y_0 = f(a)$ and $y_N = f(b)$. Then

$$R_N = \Delta x \sum_{j=1}^{N} y_j, \qquad L_N = \Delta x \sum_{j=0}^{N-1} y_j$$

When we form the sum $R_N + L_N$, each value y_j occurs twice except for the first y_0 (which only occurs in L_N) and the last y_N (which only occurs in R_N). Therefore,

$$R_N + L_N = \Delta x \big(y_0 + 2y_1 + 2y_2 + \cdots + 2y_{N-1} + y_N\big)$$

Dividing by 2, we obtain the formula for T_N.

Trapezoidal Rule   The Nth trapezoidal approximation to $\displaystyle\int_a^b f(x)\,dx$ is

$$T_N = \frac{1}{2}\,\Delta x\big(y_0 + 2y_1 + \cdots + 2y_{N-1} + y_N\big)$$

where $\Delta x = \dfrac{b-a}{N}$ and $y_j = f(a + j\,\Delta x)$.

The name "Trapezoidal Rule" is appropriate because T_N is equal to the sum of the areas of the trapezoids obtained by joining the points (x_0, y_0), (x_1, y_1), ..., (x_N, y_N) with line segments (Figure 2). To verify this, we note that the area of a typical trapezoid is equal to the average of the areas of the left- and right-hand rectangles (Figure 3).

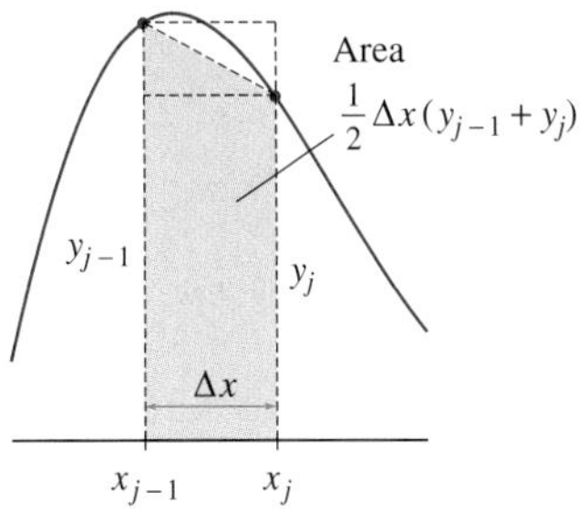

FIGURE 3 The area of a trapezoid is equal to the average of the areas of the left- and right-endpoint rectangles.

■ **EXAMPLE 1** *CAS* Calculate the tenth trapezoidal approximation to $\displaystyle\int_0^1 \sin(x^2)\,dx$. Then use a computer algebra system to calculate T_N for $N = 20, 50, 100, 500, 1{,}000$, and $5{,}000$.

Solution Divide $[0, 1]$ into 10 subintervals of length $\Delta x = \frac{1}{10} = 0.1$. To obtain T_{10}, sum the function values at the endpoints $0, 0.1, 0.2, \ldots, 1$ with the appropriate coefficients:

$$T_{10} = \frac{1}{2}(0.1)\big[\sin(0^2) + 2\sin(0.1^2) + 2\sin(0.2^2) + 2\sin(0.3^2) + 2\sin(0.4^2)$$

$$+\, 2\sin(0.5^2) + 2\sin(0.6^2) + 2\sin(0.7^2) + 2\sin(0.8^2) + 2\sin(0.9^2) + \sin(1^2)\big]$$

$$\approx 0.311$$

For general N, $\Delta x = \dfrac{1-0}{N} = \dfrac{1}{N}$ and

$$T_N = \frac{1}{2N}\Bigg[\sin(0^2) + 2\underbrace{\left(\sin\left(\left(\frac{1}{N}\right)^2\right) + \sin\left(\left(\frac{2}{N}\right)^2\right) + \cdots + \sin\left(\left(\frac{N-1}{N}\right)^2\right)\right)}_{\text{Inner sum}} + \sin(1^2)\Bigg]$$

The inner sum may be evaluated on a computer algebra system using a command such as

```
Sum[Sin[(i/N)^2], {i, 1, N − 1}]
```

We obtain the results in Table 1. Thus $\displaystyle\int_0^1 \sin(x^2)\,dx$ is approximately 0.31027. ■

TABLE 1	
N	T_N
20	0.3104936
50	0.3103043
100	0.3102773
500	0.3102687
1000	0.3102684
5000	**0.3102683**

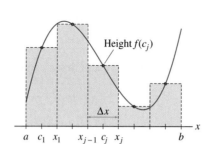
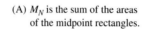
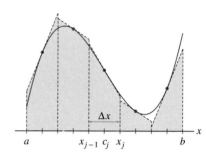

(A) M_N is the sum of the areas of the midpoint rectangles.

(B) M_N is also the sum of the areas of the tangential trapezoids.

FIGURE 4 Two interpretations of M_N.

We introduced the Midpoint Rule M_N in Section 5.1. Recall that M_N is the sum of the areas of the rectangles of height $f(c_j)$ and base Δx, where c_j is the midpoint of the interval $[x_{j-1}, x_j]$ [Figure 4(A)].

Midpoint Rule   The Nth Midpoint Approximation to $\displaystyle\int_a^b f(x)\,dx$ is

$$M_N = \Delta x\big(f(c_1) + f(c_2) + \cdots + f(c_N)\big)$$

where $\Delta x = \dfrac{b-a}{N}$ and $c_j = a + (j - \frac{1}{2})\,\Delta x$ is the midpoint of the jth interval $[x_{j-1}, x_j]$.

GRAPHICAL INSIGHT   Although M_N is defined using midpoint rectangles [Figure 4(A)], there is a second interpretation of M_N as the sum of the areas of the trapezoids whose top edges are tangent to the graph of $y = f(x)$ at the midpoints c_j [Figure 4(B)]. What allows us to replace the rectangles by trapezoids? The following general fact: A rectangle has the same area as any trapezoid of the same base whose top edge passes through the midpoint of the top of the rectangle (Figure 5). Because of this interpretation, the Midpoint Rule is sometimes called the Tangent Rule.

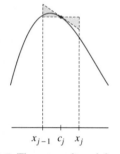

FIGURE 5 The rectangle and the trapezoid have the same area.

Error Bounds

In applications, it is important to know the accuracy of a numerical approximation. We define the error in T_N and M_N by

$$\text{Error}(T_N) = \left| T_N - \int_a^b f(x)\,dx \right|, \qquad \text{Error}(M_N) = \left| M_N - \int_a^b f(x)\,dx \right|$$

According to the following theorem, the magnitudes of these errors are related to the size of the *second* derivative $f''(x)$—here we assume $f''(x)$ exists and is continuous.

In the Error Bound, you can let K_2 be the maximum of $|f''(x)|$ on $[a, b]$, but if it is inconvenient to find this maximum exactly, take K_2 to be any number that is definitely larger than the maximum.

THEOREM 1   Error Bound for T_N and M_N   Let K_2 be a number such that $|f''(x)| \le K_2$ for all $x \in [a, b]$. Then

$$\text{Error}(T_N) \le \frac{K_2(b-a)^3}{12N^2}, \qquad \text{Error}(M_N) \le \frac{K_2(b-a)^3}{24N^2}$$

GRAPHICAL INSIGHT   The proof of the Error Bound is technical and we omit it. Intuitively, however, we can see why Error(T_N) is related to the size of $f''(x)$. The second derivative measures concavity: If $|f''(x)|$ is large, then the graph of f bends a lot (either up or down) and the trapezoids do not provide a good approximation to the area under the graph (Figure 6). Since the Midpoint Rule can be interpreted in terms of tangential trapezoids, Error(M_N) is also related to the size of $f''(x)$.

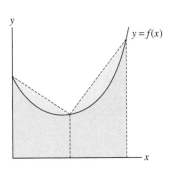

(A) Larger second derivative $f''(x)$

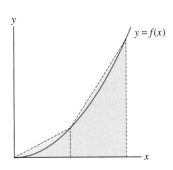

(B) Smaller second derivative $f''(x)$

FIGURE 6 The trapezoidal approximation is more accurate when $|f''(x)|$ is small.

■ **EXAMPLE 2**   Checking the Error Bound   Calculate T_6 and M_6 for $\displaystyle\int_1^4 \sqrt{x}\,dx$. Then

(a) Use the Error Bound to find bounds for the errors.

(b) Calculate the integral exactly and verify that the Error Bounds are satisfied.

Solution   We divide $[1, 4]$ into six subintervals of width $\Delta x = \frac{4-1}{6} = \frac{1}{2}$. Using the endpoints and midpoints shown in Figure 7, we obtain

$$T_6 = \frac{1}{2}\left(\frac{1}{2}\right)\left(\sqrt{1} + 2\sqrt{1.5} + 2\sqrt{2} + 2\sqrt{2.5} + 2\sqrt{3} + 2\sqrt{3.5} + \sqrt{4}\right)$$

$$\approx 4.661488$$

$$M_6 = \frac{1}{2}\left(\sqrt{1.25} + \sqrt{1.75} + \sqrt{2.25} + \sqrt{2.75} + \sqrt{3.25} + \sqrt{3.75}\right)$$

$$\approx 4.669245$$

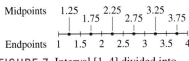

FIGURE 7 Interval $[1, 4]$ divided into $N = 6$ subintervals.

(a) Let $f(x) = \sqrt{x}$. To use the Error Bound, we must find a number K_2 such that $|f''(x)| \le K_2$ for $1 \le x \le 4$. We have $f'(x) = \frac{1}{2}x^{-1/2}$ and $f''(x) = -\frac{1}{4}x^{-3/2}$. The absolute value $|f''(x)| = \frac{1}{4}x^{-3/2}$ is decreasing on $[1, 4]$, so its maximum occurs at $x = 1$ (Figure 8). Thus, we may take $K_2 = |f''(1)| = \frac{1}{4}$. By Theorem 1,

$$\text{Error}(T_6) \le \frac{K_2(b-a)^3}{12N^2} = \frac{\frac{1}{4}(4-1)^3}{12(6)^2} = \frac{1}{64} \approx 0.0156 \qquad \boxed{1}$$

$$\text{Error}(M_6) \le \frac{K_2(b-a)^3}{24N^2} = \frac{\frac{1}{4}(4-1)^3}{24(6)^2} = \frac{1}{128} \approx 0.0078 \qquad \boxed{2}$$

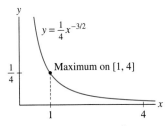

FIGURE 8 Graph of $y = |f''(x)|$ for $f(x) = \sqrt{x}$.

(b) We use the exact value $\displaystyle\int_1^4 \sqrt{x}\,dx = \frac{2}{3}x^{3/2}\Big|_1^4 = \frac{14}{3}$ to compute the actual errors:

In Example 2, the error in T_6 is approximately twice as large as the error in M_6. In practice, this is often the case. Note that the Error Bound for T_N in Theorem 1 is twice as large as the Error Bound for M_N.

$$\text{Error}(T_6) = \left| T_6 - \frac{14}{3} \right| \approx |4.661488 - 4.666667| \approx 0.00518$$

$$\text{Error}(M_6) = \left| M_6 - \frac{14}{3} \right| \approx |4.669245 - 4.666667| \approx 0.00258$$

Both of these errors lie well within the error bounds (1) and (2). ■

The Error Bound can be used to determine values of N that provide a given accuracy.

■ **EXAMPLE 3** Obtaining the Desired Accuracy Find N such that M_N approximates $\int_0^3 e^{-x^2}\, dx$ with an error of at most 0.0001.

A quick way to find a value for K_2 is to plot $f''(x)$ using a graphing utility and find a bound for $|f''(x)|$ visually, as we do in Example 3.

Solution Let $f(x) = e^{-x^2}$. To apply the Error Bound, we must find a number K_2 such that $|f''(x)| \le K_2$ for all $x \in [0, 3]$. We have $f'(x) = -2xe^{-x^2}$ and

$$f''(x) = (4x^2 - 2)e^{-x^2}$$

To find a bound for $|f''(x)|$, we use a graphing utility to plot $f''(x)$ over $[0, 3]$ (Figure 9). The graph shows that the maximum value of $|f''(x)|$ is $|f''(0)| = |-2| = 2$, so we take $K_2 = 2$ in the Error Bound:

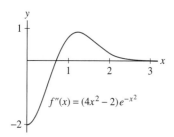

$f''(x) = (4x^2 - 2)e^{-x^2}$

FIGURE 9 Graph of the second derivative of $f(x) = e^{-x^2}$.

$$\text{Error}(M_N) \le \frac{K_2(b-a)^3}{24N^2} = \frac{2(3-0)^3}{24N^2} = \frac{9}{4N^2}$$

The error is at most 0.0001 if

$$\frac{9}{4N^2} \le 0.0001 \qquad \text{or} \qquad N^2 \ge \frac{9 \times 10^4}{4}$$

Thus, $N \ge 150$. We conclude that M_{150} has error at most 0.0001. Computing M_{150} with a computer algebra system, we find that $M_{150} \approx 0.886207$, so the value of the integral lies between 0.886107 and 0.886307 (in fact, the value to nine places is 0.886207348). ■

Can we improve on the Trapezoidal and Midpoint Rules? One clue is that when $f(x)$ is concave (up or down), then the exact value of the integral lies between T_N and M_N. In fact, we see geometrically (Figure 10) that

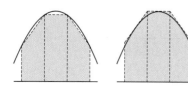

(A) Trapezoids used to compute T_N (B) Trapezoids used to compute M_N

FIGURE 10 If $f(x)$ is concave down, then T_N is smaller and M_N is larger than the integral.

- If $f(x)$ is concave down $\Rightarrow$ $T_N \le \int_a^b f(x)\, dx \le M_N$.

- If $f(x)$ is concave up $\Rightarrow$ $M_N \le \int_a^b f(x)\, dx \le T_N$

This suggests that the average of T_N and M_N may be more accurate than either T_N or M_N alone.

Simpson's Rule S_N exploits this idea, but it takes into account that M_N is roughly twice as accurate as T_N. Therefore, S_N uses a **weighted average**, which is more accurate than the ordinary average. For N even, we define

$$S_N = \frac{1}{3} T_{N/2} + \frac{2}{3} M_{N/2}$$

$\boxed{3}$

To derive a formula for S_N, let $\Delta x = \dfrac{b-a}{N}$ and, as before, set $x_j = a + j\,\Delta x$ and $y_j = f(a + j\,\Delta x)$. If we use just the even-numbered endpoints, we obtain a partition of

$[a, b]$ into $N/2$ subintervals (keep in mind that N is even):

$$[x_0, x_2], [x_2, x_4], \ldots, [x_{N-2}, x_N] \qquad \boxed{4}$$

The *endpoints* of these intervals are used to compute $T_{N/2}$, and we use the *midpoints* $x_1, x_3, \ldots, x_{N-1}$ to compute $M_{N/2}$ (see Figure 11 for the case $N = 8$):

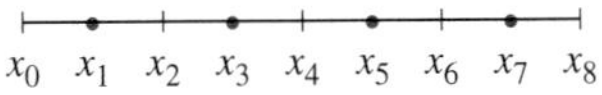

x_0 x_1 x_2 x_3 x_4 x_5 x_6 x_7 x_8

FIGURE 11 We compute S_8 using eight subintervals. The even endpoints are used for T_4, the odd endpoints for M_4, and $S_8 = \frac{1}{3}T_4 + \frac{2}{3}M_4$.

$$T_{N/2} = \left(\frac{1}{2}\right)\frac{b-a}{N/2}\left(y_0 + 2y_2 + 2y_4 + \cdots + 2y_{N-2} + y_N\right)$$

$$M_{N/2} = \frac{b-a}{N/2}\left(y_1 + y_3 + y_5 + \cdots + y_{N-1}\right) = \frac{b-a}{N}\left(2y_1 + 2y_3 + 2y_5 + \cdots + 2y_{N-1}\right)$$

Thus,

$$S_N = \frac{1}{3}T_{N/2} + \frac{2}{3}M_{N/2} = \frac{1}{3}\Delta x\left(y_0 + 2y_2 + 2y_4 + \cdots + 2y_{N-2} + y_N\right)$$

$$+ \frac{1}{3}\Delta x\left(4y_1 + 4y_3 + 4y_5 + \cdots + 4y_{N-1}\right)$$

Here is the pattern of coefficients in S_N:

$$1, 4, 2, 4, 2, 4, \ldots, 4, 2, 4, 1$$

The intermediate coefficients alternate 4, 2, 4, 2, ..., 2, 4 (beginning and ending with 4).

Simpson's Rule Assume that N is even. Let $\Delta x = \dfrac{b-a}{N}$ and $y_j = f(a + j\,\Delta x)$. The Nth approximation to $\displaystyle\int_a^b f(x)\,dx$ by Simpson's Rule is the quantity

$$S_N = \frac{1}{3}\Delta x\left[y_0 + 4y_1 + 2y_2 + \cdots + 4y_{N-3} + 2y_{N-2} + 4y_{N-1} + y_N\right] \qquad \boxed{5}$$

CONCEPTUAL INSIGHT Both T_N and M_N (for all $N \geq 1$) give the exact value of the integral when $f(x)$ is a linear function (Exercise 57). However, of all possible combinations of $T_{N/2}$ and $M_{N/2}$, only the particular combination $S_N = \frac{1}{3}T_{N/2} + \frac{2}{3}M_{N/2}$ gives the exact value of the integral for all quadratic polynomials for all N (Exercises 58 and 59). In fact, S_N is exact for all cubic polynomials (Exercise 60).

■ **EXAMPLE 4** Use Simpson's Rule with $N = 8$ to approximate $\displaystyle\int_2^4 \sqrt{1 + x^3}\,dx$.

1   4   2   4   2   4   2   4   1

2 2.25 2.5 2.75 3 3.25 3.5 3.75 4

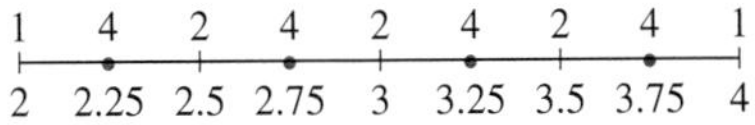

FIGURE 12 Pattern of coefficients for S_8 on $[2, 4]$.

Simpson's Rule yields impressive accuracy. Using a computer algebra system, we find that the approximation in Example 4 has an error of less than 3×10^{-6}.

Solution We have $\Delta x = \dfrac{4-2}{8} = 0.25$. Figure 12 shows the endpoints and coefficients needed to compute S_8 using Eq. (5):

$$\frac{1}{3}(0.25)\Big[\sqrt{1+2^3} + 4\sqrt{1+2.25^3} + 2\sqrt{1+2.5^3} + 4\sqrt{1+2.75^3} + 2\sqrt{1+3^3}$$

$$+ 4\sqrt{1+3.25^3} + 2\sqrt{1+3.5^3} + 4\sqrt{1+3.75^3} + \sqrt{1+4^3}\Big]$$

$$\approx \frac{1}{12}\Big[3 + 4(3.52003) + 2(4.07738) + 4(4.66871) + 2(5.2915)$$

$$+ 4(5.94375) + 2(6.62382) + 4(7.33037) + 8.06226\Big] \approx 10.74159 \qquad ■$$

We now state (without proof) a bound for the error in Simpson's Rule, where

$$\text{Error}(S_N) = \left|S_N(f) - \int_a^b f(x)\,dx\right|$$

The error involves the fourth derivative, which we assume exists and is continuous.

Although Simpson's Rule provides quite good approximations, more sophisticated techniques are implemented in computer algebra systems. These techniques are studied in the area of mathematics called "numerical analysis."

THEOREM 2 Error Bound for S_N   Let K_4 be a number such that $|f^{(4)}(x)| \le K_4$ for all $x \in [a, b]$. Then

$$\text{Error}(S_N) \le \frac{K_4(b-a)^5}{180N^4}$$

■ **EXAMPLE 5** Calculate S_8 for $\displaystyle\int_1^3 \frac{1}{x}\,dx$. Then

(a) Find a bound for the error in S_8.

(b) Find N such that S_N has an error of at most 10^{-6}.

Solution We have $\Delta x = \dfrac{3-1}{8} = 0.25$ and the endpoints in the partition of $[1, 3]$ are $1, 1.25, 1.5, \ldots, 2.75$, and 3. Therefore, using Eq. (5),

$$S_8 = \frac{1}{3}(0.25)\left[\frac{1}{1} + \frac{4}{1.25} + \frac{2}{1.5} + \frac{4}{1.75} + \frac{2}{2} + \frac{4}{2.25} + \frac{2}{2.5} + \frac{4}{2.75} + \frac{1}{3}\right]$$

$$\approx 1.09873$$

(a) Let $f(x) = x^{-1}$. The fourth derivative $f^{(4)}(x) = 24x^{-5}$ is decreasing, so its maximum on $[1, 3]$ is $f^{(4)}(1) = 24$. Therefore, we may set $K_4 = 24$, and

$$\text{Error}(S_N) \le \frac{K_4(b-a)^5}{180N^4} = \frac{24(3-1)^5}{180N^4} = \frac{64}{15N^4}$$

$$\text{Error}(S_8) \le \frac{K_4(b-a)^5}{180(8)^4} = \frac{24(3-1)^5}{180(8^4)} \approx 0.001$$

Using a CAS, we find that

$$S_{46} \approx 1.09861241$$

$$\int_1^3 \frac{1}{x}\,dx = \ln 3 \approx 1.09861229$$

The error is indeed less than 10^{-6}.

(b) To ensure that the error is at most 10^{-6}, we choose N so that

$$\text{Error}(S_N) = \frac{64}{15N^4} \le 10^{-6}$$

This gives

$$N^4 \ge 10^6 \left(\frac{64}{15}\right) \quad \text{or} \quad N \ge \left(\frac{10^6 \cdot 64}{15}\right)^{1/4} \approx 45.45$$

Thus, we may take $N = 46$ (see the comment in the margin).   ■

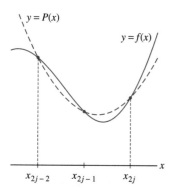

FIGURE 13 There is a unique quadratic function $P(x)$ that agrees with $f(x)$ at $x_{2j-2}, x_{2j-1}, x_{2j}$.

GRAPHICAL INSIGHT   We defined Simpson's Rule in terms of T_N and M_N, but it has another interpretation as an approximation to area under a graph using parabolas rather than rectangles. If we take three consecutive points $x_{2j-2}, x_{2j-1}, x_{2j}$, there is a unique quadratic function $P(x)$ (whose graph is a parabola) taking the same values as $f(x)$ at $x_{2j-2}, x_{2j-1}, x_{2j}$ (Figure 13). The area under the parabola over $[x_{2j-2}, x_{2j}]$ is an approximation to the area under the graph of $f(x)$ on this interval. Simpson's Rule S_N is equal to the sum of these parabolic approximations (Figure 14). This interpretation follows from the result of Exercise 58.

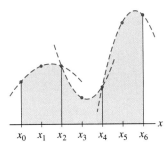

FIGURE 14 Simpson's Rule approximates the graph of $y = f(x)$ by parabolic arcs.

Graph of $y = f(x)$

Parabolic arcs used in Simpson's Rule.

■ **EXAMPLE 6** Estimating Integrals from Numerical Data   The velocity (in miles per hour) of a Piper Cub aircraft traveling due west is recorded every minute during the first 10 min after takeoff. Use the Trapezoidal Rule and Simpson's Rule to estimate the distance traveled after 10 min.

t	0	1	2	3	4	5	6	7	8	9	10
$v(t)$	0	50	60	80	90	100	95	85	80	75	85

Solution   Since the velocity is given in miles per hour, we use hours as our unit of time. The 10-min interval has length $\frac{1}{6}$ hours. We divide this interval into $N = 10$ subintervals of length $\Delta t = \frac{1}{60}$ hours. Then

$$T_{10} = \frac{1}{120}\big(0 + 2(50 + 60 + 80 + 90 + 100 + 95 + 85 + 80 + 75) + 85\big) = 12.625$$

$$S_{10} = \frac{1}{180}\big(0 + 4(50) + 2(60) + 4(80) + 2(90) + 4(100)$$
$$+ 2(95) + 4(85) + 2(80) + 4(75) + 85\big) = 12.75$$

The plane has traveled approximately 12.7 miles (Figure 15).   ■

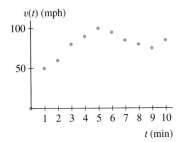

FIGURE 15 Velocity of a Piper Cub. The distance traveled is estimated by approximating the area under the curve through these data points.

7.1 SUMMARY

- We consider three numerical approximations to $\displaystyle\int_a^b f(x)\,dx$: the *Trapezoidal Rule* T_N, the *Midpoint Rule* M_N, and *Simpson's Rule* S_N (if N is even).

$$T_N = \frac{1}{2}\Delta x\big(y_0 + 2y_1 + 2y_2 + \cdots + 2y_{N-1} + y_N\big)$$

$$M_N = \Delta x\big(f(c_1) + f(c_2) + \cdots + f(c_N)\big) \qquad \left(c_j = a + \left(j - \frac{1}{2}\right)\Delta x\right)$$

$$S_N = \frac{1}{3}\Delta x\big[y_0 + 4y_1 + 2y_2 + \cdots + 4y_{N-3} + 2y_{N-2} + 4y_{N-1} + y_N\big]$$

where $\Delta x = \dfrac{b - a}{N}$ and $y_j = f(a + j\,\Delta x)$.

- T_N is equal to the sum of the areas of the trapezoids obtained by connecting the points $(x_0, y_0), (x_1, y_1), \ldots, (x_N, y_N)$ with line segments.
- M_N has two geometric interpretations, either as the sum of the areas of the midpoint rectangles or as the sum of the areas of the tangential trapezoids.
- S_N is equal to $\frac{1}{3} T_{N/2} + \frac{2}{3} M_{N/2}$.
- Error Bounds:

$$\text{Error}(T_N) \le \frac{K_2(b-a)^3}{12N^2}, \qquad \text{Error}(M_N) \le \frac{K_2(b-a)^3}{24N^2}, \qquad \text{Error}(S_N) \le \frac{K_4(b-a)^5}{180N^4}$$

where K_2 is any number such that $|f''(x)| \le K_2$ for all $x \in [a, b]$ and K_4 is any number such that $|f^{(4)}(x)| \le K_4$ for all $x \in [a, b]$.

7.1 EXERCISES

Preliminary Questions

1. What are T_1 and T_2 for a function on $[0, 2]$ such that $f(0) = 3$, $f(1) = 4$, and $f(2) = 3$?

2. For which graph in Figure 16 will T_N overestimate the integral? What about M_N?

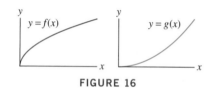

FIGURE 16

3. How large is the error when the Trapezoidal Rule is applied to a linear function? Explain graphically.

4. Suppose T_4 is used to approximate $\int_0^3 f(x) \, dx$, where $|f''(x)| \le 2$ for all x. What is the maximum possible error?

5. What are the two graphical interpretations of the Midpoint Rule?

Exercises

In Exercises 1–12, calculate T_N and M_N for the value of N indicated.

1. $\int_0^2 x^2 \, dx$, $\quad N = 4$

2. $\int_0^4 \sqrt{x} \, dx$, $\quad N = 4$

3. $\int_1^4 x^3 \, dx$, $\quad N = 6$

4. $\int_0^{\pi/2} \sqrt{\sin x} \, dx$, $\quad N = 8$

5. $\int_1^4 \frac{dx}{x}$, $\quad N = 6$

6. $\int_{-2}^{-1} \frac{dx}{x}$, $\quad N = 5$

7. $\int_0^{\pi/4} \sec x \, dx$, $\quad N = 6$

8. $\int_1^2 \ln x$, $\quad N = 5$

9. $\int_2^3 \frac{dx}{\ln x}$, $\quad N = 5$

10. $\int_0^1 e^{-x^2} \, dx$, $\quad N = 6$

11. $\int_0^2 \frac{e^x}{x+1} \, dx$, $\quad N = 8$

12. $\int_{-2}^1 e^{x^2} \, dx$, $\quad N = 6$

In Exercises 13–22, calculate S_N given by Simpson's Rule for the value of N indicated.

13. $\int_0^4 \sqrt{x} \, dx$, $\quad N = 4$

14. $\int_3^5 (9 - x^2) \, dx$, $\quad N = 4$

15. $\int_0^2 \frac{dx}{x^4 + 1}$, $\quad N = 4$

16. $\int_0^1 \cos(x^2) \, dx$, $\quad N = 6$

17. $\int_0^1 e^{-x^2} \, dx$, $\quad N = 6$

18. $\int_1^2 e^{-x} \, dx$, $\quad N = 6$

19. $\int_1^4 \ln x \, dx$, $\quad N = 8$

20. $\int_2^4 \sqrt{x^4 + 1} \, dx$, $\quad N = 8$

21. $\int_0^{\pi/4} \sec x \, dx$, $\quad N = 10$

22. $\int_2^4 \sqrt{x^4 + 1} \, dx$, $\quad N = 10$

In Exercises 23–26, calculate the approximation to the volume of the solid obtained by rotating the graph around the given axis.

23. $y = \cos x$; $\quad [0, \frac{\pi}{2}]$; $\quad x$-axis; $\quad M_8$

24. $y = \cos x$; $\quad [0, \frac{\pi}{2}]$; $\quad y$-axis; $\quad S_8$

25. $y = e^{-x^2}$; $\quad [0, 1]$; $\quad x$-axis; $\quad T_8$

26. $y = e^{-x^2}$; $\quad [0, 1]$; $\quad y$-axis; $\quad S_8$

27. Use S_8 to estimate $\int_0^{\pi/2} \dfrac{\sin x}{x} \, dx$, taking the value of $(\sin x)/x$ at $x = 0$ to be 1.

28. Calculate T_6 for the integral $I = \int_0^2 x^3 \, dx$.

(a) Is T_6 too large or too small? Explain graphically.

(b) Show that $K_2 = |f''(2)|$ may be used in the Error Bound and find a bound for the error.

(c) Evaluate I and check that the actual error is less than the bound computed in (b).

29. Calculate M_6 for the integral $I = \int_0^{\pi/2} \cos x \, dx$.

(a) Is M_6 too large or too small? Explain graphically.

(b) Show that $K_2 = 1$ may be used in the Error Bound and find a bound for the error.

(c) Evaluate I and check that the actual error is less than the bound computed in (b).

In Exercises 30–33, state whether T_N or M_N underestimates or overestimates the integral and find a bound for the error (but do not calculate T_N or M_N).

30. $\int_1^4 \dfrac{1}{x} \, dx, \quad T_{10}$

31. $\int_0^2 e^{-x/4} \, dx, \quad T_{20}$

32. $\int_1^4 \ln x \, dx, \quad M_{10}$

33. $\int_0^{\pi/4} \cos x, \quad M_{20}$

CAS *In Exercises 34–37, use the Error Bound to find a value of N for which $\mathrm{Error}(T_N) \le 10^{-6}$. If you have a computer algebra system, calculate the corresponding approximation and confirm that the error satisfies the required bound.*

34. $\int_0^1 x^4 \, dx$

35. $\int_0^{\pi/6} \cos x \, dx$

36. $\int_2^5 \dfrac{1}{x} \, dx$

37. $\int_0^3 e^{-x} \, dx$

38. Find a bound for the error in the approximations T_{10} and M_{10} to $\int_0^5 (x^3 + 1)^{-1} \, dx$ (use Figure 17 to determine a value of K_2). Then find a value of N such that the error in M_N is at most 10^{-6}.

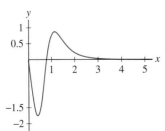

FIGURE 17 Graph of $f''(x)$, where $f(x) = (x^3 + 1)^{-1}$.

39. (a) Compute S_6 for the integral $I = \int_0^1 e^{-2x} \, dx$.

(b) Show that $K_4 = 16$ may be used in the Error Bound and find a bound for the error.

(c) Evaluate I and check that the actual error is less than the bound for the error computed in (b).

40. Calculate S_8 for the integral $\int_0^1 x^4 \, dx$. Use the Error Bound to find a bound for the error and verify that the actual error satisfies this bound.

41. Calculate S_8 for $\int_1^5 \ln x \, dx$ and find a bound for the error. Then find a value of N such that S_N has an error of at most 10^{-6}.

42. Find a bound for the error in the approximation S_{10} to $\int_0^3 e^{-x^2} \, dx$ (use Figure 18 to determine a value of K_4). Then find a value of N such that S_N has an error of at most 10^{-6}.

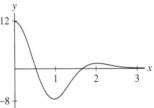

FIGURE 18 Graph of $f^{(4)}(x)$, where $f(x) = e^{-x^2}$.

43. CAS Use a computer algebra system to compute and graph $f^{(4)}(x)$ for $f(x) = \sqrt{1 + x^4}$ and find a bound for the error in the approximation S_{40} to $\int_0^5 f(x) \, dx$.

44. CAS Use a computer algebra system to compute and graph $f^{(4)}(x)$ for $f(x) = \tan x - \sec x$ and find a bound for the error in the approximation S_{40} to $\int_0^{\pi/4} f(x) \, dx$.

In Exercises 45–48, use the Error Bound to find a value of N for which $\mathrm{Error}(S_N) \le 10^{-9}$.

45. $\int_1^7 x^{3/2} \, dx$

46. $\int_0^4 x e^x \, dx$

47. $\int_0^1 e^{x^2} \, dx$

48. $\int_1^4 \sin(\ln x) \, dx$

49. CAS Show that $\int_0^1 \dfrac{dx}{1 + x^2} = \dfrac{\pi}{4}$ [use Eq. (3) in Section 5.7].

(a) Use a computer algebra system to graph $f^{(4)}(x)$ for $f(x) = (1 + x^2)^{-1}$ and find its maximum on $[0, 1]$.

(b) Find a value of N such that S_N approximates the integral with an error of at most 10^{-6}. Calculate the corresponding approximation and confirm that you have computed $\frac{\pi}{4}$ to at least four places.

50. Let $f(x) = \sin(x^2)$ and $I = \int_0^1 f(x)\,dx$.

(a) Check that $f''(x) = 2\cos(x^2) - 4x^2\sin(x^2)$. Then show that $|f''(x)| \le 6$ for $x \in [0, 1]$. *Hint:* Note that $|2\cos(x^2)| \le 2$ and $|4x^2\sin(x^2)| \le 4$ for $x \in [0, 1]$.

(b) Show that $\text{Error}(M_N)$ is at most $\dfrac{1}{4N^2}$.

(c) Find an N such that $|I - M_N| \le 10^{-3}$.

51. *CAS* 📖 The Error Bound for M_N is proportional to $1/N^2$, so the Error Bound decreases by $\frac{1}{4}$ if N is increased to $2N$. Compute the actual error in M_N for $\int_0^\pi \sin x\,dx$ for $N = 4, 8, 16, 32,$ and 64. Does the actual error seem to decrease by $\frac{1}{4}$ as N is doubled?

52. *CAS* 📖 Observe that the Error Bound for T_N (which has 12 in the denominator) is twice as large as the Error Bound for M_N (which has 24 in the denominator). Compute the actual error in T_N for $\int_0^\pi \sin x\,dx$ for $N = 4, 8, 16, 32,$ and 64 and compare with the calculations of Exercise 51. Does the actual error in T_N seem to be roughly twice as large as the error in M_N in this case?

53. *CAS* 📖 Explain why the Error Bound for S_N decreases by $\frac{1}{16}$ if N is increased to $2N$. Compute the actual error in S_N for $\int_0^\pi \sin x\,dx$ for $N = 4, 8, 16, 32,$ and 64. Does the actual error seem to decrease by $\frac{1}{16}$ as N is doubled?

54. An airplane's velocity is recorded at 5-min intervals during a 1-hour period with the following results, in mph:

550, 575, 600, 580, 610, 640, 625, 595, 590, 620, 640, 640, 630

Use Simpson's Rule to estimate the distance traveled during the hour.

55. Use Simpson's Rule to determine the average temperature in a museum over a 3-hour period, if the temperatures (in degrees Celsius), recorded at 15-min intervals, are

21, 21.3, 21.5, 21.8, 21.6, 21.2, 20.8,

20.6, 20.9, 21.2, 21.1, 21.3, 21.2

56. 📖 **Tsunami Arrival Times** Scientists estimate the arrival times of tsunamis (seismic ocean waves) based on the point of origin P and ocean depths. The speed s of a tsunami in miles per hour is approximately $s = \sqrt{15d}$, where d is the ocean depth in feet.

(a) Let $f(x)$ be the ocean depth x miles from P (in the direction of the coast). Argue using Riemann sums that the time T required for the tsunami to travel M miles toward the coast is

$$T = \int_0^M \frac{dx}{\sqrt{15f(x)}}$$

(b) Use Simpson's Rule to estimate T if $M = 1{,}000$ and the ocean depths (in feet), measured at 100-mile intervals starting from P, are

13,000, 11,500, 10,500, 9,000, 8,500, 7,000,

6,000, 4,400, 3,800, 3,200, 2,000

Further Insights and Challenges

57. Show that $T_N = \int_a^b f(x)\,dx$ for all N and all endpoints a, b if $f(x) = rx + s$ is a linear function (r, s constants).

58. Show that if $f(x) = px^2 + qx + r$ is a quadratic polynomial, then $S_2 = \int_a^b f(x)\,dx$. In other words, show that

$$\int_a^b f(x)\,dx = \frac{b-a}{6}(y_0 + 4y_1 + y_2)$$

where $y_0 = f(a)$, $y_1 = f\left(\dfrac{a+b}{2}\right)$, and $y_2 = f(b)$. *Hint:* Show this first for $f(x) = 1, x, x^2$ and use linearity.

59. For N even, divide $[a, b]$ into N subintervals of width $\Delta x = \dfrac{b-a}{N}$. Set $x_j = a + j\,\Delta x$, $y_j = f(x_j)$, and

$$S_2^{2j} = \frac{b-a}{3N}(y_{2j} + 4y_{2j+1} + y_{2j+2})$$

(a) Show that S_N is the sum of the approximations on the intervals $[x_{2j}, x_{2j+2}]$, that is, $S_N = S_2^0 + S_2^2 + \cdots + S_2^{N-2}$.

(b) By Exercise 58, $S_2^{2j} = \int_{x_{2j}}^{x_{2j+2}} f(x)\,dx$ if $f(x)$ is a quadratic polynomial. Use (a) to show that S_N is exact *for all N* if $f(x)$ is a quadratic polynomial.

60. Show that S_2 also gives the exact value for $\int_a^b x^3\,dx$ and conclude, as in Exercise 59, that S_N is exact for all cubic polynomials. Show by counterexample that S_2 is not exact for integrals of x^4.

61. Use the Error Bound for S_N to obtain another proof that Simpson's Rule is exact for all cubic polynomials.

62. 📖 **Sometimes, Simpson's Rule Performs Poorly** Calculate M_{10} and S_{10} for the integral $\int_0^1 \sqrt{1 - x^2}\,dx$, whose value we know to be $\frac{\pi}{4}$ (one-quarter of the area of the unit circle).

(a) We usually expect S_N to be more accurate than M_N. Which of M_{10} and S_{10} is more accurate in this case?

(b) How do you explain the result of part (a)? *Hint:* The Error Bounds are not valid because $|f''(x)|$ and $|f^{(4)}(x)|$ tend to ∞ as $x \to 1$, but $|f^{(4)}(x)|$ goes to infinity faster.

7.2 Integration by Parts

The Integration by Parts formula is derived from the Product Rule:

$$\big(u(x)v(x)\big)' = u(x)v'(x) + u'(x)v(x)$$

This formula states that $u(x)v(x)$ is an antiderivative of the right-hand side, so

$$u(x)v(x) = \int u(x)v'(x)\,dx + \int u'(x)v(x)\,dx$$

Combining the term on the left with the second integral on the right, we obtain

> *The Integration by Parts formula is often written using differentials:*
>
> $$\int u\,dv = uv - \int v\,du$$
>
> *Recall that $dv = v'(x)\,dx$ and $du = u'(x)\,dx$.*

Integration by Parts Formula

$$\int u(x)v'(x)\,dx = u(x)v(x) - \int u'(x)v(x)\,dx \qquad \boxed{1}$$

Note that the Integration by Parts formula applies to a product $u(x)v'(x)$, so we should consider using it when the integrand is a product of two functions.

■ **EXAMPLE 1** Evaluate $\displaystyle\int x\cos x\,dx$.

Solution The integrand is the product $x\cos x$, so we try $u(x) = x$ and $v'(x) = \cos x$. With this choice, the integral has the required form:

$$\int x\cos x\,dx = \int u(x)v'(x)\,dx$$

> *In applying Eq. (1), any antiderivative $v(x)$ of $v'(x)$ may be used.*

We apply Integration by Parts, observing that $u'(x) = 1$ and $v(x) = \sin x$ (since $v'(x) = \cos x$):

$$\int \underbrace{x\cos x}_{uv'}\,dx = \underbrace{x\sin x}_{uv} - \int \underbrace{\sin x}_{u'v}\,dx = x\sin x + \cos x + C$$

Let's check the answer by taking the derivative:

$$\frac{d}{dx}(x\sin x + \cos x + C) = x\cos x + \sin x - \sin x = x\cos x \qquad\blacksquare$$

The key step in Integration by Parts is deciding how to write the integrand as a product uv'. Keep in mind that Integration by Parts expresses $\displaystyle\int uv'\,dx$ in terms of $\displaystyle\int u'v\,dx$. *This is useful if $u'v$ is easier to integrate than uv'.* Here are two guidelines:

- Choose u so that u' is "simpler" than u itself.
- Choose v' so that $v = \displaystyle\int v'\,dx$ can be evaluated.

■ **EXAMPLE 2** Good Versus Bad Choices of u and v' Evaluate $\int xe^x\,dx$.

Solution Based on our guidelines, it makes sense to let

- $u = x$ (since $u' = 1$ is simpler)
- $v' = e^x$ (since we can evaluate $v = \int e^x\,dx = e^x + C$)

Integration by Parts gives us

$$\int xe^x\,dx = u(x)v(x) - \int u'(x)v(x)\,dx = xe^x - \int e^x\,dx = xe^x - e^x + C$$

Let's see what would have happened if we had chosen $u = e^x$, $v' = x$. Then

$$u'(x) = e^x, \qquad v(x) = \int x\,dx = \frac{1}{2}x^2 + C$$

$$\int \underbrace{xe^x}_{uv'}\,dx = \underbrace{\tfrac{1}{2}x^2 e^x}_{uv} - \boxed{\int \underbrace{\tfrac{1}{2}x^2 e^x}_{u'v}\,dx}$$

This is a poor choice because the integral on the right is more complicated than our original integral. ■

The Integration by Parts formula applies to *definite integrals*:

$$\boxed{\int_a^b u(x)v'(x)\,dx = u(x)v(x)\Big|_a^b - \int_a^b u'(x)v(x)\,dx}$$

■ **EXAMPLE 3** Taking $v' = 1$ Evaluate $\int_1^3 \ln x\,dx$.

Solution At first glance, this integral does not appear to be a candidate for Integration by Parts because the integrand is not a product. However, we are free to add a factor of 1 and write $\ln x = (\ln x)\cdot 1 = uv'$. Then

$$u = \ln x, \qquad v' = 1$$
$$u' = x^{-1}, \qquad v = x$$

Surprisingly, the choice $v' = 1$ is effective in some cases. Using it as in Example 3, we find that

$$\int \ln x\,dx = x\ln x - x + C$$

This trick also works for the inverse trigonometric functions (see Exercise 6).

Noting that $u'v = 1$, we obtain

$$\int_1^3 \ln x\,dx = \int_1^3 \underbrace{(\ln x)\cdot 1}_{uv'}\,dx = x\ln x\Big|_1^3 - \int_1^3 1\,dx = (3\ln 3 - 0) - 2 = 3\ln 3 - 2 \quad ■$$

■ **EXAMPLE 4** Integrating by Parts More Than Once Evaluate $\int x^2\cos x\,dx$.

In Example 4, it makes sense to take $u = x^2$ because Integration by Parts reduces the integration of $x^2\cos x$ to the integration of $2x\sin x$, which is easier.

Solution We use Integration by Parts with $u = x^2$ and $v' = \cos x$:

$$\int \underbrace{x^2\cos x}_{uv'}\,dx = \underbrace{x^2\sin x}_{uv} - \int \underbrace{(2x)(\sin x)}_{u'v}\,dx = x^2\sin x - 2\int x\sin x\,dx \quad \boxed{2}$$

We apply Integration by Parts a second time to evaluate $\int x \sin x \, dx$. Taking $u = x$ and $v' = \sin x$, we obtain

$$\int \underbrace{x \sin x}_{uv'} \, dx = \underbrace{-x \cos x}_{uv} - \int \underbrace{(-\cos x)}_{u'v} \, dx = -x \cos x + \sin x + C$$

Now use this result in Eq. (2):

$$\int x^2 \cos x \, dx = x^2 \sin x - 2(-x \cos x + \sin x) + C$$

$$= x^2 \sin x + 2x \cos x - 2 \sin x + C$$ ■

■ **EXAMPLE 5** Going in a Circle? Evaluate $\int e^x \cos x \, dx$.

We can also carry out the computation in Example 5 using the choice $u = e^x$, $v' = \cos x$.

Solution There are two natural ways of writing $e^x \cos x$ as uv'. Let's try the following:

$$u = \cos x, \qquad v' = e^x$$
$$u' = -\sin x, \qquad v = e^x$$

Integration by Parts yields

$$\int \underbrace{e^x \cos x}_{uv'} \, dx = \underbrace{e^x \cos x}_{uv} + \int \underbrace{e^x \sin x}_{-u'v} \, dx \qquad \boxed{3}$$

Now use Integration by Parts again to evaluate $\int e^x \sin x \, dx$:

$$u = \sin x, \qquad v' = e^x$$
$$u' = \cos x, \qquad v = e^x$$

$$\int e^x \sin x \, dx = e^x \sin x - \int e^x \cos x \, dx \qquad \boxed{4}$$

Equation (4) brings us back to our original integral of $e^x \cos x$, so it may seem that we're going in a circle. Fortunately, we can combine Eq. (4) with Eq. (3) and solve

$$\int e^x \cos x \, dx = e^x \cos x + e^x \sin x - \int e^x \cos x \, dx$$

$$2 \int e^x \cos x \, dx = e^x \cos x + e^x \sin x + C$$

$$\int e^x \cos x \, dx = \frac{1}{2} e^x (\cos x + \sin x) + C$$ ■

Integration by Parts can be used to derive **reduction formulas** for integrals that depend on a positive integer n such as $\int x^n e^x \, dx$ or $\int \ln^n x \, dx$. A reduction formula (also called a **recursive formula**) expresses the integral for a given value of n in terms of a similar integral for a smaller value of n. The desired integral is evaluated by applying the reduction formula repeatedly.

■ **EXAMPLE 6** A Reduction Formula   Verify the reduction formula

$$\int x^n e^x \, dx = x^n e^x - n \int x^{n-1} e^x \, dx \qquad \boxed{5}$$

Then use the reduction formula to evaluate $\int x^3 e^x \, dx$.

Solution We derive the reduction formula using Integration by Parts once:

$$u = x^n, \qquad v' = e^x$$
$$u' = nx^{n-1}, \qquad v = e^x$$
$$\int x^n e^x \, dx = uv - \int u'v \, dx = x^n e^x - n \int x^{n-1} e^x \, dx$$

To evaluate $\int x^3 e^x \, dx$, we'll need to use the reduction formula for $n = 3, 2, 1$:

Notice that $\int x^3 e^x \, dx$ is equal to $P(x)e^x + C$, where $P(x)$ is a polynomial of degree 3. In general, $\int x^n e^x \, dx = Q(x)e^x + C$, where $Q(x)$ is a polynomial of degree n.

$$\int x^3 e^x \, dx = x^3 e^x - 3 \int x^2 e^x \, dx$$

$$= x^3 e^x - 3\left(x^2 e^x - 2 \int x e^x \, dx \right) = x^3 e^x - 3x^2 e^x + 6 \int x e^x \, dx$$

$$= x^3 e^x - 3x^2 e^x + 6\left(x e^x - \int e^x \, dx \right) = x^3 e^x - 3x^2 e^x + 6x e^x - 6e^x + C$$

$$= (x^3 - 3x^2 + 6x - 6)e^x + C \qquad ■$$

7.2 SUMMARY

- *Integration by Parts* formula: $\int u(x)v'(x) \, dx = u(x)v(x) - \int u'(x)v(x) \, dx$.
- The key step in applying Integration by Parts is deciding how to write the integrand as a product uv'. Keep in mind that Integration by Parts is useful when $u'v$ is easier (or, at least, not more difficult) to integrate than uv'. Here are some guidelines:
  - Choose u so that u' is simpler than u itself.
  - Choose v' so that $v = \int v' \, dx$ can be evaluated.
  - Sometimes, $v' = 1$ is a good choice.

7.2 EXERCISES

Preliminary Questions

1. Which derivative rule is used to derive the Integration by Parts formula?

2. For each of the following integrals, state whether substitution or Integration by Parts should be used:

$$\int x \cos(x^2) \, dx, \qquad \int x \cos x \, dx, \qquad \int x^2 e^x \, dx, \qquad \int x e^{x^2} \, dx$$

3. Why is $u = \cos x$, $v' = x$ a poor choice for evaluating $\int x \cos x \, dx$?

Exercises

In Exercises 1–6, evaluate the integral using the Integration by Parts formula with the given choice of u and v′.

1. $\int x \sin x \, dx; \quad u = x, v' = \sin x$

2. $\int x e^{2x} \, dx; \quad u = x, v' = e^{2x}$

3. $\int (2x + 9) e^x \, dx; \quad u = 2x + 9, v' = e^x$

4. $\int x \cos(4x) \, dx; \quad u = x, v' = \cos(4x)$

5. $\int x^3 \ln x \, dx; \quad u = \ln x, v' = x^3$

6. $\int \tan^{-1} x \, dx; \quad u = \tan^{-1} x, v' = 1$

In Exercises 7–32, use Integration by Parts to evaluate the integral.

7. $\int (3x - 1) e^{-x} \, dx$

8. $\int x e^{-x} \, dx$

9. $\int x^2 e^x \, dx$

10. $\int x^2 \sin x \, dx$

11. $\int x \cos 2x \, dx$

12. $\int x^2 \sin(3x + 1) \, dx$

13. $\int e^{-x} \sin x \, dx$

14. $\int e^{4x} \cos 3x \, dx$

15. $\int x \ln x \, dx$

16. $\int \frac{\ln x}{x^2} \, dx$

17. $\int x^{-9} \ln x \, dx$

18. $\int e^x \sin(2x) \, dx$

19. $\int x \cos(2 - x) \, dx$

20. $\int x^2 \ln x \, dx$

21. $\int x \, 2^x \, dx$

22. $\int x \sec^2 x \, dx$

23. $\int (\ln x)^2 \, dx$

24. $\int \cos^{-1} x \, dx$

25. $\int \sin^{-1} x \, dx$

26. $\int \sec^{-1} x \, dx$

27. $\int x \, 5^x \, dx$

28. $\int (\sin x) 5^x \, dx$

29. $\int x \cosh 2x \, dx$

30. $\int \tanh^{-1}(4x) \, dx$

31. $\int \sinh^{-1} x \, dx$

32. $\int (\cos x)(\cosh x) \, dx$

33. Use the substitution $u = x^{1/2}$ and then Integration by Parts to evaluate $\int e^{\sqrt{x}} \, dx$.

34. Use substitution and then Integration by Parts to evaluate $\int x^3 e^{x^2} \, dx$.

In Exercises 35–44, evaluate using Integration by Parts, substitution, or both if necessary.

35. $\int x \cos 4x \, dx$

36. $\int \frac{\ln(\ln x) \, dx}{x}$

37. $\int \frac{x \, dx}{\sqrt{x + 1}}$

38. $\int \sin(\ln x) \, dx$

39. $\int \cos x \ln(\sin x) \, dx$

40. $\int x^3 (x^2 + 1)^{12} \, dx$

41. $\int \sin \sqrt{x} \, dx$

42. $\int \sqrt{x} e^{\sqrt{x}} \, dx$

43. $\int \frac{\ln(\ln x) \ln x \, dx}{x}$

44. $\int x \tan^{-1} x \, dx$

In Exercises 45–50, compute the definite integral.

45. $\int_0^2 x e^{9x} \, dx$

46. $\int_1^3 \ln x \, dx$

47. $\int_0^4 x \sqrt{4 - x} \, dx$

48. $\int_0^{\pi/4} x \sin(2x) \, dx$

49. $\int_1^4 \sqrt{x} \ln x \, dx$

50. $\int_0^1 \tan^{-1} x \, dx$

51. Use Eq. (5) to evaluate $\int x^4 e^x \, dx$.

52. Use substitution and then Eq. (5) to evaluate $\int x^4 e^{7x} \, dx$.

53. Find a reduction formula for $\int x^n e^{-x} \, dx$ similar to Eq. (5).

54. Evaluate $\int x^n \ln x \, dx$ for $n \neq -1$. Which method should be used to evaluate $\int x^{-1} \ln x \, dx$?

In Exercises 55–62, indicate a good method for evaluating the integral (but do not evaluate). Your choices are algebraic manipulation, substitution (specify u and du), and Integration by Parts (specify u and v′). If it appears that the techniques you have learned thus far are not sufficient, state this.

55. $\int \sqrt{x} \ln x \, dx$

56. $\int \frac{x^2 - \sqrt{x}}{2x} \, dx$

57. $\int \frac{x^3}{\sqrt{4 - x^2}} \, dx$

58. $\int \frac{dx}{\sqrt{4 - x^2}}$

59. $\int \frac{2x + 3}{x^2 + 3x + 6} \, dx$

60. $\int \frac{1}{x^2 + 3x + 6} \, dx$

61. $\int x \sin(3x + 4) \, dx$

62. $\int x \cos 3x \, dx$

63. Evaluate $\int (\sin^{-1} x)^2 \, dx$. *Hint:* First use Integration by Parts and then substitution.

64. Evaluate $\int \frac{(\ln x)^2 \, dx}{x^2}$. *Hint:* First use substitution and then Integration by Parts.

65. Evaluate $\int x^7 \cos(x^4) \, dx$.

66. Find $f(x)$, assuming that

$$\int f(x)e^x \, dx = f(x)e^x - \int x^{-1}e^x \, dx$$

67. Find the volume of the solid obtained by revolving $y = e^x$ for $0 \le x \le 2$ about the y-axis.

68. Find the volume of the solid obtained by revolving $y = \cos x$ for $0 \le x \le \frac{\pi}{2}$ about the y-axis.

69. Recall that the *present value* of an investment which pays out income continuously at a rate $R(t)$ for T years is $\int_0^T R(t)e^{-rt} \, dt$, where

r is the interest rate. Find the present value if income is produced at a rate $R(t) = 5,000 + 100t$ dollars/year for 10 years.

70. Prove the reduction formula

$$\int (\ln x)^k \, dx = x(\ln x)^k - k \int (\ln x)^{k-1} \, dx \qquad \boxed{6}$$

71. Use Eq. (6) to calculate $\int (\ln x)^k \, dx$ for $k = 2, 3$.

72. Prove the reduction formulas

$$\int x^n \cos x \, dx = x^n \sin x - n \int x^{n-1} \sin x \, dx$$

$$\int x^n \sin x \, dx = -x^n \cos x + n \int x^{n-1} \cos x \, dx$$

73. Prove $\int xb^x \, dx = b^x \left(\frac{x}{\ln b} - \frac{1}{\ln^2 b} \right) + C$.

Further Insights and Challenges

74. The Integration by Parts formula can be written

$$\int u(x)v(x) \, dx = u(x)V(x) - \int u'(x)V(x) \, dx \qquad \boxed{7}$$

where $V(x)$ satisfies $V'(x) = v(x)$.

(a) Show directly that the right-hand side of Eq. (7) does not change if $V(x)$ is replaced by $V(x) + C$, where C is a constant.

(b) Use $u = \tan^{-1} x$ and $v = x$ in Eq. (7) to calculate $\int x \tan^{-1} x \, dx$,

but carry out the calculation twice: first with $V(x) = \frac{1}{2}x^2$ and then with $V(x) = \frac{1}{2}x^2 + \frac{1}{2}$. Which choice of $V(x)$ results in a simpler calculation?

75. Prove in two ways

$$\int_0^a f(x) \, dx = af(a) - \int_0^a xf'(x) \, dx \qquad \boxed{8}$$

First use Integration by Parts. Then assume $f(x)$ is increasing. Use the substitution $u = f(x)$ to prove that $\int_0^a xf'(x) \, dx$ is equal to the area of the shaded region in Figure 1 and derive Eq. (8) a second time.

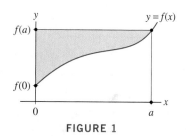

FIGURE 1

76. Assume that $f(0) = f(1) = 0$ and that f'' exists. Prove

$$\int_0^1 f''(x)f(x) \, dx = -\int_0^1 f'(x)^2 \, dx \qquad \boxed{9}$$

Use this to prove that if $f(0) = f(1) = 0$ and $f''(x) = \lambda f(x)$ for some constant λ, then $\lambda < 0$. Can you think of a function satisfying these conditions for some λ?

77. Set $I(a, b) = \int_0^1 x^a(1 - x)^b \, dx$, where a, b are whole numbers.

(a) Use substitution to show that $I(a, b) = I(b, a)$.

(b) Show that $I(a, 0) = I(0, a) = \dfrac{1}{a + 1}$.

(c) Prove that for $a \ge 1$ and $b \ge 0$,

$$I(a, b) = \frac{a}{b + 1} I(a - 1, b + 1)$$

(d) Use (b) and (c) to calculate $I(1, 1)$ and $I(3, 2)$.

(e) Show that $I(a, b) = \dfrac{a! \, b!}{(a + b + 1)!}$.

78. Show by differentiation that if $P_n(x)$ is a polynomial of degree n satisfying $P_n(x) + P_n'(x) = x^n$, then

$$\int x^n e^x \, dx = P_n(x) \, e^x + C$$

Find $P_n(x)$ for $n = 2, 3, 4$.

79. Let $I_n = \int x^n \cos(x^2) \, dx$ and $J_n = \int x^n \sin(x^2) \, dx$.

(a) Find a reduction formula that expresses I_n in terms of J_{n-2}. *Hint:* Write $x^n \cos(x^2)$ as $x^{n-1}(x \cos(x^2))$.

(b) Use the result of (a) to show that I_n can be evaluated explicitly if n is odd.

(c) Evaluate I_3.

7.3 Trigonometric Integrals

In this section, we combine substitution and Integration by Parts with the appropriate identities to integrate various trigonometric functions. First, consider

$$\int \sin^m x \cos^n x\, dx$$

where n, m are positive integers. The easier case occurs when one or both of the exponents m, n is *odd*.

■ **EXAMPLE 1** Odd Power of $\sin x$ Evaluate $\int \sin^3 x\, dx$.

Solution First use the identity $\sin^2 x = 1 - \cos^2 x$ to rewrite the integrand:

$$\sin^3 x = \sin^2 x \sin x = (1 - \cos^2 x) \sin x$$

$$\int \sin^3 x\, dx = \int (1 - \cos^2 x) \sin x\, dx$$

Then use the substitution $u = \cos x$, $du = -\sin x\, dx$:

$$\int \sin^3 x\, dx = \int (1 - \cos^2 x) \sin x\, dx = -\int (1 - u^2)\, du \qquad \boxed{1}$$

$$= \frac{u^3}{3} - u + C = \frac{\cos^3 x}{3} - \cos x + C \qquad \boxed{2}$$

■

In general, if n is odd, we may use the identity $\cos^2 x = 1 - \sin^2 x$ to write $\cos^n x$ as a power of $(1 - \sin^2 x)$ times $\cos x$ and substitute $u = \sin x$, $du = \cos x\, dx$. Alternatively, if m is odd, we write $\sin^m x$ as a power of $(1 - \cos^2 x)$ times $\sin x$.

___

Sine or Cosine to Odd Power

If $\cos x$ occurs to an odd power $2k + 1$, write $\cos^{2k+1} x$ as $(1 - \sin^2 x)^k \cos x$. Then

$$\int \sin^m x \cos^{2k+1} x\, dx \text{ becomes}$$

$$\int (\sin^m x)(1 - \sin^2 x)^k \cos x\, dx$$

Now substitute $u = \sin x$, $du = \cos x\, dx$. If $\sin x$ occurs to an odd power $2k + 1$, write $\sin^{2k+1} x$ as $(1 - \cos^2 x)^k \sin x$ and substitute $u = \cos x$, $du = -\sin x\, dx$.

___

■ **EXAMPLE 2** Odd Power of $\sin x$ or $\cos x$ Evaluate $\int \sin^4 x \cos^5 x\, dx$.

Solution Write $\cos^5 x = \cos^4 x \cdot \cos x = (1 - \sin^2 x)^2 \cos x$. Then

$$\int \sin^4 x \cos^5 x\, dx = \int (\sin^4 x)(1 - \sin^2 x)^2 \cos x\, dx$$

The factor of $\cos x$ in the integrand allows us to use the substitution $u = \sin x$. We have $(\sin^4 x)(1 - \sin^2 x)^2 = u^4(1 - u^2)^2$ and the leftover factor $\cos x\, dx$ is equal to du:

$$\int \sin^4 x \cos^5 x\, dx = \int u^4(1 - u^2)^2\, du = \int (u^4 - 2u^6 + u^8)\, du$$

$$= \frac{u^5}{5} - \frac{2u^7}{7} + \frac{u^9}{9} + C = \frac{\sin^5 x}{5} - \frac{2\sin^7 x}{7} + \frac{\sin^9 x}{9} + C \quad ■$$

Since the method of previous examples require an odd exponent, we state the following reduction formulas that can be used for any exponent, even or odd. They are derived using Integration by Parts (see Exercise 62).

Reduction Formulas for Sine and Cosine

$$\int \sin^n x \, dx = -\frac{1}{n} \sin^{n-1} x \cos x + \frac{n-1}{n} \int \sin^{n-2} x \, dx \qquad \boxed{3}$$

$$\int \cos^n x \, dx = \frac{1}{n} \cos^{n-1} x \sin x + \frac{n-1}{n} \int \cos^{n-2} x \, dx \qquad \boxed{4}$$

■ **EXAMPLE 3** Evaluate $\int \sin^4 x \, dx$.

Solution By reduction formula (3) with $n = 4$, we obtain

$$\int \sin^4 x \, dx = -\frac{1}{4} \sin^3 x \cos x + \frac{3}{4} \int \sin^2 x \, dx \qquad \boxed{5}$$

To evaluate the integral on the right, we apply the reduction formula again with $n = 2$:

$$\int \sin^2 x \, dx = -\frac{1}{2} \sin x \cos x + \frac{1}{2} \int dx = -\frac{1}{2} \sin x \cos x + \frac{1}{2} x + C \qquad \boxed{6}$$

Using Eq. (6) in Eq. (5), we obtain

$$\int \sin^4 x \, dx = -\frac{1}{4} \sin^3 x \cos x - \frac{3}{8} \sin x \cos x + \frac{3}{8} x + C \qquad ■$$

Trigonometric integrals can be expressed in more than one way because of the large number of trigonometric identities. For example, a computer algebra system may yield the following evaluation of the integral in the previous example:

$$\int \sin^4 x \, dx = \frac{1}{32}(x - 8 \sin 2x + \sin 4x) + C$$

You can check that this agrees with the result in Example 3 (Exercise 55). The following formulas are verified using the identities recalled in the margin.

◀·· *REMINDER Useful Identities:*

$$\sin^2 x = \frac{1}{2}(1 - \cos 2x)$$

$$\cos^2 x = \frac{1}{2}(1 + \cos 2x)$$

$$\sin 2x = 2 \sin x \cos x$$

$$\cos 2x = \cos^2 x - \sin^2 x$$

$$\int \sin^2 x \, dx = \frac{x}{2} - \frac{\sin 2x}{4} + C = \frac{x}{2} - \frac{1}{2} \sin x \cos x + C$$

$$\int \cos^2 x \, dx = \frac{x}{2} + \frac{\sin 2x}{4} + C = \frac{x}{2} + \frac{1}{2} \sin x \cos x + C$$

More work is required to integrate $\sin^m x \cos^n x$ when both n and m are even:

• If $m \leq n$, use the identity $\sin^2 x = 1 - \cos^2 x$ to replace $\sin^m x$ by $(1 - \cos^2 x)^{m/2}$:

$$\int \sin^m x \cos^n x \, dx = \int (1 - \cos^2 x)^{m/2} \cos^n x \, dx$$

Expand the integral on the right to obtain a sum of integrals of powers of $\cos x$ and use reduction formula (4).

- If $m \geq n$, replace $\cos^n x$ by $(1 - \sin^2 x)^{n/2}$:

$$\int \sin^m x \cos^n x \, dx = \int (\sin^m x)(1 - \sin^2 x)^{n/2} \, dx$$

Expand the integral on the right to obtain a sum of integrals of powers of $\sin x$ and again evaluate using reduction formula (3).

Another method for integrating $\sin^m x \cos^n x$ when m and n are even is discussed in Exercise 63.

■ **EXAMPLE 4** Even Powers of $\sin x$ and $\cos x$ Evaluate $\displaystyle\int \sin^2 x \cos^4 x \, dx$.

Solution Since $m < n$, we replace $\sin^2 x$ by $1 - \cos^2 x$:

$$\int \sin^2 x \cos^4 x \, dx = \int (1 - \cos^2 x) \cos^4 x \, dx = \int \cos^4 x \, dx - \int \cos^6 x \, dx \quad \boxed{7}$$

The reduction formula for $n = 6$ gives

$$\int \cos^6 x \, dx = \frac{1}{6} \cos^5 x \sin x + \frac{5}{6} \int \cos^4 x \, dx$$

Using this result in Eq. (7), we obtain

$$\int \sin^2 x \cos^4 x \, dx = -\frac{1}{6} \cos^5 x \sin x + \frac{1}{6} \int \cos^4 x \, dx$$

Next evaluate $\displaystyle\int \cos^4 x \, dx$ using the reduction formulas for $n = 4$ and $n = 2$:

$$\int \cos^4 x \, dx = \frac{1}{4} \cos^3 x \sin x + \frac{3}{4} \int \cos^2 x \, dx$$

$$= \frac{1}{4} \cos^3 x \sin x + \frac{3}{4} \left(\frac{1}{2} \cos x \sin x + \frac{1}{2} x \right) + C$$

$$= \frac{1}{4} \cos^3 x \sin x + \frac{3}{8} \cos x \sin x + \frac{3}{8} x + C$$

As noted above, trigonometric integrals can be expressed in more than one way. According to Mathematica,

$$\int \sin^2 x \cos^4 x \, dx$$

$$= \tfrac{1}{16} x + \tfrac{1}{64} \sin 2x - \tfrac{1}{64} \sin 4x - \tfrac{1}{192} \sin 6x$$

Trigonometric identities can be used to show that this agrees with Eq. (8).

Altogether,

$$\int \sin^2 x \cos^4 x \, dx = -\frac{1}{6} \cos^5 x \sin x + \frac{1}{6} \left(\frac{1}{4} \cos^3 x \sin x + \frac{3}{8} \cos x \sin x + \frac{3}{8} x \right) + C$$

$$= -\frac{1}{6} \cos^5 x \sin x + \frac{1}{24} \cos^3 x \sin x + \frac{1}{16} \cos x \sin x + \frac{1}{16} x + C$$

$$\boxed{8}$$

■

Many other trigonometric integrals can be evaluated using the appropriate trigonometric identities combined with substitution or Integration by Parts.

■ **EXAMPLE 5** Integral of the Tangent and Secant Derive the formulas

$$\int \tan x \, dx = \ln | \sec x | + C, \qquad \int \sec x \, dx = \ln \left| \sec x + \tan x \right| + C$$

The integral $\int \sec x \, dx$ was first computed numerically in the 1590s by the English mathematician Edward Wright, decades before the invention of calculus. Although he did not invent the concept of an integral, Wright realized that the sums which approximate the integral hold the key to understanding the Mercator map projection, of great importance in sea navigation because it allowed sailors to reach their destinations along lines of fixed compass direction. The formula for the integral was first proved by James Gregory in 1668.

Solution To integrate $\tan x$, use the substitution $u = \cos x$, $du = -\sin x \, dx$:

$$\int \tan x \, dx = \int \frac{\sin x}{\cos x} \, dx = -\int \frac{du}{u} = -\ln|u| + C = -\ln|\cos x| + C$$

$$= \ln \frac{1}{|\cos x|} + C = \ln|\sec x| + C$$

To integrate $\sec x$, we employ a clever substitution $u = \sec x + \tan x$. Then

$$du = (\sec x \tan x + \sec^2 x) \, dx = (\sec x) \underbrace{(\tan x + \sec x)}_{u} \, dx$$

Dividing both sides by u, we obtain $\dfrac{du}{u} = \sec x \, dx$ and thus

$$\int \sec x \, dx = \int \frac{du}{u} = \ln|u| + C = \ln\left| \sec x + \tan x \right| + C \qquad \blacksquare$$

The table of integrals at the end of this section (page 438) contains a list of additional trigonometric integrals and reduction formulas.

■ **EXAMPLE 6** Power of $\tan x$ Evaluate $\displaystyle\int_0^{\pi/4} \tan^3 x \, dx$.

Solution We use reduction formula (18) in the table with $k = 3$:

$$\int_0^{\pi/4} \tan^3 x \, dx = \frac{\tan^2 x}{2}\bigg|_0^{\pi/4} - \int_0^{\pi/4} \tan x \, dx = \left(\frac{1}{2}\tan^2 x - \ln|\sec x| \right)\bigg|_0^{\pi/4}$$

$$= \left(\frac{1}{2}\tan^2 \frac{\pi}{4} - \ln\left|\sec \frac{\pi}{4}\right| \right) - \left(\frac{1}{2}\tan^2 0 - \ln|\sec 0| \right)$$

$$= \left(\frac{1}{2}(1)^2 - \ln\sqrt{2} \right) - \left(\frac{1}{2}0^2 - \ln|1| \right) = \frac{1}{2} - \ln\sqrt{2} \qquad \blacksquare$$

In the margin, we describe a method for integrating $\tan^m x \sec^n x$.

■ **EXAMPLE 7** Using Reduction Formulas and the Table of Integrals   Evaluate

$$\int \tan^2 x \sec^3 x \, dx.$$

Solution Referring to the cases described in the marginal note, we see that this integral is covered by Case 3 since the integrand is $\tan^m x \sec^n x$, with $m = 2$ and $n = 3$. The first step is to use the identity $\tan^2 x = \sec^2 x - 1$:

$$\int \tan^2 x \sec^3 x \, dx = \int (\sec^2 x - 1)\sec^3 x \, dx = \int \sec^5 x \, dx - \int \sec^3 x \, dx \qquad \boxed{9}$$

Next we use the reduction formula (22) in the table with $m = 5$:

$$\int \sec^5 x \, dx = \frac{\tan x \sec^3 x}{4} + \frac{3}{4}\int \sec^3 x \, dx$$

$$\int \tan^2 x \sec^3 x \, dx = \left(\frac{\tan x \sec^3 x}{4} + \frac{3}{4}\int \sec^3 x \, dx \right) - \int \sec^3 x \, dx$$

$$= \frac{1}{4}\tan x \sec^3 x - \frac{1}{4}\int \sec^3 x \, dx \qquad \boxed{10}$$

Integrating $\tan^m x \sec^n x$

Case 1 $m = 2k + 1$ **odd**

Use the identity $\tan^2 x = \sec^2 x - 1$ to write $\tan^{2k+1} x \sec^n x$ as

$$(\sec^2 x - 1)^k (\sec^{n-1} x)(\sec x \tan x)$$

Then substitute $u = \sec x$, $du = \sec x \tan x \, dx$ to obtain an integral involving only powers of u.

Case 2 $n = 2k$ **even**

Use the identity $\sec^2 x = 1 + \tan^2 x$ to write $\tan^m x \sec^n x$ as

$$(\tan^m x)(1 + \tan^2 x)^{k-1} \sec^2 x$$

Then substitute $u = \tan x$, $du = \sec^2 x \, dx$ to obtain an integral involving only powers of u.

Case 3 m **even and** n **odd**

Use the identity $\tan^2 x = \sec^2 x - 1$ to write $\tan^m x \sec^n x$ as

$$(\sec^2 x - 1)^{m/2} \sec^n x$$

Expand to obtain an integral involving only powers of $\sec x$ and use the reduction formula (22).

Finally, we use the reduction formula (22) again with $m = 3$ and formula (21):

$$\int \sec^3 x \, dx = \frac{\tan x \sec x}{2} + \frac{1}{2} \int \sec x \, dx = \frac{1}{2} \tan x \sec x + \frac{1}{2} \ln |\sec x + \tan x| + C$$

Then Eq. (10) becomes

$$\int \tan^2 x \sec^3 x \, dx = \frac{1}{4} \tan x \sec^3 x - \frac{1}{4} \left(\frac{1}{2} \tan x \sec x + \frac{1}{2} \ln |\sec x + \tan x| \right) + C$$

$$= \frac{1}{4} \tan x \sec^3 x - \frac{1}{8} \tan x \sec x - \frac{1}{8} \ln |\sec x + \tan x| + C \quad \blacksquare$$

Formulas (25)–(27) in the table describe the integrals of products of the form $\sin mx \sin nx$, $\cos mx \cos nx$, and $\sin mx \cos nx$. These integrals play an important role in the theory of Fourier Series, a more advanced technique used extensively in engineering and physics.

■ **EXAMPLE 8** Integral of $\sin mx \cos nx$ Evaluate $\int_0^\pi \sin 4x \cos 3x \, dx$.

Solution We apply (26), with $m = 4$ and $n = 3$:

$$\int_0^\pi \sin 4x \cos 3x \, dx = \left(-\frac{\cos(4-3)x}{2(4-3)} - \frac{\cos(4+3)x}{2(4+3)} \right) \Big|_0^\pi$$

$$= \left(-\frac{\cos x}{2} - \frac{\cos 7x}{14} \right) \Big|_0^\pi$$

$$= \left(\frac{1}{2} + \frac{1}{14} \right) - \left(-\frac{1}{2} - \frac{1}{14} \right) = \frac{8}{7} \quad \blacksquare$$

7.3 SUMMARY

- To integrate an odd power of $\sin x$ times $\cos^n x$, write

$$\int \sin^{2k+1} x \cos^n x \, dx = \int (1 - \cos^2 x)^k \cos^n x \sin x \, dx$$

Then use the substitution $u = \cos x$, $du = -\sin x \, dx$.
- To integrate an odd power of $\cos x$ times $\sin^m x$, write

$$\int \sin^m x \cos^{2k+1} x \, dx = \int (\sin^m x)(1 - \sin^2 x)^k \cos x \, dx$$

Then use the substitution $u = \sin x$, $du = \cos x \, dx$.
- If both $\sin x$ and $\cos x$ occur to an even power, write

$$\int \sin^m x \cos^n x \, dx = \int (1 - \cos^2 x)^{m/2} \cos^n x \, dx \quad (\text{if } m \leq n)$$

$$\int \sin^m x \cos^n x \, dx = \int \sin^m x (1 - \sin^2 x)^{n/2} \, dx \quad (\text{if } m \geq n)$$

Expand the right-hand side to obtain a sum of powers of $\cos x$ or powers of $\sin x$. Then use the reduction formulas

$$\int \sin^n x \, dx = -\frac{1}{n} \sin^{n-1} x \cos x + \frac{n-1}{n} \int \sin^{n-2} x \, dx$$

$$\int \cos^n x \, dx = \frac{1}{n} \cos^{n-1} x \sin x + \frac{n-1}{n} \int \cos^{n-2} x \, dx$$

• The integral $\int \tan^m x \sec^n x \, dx$ can be evaluated by a substitution that depends on three cases: m odd, n even, or m even and n odd. See the marginal note on page 436.

TABLE OF TRIGONOMETRIC INTEGRALS

$$\int \sin^2 x \, dx = \frac{x}{2} - \frac{\sin 2x}{4} + C = \frac{x}{2} - \frac{1}{2} \sin x \cos x + C \qquad \boxed{11}$$

$$\int \cos^2 x \, dx = \frac{x}{2} + \frac{\sin 2x}{4} + C = \frac{x}{2} + \frac{1}{2} \sin x \cos x + C \qquad \boxed{12}$$

$$\int \sin^n x \, dx = -\frac{\sin^{n-1} x \cos x}{n} + \frac{n-1}{n} \int \sin^{n-2} x \, dx \qquad \boxed{13}$$

$$\int \cos^n x \, dx = \frac{\cos^{n-1} x \sin x}{n} + \frac{n-1}{n} \int \cos^{n-2} x \, dx \qquad \boxed{14}$$

$$\int \sin^m x \cos^n x \, dx = \frac{\sin^{m+1} x \cos^{n-1} x}{m+n} + \frac{n-1}{m+n} \int \sin^m x \cos^{n-2} x \, dx \qquad \boxed{15}$$

$$\int \sin^m x \cos^n x \, dx = -\frac{\sin^{m-1} x \cos^{n+1} x}{m+n} + \frac{m-1}{m+n} \int \sin^{m-2} x \cos^n x \, dx \qquad \boxed{16}$$

$$\int \tan x \, dx = \ln |\sec x| + C = -\ln |\cos x| + C \qquad \boxed{17}$$

$$\int \tan^m x \, dx = \frac{\tan^{m-1} x}{m-1} - \int \tan^{m-2} x \, dx \qquad \boxed{18}$$

$$\int \cot x \, dx = -\ln |\csc x| + C = \ln |\sin x| + C \qquad \boxed{19}$$

$$\int \cot^m x \, dx = -\frac{\cot^{m-1} x}{m-1} - \int \cot^{m-2} x \, dx \qquad \boxed{20}$$

$$\int \sec x \, dx = \ln |\sec x + \tan x| + C \qquad \boxed{21}$$

$$\int \sec^m x \, dx = \frac{\tan x \sec^{m-2} x}{m-1} + \frac{m-2}{m-1} \int \sec^{m-2} x \, dx \qquad \boxed{22}$$

$$\int \csc x \, dx = \ln |\csc x - \cot x| + C \qquad \boxed{23}$$

$$\int \csc^m x \, dx = -\frac{\cot x \csc^{m-2} x}{m-1} + \frac{m-2}{m-1} \int \csc^{m-2} x \, dx \qquad \boxed{24}$$

$$\int \sin mx \sin nx \, dx = \frac{\sin(m-n)x}{2(m-n)} - \frac{\sin(m+n)x}{2(m+n)} + C \quad (m \neq \pm n) \qquad \boxed{25}$$

$$\int \sin mx \cos nx \, dx = -\frac{\cos(m-n)x}{2(m-n)} - \frac{\cos(m+n)x}{2(m+n)} + C \quad (m \neq \pm n) \qquad \boxed{26}$$

$$\int \cos mx \cos nx \, dx = \frac{\sin(m-n)x}{2(m-n)} + \frac{\sin(m+n)x}{2(m+n)} + C \quad (m \neq \pm n) \qquad \boxed{27}$$

7.3 EXERCISES

Preliminary Questions

1. Describe the technique used to evaluate $\int \sin^5 x \, dx$.

2. Describe a way of evaluating $\int \sin^6 x \, dx$.

3. Are reduction formulas needed to evaluate $\int \sin^7 x \cos^2 x \, dx$? Why or why not?

4. Describe a way of evaluating $\int \sin^6 x \cos^2 x \, dx$.

5. Which integral requires more work to evaluate?

$$\int \sin^{798} x \cos x \, dx \quad \text{or} \quad \int \sin^4 x \cos^4 x \, dx$$

Explain your answer.

Exercises

In Exercises 1–6, use the method for odd powers to evaluate the integral.

1. $\int \cos^3 x \, dx$.

2. $\int \sin^5 x \, dx$

3. $\int \sin^3 \theta \cos^2 \theta \, d\theta$

4. $\int \cos x \sin^5 x \, dx$

5. $\int \sin^3 t \cos^3 t \, dt$

6. $\int \sin^2 x \cos^5 x \, dx$

7. Find the area of the shaded region in Figure 1.

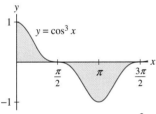

FIGURE 1 Graph of $y = \cos^3 x$.

8. Use the identity $\sin^2 x = 1 - \cos^2 x$ to write $\int \sin^2 x \cos^2 x \, dx$ as a sum of two integrals and then evaluate using the reduction formula.

In Exercises 9–12, evaluate the integral using methods employed in Examples 3 and 4.

9. $\int \cos^4 y \, dy$

10. $\int \cos^2 \theta \sin^2 \theta \, d\theta$

11. $\int \sin^4 x \cos^2 x \, dx$

12. $\int \sin^4 x \cos^6 x \, dx$

In Exercises 13–20, evaluate the integral using the reduction formulas on page 438 as necessary.

13. $\int \tan^2 x \sec^2 x \, dx$

14. $\int \tan^3 x \sec^2 x \, dx$

15. $\int \tan^4 \theta \sec^2 \theta \, d\theta$

16. $\int \tan^2 x \sec x \, dx$

17. $\int \tan^2 t \, dt$

18. $\int \cot^3 x \, dx$

19. $\int \sec^3 x \, dx$

20. $\int \csc^2 x \, dx$

In Exercises 21–54, use the techniques and reduction formulas necessary to evaluate the integral.

21. $\int \cos^5 x \sin x \, dx$

22. $\int \cos^3 2x \sin 2x \, dx$

23. $\int \cos^4(3x) \, dx$

24. $\int \cos^7 3x \, dx$

25. $\int \cos^3(\pi\theta) \sin^4(\pi\theta) \, d\theta$

26. $\int \cos^{498} y \sin^3 y \, dy$

27. $\int \sin^4(3x) \, dx$

28. $\int \sin^2 x \cos^6 x \, dx$

29. $\int \sec 7t \, dt$

30. $\int \csc^3 x \, dx$

31. $\int \tan x \sec^2 x \, dx$

32. $\int \tan^3 \theta \sec^3 \theta \, d\theta$

33. $\int \tan^5 x \sec^4 x \, dx$

34. $\int \tan^2 x \sec^4 x \, dx$

35. $\int \tan^4 x \sec x \, dx$

36. $\int \tan^6 x \sec^4 x \, dx$

37. $\int \tan^4 x \sec^3 x \, dx$

38. $\int \tan^2 x \csc^2 x \, dx$

39. $\int \sin 2x \cos 2x \, dx$

40. $\int \cos 4x \cos 6x \, dx$

41. $\int \sin 2x \cos 4x \, dx$

42. $\int t \cos^3(t^2) \, dt$

43. $\int \frac{\tan^3(\ln t)}{t} \, dt$

44. $\int \cos^2(\sin t) \cos t \, dt$

45. $\int_0^{2\pi} \sin^2 x \, dx$

46. $\int_0^{\pi/2} \cos^3 x \, dx$

47. $\int_0^{\pi/3} \sin^3 x \, dx$

48. $\int_0^{\pi/4} \frac{dx}{\cos x}$

49. $\displaystyle\int_{\pi/4}^{\pi/2} \frac{dx}{\sin x}$

50. $\displaystyle\int_{0}^{\pi/3} \tan x \, dx$

51. $\displaystyle\int_{0}^{\pi/4} \tan^5 x \, dx$

52. $\displaystyle\int_{-\pi/4}^{\pi/4} \sec^4 x \, dx$

53. $\displaystyle\int_{0}^{\pi} \sin 3x \cos 4x \, dx$

54. $\displaystyle\int_{0}^{\pi} \sin x \sin 3x \, dx$

55. Use the identities for $\sin 2x$ and $\cos 2x$ listed on page 434 to verify that the first of the following formulas is equivalent to the second:

$$\int \sin^4 x \, dx = \frac{1}{32}\left(12x - 8\sin 2x + \sin 4x\right) + C$$

$$\int \sin^4 x \, dx = -\frac{1}{4}\sin^3 x \cos x - \frac{3}{8}\sin x \cos x + \frac{3}{8}x + C$$

56. Evaluate $\displaystyle\int \sin^2 x \cos^3 x \, dx$ using the method described in the text and verify that your result is equivalent to the following result produced by a computer algebra system:

$$\int \sin^2 x \cos^3 x \, dx = \frac{1}{30}(7 + 3\cos 2x)\sin^3 x + C$$

In Exercises 57–60, evaluate using the identity $\cot^2 x + 1 = \csc^2 x$ and methods similar to those for integrating $\tan^m x \sec^n x$.

57. $\displaystyle\int \cot^3 x \csc x \, dx$

58. $\displaystyle\int \cot^4 x \csc^2 x \, dx$

59. $\displaystyle\int \cot^2 x \csc^2 x \, dx$

60. $\displaystyle\int \cot^2 x \csc^3 x \, dx$

61. Find the volume of the solid obtained by revolving $y = \sin x$ for $0 \le x \le \pi$ about the x-axis.

62. Use Integration by Parts to prove the reductions formulas Eqs. (3) and (4).

63. Here is another reduction method for evaluating the integral $J = \displaystyle\int \sin^m x \cos^n x \, dx$ when m and n are even. Use the identities

$$\sin^2 x = \frac{1}{2}(1 - \cos 2x), \qquad \cos^2 x = \frac{1}{2}(1 + \cos 2x)$$

to write $J = \dfrac{1}{4}\displaystyle\int (1 - \cos 2x)^{m/2}(1 + \cos 2x)^{n/2}\, dx$. Then expand the right-hand side as a sum of integrals involving smaller powers of sine and cosine in the variable $2x$. Use this method to evaluate $J = \displaystyle\int \sin^2 x \cos^2 x \, dx$.

64. Use the method of Exercise 63 to evaluate $\displaystyle\int \cos^4 x \, dx$.

65. Use the method of Exercise 63 to evaluate $\displaystyle\int \sin^4 x \cos^2 x \, dx$.

66. Show that for $n \ge 2$,

$$\int_{0}^{\pi/2} \sin^n x \, dx = \frac{n-1}{n}\int_{0}^{\pi/2} \sin^{n-2} x \, dx \qquad \boxed{28}$$

67. Prove the reduction formula

$$\int \tan^k x \, dx = \frac{\tan^{k-1} x}{k-1} - \int \tan^{k-2} x \, dx$$

Hint: Use the identity $\tan^2 x = (\sec^2 x - 1)$ to write

$$\tan^k x = (\sec^2 x - 1)\tan^{k-2} x.$$

68. Evaluate $I = \displaystyle\int \sin^2 x \cos^4 x \, dx$ using (15). Show:

(a) $I = \dfrac{1}{6}\sin^3 x \cos^3 x + \dfrac{1}{2}\displaystyle\int \sin^2 x \cos^2 x \, dx$

(b) $\displaystyle\int \sin^2 x \cos^2 x \, dx = \dfrac{1}{4}\sin^3 x \cos x + \dfrac{1}{4}\displaystyle\int \sin^2 x \, dx$

69. Use the substitution $u = \csc x - \cot x$ to evaluate $\displaystyle\int \csc x \, dx$ (see Example 5).

70. Total Energy A 100-W light bulb has resistance $R = 144 \ \Omega$ (ohms) when attached to household current, where the voltage varies as $V = V_0 \sin(2\pi f t)$ ($V_0 = 110$ V, $f = 60$ Hz). The power supplied to the bulb is $P = V^2/R$ (joules per second) and the total energy expended over a time period $[0, T]$ (in seconds) is $U = \displaystyle\int_{0}^{T} P(t)\, dt$. Compute U if the bulb remains on for 5 hours.

71. Let m, n be integers with $m \ne \pm n$. Use formulas (25)–(27) in the table of trigonometric integrals to prove that

$$\int_{0}^{\pi} \sin mx \sin nx \, dx = 0, \qquad \int_{0}^{\pi} \cos mx \cos nx \, dx = 0$$

$$\int_{0}^{2\pi} \sin mx \cos nx \, dx = 0$$

These formulas, known as the **orthogonality relations**, play a basic role in the theory of Fourier Series (Figure 2).

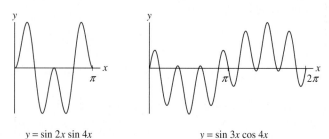

$y = \sin 2x \sin 4x$ $\qquad\qquad$ $y = \sin 3x \cos 4x$

FIGURE 2 By the orthogonality relations, the signed area under these graphs is zero.

Further Insights and Challenges

72. Use the trigonometric identity

$$\sin mx \cos nx = \frac{1}{2}\big(\sin(m-n)x + \sin(m+n)x\big)$$

to prove Eq. (26) in the table of integrals on page 438.

73. Evaluate $\displaystyle\int_0^\pi \sin^2 mx\, dx$ for m an arbitrary integer.

74. Use Integration by Parts to prove (for $m \neq 1$)

$$\int \sec^m x\, dx = \frac{\tan x \sec^{m-2} x}{m-1} + \frac{m-2}{m-1} \int \sec^{m-2} x\, dx$$

75. Set $I_m = \displaystyle\int_0^{\pi/2} \sin^m x\, dx$.

(a) Show that $I_1 = 1$, $I_2 = \left(\frac{1}{2}\right)\left(\frac{\pi}{2}\right)$ and use Eq. (28) to prove that for $m > 1$, $I_m = \left(\dfrac{m-1}{m}\right) I_{m-2}$.

(b) Show that $I_3 = \frac{2}{3}$ and $I_4 = \left(\frac{3}{4}\right)\left(\frac{1}{2}\right)\left(\frac{\pi}{2}\right)$.

(c) Show more generally:

$$I_{2m} = \frac{2m-1}{2m}\frac{2m-3}{2m-2}\cdots\frac{1}{2}\cdot\frac{\pi}{2}$$

$$I_{2m+1} = \frac{2m}{2m+1}\frac{2m-2}{2m-1}\cdots\frac{2}{3}$$

(d) Conclude that

$$\frac{\pi}{2} = \frac{2\cdot 2}{1\cdot 3}\cdot\frac{4\cdot 4}{3\cdot 5}\cdots\frac{2m\cdot 2m}{(2m-1)(2m+1)}\frac{I_{2m}}{I_{2m+1}}$$

76. This is a continuation of Exercise 75.

(a) Prove that $I_{2m+1} \leq I_{2m} \leq I_{2m-1}$. *Hint:* Observe that

$$\sin^{2m+1} x \leq \sin^{2m} x \leq \sin^{2m-1} x \quad \text{for} \quad 0 \leq x \leq \frac{\pi}{2}$$

(b) Show that $\dfrac{I_{2m-1}}{I_{2m+1}} = 1 + \dfrac{1}{2m}$.

(c) Show that

$$1 \leq \frac{I_{2m}}{I_{2m+1}} \leq 1 + \frac{1}{2m}$$

(d) Prove that $\displaystyle\lim_{m\to\infty} \frac{I_{2m}}{I_{2m+1}} = 1$.

(e) Finally, deduce the infinite product for $\frac{\pi}{2}$ discovered by English mathematician John Wallis (1616–1703):

$$\frac{\pi}{2} = \lim_{m\to\infty} \frac{2}{1}\cdot\frac{2}{3}\cdot\frac{4}{3}\cdot\frac{4}{5}\cdots\frac{2m\cdot 2m}{(2m-1)(2m+1)}$$

7.4 Trigonometric Substitution

Our next goal is to integrate functions involving one of the square root expressions:

$$\sqrt{a^2 - x^2}, \qquad \sqrt{x^2 + a^2}, \qquad \sqrt{x^2 - a^2}$$

In each case, a substitution transforms the integral into a trigonometric integral. For example, the substitution $x = a\sin\theta$ may be used when the integrand involves $\sqrt{a^2 - x^2}$.

■ **EXAMPLE 1** Evaluate $\displaystyle\int \sqrt{1-x^2}\, dx$.

Solution

Step 1. **Substitute to eliminate the square root.**
In this integral, $-1 \leq x \leq 1$. Therefore, we may use the substitution $x = \sin\theta$, $dx = \cos\theta\, d\theta$ where $-\frac{\pi}{2} \leq \theta \leq \frac{\pi}{2}$. For such θ, $\cos\theta \geq 0$, and we obtain

> When making a trigonometric substitution such as $x = \sin\theta$ or $x = \tan\theta$, we choose the angle θ in the domain of the relevant inverse trigonometric function, i.e., $\sin^{-1} x$ or $\tan^{-1} x$.

$$\sqrt{1-x^2} = \sqrt{1 - \sin^2\theta} = \cos\theta \qquad \boxed{1}$$

$$\int \sqrt{1-x^2}\, dx = \int \cos\theta\, \overbrace{\cos\theta\, d\theta}^{dx} = \int \cos^2\theta\, d\theta$$

Step 2. **Evaluate the trigonometric integral.**
Use the formula recalled in the margin:

> ◄·· **REMINDER**
> $\displaystyle\int \cos^2\theta\, d\theta = \frac{1}{2}\theta + \frac{1}{4}\sin 2\theta + C$

$$\int \sqrt{1-x^2}\, dx = \int \cos^2\theta\, d\theta = \frac{1}{2}\theta + \frac{1}{4}\sin 2\theta + C \qquad \boxed{2}$$

Step 3. Convert back to original variable.

We need to express our answer in terms of the original variable x. We use the trigonometric identity $\sin 2\theta = 2 \sin \theta \cos \theta$ and Eq. (1) to write

$$\sin 2\theta = 2 \sin \theta \cos \theta = 2x\sqrt{1 - x^2}$$

Furthermore, $\theta = \sin^{-1} x$, so Eq. (2) becomes

$$\int \sqrt{1 - x^2}\, dx = \frac{1}{2}\theta + \frac{1}{4}\sin 2\theta + C = \frac{1}{2}\sin^{-1} x + \frac{1}{2}x\sqrt{1 - x^2} + C \qquad \blacksquare$$

Integrals Involving $\sqrt{a^2 - x^2}$ If $\sqrt{a^2 - x^2}$ occurs in an integral where $a > 0$, try the substitution

$$x = a \sin \theta, \qquad dx = a \cos \theta\, d\theta, \qquad \sqrt{a^2 - x^2} = a \cos \theta$$

Note that if $x = a \sin \theta$ and $a > 0$, then

$$a^2 - x^2 = a^2(1 - \sin^2 \theta) = a^2 \cos^2 \theta$$

For $-\frac{\pi}{2} \le \theta \le \frac{\pi}{2}$, $\cos \theta \ge 0$ and

$$\sqrt{a^2 - x^2} = a \cos \theta$$

Example 2 shows that trigonometric substitution can be used with integrands involving $(a^2 - x^2)^{n/2}$, where n is any integer.

◄·· **REMINDER**

$$\int \tan^m x\, dx = \frac{\tan^{m-1} x}{m - 1} - \int \tan^{m-2} x\, dx$$

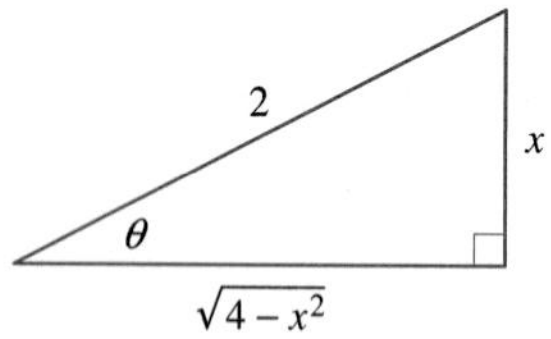

FIGURE 1 Right triangle with $\sin \theta = x/2$.

■ **EXAMPLE 2** Integrand Involving $(a^2 - x^2)^{3/2}$ Evaluate $\displaystyle\int \frac{x^2}{(4 - x^2)^{3/2}}\, dx$.

Solution

Step 1. Substitute to eliminate the square root.

In this case, $a = 2$ since $\sqrt{4 - x^2} = \sqrt{2^2 - x^2}$. Therefore, we use

$$x = 2 \sin \theta, \qquad dx = 2 \cos \theta\, d\theta, \qquad \sqrt{4 - x^2} = 2 \cos \theta$$

$$\int \frac{x^2}{(4 - x^2)^{3/2}}\, dx = \int \frac{4 \sin^2 \theta}{2^3 \cos^3 \theta} 2 \cos \theta\, d\theta = \int \frac{\sin^2 \theta}{\cos^2 \theta}\, d\theta = \int \tan^2 \theta\, d\theta$$

Step 2. Evaluate the trigonometric integral.

Use the reduction formula in the marginal note with $m = 2$:

$$\int \tan^2 \theta\, d\theta = \tan \theta - \int d\theta = \tan \theta - \theta + C$$

Step 3. Convert back to original variable.

We must express $\tan \theta$ and θ in terms of x. By definition, $x = 2 \sin \theta$, so

$$\sin \theta = \frac{x}{2}, \qquad \theta = \sin^{-1}\frac{x}{2}$$

How do we express $\tan \theta$ in terms of x? We construct a right triangle with angle θ such that $\sin \theta = x/2$ as in Figure 1. By the Pythagorean Theorem, the adjacent side has length $\sqrt{4 - x^2}$, and thus

$$\tan \theta = \frac{\text{opposite}}{\text{adjacent}} = \frac{x}{\sqrt{4 - x^2}}$$

$$\int \frac{x^2}{(4 - x^2)^{3/2}}\, dx = \tan \theta - \theta + C = \frac{x}{\sqrt{4 - x^2}} - \sin^{-1}\frac{x}{2} + C \qquad \blacksquare$$

When the integrand involves $\sqrt{a^2 + x^2}$, try the substitution $x = a \tan \theta$. Then

$$x^2 + a^2 = a^2 \tan^2 \theta + a^2 = a^2(1 + \tan^2 \theta) = a^2 \sec^2 \theta$$

and thus $\sqrt{a^2 + x^2} = a \sec \theta$.

In the substitution $x = a \tan \theta$, we choose $-\frac{\pi}{2} < \theta < \frac{\pi}{2}$. Therefore $a \sec \theta$ is the positive square root $\sqrt{a^2 + x^2}$.

> **Integrals Involving $\sqrt{a^2 + x^2}$**   If $\sqrt{a^2 + x^2}$ occurs in an integral where $a > 0$, try the substitution
>
> $$x = a \tan \theta, \qquad dx = a \sec^2 \theta \, d\theta, \qquad \sqrt{a^2 + x^2} = a \sec \theta$$

■ **EXAMPLE 3** Evaluate $\displaystyle\int \sqrt{4x^2 + 20} \, dx$.

Solution To obtain an integrand of the form $\sqrt{a^2 + x^2}$, we factor out a constant:

$$\int \sqrt{4x^2 + 20} \, dx = \int \sqrt{4(x^2 + 5)} \, dx = 2 \int \sqrt{x^2 + 5} \, dx$$

Next, let $a = \sqrt{5}$ and use the substitution

$$x = \sqrt{5} \tan \theta, \qquad dx = \sqrt{5} \sec^2 \theta \, d\theta, \qquad \sqrt{x^2 + 5} = \sqrt{5} \sec \theta$$

$$2 \int \sqrt{x^2 + 5} \, dx = 2 \int \left(\sqrt{5} \sec \theta \right) \sqrt{5} \sec^2 \theta \, d\theta = 10 \int \sec^3 \theta \, d\theta$$

Now use the reduction formula for $\sec^m x$ in the marginal note with $m = 3$:

◀·· **REMINDER**

$$\int \sec^m x \, dx = \frac{\tan x \sec^{m-2} x}{m - 1}$$
$$+ \frac{m - 2}{m - 1} \int \sec^{m-2} x \, dx$$

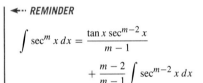

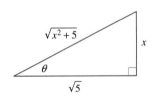

FIGURE 2

$$\int \sqrt{4x^2 + 20} \, dx = 10 \int \sec^3 \theta \, d\theta = 10 \frac{\tan \theta \sec \theta}{2} + 10 \left(\frac{1}{2} \right) \int \sec \theta \, dx$$
$$= 5 \tan \theta \sec \theta + 5 \ln \left| \sec \theta + \tan \theta \right| + C \qquad \boxed{3}$$

It remains to express the result in terms of x. Since $x = \sqrt{5} \tan \theta$, we refer to the right triangle in Figure 2 to obtain

$$\tan \theta = \frac{\text{opposite}}{\text{adjacent}} = \frac{x}{\sqrt{5}}, \qquad \sec \theta = \frac{\text{hypotenuse}}{\text{adjacent}} = \frac{\sqrt{x^2 + 5}}{\sqrt{5}}$$

Thus Eq. (3) yields

$$\int \sqrt{4x^2 + 20} \, dx = 5 \frac{x}{\sqrt{5}} \frac{\sqrt{x^2 + 5}}{\sqrt{5}} + 5 \ln \left| \frac{\sqrt{x^2 + 5}}{\sqrt{5}} + \frac{x}{\sqrt{5}} \right| + C$$
$$= x\sqrt{x^2 + 5} + 5 \ln \left| \frac{\sqrt{x^2 + 5} + x}{\sqrt{5}} \right| + C$$

The logarithmic term can be rewritten as

$$5 \ln \left| \frac{\sqrt{x^2 + 5} + x}{\sqrt{5}} \right| + C = 5 \ln \left| \sqrt{x^2 + 5} + x \right| + \underbrace{5 \ln \left| \frac{1}{\sqrt{5}} \right| + C}_{\text{Constant}}$$

Since C is an arbitrary constant, we may absorb $5 \ln \left| \dfrac{1}{\sqrt{5}} \right|$ into C and write

$$\int \sqrt{4x^2 + 20} \, dx = x\sqrt{x^2 + 5} + 5 \ln \left| \sqrt{x^2 + 5} + x \right| + C \qquad ■$$

Our last trigonometric substitution is $x = a \sec \theta$, which transforms $\sqrt{x^2 - a^2}$ into $a \tan \theta$ because

$$x^2 - a^2 = a^2 \sec^2 \theta - a^2 = a^2(\sec^2 \theta - 1) = a^2 \tan^2 \theta$$

In the substitution $x = a \sec \theta$, we choose $0 \le \theta < \frac{\pi}{2}$ if $x \ge a$ and $\pi \le \theta < \frac{3\pi}{2}$ if $x \le -a$. With these choices, $a \tan \theta$ is the positive square root $\sqrt{x^2 - a^2}$.

> **Integrals Involving $\sqrt{x^2 - a^2}$**   If $\sqrt{x^2 - a^2}$ occurs in an integral where $a > 0$, try the substitution
>
> $$x = a \sec \theta, \qquad dx = a \sec \theta \tan \theta \, d\theta, \qquad \sqrt{x^2 - a^2} = a \tan \theta$$

■ **EXAMPLE 4**   Evaluate $\displaystyle\int \frac{dx}{x^2\sqrt{x^2 - 9}}$.

Solution   In this case, make the substitution

$$x = 3 \sec \theta, \qquad dx = 3 \sec \theta \tan \theta \, d\theta, \qquad \sqrt{x^2 - 9} = 3 \tan \theta$$

Then

$$\int \frac{dx}{x^2\sqrt{x^2 - 9}} = \int \frac{3 \sec \theta \tan \theta \, d\theta}{(9 \sec^2 \theta)(3 \tan \theta)} = \frac{1}{9} \int \cos \theta \, d\theta = \frac{1}{9} \sin \theta + C$$

Since $x = 3 \sec \theta$, we construct a right triangle as in Figure 3 for which

$$\sec \theta = \frac{\text{hypotenuse}}{\text{adjacent}} = \frac{x}{3}$$

The right triangle shows that

$$\sin \theta = \frac{\text{opposite}}{\text{hypotenuse}} = \frac{\sqrt{x^2 - 9}}{x}$$

Therefore,

$$\int \frac{dx}{x^2\sqrt{x^2 - 9}} = \frac{1}{9} \sin \theta + C = \frac{\sqrt{x^2 - 9}}{9x} + C \qquad ■$$

So far we have used trigonometric substitution to deal with integrands involving the square roots $\sqrt{x^2 \pm a^2}$ or $\sqrt{a^2 - x^2}$. We treat more general square roots of the form $\sqrt{ax^2 + bx + c}$ by completing the square (Section 1.2).

■ **EXAMPLE 5**   **Completing the Square**   Evaluate $\displaystyle\int \frac{dx}{(x^2 - 6x + 11)^2}$.

Solution

***Step 1.* Complete the square.**

$$x^2 - 6x + 11 = (x^2 - 6x + 9) + 2 = (x - 3)^2 + 2$$

***Step 2.* Substitution.**
Let $u = x - 3$. Then $x^2 - 6x + 11 = u^2 + 2$ and

$$\int \frac{dx}{(x^2 - 6x + 11)^2} = \int \frac{du}{(u^2 + 2)^2} \qquad \boxed{4}$$

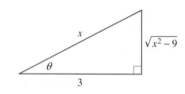

FIGURE 3

Step 3. **Trigonometric substitution.**

Evaluate the u-integral using trigonometric substitution:

$$u = \sqrt{2}\tan\theta, \qquad \sqrt{u^2 + 2} = \sqrt{2}\sec\theta, \qquad du = \sqrt{2}\sec^2\theta\,d\theta$$

$$\int \frac{du}{(u^2 + 2)^2} = \int \frac{\sqrt{2}\sec^2\theta\,d\theta}{4\sec^4\theta} = \frac{1}{2\sqrt{2}}\int \cos^2\theta\,d\theta$$

$$= \frac{1}{2\sqrt{2}}\left(\frac{\theta}{2} + \frac{\sin\theta\cos\theta}{2}\right) + C \qquad \boxed{5}$$

◀⋯ *REMINDER*

$$\int \cos^2\theta\,d\theta = \frac{\theta}{2} + \frac{\sin\theta\cos\theta}{2} + C$$

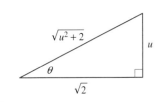

FIGURE 4

To write this result in terms of u, note that $\theta = \tan^{-1}\dfrac{u}{\sqrt{2}}$. From the right triangle in Figure 4, we obtain

$$\sin\theta\cos\theta = \left(\frac{\text{opposite}}{\text{hypotenuse}}\right)\left(\frac{\text{adjacent}}{\text{hypotenuse}}\right) = \frac{u}{\sqrt{u^2 + 2}}\cdot\frac{\sqrt{2}}{\sqrt{u^2 + 2}} = \frac{\sqrt{2}u}{u^2 + 2}$$

Thus, Eq. (5) becomes

$$\int \frac{du}{(u^2 + 2)^2} = \frac{1}{4\sqrt{2}}\left(\tan^{-1}\frac{u}{\sqrt{2}} + \frac{\sqrt{2}u}{u^2 + 2}\right) + C$$

$$= \frac{1}{4\sqrt{2}}\tan^{-1}\frac{u}{\sqrt{2}} + \frac{u}{4(u^2 + 2)} + C \qquad \boxed{6}$$

Step 4. **Convert to original variable.**

Since $u = x - 3$ and $u^2 + 2 = x^2 - 6x + 11$, Eq. (6) becomes

$$\int \frac{du}{(u^2 + 2)^2} = \frac{1}{4\sqrt{2}}\tan^{-1}\frac{x - 3}{\sqrt{2}} + \frac{x - 3}{4(x^2 - 6x + 11)} + C$$

This is our final answer by Eq. (4):

$$\int \frac{dx}{(x^2 - 6x + 11)^2} = \frac{1}{4\sqrt{2}}\tan^{-1}\frac{x - 3}{\sqrt{2}} + \frac{x - 3}{4(x^2 - 6x + 11)} + C \qquad ■$$

7.4 SUMMARY

- Trigonometric substitution:

Square Root Form in Integrand	Trigonometric Substitution		
$\sqrt{a^2 - x^2}$	$x = a\sin\theta,$	$dx = a\cos\theta\,d\theta,$	$\sqrt{a^2 - x^2} = a\cos\theta$
$\sqrt{a^2 + x^2}$	$x = a\tan\theta,$	$dx = a\sec^2\theta\,d\theta,$	$\sqrt{a^2 + x^2} = a\sec\theta$
$\sqrt{x^2 - a^2}$	$x = a\sec\theta,$	$dx = a\sec\theta\tan\theta\,d\theta,$	$\sqrt{x^2 - a^2} = a\tan\theta$

Step 1. Substitute to eliminate the square root.

Step 2. Evaluate the trigonometric integral.

Step 3. Convert back to original variable.

- The three trigonometric substitutions correspond to three right triangles (Figure 5) that we use to express the trigonometric functions of θ in terms of x.

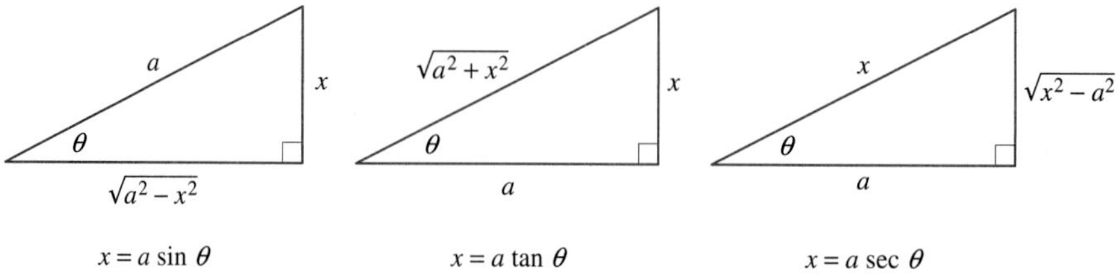

FIGURE 5

$x = a \sin \theta$ $x = a \tan \theta$ $x = a \sec \theta$

• Integrands involving $\sqrt{x^2 + bx + c}$ are treated by completing the square (see Example 5).

7.4 EXERCISES

Preliminary Questions

1. Explain why trigonometric substitution is *not needed* to evaluate $\int x\sqrt{9 - x^2}\, dx$.

2. State the trigonometric substitution appropriate to the given integral:

(a) $\int \sqrt{9 - x^2}\, dx$ **(b)** $\int x^2(x^2 - 16)^{3/2}\, dx$

(c) $\int x^2(x^2 + 16)^{3/2}\, dx$ **(d)** $\int (x^2 - 5)^{-2}\, dx$

3. Which of the triangles in Figure 6 would be used together with the substitution $x = 3 \sin \theta$?

4. Express $\tan \theta$ in terms of x for the angle in Figure 6(A).

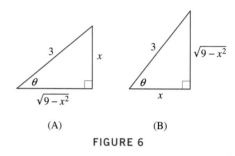

(A)   **(B)**

FIGURE 6

5. Express $\sec \theta$ in terms of x for the angle in Figure 6(B).

6. Express $\sin 2\theta$ in terms of x, where $x = \sin \theta$.

Exercises

In Exercises 1–4, evaluate the integral by following the steps.

1. $I = \int \dfrac{dx}{\sqrt{9 - x^2}}$

(a) Show that the substitution $x = 3 \sin \theta$ transforms I into $\int d\theta$ and evaluate I in terms of θ.

(b) Evaluate I in terms of x.

2. $I = \int \dfrac{dx}{x^2\sqrt{x^2 - 2}}$

(a) Show that the substitution $x = \sqrt{2}\sec \theta$ transforms the integral I into $\dfrac{1}{2}\int \cos \theta\, d\theta$ and evaluate I in terms of θ.

(b) Use a right triangle to show that with the above substitution, $\sin \theta = \sqrt{x^2 - 2}/x$.

(c) Evaluate I in terms of x.

3. $I = \int \dfrac{dx}{\sqrt{x^2 + 9}}$

(a) Show that the substitution $x = 3 \tan \theta$ transforms I into $\int \sec \theta\, d\theta$ and evaluate I in terms of θ (refer to the table of integrals in Section 7.3 if necessary).

(b) Show that if $x = 3 \tan \theta$, then $\sec \theta = \frac{1}{3}\sqrt{x^2 + 9}$.

(c) Express I in terms of x.

4. $I = \int \dfrac{dx}{(x^2 + 4)^2}$

(a) Show that the substitution $x = 2 \tan \theta$ transforms the integral I into $\dfrac{1}{8}\int \cos^2 \theta\, d\theta$.

(b) Use the formula $\int \cos^2 \theta\, d\theta = \dfrac{1}{2}\theta + \dfrac{1}{2}\sin \theta \cos \theta$ to evaluate I in terms of θ.

(c) Show that $\sin \theta = \dfrac{x}{\sqrt{x^2 + 4}}$ and $\cos \theta = \dfrac{2}{\sqrt{x^2 + 4}}$.

(d) Express I in terms of x.

In Exercises 5–10, use the indicated substitution to evaluate the integral.

5. $\int \sqrt{4 - x^2}\, dx, \quad x = 2 \sin \theta$

6. $\int_0^{1/2} \dfrac{x^2}{\sqrt{1 - x^2}}\, dx, \quad x = \sin \theta$

7. $\displaystyle\int \frac{dx}{x\sqrt{x^2-9}}$, $\quad x = 3\sec\theta$

8. $\displaystyle\int_{1/2}^{1} \frac{dx}{x^2\sqrt{x^2+4}}$, $\quad x = 2\tan\theta$

9. $\displaystyle\int \frac{dx}{(x^2-4)^{3/2}}$, $\quad x = 2\sec\theta$

10. $\displaystyle\int_{0}^{1} \frac{dx}{(16+x^2)^2}$, $\quad x = 4\tan\theta$

11. Is the substitution $u = x^2 - 4$ effective for evaluating the integral $\displaystyle\int \frac{x^2\,dx}{\sqrt{x^2-4}}$? If not, evaluate using trigonometric substitution.

12. Evaluate both $\displaystyle\int \frac{x\,dx}{\sqrt{x^2-4}}$ and $\displaystyle\int \frac{x^3\,dx}{\sqrt{x^2-4}}$ in two ways: using trigonometric substitution and using the direct substitution $u = x^2 - 4$.

In Exercises 13–30, evaluate the integral using trigonometric substitution. Refer to the table of trigonometric integrals as necessary.

13. $\displaystyle\int \frac{x^2\,dx}{\sqrt{9-x^2}}$

14. $\displaystyle\int \frac{dx}{\sqrt{x^2-9}}$

15. $\displaystyle\int \sqrt{12+4x^2}\,dx$

16. $\displaystyle\int \frac{dx}{x^2\sqrt{x^2-25}}$

17. $\displaystyle\int \frac{dt}{(4-t^2)^{3/2}}$

18. $\displaystyle\int \frac{dt}{\sqrt{t^2-5}}$

19. $\displaystyle\int \frac{dy}{y^2\sqrt{5-y^2}}$

20. $\displaystyle\int \frac{dt}{t\sqrt{t^2-4}}$

21. $\displaystyle\int \frac{dz}{z^3\sqrt{z^2-4}}$

22. $\displaystyle\int \frac{dx}{\sqrt{25+x^2}}$

23. $\displaystyle\int \frac{x^2\,dx}{\sqrt{x^2+1}}$

24. $\displaystyle\int \frac{x^2\,dx}{(x^2-4)^{3/2}}$

25. $\displaystyle\int \frac{dx}{(x^2+9)^2}$

26. $\displaystyle\int \frac{dx}{(9x^2+4)^2}$

27. $\displaystyle\int \frac{dx}{(x^2-4)^2}$

28. $\displaystyle\int \frac{dx}{(x^2+1)^3}$

29. $\displaystyle\int x^3\sqrt{9-x^2}\,dx$

30. $\displaystyle\int \frac{x^2\,dx}{(x^2+1)^{3/2}}$

31. Prove the following for $a > 0$:
$$\int \frac{dx}{x^2+a} = \frac{1}{\sqrt{a}}\tan^{-1}\frac{x}{\sqrt{a}} + C$$

32. Prove the following for $a > 0$:
$$\int \frac{dx}{(x^2+a)^2} = \frac{1}{2a}\left(\frac{x}{x^2+a} + \frac{1}{\sqrt{a}}\tan^{-1}\frac{x}{\sqrt{a}}\right) + C$$

33. Let $I = \displaystyle\int \frac{dx}{\sqrt{x^2-4x+8}}$.

(a) Complete the square to show that $x^2 - 4x + 8 = (x-2)^2 + 4$.

(b) Use the substitution $u = x - 2$ to show that $I = \displaystyle\int \frac{du}{\sqrt{u^2+2^2}}$.

Evaluate I.

34. Evaluate $\displaystyle\int \frac{dx}{\sqrt{12x-x^2}}$. First complete the square to write $12x - x^2 = 36 - (x-6)^2$.

In Exercises 35–40, evaluate the integral by completing the square and using trigonometric substitution.

35. $\displaystyle\int \frac{dx}{\sqrt{x^2+4x+13}}$

36. $\displaystyle\int \frac{dx}{\sqrt{2+x-x^2}}$

37. $\displaystyle\int \frac{dx}{\sqrt{x+x^2}}$

38. $\displaystyle\int \sqrt{x^2-4x+7}\,dx$

39. $\displaystyle\int \sqrt{x^2-4x+3}\,dx$

40. $\displaystyle\int \frac{dx}{(x^2+6x+6)^2}$

41. Evaluate $\displaystyle\int \sec^{-1}x\,dx$. *Hint:* First use Integration by Parts.

42. Evaluate $\displaystyle\int \frac{\sin^{-1}x}{x^2}\,dx$. *Hint:* First use Integration by Parts.

In Exercises 43–52, indicate a good method for evaluating the integral (but do not evaluate). Your choices are recognizing a basic integration formula, algebraic manipulation, substitution (specify u and du), Integration by Parts (specify u and v′), a trigonometric method, or trigonometric substitution (specify). If it appears that the techniques you have learned thus far are not sufficient, state this.

43. $\displaystyle\int \frac{dx}{\sqrt{12-6x-x^2}}$

44. $\displaystyle\int \sin^3 x \cos^3 x\,dx$

45. $\displaystyle\int x\sec^2 x\,dx$

46. $\displaystyle\int \sqrt{4x^2-1}\,dx$

47. $\displaystyle\int \frac{e^{2x}}{e^{4x}+1}\,dx$

48. $\displaystyle\int \frac{dx}{\sqrt{9-x^2}}$

49. $\displaystyle\int \cot x\csc x\,dx$

50. $\displaystyle\int \ln x\,dx$

51. $\displaystyle\int \frac{dx}{(x+2)^3}$

52. $\displaystyle\int \frac{dx}{(x+1)(x+2)^3}$

53. Which of the following integrals can be evaluated using the substitution $u = 1 - x^2$ and which require trigonometric substitution? Determine the integral obtained after substitution in each case.

(a) $\displaystyle\int x^3\sqrt{1-x^2}\,dx$

(b) $\displaystyle\int x^2\sqrt{1-x^2}\,dx$

(c) $\displaystyle\int \frac{x^4}{\sqrt{1-x^2}}\,dx$

(d) $\displaystyle\int \frac{x}{\sqrt{1-x^2}}\,dx$

54. Find the average height of a point on the semicircle $y = \sqrt{1 - x^2}$ for $-1 \le x \le 1$.

55. Find the volume of the solid obtained by revolving the graph of $y = x\sqrt{1 - x^2}$ over $[0, 1]$ about the y-axis.

56. Find the volume of the solid obtained by revolving the region between the graph of $y^2 - x^2 = 1$ and the line $y = 2$ about the line $y = 2$.

57. Find the volume of revolution for the region in Exercise 56, but revolve around $y = 3$.

58. A charged wire creates an electric field at a point P located at a distance D from the wire (Figure 7). The component $E_\perp$ of the field perpendicular to the wire (in volts) is

$$E_\perp = \int_{x_1}^{x_2} \frac{k\lambda D}{(x^2 + D^2)^{3/2}} \, dx$$

where $k = 8.99 \times 10^9$ N · m^2/C^2 (Coulomb constant), λ is the charge density (coulombs per meter), and x_1, x_2 are shown in the figure. Suppose that $\lambda = 6 \times 10^{-4}$ C/m, and $D = 3$ m. Find $E_\perp$ if (a) $x_1 = 0$ and $x_2 = 30$ m, and (b) $x_1 = -15$ m and $x_2 = 15$ m.

59. *CAS* Having ordered an 18-in. pizza for yourself and two friends, you want to divide it up using vertical slices as in Figure 8. Use Eq. (7) in Exercise 63 below and a computer algebra system to find the value of x that divides the pizza into equal parts.

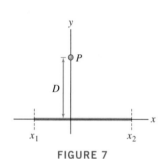

FIGURE 7

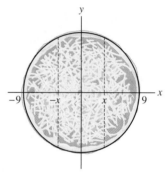

FIGURE 8 Dividing a pizza into three equal parts.

Further Insights and Challenges

60. Let $J_n = \displaystyle\int \frac{dx}{(x^2 + 1)^n}$. Prove

$$J_{n+1} = \left(1 - \frac{1}{2n}\right) J_n + \left(\frac{1}{2n}\right) \frac{x}{(x^2 + 1)^{n+1}}$$

Hint: Compute J_n using Integration by Parts with $v' = 1$. Use this recursion relation to calculate J_2 and J_3.

61. Hyperbolic Substitution Hyperbolic functions can be used instead of trigonometric substitution to treat integrals involving $\sqrt{x^2 \pm a^2}$. Let $I = \displaystyle\int \frac{dx}{\sqrt{x^2 - 1}}$.

(a) Show that the substitution $x = \cosh t$ transforms I into the integral $\displaystyle\int dt = t + C$.

(b) Show that $I = \cosh^{-1} x + C$.

(c) Trigonometric substitution with $x = \sec \theta$ leads to

$$I = \ln |x + \sqrt{x^2 - 1}| + C$$

Show that the two answers coincide.

62. Let $I = \displaystyle\int \sqrt{x^2 - 9} \, dx$.

(a) Show that the substitution $x = 3 \cosh t$ transforms I into the integral $9 \displaystyle\int \sinh^2 t \, dt$.

(b) Evaluate the hyperbolic integral using the identity

$$\sinh^2 t = \frac{1}{2}(\cosh 2t - 1)$$

(c) Express the result in terms of x. Note that $\sinh 2t = 2 \sinh t \cosh t$.

63. In Example 1, we proved the formula

$$\int \sqrt{1 - x^2} \, dx = \frac{1}{2} \sin^{-1} x + \frac{1}{2} x \sqrt{1 - x^2} + C \qquad \boxed{7}$$

Derive this formula using geometry rather than calculus by interpreting the integral as the area of part of the unit circle.

64. Compute $\displaystyle\int \frac{dx}{x^2 - 1}$ in two ways and verify that the answers agree: first via trigonometric substitution and then using the identity

$$\frac{1}{x^2 - 1} = \frac{1}{2}\left(\frac{1}{x - 1} - \frac{1}{x + 1}\right)$$

7.5 Integrals of Hyperbolic and Inverse Hyperbolic Functions

As we noted in Section 1.6, hyperbolic and trigonometric functions are similar in many ways (see the Excursion at the end of this section for an explanation of these similarities). We also saw in Section 3.10 that the formulas for their derivatives resemble each other, differing in at most a sign. The derivative formulas for the hyperbolic functions are equivalent to the following integral formulas.

Hyperbolic Integral Formulas

$$\int \sinh x \, dx = \cosh x + C, \qquad \int \cosh x \, dx = \sinh x + C$$

$$\int \operatorname{sech}^2 x \, dx = \tanh x + C, \qquad \int \operatorname{csch}^2 x \, dx = -\coth x + C$$

$$\int \operatorname{sech} x \tanh x \, dx = -\operatorname{sech} x + C, \qquad \int \operatorname{csch} x \coth x \, dx = -\operatorname{csch} x + C$$

■ **EXAMPLE 1** Calculate $\displaystyle\int x \cosh(x^2) \, dx$.

Solution The substitution $u = x^2, du = 2x \, dx$ yields

$$\int x \cosh(x^2) \, dx = \frac{1}{2} \int \cosh u \, du = \frac{1}{2} \sinh u + C = \frac{1}{2} \sinh(x^2) + C \qquad ■$$

The techniques for computing trigonometric integrals discussed in Section 7.3 apply with little change to hyperbolic integrals. In place of trigonometric identities, we use the corresponding hyperbolic identities. The following identities are useful.

$$\cosh^2 x - \sinh^2 x = 1, \qquad \cosh^2 x = 1 + \sinh^2 x$$

$$\cosh^2 x = \frac{1}{2}(\cosh 2x + 1), \qquad \sinh^2 x = \frac{1}{2}(\cosh 2x - 1)$$

$$\sinh 2x = 2 \sinh x \cosh x, \qquad \cosh 2x = \cosh^2 x + \sinh^2 x$$

■ **EXAMPLE 2** Powers of $\sinh x$ or $\cosh x$ Calculate (a) $\displaystyle\int \sinh^4 x \cosh^5 x \, dx$ and

(b) $\displaystyle\int \cosh^2 x \, dx$.

Solution

(a) Since $\cosh x$ appears to an odd power, use the identity $\cosh^2 x = 1 + \sinh^2 x$ to write

$$\cosh^5 x = \cosh^4 x \cdot \cosh x = (\sinh^2 x + 1)^2 \cosh x$$

Then

$$\int \sinh^4 x \cosh^5 x \, dx = \int (\sinh^4 x)(\sinh^2 x + 1)^2 \cosh x \, dx$$

The extra factor of $\cosh x$ in the integrand allows us to use the substitution $u = \sinh x$, $du = \cosh x \, dx$. We have

$$(\sinh^4 x)(\sinh^2 x + 1)^2 \cosh x \, dx = u^4(u^2 + 1)^2 \, du$$

$$\int \sinh^4 x \cosh^5 x \, dx = \int u^4 (u^2 + 1)^2 \, du = \int (u^8 + 2u^6 + u^4) \, du$$

$$= \frac{u^9}{9} + \frac{2u^7}{7} + \frac{u^5}{5} + C = \frac{\sinh^9 x}{9} + \frac{2 \sinh^7 x}{7} + \frac{\sinh^5 x}{5} + C$$

(b) Use the identity $\cosh^2 x = \frac{1}{2}(\cosh 2x + 1)$:

$$\int \cosh^2 x \, dx = \frac{1}{2} \int (\cosh 2x + 1) \, dx = \frac{1}{2} \left(\frac{\sinh 2x}{2} + x \right) + C$$

$$= \frac{1}{4} \sinh 2x + \frac{1}{2}x + C \qquad\blacksquare$$

Hyperbolic substitution may be used as an alternative to trigonometric substitution to integrate functions involving the following square root expressions:

In trigonometric substitution, we treat $\sqrt{x^2 + a^2}$ using the substitution $x = a \tan \theta$ and $\sqrt{x^2 - a^2}$ using $x = a \sec \theta$. Identities may be used to show that the results coincide with those obtained from hyperbolic substitution (see Exercises 31–35).

Square Root Form	Hyperbolic Substitution
$\sqrt{x^2 + a^2}$	$x = a \sinh u, \, dx = a \cosh u, \, \sqrt{x^2 + a^2} = a \cosh u$
$\sqrt{x^2 - a^2}$	$x = a \cosh u, \, dx = a \sinh u, \, \sqrt{x^2 - a^2} = a \sinh u$

When hyperbolic substitution is applied, the answer may involve inverse hyperbolic functions just as trigonometric substitution may yield inverse trigonometric functions.

■ **EXAMPLE 3** Hyperbolic Substitution Calculate $\int \sqrt{x^2 + 16} \, dx$.

Solution

Step 1. **Substitute to eliminate the square root.**
Use the hyperbolic substitution $x = 4 \sinh u, \, dx = 4 \cosh u \, du$. Then

$$x^2 + 16 = 16(\sinh^2 u + 1) = (4 \cosh u)^2$$

$$\int \sqrt{x^2 + 16} \, dx = \int (4 \cosh u) 4 \cosh u \, du = 16 \int \cosh^2 u \, du$$

Step 2. **Evaluate the hyperbolic integral.**
We evaluated the integral of $y = \cosh^2 u$ in Example 2(b):

$$\int \sqrt{x^2 + 16} \, dx = 16 \int \cosh^2 u \, du = 16 \left(\frac{1}{4} \sinh 2u + \frac{1}{2}u + C \right)$$

$$= 4 \sinh 2u + 8u + C \qquad\boxed{1}$$

Step 3. **Convert back to original variable.**
To write the answer in terms of the original variable x, we note that

$$\sinh u = \frac{x}{4}, \qquad u = \sinh^{-1} \frac{x}{4}$$

However, in Eq. (1) we must express $\sinh 2u$ in terms of x, so we use the identities recalled in the margin to write

REMINDER

$$\sinh 2u = 2 \sinh u \cosh u$$

$$\cosh u = \sqrt{\sinh^2 u + 1}$$

$$4 \sinh 2u = 4(2 \sinh u \cosh u) = 8 \sinh u \sqrt{\sinh^2 u + 1}$$

$$= 8 \left(\frac{x}{4}\right) \sqrt{\left(\frac{x}{4}\right)^2 + 1} = 2x \sqrt{\frac{x^2}{16} + 1} = \frac{1}{2} x \sqrt{x^2 + 16}$$

Then Eq. (1) becomes

$$\int \sqrt{x^2 + 16} \, dx = 4 \sinh 2u + 8u + C = \frac{1}{2} x \sqrt{x^2 + 16} + 8 \sinh^{-1} \frac{x}{4} + C \quad \blacksquare$$

In the next theorem, we state the integral formulas corresponding to the derivative formulas for the inverse hyperbolic functions recorded in Section 3.10. Each formula is valid on the domain where the integrand and inverse hyperbolic function are defined.

THEOREM 1 Integrals Involving Inverse Hyperbolic Functions

$$\int \frac{dx}{\sqrt{x^2 + 1}} = \sinh^{-1} x + C$$

$$\int \frac{dx}{\sqrt{x^2 - 1}} = \cosh^{-1} x + C \qquad \text{(for } x > 1)$$

$$\int \frac{dx}{1 - x^2} = \tanh^{-1} x + C \qquad \text{(for } |x| < 1)$$

$$\int \frac{dx}{1 - x^2} = \coth^{-1} x + C \qquad \text{(for } |x| > 1)$$

$$\int \frac{dx}{x\sqrt{1 - x^2}} = -\operatorname{sech}^{-1} x + C \qquad \text{(for } 0 < x < 1)$$

$$\int \frac{dx}{|x|\sqrt{1 + x^2}} = -\operatorname{csch}^{-1} x + C \qquad \text{(for } x \neq 0)$$

■ **EXAMPLE 4** Evaluate **(a)** $\displaystyle\int_2^4 \frac{dx}{\sqrt{x^2 - 1}}$ and **(b)** $\displaystyle\int_{0.2}^{0.6} \frac{x \, dx}{1 - x^4}$.

Solution

(a) By Theorem 1,

$$\int_2^4 \frac{dx}{\sqrt{x^2 - 1}} = \cosh^{-1} x \Big|_2^4 = \cosh^{-1} 4 - \cosh^{-1} 2 \approx 0.75$$

(b) First use the substitution $u = x^2$, $du = 2x \, du$. The new limits of integration become $u = (0.2)^2 = 0.04$ and $u = (0.6)^2 = 0.36$. We obtain

$$\int_{0.2}^{0.6} \frac{x \, dx}{1 - x^4} = \int_{0.04}^{0.36} \frac{\frac{1}{2} du}{1 - u^2} = \frac{1}{2} \int_{0.04}^{0.36} \frac{du}{1 - u^2}$$

By Theorem 1, both $\tanh^{-1} u$ and $\coth^{-1} u$ are antiderivatives of $y = (1 - u^2)^{-1}$. We use $\tanh^{-1} u$ because the interval of integration $[0.04, 0.36]$ is contained in the domain $\{u : |u| < 1\}$ of $\tanh^{-1} u$. If the limits of integration were contained in $\{u : |u| > 1\}$, we would use $\coth^{-1} u$. The result is

$$\frac{1}{2} \int_{0.04}^{0.36} \frac{du}{1-u^2} = \frac{1}{2}\left(\tanh^{-1}(0.36) - \tanh^{-1}(0.04)\right) \approx 0.1684 \qquad \blacksquare$$

Excursion: A Leap of Imagination

The terms "hyperbolic sine" and "hyperbolic cosine" suggest a connection between the hyperbolic and trigonometric functions. This excursion explores the source of this connection, which leads us to **complex numbers** and Euler's famous formula.

Recall that $y = e^t$ satisfies the differential equation $y' = y$ and *every* solution is of the form $y = Ce^t$ for some constant C. On the other hand, we can check that both $y = e^t$ and $y = e^{-t}$ satisfy the **second-order differential equation**

This differential equation is called "second-order" because it involves the second derivative y''.

$$\boxed{y'' = y} \qquad \boxed{2}$$

Indeed, $(e^t)'' = e^t$ and $(e^{-t})'' = (-e^{-t})' = e^{-t}$. It can be shown that every solution of Eq. (2) has the form $y = Ae^t + Be^{-t}$ for some constants A and B.

Now let's examine what happens when we change Eq. (2) by a minus sign:

$$\boxed{y'' = -y} \qquad \boxed{3}$$

In this case, $y = \sin t$ and $y = \cos t$ are solutions because

$$(\sin t)'' = (\cos t)' = -\sin t, \qquad (\cos t)'' = (-\sin t)' = -\cos t$$

Furthermore, it can be proved as before that every solution of Eq. (3) has the form

$$y = A \cos t + B \sin t$$

This might seem to be the end of the story. However, there is another way to write down solutions to Eq. (3). Consider the exponential functions $y = e^{it}$ and $y = e^{-it}$, where

$$i = \sqrt{-1}$$

The number i is an *imaginary* complex number satisfying $i^2 = -1$. Since i is not a real number, the exponential e^{it} is not defined without further explanation. But if we accept that e^{it} *can* be defined and that it is legitimate to apply the usual rules of calculus to it, we obtain

$$\frac{d}{dt} e^{it} = i e^{it}$$

Differentiating a second time yields

$$(e^{it})'' = (ie^{it})' = i^2 e^{it} = -e^{it}$$

In other words, $y = e^{it}$ is a solution to the equation $y'' = -y$ and must therefore be expressible in terms of $\sin t$ and $\cos t$. That is, for some constants A and B,

$$e^{it} = A \cos t + B \sin t \qquad \boxed{4}$$

We determine A and B by considering initial conditions. First, set $t = 0$ in Eq. (4):

$$1 = e^{i0} = A \cos 0 + B \sin 0 = A$$

FIGURE 1 Leonhard Euler (1707–1783). Euler (pronounced "oil-er") ranks among the greatest mathematicians of all time. His work (printed in more than 70 volumes) contains fundamental contributions to almost every aspect of the mathematics and physics of his time. The French mathematician Pierre Simon de Laplace once declared: "Read Euler, he is our master in everything."

Then take the derivative of Eq. (4) and set $t = 0$:

$$ie^{it} = \frac{d}{dt}e^{it} = A\cos' t + B\sin' t = -A\sin t + B\cos t$$

$$i = ie^{i0} = -A\sin 0 + B\cos 0 = B$$

Thus, $A = 1$ and $B = i$, and Eq. (4) yields **Euler's Formula**:

$$\boxed{e^{it} = \cos t + i\sin t}$$

Euler derived this result using power series, which allow us to define e^{it} in a rigorous fashion.

At $t = \pi$, Euler's Formula yields

$$\boxed{e^{i\pi} = -1}$$

Here we have a simple but surprising relation between the three important numbers e, π, and i.

Euler's Formula also reveals the source of the analogy between hyperbolic and trigonometric functions. Let us calculate the hyperbolic cosine at $x = it$:

$$\cosh(it) = \frac{e^{it} + e^{-it}}{2} = \frac{\cos t + i\sin t}{2} + \frac{\cos(-t) + i\sin(-t)}{2} = \cos t$$

A similar calculation shows that $\sinh(it) = i\sin(t)$. In other words, the hyperbolic and trigonometric functions are not merely analogous—once we introduce complex numbers, we see that they are very nearly the same functions.

7.5 SUMMARY

- Integrals of hyperbolic functions:

$$\int \sinh x\, dx = \cosh x + C, \qquad \int \cosh x\, dx = \sinh x + C$$

$$\int \operatorname{sech}^2 x\, dx = \tanh x + C, \qquad \int \operatorname{csch}^2 x\, dx = -\coth x + C$$

$$\int \operatorname{sech} x \tanh x\, dx = -\operatorname{sech} x + C, \qquad \int \operatorname{csch} x \coth x\, dx = -\operatorname{csch} x + C$$

- Integrals involving inverse hyperbolic functions:

$$\int \frac{dx}{\sqrt{x^2 + 1}} = \sinh^{-1} x + C$$

$$\int \frac{dx}{\sqrt{x^2 - 1}} = \cosh^{-1} x + C \qquad \text{(for } x > 1)$$

$$\int \frac{dx}{1 - x^2} = \tanh^{-1} x + C \qquad \text{(for } |x| < 1)$$

$$\int \frac{dx}{1 - x^2} = \coth^{-1} x + C \qquad \text{(for } |x| > 1)$$

$$\int \frac{dx}{x\sqrt{1-x^2}} = -\operatorname{sech}^{-1} x + C \qquad (\text{for } 0 < x < 1)$$

$$\int \frac{dx}{|x|\sqrt{1+x^2}} = -\operatorname{csch}^{-1} x + C \quad (\text{for } x \neq 0)$$

7.5 EXERCISES

Preliminary Questions

1. Which hyperbolic substitution can be used to evaluate the following integrals?

(a) $\displaystyle\int \frac{dx}{\sqrt{x^2+1}}$ 　　**(b)** $\displaystyle\int \frac{dx}{\sqrt{x^2+9}}$ 　　**(c)** $\displaystyle\int \frac{dx}{\sqrt{9x^2+1}}$

2. Which two hyperbolic integration formulas differ from their trigonometric counterparts by a minus sign?

3. Which antiderivative of $y = (1-x^2)^{-1}$ should be used to evaluate $\displaystyle\int_3^5 (1-x^2)^{-1}\,dx$?

Exercises

In Exercises 1–16, calculate the integral.

1. $\displaystyle\int \cosh(3x)\,dx$

2. $\displaystyle\int \sinh(x+1)\,dx$

3. $\displaystyle\int x \sinh(x^2+1)\,dx$

4. $\displaystyle\int \sinh^2 x \cosh x\,dx$

5. $\displaystyle\int \operatorname{sech}^2(1-2x)\,dx$

6. $\displaystyle\int \tanh(3x)\operatorname{sech}(3x)\,dx$

7. $\displaystyle\int \tanh x \operatorname{sech}^2 x\,dx$

8. $\displaystyle\int \frac{\cosh x}{3\sinh x+4}\,dx$

9. $\displaystyle\int \tanh x\,dx$

10. $\displaystyle\int x \operatorname{csch}(x^2)\coth(x^2)\,dx$

11. $\displaystyle\int \frac{\cosh x}{\sinh x}\,dx$

12. $\displaystyle\int \frac{\cosh x}{\sinh^2 x}\,dx$

13. $\displaystyle\int \sinh^2(4x-9)\,dx$

14. $\displaystyle\int \sinh^3 x \cosh^6 x\,dx$

15. $\displaystyle\int \sinh^2 x \cosh^2 x\,dx$

16. $\displaystyle\int \tanh^3 x\,dx$

In Exercises 17–30, calculate the integral in terms of the inverse hyperbolic functions.

17. $\displaystyle\int \frac{dx}{\sqrt{x^2-1}}$

18. $\displaystyle\int \frac{dx}{\sqrt{x^2-4}}$

19. $\displaystyle\int \frac{dx}{\sqrt{16+x^2}}$

20. $\displaystyle\int \frac{dx}{\sqrt{1+3x^2}}$

21. $\displaystyle\int \sqrt{x^2-1}\,dx$

22. $\displaystyle\int \frac{x^2\,dx}{\sqrt{x^2+1}}$

23. $\displaystyle\int_{-1/2}^{1/2} \frac{dx}{1-x^2}$

24. $\displaystyle\int_4^5 \frac{dx}{1-x^2}$

25. $\displaystyle\int_0^1 \frac{dx}{\sqrt{1+x^2}}$

26. $\displaystyle\int_2^{10} \frac{dx}{4x^2-1}$

27. $\displaystyle\int_{-3}^{-1} \frac{dx}{x\sqrt{x^2+16}}$

28. $\displaystyle\int_{0.2}^{0.8} \frac{dx}{x\sqrt{1-x^2}}$

29. $\displaystyle\int \frac{\sqrt{x^2-1}\,dx}{x^2}$

30. $\displaystyle\int_1^9 \frac{dx}{x\sqrt{x^4+1}}$

31. Verify the formulas

$$\sinh^{-1} x = \ln|x + \sqrt{x^2+1}|$$

$$\cosh^{-1} x = \ln|x + \sqrt{x^2-1}| \quad \text{for } x \geq 1$$

32. Verify that for $|x| < 1$,

$$\tanh^{-1} x = \frac{1}{2}\ln\left|\frac{1+x}{1-x}\right|$$

33. Evaluate $\displaystyle\int \sqrt{x^2+16}\,dx$ using trigonometric substitution. Then use Exercise 31 to verify that your answer agrees with the answer in Example 3.

34. Evaluate $\displaystyle\int \sqrt{x^2-9}\,dx$ in two ways: using trigonometric substitution and using hyperbolic substitution. Then use Exercise 31 to verify that the two answers agree.

35. Evaluate $\displaystyle\int \frac{dx}{1-x^2}$ in terms of the logarithm by writing $\dfrac{1}{1-x^2} = \dfrac{1}{2}\left(\dfrac{1}{1+x} + \dfrac{1}{1-x}\right)$. Then use Exercise 32 to verify that your result agrees with Theorem 1.

36. GU Find an equation for the tangent line to the graph of $f(x) = \tanh^{-1} x$ at $x = 0.8$. Using a graphing utility, plot the graphs of $f(x)$ and the tangent line on the same pair of axes.

37. GU Repeat Exercise 36 with $f(x) = \operatorname{csch}^{-1} x$.

38. Prove the reduction formula for $n \geq 2$:

$$\int \cosh^n x \, dx = \frac{1}{n} \cosh^{n-1} x \sinh x$$

$$+ \frac{n-1}{n} \int \cosh^{n-2} x \, dx \qquad \boxed{5}$$

39. Use Eq. (5) to compute $\int \cosh^4 x \, dx$.

40. Compute $\int \frac{\tanh^{-1} x \, dx}{x^2 - 1}$.

41. Compute $\int \sinh^{-1} x \, dx$. *Hint:* Use Integration by Parts with $v' = 1$.

42. Compute $\int \tanh^{-1} x \, dx$.

43. Compute $\int x \tanh^{-1} x \, dx$.

Further Insights and Challenges

44. Show that if $u = \tanh(x/2)$, then

$$\cosh x = \frac{1 + u^2}{1 - u^2}, \quad \sinh x = \frac{2u}{1 - u^2}, \quad dx = \frac{2 \, du}{1 - u^2}$$

Hint: For the first relation, use the identities

$$\sinh^2 \left(\frac{x}{2} \right) = \frac{1}{2} (\cosh x - 1), \quad \cosh^2 \left(\frac{x}{2} \right) = \frac{1}{2} (\cosh x + 1)$$

45. Use the substitution of Exercise 44 to evaluate $\int \operatorname{sech} x \, dx$.

46. Use the substitution of Exercise 44 to evaluate $\int \frac{dx}{1 + \cosh x}$.

47. The relations $\cosh(it) = \cos t$ and $\sinh(it) = i \sin t$ were discussed in the Excursion. Use these relations to show that the identity $\cos^2 t + \sin^2 t = 1$ results from setting $x = it$ in the identity $\cosh^2 x - \sinh^2 x = 1$.

7.6 The Method of Partial Fractions

◀┄ **REMINDER** A "rational function" is a quotient of two polynomials.

In this section, we introduce the Method of Partial Fractions, used to integrate rational functions $\frac{P(x)}{Q(x)}$, where $P(x)$ and $Q(x)$ are polynomials. As we will see, the integral of a rational function can be expressed as a sum of three types of terms: rational functions, arctangents of linear or quadratic functions, and logarithms of polynomials. Here is a typical example:

$$\int \frac{(2x^3 + x^2 - 2x + 2) \, dx}{(x^2 + 1)^2} = \frac{x + 4}{2x^2 + 2} + \frac{3}{2} \tan^{-1} x + \ln(x^2 + 1) + C$$

It is a fact from algebra that every polynomial $Q(x)$ with real coefficients can be written as a product of linear and quadratic factors with real coefficients. However, it is not always possible to find these factors explicitly.

The Method of Partial Fractions yields an explicit antiderivative of this type whenever we can factor the denominator $Q(x)$ into a product of linear and quadratic factors.

A rational function $\frac{P(x)}{Q(x)}$ is called **proper** if the degree of $P(x)$ [denoted $\deg(P)$] is *less than* the degree of $Q(x)$. For example,

$$\underbrace{\frac{x^2 - 3x + 7}{x^4 - 16}}_{\text{Proper}}, \qquad \underbrace{\frac{2x^2 + 7}{x - 5}, \quad \frac{x - 2}{x - 5}}_{\text{Not proper}}$$

Suppose first that $\frac{P(x)}{Q(x)}$ is proper and that the denominator $Q(x)$ factors as a product of *distinct linear factors*. In other words,

$$\frac{P(x)}{Q(x)} = \frac{P(x)}{(x - a_1)(x - a_2) \cdots (x - a_n)}$$

Each distinct linear factor $(x - a)$ in the denominator contributes a term of the form

$$\frac{A}{x - a}$$

to the partial fraction decomposition.

where the roots $a_1, a_2, \ldots, a_n$ are all distinct and $\deg(P) < n$. Then there is a **partial fraction decomposition**:

$$\frac{P(x)}{Q(x)} = \frac{A_1}{(x - a_1)} + \frac{A_2}{(x - a_2)} + \cdots + \frac{A_n}{(x - a_n)}$$

for suitable constants $A_1, \ldots, A_n$. For example,

$$\frac{-5x^2 - x + 28}{(x + 1)(x - 2)(x - 3)} = \frac{2}{x + 1} - \frac{2}{x - 2} - \frac{5}{x - 3}$$

Once we have found the partial fraction decomposition, we can integrate the individual terms.

■ **EXAMPLE 1** Finding the Constants $A_1, A_2, \ldots, A_n$ Evaluate $\displaystyle\int \frac{dx}{x^2 - 7x + 10}$.

Solution The denominator factors as $x^2 - 7x + 10 = (x - 2)(x - 5)$, so we look for a partial fraction decomposition:

$$\frac{1}{(x - 2)(x - 5)} = \frac{A}{x - 2} + \frac{B}{x - 5}$$

To find the constants A and B, first multiply by $(x - 2)(x - 5)$ to clear denominators:

$$1 = (x - 2)(x - 5)\left(\frac{A}{x - 2} + \frac{B}{x - 5}\right)$$

$$1 = A(x - 5) + B(x - 2) \qquad \boxed{1}$$

This equation holds for all values of x (including $x = 2$ and $x = 5$, by continuity). We determine A by setting $x = 2$ (which makes the second term disappear):

$$1 = A(2 - 5) + \underbrace{B(2 - 2)}_{\text{This is zero}} = -3A \quad \Rightarrow \quad \boxed{A = -\frac{1}{3}}$$

Similarly, to calculate B, set $x = 5$ in Eq. (1):

$$1 = A(5 - 5) + B(5 - 2) = 3B \quad \Rightarrow \quad \boxed{B = \frac{1}{3}}$$

The resulting partial fraction decomposition is

$$\boxed{\frac{1}{(x - 2)(x - 5)} = \frac{-1/3}{x - 2} + \frac{1/3}{x - 5}}$$

The integration can now be carried out:

$$\int \frac{dx}{(x - 2)(x - 5)} = -\frac{1}{3}\int \frac{dx}{x - 2} + \frac{1}{3}\int \frac{dx}{x - 5}$$

$$= -\frac{1}{3}\ln|x - 2| + \frac{1}{3}\ln|x - 5| + C \qquad ■$$

■ **EXAMPLE 2** Evaluate $\displaystyle\int \frac{x^2 + 2}{(x - 1)(2x - 8)(x + 2)}\, dx$.

Solution

Step 1. **Find the partial fraction decomposition.**

The decomposition has the form

$$\frac{x^2 + 2}{(x - 1)(2x - 8)(x + 2)} = \frac{A}{x - 1} + \frac{B}{2x - 8} + \frac{C}{x + 2} \qquad \boxed{2}$$

In Eq. (2), the linear factor $2x - 8$ does not have the form $(x - a)$ used previously, but the partial fraction decomposition can be carried out in the same way.

As before, multiply by $(x - 1)(2x - 8)(x + 2)$ to clear denominators:

$$x^2 + 2 = A(2x - 8)(x + 2) + B(x - 1)(x + 2) + C(x - 1)(2x - 8) \qquad \boxed{3}$$

Since A goes with the factor $(x - 1)$, we set $x = 1$ in Eq. (3) to compute A:

$$1^2 + 2 = A(2 - 8)(1 + 2) + \overbrace{B(1 - 1)(1 + 2) + C(1 - 1)(2 - 8)}^{\text{Zero}}$$

$$3 = -18A \quad \Rightarrow \quad \boxed{A = -\frac{1}{6}}$$

Similarly, 4 is the root of $2x - 8$, so we compute B by setting $x = 4$ in Eq. (3):

$$4^2 + 2 = A(8 - 8)(4 + 2) + B(4 - 1)(4 + 2) + C(4 - 1)(8 - 8)$$

$$18 = 18B \quad \Rightarrow \quad \boxed{B = 1}$$

Finally, C is determined by setting $x = -2$ in Eq. (3):

$$(-2)^2 + 2 = A(-4 - 8)(-2 + 2) + B(-2 - 1)(-2 + 2) + C(-2 - 1)(-4 - 8)$$

$$6 = 36C \quad \Rightarrow \quad \boxed{C = \frac{1}{6}}$$

The result is

$$\boxed{\frac{x^2 + 2}{(x - 1)(2x - 8)(x + 2)} = -\frac{1/6}{x - 1} + \frac{1}{2x - 8} + \frac{1/6}{x + 2}}$$

Step 2. **Carry out the integration.**

$$\int \frac{x^2 + 2}{(x - 1)(2x - 8)(x + 2)} \, dx = -\frac{1}{6} \int \frac{dx}{x - 1} + \int \frac{dx}{2x - 8} + \frac{1}{6} \int \frac{dx}{x + 2}$$

$$= -\frac{1}{6} \ln |x - 1| + \frac{1}{2} \ln |2x - 8| + \frac{1}{6} \ln |x + 2| + C$$

∎

If $\dfrac{P(x)}{Q(x)}$ is not proper, that is, if $\deg(P) \geq \deg(Q)$, we use long division to write

$$\frac{P(x)}{Q(x)} = g(x) + \frac{R(x)}{Q(x)}$$

where $g(x)$ is a polynomial and $\dfrac{R(x)}{Q(x)}$ is proper. We may then integrate $\dfrac{P(x)}{Q(x)}$ using the partial fraction decomposition of $\dfrac{R(x)}{Q(x)}$.

■ **EXAMPLE 3** **Long Division Necessary** Evaluate $\displaystyle\int \dfrac{x^3 + 1}{x^2 - 4}\, dx$.

Solution Using long division, we write

$$\frac{x^3 + 1}{x^2 - 4} = x + \frac{4x + 1}{x^2 - 4} = x + \frac{4x + 1}{(x - 2)(x + 2)}$$

The second term has a partial fraction decomposition:

$$\frac{4x + 1}{(x - 2)(x + 2)} = \frac{A}{x - 2} + \frac{B}{x + 2}$$

To find A and B, clear denominators:

$$4x + 1 = A(x + 2) + B(x - 2)$$

Setting $x = 2$, we get $A = 9/4$, and setting $x = -2$, we get $B = 7/4$. Therefore,

$$\frac{x^3 + 1}{x^2 - 4} = x + \frac{9/4}{x - 2} + \frac{7/4}{x + 2}$$

$$\int \frac{(x^3 + 1)\, dx}{x^2 - 4} = \int x\, dx + \frac{9}{4}\int \frac{dx}{x - 2} + \frac{7}{4}\int \frac{dx}{x + 2}$$

$$= \frac{1}{2}x^2 + \frac{9}{4}\ln|x - 2| + \frac{7}{4}\ln|x + 2| + C \qquad ■$$

Now suppose we have a rational function $\dfrac{P(x)}{Q(x)}$ whose denominator has repeated linear factors:

$$\frac{P(x)}{Q(x)} = \frac{P(x)}{(x - a_1)^{M_1}(x - a_2)^{M_2}\cdots(x - a_n)^{M_n}}$$

Each factor $(x - a_i)^{M_i}$ contributes the following sum of terms to the partial fraction decomposition:

$$\frac{B_1}{(x - a_i)} + \frac{B_2}{(x - a_i)^2} + \cdots + \frac{B_{M_i}}{(x - a_i)^{M_i}}$$

■ **EXAMPLE 4** **Repeated Linear Factors** Evaluate $\displaystyle\int \dfrac{3x - 9}{(x - 1)(x + 2)^2}\, dx$.

Solution We are looking for a partial fraction decomposition of the form

$$\frac{3x - 9}{(x - 1)(x + 2)^2} = \frac{A}{x - 1} + \frac{B_1}{x + 2} + \frac{B_2}{(x + 2)^2}$$

Let's clear denominators to obtain

$$3x - 9 = A(x + 2)^2 + B_1(x - 1)(x + 2) + B_2(x - 1) \qquad \boxed{4}$$

Long division:

$$
\begin{array}{r}
x \phantom{{}+ 1} \\
x^2 - 4 \overline{\smash{\big)}\, x^3 + 1 } \\
\underline{x^3 - 4x } \\
4x + 1
\end{array}
$$

The quotient $\dfrac{x^3 + 1}{x^2 - 4}$ is equal to x with remainder $4x + 1$.

We compute A and B_2 by substituting in Eq. (4) in the usual way:

- Set $x = 1$: This gives $-6 = 9A$ or $\boxed{A = -\dfrac{2}{3}}$.

- Set $x = -2$: This gives $-15 = -3B_2$ or $\boxed{B_2 = 5}$.

With these constants, Eq. (4) becomes

$$3x - 9 = -\frac{2}{3}(x+2)^2 + B_1(x-1)(x+2) + 5(x-1) \qquad \boxed{5}$$

We cannot determine B_1 in the same way as A and B_2. Here are two ways to proceed.

- **First method (substitution):** If we substitute $x = 1$ or $x = -2$ in Eq. (5), then the term involving B_1 drops out. But Eq. (5) is valid for all values of x, so let's plug in a value different from $x = 1, -2$, such as $x = 2$. Then Eq. (5) gives us

$$3(2) - 9 = -\frac{2}{3}(2+2)^2 + B_1(2-1)(2+2) + 5(2-1)$$

$$-3 = -\frac{32}{3} + 4B_1 + 5$$

$$B_1 = \frac{1}{4}\left(-8 + \frac{32}{3}\right) = \frac{2}{3}$$

- **Second method (undetermined coefficients):** Expand the terms in Eq. (5):

$$3x - 9 = -\frac{2}{3}(x^2 + 4x + 4) + B_1(x^2 + x - 2) + 5(x-1)$$

$$3x - 9 = \left(B_1 - \frac{2}{3}\right)x^2 + \left(B_1 + \frac{7}{3}\right)x - \left(2B_1 + \frac{23}{3}\right)$$

The coefficients of the powers of x on each side of the equation must be equal. Since x^2 does not occur on the left-hand side, $0 = B_1 - \frac{2}{3}$ or $B_1 = \frac{2}{3}$.

Either way, we have shown that

$$\boxed{\frac{3x - 9}{(x-1)(x+2)^2} = -\frac{2/3}{x-1} + \frac{2/3}{x+2} + \frac{5}{(x+2)^2}}$$

$$\int \frac{3x - 9}{(x-1)(x+2)^2}\, dx = -\frac{2}{3}\int \frac{dx}{x-1} + \frac{2}{3}\int \frac{dx}{x+2} + 5\int \frac{dx}{(x+2)^2}$$

$$= -\frac{2}{3}\ln|x-1| + \frac{2}{3}\ln|x+2| - \frac{5}{x+2} + C \qquad \blacksquare$$

Quadratic Factors

A quadratic polynomial $ax^2 + bx + c$ is called **irreducible** if it cannot be written as a product of two linear factors (without using complex numbers). A power of an irreducible quadratic factor $(ax^2 + bx + c)^M$ contributes a sum of the following type to a partial fraction decomposition:

$$\frac{A_1 x + B_1}{ax^2 + bx + c} + \frac{A_2 x + B_2}{(ax^2 + bx + c)^2} + \cdots + \frac{A_M x + B_M}{(ax^2 + bx + c)^M}$$

For example,

$$\frac{4 - x}{x(x^2 + 4x + 2)^2} = \frac{1}{x} - \frac{x + 4}{x^2 + 4x + 2} - \frac{2x + 9}{(x^2 + 4x + 2)^2}$$

You may need to use trigonometric substitution to integrate these terms. In particular, the following result may be useful (see Exercise 31 in Section 7.4).

$$\boxed{\int \frac{dx}{x^2 + a} = \frac{1}{\sqrt{a}} \tan^{-1}\left(\frac{x}{\sqrt{a}}\right) + C \quad \text{(for } a > 0\text{)}} \qquad \boxed{6}$$

←‥ REMINDER If $b > 0$, then $x^2 + b$ is irreducible, but $x^2 - b$ is reducible because

$$x^2 - b = (x + \sqrt{b})(x - \sqrt{b})$$

■ **EXAMPLE 5**   Irreducible versus Reducible Quadratic Factors   Evaluate

(a) $\displaystyle\int \frac{18}{(x + 3)(x^2 + 9)}\, dx$ **(b)** $\displaystyle\int \frac{18}{(x + 3)(x^2 - 9)}\, dx$

Solution

(a) The quadratic factor $x^2 + 9$ is irreducible, so the partial fraction decomposition has the form

$$\frac{18}{(x + 3)(x^2 + 9)} = \frac{A}{x + 3} + \frac{Bx + C}{x^2 + 9}$$

Clear denominators to obtain

$$18 = A(x^2 + 9) + (Bx + C)(x + 3) \qquad \boxed{7}$$

To find A, set $x = -3$:

$$18 = A((-3)^2 + 9) + 0 \quad \Rightarrow \quad A = 1$$

Then substitute $A = 1$ in (7) to obtain

$$18 = (x^2 + 9) + (Bx + C)(x + 3) = (B + 1)x^2 + (C + 3B)x + (9 + 3C)$$

Equating coefficients, we get $B + 1 = 0$ or $B = -1$ and $9 + 3C = 18$ or $C = 3$. Thus,

$$\int \frac{18\, dx}{(x + 3)(x^2 + 9)} = \int \frac{dx}{x + 3} - \int \frac{(x - 3)\, dx}{x^2 + 9}$$

$$= \int \frac{dx}{x + 3} - \int \frac{x\, dx}{x^2 + 9} + \int \frac{3\, dx}{x^2 + 9}$$

$$= \ln|x + 3| - \frac{1}{2} \ln(x^2 + 9) + \tan^{-1}\frac{x}{3} + C$$

Here we have used

$$\int \frac{x}{x^2 + 9}\, dx = \frac{1}{2} \int \frac{du}{u} = \frac{1}{2} \ln(x^2 + 9) + C \quad (u = x^2 + 9, du = 2x\, dx)$$

$$\int \frac{dx}{x^2 + 9} = \frac{1}{3} \tan^{-1}\frac{x}{3} + C \qquad\qquad \text{[by Eq. (6)]}$$

(b) The polynomial $x^2 - 9$ is not irreducible because $x^2 - 9 = (x - 3)(x + 3)$. Therefore, the partial fraction decomposition has the form

$$\frac{18}{(x+3)(x^2-9)} = \frac{18}{(x+3)^2(x-3)} = \frac{A}{x-3} + \frac{B}{x+3} + \frac{C}{(x+3)^2}$$

Clear denominators:

$$18 = A(x+3)^2 + B(x+3)(x-3) + C(x-3)$$

Setting $x = 3$, we obtain $18 = (6^2)A$ or $A = \frac{1}{2}$, and setting $x = -3$, we obtain $18 = -6C$ or $C = -3$. Therefore,

$$18 = \frac{1}{2}(x+3)^2 + B(x+3)(x-3) - 3(x-3)$$

To find B, we plug in any value other than ± 3, such as $x = 2$. Then $18 = \frac{1}{2}(25) - 5B + 3$ or $B = -\frac{1}{2}$, and

$$\int \frac{18}{(x+3)(x^2-9)}\,dx = \frac{1}{2}\int \frac{dx}{x-3} - \frac{1}{2}\int \frac{dx}{x+3} - 3\int \frac{dx}{(x+3)^2}$$

$$= \frac{1}{2}\ln|x-3| - \frac{1}{2}\ln|x+3| + 3(x+3)^{-1} + C \qquad \blacksquare$$

■ EXAMPLE 6 Repeated Quadratic Factor Evaluate $\displaystyle\int \frac{4-x}{x(x^2+2)^2}\,dx$.

Solution The partial fraction decomposition has the form

$$\frac{4-x}{x(x^2+2)^2} = \frac{A}{x} + \frac{Bx+C}{x^2+2} + \frac{Dx+E}{(x^2+2)^2}$$

Clear denominators by multiplying through by $x(x^2+2)^2$:

$$4 - x = A(x^2+2)^2 + (Bx+C)(x(x^2+2)) + (Dx+E)x \qquad \boxed{8}$$

We compute A directly by setting $x = 0$. Then Eq. (8) reduces to $4 = 4A$ or $A = 1$. We find the remaining coefficients by the method of undetermined coefficients. Set $A = 1$ in Eq. (8) and expand:

$$4 - x = (x^4 + 4x^2 + 4) + (Bx^4 + 2Bx^2 + Cx^3 + 2C) + (Dx^2 + Ex)$$

$$= (1+B)x^4 + Cx^3 + (4+2B+D)x^2 + Ex + 2C + 4$$

Now equate the coefficients on the two sides of the equation:

$$1 + B = 0 \qquad \text{(Coefficient of } x^4)$$
$$C = 0 \qquad \text{(Coefficient of } x^3)$$
$$4 + 2B + D = 0 \qquad \text{(Coefficient of } x^2)$$
$$E = -1 \qquad \text{(Coefficient of } x)$$
$$2C + 4 = 4 \qquad \text{(Constant term)}$$

By the first equation, $B = -1$. Setting $B = -1$ in the third equation gives $D = -2$. Thus,

$$\int \frac{(4-x)\,dx}{x(x^2+2)^2} = \int \frac{dx}{x} - \int \frac{x\,dx}{x^2+2} - \int \frac{(2x+1)\,dx}{(x^2+2)^2}$$

$$= \ln|x| - \frac{1}{2}\ln(x^2+2) - \int \frac{(2x+1)\,dx}{(x^2+2)^2}$$

The middle integral was evaluated using the substitution $u = x^2 + 2$, $du = 2x\, dx$. The third integral breaks up as a sum:

$$\int \frac{(2x+1)\,dx}{(x^2+2)^2} = \overbrace{\int \frac{2x\,dx}{(x^2+2)^2}}^{\text{Use substitution } u = x^2 + 2} + \int \frac{dx}{(x^2+2)^2}$$

$$= -(x^2+2)^{-1} + \int \frac{dx}{(x^2+2)^2} \qquad \boxed{9}$$

To evaluate the integral in Eq. (9), we use the trigonometric substitution

$$x = \sqrt{2}\tan\theta, \qquad dx = \sqrt{2}\sec^2\theta\,d\theta, \qquad x^2 + 2 = 2\tan^2\theta + 2 = 2\sec^2\theta$$

Referring to Figure 1, we obtain

$$\int \frac{dx}{(x^2+2)^2} = \int \frac{\sqrt{2}\sec^2\theta\,d\theta}{(2\tan^2\theta+2)^2} = \int \frac{\sqrt{2}\sec^2\theta\,d\theta}{4\sec^4\theta}$$

$$= \frac{\sqrt{2}}{4}\int \cos^2\theta\,d\theta = \frac{\sqrt{2}}{4}\left(\frac{1}{2}\theta + \frac{1}{2}\sin\theta\cos\theta\right) + C$$

$$= \frac{\sqrt{2}}{8}\tan^{-1}\frac{x}{\sqrt{2}} + \frac{\sqrt{2}}{8}\frac{x}{\sqrt{x^2+2}}\frac{\sqrt{2}}{\sqrt{x^2+2}} + C$$

$$= \frac{1}{4\sqrt{2}}\tan^{-1}\frac{x}{\sqrt{2}} + \frac{1}{4}\frac{x}{x^2+2} + C$$

Finally, collecting all the terms, we have

$$\int \frac{4-x}{x(x^2+2)^2}\,dx = \ln|x| - \frac{1}{2}\ln(x^2+2) + \frac{1}{x^2+2} - \frac{1}{4\sqrt{2}}\arctan\frac{x}{\sqrt{2}}$$

$$- \frac{1}{4}\frac{x}{x^2+2} + C$$

$$= \ln|x| - \frac{1}{2}\ln(x^2+2) + \frac{1-\frac{1}{4}x}{x^2+2} - \frac{1}{4\sqrt{2}}\arctan\frac{x}{\sqrt{2}} + C \qquad \blacksquare$$

Using a Computer Algebra System

Finding partial fraction decompositions often requires laborious computation. Fortunately, most computer algebra systems have a command (with names such as "Apart" or "parfrac") that produce partial fraction decompositions. For example, the command

```
Apart[(x^2 − 2)/((x + 2)(x^2 + 4)^3)]
```

produces the partial fraction decomposition

$$\frac{x^2-2}{(x+2)(x^2+4)^3} = \frac{1}{256(2+x)} + \frac{3(x-2)}{4(4+x^2)^3} + \frac{2-x}{32(4+x^2)^2} + \frac{2-x}{256(4+x^2)}$$

However, a computer algebra system cannot produce a partial fraction decomposition in cases where $Q(x)$ cannot be factored explicitly.

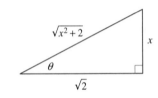

FIGURE 1

The polynomial $x^5 + 2x + 2$ cannot be factored explicitly, so the command

```
Apart[1/(x^5 + 2x + 2)]
```

returns the useless response

$$\frac{1}{x^5 + 2x + 2}$$

7.6 SUMMARY

The integral of a rational function $\int \dfrac{P(x)\,dx}{Q(x)}$ is evaluated using the Method of Partial Fractions. Assume that $P(x)/Q(x)$ is *proper* [i.e., $\deg(P) < \deg(Q)$] and that $Q(x)$ can be factored explicitly as a product of linear and irreducible quadratic terms.

- If $Q(x) = (x - a_1)(x - a_2) \cdots (x - a_n)$ where the roots a_j are distinct, then

$$\frac{P(x)}{(x - a_1)(x - a_2) \cdots (x - a_n)} = \frac{A_1}{x - a_1} + \frac{A_2}{x - a_2} + \cdots + \frac{A_n}{x - a_n}$$

To calculate the constants, clear denominators and substitute, in turn, the values $x = a_1$, $a_2, \ldots, a_n$.

- If $Q(x)$ is equal to a product of powers of linear factors $(x - a)^M$ and irreducible quadratic factors $(x^2 + b)^N$ with $b > 0$, then the partial fraction decomposition of $\dfrac{P(x)}{Q(x)}$ is a sum of terms of the following type:

$$(x - a)^M \quad \text{contributes} \quad \frac{A_1}{x - a} + \frac{A_2}{(x - a)^2} + \cdots + \frac{A_M}{(x - a)^M}$$

$$(x^2 + b)^N \quad \text{contributes} \quad \frac{A_1 x + B_1}{x^2 + b} + \frac{A_2 x + B_2}{(x^2 + b)^2} + \cdots + \frac{A_N x + B_N}{(x^2 + b)^N}$$

Substitution and trigonometric substitution may be needed to integrate the terms corresponding to $(x^2 + b)^N$ (see Example 6).

- If $P(x)/Q(x)$ is improper, use long division (see Example 3).

7.6 EXERCISES

Preliminary Questions

1. Suppose that $\int f(x)\,dx = \ln x + \sqrt{x + 1} + C$. Can $f(x)$ be a rational function? Explain.

2. Which of the following are *proper* rational functions?

(a) $\dfrac{x}{x - 3}$

(b) $\dfrac{4}{9 - x}$

(c) $\dfrac{x^2 + 12}{(x + 2)(x + 1)(x - 3)}$

(d) $\dfrac{4x^3 - 7x}{(x - 3)(2x + 5)(9 - x)}$

3. Which of the following quadratic polynomials are irreducible? To check, complete the square if necessary.

(a) $x^2 + 5$

(b) $x^2 - 5$

(c) $x^2 + 4x + 6$

(d) $x^2 + 4x + 2$

4. Let $P(x)/Q(x)$ be a proper rational function where $Q(x)$ factors as a product of distinct linear factors $(x - a_i)$. Then

$$\int \frac{P(x)\,dx}{Q(x)}$$

(choose correct answer):

(a) is a sum of logarithmic terms $A_i \ln(x - a_i)$ for some constants A_i.

(b) may contain a term involving the arctangent.

Exercises

1. Match the rational function (a)–(d) with the corresponding partial fraction decomposition (i)–(iv).

(a) $\dfrac{x^2 + 4x + 12}{(x + 2)(x^2 + 4)}$

(b) $\dfrac{2x^2 + 8x + 24}{(x + 2)^2(x^2 + 4)}$

(c) $\dfrac{x^2 - 4x + 8}{(x - 1)^2(x - 2)^2}$

(d) $\dfrac{x^4 - 4x + 8}{(x + 2)(x^2 + 4)}$

(i) $x - 2 + \dfrac{4}{x + 2} - \dfrac{4x - 4}{x^2 + 4}$

(ii) $\dfrac{-8}{x - 2} + \dfrac{4}{(x - 2)^2} + \dfrac{8}{x - 1} + \dfrac{5}{(x - 1)^2}$

(iii) $\dfrac{1}{x + 2} + \dfrac{2}{(x + 2)^2} + \dfrac{-x + 2}{x^2 + 4}$

(iv) $\dfrac{1}{x + 2} + \dfrac{4}{x^2 + 4}$

2. Determine the constants A, B:

$$\frac{2x - 3}{(x - 3)(x - 4)} = \frac{A}{x - 3} + \frac{B}{x - 4}$$

3. Clear denominators in the following partial fraction decomposition and determine the constant B (substitute a value of x or use the method of undetermined coefficients):

$$\frac{3x^2 + 11x + 12}{(x + 1)(x + 3)^2} = \frac{1}{x + 1} - \frac{B}{x + 3} - \frac{3}{(x + 3)^2}$$

4. Find the constants in the partial fraction decomposition

$$\frac{2x + 4}{(x - 2)(x^2 + 4)} = \frac{A}{x - 2} + \frac{Bx + C}{x^2 + 4}$$

In Exercises 5–8, use long division to write $f(x)$ as the sum of a polynomial and a proper rational function. Then calculate $\int f(x)\, dx$.

5. $f(x) = \dfrac{x}{3x - 9}$ **6.** $f(x) = \dfrac{x^2 + 2}{x + 3}$

7. $f(x) = \dfrac{x^3 + x + 1}{x - 2}$ **8.** $f(x) = \dfrac{x^3 - 1}{x^2 - x}$

In Exercises 9–46, evaluate the integral.

9. $\displaystyle\int \frac{dx}{(x - 2)(x - 4)}$ **10.** $\displaystyle\int \frac{dx}{(x - 3)(x + 7)}$

11. $\displaystyle\int \frac{dx}{x(2x + 1)}$ **12.** $\displaystyle\int \frac{(3x + 5)\, dx}{x^2 - 4x - 5}$

13. $\displaystyle\int \frac{(2x - 1)\, dx}{x^2 - 5x + 6}$ **14.** $\displaystyle\int \frac{dx}{(x - 2)(x - 3)(x + 2)}$

15. $\displaystyle\int \frac{(x^2 + 3x - 44)\, dx}{(x + 3)(x + 5)(3x - 2)}$ **16.** $\displaystyle\int \frac{3\, dx}{(x + 1)(x^2 + x)}$

17. $\displaystyle\int \frac{(x^2 + 11x)\, dx}{(x - 1)(x + 1)^2}$ **18.** $\displaystyle\int \frac{(4x^2 - 21x)\, dx}{(x - 3)^2(2x + 3)}$

19. $\displaystyle\int \frac{dx}{(x - 1)^2(x - 2)^2}$ **20.** $\displaystyle\int \frac{dx}{(x + 4)^3}$

21. $\displaystyle\int \frac{48\, dx}{x(x + 4)^2}$ **22.** $\displaystyle\int \frac{(x^2 + x + 3)\, dx}{(x - 1)^3}$

23. $\displaystyle\int \frac{dx}{(x - 4)^2(x - 1)}$ **24.** $\displaystyle\int \frac{x}{(3x + 7)^3}\, dx$

25. $\displaystyle\int \frac{3x + 6}{x^2(x - 1)(x - 3)}\, dx$ **26.** $\displaystyle\int \frac{dx}{x(x - 1)^3}$

27. $\displaystyle\int \frac{(3x^2 - 2)\, dx}{x - 4}$ **28.** $\displaystyle\int \frac{(x^2 - x + 1)\, dx}{x^2 + x}$

29. $\displaystyle\int \frac{dx}{x(x^2 + 1)}$ **30.** $\displaystyle\int \frac{(3x^2 - 4x + 5)\, dx}{(x - 1)(x^2 + 1)}$

31. $\displaystyle\int \frac{x^2\, dx}{x^2 + 3}$ **32.** $\displaystyle\int \frac{dx}{2x^2 - 3}$

33. $\displaystyle\int \frac{x^2}{(x + 1)(x^2 + 1)}\, dx$ **34.** $\displaystyle\int \frac{6x^2 + 7x - 6}{(x^2 - 4)(x + 2)}\, dx$

35. $\displaystyle\int \frac{x^2\, dx}{(3x + 7)^3}$ **36.** $\displaystyle\int \frac{dx}{x(x^2 + 25)}$

37. $\displaystyle\int \frac{dx}{x^2(x^2 + 25)}$ **38.** $\displaystyle\int \frac{10\, dx}{(x - 1)^2(x^2 + 9)}$

39. $\displaystyle\int \frac{10\, dx}{(x + 1)(x^2 + 9)^2}$ **40.** $\displaystyle\int \frac{dx}{x(x^2 + 8)^2}$

41. $\displaystyle\int \frac{100x\, dx}{(x - 3)(x^2 + 1)^2}$ **42.** $\displaystyle\int \frac{dx}{(x + 2)(x^2 + 4x + 10)}$

43. $\displaystyle\int \frac{9\, dx}{(x + 1)(x^2 - 2x + 6)}$ **44.** $\displaystyle\int \frac{25\, dx}{x(x^2 + 2x + 5)^2}$

45. $\displaystyle\int \frac{(x^2 + 3)\, dx}{(x^2 + 2x + 3)^2}$ **46.** $\displaystyle\int \frac{dx}{x^4 - 1}$

47. Evaluate $\displaystyle\int \frac{dx}{x^2 - 1}$ in two ways: using partial fractions and trigonometric substitution. Verify that the two answers agree.

48. ⏹GU⏹ Graph the equation $(x - 40)y^2 = 10x(x - 30)$ and find the volume of the solid obtained by revolving the region between the graph and the x-axis for $0 \le x \le 30$ around the x-axis.

49. Evaluate $\displaystyle\int \frac{\sqrt{x}\, dx}{x - 1}$. *Hint:* Use the substitution $u = \sqrt{x}$ (sometimes called a **rationalizing substitution**).

50. Evaluate $\displaystyle\int \frac{x\, dx}{x^{1/2} - x^{1/3}}$.

In Exercises 51–66, evaluate the integral using the appropriate method or combination of methods covered thus far in the text.

51. $\displaystyle\int \frac{dx}{x^2\sqrt{4 - x^2}}$ **52.** $\displaystyle\int \cos^2 4x\, dx$

53. $\displaystyle\int \frac{dx}{x(x - 1)^2}$ **54.** $\displaystyle\int x \sec^2 x\, dx$

55. $\displaystyle\int \frac{dx}{(x^2 + 9)^2}$ **56.** $\displaystyle\int \theta \sec^{-1}\theta\, d\theta$

57. $\displaystyle\int \tan^5 x \sec x\, dx$ **58.** $\displaystyle\int \frac{x\, dx}{(x^2 - 1)^{3/2}}$

59. $\displaystyle\int \frac{x^2\, dx}{(x^2 - 1)^{3/2}}$ **60.** $\displaystyle\int \frac{dx}{x(x^2 + x)}$

61. $\displaystyle\int \frac{dx}{x(x^2 - 1)}$ **62.** $\displaystyle\int \frac{(x + 1)\, dx}{(x^2 + 4x + 8)^2}$

63. $\int \dfrac{\sqrt{x}\,dx}{x^3+1}$

64. $\int \tan^{-1}3\theta\,d\theta$

65. $\int \dfrac{dx}{x^4(x^3+1)}$

66. $\int \dfrac{x^{1/2}dx}{x^{1/3}+1}$

67. Show that the substitution $\theta = 2\tan^{-1}t$ (Figure 2) yields the formulas

$$\cos\theta = \frac{1-t^2}{1+t^2}, \qquad \sin\theta = \frac{2t}{1+t^2}, \qquad d\theta = \frac{2\,dt}{1+t^2} \qquad \boxed{10}$$

This substitution transforms the integral of any rational function of $\cos\theta$ and $\sin\theta$ into an integral of a rational function of t (which

can then be evaluated using partial fractions). Use it to evaluate

$$\int \frac{d\theta}{\cos\theta + (3/4)\sin\theta}.$$

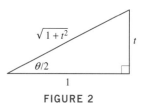

FIGURE 2

68. Use the substitution of Exercise 67 to evaluate $\displaystyle\int \frac{d\theta}{\cos\theta + \sin\theta}$.

Further Insights and Challenges

69. Prove the general formula

$$\int \frac{dx}{(x-a)(x-b)} = \frac{1}{a-b}\ln\frac{x-a}{x-b} + C$$

where a, b are constants such that $a \neq b$.

70. The method of partial fractions shows that

$$\int \frac{dx}{x^2-1} = \frac{1}{2}\ln|x-1| - \frac{1}{2}\ln|x+1| + C$$

The computer algebra system Mathematica evaluates this integral as $-\tanh^{-1}x$, where $\tanh^{-1}x$ is the inverse hyperbolic tangent function. Can you reconcile the two answers?

71. Suppose that $Q(x) = (x-a)(x-b)$, where $a \neq b$, and let $\dfrac{P(x)}{Q(x)}$ be a proper rational function so that

$$\frac{P(x)}{Q(x)} = \frac{A}{(x-a)} + \frac{B}{(x-b)}$$

(a) Show that $A = \dfrac{P(a)}{Q'(a)}$ and $B = \dfrac{P(b)}{Q'(b)}$.

(b) Use this result to find the partial fraction decomposition for $P(x) = 3x - 2$ and $Q(x) = x^2 - 4x - 12$.

72. Suppose that $Q(x) = (x-a_1)(x-a_2)\cdots(x-a_n)$ where the roots a_j are all distinct. Let $\dfrac{P(x)}{Q(x)}$ be a proper rational function so that

$$\frac{P(x)}{Q(x)} = \frac{A_1}{(x-a_1)} + \frac{A_2}{(x-a_2)} + \cdots + \frac{A_n}{(x-a_n)}$$

(a) Show that $A_j = \dfrac{P(a_j)}{Q'(a_j)}$ for $j = 1,\ldots,n$.

(b) Use this result to find the partial fraction decomposition for $P(x) = 2x^2 - 1$, $Q(x) = x^3 - 4x^2 + x + 6 = (x+1)(x-2)(x-3)$.

7.7 Improper Integrals

Sometimes, we are interested in integrating a function over an infinite interval. For example, in statistics, the probability integral

$$\int_{-\infty}^{\infty} e^{-x^2/2}\,dx$$

represents the area under the bell-shaped curve for $-\infty < x < \infty$ (Figure 1). This is an example of an **improper integral**. Although the region under the curve extends infinitely to the left and right, the total area is finite (see Exercise 62).

We consider three types of improper integrals, depending on whether one or both endpoints are infinite:

$$\int_{-\infty}^{b} f(x)\,dx, \qquad \int_{a}^{\infty} f(x)\,dx, \qquad \int_{-\infty}^{\infty} f(x)\,dx$$

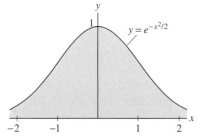

FIGURE 1 Total area under the bell-shaped curve is finite.

$y = e^{-x^2/2}$

In what sense might the area of an infinite region be finite? Consider the function $f(x) = e^{-x}$. The area under the graph over the finite interval $[0, R]$ is [Figure 2(A)]

$$\int_0^R e^{-x}\, dx = -e^{-x}\Big|_0^R = -e^{-R} + e^0 = 1 - e^{-R}$$

Now observe that this area tends to a limit as $R \to \infty$:

$$\int_0^\infty e^{-x}\, dx = \lim_{R \to \infty} \int_0^R e^{-x}\, dx = \lim_{R \to \infty} \left(1 - e^{-R}\right) = 1 \qquad \boxed{1}$$

We take this limit as the definition of the integral over the infinite interval $[0, \infty)$, and thus the infinite region in Figure 2(B) has area 1.

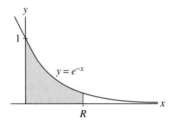
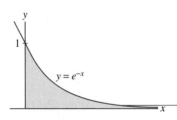

(A) Finite region has area $1 - e^{-R}$.   (B) Infinite region has area 1.

FIGURE 2 The area under $y = e^{-x}$ for $0 \le x \le R$ approaches the limit 1 as $R \to \infty$.

DEFINITION Improper Integral Fix a number a and assume that $f(x)$ is integrable over $[a, b]$ for all $b \ge a$. The *improper integral* of $f(x)$ over $[a, \infty)$ is defined as the limit (if it exists)

$$\int_a^\infty f(x)\, dx = \lim_{R \to \infty} \int_a^R f(x)\, dx$$

When the limit exists, we say that the improper integral *converges*. In a similar fashion, we define $\displaystyle\int_{-\infty}^a f(x)\, dx = \lim_{R \to -\infty} \int_R^a f(x)\, dx$.

A doubly infinite improper integral is defined as a sum (provided that both integrals on the right converge):

$$\int_{-\infty}^\infty f(x)\, dx = \int_{-\infty}^0 f(x)\, dx + \int_0^\infty f(x)\, dx \qquad \boxed{2}$$

■ **EXAMPLE 1** Show that $\displaystyle\int_2^\infty \frac{dx}{x^2}$ converges and compute its value.

Solution First we evaluate the definite integral over a finite interval $[2, R]$:

$$\int_2^R \frac{dx}{x^2} = -x^{-1}\Big|_2^R = \frac{1}{2} - \frac{1}{R}$$

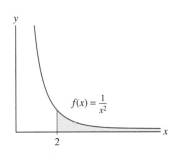

FIGURE 3 The area over $[2, \infty)$ is equal to $\frac{1}{2}$.

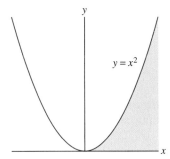

FIGURE 4 Total area under the graph of $y = x^2$ over $[0, \infty)$ is infinite.

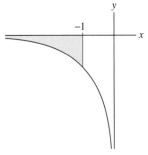

FIGURE 5 The integral of $f(x) = x^{-1}$ over $(-\infty, -1]$ is infinite.

The limit as $R \to \infty$ exists, so the improper integral converges (Figure 3):

$$\int_2^\infty \frac{dx}{x^2} = \lim_{R \to \infty} \int_2^R \frac{dx}{x^2} = \lim_{R \to \infty} \left(\frac{1}{2} - \frac{1}{R} \right) = \frac{1}{2} \qquad \blacksquare$$

When the limit defining an improper integral is either infinite or does not exist, we say that the improper integral *diverges*. For instance, the area under the parabola $y = x^2$ appears infinite (Figure 4) and indeed, the following integral diverges:

$$\int_0^\infty x^2 \, dx = \lim_{R \to \infty} \int_0^R x^2 \, dx = \lim_{R \to \infty} \frac{1}{3} R^3 = \infty$$

In general, however, we cannot tell if an improper integral converges or diverges merely by looking at the graph of the function. The functions $f(x) = x^{-2}$ and $g(x) = x^{-1}$ are both decreasing, but $f(x) = x^{-2}$ decreases more rapidly as $x \to \infty$, and the area under its graph over $[2, \infty)$ is finite, as we saw in the previous example. The next example shows that $g(x) = x^{-1}$ does not tend to zero quickly enough for the improper integral to converge.

■ **EXAMPLE 2** Determine if $\int_{-\infty}^{-1} \frac{dx}{x}$ converges.

Solution The integral over a finite interval $[R, -1]$ for $R < -1$ is

$$\int_R^{-1} \frac{dx}{x} = \ln |x| \Big|_R^{-1} = \ln |-1| - \ln |R| = -\ln |R|$$

The improper integral diverges because the limit as $R \to -\infty$ is infinite (Figure 5):

$$\lim_{R \to -\infty} \int_R^{-1} \frac{dx}{x} = \lim_{R \to -\infty} (-\ln |R|) = -\infty \qquad \blacksquare$$

■ **EXAMPLE 3** Using L'Hôpital's Rule   Determine if $\int_0^\infty x e^{-x} \, dx$ converges and, if so, find its value.

Solution First, we use Integration by Parts with $u = x$ and $v' = e^{-x}$:

$$\int x e^{-x} \, dx = -x e^{-x} + \int e^{-x} \, dx = -x e^{-x} - e^{-x} = -(x+1)e^{-x} + C$$

$$\int_0^R x e^{-x} \, dx = -(x+1)e^{-x} \Big|_0^R = -(R+1)e^{-R} + 1 = 1 - \frac{R+1}{e^R}$$

Now we can compute the improper integral as a limit using L'Hôpital's Rule:

$$\int_0^\infty x e^{-x} \, dx = 1 - \lim_{R \to \infty} \frac{R+1}{e^R} = 1 - \lim_{R \to \infty} \frac{1}{e^R} = 1 - 0 = 1 \qquad \blacksquare$$

For which exponents p does the integral $\int_1^\infty \frac{dx}{x^p}$ converge? The answer is given in the following theorem.

The integral $\int_1^\infty \frac{dx}{x^p}$ is called the "p-integral." It is often used in the Comparison Test to determine convergence or divergence of more complicated improper integrals (see Example 8).

THEOREM 1   Improper Integral of x^{-p} over $[1, \infty)$

$$\int_1^\infty \frac{dx}{x^p} = \begin{cases} \dfrac{1}{p-1} & \text{if } p > 1 \\ \infty & \text{if } p \leq 1 \end{cases}$$

Proof Let $J = \int_1^\infty \frac{dx}{x^p}$. If $p \neq 1$, then

$$J = \lim_{R \to \infty} \int_1^R x^{-p}\, dx = \lim_{R \to \infty} \frac{x^{1-p}}{1-p} \bigg|_1^R = \lim_{R \to \infty} \left(\frac{R^{1-p}}{1-p} - \frac{1}{1-p} \right)$$

If $p > 1$, then R^{1-p} tends to zero as $R \to \infty$. In this case, J converges and $J = 1/(p-1)$. If $p < 1$, then R^{1-p} tends to ∞ and J diverges. Finally, if $p = 1$, then J also diverges because $\int_1^R x^{-1}\, dx = \ln R$ and $\lim_{R \to \infty} \ln R = \infty$. ∎

Improper integrals arise in physical applications when it makes sense to treat a very large quantity as if it were infinite. A typical example is the calculation of escape velocity, the minimum initial velocity needed to launch an object into space.

In physics, we speak of moving an object "infinitely far away." In practice this means "very far way," but it is more convenient to work with an improper integral.

■ **EXAMPLE 4   Escape Velocity** The earth exerts a gravitational force of magnitude $F(r) = \dfrac{GM_e m}{r^2}$ on an object of mass m, where r is the distance to the center of the earth.

(a) Find the work required to move an object of mass m infinitely far from the earth.

(b) Calculate the escape velocity on the earth's surface.

Solution Recall that work is the integral of force as a function of distance (Section 6.5).

(a) The work required to move an object from the earth's surface ($r = r_e$) to a distance R from the center is

$$\int_{r_e}^R \frac{GM_e m}{r^2}\, dr = -\frac{GM_e m}{r} \bigg|_{r_e}^R = GM_e m \left(\frac{1}{r_e} - \frac{1}{R} \right) \text{ J}$$

The work moving the object "infinitely far away" is the improper integral

$$GM_e m \int_{r_e}^\infty \frac{dr}{r^2} = \lim_{R \to \infty} GM_e m \left(\frac{1}{r_e} - \frac{1}{R} \right) = \frac{GM_e m}{r_e} \text{ J}$$

◀··· **REMINDER** The mass of the earth is

$$M_e \approx 5.98 \cdot 10^{24} \text{ kg}$$

The radius of the earth is

$$r_e \approx 6.37 \cdot 10^6 \text{ m}$$

The universal gravitational constant is

$$G \approx 6.67 \cdot 10^{-11} \text{ N-m}^2/\text{kg}^2$$

where a newton is 1 kg-m/s^2. A joule is 1 N-m.

(b) Using conservation of energy in physics, we can show that an object launched with initial velocity v_0 will escape the earth's gravitational field if its kinetic energy $\frac{1}{2}mv_0^2$ is at least as large as the work required to move the object to infinity, that is, if

$$\frac{1}{2}mv_0^2 \geq \frac{GM_e m}{r_e} \qquad \text{or} \qquad v_0 \geq \left(\frac{2GM_e}{r_e} \right)^{1/2}$$

Escape velocity in miles per hour is approximately 25,000 mph.

Using the values of the constants recalled in the marginal note, we find that the escape velocity v_{esc} is

$$v_{\text{esc}} = \left(\frac{2GM_e}{r_e} \right)^{1/2} \approx \sqrt{\frac{2(6.67 \cdot 10^{-11})(5.98 \cdot 10^{24})}{6.37 \cdot 10^6}} \approx 11,200 \text{ m/s} \qquad ■$$

In the next example, we calculate the present value of an investment that pays out a dividend "forever."

In practice, the word "forever" means a "long but unspecified length of time." For example, if the investment pays out dividends for 100 years, then its present value is

$$\int_0^{100} 5{,}000e^{-0.5t}\,dt \approx \$99{,}326$$

The improper integral ($100,000) gives a useful approximation to this value.

■ **EXAMPLE 5** Perpetual Annuity   An investment pays a dividend continuously at a rate of $5,000/year. Compute the present value of the income stream if the interest rate is 5% and the dividends continue forever.

Solution Recall from Section 5.8 that the present value (PV) of the income stream after T years is $\int_0^T 5{,}000e^{-0.05t}\,dt$. The PV over an infinite time interval is the improper integral

$$\int_0^\infty 5{,}000e^{-0.05t}\,dt = \lim_{T\to\infty} \frac{5{,}000e^{-0.05t}}{-0.05}\bigg|_0^T = \frac{5{,}000}{0.05} = \$100{,}000$$

An infinite number of dollars are paid out during the infinite time interval, but their total present value is finite. ■

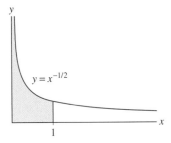

FIGURE 6 Region under the graph over $(0, 1]$ has total area 2 by Example 7.

Infinite Discontinuities at the Endpoints

Another type of improper integral occurs when the integrand $f(x)$ becomes infinite at one or both of the endpoints of the interval of integration. For example, $\int_0^1 \dfrac{dx}{\sqrt{x}}$ is an improper integral because the integrand $x^{-1/2}$ tends to ∞ as $x \to 0+$ (Figure 6). This integral represents the area of a region that is infinite in the vertical direction. Improper integrals of this type are defined as one-sided limits.

DEFINITION Integrands with Infinite Discontinuities   If $f(x)$ is continuous on $[a, b)$ but discontinuous at $x = b$, we define

$$\int_a^b f(x)\,dx = \lim_{R\to b-} \int_a^R f(x)\,dx$$

Similarly, if $f(x)$ is continuous on $(a, b]$ but discontinuous at $x = a$,

$$\int_a^b f(x)\,dx = \lim_{R\to a+} \int_R^b f(x)\,dx$$

If, in either case, the limit does not exist, the integral is said to diverge.

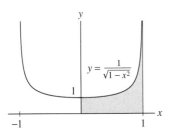

FIGURE 7 The infinite shaded region under the graph of $y = (1 - x^2)^{-1/2}$ has area $\frac{\pi}{2}$.

■ **EXAMPLE 6** Evaluate $\displaystyle\int_0^1 \frac{dx}{\sqrt{1 - x^2}}$.

Solution This is an improper integral because the integrand has an infinite discontinuity at $x = 1$ (Figure 7). Recall that $\displaystyle\int \frac{dx}{\sqrt{1 - x^2}} = \sin^{-1} x + C$. Thus,

$$\int_0^1 \frac{dx}{\sqrt{1 - x^2}} = \lim_{R\to 1-} \int_0^R \frac{dx}{\sqrt{1 - x^2}} = \lim_{R\to 1-} (\sin^{-1} R - \sin^{-1} 0)$$

$$= \sin^{-1} 1 - \sin^{-1} 0 = \frac{\pi}{2} - 0 = \frac{\pi}{2} \qquad ■$$

■ **EXAMPLE 7** Determine the convergence of **(a)** $\displaystyle\int_0^1 \frac{dx}{\sqrt{x}}$ and **(b)** $\displaystyle\int_0^1 \frac{dx}{x}$.

Solution Both integrals are improper because the integrands have infinite discontinuities at $x = 0$. The first integral converges:

$$\int_0^1 \frac{dx}{\sqrt{x}} = \lim_{R\to 0+} \int_R^1 \frac{dx}{\sqrt{x}} = \lim_{R\to 0+} 2x^{1/2} \Big|_R^1 = \lim_{R\to 0+} (2 - 2R^{1/2}) = 2$$

The second integral diverges (recall that $\ln R$ approaches $-\infty$ as $R \to 0+$):

$$\int_0^1 \frac{dx}{x} = \lim_{R\to 0+} \int_R^1 \frac{dx}{x} = \lim_{R\to 0+} (\ln 1 - \ln R) = -\lim_{R\to 0+} \ln R = \infty$$ ■

The integrals in the previous example are both of the form $\displaystyle\int_0^1 x^{-p}\,dx$. According to the next theorem, this integral converges if $p < 1$ and diverges otherwise. The proof is similar to the proof of Theorem 1 (see Exercise 56).

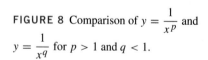

The result of Theorem 2 is valid for all exponents p. However, the integral is not improper if p < 0.

THEOREM 2 Improper Integral of x^{-p} over $[0, 1]$

$$\int_0^1 \frac{dx}{x^p} = \begin{cases} \dfrac{1}{1-p} & \text{if } p < 1 \\ \infty & \text{if } p \geq 1 \end{cases}$$

Comparing Theorems 1 and 2, we see that the integrals $\displaystyle\int_1^\infty \frac{dx}{x^p}$ and $\displaystyle\int_0^1 \frac{dx}{x^p}$ have opposite behavior for $p \neq 1$. The first integral converges only for $p > 1$ and the second integral converges only for $p < 1$ (Figure 8). They both diverge for $p = 1$.

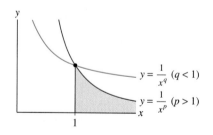

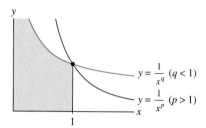

FIGURE 8 Comparison of $y = \dfrac{1}{x^p}$ and $y = \dfrac{1}{x^q}$ for $p > 1$ and $q < 1$.

Comparison of Integrals

Sometimes we are interested in determining whether an improper integral converges, even if we cannot find its exact value. For instance, the following integral cannot be evaluated explicitly:

$$\int_1^\infty \frac{e^{-x}}{x}\,dx$$

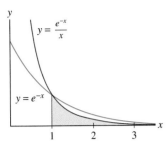

FIGURE 9 There is less area under $y = \dfrac{e^{-x}}{x}$ than under $y = e^{-x}$ in the interval $[1, \infty)$.

What the Comparison Test says (for nonnegative functions):

- *If the integral of the bigger function converges, then the integral of the smaller function also converges.*
- *If the integral of the smaller function diverges, then the integral of the larger function also diverges.*
- *The Comparison Test is also valid for improper integrals with infinite discontinuities at endpoints.*

However, we have the inequalities

$$0 \le \frac{e^{-x}}{x} \le e^{-x} \qquad \text{for } x \ge 1$$

Thus, the graph of e^{-x}/x lies *underneath* the graph of e^{-x} to the right of $x = 1$ (Figure 9) and we should be able to conclude that our integral converges:

$$\underbrace{\int_1^\infty \frac{e^{-x}}{x}\, dx}_{\text{This converges by comparison}} \le \underbrace{\int_1^\infty e^{-x}\, dx = e^{-1}}_{\text{Converges by direct computation}}$$

The next theorem (whose proof is omitted) summarizes this type of argument.

THEOREM 3 Comparison Test for Improper Integrals
Assume that $f(x) \ge g(x) \ge 0$ for $x \ge a$.

- If $\displaystyle\int_a^\infty f(x)\, dx$ converges, then $\displaystyle\int_a^\infty g(x)\, dx$ also converges.

- If $\displaystyle\int_a^\infty g(x)\, dx$ diverges, then $\displaystyle\int_a^\infty f(x)\, dx$ also diverges.

■ **EXAMPLE 8** Show that $\displaystyle\int_1^\infty \frac{dx}{\sqrt{x^3 + 1}}$ converges.

Solution To show that the integral converges, we compare the integrand $(x^3 + 1)^{-1/2}$ with a *larger* function whose integral we can compute. It makes sense to compare with $x^{-3/2}$ because, for $x \ge 1$,

$$\sqrt{x^3 + 1} \ge \sqrt{x^3} = x^{3/2} \quad \Rightarrow \quad \frac{1}{x^{3/2}} \ge \frac{1}{\sqrt{x^3 + 1}}$$

The integral $\displaystyle\int_1^\infty \frac{dx}{x^{3/2}}$ converges by Theorem 1 (the case $p = \frac{3}{2}$). Therefore, we may apply the Comparison Test:

$$\int_1^\infty \frac{dx}{x^{3/2}} \text{ converges} \quad \Rightarrow \quad \int_1^\infty \frac{dx}{\sqrt{x^3 + 1}} \text{ also converges} \qquad ■$$

If $f(x)$ is continuous, the convergence or divergence of $\displaystyle\int_a^\infty f(x)\, dx$ is not affected by a change in the lower limit a. If we change the lower limit to b, the integral changes by a finite amount:

$$\int_b^\infty f(x)\, dx = \underbrace{\int_b^a f(x)\, dx}_{\text{Ordinary integral whose value is finite}} + \int_a^\infty f(x)\, dx$$

Therefore, the integral over $[b, \infty)$ converges if and only if the integral over $[a, \infty)$ converges:

$$\int_b^\infty f(x)\, dx \text{ converges} \quad \Leftrightarrow \quad \int_a^\infty f(x)\, dx \text{ converges} \qquad \boxed{3}$$

EXAMPLE 9 Determine whether $\displaystyle\int_0^\infty \frac{dx}{\sqrt{x^2+1}}$ converges or diverges.

Solution To decide whether the integral converges or diverges, we might start by observing that if x is large, then

$$\frac{1}{\sqrt{x^2+1}} \approx \frac{1}{\sqrt{x^2}} = \frac{1}{x}$$

We know that $\displaystyle\int_1^\infty \frac{dx}{x}$ diverges (by Theorem 1 with $p=1$), so this suggests that our integral also diverges. However, $\dfrac{1}{\sqrt{x^2+1}} \le \dfrac{1}{x}$, and we need an inequality in the other direction to prove divergence. But note that if $x \ge 1$, then $x^2+1 \le x^2 + x^2 = 2x^2$ and thus $\sqrt{x^2+1} \le \sqrt{2}\,x$. This gives us the inequality

$$\frac{1}{\sqrt{x^2+1}} \ge \frac{1}{\sqrt{2}\,x} \qquad (\text{for } x \ge 1)$$

Now we may apply the Comparison Test and (3):

$$\frac{1}{\sqrt{2}}\int_1^\infty \frac{dx}{x} \text{ diverges} \Rightarrow \int_1^\infty \frac{dx}{\sqrt{x^2+1}} \text{ diverges} \Rightarrow \int_0^\infty \frac{dx}{\sqrt{x^2+1}} \text{ diverges} \quad \blacksquare$$

EXAMPLE 10 Assumptions Matter: Choosing the Right Comparison Determine whether $\displaystyle\int_1^\infty \frac{dx}{\sqrt{x}+e^{3x}}$ converges or diverges.

Solution Let's try using the inequality

$$\frac{1}{\sqrt{x}+e^{3x}} \le \frac{1}{e^{3x}} \quad (x \ge 0)$$

$$\int_1^\infty \frac{dx}{e^{3x}} = \lim_{R\to\infty} -\frac{1}{3}e^{-3x}\Big|_1^R = \lim_{R\to\infty} \frac{1}{3}\left(e^{-3} - e^{-3R}\right) = e^{-3} \quad (\text{converges})$$

The Comparison Test tells us that our integral converges:

$$\int_1^\infty \frac{dx}{e^{3x}} \quad \text{converges} \quad \Rightarrow \quad \int_1^\infty \frac{dx}{\sqrt{x}+e^{3x}} \quad \text{also converges}$$

Had we not been thinking, we might have tried to use the inequality

$$\frac{1}{\sqrt{x}+e^{3x}} \le \frac{1}{\sqrt{x}} \qquad \boxed{4}$$

The integral $\displaystyle\int_1^\infty \frac{dx}{\sqrt{x}}$ diverges by Theorem 1 (with $p=\frac{1}{2}$), but this tells us nothing about our integral which, by (4), is smaller (Figure 10). $\quad \blacksquare$

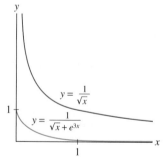

FIGURE 10 The divergence of $\displaystyle\int_1^\infty \frac{dx}{\sqrt{x}}$ says nothing about the divergence or convergence of the smaller integral $\displaystyle\int_1^\infty \frac{dx}{\sqrt{x}+e^{3x}}$.

7.7 SUMMARY

• An *improper integral* is defined as the limit of ordinary integrals. For example,

$$\int_a^\infty f(x)\,dx = \lim_{R\to\infty} \int_a^R f(x)\,dx$$

• If this limit exists, we say that the improper integral *converges*. Otherwise, we say that it *diverges*.

$$p > 1: \quad \int_1^\infty \frac{dx}{x^p} \quad \text{converges} \quad \text{and} \quad \int_0^1 \frac{dx}{x^p} \quad \text{diverges}$$

$$p < 1: \quad \int_1^\infty \frac{dx}{x^p} \quad \text{diverges} \quad \text{and} \quad \int_0^1 \frac{dx}{x^p} \quad \text{converges}$$

$$p = 1: \quad \int_1^\infty \frac{dx}{x} \quad \text{and} \quad \int_0^1 \frac{dx}{x} \quad \text{both diverge}$$

• The Comparison Test: Assume that $f(x) \geq g(x) \geq 0$ for $x \geq a$. Then

$$\text{If} \int_a^\infty f(x)\,dx \text{ converges,} \quad \text{then} \quad \int_a^\infty g(x)\,dx \text{ converges.}$$

$$\text{If} \int_a^\infty g(x)\,dx \text{ diverges,} \quad \text{then} \quad \int_a^\infty f(x)\,dx \text{ diverges.}$$

Remember: The Comparison Test provides no information if the larger integral $\int_a^\infty f(x)\,dx$ diverges or the smaller integral $\int_a^\infty g(x)\,dx$ converges.

• The Comparison Test is also valid for improper integrals with infinite discontinuities at endpoints.

7.7 EXERCISES

Preliminary Questions

1. State whether the integral converges or diverges:

(a) $\int_1^\infty x^{-3}\,dx$

(b) $\int_0^1 x^{-3}\,dx$

(c) $\int_1^\infty x^{-2/3}\,dx$

(d) $\int_0^1 x^{-2/3}\,dx$

2. Is $\int_0^{\pi/2} \cot x\,dx$ an improper integral? Explain.

3. Find a value of $b > 0$ that makes $\int_0^b \frac{1}{x^2 - 4}\,dx$ an improper integral.

4. Which comparison would show that $\int_0^\infty \frac{dx}{x + e^x}$ converges?

5. Explain why it is not possible to draw any conclusions about the convergence of $\int_1^\infty \frac{e^{-x}}{x}\,dx$ by comparing with the integral $\int_1^\infty \frac{dx}{x}$.

Exercises

1. Which of the following integrals is improper? Explain your answer, but do not evaluate the integral.

(a) $\int_0^2 \frac{dx}{x^{1/3}}\,dx$

(b) $\int_1^\infty \frac{dx}{x^{0.2}}$

(c) $\int_{-1}^\infty e^{-x}\,dx$

(d) $\int_0^1 e^{-x}$

(e) $\int_0^{\pi/2} \sec x\,dx$

(f) $\int_0^\infty \sin x\,dx$

(g) $\int_0^1 \sin x\,dx$

(h) $\int_0^1 \frac{dx}{\sqrt{3 - x^2}}$

(i) $\int_1^\infty \ln x\,dx$

(j) $\int_0^3 \ln x\,dx$

2. Let $f(x) = x^{-4/3}$.

(a) Evaluate $\int_1^R f(x)\,dx$.

(b) Evaluate $\int_1^\infty f(x)\,dx$ by computing the limit

$$\lim_{R \to \infty} \int_1^R f(x)\,dx$$

3. Prove that $\int_1^\infty x^{-2/3}\,dx$ diverges by showing that

$$\lim_{R\to\infty}\int_1^R x^{-2/3}\,dx = \infty$$

4. Determine if $\int_0^3 \dfrac{dx}{(3-x)^{3/2}}$ converges by computing

$$\lim_{R\to 3-}\int_0^R \dfrac{dx}{(3-x)^{3/2}}$$

In Exercises 5–46, determine whether the improper integral converges and, if so, evaluate it.

5. $\int_1^\infty \dfrac{dx}{x^{19/20}}$

6. $\int_1^\infty \dfrac{dx}{x^{20/19}}$

7. $\int_{-\infty}^4 e^{0.0001t}\,dt$

8. $\int_{20}^\infty \dfrac{dt}{t}$

9. $\int_0^5 \dfrac{dx}{x^{20/19}}$

10. $\int_0^5 \dfrac{dx}{x^{19/20}}$

11. $\int_0^4 \dfrac{dx}{\sqrt{4-x}}$

12. $\int_5^6 \dfrac{dx}{(x-5)^{3/2}}$

13. $\int_2^\infty x^{-3}\,dx$

14. $\int_0^\infty \dfrac{dx}{(x+1)^3}$

15. $\int_{-3}^\infty \dfrac{dx}{(x+4)^{3/2}}$

16. $\int_2^\infty e^{-2x}\,dx$

17. $\int_0^1 \dfrac{dx}{x^{0.2}}$

18. $\int_2^\infty x^{-1/3}\,dx$

19. $\int_4^\infty e^{-3x}\,dx$

20. $\int_4^\infty e^{3x}\,dx$

21. $\int_{-\infty}^0 e^{3x}\,dx$

22. $\int_0^1 \dfrac{dx}{x^{3.7}}$

23. $\int_2^\infty \dfrac{dx}{(x+3)^4}$

24. $\int_1^2 \dfrac{dx}{(x-1)^2}$

25. $\int_1^3 \dfrac{dx}{\sqrt{3-x}}$

26. $\int_{-2}^4 \dfrac{dx}{(x+2)^{1/3}}$

27. $\int_0^\infty \dfrac{dx}{1+x}$

28. $\int_{-\infty}^0 xe^{-x^2}\,dx$

29. $\int_0^\infty \dfrac{dx}{(1+x^2)^2}$

30. $\int_0^3 \dfrac{x\,dx}{\sqrt{x-3}}$

31. $\int_0^\infty e^{-x}\cos x\,dx$

32. $\int_{-\infty}^0 x^2 e^{-x}\,dx$

33. $\int_0^1 \dfrac{dx}{\sqrt{1-x^2}}$

34. $\int_0^1 \dfrac{e^{\sqrt{x}}\,dx}{\sqrt{x}}$

35. $\int_1^\infty \dfrac{e^{\sqrt{x}}\,dx}{\sqrt{x}}$

36. $\int_1^\infty xe^{-2x}\,dx$

37. $\int_0^\infty \dfrac{x\,dx}{x^2+1}$

38. $\int_0^{\pi/2} \sec\theta\,d\theta$

39. $\int_0^\infty \sin x\,dx$

40. $\int_0^{\pi/2} \tan x\,dx$

41. $\int_0^{\pi/2} \tan x\sec x\,dx$

42. $\int_0^1 \ln x\,dx$

43. $\int_0^1 x\ln x\,dx$

44. $\int_1^2 \dfrac{dx}{x\ln x}$

45. $\int_0^1 \dfrac{\ln x}{x^2}\,dx$

46. $\int_1^\infty \dfrac{\ln x}{x^2}\,dx$

47. Let $I = \int_4^\infty \dfrac{dx}{(x-2)(x-3)}$.
(a) Show that for $R > 4$,

$$\int_4^R \dfrac{dx}{(x-2)(x-3)} = \ln\left|\dfrac{R-3}{R-2}\right| - \ln\dfrac{1}{2}$$

(b) Then show that $I = \ln 2$.

48. Evaluate the integral $I = \int_1^\infty \dfrac{dx}{x(2x+5)}$.

49. Determine whether $I = \int_0^1 \dfrac{dx}{x(2x+5)}$ converges and, if so, evaluate.

50. Determine whether $I = \int_2^\infty \dfrac{dx}{(x+3)(x+1)^2}$ converges and, if so, evaluate.

In Exercises 51–54, determine if the doubly infinite improper integral converges and, if so, evaluate it. Use definition (2).

51. $\int_{-\infty}^\infty \dfrac{dx}{1+x^2}$

52. $\int_{-\infty}^\infty e^{-|x|}\,dx$

53. $\int_{-\infty}^\infty xe^{-x^2}\,dx$

54. $\int_{-\infty}^\infty \dfrac{dx}{(x^2+1)^{3/2}}$

55. For which values of a does $\int_0^\infty e^{ax}\,dx$ converge?

56. Show that $\int_0^1 \dfrac{dx}{x^p}$ converges if $p < 1$ and diverges if $p \geq 1$.

57. Sketch the region under the graph of $f(x) = \dfrac{1}{1+x^2}$ for $-\infty < x < \infty$ and show that its area is π.

58. Show that $\dfrac{1}{\sqrt{x^4+1}} \leq \dfrac{1}{x^2}$ for all x and use this to prove that

$$\int_1^\infty \dfrac{dx}{\sqrt{x^4+1}} \text{ converges.}$$

59. Show that $\displaystyle\int_1^\infty \frac{dx}{x^3+4}$ converges by comparing with $\displaystyle\int_1^\infty x^{-3}\,dx$.

60. Show that $\displaystyle\int_2^\infty \frac{dx}{x^3-4}$ converges by comparing with $\displaystyle\int_2^\infty 2x^{-3}\,dx$.

61. 📖 Show that $0 \le e^{-x^2} \le e^{-x}$ for $x \ge 1$ (Figure 11). Then use the Comparison Test and (3) to show that $\displaystyle\int_0^\infty e^{-x^2}\,dx$ converges.

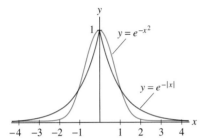

FIGURE 11 Comparison of $y = e^{-|x|}$ and $y = e^{-x^2}$.

62. Prove that $\displaystyle\int_{-\infty}^\infty e^{-x^2}\,dx$ converges by comparing with $\displaystyle\int_{-\infty}^\infty e^{-|x|}\,dx$ (Figure 11).

63. Show that $\displaystyle\int_1^\infty \frac{1-\sin x}{x^2}\,dx$ converges.

64. Let $a > 0$. Recall that $\displaystyle\lim_{x\to\infty} \frac{x^a}{\ln x} = \infty$ (by Exercise 60 in Section 4.7).

(a) Show that $x^a > 2\ln x$ for all x sufficiently large.
(b) Show that $e^{-x^a} < x^{-2}$ for all x sufficiently large.
(c) Show that $\displaystyle\int_1^\infty e^{-x^a}\,dx$ converges.

In Exercises 65–74, use the Comparison Test to determine whether or not the integral converges.

65. $\displaystyle\int_1^\infty \frac{1}{\sqrt{x^5+2}}\,dx$

66. $\displaystyle\int_1^\infty \frac{dx}{(x^3+2x+4)^{1/2}}$

67. $\displaystyle\int_3^\infty \frac{dx}{\sqrt{x-1}}$

68. $\displaystyle\int_0^5 \frac{dx}{x^{1/3}+x^3}$

69. $\displaystyle\int_1^\infty e^{-(x+x^{-1})}\,dx$

70. $\displaystyle\int_0^1 \frac{\sin x}{\sqrt{x}}\,dx$

71. $\displaystyle\int_0^1 \frac{e^x}{x^2}\,dx$

72. $\displaystyle\int_1^\infty \frac{1}{x^4+e^x}\,dx$

73. $\displaystyle\int_0^1 \frac{1}{x^4+\sqrt{x}}\,dx$

74. $\displaystyle\int_1^\infty \frac{\ln x}{\sinh x}\,dx$

75. An investment pays a dividend of \$250/year continuously forever. If the interest rate is 7%, what is the present value of the entire income stream generated by the investment?

76. An investment is expected to earn profits at a rate of $10,000e^{0.01t}$ dollars/year forever. Find the present value of the income stream if the interest rate is 4%.

77. Compute the present value of an investment that generates income at a rate of $5,000te^{0.01t}$ dollars/year forever, assuming an interest rate of 6%.

78. Find the volume of the solid obtained by rotating the region below the graph of $y = e^{-x}$ about the x-axis for $0 \le x < \infty$.

79. Let S be the solid obtained by rotating the region below the graph of $y = x^{-1}$ about the x-axis for $1 \le x < \infty$ (Figure 12).

(a) Use the Disk Method (Section 6.3) to compute the volume of S. Note that the volume is finite even though S is an infinite region.

(b) It can be shown that the surface area of S is

$$A = 2\pi \int_1^\infty x^{-1}\sqrt{1+x^{-4}}\,dx.$$

Show that A is infinite.

If S were a container, you could fill its interior with a finite amount of paint, but you could not paint its surface with a finite amount of paint.

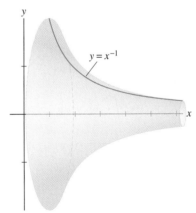

FIGURE 12

80. Compute the volume of the solid obtained by rotating the region below the graph of $y = (x^2+1)^{-1}$ about the x-axis for $-\infty < x < \infty$.

81. When a capacitor of capacitance C is charged by a source of voltage V, the power expended at time t is

$$P(t) = \frac{V^2}{R}(e^{-t/RC} - e^{-2t/RC}),$$

where R is the resistance in the circuit. The total energy stored in the capacitor is

$$W = \int_0^\infty P(t)\, dt$$

Show that $W = \frac{1}{2}CV^2$.

82. Conservation of Energy can be used to show that when a mass m oscillates at the end of a spring with spring constant k, the period of oscillation is

$$T = 4\sqrt{m} \int_0^{\sqrt{2E/k}} \frac{dx}{\sqrt{2E - kx^2}}$$

where E is the total energy of the mass. Show that this is an improper integral with value $T = 2\pi\sqrt{m/k}$.

83. When a radioactive substance decays, the fraction of atoms present at time t is $f(t) = e^{-kt}$, where $k > 0$ is the decay constant. It can be shown that the *average* life of an atom (until it decays) is $A = -\int_0^\infty t f'(t)\, dt$. Use Integration by Parts to show that $A = \int_0^\infty f(t)\, dt$ and compute A. What is the average decay time of Radon-222, whose half-life is 3.825 days?

84. According to **Maxwell's Law of velocity distribution** for a gas at temperature T, the fraction of gas molecules with speed between v and $v + \Delta v$ is approximately $f(v)\,\Delta v$, where

$$f(v) = 4\pi \left(\frac{m}{2\pi kT}\right)^{3/2} v^2 e^{-mv^2/(2kT)}$$

(m is the molecular mass of the gas and k Boltzmann's constant). Show that the average speed $\overline{v}$ is $\left(\frac{8kT}{\pi m}\right)^{1/2}$, where

$$\overline{v} = \int_0^\infty v f(v)\, dv$$

85. Let $J_n = \int_0^\infty x^n e^{-\alpha x}\, dx$, where $n \geq 1$ is an integer and $\alpha > 0$. Prove that $J_n = (n/\alpha)J_{n-1}$ and $J_0 = 1/\alpha$. Use this to compute J_4. Show that $J_n = \dfrac{n!}{\alpha^{n+1}}$.

86. Let $a > 0$ and $n > 1$. Define $f(x) = \dfrac{x^n}{e^{ax} - 1}$ for $x \neq 0$ and $f(0) = 0$.

(a) Use L'Hôpital's Rule to show that $f(x)$ is continuous at $x = 0$.

(b) Show that $\int_0^\infty f(x)\, dx$ converges. *Hint:* Show that $f(x) \leq 2x^n e^{-ax}$ if x is large enough. Then use the Comparison Test and Exercise 85.

87. There is a function $F(v)$ such that the amount of electromagnetic energy with frequency between v and $v + \Delta v$ radiated by a so-called black body at temperature T is proportional to $F(v)\,\Delta v$. The total radiated energy is $E = \int_0^\infty F(v)\, dv$. According to **Planck's Radiation Law**,

$$F(v) = \left(\frac{8\pi h}{c^3}\right) \frac{v^3}{e^{hv/kT} - 1}$$

where c, h, k are physical constants. To derive this law, Planck introduced the quantum hypothesis in 1900, which thus marked the birth of quantum mechanics. Show that E is finite (use Exercise 86).

88. Let $J = \int_0^\infty e^{-x^2}\, dx$ and $J_N = \int_0^N e^{-x^2}\, dx$. Although e^{-x^2} has no elementary antiderivative, it is known that $J = \sqrt{\pi}/2$. Let T_N be the Nth trapezoidal approximation to J_N. Calculate T_4 and show that T_4 approximates J to three decimal places.

89. A **probability density** function on $[0, \infty)$ is a function $f(x)$ defined for $x \geq 0$ such that $f(x) \geq 0$ and $\int_0^\infty f(x)\, dx = 1$. The *mean value* of $f(x)$ is the quantity $\mu = \int_0^\infty x f(x)\, dx$. For $k > 0$, find a constant C such that Ce^{-kx} is a probability density and compute μ.

*In Exercises 90–93, the **Laplace transform** of a function $f(x)$ is the function $Lf(s)$ of the variable s defined by the improper integral (if it converges):*

$$Lf(s) = \int_0^\infty f(x)e^{-sx}\, dx$$

Laplace transforms are widely used in physics and engineering.

90. Show that if $f(x) = C$ where C a constant, then $Lf(s) = C/s$ for $s > 0$.

91. Show that if $f(x) = \sin \alpha x$, then $Lf(s) = \dfrac{\alpha}{s^2 + \alpha^2}$.

92. Compute $Lf(s)$, where $f(x) = e^{\alpha x}$ and $s > \alpha$.

93. Compute $Lf(s)$, where $f(x) = \cos \alpha x$ and $s > 0$.

Further Insights and Challenges

94. Let p be an integer. Show that $\int_0^{1/2} \dfrac{dx}{x(\ln x)^p}$ converges if and only if $p > 1$.

95. Let $I = \int_0^1 x^p \ln x\, dx$.

(a) Show that I diverges for $p = -1$.

(b) Show that if $p \neq -1$, then

$$\int x^p \ln x\, dx = \frac{x^{p+1}}{p+1}\left(\ln x - \frac{1}{p+1}\right) + C$$

(c) Use L'Hôpital's Rule to show that I converges if $p > -1$ and diverges if $p < -1$.

In Exercises 96–98, an improper integral $I = \displaystyle\int_a^\infty f(x)\,dx$ is called **absolutely convergent** *if $\displaystyle\int_a^\infty |f(x)|\,dx$ converges. It can be shown that if I is absolutely convergent, then it is convergent.*

96. Show that $\displaystyle\int_1^\infty \frac{\sin x}{x^2}\,dx$ is absolutely convergent.

97. Show that $\displaystyle\int_1^\infty e^{-x^2}\cos x\,dx$ is absolutely convergent.

98. Let $f(x) = \dfrac{\sin x}{x}$ and $I = \displaystyle\int_0^\infty f(x)\,dx$. We define $f(0) = 1$. Then $f(x)$ is continuous and I is not improper at $x = 0$.

(a) Show that

$$\int_1^R \frac{\sin x}{x}\,dx = -\frac{\cos x}{x}\Big|_1^R - \int_1^R \frac{\cos x}{x^2}\,dx$$

(b) Show that $\displaystyle\int_1^\infty \frac{\cos x}{x^2}\,dx$ converges. Conclude that the limit as $R \to \infty$ of the integral in (a) exists and is finite.

(c) Show that I converges.

It is known that $I = \frac{\pi}{2}$. However, I is *not* absolutely convergent. The convergence depends on cancellation, as shown in Figure 13.

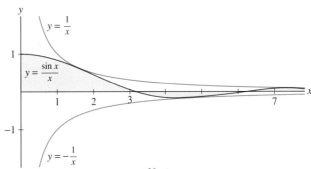

FIGURE 13 Convergence of $\displaystyle\int_1^\infty \frac{\sin x}{x}\,dx$ is due to the cancellation arising from the periodic change of sign.

99. The **gamma function**, which plays an important role in advanced applications, is defined for $n \geq 1$ by

$$\Gamma(n) = \int_0^\infty t^{n-1}e^{-t}\,dt$$

(a) Show that the integral defining $\Gamma(n)$ converges for $n \geq 1$ (it actually converges for all $n > 0$). *Hint:* Show that $t^{n-1}e^{-t} < t^{-2}$ for t sufficiently large.

(b) Show that $\Gamma(n + 1) = n\Gamma(n)$ using Integration by Parts.

(c) Show that $\Gamma(n + 1) = n!$ if $n \geq 1$ is an integer. *Hint:* Use (a) repeatedly. Thus, $\Gamma(n)$ provides a way of defining n-factorial when n is not an integer.

100. Use the results of Exercise 99 to show that the Laplace transform (see Exercises 90–93 above) of x^n is $\dfrac{n!}{s^{n+1}}$.

CHAPTER REVIEW EXERCISES

1. Estimate $\displaystyle\int_2^5 f(x)\,dx$ by computing T_2, M_3, T_6, and S_6 for a function $f(x)$ taking on the values in the table below:

x	2	2.5	3	3.5	4	4.5	5
$f(x)$	$\frac{1}{2}$	2	1	0	$-\frac{3}{2}$	-4	-2

2. State whether the approximation M_N or T_N is larger or smaller than the integral.

(a) $\displaystyle\int_0^\pi \sin x\,dx$ **(b)** $\displaystyle\int_\pi^{2\pi} \sin x\,dx$

(c) $\displaystyle\int_1^8 \frac{dx}{x^2}$ **(d)** $\displaystyle\int_2^5 \ln x\,dx$

3. The rainfall rate (in inches per hour) was measured hourly during a 10-hour thunderstorm with the following results:

0, 0.41, 0.49, 0.32, 0.3, 0.23, 0.09, 0.08, 0.05, 0.11, 0.12

Use Simpson's Rule to estimate the total rainfall during the 10-hour period.

In Exercises 4–9, compute the given approximation to the integral.

4. $\displaystyle\int_0^1 e^{-x^2}\,dx, \quad M_5$

5. $\displaystyle\int_2^4 \sqrt{6t^3 + 1}\,dt, \quad T_3$

6. $\displaystyle\int_{\pi/4}^{\pi/2} \sqrt{\sin\theta}\,d\theta, \quad M_4$

7. $\displaystyle\int_1^4 \frac{dx}{x^3 + 1}, \quad T_6$

8. $\displaystyle\int_0^1 e^{-x^2}\,dx, \quad S_4$

9. $\displaystyle\int_5^9 \cos(x^2)\,dx, \quad S_8$

10. The following table gives the area $A(h)$ of a horizontal cross section of a pond at depth h. Use the Trapezoidal Rule to estimate the volume V of the pond (Figure 1).

h (ft)	A(h) (acres)	h (ft)	A(h) (acres)
0	2.8	10	0.8
2	2.4	12	0.6
4	1.8	14	0.2
6	1.5	16	0.1
8	1.2	18	0

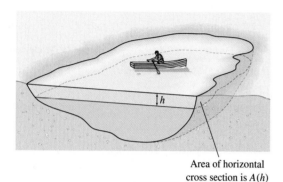

Area of horizontal
cross section is $A(h)$

FIGURE 1

11. Suppose that the second derivative of the function $A(h)$ in Exercise 10 satisfies $|A''(h)| \le 1.5$. Use the Error Bound to find the maximum possible error in your estimate of the volume V of the pond.

12. Find a bound for the error $\left| M_{16} - \int_1^3 x^3\, dx \right|$.

13. $\boxed{\text{GU}}$ Let $f(x) = \sin(x^3)$. Find a bound for the error

$$\left| T_{24} - \int_0^{\pi/2} f(x)\, dx \right|.$$

Hint: Find a bound K_2 for $|f''(x)|$ by plotting $f''(x)$ with a graphing utility.

14. Find a value of N such that

$$\left| M_N - \int_0^{\pi/4} \tan x\, dx \right| \le 10^{-4}$$

15. Find a value of N such that S_N approximates $\int_2^5 x^{-1/4}\, dx$ with an error of at most 10^{-2} (but do not calculate S_N).

16. Match the integrals (a)–(e) with their antiderivatives (i)–(v) on the basis of the general form (do not evaluate the integrals).

(a) $\displaystyle\int \frac{x\, dx}{x^2 - 4}$

(b) $\displaystyle\int \frac{(2x + 9)\, dx}{x^2 + 4}$

(c) $\displaystyle\int \sin^3 x \cos^2 x\, dx$

(d) $\displaystyle\int \frac{dx}{x\sqrt{16x^2 - 1}}$

(e) $\displaystyle\int \frac{16\, dx}{x(x - 4)^2}$

(i) $\sec^{-1} 4x + C$

(ii) $\log|x| - \log|x - 4| - \dfrac{4}{x - 4} + C$

(iii) $\dfrac{1}{30}(3\cos^5 x - 3\cos^3 x \sin^2 x - 7\cos^3 x) + C$

(iv) $\dfrac{9}{2} \tan^{-1} \dfrac{x}{2} + \ln(x^2 + 4) + C$

(v) $\sqrt{x^2 - 4} + C$

In Exercises 17–25, compute the integral using the suggested method.

17. $\displaystyle\int \cos^3 \theta \sin^8 \theta\, d\theta$ [write $\cos^3 \theta$ as $\cos\theta(1 - \sin^2 \theta)$]

18. $\displaystyle\int xe^{-12x}\, dx$ (Integration by Parts)

19. $\displaystyle\int \sec^3 \theta \tan^4 \theta\, d\theta$ (trigonometric identity, reduction formula)

20. $\displaystyle\int \frac{4x + 4}{(x - 5)(x + 3)}\, dx$ (partial fractions)

21. $\displaystyle\int \frac{dx}{x(x^2 - 1)^{3/2}}\, dx$ (trigonometric substitution)

22. $\displaystyle\int \frac{dx}{x^{3/2} + x^{1/2}}$ (substitution)

23. $\displaystyle\int \frac{dx}{x + x^{-1}}$ (rewrite integrand)

24. $\displaystyle\int x^{-2} \tan^{-1} x\, dx$ (Integration by Parts)

25. $\displaystyle\int \frac{dx}{x^2 + 4x - 5}$ (complete square, substitution, partial fractions)

In Exercises 26–69, compute the integral using the appropriate method or combination of methods.

26. $\displaystyle\int x^2 e^{4x}\, dx$

27. $\displaystyle\int \frac{x^2}{\sqrt{9 - x^2}}\, dx$

28. $\displaystyle\int \cos^9 6\theta \sin^3 6\theta\, d\theta$

29. $\displaystyle\int \sec^2 \theta \tan^4 \theta\, d\theta$

30. $\displaystyle\int \frac{(6x + 4)\, dx}{x^2 - 1}$

31. $\displaystyle\int \frac{dt}{(t^2 - 1)^2}$

32. $\displaystyle\int \frac{d\theta}{\cos^4 \theta}$

33. $\displaystyle\int \sin 2\theta \sin^2 \theta\, d\theta$

34. $\displaystyle\int \ln(9 - 2x)\, dx$

35. $\displaystyle\int (\ln(x + 1))^2\, dx$

36. $\displaystyle\int \sin^5 \theta\, d\theta$

37. $\displaystyle\int \cos^4(9x - 2)\, dx$

38. $\displaystyle\int \sin 3x \cos 5x \, dx$

39. $\displaystyle\int \sin 2x \sec^2 x \, dx$

40. $\displaystyle\int \sqrt{\tan x} \, \sec^2 x \, dx$

41. $\displaystyle\int (\sec x + \tan x)^2 \, dx$

42. $\displaystyle\int \sin^5 \theta \cos^3 \theta \, d\theta$

43. $\displaystyle\int \cot^2 \frac{\theta}{2} \, d\theta$

44. $\displaystyle\int \frac{dt}{(t-3)(t+4)}$

45. $\displaystyle\int \frac{dt}{(t-3)^2(t+4)}$

46. $\displaystyle\int \sqrt{x^2 + 9} \, dx$

47. $\displaystyle\int \frac{dx}{x\sqrt{x^2-4}}$

48. $\displaystyle\int \frac{dx}{x + x^{2/3}}$

49. $\displaystyle\int \frac{dx}{x^{3/2} + ax^{1/2}}$

50. $\displaystyle\int \frac{dx}{(x-b)^2 + 4}$

51. $\displaystyle\int \frac{(x^2 - x)\, dx}{(x+2)^3}$

52. $\displaystyle\int \frac{(7x^2 + x)\, dx}{(x-2)(2x+1)(x+1)}$

53. $\displaystyle\int \frac{16\, dx}{(x-2)^2(x^2+4)}$

54. $\displaystyle\int \frac{dx}{(x^2 + 25)^2}$

55. $\displaystyle\int \frac{dx}{x^2 + 8x + 25}$

56. $\displaystyle\int \frac{dx}{x^2 + 8x + 4}$

57. $\displaystyle\int \frac{(x^2 - x)\, dx}{(x+2)^3}$

58. $\displaystyle\int t^2 \sqrt{1 - t^2} \, dt$

59. $\displaystyle\int \frac{dx}{x^4 \sqrt{x^2 + 4}}$

60. $\displaystyle\int \frac{dx}{(x^2 + 5)^{3/2}}$

61. $\displaystyle\int (x+1) e^{4-3x} \, dx$

62. $\displaystyle\int \frac{dx}{x^2 + x^{-2}}$

63. $\displaystyle\int x^{-2} \tan^{-1} x \, dx$

64. $\displaystyle\int x^3 \cos(x^2) \, dx$

65. $\displaystyle\int x^2 (\ln x)^2 \, dx$

66. $\displaystyle\int x \tanh^{-1} x \, dx$

67. $\displaystyle\int \frac{\tan^{-1} t \, dt}{1 + t^2}$

68. $\displaystyle\int \ln(x^2 + 9) \, dx$

69. $\displaystyle\int (\sin x)(\cosh x) \, dx$

In Exercises 70–77, evaluate the integral.

70. $\displaystyle\int_0^1 \cosh 2t \, dt$

71. $\displaystyle\int \tanh 5x \, dx$

72. $\displaystyle\int \sinh^3 x \cosh x \, dx$

73. $\displaystyle\int \coth^2 (1 - 4t) \, dt$

74. $\displaystyle\int_{-0.3}^{0.3} \frac{dx}{1 - x^2}$

75. $\displaystyle\int_0^{3\sqrt{3}/2} \frac{dx}{\sqrt{9 - x^2}}$

76. $\displaystyle\int \frac{\sqrt{x^2 + 1}\, dx}{x^2}$

77. $\displaystyle\int \frac{dt}{\cosh^2 t + \sinh^2 t}$ *Hint: Use the substitution $u = \tanh t$.*

*In Exercises 78–81, let $gd(y) = \tan^{-1}(\sinh y)$ be the so-called **gudermannian**, which arises in cartography. In a map of the earth constructed by Mercator projection, points located y radial units from the equator correspond to points on the globe of latitude $gd(y)$.*

78. Prove that $\dfrac{d}{dy} gd(y) = \operatorname{sech} y$.

79. Let $f(y) = 2 \tan^{-1}(e^y) - \pi/2$. Prove that $gd(y) = f(y)$. *Hint:* Show that $gd'(y) = f'(y)$ and $f(0) = g(0)$.

80. Show that $t(y) = \sinh^{-1}(\tan y)$ is the inverse of $gd(y)$ for $0 \le y < \pi/2$.

81. Verify that $t(y)$ in Exercise 80 satisfies $t'(y) = \sec y$ and find a value of a such that

$$t(y) = \int_a^y \frac{dt}{\cos t}$$

In Exercises 82–91, determine whether the improper integral converges and, if so, evaluate it.

82. $\displaystyle\int_0^\infty \frac{dx}{(x+2)^2}$

83. $\displaystyle\int_4^\infty \frac{dx}{x^{2/3}}$

84. $\displaystyle\int_0^4 \frac{dx}{x^{2/3}}$

85. $\displaystyle\int_9^\infty \frac{dx}{x^{12/5}}$

86. $\displaystyle\int_{-\infty}^0 \frac{dx}{x^2 + 1}$

87. $\displaystyle\int_{-\infty}^9 e^{4x} \, dx$

88. $\displaystyle\int_0^{\pi/2} \cot \theta \, d\theta$

89. $\displaystyle\int_1^\infty \frac{dx}{(x+2)(2x+3)}$

90. $\displaystyle\int_0^\infty (5 + x)^{-1/3} \, dx$

91. $\displaystyle\int_2^5 (5 - x)^{-1/3} \, dx$

In Exercises 92–97, use the Comparison Test to determine if the improper integral converges or diverges.

92. $\displaystyle\int_8^\infty \frac{dx}{x^2 - 4}$

93. $\displaystyle\int_8^\infty (\sin^2 x) e^{-x} \, dx$

94. $\displaystyle\int_3^\infty \frac{dx}{x^4 + \cos^2 x}$

95. $\displaystyle\int_1^\infty \frac{dx}{x^{1/3} + x^{2/3}}$

96. $\displaystyle\int_0^1 \frac{dx}{x^{1/3} + x^{2/3}}$

97. $\displaystyle\int_0^\infty e^{-x^3} \, dx$

98. Calculate the volume of the infinite solid obtained by rotating the region under $y = (x^2 + 1)^{-2}$ for $0 \le x < \infty$ about the y-axis.

99. Let R be the region under the graph of $y = (x + 1)^{-1}$ for $0 \le x < \infty$. Which of the following quantities is finite?

(a) The area of R

(b) The volume of the solid obtained by rotating R about the x-axis

(c) The volume of the solid obtained by rotating R about the y-axis

100. Show that $\displaystyle\int_0^\infty x^n e^{-x^2}\,dx$ converges for all $n > 0$. *Hint:* First observe that $x^n e^{-x^2} < x^n e^{-x}$ for $x > 1$. Then show that $x^n e^{-x} < x^{-2}$ for x sufficiently large.

101. According to kinetic theory, the molecules of ordinary matter are in constant random motion. The probability that a molecule has kinetic energy in a small interval $[E, E + \Delta E]$ is approximately $\dfrac{1}{kT} e^{-E/(kT)}$, where T is the temperature (in kelvins) and k Boltzmann's constant. Compute the *average* kinetic energy $\overline{E}$ in terms of k and T, where

$$\overline{E} = \frac{1}{kT} \int_0^\infty E e^{-E/kT}\,dE$$

102. Compute the Laplace transform $Lf(s)$ of the function $f(x) = x$ for $s > 0$. See Exercises 90–93 in Section 7.7 for the definition of $Lf(s)$.

103. Compute the Laplace transform $Lf(s)$ of the function $f(x) = x^2 e^{\alpha x}$ for $s > \alpha$.

104. Let $I_n = \displaystyle\int \frac{x^n\,dx}{x^2 + 1}$.

(a) Prove that $I_n = \dfrac{x^{n-1}}{n-1} - I_{n-2}$.

(b) Use (a) to calculate I_n for $0 \le n \le 5$.

(c) Show that, in general,

$$I_{2n+1} = \frac{x^{2n}}{2n} - \frac{x^{2n-2}}{2n-2} + \cdots$$

$$+ (-1)^{n-1}\frac{x^2}{2} + (-1)^n \frac{1}{2}\ln(x^2 + 1) + C$$

$$I_{2n} = \frac{x^{2n-1}}{2n-1} - \frac{x^{2n-3}}{2n-3} + \cdots$$

$$+ (-1)^{n-1}x + (-1)^n \tan^{-1} x + C$$

105. Let $J_n = \displaystyle\int x^n e^{-x^2/2}\,dx$.

(a) Show that $J_1 = -e^{-x^2/2}$.

(b) Prove $J_n = -x^{n-1}e^{-x^2/2} + (n-1)J_{n-2}$.

(c) Use (a) and (b) to compute J_3 and J_5.

8 | FURTHER APPLICATIONS OF THE INTEGRAL AND TAYLOR POLYNOMIALS

In the first three sections of this chapter, we discuss some additional uses of integration, including two applications to physics. The last section introduces Taylor polynomials, which are higher-order generalizations of the linear approximation used to approximate algebraic and transcendental functions to arbitrary accuracy. Taylor polynomials are a high point of the theory we have developed so far and provide a beautiful illustration of the power and utility of calculus.

This NASA simulation, depicting streamlines of hot gas from the nozzles of a Harrier Jet during vertical takeoff, is based on a branch of mathematics called computational fluid dynamics.

8.1 Arc Length and Surface Area

One of the first problems we solved using integration was finding the area under the parabola $y = x^2$. What about the length of the parabola (Figure 1) or more generally, the length of a curve $y = f(x)$? In this section, we show that the length of a curve (often called **arc length**) can also be computed as an integral.

The arc length of $y = f(x)$ over an interval $[a, b]$ is defined using **polygonal approximations** to the graph. To construct a polygonal approximation L, choose a partition of $[a, b]$ into N subintervals with endpoints

$$a = x_0 < x_1 < \cdots < x_N = b$$

and join the corresponding points $P_0, P_1, \ldots, P_N$ on the graph by line segments (Figure 2). Let $|L|$ be the length of L.

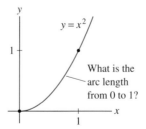

FIGURE 1 The arc length of $y = x^2$ over $[0, 1]$ is

$$\tfrac{1}{2}\sqrt{5} + \tfrac{1}{4}\ln(2 + \sqrt{5}) \approx 1.148$$

(see Exercise 22).

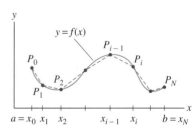

FIGURE 2 A polygonal approximation L.

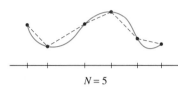 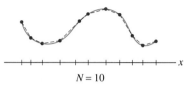

$N = 3$ $N = 5$ $N = 10$

FIGURE 3 The polygonal approximations L improve as the widths of the subintervals decrease.

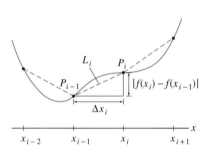

FIGURE 4

Figure 3 suggests that L approximates the graph more closely as the width of the partition decreases. We define the arc length of the graph to be the limit $\lim |L|$ as the width of the partition tends to zero, provided that the limit exists.

Now assume that $f(x)$ is differentiable and that $f'(x)$ is continuous on $[a, b]$. In this case, the limit $\lim |L|$ exists and we can compute it as an integral. Let $|L_i|$ be the length of the ith segment $L_i = \overline{P_{i-1} P_i}$. Figure 4 shows that $|L_i|$ is the hypotenuse of a right triangle of base $\Delta x_i = x_i - x_{i-1}$ and height $|f(x_i) - f(x_{i-1})|$. By the Pythagorean Theorem,

$$|L_i| = \sqrt{\Delta x_i^2 + (f(x_i) - f(x_{i-1}))^2}$$

The Mean Value Theorem tells us that for some point c_i in $[x_{i-1}, x_i]$, we have

$$f(x_i) - f(x_{i-1}) = f'(c_i)(x_i - x_{i-1}) = f'(c_i)\Delta x_i$$

Therefore,

$$|L_i| = \sqrt{(\Delta x_i)^2 + (f'(c_i)\Delta x_i)^2} = \sqrt{(\Delta x_i)^2(1 + f'(c_i)^2)} = \sqrt{1 + f'(c_i)^2}\, \Delta x_i$$

The total length $|L|$ is the sum of the lengths of the segments L_i:

$$|L| = |L_1| + |L_2| + \cdots + |L_N| = \sum_{i=1}^{N} \sqrt{1 + f'(c_i)^2}\, \Delta x_i \qquad \boxed{1}$$

◄┅ **REMINDER** A Riemann sum for the
integral $\displaystyle\int_a^b g(x)\,dx$ is a sum of the form

$$\sum_{i=1}^{N} g(c_i)\,\Delta x_i$$

where $x_0, x_1, \ldots, x_N$ is any partition of
$[a, b]$, $\Delta x_i = x_i - x_{i-1}$, and c_i is any
number in $[x_{i-1}, x_i]$.

We recognize the sum in Eq. (1) as a Riemann sum for the integral $\displaystyle\int_a^b \sqrt{1 + f'(x)^2}\,dx$. Since $f'(x)$ is continuous, the integrand $\sqrt{1 + f'(x)^2}$ is also continuous and hence integrable, so the limit $\lim |L|$ exists and is equal to the integral.

In Exercises 19–21, we verify that Eq. (2)
gives the correct result for the length of a
line and circumference of a circle.

THEOREM 1 Formula for Arc Length   If $f'(x)$ exists and is continuous on $[a, b]$, then the arc length of the graph of $y = f(x)$ over $[a, b]$ is equal to

$$\boxed{\text{Arc length over } [a, b] = \int_a^b \sqrt{1 + f'(x)^2}\,dx} \qquad \boxed{2}$$

■ **EXAMPLE 1** Calculate the arc length of the graph of $f(x) = \dfrac{1}{12}x^3 + x^{-1}$ over $[1, 3]$.

Solution First, let's calculate $1 + f'(x)^2$. Since $f'(x) = \frac{1}{4}x^2 - x^{-2}$,

$$1 + f'(x)^2 = 1 + \left(\frac{1}{4}x^2 - x^{-2}\right)^2 = 1 + \left(\frac{1}{16}x^4 - \frac{1}{2} + x^{-4}\right)$$

$$= \frac{1}{16}x^4 + \frac{1}{2} + x^{-4} = \left(\frac{1}{4}x^2 + x^{-2}\right)^2$$

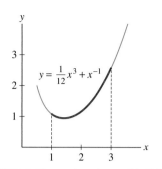

FIGURE 5 The arc length over [1, 3] is $\frac{17}{6}$.

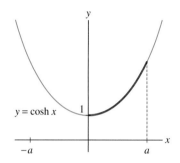

FIGURE 6

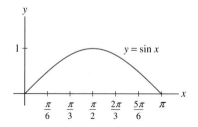

FIGURE 7 The arc length from 0 to π is approximately 3.82.

Fortunately, $1 + f'(x)^2$ is a square, so we can easily compute the arc length (Figure 5)

$$\int_1^3 \sqrt{1 + f'(x)^2}\,dx = \int_1^3 \left(\frac{1}{4}x^2 + x^{-2}\right)dx = \left(\frac{1}{12}x^3 - x^{-1}\right)\Big|_1^3$$

$$= \left(\frac{9}{4} - \frac{1}{3}\right) - \left(\frac{1}{12} - 1\right) = \frac{17}{6} \qquad \blacksquare$$

■ **EXAMPLE 2** Arc Length as a Function of the Upper Limit Find the arc length of $y = \cosh x$ over $[0, a]$ (Figure 6). Then find the arc length over [0, 2].

Solution Recall that $y' = (\cosh x)' = \sinh x$. Using Eq. (3) in the margin, we obtain

$$1 + (y')^2 = 1 + \sinh^2 x = \cosh^2 x$$

$$\int_0^a \sqrt{1 + (y')^2}\,dx = \int_0^a \cosh x\,dx = \sinh x\Big|_0^a = \sinh a$$

Therefore, the arc length over $[0, a]$ is $\sinh a$. Setting $a = 2$, we find that the arc length over [0, 2] is $\sinh 2 \approx 3.63$. ■

In Examples 1 and 2, the quantity $1 + f'(x)^2$ turned out to be a perfect square and we were able to evaluate the arc length integral. When $\sqrt{1 + f'(x)^2}$ does not have an elementary antiderivative, as is often the case, we may approximate the arc length using numerical integration.

■ **EXAMPLE 3** When Arc Length Cannot Be Calculated Exactly $\mathcal{CAS}$ Express the arc length L of $y = \sin x$ over $[0, \pi]$ as an integral. Then approximate L using Simpson's Rule S_N with $N = 6$ and using a computer algebra system.

Solution We have $y' = \cos x$. Setting $g(x) = \sqrt{1 + y'^2} = \sqrt{1 + \cos^2 x}$,

$$\text{Arc length} = \int_0^\pi \sqrt{1 + \cos^2 x}\,dx = \int_0^\pi g(x)\,dx$$

This integral cannot be evaluated explicitly, so we approximate it using Simpson's Rule (Section 7.1). We divide $[0, \pi]$ into $N = 6$ subintervals of width $\Delta x = \pi/6$ and endpoints $j\,\Delta x = j\pi/6$ for $j = 0, 1, \ldots, 6$. Then

$$S_6 = \frac{\Delta x}{3}\left(g(0) + 4g\left(\frac{\pi}{6}\right) + 2g\left(\frac{2\pi}{6}\right) + 4g\left(\frac{3\pi}{6}\right) + 2g\left(\frac{4\pi}{6}\right) + 4g\left(\frac{5\pi}{6}\right) + g(\pi)\right)$$

$$\approx \frac{\pi}{18}(1.4142 + 5.2915 + 2.2361 + 4 + 2.2361 + 5.2915 + 1.4142)$$

$$\approx 3.82$$

Using a computer algebra system to evaluate the integral numerically, we obtain the approximation 3.820198 (Figure 7). ■

The surface area of a surface of revolution can be computed by an integral that is closely related to the arc length integral (Figure 8). This formula can be justified using polynomial approximations to $y = f(x)$ as discussed above for arc length.

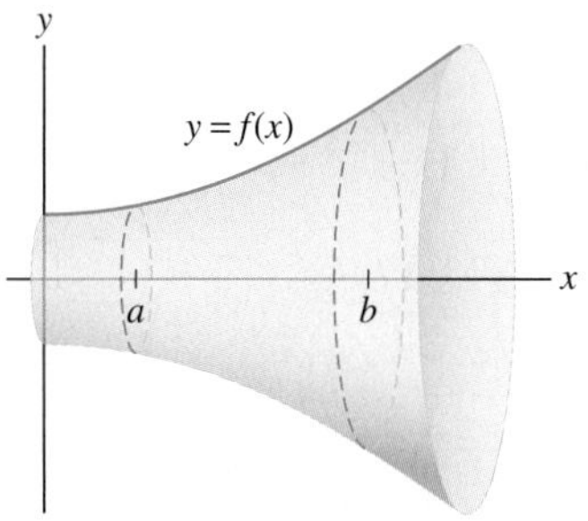

FIGURE 8 Surface obtained by revolving $y = f(x)$ about the x-axis.

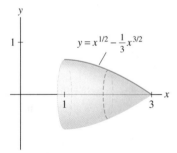

FIGURE 9 Surface obtained by revolving $y = 2x$ about the x-axis.

> **Area of a Surface of Revolution**   If $f'(x)$ exists and is continuous on $[a, b]$, then the area of the surface obtained by rotating the graph of $f(x)$ about the x-axis for $a \leq x \leq b$ is equal to
>
> $$\text{Surface area over } [a, b] = 2\pi \int_a^b f(x)\sqrt{1 + f'(x)^2}\, dx$$

■ **EXAMPLE 4**   Find the surface area S of the cone obtained by rotating the line $y = 2x$ about the x-axis for $0 \leq x \leq 4$.

Solution   We have $f(x) = 2x$ and $\sqrt{1 + f'(x)^2} = \sqrt{1 + 4} = \sqrt{5}$, so the surface area is equal to (Figure 9)

$$S = 2\pi \int_0^4 f(x)\sqrt{1 + f'(x)^2}\, dx = 2\pi \int_0^4 2x\sqrt{5}\, dx = 2\pi\sqrt{5}\, x^2 \Big|_0^4 = 32\pi\sqrt{5} \quad ■$$

■ **EXAMPLE 5**   Find the area S of the surface obtained by rotating $y = x^{1/2} - \frac{1}{3}x^{3/2}$ about the x-axis for $1 \leq x \leq 3$.

Solution   Let $f(x) = x^{1/2} - \frac{1}{3}x^{3/2}$. Then $f'(x) = \frac{1}{2}(x^{-1/2} - x^{1/2})$ and

$$1 + f'(x)^2 = 1 + \left(\frac{x^{-1/2} - x^{1/2}}{2}\right)^2 = 1 + \frac{x^{-1} - 2 + x}{4}$$

$$= \frac{x^{-1} + 2 + x}{4} = \left(\frac{x^{1/2} + x^{-1/2}}{2}\right)^2$$

The surface area (Figure 10) is equal to

$$S = 2\pi \int_1^3 f(x)\sqrt{1 + f'(x)^2}\, dx$$

$$= 2\pi \int_1^3 \left(x^{1/2} - \frac{1}{3}x^{3/2}\right)\left(\frac{x^{1/2} + x^{-1/2}}{2}\right) dx$$

$$= \pi \int_1^3 \left(1 + \frac{2}{3}x - \frac{1}{3}x^2\right) dx = \pi\left(x + \frac{1}{3}x^2 - \frac{1}{9}x^3\right)\Big|_1^3 = \frac{16\pi}{9} \quad ■$$

FIGURE 10 Surface obtained by revolving $y = x^{1/2} - \frac{1}{3}x^{3/2}$ about the x-axis for $1 \leq x \leq 3$.

8.1 SUMMARY

- The *arc length* of the curve $y = f(x)$ for $a \leq x \leq b$ is $\displaystyle\int_a^b \sqrt{1 + f'(x)^2}\, dx$.

- When the arc length integral cannot be evaluated directly, we may use numerical integration to obtain an approximation.

- The *surface area* of the surface obtained by rotating the graph of $f(x)$ about the x-axis for $a \leq x \leq b$ is

$$\text{Surface area} = 2\pi \int_a^b f(x)\sqrt{1 + f'(x)^2}\, dx$$

8.1 EXERCISES

Preliminary Questions

1. Which integral represents the length of the curve $y = \cos x$ between 0 and π?

$$\int_0^\pi \sqrt{1 + \cos^2 x}\, dx, \qquad \int_0^\pi \sqrt{1 + \sin^2 x}\, dx$$

2. How do the arc lengths of the curves $y = f(x)$ and $y = f(x) + C$ over an interval $[a, b]$ differ (C is a constant)? Explain geometrically and then justify using the arc length formula.

Exercises

1. Express the arc length of the curve $y = x^4$ between $x = 2$ and $x = 6$ as an integral (but do not evaluate).

2. Express the arc length of the curve $y = \tan x$ for $0 \le x \le \frac{\pi}{4}$ as an integral (but do not evaluate).

3. Find the arc length of $y = \dfrac{1}{12}x^3 + x^{-1}$ for $1 \le x \le 2$. *Hint:* Show that $1 + (y')^2 = \left(\dfrac{1}{4}x^2 + x^{-2}\right)^2$.

4. Find the arc length of $y = \left(\dfrac{x}{2}\right)^4 + \dfrac{1}{2x^2}$ over $[1, 4]$. *Hint:* Show that $1 + (y')^2$ is a perfect square.

In Exercises 5–10, calculate the arc length over the given interval.

5. $y = 3x + 1$, $[0, 3]$

6. $y = 9 - 3x$, $[1, 3]$

7. $y = x^{3/2}$, $[1, 2]$

8. $y = \frac{1}{3}x^{3/2} - x^{1/2}$, $[2, 8]$

9. $y = \frac{1}{4}x^2 - \frac{1}{2}\ln x$, $[1, 2e]$

10. $y = \ln(\cos x)$, $[0, \frac{\pi}{4}]$

In Exercises 11–14, approximate the arc length of the curve over the interval using the Trapezoidal Rule T_N, the Midpoint Rule M_N, or Simpson's Rule S_N as indicated.

11. $y = \frac{1}{4}x^4$, $[1, 2]$, T_5

12. $y = \sin x$, $[0, \frac{\pi}{2}]$, M_8

13. $y = x^{-1}$, $[1, 2]$, S_8

14. $y = e^{-x^2}$, $[0, 2]$, S_8

15. Calculate the length of the astroid $x^{2/3} + y^{2/3} = 1$ (Figure 11).

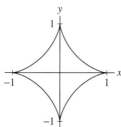

FIGURE 11 Graph of $x^{2/3} + y^{2/3} = 1$.

16. Show that the arc length of the astroid $x^{2/3} + y^{2/3} = a^{2/3}$ (for $a > 0$) is proportional to a.

17. Find the arc length of the curve shown in Figure 12.

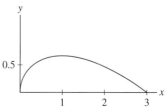

FIGURE 12 Graph of $9y^2 = x(x - 3)^2$.

18. Find the value of a such that the arc length of the *catenary* $y = \cosh x$ for $-a \le x \le a$ equals 10.

19. Let $f(x) = mx + r$ be a linear function (Figure 13). Use the Pythagorean Theorem to verify that the arc length over $[a, b]$ is equal to

$$\int_a^b \sqrt{1 + f'(x)^2}\, dx$$

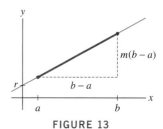

FIGURE 13

20. Show that the circumference of the unit circle is equal to

$$2\int_{-1}^{1} \frac{dx}{\sqrt{1 - x^2}} \quad \text{(an improper integral).}$$

Evaluate, thus verifying that the circumference is 2π.

21. Show that the circumference C of the circle of radius r is $C = 2\int_{-r}^{r} \dfrac{dx}{\sqrt{1 - x^2/r^2}}$. Use a substitution to show that C is proportional to r.

22. Calculate the arc length of $y = x^2$ over $[0, a]$. *Hint:* Use trigonometric substitution. Evaluate for $a = 1$.

23. Express the arc length of $g(x) = \sqrt{x}$ over $[0, 1]$ as a definite integral. Then use the substitution $u = \sqrt{x}$ to show that this arc length is equal to the arc length of x^2 over $[0, 1]$ (but do not evaluate the integrals). Explain this result graphically.

24. $\boxed{\text{GU}}$ Use a graphing utility to sketch the graphs of $f(x) = \sqrt{x}$ and $g(x) = \ln x$ over $[1, 4]$. Then, by applying the Comparison Test to the arc length integrals, show that the arc length of $y = f(x)$ is:
(a) Less than the arc length of $y = g(x)$ over $[1, 4]$.
(b) Greater than the arc length of $y = g(x)$ over $[4, 8]$.

25. Find the arc length of $y = e^x$ over $[0, a]$. *Hint:* Try the substitution $u = \sqrt{1 + e^{2x}}$ followed by partial fractions.

26. Show that the arc length of $y = \ln(f(x))$ for $a \le x \le b$ is

$$\int_a^b \frac{\sqrt{f(x)^2 + f'(x)^2}}{f(x)} \, dx \qquad \boxed{4}$$

27. Use Eq. (4) to compute the arc length of $y = \ln(\sin x)$ for $\frac{\pi}{4} \le x \le \frac{\pi}{2}$.

28. Use Eq. (4) to compute the arc length of $y = \ln\left(\dfrac{e^x + 1}{e^x - 1}\right)$ over $[1, 3]$.

29. Show that if $0 \le f'(x) \le 1$ for all x, then the arc length of $y = f(x)$ over $[a, b]$ is at most $\sqrt{2}(b - a)$. Show that for $f(x) = x$, the arc length equals $\sqrt{2}(b - a)$.

30. Let $a, r > 0$. Show that the arc length of the curve $x^r + y^r = a^r$ for $0 \le x \le a$ is proportional to a.

31. Approximate the arc length of one-quarter of the unit circle (which we know is $\frac{\pi}{2}$) by computing the length of the polygonal approximation with $N = 4$ segments (Figure 14).

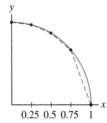

FIGURE 14 One-quarter of the unit circle

In Exercises 32–39, compute the surface area of revolution about the x-axis over the interval.

32. $y = 4x + 3$, $[0, 1]$

33. $y = 4x + 3$, $[2, 4]$

34. $y = \frac{1}{12}x^3 + x^{-1}$, $[1, 4]$

35. $y = (4 - x^{2/3})^{3/2}$, $[0, 8]$

36. $y = e^{-x}$, $[0, 1]$

37. $y = e^x$, $[0, 1]$

38. $y = x^2$, $[0, 4]$

39. $y = \sin x$, $[0, \pi]$

40. Find the area of the surface obtained by rotating $y = \cosh x$ over $[-\ln 2, \ln 2]$ around the x-axis.

41. Prove that the surface area of a sphere of radius r is $4\pi r^2$ by rotating the top half of the unit circle $x^2 + y^2 = r^2$ about the x-axis.

$\mathcal{CAS}$ *In Exercises 42–45, use a computer algebra system to find the exact or approximate surface area of the solid generated by rotating the curve about the x-axis.*

42. $y = \frac{1}{4}x^2 - \frac{1}{2}\ln x$ for $1 \le x \le e$, exact area

43. $y = x^2$ for $0 \le x \le 4$, exact area

44. $y = x^3$ for $0 \le x \le 4$, approximate area

45. $y = \tan x$ for $0 \le x \le \frac{\pi}{4}$, approximate area

46. Find the surface area of the torus obtained by rotating the circle $x^2 + (y - b)^2 = r^2$ about the x-axis.

47. $\mathcal{CAS}$ A merchant intends to produce specialty carpets in the shape of the region in Figure 15, bounded by the axes and graph of $y = 1 - x^n$ (units in yards). Assume that material costs \$50/yd^2 and that it costs $50L$ dollars to cut the carpet, where L is the length of the curved side of the carpet. The carpet can be sold for $150A$ dollars, where A is the carpet's area. Using numerical integration with a computer algebra system, find the whole number n for which the merchant's profits are maximal.

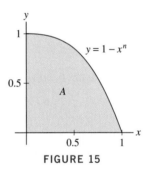

FIGURE 15

48. Prove that the portion of a sphere of radius R seen by an observer located at a distance d above the North Pole has area $A = \dfrac{2\pi d R^2}{d + R}$.
Hint: First show that if the cap has height h, then its surface area is $2\pi R h$. Then show that $h = \dfrac{dR}{d + R}$ by applying the Pythagorean Theorem to three right triangles in Figure 16.

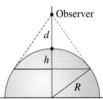

FIGURE 16 Spherical cap observed from a distance d above the North Pole.

49. $\boxed{\text{📖}}$ Suppose that the observer in Exercise 48 moves off to infinity, that is, $d \to \infty$. What do you expect the limiting value of the observed area to be? Check your guess by calculating the limit using the formula for the area in the previous exercise.

Further Insights and Challenges

50. Show that the surface area of a right circular cone of radius r and height h is $\pi r \sqrt{r^2 + h^2}$. *Hint:* Rotate a line $y = mx$ about the x-axis for $0 \le x \le h$, where m is determined suitably by the radius r.

51. Find the surface area of the ellipsoid obtained by rotating the ellipse $\left(\dfrac{x}{a}\right)^2 + \left(\dfrac{y}{b}\right)^2 = 1$ about the x-axis.

52. Show that if the arc length of $f(x)$ over $[0, a]$ is proportional to a, then $f(x)$ must be a linear function

53. *CAS* Let L be the arc length of the upper half of the ellipse with equation $y = \left(\dfrac{b}{a}\right)\sqrt{a^2 - x^2}$ (Figure 17) and let $\eta = \sqrt{1 - \dfrac{b^2}{a^2}}$.

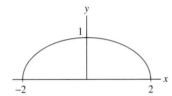

FIGURE 17 Graph of the ellipse $y = \frac{1}{2}\sqrt{4 - x^2}$.

Use substitution to show that

$$L = a \int_{-\pi/2}^{\pi/2} \sqrt{1 - \eta^2 \sin^2 \theta}\, d\theta$$

Use a computer algebra system to approximate L for $a = 2, b = 1$.

54. Let M be the total mass of a metal rod in the shape of the curve $y = f(x)$ over $[a, b]$ whose mass density $\rho(x)$ varies as a function of x. Use Riemann sums to justify the formula

$$M = \int_a^b \rho(x)\sqrt{1 + f'(x)^2}\, dx$$

55. Let $f(x)$ be an increasing function on $[a, b]$ and let $g(x)$ be its inverse. Argue on the basis of arc length that the following equality holds:

$$\int_a^b \sqrt{1 + f'(x)^2}\, dx = \int_{f(a)}^{f(b)} \sqrt{1 + g'(y)^2}\, dy \qquad \boxed{5}$$

Then use the substitution $u = f(x)$ to prove Eq. (5).

8.2 Fluid Pressure and Force

FIGURE 1 Since water pressure is proportional to depth, a diver's lungs are compressed and may be injured during a deep dive.

In this section, we use integration to compute the force exerted on an object submerged in a fluid. This is the force you feel when you try to push a basketball underwater in a swimming pool. Our computations are based on a law from physics which states that for a fluid at rest (such as water in a tank as opposed to water flowing in a stream), fluid pressure p is proportional to depth h.

Pressure is defined as force per unit area. However, there is an important difference between fluid pressure and the pressure exerted by one solid object on another. Fluid pressure does not act in any specific direction. Instead, a fluid exerts pressure on each side of an object in the direction perpendicular to that side (Figure 2). Recall that the density of a fluid is measured in units of mass per volume or weight per volume.

Pressure as a Function of Depth The fluid pressure at depth h in a fluid of density w (in weight per volume) is

$$\boxed{p = \text{fluid pressure} = wh} \qquad \boxed{1}$$

The pressure acts at each point on an object in the direction perpendicular to the object's surface at that point.

In our first example, we calculate the fluid force (also called the "total" or "resultant" force) in a situation where the pressure p is constant. In this case, the total force acting

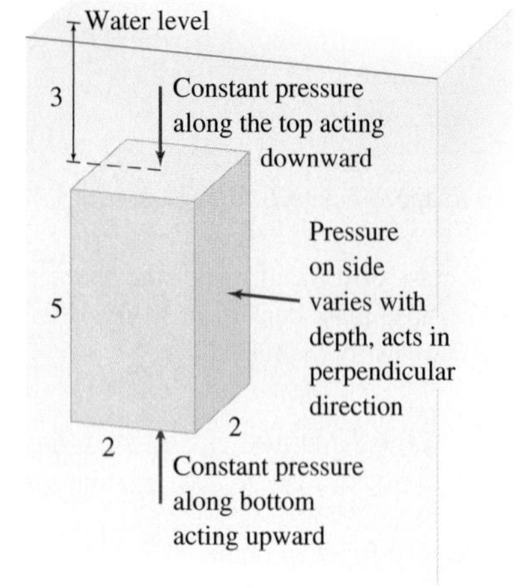

FIGURE 2 The pressure acts on each side in the perpendicular direction.

on a surface of area A is

$$\boxed{\text{Force} = \text{pressure} \times \text{area} = pA}$$

The density of water is $w = 62.5$ lb/ft^3.

■ **EXAMPLE 1** Calculate the force on the top and bottom of a box of dimensions $2 \times 2 \times 5$ ft, submerged in a pool of water with its top 3 ft below the water surface (Figure 2).

Solution The top of the box is located at depth $h = 3$ ft, so by Eq. (1),

$$\text{Pressure on top} = wh = 62.5(3) = 187.5 \text{ lb/ft}^2$$

Since the pressure along the top is constant and the top has area $A = 4$ ft^2, the total force on the top, acting downward, is

$$\text{Force on top} = \text{pressure} \times \text{area} = 187.5 \times 4 = 750 \text{ lb}$$

The bottom of the box is at depth $h = 8$ ft, so the total force on the bottom is

$$\text{Force on bottom} = \text{pressure} \times \text{area} = 62.5(8) \times 4 = 2{,}000 \text{ lb}$$

The force on the bottom acts in the upward direction. ■

In the next example, we calculate the total force on the side of a box. Since the pressure varies with depth, we use integration to find the total force.

■ **EXAMPLE 2** Calculating Force on a Side Using Integration Calculate the force F on the side of a box submerged with its top 3 ft below the surface of the water. As before, the box has height 5 and a square base with a side of 2 ft.

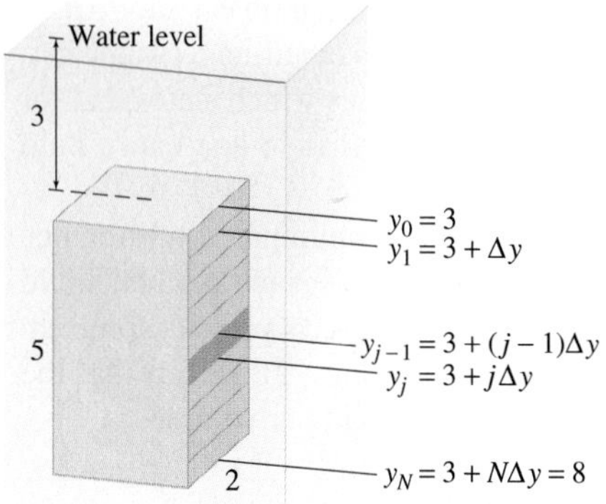

Solution Since the pressure varies with depth, we divide the side into N thin horizontal strips (Figure 3). If F_j is the force on the jth strip, then the total force F is

$$F = F_1 + F_2 + \cdots + F_N$$

Step 1. **Approximate the force on a strip.**

We'll use the variable y to denote depth. Thus, a larger value of y denotes greater depth. Each strip has height $\Delta y = 5/N$ and width 2, so the area of a strip is $2\Delta y$. The bottom edge of the jth strip has depth $y_j = 3 + j\Delta y$ (Figure 3).

If Δy is small, then the pressure on the jth strip is nearly constant with value wy_j (since the strip is thin, all points on it lie at nearly the same depth y_j), so we can approximate the force on the jth strip:

$$F_j = \text{force on the } j\text{th strip at depth } y_j \approx \underbrace{wy_j}_{\text{Pressure}} \times \underbrace{(2\Delta y)}_{\text{Area}} = 2wy_j\,\Delta y$$

Step 2. **Approximate total force as a Riemann sum.**

The total force on the side is

$$F = F_1 + F_2 + \cdots + F_N \approx \sum_{j=1}^{N} 2wy_j\,\Delta y$$

FIGURE 3 Each strip has area $2\Delta y$.

We recognize the sum on the right as a Riemann sum for the integral $w\displaystyle\int_3^8 2y\,dy$. The interval of integration is $[3, 8]$ because the box extends from $y = 3$ to $y = 8$ (the Riemann sum has been set up with $y_0 = 3$ and $y_N = 8$).

The computations of fluid force and work
(Example 2 in Section 6.5) are similar. In
both cases, we estimate the force or work
as a sum (corresponding to thin strips,
slices, etc.), and express the quantity
precisely as an integral.

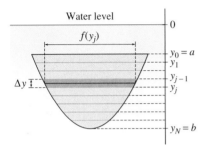

Water level

$f(y_j)$

Δy

$y_0 = a$
y_1
y_{j-1}
y_j
$y_N = b$

0

FIGURE 4 The area of the shaded strip is
approximately $f(y_j)\,\Delta y$.

Step 3. Evaluate total force as an integral.

As Δy tends to zero, the Riemann sum approaches the integral and we obtain

$$F = w \int_3^8 2y\,dy = (62.5)y^2 \Big|_3^8 = 62.5(8^2 - 3^2) = 3{,}437.5 \text{ lb} \qquad \blacksquare$$

The method of the previous example can be used to find the fluid force on a flat side
of any object submerged vertically (not at an incline) in a fluid. Assume that the object
extends from depth $y = a$ to $y = b$ and, as before, divide the side into N horizontal strips
of thickness $\Delta y = (b - a)/N$ (Figure 4). Let

$$f(y) = \text{width of the side at depth } y$$

If Δy is small, then the jth strip is nearly rectangular with height Δy, width $f(y)$ for
some continuous function $f(y)$, and area $f(y)\Delta y$. Since the strip lies at depth $y_j =
a + j\Delta y$, the force F_j on the jth strip can be approximated:

$$F_j = \text{force on } j\text{th strip at depth } y_j \approx \underbrace{wy_j}_{\text{Pressure}} \times \underbrace{f(y_j)\Delta y}_{\text{Area}} = wy_j f(y_j)\,\Delta y$$

The total force F on the side is approximated by a Riemann sum that converges to an
integral as $N \to \infty$:

$$F = F_1 + \cdots + F_N = \sum_{j=1}^{N} wy_j f(y_j)\,\Delta y \quad \to \quad w \int_a^b yf(y)\,dy$$

THEOREM 1 Fluid Force on a Flat Surface Submerged Vertically   The force on a flat
side of an object submerged vertically in a fluid is

$$\text{Force on the side} = w \int_a^b \big(\text{depth} \times \text{width}\big)d(\text{depth}) = w \int_a^b yf(y)\,dy \qquad \boxed{2}$$

where $f(y)$ is the horizontal width of the side at depth y, w is the fluid density (in
weight per unit volume), and the object extends from depth $y = a$ to depth $y = b$.

■ EXAMPLE 3 A plate in the shape of an equilateral triangle of side 2 ft is submerged
vertically in a tank of oil of mass density $\rho = 900$ kg/m^3 as in Figure 5. Calculate the
total force F on one side of the plate.

Solution To use Eq. (2), we calculate the horizontal width $f(y)$ of the plate at depth y.
Recall that an equilateral triangle of side s has height $\dfrac{\sqrt{3}}{2}s$. Since $s = 2$, our triangle has
height $\sqrt{3}$. By similar triangles (Figure 5), $y/f(y) = \sqrt{3}/2$ and thus $f(y) = 2y/\sqrt{3}$.
The fluid density in weight per volume is $w = \rho g = (900)(9.8) = 8{,}820$ N/m^3. By
Eq. (2),

$$F = w \int_0^{\sqrt{3}} y\,f(y)\,dy = 8{,}820 \int_0^{\sqrt{3}} \frac{2}{\sqrt{3}}y^2\,dy = \left(\frac{17{,}640}{\sqrt{3}}\right)\frac{y^3}{3}\Big|_0^{\sqrt{3}} = 17{,}640 \text{ N} \quad \blacksquare$$

If the side of the submerged object is inclined at an angle, we must modify the force
calculation, as explained in the next example.

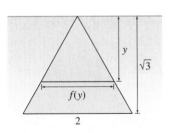

y

$\sqrt{3}$

$f(y)$

2

FIGURE 5 Triangular plate submerged in
a tank of oil.

Hoover Dam, located on the Nevada–
Arizona border.

FIGURE 6 Water exerts pressure on the
side of a dam.

■ EXAMPLE 4 Force on an Inclined Surface   Calculate the fluid force F on the side of a dam, which is inclined at an angle of $45°$. The height of the dam is 700 ft and the width of the base is 1,500 ft as in Figure 6. Assume the reservoir is filled so that water pressure acts along the full slanted side of the dam.

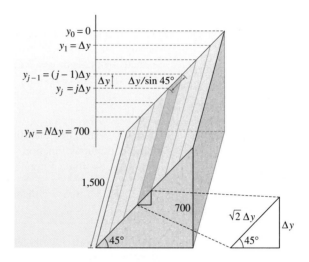

Solution The vertical height of the dam is 700 ft. We divide the vertical axis from 0 to 700 into N subintervals of length $\Delta y = 700/N$. This divides the face of the dam into N strips as in Figure 6. Note that the width of each strip is the length of the hypotenuse of a triangle that, by trigonometry, has length $\Delta y / \sin(45°) = \sqrt{2}\Delta y$. Therefore,

$$\text{Area of each strip} = \text{length} \times \text{width} = 1{,}500(\sqrt{2}\,\Delta y)$$

Now we may proceed as in the previous example. We approximate the force F_j on the jth strip and write the total force F as the sum of the forces on the strips:

$$F_j \approx \underbrace{wy_j}_{\text{Pressure}} \times \underbrace{1{,}500\sqrt{2}\Delta y}_{\text{Area of strip}} = wy_j \times 1{,}500\sqrt{2}\,\Delta y$$

$$F = \sum_{j=1}^{N} F_j \approx \sum_{j=1}^{N} wy_j\left(1{,}500\sqrt{2}\,\Delta y\right) = 1{,}500\sqrt{2}\,w \sum_{j=1}^{N} y_j\,\Delta y$$

The sum on the right is a Riemann sum for the integral $1{,}500\sqrt{2}w \int_0^{700} y\,dy$. Since water has density $w = 62.5$ lb/ft^3,

$$F = 1{,}500\sqrt{2}w \int_0^{700} y\,dy = 1{,}500\sqrt{2}(62.5)\frac{700^2}{2} \approx 3.25 \times 10^{10} \text{ lb} \quad ■$$

8.2 SUMMARY

• The fluid pressure at depth h is equal to wh, where w is the density of the fluid (weight per volume). Fluid pressure acts on a surface in the direction perpendicular to the surface. Water has density 62.5 lb/ft^3 and mass density 1,000 kg/m^3.

- If an object is submerged vertically in a fluid and extends from depth $y = a$ to $y = b$, then the total force on a side of the object is

$$F = w \int_a^b yf(y)\,dy$$

where $f(y)$ is the horizontal width of the side at depth y.

8.2 EXERCISES

Preliminary Questions

1. How is pressure defined?

2. Fluid pressure is proportional to depth. What is the factor of proportionality?

3. When fluid force acts on the side of a submerged object, in which direction does it act?

4. Why is fluid pressure on a surface calculated using thin horizontal strips rather than thin vertical strips?

5. If a thin plate is submerged horizontally, then the fluid force on one side of the plate is equal to pressure times area. Is this true if the plate is submerged vertically?

Exercises

1. A box of height 6 ft and square base of side 3 ft is submerged in a pool of water. The top of the box is 2 ft below the surface of the water.

(a) Calculate the fluid force on the top and bottom of the box.

(b) Write a Riemann sum that approximates the fluid force on a side of the box by dividing the side into N horizontal strips of thickness $\Delta y = 6/N$.

(c) To which integral does the Riemann sum converge?

(d) Compute the fluid force on a side of the box.

2. A plate in the shape of an isosceles triangle with base 1 ft and height 2 ft is submerged vertically in a tank of water so that its vertex touches the surface of the water (Figure 7).

(a) Show that the width of the triangle at depth y is $f(y) = \frac{1}{2}y$.

(b) Consider a thin strip of thickness Δy at depth y. Explain why the fluid force on a side of this strip is approximately equal to $w\frac{1}{2}y^2\Delta y$, where $w = 62.5$ lb/ft^3.

(c) Write an approximation for the total fluid force F on a side of the plate as a Riemann sum and indicate the integral to which it converges.

(d) Calculate F.

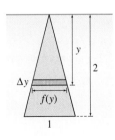

FIGURE 7

3. Repeat Exercise 2, but assume that the top of the triangle is located 3 ft below the surface of the water.

4. The thin plate R in Figure 8, bounded by the parabola $y = x^2$ and $y = 1$, is submerged vertically in water. Let F be the fluid force on one side of R.

(a) Show that the width of R at height y is $f(y) = 2\sqrt{y}$ and the fluid force on a side of a horizontal strip of thickness Δy at height y is approximately $2wy^{1/2}(1 - y)\Delta y$.

(b) Write a Riemann sum that approximates F and use it to explain why $F = 2w \int_0^1 y^{1/2}(1 - y)\,dy$.

(c) Calculate F.

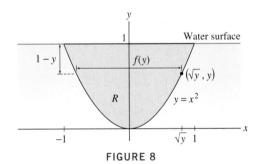

FIGURE 8

5. Let F be the fluid force (in Newtons) on a side of a semicircular plate of radius r meters, submerged in water so that its diameter is level with the water's surface (Figure 9).

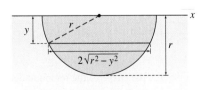

FIGURE 9

(a) Show that the width of the plate at depth y is $2\sqrt{r^2 - y^2}$.

(b) Calculate F using Eq. (2).

6. Calculate the force on one side of a circular plate with radius 2 ft, submerged vertically in a tank of water so that the top of the circle is tangent to the water surface.

7. A semicircular plate of radius r, oriented as in Figure 9, is submerged in water so that its diameter is located at a depth of m feet. Calculate the force on one side of the plate in terms of m and r.

8. Figure 10 shows the wall of a dam on a water reservoir. Use the Trapezoidal Rule and the width and depth measurements in the figure to estimate the total force on the wall.

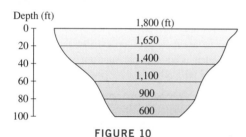

FIGURE 10

9. Calculate the total force (in Newtons) on a side of the plate in Figure 11(A), submerged in water.

10. Calculate the total force (in Newtons) on a side of the plate in Figure 11(B), submerged in a fluid of mass density $\rho = 800 \text{ kg/m}^3$.

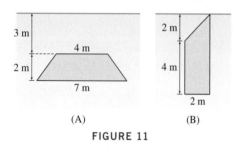

(A)   (B)

FIGURE 11

11. The plate in Figure 12 is submerged in water with its top level with the surface of the water. The left and right edges of the plate are the curves $y = x^{1/3}$ and $y = -x^{1/3}$. Find the fluid force on a side of the plate.

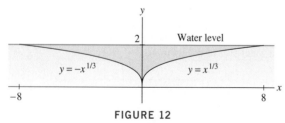

FIGURE 12

12. Let R be the plate in the shape of the region under $y = \sin x$ for $0 \le x \le \frac{\pi}{2}$ in Figure 13(A). Find the fluid force on a side of R if it is rotated counterclockwise by 90° and submerged in a fluid of density 140 lb/ft^3 with its top edge level with the surface of the fluid as in (B).

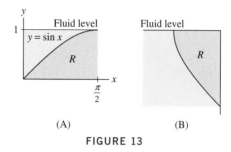

(A)   (B)

FIGURE 13

13. In the notation of Exercise 12, calculate the fluid force on a side of the plate R if it is oriented as in Figure 13(A). You may need to use Integration by Parts and trigonometric substitution.

14. Let A be the region under the graph of $y = \ln x$ for $1 \le x \le e$ (Figure 14). Calculate the fluid force on one side of a plate in the shape of region A if the water surface is at $y = 1$.

15. Calculate the fluid force on one side of the "infinite" plate B in Figure 14.

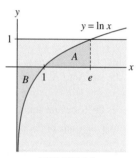

FIGURE 14

16. A square plate of side 3 m is submerged in water at an incline of 30° with the horizontal. Its top edge is located at the surface of the water. Calculate the fluid force (in Newtons) on one side of the plate.

17. Repeat Exercise 16, but assume that the top edge of the plate lies at a depth of 6 m.

18. Figure 15(A) shows a ramp inclined at 30° leading into a swimming pool. Calculate the fluid force on the ramp.

19. Calculate the fluid force on one side of the plate (an isosceles triangle) shown in Figure 15(B).

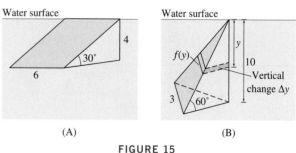

(A)   (B)

FIGURE 15

20. The trough in Figure 16 is filled with corn syrup, whose density is 90 lb/ft^3. Calculate the force on the front side of the trough.

21. Calculate the fluid pressure on one of the slanted sides of the trough in Figure 16, filled with corn syrup as in Exercise 20.

22. Figure 17 shows an object whose face is an equilateral triangle with 5-ft sides. The object is 2 ft thick and is submerged in water with its vertex 3 ft below the water surface. Calculate the fluid force on both a triangular face and a slanted rectangular edge of the object.

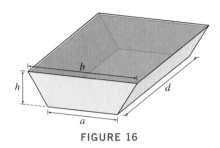

FIGURE 16

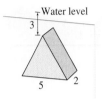

FIGURE 17

Further Insights and Challenges

23. The end of the trough in Figure 18 is an equilateral triangle of side 3. Assume that the trough is filled with water to height y. Calculate the fluid force on each side of the trough as a function of the level y and the length l of the trough.

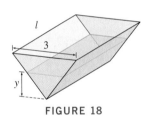

FIGURE 18

24. A rectangular plate of side ℓ is submerged vertically in a fluid of density w, with its top edge at depth h. Show that if the depth is increased by an amount Δh, then the force on a side of the plate increases by $wA\Delta h$, where A is the area of the plate.

25. Prove that the force on the side of a rectangular plate of area A submerged vertically in a fluid is equal to $p_0 A$, where p_0 is the fluid pressure at the center point of the rectangle.

26. If the density of a fluid varies with depth, then the pressure at depth y is a function $p(y)$ (which need not equal wy as in the case of constant density). Use Riemann sums to argue that the total force F on the flat side of a submerged object submerged vertically is
$$F = \int_a^b f(y) p(y)\, dy,$$ where $f(y)$ is the width of the side at depth y.

8.3 Center of Mass

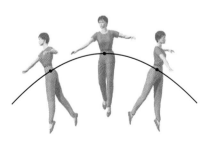

FIGURE 1 An "extended object" such as a dancer may rotate in a complicated fashion as it moves through the air under the influence of gravity, but its center of mass traces a simple parabolic trajectory (ignoring air friction).

Every object has a balance point called the *center of mass* (COM) (Figure 1). You can balance an object such as a tray or ball on the tip of one finger if you place your finger at a point directly below the COM. More generally, an object does not rotate when a force is applied along a line passing through the COM.

In this section, we review the formula for the COM of a system of particles (or "point masses") and then show how integration may be used to compute the centers of mass of thin plates of uniform density.

The center of mass is expressed in terms of quantities called **moments**. For a single particle of mass m, the moment with respect to a line L is defined as m times the directed distance (positive or negative) from the particle to the line:

Moment with respect to line $L = m \times$ directed distance to L

The moments with respect to the x- and y-axes, denoted M_x and M_y, are called the x- and y-moments. If the mass m is located at the point (x, y) in the plane (Figure 2),

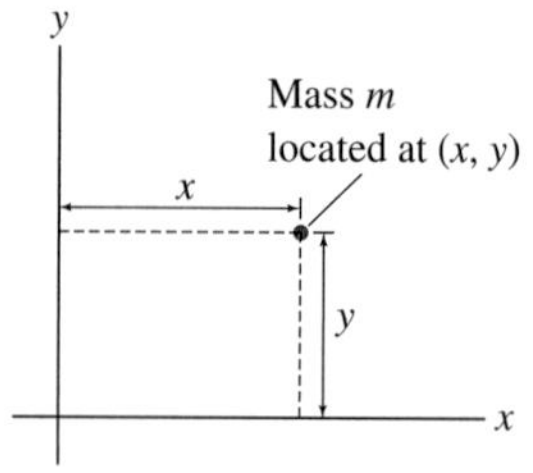

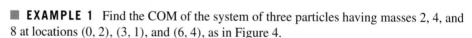

FIGURE 2 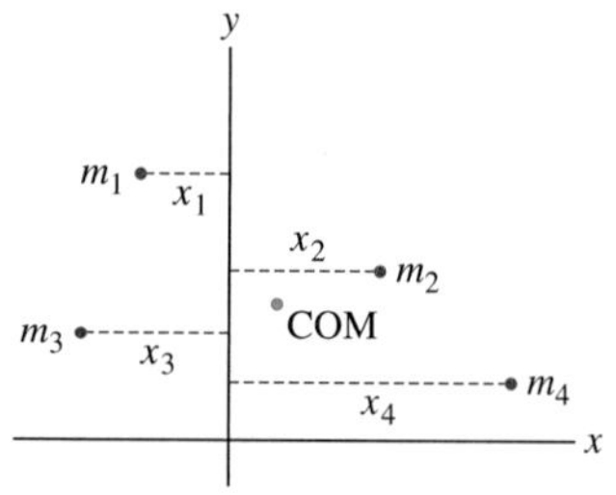   FIGURE 3

$$M_x = \text{moment with respect to } x\text{-axis} = \underbrace{m}_{\text{Mass}} \times \underbrace{y}_{\text{Directed distance to } x\text{-axis}}$$

$$M_y = \text{moment with respect to } y\text{-axis} = \underbrace{m}_{\text{Mass}} \times \underbrace{x}_{\text{Directed distance to } y\text{-axis}}$$

CAUTION *The notation is potentially confusing: M_x is defined in terms of y-coordinates and M_y in terms of x-coordinates. However, $x_{\text{CM}} = M_y/M$ and $y_{\text{CM}} = M_x/M$.*

Given n particles with coordinates (x_j, y_j) and mass m_j, we define the moment of the system as the sum of the moments of the individual particles (Figure 3):

$$M_x = m_1 y_1 + m_2 y_2 + \cdots + m_n y_n$$
$$M_y = m_1 x_1 + m_2 x_2 + \cdots + m_n x_n$$

The **center of mass** of the system is the point $P_{\text{CM}} = (x_{\text{CM}}, y_{\text{CM}})$ with coordinates

$$x_{\text{CM}} = \frac{M_y}{M}, \qquad y_{\text{CM}} = \frac{M_x}{M}$$

where $M = m_1 + m_2 + \cdots + m_n$ is the total mass of the system.

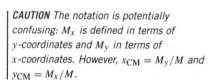

■ **EXAMPLE 1** Find the COM of the system of three particles having masses 2, 4, and 8 at locations $(0, 2)$, $(3, 1)$, and $(6, 4)$, as in Figure 4.

Solution The total mass is $M = 2 + 4 + 8 = 14$ and the moments are

$$M_x = m_1 y_1 + m_2 y_2 + m_3 y_3 = 2 \cdot 2 + 4 \cdot 1 + 8 \cdot 4 = 40$$
$$M_y = m_1 x_1 + m_2 x_2 + m_3 x_3 = 2 \cdot 0 + 4 \cdot 3 + 8 \cdot 6 = 60$$

Therefore, $x_{\text{CM}} = \frac{60}{14} = \frac{30}{7}$ and $y_{\text{CM}} = \frac{40}{14} = \frac{20}{7}$. The COM is $\left(\frac{30}{7}, \frac{20}{7}\right)$. ■

FIGURE 4 Centers of mass for Example 1.

Laminas (Thin Plates)

A lamina is a thin plate whose mass is distributed continuously throughout a region in the plane. We shall compute the moments and the center of mass of a lamina assuming that the mass density ρ is constant. Suppose first that the lamina occupies a region in the plane under the graph of a continuous function $f(x)$ over an interval $[a, b]$, where $f(x) \geq 0$ (Figure 5).

We follow our usual procedure to compute M_y. That is, we divide $[a, b]$ into N subintervals of width $\Delta x = \dfrac{b - a}{N}$ with endpoints $x_j = a + j\Delta x$. This divides the lamina into N vertical strips (Figure 6). If Δx is small, the jth strip is nearly rectangular with base Δx and height $f(x_j)$. Hence, the strip has area $f(x_j)\,\Delta x$ and mass $\rho f(x_j)\,\Delta x$. Since the strip is thin, all points in the strip lie at approximately the same distance x_j from the y-axis (Figure 6). Therefore, we can approximate the moment of the jth strip, which

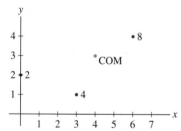

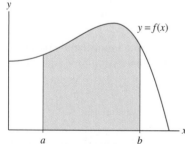

FIGURE 5 Lamina occupying the region under the graph of $f(x)$ over $[a, b]$.

In this section, we restrict our attention to thin plates of constant mass density (also called "uniform density"). COM computations when mass density is not constant require multiple integration, covered in Chapter 15.

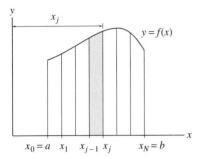

FIGURE 6 The shaded strip is nearly rectangular of area $f(x_j)\,\Delta x$.

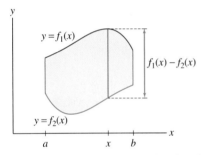

FIGURE 7 The vertical cut at x has length $f_1(x) - f_2(x)$.

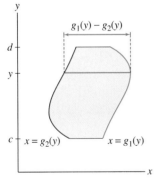

FIGURE 8 The horizontal cut at y has length $g_1(y) - g_2(y)$.

we call $M_{y,j}$:

$$M_{y,j} \approx (\text{mass}) \times (\text{directed distance to } y\text{-axis}) = (\rho f(x_j)\,\Delta x) x_j$$

The moment M_y is the sum of the $M_{y,j}$ and thus may be approximated:

$$M_y = \sum_{j=1}^{N} M_{y,j} \approx \rho \sum_{j=1}^{N} x_j f(x_j)\,\Delta x$$

The sum on the right is a Riemann sum whose value approaches $\int_a^b \rho x f(x)\,dx$ as $N \to \infty$, and the moment itself is given by the integral:

$$M_y = \rho \int_a^b x f(x)\,dx$$

We argue similarly if the lamina occupies the region *between* the graphs of two functions $f_1(x)$ and $f_2(x)$ over $[a, b]$, where $f_1(x) \geq f_2(x)$. The length of the ith vertical strip is then $f_1(x_i) - f_2(x_i)$, so we replace $f(x)$ by $f_1(x) - f_2(x)$ in the formula for M_y. We may think of $f_1(x) - f_2(x)$ as the length of a *vertical* cut or strip of "width zero" (Figure 7):

$$M_y = \rho \int_a^b x\,(\text{length of vertical cut})\,dx = \rho \int_a^b x\big(f_1(x) - f_2(x)\big)\,dx \qquad \boxed{1}$$

We may compute the x-moment in a similar way by dividing the lamina into thick *horizontal* strips. To do so, we must describe the lamina as a region between two curves $x = g_1(y)$ and $x = g_2(y)$ with $g_1(y) \geq g_2(y)$ over an interval $[c, d]$ along the y-axis (Figure 8). We obtain the formula

$$M_x = \rho \int_c^d y\,(\text{length of horizontal cut})\,dy = \rho \int_c^d y\big(g_1(y) - g_2(y)\big)\,dy \qquad \boxed{2}$$

The coordinates of the center of mass are the moments divided by the total mass M of the lamina:

$$x_{\text{CM}} = \frac{M_y}{M}, \qquad y_{\text{CM}} = \frac{M_x}{M}$$

The total mass of the lamina is $M = \rho A$ where A is the area of the lamina:

$$M = \rho A = \rho \int_a^b \big(f_1(x) - f_2(x)\big)\,dx \quad \text{or} \quad \rho \int_c^d \big(g_1(y) - g_2(y)\big)\,dy$$

The COM of a lamina of constant mass density ρ occupying a given region is also called the **centroid** of the lamina or region. Notice that the moments M_x and M_y and the total mass M all contain the factor ρ. Since this factor cancels in the ratios M_y/M and M_x/M, the centroid depends on only the shape of the lamina, not on its mass density. Therefore, *in calculating the centroid, we are free to take $\rho = 1$*. In the more general situation where the mass density is not constant (not covered here), the COM depends on both shape and mass density.

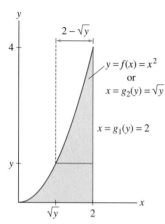

FIGURE 9 Lamina occupying the region under the graph of $f(x) = x^2$ over $[0, 2]$.

EXAMPLE 2 Find the moments and COM of the lamina of uniform density ρ occupying the region underneath the curve $y = x^2$ for $0 \le x \le 2$.

Solution First, we compute M_y using Eq. (1):

$$M_y = \rho \int_0^2 x f(x)\, dx = \rho \int_0^2 x(x^2)\, dx = \rho\, \frac{x^4}{4}\Big|_0^2 = 4\rho$$

To compute M_x using Eq. (2), we must describe our lamina as the region between the graph of $x = g_2(y) = \sqrt{y}$ and the vertical line $x = g_1(y) = 2$ over the interval $[0, 4]$ along the y-axis (Figure 9). Then we obtain

$$M_x = \rho \int_0^4 y\big(g_1(y) - g_2(y)\big)\, dy = \rho \int_0^4 y(2 - \sqrt{y})\, dy$$

$$= \rho\left(y^2 - \frac{2}{5}y^{5/2}\right)\Big|_0^4 = \rho\left(16 - \frac{2}{5}\cdot 32\right) = \frac{16}{5}\rho$$

The area of the plate is $A = \int_0^2 x^2\, dx = \frac{8}{3}$ and its total mass is $M = \frac{8}{3}\rho$. Therefore, the coordinates of the COM are

$$x_{\mathrm{CM}} = \frac{M_y}{M} = \frac{4\rho}{(8/3)\rho} = \frac{3}{2}, \qquad y_{\mathrm{CM}} = \frac{M_x}{M} = \frac{(16/5)\rho}{(8/3)\rho} = \frac{6}{5} \qquad \blacksquare$$

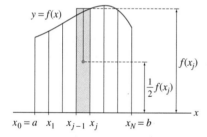

FIGURE 10 The shaded strip is nearly rectangular and its COM has an approximate height of $\frac{1}{2}f(x_j)$.

A drawback of Eq. (2) for M_x is that it requires integration along the y-axis. Fortunately, there is a second formula for M_x as an integral along the x-axis. As before, divide the region into N thin vertical strips of width Δx (see Figure 10). Let $M_{x,j}$ be the x-moment of the jth strip and, in addition, let m_j be its mass and y_j the y-coordinate of its COM. Then $y_j = M_{x,j}/m_j$ and we can use the following trick to approximate $M_{x,j}$. The strip is nearly rectangular with height $f(x_j)$ and width Δx, so $m_j \approx \rho f(x_j)\,\Delta x$. Furthermore, $y_j \approx \frac{1}{2}f(x_j)$ because the COM of a rectangle is located at its center. Thus,

$$M_{x,j} = m_j y_j \approx \rho f(x_j)\Delta x \cdot \frac{1}{2}f(x_j) = \frac{1}{2}\rho f(x_j)^2 \Delta x$$

$$M_x = \sum_{j=1}^{N} M_{x,j} \approx \frac{1}{2}\rho \sum_{j=1}^{N} f(x_j)^2 \Delta x$$

The sum on the right is a Riemann sum whose value approaches $\dfrac{1}{2}\rho \displaystyle\int_a^b f(x)^2\, dx$ as $N \to \infty$. The region *between* the graphs of two functions $f_1(x)$ and $f_2(x)$ where $f_1(x) \ge f_2(x) \ge 0$ is treated similarly, so we obtain the following alternate formulas for M_x:

$$\boxed{\; M_x = \frac{1}{2}\rho \int_a^b f(x)^2\, dx \qquad \text{or} \qquad \frac{1}{2}\rho \int_a^b \big(f_1(x)^2 - f_2(x)^2\big)\, dx \;} \qquad \boxed{3}$$

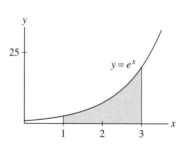

FIGURE 11 Region under the curve $y = e^x$ between $x = 1$ and $x = 3$.

EXAMPLE 3 Find the centroid of the region under the curve $y = e^x$ over $[1, 3]$ (Figure 11).

Solution The centroid does not depend on ρ, so we may set $\rho = 1$. By Eqs. (1) and (3),

$$M_x = \frac{1}{2} \int_1^3 f(x)^2 \, dx = \frac{1}{2} \int_1^3 e^{2x} \, dx = \frac{1}{4} e^{2x} \Big|_1^3 = \frac{e^6 - e^2}{4}$$

$$M_y = \int_1^3 x f(x) \, dx = \int_1^3 x e^x \, dx = (x - 1) e^x \Big|_1^3 = 2e^3$$

The total mass is $M = \int_1^3 e^x \, dx = (e^3 - e)$. Therefore, the centroid has coordinates

$$x_{\text{CM}} = \frac{M_y}{M} = \frac{2e^3}{e^3 - e} \approx 2.313, \qquad y_{\text{CM}} = \frac{M_x}{M} = \frac{e^6 - e^2}{4(e^3 - e)} \approx 5.701 \qquad \blacksquare$$

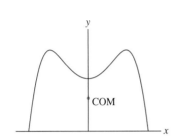

FIGURE 12 The COM of a symmetrical plate lies on the axis of symmetry.

The symmetry properties of an object give information about its centroid (Figure 12). For instance, the centroid of a square or circular plate is located at the center of the plate. Here is a precise formulation of the Symmetry Principle (see Exercise 42).

THEOREM 1 Symmetry Principle If a lamina of constant density is symmetrical with respect to a line, then its centroid lies on that line.

■ **EXAMPLE 4** Using Symmetry Find the centroid of a semicircle of radius 3.

Solution We use symmetry to cut our work in half. Set $\rho = 1$. Since the semicircle is symmetrical with respect to y-axis, the centroid lies on the y-axis and hence $x_{\text{CM}} = 0$. It remains to calculate M_x and y_{CM}. The semicircle is the graph of $f(x) = \sqrt{9 - x^2}$ (Figure 13). By Eq. (3),

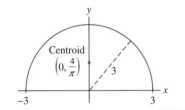

FIGURE 13 The semicircle $y = \sqrt{9 - x^2}$.

$$M_x = \frac{1}{2} \int_{-3}^3 f(x)^2 \, dx = \frac{1}{2} \int_{-3}^3 (9 - x^2) \, dx = \frac{1}{2} \left(9x - \frac{1}{3} x^3 \right) \Big|_{-3}^3 = 9 - (-9) = 18$$

Furthermore, the semicircle has area (and mass) equal to $A = \frac{1}{2} \pi (3^2) = 9\pi/2$, so

$$y_{\text{CM}} = \frac{M_x}{M} = \frac{18}{9\pi/2} = \frac{4}{\pi} \approx 1.27 \qquad \blacksquare$$

Moments satisfy a key additivity property: If a region S consists of two or more disjoint smaller regions, then the moment of S (with respect to any axis or line) is the *sum* of the moments of the smaller regions. This follows from the additivity of the integral. The next example shows how the additivity property can be used to simplify COM calculations.

■ **EXAMPLE 5** Using Additivity of Moments Find the centroid of the region R in Figure 14.

Solution We set $\rho = 1$ because we are computing a centroid. The region R is symmetrical with respect to the y-axis, so we know that the centroid lies on the y-axis. Thus $x_{\text{CM}} = 0$. It remains to compute y_{CM}.

Step 1. **Use additivity of moments.**
To find y_{CM}, we use additivity of the x-moment:

$$M_x^R = M_x^{\text{triangle}} + M_x^{\text{circle}}$$

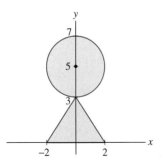

FIGURE 14 The moment of region R is the sum of the moments of the triangle and circle.

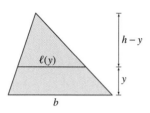

FIGURE 15 By similar triangles, $\dfrac{\ell(y)}{h-y} = \dfrac{b}{h}$.

where M_x^R, M_x^{triangle}, and M_x^{circle} are the x-moments of R, the triangle, and circle, respectively.

Step 2. Moment of a triangle.

Let's compute M_x^{triangle} for an arbitrary triangle of height h and base b. If $\ell(y)$ is the width of the triangle at height y (Figure 15), then by similar triangles,

$$\frac{\ell(y)}{h-y} = \frac{b}{h} \qquad \Rightarrow \qquad \ell(y) = b - \frac{b}{h}y$$

Therefore,

$$M_x^{\text{triangle}} = \int_0^h y\,\ell(y)\,dy = \int_0^h y\left(b - \frac{b}{h}y\right)dy = \left(\frac{by^2}{2} - \frac{by^3}{3h}\right)\Bigg|_0^h = \frac{bh^2}{6}$$

In our case, $b = 4$, $h = 3$, and $M_x^{\text{triangle}} = \dfrac{4 \cdot 3^2}{6} = 6$.

Step 3. Moment of the circle.

To compute M_x^{circle}, we exploit the fact that, by symmetry, the centroid of the circle is located at the center $(0, 5)$. Thus $y_{\text{CM}}^{\text{circle}} = 5$. The circle has radius 2, hence it has area 4π and also mass $M^{\text{circle}} = 4\pi$. Now we can solve for M_x^{circle}:

$$y_{\text{CM}}^{\text{circle}} = \frac{M_x^{\text{circle}}}{M^{\text{circle}}} = \frac{M_x^{\text{circle}}}{4\pi} = 5 \qquad \Rightarrow \qquad M_x^{\text{circle}} = 20\pi$$

Step 4. Computation of y_{CM}.

We may now conclude that

$$M_x^R = M_x^{\text{triangle}} + M_x^{\text{circle}} = 6 + 20\pi$$

The triangle has mass $\frac{1}{2} \cdot 4 \cdot 3 = 6$ and the circle has mass 4π, so R has mass $M = 6 + 4\pi$, and

$$y_{\text{CM}} = \frac{M_x^R}{M} = \frac{6 + 20\pi}{6 + 4\pi} \approx 3.71 \qquad \blacksquare$$

HISTORICAL PERSPECTIVE

FIGURE 16 Archimedes's Law of the Lever: $L_1m_1 = L_2m_2$.

We take it for granted that the basic laws of physics are best expressed as mathematical relationships. Think of $F = ma$ or the universal law of gravitation. However, the fundamental insight that mathematics could be used to formulate laws of nature (and not just for counting or measuring) developed gradually, beginning with the philosophers of ancient Greece and culminating some 2,000 years later in the discoveries of Galileo and Newton. Archimedes

(287–212 BCE) was one of the first scientists (perhaps the first) to formulate a precise physical law. He considered the following question: If weights of mass m_1 and m_2 are placed at the ends of a weightless lever, where should the fulcrum P be located so that the lever does not tip to one side? Suppose that the distance from P to m_1 and m_2 is L_1 and L_2, respectively (Figure 16). Archimedes's Law states that the lever will balance if

$$\boxed{L_1m_1 = L_2m_2}$$

In our terminology, what Archimedes had discovered was the center of mass P of the system of weights (see Exercises 40 and 41).

8.3 SUMMARY

- The *moments* of a system of particles of mass m_j located at (x_j, y_j) are

$$M_x = m_1 y_1 + \cdots + m_n y_n, \qquad M_y = m_1 x_1 + \cdots + m_n x_n$$

The *center of mass* (COM) has coordinates

$$x_{CM} = \frac{M_y}{M} \quad \text{and} \quad y_{CM} = \frac{M_x}{M},$$

where $M = m_1 + \cdots + m_n$.

- The *y-moment* M_y of a lamina (thin plate) of constant mass density ρ occupying the region under the graph of $y = f(x)$ where $f(x) \geq 0$, or *between* the graphs of $f_1(x)$ and $f_2(x)$ where $f_1(x) \geq f_2(x)$ is equal to

$$M_y = \rho \int_a^b x f(x)\, dx \quad \text{or} \quad \rho \int_a^b x \big(f_1(x) - f_2(x)\big)\, dx$$

- There are two ways to compute the *x-moment* M_x. If the lamina occupies the region between the graph of $x = g(y)$ and the y-axis where $g(y) \geq 0$, or *between* the graphs of $g_1(y)$ and $g_2(y)$ where $g_1(y) \geq g_2(y)$, then

$$M_x = \rho \int_c^d y g(y)\, dy \quad \text{or} \quad \rho \int_c^d y \big(g_1(y) - g_2(y)\big)\, dy$$

- Alternate (often more convenient) formula for M_x:

$$M_x = \frac{1}{2}\rho \int_a^b f(x)^2\, dx \quad \text{or} \quad \frac{1}{2}\rho \int_a^b \big(f_1(x)^2 - f_2(x)^2\big)\, dx$$

- The total mass of the lamina is $M = \rho \displaystyle\int_a^b \big(f_1(x) - f_2(x)\big)\, dx$. The coordinates of the center of mass (also called the *centroid*) are

$$x_{CM} = \frac{M_y}{M}, \qquad y_{CM} = \frac{M_x}{M}$$

- Symmetry Principle: If a lamina of constant mass density is symmetrical with respect to a given line, then the center of mass (centroid) lies on that line.
- Additivity property: If a region S consists of two or more disjoint smaller regions, then each moment of S is the sum of the corresponding moments of the smaller regions.

8.3 EXERCISES

Preliminary Questions

1. What are the *x*- and *y*-moments of a lamina whose center of mass is located at the origin?

2. A thin plate has mass 3. What is the *x*-moment of the plate if its center of mass has coordinates $(2, 7)$?

3. The center of mass of a lamina of total mass 5 has coordinates $(2, 1)$. What are the lamina's *x*- and *y*-moments?

4. Explain how the Symmetry Principle is used to conclude that the centroid of a rectangle is the center of the rectangle.

Exercises

1. Four particles are located at points

$$(1, 1) \quad (1, 2) \quad (4, 0) \quad (3, 1)$$

(a) Find the moments M_x and M_y and the center of mass of the system, assuming that the particles have equal mass m.

(b) Find the center of mass of the system, assuming the particles have mass 3, 2, 5, and 7, respectively.

2. Find the center of mass for the system of particles of mass 4, 2, 5, 1 located at $(1, 2)$, $(-3, 2)$, $(2, -1)$, $(4, 0)$.

3. Point masses of equal size are placed at the vertices of the triangle with coordinates $(a, 0)$, $(b, 0)$, and $(0, c)$. Show that the center of mass of the system of masses has coordinates $(\frac{1}{3}(a + b), \frac{1}{3}c)$.

4. Point masses of mass m_1, m_2, and m_3 are placed at the points $(-1, 0)$, $(3, 0)$, and $(0, 4)$.
(a) Suppose that $m_1 = 6$. Show that there is a unique value of m_2 such that the center of mass lies on the y-axis.
(b) Suppose that $m_1 = 6$ and $m_2 = 4$. Find the value of m_3 such that $y_{CM} = 2$.

5. Sketch the lamina S of constant density $\rho = 3$ g/cm^2 occupying the region beneath the graph of $y = x^2$ for $0 \le x \le 3$.
(a) Use formulas (1) and (2) to compute M_x and M_y.
(b) Find the area and the center of mass of S.

6. Use Eqs. (1) and (3) to find the moments and center of mass of the lamina S of constant density $\rho = 2$ g/cm^2 occupying the region between $y = x^2$ and $y = 9x$ over $[0, 3]$. Sketch S, indicating the location of the center of mass.

7. Find the moments and center of mass of the lamina of uniform density ρ occupying the region underneath $y = x^3$ for $0 \le x \le 2$.

8. Calculate M_x (assuming $\rho = 1$) for the region underneath the graph of $y = 1 - x^2$ for $0 \le x \le 1$ in two ways, first using Eq. (2) and then using Eq. (3).

9. Let T be the triangular lamina in Figure 17.
(a) Show that the horizontal cut at height y has length $4 - \frac{2}{3}y$ and use Eq. (2) to compute M_x (with $\rho = 1$).
(b) Use the Symmetry Principle to show that $M_y = 0$ and find the center of mass.

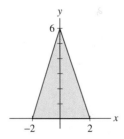

FIGURE 17 Isosceles triangle.

In Exercises 10–17, find the centroid of the region lying underneath the graph of the function over the given interval.

10. $f(x) = 6 - 2x$, $[0, 3]$

11. $f(x) = \sqrt{x}$, $[4, 9]$

12. $f(x) = x^3$, $[0, 1]$

13. $f(x) = 9 - x^2$, $[0, 3]$

14. $f(x) = (1 + x^2)^{-1/2}$, $[0, 1]$

15. $f(x) = e^{-x}$, $[0, 4]$

16. $f(x) = \ln x$, $[1, 2]$

17. $f(x) = \sin x$, $[0, \pi]$

18. Calculate the moments and center of mass of the lamina occupying the region between the curves $y = x$ and $y = x^2$ for $0 \le x \le 1$.

19. Sketch the region between $y = x + 4$ and $y = 2 - x$ for $0 \le x \le 2$. Using symmetry, explain why the centroid of the region lies on the line $y = 3$. Verify this by computing the moments and the centroid.

In Exercises 20–25, find the centroid of the region lying between the graphs of the functions over the given interval.

20. $y = x$, $y = \sqrt{x}$, $[0, 1]$

21. $y = x^2$, $y = \sqrt{x}$, $[0, 1]$

22. $y = x^{-1}$, $y = 2 - x$, $[1, 2]$

23. $y = e^x$, $y = 1$, $[0, 1]$

24. $y = \ln x$, $y = x - 1$, $[1, 3]$

25. $y = \sin x$, $y = \cos x$, $[0, \pi/4]$

26. Sketch the region enclosed by $y = 0$, $y = x + 1$, and $y = (x - 1)^2$ and find its centroid.

27. Sketch the region enclosed by $y = 0$, $y = (x + 1)^3$, and $y = (1 - x)^3$ and find its centroid.

In Exercises 28–32, find the centroid of the region.

28. Top half of the ellipse $\left(\frac{x}{2}\right)^2 + \left(\frac{y}{4}\right)^2 = 1$

29. Top half of the ellipse $\left(\frac{x}{a}\right)^2 + \left(\frac{y}{b}\right)^2 = 1$ for arbitrary $a, b > 0$

30. Semicircle of radius r with center at the origin

31. Quarter of the unit circle lying in the first quadrant

32. Triangular plate with vertices $(-c, 0)$, $(0, c)$, (a, b), where $a, b, c > 0$, and $b < c$

33. Find the centroid for the shaded region of the semicircle of radius r in Figure 18. What is the centroid when $r = 1$ and $h = \frac{1}{2}$? *Hint:* Use geometry rather than integration to show that the *area* of the region is $r^2 \sin^{-1}(\sqrt{1 - h^2/r^2}) - h\sqrt{r^2 - h^2}$.

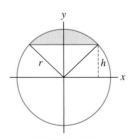

FIGURE 18

In Exercises 34–36, use the additivity of moments to find the COM of the region.

34. Isosceles triangle of height 2 on top of a rectangle of base 4 and height 3 (Figure 19)

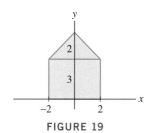

FIGURE 19

35. An ice cream cone consisting of a semicircle on top of an equilateral triangle of side 6 (Figure 20)

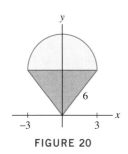

FIGURE 20

36. Three-quarters of the unit circle (remove the part in the fourth quadrant)

37. Let S be the lamina of mass density $\rho = 1$ obtained by removing a circle of radius r from the circle of radius $2r$ shown in Figure 21. Let

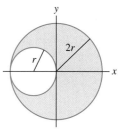

FIGURE 21

M_x^S and M_y^S denote the moments of S. Similarly, let M_y^{big} and M_y^{small} be the y-moments of the larger and smaller circles.

(a) Use the Symmetry Principle to show that $M_x^S = 0$.

(b) Show that $M_y^S = M_y^{\text{big}} - M_y^{\text{small}}$ using the additivity of moments.

(c) Find M_y^{big} and M_y^{small} using the fact that the COM of a circle is its center. Then compute M_y^S using (b).

(d) Determine the COM of S.

38. Find the COM of the laminas in Figure 22 obtained by removing squares of side 2 from a square of side 8.

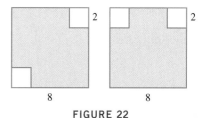

FIGURE 22

Further Insights and Challenges

39. A **median** of a triangle is a segment joining a vertex to the midpoint of the opposite side. Show that the centroid of a triangle lies on each of its medians, at a distance two-thirds down from the vertex. Then use this fact to prove that the three medians intersect at a single point. *Hint:* Simplify the calculation by assuming that one vertex lies at the origin and another on the x-axis.

40. Let P be the COM of a system of two weights with masses m_1 and m_2 separated by a distance d. Prove Archimedes's Law of the (weightless) Lever: P is the point on line between the two weights such that $\ell_1 m_1 = \ell_2 m_2$, where ℓ_j is the distance from mass j to P.

41. Find the COM of a system of two weights of masses m_1 and m_2 connected by a lever of length d whose mass density ρ is uniform. *Hint:* The moment of the system is the sum of the moments of the weights and the lever.

42. **Symmetry Principle** Let $\mathcal{R}$ be the region under the graph of $f(x)$ over the interval $[-a, a]$, where $f(x) \geq 0$. Assume that $\mathcal{R}$ is symmetrical with respect to the y-axis.

(a) Explain why $f(x)$ is even, that is, $f(x) = f(-x)$.

(b) Show that $xf(x)$ is an *odd* function.

(c) Use (b) to prove that $M_y = 0$.

(d) Prove that the COM of $\mathcal{R}$ lies on the y-axis (a similar argument applies to symmetry with respect to the x-axis).

43. Prove directly that Eqs. (2) and (3) are equivalent in the following situation. Let $f(x)$ be a positive decreasing function on $[0, b]$ such that $f(b) = 0$. Set $d = f(0)$ and $g(y) = f^{-1}(y)$. Show that

$$\frac{1}{2} \int_0^b f(x)^2 \, dx = \int_0^d y g(y) \, dy$$

Hint: First apply the substitution $y = f(x)$ to the integral on the left and observe that $dx = g'(y) \, dy$. Then apply Integration by Parts.

44. Let R be a lamina of uniform density submerged in a fluid of density w (Figure 23). Prove the following law: The fluid force on one side of R is equal to the area of R times the fluid pressure on the centroid. *Hint:* Let $g(y)$ be the horizontal width of R at depth y. Express both the fluid pressure [Eq. (2) in Section 8.2] and y-coordinate of the centroid in terms of $g(y)$.

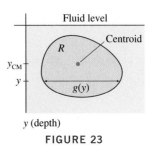

FIGURE 23

8.4 Taylor Polynomials

English mathematician Brook Taylor (1685–1731) made important contributions to calculus and physics, as well as to the theory of linear perspective used in drawing.

We have seen that approximation is a basic theme in calculus. For example, we may approximate a differentiable function $f(x)$ at $x = a$ by the linear function (called the linearization)

$$L(x) = f(a) + f'(a)(x - a)$$

However, a drawback of the linearization is that it is accurate only in a small interval around $x = a$. Taylor polynomials are higher-degree approximations that generalize the linearization using the higher derivatives $f^{(k)}(a)$. They are useful because, by taking sufficiently high degree, we can approximate transcendental functions such as $\sin x$ and e^x to arbitrary accuracy on any given interval.

Assume that $f(x)$ is defined on an open interval I and that all higher derivatives $f^{(k)}(x)$ exist on I. Fix a number $a \in I$. The nth **Taylor polynomial** for f centered at $x = a$ is the polynomial

$$\boxed{T_n(x) = f(a) + \frac{f'(a)}{1!}(x - a) + \frac{f''(a)}{2!}(x - a)^2 + \cdots + \frac{f^{(n)}(a)}{n!}(x - a)^n}$$

> **←·· REMINDER** k-*factorial is the number* $k! = k(k-1)(k-2)\cdots(2)(1)$. *Thus,*
>
> $1! = 1, \quad 2! = (2)1 = 2, \quad 3! = (3)(2)1 = 6$
>
> *By convention, we define* $0! = 1$.

It is convenient to regard $f(x)$ itself as the *zeroth* derivative $f^{(0)}(x)$. Then we may write the Taylor polynomial in summation notation,

$$T_n(x) = \sum_{j=0}^{n} \frac{f^{(j)}(a)}{j!} (x - a)^j$$

When $a = 0$, $T_n(x)$ is also called the nth **Maclaurin polynomial**. The first few Taylor polynomials are

$$T_0(x) = f(a)$$
$$T_1(x) = f(a) + f'(a)(x - a)$$
$$T_2(x) = f(a) + f'(a)(x - a) + \frac{1}{2}f''(a)(x - a)^2$$
$$T_3(x) = f(a) + f'(a)(x - a) + \frac{1}{2}f''(a)(x - a)^2 + \frac{1}{6}f'''(a)(x - a)^3$$

Scottish mathematician Colin Maclaurin (1698–1746) was a professor in Edinburgh. Newton was so impressed by his work that he once offered to pay part of Maclaurin's salary.

Note that $T_1(x)$ is the linearization of $f(x)$ at a. In most cases, the higher-degree Taylor polynomials provide increasingly better approximations to $f(x)$. Before computing some Taylor polynomials, we record two important properties that follow from the definition:

- $T_n(a) = f(a)$ [since all terms in $T_n(x)$ after the first are zero at $x = a$].
- $T_n(x)$ is obtained from $T_{n-1}(x)$ by adding on a term of degree n:

$$T_n(x) = T_{n-1}(x) + \frac{f^{(n)}(a)}{n!}(x - a)^n$$

■ **EXAMPLE 1** Computing Taylor Polynomials Let $f(x) = \sqrt{x + 1}$. Compute $T_n(x)$ at $a = 3$ for $n = 0, 1, 2, 3,$ and 4.

Solution First evaluate the derivatives $f^{(j)}(3)$:

$$f(x) = (x + 1)^{1/2}, \qquad f(3) = 2$$

$$f'(x) = \frac{1}{2}(x + 1)^{-1/2} \qquad f'(3) = \frac{1}{4}$$

$$f''(x) = -\frac{1}{4}(x + 1)^{-3/2} \qquad f''(3) = -\frac{1}{32}$$

$$f'''(x) = \frac{3}{8}(x + 1)^{-5/2} \qquad f'''(3) = \frac{3}{256}$$

$$f^{(4)}(x) = -\frac{15}{16}(x + 1)^{-7/2} \quad f^{(4)}(3) = -\frac{15}{2{,}048}$$

Then compute the coefficients $\dfrac{f^{(j)}(3)}{j!}$:

The first term $f(a)$ in the Taylor polynomial $T_n(x)$ is called the constant term.

$$\text{Constant term} \quad = f(3) = 2$$

$$\text{Coefficient of } (x - 3) \ = f'(3) = \frac{1}{4}$$

$$\text{Coefficient of } (x - 3)^2 = \frac{f''(3)}{2!} = -\frac{1}{32} \cdot \frac{1}{2!} = -\frac{1}{64}$$

$$\text{Coefficient of } (x - 3)^3 = \frac{f'''(3)}{3!} = \frac{3}{256} \cdot \frac{1}{3!} = \frac{1}{512}$$

$$\text{Coefficient of } (x - 3)^4 = \frac{f^{(4)}(3)}{4!} = -\frac{15}{2{,}048} \cdot \frac{1}{4!} = -\frac{5}{16{,}384}$$

The first four Taylor polynomials centered at $a = 3$ are (see Figure 1):

$$T_0(x) = 2$$

$$T_1(x) = 2 + \frac{1}{4}(x - 3)$$

$$T_2(x) = 2 + \frac{1}{4}(x - 3) - \frac{1}{64}(x - 3)^2$$

$$T_3(x) = 2 + \frac{1}{4}(x - 3) - \frac{1}{64}(x - 3)^2 + \frac{1}{512}(x - 3)^3$$

$$T_4(x) = 2 + \frac{1}{4}(x - 3) - \frac{1}{64}(x - 3)^2 + \frac{1}{512}(x - 3)^3 - \frac{5}{16{,}384}(x - 3)^4 \qquad ■$$

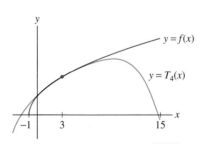

FIGURE 1 Graph of $f(x) = \sqrt{x + 1}$ and $T_4(x)$ centered at $x = 3$.

■ **EXAMPLE 2** Finding a General Formula for T_n Find the Taylor polynomials $T_n(x)$ of $f(x) = \ln x$ at $a = 1$.

After computing several derivatives of $f(x) = \ln x$, we begin to discern the pattern. Often, however, the derivatives follow no simple pattern and there is no simple formula for the general Taylor polynomial.

Solution For $f(x) = \ln x$, the constant term of $T_n(x)$ at $a = 1$ is zero since $f(1) = \ln 1 = 0$. Next, we compute the derivatives:

$$f'(x) = x^{-1}, \qquad f''(x) = -x^{-2}, \qquad f'''(x) = 2x^{-3}, \qquad f^{(4)}(x) = -3 \cdot 2x^{-4}$$

Similarly, $f^{(5)}(x) = 4 \cdot 3 \cdot 2x^{-5}$. The general pattern is that $f^{(k)}(x)$ is a multiple of x^{-k}, with a coefficient $(k-1)!$ that alternates in sign:

$$f^{(k)}(x) = (-1)^{k-1}(k-1)! \, x^{-k} \qquad \boxed{1}$$

Therefore, the coefficient of $(x-1)^k$ in $T_n(x)$ is

$$\frac{f^{(k)}(1)}{k!} = \frac{(-1)^{k-1}(k-1)!}{k!} = \frac{(-1)^{k-1}}{k} \qquad \text{(for } k \geq 1\text{)}$$

Taylor polynomials for $\ln x$ at $a = 1$:

$T_1(x) = (x-1)$

$T_2(x) = (x-1) - \dfrac{1}{2}(x-1)^2$

$T_3(x) = (x-1) - \dfrac{1}{2}(x-1)^2 + \dfrac{1}{3}(x-1)^3$

In other words, the coefficients for $k \geq 1$ are $1, -\frac{1}{2}, \frac{1}{3}, -\frac{1}{4}, \dots$, and

$$T_n(x) = (x-1) - \frac{1}{2}(x-1)^2 + \frac{1}{3}(x-1)^3 - \cdots + (-1)^{n-1}\frac{1}{n}(x-1)^n \qquad \blacksquare$$

■ **EXAMPLE 3** Maclaurin Polynomials for $f(x) = \cos x$ Find the Maclaurin polynomials of $f(x) = \cos x$.

Solution Recall that the Maclaurin polynomials are the Taylor polynomials centered at $a = 0$. The key observation is that the derivatives of $f(x) = \cos x$ form a pattern that repeats with period 4:

$$f'(x) = -\sin x, \qquad f''(x) = -\cos x, \qquad f'''(x) = \sin x, \qquad f^{(4)}(x) = \cos x,$$

and, in general, $f^{(j+4)}(x) = f^{(j)}(x)$. At $x = 0$, the derivatives form the repeating pattern $1, 0, -1,$ and 0:

$f(0)$	$f'(0)$	$f''(0)$	$f'''(0)$	$f^{(4)}(0)$	$f^{(5)}(0)$	$f^{(6)}(0)$	$f^{(7)}(0)$	$\cdots$
1	0	-1	0	1	0	-1	0	$\cdots$

In other words, the even derivatives are $f^{(2k)}(0) = (-1)^k$ and the odd derivatives are zero: $f^{(2k+1)}(0) = 0$. Therefore, the coefficient of x^{2k} is $(-1)^k/(2k)!$ and the coefficient of x^{2k+1} is zero. We have

$$T_0(x) = T_1(x) = 1$$

$$T_2(x) = T_3(x) = 1 - \frac{1}{2!}x^2$$

$$T_4(x) = T_5(x) = 1 - \frac{x^2}{2} + \frac{x^4}{4!}$$

and, in general,

$$T_{2n}(x) = T_{2n+1}(x) = 1 - \frac{1}{2}x^2 + \frac{1}{4!}x^4 - \frac{1}{6!}x^6 + \cdots + (-1)^n \frac{1}{(2n)!}x^{2n} \qquad \blacksquare$$

Taylor polynomials $T_n(x)$ are designed to approximate $f(x)$ in an interval around $x = a$. Figure 2 shows the first few Maclaurin polynomials for $f(x) = \cos x$. Observe that as n gets larger, $T_n(x)$ approximates $f(x) = \cos x$ well over larger and larger intervals. Outside this interval, the approximation fails.

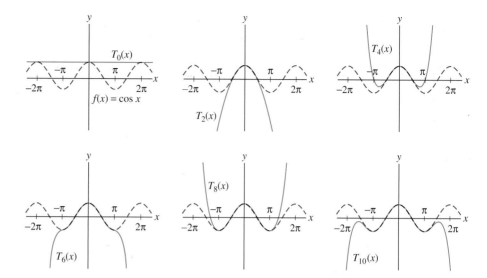

FIGURE 2 Graphs of Taylor polynomials for $f(x) = \cos x$. The graph of $f(x)$ is shown as a dashed curve.

CONCEPTUAL INSIGHT   To understand how the formula for T_n arises, it is useful to introduce a notion of closeness for two functions. We say that f and g *agree to order n* at $x = a$ if

$$f(a) = g(a), \quad f'(a) = g'(a), \quad f''(a) = g''(a), \dots, \quad f^{(n)}(a) = g^{(n)}(a)$$

Taylor polynomials are defined so that $T_n(x)$ and $f(x)$ agree to order n at $x = a$ (see Exercise 59) and, in fact, $T_n(x)$ is the only polynomial of degree at most n with this property. For example, we can check that $T_2(x)$ agrees with $f(x)$ to order $n = 2$:

$$T_2(x) = f(a) + f'(a)(x - a) + \frac{1}{2}f''(a)(x - a)^2, \quad T_2(a) = f(a)$$

$$T_2'(x) = f'(a) + f''(a)(x - a), \qquad\qquad\qquad T_2'(a) = f'(a)$$

$$T_2''(x) = f''(a), \qquad\qquad\qquad\qquad\qquad\qquad T_2''(a) = f''(a)$$

The Remainder Term

Our next goal is to study the error $|T_n(x) - f(x)|$ in the approximation provided by the nth Taylor polynomial centered at $x = a$. Define the **nth remainder** for $f(x)$ at $x = a$ by

$$\boxed{R_n(x) = f(x) - T_n(x)}$$

The error is the absolute value of the remainder. Also, $f(x) = T_n(x) + R_n(x)$, so

$$f(x) = f(a) + \frac{f'(a)}{1!}(x - a)^1 + \frac{f''(a)}{2!}(x - a)^2 + \cdots + \frac{f^{(n)}(a)}{n!}(x - a)^n + R_n(x)$$

Taylor's Theorem   Assume that $f^{(n+1)}(x)$ exists and is continuous. Then

$$\boxed{R_n(x) = \frac{1}{n!}\int_a^x (x - u)^n f^{(n+1)}(u)\, du}$$ $\boxed{2}$

Proof　Set

$$I_n(x) = \frac{1}{n!} \int_a^x (x-u)^n f^{(n+1)}(u)\, du$$

Our goal is to show that $R_n(x) = I_n(x)$. For $n = 0$, $R_0(x) = f(x) - f(a)$ and the desired result is just a restatement of the Fundamental Theorem of Calculus:

$$I_0(x) = \int_a^x f'(u)\, du = f(x) - f(a) = R_0(x)$$

Exercise 55 reviews this proof for the special case $n = 2$.

To prove the formula for $n > 0$, we apply Integration by Parts to $I_n(x)$ with

$$h(u) = \frac{(x-u)^n}{n!}, \qquad g(u) = f^{(n)}(u)$$

$$I_n(x) = \int_a^x h(u)\, g'(u)\, du = h(u)g(u)\Big|_a^x - \int_a^x h'(u)g(u)\, du$$

$$= \frac{1}{n!}(x-u)^n f^{(n)}(u)\Big|_a^x - \frac{1}{n!}\int_a^x (-n)(x-u)^{n-1} f^{(n)}(u)\, du$$

$$= -\frac{1}{n!}(x-a)^n f^{(n)}(a) + I_{n-1}(x)$$

This result can be rewritten as

$$I_{n-1}(x) = \frac{f^{(n)}(a)}{n!}(x-a)^n + I_n(x)$$

Now apply this relation n times:

$$f(x) = f(a) + I_0(x)$$

$$= f(a) + \frac{f'(a)}{1!}(x-a) + I_1(x)$$

$$= f(a) + \frac{f'(a)}{1!}(x-a) + \frac{f''(a)}{2!}(x-a)^2 + I_2(x)$$

$$\vdots$$

$$= f(a) + \frac{f'(a)}{1!}(x-a) + \cdots + \frac{f^{(n)}(a)}{n!}(x-a)^n + I_n(x)$$

This shows that $f(x) = T_n(x) + I_n(x)$ and hence $I_n(x) = R_n(x)$ as desired.　■

Although Taylor's Theorem gives us an explicit formula for the remainder, we will not use this formula directly. Instead, we will use it to *estimate* the size of the error.

THEOREM 1　Error Bound　Assume that $f^{(n+1)}(x)$ exists and is continuous. Let K be a number such that $|f^{(n+1)}(u)| \le K$ for all u between a and x. Then

$$|T_n(x) - f(x)| \le K \frac{|x-a|^{n+1}}{(n+1)!}$$

Proof Assume that $x \geq a$ (the case $x \leq a$ is similar). Then, since $|f^{(n+1)}(u)| \leq K$ for $a \leq u \leq x$,

$$|T_n(x) - f(x)| = |R_n(x)| = \left| \frac{1}{n!} \int_a^x (x - u)^n f^{(n+1)}(u)\, du \right|$$

$$\leq \frac{1}{n!} \int_a^x \left| (x - u)^n f^{(n+1)}(u) \right|\, du \qquad \boxed{3}$$

$$\leq \frac{K}{n!} \int_a^x (x - u)^n\, du$$

$$= \frac{K}{n!} \left. \frac{-(x - u)^{n+1}}{n + 1} \right|_{u=a}^x = K \frac{|x - a|^{n+1}}{(n + 1)!} \qquad \blacksquare$$

In (3), we use the inequality

$$\left| \int_a^b f(x)\, dx \right| \leq \int_a^b |f(x)|\, dx$$

which is valid for all integrable functions.

■ **EXAMPLE 4** Using the Error Bound   Use the Error Bound to find a bound for the error $|T_3(1.2) - \ln 1.2|$, where $T_3(x)$ is the third Taylor polynomial for $f(x) = \ln x$ at $a = 1$. Check your result using a calculator.

Solution To use the Error Bound with $n = 3$, we must find a value of K such that $|f^{(4)}(u)| \leq K$ for $1 \leq u \leq 1.2$. By Eq. (1) in Example 2, $|f^{(4)}(x)| = 6x^{-4}$. Since $|f^{(4)}(x)|$ is decreasing for $x \geq 1$, its maximum value on $[1, 1.2]$ is $|f^{(4)}(1)| = 6$. Therefore, we may apply the Error Bound with $K = 6$ to obtain

$$|T_3(1.2) - \ln 1.2| \leq K \frac{|x - a|^{n+1}}{(n + 1)!} = 6 \frac{|1.2 - 1|^4}{4!} \approx 0.0004$$

By Example 2,

$$T_3(x) = (x - 1) - \frac{1}{2}(x - 1)^2 + \frac{1}{3}(x - 1)^3$$

The following values from a calculator confirm that the error is at most 0.0004:

$$|T_3(1.2) - \ln 1.2| \approx |0.182667 - 0.182322| \approx 0.00035 < 0.0004 \qquad \blacksquare$$

■ **EXAMPLE 5** Approximating with a Given Accuracy   Let $T_n(x)$ be the nth Maclaurin polynomial for $f(x) = \cos x$. Find a value of n such that

$$|T_n(0.2) - \cos(0.2)| < 10^{-5}$$

Use a calculator to verify that this value of n works.

Solution Since $|f^{(n)}(x)|$ is $|\cos x|$ or $|\sin x|$, depending on whether n is even or odd, we have $|f^{(n)}(u)| \leq 1$ for all u. Thus, we may apply the Error Bound with $K = 1$:

$$|T_n(0.2) - \cos(0.2)| \leq K \frac{|0.2 - 0|^{n+1}}{(n + 1)!} = \frac{|0.2|^{n+1}}{(n + 1)!}$$

To use the Error Bound, it is not necessary to find the smallest possible value of K. In this example, we take $K = 1$. This works for all n, but for odd n we could have used the smaller value $K = \sin(0.2) \approx 0.2$.

To make the error less than 10^{-5}, we must choose n so that

$$\frac{|0.2|^{n+1}}{(n + 1)!} < 10^{-5}$$

We find a suitable n by checking several values:

n	2	3	4
$\dfrac{\|0.2\|^{n+1}}{(n+1)!}$	$\dfrac{0.2^3}{3!} \approx 0.0013$	$\dfrac{0.2^4}{4!} \approx 6.67 \times 10^{-5}$	$\dfrac{0.2^5}{5!} \approx 2.67 \times 10^{-6} < 10^{-5}$

We see that the error is less than 10^{-5} for $n = 4$. To verify this, recall that $T_4(x) = 1 - \frac{1}{2}x^2 + \frac{1}{4!}x^4$ by Example 3. The following values from a calculator confirm that the error is significantly less than 10^{-5} as required:

$$\text{Actual error} = |T_4(0.2) - \cos(0.2)| \approx |0.98006667 - 0.98006657| = 10^{-7} \qquad ∎$$

CONCEPTUAL INSIGHT Recall that functions $g(x)$ and $f(x)$ agree at $x = a$ to order n if $g^{(k)}(a) = f^{(k)}(a)$ for $0 \le k \le n$. The defining property of the Taylor polynomial $T_n(x)$ is that it has at most degree n and agrees with $f(x)$ at $x = a$ to order n. Now, let us say that a function $g(x)$ *approximates* $f(x)$ *at* $x = a$ *to order* n if the error $E(x) = |f(x) - g(x)|$ satisfies

$$\lim_{x \to a} \frac{E(x)}{|x - a|^n} = 0 \qquad \boxed{4}$$

Thus, in an nth-order approximation, the error tends to zero faster than $|x - a|^n$. The Error Bound tells us that $T_n(x)$ is an nth-order approximation to $f(x)$ at $x = a$, provided we assume that $f^{(n+1)}(x)$ exists and is continuous and that $|f^{(n+1)}(u)| \le K$ for u in some open interval I around a. In fact, the Error Bound gives

$$|f(x) - T_n(x)| \le C|x - a|^{n+1} \qquad \text{for } x \in I$$

where $C = \dfrac{K}{(n+1)!}$, so

$$\frac{|f(x) - T_n(x)|}{|x - a|^n} \le C|x - a| \to 0 \qquad (\text{as } x \to a)$$

If $f^{(k)}(x)$ exists for all k [as is the case for transcendental functions such as $f(x) = \sin x$ or $f(x) = e^x$], then we can approximate $f(x)$ to arbitrarily high order using Taylor polynomials.

8.4 SUMMARY

- The nth *Taylor polynomial* centered at $x = a$ for the function $f(x)$ is

$$T_n(x) = f(a) + \frac{f'(a)}{1!}(x - a)^1 + \frac{f''(a)}{2!}(x - a)^2 + \cdots + \frac{f^{(n)}(a)}{n!}(x - a)^n$$

When $a = 0$, $T_n(x)$ is also called the nth *Maclaurin polynomial*.
- If $f^{(n+1)}(x)$ exists and is continuous, then we have the *Error Bound*

$$|T_n(x) - f(x)| \le K \frac{|x - a|^{n+1}}{(n+1)!}$$

where K is a number such that $|f^{(n+1)}(u)| \le K$ for all u between a and x.

• For reference, we include a table of standard Maclaurin and Taylor polynomials.

$f(x)$	a	Maclaurin or Taylor Polynomial
e^x	0	$T_n(x) = 1 + x + \dfrac{x^2}{2!} + \dfrac{x^3}{3!} + \cdots + \dfrac{x^n}{n!}$
$\sin x$	0	$T_{2n+1}(x) = T_{2n+2}(x) = x - \dfrac{x^3}{3!} + \cdots + (-1)^n \dfrac{x^{2n+1}}{(2n+1)!}$
$\cos x$	0	$T_{2n}(x) = T_{2n+1}(x) = 1 - \dfrac{x^2}{2!} + \dfrac{x^4}{4!} - \cdots + (-1)^n \dfrac{x^{2n}}{(2n)!}$
$\ln x$	1	$T_n(x) = (x-1) - \dfrac{1}{2}(x-1)^2 + \cdots + \dfrac{(-1)^{n-1}}{n}(x-1)^n$
$\dfrac{1}{1-x}$	0	$T_n(x) = 1 + x + x^2 + \cdots + x^n$

8.4 EXERCISES

Preliminary Questions

1. What is $T_3(x)$ centered at $a = 3$ for a function $f(x)$ such that $f(3) = 9$, $f'(3) = 8$, $f''(3) = 4$, and $f'''(3) = 12$.

2. The dashed graphs in Figure 3 are Taylor polynomials for a function $f(x)$. Which of the two is a Maclaurin polynomial?

3. For which value of x does the Maclaurin polynomial $T_n(x)$ satisfy $T_n(x) = f(x)$, no matter what $f(x)$ is?

4. Let $T_n(x)$ be the Maclaurin polynomial of a function $f(x)$ satisfying $|f^{(4)}(x)| \le 1$ for all x. Which of the following statements follow from the Error Bound?

(a) $|T_4(2) - f(2)| \le 2^4/24$

(b) $|T_3(2) - f(2)| \le 2^3/6$

(c) $|T_3(2) - f(2)| \le 1/3$

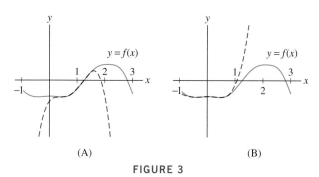

(A)   (B)

FIGURE 3

Exercises

In Exercises 1–14, calculate the Taylor polynomials $T_2(x)$ and $T_3(x)$ centered at $x = a$ for the given function and value of a.

1. $f(x) = \sin x$, $\quad a = 0$

2. $f(x) = \sin x$, $\quad a = \pi/2$

3. $f(x) = \dfrac{1}{1+x}$, $\quad a = 0$

4. $f(x) = \dfrac{1}{1+x}$, $\quad a = 1$

5. $f(x) = \tan x$, $\quad a = 0$

6. $f(x) = \tan x$, $\quad a = \frac{\pi}{4}$

7. $f(x) = \dfrac{1}{1+x^2}$, $\quad a = 0$

8. $f(x) = \dfrac{1}{1+x^2}$, $\quad a = -1$

9. $f(x) = e^x$, $\quad a = 0$

10. $f(x) = e^x$, $\quad a = \ln 2$

11. $f(x) = e^{-x} + e^{-2x}$, $\quad a = 0$

12. $f(x) = x^2 e^{-x}$, $\quad a = 1$

13. $f(x) = \ln(x + 1)$, $\quad a = 0$

14. $f(x) = \cosh x$, $\quad a = 0$

In Exercises 15–18, compute $T_2(x)$ at $x = a$ and use a calculator to compute the error $|f(x) - T_2(x)|$ at the given value of x.

15. $y = e^x$, $\quad x = -0.5$, $\quad a = 0$

16. $y = \cos x$, $\quad x = \frac{\pi}{12}$, $\quad a = 0$

17. $y = x^{-3/2}$, $\quad x = 0.3$, $\quad a = 1$

18. $y = e^{\sin x}$, $\quad x = 1.5$, $\quad a = \frac{\pi}{2}$

19. Show that the nth Maclaurin polynomial for $f(x) = e^x$ is

$$T_n(x) = 1 + \frac{x}{1!} + \frac{x^2}{2!} + \cdots + \frac{x^n}{n!}$$

20. Show that the nth Taylor polynomial for $\dfrac{1}{x+1}$ at $a = 1$ is

$$T_n(x) = \frac{1}{2} - \frac{x-1}{4} + \frac{(x-1)^2}{8} + \cdots + (-1)^n \frac{(x-1)^n}{2^{n+1}}$$

In Exercises 21–26, find $T_n(x)$ at $x = a$ for all n.

21. $f(x) = \dfrac{1}{1-x}, \quad a = 0$

22. $f(x) = \dfrac{1}{x-1}, \quad a = 4$

23. $f(x) = e^x, \quad a = 1$

24. $f(x) = \sqrt{x}, \quad a = 1$

25. $f(x) = x^{5/2}, \quad a = 2$

26. $f(x) = \cos x, \quad a = \pi/4$

27. *CAS* Plot $y = e^x$ together with the Maclaurin polynomials $T_n(x)$ for $n = 1, 3, 5$ and then for $n = 2, 4, 6$ on the interval $[-3, 3]$. What difference do you notice between the even and odd Maclaurin polynomials?

28. *CAS* Plot $f(x) = \dfrac{1}{1+x}$ together with the Taylor polynomials $T_n(x)$ at $a = 1$ for $1 \le n \le 4$ on the interval $[-2, 8]$ (be sure to limit the upper plot range).

(a) Over which interval does T_4 appear to give a close approximation to $f(x)$?

(b) What happens with $x < -1$?

(c) Use your computer algebra system to produce T_{30} and plot it together with $f(x)$ on $[-2, 8]$. Over which interval does T_{30} appear to give a close approximation?

29. Use the Error Bound to find the maximum possible size of $|\cos 0.3 - T_5(0.3)|$, where $T_5(x)$ is the Maclaurin polynomial. Verify your result with a calculator.

30. Calculate $T_4(x)$ at $a = 1$ for $f(x) = x^{11/2}$ and use the Error Bound to find the maximum possible size of $|T_4(1.2) - f(1.2)|$. *Hint:* Show that the max $|f^{(5)}(x)|$ on $[1, c]$ with $c > 1$ is $f^{(5)}(c)$.

31. Let T_n be the Maclaurin polynomial of $f(x) = e^x$ and let $c > 0$. Show that we may take $K = e^c$ in the Error Bound for $|T_n(c) - f(c)|$. Then use the Error Bound to determine a number n such that $|T_n(0.1) - e^{0.1}| \le 10^{-5}$.

32. Let $f(x) = \sqrt{1+x}$ and let $T_n(x)$ be the Taylor polynomial centered at $a = 8$.

(a) Find $T_3(x)$ and calculate $T_3(8.02)$.

(b) Use the Error Bound to find a bound for $|T_3(8.02) - \sqrt{9.02}|$.

33. Calculate $T_3(x)$ at $a = 0$ for $f(x) = \tan^{-1} x$. Then compute $T_3(\frac{1}{2})$ and use the Error Bound to find a bound for $|T_3(\frac{1}{2}) - \tan^{-1}(\frac{1}{2})|$. Refer to the graph in Figure 4 to find an acceptable value of K.

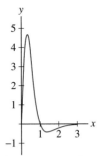

FIGURE 4 Graph of $f^{(4)}(x) = -24x(x^2 - 1)/(x^2 + 1)^4$, where $f(x) = \tan^{-1} x$.

34. ⬛GU Calculate $T_2(x)$ for $f(x) = \operatorname{sech} x$ at $a = 0$. Then compute $T_2(\frac{1}{2})$ and use the Error Bound to find a bound for

$$\left| T_2\left(\frac{1}{2}\right) - f\left(\frac{1}{2}\right) \right|.$$

Hint: Plot $f'''(x)$ to find an acceptable value of K.

35. Show that the Maclaurin polynomials for $f(x) = \sin x$ are

$$T_{2n-1}(x) = T_{2n} = x - \frac{x^3}{3!} + \frac{x^5}{5!} - \cdots + (-1)^{n-1} \frac{x^{2n-1}}{(2n-1)!}$$

Use the Error Bound with $n = 4$ to show that

$$\left| \sin x - \left(x - \frac{x^3}{6} \right) \right| \le \frac{|x|^5}{120} \quad \text{(for all } x)$$

36. Find n such that $|T_n(0.1) - \cos(0.1)| \le 10^{-7}$, where T_n is the Maclaurin polynomial for $f(x) = \cos x$ (see Example 5). Calculate $|T_n(0.1) - \cos(0.1)|$ for this value of n and verify the bound on the error.

37. Find n such that $|T_n(1.3) - \ln(1.3)| \le 10^{-4}$, where T_n is the Taylor polynomial for $f(x) = \ln x$ at $a = 1$.

38. Find n such that $|T_n(1) - e| \le 10^{-6}$, where T_n is the Maclaurin polynomial for $f(x) = e^x$.

39. Find n such that $|T_n(1.3) - \sqrt{1.3}| \le 10^{-6}$, where $T_n(x)$ is the Taylor polynomial for $f(x) = \sqrt{x}$ at $a = 1$.

40. Let $T_n(x)$ be the Taylor polynomial for $f(x) = \ln x$ at $a = 1$ and let $c > 1$.

(a) Show that the maximum of $f^{(k+1)}(x)$ on $[1, c]$ is $f^{(k+1)}(1)$ (see Example 4).

(b) Prove $|T_n(c) - \ln c| \le \dfrac{|c-1|^{n+1}}{n+1}$.

(c) Find n such that $|T_n(1.5) - \ln 1.5| \le 10^{-2}$.

41. Let $n \ge 1$. Show that if $|x|$ is small, then

$$(x+1)^{1/n} \approx 1 + \frac{x}{n} + \frac{1-n}{2n^2} x^2$$

Use this approximation with $n = 6$ to estimate $1.5^{1/6}$.

42. Verify that the third Maclaurin polynomial for $f(x) = e^x \sin x$ is equal to the product of the third Maclaurin polynomials of e^x and $\sin x$ (after discarding terms of degree greater than 3 in the product).

43. Find the fourth Maclaurin polynomial for $f(x) = \sin x \cos x$ by multiplying the fourth Maclaurin polynomials for $f(x) = \sin x$ and $f(x) = \cos x$.

44. Find the Maclaurin polynomials $T_n(x)$ for $f(x) = \cos(x^2)$. You may use the fact that $T_n(x)$ is equal to the sum of the terms up to degree n obtained by substituting x^2 for x in the nth Maclaurin polynomial of $\cos x$.

45. Find the Maclaurin polynomials of $\dfrac{1}{1+x^2}$ by substituting $-x^2$ for x in the Maclaurin polynomials of $\dfrac{1}{1-x}$ (see Exercise 21).

46. The seventeenth-century Dutch scientist Christian Huygens used the approximation $\theta \approx \dfrac{8b-a}{3}$ for the length θ of a circular arc of the unit circle, where a is the length of the chord $\overline{AC}$ of angle θ and b is length of the chord $\overline{AB}$ of angle $\theta/2$ (Figure 5).
(a) Prove that $a = 2\sin(\theta/2)$ and $b = 2\sin(\theta/4)$, and show that the Huygens approximation amounts to the approximation

$$\theta \approx \frac{16}{3}\sin\frac{\theta}{4} - \frac{2}{3}\sin\frac{\theta}{2}$$

(b) Compute the fifth MacLaurin polynomial of the function on the right.
(c) Use the Error Bound to show that the error in the Huygens approximation is less than $0.00022|\theta|^5$.

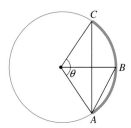

FIGURE 5 Unit circle.

47. Let $f(x) = 3x^3 + 2x^2 - x - 4$. Calculate $T_j(x)$ for $j = 1, 2, 3, 4, 5$ at both $a = 0$ and $a = 1$. Show that $T_3(x) = f(x)$ in both cases.

48. Let $T_n(x)$ be the nth Taylor polynomial at $x = a$ for a polynomial $f(x)$ of degree n. Based on the result of Exercise 47, guess the value of $|f(x) - T_n(x)|$. Prove that your guess is correct using the Error Bound.

49. Taylor polynomials can be used instead of L'Hôpital's Rule to evaluate limits. Consider $L = \lim\limits_{x \to 0} \dfrac{e^x + e^{-x} - 2}{1 - \cos x}$.
(a) Show that the second Maclaurin polynomial for

$$f(x) = e^x + e^{-x} - 2$$

is $T_2(x) = x^2$. Use the Error Bound with $n = 2$ to show that

$$e^x + e^{-x} - 2 = x^2 + g_1(x)$$

where $\lim\limits_{x \to 0} \dfrac{g_1(x)}{x^2} = 0$. Similarly, prove that

$$1 - \cos x = \frac{1}{2}x^2 + g_2(x)$$

where $\lim\limits_{x \to 0} \dfrac{g_2(x)}{x^2} = 0$.

(b) Evaluate L by using (a) to show that $L = \lim\limits_{x \to 0} \dfrac{1 + \dfrac{g_1(x)}{x^2}}{\dfrac{1}{2} + \dfrac{g_2(x)}{x^2}}$.

(c) Evaluate L again using L'Hôpital's Rule.

50. Use the method of Exercise 49 to evaluate

$$\lim_{x \to 0} \frac{1 - \cos x}{x^2} \qquad \text{and} \qquad \lim_{x \to 0} \frac{\sin x - x}{x^3}.$$

51. A light wave of wavelength λ travels from A to B by passing through an aperture (see Figure 6 for the notation). The aperture (circular region) is located in a plane that is perpendicular to $\overline{AB}$. Let $f(r) = d' + h'$, that is, $f(r)$ is the distance $AC + CB$ as a function of r. The **Fresnel zones**, used to determine the optical disturbance at B, are the concentric bands bounded by the circles of radius R_n such that $f(R_n) = AB + n\lambda/2 = d + h + n\lambda/2$.
(a) Show that $f(r) = \sqrt{d^2 + r^2} + \sqrt{h^2 + r^2}$, and use the Maclaurin polynomial of order 2 to show

$$f(r) \approx d + h + \frac{1}{2}\left(\frac{1}{d} + \frac{1}{h}\right)r^2$$

(b) Deduce that $R_n \approx \sqrt{n\lambda L}$, where $L = (d^{-1} + h^{-1})^{-1}$.
(c) Estimate the radii R_1 and R_{100} for blue light ($\lambda = 475 \times 10^{-7}$ cm) if $d = h = 100$ cm.

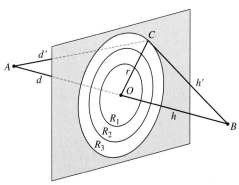

FIGURE 6 The Fresnel zones are the regions between the circles of radius R_n.

Further Insights and Challenges

52. Show that the nth Maclaurin polynomial of $f(x) = \arcsin x$ for n odd is

$$T_n(x) = x + \frac{1}{2}\frac{x^3}{3} + \frac{1 \cdot 3}{2 \cdot 4}\frac{x^5}{5} + \frac{1 \cdot 3 \cdot 5}{2 \cdot 4 \cdot 6}\frac{x^7}{7} + \cdots$$
$$+ \frac{1 \cdot 3 \cdot 5 \cdots (n-2)}{2 \cdot 4 \cdot 6 \cdots (n-1)}\frac{x^n}{n}$$

53. Use Taylor's Theorem to show that if $f^{(n+1)}(t) \geq 0$ for all t, then the nth Maclaurin polynomial $T_n(x)$ satisfies $T_n(x) \leq f(x)$ for all $x \geq 0$.

54. Use Exercise 53 to show that for $x \geq 0$ and all n,

$$e^x \geq 1 + x + \frac{x^2}{2!} + \cdots + \frac{x^n}{n!}$$

Sketch the graphs of e^x, $T_1(x)$, and $T_2(x)$ on the same coordinate axes. Does this inequality remain true for $x < 0$?

55. This exercise is intended to reinforce the proof of Taylor's Theorem.

(a) Show that $f(x) = T_0(x) + \int_a^x f'(u)\,du$.

(b) Use Integration by Parts to prove the formula

$$\int_a^x (x-u)f^{(2)}(u)\,du = -f'(a)(x-a) + \int_a^x f'(u)\,du$$

(c) Prove the case $n = 2$ of Taylor's Theorem:

$$f(x) = T_1(x) + \int_a^x (x-u)f^{(2)}(u)\,du.$$

56. Approximating Integrals Using Taylor Polynomials   Taylor polynomials can be used to obtain numerical approximations to integrals.

(a) Let $L > 0$. Show that if two functions $f(x)$ and $g(x)$ satisfy $|f(x) - g(x)| < L$ for all $x \in [a, b]$, then

$$\int_a^b \left(f(x) - g(x) \right) dx < L(b-a)$$

(b) Show that $|T_3(x) - \sin x| \leq (\frac{1}{2})^5/5!$ for all $x \in [0, \frac{1}{2}]$.

(c) Evaluate $\int_0^{1/2} T_3(x)\,dx$ as an approximation to $\int_0^{1/2} \sin x\,dx$. Use (a) and (b) to find a bound for the size of the error.

57. Use the fourth Maclaurin polynomial and the method of Exercise 56 to approximate $\int_0^{1/2} \sin(x^2)\,dx$. Find a bound for the error.

58. The Second Derivative Test for local extrema fails when $f''(a) = 0$. This exercise shows us how to extend the test under the following assumption:

$$f'(a) = f''(a) = f'''(a) = 0 \quad \text{but} \quad f^{(4)}(a) > 0$$

(a) Show that $T_4(x) = f(a) + \frac{1}{24}f^{(4)}(a)(x-a)^4$.

(b) Show that $T_4(x) > f(a)$ for $x \neq a$.

(c) By the Error Bound, there is a constant C such that $|T_4(x) - f(x)| \leq C|x - a|^5$ for all x in an interval containing a. Use this and (b) to show that $f(x) > f(a)$ for x sufficiently close to but not equal to a.

(d) Conclude that $f(a)$ is a local minimum. A similar argument shows that $f(a)$ is a local maximum if $f^{(4)}(a) < 0$.

59. Prove by induction that for all k,

$$\frac{d^j}{dx^j}\left(\frac{(x-a)^k}{k!} \right) = \frac{k(k-1)\cdots(k-j+1)(x-a)^{k-j}}{k!}$$

$$\frac{d^j}{dx^j}\left(\frac{(x-a)^k}{k!} \right)\bigg|_{x=a} = \begin{cases} 1 & \text{for } k = j \\ 0 & \text{for } k \neq j \end{cases}$$

Use this to prove that $T_n(x)$ agrees with $f(x)$ at $x = a$ to order n.

60. The following equation arises in the description of Bose–Einstein condensation (the quantum theory of gases cooled to near absolute zero):

$$A_0 = \frac{4A}{\sqrt{\pi}}\int_0^\infty \frac{x^2 e^{-x^2}\,dx}{1 - Ae^{-x^2}}$$

It is necessary to derive an approximate expression for A_0 in terms of A for $|A|$ small.

(a) Show that the second Maclaurin polynomial for the function $f(A) = \dfrac{x^2 e^{-x^2}}{1 - Ae^{-x^2}}$ (where A is the variable and x is treated as a constant) is

$$T_2(A) = x^2 e^{-x^2} + x^2 e^{-2x^2}A + x^2 e^{-3x^2}A^2$$

(b) Use the approximation $A_0 \approx \dfrac{4A}{\sqrt{\pi}}\int_0^\infty T_2(A)\,dx$ to show

$$A_0 \approx A + \frac{1}{2\sqrt{2}}A^2 + \frac{1}{3\sqrt{3}}A^3$$

You may use the formula (valid for $\lambda > 0$)

$$\int_0^\infty x^2 e^{-\lambda x^2}\,dx = \frac{1}{4}\sqrt{\frac{\pi}{\lambda^3}}$$

CHAPTER REVIEW EXERCISES

In Exercises 1–4, calculate the arc length over the given interval.

1. $y = \dfrac{x^5}{10} + \dfrac{x^{-3}}{6}$, $[3, 5]$

2. $y = e^{x/2} + e^{-x/2}$, $[0, 2]$

3. $y = 4x - 2$, $[-2, 2]$

4. $y = x^{2/3}$, $[1, 2]$

5. Show that the arc length of $y = 2\sqrt{x}$ over $[0, a]$ is equal to $\sqrt{a(a+1)} + \ln(\sqrt{a} + \sqrt{a+1})$. *Hint:* Apply the substitution $x = \tan^2 \theta$ to the arc length integral.

6. *CAS*   Find a numerical approximation to the arc length of $y = \tan x$ over $[0, \pi/4]$.

In Exercises 7–10, calculate the surface area of the solid obtained by rotating the curve over the given interval about the x-axis.

7. $y = x + 1$, $[0, 4]$

8. $y = \dfrac{2}{3}x^{3/4} - \dfrac{2}{5}x^{5/4}$, $[0, 1]$

9. $y = \dfrac{2}{3}x^{3/2} - \dfrac{1}{2}x^{1/2}$, $[1, 2]$

10. $y = \dfrac{1}{2}x^2$, $[0, 2]$

11. Compute the total surface area of the coin obtained by rotating the region in Figure 1 about the x-axis. The top and bottom parts of the region are semicircles with a radius of 1 mm.

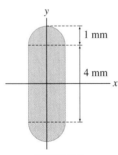

FIGURE 1

12. Calculate the fluid force on the side of a right triangle of height 3 ft and base 2 ft submerged in water vertically, with its upper vertex at the surface of the water.

13. Calculate the fluid force on the side of a right triangle of height 3 ft and base 2 ft submerged in water vertically, with its upper vertex located at a depth of 4 ft.

14. The end of a horizontal oil tank is an ellipse (Figure 2) with equation $(x/4)^2 + (y/3)^2 = 1$ (length in feet). Assume that the tank is filled with oil of density 55 lb/ft³.

(a) Calculate the total force F on the end of the tank when the tank is full.

(b) Would you expect the total force on the lower half of the tank to be greater than, less than, or equal to $\frac{1}{2}F$? Explain. Then compute the force on the lower half exactly and confirm (or refute) your expectation.

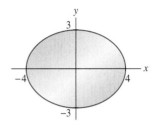

FIGURE 2

15. Calculate the moments and COM of the lamina occupying the region under $y = x(4 - x)$ for $0 \le x \le 4$, assuming a density of $\rho = 5$ lb/ft².

16. Sketch the region between $y = 4(x + 1)^{-1}$ and $y = 1$ for $0 \le x \le 3$ and find its centroid.

17. Find the centroid of the region between the semicircle $y = \sqrt{1 - x^2}$ and the top half of the ellipse $y = \frac{1}{2}\sqrt{1 - x^2}$ (Figure 3).

18. A plate in the shape of the shaded region in Figure 3 is submerged in water. Calculate the fluid pressure on a side of the plate if the water surface is $y = 1$.

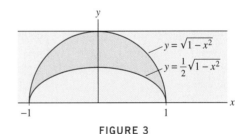

FIGURE 3

19. Find the centroid of the shaded region in Figure 4 bounded on the left by $x = 2y^2 - 2$ and on the right by a semicircle of radius 1. *Hint:* Use symmetry and additivity of moments.

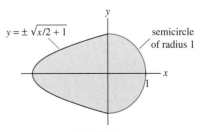

FIGURE 4

In Exercises 20–25, find the Taylor polynomial at $x = a$ for the given function.

20. $f(x) = x^3$, $T_3(x)$, $a = 1$

21. $f(x) = 3(x+2)^3 - 5(x+2)$, $T_3(x)$, $a = -2$

22. $f(x) = x \ln(x)$, $T_4(x)$, $a = 1$

23. $f(x) = (3x+2)^{1/3}$, $T_3(x)$, $a = 2$

24. $f(x) = xe^{-x^2}$, $T_4(x)$, $a = 0$

25. $f(x) = \ln(\cos x)$, $T_3(x)$, $a = 0$

26. Find the nth Maclaurin polynomial for $f(x) = e^{3x}$.

27. Use the fifth Maclaurin polynomial of $f(x) = e^x$ to approximate $\sqrt{e}$. Use a calculator to determine the error.

28. Use the third Taylor polynomial of $f(x) = \tan^{-1} x$ at $a = 1$ to approximate $f(1.1)$. Use a calculator to determine the error.

29. Let $T_4(x)$ be the Taylor polynomial for $f(x) = \sqrt{x}$ at $a = 16$. Use the Error Bound to find the maximum possible size of $|f(17) - T_4(17)|$.

30. Find n such that $|T_n(1) - e| < 10^{-8}$, where $T_n(x)$ is the nth Maclaurin polynomial for $f(x) = e^x$.

31. Let $T_4(x)$ be the Taylor polynomial for $f(x) = x \ln x$ at $x = 1$ computed in Exercise 22. Use the Error Bound to find a bound for $|f(1.2) - T_4(1.2)|$.

32. Let $T_n(x)$ be the nth Maclaurin polynomial for $f(x) = \sin x + \sinh x$.

(a) Show that $T_5(x) = T_6(x) = T_7(x) = T_8(x)$. *Hint:* $T_n(x)$ is the sum of the Maclaurin polynomials of $\sin x$ and $\sinh x$.

(b) Show that $|f^n(x)| \le 1 + \cosh x$ for all n. *Hint:* Note that $|\sinh x| \le |\cosh x|$ for all x.

(c) Show that $|T_8(x) - f(x)| \le \dfrac{2.6}{9!}|x|^9$ for $-1 \le x \le 1$.

33. Show that the nth Maclaurin polynomial for $f(x) = \dfrac{1}{1-x}$ is $T_n(x) = 1 + x + x^2 + \cdots + x^n$. Conclude by substituting $x/4$ for x so that the nth Maclaurin polynomial for $f(x) = \dfrac{1}{1+x/4}$ is

$$T_n(x) = 1 + \frac{1}{4}x + \frac{1}{4^2}x^2 + \cdots + \frac{1}{4^n}x^n$$

What is the nth Maclaurin polynomial for $g(x) = \dfrac{1}{1+x}$? *Hint:* $g(x) = f(-x)$.

34. Let $T_n(x)$ be the Maclaurin polynomial of $f(x) = \dfrac{5}{4+3x-x^2}$.

(a) Show that $f(x) = 5\left(\dfrac{1/4}{1-x/4} + \dfrac{1}{1+x}\right)$.

(b) Let a_k be the coefficient of x^k in $T_n(x)$ for $k \le n$. Use the result of Exercise 33 to show that

$$a_k = \frac{1}{4^{k+1}} + (-1)^k$$

9 | INTRODUCTION TO DIFFERENTIAL EQUATIONS

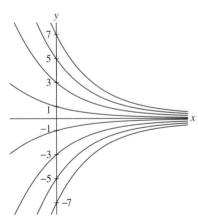

NASA technicians preparing a ground vibration test of an F/A-18 aircraft with aeroelastic wings.

Will this airplane fly? ... How can we create an image of the interior of the human body using very weak X-rays? ... What is a design of a bicycle frame that combines low weight with rigidity? ... How much would the mean temperature of the earth increase if the amount of carbon dioxide in the atmosphere increased by 20 percent?

—An overview of applications of differential equations in *Computational Differential Equations*, K. Eriksson, D. Estep, P. Hansbo, and C. Johnson, Cambridge University Press, New York, 1996

Differential equations are among the most powerful tools we have for analyzing the world mathematically. They are used to formulate the fundamental laws of nature (from Newton's Laws to Maxwell's equations and the laws of quantum mechanics) and to model the most diverse physical phenomena. The above quote lists just a few of the myriad applications. In this chapter, we study some elementary techniques and basic applications of differential equations and give a few indications of where the more general theory leads.

9.1 Solving Differential Equations

A differential equation is an equation that involves an unknown function $y = y(x)$ and its first or higher derivatives. A **solution** is a function $y = f(x)$ satisfying the given equation. As we have seen in previous chapters, solutions usually depend on one or more arbitrary constants (denoted A, B, and C in the following examples):

Differential Equation	General Solution
$y' = -2y$	$y = Ce^{-2x}$
$\dfrac{dy}{dt} = t$	$y = \dfrac{1}{2}t^2 + C$
$y'' + y = 0$	$y = A\sin x + B\cos x$

An expression such as $y = Ce^{-2x}$ is called a **general solution**. For each value of C, we obtain a particular solution. The graphs of the solutions as C varies form a family of curves in the xy-plane (Figure 1).

FIGURE 1 Family of solutions of $y' = -2y$.

The first step in any study of differential equations is to classify the equations according to various properties. The most important attributes of a differential equation are its order and whether or not it is linear. The **order** of a differential equation is the order of the highest derivative appearing in the equation. The general solution of an equation of order n usually involves n arbitrary constants. For example, $y'' + y = 0$ has order two and its general solution has two arbitrary constants A and B as listed above.

A differential equation is called **linear** if it can be written in the form

$$a_n(x)y^{(n)} + a_{n-1}(x)y^{(n-1)} + \cdots + a_1(x)y' + a_0(x)y = b(x)$$

The coefficients $a_j(x)$ and $b(x)$ are functions of x alone. A linear equation cannot have terms such as y^3, yy', or $\sin y$, but the coefficients $a_j(x)$ may be arbitrary.

Differential Equation	Type
$x^2 y' + e^x y = 4$	First-order, linear
$x(y')^2 = y + x$	First-order, nonlinear [since $(y')^2$ appears]
$y'' = (\sin x)y'$	Second-order, linear
$y'' = x(\sin y)$	Second-order, nonlinear (since $\sin y$ appears)

Separation of Variables

The simplest differential equations are those of the form $y' = f(x)$. A solution is an antiderivative of $f(x)$, and thus we may write the general solution as $y = \int f(x)\,dx$. A more general class of first-order equations that can be solved directly by integration are the **separable equations**, which have the form

$$\frac{dy}{dx} = f(x)g(y) \qquad \boxed{1}$$

For example,

- $\dfrac{dy}{dx} = (\sin x)y$ is separable.

- $\dfrac{dy}{dx} = x + y$ is not separable because $x + y$ is not a *product* $f(x)g(y)$.

In separation of variables, we manipulate dx and dy symbolically, just as in the Substitution Rule.

To solve a separable equation, we use the method of **separation of variables:** Move the terms involving y and dy to the left and those involving x and dx to the right. Then integrate both sides:

$$\frac{dy}{dx} = f(x)g(y) \qquad \text{(separable equation)}$$

$$\frac{dy}{g(y)} = f(x)\,dx \qquad \text{(separate the variables)}$$

$$\int \frac{dy}{g(y)} = \int f(x)\,dx \qquad \text{(integrate)}$$

If these integrals can be evaluated, we can try to solve for y as a function of x.

■ **EXAMPLE 1** Show that $y\dfrac{dy}{dx} - x = 0$ is separable but not linear. Then find the general solution and plot the family of solutions.

Solution This differential equation is nonlinear because it contains the term yy'. To show that it is separable, rewrite the equation:

$$y\frac{dy}{dx} - x = 0 \quad \Rightarrow \quad \frac{dy}{dx} = xy^{-1} \quad \text{(separable)}$$

Now use separation of variables:

$$y\,dy = x\,dx \qquad \text{(separate variables)}$$

$$\int y\,dy = \int x\,dx \qquad \text{(integrate)}$$

$$\frac{1}{2}y^2 = \frac{1}{2}x^2 + C \qquad \boxed{2}$$

$$y = \pm\sqrt{x^2 + 2C} \quad \text{(solve for } y\text{)}$$

Note that one constant of integration is sufficient in Eq. (2). An additional constant for the integral on the left is not needed.

Since C is arbitrary, we may replace $2C$ by C to obtain (Figure 2)

$$\boxed{y = \pm\sqrt{x^2 + C}}$$

Each choice of sign yields a solution. It is a good idea to check the solution by verifying that it satisfies the differential equation. For the positive square root (the negative square root is similar),

$$\frac{dy}{dx} = \frac{d}{dx}\sqrt{x^2 + C} = \frac{x}{\sqrt{x^2 + C}}$$

$$y\frac{dy}{dx} = \sqrt{x^2 + C}\left(\frac{x}{\sqrt{x^2 + C}}\right) = x$$

Thus, $y\dfrac{dy}{dx} - x = 0$ as required. ∎

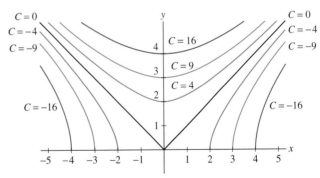

FIGURE 2 Solutions $y = \sqrt{x^2 + C}$ to $y\dfrac{dy}{dx} - x = 0$.

Most of the differential equations arising in applications have an existence and uniqueness property: There exists one and only one solution satisfying a given initial condition. General existence and uniqueness theorems are discussed in more advanced treatments of differential equations.

Although it is useful to find general solutions, in applications we are usually interested in the solution that describes a particular physical situation. The general solution to a first-order equation generally depends on one arbitrary constant, so we can pick out a particular solution $y(x)$ by specifying the value $y(x_0)$ for some fixed x_0 (Figure 3). This specification is called an **initial condition**. A differential equation together with an initial condition is called an **initial value problem.**

■ **EXAMPLE 2** **Initial Value Problem** Solve the initial value problem $y' = -ty$ with $y(0) = 3$.

Solution   First we find the general solution by separation of variables:

$$\frac{dy}{dt} = -ty$$

$$\frac{dy}{y} = -t\,dt$$

$$\int \frac{dy}{y} = -\int t\,dt$$

$$\ln|y| = -\frac{1}{2}t^2 + C$$

$$|y| = e^{-t^2/2+C} = e^C\,e^{-t^2/2}$$

Thus, $y = \pm e^C\,e^{-t^2/2}$. Since C is arbitrary, e^C represents an arbitrary *positive number* and $\pm e^C$ an arbitrary nonzero number. We replace $\pm e^C$ by C and write the general solution as

$$\boxed{y = Ce^{-t^2/2}} \qquad \boxed{3}$$

If we set $C = 0$ in Eq. (3), we obtain the solution $y = 0$. The separation of variables procedure did not directly yield this solution because we divided by y (and thus assumed implicitly that $y \neq 0$).

Now use the initial condition $y(0) = Ce^{-0^2/2} = 3$. Thus, $C = 3$ and $y = 3e^{-t^2/2}$ is the solution to the initial value problem (Figure 3).   ∎

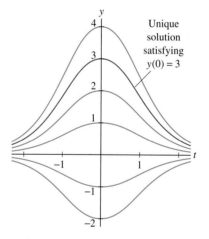

FIGURE 3   The initial condition $y(0) = 3$ determines one curve in the family of solutions to $y' = -ty$.

In the context of differential equations, the term "modeling" means finding a differential equation that describes a given physical situation. As an example, we consider water leaking through a hole at the bottom of a tank (Figure 4). The problem is to find the water level $y(t)$ at time t.

We solve this problem by showing that $y(t)$ satisfies a differential equation. The key observation is that the water lost during the interval from t to $t + \Delta t$ can be computed in two ways. Let

$$v(y) = \text{velocity of the water flowing through the hole}$$
$$\text{when the tank is filled to height } y$$

$$B = \text{area of the hole}$$

$$A(y) = \text{area of horizontal cross section of the tank at height } y$$

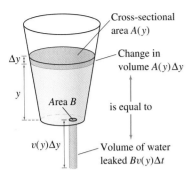

FIGURE 4 Water leaks out of a tank through a hole of area B at the bottom.

Like most if not all mathematical models, our model of water draining from a tank is at best an approximation. The differential equation (4) does not take into account viscosity (resistance of a fluid to flow). This can be remedied by using the differential equation

$$\frac{dy}{dt} = k\frac{Bv(y)}{A(y)}$$

where $k < 1$ is a viscosity constant. Furthermore, Torricelli's Law is valid only when the hole size B is small relative to the cross-sectional areas $A(y)$.

First we observe that the water exiting through the hole during a time interval Δt forms a cylinder of base B and height $v(y)\,\Delta t$ (because the water travels a distance $v(y)\,\Delta t$—see Figure 4). The volume of this cylinder is approximately $Bv(y)\,\Delta t$ [approximately but not exactly, because $v(y)$ may not be constant]. Thus,

$$\text{Water lost between } t \text{ and } t + \Delta t \approx Bv(y)\,\Delta t$$

Second, we note that if the water level drops by an amount Δy during the interval Δt, then the volume of water lost is approximately $A(y)\,\Delta y$ (Figure 4). Therefore,

$$\text{Water lost between } t \text{ and } t + \Delta t \approx A(y)\,\Delta y$$

This is also an approximation because the cross-sectional area may not be constant. Comparing the two results, we obtain $A(y)\Delta y \approx Bv(y)\,\Delta t$, or

$$\frac{\Delta y}{\Delta t} \approx \frac{Bv(y)}{A(y)}$$

Now take the limit as $\Delta t \to 0$ to obtain the differential equation

$$\boxed{\frac{dy}{dt} = \frac{Bv(y)}{A(y)}} \qquad \boxed{4}$$

To use Eq. (4), we need to know the velocity of the water leaving the hole. This is given by **Torricelli's Law**, which states that

$$v(y) = -\sqrt{2gy} = -\sqrt{2(32)y} = -8\sqrt{y} \text{ ft/s} \qquad \boxed{5}$$

where $g = 32$ ft/s^2 is the acceleration due to gravity on the earth's surface.

■ **EXAMPLE 3** Application of Torricelli's Law   A cylindrical tank of height 9 ft and radius 2 ft is filled with water. Water drains through a square hole of side 1 inch in the bottom. Determine the water level $y(t)$ at time t (seconds). How long does it take for the tank to go from full to empty?

Solution The horizontal cross section of the cylinder is a circle of radius $r = 2$ ft and area $A(y) = \pi r^2 = 4\pi$ ft^2 (Figure 5). The hole is a square of side $\frac{1}{12}$ ft and area $B = \frac{1}{144}$ ft^2. By Torricelli's Law [Eq. (5)], $v(y) = -8\sqrt{y}$, so Eq. (4) becomes

$$\frac{dy}{dt} = \frac{Bv(y)}{A(y)} = -\frac{\frac{1}{144}(8\sqrt{y})}{4\pi} = -\frac{\sqrt{y}}{72\pi} \qquad \boxed{6}$$

We solve this equation using separation of variables:

$$\int \frac{dy}{\sqrt{y}} = -\int \frac{dt}{72\pi}$$

$$2y^{1/2} = -\frac{1}{72\pi}t + C \qquad \boxed{7}$$

$$y = \left(-\frac{1}{144\pi}t + \frac{1}{2}C\right)^2$$

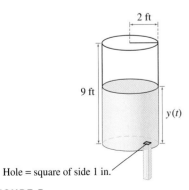

2 ft

9 ft

$y(t)$

Hole = square of side 1 in.

FIGURE 5

Since C is arbitrary, we may replace $\frac{1}{2}C$ by C and write the general solution as

$$y = \left(-\frac{t}{144\pi} + C\right)^2$$

The tank is full at $t = 0$ so we have the initial condition $y(0) = 9$. Thus

$$y(0) = \left(-\frac{0}{144\pi} + C\right)^2 = C^2 = 9 \quad \Rightarrow \quad C = \pm 3$$

Which sign is correct? You might think that both sign choices are possible, but notice that the *positive* square root $y^{1/2}$ appears on the left in Eq. (7). Therefore, the right-hand side of Eq. (7) must be nonnegative for $t \geq 0$ and this occurs only if $C \geq 0$. Thus, $C = 3$ is the only possible choice (Figure 6):

$$y(t) = \left(3 - \frac{t}{144\pi}\right)^2$$

Suppose that the tank empties at time t_e. Then $y(t_e) = 0$ and we may solve

$$y(t_e) = \left(3 - \frac{t_e}{144\pi}\right)^2 = 0 \quad \Rightarrow \quad t_e = (3)144\pi \approx 1{,}357 \text{ s}$$

Thus, the tank is empty after $t_e = 1{,}357$ s, or approximately 23 min.   ∎

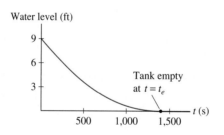

(A) Water level $y(t) = \left(3 - \frac{t}{144\pi}\right)^2$ $(C = 3)$

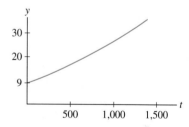

(B) Graph of $y(t) = \left(3 + \frac{t}{144\pi}\right)^2$ $(C = -3)$

FIGURE 6

CONCEPTUAL INSIGHT The previous example highlights the need to check and analyze solutions to a differential equation rather than rely on algebra alone. The algebra seemed to suggest that $C = \pm 3$, but further analysis of the differential equation showed that $C = -3$ does not yield a solution for $t \geq 0$. Furthermore, our formula for the solution is valid only for $t \leq t_e$. The function

$$y(t) = \left(3 - \frac{t}{144\pi}\right)^2$$

does not satisfy Eq. (6) for $t > t_e$ because $y'(t)$ is positive for $t > t_e$, but by Eq. (6) the derivative of a solution is ≤ 0. In fact, the tank is empty for $t \geq t_e$ so the full solution, valid for all $t \geq 0$, is

$$
y(t) = \begin{cases} \left(3 - \dfrac{t}{144\pi}\right)^2 & \text{for } 0 \leq t \leq t_e \\ 0 & \text{for } t > t_e \end{cases}
$$

$$\boxed{8}$$

9.1 SUMMARY

- A differential equation has order n if $y^{(n)}$ is the highest-order derivative appearing in the equation.
- A differential equation is *linear* if it can be written as

$$
a_n(x)y^{(n)} + a_{n-1}(x)y^{(n-1)} + \cdots + a_1(x)y' + a_0(x)y = b(x)
$$

- A first-order equation is called *separable* if it can be written in the form

$$
\frac{dy}{dx} = f(x)g(y)
$$

To solve a separable equation, move all terms involving y to the left and all terms involving x to the right and integrate:

$$
\frac{dy}{g(y)} = f(x)\,dx
$$

$$
\int \frac{dy}{g(y)} = \int f(x)\,dx
$$

9.1 EXERCISES

Preliminary Questions

1. Determine the order of the following differential equations:

(a) $x^5 y' = 1$

(b) $(y')^3 + x = 1$

(c) $y''' + x^4 y' = 2$

2. Is $y' = \sin x$ a linear differential equation?

3. Give an example of a nonlinear differential equation of the form $y' = f(y)$.

4. Can a nonlinear differential equation be separable? If so, give an example.

5. Give an example of a linear, nonseparable differential equation.

Exercises

1. Which of the following differential equations are first-order?

(a) $y' = x^2$

(b) $y'' = y^2$

(c) $(y')^3 + yy' = \sin x$

(d) $x^2 y' - e^x y = \sin y$

(e) $y'' + 3y' = \dfrac{y}{x}$

(f) $yy' + x + y = 0$

2. Which of the equations in Exercise 1 are linear?

In Exercises 3–9, verify that the given function is a solution of the differential equation.

3. $y' + 8x = 0, \quad y = 4x^2$

4. $y' + 8y = 0, \quad y = 4e^{-8x}$

5. $yy' + 4x = 0, \quad y = \sqrt{12 - 4x^2}$

6. $y' + 4xy = 0, \quad y = 3e^{-2x^2}$

7. $(x^2 - 1)y' + xy = 0, \quad y = 4(x^2 - 1)^{-1/2}$

8. $y'' - 2xy' + 8y = 0, \quad y = 4x^4 - 12x^2 + 3$

9. $y'' - 2y' + 5y = 0, \quad y = e^x \sin 2x$

10. Which of the following equations are separable? Write those that are separable in the form $y' = f(x)g(y)$ (but do not solve).

(a) $xy' + y^2 = 0$

(b) $\sqrt{1 - x^2}\,y' = e^y \sin x$

(c) $y' = x^2 + y^2$

(d) $y' = 9 - y^2$

11. Consider the differential equation $y' = e^y \cos x$.

(a) Write it as $e^{-y}\,dy = \cos x\,dx$.

(b) Integrate both sides of $\int e^{-y}\,dy = \int \cos x\,dx$.

(c) Show that $y = -\ln|C - \sin x|$ is the general solution.

(d) Find the particular solution satisfying $y(0) = 0$.

12. Verify that $x^2 y' + e^y = 0$ is separable by rewriting it in the form $y' = f(x)g(y)$.

(a) Show that $e^{-y}\,dy = -x^{-2}\,dx$.

(b) Show that $e^{-y} = x^{-1} + C$.

(c) Verify that $y = -\ln(C - x^{-1})$ is the general solution.

In Exercises 13–28, solve using separation of variables.

13. $y' = xy^2$

14. $y' = \dfrac{1}{2}xy$

15. $y' = 9y$

16. $y' = 2(4 - y)$

17. $2\dfrac{dy}{dx} + 6y + 4 = 0$

18. $\dfrac{dy}{dt} = 2\sqrt{y}$

19. $\dfrac{dy}{dt} - te^y = 0$

20. $\sqrt{1 - x^2}\,y' = xy$

21. $y' = y^2(1 - x^2)$

22. $yy' = x$

23. $(t^2 + 1)\dfrac{dx}{dt} = x^2 + 1$

24. $(1 + x^2)y' = x^3 y$

25. $y' = x \sec y$

26. $\dfrac{dy}{dt} = \tan y$

27. $\dfrac{dy}{dt} = y \tan t$

28. $\dfrac{dx}{dt} = t \tan x$

In Exercises 29–41, solve the initial value problem.

29. $y' + 2y = 0, \quad y(\ln 2) = 3$

30. $y' - 2y + 4 = 0, \quad y(1) = 4$

31. $yy' = xe^{-y^2}, \quad y(0) = -1$

32. $y^2\dfrac{dy}{dx} = x^{-3}, \quad y(2) = 0$

33. $y' = (x - 1)(y - 2), \quad y(0) = 3$

34. $(1 - t)\dfrac{dy}{dt} - y = 0, \quad y(2) = -4$

35. $\dfrac{dy}{dt} = ye^{-t}, \quad y(0) = 1$

36. $\dfrac{dy}{dt} = te^{-y}, \quad y(1) = 0$

37. $t^2\dfrac{dy}{dt} - t = 1 + y + ty, \quad y(1) = 0$

38. $\sqrt{1 - x^2}\,y' = y^2 + 1, \quad y(0) = 0$

39. $\sqrt{1 - x^2}\,y' = y^2, \quad y(0) = 1$

40. $y' = \tan y, \quad y(\ln 2) = \dfrac{\pi}{2}$

41. $y' = y^2 \sin x, \quad y(0) = 3$

42. Find all values of a such that $y = x^a$ is a solution of

$$y'' - 6x^{-2}y = 0$$

43. Find all values of a such that $y = e^{ax}$ is a solution of

$$y'' + 2y' - 8y = 0$$

44. Show that if $y(t)$ is a solution of $t(y - 1)y' = 2y$, then $t^2 y = Ce^y$ for some constant C [we cannot solve for $y(t)$ explicitly]. Find t such that $y(t) = 2$, assuming that $y(1) = 1$.

In Exercises 45–48, use Eq. (4) and Torricelli's Law [Eq. (5)].

45. A cylindrical tank filled with water has height 10 ft and a base of area 30 ft^2. Water leaks through a hole in the bottom of area $\frac{1}{3}$ ft^2. How long does it take (a) for half of the water to leak out and (b) for the tank to empty?

46. A conical tank filled with water has height 12 ft [Figure 7(A)]. Assume that the top is a circle of radius 4 ft and that water leaks through a hole in the bottom of area 2 in.^{2}. Let $y(t)$ be the water level at time t.

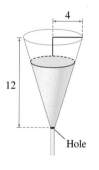
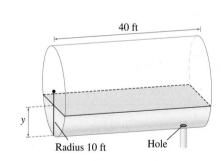

(A) Conical tank   (B) Horizontal tank

FIGURE 7

(a) Show that the cross-sectional area of the tank at height y is $A(y) = \frac{\pi}{9}y^2$.

(b) Find the differential equation satisfied by $y(t)$ and solve for $y(t)$. Use the initial condition $y(0) = 12$.

(c) How long does it take for the tank to empty?

47. The tank in Figure 7(B) is a cylinder of radius 10 ft and length 40 ft. Assume that the tank is half-filled with water and that water leaks through a hole in the bottom of area $B = 3$ in.^{2}. Determine the water level $y(t)$ and the time t_e when the tank is empty.

48. 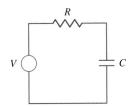 A cylindrical tank filled with water has height h ft and a base of area A ft^2. Water leaks through a hole in the bottom of area B ft^2.

(a) Show that the time required for the tank to empty is proportional to $A\sqrt{h}/B$.

(b) Show that the emptying time is proportional to $Vh^{-1/2}$, where V is the volume of the tank.

(c) Two tanks have the same volume and same-sized hole, but different heights and bases. Which tank empties first: the taller or shorter tank?

49. Figure 8 shows a circuit consisting of a resistor of R ohms, a capacitor of C farads, and a battery of voltage V. When the circuit is completed, the amount of charge $q(t)$ (in coulombs) on the plates of the capacitor varies according to the differential equation (t in seconds)

$$R\frac{dq}{dt} + \frac{1}{C}q = V$$

(a) Solve for $q(t)$.

(b) Show that $\lim\limits_{t \to \infty} q(t) = CV$.

(c) Find $q(t)$, assuming that $q(0) = 0$. Show that the capacitor charges to approximately 63% of its final value CV after a time period of length $\tau = RC$ (τ is called the time constant of the capacitor).

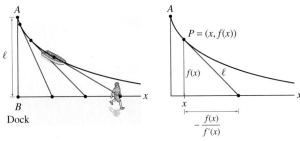

FIGURE 8 An RC circuit.

50. Assume in the circuit of Figure 8 that $R = 100\ \Omega$, $C = 0.01$ F, and $V = 10$ V. How many seconds does it take for the charge on the capacitor plates to reach half of its limiting value?

51. 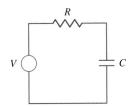 One hypothesis for the growth rate of the volume V of a cell is that $\dfrac{dV}{dt}$ is proportional to the cell's surface area A. Since V has cubic units such as cm^3 and A has square units such as cm^2, we may assume roughly that $A \propto V^{2/3}$, and hence $\dfrac{dV}{dt} = kV^{2/3}$ for some constant k. If this hypothesis is correct, which dependence of volume on time would we expect to see (again, roughly speaking) in the laboratory?

(a) Linear   **(b)** Quadratic   **(c)** Cubic

52. We might also guess that the rate at which a snowball melts is proportional to its surface area. What is the differential equation satisfied by the volume V of a spherical snowball at time t? Suppose the snowball has radius 4 cm and that it loses half of its volume after 10 min. According to this model, when will the snowball disappear?

53. In general, $(fg)'$ is not equal to $f'g'$, but let $f(x) = e^{2x}$ and find a function $g(x)$ such that $(fg)' = f'g'$. Do the same for $f(x) = x$.

54. A boy standing at point B on a dock holds a rope of length ℓ attached to a boat at point A [Figure 9(A)]. As the boy walks along the dock, holding the rope taut, the boat moves along a curve called a **tractrix** (from the Latin *tractus* meaning "to pull"). The segment from a point P on the curve to the x-axis along the tangent line has constant length ℓ. Let $y = f(x)$ be the equation of the tractrix.

(a) Show that $y^2 + (y/y')^2 = \ell^2$ and conclude $y' = -\dfrac{y}{\sqrt{\ell^2 - y^2}}$. Why must we choose the negative square root?

(b) Prove that the tractrix is the graph of

$$x = \ell \ln\left(\frac{\ell + \sqrt{\ell^2 - y^2}}{y}\right) - \sqrt{\ell^2 - y^2}$$

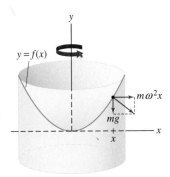

FIGURE 9

55. If a bucket of water spins about a vertical axis with constant angular velocity ω (in radians per second), the water climbs up the side of the bucket until it reaches an equilibrium position (Figure 10). Two forces act on a particle located at a distance x from the vertical axis: the

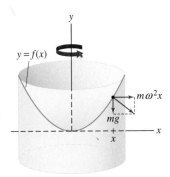

FIGURE 10

gravitational force $-mg$ acting downward and the force of the bucket on the particle (transmitted indirectly through the liquid) in the direction perpendicular to the surface of the water. These two forces must combine to supply a centripetal force $m\omega^2 x$, and this occurs if the diagonal of the rectangle in Figure 10 is normal to the water's surface (that is, perpendicular to the tangent line). Prove that if $y = f(x)$ is the equation of the curve obtained by taking a vertical cross section through the axis, then $-1/y' = -g/(\omega^2 x)$. Show that $y = f(x)$ is a parabola.

56. Show that the differential equations $y' = \dfrac{3y}{x}$ and $y' = -\dfrac{x}{3y}$ define **orthogonal families** of curves; that is, the graphs of solutions to the first equation intersect the graphs of the solutions to the second equation in right angles (Figure 11). Find these curves explicitly.

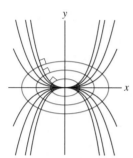

FIGURE 11 Two orthogonal families of curves.

57. Find the family of curves satisfying $y' = x/y$ and sketch several members of the family. Then find the differential equation for the orthogonal family (see Exercise 56), find its general solution, and add some members of this orthogonal family to your plot.

Further Insights and Challenges

60. In Section 6.2, we computed the volume V of a solid as the integral of cross-sectional area. Explain this formula in terms of differential equations. Let $V(y)$ be the volume of the solid up to height y and let $A(y)$ be the cross-sectional area at height y as in Figure 12.

(a) Explain the following approximation for small Δy:

$$V(y + \Delta y) - V(y) \approx A(y)\,\Delta y \qquad \boxed{9}$$

(b) Use Eq. (9) to justify the differential equation $\dfrac{dV}{dy} = A(y)$. Then derive the formula $V = \displaystyle\int_a^b A(y)\,dy$.

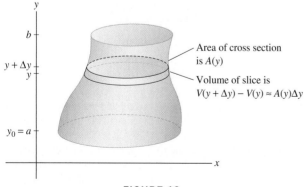

FIGURE 12

58. A 50-kg model rocket lifts off by expelling fuel at a rate of $k = 4.75$ kg/s for 10 s. The fuel leaves the end of the rocket with an exhaust velocity of $b = 100$ m/s. Let $m(t)$ be the mass of the rocket at time t. From the law of conservation of momentum, we find the following differential equation for the rocket's velocity $v(t)$ (in meters per second):

$$m(t)v'(t) = -9.8m(t) + b\frac{dm}{dt}$$

(a) Show that $m(t) = 50 - 4.75t$ kg.

(b) Solve for $v(t)$ and compute the rocket's velocity at rocket burnout (after 10 s).

59. Let $v(t)$ be the velocity of an object of mass m in free fall near the earth's surface. If we assume that air resistance is proportional to v^2, then v satisfies the differential equation $m\dfrac{dv}{dt} = -g + kv^2$ for some constant $k > 0$.

(a) Set $\alpha = (g/k)^{1/2}$ and rewrite the differential equation as

$$\frac{dv}{dt} = -\frac{k}{m}(\alpha^2 - v^2)$$

Then solve using separation of variables with initial condition $v(0) = 0$.

(b) Show that the terminal velocity $\lim\limits_{t\to\infty} v(t)$ is equal to $-\alpha$.

61. In most cases of interest, the general solution of a differential equation of order n depends on n arbitrary constants. This exercise shows there are exceptions.

(a) Show that $(y')^2 + y^2 = 0$ is a first-order equation with only one solution $y = 0$.

(b) Show that $(y')^2 + y^2 + 1 = 0$ is a first-order equation with no solutions.

62. (a) Let $P(x) = x^2 + ax + b$ with a, b constants. Show that $y = Ce^{\lambda x}$ (for any constant C) is a solution of $y'' + ay' + by = 0$ if and only if $P(\lambda) = 0$.

(b) Show that $y = C_1 e^{3x} + C_2 e^{-x}$ is a solution of $y'' - 2y' - 3y = 0$ for any constants C_1, C_2.

63. A spherical tank of radius R is half-filled with water. Suppose that water leaks through a hole in the bottom of area B ft^2. Let $y(t)$ be the water level at time t (seconds).

(a) Show that $\dfrac{dy}{dt} = \dfrac{-8B\sqrt{y}}{\pi(Ry - y^2)}$.

(b) Show that for some constant C,

$$\frac{\pi}{8B}\left(\frac{2}{3}Ry^{3/2} - \frac{2}{5}y^{5/2}\right) = C - t$$

(c) Use the initial condition $y(0) = R$ to compute C and show that $C = t_e$, the time at which the tank is empty.

(d) Show that t_e is proportional to $R^{5/2}$ and inversely proportional to B.

9.2 Models Involving $y' = k(y - b)$

We have seen that a quantity grows or decays exponentially if its *rate of change* is proportional to the amount present. This characteristic property is expressed by the differential equation $y' = ky$. We now study a closely related differential equation:

$$\frac{dy}{dt} = k(y - b) \qquad \boxed{1}$$

where k and b are constants and $k \neq 0$. This differential equation describes a quantity y whose *rate of change is proportional to the size of the difference $y - b$*. The general solution is

$$y(t) = b + Ce^{kt} \qquad \boxed{2}$$

where C is a constant determined by the initial condition. To verify this formula, we apply separation of variables to Eq. (1). Alternatively, we may observe that $(y - b)' = y'$ since b is a constant, so Eq. (1) may be rewritten

$$\frac{d}{dt}(y - b) = k(y - b)$$

In other words, $y - b$ satisfies the differential equation of an exponential function. Thus, $y - b = Ce^{kt}$ or $y = b + Ce^{kt}$, as claimed.

> **GRAPHICAL INSIGHT**   The behavior of the solution $y(t)$ depends on whether C and k are positive or negative. When $k < 0$, we usually rewrite the differential equation as $y' = -k(y - b)$ with $k > 0$. In this case, $y(t) = b + Ce^{-kt}$ and $y(t)$ approaches the horizontal asymptote $y = b$ since Ce^{-kt} tends to zero as $t \to \infty$ (Figure 1). However, $y(t)$ approaches the asymptote from above or below, depending on whether $C > 0$ or $C < 0$.

Every first-order, linear differential equation with constant coefficients may be written in the form of Eq. (1). This equation is used to model a variety of phenomena such as the cooling process, free-fall with air resistance, and current in a circuit.

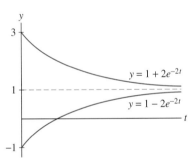

$y = 1 + 2e^{-2t}$

$y = 1 - 2e^{-2t}$

FIGURE 1 Two solutions to $y' = -2(y - 1)$ corresponding to $C = 2$ and $C = -2$.

We now consider several applications of Eq. (1), beginning with Newton's Law of Cooling. Let $y(t)$ be the temperature of a hot object cooling down in an environment where the ambient temperature is T_0. Newton assumed that the *rate of cooling* is proportional to the temperature difference $y - T_0$. We express this hypothesis in a precise way by the differential equation

$$y' = -k(y - T_0)$$

The constant k [in units of $(\text{time})^{-1}$] is called the **cooling constant** and depends on the physical properties of the object.

Newton's Law of Cooling implies that the object cools quickly when it is very hot (when $y - T_0$ is large). The rate of cooling slows as y approaches T_0. When the object's initial temperature is less than T_0, y' is positive and Newton's Law models warming.

■ **EXAMPLE 1**   **Newton's Law of Cooling**   At $t = 0$, we submerge a hot metal bar with cooling constant $k = 2.1 \text{ min}^{-1}$ in a large tank of water at temperature $T_0 = 55°F$. Let $y(t)$ be the bar's temperature at time t.

(a) Find the differential equation satisfied by $y(t)$ and determine its general solution.

(b) What is the bar's temperature after 1 min if its initial temperature is $400°F$?

(c) What was the bar's initial temperature if, after half a minute, it had cooled to $120°F$?

Solution

(a) Since $k = 2.1 \text{ min}^{-1}$, $y(t)$ (with t in minutes) satisfies

$$y' = -2.1(y - 55)$$

By Eq. (2), the general solution is $y(t) = 55 + Ce^{-2.1t}$ for some constant C.

(b) If the initial temperature is 400°F, then $y(0) = 55 + C = 400$. Thus, $C = 345$ and $y(t) = 55 + 345e^{-2.1t}$ (Figure 2). After 1 min,

$$y(1) = 55 + 345e^{-2.1(1)} \approx 97.2°\text{F}$$

(c) If the temperature after half a minute is 120°F, then $y(t) = 55 + Ce^{-2.1t}$ where $y(0.5) = 120$. We solve for C:

$$55 + Ce^{-2.1(0.5)} = 120$$

$$Ce^{-1.05} = 65 \quad \Rightarrow \quad C = 65e^{1.05} \approx 185.7$$

It follows that $y(t) = 55 + 185.7e^{-2.1t}$ and the initial temperature is

$$y(0) = 55 + 185.7e^{-2.1(0)} = 55 + 185.7 = 240.7°\text{F}$$ ■

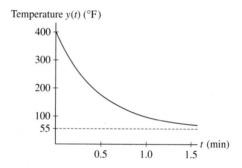

Temperature $y(t)$ (°F)

FIGURE 2 Temperature of metal bar as it cools.

The effect of air resistance depends on the physical situation. A high-speed bullet is affected differently than a skydiver. Our model is fairly realistic for a large object such as a skydiver falling from high altitudes.

The differential equation $y' = k(y - b)$ is also used to model free fall when air resistance is taken into account. Assume that the force due to air resistance is proportional to the velocity v and acts opposite to the direction of the fall. We write this force as $-kv$ where $k > 0$. The force due to gravity on a falling object of mass m is $-mg$, where g is the acceleration due to gravity, so the total force is $F = -mg - kv$. By Newton's Law, $F = ma = mv'$ ($a = v'$ is the acceleration). Thus $mv' = -mg - kv$, which can be written

In this model of free fall, k has units of mass per time, such as kg/s.

$$v' = -\frac{k}{m}\left(v + \frac{gm}{k}\right)$$

This equation has the form $v' = -k(v - b)$ with k/m instead of k and $b = -gm/k$. By Eq. (2) the general solution is

$$v(t) = -\frac{gm}{k} + Ce^{-(k/m)t} \qquad \boxed{3}$$

Since $Ce^{-(k/m)t}$ tends to zero as $t \to \infty$, $v(t)$ tends to a limiting terminal velocity:

$$\text{Terminal velocity} = \lim_{t \to \infty} v(t) = -\frac{gm}{k} \qquad \boxed{4}$$

Without air resistance the velocity would increase indefinitely.

■ **EXAMPLE 2** An 80-kg skydiver jumps out of an airplane (with zero initial velocity).

(a) What is his terminal velocity if $k = 8$ kg/s?

(b) What is his velocity after 30 s?

Solution

(a) By Eq. (4), with $k = 8$ kg/s and $g = 9.8$ m/s^2, the terminal velocity is

$$-\frac{gm}{k} = -\frac{9.8(80)}{8} = -98 \text{ m/s}$$

(b) By Eq. (3),

$$v(t) = -98 + Ce^{-(k/m)t} = -98 + Ce^{-(8/80)t} = -98 + Ce^{-0.1t}$$

Since the initial velocity is zero, $v(0) = -98 + C = 0$ or $C = 98$. Therefore, $v(t) = -98(1 - e^{-0.1t})$ (Figure 3). The skydiver's velocity after 30 s is

$$v(30) = -98(1 - e^{-0.1(30)}) \approx -93.1 \text{ m/s} \qquad ■$$

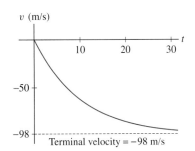

FIGURE 3 Velocity of 80-kg skydiver in free fall with air resistance ($k = 8$).

An **annuity** is an investment in which a principal P_0 is placed in an account earning interest at a rate r and money is withdrawn at regular intervals. To model an annuity by a differential equation, we assume that the money is withdrawn continuously at a rate of N dollars/year. The balance $P(t)$ in the annuity at time t satisfies

$$\underbrace{P'(t)}_{\substack{\text{Rate of} \\ \text{change}}} = \underbrace{rP(t)}_{\substack{\text{Growth due} \\ \text{to interest}}} - \underbrace{N}_{\substack{\text{Withdrawal} \\ \text{rate}}} = r\left(P(t) - \frac{N}{r}\right) \qquad \boxed{5}$$

Notice in Eq. (5) that $P'(t)$ is determined by the growth rate and the withdrawal rate. If no withdrawals occurred, $P(t)$ would grow with compound interest and would satisfy $P'(t) = rP(t)$.

This equation has the form $y' = k(y - b)$ with $k = r$ and $b = N/r$, so by Eq. (2), the general solution is

$$P(t) = \frac{N}{r} + Ce^{rt} \qquad \boxed{6}$$

Since e^{rt} tends to infinity as $t \to \infty$, the balance $P(t)$ tends to ∞ if $C > 0$. If $C < 0$, then $P(t)$ tends to $-\infty$ (i.e., the annuity eventually runs out of money).

■ **EXAMPLE 3** **Does an Annuity Pay Out Forever?** Let $P(t)$ be the balance in an annuity earning interest at the rate $r = 0.07$, with withdrawals made continuously at a rate of $N = \$500$/year.

(a) Assume that $P(0) = \$5,000$. Find $P(t)$ and determine when the annuity runs out of money.

(b) Assume that $P(0) = \$9,000$. Find $P(t)$ and show that the balance increases indefinitely.

Solution We have $\dfrac{N}{r} = \dfrac{500}{0.07} \approx 7,143$, so $P(t) = 7,143 + Ce^{0.07t}$ by Eq. (6).

(a) If $P(0) = 5,000 = 7,143 + Ce^0$, then $C = -2,143$ and

$$P(t) = 7,143 - 2,143e^{0.07t}$$

The account runs out of money when $P(t) = 7,143 - 2,143e^{0.07t} = 0$. We solve for t:

$$e^{0.07t} = \frac{7,143}{2,143} \quad \Longrightarrow \quad 0.07t = \ln\left(\frac{7,143}{2,143}\right) \approx 1.2$$

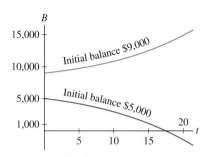

FIGURE 4 The balance in an annuity may increase indefinitely or decrease to zero (eventually becoming negative), depending on the size of initial deposit P_0.

The money runs out at time $t = \dfrac{1.2}{0.07} \approx 17$ years.

(b) If $P(0) = 9,000 = 7,143 + Ce^0$, then $C = 1,857$ and

$$P(t) = 7,143 + 1,857e^{0.07t}$$

Since the coefficient $C = 1,857$ is positive, the account never runs out of money. In fact, $P(t)$ increases indefinitely as $t \to \infty$. Figure 4 illustrates the two cases. ∎

9.2 SUMMARY

- The general solution of $y' = k(y - b)$ is $y = b + Ce^{kt}$, where C is a constant.

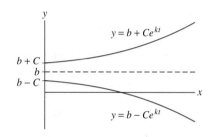
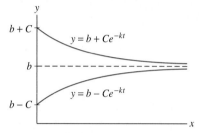

Solutions to $y' = k(y - b)$ with $k, C > 0$ Solutions to $y' = -k(y - b)$ with $k, C > 0$

FIGURE 5

- The following tables describe the solutions to $y' = k(y - b)$ (see Figure 5):

Equation ($k > 0$)	Solution	Behavior as $t \to \infty$
$y' = k(y - b)$	$y(t) = b + Ce^{kt}$	$\displaystyle\lim_{t\to\infty} y(t) = \begin{cases} \infty & \text{if } C > 0 \\ -\infty & \text{if } C < 0 \end{cases}$
$y' = -k(y - b)$	$y(t) = b + Ce^{-kt}$	$\displaystyle\lim_{t\to\infty} y(t) = b$

- The equation $P' = r(P - N/r)$ models the balance $P(t)$ of a continuous annuity with interest rate r and withdrawal rate N.

9.2 EXERCISES

Preliminary Questions

1. What is the general solution to $y' = -k(y - b)$?

2. Write down a solution to $y' = 4(y - 5)$ that tends to $-\infty$ as $t \to \infty$.

3. Does there exist a solution of $y' = -4(y - 5)$ that tends to ∞ as $t \to \infty$?

4. True or false? If $k > 0$, then all solutions of $y' = -k(y - b)$ approach the same limit as $t \to \infty$.

5. Suppose that material A cools more rapidly than material B. Which material has a larger cooling constant k in Newton's Law of Cooling, $y' = -k(y - T_0)$?

Exercises

1. Find the general solution of $y' = 2(y - 10)$. Then find the two solutions satisfying $y(0) = 25$ and $y(0) = 5$, and sketch their graphs.

2. Find the general solution of $y' = -3(y - 12)$. Then find the two solutions satisfying $y(0) = 20$ and $y(0) = 0$, and sketch their graphs.

3. Verify directly that $y = b + Ce^{kt}$ satisfies $y' = k(y - b)$ for any constant C.

4. Let $F(t)$ be the temperature of a hot object submerged in a large pool of water whose temperature is $70°F$.
(a) What is the differential equation satisfied by $F(t)$ if the cooling constant is $k = 1.5$?
(b) Find a formula for $F(t)$ if the object's initial temperature is $250°F$.

5. A hot metal bar is submerged in a large reservoir of water whose temperature is $60°F$. The temperature of the bar 20 s after submersion is $100°F$. After 1 min, the temperature has cooled to $80°F$.
(a) Determine the cooling constant k.
(b) What is the differential equation satisfied by the temperature $F(t)$ of the bar?
(c) What is the formula for $F(t)$?
(d) Determine the temperature of the bar at the moment it is submerged.

6. A hot metal rod is placed in a water bath whose temperature is $40°F$. The rod cools from 300 to $200°F$ in 1 min. How long will it take for the rod to cool to $150°F$?

7. When a hot object is placed in a water bath whose temperature is $25°C$, it cools from 100 to $50°C$ in 150 s. In another bath, the same cooling occurs in 120 s. Find the temperature of the second bath.

8. A cold metal bar at $-30°C$ is submerged in a pool maintained at a temperature of $40°C$. Half a minute later, the temperature of the bar is $20°C$. How long will it take for the bar to attain a temperature of $30°C$?

9. A cup of coffee, cooling off in a room at temperature $20°C$, has cooling constant $k = 0.09 \text{ min}^{-1}$.
(a) How fast is the coffee cooling (in degrees per minute) when its temperature is $T = 80°C$?
(b) Use the Linear Approximation to estimate the change in temperature over the next 6 s when $T = 80°C$.
(c) The coffee is served at a temperature of $90°C$. How long should you wait before drinking it if the optimal temperature is $65°C$?

10. Two identical objects are heated to different temperatures T_1 and T_2. Both are submerged in a cold bath of temperature $T_0 = 40°C$ at $t = 0$. Measurements show that $T_2 = 400°C$ and the cooling rate of object 1 is twice as large as the cooling rate of object 2 (at each time t). Find T_1.

In Exercises 11–14, use the model for free-fall with air resistance discussed in this section. If the weight w of an object (in pounds) is given rather than its mass (as in Example 2), then the differential equation for free fall with air resistance is $v' = -(kg/w)(v + w/k)$, where $g = 32 \text{ ft/s}^2$.

11. A 60-kg skydiver jumps out of an airplane. What is her terminal velocity in miles per hour, assuming that $k = 10$ kg/s for free-fall (no parachute)?

12. Find the terminal velocity of a 192-lb skydiver if $k = 1.2$ lb-s/ft. How long does it take him to reach half of his terminal velocity if his initial velocity is zero?

13. A 175-lb skydiver jumps out of an airplane (with zero initial velocity). Assume that $k = 0.7$ lb-s/ft with a closed parachute and $k = 5$ lb-s/ft with an open parachute. What is the skydiver's velocity at $t = 25$ s if the parachute opens after 20 seconds of free fall?

14. Does a heavier or lighter skydiver reach terminal velocity faster?

15. A continuous annuity with withdrawal rate $N = \$1,000$/year and interest rate $r = 5\%$ is funded by an initial deposit P_0.
(a) When will the annuity run out of funds if $P_0 = \$15,000$?
(b) Which initial deposit P_0 yields a constant balance?

16. Show that a continuous annuity with withdrawal rate $N = \$5,000$/year and interest rate $r = 8\%$, funded by an initial deposit of $P_0 = \$75,000$, never runs out of money.

17. Find the minimum initial deposit that will allow an annuity to pay out $\$500$/year indefinitely if it earns interest at a rate of 5%.

18. What is the minimum initial deposit necessary to fund an annuity for 30 years if withdrawals are made at a rate of $\$2,000$/year at an interest rate of 7%?

19. An initial deposit of $\$5,000$ is placed in a bank account. What is the minimum interest rate the bank must pay to allow continuous withdrawals at a rate of $\$500$/year to continue indefinitely?

20. Show that a continuous annuity never runs out of money if the initial balance is greater than or equal to N/r, where N is the withdrawal rate and r the interest rate.

21. Sam borrows $\$10,000$ from a bank at an interest rate of 9% and pays back the loan continuously at a rate of N dollars/year. Let $P(t)$ denote the amount still owed at time t.
(a) Explain why $P(t)$ satisfies the differential equation

$$y' = 0.09y - N$$

(b) How long will it take Sam to pay back the loan if $N = \$1,200$?
(c) Will he ever be able to pay back the loan if $N = \$800$?

22. Let $N(t)$ be the fraction of the population who have heard a given piece of news t hours after its initial release. According to one model, the rate $N'(t)$ at which the news spreads is equal to k times the fraction of the population that has not yet heard the news, for some constant k.
(a) Determine the differential equation satisfied by $N(t)$.
(b) Find the solution of this differential equation with the initial condition $N(0) = 0$ in terms of k.

(c) Suppose that half of the population is aware of an earthquake 8 hours after it occurs. Use the model to calculate k and estimate the percentage that will know about the earthquake 12 hours after it occurs.

23. Current in a Circuit The electric current flowing in the circuit in Figure 6 (consisting of a battery of V volts, a resistor of R ohms, and an inductor) satisfies

$$\frac{dI}{dt} = -k(I - b)$$

for some constants k and b with $k > 0$. Initially, $I(0) = 0$ and $I(t)$ approaches a maximum level V/R as $t \to \infty$.

(a) What is the value of b?

(b) Find a formula for $I(t)$ in terms of k, V, and R.

(c) Show that $I(t)$ reaches approximately 63% of its maximum value at time $t = 1/k$.

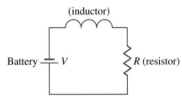

FIGURE 6 Current flow approaches the level $I_{\max} = V/R$.

Further Insights and Challenges

24. Show that the cooling constant of an object can be determined from two temperature readings $y(t_1)$ and $y(t_2)$ at times $t_1 \neq t_2$ by the formula

$$k = \frac{1}{t_1 - t_2} \ln\left(\frac{y(t_2) - T_0}{y(t_1) - T_0}\right)$$

25. Show that by Newton's Law of Cooling, the time required to cool an object from temperature A to temperature B is

$$t = \frac{1}{k} \ln\left(\frac{A - T_0}{B - T_0}\right)$$

where T_0 is the ambient temperature.

26. Air Resistance A projectile of mass $m = 1$ travels straight up from ground level with initial velocity v_0. Suppose that the velocity v satisfies $v' = -g - kv$.

(a) Find a formula for $v(t)$.

(b) Show that the projectile's height $h(t)$ is given by

$$h(t) = C(1 - e^{-kt}) - \frac{g}{k}t$$

where $C = k^{-2}(g + kv_0)$.

(c) Show that the projectile reaches its maximum height at time $t_{\max} = k^{-1} \ln(1 + kv_0/g)$.

(d) In the absence of air resistance, the maximum height is reached at time $t = v_0/g$. In view of this, explain why we should expect that

$$\lim_{k \to 0} \frac{\ln(1 + \frac{kv_0}{g})}{k} = \frac{v_0}{g} \qquad \boxed{7}$$

(e) Verify Eq. (7). *Hint:* Use Theorem 2 in Section 5.8 to show that

$$\lim_{k \to 0} \left(1 + \frac{kv_0}{g}\right)^{1/k} = e^{v_0/g} \text{ or use L'Hôpital's Rule.}$$

9.3 Graphical and Numerical Methods

"To imagine yourself subject to a differential equation, start somewhere. There you are tugged in some direction, so you move that way ... as you move, the tugging forces change, pulling you in a new direction; for your motion to solve the differential equation you must keep drifting with and responding to the ambient forces."

——From the introduction to *Differential Equations*, J. H. Hubbard and Beverly West, Springer-Verlag, New York, 1991

In the previous two sections, we focused on finding solutions to differential equations. However, most differential equations cannot be solved explicitly. Fortunately, there are techniques for analyzing the solutions that do not rely on explicit formulas. In this section, we discuss the method of slope fields, which provide us with a good visual understanding of first-order equations. We also discuss Euler's Method for finding numerical approximations to solutions.

We use t as the independent variable and write $\dot{y}$ for $\dfrac{dy}{dt}$. The notation $\dot{y}$, often used for time derivatives in physics and engineering, was introduced by Isaac Newton. A first-order differential equation can then be written in the form

$$\boxed{\dot{y} = F(t, y)} \qquad \boxed{1}$$

where $F(t, y)$ is a function of t and y. For example, $\dfrac{dy}{dt} = ty$ becomes $\dot{y} = ty$.

It is useful to think of Eq. (1) as a set of instructions that "tells a solution" which direction to go in. Thus, a solution passing through a point (t, y) is "instructed" to continue in the direction of slope $F(t, y)$. To visualize this set of instructions, we draw a **slope field**, which is an array of small segments of slope $F(t, y)$ at points (t, y) lying on a rectangular grid in the plane.

To illustrate, let's return to the differential equation:

$$\dot{y} = -ty$$

In this case, $F(t, y) = -ty$. According to Example 2 of Section 9.1, the general solution is $y = Ce^{-t^2/2}$. Figure 1(A) shows segments of slope $-ty$ at points (t, y) along the graph of a particular solution $y(t)$. This particular solution passes through $(-1, 3)$ and, according to the differential equation, $\dot{y}(-1) = -ty = -(-1)3 = 3$. Thus, the segment located at the point $(-1, 3)$ has slope 3. The graph of the solution is tangent to each segment [Figure 1(B)].

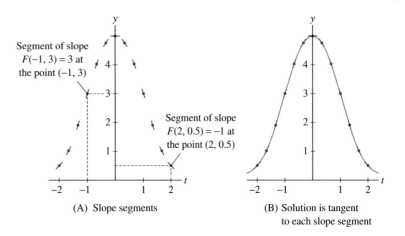

Segment of slope $F(-1, 3) = 3$ at the point $(-1, 3)$

Segment of slope $F(2, 0.5) = -1$ at the point $(2, 0.5)$

(A) Slope segments

(B) Solution is tangent to each slope segment

FIGURE 1 The solution of $\dot{y} = -ty$ satisfying $y(-1) = 3$.

To sketch the slope field for $\dot{y} = -ty$, we draw small segments of slope $-ty$ at an array of points in the plane, as in Figure 2(A). *The slope field allows us to visualize all*

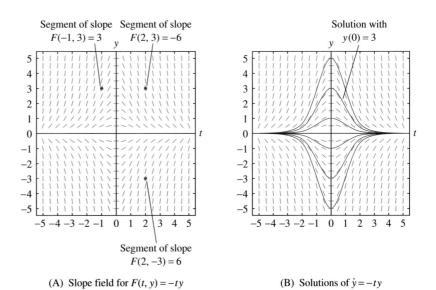

Segment of slope $F(-1, 3) = 3$ Segment of slope $F(2, 3) = -6$

Solution with $y(0) = 3$

Segment of slope $F(2, -3) = 6$

(A) Slope field for $F(t, y) = -ty$

(B) Solutions of $\dot{y} = -ty$

FIGURE 2 Slope field for $F(t, y) = -ty$.

of the solutions at a glance. Starting at any point, we may sketch a solution by drawing a curve that runs tangent to the slope segments at each point [Figure 2(B)]. The graph of a solution is also called an **integral curve**.

■ **EXAMPLE 1** Using Isoclines   Draw the slope field for $\dot{y} = y - t$ and sketch the integral curves satisfying $y(0) = 1$ and $y(1) = -2$.

Solution   A good way to sketch the slope field of $\dot{y} = F(t, y)$ is to choose several values c and identify the points (t, y) where the slope has the given value c, that is, where $F(t, y) = c$. The points where $F(t, y) = c$ form a curve called the **isocline** of slope c (Figure 3).

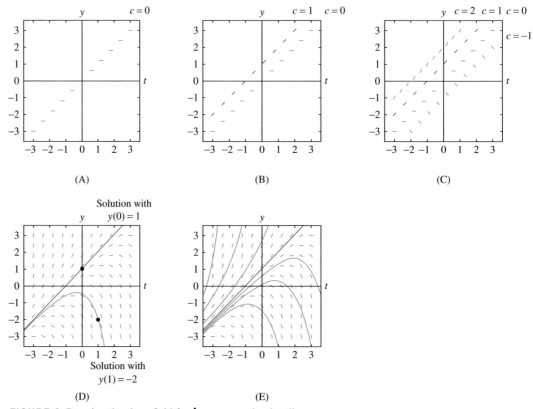

FIGURE 3 Drawing the slope field for $\dot{y} = y - t$ using isoclines.

In our case, $F(t, y) = y - t$, so the isocline of slope c has equation $y - t = c$ or $y = t + c$, which is a line. Consider the following values:

- $c = 0$: This isocline is $y - t = 0$ or $y = t$. We draw segments of slope $c = 0$ at points along the line $y = t$, as in Figure 3(A).
- $c = 1$: This isocline is $y - t = 1$ or $y = t + 1$. We draw segments of slope 1 at points along $y = t + 1$, as in Figure 3(B).
- $c = 2$: This isocline is $y - t = 2$ or $y = t + 2$. We draw segments of slope 2 at points along $y = t + 2$, as in Figure 3(C).
- $c = -1$: This isocline is $y - t = -1$ or $y = t - 1$ [Figure 3(C)].

A more detailed slope field is shown in Figure 3(D). To sketch the solution satisfying $y(0) = 1$, begin at the point $(t_0, y_0) = (0, 1)$ and draw the integral curve that follows the directions indicated by the slope field. Similarly, the graph of the solution satisfying $y(1) = -2$ is the integral curve obtained by starting at $(t_0, y_0) = (1, -2)$ and moving along the slope field. ■

> **GRAPHICAL INSIGHT**   Often, we are interested in the *asymptotic* behavior of solutions, that is, behavior as $t \to \infty$. The slope field in Figure 3(E) suggests that the asymptotic behavior depends on the initial value: If $y(0) > 1$, then $y(t)$ tends to ∞, and if $y(0) < 1$, then $y(t)$ tends to $-\infty$. We can check this using the general solution $y(t) = 1 + t + Ce^t$, where $y(0) = 1 + C$. If $y(0) > 1$, then $C > 0$ and $y(t)$ tends to ∞, but if $y(0) < 1$, then $C < 0$ and $y(t)$ tends to $-\infty$. The solution $y = 1 + t$ with initial condition $y(0) = 1$ is the straight line shown in Figure 3(D).

■ **EXAMPLE 2** Newton's Law of Cooling Revisited   The temperature $y(t)$ (°F) of an object placed in a refrigerator satisfies $\dot{y} = -3(y - 40)$. Draw the slope field and describe the behavior of the solutions.

Solution   The function $F(t, y) = -3(y - 40)$ depends only on y, so slopes of the segments in the slope field do not vary in the t-direction. The slope $F(t, y)$ is positive for $y < 40$ and negative for $y > 40$. Furthermore, the slope at height y is equal to $-3(y - 40) = -3y + 120$, so the segments grow steeper as y tends to ∞ or $-\infty$ (Figure 4).

The slope field shows that if the initial temperature satisfies $y_0 > 40$, then $y(t)$ decreases to 40 as $t \to \infty$. The object cools down to 40° when placed in the refrigerator. If $y_0 < 40$, then $y(t)$ increases to 40 as $t \to \infty$. In this case, the object warms up when placed in the refrigerator. If $y_0 = 40$, then y remains 40 for all time t. ■

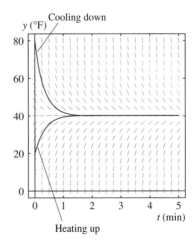

FIGURE 4 Slope field for $\dot{y} = -3(y - 40)$.

> **CONCEPTUAL INSIGHT**   As we pointed out above, most first-order equations arising in applications have a uniqueness property: There is precisely one solution $y(t)$ satisfying a given initial condition $y(t_0) = y_0$. Graphically, this means that precisely one integral

curve (solution) passes through the point (t_0, y_0). Thus, when uniqueness holds, distinct integral curves never cross as in Figure 3(E). Figure 5 shows the slope field of $\dot{y} = -\sqrt{|y|}$, where uniqueness fails. We can show that when a solution touches the t-axis, it can either remain on the t-axis or continue to the region below the t-axis. However, the slope field does not show this clearly, and we see again that it is necessary to analyze solutions rather than rely on visual impressions alone.

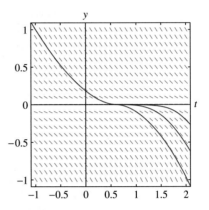

FIGURE 5 Overlapping integral curves for $\dot{y} = -\sqrt{|y|}$ (uniqueness fails for this differential equation).

Euler's Method

Euler's Method is the simplest method for solving initial value problems numerically, but it is not very efficient. More sophisticated versions are implemented on computer systems to plot and analyze solutions to the complex systems of differential equations arising in areas such as weather prediction, aerodynamic modeling, and economic forecasting.

Euler's Method produces numerical approximations to the solution of a first-order initial value problem:

$$\dot{y} = F(t, y), \qquad y(t_0) = y_0 \qquad \boxed{2}$$

We begin by choosing a small number h, called the **time step**, and consider the sequence of times spaced at intervals of size h:

$$t_0, \qquad t_1 = t_0 + h, \qquad t_2 = t_0 + 2h, \qquad t_3 = t_0 + 3h, \qquad \ldots$$

In general, $t_k = t_0 + kh$. Euler's Method consists of computing a sequence of values $y_1, y_2, y_3, \ldots, y_n$ successively using the formula

$$\boxed{y_k = y_{k-1} + hF(t_{k-1}, y_{k-1})} \qquad \boxed{3}$$

Starting with the initial value $y_0 = y(t_0)$, we compute $y_1 = y_0 + hF(t_0, y_0)$, etc. The value y_k is the Euler approximation to $y(t_k)$. We connect the points $P_k = (t_k, y_k)$ by segments to obtain an approximation to the graph of $y(t)$ (Figure 6).

GRAPHICAL INSIGHT The values y_k are defined so that the segment joining P_{k-1} to P_k has slope

$$\frac{y_k - y_{k-1}}{t_k - t_{k-1}} = \frac{(y_{k-1} + hF(t_{k-1}, y_{k-1})) - y_{k-1}}{h} = F(t_{k-1}, y_{k-1})$$

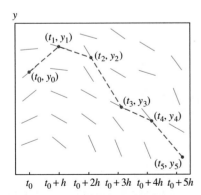

FIGURE 6 In Euler's Method, we move from one point to the next by traveling along the line indicated by the slope field.

Thus, in Euler's method we move from P_{k-1} to P_k by traveling in the direction specified by the slope field at P_{k-1} for a time interval of length h (Figure 6).

■ **EXAMPLE 3** Use Euler's Method with time step $h = 0.2$ and $n = 4$ steps to approximate the solution of $\dot{y} = y - t^2$, $y(0) = 3$.

Solution Our initial value at $t_0 = 0$ is $y_0 = 3$. Since $h = 0.2$, the time values are $t_1 = 0.2$, $t_2 = 0.4$, $t_3 = 0.6$, and $t_4 = 0.8$. We use Eq. (3) with $F(t, y) = y - t^2$ to calculate

$$y_1 = y_0 + hF(t_0, y_0) = 3 + 0.2(3 - (0)^2) = 3.6$$

$$y_2 = y_1 + hF(t_1, y_1) = 3.6 + 0.2(3.6 - (0.2)^2) \approx 4.3$$

$$y_3 = y_2 + hF(t_2, y_2) = 4.3 + 0.2(4.3 - (0.4)^2) \approx 5.14$$

$$y_4 = y_3 + hF(t_3, y_3) = 5.14 + 0.2(5.14 - (0.6)^2) \approx 6.1$$

Figure 7(A) shows the exact solution $y(t) = 2 + 2t + t^2 + e^t$ together with a plot of the points (t_k, y_k) for $k = 0, 1, 2, 3, 4$ connected by line segments.   ■

FIGURE 7 Euler's Method applied to $\dot{y} = y - t^2$, $y(0) = 3$.

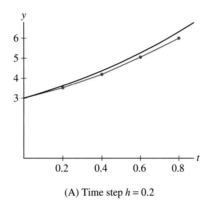

(A) Time step $h = 0.2$

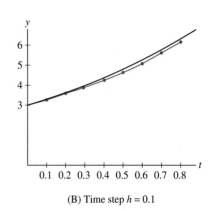

(B) Time step $h = 0.1$

CONCEPTUAL INSIGHT   Figure 7(B) shows that the timestep $h = 0.1$ gives a better approximation than $h = 0.2$. In general, the smaller the time step, the better the approximation. In fact, if we start at a point $(a, y(a))$ and use Euler's Method to approximate $(b, y(b))$ using N steps (i.e., $h = \dfrac{b - a}{N}$), then the error is roughly proportional to $1/N$ [provided that $F(t, y)$ is a well-behaved function]. This is similar to the error size in the Nth left- and right-endpoint approximations to an integral. What this means, however, is that Euler's Method is quite inefficient; to cut the error in half, it is necessary to double the number of steps and to achieve n-digit accuracy requires roughly 10^n steps. Fortunately, there are several methods that improve on Euler's Method in much the same way that the Midpoint Rule and Simpson's Rule improve on the endpoint approximations (see Exercises 21–26).

■ **EXAMPLE 4** *CAS*   Let $y(t)$ be the solution of $\dot{y} = \sin t \cos y$, $y(0) = 0$.

(a) Use Euler's Method with time step $h = 0.1$ to approximate $y(0.5)$.

(b) Use a computer algebra system to implement Euler's Method with time steps $h = 0.01, 0.001$, and 0.0001 to approximate $y(0.5)$.

Euler's Method:

$$y_k = y_{k-1} + hF(t_{k-1}, y_{k-1})$$

Solution

(a) When $h = 0.1$, y_k is an approximation to $y(0 + k(0.1)) = y(0.1k)$. So y_5 is an approximation to $y(0.5)$. It is convenient to organize calculations in the following table. Note that the value y_{k+1} computed in the last column of each line is used in the next line to continue the process.

t_k	y_k	$F(t_k, y_k) = \sin t_k \cos y_k$	$y_{k+1} = y_k + hF(t_k, y_k)$
$t_0 = 0$	$y_0 = 0$	$(\sin 0) \cos 0 = 0$	$y_1 = 0 + 0.1(0) = 0$
$t_1 = 0.1$	$y_1 = 0$	$(\sin 0.1) \cos 0 \approx 0.1$	$y_2 \approx 0 + 0.1(0.1) = 0.01$
$t_2 = 0.2$	$y_2 \approx 0.01$	$(\sin 0.2) \cos(0.01) \approx 0.2$	$y_3 \approx 0.01 + 0.1(0.2) = 0.03$
$t_3 = 0.3$	$y_3 \approx 0.03$	$(\sin 0.3) \cos(0.03) \approx 0.3$	$y_4 \approx 0.03 + 0.1(0.3) = 0.06$
$t_4 = 0.4$	$y_4 \approx 0.06$	$(\sin 0.4) \cos(0.06) \approx 0.4$	$y_5 \approx 0.06 + 0.1(0.4) = 0.10$

Thus, Euler's Method yields the approximation $y(0.5) \approx y_5 \approx 0.1$.

A typical CAS command to implement Euler's Method with time step $h = 0.01$ reads:

```
>> For[n = 0; y = 0, n < 50, n++,
>> y = y + (.01) * (Sin[.01 * n] * Cos[y])]
>> y
>> 0.119746
```

The command For[...] updates the variable y successively through the values $y_1, y_2, \ldots, y_{50}$ according to Euler's Method.

(b) When the number of steps is large, the calculations are too lengthy to do by hand, but they are easily carried out using a CAS. Note that for $h = 0.01$, the kth value y_k is an approximation to $y(0 + k(0.01)) = y(0.01k)$, and y_{50} gives an approximation to $y(0.5)$. Similarly, when $h = 0.001$, y_{500} is an approximation to $y(0.5)$, and when $h = 0.0001$, $y_{5,000}$ is an approximation to $y(0.5)$. Here are the results obtained using a CAS:

Time step $h = 0.01$	$y_{50} \approx 0.120$
Time step $h = 0.001$	$y_{500} \approx 0.1219$
Time step $h = 0.0001$	$y_{5,000} \approx 0.1221$

The values appear to converge and we may assume that $y(0.5) \approx 0.12$. However, we see here that Euler's Method converges quite slowly. ∎

9.3 SUMMARY

- The *slope field* for a first-order differential equation $\dot{y} = F(t, y)$ is obtained by drawing small segments of slope $F(t, y)$ at points (t, y) lying on a rectangular grid in the plane.
- The graph of a solution (also called an *integral curve*) satisfying $y(t_0) = y_0$ is a curve through (t_0, y_0) that runs tangent to the segments of the slope field at each point.
- *Euler's Method* is used to approximate a solution to $\dot{y} = F(t, y)$ with initial condition $y(t_0) = y_0$. Fix a time step h and set $t_k = t_0 + kh$. Define $y_1, y_2, \ldots$ successively by the formula

$$\boxed{y_k = y_{k-1} + hF(t_{k-1}, y_{k-1})} \qquad \boxed{4}$$

The values $y_0, y_1, y_2, \ldots$ are approximations to the values $y(t_0), y(t_1), y(t_2), \ldots$.

9.3 EXERCISES

Preliminary Questions

1. What is the slope of the segment in the slope field for $\dot{y} = ty + 1$ at the point $(2, 3)$?

2. What is the equation of the isocline of slope $c = 1$ for $\dot{y} = y^2 - t$?

3. True or false? In the slope field for $\dot{y} = \ln y$, the slopes at points on a vertical line $t = C$ are all equal.

4. What about the slope field for $\dot{y} = \ln t$? Are the slopes at points on a vertical line $t = C$ all equal?

5. Let $y(t)$ be the solution to $\dot{y} = F(t, y)$ with $y(1) = 3$. How many iterations of Euler's Method are required to approximate $y(3)$ if the time step is $h = 0.1$?

Exercises

1. Figure 8 shows the slope field for $\dot{y} = \sin y \sin t$. Sketch the graphs of the solutions with initial conditions $y(0) = 1$ and $y(0) = -1$. Show that $y(t) = 0$ is a solution and add its graph to the plot.

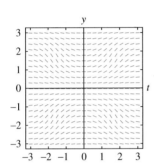

FIGURE 8 Slope field for $\dot{y} = \sin y \sin t$.

2. Figure 9 shows the slope field of $\dot{y} = y^2 - t^2$. Sketch the integral curve passing through the point $(0, -1)$, the curve through $(0, 0)$, and the curve through $(0, 2)$. Is $y(t) = 0$ a solution?

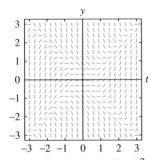

FIGURE 9 Slope field for $\dot{y} = y^2 - t^2$.

3. Show that $f(t) = \frac{1}{2}\left(t - \frac{1}{2}\right)$ is a solution to $\dot{y} = t - 2y$. Sketch the four solutions with $y(0) = \pm 0.5, \pm 1$ on the slope field in Fig-

ure 10. The slope field suggests that every solution approaches $f(t)$ as $t \to \infty$. Confirm this by showing that $y = f(t) + Ce^{-2t}$ is the general solution.

4. Consider the differential equation $\dot{y} = t - y$.

(a) Sketch the slope field of the differential equation $\dot{y} = t - y$ in the range $-1 \le t \le 3, -1 \le y \le 3$. As an aid, observe that the isocline of slope c is the line $t - y = c$, so the segments have slope c at points on the line $y = t - c$.

(b) Show that $y = t - 1 + Ce^{-t}$ is a solution for all C. Since $\lim_{t \to \infty} e^{-t} = 0$, these solutions approach the particular solution $y = t - 1$ as $t \to \infty$. Explain how this behavior is reflected in your slope field.

5. Show that the isoclines of $\dot{y} = 1/y$ are horizontal lines. Sketch the slope field for $-2 \le t, y \le 2$ and plot the solutions with initial conditions $y(0) = 0$ and $y(0) = 1$.

6. Show that the isoclines of $\dot{y} = t$ are vertical lines. Sketch the slope field for $-2 \le t, y \le 2$ and plot the integral curves passing through $(0, -1)$ and $(0, 1)$.

7. Sketch the slope field of $\dot{y} = ty$ for $-2 \le t, y \le 2$. Based on the sketch, determine $\lim_{t \to \infty} y(t)$, where $y(t)$ is a solution with $y(0) > 0$. What is $\lim_{t \to \infty} y(t)$ if $y(0) < 0$?

8. Match the differential equation with its slope field in Figures 11(A)–(F).

(i) $\dot{y} = -1$ | **(ii)** $\dot{y} = \dfrac{y}{t}$

(iii) $\dot{y} = t^2 y$ | **(iv)** $\dot{y} = t y^2$

(v) $\dot{y} = t^2 + y^2$ | **(vi)** $\dot{y} = t$

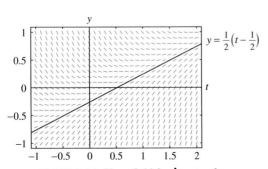

FIGURE 10 Slope field for $\dot{y} = t - 2y$.

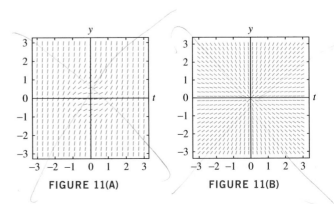

FIGURE 11(A)   **FIGURE 11(B)**

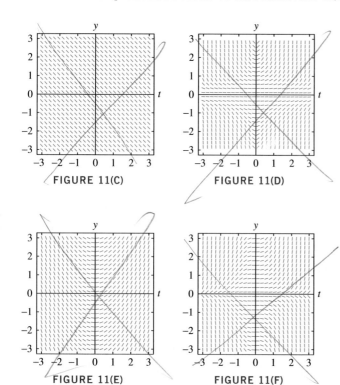

FIGURE 11(C)

FIGURE 11(D)

FIGURE 11(E)

FIGURE 11(F)

9. One of the slope fields in Figures 12(A) and (B) is the slope field for $\dot{y} = t^2$. The other is for $\dot{y} = y^2$. Identify which is which. In each case, sketch the solutions with initial conditions $y(0) = 1$, $y(0) = 0$, and $y(0) = -1$.

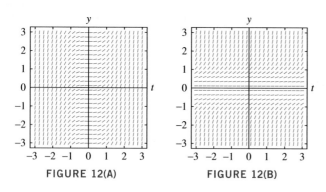

FIGURE 12(A)

FIGURE 12(B)

10. (a) Sketch the slope field of $\dot{y} = t/y$ in the region $-2 \le t \le 2$, $-2 \le y \le 2$.

(b) Check that $y = \pm\sqrt{t^2 + C}$ is the general solution.

(c) Sketch the solutions on the slope field with initial conditions $y(0) = 1$ and $y(0) = -1$.

11. Sketch the slope field of $\dot{y} = t^2 - y$ in the region $-3 \le t \le 3$, $-3 \le y \le 3$ and sketch the solutions satisfying $y(1) = 0$, $y(1) = 1$, and $y(1) = -1$.

12. Let $F(t, y) = t^2 - y$ and let $y(t)$ be the solution of $\dot{y} = F(t, y)$ satisfying $y(2) = 3$. Let $h = 0.1$ be the time step in Euler's Method and set $y_0 = y(2) = 3$.

(a) Calculate $y_1 = y_0 + hF(2, 3)$.

(b) Calculate $y_2 = y_1 + hF(2.1, y_1)$.

(c) Calculate $y_3 = y_2 + hF(2.2, y_2)$ and continue computing y_4, y_5, and y_6.

(d) Find approximations to $y(2.2)$ and $y(2.5)$.

13. Let $y(t)$ be the solution to $\dot{y} = te^{-y}$ satisfying $y(0) = 0$.

(a) Use Euler's Method with time step $h = 0.1$ to approximate $y(0.1), y(0.2), \dots, y(1)$.

(b) Use separation of variables to find $y(t)$ exactly.

(c) Compute the error in the approximations to $y(0.1)$, $y(0.5)$, and $y(1)$.

In Exercises 14–19, use Euler's Method with $h = 0.1$ to approximate the given value of $y(t)$.

14. $y(0.5)$; $\dot{y} = y^2$, $y(0) = 0$

15. $y(1)$; $\dot{y} = y$, $y(0) = 0$

16. $y(0.7)$; $\dot{y} = -yt$, $y(0) = 1$

17. $y(1.5)$; $\dot{y} = t \sin y$, $y(1) = 2$

18. $y(2.6)$; $\dot{y} = t/y$, $y(0) = 2$

19. $y(0.5)$; $\dot{y} = t - y$, $y(0) = 1$

Further Insights and Challenges

20. If $f(t)$ is continuous on $[a, b]$, then the solution to $\dot{y} = f(t)$ with initial condition $y(a) = 0$ is $y(t) = \int_a^t f(u)\, du$. Show that Euler's Method with time step $h = \dfrac{b - a}{N}$ for N steps yields the Nth left-endpoint approximation to $y(b) = \int_a^b f(u)\, du$.

In Exercises 21–22, use a modification of Euler's Method, Euler's Midpoint Method, that gives a significant improvement in accuracy. With

time step h and initial value $y_0 = y(t_0)$, the values y_k are defined successively by the equation

$$y_k = y_{k-1} + hm_{k-1}$$

where $m_{k-1} = F\left(t_{k-1} + \dfrac{h}{2}, y_{k-1} + \dfrac{h}{2}F(t_{k-1}, y_{k-1})\right).$

21. Apply both Euler's Method and the Euler Midpoint Method with $h = 0.1$ to estimate $y(1.5)$, where $y(t)$ satisfies $\dot{y} = y$ with $y(0) = 1$. Find $y(t)$ exactly and compute the errors in these two approximations.

22. If $f(t)$ is continuous on $[a, b]$, then the solution to $\dot{y} = f(t)$ with initial condition $y(a) = 0$ is $y(t) = \int_a^t f(u)\, du$. Show that the Euler Midpoint Method with time step $h = \dfrac{b-a}{N}$ for N steps yields the Nth midpoint approximation to $y(b) = \int_a^b f(u)\, du$.

In Exercises 23–26, use Euler's Midpoint Method with $h = 0.1$ to approximate the given value of $y(t)$.

23. $y(0.5)$; $\quad \dot{y} = y^2$, $\quad y(0) = 0$

24. $y(1)$; $\quad \dot{y} = y^2$, $\quad y(0) = 0$

25. $y(1)$; $\quad \dot{y} = t + y$, $\quad y(0) = 0$

26. $y(0.5)$; $\quad \dot{y} = y^2 - t$, $\quad y(0) = 1$

9.4 The Logistic Equation

The logistic equation was first introduced in 1838 by the Belgian mathematician Pierre-François Verhulst (1804–1849). Based on the population of Belgium for three years (1815, 1830, and 1845), which was then between 4 and 4.5 million, Verhulst predicted that the population would never exceed 9.4 million. This prediction has held up reasonably well. Belgium's current population is around 10.4 million.

The first and simplest model of population growth is $dy/dt = ky$, which predicts that populations grow exponentially. This may be true over short periods of time, but it is obvious that no population can increase without limit. Therefore, population biologists use a variety of other differential equations that take into account factors which limit growth, such as food scarcity, competition between species, and environmental limitations. One widely used model is based on the **logistic differential equation**:

$$\boxed{\frac{dy}{dt} = ky\left(1 - \frac{y}{A}\right)} \qquad \boxed{1}$$

Here $k > 0$ is the growth constant (in units of time^{-1}) and A is a nonzero constant called the **carrying capacity**. Figure 1 shows a typical S-shaped solution of Eq. (1).

As in the previous section, we also denote dy/dt by $\dot{y}$.

CONCEPTUAL INSIGHT The logistic equation $\dot{y} = ky(1 - y/A)$ is obtained by inserting the factor $(1 - y/A)$ in the exponential differential equation $\dot{y} = ky$. As long as $y = y(t)$ is small relative to A, $(1 - y/A)$ is close to 1 and can be ignored, yielding $\dot{y} \approx ky$. Thus, $y(t)$ grows nearly exponentially when the population is small (Figure 1). As $y(t)$ approaches A, the factor $(1 - y/A)$ tends to zero. This causes the growth rate to decrease to zero and prevents $y(t)$ from exceeding the carrying capacity A.

The slope field in Figure 2 provides a more complete picture of the solutions. Let's write Eq. (1) as $\dot{y} = F(y)$ with $F(y) = ky(1 - y/A)$. Since $F(y)$ does not depend on

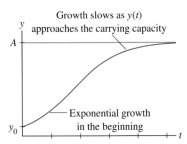

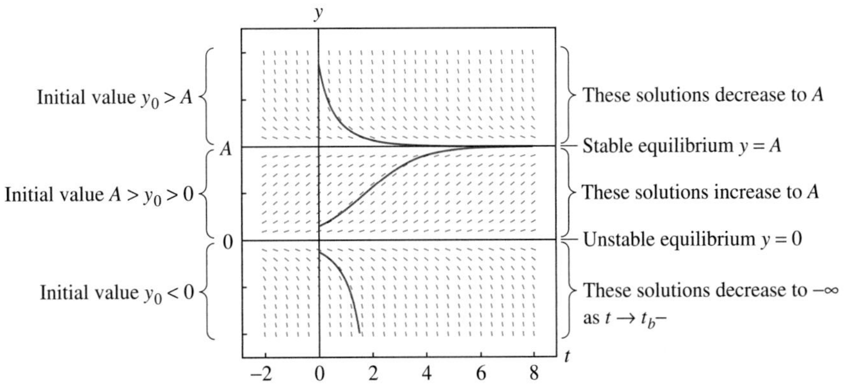

Growth slows as $y(t)$ approaches the carrying capacity

Exponential growth in the beginning

FIGURE 1 Solution of the logistic equation.

Initial value $y_0 > A$ { These solutions decrease to A

Stable equilibrium $y = A$

Initial value $A > y_0 > 0$ { These solutions increase to A

Unstable equilibrium $y = 0$

Initial value $y_0 < 0$ { These solutions decrease to $-\infty$ as $t \to t_b-$

FIGURE 2 Slope field for $\dfrac{dy}{dt} = ky\left(1 - \dfrac{y}{A}\right)$.

t, the slope field varies only in the y-direction. We see that the solutions come in three families plus two special cases, depending on the initial value $y_0 = y(0)$:

- If $y_0 > A$, then $y(t)$ is decreasing and approaches A as $t \to \infty$.
- If $0 < y_0 < A$, then $y(t)$ is increasing and approaches A as $t \to \infty$.
- If $y_0 < 0$, then $y(t)$ is decreasing and $\lim\limits_{t \to t_b-} y(t) = -\infty$ for some time t_b.

Solutions of the logistic equation with $y_0 < 0$ are not relevant to populations because a population cannot be negative (see Exercise 18).

Equation (1) also has two constant solutions, namely $y = 0$ and $y = A$ [because both sides of $\dot{y} = F(y)$ are zero for $y = 0$ and $y = A$]. These are called **equilibrium** or **steady-state** solutions. The equilibrium solution $y = A$ is called a **stable equilibrium** because every solution with initial condition $y_0 > 0$ approaches the equilibrium $y = A$ as $t \to \infty$. By contrast, $y = 0$ is an **unstable equilibrium**. Nonequilibrium solutions that start near $y = 0$ move away from $y = 0$, either increasing to A or decreasing to $-\infty$.

Having described the solutions of the logistic equation qualitatively, we now find the nonequilibrium solutions explicitly using separation of variables (assume $y \ne 0, A$):

In Eq. (2), we use the the partial fraction decomposition

$$\frac{1}{y(1 - y/A)} = \frac{1}{y} - \frac{1}{y - A}$$

$$\frac{dy}{dt} = ky\left(1 - \frac{y}{A}\right)$$

$$\frac{dy}{y(1 - y/A)} = k\, dt$$

$$\int\left(\frac{1}{y} - \frac{1}{y - A}\right) dy = \int k\, dt \qquad \boxed{2}$$

$$\ln|y| - \ln|y - A| = kt + C$$

$$\left|\frac{y}{y - A}\right| = e^{kt+C} \quad \Rightarrow \quad \frac{y}{y - A} = \pm e^C e^{kt}$$

Since $\pm e^C$ takes on arbitrary nonzero values, we replace $\pm e^C$ with C (nonzero):

$$\frac{y}{y - A} = Ce^{kt} \qquad \boxed{3}$$

For $t = 0$, this gives a useful relation between C and the initial value $y_0 = y(0)$:

$$\frac{y_0}{y_0 - A} = C \qquad \boxed{4}$$

To solve for y, multiply each side of Eq. (3) by $(y - A)$:

$$y = (y - A)Ce^{kt}$$

$$y(1 - Ce^{kt}) = -ACe^{kt}$$

$$y = \frac{ACe^{kt}}{Ce^{kt} - 1}$$

As $C \ne 0$, we may divide by Ce^{kt} to obtain the general nonequilibrium solution:

$$\boxed{\frac{dy}{dt} = ky\left(1 - \frac{y}{A}\right), \qquad y = \frac{A}{1 - e^{-kt}/C}} \qquad \boxed{5}$$

■ **EXAMPLE 1** Solve $\dot{y} = 0.3y(4 - y)$ with initial condition $y(0) = 1$.

Solution   To apply (5), we must rewrite the equation in the form

$$\dot{y} = 1.2y\left(1 - \frac{y}{4}\right)$$

Thus, $k = 1.2$ and $A = 4$, and the general solution is

$$y = \frac{4}{1 - e^{-1.2t}/C}$$

There are two ways to find C. One way is to solve $y(0) = 1$ for C directly. An easier way is to use Eq. (4):

$$C = \frac{y_0}{y_0 - A} = \frac{1}{1 - 4} = -\frac{1}{3}$$

We find that the particular solution is $y = \dfrac{4}{1 + 3e^{-1.2t}}$ (Figure 3). ■

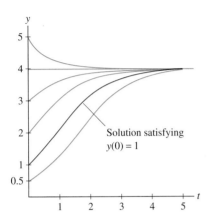

FIGURE 3 Several solutions of $\dot{y} = 0.3y(4 - y)$.

In the figure: Solution satisfying $y(0) = 1$

The logistic equation may be too simple to describe a real deer population accurately, but it serves as a starting point for more sophisticated models used by ecologists, population biologists, and forestry professionals.

■ **EXAMPLE 2** A 10,000-acre forest has a carrying capacity of 1,000 deer. Assume that the deer population grows logistically with growth constant $k = 0.4 \text{ yr}^{-1}$.

(a) Find the deer population $P(t)$ if the initial population is $P_0 = 100$.

(b) How long does it take for the deer population to reach 500?

Solution

(a) Since $k = 0.4$ and $A = 1,000$, $P(t)$ satisfies the differential equation

$$\frac{dP}{dt} = 0.4P\left(1 - \frac{P}{1,000}\right)$$

The general solution is given by Eq. (5):

$$P(t) = \frac{1,000}{1 - e^{-0.4t}/C} \qquad \boxed{6}$$

According to Eq. (4) (see Figure 4),

$$C = \frac{P_0}{P_0 - A} = \frac{100}{100 - 1,000} = -\frac{1}{9} \quad \Rightarrow \quad P(t) = \frac{1,000}{1 + 9e^{-0.4t}}$$

Logistic Growth of Deer Population

Year	Deer Population	Year	Deer Population
0	100	8	732
1	142	9	803
2	198	10	858
3	269	11	900
4	355	12	931
5	451	13	953
6	551	14	968
7	646	15	978

(b) To find the time t when $P(t) = 500$, we could solve the equation

$$P(t) = \frac{1{,}000}{1 + 9e^{-0.4t}} = 500$$

But it is easier to use Eq. (3):

$$\frac{P}{P - A} = Ce^{kt}$$

$$\frac{P}{P - 1{,}000} = -\frac{1}{9}e^{0.4t}$$

Set $P = 500$ and solve for t:

$$-\frac{1}{9}e^{0.4t} = \frac{500}{500 - 1{,}000} = -1 \quad \Rightarrow \quad e^{0.4t} = 9 \quad \Rightarrow \quad 0.4t = \ln 9$$

This gives $t = (\ln 9)/0.4 \approx 5.5$ years.   ■

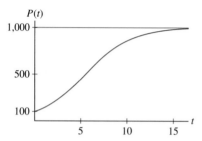

FIGURE 4 Deer population as a function of t (in years).

9.4 SUMMARY

- The *logistic equation* and its general nonequilibrium solution are

$$\frac{dy}{dt} = ky\left(1 - \frac{y}{A}\right), \qquad y = \frac{A}{1 - e^{-kt}/C}$$

- The logistic equation has two equilibrium (constant) solutions: $y = 0$ is an unstable equilibrium and $y = A$ is a stable equilibrium.
- Assume that $k > 0$. If the initial value $y_0 = y(0)$ satisfies $y_0 > 0$, then $y(t)$ approaches the stable equilibrium $y = A$, that is,

$$\lim_{t \to \infty} y(t) = A$$

9.4 EXERCISES

Preliminary Questions

1. Which of the following is a logistic differential equation?

(a) $\dot{y} = 2y(1 - y^2)$ **(b)** $\dot{y} = 2y\left(1 - \frac{y}{3}\right)$

(c) $\dot{y} = 2y\left(1 - \frac{x}{4}\right)$ **(d)** $\dot{y} = 2y(1 - 3y)$

2. True or false? The logistic equation is linear.

3. True or false? The logistic equation is separable.

4. Let $y(t)$ be a solution to $\dot{y} = 4y(3 - y)$. What is $\lim_{t \to \infty} y(t)$ in the following three cases:

(a) $y(0) = 3$ **(b)** $y(0) = 4$ **(c)** $y(0) = -2$

Exercises

1. Find the general solution of the logistic equation

$$\dot{y} = 3y(1 - y/5)$$

Then find the particular solution satisfying $y(0) = 2$.

2. Find the solution of $\dot{y} = 2y(3 - y)$, $y(0) = 10$.

3. Let $y(t)$ be a solution of $\dot{y} = 0.5y(1 - 0.5y)$ such that $y(0) = 4$. Determine $\lim_{t \to \infty} y(t)$ without finding $y(t)$ explicitly.

4. Let $y(t)$ be a solution of $\dot{y} = 5y(1 - y/A)$ satisfying $y(0) = 10$. Find $\lim_{t \to \infty} y(t)$, assuming that

(a) $A = 15$ **(b)** $A = 5$ **(c)** $A = 10$

5. A population of squirrels lives in a forest with a carrying capacity of 2,000. Assume logistic growth with growth constant $k = 0.6$ yr^{-1}.

(a) Find a formula for the squirrel population $P(t)$, assuming an initial population of 500 squirrels.

(b) How long will it take for the squirrel population to double?

6. The population $P(t)$ of mosquito larvae growing in a tree hole increases according to the logistic equation with growth constant $k = 0.3$ days^{-1} and carrying capacity $A = 500$.

(a) Find a formula for the larvae population $P(t)$, assuming an initial population of $P_0 = 50$ larvae.

(b) After how many days will the larvae population reach 200?

7. Sunset Lake is stocked with 2,000 rainbow trout and after 1 year the population has grown to 4,500. Assuming logistic growth with a carrying capacity of 20,000, find the growth constant k (specify the units) and determine when the population will increase to 10,000.

8. Spread of a Rumor   A rumor spreads through a small town. Let $y(t)$ be the fraction of the population that has heard the rumor at time t and assume that the rate at which the rumor spreads is proportional to the product of the fraction y of the population that has heard the rumor and the fraction $1 - y$ that has not yet heard the rumor.

(a) Write down the differential equation satisfied by y in terms of a proportionality factor k.

(b) Find k (in units of days^{-1}), assuming that 10% of the population knows the rumor at $t = 0$ and 40% knows it at $t = 2$ days.

(c) Using the assumptions of part (b), determine when 75% of the population will know the rumor.

9. A rumor spreads through a school with 1,000 students. At 8 AM, 80 students have heard the rumor, and by noon, half the school has heard it. Using the logistic model of Exercise 8, determine when 90% of the students will have heard the rumor.

10. $\boxed{\text{GU}}$ A simpler model for the spread of a rumor assumes that the rate at which the rumor spreads is proportional (with factor k) to the fraction of the population that has not yet heard the rumor.

(a) Compute the solutions to this model and the model of Exercise 8 with the values $k = 0.9$ and $y_0 = 0.1$.

(b) Graph the two solutions on the same axis.

(c) Which model seems more realistic? Why?

11. Let $k = 1$ and $A = 1$ in the logistic equation.

(a) Find the solutions satisfying $y_1(0) = 10$ and $y_2(0) = -1$.

(b) Find the time t when $y_1(t) = 5$.

(c) When does $y_2(t)$ become infinite?

12. ✎   **Reverse Logistic Equation**   Consider the logistic equation (with k, $B > 0$)

$$\frac{dP}{dt} = -kP\left(1 - \frac{P}{B}\right) \qquad \boxed{7}$$

(a) Sketch the slope field of this equation.

(b) The general solution is $P(t) = B/(1 - e^{kt}/C)$, where C is a nonzero constant. Show that $P(0) > B$ if $C > 1$ and $0 < P(0) < B$ if $C < 0$.

(c) Show that Eq. (7) models an "extinction–explosion" population. That is, $P(t)$ tends to zero if the initial population satisfies $0 < P(0) < B$ and it tends to ∞ after a finite amount of time if $P(0) > B$.

(d) Show that $P = 0$ is a stable equilibrium and $P = B$ an unstable equilibrium.

13. A tissue culture grows until it has a maximum area of M cm^2. The area $A(t)$ of the culture at time t may be modeled by the differential equation

$$\dot{A} = k\sqrt{A}\left(1 - \frac{A}{M}\right) \qquad \boxed{8}$$

where k is a growth constant.

(a) By setting $A = u^2$, show that the equation can be rewritten

$$\dot{u} = \frac{1}{2}k\left(1 - \frac{u^2}{M}\right)$$

Then find the general solution using separation of variables.

(b) Show that the general solution to Eq. (8) is

$$A(t) = M\left(\frac{Ce^{(k/\sqrt{M})t} - 1}{Ce^{(k/\sqrt{M})t} + 1}\right)^2 \qquad \boxed{9}$$

14. $\boxed{\text{GU}}$ Use the model of Exercise 13 to determine the area $A(t)$ (t in hours) of a tissue culture with initial size $A(0) = 0.2$ cm^2, assuming that the maximum area is $M = 5$ cm^2 and the growth constant is $k = 0.06$. Graph the solution using a graphing utility.

15. Show that if a tissue culture grows according to Eq. (8), then the growth rate reaches a maximum when $A = M/3$.

16. In 1751, Benjamin Franklin predicted that the U.S. population $P(t)$ would increase with growth constant $k = 0.028$ yr^{-1}. According to the census, the U.S. population was 5 million in 1800 and 76 million in 1900. Assuming logistic growth with $k = 0.028$, what is the predicted carrying capacity for the U.S. population? *Hint:* Use Eqs. (3) and (4) to show that

$$\frac{P(t)}{P(t) - A} = \frac{P_0}{P_0 - A}e^{kt}$$

Further Insights and Challenges

17. Let $y(t)$ be a solution of the logistic equation

$$\frac{dy}{dt} = ky\left(1 - \frac{y}{A}\right) \qquad \boxed{10}$$

(a) Differentiate Eq. (10) with respect to t and use the Chain Rule to show that

$$\frac{d^2y}{dt^2} = k^2y\left(1 - \frac{y}{A}\right)\left(1 - \frac{2y}{A}\right)$$

(b) Show that $y(t)$ is concave up if $0 < y < A/2$ and concave down if $A/2 < y < A$.

(c) Show that if $0 < y(0) < A/2$, then $y(t)$ has a point of inflection at $y = A/2$ (Figure 5).

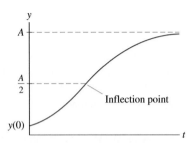

FIGURE 5 Inflection point in a logistic curve occurs at $y = A/2$.

(d) Assume that $0 < y(0) < A/2$. Find the time t when $y(t)$ reaches the inflection point.

18. Let

$$y = \frac{A}{1 - e^{-kt}/C}$$

be the general nonequilibrium solution of $\dot{y} = ky(1 - y/A)$ with $k > 0$. If $y(t)$ has a vertical asymptote at $t = t_b$, that is, if $\lim\limits_{t \to t_b-} y(t) = \pm\infty$, we say that the solution "blows up" at $t = t_b$.

(a) Show that if $0 < y(0) < A$, then y does not blow up at any time t_b.

(b) Show that if $y(0) > A$, then y blows up at a time t_b, which is negative (and hence does not correspond to a real time).

(c) Show that y blows up at some positive time t_b if and only if $y(0) < 0$ (and hence does not correspond to a real population).

9.5 First-Order Linear Equations

The separation of variables method applies to all separable first-order equations, whether linear or not. In this section, we introduce a method for solving all linear first-order equations whether separable or not.

A first-order linear equation has the form $a(x)y' + b(x)y = c(x)$, where $a(x)$ is not identically equal to zero. We divide by $a(x)$ and write the equation in the standard form

$$\boxed{y' + A(x)y = B(x)} \qquad \boxed{1}$$

The idea in solving Eq. (1) is to multiply through by a function $\alpha(x)$, called an **integrating factor**, which turns the left-hand side into the derivative of $\alpha(x)y$:

$$\alpha(x)\big(y' + A(x)y\big) = \big(\alpha(x)y\big)' \qquad \boxed{2}$$

Suppose we can find a function $\alpha(x)$ satisfying Eq. (2). Then Eq. (1) yields

$$\alpha(x)\big(y' + A(x)y\big) = \alpha(x)B(x)$$

$$\big(\alpha(x)y\big)' = \alpha(x)B(x)$$

We can solve this equation by integration:

$$\alpha(x)y = \int \alpha(x)B(x)\,dx + C \qquad \text{or} \qquad y = \frac{1}{\alpha(x)}\left(\int \alpha(x)B(x)\,dx + C\right)$$

To find $\alpha(x)$, expand Eq. (2), using the Product Rule on the right-hand side:

$$\alpha(x)y' + \alpha(x)A(x)y = \alpha(x)y' + \alpha'(x)y$$

Then subtract $\alpha(x)y'$ from both sides and divide by y to obtain $\alpha'(x) = \alpha(x)A(x)$, or

$$\frac{d\alpha}{dx} = \alpha(x)A(x) \qquad \boxed{3}$$

We solve this equation using separation of variables:

$$\frac{d\alpha}{\alpha} = A(x)\,dx \quad \Rightarrow \quad \int \frac{d\alpha}{\alpha} = \int A(x)\,dx$$

Therefore, $\ln|\alpha(x)| = \displaystyle\int A(x)\,dx$, and by exponentiation, $\alpha(x) = \pm e^{\int A(x)\,dx}$. Since we need just one solution of Eq. (3), we choose the positive solution.

In the formula for the integrating factor $\alpha(x)$, the integral $\displaystyle\int A(x)\,dx$ denotes any antiderivative of $A(x)$.

THEOREM 1 The General Solution   The general solution of $y' + A(x)y = B(x)$ is

$$y = \alpha(x)^{-1}\left(\int \alpha(x)B(x)\,dx + C\right) \qquad \boxed{4}$$

where $\alpha(x)$ is an integrating factor:

$$\alpha(x) = e^{\int A(x)\,dx} \qquad \boxed{5}$$

■ **EXAMPLE 1**   Solve $xy' - 3y = x^2$, $y(1) = 2$.

Solution   First put the equation in the form $y' + A(x)y = B(x)$ by dividing by x:

$$y' - \frac{3}{x}y = x \qquad \boxed{6}$$

***Step 1.* Find an integrating factor.**

In our case, $A(x) = -\dfrac{3}{x}$ and by Eq. (5)

$$\alpha(x) = e^{\int A(x)\,dx} = e^{\int (-3/x)\,dx} = e^{-3\ln x} = e^{\ln(x^{-3})} = x^{-3}$$

***Step 2.* Find the general solution.**

We have found $\alpha(x)$, so we can use Eq. (4) to write down the general solution. However, to illustrate the ideas, let's go through the steps leading to Eq. (4). First, multiply Eq. (6) by $\alpha(x) = x^{-3}$ to rewrite the equation:

$$x^{-3}\left(y' - \frac{3}{x}y\right) = x^{-3}x \quad \Rightarrow \quad x^{-3}y' - 3x^{-4}y = x^{-2} \quad \Rightarrow \quad \frac{d}{dx}\left(x^{-3}y\right) = x^{-2}$$

Thus, $x^{-3}y$ is an antiderivative of x^{-2}, and we obtain the general solution (see Figure 1)

$$x^{-3}y = \int x^{-2}\,dx = -x^{-1} + C \qquad \boxed{7}$$

$$\boxed{y = -x^2 + Cx^3}$$

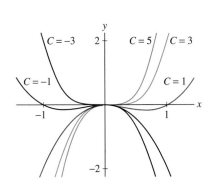

FIGURE 1 Family of solutions to $xy' - 3y = x^2$.

CAUTION The constant of integration C in Eq. (7) is essential, but note that in the general solution, C does not appear as an additive constant. The general solution is $y = -x^2 + Cx^3$, not $-x^2 + C$, a common mistake.

Step 3. **Solve the initial value problem.**

The initial condition gives

$$y(1) = -1^2 + C \cdot 1^3 = 2 \qquad \text{or} \qquad C = 3$$

Therefore, the solution of the initial value problem is $y = -x^2 + 3x^3$.

Finally, let's check that $y = -x^2 + 3x^3$ satisfies our equation $xy' - 3y = x^2$:

$$xy' - 3y = x(-2x + 9x^2) - 3(-x^2 + 3x^3)$$
$$= (-2x^2 + 9x^3) + (3x^2 - 9x^3) = x^2 \qquad \blacksquare$$

■ **EXAMPLE 2** Solve the initial value problem: $y' + (1 - x^{-1})y = x^2$, $y(1) = 2$.

Solution This equation has the form $y' + A(x)y = B(x)$ with $A(x) = (1 - x^{-1})$. By Eq. (5), an integrating factor is

$$\alpha(x) = e^{\int (1 - x^{-1})\,dx} = e^{x - \ln x} = e^x e^{\ln x^{-1}} = x^{-1}e^x$$

Using Eq. (4) with $B(x) = x^2$, we obtain the general solution:

$$y = \alpha(x)^{-1}\left(\int \alpha(x)B(x)\,dx + C\right) = xe^{-x}\left(\int (x^{-1}e^x)x^2\,dx + C\right)$$
$$= xe^{-x}\left(\int xe^x\,dx + C\right)$$

Integration by Parts shows that $\int xe^x\,dx = (x - 1)e^x + C$, so we obtain

$$y = xe^{-x}\big((x - 1)e^x + C\big) = x(x - 1) + Cxe^{-x}$$

The initial condition $y(1) = 2$ gives

$$y(1) = 1(1 - 1) + Ce^{-1} = Ce^{-1} = 2 \quad \Rightarrow \quad C = 2e$$

The desired particular solution is

$$y = x(x - 1) + (2e)xe^{-x} = x(x - 1) + 2xe^{1-x} \qquad \blacksquare$$

> *Summary: The general solution of $y' + A(x)y = B(x)$ is*
>
> $$y = \alpha(x)^{-1}\left(\int \alpha(x)B(x) + C\right)$$
>
> *where*
>
> $$\alpha(x) = e^{\int A(x)\,dx}$$

CONCEPTUAL INSIGHT Our method expresses the general solution in terms of the integrals in Eqs. (4) and (5). In this sense, we have an explicit general solution of any first-order linear equation. However, it is not always possible to evaluate the integrals in elementary terms. For example, the general solution of $y' + xy = 1$ is

$$y = e^{-x^2/2}\left(\int e^{x^2/2}\,dx + C\right)$$

and $\int e^{x^2/2}\,dx$ cannot be evaluated in elementary terms. However, we may approximate the integral numerically and plot the solutions by computer (Figure 2).

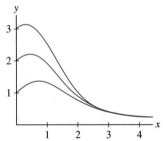

FIGURE 2 Solutions to $y' + xy = 1$ solved numerically and plotted by computer.

In the next example, we use a differential equation to model the "mixing problem," which has applications in biology, chemistry, and medicine.

R_{in} (gal/min)

Water level

R_{out} (gal/min)

FIGURE 3

■ **EXAMPLE 3** A Mixing Problem   A tank contains 100 gal of water with a sucrose concentration of 0.2 lb/gal. We begin adding water with a sucrose concentration of 0.4 lb/gal at a rate of $R_{in} = 10$ gal/min (Figure 3). The water is mixed instantaneously and exits the bottom of the tank at a rate of $R_{out} = 5$ gal/min. Let $y(t)$ be the quantity of sucrose in the tank at time t. Set up a differential equation for $y(t)$ and solve for $y(t)$.

Solution

Step 1. **Set up the differential equation.**

The derivative dy/dt is the difference of two rates of change, namely the rate at which sucrose enters the tank and the rate at which it leaves:

$$\frac{dy}{dt} = \text{sucrose rate in} - \text{sucrose rate out} \qquad \boxed{8}$$

The rate at which sucrose enters the tank is

$$\text{Sucrose rate in} = \underbrace{(0.4 \text{ lb/gal})(10 \text{ gal/min})}_{\text{Concentration times water rate in}} = 4 \text{ lb/min}$$

Next, we compute the sucrose concentration in the tank at time t. Water flows in at 10 gal/min and out at 5 gal/min, so there is a net inflow of 5 gal/min. Since the tank has 100 gal at time $t = 0$, it has $100 + 5t$ gallons at time t minutes and

$$\text{Sucrose concentration at time } t = \frac{\text{lb of sucrose in tank}}{\text{gal of water in tank}} = \frac{y(t)}{100 + 5t} \text{ lb/gal}$$

The rate at which sucrose leaves the tank is the product of the concentration and the rate at which water flows out:

$$\text{Sucrose rate out} = \underbrace{\left(\frac{y}{100 + 5t} \frac{\text{lb}}{\text{gal}} \right) \left(5 \frac{\text{gal}}{\text{min}} \right)}_{\text{Concentration times water rate out}} = \frac{5y}{100 + 5t} = \frac{y}{t + 20} \text{ lb/min}$$

Now Eq. (8) gives us the differential equation

$$\frac{dy}{dt} = 4 - \frac{y}{t + 20} \qquad \boxed{9}$$

Step 2. **Find the general solution.**

We write Eq. (9) in standard form:

$$\frac{dy}{dt} + \underbrace{\frac{1}{t + 20}}_{A(t)} y = \underbrace{4}_{B(t)} \qquad \boxed{10}$$

An integrating factor is

$$\alpha(t) = e^{\int A(t)\, dt} = e^{\int dt/(t+20)} = e^{\ln(t+20)} = t + 20$$

The general solution is

$$y(t) = \alpha(t)^{-1} \left(\int \alpha(t) B(t)\, dt + C \right) = \frac{1}{t + 20} \left(\int (t + 20)(4)\, dt + C \right)$$

$$= \frac{1}{t + 20} \left(2(t + 20)^2 + C \right) = 2t + 40 + \frac{C}{t + 20}$$

Summary:

sucrose rate in = 4 lb/min

sucrose rate out = $\dfrac{y}{t + 20}$ lb/min

$$\frac{dy}{dt} = 4 - \frac{y}{t + 20}$$

$$y(t) = 2t + 40 + \frac{C}{t + 20}$$

Step 3. **Solve the initial value problem.**

At $t = 0$, the tank contains 100 gal of water with a sucrose concentration of 0.2 lb/gal. Thus, the total sucrose at $t = 0$ is $y(0) = (100)(0.2) = 20$ lb, and

$$y(0) = 2(0) + 40 + \frac{C}{0 + 20} = 40 + \frac{C}{20} = 20 \quad \Rightarrow \quad C = -400$$

We obtain the following formula (for t in minutes), valid until the tank overflows:

$$y(t) = 2t + 40 - \frac{400}{t + 20} \text{ lb sucrose}$$

■

9.5 SUMMARY

- A *first-order linear differential equation* can always be written in the form

$$y' + A(x)y = B(x) \qquad \boxed{11}$$

- The general solution of Eq. (11) is

$$y = \alpha(x)^{-1} \left(\int \alpha(x)B(x)\, dx + C \right)$$

where $\alpha(x)$ is an *integrating factor*: $\alpha(x) = e^{\int A(x)\, dx}$.

9.5 EXERCISES

Preliminary Questions

1. Which of the following are first-order linear equations?

(a) $y' + x^2 y = 1$

(b) $y' + xy^2 = 1$

(c) $x^5 y' + y = e^x$

(d) $x^5 y' + y = e^y$

2. If $\alpha(x)$ is an integrating factor for $y' + A(x)y = B(x)$, then $\alpha'(x)$ is equal to (choose the correct answer):

(a) $B(x)$

(b) $\alpha(x)A(x)$

(c) $\alpha(x)A'(x)$

(d) $\alpha(x)B(x)$

Exercises

1. Consider $y' + x^{-1}y = x^3$.

(a) Verify that $\alpha(x) = x$ is an integrating factor.

(b) Show that when multiplied by $\alpha(x)$, the differential equation can be written $(xy)' = x^4$.

(c) Conclude that xy is an antiderivative of x^4 and use this information to find the general solution.

(d) Find the particular solution satisfying $y(1) = 0$.

2. Consider $\dfrac{dy}{dt} + 2y = e^{-3t}$.

(a) Verify that $\alpha(t) = e^{2t}$ is an integrating factor.

(b) Use Eq. (4) to find the general solution.

(c) Find the particular solution with initial condition $y(0) = 1$.

3. Let $\alpha(x) = e^{x^2}$. Verify the identity

$$(\alpha(x)y)' = \alpha(x)(y' + 2xy)$$

and explain how it is used to find the general solution of

$$y' + 2xy = x.$$

4. Find the solution of $y' - y = e^{2x}$, $y(0) = 1$.

In Exercises 5–18, find the general solution of the first-order linear differential equation.

5. $xy' + y = x$

6. $xy' - y = x^2 - x$

7. $3xy' - y = x^{-1}$

8. $y' + xy = x$

9. $y' + 3x^{-1}y = x + x^{-1}$

10. $y' + x^{-1}y = \cos(x^2)$

11. $xy' = y - x$

12. $xy' = x^3 - \dfrac{3y}{x}$

13. $y' + y = e^x$

14. $y' + (\sec x)y = \cos x$

15. $y' + (\tan x)y = \cos x$

16. $e^{2x}y' = 1 - e^x y$

17. $y' - (\ln x)y = x^x$

18. $y' + y = \cos x$

In Exercises 19–26, solve the initial value problem.

19. $y' + 3y = e^{2x}$, $y(0) = -1$

20. $xy' + y = e^x$, $y(1) = 3$

21. $y' + \dfrac{1}{x+1}y = x^{-2}$, $y(1) = 2$

22. $y' + y = \sin x$, $y(0) = 1$

23. $(\sin x)y' = (\cos x)y + 1$, $y(\frac{\pi}{4}) = 0$

24. $y' + (\sec t)y = \sec t$, $y(0) = 1$ and $y(\frac{\pi}{4}) = 1$

25. $y' + (\tanh x)y = 1$, $y(0) = 3$

26. $y' + \dfrac{x}{1+x^2}y = \dfrac{1}{(1+x^2)^{3/2}}$, $y(1) = 0$

27. Find the general solution of $y' + ny = e^{mx}$ for all m, n. *Note:* The case $m = -n$ must be treated separately.

28. A 200-gal tank contains 100 gal of water with a salt concentration of 0.1 lb/gal. Water with a salt concentration of 0.4 lb/gal flows into the tank at a rate of 20 gal/min. The fluid is mixed instantaneously, and water is pumped out at a rate of 10 gal/min. Let $y(t)$ be the amount of salt in the tank at time t.

(a) Set up and solve the differential equation for $y(t)$.

(b) What is the salt concentration when the tank overflows?

29. Repeat Exercise 28(a), assuming that water is pumped out at a rate of 20 gal/min. What is the limiting salt concentration for large t?

30. Repeat Exercise 28(a), assuming that water is pumped out at a rate of 25 gal/min. What is the limiting salt concentration for t large?

31. Water flows into a tank at the variable rate $R_{in} = \dfrac{20}{1+t}$ gal/min and out at the constant rate $R_{out} = 5$ gal/min. Let $V(t)$ be the volume of water in the tank at time t.

(a) Set up a differential equation for $V(t)$ and solve it with the initial condition $V(0) = 100$.

(b) Find the maximum value of V.

(c) *CAS* Plot $V(t)$ and estimate the time t when the tank is empty.

32. A stream feeds into a lake at a rate of 1,000 m³/day. The stream is polluted with a toxin whose concentration is 5 g/m³. Assume that the lake has volume 10^6 m³ and that water flows out of the lake at the same rate of 1,000 m³/day. Set up a differential equation for the concentration $c(t)$ of toxin in the lake and solve for $c(t)$, assuming that $c(0) = 0$. What is the limiting concentration for large t?

In Exercises 33–35, consider a series circuit (Figure 4) consisting of a resistor of R ohms, an inductor of L henries, and variable voltage source of V(t) volts (time t in seconds). The current through the circuit I(t) (in amperes) satisfies the differential equation:

$$\frac{dI}{dt} + \frac{R}{L}I = \frac{1}{L}V(t) \qquad \boxed{12}$$

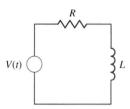

FIGURE 4 *RL* circuit.

33. Find the solution to Eq. (12) with initial condition $I(0) = 0$, assuming that $R = 100\ \Omega$, $L = 5$ H, and $V(t)$ is constant with $V(t) = 10$ V.

34. ⟨GU⟩ Assume that $R = 110\ \Omega$, $L = 10$ H, and $V(t) = e^{-t}$.

(a) Solve Eq. (12) with initial condition $I(0) = 0$.

(b) Use a computer algebra system to sketch the graph of the solution for $0 \le t \le 3$.

(c) Calculate t_m and $I(t_m)$, where t_m is the time at which $I(t)$ has a maximum value.

35. Assume that $V(t) = V$ is constant and $I(0) = 0$.

(a) Solve for $I(t)$.

(b) Show that $\lim_{t\to\infty} I(t) = V/R$ and that $I(t)$ reaches approximately 63% of its limiting value after L/R seconds.

(c) How long does it take for $I(t)$ to reach 90% of its limiting value if $R = 500\ \Omega$, $L = 4$ H, and $V = 20$ V?

36. ✎ Tank 1 in Figure 5 is filled with V_1 gallons of water. Water flows into the tank at a rate of R gal/min and out through the bottom at the same rate R. Suppose that I gallons of blue ink are dumped into the tank at time $t = 0$ and mixed instantaneously. Let $y_1(t)$ be the quantity of ink in the tank at time t.

(a) Explain why y_1 satisfies the differential equation $\dfrac{dy_1}{dt} = -\dfrac{R}{V_1}y_1$.

(b) Solve for $y_1(t)$ with $V_1 = 100$, $R = 10$, and $I = 2$.

37. ✎ Continuing with the previous exercise, let Tank 2 be another tank filled with V_2 gallons of water. Assume that the inky water from Tank 1 empties into Tank 2 as in Figure 5, mixes instantaneously, and leaves Tank 2 at the same rate R. Let $y_2(t)$ be the amount of ink in Tank 2 at time t.

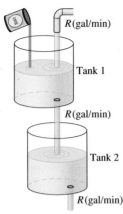

R(gal/min)

Tank 1

R(gal/min)

Tank 2

R(gal/min)

FIGURE 5

(a) Explain why y_2 satisfies the differential equation

$$\frac{dy_2}{dt} = R\left(\frac{y_1}{V_1} - \frac{y_2}{V_2}\right)$$

(b) Use the solution to Exercise 36 to solve for $y_2(t)$ if $V_1 = 100$, $V_2 = 200$, $R = 10$, $I = 2$, and $y_2(0) = 0$.

(c) [GU]   Plot the solution for $0 \le t \le 120$.

(d) Find the maximum ink concentration in Tank 2.

Further Insights and Challenges

38. Let $\alpha(x)$ be an integrating factor for $y' + A(x)y = B(x)$. The differential equation $y' + A(x)y = 0$ is called the associated **homogeneous equation**.

(a) Show that $\dfrac{1}{\alpha(x)}$ is a solution of the associated homogeneous equation.

(b) Show that if $y = f(x)$ is a particular solution of $y' + A(x)y = B(x)$, then $f(x) + \dfrac{C}{\alpha(x)}$ is also a solution for any constant C.

39. Use the Fundamental Theorem of Calculus and the Product Rule to verify directly that for any x_0, the function

$$f(x) = \alpha(x)^{-1} \int_{x_0}^{x} \alpha(t) B(t)\, dt$$

is a solution of the initial value problem

$$y' + A(x)y = B(x), \qquad y(x_0) = 0,$$

where $\alpha(x)$ is an integrating factor [a solution to Eq. (3)].

40. Transient Currents   Suppose the circuit described by Eq. (12) is driven by a sinusoidal voltage source $V(t) = V \sin \omega t$ (where V and ω are constant).

(a) Show that

$$I(t) = \frac{V}{R^2 + L^2\omega^2}(R \sin \omega t - L\omega \cos \omega t) + Ce^{-(R/L)t}$$

(b) Let $Z = \sqrt{R^2 + L^2\omega^2}$. Choose θ so that $Z \cos \theta = R$ and $Z \sin \theta = L\omega$. Use the addition formula for the sine function to show that

$$I(t) = \frac{V}{Z} \sin(\omega t - \theta) + +Ce^{-(R/L)t}$$

This shows that the current in the circuit varies sinusoidally apart from a DC term (called the **transient current** in electronics) that decreases exponentially.

CHAPTER REVIEW EXERCISES

1. Which of the following differential equations are linear? Determine the order of each equation.

(a) $y' = y^5 - 3x^4 y$

(b) $y' = x^5 - 3x^4 y$

(c) $y = y''' - 3x\sqrt{y}$

(d) $\sin x \cdot y'' = y - 1$

2. Find a value of c such that $y = x - 2 + e^{cx}$ is a solution of $2y' + y = x$.

In Exercises 3–6, solve using separation of variables.

3. $\dfrac{dy}{dt} = t^2 y^{-3}$

4. $xyy' = 1 - x^2$

5. $x\dfrac{dy}{dx} - y = 1$

6. $y' = \dfrac{xy^2}{x^2 + 1}$

In Exercises 7–10, solve the initial value problem using separation of variables.

7. $y' = \cos^2 x$, $y(0) = \dfrac{\pi}{4}$

8. $y' = \cos^2 y$, $y(0) = \dfrac{\pi}{4}$

9. $y' = xy^2$, $y(1) = 2$

10. $xyy' = 1$, $y(3) = 2$

11. Figure 1 shows the slope field for $\dot{y} = \sin y + ty$. Sketch the graphs of the solutions with the initial conditions $y(0) = 1$, $y(0) = 0$, and $y(0) = -1$.

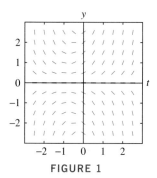

FIGURE 1

12. Which of the equations (i)–(iii) corresponds to the slope field in Figure 2?

(i) $\dot{y} = 1 - y^2$ **(ii)** $\dot{y} = 1 + y^2$ **(iii)** $\dot{y} = y^2$

13. Let $y(t)$ be the solution to the differential equation with slope field as shown in Figure 2, satisfying $y(0) = 0$. Sketch the graph of $y(t)$. Then use your answer to Exercise 12 to solve for $y(t)$.

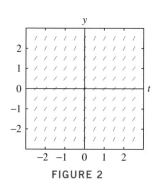

FIGURE 2

14. Let $y(t)$ be the solution of $4\dot{y} = y^2 + t$ satisfying $y(2) = 1$. Carry out Euler's Method with time step $h = 0.05$ for $n = 6$ steps.

15. Let $y(t)$ be the solution of $(x^3 + 1)\dot{y} = y$ satisfying $y(0) = 1$. Compute approximations to $y(0.1)$, $y(0.2)$, and $y(0.3)$ using Euler's Method with time step $h = 0.1$.

In Exercises 16–19, solve using the method of integrating factors.

16. $\dfrac{dy}{dt} = y + t^2$, $y(0) = 4$

17. $\dfrac{dy}{dx} = \dfrac{y}{x} + x$, $y(1) = 3$

18. $\dfrac{dy}{dt} = y - 3t$, $y(-1) = 2$

19. $y' + 2y = 1 + e^{-x}$, $y(0) = -4$

In Exercises 20–27, solve using the appropriate method.

20. $x^2 y' = x^2 + 1$, $y(1) = 10$

21. $y' + (\tan x)y = \cos^2 x$, $y(\pi) = 2$

22. $xy' = 2y + x - 1$, $y(\tfrac{3}{2}) = 9$

23. $(y - 1)y' = t$, $y(1) = -3$

24. $(\sqrt{y} + 1)y' = yte^{t^2}$, $y(0) = 1$.

25. $\dfrac{dw}{dx} = k\dfrac{1 + w^2}{x}$, $y(1) = 1$

26. $y' + \dfrac{3y - 1}{t} = t + 2$

27. $y' + \dfrac{y}{x} = \sin x$

28. Find the solutions to $y' = 4(y - 12)$ satisfying $y(0) = 20$ and $y(0) = 0$, and sketch their graphs.

29. Find the solutions to $y' = -2y + 8$ satisfying $y(0) = 3$ and $y(0) = 4$, and sketch their graphs.

30. Show that $y = \sin^{-1} x$ satisfies the differential equation $y' = \sec y$ with initial condition $y(0) = 0$.

31. What is the limit $\lim\limits_{t \to \infty} y(t)$ if $y(t)$ is a solution of:

(a) $\dfrac{dy}{dt} = -4(y - 12)$?

(b) $\dfrac{dy}{dt} = 4(y - 12)$?

(c) $\dfrac{dy}{dt} = -4y - 12$?

In Exercises 32–35, let $P(t)$ denote the balance at time t (years) of an annuity that earns 5% interest continuously compounded and pays out \$2000/year continuously.

32. Find the differential equation satisfied by $P(t)$.

33. Determine $P(2)$ if $P(0) = \$5{,}000$.

34. When does the annuity run out of money if $P(0) = \$2{,}000$?

35. What is the minimum initial balance that will allow the annuity to make payments indefinitely?

36. State whether the differential equation can be solved using separation of variables, the method of integrating factors, both, or neither.

(a) $y' = y + x^2$ **(b)** $xy' = y + 1$

(c) $y' = y^2 + x^2$ **(d)** $xy' = y^2$

37. Let A and B be constants. Prove that if $A > 0$, then all solution of $\dfrac{dy}{dt} + Ay = B$ approach the same limit as $t \to \infty$.

38. The trough in Figure 3 is filled with water. At time $t = 0$ (in seconds), water begins leaking through a circular hole at the bottom of radius 3 in. Let $y(t)$ be the water height at time t. Find a differential

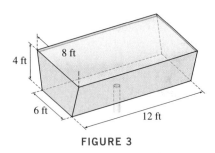

FIGURE 3

equation for $y(t)$ and solve it to determine when the trough will be half empty.

39. Find the solution of the logistic equation $\dot{y} = 0.4y(4 - y)$ satisfying $y(0) = 8$.

40. Let $y(t)$ be the solution of $\dot{y} = 0.3y(2 - y)$ with $y(0) = 1$. Determine $\lim_{t\to\infty} y(t)$ without solving for y explicitly.

41. Suppose that $y = ky(1 - y/8)$ has a solution satisfying $y(0) = 12$ and $y(10) = 24$. Find k.

42. A lake has a carrying capacity of 1,000 fish. Assume that the fish population grows logistically with growth constant $k = 0.2$ days^{-1}. How many days will it take for the population to reach 900 fish if the initial population is 20 fish?

43.   A rabbit population on a deserted island increases exponentially with growth rate $k = 0.12$ months^{-1}. When the population reaches 150 rabbits (say, at time $t = 0$), hunters begin killing the rabbits at a rate of r rabbits per month.

(a) Find a differential equation satisfied by the rabbit population $P(t)$.

(b) How large can r be without the rabbit population becoming extinct?

44. Show that $y = \sin(\tan^{-1} x + C)$ is the general solution of $y' = \dfrac{\sqrt{1 - y^2}}{1 + x^2}$. Then use the addition formula for the sine function to show that the general solution may be written $y = \dfrac{(\cos C)x + \sin C}{\sqrt{1 + x^2}}$.

45.   A tank contains 100 gal of pure water. Water is pumped out at a rate of 20 gal/min, and polluted water with a toxin concentration of 0.1 lb/gal is pumped in at a rate of 15 gal/min. Let $y(t)$ be the amount of toxin present in the tank at time t.

(a) Find a differential equation satisfied by $y(t)$.

(b) Solve for $y(t)$.

(c) *CAS*   Plot $y(t)$ and find the time t at which the amount of toxin is maximal.

10 | INFINITE SERIES

Our knowledge of what stars are made of is based on the study of absorption spectra, the sequences of wavelengths absorbed by gases in the star's atmosphere.

The theory of infinite series is a third branch of calculus, in addition to differential and integral calculus. Infinite series provide us with a new perspective on functions and on many interesting numbers. Two examples are the Gregory–Leibniz series

$$\frac{\pi}{4} = 1 - \frac{1}{3} + \frac{1}{5} - \frac{1}{7} + \frac{1}{9} - \cdots$$

and the infinite series for the exponential function

$$e^x = 1 + x + \frac{x^2}{2!} + \frac{x^3}{3!} + \frac{x^4}{4!} + \cdots$$

The first reveals that π is related to the reciprocals of the odd integers in an unexpected way, whereas the second shows that e^x can be expressed as an "infinite polynomial." Series of this type are widely used in applications, both in computations and in the analysis of functions. To make sense of infinite series, we need to define precisely what it means to add up infinitely many terms. Limits play a key role here, just as they do in differential and integral calculus.

10.1 Sequences

Sequences of numbers appear in diverse situations. If you divide a cake in half, and then divide the remaining half in half, and continue dividing in half indefinitely (Figure 1), then the fraction of cake left at each step forms the sequence

$$1, \quad \frac{1}{2}, \quad \frac{1}{4}, \quad \frac{1}{8}, \quad \cdots$$

This is the sequence of values of $f(n) = \dfrac{1}{2^n}$ for $n = 0, 1, 2, \ldots$.

Formally, a **sequence** is a function $f(n)$ whose domain is a subset of the integers. The values $a_n = f(n)$ are called the **terms** of the sequence and n is called the **index**. We usually think of a sequence informally as a collection of values $\{a_n\}$, or a list of terms:

$$a_1, \quad a_2, \quad a_3, \quad a_4, \quad \ldots$$

When a_n is given by a formula, we refer to a_n as the **general term**.

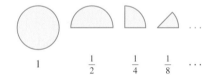

General Term	Domain	Sequence
$a_n = 1 - \dfrac{1}{n}$	$n \geq 1$	$0, \ \dfrac{1}{2}, \ \dfrac{2}{3}, \ \dfrac{3}{4}, \ \dfrac{4}{5}, \ \ldots$
$a_n = (-1)^n n$	$n \geq 0$	$0, \ -1, \ 2, \ -3, \ 4, \ \ldots$
$b_n = \dfrac{364.5n^2}{n^2 - 4}$	$n \geq 3$	$656.1, \ 486, \ 433.9, \ 410.1, \ 396.9, \ \ldots$

FIGURE 1

The sequence $b_n = \dfrac{364.5n^2}{n^2 - 4}$, known as the "Balmer series" in physics and chemistry, plays a key role in spectroscopy. The terms of this sequence are the absorption wavelengths of the hydrogen atom in nanometers.

In the following example, we consider a sequence whose terms are defined *recursively*. The first term is given and the nth term a_n is computed in terms of the preceding term a_{n-1}.

■ **EXAMPLE 1** Recursively Defined Sequence   Compute a_2, a_3, a_4 for the sequence defined recursively by

$$a_1 = 1, \qquad a_n = \frac{1}{2}\left(a_{n-1} + \frac{2}{a_{n-1}}\right)$$

Solution

$$a_2 = \frac{1}{2}\left(a_1 + \frac{2}{a_1}\right) = \frac{1}{2}\left(1 + \frac{2}{1}\right) = \frac{3}{2} = 1.5$$

$$a_3 = \frac{1}{2}\left(a_2 + \frac{2}{a_2}\right) = \frac{1}{2}\left(\frac{3}{2} + \frac{2}{3/2}\right) = \frac{17}{12} \approx 1.4167$$

$$a_4 = \frac{1}{2}\left(a_3 + \frac{2}{a_3}\right) = \frac{1}{2}\left(\frac{17}{12} + \frac{2}{17/12}\right) = \frac{577}{408} \approx 1.414216 \qquad ■$$

You may recognize the sequence in Example 1 as the sequence of approximations to $\sqrt{2} \approx 1.4142136$ produced by Newton's method with starting value $a_1 = 1$. As n tends to infinity, a_n approaches $\sqrt{2}$.

Our main goal is to study convergence of sequences. A sequence $\{a_n\}$ converges to a limit L if the terms a_n get closer and closer to L as $n \to \infty$.

DEFINITION Limit of a Sequence   A sequence $\{a_n\}$ **converges** to a limit L, and we write

$$\lim_{n\to\infty} a_n = L \qquad \text{or} \qquad a_n \to L$$

if, for every $\epsilon > 0$, there is a number M such that $|a_n - L| < \epsilon$ for all $n > M$. If no limit exists, we say that $\{a_n\}$ **diverges**.

■ **EXAMPLE 2** Proving Convergence of a Sequence   Let $a_n = \dfrac{n+4}{n+1}$. Prove formally that $\lim\limits_{n\to\infty} a_n = 1$.

Solution The definition requires us to find, for every $\epsilon > 0$, a number M such that

$$|a_n - 1| < \epsilon \qquad \text{for all } n > M \qquad \boxed{1}$$

We have

$$|a_n - 1| = \left|\frac{n+4}{n+1} - 1\right| = \frac{3}{n+1}$$

Therefore, $|a_n - 1| < \epsilon$ if

$$\frac{3}{n+1} < \epsilon \qquad \text{or} \qquad n > \frac{3}{\epsilon} - 1$$

It follows that (1) is valid with $M = \dfrac{3}{\epsilon} - 1$. For example, if $\epsilon = 0.01$, then we may take

$$M = \frac{3}{0.01} - 1 = 299. \text{ Thus, } |a_n - 1| < 0.01 \text{ for } n = 300, 301, 302, \ldots . \qquad ■$$

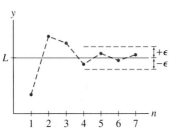

FIGURE 2 Plot of a sequence with limit L. For any ϵ, the dots eventually remain within an ϵ-band around L.

We may visualize a sequence by plotting its "graph," that is, by plotting the points $(1, a_1), (2, a_2), (3, a_3), \ldots$ (Figure 2). The sequence converges to a limit L if, for every $\epsilon > 0$, the plotted points eventually remain within an ϵ-band around the horizontal line $y = L$ (Figure 2). Figure 3 shows the plot of a sequence converging to $L = 1$. On the other hand, it can be shown that the sequence $a_n = \cos n$ shown in Figure 4 has no limit.

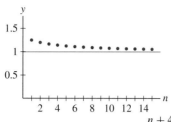

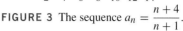

FIGURE 3 The sequence $a_n = \dfrac{n+4}{n+1}$.

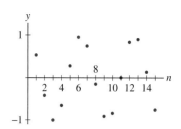

FIGURE 4 The sequence $a_n = \cos n$ has no limit.

We note the following:

- The limit does not change if we modify or drop finitely many terms of the sequence.
- If C is a constant and $a_n = C$ for all n sufficiently large, then $\lim\limits_{n\to\infty} a_n = C$.

Suppose that $f(x)$ is a function and that $f(x)$ approaches a limit L as $x \to \infty$. In this case, the sequence $a_n = f(n)$ approaches the same limit L (Figure 5). Indeed, in this case, for all $\epsilon > 0$, we can find M so that $|f(x) - L| < \epsilon$ for all $x > M$. It follows automatically that $|f(n) - L| < \epsilon$ for all integers $n > M$.

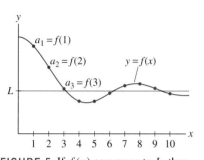

FIGURE 5 If $f(x)$ converges to L, then the sequence $a_n = f(n)$ also converges to L.

> **THEOREM 1 Sequence Defined by a Function** Let $f(x)$ be a function defined on $[c, \infty)$ for some constant c. If $\lim\limits_{x\to\infty} f(x)$ exists, then the sequence $a_n = f(n)$, defined for $n \geq c$, converges and
>
> $$\lim_{n\to\infty} a_n = \lim_{x\to\infty} f(x)$$

■ **EXAMPLE 3** Find the limit of the sequence $0, \dfrac{1}{2}, \dfrac{2}{3}, \dfrac{3}{4}, \dfrac{4}{5} \cdots$.

Solution This is the sequence with general term

$$a_n = \frac{n-1}{n} = 1 - \frac{1}{n}$$

Let $f(x) = 1 - \dfrac{1}{x}$. Then $a_n = f(n)$ and by Theorem 1,

$$\lim_{n\to\infty} a_n = \lim_{x\to\infty} f(x) = \lim_{x\to\infty}\left(1 - \frac{1}{x}\right) = 1 - \lim_{x\to\infty}\frac{1}{x} = 1 \qquad ■$$

■ **EXAMPLE 4** Calculate $\lim\limits_{n\to\infty} a_n$, where $a_n = \dfrac{n + \ln n}{n^2}$.

Solution The limit of the sequence is equal to the limit of the function $f(x) = \dfrac{x + \ln x}{x^2}$, which we calculate using L'Hôpital's Rule:

$$\lim_{n\to\infty} a_n = \lim_{x\to\infty} f(x) = \lim_{x\to\infty}\frac{x + \ln x}{x^2} = \lim_{x\to\infty}\frac{1 + (1/x)}{2x} = 0 \qquad ■$$

The limit of the Balmer wavelengths b_n defined in the margin on page 553 is of interest in physics and chemistry because it determines the ionization energy of the hydrogen

TABLE 1   The Wavelengths in the Balmer Series Approach the Limit $L = 364.5$

**TABLE 1   The Wavelengths
in the Balmer Series
Approach the Limit $L = 364.5$**

n	b_n
3	656.1
4	486
5	433.9
6	410.1
7	396.9
10	379.7
20	368.2
40	365.4
60	364.9
80	364.7
100	364.6

atom. Table 1 suggests that b_n approaches 364.5 as $n \to \infty$. Figure 6 shows the graph of b_n, and in Figure 7, the wavelengths are shown "crowding in" toward their limiting value.

■ **EXAMPLE 5** Limit of Balmer Wavelengths   Calculate the limit of the Balmer wavelengths $b_n = \dfrac{364.5n^2}{n^2 - 4}$, where $n \geq 3$.

Solution Observe that $b_n = f(n)$, where $f(x) = \dfrac{364.5x^2}{x^2 - 4}$. We compute the limit by dividing the numerator and denominator by x^2:

$$\lim_{n \to \infty} b_n = \lim_{x \to \infty} f(x) = 364.5 \lim_{x \to \infty} \frac{x^2}{x^2 - 4} = 364.5 \lim_{x \to \infty} \frac{1}{1 - 4/x^2}$$

$$= \frac{364.5}{\lim_{x \to \infty} (1 - 4/x^2)} = 364.5 \qquad ■$$

The geometric sequence $a_n = cr^n$ is the sequence defined by the exponential function $f(x) = cr^x$ whose base r is the common ratio.

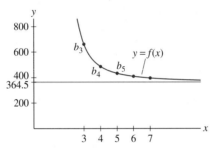

FIGURE 6 The sequence b_n and the function approach the same limit.

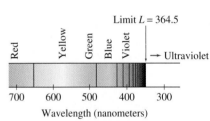

FIGURE 7

A **geometric sequence** is a sequence of the form $a_n = cr^n$, where c and r are nonzero constants. For instance, if $c = 2$ and $r = 3$, we obtain the geometric sequence

$$2, \quad 2 \cdot 3, \quad 2 \cdot 3^2, \quad 2 \cdot 3^3, \quad 2 \cdot 3^4, \quad 2 \cdot 3^5, \quad \ldots$$

The number r is called the **common ratio**. Each term a_n is r times the previous term a_{n-1}, that is, $a_n/a_{n-1} = r$.

We say that $\{a_n\}$ *diverges to* ∞, and we write $\lim_{n \to \infty} a_n = \infty$, if the terms a_n increase beyond all bounds, that is, if, for every $N > 0$, we have $a_n > N$ for all sufficiently large n (Figure 8).

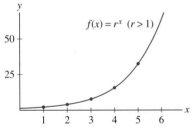

FIGURE 8 If $r > 1$, the geometric sequence $a_n = r^n$ diverges to ∞.

■ **EXAMPLE 6** Limit of a Geometric Sequence   Prove that:

$$\lim_{n \to \infty} r^n = \begin{cases} 0 & \text{if} \quad 0 < r < 1 \\ 1 & \text{if} \quad r = 1 \\ \text{diverges to } \infty & \text{if} \quad r > 1 \end{cases}$$

Solution We apply Theorem 1 to the exponential function $f(x) = r^x$. If $0 < r < 1$, then (Figure 9)

$$\lim_{n \to \infty} r^n = \lim_{x \to \infty} r^x = 0$$

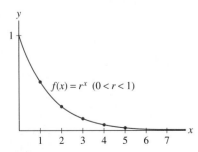

FIGURE 9 If $0 < r < 1$, the geometric sequence $a_n = r^n$ converges to 0.

Similarly, if $r > 1$, then $f(x)$ diverges to ∞ as $x \to \infty$, so $\{r^n\}$ also diverges to ∞ (Figure 8). If $r = 1$, $r^n = 1$ for all n and the limit is 1.   ■

Most of the limit laws for functions also apply to sequences. The proofs are similar and are omitted.

THEOREM 2 Limit Laws for Sequences Assume that $\{a_n\}$ and $\{b_n\}$ are convergent sequences with

$$\lim_{n \to \infty} a_n = L, \qquad \lim_{n \to \infty} b_n = M$$

Then:

(i) $\displaystyle \lim_{n \to \infty} (a_n \pm b_n) = \lim_{n \to \infty} a_n \pm \lim_{n \to \infty} b_n = L \pm M$

(ii) $\displaystyle \lim_{n \to \infty} a_n b_n = \left(\lim_{n \to \infty} a_n \right) \left(\lim_{n \to \infty} b_n \right) = LM$

(iii) $\displaystyle \lim_{n \to \infty} \frac{a_n}{b_n} = \frac{\lim_{n \to \infty} a_n}{\lim_{n \to \infty} b_n} = \frac{L}{M}$ if $M \neq 0$

(iv) $\displaystyle \lim_{n \to \infty} c a_n = c \lim_{n \to \infty} a_n = cL$ for any constant c

---

THEOREM 3 Squeeze Theorem for Sequences Let $\{a_n\}$, $\{b_n\}$, $\{c_n\}$ be sequences such that for some number M,

$$b_n \leq a_n \leq c_n \quad \text{for } n > M \qquad \text{and} \qquad \lim_{n \to \infty} b_n = \lim_{n \to \infty} c_n = L$$

Then $\displaystyle \lim_{n \to \infty} a_n = L$.

■ **EXAMPLE 7** Show that if $\displaystyle \lim_{n \to \infty} |a_n| = 0$, then $\displaystyle \lim_{n \to \infty} a_n = 0$.

Solution We have

$$-|a_n| \leq a_n \leq |a_n|$$

Since $|a_n|$ tends to zero, $-|a_n|$ also tends to zero and the Squeeze Theorem implies that $\displaystyle \lim_{n \to \infty} a_n = 0$. ■

As another application of the Squeeze Theorem, consider the sequence

$$a_n = \frac{5^n}{n!}$$

Both the numerator and denominator tend to infinity, so it is not clear in advance whether $\{a_n\}$ converges. Figure 10 and Table 2 suggest that a_n increases initially and then tends to zero. In the next example, we prove that, indeed, $a_n = \dfrac{R^n}{n!}$ converges to zero for all R. We will use this fact in the discussion of Taylor series in Section 10.7.

■ **EXAMPLE 8** Prove that $\displaystyle \lim_{n \to \infty} \frac{R^n}{n!} = 0$ for all R.

Solution We may assume without loss of generality that $R > 0$ by the result of Example 7. Then there is a unique integer $M \geq 0$ such that

$$M \leq R < M + 1$$

◄⋯ **REMINDER** $n!$ *(n-factorial) is the number*

$$n! = n(n-1)(n-2)\cdots 2 \cdot 1$$

For example, $4! = 4 \cdot 3 \cdot 2 \cdot 1 = 24$.

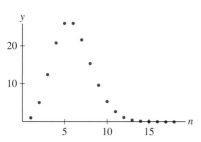

FIGURE 10 Graph of the sequence $a_n = \dfrac{5^n}{n!}$.

TABLE 2

n	$a_n = \dfrac{5^n}{n!}$
1	5
2	12.5
3	20.83
4	26.04
10	2.69
15	0.023
20	0.00004

For $n > M$, we write $R^n/n!$ as a product of n factors:

$$\frac{R^n}{n!} = \underbrace{\left(\frac{R}{1}\frac{R}{2}\cdots\frac{R}{M}\right)}_{\text{Call this constant } C}\underbrace{\left(\frac{R}{M+1}\right)\left(\frac{R}{M+2}\right)\cdots\left(\frac{R}{n}\right)}_{\text{Each factor is less than 1}} \leq C\left(\frac{R}{n}\right)$$

The first M factors are ≥ 1 and the last $n - M$ factors are < 1. If we lump together the first M factors and call the product C, and drop all the remaining factors except the last factor R/n, we obtain

$$0 \leq \frac{R^n}{n!} \leq \frac{CR}{n}$$

Since $\dfrac{CR}{n} \to 0$, the Squeeze Theorem implies that $\displaystyle\lim_{n\to\infty} \frac{R^n}{n!} = 0$. ∎

We can apply a function $f(x)$ to a sequence $\{a_n\}$ to obtain a new sequence $\{f(a_n)\}$. It is useful to know that if $f(x)$ is continuous and $a_n \to L$, then $f(a_n) \to f(L)$. We state this result in the next theorem. See Appendix D for a proof.

THEOREM 4 If $f(x)$ is continuous and the limit $\displaystyle\lim_{n\to\infty} a_n = L$ exists, then

$$\lim_{n\to\infty} f(a_n) = f\left(\lim_{n\to\infty} a_n\right) = f(L)$$

■ **EXAMPLE 9** Calculate $\displaystyle\lim_{n\to\infty} e^{3n/(n+1)}$.

Solution We have $e^{3n/(n+1)} = f(a_n)$, where $a_n = \dfrac{3n}{n+1}$ and $f(x) = e^x$. Furthermore,

$$\lim_{n\to\infty} a_n = \lim_{n\to\infty} \frac{3n}{n+1} = 3$$

By Theorem 4, $\displaystyle\lim_{n\to\infty} f(a_n) = f\left(\lim_{n\to\infty} a_n\right) = f(3)$, that is,

$$\lim_{n\to\infty} e^{3n/(n+1)} = e^{\lim_{n\to\infty} 3n/(n+1)} = e^3$$ ∎

We now introduce two concepts that are important for understanding convergence: the concepts of a bounded sequence and a monotonic sequence.

DEFINITION Bounded Sequences A sequence $\{a_n\}$ is:

- **Bounded from above** if there is a number M such that $a_n \leq M$ for all n. The number M is called an *upper bound*.
- **Bounded from below** if there exists m such that $a_n \geq m$ for all n. The number m is called a *lower bound*.

If $\{a_n\}$ is bounded from above and below, we say that $\{a_n\}$ is *bounded*. If $\{a_n\}$ is not bounded, we call $\{a_n\}$ an *unbounded sequence*.

Upper and lower bounds are not unique. If M is an upper bound, then any number larger than M is also an upper bound (Figure 11). Similarly, if m is a lower bound, then any number smaller than m is also a lower bound.

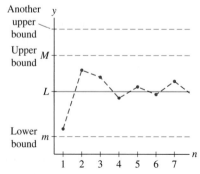

FIGURE 11 A convergent sequence is bounded.

It would seem that a convergent sequence $\{a_n\}$ must be bounded because the terms a_n get closer and closer to the limit (Figure 11). This leads to the next theorem.

THEOREM 5 Convergent Sequences Are Bounded If $\{a_n\}$ converges, then $\{a_n\}$ is bounded.

Proof Let $L = \lim\limits_{n\to\infty} a_n$. Then there exists $N > 0$ such that $|a_n - L| < 1$ for $n > N$. In other words,

$$L - 1 < a_n < L + 1 \qquad \text{for } n > N$$

If M is any number larger than $L + 1$ and also larger than the numbers $a_1, a_2, \ldots, a_N$, then $a_n < M$ for all n. Thus, M is an upper bound. Similarly, any number m smaller than $L - 1$ and $a_1, a_2, \ldots, a_N$ is a lower bound. ∎

There are two ways for a sequence $\{a_n\}$ to be divergent. First, if $\{a_n\}$ is unbounded, then it certainly diverges by Theorem 5. For example, the following sequence diverges:

$$-1, \quad 2, \quad -3, \quad 4, \quad -5, \quad 6, \quad \ldots$$

On the other hand, a sequence may diverge even if it is bounded if the terms a_n bounce around and never settle down to approach a limit. For example, the sequence $a_n = (-1)^n$ is bounded but does not converge:

$$1, \quad -1, \quad 1, \quad -1, \quad 1, \quad -1, \quad \ldots$$

When can we be sure that a sequence converges? One situation is when $\{a_n\}$ is both bounded and **monotonic** increasing or decreasing. The reason, intuitively, is that if $\{a_n\}$ is increasing and bounded above by M, then the terms must eventually bunch up near some limiting value L that is not greater than M (Figure 12). We state this formally in the next theorem, whose proof is given in Appendix B.

THEOREM 6 Bounded Monotonic Sequences Converge

- If $\{a_n\}$ is increasing and $a_n \le M$ for all n, then $\{a_n\}$ converges and $\lim\limits_{n\to\infty} a_n \le M$.
- If $\{a_n\}$ is decreasing and $a_n \ge m$ for all n, then $\{a_n\}$ converges and $\lim\limits_{n\to\infty} a_n \ge m$.

■ **EXAMPLE 10** Verify that $a_n = \sqrt{n+1} - \sqrt{n}$ is decreasing and bounded below. Does $\lim\limits_{n\to\infty} a_n$ exist?

Solution The function $f(x) = \sqrt{x+1} - \sqrt{x}$ is decreasing because its derivative is negative:

$$f'(x) = \frac{1}{2\sqrt{x+1}} - \frac{1}{2\sqrt{x}} < 0 \qquad \text{for } x > 0$$

It follows that $a_n = f(n)$ is also decreasing (see Table 3). The sequence is bounded below by $m = 0$ since $a_n > 0$ for all n. Theorem 6 guarantees that the limit $L = \lim\limits_{n\to\infty} a_n$ exists and $L \ge 0$ (it can be shown that $L = 0$). ■

A sequence $\{a_n\}$ is monotonic

- *Increasing if $a_j \le a_{j+1}$ for all j*
- *Decreasing if $a_j \ge a_{j+1}$ for all j.*

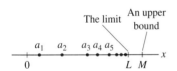

FIGURE 12 An increasing sequence with upper bound M approaches a limit L.

TABLE 3 The Sequence
$a_n = \sqrt{n+1} - \sqrt{n}$ **Is Decreasing**

$a_n = \sqrt{n+1} - \sqrt{n}$
$a_1 \approx 0.414$
$a_2 \approx 0.318$
$a_3 \approx 0.268$
$a_4 \approx 0.213$
$a_5 \approx 0.196$

■ **EXAMPLE 11** Show that the following sequence is bounded and increasing:

$$a_0 = 0, \quad a_1 = \sqrt{2}, \quad a_2 = \sqrt{2 + \sqrt{2}}, \quad a_3 = \sqrt{2 + \sqrt{2 + \sqrt{2}}}, \quad \dots$$

Prove that $L = \lim_{n \to \infty} a_n$ exists and compute its value.

Solution This sequence is defined recursively by

$$a_0 = 0, \qquad a_{n+1} = \sqrt{2 + a_n}$$

It would not be hard to find the limit L if we knew in advance that the limit exists. We could proceed as follows. The sequence $b_n = a_{n+1}$ (the same sequence $\{a_n\}$ but starting with a_1) would converge to the same limit L and we would have, using Theorem 4,

$$L = \lim_{n \to \infty} a_{n+1} = \lim_{n \to \infty} \sqrt{2 + a_n} = \sqrt{2 + \lim_{n \to \infty} a_n} = \sqrt{2 + L}$$

Thus, $L = \sqrt{2 + L}$ and hence

$$L^2 = 2 + L \quad \Rightarrow \quad L^2 - L - 2 = (L - 2)(L + 1) = 0$$

It follows that $L = -1$ or $L = 2$, and since $L \geq 0$, we conclude that $L = 2$ (see Table 4). To justify this conclusion, we must prove that the limit L exists. By Theorem 6, it suffices to show that $\{a_n\}$ is bounded above and increasing.

Step 1. **Show that $\{a_n\}$ is bounded above by $M = 2$.**

First, we observe that

$$\text{if } a_n < 2, \quad \text{then} \quad a_{n+1} = \sqrt{a_n + 2} < \sqrt{2 + 2} = 2 \qquad \boxed{2}$$

Now we can prove that $a_n < 2$ for all n. Since $a_0 = 0$, (2) implies that $a_1 < 2$. But then $a_1 < 2$ implies that $a_2 < 2$ by (2), and $a_2 < 2$ implies $a_3 < 2$, etc. for all n (formally speaking, this is a proof by induction).

Step 2. **Show that $\{a_n\}$ is increasing.**

Since a_n is positive and $a_n < 2$,

$$a_{n+1} = \sqrt{a_n + 2} > \sqrt{a_n + a_n} = \sqrt{2a_n} > \sqrt{a_n \cdot a_n} = a_n$$

Thus, $a_{n+1} > a_n$ for all n and $\{a_n\}$ is increasing.   ■

TABLE 4   Terms of the Recursive Sequence $a_{n+1} = \sqrt{2 + a_n}$

a_0	0
a_1	1.4142
a_2	1.8478
a_3	1.9616
a_4	1.9904
a_5	1.9976

10.1 SUMMARY

• A *sequence* is a function $f(n)$ whose domain is a subset of the integers. We write $a_n = f(n)$ for the nth term and denote the sequence itself by $\{a_n\}$ or simply a_n.

• We say that $\{a_n\}$ *converges* to a limit L, and we write $\lim_{n \to \infty} a_n = L$ or $a_n \to L$ if, for every $\epsilon > 0$, there is a number M such that

$$|a_n - L| < \epsilon \qquad \text{for all } n > M$$

If no limit exists, we say that $\{a_n\}$ *diverges*.

• Let $f(x)$ be a function on $[c, \infty)$ for some number c and let $a_n = f(n)$ for $n \geq c$. If $\lim_{x \to 0} f(x) = L$, then $\lim_{n \to \infty} a_n = L$.

- A *geometric sequence* is a sequence of the form $a_n = cr^n$, where c and r are nonzero.
- The basic Limit Laws and the Squeeze Theorem apply to sequences.
- If $f(x)$ is continuous and $\lim_{n\to\infty} a_n = L$, then $\lim_{n\to\infty} f(a_n) = f(L)$.
- We say that $\{a_n\}$ is *bounded above* by M if $a_n \le M$ for all n and *bounded below* by m if $a_n \ge m$ for all n. If $\{a_n\}$ is bounded above and below, $\{a_n\}$ is called *bounded*.
- A sequence $\{a_n\}$ is *monotonic* if it is increasing ($a_j \le a_{j+1}$) or decreasing ($a_j \ge a_{j+1}$) for all j.
- Theorem 6 states that every increasing sequence that is bounded above and every decreasing sequence that is bounded below converges.

10.1 EXERCISES

Preliminary Questions

1. What is a_4 for the sequence $a_n = n^2 - n$?

2. Which of the following sequences converge to zero?

(a) $\dfrac{n^2}{n^2 + 1}$ **(b)** 2^n **(c)** $\left(\dfrac{-1}{2}\right)^n$

3. Let a_n be the nth decimal approximation to $\sqrt{2}$. That is, $a_1 = 1$, $a_2 = 1.4$, $a_3 = 1.41$, etc. What is $\lim_{n\to\infty} a_n$?

4. Which sequence is defined recursively?

(a) $a_n = \sqrt{2 + n^{-1}}$ **(b)** $b_n = \sqrt{4 + b_{n-1}}$

5. Theorem 5 says that every convergent sequence is bounded. Which of the following statements follow from Theorem 5 and which are false? If false, give a counterexample.

(a) If $\{a_n\}$ is bounded, then it converges.

(b) If $\{a_n\}$ is not bounded, then it diverges.

(c) If $\{a_n\}$ diverges, then it is not bounded.

Exercises

1. Match the sequence with the general term:

$a_1, a_2, a_3, a_4, \ldots$	**General term**
(a) $\frac{1}{2}, \frac{2}{3}, \frac{3}{4}, \frac{4}{5}, \ldots$	(i) $\cos \pi n$
(b) $-1, 1, -1, 1, \ldots$	(ii) $\dfrac{n!}{2^n}$
(c) $1, -1, 1, -1, \ldots$	(iii) $(-1)^{n+1}$
(d) $\frac{1}{2}, \frac{2}{4}, \frac{6}{8}, \frac{24}{16} \cdots$	(iv) $\dfrac{n}{n+1}$

2. Let $a_n = \dfrac{1}{2n - 1}$ for $n = 1, 2, 3, \ldots$. Write out the first three terms of the following sequences.

(a) $b_n = a_{n+1}$ **(b)** $c_n = a_{n+3}$

(c) $d_n = a_n^2$ **(d)** $e_n = 2a_n - a_{n+1}$

In Exercises 3–10, calculate the first four terms of the following sequences, starting with $n = 1$.

3. $c_n = \dfrac{2^n}{n!}$ **4.** $b_n = \cos \pi n$

5. $a_1 = 3$, $a_{n+1} = 1 + a_n^2$ **6.** $b_n = 2 + (-1)^n$

7. $c_n = 1 + \dfrac{1}{2} + \dfrac{1}{3} + \cdots + \dfrac{1}{n}$

8. $a_n = n + (n + 1) + (n + 2) + \cdots + (2n)$

9. $b_1 = 2$, $b_2 = 5$, $b_n = b_{n-1} + 2b_{n-2}$

10. $c_n = n$th decimal approximation to e^{-1}

11. Find a formula for the nth term of the following sequence:

(a) $\dfrac{1}{1}, \dfrac{-1}{8}, \dfrac{1}{27}, \ldots$ **(b)** $\dfrac{2}{6}, \dfrac{3}{7}, \dfrac{4}{8}, \ldots$

12. Suppose that $\lim_{n\to\infty} a_n = 4$ and $\lim_{n\to\infty} b_n = 7$. Determine:

(a) $\lim_{n\to\infty} (a_n + b_n)$ **(b)** $\lim_{n\to\infty} a_n^3$

(c) $\lim_{n\to\infty} 4b_n$ **(d)** $\lim_{n\to\infty} (a_n^2 - 2a_nb_n)$

In Exercises 13–26, use Theorem 1 to determine the limit of the sequence or state that the sequence diverges.

13. $a_n = 4$ **14.** $b_n = \dfrac{3n + 1}{2n + 4}$

15. $a_n = 5 - \dfrac{9}{n^2}$ **16.** $b_n = (-1)^{2n+1}$

17. $c_n = -2^{-n}$ **18.** $c_n = 4(2^n)$

19. $z_n = \left(\dfrac{1}{3}\right)^n$ **20.** $z_n = (0.1)^{-1/n}$

21. $a_n = \dfrac{(-1)^n n^2 + n}{4n^2 + 1}$ **22.** $a_n = \dfrac{n}{\sqrt{n^2 + 1}}$

23. $a_n = \dfrac{n}{\sqrt{n^3 + 1}}$ **24.** $a_n = \sin \pi n$

25. $a_n = \cos \pi n$

26. $a_n = n((1 + n^{-1})^2 - 1)$

27. Let $a_n = \dfrac{n}{n+1}$. Find a number M such that:

(a) $|a_n - 1| \le 0.001$ for $n \ge M$.

(b) $|a_n - 1| \le 0.00001$ for $n \ge M$.

Then use the limit definition to prove that $\lim\limits_{n \to \infty} a_n = 1$.

28. Let $b_n = (\frac{1}{3})^n$.

(a) Find a value of M such that $|b_n| \le 10^{-5}$ for $n \ge M$.

(b) Use the limit definition to prove that $\lim\limits_{n \to \infty} b_n = 0$.

29. Use the limit definition to prove that $\lim\limits_{n \to \infty} n^{-2} = 0$.

30. Find the limit of $d_n = \sqrt{n+3} - \sqrt{n}$.

31. Find $\lim\limits_{n \to \infty} 2^{1/n}$.

32. Show that $\lim\limits_{n \to \infty} b^{1/n}$ is independent of b for $b > 0$.

33. Find $\lim\limits_{n \to \infty} n^{1/n}$.

34. Find the limit of $a_n = n^2(\sqrt[3]{n^3 + 1} - n)$. *Hint:* Write $a_n = \dfrac{(1 + n^{-3})^{1/3} - 1}{n^{-3}}$ and apply L'Hôpital's Rule.

35. Find $\lim\limits_{n \to \infty} \left(1 + \dfrac{1}{n}\right)^n$.

36. Find $\lim\limits_{n \to \infty} \left(1 + \dfrac{1}{n^2}\right)^n$.

37. Use the Squeeze Theorem to find $\lim\limits_{n \to \infty} a_n$, where $a_n = \dfrac{1}{\sqrt{n^4 + n^8}}$ by proving that
$$\frac{1}{\sqrt{2}n^4} \le a_n \le \frac{1}{\sqrt{2}n^2}$$

38. Evaluate $\lim\limits_{n \to \infty} \dfrac{\cos n}{n}$.

39. Evaluate $\lim\limits_{n \to \infty} n \sin \dfrac{1}{n}$.

40. Evaluate $\lim\limits_{n \to \infty} (2^n + 3^n)^{1/n}$. *Hint:* Show that
$$3 \le a_n \le (2 \cdot 3^n)^{1/n}$$

41. [✎] Which statement is equivalent to the assertion $\lim\limits_{n \to \infty} a_n = L$? Explain.

(a) For every $\epsilon > 0$, the interval $(L - \epsilon, L + \epsilon)$ contains at least one element of the sequence $\{a_n\}$.

(b) For every $\epsilon > 0$, the interval $(L - \epsilon, L + \epsilon)$ contains all but at most finitely many elements of the sequence $\{a_n\}$.

42. [✎] Which statement is equivalent to the assertion that $\{a_n\}$ is bounded? Explain.

(a) There exists a finite interval $[m, M]$ containing every element of the sequence $\{a_n\}$.

(b) There exists a finite interval $[m, M]$ containing an element of the sequence $\{a_n\}$.

In Exercises 43–63, determine the limit of the sequence or show that the sequence diverges by using the appropriate Limit Laws or theorems.

43. $a_n = \dfrac{3n^2 + n + 2}{2n^2 - 3}$

44. $a_n = \dfrac{\sqrt{n}}{\sqrt{n} + 4}$

45. $a_n = 3 + \left(-\dfrac{1}{2}\right)^n$

46. $a_n = \left(2 + \dfrac{4}{n^2}\right)^{1/3}$

47. $b_n = \tan^{-1}\left(1 - \dfrac{2}{n}\right)$

48. $b_n = e^{n^2 - n}$

49. $c_n = \ln\left(\dfrac{2n+1}{3n+4}\right)$

50. $c_n = \dfrac{n}{n + n^{1/n}}$

51. $y_n = \dfrac{e^n + 3^n}{5^n}$

52. $y_n = \dfrac{e^n}{2^n}$

53. $a_n = \dfrac{e^n}{2^{n^2}}$

54. $a_n = \dfrac{n}{2^n}$

55. $b_n = \dfrac{n^3 + 2e^{-n}}{3n^3 + 4e^{-n}}$

56. $b_n = \dfrac{3 - 4^n}{2 + 7 \cdot 4^n}$

57. $A_n = \dfrac{3 - 4^n}{2 + 7 \cdot 3^n}$

58. $B_n = \dfrac{10^n}{n!}$

59. $A_n = \dfrac{(-4{,}000)^n}{n!}$

60. $B_n = \dfrac{n!}{2^n}$

61. $a_n = \cos \dfrac{\pi}{n}$

62. $a_n = n \sin \dfrac{\pi}{n}$

63. $a_n = \sqrt[n]{n}$

64. Show that $a_n = \dfrac{1}{2n+1}$ is strictly decreasing.

65. Show that $a_n = \dfrac{3n^2}{n^2 + 2}$ is strictly increasing. Find an upper bound.

66. Show that $a_n = \sqrt[3]{n+1} - n$ is decreasing.

67. Use the limit definition to prove that the limit does not change if a finite number of terms are added or removed from a convergent sequence.

68. Let $b_n = a_{n+1}$. Prove that if $\{a_n\}$ converges, then $\{b_n\}$ also converges and $\lim\limits_{n \to \infty} a_n = \lim\limits_{n \to \infty} b_n$.

69. Let $\{a_n\}$ be a sequence such that $\lim\limits_{n \to \infty} |a_n|$ exists and is nonzero. Show that $\lim\limits_{n \to \infty} a_n$ exists if and only if there exists an integer M such that the sign of a_n does not change for $n > M$.

70. Give an example of a divergent sequence $\{a_n\}$ such that $\lim\limits_{n \to \infty} |a_n|$ converges.

71. Show, by giving an example, that there exist *divergent* sequences $\{a_n\}$ and $\{b_n\}$ such that $\{a_n + b_n\}$ converges.

72. Using the limit definition, prove that if $\{a_n\}$ converges and $\{b_n\}$ diverges, then $\{a_n + b_n\}$ diverges.

73. Use the limit definition to prove that if $\{a_n\}$ is a convergent sequence of integers with limit L, then there exists a number M such that $a_n = L$ for all $n \geq M$.

74. Theorem 1 states that if $\lim\limits_{x \to \infty} f(x) = L$, then the sequence $a_n = f(n)$ converges and $\lim\limits_{n \to \infty} a_n = L$. Show that the *converse* is false. In other words, find a function $f(x)$ such that $a_n = f(n)$ converges but $\lim\limits_{x \to \infty} f(x)$ does not exist.

75. Prove that the following sequence is bounded and increasing. Then find its limit:

$$a_1 = \sqrt{5}, \quad a_2 = \sqrt{5 + \sqrt{5}}, \quad a_3 = \sqrt{5 + \sqrt{5 + \sqrt{5}}}, \ldots$$

76. Let $\{a_n\}$ be the sequence

$$\sqrt{2}, \quad \sqrt{2\sqrt{2}}, \quad \sqrt{2\sqrt{2\sqrt{2}}}, \ldots$$

Show that $\{a_n\}$ is increasing and $0 \leq a_n \leq 2$. Then prove that $\{a_n\}$ converges and find the limit.

77. Find the limit of the sequence

$$c_n = \frac{1}{\sqrt{n^2 + 1}} + \frac{1}{\sqrt{n^2 + 2}} + \cdots + \frac{1}{\sqrt{n^2 + n}}$$

Hint: Show that

$$\frac{n}{\sqrt{n^2 + n}} \leq c_n \leq \frac{n}{\sqrt{n^2 + 1}}$$

Further Insights and Challenges

78. Show that $a_n = \sqrt[n]{n!}$ diverges. *Hint:* Show that $n! \geq (n/2)^{n/2}$ by observing that half of the factors of $n!$ are greater than or equal to $n/2$.

79. Let $b_n = \dfrac{\sqrt[n]{n!}}{n}$.

(a) Show that $\ln b_n = \dfrac{\ln(n!) - n \ln n}{n} = \dfrac{1}{n} \sum\limits_{k=1}^{n} \ln \dfrac{k}{n}$.

(b) Show that $\ln b_n$ converges to $\displaystyle\int_0^1 \ln x \, dx$ and conclude that $b_n \to e^{-1}$.

80. Given positive numbers $a_1 < b_1$, define two sequences recursively by

$$a_{n+1} = \sqrt{a_n b_n}, \qquad b_n = \frac{a_n + b_n}{2}$$

(a) Show that $a_n \leq b_n$ for all n (Figure 13).

(b) Show that $\{a_n\}$ is increasing and $\{b_n\}$ is decreasing.

(c) Show that

$$b_{n+1} - a_{n+1} \leq \frac{b_n - a_n}{2}$$

Prove that both $\{a_n\}$ and $\{b_n\}$ converge and have the same limit. This limit, denoted $\text{AGM}(a_1, b_1)$, is called the **arithmetic-geometric mean** of a_1 and b_1. See Figure 13.

(d) Estimate $\text{AGM}(1, \sqrt{2})$ to three decimal places.

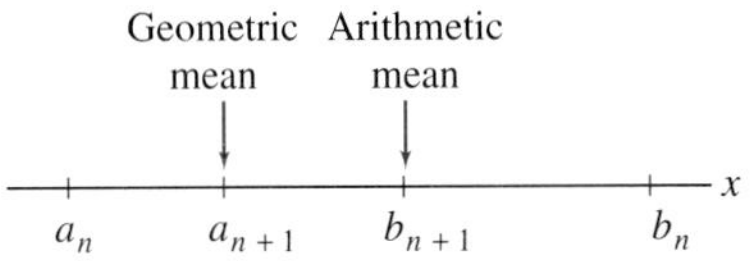

81. Let $c_n = \dfrac{1}{n} + \dfrac{1}{n+1} + \dfrac{1}{n+2} + \cdots + \dfrac{1}{2n}$.

Geometric mean   Arithmetic mean

FIGURE 13

(a) Calculate c_1, c_2, c_3, c_4.

(b) Use a comparison of rectangles with the area under $y = x^{-1}$ over the interval $[n, 2n]$ to prove that

$$\int_n^{2n} \frac{dx}{x} + \frac{1}{2n} \leq c_n \leq \int_n^{2n} \frac{dx}{x} + \frac{1}{n}$$

(c) Use the Squeeze Theorem to determine $\lim\limits_{n \to \infty} c_n$.

82. The nth harmonic number is the number

$$H_n = 1 + \frac{1}{2} + \frac{1}{3} + \cdots + \frac{1}{n}$$

Let $a_n = H_n - \ln n$.

(a) Show that $a_n \geq 0$ for $n \geq 1$. *Hint:* Show that $H_n \geq \displaystyle\int_1^{n+1} \frac{dx}{x}$.

(b) Show that $\{a_n\}$ is decreasing by interpreting $a_n - a_{n+1}$ as an area.

(c) Prove that $\lim\limits_{n \to \infty} a_n$ exists. This limit, denoted γ and known as *Euler's Constant*, appears in many areas of mathematics, including analysis and number theory. It has been calculated to more than 100 million decimal places, but it is still not known if γ is an irrational number. The first 10 digits are $\gamma \approx 0.5772156649$.

10.2 Summing an Infinite Series

Quantities that arise in applications often cannot be computed exactly. We cannot write down an exact decimal expression for the number π or for most values of the sine function such as $\sin 1$. Sometimes these quantities can be represented as infinite sums. For example,

$$\sin 1 = 1 - \frac{1}{3!} + \frac{1}{5!} - \frac{1}{7!} + \frac{1}{9!} - \frac{1}{11!} + \cdots \qquad \boxed{1}$$

Infinite sums of this type are called **infinite series**.

What precisely does Eq. (1) mean? Although it is impossible to add up infinitely many numbers, we can compute the **partial sums** S_N, defined as the first N terms of the infinite series. Let's compare the first few partial sums of the series above with $\sin 1$:

$$S_1 = 1$$

$$S_2 = 1 - \frac{1}{3!} = 1 - \frac{1}{6} \qquad\qquad \approx 0.833$$

$$S_3 = 1 - \frac{1}{3!} + \frac{1}{5!} = 1 - \frac{1}{6} + \frac{1}{120} \qquad \approx 0.841667$$

$$S_4 = 1 - \frac{1}{6} + \frac{1}{120} - \frac{1}{5,040} \qquad \approx 0.841468$$

$$S_5 = 1 - \frac{1}{6} + \frac{1}{120} - \frac{1}{5,040} + \frac{1}{362,880} \approx \mathbf{0.8414709846}$$

$$\sin 1 \approx \mathbf{0.8414709848079}$$

The partial sums appear to converge to $\sin 1$ and, indeed, we will prove that $\lim\limits_{N \to \infty} S_N = \sin 1$ in Section 10.7. This is the precise meaning of Eq. (1).

In general, an infinite series is an expression of the form

$$\sum_{n=1}^{\infty} a_n = a_1 + a_2 + a_3 + \cdots$$

where $\{a_n\}$ is any sequence. For example,

Infinite series may begin with any index. For example,

$$\sum_{n=3}^{\infty} \frac{1}{n} = \frac{1}{3} + \frac{1}{4} + \frac{1}{5} + \cdots$$

When it is not necessary to specify the starting point, we write simply $\sum a_n$. Any letter may be used for the index. Thus, we may write a_m, a_k, a_i, etc.

Sequence	General Term	Infinite Series
$\dfrac{1}{3}, \dfrac{1}{9}, \dfrac{1}{27}, \dots$	$a_n = \dfrac{1}{3^n}$	$\displaystyle\sum_{n=1}^{\infty} \frac{1}{3^n} = \frac{1}{3} + \frac{1}{9} + \frac{1}{27} + \frac{1}{81} + \cdots$
$\dfrac{1}{1}, \dfrac{1}{4}, \dfrac{1}{9}, \dfrac{1}{16}, \dots$	$a_n = \dfrac{1}{n^2}$	$\displaystyle\sum_{n=1}^{\infty} \frac{1}{n^2} = \frac{1}{1^2} + \frac{1}{2^2} + \frac{1}{3^2} + \frac{1}{4^2} + \cdots$

The Nth partial sum S_N is the finite sum of the first N terms of the series:

$$S_N = \sum_{n=1}^{N} a_n = a_1 + a_2 + a_3 + \cdots + a_N$$

The sum of the infinite series $\displaystyle\sum_{n=1}^{\infty} a_n$ is defined as the limit of the sequence of partial sums S_N, if this limit exists.

> **DEFINITION Convergence of an Infinite Series** An infinite series $\sum_{n=1}^{\infty} a_n$ *converges* to S if $\lim_{N \to \infty} S_N = S$. The limit S is called the *sum* of the infinite series, and we write $S = \sum_{n=1}^{\infty} a_n$. If the limit does not exist, the infinite series is said to *diverge*.

It is easy to give examples of series that diverge. For example, $\sum_{n=1}^{\infty} 1$ diverges because the partial sums S_N diverge to ∞:

$$S_N = \sum_{n=1}^{N} 1 = \overbrace{1 + 1 + 1 + 1 + \cdots + 1}^{N \text{ times}} = N$$

Similarly, $\sum_{n=1}^{\infty} (-1)^{n-1}$ diverges because the partial sums jump back and forth between 1 and 0:

$$S_1 = 1, \quad S_2 = 1 - 1 = 0, \quad S_3 = 1 - 1 + 1 = 1, \quad S_4 = 1 - 1 + 1 - 1 = 0, \quad \ldots$$

We may investigate series numerically by computing several partial sums S_N. If the partial sums show a trend of convergence to some number S, then we have evidence (but not proof) that the series converges to S. The next example treats a convergent **telescoping series**, where the partial sums are particularly easy to evaluate.

■ **EXAMPLE 1** Telescoping Series Investigate the following series numerically:

$$S = \sum_{n=1}^{\infty} \frac{1}{n(n+1)} = \frac{1}{1(2)} + \frac{1}{2(3)} + \frac{1}{3(4)} + \frac{1}{4(5)} + \cdots$$

Then compute the sum S using the identity:

$$\frac{1}{n(n+1)} = \frac{1}{n} - \frac{1}{n+1}$$

Although there is a simple formula for the partial sums in Example 1, this is the exception rather than the rule. Apart from telescoping series and the geometric series introduced below, there is usually no formula for S_N, and we must develop techniques for studying infinite series that do not rely on formulas.

Solution Table 1 lists some partial sums, computed with the help of a CAS. These numerics suggest convergence to $S = 1$. To evaluate the limit exactly, we use the identity to rewrite the terms of the series. We find that each partial sum collapses down to two terms due to cancellation:

$$S_1 = \frac{1}{1(2)} = \frac{1}{1} - \frac{1}{2}$$

$$S_2 = \frac{1}{1(2)} + \frac{1}{2(3)} = \left(\frac{1}{1} - \frac{1}{2} \right) + \left(\frac{1}{2} - \frac{1}{3} \right) = 1 - \frac{1}{3}$$

$$S_3 = \frac{1}{1(2)} + \frac{1}{2(3)} + \frac{1}{3(4)} = \left(\frac{1}{1} - \frac{1}{2} \right) + \left(\frac{1}{2} - \frac{1}{3} \right) + \left(\frac{1}{3} - \frac{1}{4} \right) = 1 - \frac{1}{4}$$

In general,

$$S_N = \left(\frac{1}{1} - \frac{1}{2} \right) + \left(\frac{1}{2} - \frac{1}{3} \right) + \cdots + \left(\frac{1}{N} - \frac{1}{N+1} \right) = 1 - \frac{1}{N+1}$$

TABLE 1   Partial Sums
for $\sum_{n=1}^{\infty} \dfrac{1}{n(n+1)}$

N	S_N
10	0.90909
50	0.98039
100	0.990099
200	0.995025
300	0.996678

Now we may compute the sum S as the limit of the partial sums:

$$S = \lim_{N \to \infty} S_N = \lim_{N \to \infty} \left(1 - \frac{1}{N+1} \right) = 1$$

It is important to keep in mind the difference between a sequence $\{a_n\}$ and an infinite series $\sum_{n=1}^{\infty} a_n$, which is a sum of terms of a sequence.

■ **EXAMPLE 2** Difference between a Sequence and a Series   Discuss the difference between $\{a_n\}$ and $\sum_{n=0}^{\infty} a_n$, where $a_n = 3^{-n}$.

Solution The sequence $a_n = 3^{-n}$ converges to zero:

$$1, \quad \frac{1}{3}, \quad \frac{1}{3^2}, \quad \frac{1}{3^3}, \quad \to 0$$

The infinite series defined by this sequence is an infinite sum:

$$\sum_{n=0}^{\infty} 3^{-n} = 1 + \frac{1}{3} + \frac{1}{3^2} + \cdots$$

The value of this sum is nonzero. In fact, the partial sum S_5 gives an approximation to this sum:

$$S_5 = 1 + \frac{1}{3} + \frac{1}{9} + \frac{1}{27} + \frac{1}{81} \approx 1.494$$

One of the most important types of infinite series is the **geometric series**, defined as the sum of terms cr^n, where c and r are fixed nonzero numbers:

$$\sum_{n=0}^{\infty} cr^n = c + cr + cr^2 + cr^3 + cr^4 + cr^5 + \cdots$$

The nonzero number r is called the **common ratio**.

For $r = \frac{1}{2}$, we can visualize the sum of the geometric series (Figure 1):

$$\sum_{n=1}^{\infty} \frac{1}{2^n} = \frac{1}{2} + \frac{1}{4} + \frac{1}{8} + \frac{1}{16} + \cdots = 1$$

The sum is 1 because adding terms in the series corresponds to moving stepwise from 0 to 1, where each step consists of a move to the right by half of the remaining distance.

A simple device exists for computing the partial sums of a geometric series:

$$S_N = c + cr + cr^2 + cr^3 + \cdots + cr^N$$
$$r S_N = cr + cr^2 + cr^3 + \cdots + cr^N + cr^{N+1}$$
$$S_N - r S_N = S_N(1 - r) = c - cr^{N+1} = c(1 - r^{N+1})$$

If $r \neq 1$, we may divide by $(1 - r)$ to obtain

$$\boxed{S_N = c + cr + cr^2 + cr^3 + \cdots + cr^N = \frac{c(1 - r^{N+1})}{1 - r}}$$ $\boxed{2}$

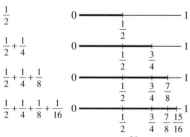

FIGURE 1 Partial sums of $\sum_{n=1}^{\infty} \frac{1}{2^n}$.

Geometric series are important because they

- *Arise often in applications.*
- *Can be evaluated explicitly.*
- *Are used to study other nongeometric series (by comparison).*

> **THEOREM 1 Sum of a Geometric Series** A geometric series with common ratio r converges if $|r| < 1$ and diverges if $|r| \geq 1$. Furthermore,
>
> $$\sum_{n=0}^{\infty} cr^n = \frac{c}{1-r}, \qquad |r| < 1 \qquad \boxed{3}$$
>
> $$\sum_{n=M}^{\infty} cr^n = \frac{cr^M}{1-r}, \qquad |r| < 1 \qquad \boxed{4}$$

Proof If $r \neq 1$, then by Eq. (2),

$$\lim_{N \to \infty} S_N = \lim_{N \to \infty} \frac{c(1 - r^{N+1})}{1-r} = \frac{c}{1-r}\left(1 - \lim_{N \to \infty} r^{N+1}\right)$$

If $|r| < 1$, then $\displaystyle\lim_{N \to \infty} r^{N+1} = 0$ and we obtain Eq. (3):

$$\sum_{n=0}^{\infty} cr^n = \lim_{N \to \infty} S_N = \frac{c}{1-r}$$

If $|r| > 1$, $\displaystyle\lim_{N \to \infty} r^{N+1}$ diverges and thus the geometric series diverges. It also diverges in the borderline cases $r = \pm 1$, as we saw in the discussion before Example 1. If the geometric series starts with the term cr^M rather than cr^0, then

$$\sum_{n=M}^{\infty} cr^n = \sum_{n=0}^{\infty} cr^{M+n} = \sum_{n=0}^{\infty} cr^M r^n = r^M \sum_{n=0}^{\infty} cr^n = \frac{cr^M}{1-r} \qquad \blacksquare$$

■ **EXAMPLE 3** Evaluate $\displaystyle\sum_{n=3}^{\infty} 7\left(-\frac{3}{4}\right)^n$.

Solution This is the geometric series with $c = 7$ and $r = -\dfrac{3}{4}$. The general term is $cr^n = 7\left(-\dfrac{3}{4}\right)^n$ and the sum starts at $n = 3$. By Eq. (4), the sum is

$$S = \sum_{n=3}^{\infty} 7\left(-\frac{3}{4}\right)^n = \frac{7\left(-\frac{3}{4}\right)^3}{1 - \left(-\frac{3}{4}\right)} = -\frac{27}{16} \qquad \blacksquare$$

A main goal in this chapter is to develop techniques for determining whether a series converges or diverges. Sometimes, it is obvious that an infinite series diverges. For example, $\displaystyle\sum_{k=1}^{\infty} 1$ diverges because its Nth partial sum is $S_N = N$. It is less clear if the following series converges or diverges:

$$\sum_{n=1}^{\infty} \frac{(-1)^{n+1} n}{n+1} = \frac{1}{2} - \frac{2}{3} + \frac{3}{4} - \frac{4}{5} + \frac{5}{6} - \cdots$$

We show that this series diverges in Example 4, using the next theorem.

> **THEOREM 2 Divergence Test** If $\{a_n\}$ does not converge to zero, then $\displaystyle\sum_{n=1}^{\infty} a_n$ diverges.

Proof We use the relation

$$S_n = a_1 + a_2 + \cdots + a_{n-1} + a_n = S_{n-1} + a_n$$

to write $a_n = S_n - S_{n-1}$. If $\displaystyle\sum_{n=1}^{\infty} a_n$ is convergent with sum S, then

$$\lim_{n\to\infty} a_n = \lim_{n\to\infty} (S_n - S_{n-1}) = \lim_{n\to\infty} S_n - \lim_{n\to\infty} S_{n-1} = S - S = 0$$

Therefore, if $\{a_n\}$ does not converge to zero, $\displaystyle\sum_{n=1}^{\infty} a_n$ must diverge. ∎

■ **EXAMPLE 4** Using the Divergence Test Does

$$\sum_{n=1}^{\infty} (-1)^n \frac{n}{n+1} = -\frac{1}{2} + \frac{2}{3} - \frac{3}{4} + \frac{4}{5} - \cdots$$

converge?

Solution The general term $a_n = (-1)^n \dfrac{n}{n+1}$ does not tend to zero. In fact, $\dfrac{n}{n+1}$ tends to 1, so the even terms a_{2n} tend to 1 and the odd terms a_{2n+1} to -1. Therefore, the infinite series diverges by Theorem 2. ∎

The Divergence Test only tells part of the story. If $\{a_n\}$ does not tend to zero, then $\sum a_n$ certainly diverges. But what if $\{a_n\}$ does tend to zero? In this case, the series may or may not converge. Here is an example of a series that diverges even though its terms tend to zero.

■ **EXAMPLE 5** A Divergent Series Where $\{a_n\}$ Tends to Zero Show that

$$\sum_{n=1}^{\infty} \frac{1}{\sqrt{n}} = \frac{1}{\sqrt{1}} + \frac{1}{\sqrt{2}} + \frac{1}{\sqrt{3}} + \cdots$$

is divergent.

Solution Each term in the Nth partial sum is greater than or equal to $1/\sqrt{N}$:

$$S_N = \overbrace{\frac{1}{\sqrt{1}} + \frac{1}{\sqrt{2}} + \cdots + \frac{1}{\sqrt{N}}}^{N \text{ terms}} \geq \frac{1}{\sqrt{N}} + \frac{1}{\sqrt{N}} + \cdots + \frac{1}{\sqrt{N}} = N\left(\frac{1}{\sqrt{N}}\right)$$

Therefore,

$$S_N \geq N\left(\frac{1}{\sqrt{N}}\right) = \sqrt{N}$$

Since $S_N \geq \sqrt{N}$, we have $\displaystyle\lim_{N\to\infty} S_N = \infty$ and the series diverges. ∎

The next theorem shows that infinite series may be added or subtracted like ordinary sums, *provided that the series are convergent*.

> **THEOREM 3 Linearity of Infinite Series**   If $\sum a_n$ and $\sum b_n$ are both convergent,
> then $\sum (a_n \pm b_n)$ and $\sum c a_n$ are convergent (c any constant), and
>
> $$\sum a_n + \sum b_n = \sum (a_n + b_n)$$
> $$\sum a_n - \sum b_n = \sum (a_n - b_n)$$
> $$\sum c a_n = c \sum a_n \qquad (c \text{ any constant})$$

Proof   These rules follow from the corresponding linearity rules for limits. For the first rule, we have

$$\sum_{n=1}^{\infty} (a_n + b_n) = \lim_{N \to \infty} \sum_{n=1}^{N} (a_n + b_n) = \lim_{N \to \infty} \left(\sum_{n=1}^{N} a_n + \sum_{n=1}^{N} b_n \right)$$

$$= \lim_{N \to \infty} \sum_{n=1}^{N} a_n + \lim_{N \to \infty} \sum_{n=1}^{\infty} b_n = \sum_{n=1}^{\infty} a_n + \sum_{n=1}^{\infty} b_n$$

The remaining statements are proved similarly.   ∎

■ **EXAMPLE 6**   Evaluate $S = \displaystyle\sum_{n=0}^{\infty} \frac{2 + 3^n}{5^n}$.

Solution   We write the series as a sum of two geometric series. This is valid by Theorem 3 because both geometric series are convergent:

$$\sum_{n=0}^{\infty} \frac{2 + 3^n}{5^n} = \sum_{n=0}^{\infty} \frac{2}{5^n} + \sum_{n=0}^{\infty} \frac{3^n}{5^n} = \overbrace{2 \sum_{n=0}^{\infty} \frac{1}{5^n} + \sum_{n=0}^{\infty} \left(\frac{3}{5} \right)^n}^{\text{Both convergent geometric series}}$$

$$= 2 \cdot \frac{1}{1 - \frac{1}{5}} + \frac{1}{1 - \frac{3}{5}} = 5$$ ■

CONCEPTUAL INSIGHT   Sometimes, the following *incorrect argument* is given for summing a geometric series:

$$S = \frac{1}{2} + \frac{1}{4} + \frac{1}{8} + \cdots$$

$$2S = 1 + \frac{1}{2} + \frac{1}{4} + \frac{1}{8} + \cdots = 1 + S$$

Thus, $2S = 1 + S$ or $S = 1$. The answer is correct, so why is the argument wrong? It is wrong because we do not know in advance that the geometric series converges. Observe what happens when this argument is applied to a divergent series:

$$S = 1 - 1 + 1 - 1 + \cdots$$

$$-S = -1 + (1 - 1 + \cdots) = -1 + S$$

This would yield $-S = -1 + S$ or $S = \frac{1}{2}$, which is clearly incorrect since S diverges. Mathematicians developed the formal definition of the sum of an infinite series as the limit of partial sums in order to avoid erroneous conclusions of this type.

Archimedes (287 BC–212 BC), who discovered the law of the lever, said "Give me a place to stand on, and I can move the earth" (quoted by Pappus of Alexandria c. AD 340).

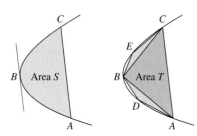

FIGURE 2 Archimedes showed that the area S of the parabolic segment is $\frac{4}{3}T$, where T is the area of $\triangle ABC$.

HISTORICAL PERSPECTIVE

Infinite series have been a part of calculus since the beginning of the subject and they have remained an indispensable tool of mathematical analysis ever since. Geometric series were used as early as the third century BC by Archimedes in a brilliant argument for determining the area S of a parabolic segment (shaded region in Figure 2). Archimedes's result is equivalent to our formula for the integral of $f(x) = x^2$, but he discovered it 2,000 years before the invention of calculus. Archimedes expressed his result geometrically rather than in terms of functions (which had not yet been invented). Given any two points A and C on a parabola, we may choose B between A and C so that the tangent at B is parallel to $\overline{AC}$. Let T be the area of triangle $\triangle ABC$. Archimedes proved that if D is chosen in a similar fashion relative to $\overline{AB}$ and E relative to $\overline{BC}$, then

$$\frac{1}{4}T = \frac{1}{4}\text{Area}(\triangle ABC)$$

$$= \text{Area}(\triangle ADB) + \text{Area}(\triangle BEC) \quad \boxed{5}$$

This construction of triangles can be continued. The next step would be to construct the four triangles on the segments $\overline{AD}$, $\overline{DB}$, $\overline{BE}$, $\overline{EC}$, of total area $(1/4)^2 T$, etc. In this way, we obtain infinitely many triangles that completely fill up the parabolic segment. By Eq. (5) and the formula for the sum of a geometric series,

$$S = T + \frac{1}{4}T + \frac{1}{16}T + \cdots = T\sum_{n=0}^{\infty}\frac{1}{4^n} = \frac{4}{3}T$$

For this and many other achievements, Archimedes is ranked as one of the greatest scientists of all time, in the same league as Newton and Gauss.

The modern study of infinite series began in the seventeenth century with Newton, Leibniz, and their contemporaries. The divergence of $\sum_{n=1}^{\infty}\frac{1}{n}$ (called the **harmonic series**) was known to the medieval scholar Nicole d'Oresme (1323–1382), but his proof was lost for centuries and the result was rediscovered on more than one occasion. It was also known that the sum of the reciprocal squares $\sum_{n=1}^{\infty}\frac{1}{n^2}$ converges and, in the 1640s, the Italian Pietro Mengoli put forward the challenge of finding its sum. Despite the efforts of the best mathematicians of the day, including Leibniz and the Bernoulli brothers Jakob and Johann, the problem resisted solution for nearly a century. In 1735, the great master Leonhard Euler astonished his contemporaries by proving that

$$\frac{1}{1^2} + \frac{1}{2^2} + \frac{1}{3^2} + \frac{1}{4^2} + \frac{1}{5^2} + \frac{1}{6^2} + \cdots = \frac{\pi^2}{6}$$

$$\boxed{6}$$

This formula is used in a variety of ways in number theory. For example, the probability p that two whole numbers, chosen randomly, have no common factor is $p = 6/\pi^2 \approx 0.6$ (the reciprocal of Euler's result). This application and others like it lie in the realm of "pure mathematics," and for hundreds of years, it seemed that Euler's result had no real-world applications. Surprisingly, there is now evidence that this result and its generalizations may play a role in the area of advanced physics called quantum field theory. History seems to show that even the "purest" branches of mathematics are connected to the real world.

10.2 SUMMARY

• An *infinite series* is an expression

$$\sum_{n=1}^{\infty} a_n = a_1 + a_2 + a_3 + a_4 + \cdots$$

We call a_n the *general term* of the series.

- The Nth *partial sum* is the finite sum:

$$S_N = \sum_{n=1}^{N} a_n = a_1 + a_2 + a_3 + \cdots + a_N$$

If the limit $S = \lim_{N \to \infty} S_N$ exists, we say that the infinite series is *convergent* or *converges* to the sum S. If the limit does not exist, the infinite series is called *divergent*.

- *Divergence Test:* If $\{a_n\}$ does not tend to zero, then $\sum_{n=1}^{\infty} a_n$ diverges. However, a series may diverge, even if its general term $\{a_n\}$ tends to zero.

- A *geometric series* with common ratio r satisfying $|r| < 1$ converges:

$$\sum_{n=M}^{\infty} cr^n = cr^M + cr^{M+1} + cr^{M+2} + \cdots = \frac{cr^M}{1-r}$$

The geometric series diverges for $|r| \geq 1$. There is a formula for the partial sum:

$$c + cr + cr^2 + cr^3 + \cdots + cr^N = \frac{c(1 - r^{N+1})}{1-r}$$

10.2 EXERCISES

Preliminary Questions

1. What role do partial sums play in defining the sum of an infinite series?

2. What is the sum of the following infinite series?

$$\frac{1}{4} + \frac{1}{8} + \frac{1}{16} + \frac{1}{32} + \frac{1}{64} + \cdots$$

3. What happens if you apply the formula for the sum of a geometric series to the following series? Is the formula valid?

$$1 + 3 + 3^2 + 3^3 + 3^4 + \cdots$$

4. Arvind asserts that $\sum_{n=1}^{\infty} \frac{1}{n^2} = 0$ because $\frac{1}{n^2}$ tends to zero. Is this valid reasoning?

5. Colleen claims that $\sum_{n=1}^{\infty} \frac{1}{\sqrt{n}}$ converges because $\lim_{n \to \infty} \frac{1}{\sqrt{n}} = 0$. Is this valid reasoning?

6. Find an N such that $S_N > 25$ for the series $\sum_{n=1}^{\infty} 2$.

7. Does there exist an N such that $S_N > 25$ for the series $\sum_{n=1}^{\infty} 2^{-n}$? Explain.

8. Give an example of a divergent infinite series whose general term tends to zero.

Exercises

1. Find a formula for the general term a_n (not the partial sum) of the infinite series.

(a) $\dfrac{1}{3} + \dfrac{1}{9} + \dfrac{1}{27} + \dfrac{1}{81} + \cdots$

(b) $\dfrac{1}{1} + \dfrac{5}{2} + \dfrac{25}{4} + \dfrac{125}{8} + \cdots$

(c) $\dfrac{1}{1} - \dfrac{2^2}{2 \cdot 1} + \dfrac{3^3}{3 \cdot 2 \cdot 1} - \dfrac{4^4}{4 \cdot 3 \cdot 2 \cdot 1} + \cdots$

(d) $\dfrac{2}{1^2 + 1} + \dfrac{1}{2^2 + 1} + \dfrac{2}{3^2 + 1} + \dfrac{1}{4^2 + 1} + \cdots$

2. Write in summation notation:

(a) $1 + \dfrac{1}{4} + \dfrac{1}{9} + \dfrac{1}{16} + \cdots$

(b) $\dfrac{1}{9} + \dfrac{1}{16} + \dfrac{1}{25} + \dfrac{1}{36} + \cdots$

(c) $1 - \dfrac{1}{3} + \dfrac{1}{5} - \dfrac{1}{7} + \cdots$

(d) $\dfrac{125}{9} + \dfrac{625}{16} + \dfrac{3{,}125}{25} + \dfrac{15{,}625}{36} + \cdots$

In Exercises 3–6, compute the partial sums S_2, S_4, and S_6.

3. $1 + \dfrac{1}{2^2} + \dfrac{1}{3^2} + \dfrac{1}{4^2} + \cdots$

4. $\displaystyle\sum_{k=1}^{\infty} (-1)^k k^{-1}$

5. $\dfrac{1}{1 \cdot 2} + \dfrac{1}{2 \cdot 3} + \dfrac{1}{3 \cdot 4} + \cdots$

6. $\displaystyle\sum_{j=1}^{\infty} \dfrac{1}{j!}$

7. Compute S_5, S_{10}, and S_{15} for the series

$$S = \dfrac{1}{2 \cdot 3 \cdot 4} - \dfrac{1}{4 \cdot 5 \cdot 6} + \dfrac{1}{6 \cdot 7 \cdot 8} - \dfrac{1}{8 \cdot 9 \cdot 10} + \cdots$$

This series S is known to converge to $\dfrac{\pi - 3}{4}$. Do your calculations support this conclusion?

8. The series

$$S = \dfrac{1}{0!} - \dfrac{1}{1!} + \dfrac{1}{2!} - \dfrac{1}{3!} + \cdots$$

is known to converge to e^{-1} (recall that $0! = 1$). Find a partial sum that approximates e^{-1} with an error at most 10^{-3}.

9. Calculate S_3, S_4, and S_5 and then find the sum of the telescoping series

$$S = \sum_{n=1}^{\infty} \left(\dfrac{1}{n+1} - \dfrac{1}{n+2} \right)$$

10. Calculate S_3, S_4, and S_5 and then find the sum $S = \displaystyle\sum_{n=1}^{\infty} \dfrac{1}{4n^2 - 1}$ using the identity

$$\dfrac{1}{4n^2 - 1} = \dfrac{1}{2} \left(\dfrac{1}{2n - 1} - \dfrac{1}{2n + 1} \right)$$

11. Write $\displaystyle\sum_{n=3}^{\infty} \dfrac{1}{n(n-1)}$ as a telescoping series and find its sum.

12. Find a formula for the partial sums S_N of $\displaystyle\sum_{n=1}^{\infty} (-1)^{n-1}$ and show that the series diverges.

In Exercises 13–16, use Theorem 2 to prove that the following series diverge.

13. $\displaystyle\sum_{n=1}^{\infty} (-1)^n n^2$

14. $\dfrac{0}{1} - \dfrac{1}{2} + \dfrac{2}{3} - \dfrac{3}{4} + \cdots$

15. $\displaystyle\sum_{n=1}^{\infty} \left(\sqrt{n+1} - \sqrt{n} \right)$

16. $\cos \frac{1}{2} + \cos \frac{1}{3} + \cos \frac{1}{4} + \cdots$

17. Which of these series converge?

(a) $\displaystyle\sum_{n=1}^{\infty} \left(\dfrac{1}{\sqrt{n}} - \dfrac{1}{\sqrt{n+1}} \right)$

(b) $\displaystyle\sum_{n=1}^{\infty} (\ln n - \ln(n+1))$

In Exercises 18–31, use the formula for the sum of a geometric series to find the sum or state that the series diverges.

18. $1 + \dfrac{1}{5} + \dfrac{1}{5^2} + \dfrac{1}{5^3} + \cdots$

19. $\dfrac{1}{3^3} + \dfrac{1}{3^4} + \dfrac{1}{3^5} + \cdots$

20. $\displaystyle\sum_{n=0}^{\infty} \dfrac{3^n}{11^n}$

21. $\displaystyle\sum_{n=3}^{\infty} \dfrac{3^n}{11^n}$

22. $\displaystyle\sum_{n=0}^{\infty} \dfrac{7 \cdot 3^n}{11^n}$

23. $1 + \dfrac{2}{7} + \dfrac{2^2}{7^2} + \dfrac{2^3}{7^3} + \cdots$

24. $\displaystyle\sum_{n=1}^{\infty} e^{-n}$

25. $\displaystyle\sum_{n=2}^{\infty} e^{3-2n}$

26. $\displaystyle\sum_{n=0}^{\infty} \dfrac{8 + 2^n}{5^n}$

27. $\displaystyle\sum_{n=0}^{\infty} \dfrac{93^n + 4^{n-2}}{5^n}$

28. $5 - \dfrac{5}{4} + \dfrac{5}{4^2} - \dfrac{5}{4^3} + \cdots$

29. $\dfrac{2^3}{7} + \dfrac{2^4}{7^2} + \dfrac{2^5}{7^3} + \dfrac{2^6}{7^4} + \cdots$

30. $\dfrac{7}{8} - \dfrac{49}{64} + \dfrac{343}{512} - \dfrac{2{,}401}{4{,}096} + \cdots$

31. $\dfrac{64}{49} + \dfrac{8}{7} + 1 + \dfrac{7}{8} + \dfrac{49}{64} + \dfrac{343}{512} + \cdots$

32. Which of the following are *not* geometric series?

(a) $\displaystyle\sum_{n=0}^{\infty} \dfrac{7^n}{29^n}$

(b) $\displaystyle\sum_{n=3}^{\infty} \dfrac{1}{n^4}$

(c) $\displaystyle\sum_{n=0}^{\infty} \dfrac{n^2}{2^n}$

(d) $\displaystyle\sum_{n=5}^{\infty} \pi^{-n}$

33. Which of the following series are divergent?

(a) $\displaystyle\sum_{n=0}^{\infty} \dfrac{2^n}{5^n}$

(b) $\displaystyle\sum_{n=3}^{\infty} 1.5^n$

(c) $\displaystyle\sum_{n=0}^{\infty} \dfrac{5^n}{2^n}$

(d) $\displaystyle\sum_{n=0}^{\infty} (0.4)^n$

34. ✏ Explain why each of the following statements is incorrect.

(a) If the general term a_n tends to zero, $\displaystyle\sum_{n=1}^{\infty} a_n = 0$.

(b) The Nth partial sum of the infinite series defined by $\{a_n\}$ is a_N.

(c) If a_n tends to zero, then $\displaystyle\sum_{n=1}^{\infty} a_n$ converges.

(d) If a_n tends to L, then $\displaystyle\sum_{n=1}^{\infty} a_n = L$.

35. Let $S = \displaystyle\sum_{n=1}^{\infty} a_n$ be an infinite series such that $S_N = 5 - \dfrac{2}{N^2}$.

(a) What are the values of $\displaystyle\sum_{n=1}^{10} a_n$ and $\displaystyle\sum_{n=4}^{16} a_n$?

(b) What is the value of a_3?

(c) Find a general formula for a_n.

(d) Find the sum $\sum_{n=1}^{\infty} a_n$.

36. Compute the total area of the (infinitely many) triangles in Figure 3.

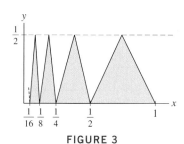

FIGURE 3

37. Use the method of Example 5 to show that $\sum_{k=1}^{\infty} \dfrac{1}{k^{1/3}}$ diverges.

38. Let S_N be the Nth partial sum of the **harmonic series** $S = \sum_{n=1}^{\infty} \dfrac{1}{n}$.

(a) Verify the following inequality for $n = 1, 2, 3$. Then prove it for general n.

$$\frac{1}{2^{n-1}+1} + \frac{1}{2^{n-1}+2} + \frac{1}{2^{n-1}+3} + \cdots + \frac{1}{2^n} \geq \frac{1}{2}$$

(b) Prove that S diverges by showing that $S_N \geq 1 + \dfrac{n}{2}$ for $N = 2^n$.

Hint: Break up S_N into $n + 1$ sums of length $1, 2, 4, 8 \ldots$, as in the following:

$$S_{2^3} = 1 + \left(\frac{1}{2}\right) + \left(\frac{1}{3} + \frac{1}{4}\right) + \left(\frac{1}{5} + \frac{1}{6} + \frac{1}{7} + \frac{1}{8}\right)$$

39. A ball dropped from a height of 10 ft begins to bounce. Each time it strikes the ground, it returns to two-thirds of its previous height. What is the total distance traveled by the ball if it bounces infinitely many times?

40. Use partial fractions to rewrite $\sum_{n=1}^{\infty} \dfrac{1}{n(n+3)}$ as a telescoping series and find its sum.

41. Find the sum of $\dfrac{1}{1 \cdot 3} + \dfrac{1}{3 \cdot 5} + \dfrac{1}{5 \cdot 7} + \cdots$.

42. Let $S = \sum_{n=1}^{\infty} \left(\dfrac{1}{n} - \dfrac{1}{n+2}\right)$. Compute S_N for $N = 1, 2, 3, 4$. Find S by showing that

$$S_N = \frac{3}{2} - \frac{1}{N+1} - \frac{1}{N+2}$$

43. Let $\{b_n\}$ be a sequence and let $a_n = b_n - b_{n-1}$. Show that $\sum_{n=1}^{\infty} a_n$ converges if and only if $\lim_{n \to \infty} b_n$ exists.

44. The winner of a lottery receives m dollars at the end of each year for N years. The present value (PV) of this prize in today's dollars is $\text{PV} = \sum_{i=1}^{N} m(1+r)^{-i}$, where r is the interest rate. Calculate PV if $m = \$50,000$, $r = 0.06$, and $N = 20$. What is the PV if $N = \infty$?

45. Find the total length of the infinite zigzag path in Figure 4 (each zag occurs at an angle of $\frac{\pi}{4}$).

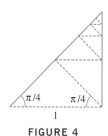

FIGURE 4

46. Evaluate $\sum_{n=1}^{\infty} \dfrac{1}{n(n+1)(n+2)}$. *Hint:* Find constants A, B, and C such that

$$\frac{1}{n(n+1)(n+2)} = \frac{A}{n} + \frac{B}{(n+1)} + \frac{C}{n+2}$$

47. Show that if a is a positive integer, then

$$\sum_{n=1}^{\infty} \frac{1}{n(n+a)} = \frac{1}{a}\left(1 + \frac{1}{2} + \cdots + \frac{1}{a}\right)$$

48. Assumptions Matter Show, by giving counterexamples, that the assertions of Theorem 3 is not valid if the series $\sum_{n=0}^{\infty} a_n$ and $\sum_{n=0}^{\infty} b_n$ are not convergent.

Further Insights and Challenges

49. Professor George Andrews of Pennsylvania State University observed that geometric sums can be used to calculate the derivative of $f(x) = x^N$ in a new way. By Eq. (2),

$$1 + r + r^2 + \cdots + r^{N-1} = \frac{1 - r^N}{1 - r} \qquad \boxed{7}$$

Assume that $a \neq 0$ and let $x = ra$. Show that

$$f'(a) = \lim_{x \to a} \frac{x^N - a^N}{x - a} = a^{N-1} \lim_{r \to 1} \frac{r^N - 1}{r - 1}$$

Then use Eq. (7) to evaluate the limit on the right.

50. Pierre de Fermat used geometric series to compute the area under the graph of $f(x) = x^N$ over $[0, A]$. For $0 < r < 1$, let $F(r)$ be the sum of the areas of the infinitely many right-endpoint rectangles with endpoints Ar^n, as in Figure 5. As r tends to 1, the rectangles become narrower and $F(r)$ tends to the area under the graph.

(a) Show that $F(r) = A^{N+1} \dfrac{1 - r}{1 - r^{N+1}}$.

(b) Use Eq. (7) to evaluate $\displaystyle\int_0^A x^N \, dx = \lim_{r \to 1} F(r)$.

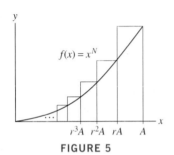

$f(x) = x^N$

r^3A r^2A rA A

FIGURE 5

51. Cantor's Disappearing Table (following Larry Knop of Hamilton College)   Take a table of length L (Figure 6). At stage 1, remove the section of length $L/4$ centered at the midpoint. Two sections remain, each with a length less than $L/2$. At stage 2, remove sections of length $L/4^2$ from each of these two sections (this stage removes $L/8$ of the table). Now four sections remain, each of length less than $L/4$. At stage 3, remove the four central sections of length $L/4^3$, etc.

(a) Show that at the Nth stage, each remaining section has length less than $L/2^N$ and that the total amount of table removed is

$$L \left(\frac{1}{4} + \frac{1}{8} + \frac{1}{16} + \cdots + \frac{1}{2^{N+1}} \right)$$

(b) Show that in the limit as $N \to \infty$, precisely one-half of the table remains.

This result is curious, because there are no nonzero intervals of table left (at each stage, the remaining sections have a length less than $L/2^N$). So the table has "disappeared." However, we can place any object longer than $L/4$ on the table and it will not fall through since it will not fit through any of the removed sections.

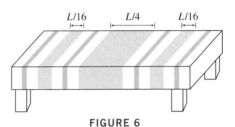

$L/16$ $L/4$ $L/16$

FIGURE 6

52. The **Koch snowflake** (described in 1904 by Swedish mathematician Helge von Koch) is an infinitely jagged "fractal" curve obtained as a limit of polygonal curves (it is continuous, but has no tangent line at any point). Begin with an equilateral triangle (stage 0) and produce stage 1 by replacing each edge with four edges of one-third the length, arranged as in Figure 7. Continue the process: At the nth stage, replace each edge with four edges of one-third the length.

(a) Show that the perimeter P_n of the polygon at the nth stage satisfies $P_n = \frac{4}{3} P_{n-1}$. Prove that $\lim\limits_{n \to \infty} P_n = \infty$. The snowflake has infinite length.

(b) Let A_0 be the area of the original equilateral triangle. Show that $(3)4^{n-1}$ new triangles are added at the nth stage, each with area $A_0/9^n$ (for $n \geq 1$). Show that the total area of the Koch snowflake is $\frac{8}{5} A_0$.

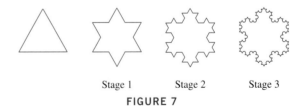

Stage 1   Stage 2   Stage 3

FIGURE 7

10.3 Convergence of Series with Positive Terms

In the next three sections, we focus on the problem of determining whether an infinite series converges or diverges. This is easier than finding the sum of an infinite series, which is possible only in special cases.

In this section, we consider **positive series** $\sum a_n$, defined as a series such that $a_n \geq 0$ for all n (thus the terms of the series are nonnegative). The terms of a positive series may be visualized as rectangles of width 1 and height a_n (Figure 1). The partial sum

$$S_N = a_1 + a_2 + \cdots + a_N$$

is equal to the area of the first N rectangles.

There are powerful numerical methods for finding approximations to infinite series. When implemented on a computer, these methods can be used to compute sums to millions of digits (see Exercises 75–77).

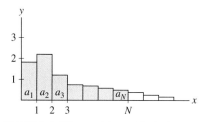

FIGURE 1 The partial sum S_N is the sum of the areas of the N shaded rectangles.

A key property of positive series is that the partial sums S_N form an increasing sequence. Each partial sum is obtained from the previous one by adding a nonnegative number:

$$S_{N+1} = \left(a_1 + a_2 + \cdots + a_N\right) + a_{N+1} = S_N + \underbrace{a_{N+1}}_{\text{Nonnegative}}$$

and thus $S_N \leq S_{N+1}$. Recall that an increasing sequence converges if it is bounded above and diverges otherwise (Theorem 6, Section 10.1). It follows that there are just two ways a positive series can behave (we refer to this as the "dichotomy").

THEOREM 1 Dichotomy Theorem for Positive Series   If $S = \displaystyle\sum_{n=1}^{\infty} a_n$ is a positive series, then there are two possibilities:

  (i) The partial sums S_N are bounded above. In this case, S converges.
  (ii) The partial sums S_N are not bounded above. In this case, S diverges.

Assumptions Matter   This dichotomy does not hold in general for nonpositive series. The partial sums of the nonpositive series

$$S = \sum_{n=1}^{\infty} (-1)^n = 1 - 1 + 1 - 1 + 1 - 1 + \cdots$$

are bounded since $S_N = 1$ or 0, but S diverges.

One of the most important applications of Theorem 1 is the following Integral Test, which is useful because integrals are often easier to evaluate than series.

The Integral Test is valid for any series $\displaystyle\sum_{n=k}^{\infty} f(n)$, provided that $f(x)$ is positive, decreasing, and continuous for $x \geq M$ for some M. The convergence of the series is determined by the convergence of

$$\int_M^{\infty} f(x)\, dx$$

THEOREM 2 Integral Test   Let $a_n = f(n)$, where $f(x)$ is positive, decreasing, and continuous for $x \geq 1$.

  (i) If $\displaystyle\int_1^{\infty} f(x)\, dx$ converges, then $\displaystyle\sum_{n=1}^{\infty} a_n$ converges.
  (ii) If $\displaystyle\int_1^{\infty} f(x)\, dx$ diverges, then $\displaystyle\sum_{n=1}^{\infty} a_n$ diverges.

Proof We compare S_N with the area under the graph of $f(x)$ over the interval $[1, N]$. Since $f(x)$ is decreasing (Figure 2),

$$\underbrace{a_2 + \cdots + a_N}_{\text{Area of shaded rectangles in Figure 2}} \leq \int_1^N f(x)\, dx \leq \int_1^{\infty} f(x)\, dx$$

If the improper integral on the right converges, then the sums $a_2 + \cdots + a_N$ remain bounded. In this case, S_N also remains bounded and the infinite series converges by the Dichotomy Theorem (Theorem 1). This proves (i).

On the other hand (Figure 3),

$$\int_1^N f(x)\, dx \leq \underbrace{a_1 + a_2 + \cdots + a_{N-1}}_{\text{Area of shaded rectangles in Figure 3}} \qquad \boxed{1}$$

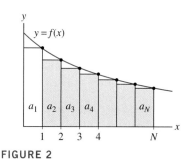

FIGURE 2

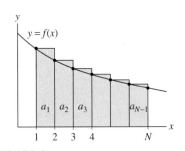

FIGURE 3

The infinite series

$$\sum_{n=1}^{\infty} \frac{1}{n}$$

is called the "harmonic series."

If $\int_{1}^{\infty} f(x)\,dx$ diverges, then $\int_{1}^{N} f(x)\,dx$ tends to ∞ and (1) shows that S_N also tends to ∞. This proves (ii). ∎

■ **EXAMPLE 1** **Divergence of the Harmonic Series** Show that $\displaystyle\sum_{n=1}^{\infty} \frac{1}{n}$ diverges.

Solution The function $f(x) = \dfrac{1}{x}$ is positive, decreasing, and continuous for $x \geq 1$, so we may apply the Integral Test:

$$\int_{1}^{\infty} \frac{dx}{x} = \lim_{R \to \infty} \int_{1}^{R} \frac{dx}{x} = \lim_{R \to \infty} \ln R = \infty$$

The integral diverges, and hence the sum $\displaystyle\sum_{n=1}^{\infty} \frac{1}{n}$ also diverges. ∎

■ **EXAMPLE 2** Determine whether $\displaystyle\sum_{n=1}^{\infty} \frac{n}{(n^2+1)^2} = \frac{1}{2^2} + \frac{2}{5^2} + \frac{3}{10^2} + \cdots$ converges.

Solution The function $f(x) = \dfrac{x}{(x^2+1)^2}$ is positive and continuous for $x \geq 1$, and it is decreasing since $f'(x)$ is negative:

$$f'(x) = \frac{1 - 3x^2}{(x^2+1)^3} < 0 \qquad \text{for } x \geq 1$$

Therefore, we may apply the Integral Test. We use the substitution $u = x^2 + 1$, $du = 2x\,dx$ to evaluate the improper integral:

$$\int_{1}^{\infty} \frac{x}{(x^2+1)^2}\,dx = \lim_{R \to \infty} \int_{1}^{R} \frac{x}{(x^2+1)^2}\,dx = \lim_{R \to \infty} \frac{1}{2} \int_{2}^{R} \frac{du}{u^2}$$

$$= \lim_{R \to \infty} \frac{-1}{2u}\bigg|_{2}^{R} = \lim_{R \to \infty} \left(\frac{1}{4} - \frac{1}{2R} \right) = \frac{1}{4}$$

The integral converges and hence $\displaystyle\sum_{n=1}^{\infty} \frac{n}{(n^2+1)^2}$ also converges. ∎

The Integral Test applies to the sums of reciprocal powers n^{-p}, called ***p*-series**.

THEOREM 3 Convergence of *p*-Series The infinite series $\displaystyle\sum_{n=1}^{\infty} \frac{1}{n^p}$ converges if $p > 1$ and diverges otherwise.

Proof If $p \neq 1$, we have

$$\int_{1}^{\infty} \frac{1}{x^p}\,dx = \lim_{R \to \infty} \int_{1}^{R} \frac{1}{x^p}\,dx = \lim_{R \to \infty} \frac{x^{1-p}}{1-p}\bigg|_{1}^{R} = \lim_{R \to \infty} \frac{R^{1-p} - 1}{1-p}$$

Since R^{1-p} tends to zero if $p > 1$ and to ∞ if $p < 1$, the improper integral converges for $p > 1$ and diverges for $p < 1$. The same is true of the *p*-series by the Integral Test. For $p = 1$, the series diverges, as shown in Example 1. ∎

Here are two examples of p-series:

$$p = \frac{1}{2}: \qquad \sum_{n=1}^{\infty} \frac{1}{\sqrt{n}} = \frac{1}{\sqrt{1}} + \frac{1}{\sqrt{2}} + \frac{1}{\sqrt{3}} + \frac{1}{\sqrt{4}} + \cdots = \infty \quad \text{diverges}$$

$$p = 2: \qquad \sum_{n=1}^{\infty} \frac{1}{n^2} = \frac{1}{1} + \frac{1}{2^2} + \frac{1}{3^2} + \frac{1}{4^2} + \cdots \qquad \text{converges}$$

Another powerful method for determining convergence of positive series is comparison. Suppose that $0 \le a_n \le b_n$. Figure 4 suggests that if the larger sum $\sum b_n$ *converges*, then the smaller sum $\sum a_n$ also converges and, similarly, if the smaller sum *diverges*, then the larger sum also diverges.

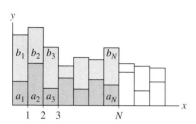

FIGURE 4 The series $\sum a_n$ is dominated by the series $\sum b_n$.

THEOREM 4 Comparison Test
Assume that there exists $M > 0$ such that $0 \le a_n \le b_n$ for $n \ge M$.

(i) If $\displaystyle\sum_{n=1}^{\infty} b_n$ converges, then $\displaystyle\sum_{n=1}^{\infty} a_n$ also converges.

(ii) If $\displaystyle\sum_{n=1}^{\infty} a_n$ diverges, then $\displaystyle\sum_{n=1}^{\infty} b_n$ also diverges.

The convergence of an infinite series does not depend on where the series begins. Therefore, the Comparison Test remains valid even if the series does not begin with $n = 1$.

Proof Assume without loss of generality that $M = 1$. If $S = \displaystyle\sum_{n=1}^{\infty} b_n$ converges, then

$$a_1 + a_2 + \cdots + a_N \le b_1 + b_2 + \cdots + b_N \le \sum_{n=1}^{\infty} b_n = S \qquad \boxed{2}$$

Thus, the partial sums of $\displaystyle\sum_{n=1}^{\infty} a_n$ are bounded above by S, and $\displaystyle\sum_{n=1}^{\infty} a_n$ converges by the Dichotomy Theorem (Theorem 1). On the other hand, if $\displaystyle\sum_{n=1}^{\infty} a_n$ diverges, then its partial sums increase beyond bound and (2) shows that $\displaystyle\sum_{n=1}^{\infty} b_n$ also diverges. ∎

In words, the Comparison Test states that for positive series:

- *Convergence of larger series forces convergence of smaller series.*
- *Divergence of smaller series forces divergence of larger series.*

■ **EXAMPLE 3** Show that $S = \displaystyle\sum_{n=1}^{\infty} 2^{-n!} = 2^{-1} + 2^{-2} + 2^{-6} + 2^{-24} + \cdots$ converges.

Solution We apply the Comparison Test with $a_n = 2^{-n!}$ and $b_n = 2^{-n}$. This is valid because $n! \ge n$ and thus $0 \le 2^{-n!} \le 2^{-n}$ for $n \ge 1$. The geometric series converges:

$$\sum_{n=1}^{\infty} b_n = \sum_{n=1}^{\infty} 2^{-n} = 1 \quad \text{(converges)}$$

Therefore, the smaller series $\displaystyle\sum_{n=1}^{\infty} a_n = \sum_{n=1}^{\infty} 2^{-n!}$ converges. ■

■ **EXAMPLE 4** Show that $\displaystyle\sum_{n=1}^{\infty} \frac{1}{\sqrt{n}\, 3^n}$ converges.

Solution We compare $\sum_{n=1}^{\infty} \dfrac{1}{\sqrt{n}\,3^n}$ with the geometric series $\sum_{n=1}^{\infty} \dfrac{1}{3^n}$. For $n \geq 1$,

$$\frac{1}{\sqrt{n}\,3^n} \leq \frac{1}{3^n}$$

The geometric series $\sum_{n=1}^{\infty} \dfrac{1}{3^n}$ converges, so $\sum_{n=1}^{\infty} \dfrac{1}{\sqrt{n}\,3^n}$ also converges. ∎

In Example 5, the series begins with $n = 2$ because $1/\ln n$ is undefined for $n = 1$.

■ **EXAMPLE 5** Determine whether $\sum_{n=2}^{\infty} \dfrac{1}{\ln n}$ converges.

Solution We compare $\sum_{n=2}^{\infty} \dfrac{1}{\ln n}$ with the harmonic series $\sum_{n=2}^{\infty} \dfrac{1}{n}$ by showing that for $n \geq 2$,

$$\frac{1}{\ln n} \geq \frac{1}{n} \qquad \text{or equivalently,} \qquad n \geq \ln n$$

It suffices to show that $f(x) = x - \ln x$ is positive for $x \geq 2$. However, $f(1) = 1$ and $f(x)$ is increasing since $f'(x) = 1 - x^{-1} > 0$ for $x > 1$. Therefore, $f(x) > 1$ for $x > 1$ as required. Since the harmonic series diverges, the larger series $\sum_{n=2}^{\infty} \dfrac{1}{\ln n}$ also diverges. ∎

■ **EXAMPLE 6** Using the Comparison Test Correctly   Study the convergence of

$$\sum_{n=2}^{\infty} \frac{1}{n(\ln n)^2}$$

Solution We might be tempted to compare $\sum_{n=2}^{\infty} \dfrac{1}{n(\ln n)^2}$ to the harmonic series $\sum_{n=2}^{\infty} \dfrac{1}{n}$ using the inequality (valid for $n \geq 3$)

$$\frac{1}{n(\ln n)^2} \leq \frac{1}{n}$$

However, $\sum_{n=2}^{\infty} \dfrac{1}{n}$ diverges, so this inequality gives us no information about the *smaller* series $\sum \dfrac{1}{n(\ln n)^2}$. Fortunately, in this case we can use the Integral Test. The substitution $u = \ln x$ yields

$$\int_2^{\infty} \frac{dx}{x(\ln x)^2} = \int_{\ln 2}^{\infty} \frac{du}{u^2} = \lim_{R \to \infty} \left(\frac{1}{\ln 2} - \frac{1}{\ln R} \right) = \frac{1}{\ln 2} < \infty$$

The Integral Test shows that $\sum_{n=2}^{\infty} \dfrac{1}{n(\ln n)^2}$ converges. ∎

Suppose we wish to study the convergence of

$$S = \sum_{n=2}^{\infty} \frac{n^2}{n^4 - n} \qquad \boxed{3}$$

For large n, the general term is very close to $\dfrac{1}{n^2}$:

$$\frac{n^2}{n^4 - n} = \frac{1}{n^2 - n^{-1}} \approx \frac{1}{n^2}$$

so we might try to compare S with the convergent series $\displaystyle\sum_{n=2}^{\infty} \frac{1}{n^2}$. However, the Comparison Test cannot be used directly because the precise inequality goes in the wrong direction:

$$\frac{n^2}{n^4 - n} > \frac{n^2}{n^4} = \frac{1}{n^2}$$

In this case, we may apply the following variation of the Comparison Test.

THEOREM 5 Limit Comparison Test Let $\{a_n\}$ and $\{b_n\}$ be *positive* sequences. Assume that the following limit exists:

$$L = \lim_{n \to \infty} \frac{a_n}{b_n}$$

- If $L > 0$, then $\displaystyle\sum_{n=1}^{\infty} a_n$ converges if and only if $\displaystyle\sum_{n=1}^{\infty} b_n$ converges.
- If $L = 0$ and $\displaystyle\sum_{n=1}^{\infty} b_n$ converges, then $\displaystyle\sum_{n=1}^{\infty} a_n$ converges.

CAUTION The Limit Comparison Test may not be applied if the series are not positive. See Exercise 38 in Section 10.4.

Proof First we show that if $\displaystyle\sum_{n=1}^{\infty} b_n$ converges, then $\displaystyle\sum_{n=1}^{\infty} a_n$ converges for $L > 0$ or $L = 0$. Choose a positive number $R > L$. Since the sequences are positive and a_n/b_n approaches L, we have $0 \le a_n/b_n \le R$ and thus $a_n \le Rb_n$ for all n sufficiently large. Since $\displaystyle\sum_{n=1}^{\infty} Rb_n$ also converges, $\displaystyle\sum_{n=1}^{\infty} a_n$ converges by the Comparison Test.

Now suppose that $L > 0$ and $\displaystyle\sum_{n=1}^{\infty} a_n$ converges. We may choose r such that $0 < r < L$. Since a_n/b_n approaches L, we have $a_n/b_n \ge r$ and thus $a_n \ge rb_n$ for all n sufficiently large. Therefore, $\displaystyle\sum_{n=1}^{\infty} rb_n$ converges by the Comparison Test and hence $\displaystyle\sum_{n=1}^{\infty} b_n$ converges. ∎

■ EXAMPLE 7 Show that $\displaystyle\sum_{n=2}^{\infty} \frac{n^2}{n^4 - n}$ converges.

Solution Let $a_n = \dfrac{n^2}{n^4 - n}$. We observed above that $a_n \approx n^{-2}$ for large n, so it makes sense to apply the Limit Comparison Test with $b_n = n^{-2}$:

$$\frac{a_n}{b_n} = \left(\frac{n^2}{n^4 - n}\right)n^2 = \frac{n^4}{n^4 - n} = \frac{1}{1 - n^{-3}}$$

$$L = \lim_{n \to \infty} \frac{a_n}{b_n} = \lim_{n \to \infty} \frac{1}{1 - n^{-3}} = 1$$

Since L exists and $\displaystyle\sum_{n=2}^{\infty} b_n = \sum_{n=2}^{\infty} n^{-2}$ converges, $\displaystyle\sum_{n=2}^{\infty} a_n$ also converges. ■

■ **EXAMPLE 8** Determine whether $\displaystyle\sum_{n=3}^{\infty} \frac{1}{\sqrt{n^2-4}}$ converges.

Solution Let $a_n = \dfrac{1}{\sqrt{n^2-4}}$ and $b_n = \dfrac{1}{n}$. Then

$$L = \lim_{n\to\infty} \frac{a_n}{b_n} = \lim_{n\to\infty} \frac{n}{\sqrt{n^2-4}} = \lim_{n\to\infty} \frac{1}{\sqrt{1-4/n^2}} = 1$$

Since $\displaystyle\sum_{n=3}^{\infty} \frac{1}{n}$ diverges and $L > 0$, the series $\displaystyle\sum_{n=3}^{\infty} \frac{1}{\sqrt{n^2-4}}$ also diverges.   ■

10.3 SUMMARY

- The partial sums S_N of a positive series $S = \sum a_n$ form an increasing sequence.
- *Dichotomy Theorem:* A positive series S converges if its partial sums S_N remain bounded. Otherwise, it diverges.
- *Integral Test:* If f is positive, decreasing, and continuous, then $S = \sum f(n)$ converges (or diverges) if $\displaystyle\int_M^{\infty} f(x)\,dx$ converges (or diverges) for some $M > 0$.
- *p-Series:* The series $\displaystyle\sum_{n=1}^{\infty} \frac{1}{n^p}$ converges if $p > 1$ and diverges if $p \le 1$.
- *Comparison Test:* Assume that there exists $M > 0$ such that $0 \le a_n \le b_n$ for $n \ge M$. If $\displaystyle\sum_{n=1}^{\infty} b_n$ converges, then $\displaystyle\sum_{n=1}^{\infty} a_n$ converges; if $\displaystyle\sum_{n=1}^{\infty} a_n$ diverges, then $\displaystyle\sum_{n=1}^{\infty} b_n$ diverges.
- *Limit Comparison Test:* Let $\{a_n\}$ and $\{b_n\}$ be positive sequences and assume that the following limit exists:

$$L = \lim_{n\to\infty} \frac{a_n}{b_n}$$

  – If $L > 0$, then $\displaystyle\sum_{n=1}^{\infty} a_n$ converges if and only if $\displaystyle\sum_{n=1}^{\infty} b_n$ converges.

  – If $L = 0$ and $\displaystyle\sum_{n=1}^{\infty} b_n$ converges, then $\displaystyle\sum_{n=1}^{\infty} a_n$ converges.

10.3 EXERCISES

Preliminary Questions

1. Let $S = \displaystyle\sum_{n=1}^{\infty} a_n$. If the partial sums S_N are increasing, then (choose correct conclusion)

(a) $\{a_n\}$ is an increasing sequence.

(b) $\{a_n\}$ is a positive sequence.

2. What are the hypotheses of the Integral Test?

3. Which test would you use to determine whether $\displaystyle\sum_{n=1}^{\infty} n^{-3.2}$ converges?

4. Which test would you use to determine whether $\displaystyle\sum_{n=1}^{\infty} \frac{1}{2^n + \sqrt{n}}$ converges?

5. Ralph hopes to investigate the convergence of $\displaystyle\sum_{n=1}^{\infty} \frac{e^{-n}}{n}$ by comparing it with $\displaystyle\sum_{n=1}^{\infty} \frac{1}{n}$. Is Ralph on the right track?

Exercises

In Exercises 1–14, use the Integral Test to determine whether the infinite series is convergent.

1. $\displaystyle\sum_{n=1}^{\infty} \frac{1}{n^4}$

2. $\displaystyle\sum_{n=1}^{\infty} \frac{1}{n+3}$

3. $\displaystyle\sum_{n=1}^{\infty} n^{-1/3}$

4. $\displaystyle\sum_{n=5}^{\infty} \frac{1}{\sqrt{n-4}}$

5. $\displaystyle\sum_{n=25}^{\infty} \frac{n^2}{(n^3+9)^{5/2}}$

6. $\displaystyle\sum_{n=1}^{\infty} \frac{n}{(n^2+1)^{3/5}}$

7. $\displaystyle\sum_{n=1}^{\infty} \frac{1}{n^2+1}$

8. $\displaystyle\sum_{n=1}^{\infty} \frac{1}{n(n+1)}$

9. $\displaystyle\sum_{n=1}^{\infty} ne^{-n^2}$

10. $\displaystyle\sum_{n=2}^{\infty} \frac{1}{n(\ln n)^2}$

11. $\displaystyle\sum_{n=1}^{\infty} \frac{1}{2^{\ln n}}$

12. $\displaystyle\sum_{n=4}^{\infty} \frac{1}{n^2-1}$

13. $\displaystyle\sum_{n=1}^{\infty} \frac{\ln n}{n^2}$

14. $\displaystyle\sum_{n=1}^{\infty} \frac{n}{2^n}$

15. Use the Comparison Test to show that $\displaystyle\sum_{n=1}^{\infty} \frac{1}{n^3+8n}$ converges.
Hint: Compare with $\displaystyle\sum_{n=1}^{\infty} n^{-3}$.

16. Show that $\displaystyle\sum_{n=2}^{\infty} \frac{1}{\sqrt{n^2-3}}$ diverges by comparing with $\displaystyle\sum_{n=2}^{\infty} n^{-1}$.

17. Let $S = \displaystyle\sum_{n=1}^{\infty} \frac{1}{n+\sqrt{n}}$. Verify that for $n \geq 1$

$$\frac{1}{n+\sqrt{n}} \leq \frac{1}{n}, \qquad \frac{1}{n+\sqrt{n}} \leq \frac{1}{\sqrt{n}}$$

Can either inequality be used to show that S diverges? Show that $\dfrac{1}{n+\sqrt{n}} \geq \dfrac{1}{2n}$ and conclude that S diverges.

18. Which of the following inequalities can be used to study the convergence of $\displaystyle\sum_{n=2}^{\infty} \frac{1}{n^2+\sqrt{n}}$? Explain.

$$\frac{1}{n^2+\sqrt{n}} \leq \frac{1}{\sqrt{n}}, \qquad \frac{1}{n^2+\sqrt{n}} \leq \frac{1}{n^2}$$

In Exercises 19–31, use the Comparison Test to determine whether the infinite series is convergent.

19. $\displaystyle\sum_{n=1}^{\infty} \frac{1}{n2^n}$

20. $\displaystyle\sum_{n=1}^{\infty} \frac{1}{\sqrt{n}+2^n}$

21. $\displaystyle\sum_{k=1}^{\infty} \frac{k^{1/3}}{k^2+k}$

22. $\displaystyle\sum_{n=4}^{\infty} \frac{\sqrt{n}}{n-3}$

23. $\displaystyle\sum_{n=1}^{\infty} \frac{1}{\sqrt{n^3+1}}$

24. $\displaystyle\sum_{n=1}^{\infty} \frac{n^3}{n^5+4n+1}$

25. $\displaystyle\sum_{k=1}^{\infty} \frac{\sin^2 k}{k^2}$

26. $\displaystyle\sum_{n=2}^{\infty} \frac{1}{(\ln n)2^n}$

27. $\displaystyle\sum_{m=1}^{\infty} \frac{4}{m!+4^m}$

28. $\displaystyle\sum_{n=1}^{\infty} \frac{2}{3^n+3^{-n}}$

29. $\displaystyle\sum_{k=1}^{\infty} 2^{-k^2}$

30. $\displaystyle\sum_{n=1}^{\infty} \frac{\ln n}{n^3}$

31. $\displaystyle\sum_{n=1}^{\infty} \frac{\ln n}{n^3+3\ln n}$

32. Show that $\displaystyle\sum_{n=1}^{\infty} \sin \frac{1}{n^2}$ is a positive, convergent series. *Hint:* Use the inequality $\sin x \leq x$ for $x \geq 0$.

33. Does $\displaystyle\sum_{n=1}^{\infty} \frac{n}{\sqrt{n^2+c}}$ converge for any c?

In Exercises 34–42, use the Limit Comparison Test to prove convergence or divergence of the infinite series.

34. $\displaystyle\sum_{n=2}^{\infty} \frac{n^2}{n^4-1}$

35. $\displaystyle\sum_{n=2}^{\infty} \frac{1}{n^2-\sqrt{n}}$

36. $\displaystyle\sum_{n=2}^{\infty} \frac{n}{\sqrt{n^3-1}}$

37. $\displaystyle\sum_{n=3}^{\infty} \frac{n^3}{\sqrt{n^4-2n^2+1}}$

38. $\displaystyle\sum_{n=3}^{\infty} \frac{3n+5}{n(n-1)(n-2)}$

39. $\displaystyle\sum_{n=1}^{\infty} \frac{e^n+n}{e^{2n}-n^2}$

40. $\displaystyle\sum_{n=1}^{\infty} \frac{\ln n}{n^2}$. *Hint:* Use L'Hôpital's Rule to compare with $\displaystyle\sum_{n=1}^{\infty} \frac{1}{n^{3/2}}$.

41. $\displaystyle\sum_{n=1}^{\infty} \left(1-\cos\frac{1}{n}\right)$ *Hint:* Compare with $\displaystyle\sum_{n=1}^{\infty} n^{-2}$.

42. $\displaystyle\sum_{n=1}^{\infty} (1-2^{-1/n})$ *Hint:* Compare with the harmonic series.

43. Show that if $a_n \geq 0$ and $\displaystyle\lim_{n\to\infty} n^2 a_n$ exists, then $\displaystyle\sum_{n=1}^{\infty} a_n$ converges.
Hint: Show that if M is larger than $\displaystyle\lim_{n\to\infty} n^2 a_n$, then $a_n \leq M/n^2$ for n sufficiently large.

44. Show that $\displaystyle\sum_{n=2}^{\infty} \frac{1}{n^{\ln n}}$ converges. *Hint:* Show that $n^{\ln n} \geq n^2$ for $n > e^2$.

45. Show that $\displaystyle\sum_{n=2}^{\infty} (\ln n)^{-2}$ diverges. *Hint:* Show that for x sufficiently large, $\ln x < x^{1/2}$.

46. For which a does $\displaystyle\sum_{n=2}^{\infty} \frac{1}{n(\ln n)^a}$ converge?

47. For which a does $\displaystyle\sum_{n=2}^{\infty} \frac{1}{n^a \ln n}$ converge?

48. Use the Integral Test to show that $\displaystyle\sum_{n=2}^{\infty} \frac{(\ln n)^k}{n^2}$ converges for all exponents k. You may use that $\displaystyle\int_1^{\infty} u^k e^{-u}\, du$ converges for all k.

In Exercises 49–74, determine convergence or divergence using any method covered so far.

49. $\displaystyle\sum_{n=4}^{\infty} \frac{1}{n^2 - 9}$

50. $\displaystyle\sum_{n=1}^{\infty} \frac{\cos^2 n}{n^2}$

51. $\displaystyle\sum_{n=1}^{\infty} \frac{\sqrt{n}}{4n + 9}$

52. $\displaystyle\sum_{n=1}^{\infty} e^{-n}$

53. $\displaystyle\sum_{n=1}^{\infty} \frac{1}{3^{n^2}}$

54. $\displaystyle\sum_{n=1}^{\infty} \frac{1}{n^2 + \sin n}$

55. $\displaystyle\sum_{n=2}^{\infty} \frac{1}{n^{3/2} \ln n}$

56. $\displaystyle\sum_{k=1}^{\infty} 2^{1/k}$

57. $\displaystyle\sum_{n=2}^{\infty} \frac{1}{n^{1/2} \ln n}$

58. $\displaystyle\sum_{n=1}^{\infty} \frac{4^n}{5^n - 2n}$

59. $\displaystyle\sum_{n=2}^{\infty} \frac{n}{e^{n^2}}$

60. $\displaystyle\sum_{n=1}^{\infty} \frac{n - \sin n}{n^3}$

61. $\displaystyle\sum_{n=1}^{\infty} \frac{2^n}{3^n - n}$

62. $\displaystyle\sum_{n=2}^{\infty} \frac{1}{n \ln n - n}$

63. $\displaystyle\sum_{n=1}^{\infty} \frac{\tan^{-1} n}{n^2}$

64. $\displaystyle\sum_{n=1}^{\infty} \frac{1}{n^n}$

65. $\displaystyle\sum_{n=1}^{\infty} \frac{\ln n}{n^3}$

66. $\displaystyle\sum_{n=1}^{\infty} \frac{2 + (-1)^n}{n}$

67. $\displaystyle\sum_{n=1}^{\infty} \frac{2 + (-1)^n}{n^{3/2}}$

68. $\displaystyle\sum_{n=1}^{\infty} \sin \frac{1}{n}$

69. $\displaystyle\sum_{n=1}^{\infty} \frac{2n + 1}{4^n}$

70. $\displaystyle\sum_{n=3}^{\infty} \frac{1}{n^4 - \sqrt{n}}$

71. $\displaystyle\sum_{n=1}^{\infty} \frac{n^2 - n}{n^5 + n}$

72. $\displaystyle\sum_{n=2}^{\infty} \frac{n^2 + n}{n^5 - n}$

73. $\displaystyle\sum_{n=1}^{\infty} \frac{1}{n^{1.2} \ln n}$

74. $\displaystyle\sum_{n=1}^{\infty} \frac{\ln n}{n^{1.2}}$

Approximating Infinite Sums *In Exercises 75–77, let $a_n = f(n)$, where $f(x)$ is a continuous, decreasing function such that*

$$\int_1^{\infty} f(x)\, dx$$

converges.

75. Show that

$$\int_1^{\infty} f(x)\, dx \le \sum_{n=1}^{\infty} a_n \le a_1 + \int_1^{\infty} f(x)\, dx \qquad \boxed{4}$$

76. $\mathcal{CAS}$ Using Eq. (4), show that

$$5 \le \sum_{n=1}^{\infty} \frac{1}{n^{1.2}} \le 6$$

This series converges slowly. Use a computer algebra system to verify that $S_N < 5$ for $N \le 43{,}128$ and $S_{43{,}129} \approx 5.00000021$.

77. Let $S = \displaystyle\sum_{n=1}^{\infty} a_n$. Arguing as in Exercise 75, show that

$$\sum_{n=1}^{M} a_n + \int_{M+1}^{\infty} f(x)\, dx \le S \le \sum_{n=1}^{M+1} a_n + \int_{M+1}^{\infty} f(x)\, dx \qquad \boxed{5}$$

Conclude that

$$0 \le S - \left(\sum_{n=1}^{M} a_n + \int_{M+1}^{\infty} f(x)\, dx \right) \le a_{M+1} \qquad \boxed{6}$$

This yields a method for approximating S with an error of at most a_{M+1}.

78. $\mathcal{CAS}$ Use Eq. (5) with $M = 43{,}129$ to prove that

$$5.5915810 \le \sum_{n=1}^{\infty} \frac{1}{n^{1.2}} \le 5.5915839$$

79. $\mathcal{CAS}$ Apply Eq. (5) with $M = 40{,}000$ to show that

$$1.644934066 \le \sum_{n=1}^{\infty} \frac{1}{n^2} \le 1.644934068$$

Is this consistent with Euler's result, according to which this infinite series has sum $\pi^2/6$?

80. $\mathcal{CAS}$ Using a CAS and Eq. (6), determine the value of $\displaystyle\sum_{n=1}^{\infty} n^{-4}$ to within an error less than 10^{-4}. Check that your result is consistent with that of Euler, who proved that the sum is equal to $\pi^4/90$.

81. $\mathcal{CAS}$ Using a CAS and Eq. (6), determine the value of $\displaystyle\sum_{n=1}^{\infty} n^{-5}$ to within an error less than 10^{-4}.

82. The harmonic series diverges, but it does so very slowly. Show that the partial sum S_N satisfies

$$\ln N \le 1 + \frac{1}{2} + \frac{1}{3} + \cdots + \frac{1}{N} \le 1 + \ln N$$

Verify that $S_N \le 10$ for $N \le 8{,}200$. Find an N such that $S_N \ge 100$.

83. Let p_n denote the nth prime number ($p_1 = 2$, $p_2 = 3$, etc.). It is known that there is a constant C such that $p_n \le Cn \ln n$. Prove the divergence of

$$\sum_{n=1}^{\infty} \frac{1}{p_n} = \frac{1}{2} + \frac{1}{3} + \frac{1}{5} + \frac{1}{7} + \frac{1}{11} + \cdots$$

84. How far can a stack of identical books (each of unit length) extend without tipping over? The stack will not tip over if the $(n+1)$st book is placed at the bottom of the stack with its left edge located at the center of mass of the first n books (Figure 5). Let c_n be the center of mass of the first n books, measured along the x-axis.

(a) Prove that $c_{n+1} = c_n + \dfrac{1}{2(n+1)}$. Recall that if objects of mass $m_1, \ldots, m_n$ are placed along the x-axis with their centers of mass at $x_1, \ldots, x_n$, then the center of mass of the system is located at

$$\frac{m_1 x_1 + \cdots + m_n x_n}{x_1 + \cdots + x_n}$$

(b) Prove that $\lim\limits_{n \to \infty} c_n = \infty$. Thus, by using enough books, the stack can be extended as far as desired without tipping over.

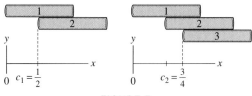

FIGURE 5

Further Insights and Challenges

85. Use the Integral Test to prove again that the geometric series $\sum\limits_{n=1}^{\infty} r^n$ converges if $0 < r < 1$ and diverges if $r > 1$.

86. Let $S = \sum\limits_{n=2}^{\infty} a_n$, where $a_n = (\ln(\ln n))^{-\ln n}$.

(a) Show, by taking logarithms, that $a_n = n^{-\ln(\ln(\ln n))}$.
(b) Show that $\ln(\ln(\ln n)) \ge 2$ if $n > C$, where $C = e^{e^{e^2}}$.
(c) Show that S converges.

87. Kummer's Acceleration Method Suppose we wish to approximate $S = \sum\limits_{n=1}^{\infty} \frac{1}{n^2}$. There is a similar telescoping series whose value can be computed exactly (see Example 1 in Section 10.2):

$$\sum_{n=1}^{\infty} \frac{1}{n(n+1)} = 1$$

(a) Verify that

$$S = \sum_{n=1}^{\infty} \frac{1}{n(n+1)} + \sum_{n=1}^{\infty} \left(\frac{1}{n^2} - \frac{1}{n(n+1)} \right)$$

Thus for M large,

$$S \approx 1 + \sum_{n=1}^{M} \frac{1}{n^2(n+1)} \qquad \boxed{7}$$

(b) Explain what has been gained. Why is (7) a better approximation to S than $\sum\limits_{n=1}^{M} \frac{1}{n^2}$?

(c) ⌊*CAS*⌋ Compute

$$\sum_{n=1}^{1{,}000} \frac{1}{n^2}, \qquad 1 + \sum_{n=1}^{100} \frac{1}{n^2(n+1)}$$

Which is a better approximation to S, whose exact value is $\pi^2/6$?

88. ⌊*CAS*⌋ The series $S = \sum\limits_{k=1}^{\infty} k^{-3}$ has been computed to more than 100 million digits. The first 30 digits are

$$S = 1.202056903159594285399738161511$$

Approximate S using the Acceleration Method of Exercise 87 with $M = 100$ and auxiliary series $R = \sum\limits_{n=1}^{\infty} (n(n+1)(n+2))^{-3}$. According to Exercise 46 in Section 10.2, R is a telescoping series with the sum $R = \frac{1}{4}$.

10.4 Absolute and Conditional Convergence

In the previous section, we studied the convergence of positive series, but we still lack the tools to analyze series with both positive and negative terms such as

$$\frac{1}{1^2} - \frac{1}{2^2} + \frac{1}{3^2} - \frac{1}{4^2} + \cdots \qquad \boxed{1}$$

One of the keys to understanding these more general series is the concept of absolute convergence.

DEFINITION Absolute Convergence $\sum a_n$ is called **absolutely convergent** if $\sum |a_n|$ converges.

As an example, consider the series (1). It is absolutely convergent because the sequence of absolute values is a convergent p-series:

$$\frac{1}{1^2} + \frac{1}{2^2} + \frac{1}{3^2} + \frac{1}{4^2} + \cdots$$

The following theorem states that if the series of absolute values converges, then the original series also converges.

THEOREM 1 Absolute Convergence Implies Convergence If $\sum a_n$ is absolutely convergent, then $\sum a_n$ converges.

Proof We have

$$0 \le a_n + |a_n| \le 2|a_n|$$

and thus $\sum (a_n + |a_n|)$ is a positive series. If $\sum |a_n|$ converges, then $\sum 2|a_n|$ also converges and hence $\sum (a_n + |a_n|)$ converges by the Comparison Test. Our original series is the difference of two convergent series and hence it converges:

$$\sum a_n = \sum (a_n + |a_n|) - \sum |a_n| \qquad \blacksquare$$

■ **EXAMPLE 1** Verify the convergence of $S = \dfrac{1}{1^2} - \dfrac{1}{2^2} + \dfrac{1}{3^2} - \dfrac{1}{4^2} + \cdots$.

Solution The positive series $\displaystyle\sum_{n=1}^{\infty} \frac{1}{n^2}$ converges because it is a p-series with $p = 2$. Thus, S converges absolutely. By Theorem 1, S converges. ■

■ **EXAMPLE 2** Does $S = \displaystyle\sum_{n=2}^{\infty} \frac{(-1)^n}{n \ln n}$ converge absolutely?

Solution We apply the Integral Test to the positive series $\displaystyle\sum_{n=2}^{\infty} \frac{1}{n \ln n}$. Using the substitution $u = \ln x, du = x^{-1} dx$, we find that the improper integral diverges:

$$\int_2^{\infty} \frac{dx}{x \ln x} = \int_{\ln 2}^{\infty} \frac{du}{u} = \lim_{R \to \infty} \int_{\ln 2}^{R} \frac{du}{u} = \lim_{R \to \infty} \left(\ln R - \ln(\ln 2) \right) = \infty$$

Therefore, $\displaystyle\sum_{n=2}^{\infty} \frac{1}{n \ln n}$ diverges and S does not converge absolutely. ■

In the previous example, we showed that the series $\sum_{n=2}^{\infty} \dfrac{(-1)^n}{n \ln n}$ does not converge *absolutely*, but we still do not know whether or not it converges. A series $\sum a_n$ may converge without converging absolutely. In this case, we say that $\sum a_n$ is conditionally convergent.

DEFINITION Conditional Convergence   An infinite series $\sum a_n$ is called **conditionally convergent** if it converges but $\sum |a_n|$ diverges.

If we encounter a series that is not absolutely convergent, how can we determine whether it is conditionally convergent? The Integral Test and Comparison Test cannot be used because they apply only to positive series. Our next test applies to **alternating series**, that is, series in which the terms alternate in sign. Such a series has the following form, where $\{a_n\}$ is a *positive* sequence (Figure 1):

$$\sum_{n=1}^{\infty} (-1)^{n-1} a_n = a_1 - a_2 + a_3 - a_4 + a_5 + \cdots$$

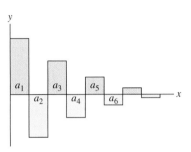

FIGURE 1 An alternating series with decreasing terms. The sum of the signed areas is positive and at most a_1.

Assumptions Matter   The Leibniz Test is not valid if we drop the assumption that a_n is decreasing (see Exercise 35).

THEOREM 2 Leibniz Test for Alternating Series   Let $\{a_n\}$ be a decreasing positive sequence that converges to 0:

$$a_1 \geq a_2 \geq a_3 \geq a_4 \geq \cdots \geq 0, \qquad \lim_{n \to \infty} a_n = 0$$

Then the following alternating series converges:

$$S = \sum_{n=1}^{\infty} (-1)^{n-1} a_n = a_1 - a_2 + a_3 - a_4 + \cdots$$

Furthermore, $0 \leq S \leq a_1$ and $S_{2N} \leq S \leq S_{2N+1}$ for all N.

Proof We analyze the even and odd partial sums separately. First, we observe that the even partial sums form an increasing sequence:

$$S_{2N+2} = \underbrace{(a_1 - a_2) + \cdots + (a_{2N-1} - a_{2N})}_{S_{2N}} + (a_{2N+1} - a_{2N+2})$$

$$= S_{2N} + (a_{2N+1} - a_{2N+2})$$

Since $\{a_n\}$ is decreasing, the quantity $a_{2N+1} - a_{2N+2}$ is positive and thus S_{2N+2} is obtained by adding the positive quantity $(a_{2N+1} - a_{2N+2})$ to S_{2N}. On the other hand, we may regroup the terms:

$$S_{2N} = a_1 - (a_2 - a_3) - (a_4 - a_5) - \cdots - (a_{2N-2} - a_{2N-1}) - a_{2N}$$

We conclude that $S_{2N} \leq a_1$ because each term $(a_{2j} - a_{2j+1})$ is nonnegative. This shows that $\{S_{2N}\}$ is a positive increasing sequence with upper bound a_1. Consequently, the limit $S = \lim S_{2N}$ exists and $0 \leq S_{2N} \leq S \leq a_1$ by Theorem 6 in Section 10.1.

The odd partial sums form a decreasing sequence:

$$S_{2N+1} = S_{2N-1} - (a_{2N} - a_{2N+1}) \leq S_{2N-1}$$

Furthermore, the odd partial sums converge to the same limit:

$$\lim_{N \to \infty} S_{2N+1} = \lim_{N \to \infty} (S_{2N} + a_{2N+1}) = \lim_{N \to \infty} S_{2N} + \lim_{N \to \infty} a_{2N+1} = S + 0 = S$$

This proves that the alternating series converges to S and that $S_{2N+1} \geq S$ for all N.   ∎

■ **EXAMPLE 3**   Show that $S = \sum_{n=1}^{\infty} \dfrac{(-1)^{n-1}}{\sqrt{n}}$ is conditionally convergent and that $0 \leq S \leq 1$.

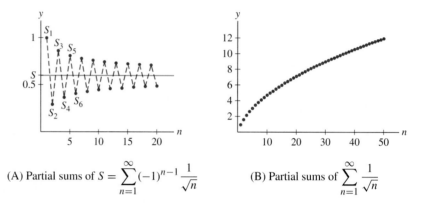

(A) Partial sums of $S = \sum_{n=1}^{\infty}(-1)^{n-1}\dfrac{1}{\sqrt{n}}$ (B) Partial sums of $\sum_{n=1}^{\infty}\dfrac{1}{\sqrt{n}}$

FIGURE 2 The series $S = \sum_{n=1}^{\infty} \dfrac{(-1)^{n-1}}{\sqrt{n}}$ converges conditionally but not absolutely.

Solution   The terms $a_n = \dfrac{1}{\sqrt{n}}$ form a decreasing sequence that tends to zero. The Leibniz Test implies that $S = \sum_{n=1}^{\infty} \dfrac{(-1)^{n-1}}{\sqrt{n}}$ converges and that $0 \leq S \leq a_1 = 1$ [Figure 2(A)]. However, S is only conditionally convergent because the positive series $\sum_{n=1}^{\infty} \dfrac{1}{\sqrt{n}}$ is a divergent p-series [Figure 2(B)].   ∎

According to Theorem 2, if $\{a_n\}$ is decreasing and positive, then the partial sums S_N of the alternating series satisfy

$$S_{2N} \leq S \leq S_{2N+1}$$

Therefore, the partial sums zigzag above and below the limit (Figure 3). It follows that the error $|S_N - S|$ is not greater than $|S_N - S_{N+1}| = a_{N+1}$:

$$\boxed{|S_N - S| \leq a_{N+1}}$$ $\boxed{2}$

In other words, for an alternating series, *the error committed when we approximate S by S_N is at most the size of the first omitted term a_{N+1}*.

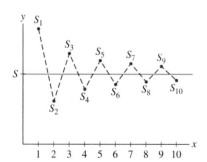

FIGURE 3 The partial sums of an alternating series zigzag, above and below the limit.

■ **EXAMPLE 4**   **The Alternating Harmonic Series**   Show that $S = \sum_{n=1}^{\infty} \dfrac{(-1)^{n+1}}{n}$ converges conditionally. Then

(a) Show that $|S_6 - S| \leq \frac{1}{7}$.

(b) Find an N such that S_N approximates S with an error less than 10^{-3}.

Solution The harmonic series $\sum_{n=1}^{\infty} n^{-1}$ diverges, but S converges conditionally by the Leibniz Test, since $a_n = n^{-1}$ is a positive, decreasing sequence.

(a) By Eq. (2), $|S_6 - S| \le a_7 = \frac{1}{7}$.

(b) The error in the Nth partial sum is at most a_{N+1}:

$$\left| \sum_{n=1}^{N} \frac{(-1)^{n+1}}{n} - S \right| \le a_{N+1} = \frac{1}{N+1}$$

To make the error less than 10^{-3}, we choose N so that $\dfrac{1}{N+1} < 10^{-3}$ or $N + 1 > 10^3$. This gives $N > 999$. Using a computer algebra system, we find that $S_{1,000} \approx 0.6926$. ∎

CONCEPTUAL INSIGHT   The convergence of an infinite series $\sum a_n$ depends on two factors: (1) how quickly a_n tends to zero, and (2) how much cancellation takes place among the terms. Consider the series

Harmonic series (diverges):	$1 + \dfrac{1}{2} + \dfrac{1}{3} + \dfrac{1}{4} + \dfrac{1}{5} + \cdots$
p-Series with $p = 2$ (converges):	$1 + \dfrac{1}{2^2} + \dfrac{1}{3^2} + \dfrac{1}{4^2} + \dfrac{1}{5^2} + \cdots$
Alternating harmonic series (converges):	$1 - \dfrac{1}{2} + \dfrac{1}{3} - \dfrac{1}{4} + \dfrac{1}{5} - \cdots$

The harmonic series diverges because reciprocals $\dfrac{1}{n}$ do not tend to zero quickly enough. By contrast, the reciprocal squares $\dfrac{1}{n^2}$ tend to zero quickly enough for the series $\sum_{n=1}^{\infty} \dfrac{1}{n^2}$ to converge. The alternating harmonic series converges as a result of cancellation among the terms.

10.4 SUMMARY

- An infinite series $\sum a_n$ is called *absolutely convergent* if the positive series $\sum |a_n|$ converges.
- Theorem: Absolute convergence implies convergence. Namely, if $\sum |a_n|$ converges, then $\sum a_n$ also converges.
- An infinite series $\sum a_n$ is *conditionally convergent* if it converges but $\sum |a_n|$ diverges.
- *Leibniz Test:* If $\{a_n\}$ is a positive decreasing sequence such that $\lim_{n\to\infty} a_n = 0$, then the following alternating series converges:

$$S = \sum_{n=1}^{\infty} (-1)^{n-1} a_n = a_1 - a_2 + a_3 - a_4 + a_5 - \cdots$$

Furthermore, $|S - S_N| \le a_{N+1}$.

- We have developed two ways to handle nonpositive series: Either show absolute convergence or use the Leibniz Test, if applicable.

10.4 EXERCISES

Preliminary Questions

1. Suppose that $S = \sum_{n=0}^{\infty} a_n$ is conditionally convergent. Which of the following statements are correct?

(a) $\sum_{n=0}^{\infty} |a_n|$ may or may not converge.

(b) S may or may not converge.

(c) $\sum_{n=0}^{\infty} |a_n|$ diverges.

2. Which of the following statements is equivalent to Theorem 1?

(a) If $\sum_{n=0}^{\infty} |a_n|$ diverges, then $\sum_{n=0}^{\infty} a_n$ also diverges.

(b) If $\sum_{n=0}^{\infty} a_n$ diverges, then $\sum_{n=0}^{\infty} |a_n|$ also diverges.

(c) If $\sum_{n=0}^{\infty} a_n$ converges, then $\sum_{n=0}^{\infty} |a_n|$ also converges.

3. Lathika argues that $\sum_{n=1}^{\infty} (-1)^n \sqrt{n}$ is an alternating series and therefore converges. Is Lathika right?

4. Give an example of a series such that $\sum a_n$ converges but $\sum |a_n|$ diverges.

Exercises

1. Show that $\sum_{n=0}^{\infty} \dfrac{(-1)^n}{2^n}$ converges absolutely.

2. Show that the following series converges conditionally:

$$\sum_{n=1}^{\infty} (-1)^{n-1} \frac{1}{n^{2/3}} = \frac{1}{1^{2/3}} - \frac{1}{2^{2/3}} + \frac{1}{3^{2/3}} - \frac{1}{4^{2/3}} + \cdots$$

In Exercises 3–12, determine whether the series converges absolutely, conditionally, or not at all.

3. $\sum_{n=1}^{\infty} \dfrac{(-1)^n}{\sqrt{n}}$

4. $\sum_{n=1}^{\infty} \dfrac{(-1)^n n^4}{n^3 + 1}$

5. $\sum_{n=1}^{\infty} \dfrac{(-1)^{n-1}}{(1.1)^n}$

6. $\sum_{n=1}^{\infty} \dfrac{\sin n}{n^2}$

7. $\sum_{n=2}^{\infty} \dfrac{(-1)^{n+1}}{n \ln n}$

8. $\sum_{n=1}^{\infty} \dfrac{(-1)^n}{1 + \frac{1}{n}}$

9. $\sum_{n=1}^{\infty} \dfrac{\sin n\pi}{\sqrt{n}}$

10. $\sum_{n=4}^{\infty} (-1)^n \tan \dfrac{1}{n}$

11. $\sum_{n=1}^{\infty} \dfrac{\cos \frac{1}{n}}{n^2}$

12. $\sum_{n=1}^{\infty} \dfrac{\cos \frac{1}{n}}{n}$

13. Let $S = \sum_{n=1}^{\infty} (-1)^{n+1} \dfrac{1}{n^3}$.

(a) Calculate S_n for $1 \le n \le 10$.

(b) Use Eq. (2) to show that $0.9 \le S \le 0.902$.

14. Use Eq. (2) to approximate $\sum_{n=1}^{\infty} \dfrac{(-1)^{n+1}}{n!}$ to four decimal places.

15. Approximate $\sum_{n=1}^{\infty} \dfrac{(-1)^{n+1}}{n^4}$ to three decimal places.

16. *CAS* Let $S = \sum_{n=1}^{\infty} (-1)^{n-1} \dfrac{n}{n^2 + 1}$. Use a computer algebra system to calculate and plot the partial sums S_n for $1 \le n \le 100$. Observe that the partial sums zigzag above and below the limit.

In Exercises 17–18, use Eq. (2) to approximate the value of the series to within an error of at most 10^{-5}.

17. $\sum_{n=1}^{\infty} \dfrac{(-1)^{n+1}}{n(n+2)(n+3)}$

18. $\sum_{n=1}^{\infty} \dfrac{(-1)^{n+1} \ln n}{n!}$

In Exercises 19–26, determine convergence or divergence by any method.

19. $\sum_{n=1}^{\infty} \dfrac{1}{3^n + 5^n}$

20. $\sum_{n=2}^{\infty} \dfrac{n}{n^2 - n}$

21. $\sum_{n=1}^{\infty} \dfrac{(-1)^n}{\sqrt{n^2 + 1}}$

22. $\sum_{n=1}^{\infty} \dfrac{1}{\sqrt{n^2 + 1}}$

23. $\sum_{n=1}^{\infty} \dfrac{3^n + (-1)^n 2^n}{5^n}$

24. $\sum_{n=1}^{\infty} \dfrac{(-1)^{n+1}}{(2n+1)!}$

25. $\sum_{n=1}^{\infty} (-1)^n n e^{-n}$

26. $\sum_{n=1}^{\infty} (-1)^n n^4 2^{-n}$

27. Show that

$$S = \frac{1}{2} - \frac{1}{2} + \frac{1}{3} - \frac{1}{3} + \frac{1}{4} - \frac{1}{4}$$

converges by computing the partial sums. Does it converge absolutely?

28. The Leibniz Test cannot be applied to

$$\frac{1}{2} - \frac{1}{3} + \frac{1}{2^2} - \frac{1}{3^2} + \frac{1}{2^3} - \frac{1}{3^3} + \cdots$$

Why not? Show that it converges by another method.

29. Determine whether the following series converges conditionally:

$$1 - \frac{1}{3} + \frac{1}{2} - \frac{1}{5} + \frac{1}{3} - \frac{1}{7} + \frac{1}{4} - \frac{1}{9} + \frac{1}{5} - \frac{1}{11} + \cdots$$

30. Prove that if $\sum a_n$ converges absolutely, then $\sum a_n^2$ also converges. Then show by giving a counterexample that $\sum a_n^2$ need not converge if $\sum a_n$ is only conditionally convergent.

Further Insights and Challenges

31. Prove the following variant of the Leibniz Test: If $\{a_n\}$ is a positive, decreasing sequence with $\lim_{n \to \infty} a_n = 0$, then the series

$$a_1 + a_2 - 2a_3 + a_4 + a_5 - 2a_6 + \cdots$$

converges. *Hint:* Show that S_{3N} is increasing and bounded by $a_1 + a_2$, and continue as in the proof of the Leibniz Test.

32. Use Exercise 31 to show that the following series converges:

$$S = \frac{1}{\ln 2} + \frac{1}{\ln 3} - \frac{2}{\ln 4} + \frac{1}{\ln 5} + \frac{1}{\ln 6} - \frac{2}{\ln 7} + \cdots$$

33. Prove the conditional convergence of

$$R = 1 + \frac{1}{2} + \frac{1}{3} - \frac{3}{4} + \frac{1}{5} + \frac{1}{6} + \frac{1}{7} - \frac{3}{8} + \cdots$$

34. Show that the following series diverges:

$$S = 1 + \frac{1}{2} + \frac{1}{3} - \frac{2}{4} + \frac{1}{5} + \frac{1}{6} + \frac{1}{7} - \frac{2}{8} + \cdots$$

Hint: Use the result of Exercise 33 to write S as the sum of a convergent and a divergent series.

35. **Assumptions Matter**   Show by counterexample that the Leibniz Test does not remain true if $\{a_n\}$ tends to zero but we drop the assumption that the sequence a_n is decreasing. *Hint:* Consider

$$R = \frac{1}{2} - \frac{1}{4} + \frac{1}{3} - \frac{1}{8} + \frac{1}{4} - \frac{1}{16} + \cdots + \left(\frac{1}{n} - \frac{1}{2^n}\right) + \cdots$$

36. Prove that $\sum_{n=1}^{\infty} (-1)^{n+1} \frac{(\ln n)^a}{n}$ converges for all exponents a.

Hint: Show that $f(x) = \frac{(\ln x)^a}{x}$ is decreasing for x sufficiently large.

37. We say that $\{b_n\}$ is a rearrangement of $\{a_n\}$ if $\{b_n\}$ has the same terms as $\{a_n\}$ but occurring in a different order. Show that if $\{b_n\}$ is a rearrangement of $\{a_n\}$ and $S = \sum_{n=1}^{\infty} a_n$ converges absolutely, then

$$T = \sum_{n=1}^{\infty} b_n$$ also converges absolutely. (This result does not hold if S is only conditionally convergent.) *Hint:* Prove that the partial sums $\sum_{n=1}^{N} |b_n|$ are bounded. It can be shown further that $S = T$.

38. **Assumptions Matter**   In 1829, Lejeune Dirichlet pointed out that the great French mathematician Augustin Louis Cauchy made a mistake in a published paper by improperly assuming the Limit Comparison Test to be valid for nonpositive series. Here are Dirichlet's two series:

$$\sum_{n=1}^{\infty} \frac{(-1)^n}{\sqrt{n}}, \qquad \sum_{n=1}^{\infty} \frac{(-1)^n}{\sqrt{n}}\left(1 + \frac{(-1)^n}{\sqrt{n}}\right)$$

Explain how they provide a counterexample to the Limit Comparison Test when the series are not assumed to be positive.

10.5 The Ratio and Root Tests

As we will show in Section 10.7, the number e has a well-known expression as an infinite series:

$$e = 1 + \frac{1}{1!} + \frac{1}{2!} + \frac{1}{3!} + \cdots$$

However, the convergence tests developed so far cannot be easily applied to this series. This points to the need for the following test, which is also of key importance in the study of power series (Section 10.6).

The symbol ρ, pronounced "rho," is the seventeenth letter of the Greek alphabet.

THEOREM 1 Ratio Test   Let $\{a_n\}$ be a sequence and assume that the following limit exists:

$$\rho = \lim_{n \to \infty} \left| \frac{a_{n+1}}{a_n} \right|$$

(i) If $\rho < 1$, then $\sum a_n$ converges absolutely.

(ii) If $\rho > 1$, then $\sum a_n$ diverges.

(iii) If $\rho = 1$, the Ratio Test is inconclusive (the series may converge or diverge).

Proof If $\rho < 1$, we may choose a number r such that $\rho < r < 1$. Since $\left| \frac{a_{n+1}}{a_n} \right|$ converges to ρ, there exists a number M such that $\left| \frac{a_{n+1}}{a_n} \right| < r$ for $n \geq M$. Therefore,

$$|a_{M+1}| < r|a_M|$$

$$|a_{M+2}| < r|a_{M+1}| < r^2|a_M|$$

$$|a_{M+3}| < r|a_{M+2}| < r^3|a_M|$$

In general, $|a_{M+n}| < r^n|a_M|$ and therefore

$$\sum_{n=M}^{\infty} |a_n| = \sum_{n=0}^{\infty} |a_{M+n}| \leq \sum_{n=0}^{\infty} |a_M| \, r^n = |a_M| \sum_{n=0}^{\infty} r^n$$

The geometric series on the right converges because $0 < r < 1$, so $\sum_{n=M}^{\infty} |a_n|$ converges by the Comparison Test. Thus, $\sum_{n=1}^{\infty} a_n$ converges absolutely.

If $\rho > 1$, choose a number r such that $1 < r < \rho$. Since $\left| \frac{a_{n+1}}{a_n} \right|$ converges to ρ, there exists a number M such that $\left| \frac{a_{n+1}}{a_n} \right| > r$ for $n \geq M$. Arguing as before with the inequalities reversed, we find that $|a_{M+n}| \geq r^n|a_M|$. Since r^n tends to ∞, we see that the terms a_{M+n} do not tend to zero and, consequently, $\sum_{n=1}^{\infty} a_n$ diverges. Finally, Example 4 below shows that when $\rho = 1$, both convergence and divergence are possible, so the test is inconclusive.   ∎

■ **EXAMPLE 1** Prove that $S = \sum_{n=1}^{\infty} \frac{1}{n!}$ converges.

Solution We compute the limit ρ. Let $a_n = \frac{1}{n!}$. Then

$$\frac{a_{n+1}}{a_n} = \frac{1}{(n+1)!} \frac{n!}{1} = \frac{n!}{(n+1)!} = \frac{1}{n+1}$$

$$\rho = \lim_{n \to \infty} \left| \frac{a_{n+1}}{a_n} \right| = \lim_{n \to \infty} \frac{1}{n+1} = 0$$

Since $\rho < 1$, S converges by the Ratio Test.   ■

■ **EXAMPLE 2** Apply the Ratio Test to determine if $\displaystyle\sum_{n=1}^{\infty} \frac{n^2}{2^n}$ converges.

Solution Let $a_n = \dfrac{n^2}{2^n}$. We have

$$\left|\frac{a_{n+1}}{a_n}\right| = \frac{(n+1)^2}{2^{n+1}} \frac{2^n}{n^2} = \frac{1}{2}\left(\frac{n^2+2n+1}{n^2}\right) = \frac{1}{2}\left(1 + \frac{2}{n} + \frac{1}{n^2}\right)$$

$$\rho = \lim_{n\to\infty}\left|\frac{a_{n+1}}{a_n}\right| = \frac{1}{2}\lim_{n\to\infty}\left(1 + \frac{2}{n} + \frac{1}{n^2}\right) = \frac{1}{2}$$

Since $\rho < 1$, the series converges by the Ratio Test.   ■

■ **EXAMPLE 3** Determine whether $\displaystyle\sum_{n=0}^{\infty}(-1)^n \frac{n!}{100^n}$ converges.

Solution Let $a_n = (-1)^n \dfrac{n!}{100^n}$. Then

$$\left|\frac{a_{n+1}}{a_n}\right| = \frac{(n+1)!}{100^{n+1}} \frac{100^n}{n!} = \frac{n+1}{100}$$

We see that the ratio of the coefficients tends to infinity:

$$\rho = \lim_{n\to\infty}\left|\frac{a_{n+1}}{a_n}\right| = \lim_{n\to\infty}\frac{n+1}{100} = \infty$$

Since $\rho > 1$, $\displaystyle\sum_{n=0}^{\infty}(-1)^n \frac{n!}{100^n}$ diverges by the Ratio Test.   ■

■ **EXAMPLE 4** Ratio Test Inconclusive Show that $\rho = 1$ for both $\displaystyle\sum_{n=1}^{\infty} n^2$ and $\displaystyle\sum_{n=1}^{\infty} n^{-2}$.
Conclude that the Ratio Test is inconclusive when $\rho = 1$.

Solution For $a_n = n^2$, we have

$$\rho = \lim_{n\to\infty}\left|\frac{a_{n+1}}{a_n}\right| = \lim_{n\to\infty}\frac{(n+1)^2}{n^2} = \lim_{n\to\infty}\frac{n^2+2n+1}{n^2} = \lim_{n\to\infty}\left(1 + \frac{2}{n} + \frac{1}{n^2}\right) = 1$$

On the other hand, for $b_n = n^{-2}$,

$$\rho = \lim_{n\to\infty}\left|\frac{b_{n+1}}{b_n}\right| = \lim_{n\to\infty}\left|\frac{a_n}{a_{n+1}}\right| = \frac{1}{\displaystyle\lim_{n\to\infty}\frac{a_{n+1}}{a_n}} = 1$$

Thus, $\rho = 1$ in both cases, but $\displaystyle\sum_{n=1}^{\infty} n^2$ diverges and $\displaystyle\sum_{n=1}^{\infty} n^{-2}$ converges (a p-series with $p = 2$). This shows that both convergence and divergence are possible when $\rho = 1$.   ■

For some series, it is more convenient to use the following Root Test, based on the limit of the nth roots $\sqrt[n]{a_n}$ rather than the ratios a_{n+1}/a_n. The proof of the Root Test, like that of the Ratio Test, is based on a comparison with geometric series (see Exercise 53).

> **THEOREM 2 Root Test**   Let $\{a_n\}$ be a sequence and assume that the following limit exists:
>
> $$L = \lim_{n \to \infty} \sqrt[n]{|a_n|}$$
>
> (i) If $L < 1$, then $\sum a_n$ converges absolutely.
> (ii) If $L > 1$, then $\sum a_n$ diverges.
> (iii) If $L = 1$, the Root Test is inconclusive: The series may converge or diverge.

■ **EXAMPLE 5**   Determine whether $\displaystyle\sum_{n=1}^{\infty} \left(\frac{n}{2n+3}\right)^n$ converges.

Solution   Let $a_n = \left(\dfrac{n}{2n+3}\right)^n$. Then

$$L = \lim_{n \to \infty} \sqrt[n]{a_n} = \lim_{n \to \infty} \frac{n}{2n+3} = \frac{1}{2}$$

Since $L < \frac{1}{2}$, the series $\displaystyle\sum_{n=1}^{\infty} \left(\frac{n}{2n+3}\right)^n$ converges.   ■

10.5 SUMMARY

- *Ratio Test:* Assume that the following limit exists:

$$\rho = \lim_{n \to \infty} \left|\frac{a_{n+1}}{a_n}\right|$$

Then $\sum a_n$ converges absolutely if $\rho < 1$ and it diverges if $\rho > 1$. The test is inconclusive if $\rho = 1$.
- *Root Test:* Assume that the following limit exists: $L = \lim_{n \to \infty} \sqrt[n]{|a_n|}$. Then $\sum a_n$ converges if $L < 1$ and it diverges if $L > 1$. The test is inconclusive if $L = 1$.

10.5 EXERCISES

Preliminary Questions

1. In the Ratio Test, is ρ equal to $\displaystyle\lim_{n \to \infty} \left|\frac{a_{n+1}}{a_n}\right|$ or $\displaystyle\lim_{n \to \infty} \left|\frac{a_n}{a_{n+1}}\right|$?

2. Is the Ratio Test conclusive for $\displaystyle\sum_{n=1}^{\infty} \frac{1}{2^n}$? Is it conclusive for $\displaystyle\sum_{n=1}^{\infty} \frac{1}{n}$?

3. Can the Ratio Test be used to show convergence if the series is only conditionally convergent?

Exercises

In Exercises 1–18, apply the Ratio Test to determine convergence or divergence, or state that the Ratio Test is inconclusive.

1. $\displaystyle\sum_{n=1}^{\infty} \frac{1}{5^n}$

2. $\displaystyle\sum_{n=1}^{\infty} \frac{(-1)^{n-1}n}{5^n}$

3. $\displaystyle\sum_{n=1}^{\infty} \frac{(-1)^{n-1}}{n^n}$

4. $\displaystyle\sum_{n=0}^{\infty} \frac{3n+2}{5n^3+1}$

5. $\displaystyle\sum_{n=1}^{\infty} \frac{n}{n^2+1}$

6. $\displaystyle\sum_{n=1}^{\infty} \frac{2^n}{n}$

7. $\displaystyle\sum_{n=1}^{\infty} \frac{2^n}{n^{100}}$

8. $\displaystyle\sum_{n=1}^{\infty} \frac{n^3}{3^{n^2}}$

9. $\displaystyle\sum_{n=1}^{\infty} \frac{10^n}{2^{n^2}}$

10. $\displaystyle\sum_{n=1}^{\infty} \frac{e^n}{n!}$

11. $\displaystyle\sum_{n=1}^{\infty} \frac{e^n}{n^n}$

12. $\displaystyle\sum_{n=1}^{\infty} \frac{n^{50}}{n!}$

13. $\displaystyle\sum_{n=0}^{\infty} (-1)^n \frac{n!}{4^n}$

14. $\displaystyle\sum_{n=1}^{\infty} \frac{n!}{n^4}$

15. $\displaystyle\sum_{n=2}^{\infty} \frac{1}{n \ln n}$

16. $\displaystyle\sum_{n=1}^{\infty} \frac{1}{(2n)!}$

17. $\displaystyle\sum_{n=1}^{\infty} \frac{n^2}{(2n+1)!}$

18. $\displaystyle\sum_{n=1}^{\infty} \frac{(n!)^2}{(2n)!}$

19. Show that $\displaystyle\sum_{n=1}^{\infty} n^k \, 3^{-n}$ converges for all exponents k.

20. Show that $\displaystyle\sum_{n=1}^{\infty} n^2 x^n$ converges if $|x| < 1$.

21. Show that $\displaystyle\sum_{n=1}^{\infty} 2^n x^n$ converges if $|x| < \frac{1}{2}$.

22. Show that $\displaystyle\sum_{n=1}^{\infty} \frac{r^n}{n!}$ converges for all r.

23. Show that $\displaystyle\sum_{n=1}^{\infty} \frac{r^n}{n}$ converges if $|r| < 1$.

24. Is there any value of k such that $\displaystyle\sum_{n=1}^{\infty} \frac{2^n}{n^k}$ converges?

25. Show that $\displaystyle\sum_{n=1}^{\infty} \frac{n!}{n^n}$ converges. *Hint:* Use $\displaystyle\lim_{n\to\infty} \left(1 + \frac{1}{n}\right)^n = e$.

In Exercises 26–31, assume that $|a_{n+1}/a_n|$ converges to $\rho = \frac{1}{3}$. What can you say about the convergence of the given series?

26. $\displaystyle\sum_{n=1}^{\infty} n a_n$

27. $\displaystyle\sum_{n=1}^{\infty} n^3 a_n$

28. $\displaystyle\sum_{n=1}^{\infty} 2^n a_n$

29. $\displaystyle\sum_{n=1}^{\infty} 3^n a_n$

30. $\displaystyle\sum_{n=1}^{\infty} 4^n a_n$

31. $\displaystyle\sum_{n=1}^{\infty} a_n^2$

32. Assume that $\left|\dfrac{a_{n+1}}{a_n}\right|$ converges to $\rho = 4$. Does $\displaystyle\sum_{n=1}^{\infty} a_n^{-1}$ converge (assume that $a_n \neq 0$ for all n)?

33. Is the Ratio Test conclusive for the p-series $\displaystyle\sum_{n=1}^{\infty} \frac{1}{n^p}$?

In Exercises 34–39, use the Root Test to determine convergence or divergence (or state that the test is inconclusive).

34. $\displaystyle\sum_{n=0}^{\infty} \frac{1}{10^n}$

35. $\displaystyle\sum_{n=1}^{\infty} \frac{1}{n^n}$

36. $\displaystyle\sum_{k=0}^{\infty} \left(\frac{k}{k+10}\right)^k$

37. $\displaystyle\sum_{k=0}^{\infty} \left(\frac{k}{3k+1}\right)^k$

38. $\displaystyle\sum_{n=1}^{\infty} \left(1 + \frac{1}{n}\right)^{-n}$

39. $\displaystyle\sum_{n=4}^{\infty} \left(1 + \frac{1}{n}\right)^{-n^2}$

40. Prove that $\displaystyle\sum_{n=1}^{\infty} \frac{2^{n^2}}{n!}$ diverges. *Hint:* Use $2^{n^2} = (2^n)^n$ and $n! \le n^n$.

In Exercises 41–52, determine convergence or divergence using any method covered in the text so far.

41. $\displaystyle\sum_{n=1}^{\infty} \frac{2^n + 4^n}{7^n}$

42. $\displaystyle\sum_{n=1}^{\infty} \frac{n^3}{n!}$

43. $\displaystyle\sum_{n=1}^{\infty} \frac{n^3}{5^n}$

44. $\displaystyle\sum_{n=2}^{\infty} \frac{1}{n(\ln n)^3}$

45. $\displaystyle\sum_{n=2}^{\infty} \frac{1}{\sqrt{n^3 - n^2}}$

46. $\displaystyle\sum_{k=1}^{\infty} 4^{-2k+1}$

47. $\displaystyle\sum_{n=1}^{\infty} \frac{n^2 + 4n}{3n^4 + 9}$

48. $\displaystyle\sum_{n=1}^{\infty} (-1)^n \cos \frac{1}{n}$

49. $\displaystyle\sum_{n=1}^{\infty} \sin \frac{1}{n^2}$

50. $\displaystyle\sum_{n=1}^{\infty} \frac{(-1)^{n-1}}{\sqrt{n}}$

51. $\displaystyle\sum_{n=1}^{\infty} \left(\frac{n}{n+12}\right)^n$

52. $\displaystyle\sum_{n=1}^{\infty} \frac{(-2)^n}{\sqrt{n}}$

Further Insights and Challenges

53. ✎ **Proof of the Root Test** Let $S = \displaystyle\sum_{n=0}^{\infty} a_n$ be a positive series and assume that $L = \displaystyle\lim_{n\to\infty} \sqrt[n]{a_n}$ exists.

(a) Show that S converges if $L < 1$. *Hint:* Choose R with $\rho < R < 1$ and show that $a_n \le R^n$ for n sufficiently large. Then compare with the geometric series $\sum R^n$.

(b) Show that S diverges if $L > 1$.

54. Show that the Ratio Test is inconclusive but the Root Test indicates convergence for the series

$$\frac{1}{2} + \frac{1}{3^2} + \frac{1}{2^3} + \frac{1}{3^4} + \frac{1}{2^5} + \cdots$$

55. Let $S = \sum_{n=1}^{\infty} \dfrac{c^n n!}{n^n}$, where c is a constant.

(a) Prove that S converges absolutely if $|c| < e$ and diverges if $|c| > e$.

(b) It is known that $\lim\limits_{n \to \infty} \dfrac{e^n n!}{n^{n+1/2}} = \sqrt{2\pi}$. Verify this numerically.

(c) Use the Limit Comparison Test to prove that S diverges for $c = e$.

10.6 Power Series

Most functions that arise in applications can be represented as power series. This includes not only the familiar trigonometric, exponential, logarithm, and root functions, but also the host of more advanced "special functions" of physics and engineering such as Bessel functions and elliptic functions.

In the introduction to this chapter, we mentioned that e^x can be expressed as an "infinite polynomial" called a power series:

$$e^x = 1 + \frac{x}{1!} + \frac{x^2}{2!} + \frac{x^3}{3!} + \frac{x^4}{4!} + \cdots$$

In this section, we develop the basic properties of power series, especially the key concept of *radius of convergence*.

A **power series** centered at the point $x = c$ is an infinite series of the form

$$F(x) = \sum_{n=0}^{\infty} a_n(x - c)^n = a_0 + a_1(x - c) + a_2(x - c)^2 + a_3(x - c)^3 + \cdots$$

To make use of a power series, we must determine the values of x for which the series converges. It certainly converges at its center $x = c$:

$$F(c) = a_0 + a_1(c - c) + a_2(c - c)^2 + a_3(c - c)^3 + \cdots = a_0$$

Where else does it converge? The following basic theorem states that every power series converges absolutely on an interval that is symmetric around the center $x = c$ (the interval may be infinite or possibly reduced to the single point c).

THEOREM 1 Radius of Convergence   Let $F(x) = \sum_{n=0}^{\infty} a_n(x - c)^n$. There are three possibilities:

(i) $F(x)$ converges only for $x = c$, or

(ii) $F(x)$ converges for all x, or

(iii) There is a number $R > 0$ such that $F(x)$ converges absolutely if $|x - c| < R$ and diverges if $|x - c| > R$. It may or may not converge at the endpoints $|x - c| = R$.

In Case (i), set $R = 0$, and in Case (ii), set $R = \infty$. We call R the **radius of convergence** of $F(x)$.

Proof   We assume that $c = 0$ to simplify the notation. The key observation is that if $F(x)$ converges for some nonzero value $x = B$, then it converges absolutely for all $|x| < |B|$.

To prove this, note that if $F(B) = \sum_{n=0}^{\infty} a_n B^n$ converges, then the general term $a_n B^n$ must tend to zero. In particular, there exists $M > 0$ such that $|a_n B^n| < M$ for all n, and therefore,

$$\sum_{n=0}^{\infty} |a_n x^n| = \sum_{n=0}^{\infty} |a_n B^n| \left| \frac{x}{B} \right|^n < M \sum_{n=0}^{\infty} \left| \frac{x}{B} \right|^n$$

If $|x| < |B|$, then $|x/B| < 1$ and the series on the right is a convergent geometric series. By the Comparison Test, the series on the left also converges and thus $F(x)$ converges absolutely if $|x| < |B|$.

Let S be the set of numbers x such that $F(x)$ converges. Then S contains 0. If $S = \{0\}$, then $F(x)$ converges only for $x = 0$ and Case (i) holds. Otherwise, S contains a number $B \neq 0$. In this case, S contains the open interval $(-|B|, |B|)$ by the previous paragraph. If S is bounded, then S has a least upper bound $L > 0$ (see marginal note). Since there exist numbers $B \in S$ smaller than but arbitrarily close to L, S contains $(-B, B)$ for all $0 < B < L$. It follows that S contains the open interval $(-L, L)$. S cannot contain any number x with $|x| > L$, but S may contain one or both of the endpoints $x = \pm L$. This is Case (iii). If S is not bounded, then S contains intervals $(-B, B)$ for B arbitrarily large. Thus $S = \mathbf{R}$ and we are in Case (ii). ∎

Least Upper Bound Property: If S is a set of real numbers with an upper bound M (that is, $x \leq M$ for all $x \in S$), then S has a least upper bound L. See Appendix B.

According to Theorem 1, if the radius of convergence R is nonzero and finite, then $F(x)$ converges absolutely on an open interval around c of radius R (Figure 1). *It is necessary to check convergence at the endpoints of the interval separately.*

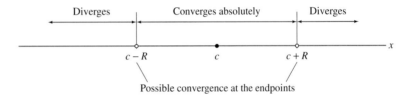

FIGURE 1 Interval of convergence of a power series.

Possible convergence at the endpoints

■ **EXAMPLE 1** Using the Ratio Test For which values of x does $F(x) = \sum_{n=0}^{\infty} \dfrac{x^n}{2^n}$ converge?

Solution Let $b_n = \dfrac{x^n}{2^n}$ and let us compute the ratio ρ of the Ratio Test:

$$\rho = \lim_{n \to \infty} \left| \frac{b_{n+1}}{b_n} \right| = \lim_{n \to \infty} \left| \frac{x^{n+1}}{2^{n+1}} \cdot \frac{2^n}{x^n} \right| = \lim_{n \to \infty} \left| \frac{2^{-(n+1)} x^{n+1}}{2^{-n} x^n} \right| = \lim_{n \to \infty} \frac{1}{2} |x| = \frac{1}{2} |x|$$

By the Ratio Test, $F(x)$ converges if $\rho = \frac{1}{2}|x| < 1$, that is, for $|x| < 2$. Similarly, $F(x)$ diverges if $\rho = \frac{1}{2}|x| > 1$ or $|x| > 2$. Therefore, the radius of convergence is $R = 2$.

What about the endpoints? The Ratio Test is inconclusive for $x = \pm 2$, so we must check these cases directly. Both series diverge:

$$F(2) = \sum_{n=0}^{\infty} \frac{2^n}{2^n} = 1 + 1 + 1 + 1 + 1 + 1 \cdots$$

$$F(-2) = \sum_{n=0}^{\infty} \frac{(-2)^n}{2^n} = 1 - 1 + 1 - 1 + 1 - 1 \cdots$$

Therefore, $F(x)$ converges only for $|x| < 2$ (Figure 2). ■

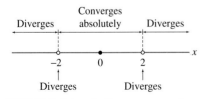

FIGURE 2 Interval of convergence of $\sum_{n=0}^{\infty} \dfrac{x^n}{2^n}$.

The method of the previous example may be applied more generally to any power series $F(x) = \sum b_n(x - c)^n$ for which the following limit exists:

$$r = \lim_{n \to \infty} \left| \frac{b_{n+1}}{b_n} \right|$$

We compute the ratio ρ of the Ratio Test applied to $F(x)$:

$$\rho = \lim_{n \to \infty} \left| \frac{b_{n+1}(x - c)^{n+1}}{b_n(x - c)^n} \right| = |x - c| \left(\lim_{n \to \infty} \left| \frac{b_{n+1}}{b_n} \right| \right) = r|x - c|$$

By the Ratio Test, $F(x)$ converges if $\rho = r|x - c| < 1$ and diverges if $\rho = r|x - c| > 1$. Thus, if r is finite and nonzero, then $F(x)$ converges if $|x - c| < r^{-1}$ and the radius of convergence is $R = r^{-1}$. If $r = 0$, then $F(x)$ converges for all x and the radius of convergence is $R = \infty$. If $r = \infty$, then $F(x)$ diverges for all $x \neq c$ and $R = 0$.

THEOREM 2 Finding the Radius of Convergence   Let $F(x) = \sum b_n(x - c)^n$ and assume that the following limit exists:

$$r = \lim_{n \to \infty} \left| \frac{b_{n+1}}{b_n} \right|$$

Then $F(x)$ has radius of convergence $R = r^{-1}$ (where we set $R = \infty$ if $r = 0$ and $R = 0$ if $r = \infty$).

■ **EXAMPLE 2** Determine the convergence of $F(x) = \sum_{n=1}^{\infty} \frac{(-1)^n}{n}(x - 5)^n$.

Solution Let $b_n = \dfrac{(-1)^n}{n}$. Then

$$r = \lim_{n \to \infty} \left| \frac{b_{n+1}}{b_n} \right| = \lim_{n \to \infty} \left| \frac{1}{n+1} \frac{n}{1} \right| = \lim_{n \to \infty} \left| \frac{n}{n+1} \right| = 1$$

The radius of convergence is $R = r^{-1} = 1$. Therefore, the power series converges absolutely if $|x - 5| < 1$ and diverges if $|x - 5| > 1$. In other words, $F(x)$ converges absolutely on the open interval $(4, 6)$. At the endpoints we have

$$x = 6: \quad \sum_{n=1}^{\infty} \frac{(-1)^n}{n}(6 - 5)^n = \sum_{n=1}^{\infty} \frac{(-1)^n}{n} \quad \text{convergent by the Leibniz Test}$$

$$x = 4: \quad \sum_{n=1}^{\infty} \frac{(-1)^n}{n}(4 - 5)^n = \sum_{n=1}^{\infty} \frac{1}{n} \quad \text{divergent (harmonic series)}$$

Therefore, the power series converges on the half-open interval $(4, 6]$. ■

■ **EXAMPLE 3** Infinite Radius of Convergence Show that $\displaystyle\sum_{n=0}^{\infty} \frac{x^n}{n!}$ converges for all x.

Solution Let $b_n = \dfrac{1}{n!}$. Then

$$r = \lim_{n \to \infty} \left| \frac{b_{n+1}}{b_n} \right| = \lim_{n \to \infty} \frac{1}{(n+1)!} \frac{n!}{1} = \lim_{n \to \infty} \frac{1}{n+1} = 0$$

Thus, $R = r^{-1} = \infty$ and the series converges for all x by Theorem 2. ■

When a function $f(x)$ is represented by a power series on an interval I, we refer to the power series as the "power series expansion" of $f(x)$ on I. In the next section, we show that a function has at most one power series expansion with a given center c on an interval.

An important example of a power series is provided by the geometric series. Recall that $\displaystyle\sum_{n=0}^{\infty} r^n = \frac{1}{1-r}$ for $|r| < 1$. Writing x instead of r, we view this formula as a power series expansion:

$$\boxed{\frac{1}{1-x} = \sum_{n=0}^{\infty} x^n \qquad \text{for } |x| < 1} \qquad \boxed{1}$$

The next two examples show how this formula may be adapted to find the power series representations of other functions.

■ **EXAMPLE 4** Using the Formula for Geometric Series Prove that $\dfrac{1}{1-2x} = \displaystyle\sum_{n=0}^{\infty} 2^n x^n$ for $|x| < \frac{1}{2}$.

Solution Substitute $2x$ for x in Eq. (1):

$$\frac{1}{1-2x} = \sum_{n=0}^{\infty} (2x)^n = \sum_{n=0}^{\infty} 2^n x^n \qquad \boxed{2}$$

Expansion (1) is valid for $|x| < 1$ and thus expansion (2) is valid for $|2x| < 1$ or $|x| < \frac{1}{2}$. ■

■ **EXAMPLE 5** Prove that $\dfrac{1}{2+x^2} = \displaystyle\sum_{n=0}^{\infty} \frac{(-1)^n x^{2n}}{2^{n+1}}$. For which x is this formula valid?

Solution We first rewrite $\dfrac{1}{2+x^2}$ in the form $\dfrac{1}{1-u}$ so we can use Eq. (1):

$$\frac{1}{2+x^2} = \frac{1}{2} \frac{1}{1+\frac{1}{2}x^2} = \frac{1}{2} \frac{1}{1-(-\frac{1}{2}x^2)}$$

We may substitute $-\frac{1}{2}x^2$ for x in Eq. (1), provided that $\left|\frac{1}{2}x^2\right| < 1$, to obtain

$$\frac{1}{2+x^2} = \frac{1}{2} \sum_{n=0}^{\infty} \left(-\frac{x^2}{2}\right)^n = \sum_{n=0}^{\infty} \frac{(-1)^n x^{2n}}{2^{n+1}}$$

This expansion is valid if $\left|-x^2/2\right| < 1$ or $|x| < \sqrt{2}$. ■

Our next theorem states, in essence, that power series are well-behaved functions in the following sense: A power series $F(x)$ is differentiable within its interval of convergence and we may differentiate and integrate $F(x)$ as if it were a polynomial.

THEOREM 3 Term-by-Term Differentiation and Integration Suppose that

$$F(x) = \sum_{n=0}^{\infty} a_n (x - c)^n$$

has radius of convergence $R > 0$. Then $F(x)$ is differentiable on $(c - R, c + R)$ and its derivative and antiderivative may be computed term by term. More precisely, for $x \in (c - R, c + R)$ we have

$$F'(x) = \sum_{n=1}^{\infty} n a_n (x - c)^{n-1}$$

$$\int F(x)\, dx = A + \sum_{n=0}^{\infty} \frac{a_n}{n+1} (x - c)^{n+1} \quad (A \text{ any constant})$$

These series have the same radius of convergence R.

See Exercise 58 for a proof that $F(x)$ is continuous. The proofs of the remaining statements are omitted.

■ **EXAMPLE 6** Differentiating a Power Series Prove that

$$\frac{1}{(1 - x)^2} = 1 + 2x + 3x^2 + 4x^3 + 5x^4 + \cdots$$

for $-1 < x < 1$.

Solution Noting that

$$\frac{d}{dx} \frac{1}{1 - x} = \frac{1}{(1 - x)^2}$$

we obtain the result by differentiating the geometric series term by term for $|x| < 1$:

$$\frac{d}{dx} \frac{1}{1 - x} = \frac{d}{dx}\left(1 + x + x^2 + x^3 + x^4 + \cdots\right)$$

$$\frac{1}{(1 - x)^2} = 1 + 2x + 3x^2 + 4x^3 + 5x^4 + \cdots \qquad \boxed{3}$$

Expansion (3) is valid for $|x| < 1$ because the geometric series has radius of convergence $R = 1$. ■

■ **EXAMPLE 7** The Power Series for $f(x) = \tan^{-1} x$ via Integration Prove that for $-1 < x < 1$,

$$\tan^{-1} x = \sum_{n=0}^{\infty} \frac{(-1)^n x^{2n+1}}{2n + 1} = x - \frac{x^3}{3} + \frac{x^5}{5} - \frac{x^7}{7} + \cdots \qquad \boxed{4}$$

Solution First, substitute $-x^2$ for x in (1) to obtain

$$\frac{1}{1 + x^2} = 1 - x^2 + x^4 - x^6 + \cdots$$

Since the geometric series has radius of convergence $R = 1$, this expansion is valid for $|x^2| < 1$, that is, $|x| < 1$. Now apply Theorem 3 to integrate this series term by term, recalling that $\tan^{-1} x$ is an antiderivative $(1 + x^2)^{-1}$:

$$\tan^{-1} x = \int \frac{dx}{1 + x^2} = \int \left(1 - x^2 + x^4 - x^6 + \cdots\right) dx = A + x - \frac{x^3}{3} + \frac{x^5}{5} - \frac{x^7}{7} + \cdots$$

To determine the constant A, set $x = 0$. We obtain $\tan^{-1} 0 = 0 = A$ and therefore $A = 0$. This proves Eq. (4) for $-1 < x < 1$. ∎

GRAPHICAL INSIGHT Let's examine the expansion of the previous example graphically. The partial sums of the power series for $f(x) = \tan^{-1} x$ are

$$S_N(x) = \sum_{n=0}^{N} (-1)^n \frac{x^{2n+1}}{2n+1} = x - \frac{x^3}{3} + \frac{x^5}{5} - \cdots + (-1)^N \frac{x^{2N+1}}{2N+1}$$

We can expect $S_N(x)$ to provide a good approximation to $f(x) = \tan^{-1} x$ on the interval $(-1, 1)$, where the power series expansion is valid. Figure 3 confirms this: The graphs of the partial sums $S_{50}(x)$ and $S_{51}(x)$ are nearly indistinguishable from the graph of $\tan^{-1} x$ on $(-1, 1)$. Thus we may use the partial sums to approximate values of the arctangent. For example, an approximation to $\tan^{-1}(0.3)$ is given by

$$S_4(0.3) = 0.3 - \frac{(0.3)^3}{3} + \frac{(0.3)^5}{5} - \frac{(0.3)^7}{7} + \frac{(0.3)^9}{9} \approx 0.2914569$$

Since the power series is an alternating series, the error is not greater than the first omitted term:

$$|\tan^{-1}(0.3) - S_4(0.3)| \leq \frac{(0.3)^{11}}{11} \approx 1.61 \times 10^{-7}$$

The situation changes drastically in the region $|x| > 1$, where the power series diverges. The partial sums $S_N(x)$ deviate sharply from $\tan^{-1} x$ outside $(-1, 1)$.

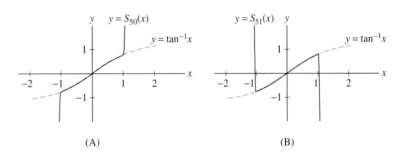

FIGURE 3 $S_{50}(x)$ and $S_{51}(x)$ are nearly indistinguishable from $\tan^{-1} x$ on $(-1, 1)$.

(A)   (B)

Power Series Solutions of Differential Equations

In the next section, we use the theory of Taylor series to prove that the exponential function $f(x) = e^x$ is represented by a power series. However, we can already show this with the tools at our disposal by making use of the differential equation satisfied by $f(x) = e^x$. Recall that by Theorem 1 in Section 5.8, $y = e^x$ is the unique function satisfying the differential equation $y' = y$ with initial condition $y(0) = 1$. Let's try to find a power series

$$P(x) = \sum_{n=0}^{\infty} a_n x^n \text{ that also satisfies } P'(x) = P(x) \text{ and } P(0) = 1.$$

We have

$$P(x) = \sum_{n=0}^{\infty} a_n x^n = a_0 + a_1 x + a_2 x^2 + a_3 x^3 + \cdots$$

$$P'(x) = \sum_{n=0}^{\infty} n a_n x^{n-1} = a_1 + 2a_2 x + 3a_3 x^2 + 4a_4 x^3 + \cdots$$

We see that $P(x)$ satisfies $P'(x) = P(x)$ if

$$a_0 = a_1, \quad a_1 = 2a_2, \quad a_2 = 3a_3, \quad a_3 = 4a_4, \quad \ldots$$

In general, $a_{n-1} = n a_n$, or

$$\boxed{a_n = \frac{a_{n-1}}{n}}$$

This equation is called a *recursion relation*. It allows us to successively determine all of the coefficients a_n from the first coefficient a_0, which may be chosen arbitrarily. For example, the recursion relation yields

$$n = 1: \qquad a_1 = \frac{a_0}{1}$$

$$n = 2: \qquad a_2 = \frac{a_1}{2} = \frac{a_0}{2 \cdot 1} = \frac{a_0}{2!}$$

$$n = 3: \qquad a_3 = \frac{a_2}{3} = \frac{a_1}{3 \cdot 2} = \frac{a_0}{3 \cdot 2 \cdot 1} = \frac{a_0}{3!}$$

To obtain a general formula for a_n, apply the recursion relation n times:

$$a_n = \frac{a_{n-1}}{n} = \frac{a_{n-2}}{n(n-1)} = \frac{a_{n-3}}{n(n-1)(n-2)} = \cdots = \frac{a_0}{n!}$$

We conclude that $P(x) = a_0 \sum_{n=0}^{\infty} \frac{x^n}{n!}$. As we showed in Example 3, this power series has an infinite radius of convergence and thus $P(x)$ is a solution of $y' = y$ for all x.

Now observe that $P(0) = a_0$, so we set $a_0 = 1$ to obtain a solution satisfying the initial condition $y(0) = 1$. Now, since $f(x) = e^x$ and $P(x)$ both satisfy the differential condition with initial condition, they are equal. Thus we have proven that for all x,

$$e^x = \sum_{n=0}^{\infty} \frac{x^n}{n!} = 1 + x + \frac{x^2}{2!} + \frac{x^3}{3!} + \frac{x^4}{4!} + \cdots \qquad \boxed{5}$$

The method just employed is a powerful tool in the study of differential equations. We knew in advance that $y = e^x$ is a solution of $y' = y$, but suppose we are given a differential equation whose solution is unknown. We may try to find a solution in the form of a power series $P(x) = \sum_{n=0}^{\infty} a_n x^n$. In favorable cases, the differential equation leads to a recursion relation that enables us to determine the coefficients a_n.

The solution of Eq. (6) satisfying $y'(0) = 1$ is called the "Bessel function of order one." The Bessel function of order n is a solution of

$$x^2 y'' + x y' + (x^2 - n^2) y = 0$$

Bessel functions appear in many areas of physics and engineering.

■ **EXAMPLE 8** Find a power series solution to the differential equation

$$x^2 y'' + x y' + (x^2 - 1) y = 0 \qquad \boxed{6}$$

and the initial condition $y'(0) = 1$.

Solution Assume that Eq. (6) has a power series solution $P(x) = \sum\limits_{n=0}^{\infty} a_n x^n$. Then

$$y' = P'(x) = \sum_{n=0}^{\infty} n a_n x^{n-1} = a_1 + 2a_2 x + 3a_3 x^2 + \cdots$$

$$y'' = P''(x) = \sum_{n=0}^{\infty} n(n-1) a_n x^{n-2} = 2a_2 + 6a_3 x + 12a_4 x^2 + \cdots$$

Now substitute the series for y, y', and y'' into the differential equation (6) to determine the recursion relation satisfied by the coefficients a_n:

$$x^2 y'' + xy' + (x^2 - 1)y$$

$$= x^2 \sum_{n=0}^{\infty} n(n-1) a_n x^{n-2} + x \sum_{n=0}^{\infty} n a_n x^{n-1} + (x^2 - 1) \sum_{n=0}^{\infty} a_n x^n$$

$$= \sum_{n=0}^{\infty} n(n-1) a_n x^n + \sum_{n=0}^{\infty} n a_n x^n - \sum_{n=0}^{\infty} a_n x^n + \sum_{n=0}^{\infty} a_n x^{n+2}$$

$$= \sum_{n=0}^{\infty} (n^2 - 1) a_n x^n + \sum_{n=2}^{\infty} a_{n-2} x^n$$

We see that the equation $x^2 y'' + xy' + (x^2 - 1)y = 0$ is satisfied if

$$\sum_{n=0}^{\infty} (n^2 - 1) a_n x^n = -\sum_{n=2}^{\infty} a_{n-2} x^n \qquad \boxed{7}$$

The first few terms on each side of this equation are

$$-a_0 + 0 \cdot x + 3a_2 x^2 + 8a_3 x^3 + 15a_4 x^4 + \cdots = 0 + 0 \cdot x - a_0 x^2 - a_1 x^3 - a_2 x^4 - \cdots$$

Matching up the coefficients of x^n, we find that

$$-a_0 = 0, \qquad 3a_2 = -a_0, \qquad 8a_3 = -a_1, \qquad 15a_4 = -a_2 \qquad \boxed{8}$$

In general, $(n^2 - 1)a_n = -a_{n-2}$, and this yields the recursion relation

$$\boxed{a_n = -\frac{a_{n-2}}{n^2 - 1} \qquad \text{for } n \geq 2} \qquad \boxed{9}$$

Note that $a_0 = 0$ by (8). The recursion relation implies that all of the even coefficients $a_2, a_4, a_6, \ldots$ are zero:

$$a_2 = \frac{a_0}{2^2 - 1} \quad \text{so } a_2 = 0, \qquad \text{and then} \qquad a_4 = \frac{a_2}{4^2 - 1} = 0 \text{ so } a_4 = 0, \qquad \text{etc.}$$

As for the odd coefficients, note that we may choose a_1 arbitrarily. Observe that $P'(0) = a_1$. Thus, we set $a_1 = 1$ so that $y = P(x)$ satisfies $y'(0) = 1$. Now apply Eq. (9):

$$n = 3 \qquad a_3 = -\frac{a_1}{3^2 - 1} = -\frac{1}{3^2 - 1}$$

$$n = 5: \qquad a_5 = -\frac{a_3}{5^2 - 1} = \frac{1}{(5^2 - 1)(3^2 - 1)}$$

$$n = 7: \qquad a_7 = -\frac{a_5}{7^2 - 1} = -\frac{1}{(7^2 - 1)(3^2 - 1)(5^2 - 1)}$$

This shows the general pattern of coefficients. To express the coefficients in a compact form, let $n = 2k + 1$. Then

$$n^2 - 1 = (2k + 1)^2 - 1 = 4k^2 + 4k = 4k(k + 1)$$

and the recursion relation may be written

$$a_{2k+1} = -\frac{a_{2k-1}}{4k(k + 1)}$$

Applying this recursion relation k times, we obtain the closed formula

$$a_{2k+1} = (-1)^k \left(\frac{1}{4k(k + 1)} \right) \left(\frac{1}{4(k - 1)k} \right) \cdots \left(\frac{1}{4(1)(2)} \right) = \frac{(-1)^k}{4^k \, k! \, (k + 1)!}$$

Thus we obtain a power series representation of our solution:

$$P(x) = \sum_{k=0}^{\infty} \frac{(-1)^k}{4^k k!(k + 1)!} x^{2k+1}$$

A straightforward application of the Ratio Test shows that $P(x)$ has an infinite radius of convergence. Therefore $P(x)$ is a solution of the initial value problem for all x. ∎

10.6 SUMMARY

- A *power series* is an infinite series of the form

$$F(x) = \sum_{n=0}^{\infty} a_n (x - c)^n$$

The constant c is called the *center* of $F(x)$.
- A power series behaves in one of three ways:

(i) $F(x)$ converges only for $x = c$, or

(ii) $F(x)$ converges for all x, or

(iii) There exists $R > 0$ such that $F(x)$ converges absolutely for $|x - c| < R$ and diverges for $|x - c| > R$.

The number R is called the *radius of convergence* of $F(x)$. Convergence at the endpoints $x = c \pm R$ must be checked separately. We set $R = 0$ in Case (i), and $R = \infty$ in Case (ii).
- If $r = \lim\limits_{n \to \infty} \left| \dfrac{a_{n+1}}{a_n} \right|$ exists, then $F(x)$ has radius of convergence $R = r^{-1}$ (where $R = 0$ if $r = \infty$ and $R = \infty$ if $r = 0$).

- If $R > 0$, then $F(x)$ is differentiable on $(c - R, c + R)$ and may be differentiated and integrated term by term:

$$F'(x) = \sum_{n=1}^{\infty} n a_n (x - c)^{n-1}, \qquad \int F(x)\, dx = A + \sum_{n=0}^{\infty} \frac{a_n}{n+1} (x - c)^{n+1}$$

(A is any constant). The power series for $F'(x)$ and $\int F(x)\, dx$ have the same radius of convergence R.

- The power series expansion $\dfrac{1}{1-x} = \sum_{n=0}^{\infty} x^n$ is valid for $|x| < 1$. It may be used to derive expansions of other related functions by substitution, integration, or differentiation.

10.6 EXERCISES

Preliminary Questions

1. Suppose that $\sum a_n x^n$ converges for $x = 5$. Must it also converge for $x = 4$? What about $x = -3$?

2. Suppose that $\sum a_n (x - 6)^n$ converges for $x = 10$. At which of the points (a)–(d) must it also converge?

(a) $x = 8$ **(b)** $x = 12$ **(c)** $x = 2$ **(d)** $x = 0$

3. Suppose that $F(x)$ is a power series with radius of convergence $R = 12$. What is the radius of convergence of $F(3x)$?

4. The power series $F(x) = \sum_{n=1}^{\infty} n x^n$ has radius of convergence $R = 1$. What is the power series expansion of $F'(x)$ and what is its radius of convergence?

Exercises

1. Use the Ratio Test to determine the radius of convergence of

$$\sum_{n=0}^{\infty} \frac{x^n}{2^n}.$$

2. Use the Ratio Test to show that $\sum_{n=1}^{\infty} \dfrac{x^n}{\sqrt{n}\, 2^n}$ has radius of convergence $R = 2$. Then determine whether it converges absolutely or conditionally at the endpoints $R = \pm 2$.

3. Show that the following three power series have the same radius of convergence. Then show that (a) diverges at both endpoints, (b) converges at one endpoint but diverges at the other, and (c) converges at both endpoints.

(a) $\displaystyle\sum_{n=1}^{\infty} \frac{x^n}{3^n}$ **(b)** $\displaystyle\sum_{n=1}^{\infty} \frac{x^n}{n 3^n}$ **(c)** $\displaystyle\sum_{n=1}^{\infty} \frac{x^n}{n^2 3^n}$

4. Repeat Exercise 3 for the following series:

(a) $\displaystyle\sum_{n=1}^{\infty} \frac{(x - 5)^n}{9^n}$ **(b)** $\displaystyle\sum_{n=1}^{\infty} \frac{(x - 5)^n}{n 9^n}$ **(c)** $\displaystyle\sum_{n=1}^{\infty} \frac{(x - 5)^n}{n^2 9^n}$

5. Show that $\displaystyle\sum_{n=0}^{\infty} n^n x^n$ diverges for all $x \neq 0$.

6. (a) Find the radius of convergence of $\displaystyle\sum_{n=1}^{\infty} \frac{x^n}{n^2}$.

(b) Determine whether the series converges at the endpoints of the interval of convergence.

In Exercises 7–26, find the values of x for which the following power series converge.

7. $\displaystyle\sum_{n=1}^{\infty} n x^n$

8. $\displaystyle\sum_{n=1}^{\infty} n(x - 3)^n$

9. $\displaystyle\sum_{n=1}^{\infty} \frac{2^n x^n}{n}$

10. $\displaystyle\sum_{n=1}^{\infty} \frac{(-5)^n (x - 3)^n}{n^2}$

11. $\displaystyle\sum_{n=2}^{\infty} \frac{x^n}{\ln n}$

12. $\displaystyle\sum_{n=1}^{\infty} \frac{x^n}{n 2^n}$

13. $\displaystyle\sum_{n=1}^{\infty} \frac{x^n}{(n!)^2}$

14. $\displaystyle\sum_{n=4}^{\infty} \frac{x^n}{n^5}$

15. $\displaystyle\sum_{n=1}^{\infty} (-1)^n n^4 (x + 4)^n$

16. $\displaystyle\sum_{n=0}^{\infty} \frac{(-1)^n x^n}{\sqrt{n^2 + 1}}$

17. $\displaystyle\sum_{n=0}^{\infty} \frac{n}{2^n} x^n$

18. $\displaystyle\sum_{n=0}^{\infty} 4^n x^n$

19. $\displaystyle\sum_{n=1}^{\infty} \frac{(x - 4)^n}{n^4}$

20. $\displaystyle\sum_{n=1}^{\infty} \frac{2^n}{3n} (x + 3)^n$

21. $\displaystyle\sum_{n=10}^{\infty} n! \, (x+5)^n$

22. $\displaystyle\sum_{n=15}^{\infty} \frac{x^{2n+1}}{3n+1}$

23. $\displaystyle\sum_{n=12}^{\infty} e^n (x-2)^n$

24. $\displaystyle\sum_{n=0}^{\infty} \frac{x^n}{n^4 + 2}$

25. $\displaystyle\sum_{n=1}^{\infty} \frac{x^n}{n - 4\ln n}$

26. $\displaystyle\sum_{n=2}^{\infty} \frac{(x-2)^n}{(n \ln n)^2}$

In Exercises 27–34, use Eq. (1) to expand the function in a power series with center $c = 0$ and determine the set of x for which the expansion is valid.

27. $f(x) = \dfrac{1}{1 - 3x}$

28. $f(x) = \dfrac{1}{1 + 3x}$

29. $f(x) = \dfrac{1}{3 - x}$

30. $f(x) = \dfrac{1}{4 + 3x}$

31. $f(x) = \dfrac{1}{1 + x^9}$

32. $f(x) = \dfrac{1}{5 - x^2}$

33. $f(x) = \dfrac{1}{1 + 3x^7}$

34. $f(x) = \dfrac{1}{4 - 2x^3}$

35. Use the equalities

$$\frac{1}{1 - x} = \frac{1}{-3 - (x-4)} = \frac{-\frac{1}{3}}{1 + \left(\frac{x-4}{3}\right)}$$

to show that for $|x - 4| < 3$

$$\frac{1}{1 - x} = \sum_{n=0}^{\infty} (-1)^{n+1} \frac{(x-4)^n}{3^{n+1}}$$

36. Use the method of Exercise 35 to expand $\dfrac{1}{1 - x}$ in power series with centers $c = 2$ and $c = -2$. Determine the set of x for which the expansions are valid.

37. Use the method of Exercise 35 to expand $\dfrac{1}{4 - x}$ in a power series with center $c = 5$. Determine the set of x for which the expansion is valid.

38. Evaluate $\displaystyle\sum_{n=1}^{\infty} \frac{n}{2^n}$. *Hint:* Show that

$$(1 - x)^{-2} = \sum_{n=1}^{\infty} n x^{n-1}$$

39. Give an example of a power series that converges for x in $[2, 6)$.

40. Prove that for $-1 < x < 1$

$$\frac{1}{1 + x} = \sum_{n=0}^{\infty} (-1)^n x^n = 1 - x + x^2 - x^3 + \cdots$$

$$\ln(1 + x) = \sum_{n=1}^{\infty} \frac{(-1)^{n-1} x^n}{n} = x - \frac{x^2}{2} + \frac{x^3}{3} - \frac{x^4}{4} + \cdots$$

41. Use Exercise 40 to prove that

$$\ln \frac{3}{2} = \frac{1}{2} - \frac{1}{2 \cdot 2^2} + \frac{1}{3 \cdot 2^3} - \frac{1}{4 \cdot 2^4} + \cdots$$

Use your knowledge of alternating series to find an N such that the partial sum S_N approximates $\ln \frac{3}{2}$ to within an error of at most 10^{-3}. Confirm this using a calculator to compute both S_N and $\ln \frac{3}{2}$.

42. Show that the following series converges absolutely for $|x| < 1$ and compute its sum:

$$F(x) = 1 - x - x^2 + x^3 - x^4 - x^5 + x^6 - x^7 - x^8 + \cdots$$

Hint: Write $F(x)$ as a sum of three geometric series with common ratio x^3.

43. Show that for $|x| < 1$

$$\frac{1 + 2x}{1 + x + x^2} = 1 + x - 2x^2 + x^3 + x^4 - 2x^5 + x^6 + x^7 - 2x^8 + \cdots$$

Hint: Use the hint from Exercise 42.

44. Find a power series $P(x) = \displaystyle\sum_{n=0}^{\infty} a_n x^n$ satisfying the differential equation $y' = -y$ with initial condition $y(0) = 1$. Then use Theorem 1 of Section 5.8 to conclude that $P(x) = e^{-x}$.

45. Use the power series for $y = e^x$ to show that

$$\frac{1}{e} = \frac{1}{2!} - \frac{1}{3!} + \frac{1}{4!} - \cdots$$

Use your knowledge of alternating series to find an N such that the partial sum S_N approximates e^{-1} to within an error of at most 10^{-3}. Confirm this using a calculator to compute both S_N and e^{-1}.

46. Let $P(x) = \displaystyle\sum_{n=0}^{\infty} a_n x^n$ be a power series solution to $y' = 2xy$ with initial condition $y(0) = 1$.

(a) Show that the odd coefficients a_{2k+1} are all zero.

(b) Prove that $a_{2k} = \dfrac{a_{2k-2}}{k}$ and use this result to determine the coefficients a_{2k}.

47. Find a power series $P(x)$ satifying the differential equation:

$$y'' - xy' + y = 0 \qquad \boxed{10}$$

with initial condition $y(0) = 1$, $y'(0) = 0$. What is the radius of convergence of the power series?

48. Find a power series satisfying Eq. (10) with initial condition $y(0) = 0$, $y'(0) = 1$.

49. Prove that $J_2(x) = \displaystyle\sum_{k=0}^{\infty} \frac{(-1)^k}{2^{2k+2} \, k! \, (k+3)!} x^{2k+2}$ is a solution of the Bessel differential equation of order two:

$$x^2 y'' + xy' + (x^2 - 4)y = 0$$

50. Use Eq. (4) to approximate $\tan^{-1}(0.5)$ to three decimal places.

51. Let $C(x) = 1 - \dfrac{x^2}{2!} + \dfrac{x^4}{4!} - \dfrac{x^6}{6!} + \cdots$.

(a) Show that $C(x)$ has an infinite radius of convergence.

(b) Prove that $C(x)$ and $f(x) = \cos x$ are both solutions of $y'' = -y$ with initial conditions $y(0) = 1$, $y'(0) = 0$. This initial value problem has a unique solution, so it follows that $C(x) = \cos x$ for all x.

52. Find all values of x such that $\displaystyle\sum_{n=1}^{\infty} \frac{x^{n^2}}{n!}$ converges.

53. Find all values of x such that the following series converges:

$$F(x) = 1 + 3x + x^2 + 27x^3 + x^4 + 243x^5 + \cdots$$

54. Explain why Theorem 2 cannot be applied directly to find the radius of convergence of $\displaystyle\sum_{n=1}^{\infty} \frac{x^{3n}}{5^n}$. What is the radius of convergence of this series?

55. Why is it impossible to expand $f(x) = |x|$ as a power series that converges in an interval around $x = 0$? Explain this using Theorem 3.

Further Insights and Challenges

56. Prove that for $b \neq 0$

$$\sum_{n=1}^{\infty} \frac{1}{n(n+b)} = \frac{1}{b} \int_0^1 \frac{1 - x^b}{1 - x}\, dx$$

Conclude that the sum on the left has the value

$$\frac{1}{b}\left(1 + \frac{1}{2} + \frac{1}{3} + \cdots + \frac{1}{b}\right)$$

57. Suppose that the coefficients of $F(x) = \displaystyle\sum_{n=0}^{\infty} a_n x^n$ are *periodic*, that is, for some whole number $M > 0$, we have $a_{M+n} = a_n$. Prove that $F(x)$ converges absolutely for $|x| < 1$ and that

$$F(x) = \frac{a_0 + a_1 x + \cdots + a_{M-1} x^{M-1}}{1 - x^M}$$

Hint: Use the hint for Exercise 42.

58. Continuity of Power Series Let $F(x) = \displaystyle\sum_{n=0}^{\infty} a_n x^n$ be a power series with radius of convergence $R > 0$.

(a) Prove the inequality

$$|x^n - y^n| \le n|x - y|(|x|^{n-1} + |y|^{n-1}) \qquad \boxed{11}$$

Hint: $x^n - y^n = (x - y)(x^{n-1} + x^{n-2}y + \cdots + y^{n-1})$.

(b) Choose R_1 with $0 < R_1 < R$. Use the Ratio Test to show that the infinite series $M = \displaystyle\sum_{n=0}^{\infty} 2n|a_n|R_1^n$ converges.

(c) Use Eq. (11) to show that if $|x| < R_1$ and $|y| < R_1$, then $|F(x) - F(y)| \le M|x - y|$.

(d) Prove that if $|x| < R$, then $F(x)$ is continuous at x. *Hint:* Choose R_1 such that $|x| < R_1 < R$. Show that if $\epsilon > 0$ is given, then $|F(x) - F(y)| \le \epsilon$ for all y such that $|x - y| < \delta$, where δ is any positive number that is less than ϵ/M and $R_1 - |x|$ (see Figure 4).

FIGURE 4 If $x > 0$, choose $\delta > 0$ less than ϵ/M and $R_1 - x$.

10.7 Taylor Series

We saw in the previous section that functions such as $f(x) = e^x$ and $f(x) = \tan^{-1} x$ can be represented as power series. These power series give us a certain tangible insight into the function represented and they allow us to approximate the values of $f(x)$ to any desired degree of accuracy. Thus, it is desirable to develop general methods for finding power series representations.

Suppose that $f(x)$ has a power series expansion centered at $x = c$ that is valid for all x in an interval $(c - R, c + R)$ with $R > 0$:

$$f(x) = \sum_{n=0}^{\infty} a_n(x - c)^n = a_0 + a_1(x - c) + a_2(x - c)^2 + \cdots$$

Then we may differentiate the series term by term (Theorem 3 in Section 10.6) to obtain

$$f(x) = a_0 + a_1(x - c) + a_2(x - c)^2 + a_3(x - c)^3 + \cdots$$

$$f'(x) = a_1 + 2a_2(x - c) + 3a_3(x - c)^2 + 4a_4(x - c)^3 + \cdots$$

$$f''(x) = 2a_2 + 2 \cdot 3a_3(x - c) + 3 \cdot 4a_4(x - c)^2 + \cdots$$

$$\vdots$$

$$f^{(k)}(x) = k!a_k + \left(2 \cdot 3 \cdots (k + 1)\right)a_{k+1}(x - c) + \cdots$$

Setting $x = c$ in each of these series, we find that

$$f(c) = a_0, \quad f'(c) = a_1, \quad f''(c) = 2a_2, \quad \ldots, \quad f^{(k)}(c) = k!a_k$$

This shows that the coefficients are given by the formula (and proves Theorem 1 below):

$$\boxed{a_k = \frac{f^{(k)}(c)}{k!}}$$ **1**

Recall that these are the coefficients of the Taylor polynomials. In summary:

$$f(x) = \sum_{n=0}^{\infty} \frac{f^{(n)}(c)}{n!}(x - c)^n$$

$$= f(c) + f'(c)(x - c) + \frac{f''(c)}{2!}(x - c)^2 + \frac{f'''(c)}{3!}(x - c)^3 + \cdots$$

The power series on the right is called the **Taylor series** of $f(x)$ centered at $x = c$. In the special case $c = 0$, the Taylor series is also called the **Maclaurin series**:

$$f(x) = f(0) + f'(0)x + \frac{f''(0)}{2!}x^2 + \frac{f'''(0)}{3!}x^3 + \frac{f^{(4)}(0)}{4!}x^4 + \cdots$$

THEOREM 1 Uniqueness of the Power Series Expansion If $f(x)$ is represented by a power series $F(x)$ centered at c on an interval $(c - R, c + R)$ with $R > 0$, then $F(x)$ is the Taylor series of $f(x)$ centered at $x = c$.

■ **EXAMPLE 1** Find the Maclaurin series for $f(x) = e^x$.

Solution The nth derivative $f^{(n)}(x)$ is $f^{(n)}(x) = e^x$ for all n and thus

$$f(0) = f'(0) = f''(0) = \cdots = e^0 = 1$$

Therefore, the coefficients of the Maclaurin series are $a_k = \dfrac{f^{(k)}(0)}{k!} = \dfrac{1}{k!}$ and the Maclaurin series is

$$1 + \frac{x}{1} + \frac{x^2}{2!} + \frac{x^3}{3!} + \cdots$$ ■

Theorem 1 tells us that if we want to represent $f(x)$ by a power series centered at c, the only candidate for the job is the Taylor series:

$$T(x) = \sum_{n=0}^{\infty} \frac{f^{(n)}(c)}{n!}(x - c)^n$$

However, there is no guarantee that $T(x)$ converges to $f(x)$. To study convergence, we consider the kth partial sum, which is the Taylor polynomial of degree k:

$$T_k(x) = f(c) + f'(c)(x - c) + \frac{f''(c)}{2!}(x - c)^2 + \cdots + \frac{f^{(k)}(c)}{k!}(x - c)^k$$

Recall that the remainder is defined by

$$R_k(x) = f(x) - T_k(x)$$

Since $T(x)$ is the limit of the partial sums $T_k(x)$, we see that

The Taylor series converges to $f(x)$ if and only if $\lim_{k \to \infty} R_k(x) = 0$

Although there is no general method for determining whether $R_k(x)$ tends to zero, the following theorem can often be applied.

← **REMINDER** $f(x)$ *is called "infinitely differentiable" if* $f^{(n)}(x)$ *exists for all n.*

THEOREM 2 Let $f(x)$ be an infinitely differentiable function on the open interval $I = (c - R, c + R)$ with $R > 0$. Assume there exists $K \geq 0$ such that for all $k \geq 0$,

$$|f^{(k)}(x)| \leq K \qquad \text{for all} \quad x \in I$$

Then $f(x)$ is represented by its Taylor series in I:

$$f(x) = \sum_{n=0}^{\infty} \frac{f^{(n)}(c)}{n!}(x - c)^n \qquad \text{for all} \quad x \in I$$

Proof We apply the Error Bound for Taylor polynomials:

$$|R_k(x)| = |f(x) - T_k(x)| \leq K \frac{|x - c|^{k+1}}{(k + 1)!}$$

If $x \in I$, then $|x - c| < R$ and

$$|R_k(x)| \leq \underbrace{K \frac{R^{k+1}}{(k + 1)!}}_{\text{This tends to zero as } k \to \infty}$$

As shown in Example 8 of Section 10.1, the quantity $R^k/k!$ tends to zero as $k \to \infty$ for every number R. We conclude that $\lim_{k \to \infty} R_k(x) = 0$ for all $x \in (c - R, c + R)$ and Theorem 2 follows. ∎

Taylor expansions were studied throughout the seventeenth and eighteenth centuries by Euler, Gregory, Leibniz, Maclaurin, Newton, Taylor, and others. These developments in Europe and England were anticipated by the great Indian mathematician Madhava (c. 1340–1425), who discovered the expansions of sine and cosine and many other results two centuries earlier.

■ **EXAMPLE 2** **Maclaurin Expansions of Sine and Cosine** Show that the following Taylor expansions are valid for all x:

$$\sin x = \sum_{n=0}^{\infty} (-1)^n \frac{x^{2n+1}}{(2n + 1)!} = x - \frac{x^3}{3!} + \frac{x^5}{5!} - \frac{x^7}{7!} + \cdots$$

$$\cos x = \sum_{n=0}^{\infty} (-1)^n \frac{x^{2n}}{(2n)!} = 1 - \frac{x^2}{2!} + \frac{x^4}{4!} - \frac{x^6}{6!} + \cdots$$

←·· REMINDER *The derivatives of*
$f(x) = \sin x$ *form a repeating pattern of*
length 4:

$f(x)$	$f'(x)$	$f''(x)$	$f'''(x)$	$\cdots$
$\sin x$	$\cos x$	$-\sin x$	$-\cos x$	$\cdots$

Solution For $f(x) = \sin x$, we have

$$f^{(2n)}(x) = (-1)^n \sin x, \qquad f^{(2n+1)}(x) = (-1)^n \cos x$$

Therefore, $f^{(2n)}(0) = 0$ and $f^{(2n+1)}(0) = (-1)^n$. The nonzero Taylor coefficients for $\sin x$ are $a_{2n+1} = \dfrac{(-1)^n}{(2n+1)!}$. Similarly, for $f(x) = \cos x$,

$$f^{(2n)}(x) = (-1)^n \cos x, \qquad f^{(2n+1)}(x) = (-1)^{n+1} \sin x$$

Therefore, $f^{(2n)}(0) = (-1)^n$ and $f^{(2n+1)}(0) = 0$. The nonzero Taylor coefficients for $\cos x$ are $a_{2n} = \dfrac{(-1)^n}{(2n)!}$.

In both cases, $|f^{(n)}(x)| \leq 1$ for all x and n. Thus, we may apply Theorem 2 with $M = 1$ and any R to conclude that the Taylor series converges to $f(x)$ for $|x| < R$. Since R is arbitrary, the Taylor expansions hold for all x. ∎

■ EXAMPLE 3 Taylor Expansion $f(x) = e^x$ **at** $x = c$ Find the Taylor series $T(x)$ of $f(x) = e^x$ at $x = c$.

Solution We have $f^{(n)}(c) = e^c$ for all x and thus

$$T(x) = \sum_{n=0}^{\infty} \frac{e^c}{n!}(x - c)^n$$

To prove convergence, we note that e^x is increasing and therefore, for any R, $|f^{(k)}(x)| \leq e^{c+R}$ for $x \in (c - R, c + R)$. Applying Theorem 2 with $M = e^{c+R}$, we conclude that $T(x)$ converges to $f(x)$ for all $x \in (c - R, c + R)$. Since R is arbitrary, the Taylor expansion holds for all x. For $c = 0$ we obtain the standard Taylor series

$$e^x = 1 + x + \frac{x^2}{2!} + \frac{x^3}{3!} + \cdots$$ ∎

Shortcuts to Finding Taylor Series

Since a Taylor series is a power series, we may differentiate and integrate a Taylor series term by term within its interval of convergence by Theorem 3 in Section 10.6. We may also multiply two Taylor series or substitute one Taylor series into another (we omit the proofs of these facts). This leads to shortcuts for generating new Taylor series from known ones.

■ EXAMPLE 4 Find the Maclaurin series for $f(x) = x^2 e^x$.

Solution We obtain the Maclaurin series of $f(x)$ by multiplying the known Maclaurin series for e^x by x^2:

$$x^2 e^x = x^2 \left(1 + x + \frac{x^2}{2!} + \frac{x^3}{3!} + \frac{x^4}{4!} + \frac{x^5}{5!} + \cdots \right)$$

$$= x^2 + x^3 + \frac{x^4}{2!} + \frac{x^5}{3!} + \frac{x^6}{4!} + \frac{x^7}{5!} + \cdots = \sum_{n=2}^{\infty} \frac{x^n}{(n-2)!}$$ ∎

In some cases, there is no convenient formula for the Taylor coefficients of a product, but we can compute as many coefficients as desired numerically.

■ **EXAMPLE 5** Multiplying Taylor Series Write out the first five terms in the Maclaurin series for $f(x) = e^x \cos x$.

Solution We multiply the fifth-order Taylor polynomials of e^x and $\cos x$ together, dropping the terms of degree greater than 5:

$$\left(1 + x + \frac{x^2}{2} + \frac{x^3}{6} + \frac{x^4}{24} + \frac{x^5}{120}\right)\left(1 - \frac{x^2}{2} + \frac{x^4}{24}\right)$$

Distributing the term on the left (and ignoring terms of degree greater than 5), we obtain

$$\left(1 + x + \frac{x^2}{2} + \frac{x^3}{6} + \frac{x^4}{24} + \frac{x^5}{120}\right) - \left(\frac{x^2}{2}\right)\left(1 + x + \frac{x^2}{2} + \frac{x^3}{6}\right) + \left(\frac{x^4}{24}\right)(1 + x)$$

$$= \underbrace{1 + x - \frac{x^3}{3} - \frac{x^4}{6} - \frac{x^5}{30}}_{\text{Retain terms of degree} \le 5} + \text{ higher-order terms}$$

We conclude that the first five terms of the Taylor series for $f(x) = e^x \cos x$ are

$$T_5(x) = 1 + x - \frac{x^3}{3} - \frac{x^4}{6} - \frac{x^5}{30} \qquad ■$$

■ **EXAMPLE 6** Substitution Use substitution to determine the Maclaurin series for e^{-x^2}.

Solution The Maclaurin series for e^{-x^2} is obtained by substituting $-x^2$ in the Maclaurin series for e^x:

$$e^{-x^2} = \sum_{n=0}^{\infty} \frac{(-1)^n x^{2n}}{n!} = 1 - x^2 + \frac{x^4}{2!} - \frac{x^6}{3!} + \frac{x^8}{4!} - \cdots \qquad \boxed{2}$$

Since the Taylor expansion of e^x is valid for all x, this expansion is also valid for all x.

■

■ **EXAMPLE 7** Integrating a Taylor Series Find the Maclaurin series for $f(x) = \ln(1 + x)$.

Solution In our study of Taylor polynomials, we computed the Maclaurin polynomials of $\ln(1 + x)$ directly. Here we obtain the same result by integrating the geometric series with common ratio $-x$:

$$\frac{1}{1 + x} = 1 - x + x^2 - x^3 + \cdots \qquad \boxed{3}$$

$$\ln(1 + x) = \int \frac{dx}{1 + x} = x - \frac{x^2}{2} + \frac{x^3}{3} - \frac{x^4}{4} + \cdots \qquad \boxed{4}$$

In principle, we might need a constant of integration on the right-hand side of Eq. (4). However, the constant of integration is zero because both $\ln(1 + x)$ and the power series take the value 0 at $x = 0$.

Furthermore, (3) is valid for $|x| < 1$, so the expansion of $\ln(1 + x)$ is also valid for $|x| < 1$. It also holds for $x = 1$ (see Exercise 83). ■

Taylor series may be used to express definite integrals as infinite series. This is useful when the integrand does not have an explicit antiderivative and the FTC cannot be applied. To justify this use of Taylor series, we appeal to Theorem 3 in Section 10.6, which implies the following: If a power series $P(x)$ centered at c has radius of convergence R, then the definite integral $\int_a^b P(x)\,dx$ over an interval $[a, b]$ contained in $(c - R, c + R)$ may be evaluated term by term.

■ **EXAMPLE 8** Let $J = \int_0^1 \sin(x^2)\,dx$.

(a) Express J as an infinite series.

(b) Determine J to within an error less than 10^{-4}.

Solution

(a) The Maclaurin expansion for $\sin x$ is valid for all x, so we have

$$\sin x = \sum_{n=0}^{\infty} \frac{(-1)^n}{(2n+1)!}x^{2n+1} \quad \Rightarrow \quad \sin(x^2) = \sum_{n=0}^{\infty} \frac{(-1)^n}{(2n+1)!}x^{4n+2}$$

We obtain an infinite series for J by integration:

$$J = \int_0^1 \sin(x^2)\,dx = \sum_{n=0}^{\infty} \frac{(-1)^n}{(2n+1)!} \int_0^1 x^{4n+2}\,dx$$

$$= \sum_{n=0}^{\infty} \frac{(-1)^n}{(2n+1)!} \left(\frac{1}{4n+3} \right)$$

$$= \frac{1}{3} - \frac{1}{42} + \frac{1}{1,320} - \frac{1}{75,600} + \cdots \qquad \boxed{5}$$

(b) The infinite series for J is an alternating series with decreasing terms, so the sum of the first N terms is accurate to within an error not greater than the $(N + 1)$st term. In other words,

$$\left| J - \sum_{n=0}^{N-1} \frac{(-1)^n}{(4n+3)(2n+1)!} \right| \le \frac{1}{(4N+3)(2N+1)!}$$

For $N = 3$, we obtain

$$J \approx \frac{1}{3} - \frac{1}{42} + \frac{1}{1,320} \approx 0.31028$$

with an error

$$\left| J - \left(\frac{1}{3} - \frac{1}{42} + \frac{1}{1,320} \right) \right| \le \frac{1}{(4(3)+3)(2(3)+1)!} = \frac{1}{75,600} \approx 1.3 \times 10^{-5}$$

We see that three terms of the series suffice to compute the integral with an error less than 10^{-4}.   ■

Binomial Series

Taylor series yield a generalization of the Binomial Theorem that was first discovered by Isaac Newton around 1665. For any number a (integer or not) and integer $n \ge 0$, we define the **binomial coefficient**:

$$\binom{a}{n} = \frac{a(a-1)(a-2)\cdots(a-n+1)}{n!}, \qquad \binom{a}{0} = 1$$

For example,

$$\binom{6}{3} = \frac{6 \cdot 5 \cdot 4}{3 \cdot 2 \cdot 1} = 20, \qquad \binom{\frac{4}{3}}{3} = \frac{\frac{4}{3} \cdot \frac{1}{3} \cdot \left(-\frac{2}{3}\right)}{3 \cdot 2 \cdot 1} = -\frac{4}{81}$$

The **Binomial Theorem** of algebra (see Appendix C) states that for any whole number a,

$$(r+s)^a = r^a + \binom{a}{1}r^{a-1}s + \binom{a}{2}r^{a-2}s^2 + \cdots + \binom{a}{a-1}rs^{a-1} + s^a$$

Setting $r = 1$ and $s = x$, we obtain an expansion of $f(x) = (1+x)^a$:

$$f(x) = (1+x)^a = 1 + \binom{a}{1}x + \binom{a}{2}x^2 + \cdots + \binom{a}{a-1}x^{a-1} + x^a$$

To derive Newton's generalization, we compute the Taylor series of $f(x) = (1+x)^a$ without assuming that a is a whole number. Observe that the derivatives of $f^{(k)}(0)$ follow a pattern:

$$f(x) = (1+x)^a \qquad\qquad f(0) = 1$$
$$f'(x) = a(1+x)^{a-1} \qquad\qquad f'(0) = a$$
$$f''(x) = a(a-1)(1+x)^{a-2} \qquad f''(0) = a(a-1)$$
$$f'''(x) = a(a-1)(a-2)(1+x)^{a-3} \quad f'''(0) = a(a-1)(a-2)$$

In general, $f^{(n)}(0) = a(a-1)(a-2)\cdots(a-n+1)$ and

$$\frac{f^{(n)}(0)}{n!} = \frac{a(a-1)(a-2)\cdots(a-n+1)}{n!} = \binom{a}{n}$$

Hence the Taylor series for $(1+x)^a$ is the binomial series

When a is a whole number, $\binom{a}{n}$ is zero for $n > a$, and in this case, the binomial series breaks off at degree n. The binomial series is infinite when a is not a whole number.

$$\sum_{n=0}^{\infty}\binom{a}{n}x^n = 1 + ax + \frac{a(a-1)}{2!}x^2 + \frac{a(a-1)(a-2)}{3!}x^3 + \cdots + \binom{a}{n}x^n + \cdots$$

The Ratio Test shows that this series has radius of convergence $R = 1$ (see Exercise 84). Furthermore, the binomial series converges to $(1+x)^a$ for $|x| < 1$ (see Exercise 85).

THEOREM 3 The Binomial Series For any exponent a, the following expansion is valid for $|x| < 1$:

$$(1+x)^a = 1 + ax + \frac{a(a-1)}{2!}x^2 + \frac{a(a-1)(a-2)}{3!}x^3 + \cdots + \binom{a}{n}x^n + \cdots \quad \boxed{6}$$

■ **EXAMPLE 9** Find the first five terms in the Maclaurin expansion of

$$f(x) = (1+x)^{1/3}$$

Solution The binomial coefficients $\binom{a}{n}$ for $a = \dfrac{1}{3}$ are

$$1, \quad \frac{1}{3}, \quad \frac{\frac{1}{3}(-\frac{2}{3})}{2!} = -\frac{1}{9}, \quad \frac{\frac{1}{3}(-\frac{2}{3})(-\frac{5}{3})}{3!} = \frac{5}{81}, \quad \frac{\frac{1}{3}(-\frac{2}{3})(-\frac{5}{3})(-\frac{8}{3})}{4!} = -\frac{10}{243}$$

Therefore, $(1 + x)^{1/3} \approx 1 + \dfrac{1}{3}x - \dfrac{1}{9}x^2 + \dfrac{5}{81}x^3 - \dfrac{10}{243}x^4 + \cdots$. ■

■ **EXAMPLE 10** Find the Maclaurin series for $f(x) = \dfrac{1}{\sqrt{1 - x^2}}$.

Solution First, we compute the coefficients $\binom{-\frac{1}{2}}{n}$ in the binomial series for $(1 + x)^{-1/2}$. The coefficients for $n = 0, 1, 2, 3$ are

$$1, \quad \frac{(-\frac{1}{2})}{1} = -\frac{1}{2}, \quad \frac{(-\frac{1}{2})(-\frac{3}{2})}{2!} = \frac{1 \cdot 3}{2 \cdot 4}, \quad \frac{(-\frac{1}{2})(-\frac{3}{2})(-\frac{5}{2})}{3!} = -\frac{1 \cdot 3 \cdot 5}{2 \cdot 4 \cdot 6}$$

The general pattern is

$$\binom{-\frac{1}{2}}{n} = (-1)^n \frac{1 \cdot 3 \cdot 5 \cdots (2n - 1)}{2^n \, n!} = (-1)^n \frac{1 \cdot 3 \cdot 5 \cdots (2n - 1)}{2 \cdot 4 \cdot 6 \cdots 2n}$$

Thus, the following binomial expansion is valid for $|x| < 1$:

$$\frac{1}{\sqrt{1 + x}} = 1 + \sum_{n=1}^{\infty} (-1)^n \frac{1 \cdot 3 \cdot 5 \cdots (2n - 1)}{2 \cdot 4 \cdot 6 \cdots (2n)} x^n = 1 - \frac{1}{2}x + \frac{1 \cdot 3}{2 \cdot 4}x^2 - \cdots$$

If $|x| < 1$, then $|x|^2 < 1$ and we may substitute $-x^2$ for x. We obtain for $|x| < 1$,

$$\boxed{\frac{1}{\sqrt{1 - x^2}} = 1 + \sum_{n=1}^{\infty} \frac{1 \cdot 3 \cdot 5 \cdots (2n - 1)}{2 \cdot 4 \cdot 6 \cdots 2n} x^{2n} = 1 + \frac{1}{2}x^2 + \frac{1 \cdot 3}{2 \cdot 4}x^4 + \cdots} \quad \boxed{7}$$

■

FIGURE 1 Pendulum released at an angle θ.

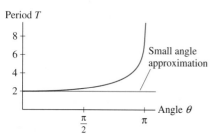

FIGURE 2 The period T of a 1-m pendulum as a function of the angle θ at which it is released.

Taylor series are used to study the transcendental functions occurring in physics and engineering, such as Bessel functions and hypergeometric functions. Though less familiar than the trigonometric and exponential functions, these so-called *special functions* appear in a wide range of applications. An example is the following **elliptic function of the first kind**, defined for $|k| < 1$:

$$E(k) = \int_0^{\pi/2} \frac{dt}{\sqrt{1 - k^2 \sin^2 t}}$$

In physics, one shows that the period T of pendulum of length L released from an angle θ is equal to $T = 4\sqrt{L/g}\, E(k)$, where $k = \sin \frac{1}{2}\theta$ (Figure 1). For small θ, we have the "small angle approximation" $T \approx 2\pi\sqrt{L/g}$, but this approximation breaks down for large angles (as we see in Figure 2).

■ **EXAMPLE 11** Elliptic Functions Find the Maclaurin series for $E(k)$ and estimate $E(k)$ for $k = \sin \dfrac{\pi}{6}$.

Solution Assume that $|k| < 1$ and substitute $x = k \sin t$ in the Taylor expansion (7):

$$\frac{1}{\sqrt{1 - k^2 \sin^2 t}} = 1 + \frac{1}{2}k^2 \sin^2 t + \frac{1 \cdot 3}{2 \cdot 4}k^4 \sin^4 t + \frac{1 \cdot 3 \cdot 5}{2 \cdot 4 \cdot 6}k^6 \sin^6 t + \cdots$$

This expansion is valid since $|x| = |k \sin t| < 1$. Thus $E(k)$ is equal to

$$\int_0^{\pi/2} \frac{dt}{\sqrt{1 - k^2 \sin^2 t}} = \int_0^{\pi/2} dt + \sum_{n=1}^{\infty} \frac{1 \cdot 3 \cdots (2n - 1)}{2 \cdot 4 \cdot (2n)} \left(\int_0^{\pi/2} \sin^{2n} t \, dt \right) k^{2n}$$

According to Exercise 75 in Section 7.3,

$$\int_0^{\pi/2} \sin^{2n} t \, dt = \left(\frac{1 \cdot 3 \cdots (2n - 1)}{2 \cdot 4 \cdot (2n)} \right) \frac{\pi}{2}$$

This yields

$$E(k) = \frac{\pi}{2} + \frac{\pi}{2} \sum_{n=1}^{\infty} \left(\frac{1 \cdot 3 \cdots (2n - 1)^2}{2 \cdot 4 \cdots (2n)} \right)^2 k^{2n}$$

We approximate $E(k)$ for $k = \sin(\frac{\pi}{6}) = \frac{1}{2}$ using the first five terms:

$$F\left(\frac{1}{2}\right) \approx \frac{\pi}{2} \left(1 + \left(\frac{1}{2}\right)^2 \left(\frac{1}{2}\right)^2 + \left(\frac{1 \cdot 3}{2 \cdot 4}\right)^2 \left(\frac{1}{2}\right)^4 \right.$$

$$\left. + \left(\frac{1 \cdot 3 \cdot 5}{2 \cdot 4 \cdot 6}\right)^2 \left(\frac{1}{2}\right)^6 + \left(\frac{1 \cdot 3 \cdot 5 \cdot 7}{2 \cdot 4 \cdot 6 \cdot 8}\right)^2 \left(\frac{1}{2}\right)^8 \right)$$

$$\approx 1.68517$$

The value given by a computer algebra system to seven places is $F(\frac{1}{2}) \approx 1.6856325$. ∎

10.7 SUMMARY

- The *Taylor series* of $f(x)$ centered at $x = c$ is

$$T(x) = \sum_{n=0}^{\infty} \frac{f^{(n)}(c)}{n!}(x - c)^n$$

The partial sum $T_k(x)$ of $T(x)$ is the kth Taylor polynomial.

- If $f(x)$ is represented as a power series $\sum_{n=0}^{\infty} a_n(x - c)^n$ on an interval $(c - R, c + R)$ with $R > 0$, then this power series is the Taylor series centered at $x = c$.
- When $c = 0$, $T(x)$ is called the *Maclaurin series* of $f(x)$.
- The equality $f(x) = T(x)$ holds if and only if the remainder, defined as $R_k(x) = f(x) - T_k(x)$, tends to zero as $k \to \infty$.
- Suppose that $f(x)$ is infinitely differentiable on an interval $I = (c - R, c + R)$ with $R > 0$ and assume there exists $K > 0$ such that $|f^{(k)}(x)| < K$ for $x \in I$. Then $f(x)$ is represented by its Taylor series on I, that is, $f(x) = T(x)$ for $x \in I$.

TABLE 1

Function $f(x)$	Maclaurin Series	Converges to $f(x)$ for
e^x	$\sum_{n=0}^{\infty} \frac{x^n}{n!} = 1 + x + \frac{x^2}{2!} + \frac{x^3}{3!} + \frac{x^4}{4!} + \cdots$	All x
$\sin x$	$\sum_{n=0}^{\infty} \frac{(-1)^n x^{2n+1}}{(2n+1)!} = x - \frac{x^3}{3!} + \frac{x^5}{5!} - \frac{x^7}{7!} + \cdots$	All x
$\cos x$	$\sum_{n=0}^{\infty} \frac{(-1)^n x^{2n}}{(2n)!} = 1 - \frac{x^2}{2!} + \frac{x^4}{4!} - \frac{x^6}{6!} + \cdots$	All x
$\dfrac{1}{1-x}$	$\sum_{n=0}^{\infty} x^n = 1 + x + x^2 + x^3 + x^4 + \cdots$	$|x| < 1$
$\dfrac{1}{1+x}$	$\sum_{n=0}^{\infty} (-1)^n x^n = 1 - x + x^2 - x^3 + x^4 - \cdots$	$|x| < 1$
$\ln(1+x)$	$\sum_{n=1}^{\infty} \frac{(-1)^{n-1} x^n}{n} = x - \frac{x^2}{2} + \frac{x^3}{3} - \frac{x^4}{4} + \cdots$	$|x| < 1$ and $x = 1$
$\tan^{-1} x$	$\sum_{n=0}^{\infty} \frac{(-1)^n x^{2n+1}}{2n+1} = x - \frac{x^3}{3} + \frac{x^5}{5} - \frac{x^7}{7} + \cdots$	$|x| \le 1$
$(1+x)^a$	$\sum_{n=0}^{\infty} \binom{a}{n} x^n = 1 + ax + \frac{a(a-1)}{2!} x^2 + \frac{a(a-1)(a-2)}{3!} x^3 + \cdots$	$|x| < 1$

- A good way to find the Taylor series of a function is to start with known Taylor expansions and apply one of the following operations: multiplication, substitution, differentiation, or integration.

10.7 EXERCISES

Preliminary Questions

1. Determine $f(0)$ and $f'''(0)$ for a function $f(x)$ with Maclaurin series

$$T(x) = 3 + 2x + 12x^2 + 5x^3 + \cdots$$

2. Determine $f(-2)$ and $f^{(4)}(-2)$ for a function with Taylor series

$$T(x) = 3(x+2) + (x+2)^2 - 4(x+2)^3 + 2(x+2)^4 + \cdots$$

3. What is the easiest way to find the Maclaurin series for the function $f(x) = \sin(x^2)$?

4. What is the Taylor series for $f(x)$ centered at $c = 3$ if $f(3) = 4$ and $f'(x)$ has a Taylor expansion

$$f'(x) = \sum_{n=1}^{\infty} \frac{(x-3)^n}{n}$$

5. Let $T(x)$ be the Maclaurin series of $f(x)$. Which of the following guarantees that $f(2) = T(2)$?

(a) $T(x)$ converges for $x = 2$.

(b) The remainder $R_k(2)$ approaches a limit as $k \to \infty$.

(c) The remainder $R_k(2)$ approaches zero as $k \to \infty$.

Exercises

1. Write out the first four terms of the Maclaurin of $f(x)$ if

$$f(0) = 2, \quad f'(0) = 3, \quad f''(0) = 4, \quad f'''(0) = 12$$

2. Write out the first four terms of the Taylor series of $f(x)$ centered at $c = 3$ if

$$f(3) = 1, \quad f'(3) = 2, \quad f''(3) = 12, \quad f'''(3) = 3$$

In Exercises 3–20, find the Maclaurin series.

3. $f(x) = \dfrac{1}{1-2x}$

4. $f(x) = \dfrac{x}{1-x^4}$

5. $f(x) = \cos 3x$

6. $f(x) = \sin(2x)$

7. $f(x) = \sin(x^2)$

8. $f(x) = e^{4x}$

9. $f(x) = \ln(1-x^2)$

10. $f(x) = (1-x)^{-1/2}$

11. $f(x) = \tan^{-1}(x^2)$

12. $f(x) = x^2 e^{x^2}$

13. $f(x) = e^{x-2}$

14. $f(x) = \cos\sqrt{x}$

15. $f(x) = \ln(1-5x)$

16. $f(x) = (x^2+2x)e^x$

17. $f(x) = \sinh x$

18. $f(x) = \cosh x$

19. $f(x) = \dfrac{1-\cos(x^2)}{x}$

20. $f(x) = \dfrac{e^x - \cos x}{x}$

21. Use multiplication to find the first four terms in the Maclaurin series for $f(x) = e^x \sin x$.

22. Find the first five terms of the Maclaurin series for $f(x) = \dfrac{\sin x}{1-x}$.

23. Find the first four terms of the Maclaurin series for $f(x) = e^x \ln(1-x)$.

24. Write out the first five terms of the binomial series for $f(x) = (1+x)^{1/3}$.

25. Write out the first five terms of the binomial series for $f(x) = (1+x)^{-3/2}$.

26. Differentiate the Maclaurin series for $\dfrac{1}{1-x}$ twice to find the Maclaurin series of $\dfrac{1}{(1-x)^3}$.

27. Find the first four terms of the Maclaurin for $f(x) = e^{(e^x)}$.

28. Find the first three terms of the Maclaurin series for $f(x) = \dfrac{1}{1+\sin x}$. *Hint:* First expand $f(x)$ as a geometric series.

29. Find the Taylor series for $\sin x$ at $c = \dfrac{\pi}{2}$.

30. What is the Maclaurin series for $f(x) = x^4 - 2x^2 + 3$? What is the Taylor series centered at $c = 2$?

In Exercises 31–40, find the Taylor series centered at c.

31. $f(x) = \dfrac{1}{x}, \quad c = 1$

32. $f(x) = \sqrt{x}, \quad c = 4$

33. $f(x) = \dfrac{1}{1-x}, \quad c = 5$

34. $f(x) = x^4 + 3x - 1, \quad c = 0$

35. $f(x) = x^4 + 3x - 1, \quad c = 2$

36. $f(x) = \dfrac{1}{x^2}, \quad c = 4$

37. $f(x) = e^{3x}, \quad c = -1$

38. $f(x) = \dfrac{1}{1-4x}, \quad c = -2$

39. $f(x) = \dfrac{1}{1-x^2}, \quad c = 3$

40. $f(x) = \dfrac{1}{3x-2}, \quad c = -1$

41. Find the Maclaurin series for $f(x) = \dfrac{1}{\sqrt{1-9x^2}}$ (see Example 10).

42. Show, by integrating the Maclaurin series for $f(x) = \dfrac{1}{\sqrt{1-x^2}}$, that for $|x| < 1$,

$$\sin^{-1} x = x + \sum_{n=1}^{\infty} \frac{1\cdot 3\cdot 5\cdots(2n-1)}{2\cdot 4\cdot 6\cdots(2n)}\frac{x^{2n+1}}{2n+1}$$

43. Use the first five terms of the Maclaurin series in Exercise 42 to approximate $\sin^{-1}\frac{1}{2}$. Compare the result with the calculator value.

44. Show that for $|x| < 1$

$$\tanh^{-1} x = x + \frac{x^3}{3} + \frac{x^5}{5} + \cdots$$

Hint: Recall that $\dfrac{d}{dx}\tanh^{-1} x = \dfrac{1}{1-x^2}$.

45. Use the Maclaurin series for $\ln(1+x)$ and $\ln(1-x)$ to show that

$$\frac{1}{2}\ln\left(\frac{1+x}{1-x}\right) = x + \frac{x^3}{3} + \frac{x^5}{5} + \cdots$$

What can you conclude by comparing this result with that of Exercise 44?

46. Use the Taylor series for $\cos x$ to compute $\cos 1$ to within an error of at most 10^{-6}. Use the fact that $\cos x$ is an alternating series with decreasing terms to estimate the error.

47. Use the Maclaurin expansion for e^{-t^2} to express $\displaystyle\int_0^x e^{-t^2}\,dt$ as an alternating power series in t.

(a) How many terms of the infinite series are needed to approximate the integral for $x = 1$ to within an error of at most 0.001?

(b) *CAS* Carry out the computation and check your answer using a computer algebra system.

48. Let $F(x) = \displaystyle\int_0^x \frac{\sin t\,dt}{t}$. Show that

$$F(x) = x - \frac{x^3}{3\cdot 3!} + \frac{x^5}{5\cdot 5!} - \frac{x^7}{7\cdot 7!} + \cdots$$

Evaluate $F(1)$ to three decimal places.

49. Which function has the MacLaurin series $\sum_{n=0}^{\infty} (-1)^n 2^n x^n$?

In Exercises 50–53, express the definite integral as an infinite series and find its value to within an error of at most 10^{-4}.

50. $\int_0^1 \cos(x^2)\, dx$

51. $\int_0^1 \tan^{-1}(x^2)\, dx$

52. $\int_0^2 e^{-x^3}\, dx$

53. $\int_0^1 \frac{dx}{\sqrt{x^4 + 1}}$

In Exercises 54–57, express the integral as an infinite series.

54. $\int_0^x \frac{1 - \cos(t)}{t}\, dt$,   for all x

55. $\int_0^x \frac{t - \sin t}{t}\, dt$,   for all x

56. $\int_0^x \ln(1 + t^2)\, dt$,   for $|x| < 1$

57. $\int_0^x \frac{dt}{\sqrt{1 - t^4}}$,   for $|x| < 1$

58. Which function has Maclaurin series $\sum_{n=0}^{\infty} (-1)^n 2^n x^n$?

59. Which function has Maclaurin series

$$\sum_{k=0}^{\infty} \frac{(-1)^k}{3^{k+1}} (x - 3)^k?$$

For which values of x is the expansion valid?

In Exercises 60–63, find the first four terms of the Taylor series.

60. $f(x) = \sin(x^2)\cos(x^2)$

61. $f(x) = e^x \tan^{-1} x$

62. $f(x) = e^{\sin x}$

63. $f(x) = \sin(x^3 + 2x)$

In Exercises 64–67, find the functions with the following Maclaurin series (refer to Table 1).

64. $1 + x^3 + \frac{x^6}{2!} + \frac{x^9}{3!} + \frac{x^{12}}{4!} + \cdots$

65. $1 - 4x + 4^2 x^2 - 4^3 x^3 + 4^4 x^5 - 4^5 x^5 + \cdots$

66. $1 - \frac{5^3 x^3}{3!} + \frac{5^5 x^5}{5!} - \frac{5^7 x^7}{7!} + \cdots$

67. $x^4 - \frac{x^{12}}{3} + \frac{x^{20}}{5} - \frac{x^{28}}{7} + \cdots$

68. When a voltage V is applied to a series circuit consisting of a resistor R and an inductor L, the current at time t is

$$I(t) = \left(\frac{V}{R}\right)\left(1 - e^{-Rt/L}\right)$$

Expand $I(t)$ in a Maclaurin series. Show that $I(t) \approx Vt/L$ if R is small.

69. ✎   Use substitution to write out the first three terms of the Maclaurin series for $f(x) = e^{x^{20}}$. Explain how the result implies that $f^{(k)}(0) = 0$ for $1 \le k \le 19$.

70. Find the Maclaurin series for $f(x) = \cos(\sqrt{x})$ and use it to determine $f^{(5)}(0)$.

71. Find $f^{(7)}(0)$ and $f^{(8)}(0)$ for $f(x) = \tan^{-1} x$.

72. Use the binomial series to find $f^{(8)}(0)$ for $f(x) = \sqrt{1 - x^2}$.

73. Show that $\pi - \frac{\pi^3}{3!} + \frac{\pi^5}{5!} - \frac{\pi^7}{7!} + \cdots$ converges to zero. How many terms must be computed to get within 0.01 of zero?

74. Does the Taylor series for $f(x) = (1 + x)^{3/4}$ converge to $f(x)$ at $x = 2$? Give numerical evidence to support your answer.

75. ✎   Explain the steps required to verify that the Maclaurin series for $f(x) = \sin x$ converges to $f(x)$ at $x = 1$.

76. Explain the steps required to verify that the Maclaurin series for $f(x) = \tan^{-1} x$ converges to $f(x)$ at $x = 0.5$.

77. GU Let $f(x) = \sqrt{1 + x}$.
(a) Use a graphing calculator to compare the graph of f with the graphs of the first five Taylor polynomials for f. What do they suggest about the interval of convergence of the Taylor series?
(b) Investigate numerically whether or not the Taylor expansion for f is valid for $x = 1$ and $x = -1$.

78. How many terms of the Maclaurin series of $f(x) = \ln(1 + x)$ are needed to compute $\ln 1.2$ to within an error of at most 0.0001? Make the computation and compare the result with the calculator value.

In Exercises 79–80, let

$$f(x) = \frac{1}{(1 - x)(1 - 2x)}$$

79. Find the Maclaurin series of $f(x)$ using the identity

$$f(x) = \frac{2}{1 - 2x} - \frac{1}{1 - x}$$

80. Find the Taylor series for $f(x)$ at $c = 2$. *Hint:* Rewrite the identity of Exercise 79 as

$$f(x) = \frac{2}{-3 - 2(x - 2)} - \frac{1}{-1 - (x - 2)}$$

81. Use the first five terms of the Maclaurin series for the elliptic function $E(k)$ to estimate the period T of a 1-m pendulum released at an angle $\theta = \frac{\pi}{4}$ (see Example 11).

82. Use Example 11 and the approximation $\sin x \approx x$ to show that the period T of a pendulum released at an angle θ has the following second-order approximation:

$$T \approx 2\pi \sqrt{\frac{L}{g}} \left(1 + \frac{\theta^2}{16}\right)$$

Further Insights and Challenges

83. In this exercise we show that the Maclaurin expansion of the function $f(x) = \ln(1 + x)$ is valid for $x = 1$.

(a) Show that for all $x \neq -1$,

$$\frac{1}{1+x} = \sum_{n=0}^{N} (-1)^n x^n + \frac{(-1)^{N+1} x^{N+1}}{1+x}$$

(b) Integrate from 0 to 1 to obtain

$$\ln 2 = \sum_{n=1}^{N} \frac{(-1)^{n-1}}{n} + (-1)^{N+1} \int_0^1 \frac{x^{N+1} \, dx}{1+x}$$

(c) Verify that the integral on the right tends to zero as $N \to \infty$ by showing that it is smaller than $\int_0^1 x^{N+1} dx$.

(d) Prove Leibniz's formula

$$\ln 2 = 1 - \frac{1}{2} + \frac{1}{3} - \frac{1}{4} + \cdots$$

In Exercises 84–85, we investigate the convergence of the binomial series

$$T_a(x) = \sum_{n=0}^{\infty} \binom{a}{n} x^n$$

84. Prove that $T_a(x)$ has radius of convergence $R = 1$ if a is not a whole number. What is the radius of convergence if a is a whole number?

85. By Exercise 84, $T_a(x)$ converges for $|x| < 1$, but we do not yet know whether $T_a(x) = (1 + x)^a$.

(a) Verify the identity

$$a \binom{a}{n} = n \binom{a}{n} + (n+1) \binom{a}{n+1}$$

(b) Use (a) to show that $y = T_a(x)$ satisfies the differential equation $(1 + x)y' = ay$ with initial condition $y(0) = 1$.

(c) Prove that $T_a(x) = (1 + x)^a$ for $|x| < 1$ by showing that the derivative of the ratio $\dfrac{T_a(x)}{(1+x)^a}$ is constant.

86. The function $G(k) = \displaystyle\int_0^{\pi/2} \sqrt{1 - k^2 \sin^2 t} \, dt$ is called an **elliptic function of the second kind**. Prove that for $|k| < 1$

$$G(k) = \frac{\pi}{2} - \frac{\pi}{2} \sum_{n=1}^{\infty} \left(\frac{1 \cdot 3 \cdots (2n-1)}{2 \cdots 4 \cdot (2n)} \right)^2 \frac{k^{2n}}{2n - 1}$$

87. Assume that $a < b$ and let L be the arc length (circumference) of the ellipse $\left(\dfrac{x}{a}\right)^2 + \left(\dfrac{y}{b}\right)^2 = 1$ shown in Figure 3. There is no explicit formula for L, but it is known that $L = 4bG(k)$, where $k = \sqrt{1 - a^2/b^2}$. Use the first three terms of the expansion of Exercise 86 to estimate the arc length when $a = 4$ and $b = 5$.

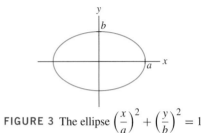

FIGURE 3 The ellipse $\left(\dfrac{x}{a}\right)^2 + \left(\dfrac{y}{b}\right)^2 = 1$

88. Use Exercise 86 to prove that if $a < b$ and a/b is near 1 (a nearly circular ellipse), then

$$L \approx \frac{\pi}{2} \left(3b + \frac{a^2}{b} \right)$$

Hint: Use the first two terms of the series for $G(k)$.

89. Irrationality of e   Prove that e is an irrational number using the following argument by contradiction. Suppose that $e = M/N$, where M, N are nonzero integers.

(a) Show that $M! \, e^{-1}$ is a whole number.

(b) Use the power series for e^x at $x = -1$ to show that there is an integer B such that $M! \, e^{-1}$ equals

$$B + (-1)^{M+1} \left(\frac{1}{M+1} - \frac{1}{(M+1)(M+2)} + \cdots \right)$$

(c) Use your knowledge of alternating series with decreasing terms to conclude that $0 < |M! \, e^{-1} - B| < 1$. However, this contradicts (a). Hence, e is not equal to M/N.

CHAPTER REVIEW EXERCISES

1. Let $a_n = \dfrac{n - 3}{n!}$ and $b_n = a_{n+3}$. Calculate the first three terms in the sequence:

(a) a_n^2

(b) b_n

(c) $a_n b_n$

(d) $2a_{n+1} - 3a_n$

2. Prove that $\displaystyle\lim_{n \to \infty} \frac{2n - 1}{3n + 2} = \frac{2}{3}$ using the limit definition.

In Exercises 3–8, compute the limit (or state that it does not exist) assuming that $\lim\limits_{n\to\infty} a_n = 2$.

3. $\lim\limits_{n\to\infty} (5a_n - 2a_n^2)$

4. $\lim\limits_{n\to\infty} \dfrac{1}{a_n}$

5. $\lim\limits_{n\to\infty} e^{a_n}$

6. $\lim\limits_{n\to\infty} \cos(\pi a_n)$

7. $\lim\limits_{n\to\infty} (-1)^n a_n$

8. $\lim\limits_{n\to\infty} \dfrac{a_n + n}{a_n + n^2}$

In Exercises 9–22, determine the limit of the sequence or show that the sequence diverges.

9. $a_n = \sqrt{n+5} - \sqrt{n+2}$

10. $a_n = \dfrac{3n^3 - n}{1 - 2n^3}$

11. $a_n = 2^{1/n^2}$

12. $a_n = \dfrac{10^n}{n!}$

13. $b_m = 1 + (-1)^m$

14. $b_m = \dfrac{1 + (-1)^m}{m}$

15. $b_n = \tan^{-1}\left(\dfrac{n+2}{n+5}\right)$

16. $a_n = \ln(n+1) - \ln n$

17. $b_n = \sqrt{n^2 + n} - \sqrt{n^2 + 1}$

18. $c_n = \sqrt{n^2 + n} - \sqrt{n^2 - n}$

19. $a_n = \dfrac{100^n}{n!} - \dfrac{3 + \pi^n}{5^n}$

20. $b_m = \left(1 + \dfrac{1}{m}\right)^m$

21. $c_n = \left(1 + \dfrac{3}{n}\right)^n$

22. $b_n = n(\ln(n+1) - \ln n)$

23. Use the Squeeze Theorem to show that $\lim\limits_{n\to\infty} \dfrac{\arctan(n^2)}{\sqrt{n}} = 0$.

24. Give an example of a divergent sequence $\{a_n\}$ such that $\{\sin a_n\}$ is convergent.

25. Given $a_n = \dfrac{1}{2}3^n - \dfrac{1}{3}2^n$,

(a) Calculate $\lim\limits_{n\to\infty} a_n$.

(b) Calculate $\lim\limits_{n\to\infty} \dfrac{a_{n+1}}{a_n}$.

26. Define $a_{n+1} = \sqrt{a_n + 6}$ with $a_1 = 2$.

(a) Compute a_n for $n = 2, 3, 4, 5$.

(b) Show that $\{a_n\}$ is increasing and bounded by 3.

(c) Prove that $\lim\limits_{n\to\infty} a_n$ exists and find its value.

27. Calculate the partial sums S_4 and S_7 of the series $\sum\limits_{n=1}^{\infty} \dfrac{n-2}{n^2 + 2n}$.

28. Find the sum $1 - \dfrac{1}{4} + \dfrac{1}{4^2} - \dfrac{1}{4^3} + \cdots$.

29. Find the sum $\dfrac{4}{9} + \dfrac{8}{27} + \dfrac{16}{81} + \dfrac{32}{243} + \cdots$.

30. Find the sum $\sum\limits_{n=2}^{\infty} \left(\dfrac{2}{e}\right)^n$.

31. Find the sum $\sum\limits_{n=0}^{\infty} \dfrac{2^{n+1}}{3^n}$.

32. Show that $\sum\limits_{n=1}^{\infty} \left(b - \tan^{-1} n^2\right)$ diverges if $b \neq \dfrac{\pi}{2}$.

33. Give an example of divergent series $\sum\limits_{n=1}^{\infty} a_n$, $\sum\limits_{n=1}^{\infty} b_n$ such that

$$\sum\limits_{n=1}^{\infty} (a_n + b_n) = 1.$$

34. Find the total area of the infinitely many circles on the interval $[0, 1]$ in Figure 1.

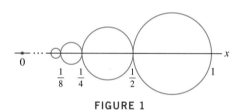

FIGURE 1

In Exercises 35–38, use the Integral Test to determine if the infinite series converges.

35. $\sum\limits_{n=1}^{\infty} \dfrac{n^2}{n^3 + 1}$

36. $\sum\limits_{n=1}^{\infty} \dfrac{n^2}{(n^3 + 1)^{1.01}}$

37. $\sum\limits_{n=1}^{\infty} \dfrac{n^3}{e^{n^4}}$

38. $\sum\limits_{n=1}^{\infty} \dfrac{1}{(2n+1)(\ln(2n+1))^2}$

In Exercises 39–46, use the Comparison or Limit Comparison Test to determine whether the infinite series converges.

39. $\displaystyle\sum_{n=1}^{\infty} \frac{1}{(n+1)^2}$

40. $\displaystyle\sum_{n=1}^{\infty} \frac{1}{\sqrt{n}+n}$

41. $\displaystyle\sum_{n=2}^{\infty} \frac{n^2+1}{n^{3.5}-2}$

42. $\displaystyle\sum_{n=1}^{\infty} \frac{1}{n-\ln n}$

43. $\displaystyle\sum_{n=2}^{\infty} \frac{\ln n}{1.5^n}$

44. $\displaystyle\sum_{n=2}^{\infty} \frac{n}{\sqrt{n^5+5}}$

45. $\displaystyle\sum_{n=1}^{\infty} \frac{1}{3^n-2^n}$

46. $\displaystyle\sum_{n=1}^{\infty} \frac{n^{10}+10^n}{n^{11}+11^n}$

47. Show that $\displaystyle\sum_{n=2}^{\infty} \left(1-\sqrt{1-\frac{1}{n}}\right)$ diverges. *Hint:* Show that

$$1-\sqrt{1-\frac{1}{n}} \geq \frac{1}{2n}$$

48. Determine whether $\displaystyle\sum_{n=2}^{\infty} \left(1-\sqrt{1-\frac{1}{n^2}}\right)$ converges.

49. Let $\displaystyle S = \sum_{n=1}^{\infty} \frac{n}{(n^2+1)^2}$.

(a) Show that S converges.

(b) $\mathcal{CAS}$ Use Eq. (5) in Exercise 77 of Section 10.3 with $M=99$ to approximate S. What is the maximum size of the error?

In Exercises 50–53, determine whether the series converges absolutely. If not, determine whether it converges conditionally.

50. $\displaystyle\sum_{n=1}^{\infty} \frac{(-1)^n}{\sqrt[3]{n}+2n}$

51. $\displaystyle\sum_{n=1}^{\infty} \frac{(-1)^n}{n^{1.1}\ln(n+1)}$

52. $\displaystyle\sum_{n=1}^{\infty} \frac{\cos\left(\frac{\pi}{4}+\pi n\right)}{\sqrt{n}}$

53. $\displaystyle\sum_{n=1}^{\infty} \frac{\cos\left(\frac{\pi}{4}+2\pi n\right)}{\sqrt{n}}$

54. $\mathcal{CAS}$ Use a computer algebra system to approximate $\displaystyle\sum_{n=1}^{\infty} \frac{(-1)^n}{n^3+\sqrt{n}}$ to within an error of at most 10^{-5}.

55. How many terms of the series are needed to calculate Catalan's constant $\displaystyle K = \sum_{k=0}^{\infty} \frac{(-1)^k}{(2k+1)^2}$ to three decimal places? Carry out the calculation.

56. Give an example of conditionally convergent series $\displaystyle\sum_{n=1}^{\infty} a_n$, $\displaystyle\sum_{n=1}^{\infty} b_n$ such that $\displaystyle\sum_{n=1}^{\infty}(a_n+b_n)$ converges absolutely.

57. Let $\displaystyle\sum_{n=1}^{\infty} a_n$ be an absolutely convergent series. Determine whether the following series are convergent or divergent:

(a) $\displaystyle\sum_{n=1}^{\infty} \left(a_n+\frac{1}{n^2}\right)$

(b) $\displaystyle\sum_{n=1}^{\infty} (-1)^n a_n$

(c) $\displaystyle\sum_{n=1}^{\infty} \frac{1}{1+a_n^2}$

(d) $\displaystyle\sum_{n=1}^{\infty} \frac{|a_n|}{n}$

In Exercises 58–65, apply the Ratio Test to determine convergence or divergence, or state that the Ratio Test is inconclusive.

58. $\displaystyle\sum_{n=1}^{\infty} \frac{n^5}{5^n}$

59. $\displaystyle\sum_{n=1}^{\infty} \frac{\sqrt{n+1}}{n^8}$

60. $\displaystyle\sum_{n=1}^{\infty} \frac{1}{n2^n+n^3}$

61. $\displaystyle\sum_{n=1}^{\infty} \frac{n^4}{n!}$

62. $\displaystyle\sum_{n=1}^{\infty} \frac{2^{n^2}}{n!}$

63. $\displaystyle\sum_{n=4}^{\infty} \frac{\ln n}{n^{3/2}}$

64. $\displaystyle\sum_{n=1}^{\infty} \left(\frac{n}{2}\right)^n \frac{1}{n!}$

65. $\displaystyle\sum_{n=1}^{\infty} \left(\frac{n}{4}\right)^n \frac{1}{n!}$

In Exercises 66–69, apply the Root Test to determine convergence or divergence, or state that the Root Test is inconclusive.

66. $\displaystyle\sum_{n=1}^{\infty} \frac{1}{4^n}$

67. $\displaystyle\sum_{n=1}^{\infty} \left(\frac{2}{n}\right)^n$

68. $\displaystyle\sum_{n=1}^{\infty} \left(\frac{3}{4n}\right)^n$

69. $\displaystyle\sum_{n=1}^{\infty} \left(\cos\frac{1}{n}\right)^{n^3}$

70. Let $\{a_n\}$ be a positive sequence such that $\displaystyle\lim_{n\to\infty}\sqrt[n]{a_n}=\frac{1}{2}$. Determine whether the following series converge or diverge:

(a) $\displaystyle\sum_{n=1}^{\infty} 2a_n$

(b) $\displaystyle\sum_{n=1}^{\infty} 3^n a_n$

(c) $\displaystyle\sum_{n=1}^{\infty} \sqrt{a_n}$

In Exercises 71–84, determine convergence or divergence using any method covered in the text.

71. $\displaystyle\sum_{n=1}^{\infty} \left(\frac{2}{3}\right)^n$

72. $\displaystyle\sum_{n=1}^{\infty} \frac{n+2^n}{3^n-1}$

73. $\displaystyle\sum_{n=1}^{\infty} e^{-0.02n}$

74. $\displaystyle\sum_{n=1}^{\infty} n e^{-0.02n}$

75. $\displaystyle\sum_{n=1}^{\infty} \frac{(-1)^{n-1}}{\sqrt{n}+\sqrt{n+1}}$

76. $\displaystyle\sum_{n=10}^{\infty} \frac{1}{n(\log n)^{3/2}}$

77. $\displaystyle\sum_{n=10}^{\infty} \frac{(-1)^n}{\log n}$

78. $\displaystyle\sum_{n=1}^{\infty} \frac{1}{n\sqrt{n}+\ln n}$

79. $\displaystyle\sum_{n=1}^{\infty} \frac{e^n}{n!}$

80. $\displaystyle\sum_{n=1}^{\infty} \frac{1}{\sqrt[3]{n}(1+\sqrt{n})}$

81. $\displaystyle\sum_{n=1}^{\infty} \frac{1}{n - 100.1}$

82. $\displaystyle\sum_{n=2}^{\infty} \frac{\cos(\pi n)}{n^{2/3}}$

83. $\displaystyle\sum_{n=1}^{\infty} \sin^2 \frac{\pi}{n}$

84. $\displaystyle\sum_{n=0}^{\infty} \frac{2^{2n}}{n!}$

In Exercises 85–90, find the values of x for which the power series converges.

85. $\displaystyle\sum_{n=0}^{\infty} \frac{2^n x^n}{n!}$

86. $\displaystyle\sum_{n=0}^{\infty} \frac{x^n}{n + 1}$

87. $\displaystyle\sum_{n=0}^{\infty} \frac{n^6 (x - 3)^n}{n^8 + 1}$

88. $\displaystyle\sum_{n=0}^{\infty} n x^n$

89. $\displaystyle\sum_{n=0}^{\infty} (nx)^n$

90. $\displaystyle\sum_{n=0}^{\infty} \frac{(2x - 3)^n}{n \ln n}$

91. Expand the function $f(x) = \dfrac{2}{4 - 3x}$ as a power series centered at $c = 0$. Determine the values of x for which the series converges.

92. Prove that $\displaystyle\sum_{n=0}^{\infty} n e^{-nx} = \dfrac{e^{-x}}{(1 - e^{-x})^2}$. *Hint:* Express the left-hand side as the derivative of a geometric series.

93. Let $F(x) = \displaystyle\sum_{k=0}^{\infty} \frac{x^{2k}}{2^k \cdot k!}$.

(a) Show that $F(x)$ has infinite radius of convergence.

(b) Show that $y = F(x)$ is a solution to the differential equation

$$y'' = xy' + y$$

satisfying $y(0) = 1$, $y'(0) = 0$.

(c) ⊂ℝ5 Plot the partial sums S_N for $N = 1, 3, 5, 7$ on the same set of axes.

94. Find a power series solution $P(x) = \displaystyle\sum_{n=0}^{\infty} a_n x^n$ to the Laguerre differential equation

$$xy'' + (1 - x)y' - y = 0$$

satisfying $P(0) = 1$.

95. Use the Maclaurin series for $f(x) = \cos x$ to calculate the limit

$$\lim_{x \to 0} \frac{1 - \frac{x^2}{2} - \cos x}{x^4}.$$

In Exercises 96–103, find the Taylor series centered at c.

96. $f(x) = e^{4x}, \quad c = 0$

97. $f(x) = e^{2x}, \quad c = -1$

98. $f(x) = x \sin x, \quad c = \pi$

99. $f(x) = \ln \frac{x}{2}, \quad c = 2$

100. $f(x) = x \ln\left(1 + \frac{x}{2}\right), \quad c = 0$

101. $f(x) = \sqrt{x} \arctan \sqrt{x}, \quad c = 0$

102. $f(x) = \dfrac{1}{1 - 2x}, \quad c = -2$

103. $f(x) = e^{x-1}, \quad c = -1$

104. Find the Maclaurin series of $f(x) = x^5 - 4x^2 + 10$. Determine its Taylor series centered at $c = 2$.

105. Use the Maclaurin series of $\sin x$ and $\sqrt{1 + x}$ to calculate $f^{(4)}(0)$, where $f(x) = (\sin x)\sqrt{1 + x}$.

106. Calculate $\dfrac{\pi}{2} - \dfrac{\pi^3}{2^3 3!} + \dfrac{\pi^5}{2^5 5!} - \dfrac{\pi^7}{2^7 7!} + \cdots$.

107. Find the Maclaurin series of the function $F(x) = \displaystyle\int_0^x \frac{e^t - 1}{t}\, dt$.

11 PARAMETRIC EQUATIONS, POLAR COORDINATES, AND CONIC SECTIONS

This chapter develops several new tools of calculus. First, we introduce parametric equations, which are used to analyze motion and to describe curves that cannot be represented as graphs of functions or equations. We then discuss polar coordinates, which are an alternative to rectangular coordinates and are useful in many applications. The chapter closes with a discussion of conic sections (ellipses, hyperbolas, and parabolas). The conic sections were first defined and studied by the ancient Greek mathematicians Menaechmus (c. 380–320 BC) and Apollonius (c. 262–190 BC) and they have played an important role in mathematics ever since.

The beautiful shell of a chambered nautilus grows in the shape of an equiangular spiral, a curve described in polar coordinates by an equation $r = e^{a\theta}$.

Parametric equations are useful in multivariable calculus, especially in three dimensions, where it is no longer possible to describe a curve as a graph of a function (the graph of a function of two variables is a surface in $\mathbf{R}^3$ rather than a curve).

11.1 Parametric Equations

In previous chapters, we have studied curves that are graphs of functions or equations. In this section, we introduce a new and important way of describing a curve, via parametric equations. Imagine a particle P moving along a path $\mathcal{C}$ as a function of time t (Figure 1). Then the coordinates of P are functions of t:

$$\boxed{x = f(t), \qquad y = g(t)} \qquad \boxed{1}$$

The equations (1) are called **parametric equations** of the path $\mathcal{C}$ with **parameter** t. We also write

$$c(t) = (f(t), g(t))$$

As t varies in a domain such as an interval or the real line, $c(t)$ represents a point moving along the path $\mathcal{C}$. We refer to $\mathcal{C}$ as a **parametrized** or **parametric** curve. The direction of motion is often indicated by an arrow next to the plot of $c(t)$ as in Figure 1. Finally, since x and y are functions of t, we often write $c(t) = (x(t), y(t))$ instead of $(f(t), g(t))$.

In physical problems, t often represents time, but we are free to use other variables such as s or θ for the parameter.

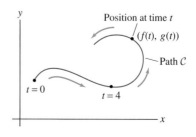

FIGURE 1 Path of a particle moving along a path $\mathcal{C}$ in the plane.

t	$x = 2t - 4$	$y = 3 + t^2$
-2	-8	7
0	-4	3
2	0	7
4	4	19

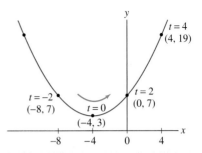

FIGURE 2 The parametric curve $x = 2t - 4$, $y = 3 + t^2$.

Consider the parametric equations

$$x = 2t - 4, \qquad y = 3 + t^2 \qquad \boxed{2}$$

The table above gives the x and y coordinates for several values of t. In Figure 2, we plot these points and join them by a smooth curve, indicating the direction of motion with an arrow.

In some cases, a parametric curve can be expressed as the graph of a function.

■ **EXAMPLE 1** Eliminating the Parameter   Describe the parametric curve

$$c(t) = (2t - 4, 3 + t^2)$$

in the form $y = f(x)$.

Solution   We "eliminate the parameter" by solving for y as a function of x. First, express t in terms of x: Since $x = 2t - 4$, we have $t = \frac{1}{2}x + 2$. Then substitute

$$y = 3 + t^2 = 3 + \left(\frac{1}{2}x + 2\right)^2 = 7 + 2x + \frac{1}{4}x^2$$

Thus, $c(t)$ traces out the parabola $y = 7 + 2x + \frac{1}{4}x^2$ (Figure 2).   ■

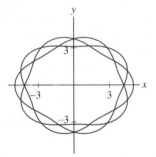

FIGURE 3 The parametric curve:
$x = 5\cos(3t)\cos\left(\frac{2}{3}\sin(5t)\right)$
$y = 4\sin(3t)\cos\left(\frac{2}{3}\sin(5t)\right)$.

CONCEPTUAL INSIGHT   The graph of a function $y = f(x)$ can always be parametrized as $c(t) = (t, f(t))$. For example, the parabola $y = x^2$ is parametrized by $c(t) = (t, t^2)$ and $y = e^t$ by $c(t) = (t, e^t)$. An advantage of parametric equations is that they allow us to describe curves that are not graphs of functions, such as the curve in Figure 3.

■ **EXAMPLE 2**   A bullet follows the trajectory (t in seconds, distance in feet):

$$x(t) = 200t, \qquad y(t) = 400t - 16t^2 \qquad 0 \le t \le 25$$

Find the bullet's height at $t = 5$ and its maximum height.

Solution   The bullet's height is $y(t) = 400t - 16t^2$. At $t = 5$, the height is

$$y(5) = 400(5) - 16(5^2) = 1,600 \text{ ft}$$

The function $y(t)$ is quadratic and takes on a maximum value where $y'(t) = 0$:

$$y'(t) = \frac{d}{dt}(400t - 16t^2) = 400 - 32t = 0 \qquad \Rightarrow \qquad t = \frac{400}{32} = 12.5$$

Thus, the maximum height is $y(12.5) = 400(12.5) - 16(12.5)^2 = 2,500 \text{ ft}$ (Figure 4).   ■

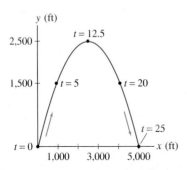

FIGURE 4 Parabolic path of a bullet.

We now discuss parametrizations of straight lines and circles. These are used frequently in multivariable calculus.

■ **EXAMPLE 3** Parametric Representation of a Line Show that the line through the point $P = (a, b)$ with slope m has parametrization

$$\boxed{c(t) = (a + t, b + mt) \qquad -\infty < t < \infty} \qquad \boxed{3}$$

Find parametric equations for the line through $P = (3, -1)$ of slope $m = 4$.

Solution We eliminate the parameter $x = a + t$. Thus, $t = x - a$ and

$$y = b + mt = b + m(x - a) \qquad \text{or} \qquad y - b = m(x - a)$$

This is the equation of the line through $P = (a, b)$ of slope m as claimed (Figure 5).

The line through $P = (3, -1)$ of slope $m = 4$ has parametrization

$$c(t) = (3 + t, -1 + 4t) \qquad \blacksquare$$

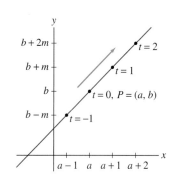

FIGURE 5 The parametric line $c(t) = (a + t, b + mt)$.

A circle of radius R centered at the origin has the parametrization

$$c(\theta) = (R \cos \theta, R \sin \theta)$$

The parameter θ represents the angle corresponding to the point (x, y) on the circle (Figure 6). The circle is traversed once in the counterclockwise direction as θ varies from 0 to 2π.

To move or "translate" a parametric curve a units horizontally and b units vertically, we replace $c(t) = (x(t), y(t))$ by $c(t) = (a + x(t), b + y(t))$. In particular, the circle of radius R with center (a, b) has parametrization (Figure 6)

$$\boxed{x = a + R \cos \theta, \qquad y = b + R \sin \theta} \qquad \boxed{4}$$

As a check, let's verify that the point (x, y) satisfies the equation of the circle of radius R centered at (a, b):

$$(x - a)^2 + (y - b)^2 = (a + R \cos \theta - a)^2 + (b + R \sin \theta - b)^2$$
$$= R^2 \cos^2 \theta + R^2 \sin^2 \theta = R^2$$

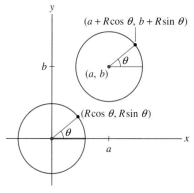

FIGURE 6 Parametrization of a circle of radius R with center (a, b).

A modification of these equations may be used to parametrize ellipses.

■ **EXAMPLE 4** Parametrization of an Ellipse Show that the ellipse $\left(\dfrac{x}{a}\right)^2 + \left(\dfrac{y}{b}\right)^2 = 1$ is parametrized by

$$c(t) = (a \cos t, b \sin t) \qquad 0 \le t < 2\pi$$

Graph the case $a = 4$, $b = 2$, and indicate the points corresponding to $t = 0, \dfrac{\pi}{6}, \dfrac{\pi}{3}, \dfrac{\pi}{2}$.

Solution We check that $x = a \cos t$, $y = b \sin t$ satisfies the equation of the ellipse:

$$\left(\frac{x}{a}\right)^2 + \left(\frac{y}{b}\right)^2 = \left(\frac{a \cos t}{a}\right)^2 + \left(\frac{b \sin t}{b}\right)^2 = \cos^2 t + \sin^2 t = 1$$

As t varies from 0 to π, $c(t)$ traces the top half of the ellipse because x decreases from a to $-a$ and y varies from 0 to b and back again to 0. Similarly, $c(t)$ traces the bottom half as t varies from π to 2π. For $a = 4$, $b = 2$, we obtain $c(t) = (4\cos t, 2\sin t)$, which parametrizes the ellipse $\left(\dfrac{x}{4}\right)^2 + \left(\dfrac{y}{2}\right)^2 = 1$. Figure 7 shows the ellipse and points indicated. ∎

t	$x(t) = 4\cos t$	$y(t) = 2\sin t$
0	4	0
$\dfrac{\pi}{6}$	$2\sqrt{3}$	1
$\dfrac{\pi}{3}$	2	$\sqrt{3}$
$\dfrac{\pi}{2}$	0	2

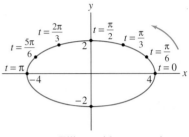

FIGURE 7 Ellipse with parametric equations $x = 4\cos t$, $y = 2\sin t$.

Before continuing, we make two important remarks:

- There is a difference between a path $c(t)$—such as the orbit of a moon around a planet—and the underlying curve $\mathcal{C}$. The curve $\mathcal{C}$ is a set of points in the plane. The path $c(t)$ (the orbit) describes not just the curve, but also the location of a point along the curve as a function of the parameter. The path may traverse all or part of $\mathcal{C}$ several times. For example, $c(t) = (\cos t, \sin t)$ describes a *path* that goes around the unit circle twice as t varies from 0 to 4π.
- Parametrizations are not unique, and in fact, every curve may be parametrized in infinitely many different ways. For instance, the parabola $y = x^2$ is parametrized by (t, t^2), but also (t^3, t^6), or (t^5, t^{10}), etc.

■ **EXAMPLE 5** **Parametrizations and Paths Versus Curves** Describe the motion of a particle moving along the paths (for $-\infty < t < \infty$):

(a) $c_1(t) = (t^3, t^6)$ **(b)** $c_2(t) = (t^2, t^4)$ **(c)** $c_3(t) = (\cos t, \cos^2 t)$

Solution The relation $y = x^2$ holds for each of these parametrizations, so all three parametrize portions of the parabola $y = x^2$.

(a) As t varies from $-\infty$ to ∞, the function t^3 also varies from $-\infty$ to ∞. Therefore, a particle following the path $c_1(t) = (t^3, t^6)$ traces the entire parabola $y = x^2$, moving from left to right and passing through each point once [Figure 8(A)].

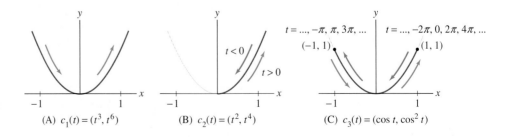

FIGURE 8 Three parametrizations of portions of the parabola.

(A) $c_1(t) = (t^3, t^6)$ (B) $c_2(t) = (t^2, t^4)$ (C) $c_3(t) = (\cos t, \cos^2 t)$

(b) In this parametrization, $x = t^2 \geq 0$, so the path $c_2(t) = (t^2, t^4)$ only traces the right half of the parabola. The particle comes in toward the origin as t varies from $-\infty$ to 0, and goes back out to the right as t varies from 0 to ∞ [Figure 8(B)].

(c) The function $\cos t$ oscillates between 1 to -1. Therefore, a particle following the path $c_3(t) = (\cos t, \cos^2 t)$ oscillates back and forth between the points $(1, 1)$ and $(-1, 1)$ on the parabola [Figure 8(C)]. The motion is repeated as t varies on intervals of length 2π. ∎

■ **EXAMPLE 6** Using Symmetry to Sketch a Loop Sketch the curve parametrized by

$$c(t) = (t^2 + 1, t^3 - 4t)$$

Label the points corresponding to $t = 0, \pm 1, \pm 2, \pm 2.5$.

Solution We note that $x(t) = t^2 + 1$ is an even function and $y(t) = t^3 - 4t$ is an odd function:

$$x(-t) = x(t), \qquad y(-t) = -y(t)$$

It follows that if a point $P = c(t)$ lies on the curve, then its reflection across the x-axis also lies on the curve and corresponds to "time" $-t$:

$$c(-t) = (x(t), -y(t)) = \text{reflection of } c(t) \text{ across the } x\text{-axis}$$

In other words, the curve is symmetric with respect to the x-axis. Therefore, it suffices to sketch the curve for $t \geq 0$ and then use symmetry.

Next, we observe that the coordinate $x(t) = t^2 + 1$ tends to ∞ as $t \to \infty$. To analyze the y-coordinate, we graph $y(t) = t^3 - 4t$ as a function of t (*not* as a function of x). Figure 9(A) shows that

$$y(t) < 0 \qquad \text{for} \qquad 0 < t < 2,$$
$$y(t) > 0 \qquad \text{for} \qquad t > 2,$$
$$y(t) \to \infty \qquad \text{as} \qquad t \to \infty$$

Therefore, starting at $c(0) = (1, 0)$, the curve first dips below the x-axis. Then it turns up and tends to ∞. The points $c(0)$, $c(1)$, $c(2)$, and $c(2.5)$ are tabulated in Table 1 and plotted in Figure 9(B). The part of the path for $t \leq 0$ is obtained by reflecting across the x-axis as shown in Figure 9(C). ∎

TABLE 1

t	$x = t^2 + 1$	$y = t^3 - 4t$
0	1	0
1	2	-3
2	5	0
2.5	7.25	5.625

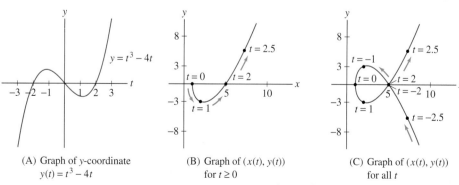

(A) Graph of y-coordinate
$y(t) = t^3 - 4t$

(B) Graph of $(x(t), y(t))$
for $t \geq 0$

(C) Graph of $(x(t), y(t))$
for all t

FIGURE 9 The curve $c(t) = (t^2 + 1, t^3 - 4t)$.

The **cycloid** is the curve traced by a point on the circumference of a rolling wheel (Figure 10).

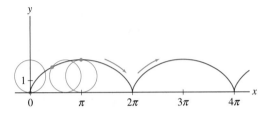

FIGURE 10 A cycloid.

The cycloid was studied intensively in the sixteenth and seventeenth centuries by a stellar cast of mathematicians (including Galileo, Pascal, Newton, Leibniz, Huygens, and Bernoulli) who discovered many of its remarkable properties. For example, to build a slide with the property that an object sliding down (without friction) reaches the bottom in the least amount of time, the slide must have the shape of an inverted cycloid. This is called the "brachistochrone property" from the Greek brachistos, *"the shortest," and* chronos, *"time."*

■ **EXAMPLE 7** **Parametrizing the Cycloid** Find parametric equations for the cycloid generated by a point P on the unit circle.

Solution We refer to Figure 11. Let P be the point on the rolling unit circle located at the origin at $t = 0$. Now let the circle roll a distance t along the x-axis. Then the length of the circular arc $\overparen{QP}$ is also t and the angle $\angle PCQ$ has radian measure t.

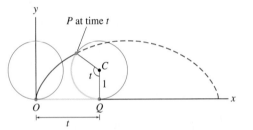
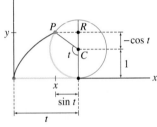

(A) The segment $\overline{OQ}$ and the arc $\overparen{QP}$ have the same length.

(B) The height of P is $1 - \cos t$. In this diagram, $\cos t$ is negative since t is obtuse.

FIGURE 11 Parametric representation of the cycloid.

To calculate the coordinates of P, observe in Figure 11(B) that the height y of P is $y = 1 - \cos t$. Note that in the figure, $\cos t$ is negative since t is obtuse and the segment $\overline{CR}$ has length $-\cos t$. The x-coordinate of P is $x = t - PR = t - \sin t$, so we obtain

$$x(t) = t - \sin t, \qquad y(t) = 1 - \cos t \qquad \boxed{5}$$

■

The argument in Example 7 may be modified to show that the cycloid generated by a circle of radius R has parametric equations

$$x = Rt - R\sin t, \qquad y = R - R\cos t \qquad \boxed{6}$$

Next, we address the problem of finding the tangent line for a curve in parametric form. For any curve in the xy-plane, the slope of the tangent line is the derivative dy/dx (whether described parametrically or not), but for a parametric curve, we must use the Chain Rule to compute dy/dx because y is not given explicitly as a function of x. If $x = f(t)$, $y = g(t)$, then

$$g'(t) = \frac{dy}{dt} = \frac{dy}{dx}\frac{dx}{dt} = \frac{dy}{dx}f'(t)$$

NOTATION In this section, we write $f'(t), x'(t), y'(t)$, etc. to denote the derivative with respect to t.

If $f'(t) \neq 0$, we may divide by $f'(t)$ to obtain

$$\frac{dy}{dx} = \frac{g'(t)}{f'(t)}$$

This calculation is valid on an interval I where $f(t)$ and $g(t)$ are differentiable, $f'(t)$ is continuous, and $f'(t) \neq 0$. In this case, $t = f^{-1}(x)$ exists and the composite $y = g(f^{-1}(x))$ is a differentiable function of x.

CAUTION Do not confuse dy/dx (the slope of the tangent line) with the derivatives dx/dt and dy/dt, which are derivatives with respect to the parameter t.

THEOREM 1 Slope of the Tangent Line   Let $c(t) = (x(t), y(t))$, where $x(t)$ and $y(t)$ are differentiable. Assume that $x'(t)$ is continuous and $x'(t) \neq 0$. Then

$$\frac{dy}{dx} = \frac{dy/dt}{dx/dt} = \frac{y'(t)}{x'(t)}$$

$\boxed{7}$

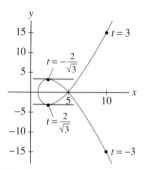

FIGURE 12 Horizontal tangent lines on $c(t) = (t^2 + 1, t^3 - 4t)$.

■ **EXAMPLE 8**   Let $c(t) = (t^2 + 1, t^3 - 4t)$. Find the equation of the tangent line at $t = 3$ and find the points where the tangent is horizontal.

Solution   We have

$$\frac{dy}{dx} = \frac{y'(t)}{x'(t)} = \frac{(t^3 - 4t)'}{(t^2 + 1)'} = \frac{3t^2 - 4}{2t}$$

The slope at $t = 3$ is

$$\frac{dy}{dx} = \left.\frac{3t^2 - 4}{2t}\right|_{t=3} = \frac{3(3)^2 - 4}{2(3)} = \frac{23}{6}$$

Since $c(3) = (10, 15)$, the equation of the tangent line in point-slope form is

$$y - 15 = \frac{23}{6}(x - 10)$$

The slope $\dfrac{dy}{dx}$ is zero if $3t^2 - 4 = 0$ and $2t \neq 0$. This gives $t = \pm 2/\sqrt{3}$ (Figure 12). Therefore, the tangent line is horizontal at the points

$$c\left(-\frac{2}{\sqrt{3}}\right) = \left(\frac{7}{3}, \frac{16}{3\sqrt{3}}\right), \qquad c\left(\frac{2}{\sqrt{3}}\right) = \left(\frac{7}{3}, -\frac{16}{3\sqrt{3}}\right) \qquad ■$$

Bézier curves were invented in the 1960s by Pierre Bézier (1910–1999), a French engineer who worked for the Renault car company. They are based on the properties of Bernstein polynomials, which were introduced 50 years earlier (in 1911) by the Russian mathematician Sergei Bernstein to study the approximation of continuous functions by polynomials. Today, Bézier curves are used in standard graphics programs such as Adobe Illustrator™ and Corel Draw™, and in the construction and storage of computer fonts such as TrueType™ and PostScript™ fonts.

Parametric curves are widely used in the field of computer graphics. A particularly important class of curves are **Bézier curves**, which we discuss here briefly in the cubic case. Given four "control points" (Figure 13):

$$P_0 = (a_0, b_0), \qquad P_1 = (a_1, b_1), \qquad P_2 = (a_2, b_2), \qquad P_3 = (a_3, b_3)$$

we define the Bézier curve $c(t) = (x(t), y(t))$ (for $0 \leq t \leq 1$), where

$$x(t) = a_0(1 - t)^3 + 3a_1 t(1 - t)^2 + 3a_2 t^2(1 - t) + a_3 t^3 \qquad \boxed{8}$$

$$y(t) = b_0(1 - t)^3 + 3b_1 t(1 - t)^2 + 3b_2 t^2(1 - t) + b_3 t^3 \qquad \boxed{9}$$

Note that $c(0) = (a_0, b_0)$ and $c(1) = (a_3, b_3)$, so the Bézier curve begins at P_0 and ends at P_3 (Figure 13). It can also be shown that the Bézier curve is contained within the quadrilateral (shown in yellow) with vertices P_0, P_1, P_2, P_3. However, $c(t)$ does not pass

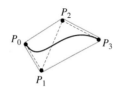

FIGURE 13 Cubic Bézier curves specifying four control points.

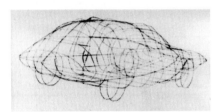

Hand sketch by Pierre Bézier for the French automobile manufacturer Renault, made in 1964.

through P_1 and P_2. Instead, these intermediate control points determine the slopes of the tangent lines at P_0 and P_3, as we show in the next example (also, see Exercises 58–60).

■ **EXAMPLE 9**   Show that the Bézier curve is tangent to the segment $\overline{P_0P_1}$ at P_0.

Solution   The Bézier curve passes through P_0 at $t = 0$, so we must show that the slope of the tangent line at $c(0) = P_0 = (a_0, b_0)$ is equal to the slope of $\overline{P_0P_1}$. To find the slope, we compute the derivatives:

$$x'(t) = -3a_0(1 - t)^2 + 3a_1(1 - 4t + 3t^2) + a_2(2t - 3t^2) + 3a_3t^2$$

$$y'(t) = -3b_0(1 - t)^2 + 3b_1(1 - 4t + 3t^2) + b_2(2t - 3t^2) + 3b_3t^2$$

Evaluating at $t = 0$, we obtain $x'(0) = 3(a_1 - a_0)$, $y'(0) = 3(b_1 - b_0)$, and

$$\left.\frac{dy}{dx}\right|_{t=0} = \frac{y'(0)}{x'(0)} = \frac{3(b_1 - b_0)}{3(a_1 - a_0)} = \frac{b_1 - b_0}{a_1 - a_0}$$

This is equal to the slope of the line segment through $P_0 = (a_0, b_0)$ and $P_1 = (a_1, b_1)$ as claimed (provided that $a_1 \neq a_0$).   ■

11.1 SUMMARY

- A path traced by a point $P = (x, y)$, where x and y are functions of a parameter t, is called a *parametric* or *parametrized* path or curve. We write $c(t) = (f(t), g(t))$ or $c(t) = (x(t), y(t))$.
- Keep in mind that the path $c(t) = (x(t), y(t))$ and the curve that it traces are different. The path $c(t)$ describes the location of a point as a function of t, whereas the underlying curve is a set of points in the plane. For example, the path $(\cos t, \sin t)$ moves around the unit circle infinitely many times as t varies from 0 to ∞.
- Parametrizations are not unique: Every path may be parametrized in infinitely many ways.
- Let $c(t) = (x(t), y(t))$, where $x(t)$ and $y(t)$ are differentiable and $x'(t)$ is continuous. Then the slope of the tangent line at $c(t)$ is the derivative

$$\frac{dy}{dx} = \frac{dy/dt}{dx/dt} = \frac{y'(t)}{x'(t)} \qquad \text{[valid if } x'(t) \neq 0\text{]}$$

- Do not confuse the derivatives with respect to t, dy/dt, and dx/dt, with the derivative dy/dx (the slope of the tangent line).

11.1 EXERCISES

Preliminary Questions

1. Describe the shape of the curve $x = 3\cos t$, $y = 3\sin t$.

2. How does $x = 4 + 3\cos t$, $y = 5 + 3\sin t$ differ from the curve in the previous question?

3. What is the maximum height of a particle whose path has parametric equations $x = t^9$, $y = 4 - t^2$?

4. Can the parametric curve $(t, \sin t)$ be represented as a graph $y = f(x)$? What about $(\sin t, t)$?

5. Match the derivatives with a verbal description:

(a) $\dfrac{dx}{dt}$ **(b)** $\dfrac{dy}{dt}$ **(c)** $\dfrac{dy}{dx}$

(i) Slope of the tangent line to the curve

(ii) Vertical rate of change with respect to time

(iii) Horizontal rate of change with respect to time

Exercises

1. Find the coordinates at times $t = 0, 2, 4$ of a particle following the path $x = 1 + t^3$, $y = 9 - 3t^2$.

2. Find the coordinates at $t = 0, \frac{\pi}{4}, \pi$ of a particle moving along the path $c(t) = (\cos 2t, \sin^2 t)$.

3. Show that the path traced by the bullet in Example 2 is a parabola by eliminating the parameter.

4. Use the table of values to sketch the parametric curve $(x(t), y(t))$, indicating the direction of motion.

t	-3	-2	-1	0	1	2	3
x	-15	0	3	0	-3	0	15
y	5	0	-3	-4	-3	0	5

5. Graph the parametric curves. Include arrows indicating the direction of motion.

(a) (t, t), $-\infty < t < \infty$ **(b)** $(\sin t, \sin t)$, $0 \le t \le 2\pi$

(c) (e^t, e^t), $-\infty < t < \infty$ **(d)** (t^3, t^3), $-1 \le t \le 1$

6. Give two different parametrizations of the line through $(4, 1)$ with slope 2.

In Exercises 7–14, express in the form $y = f(x)$ by eliminating the parameter.

7. $x = t + 3$, $y = 4t$ **8.** $x = t^{-1}$, $y = t^{-2}$

9. $x = t$, $y = \tan^{-1}(t^3 + e^t)$ **10.** $x = t^2$, $y = t^3 + 1$

11. $x = e^{-2t}$, $y = 6e^{4t}$ **12.** $x = 1 + t^{-1}$, $y = t^2$

13. $x = \ln t$, $y = 2 - t$ **14.** $x = \cos t$, $y = \tan t$

In Exercises 15–18, graph the curve and draw an arrow specifying the direction corresponding to motion.

15. $x = \frac{1}{2}t$, $y = 2t^2$ **16.** $x = 2 + 4t$, $y = 3 + 2t$

17. $x = \pi t$, $y = \sin t$ **18.** $x = t^2$, $y = t^3$

19. Match the parametrizations (a)–(d) below with their plots in Figure 14 and draw an arrow indicating the direction of motion.

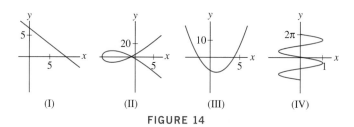

(I)	(II)	(III)	(IV)

FIGURE 14

(a) $c(t) = (\sin t, -t)$ **(b)** $c(t) = (t^2 - 9, -t^3 - 8)$

(c) $c(t) = (1 - t, t^2 - 9)$ **(d)** $c(t) = (4t + 2, 5 - 3t)$

20. The graphs of $x(t)$ and $y(t)$ as functions of t are shown in Figure 15(A). Which of (I)–(III) is the plot of $c(t) = (x(t), y(t))$? Explain.

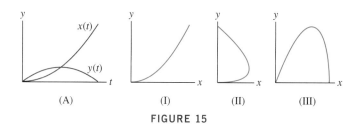

(A)	(I)	(II)	(III)

FIGURE 15

21. Find an interval of t-values such that $c(t) = (\cos t, \sin t)$ traces the lower half of the unit circle.

22. Find an interval of t-values such that $c(t) = (2t + 1, 4t - 5)$ parametrizes the segment from $(0, -7)$ to $(7, 7)$.

In Exercises 23–34, find parametric equations for the given curve.

23. $y = 9 - 4x$ **24.** $y = 8x^2 - 3x$

25. $4x - y^2 = 5$ **26.** $x^2 + y^2 = 49$

27. $(x + 9)^2 + (y - 4)^2 = 49$

28. Line of slope 8 through $(-4, 9)$

29. Line through $(2, 5)$ perpendicular to $y = 3x$

30. Circle of radius 4 with center $(3, 9)$

31. $\left(\dfrac{x}{4}\right)^2 + \left(\dfrac{y}{9}\right)^2 = 1$

32. Ellipse of Exercise 31, with its center translated to $(2, 11)$

33. The parabola $y = x^2$ translated so that its minimum occurs at $(2, 3)$

34. The curve $y = \cos x$ translated so that a maximum occurs at $(3, 5)$

35. Describe the parametrized curve $c(t) = (\sin^2 t, \cos^2 t)$ for $0 \le t \le \pi$.

36. Find the graph $y = f(x)$ traced by the path $x = \sec t$, $y = \tan t$. Which intervals of t-values trace the graph exactly once?

37. Find a parametrization $c(t)$ of the line $y = 3x - 4$ such that $c(0) = (2, 2)$.

38. Find a parametrization $c(t)$ of $y = x^2$ such that $c(0) = (3, 9)$.

39. Show that $x = \cosh t$, $y = \sinh t$ parametrizes the hyperbola $x^2 - y^2 = 1$. Calculate $\dfrac{dy}{dx}$ as a function of t. Generalize to obtain a parametrization of $\left(\dfrac{x}{a}\right)^2 - \left(\dfrac{y}{b}\right)^2 = 1$.

40. A particle moves along the path $x(t) = \dfrac{1}{4}t^3 + 2t$, $y(t) = 20t - t^2$ (in centimeters).

(a) What is the maximum height attained by the object?

(b) At what time does the object hit the ground?

(c) How far is the object from the origin when it hits the ground?

41. Which of (I) or (II) is the graph of $x(t)$ for the parametric curve in Figure 16(A)? Which represents $y(t)$?

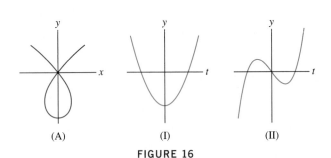

(A) (I) (II)

FIGURE 16

42. Sketch the graph of $c(t) = (t^3 - 4t, t^2)$ following the steps in Example 6.

43. Sketch $c(t) = (t^2, \sin t)$ for $-2\pi \le t \le 2\pi$.

44. Sketch $c(t) = (t^2 - 4t, 9 - t^2)$ for $-4 \le t \le 10$.

In Exercises 45–48, use Eq. (7) to find $\dfrac{dy}{dx}$ at the given point.

45. $(t^3, t^2 - 1)$, $t = -4$

46. $(2t + 9, 7t - 9)$, $t = 1$

47. $(s^{-1} - 3s, s^3)$, $s = -1$

48. $(\sin 2\theta, \cos 3\theta)$, $\theta = \frac{\pi}{4}$

In Exercises 49–52, find an equation $y = f(x)$ for the parametric curve and compute $\dfrac{dy}{dx}$ in two ways: using Eq. (7) and by differentiating $f(x)$.

49. $c(t) = (2t + 1, 1 - 9t)$

50. $c(t) = \left(\frac{1}{2}t, \frac{1}{4}t^2 - t\right)$

51. $x = s^3$, $y = s^6 + s^{-3}$

52. $x = \cos\theta$, $y = \cos\theta + \sin^2\theta$

In Exercises 53–56, let $c(t) = (t^2 - 9, t^2 - 8t)$ (see Figure 17).

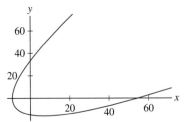

FIGURE 17 Plot of $c(t) = (t^2 - 9, t^2 - 8t)$.

53. Draw an arrow indicating the direction of motion and determine the interval of t-values corresponding to the portion of the curve in each of the four quadrants.

54. Find the equation of the tangent line at $t = 4$.

55. Find the points where the tangent has slope $\frac{1}{2}$.

56. Find the points where the tangent is horizontal or vertical.

57. Find the equation of the ellipse represented parametrically by $x = 4\cos t$, $y = 7\sin t$. Calculate the slope of the tangent line at the point $(2\sqrt{2}, 7\sqrt{2}/2)$.

In Exercises 58–60, refer to the Bézier curve defined by Eqs. (8) and (9).

58. Show that the Bézier curve with control points

$$P_0 = (1, 4), \quad P_1 = (3, 12), \quad P_2 = (6, 15), \quad P_3 = (7, 4)$$

has parametrization

$$c(t) = (1 + 6t + 3t^2 - 3t^3, 4 + 24t - 15t^2 - 9t^3)$$

Verify that the slope at $t = 0$ is equal to the slope of the segment $\overline{P_0 P_1}$.

59. Find and plot the Bézier curve $c(t)$ passing through the control points

$$P_0 = (3, 2), \quad P_1 = (0, 2), \quad P_2 = (5, 4), \quad P_3 = (2, 4)$$

60. Show that a cubic Bézier curve is tangent to the segment $\overline{P_2 P_3}$ at P_3.

61. A bullet fired from a gun follows the trajectory

$$x = at, \quad y = bt - 16t^2 \quad (a, b > 0)$$

Show that the bullet leaves the gun at an angle $\theta = \tan^{-1}\left(\dfrac{b}{a}\right)$ and lands at a distance $\dfrac{ab}{16}$ from the origin.

62. *CAS* Plot $c(t) = (t^3 - 4t, t^4 - 12t^2 + 48)$ for $-3 \le t \le 3$. Find the points where the tangent line is horizontal and vertical.

63. $\mathcal{CRS}$ Plot the astroid $x = \cos^3 \theta$, $y = \sin^3 \theta$ and find the equation of the tangent line at $\theta = \frac{\pi}{3}$.

64. Find the equation of the tangent line at $t = \frac{\pi}{4}$ to the cycloid generated by the unit circle with parametric equation (5).

65. Find the points with horizontal tangent line on the cycloid with parametric equation (5).

66. Property of the Cycloid Prove that the tangent line at a point P on the cycloid always passes through the top point on the rolling circle as indicated in Figure 18. Use the parametrization (6).

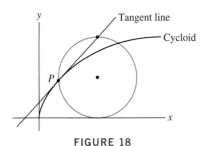

FIGURE 18

67. A *curtate cycloid* (Figure 19) is the curve traced by a point at a distance h from the center of a circle of radius R rolling along the x-axis where $h < R$. Show that this curve has parametric equations $x = Rt - h \sin t$, $y = R - h \cos t$.

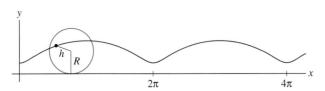

FIGURE 19 Curtate cycloid.

68. $\mathcal{CRS}$ Use a computer algebra system to explore what happens when $h > R$ in the parametric equations of Exercise 67. Describe the result.

69. Show that the line of slope t through $(-1, 0)$ intersects the unit circle in the point with coordinates

$$x = \frac{1 - t^2}{t^2 + 1}, \quad y = \frac{2t}{t^2 + 1} \qquad \boxed{10}$$

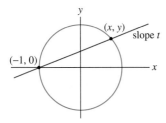

FIGURE 20 Unit circle.

Conclude that these equations parametrize the unit circle with the point $(-1, 0)$ excluded (Figure 20). Show further that $t = \dfrac{y}{x + 1}$.

70. The **folium of Descartes** is the curve with equation $x^3 + y^3 = 3axy$, where $a \neq 0$ is a constant (Figure 21).

(a) Show that for $t \neq -1, 0$, the line $y = tx$ intersects the folium at the origin and at one other point P. Express the coordinates of P in terms of t to obtain a parametrization of the folium. Indicate the direction of the parametrization on the graph.

(b) Describe the interval of t values parametrizing the parts of the curve in quadrants I, II, and IV. Note that $t = -1$ is a point of discontinuity of the parametrization.

(c) Calculate dy/dx as a function of t and find the points with horizontal or vertical tangent.

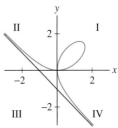

FIGURE 21 Folium $x^3 + y^3 = 3axy$.

71. Use the results of Exercise 70 to show that the asymptote of the folium is the line $x + y = -a$. Hint: show that $\lim\limits_{t \to -1} (x + y) = -a$.

72. Find a parametrization of $x^{2n+1} + y^{2n+1} = ax^n y^n$, where a and n are constants.

73. Verify that the **tractrix** curve ($\ell > 0$)

$$c(t) = \left(t - \ell \tanh \frac{t}{\ell}, \ell \operatorname{sech} \frac{t}{\ell} \right)$$

has the following property: For all t, the segment from $c(t)$ to $(0, t)$ is tangent to the curve and has length ℓ (Figure 22).

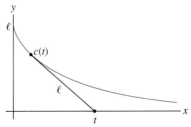

FIGURE 22 The tractrix $c(t) = \left(t - \ell \tanh \dfrac{t}{\ell}, \ell \operatorname{sech} \dfrac{t}{\ell} \right)$.

74. In Exercise 54 of Section 9.1, we described the tractrix by the differential equation

$$\frac{dy}{dx} = -\frac{y}{\sqrt{\ell^2 - y^2}}$$

Show that the curve $c(t)$ identified as the tractrix in Exercise 73 satisfies this differential equation. Note that the derivative on the left is taken with respect to x, not t.

75. Let A and B be the points where the ray of angle θ intersects the two concentric circles of radii $r < R$ centered at the origin (Figure 23). Let P be the point of intersection of the horizontal line through A and the vertical line through B. Express the coordinates of P as a function of θ and describe the curve traced by P for $0 \le \theta \le 2\pi$.

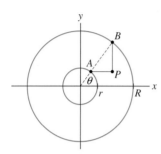

FIGURE 23

76. A 10-ft ladder slides down a wall as its bottom B is pulled away from the wall (Figure 24). Using the angle θ as parameter, find the parametric equations for the path followed by (a) the top of the ladder A, (b) the bottom of the ladder B, and (c) the point P located 4 ft from the top of the ladder. Show that P describes an ellipse.

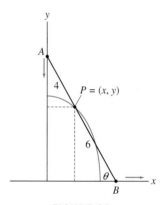

FIGURE 24

Further Insights and Challenges

77. Derive the formula for the slope of the tangent line to a parametric curve $c(t) = (x(t), y(t))$ using a different method than that presented in the text. Assume that $x'(t_0)$ and $y'(t_0)$ exist and that $x'(t_0) \ne 0$. Show that

$$\lim_{h \to 0} \frac{y(t_0 + h) - y(t_0)}{x(t_0 + h) - x(t_0)} = \frac{y'(t_0)}{x'(t_0)}$$

Then explain why this limit is equal to the slope dy/dx. Draw a diagram showing that the ratio in the limit is the slope of a secant line.

78. Second Derivative for a Parametrized Curve Given a parametrized curve $c(t) = (x(t), y(t))$, show that

$$\frac{d}{dt}\left(\frac{dy}{dx}\right) = \frac{x'(t)y''(t) - y'(t)x''(t)}{x'(t)^2}$$

Use this to prove the formula

$$\boxed{\frac{d^2y}{dx^2} = \frac{x'(t)y''(t) - y'(t)x''(t)}{x'(t)^3}} \qquad \boxed{11}$$

In Exercises 79–82, use Eq. (11) to find $\dfrac{d^2y}{dx^2}$.

79. $x = t^3 + t^2$, $y = 7t^2 - 4$, $t = 2$

80. $x = s^{-1} + s$, $y = 4 - s^{-2}$, $s = 1$

81. $x = 8t + 9$, $y = 1 - 4t$, $t = -3$

82. $x = \cos\theta$, $y = \sin\theta$, $\theta = \frac{\pi}{4}$

83. Use Eq. (11) to find the t-intervals on which $c(t) = (t^2, t^3 - 4t)$ is concave up.

84. Use Eq. (11) to find the t-intervals on which $c(t) = (t^2, t^4 - 4t)$ is concave up.

85. Area under a Parametrized Curve Let $c(t) = (x(t), y(t))$ be a parametrized curve such that $x'(t) > 0$ and $y(t) > 0$ (Figure 25). Show that the area A under $c(t)$ for $t_0 \le t \le t_1$ is

$$A = \int_{t_0}^{t_1} y(t)x'(t)\, dt \qquad \boxed{12}$$

Hint: $x(t)$ is increasing and therefore has an inverse, say, $t = g(x)$. Observe that $c(t)$ is the graph of the function $y(g(x))$ and apply the Change of Variables formula to $A = \displaystyle\int_{x(t_0)}^{x(t_1)} y(g(x))\, dx$.

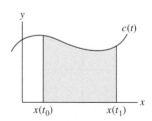

FIGURE 25

86. Calculate the area under $y = x^2$ over $[0, 1]$ using Eq. (12) with the parametrizations (t^3, t^6) and (t^2, t^4).

87. What does Eq. (12) say if $c(t) = (t, f(t))$?

88. Sketch the graph of $c(t) = (\ln t, 2 - t)$ for $1 \le t \le 2$ and compute the area under the graph using Eq. (12).

89. Use Eq. (12) to show that the area under one arch of the cycloid $c(t)$ (Figure 26) generated by a circle of radius R is equal to three times the area of the circle. Recall that

$$c(t) = (Rt - R \sin t, R - R \cos t)$$

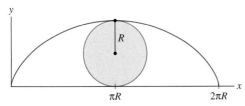

FIGURE 26 The area of the generating circle is one-third the area of one arch of the cycloid.

In Exercises 90–91, refer to Figure 27.

90. The parameter t in the standard parametrization of the ellipse $c(t) = (a \cos t, b \sin t)$ is *not* an angular parameter (unless $a = b$), but it can be interpreted as an area parameter. Show that if $c(t) = (x, y)$, then $t = 2A/ab$, where A is the area of the shaded region in Figure 27. *Hint:* Use Eq. (12).

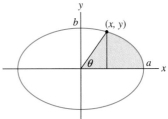

FIGURE 27 The parameter θ on the ellipse $\left(\dfrac{x}{a}\right)^2 + \left(\dfrac{y}{b}\right)^2 = 1$.

91. Show that the parametrization of the ellipse by the angle θ is

$$x = \frac{ab \cos \theta}{\sqrt{a^2 \sin^2 \theta + b^2 \cos^2 \theta}}$$

$$y = \frac{ab \sin \theta}{\sqrt{a^2 \sin^2 \theta + b^2 \cos^2 \theta}}$$

11.2 Arc Length and Speed

In Section 8.1, the length (or "arc length") s of a curve was defined as the limit of the lengths of polygonal approximations. We showed that if $f(x)$ has a continuous derivative, then the arc length of the graph of $y = f(x)$ over $[a, b]$ is equal to

$$s = \text{arc length} = \int_a^b \sqrt{1 + f'(x)^2}\, dx \qquad \boxed{1}$$

There is a more general formula for the arc length of a path $c(t) = (x(t), y(t))$ for $a \le t \le b$. To derive this formula, choose a partition P of $[a, b]$ by an increasing sequence of parameter values:

$$t_0 = a < t_1 < t_2 < \cdots < t_N = b$$

and let L be the polygonal path obtained by joining the points

$$P_0 = c(t_0), \quad P_1 = c(t_1), \ldots, P_N = c(t_N)$$

by segments as in Figure 1. We define the length s of the path to be the limit of the lengths $|L|$ of the polygonal approximations L as the norm $\|P\|$ tends to zero (recall that $\|P\|$ is the maximum of the widths $\Delta t_i = t_i - t_{i-1}$). Note that N tends to ∞ as $\|P\| \to 0$.

By the distance formula, the length of the ith segment in a polygonal approximation L is

$$P_{i-1}P_i = \sqrt{(x(t_i) - x(t_{i-1}))^2 + (y(t_i) - y(t_{i-1}))^2} \qquad \boxed{2}$$

We rewrite this expression by applying the Mean Value Theorem (MVT) to both $x(t)$ and $y(t)$. Assuming that $x(t)$ and $y(t)$ are differentiable, the MVT states that there exist

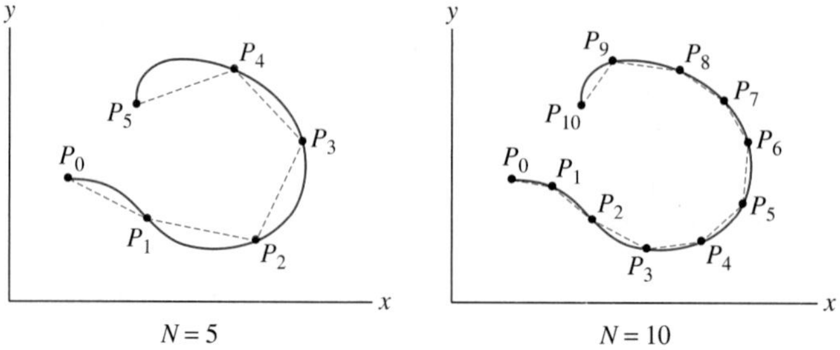

FIGURE 1 Polygonal approximations for $N = 5$ and $N = 10$.

intermediate values t_i^* and t_i^{**} in the interval $[t_{i-1}, t_i]$ such that

$$x(t_i) - x(t_{i-1}) = x'(t_i^*)\Delta t_i, \qquad y(t_i) - y(t_{i-1}) = y'(t_i^{**})\Delta t_i$$

where $\Delta t_i = t_i - t_{i-1}$. Therefore,

$$P_{i-1}P_i = \sqrt{x'(t_i^*)^2 \Delta t_i^2 + y'(t_i^{**})^2 \Delta t_i^2} = \sqrt{x'(t_i^*)^2 + y'(t_i^{**})^2}\,\Delta t_i$$

The length $|L|$ of L is the sum of the lengths of the segments:

$$|L| = \sum_{i=1}^{N} P_{i-1}P_i = \sum_{i=1}^{N} \sqrt{x'(t_i^*)^2 + y'(t_i^{**})^2}\,\Delta t_i \qquad \boxed{3}$$

This is *nearly* a Riemann sum for the function $\sqrt{x'(t)^2 + y'(t)^2}$. It would be a true Riemann sum if the two intermediate values t_i^* and t_i^{**} were equal. Although they need not be equal, it can be shown (and we will take it for granted) that if $x'(t)$ and $y'(t)$ are continuous, then the sum in Eq. (3) still approaches the integral as $\|P\| \to 0$. Thus,

$$\lim_{\|P\| \to 0} \sum_{i=1}^{N} P_{i-1}P_i = \int_a^b \sqrt{x'(t)^2 + y'(t)^2}\,dt$$

THEOREM 1 Arc Length   Let $c(t) = (x(t), y(t))$ and assume that $x'(t)$ and $y'(t)$ exist and are continuous. Then the length s of $c(t)$ for $a \le t \le b$ is equal to

$$s = \int_a^b \sqrt{x'(t)^2 + y'(t)^2}\,dt \qquad \boxed{4}$$

It is often impossible to evaluate the arc length integral (4) explicitly, but we can always approximate it numerically.

Keep in mind that the length of a path $c(t)$ is the distance traveled by a particle following the path. This is greater than the length of the underlying curve if parts of the curve are traversed more than once.

■ **EXAMPLE 1**   Length of a Circular Arc   Calculate the length s of the arc $0 \le \theta \le \theta_0$ of a circle of radius R.

Solution   We use the parametrization $x = R\cos\theta$, $y = R\sin\theta$:

$$x'(\theta)^2 + y'(\theta)^2 = (-R\sin\theta)^2 + (R\cos\theta)^2 = R^2(\sin^2\theta + \cos^2\theta) = R^2$$

$$s = \int_0^{\theta_0} \sqrt{x'(\theta)^2 + y'(\theta)^2}\,d\theta = \int_0^{\theta_0} R\,d\theta = R\theta_0$$

This agrees with the standard formula for the length of a circular arc (Figure 2).   ■

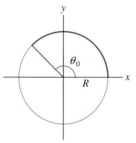

FIGURE 2 An arc of angle θ_0 on a circle of radius R has length $R\theta_0$.

It turns out that the cycloid is a curve whose arc length can be computed explicitly.

■ **EXAMPLE 2** Length of the Cycloid   Calculate the length s of one arch of the cycloid generated by a circle of radius $R = 2$ (Figure 3).

Solution By Eq. (6) in Section 1, the cycloid generated by a circle of radius $R = 2$ has parametric equations

$$x = 2(t - \sin t), \qquad y = 2(1 - \cos t)$$

Using the identity $\dfrac{1 - \cos t}{2} = \sin^2 \dfrac{t}{2}$, we find

$$x'(t)^2 + y'(t)^2 = 2^2 (1 - \cos t)^2 + 2^2 \sin^2 t$$

$$= 4 - 8 \cos t + 4 \cos^2 t + 4 \sin^2 t$$

$$= 8 - 8 \cos t = 16 \left(\frac{1 - \cos t}{2} \right) = 16 \sin^2 \frac{t}{2}$$

and thus,

$$\sqrt{x'(t)^2 + y'(t)^2} = 4 \left| \sin \frac{t}{2} \right|$$

One arch of the cycloid is traced as t varies from 0 to 2π, so

$$s = \int_0^{2\pi} \sqrt{x'(t)^2 + y'(t)^2}\, dt = 4 \int_0^{2\pi} \sin \frac{t}{2}\, dt = -8 \cos \frac{t}{2} \Big|_0^{2\pi} = -8(-1) + 8 = 16$$

Note that we may drop the absolute value in the integral over $[0, 2\pi]$ because $\sin \dfrac{t}{2} \geq 0$ for $0 \leq t \leq 2\pi$.   ■

The arc length integral leads to an expression for the **speed** of a particle moving along a path $c(t) = (x(t), y(t))$. By definition, speed is the rate of change of distance traveled with respect to time. The distance traveled over the time interval $[t_0, t]$ is given by the arc length integral:

$$s(t) = \text{distance traveled} = \int_{t_0}^{t} \sqrt{x'(u)^2 + y'(u)^2}\, du$$

Therefore, by the Fundamental Theorem of Calculus,

$$\text{Speed} = \frac{ds}{dt} = \frac{d}{dt} \int_0^{t} \sqrt{x'(u)^2 + y'(u)^2}\, du = \sqrt{x'(t)^2 + y'(t)^2}$$

In Chapter 13, we will discuss not just the speed but also the velocity of a particle moving along a curved path. Velocity is "speed plus direction" and is represented by a "vector."

THEOREM 2 Speed along a Parametrized Path The speed at time t of a particle with trajectory $c(t) = (x(t), y(t))$ is the derivative of the arc length integral

$$s(t) = \int_{t_0}^{t} \sqrt{x'(u)^2 + y'(u)^2}\, du$$

Namely,

$$\text{Speed} = \frac{ds}{dt} = \sqrt{x'(t)^2 + y'(t)^2}$$

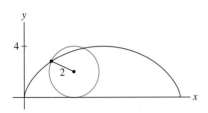

FIGURE 3 One arch of the cycloid generated by a circle of radius 2.

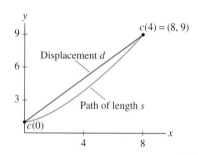

FIGURE 4 The distance traveled is greater than or equal to the displacement.

Before continuing, let us emphasize the distinction between distance traveled and **displacement** (also called net change in position). The displacement over a time interval $[t_0, t_1]$ is the distance between the initial point $c(t_0)$ and the endpoint $c(t_1)$. The distance traveled is greater than the displacement unless the particle happens to move in a straight line (Figure 4).

■ **EXAMPLE 3** A particle travels along the path $c(t) = (2t, 1 + t^{3/2})$ (t in minutes, distance in feet).

(a) Find the speed at $t = 1$.

(b) Compute distance traveled s and displacement d during the first 4 min.

Solution We have $x'(t) = 2$, $y'(t) = \frac{3}{2}t^{1/2}$, so the particle's speed at time t is

$$s'(t) = \sqrt{x'(t)^2 + y'(t)^2} = \sqrt{4 + \frac{9}{4}t} \quad \text{ft/min}$$

The speed at $t = 1$ is $s'(1) = \sqrt{4 + \frac{9}{4}} = 2.5$ ft/min. The distance traveled during the first 4 min is equal to

$$s = \int_0^4 \sqrt{4 + \frac{9}{4}t}\, dt = \frac{8}{27}\left(4 + \frac{9}{4}t\right)^{3/2}\Big|_0^4 = \frac{8}{27}\left(13^{3/2} - 8\right) \approx 11.52 \text{ ft}$$

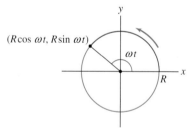

FIGURE 5 The path $c(t) = (2t, 1 + t^{3/2})$ for $0 \le t \le 4$.

The displacement d is the distance from the initial point $c(0) = (0, 1)$ to the endpoint $c(4) = (8, 1 + 4^{3/2}) = (8, 9)$ (see Figure 5):

$$d = \sqrt{(8-0)^2 + (9-1)^2} = 8\sqrt{2} \approx 11.31 \text{ ft} \qquad ■$$

■ **EXAMPLE 4** **Angular Velocity** Let $c(t) = (R \cos \omega t, R \sin \omega t)$ be a circular path, where ω is a constant (called the **angular velocity**, in units of radians per unit time). Show that $c(t)$ is a path of constant speed. Find the angular velocity ω of a counterclockwise path if $R = 3$ m and the speed is 12 m/s.

Solution We have $x = R \cos \omega t$ and $y = R \sin \omega t$, and

$$x'(t) = -\omega R \sin \omega t, \qquad y'(t) = \omega R \cos \omega t$$

We see that the speed is a constant independent of t:

$$\frac{ds}{dt} = \sqrt{x'(t)^2 + y'(t)^2} = \sqrt{(-\omega R \sin \omega t)^2 + (\omega R \cos \omega t)^2}$$

$$= \sqrt{\omega^2 R^2 (\sin^2 \omega t + \cos^2 \omega t)} = |\omega| R$$

If $R = 3$ and the speed is 12 m/s, then $|\omega| R = 3|\omega| = 12$. Therefore $|\omega| = 4$ and $\omega = 4$ since the path is counterclockwise (Figure 6). ■

FIGURE 6 A particle rotating on a circular path with angular velocity ω has speed $|\omega R|$.

The surface area S of a surface of revolution obtained by rotating a parametric curve $c(t) = (x(t), y(t))$ about the x-axis may be computed by an integral that is closely related to the arc length integral. The formula, stated in the next theorem (without proof), may be justified using polygonal approximations as in the case of the arc length formula. Note that we assume $y(t) \ge 0$ so that the curve $c(t)$ lies above the x-axis.

> **THEOREM 3 Surface Area** Let $c(t) = (x(t), y(t))$ and assume that $x'(t)$ and $y'(t)$ exist and are continuous. Assume further that $y(t) \geq 0$. Then the surface area S of the surface obtained by rotating $c(t)$ about the x-axis for $a \leq t \leq b$ is equal to
>
> $$S = 2\pi \int_a^b y(t)\sqrt{x'(t)^2 + y'(t)^2}\, dt \qquad \boxed{5}$$

■ **EXAMPLE 5** The curve $c(t) = (t - \tanh t, \operatorname{sech} t)$ is called a *tractrix*. Calculate the surface area of the infinite surface generated by revolving the tractrix about the x-axis for $0 \leq t < \infty$ (Figure 7).

Solution We have

$$x'(t) = \frac{d}{dt}(t - \tanh t) = 1 - \operatorname{sech}^2 t, \quad y'(t) = \frac{d}{dt}\operatorname{sech} t = -\operatorname{sech} t \tanh t$$

Using the identities $1 - \operatorname{sech}^2 t = \tanh^2 t$ and $\operatorname{sech}^2 t = 1 - \tanh^2 t$, we obtain

$$x'(t)^2 + y'(t)^2 = (1 - \operatorname{sech}^2 t)^2 + (-\operatorname{sech} t \tanh t)^2$$
$$= (\tanh^2 t)^2 + (1 - \tanh^2 t)\tanh^2 t = \tanh^2 t$$

The surface area is given by an improper integral, which we evaluate using the integral formula recalled in the margin:

$$S = 2\pi \int_0^\infty \operatorname{sech} t \sqrt{\tanh^2 t}\, dt = 2\pi \int_0^\infty \operatorname{sech} t \tanh t\, dt = 2\pi \lim_{R \to \infty} \int_0^R \operatorname{sech} t \tanh t\, dt$$

$$= 2\pi \lim_{R \to \infty} (-\operatorname{sech} t)\Big|_0^R = 2\pi \lim_{R \to \infty} (\operatorname{sech} 0 - \operatorname{sech} R) = 2\pi \operatorname{sech} 0 = 2\pi$$

Here we use that $\operatorname{sech} t = \dfrac{2}{e^t + e^{-t}}$ tends to 0 as $t \to \infty$ because the term e^t in the denominator tends to ∞. ■

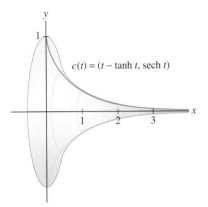

y

1

$c(t) = (t - \tanh t, \operatorname{sech} t)$

1   2   3 x

FIGURE 7 Surface generated by revolving the tractrix about the x-axis.

◄·· *REMINDER*

$$\operatorname{sech} t = \frac{1}{\cosh t} = \frac{2}{e^t + e^{-t}}$$

$$1 - \operatorname{sech}^2 t = \tanh^2 t$$

$$\frac{d}{dt}\tanh t = \operatorname{sech}^2 t$$

$$\frac{d}{dt}\operatorname{sech} t = -\operatorname{sech} t \tanh t$$

$$\int \operatorname{sech} t \tanh t\, dt = -\operatorname{sech} t + C$$

11.2 SUMMARY

- The *arc length* s of a path $c(t) = (x(t), y(t))$ for $a \leq t \leq b$ is

$$s = \text{arc length} = \int_a^b \sqrt{x'(t)^2 + y'(t)^2}\, dt$$

- The *arc length integral* $s(t) = \displaystyle\int_{t_0}^t \sqrt{x'(u)^2 + y'(u)^2}\, du$ is equal to the *distance traveled* over the time interval $[t_0, t]$. The *displacement* is the distance between the starting point $c(t_0)$ and endpoint $c(t)$.
- The speed at time t of a particle with trajectory $c(t) = (x(t), y(t))$ is

$$\frac{ds}{dt} = \sqrt{x'(t)^2 + y'(t)^2}$$

11.2 EXERCISES

Preliminary Questions

1. What is the definition of arc length?

2. What is the interpretation of $\sqrt{x'(t)^2 + y'(t)^2}$ for a particle following the trajectory $(x(t), y(t))$?

3. A particle travels along a path from $(0, 0)$ to $(3, 4)$. What is the displacement? Can the distance traveled be determined from the information given?

4. A particle traverses the parabola $y = x^2$ with constant speed 3 cm/s. What is the distance traveled during the first minute? *Hint:* No computation is necessary.

Exercises

1. Use Eq. (4) to calculate the length of the semicircle

$$x = 3 \sin t, \quad y = 3 \cos t, \quad 0 \le t \le \pi$$

2. Find the speed at $t = 4$ s of a particle whose position at time t seconds is $c(t) = (4 - t^2, t^3)$.

In Exercises 3–12, use Eq. (4) to find the length of the path over the given interval.

3. $(3t + 1, 9 - 4t), \quad 0 \le t \le 2$

4. $(1 + 2t, 2 + 4t), \quad 1 \le t \le 4$

5. $(2t^2, 3t^2 - 1), \quad 0 \le t \le 4$

6. $(3t, 4t^{3/2}), \quad 0 \le t \le 1$

7. $(3t^2, 4t^3), \quad 1 \le t \le 4$

8. $(t^3 + 1, t^2 - 3), \quad 0 \le t \le 1$

9. $(\sin 3t, \cos 3t), \quad 0 \le t \le \pi$

10. $(2 \cos t - \cos 2t, 2 \sin t - \sin 2t), \quad 0 \le t \le \frac{\pi}{2}$

11. $(\sin \theta - \theta \cos \theta, \cos \theta + \theta \sin \theta), \quad 0 \le \theta \le 2$

12. $(5(\theta - \sin \theta), 5(1 - \cos \theta)), \quad 0 \le \theta \le 2\pi$

13. Show that one arch of a cycloid generated by a circle of radius R has length $8R$.

14. Find the length of the tractrix (Figure 22 in Section 11.1) with parametrization

$$c(t) = (t - \tanh(t), \operatorname{sech}(t)), \quad 0 \le t \le A$$

15. Find the length of the spiral $c(t) = (t \cos t, t \sin t)$ for $0 \le t \le 2\pi$ to three decimal places (Figure 8). *Hint:* Use the formula

$$\int \sqrt{1 + t^2}\, dt = \frac{1}{2} t \sqrt{1 + t^2} + \frac{1}{2} \ln\left(t + \sqrt{1 + t^2}\right)$$

In Exercises 16–19, determine the speed $s(t)$ of a particle with a given trajectory at time t_0 (in units of meters and seconds).

16. $(t^3, t^2), \quad t = 2$

17. $(3 \sin 5t, 8 \cos 5t), \quad t = \frac{\pi}{4}$

18. $(5t + 1, 4t - 3), \quad t = 9$

19. $(\ln(t^2 + 1), t^3), \quad t = 1$

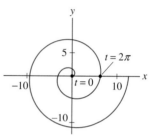

FIGURE 8 The spiral $c(t) = (t \cos t, t \sin t)$.

20. Find the speed of a particle whose path around a circle of radius r is described by the parametric curve $c(t) = (r \cos \omega t, r \sin \omega t)$.

21. Find the minimum speed of a particle with trajectory $c(t) = (t^3 - 4t, t^2 + 1)$ for $t \ge 0$. *Hint:* It is easier to find the minimum of the square of the speed.

22. Find the minimum speed of a particle with trajectory $c(t) = (t^3, t^{-2})$ for $t \ge 0.5$.

23. Find the speed of the cycloid $c(t) = (4t - 4 \sin t, 4 - 4 \cos t)$ at points where the tangent line is horizontal.

24. If you unwind thread from a stationary circular spool, keeping the thread taut at all times, then the endpoint traces a curve $\mathcal{C}$ called the **involute** of the circle (Figure 9). Observe that $\overline{PQ}$ has length $R\theta$. Show that $\mathcal{C}$ is parametrized by

$$c(\theta) = \big(R(\cos \theta + \theta \sin \theta), R(\sin \theta - \theta \cos \theta)\big)$$

Then find the length of the involute for $0 \le \theta \le 2\pi$.

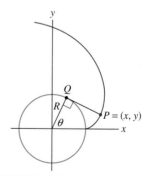

FIGURE 9 Involute of a circle.

$⨍⨍⨍$ *In Exercises 25–28, plot the curve and use the Midpoint Rule with $N = 10, 20, 30$, and 50 to approximate its length.*

25. $c(t) = (\cos t, e^{\sin t})$ for $0 \le t \le 2\pi$

26. $c(t) = (t - \sin 2t, 1 - \cos 2t)$ for $0 \le t \le 2\pi$

27. The ellipse $\left(\frac{x}{5}\right)^2 + \left(\frac{y}{3}\right)^2 = 1$

28. $x = \sin 2t, \quad y = \sin 3t$ for $0 \le t \le 2\pi$

29. Let $a > b$ and set $k = \sqrt{1 - \dfrac{b^2}{a^2}}$. Use a parametric representation to show that the ellipse $\left(\frac{x}{a}\right)^2 + \left(\frac{y}{b}\right)^2 = 1$ has length $L = 4aG\left(\frac{\pi}{2}, k\right)$, where

$$G(\theta, k) = \int_0^\theta \sqrt{1 - k^2 \sin^2 t}\, dt$$

is the elliptic integral of the second kind.

In Exercises 30–33, use Eq. (5) to compute the surface area of the given surface.

30. The cone generated by revolving $c(t) = (t, mt)$ about the x-axis for $0 \le t \le A$

31. The surface generated by revolving the astroid with parametrization $c(t) = (\cos^3 t, \sin^3 t)$ about the x-axis for $0 \le t \le \frac{\pi}{2}$

32. A sphere of radius R

33. The surface generated by revolving one arch of the cycloid $c(t) = (t - \sin t, 1 - \cos t)$ about the x-axis

Further Insights and Challenges

34. $⨍⨍⨍$ Let $b(t)$ be the "Butterfly Curve":

$$x(t) = \sin t\left(e^{\cos t} - 2\cos 4t - \sin\left(\frac{t}{12}\right)^5\right)$$

$$y(t) = \cos t\left(e^{\cos t} - 2\cos 4t - \sin\left(\frac{t}{12}\right)^5\right)$$

(a) Use a computer algebra system to plot $b(t)$ and the speed $s(t)$ for $0 \le t \le 12\pi$.

(b) Approximate the length $b(t)$ for $0 \le t \le 10\pi$.

35. $⨍⨍⨍$ Let $a \ge b > 0$ and set $k = \dfrac{2\sqrt{ab}}{a-b}$. Show that the **trochoid**

$$x = at - b\sin t, \qquad y = a - b\cos t, \qquad 0 \le t \le T$$

has length $2(a - b)G\left(\dfrac{T}{2}, k\right)$ with $G(\theta, k)$ as in Exercise 29.

36. The path of a satellite orbiting at a distance R from the center of the earth is parametrized by $x = R\cos \omega t$, $y = R\sin \omega t$.

(a) Show that the period T (the time of one revolution) is $T = 2\pi/\omega$.

(b) According to Newton's laws of motion and gravity,

$$x''(t) = -Gm_e\frac{x}{R^3}, \qquad y''(t) = -Gm_e\frac{y}{R^3}$$

where G is the universal gravitational constant and m_e is the mass of the earth. Prove that $\dfrac{R^3}{T^2} = \dfrac{Gm_e}{4\pi^2}$. Thus, $\dfrac{R^3}{T^2}$ has the same value for all orbits (a special case of Kepler's Third Law).

37. The acceleration due to gravity on the surface of the earth is
$$g = \frac{Gm_e}{R_e^2} = 9.8 \text{ m/s}^2,$$ where $R_e = 6{,}378$ km. Use Exercise 36(b) to show that a satellite orbiting at the earth's surface would have period $T_e = 2\pi\sqrt{R_e/g} \approx 84.5$ min. Then estimate the distance R_m from the moon to the center of the earth. Assume that the period of the moon (sidereal month) is $T_m \approx 27.43$ days.

11.3 Polar Coordinates

Polar coordinates are an alternative system of labeling points P in the plane. Instead of specifying the x- and y-coordinates, we specify coordinates (r, θ), where r is the distance from P to the origin O and θ is the angle between $\overline{OP}$ and the positive x-axis (Figure 1). By convention, an angle is positive if the corresponding rotation is counterclockwise. We call r the **radial coordinate** and θ the **angular coordinate**.

For example, the point P in Figure 2 has polar coordinates $(4, \frac{2\pi}{3})$. It is located at a distance 4 from the origin (so it lies on the circle of radius 4) and it lies on the ray of angle $\frac{2\pi}{3}$.

To relate polar and rectangular coordinates, we refer to Figure 1. We see that if P has polar coordinates (r, θ), then its rectangular coordinates are $x = r\cos\theta$ and $y = r\sin\theta$. On the other hand, $r^2 = x^2 + y^2$ by the distance formula and $\tan\theta = \dfrac{y}{x}$ if $x \ne 0$.

Polar coordinates are useful for problems in which the distance from the origin or angle plays a role, especially if the problem involves rotational symmetry. They are a natural choice for describing the gravitational force exerted by an object such as the sun, since the magnitude of the force depends only on distance from the sun.

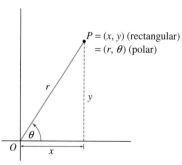

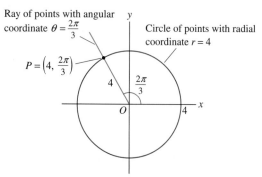

FIGURE 1 The polar coordinates of P satisfy $r = \sqrt{x^2 + y^2}$, $\tan \theta = y/x$.

FIGURE 2 The point P lies at the intersection of the circle $r = 4$ and the ray $\theta = \frac{2\pi}{3}$.

Polar to Rectangular	Rectangular to Polar
$x = r \cos \theta$	$r = \sqrt{x^2 + y^2}$
$y = r \sin \theta$	$\tan \theta = \dfrac{y}{x} \quad (x \neq 0)$

A few remarks are in order before proceeding.

- The angular coordinate θ is not uniquely determined because the polar coordinates (r, θ) and $(r, \theta + 2\pi n)$ *label the same point* for any integer n. For instance, the point $(x, y) = (0, 2)$ has polar coordinates $\left(2, \frac{\pi}{2}\right)$ and $\left(2, \frac{5\pi}{2}\right)$ and $\left(2, \frac{9\pi}{2}\right)$, etc. (see Figure 3).
- The origin O does not have a well-defined angular coordinate, so we assign to O the polar coordinates $(0, \theta)$ for any angle θ.
- By convention, we also allow *negative* radial coordinates. For $r > 0$, the point $(-r, \theta)$ is defined to be the reflection of (r, θ) through the origin (see Figure 4). With this convention, $(-r, \theta)$ and $(r, \theta + \pi)$ represent the same point.
- We may specify unique polar coordinates for points other than the origin by placing restrictions on r and θ. We commonly choose $r > 0$ and $0 \leq \theta < 2\pi$.

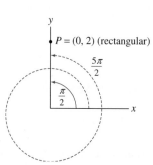

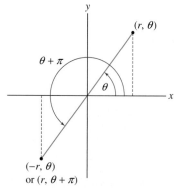

FIGURE 3 The angular coordinate of P can take on the value $\frac{\pi}{2} + 2\pi n$ for any integer n.

FIGURE 4 Relation between (r, θ) and $(-r, \theta)$.

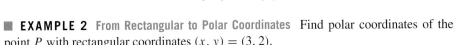

FIGURE 5 Grid lines in polar coordinates.

Figure 5 shows the two families of **grid lines** in polar coordinates:

$$\text{Circle centered at } O \quad \longleftrightarrow \quad r = \text{constant}$$

$$\text{Ray starting at } O \quad \longleftrightarrow \quad \theta = \text{constant}$$

Every point in the plane other than the origin lies at the intersection of two grid lines that determine its polar coordinates. For example, the point Q in Figure 5 lies on the circle $r = 3$ and the ray $\theta = \frac{5\pi}{6}$ so $Q = (3, \frac{5\pi}{6})$ in polar coordinates.

■ **EXAMPLE 1** From Polar to Rectangular Coordinates Find the rectangular coordinates of the point Q with polar coordinates $(r, \theta) = (3, \frac{5\pi}{6})$.

Solution The point $Q = (r, \theta) = (3, \frac{5\pi}{6})$ has rectangular coordinates:

$$x = r \cos \theta = 3 \cos \left(\frac{5\pi}{6}\right) = 3 \left(-\frac{\sqrt{3}}{2}\right) = -\frac{3\sqrt{3}}{2}$$

$$y = r \sin \theta = 3 \sin \left(\frac{5\pi}{6}\right) = 3 \left(\frac{1}{2}\right) = \frac{3}{2}$$
■

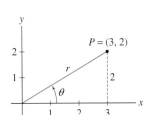

FIGURE 6 The polar coordinates of P satisfy $r = \sqrt{3^2 + 2^2}$ and $\tan \theta = \frac{2}{3}$.

■ **EXAMPLE 2** From Rectangular to Polar Coordinates Find polar coordinates of the point P with rectangular coordinates $(x, y) = (3, 2)$.

Solution Since $P = (x, y) = (3, 2)$,

$$r = \sqrt{x^2 + y^2} = \sqrt{3^2 + 2^2} = \sqrt{13} \approx 3.6$$

$$\tan \theta = \frac{y}{x} = \frac{2}{3}$$

$$\theta = \tan^{-1} \frac{y}{x} = \tan^{-1} \frac{2}{3} \approx 0.588$$

Thus, P has polar coordinates $(r, \theta) \approx (3.6, 0.588)$ (see Figure 6).   ■

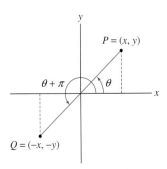

FIGURE 7 $\tan(\theta + \pi) = \tan \theta$.

By definition, the arctangent satisfies

$$-\frac{\pi}{2} < \tan^{-1} x < \frac{\pi}{2}$$

An angular coordinate θ of $P = (x, y)$ is

$$\theta = \begin{cases} \tan^{-1} \dfrac{y}{x} & \text{if } x > 0 \\[2mm] \tan^{-1} \dfrac{y}{x} + \pi & \text{if } x < 0 \\[2mm] \pm \dfrac{\pi}{2} & \text{if } x = 0 \end{cases}$$

The next example calls attention to a subtlety in determining the angular coordinate θ of a point $P = (x, y)$. There are two angles between 0 and 2π satisfying $\tan \theta = \frac{y}{x}$ because $\tan(\theta + \pi) = \tan \theta$. We must choose θ so that (r, θ) lies in the quadrant containing P and not its reflection through the origin Q (Figure 7).

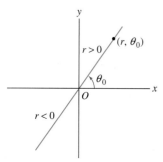

FIGURE 8

■ EXAMPLE 3 Choosing θ Correctly Find two polar representations of $P = (-1, 1)$, one using a positive radial coordinate and one using a negative radial coordinate.

Solution The positive radial coordinate of $P = (x, y) = (-1, 1)$ is

$$r = \sqrt{(-1)^2 + 1^2} = \sqrt{2}$$

To find the angular coordinate, we solve

$$\tan \theta = \frac{y}{x} = -1 \quad \Rightarrow \quad \theta = \frac{3\pi}{4}, \ \frac{3\pi}{4} + \pi = \frac{7\pi}{4}$$

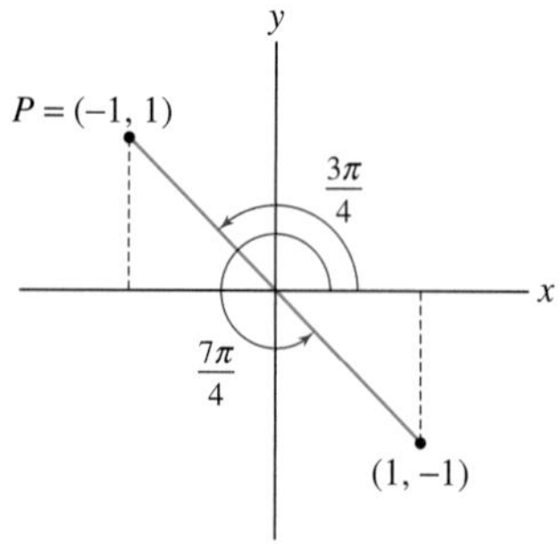

Since P lies in the second quadrant, the appropriate angular coordinate of P is $\theta = \frac{3\pi}{4}$ (Figure 8). Alternatively, P may be represented with the negative radial coordinate $r = -\sqrt{2}$ and angle $\theta = \frac{7\pi}{4}$. Thus,

$$P = \left(\sqrt{2}, \frac{3\pi}{4}\right) \qquad \text{or} \qquad \left(-\sqrt{2}, \frac{7\pi}{4}\right) \qquad \blacksquare$$

Curves are described in polar coordinates by equations relating r and θ. We refer to an equation in polar coordinates as a **polar equation** to avoid confusion with the usual equation in rectangular coordinates. By convention, we allow solutions with negative r when considering equations in polar coordinates.

The polar equation of a line through the origin O has the simple form $\theta = \theta_0$, where θ_0 is the angle between the line and the x-axis (Figure 9). Indeed, the solutions of $\theta = \theta_0$ are (r, θ_0), where r is arbitrary (positive, negative, or zero).

■ EXAMPLE 4 Line Through the Origin Find the polar equation of the line through the origin of slope $\frac{3}{2}$.

Solution A line of slope m makes an angle θ_0 with the x-axis, where $m = \tan \theta_0$. In our case, Figure 10 shows that $\tan \theta_0 = \frac{3}{2}$ and so $\theta_0 = \tan^{-1} \frac{3}{2} \approx 0.98$. The equation of the line is $\theta = \tan^{-1} \frac{3}{2}$. $\blacksquare$

FIGURE 9 Lines through O with polar equation $\theta = \theta_0$.

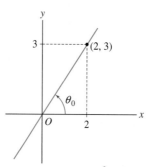

FIGURE 10 Line of slope $\frac{3}{2}$ through the origin.

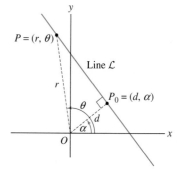

FIGURE 11 Line $\mathcal{L}$ not passing through the origin.

To describe a line not passing through the origin in polar coordinates, the key is to consider the point P_0 on the line that is closest to the origin (Figure 11).

■ **EXAMPLE 5** Line Not Passing Through the Origin Let $P_0 = (d, \alpha)$ be polar coordinates of the point on a line $\mathcal{L}$ closest to the origin. Show that $\mathcal{L}$ has polar equation

$$\boxed{r = d \sec(\theta - \alpha)} \qquad \boxed{1}$$

Solution The point P_0 on $\mathcal{L}$ closest to the origin is obtained by dropping a perpendicular from the origin to $\mathcal{L}$ (Figure 11). Therefore, if $P = (r, \theta)$ is any point on $\mathcal{L}$ other than P_0, then $\triangle OPP_0$ is a right triangle and as we see in Figure 11, $\dfrac{d}{r} = \cos(\theta - \alpha)$. We obtain the equation $r = d \sec(\theta - \alpha)$ as claimed. ■

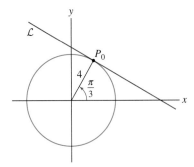

FIGURE 12 The tangent line has equation $r = 4 \sec\left(\theta - \dfrac{\pi}{3}\right)$.

■ **EXAMPLE 6** Find the polar equation of the line $\mathcal{L}$ tangent to the circle $r = 4$ at the point with polar coordinates $P_0 = (4, \frac{\pi}{3})$.

Solution Since $\mathcal{L}$ is tangent to the circle at P_0, all points $P \neq P_0$ on $\mathcal{L}$ lie outside the circle. Therefore, P_0 is the point on $\mathcal{L}$ closest to the origin (Figure 12). We take $(d, \alpha) = (4, \frac{\pi}{3})$ in Eq. (1) to obtain the equation $r = 4 \sec(\theta - \frac{\pi}{3})$. ■

It is often difficult to guess the shape of a graph of a polar equation. One way of sketching the graph of an equation $r = f(\theta)$ is to rewrite the equation in rectangular coordinates.

■ **EXAMPLE 7** Converting to Rectangular Coordinates Identify the curve with polar equation $r = 2a \cos \theta$ (a a constant).

Solution We have $r = \pm\sqrt{x^2 + y^2}$ and $\cos \theta = \dfrac{x}{r} = \dfrac{x}{\pm\sqrt{x^2 + y^2}}$. Thus

$$r = 2a \cos \theta \quad \Rightarrow \quad \pm\sqrt{x^2 + y^2} = 2a \frac{x}{\pm\sqrt{x^2 + y^2}} \quad \Rightarrow \quad x^2 + y^2 = 2ax$$

Now complete the square:

$$x^2 - 2ax + y^2 = 0$$

$$(x - a)^2 - a^2 + y^2 = 0 \qquad \text{or} \qquad \boxed{(x - a)^2 + y^2 = a^2}$$

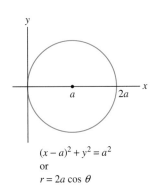

$$(x - a)^2 + y^2 = a^2$$
or
$$r = 2a \cos \theta$$

FIGURE 13 Circle with polar equation $r = 2a \cos \theta$.

Thus, $r = 2a \cos \theta$ is the circle of radius a and center $(a, 0)$ (Figure 13). ■

A similar calculation shows that $x^2 + (y - a)^2 = a^2$ has polar equation $r = 2a \sin \theta$. In the next example, we make use of symmetry. Note that the points (r, θ) and $(r, -\theta)$ are symmetric about the x-axis (Figure 14).

■ **EXAMPLE 8** The *Limaçon*: Symmetry About the x-Axis Sketch the graph of $r = 2 \cos \theta - 1$.

Solution To get started, we plot the points A–G on a grid and join them by a smooth curve (Figure 15). However, for a better understanding, it is helpful to graph r as a function of θ in rectangular coordinates. Figure 16(A) shows that

As θ varies from 0 to $\frac{\pi}{3}$, r varies from 1 to 0

As θ varies from $\frac{\pi}{3}$ to π, r is *negative* and varies from 0 to -3

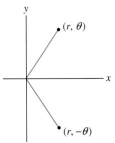

FIGURE 14 The points (r, θ) and $(r, -\theta)$ are symmetric with respect to the x-axis.

	A	B	C	D	E	F	G
θ	0	$\pi/6$	$\pi/3$	$\pi/2$	$2\pi/3$	$5\pi/6$	π
r	1	0.73	0	-1	-2	-2.73	-3

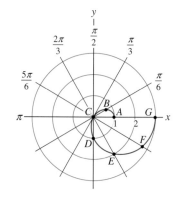

FIGURE 15 Plotting $r = 2\cos\theta - 1$ using a grid.

Since $r(\theta) = 2\cos\theta - 1$ is periodic, it suffices to plot points for $-\pi \le \theta \le \pi$.

- The graph begins at point A with polar coordinates $(r, \theta) = (1, 0)$ and moves in toward the origin as θ varies from 0 to $\pi/3$ [Figure 16(B)].
- For $\pi/3 \le \theta \le \pi$, r is negative. Therefore the curve continues into the third and fourth quadrants (rather than into the first and second quadrants), moving toward the point G with polar coordinates $(r, \theta) = (-3, \pi)$ as in Figure 16(C).
- Since $r(\theta) = r(-\theta)$, the curve is symmetric with respect to the x-axis. As θ varies from 0 to $-\pi$, we obtain the reflection through the x-axis of the first half of the curve, where $0 \le \theta \le \pi$, as in Figure 16(D). ■

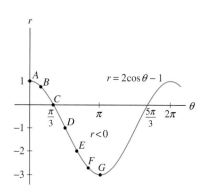

(A) Variation of r as a function of θ.

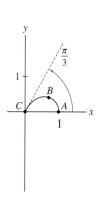

(B) As θ varies from 0 to $\pi/3$, r varies from 1 to 0.

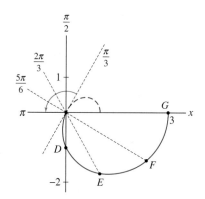

(C) As θ varies from $\pi/3$ to π, r is negative and varies from 0 to -3.

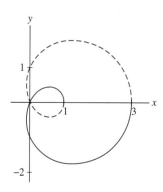

(D) The entire limaçon.

FIGURE 16 The curve $r = 2\cos\theta - 1$ is called the *limaçon*, from the Latin word for "snail." It was first described in 1525 by the German artist Albrecht Dürer.

11.3 SUMMARY

- A point P in the plane has polar coordinates (r, θ), where r is the distance from P to the origin and θ is the angle between the positive x-axis and the segment $\overline{OP}$, measured in the counterclockwise direction. The rectangular coordinates of P are

$$x = r\cos\theta, \qquad y = r\sin\theta$$

- If P has rectangular coordinates (x, y), then

$$r = \sqrt{x^2 + y^2}, \qquad \tan \theta = \frac{y}{x}$$

The angular coordinate θ must be chosen so that (r, θ) lies in the proper quadrant. We have

$$\theta = \begin{cases} \tan^{-1} \dfrac{y}{x} & \text{if } x > 0 \\ \tan^{-1} \dfrac{y}{x} + \pi & \text{if } x < 0 \\ \pm \dfrac{\pi}{2} & \text{if } x = 0 \end{cases}$$

- Nonuniqueness: (r, θ) and $(r, \theta + 2n\pi)$ represent the same point for all integers n. The origin O has polar coordinates $(0, \theta)$ for any θ.
- Negative radial coordinates: $(-r, \theta)$ and $(r, \theta + \pi)$ represent the same point.
- Polar equations:

Curve	Polar Equation
Circle of radius R, center at the origin	$r = R$
Line through origin of slope $m = \tan \theta_0$	$\theta = \theta_0$
Line on which $P_0 = (d, \alpha)$ is the point closest to the origin	$r = d \sec(\theta - \alpha)$
Circle of radius a, center at $(a, 0)$ $(x - a)^2 + y^2 = a^2$	$r = 2a \cos \theta$
Circle of radius a, center at $(0, a)$ $x^2 + (y - a)^2 = a^2$	$r = 2a \sin \theta$

11.3 EXERCISES

Preliminary Questions

1. If P and Q have the same radial coordinate, then (choose the correct answer):

(a) P and Q lie on the same circle with the center at the origin.

(b) P and Q lie on the same ray based at the origin.

2. Give two polar coordinate representations for the point $(x, y) = (0, 1)$, one with negative r and one with positive r.

3. Does a point (r, θ) have more than one representation in rectangular coordinates?

4. Describe the curves with polar equations

(a) $r = 2$ **(b)** $r^2 = 2$ **(c)** $r \cos \theta = 2$

5. If $f(-\theta) = f(\theta)$, then the curve $r = f(\theta)$ is symmetric with respect to the (choose the correct answer):

(a) x-axis   **(b)** y-axis   **(c)** origin

Exercises

1. Find polar coordinates for each of the seven points plotted in Figure 17.

2. Plot the points with polar coordinates

(a) $(2, \frac{\pi}{6})$ **(b)** $(4, \frac{3\pi}{4})$ **(c)** $(3, -\frac{\pi}{2})$ **(d)** $(0, \frac{\pi}{6})$

3. Convert from rectangular to polar coordinates:

(a) $(1, 0)$ **(b)** $(3, \sqrt{3})$

(c) $(-2, 2)$ **(d)** $(-1, \sqrt{3})$

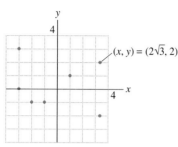

FIGURE 17

4. Use a calculator to convert from rectangular to polar coordinates (make sure your choice of θ gives the correct quadrant):

(a) $(2, 3)$ **(b)** $(4, -7)$
(c) $(-3, -8)$ **(d)** $(-5, 2)$

5. Convert from polar to rectangular coordinates:

(a) $(3, \frac{\pi}{6})$ **(b)** $(6, \frac{3\pi}{4})$ **(c)** $(5, -\frac{\pi}{2})$

6. Convert from polar to rectangular coordinates:

(a) $(0, 0)$ **(b)** $(-4, \frac{\pi}{3})$ **(c)** $(0, \frac{\pi}{6})$

7. Which of the following are possible polar coordinates for the point P with rectangular coordinates $(0, -2)$?

(a) $\left(2, \frac{\pi}{2}\right)$ **(b)** $\left(2, \frac{7\pi}{2}\right)$
(c) $\left(-2, -\frac{3\pi}{2}\right)$ **(d)** $\left(-2, \frac{7\pi}{2}\right)$
(e) $\left(-2, -\frac{\pi}{2}\right)$ **(f)** $\left(2, -\frac{7\pi}{2}\right)$

8. Describe each shaded sector in Figure 18 by inequalities in r and θ.

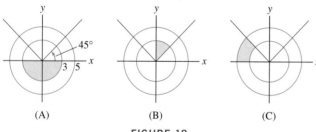

(A) **(B)** **(C)**

FIGURE 18

9. Find the equation in polar coordinates of the line through the origin with slope $\frac{1}{2}$.

10. What is the slope of the line $\theta = \frac{3\pi}{5}$?

11. Which of the two equations, $r = 2 \sec \theta$ and $r = 2 \csc \theta$, defines a horizontal line?

In Exercises 12–17, convert to an equation in rectangular coordinates.

12. $r = 7$ **13.** $r = \sin \theta$

14. $r = 2 \sin \theta$ **15.** $r = 2 \csc \theta$

16. $r = \dfrac{1}{\cos \theta - \sin \theta}$ **17.** $r = \dfrac{1}{2 - \cos \theta}$

In Exercises 18–21, convert to an equation in polar coordinates.

18. $x^2 + y^2 = 5$ **19.** $x = 5$

20. $y = x^2$ **21.** $xy = 1$

22. Match the equation with its description:

(a) $r = 2$ **(i)** Vertical line
(b) $\theta = 2$ **(ii)** Horizontal line
(c) $r = 2 \sec \theta$ **(iii)** Circle
(d) $r = 2 \csc \theta$ **(iv)** Line through origin

23. Find the values of θ in the plot of $r = 4 \cos \theta$ corresponding to points A, B, C, D in Figure 19. Then indicate the portion of the graph traced out as θ varies in the following intervals:

(a) $0 \le \theta \le \frac{\pi}{2}$ **(b)** $\frac{\pi}{2} \le \theta \le \pi$ **(c)** $\pi \le \theta \le \frac{3\pi}{2}$

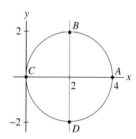

FIGURE 19 Plot of $r = 4 \cos \theta$.

24. Suppose that (x, y) has polar coordinates (r, θ). Find the polar coordinates of the following:

(a) $(x, -y)$ **(b)** $(-x, -y)$ **(c)** $(-x, y)$ **(d)** (y, x)

25. Match each equation in rectangular coordinates with its equation in polar coordinates.

(a) $x^2 + y^2 = 2$ **(i)** $r^2(1 + 2 \sin^2 \theta) = 4$
(b) $x^2 + (y - 1)^2 = 1$ **(ii)** $r(\cos \theta + \sin \theta) = 4$
(c) $x^2 - y^2 = 4$ **(iii)** $r = 2 \sin \theta$
(d) $x + y = 4$ **(iv)** $r = 2$

26. What are the polar equations of the lines parallel to the line $r \cos(\theta - \frac{\pi}{3}) = 1$?

27. Show that $r = \sin \theta + \cos \theta$ is the equation of the circle of radius $1/\sqrt{2}$ whose center in rectangular coordinates is $(\frac{1}{2}, \frac{1}{2})$. Then find the values of θ between 0 and π such that $(\theta, r(\theta))$ yields the points $A, B, C,$ and D in Figure 20.

28. Sketch the curve $r = \frac{1}{2}\theta$ (the spiral of Archimedes) for θ between 0 and 2π by plotting the points for $\theta = 0, \frac{\pi}{4}, \frac{\pi}{2}, \ldots, 2\pi$.

29. Sketch the graph of $r = 3 \cos \theta - 1$ (see Example 8).

30. Sketch the graph of $r = \cos \theta - 1$.

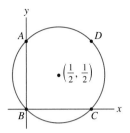

FIGURE 20 Plot of $r = \sin\theta + \cos\theta$.

31. Figure 21 displays the graphs of $r = \sin 2\theta$ in rectangular coordinates and in polar coordinates, where it is a "rose with four petals." Identify (a) the points in (B) corresponding to the points labeled A–I in (A), and (b) the parts of the curve in (B) corresponding to the angle intervals $\left[0, \frac{\pi}{2}\right]$, $\left[\frac{\pi}{2}, \pi\right]$, $\left[\pi, \frac{3\pi}{2}\right]$, and $\left[\frac{3\pi}{2}, 2\pi\right]$.

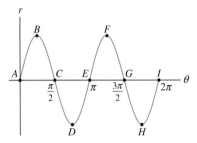 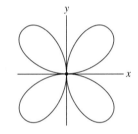

(A) Graph of r as a function of θ, where $r = \sin 2\theta$.

(B) Graph of $r = \sin 2\theta$ in polar coordinates.

FIGURE 21 Rose with four petals.

32. Sketch the curve $r = \sin 3\theta$. First fill in the table of r-values below and plot the corresponding points of the curve. Notice that the three petals of the curve correspond to the angle intervals $\left[0, \frac{\pi}{3}\right]$, $\left[\frac{\pi}{3}, \frac{2\pi}{3}\right]$, and $\left[\frac{\pi}{3}, \pi\right]$. Then plot $r = \sin 3\theta$ in rectangular coordinates and label the points on this graph corresponding to (r, θ) in the table.

θ	0	$\frac{\pi}{12}$	$\frac{\pi}{6}$	$\frac{\pi}{4}$	$\frac{\pi}{3}$	$\frac{5\pi}{12}$	$\cdots$	$\frac{11\pi}{12}$	π
r									

33. ⌐∃⌐ Plot the **cissoid** $r = 2\sin\theta\tan\theta$ and show that its equation in rectangular coordinates is $y^2 = \dfrac{x^3}{2-x}$.

34. Prove that $r = 2a\cos\theta$ is the equation of the circle in Figure 22 using only the fact that a triangle inscribed in a circle with one side a diameter is a right triangle.

35. Show that $r = a\cos\theta + b\sin\theta$ is the equation of a circle passing through the origin. Express the radius and center (in rectangular coordinates) in terms of a and b.

36. Use the previous exercise to write the equation of the circle of radius 5 and center $(3, 4)$ in the form $r = a\cos\theta + b\sin\theta$.

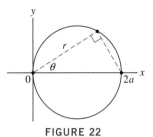

FIGURE 22

37. Use the identity $\cos 2\theta = \cos^2\theta - \sin^2\theta$ to find a polar equation of the hyperbola $x^2 - y^2 = 1$.

38. Find an equation in rectangular coordinates for the curve $r^2 = \cos 2\theta$.

39. Show that $\cos 3\theta = \cos^3\theta - 3\cos\theta\sin^2\theta$ and use this identity to find an equation in rectangular coordinates for the curve $r = \cos 3\theta$.

40. Use the addition formula for the cosine to show that the line $\mathcal{L}$ with polar equation $r\cos(\theta - \alpha) = d$ has the equation in rectangular coordinates $(\cos\alpha)x + (\sin\alpha)y = d$. Show that $\mathcal{L}$ has slope $m = -\cot\alpha$ and y-intercept $\dfrac{d}{\sin\alpha}$.

In Exercises 41–45, find an equation in polar coordinates of the line $\mathcal{L}$ with given description.

41. The point on $\mathcal{L}$ closest to the origin has polar coordinates $\left(2, \frac{\pi}{9}\right)$.

42. The point on $\mathcal{L}$ closest to the origin has rectangular coordinates $(-2, 2)$.

43. $\mathcal{L}$ is tangent to the circle $r = 2\sqrt{10}$ at the point with rectangular coordinates $(-2, -6)$.

44. $\mathcal{L}$ has slope 3 and is tangent to the unit circle in the fourth quadrant.

45. $y = 4x - 9$.

46. Show that the polar equation of the line $y = ax + b$ can be written in the form

$$r = \frac{b}{\sin\theta - a\cos\theta}$$

47. Distance Formula Use the Law of Cosines (Figure 23) to show that the distance d between two points with polar coordinates (r, θ) and (r_0, θ_0) is

$$d^2 = r^2 + r_0^2 - 2rr_0\cos(\theta - \theta_0) \qquad \boxed{2}$$

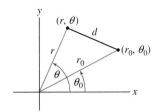

FIGURE 23

48. Use the distance formula (2) to show that the circle of radius 9 whose center has polar coordinates $\left(5, \frac{\pi}{4}\right)$ has equation:

$$r^2 - 10r \cos\left(\theta - \frac{\pi}{4}\right) = 56$$

49. Show that the cardiod $r = 1 + \sin\theta$ has equation

$$x^2 + y^2 = (x^2 + y^2 - x)^2$$

50. For $a > 0$, a **lemniscate curve** is the set of points P such that the product of the distances from P to $(a, 0)$ and $(-a, 0)$ is a^2. Show that the equation of the lemniscate is:

$$(x^2 + y^2)^2 = 2a^2(x^2 - y^2)$$

Then find the equation in polar coordinates. To obtain the simplest form of the equation, use the identity $\cos 2\theta = \cos^2\theta - \sin^2\theta$. Plot the lemniscate for $a = 2$ if you have a computer algebra system.

51. The Derivative in Polar Coordinates A polar curve $r = f(\theta)$ has parametric equations (since $x = r\cos\theta$ and $y = r\sin\theta$):

$$x = f(\theta)\cos\theta, \quad y = f(\theta)\sin\theta$$

Apply Theorem 1 of Section 11.1 to prove the formula

$$\frac{dy}{dx} = \frac{f(\theta)\cos\theta + f'(\theta)\sin\theta}{-f(\theta)\sin\theta + f'(\theta)\cos\theta} \qquad \boxed{3}$$

where $f'(\theta) = df/d\theta$.

52. Use Eq. (3) to find the slope of the tangent line to $r = \theta$ at $\theta = \frac{\pi}{2}$ and $\theta = \pi$.

53. Find the equation in rectangular coordinates of the tangent line to $r = 4\cos 3\theta$ at $\theta = \frac{\pi}{6}$.

54. Show that for the circle $r = \sin\theta + \cos\theta$,

$$\frac{dy}{dx} = \frac{\cos 2\theta + \sin 2\theta}{\cos 2\theta - \sin 2\theta}$$

Calculate the slopes of the tangent lines at points A, B, C in Figure 20 and find the polar coordinates of the points at which the tangent line is horizontal.

Further Insights and Challenges

55. 🖼 Let c be a fixed constant. Explain the relationship between the graphs of:

(a) $y = f(x + c)$ and $y = f(x)$ (rectangular)

(b) $r = f(\theta + c)$ and $r = f(\theta)$ (polar)

(c) $y = f(x) + c$ and $y = f(x)$ (rectangular)

(d) $r = f(\theta) + c$ and $r = f(\theta)$ (polar)

56. 🖼 Let $f(x)$ be a periodic function of period 2π, that is, $f(x) = f(x + 2\pi)$. Explain how this periodicity is reflected in the graph of:

(a) $y = f(x)$ in rectangular coordinates

(b) $r = f(\theta)$ in polar coordinates

57. 𝖦𝖴 Use a graphing utility to convince yourself that graphs of the polar equations $r = f_1(\theta) = 2\cos\theta - 1$ and $r = f_2(\theta) = 2\cos\theta + 1$ have the same graph. Then explain why. *Hint:* Show that the points $(f_1(\theta + \pi), \theta + \pi)$ and $(f_2(\theta), \theta)$ coincide.

58. 𝖢𝖠𝖲 Plot the limaçon curves $r = a + \cos\theta$ for several values of a.

(a) Describe how the shape changes as a increases. What happens for $a < 0$?

(b) A closed curve is called convex if, for any two points P and Q in the interior of the curve, the segment $\overline{PQ}$ is also contained in the interior. Figure 24 shows that the limaçon is convex if a is large and is not convex if a is small. Experiment with a computer algebra system to estimate the smallest value a for which it is convex.

(c) Use Eq. (3) of Exercise 51 to show that if the tangent line is vertical, then either $\theta = 0$, π or $\cos\theta = -a/2$.

(d) Show that the limaçon has three vertical tangent lines to the left of the y-axis if $1 < a < 2$ and a unique vertical tangent line to the left of the y-axis there if $a \geq 2$. What does this suggest about the convexity of the limaçon for $a > 2$?

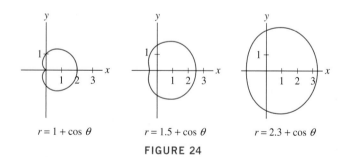

$r = 1 + \cos\theta$ $r = 1.5 + \cos\theta$ $r = 2.3 + \cos\theta$

FIGURE 24

11.4 Area and Arc Length in Polar Coordinates

In this section, we derive the formulas for area and arc length in polar coordinates. First, we compute the area of a sector bounded by a curve $r = f(\theta)$ and two rays $\theta = \alpha$ and $\theta = \beta$ [shaded region in Figure 1(A)]. Divide the region into N narrow sectors of angle

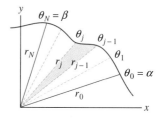

(A) Region defined by $\alpha \leq \theta \leq \beta$ (B) Region divided into narrow sectors

FIGURE 1 Area bounded by the curve $r = f(\theta)$ and the two rays $\theta = \alpha$ and $\theta = \beta$.

Rectangular and polar coordinates are suited for different kinds of area calculations. Rectangular coordinates should be used to compute the signed area under the graph $y = f(x)$ over an interval $[a, b]$. Polar coordinates should be used to calculate the area bounded by a curve in an angular sector $\alpha \leq \theta \leq \beta$.

$\Delta \theta = \frac{\beta - \alpha}{N}$ corresponding to a partition of the interval $[\alpha, \beta]$:

$$\theta_0 = \alpha \leq \theta_1 \leq \theta_2 \cdots \leq \theta_N = \beta$$

Recall that a circular sector of width $\Delta \theta$ and radius r has area $\frac{1}{2}r^2 \Delta \theta$ (Figure 2). Each narrow sector in our region is nearly circular, of radius $r_j = f(\theta_j)$, and therefore has area approximately equal to $\frac{1}{2}r_j^2 \Delta \theta$ (Figure 3). The total area is approximated by the sum:

$$\text{Area of region} \approx \sum_{j=1}^{N} \frac{1}{2}r_j^2 \Delta \theta = \sum_{j=1}^{N} \frac{1}{2}f(\theta_j)^2 \Delta \theta \qquad \boxed{1}$$

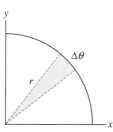

FIGURE 2 The area of a circular sector is *exactly* $\frac{1}{2}r^2 \Delta \theta$.

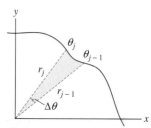

FIGURE 3 The area of the jth sector is *approximately* $\frac{1}{2}r_j^2 \Delta \theta$.

This is a Riemann sum for the integral $\int_{\alpha}^{\beta} \frac{1}{2}f(\theta)^2 \, d\theta$. If $f(\theta)$ is continuous, the sum approaches the integral as $N \to \infty$, and we obtain the following formula.

CAUTION In rectangular coordinates, the integral computes signed area, but in polar coordinates, formula (2) gives the actual (positive) area, not signed area. The formula remains valid if $f(\theta)$ takes on negative values.

THEOREM 1 Area in Polar Coordinates If $f(\theta)$ is continuous function, then the area bounded by a curve in polar form $r = f(\theta)$ and the rays $\theta = \alpha$ and $\theta = \beta$ is equal to

$$\frac{1}{2} \int_{\alpha}^{\beta} r^2 \, d\theta = \frac{1}{2} \int_{\alpha}^{\beta} f(\theta)^2 \, d\theta \qquad \boxed{2}$$

As a check on this formula, note that the graph of $f(r) = R$ is a circle of radius R. According to Eq. (2), the area of the circle is $\frac{1}{2} \int_{0}^{2\pi} R^2 \, d\theta = \frac{1}{2}R^2(2\pi) = \pi R^2$, as expected.

■ **EXAMPLE 1** Find the area of the right semicircle with equation $r = 4 \sin \theta$.

Solution The equation $r = 4 \sin \theta$ defines a circle of radius 2 tangent to the x-axis at the origin (Figure 4). The right semicircle is "swept out" as θ varies from 0 to $\frac{\pi}{2}$. By Eq. (2), the area of the semicircle is

$$\frac{1}{2} \int_0^{\pi/2} r^2 \, d\theta = \frac{1}{2} \int_0^{\pi/2} (4 \sin \theta)^2 \, d\theta = 8 \int_0^{\pi/2} \sin^2 \theta \, d\theta \qquad \boxed{4}$$

$$= 8 \int_0^{\pi/2} \frac{1}{2}(1 - \cos 2\theta) \, d\theta$$

$$= (4\theta - 2 \sin 2\theta) \Big|_0^{\pi/2} = 4 \left(\frac{\pi}{2} \right) - 0 = 2\pi$$

This is the expected result since a circle of radius $R = 2$ has total area 4π. ∎

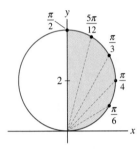

REMINDER In Eq. (4), we use the identity

$$\sin^2 \theta = \frac{1}{2}(1 - \cos 2\theta) \qquad \boxed{3}$$

to integrate $y = \sin^2 \theta$.

FIGURE 4 The circle with equation $r = 4 \sin \theta$.

CONCEPTUAL INSIGHT Keep in mind that the integral $\frac{1}{2} \int_\alpha^\beta r^2 \, d\theta$ does not compute the area *under* a curve, but rather the area "swept out" by a radial segment as θ varies from α to β. For example, integral $\frac{1}{2} \int_0^{\pi/3} (4 \sin \theta)^2 \, d\theta$ is equal to the area of the shaded region in Figure 5(A), not the area under the curve in Figure 5(B).

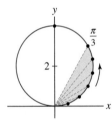 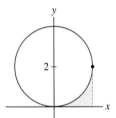

(A) Region swept out by radial segment has area
$$\frac{1}{2} \int_0^{\pi/3} r^2 \, d\theta$$

(B) Region under the curve.

FIGURE 5

■ **EXAMPLE 2** Sketch the graph of $r = \sin 3\theta$ and compute the area of one "petal."

Solution To sketch the curve, we first graph $r = \sin 3\theta$ in rectangular coordinates. Figure 6 shows that as θ varies from 0 to $\frac{\pi}{3}$, the radius r varies from 0 to 1, and then back to 0. This gives petal A (Figure 7). Petal B is traced as θ varies from $\frac{\pi}{3}$ to $\frac{2\pi}{3}$ (with $r \leq 0$), and petal C for $\frac{2\pi}{3} \leq \theta \leq \pi$. We compute the area of petal A using identity (3):

$$\frac{1}{2} \int_0^{\pi/3} (\sin 3\theta)^2 \, d\theta = \frac{1}{4} \int_0^{\pi/3} (1 - \cos 6\theta) \, d\theta = \left(\frac{1}{4}\theta - \frac{1}{24} \sin 6\theta \right) \Big|_0^{\pi/3} = \frac{\pi}{12} \quad ■$$

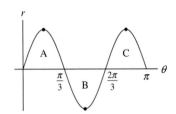

FIGURE 6 Graph of $r = \sin 3\theta$ as a function of θ.

If $r = f_1(\theta)$ and $r = f_2(\theta)$ are two polar curves with $f_2(\theta) \geq f_1(\theta)$, then we may compute the area of the region between the two curves within the sector $\alpha \leq \theta \leq \beta$ by the integral (Figure 8):

$$\boxed{\text{Area between two curves} = \frac{1}{2} \int_\alpha^\beta (f_2(\theta)^2 - f_1(\theta)^2) \, d\theta} \qquad \boxed{5}$$

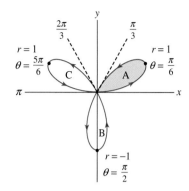

FIGURE 7 The "rose with three petals" $r = \sin 3\theta$.

FIGURE 8 Area between two polar graphs in a sector.

■ **EXAMPLE 3** Area Between Two Curves Find the area of the region inside the circle $r = 2\cos\theta$ but outside the circle $r = 1$ [Figure 9A].

Solution Recall that $r = 2a\cos\theta$ is the polar equation of the circle of radius a with center $(a, 0)$ (Section 11.3, Example 7). Thus $r = 2\cos\theta$ is a circle of radius 1 with center $(1, 0)$. We wish to compute the area of the region within this circle and outside the circle $r = 1$ of radius 1 centered at the origin. This is the region labeled (I) in Figure 9(A). The two circles intersect at the points where $r = 2\cos\theta = 1$ or $\cos\theta = \frac{1}{2}$. Thus the intersection points occur at $\theta = \pm\frac{\pi}{3}$. The area of region (I) is equal to

$$\text{Area of (I)} = \text{area of (II)} - \text{area of (III)}$$

⬅··· **REMINDER** In (6), we use the identity

$$\cos^2\theta = \frac{1}{2}(1 + \cos 2\theta)$$

$$= \frac{1}{2}\int_{-\pi/3}^{\pi/3}(2\cos\theta)^2\,d\theta - \frac{1}{2}\int_{-\pi/3}^{\pi/3}(1)^2\,d\theta$$

$$= \frac{1}{2}\int_{-\pi/3}^{\pi/3}(4\cos^2\theta - 1)\,d\theta = \frac{1}{2}\int_{-\pi/3}^{\pi/3}(2\cos 2\theta + 1)\,d\theta \qquad \boxed{6}$$

$$= \frac{1}{2}(\sin 2\theta + \theta)\Big|_{-\pi/3}^{\pi/3} = \frac{\sqrt{3}}{2} + \frac{\pi}{3} \approx 1.91 \qquad ■$$

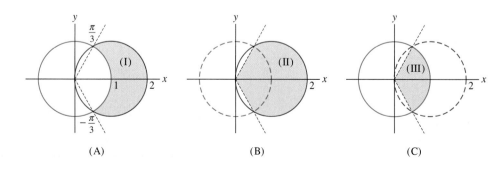

FIGURE 9 Region I is the difference of regions II and III.

(A)   (B)   (C)

To derive a formula for arc length in polar coordinates, we observe that a polar curve $r = f(\theta)$ has a natural parametrization with θ as a parameter:

$$x = r \cos \theta = f(\theta) \cos \theta, \qquad y = r \sin \theta = f(\theta) \sin \theta$$

Using a prime to denote the derivative with respect to θ, we have

$$x'(\theta) = \frac{dx}{d\theta} = -f(\theta) \sin \theta + f'(\theta) \cos \theta$$

$$y'(\theta) = \frac{dy}{d\theta} = f(\theta) \cos \theta + f'(\theta) \sin \theta$$

We check using algebra that $x'(\theta)^2 + y'(\theta)^2 = f(\theta)^2 + f'(\theta)^2$. Now recall from Section 11.2 that the length s of the arc traced out for $\alpha \leq \theta \leq \beta$ is $\int_\alpha^\beta \sqrt{x'(\theta)^2 + y'(\theta)^2}\, d\theta$. It follows that

$$\boxed{\text{Arc length } s = \int_\alpha^\beta \sqrt{f(\theta)^2 + f'(\theta)^2}\, d\theta} \qquad \boxed{7}$$

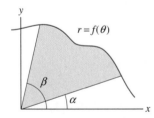

FIGURE 10 Graph of $r = 2a \cos \theta$.

■ **EXAMPLE 4** Find the total length of the circle with equation $r = 2a \cos \theta$ for $a > 0$.

Solution In this case, $f(\theta) = 2a \cos \theta$ and

$$f(\theta)^2 + f'(\theta)^2 = 4a^2 \cos^2 \theta + 4a^2 \sin^2 \theta = 4a^2$$

The total length of this circle of radius a has the expected value:

$$\int_0^\pi \sqrt{f(\theta)^2 + f'(\theta)^2}\, d\theta = \int_0^\pi (2a)\, d\theta = 2\pi a$$

Note that the upper limit of integration is π rather than 2π because the entire circle is traced out as θ varies from 0 to π (see Figure 10).   ■

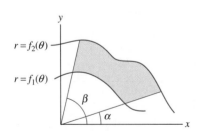

FIGURE 11 Region bounded by the polar curve $r = f(\theta)$ and the rays $\theta = \alpha$, $\theta = \beta$.

11.4 SUMMARY

• Polar coordinates are suited to calculating areas bounded by a polar curve $r = f(\theta)$ and two rays $\theta = \alpha$ and $\theta = \beta$ (Figure 11):

$$\text{Area} = \frac{1}{2} \int_\alpha^\beta f(\theta)^2\, d\theta$$

• The area between the polar curves $r = f_1(\theta)$ and $r = f_2(\theta)$, where $f_2(\theta) \geq f_1(\theta)$, is equal to (Figure 12)

$$\text{Area} = \frac{1}{2} \int_\alpha^\beta \left(f_2(\theta)^2 - f_1(\theta)^2 \right) d\theta$$

• The arc length of the polar curve $r = f(\theta)$ for $\alpha \leq \theta \leq \beta$ is

$$\text{Arc length} = \int_\alpha^\beta \sqrt{f(\theta)^2 + f'(\theta)^2}\, d\theta$$

FIGURE 12 Region between two polar curves.

11.4 EXERCISES

Preliminary Questions

1. True or False: The area under the curve with polar equation $r = f(\theta)$ is equal to the integral of $f(\theta)$.

2. Polar coordinates are best suited to finding the area (choose one):

(a) Under a curve between $x = a$ and $x = b$.

(b) Bounded by a curve and two rays through the origin.

3. True or False: The formula for area in polar coordinates is valid only if $f(\theta) \geq 0$.

4. The horizontal line $y = 1$ has polar equation $r = \csc \theta$. Which area is represented by the integral $\dfrac{1}{2} \displaystyle\int_{\pi/6}^{\pi/2} \csc^2 \theta \, d\theta$ (Figure 13)?

(a) $\square ABCD$ **(b)** $\triangle ABC$ **(c)** $\triangle ACD$

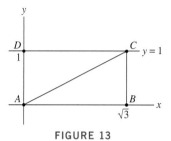

FIGURE 13

Exercises

1. Sketch the area bounded by the circle $r = 5$ and the rays $\theta = \frac{\pi}{2}$ and $\theta = \pi$, and compute its area as an integral in polar coordinates.

2. Sketch the region bounded by the line $r = \sec \theta$ and the rays $\theta = 0$ and $\theta = \frac{\pi}{3}$. Compute its area in two ways: as an integral in polar coordinates and using geometry.

3. Calculate the area of the circle $r = 4 \sin \theta$ as an integral in polar coordinates (see Figure 4). Be careful to choose the correct limits of integration.

4. Compute the area of the shaded region in Figure 14 as an integral in polar coordinates.

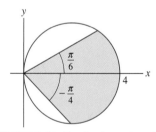

FIGURE 14 Graph of $r = 4 \cos \theta$.

5. Find the total area enclosed by the cardioid $r = 1 - \cos \theta$ (Figure 15).

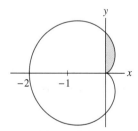

FIGURE 15 The cardioid $r = 1 - \cos \theta$.

6. Find the area of the shaded region in Figure 15.

7. Find the area of one leaf of the "four-petaled rose" $r = \sin 2\theta$ (Figure 16).

8. Prove that the total area of the four-petaled rose $r = \sin 2\theta$ is equal to one-half the area of the circumscribed circle (Figure 16).

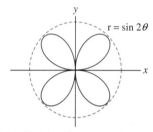

FIGURE 16 Graph of four-petaled rose $r = \sin 2\theta$.

9. Find the area enclosed by one loop of the lemniscate with equation $r^2 = \cos 2\theta$ (Figure 17). Choose your limits of integration carefully.

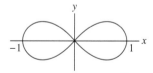

FIGURE 17 The lemniscate $r^2 = \cos 2\theta$.

10. Sketch the spiral $r = \theta$ for $0 \leq \theta \leq 2\pi$ and find the area bounded by the curve and the first quadrant.

11. Find the area enclosed by the cardioid $r = a(1 + \cos \theta)$, where $a > 0$.

12. Find the area of the intersection of the circles $r = \sin \theta$ and $r = \cos \theta$.

13. Find the area of region A in Figure 18.

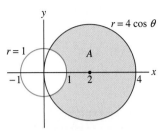

FIGURE 18

14. Find the area of the shaded region in Figure 19, enclosed by the circle $r = \frac{1}{2}$ and a petal of the curve $r = \cos 3\theta$. *Hint:* Compute the area of both the petal and the region inside the petal and outside the circle.

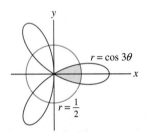

FIGURE 19

15. Find the area of the inner loop of the limaçon with polar equation $r = 2\cos\theta - 1$ (Figure 20).

16. Find the area of the region between the inner and outer loop of the limaçon $r = 2\cos\theta - 1$.

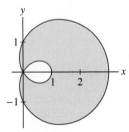

FIGURE 20   The limaçon $r = 2\cos\theta - 1$.

17. Find the area of the part of the circle $r = \sin\theta + \cos\theta$ in the fourth quadrant (see Exercise 27 in Section 11.3).

18. Compute the area of the shaded region in Figure 21.

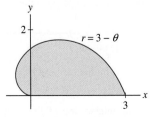

FIGURE 21

19. Find the area between the two curves in Figure 22(A).

20. Find the area between the two curves in Figure 22(B).

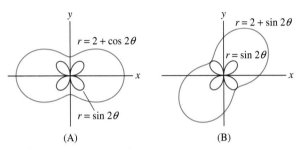

(A)   (B)

FIGURE 22

21. Find the area inside both curves in Figure 23.

22. Find the area of the region that lies inside one but not both of the curves in Figure 23.

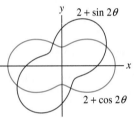

FIGURE 23

23. Figure 24 suggests that the circle $r = \sin\theta$ lies inside the spiral $r = \theta$. Which inequality from Chapter 2 assures us that this is the case? Find the area between the curves $r = \theta$ and $r = \sin\theta$ in the first quadrant.

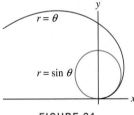

FIGURE 24

24. Calculate the total length of the circle $r = 4\sin\theta$ as an integral in polar coordinates.

25. Find the length of the spiral $r = \theta$ for $0 \le \theta \le A$.

26. Find the length of $r = \theta^2$ for $0 \le \theta \le \pi$.

27. Sketch the segment $r = \sec\theta$ for $0 \le \theta \le A$. Then compute its length in two ways: as an integral in polar coordinates and using trigonometry.

28. Sketch the circle $r = \sin\theta + \cos\theta$. Then compute its length in two ways: as an integral in polar coordinates and using trigonometry.

29. Find the length of the cardioid $r = 1 + \cos\theta$. *Hint:* Use the identity $1 + \cos\theta = 2\cos^2\left(\dfrac{\theta}{2}\right)$ to evaluate the arc length integral.

30. Find the length of the cardioid with equation $r = 1 + \cos\theta$ located in the first quadrant.

31. Find the length of the *equiangular spiral $r = e^\theta$* for $0 \le \theta \le 2\pi$.

32. Find the length of the curve $r = \cos^2\theta$.

In Exercises 33–36, express the length of the curve as an integral but do not evaluate it.

33. $r = e^{a\theta}, \quad 0 \le \theta \le \pi$

34. $r = \sin 2\theta, \quad 0 \le \theta \le \pi$

35. $r = (2 - \cos\theta)^{-1}, \quad 0 \le \theta \le 2\pi$

36. The inner loop of $r = 2\cos\theta - 1$ (see Exercise 15)

Further Insights and Challenges

37. Suppose that the polar coordinates of a moving particle at time t are $(r(t), \theta(t))$. Prove that the particle's speed is equal to
$$\sqrt{(dr/dt)^2 + r^2(d\theta/dt)^2}.$$

38. Compute the speed at time $t = 1$ of a particle whose polar coordinates at time t are $r = t$, $\theta = t$ (use Exercise 37). What would the speed be if the particle's rectangular coordinates are $x = t$, $y = t$? Why is the speed increasing in one case and constant in the other?

11.5 Conic Sections

The three familiar families of curves—ellipses, hyperbolas, and parabolas—appear throughout mathematics and its applications. They were first studied by the ancient Greek mathematicians, who recognized that they are obtained by intersecting a plane with a cone (Figure 1) and hence referred to them as "conic sections." Our main goal in this section is to derive the equations of the conic sections from their geometric definitions as curves in the plane.

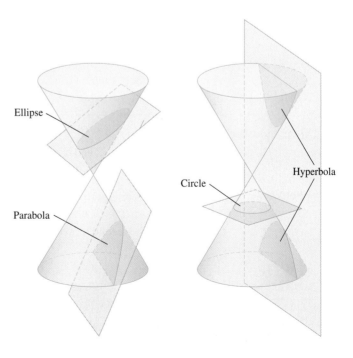

FIGURE 1 The conic sections are obtained by intersecting a plane and a cone.

An ellipse is the set of all points P such that the sum of the distances to two fixed points F_1 and F_2 is a constant $K > 0$:

$$PF_1 + PF_2 = K \qquad \boxed{1}$$

The points F_1 and F_2 are called the **foci** (plural of "focus") of the ellipse. We assume that K is greater than the distance $F_1 F_2$ between the foci (if $K = F_1 F_2$, the ellipse reduces to the line segment $\overline{F_1 F_2}$; if $K < F_1 F_2$, the ellipse has no points). Figure 2(A) shows an ellipse with foci at F_1 and F_2.

We use the following terminology:

- The midpoint of $\overline{F_1 F_2}$ is the **center** of the ellipse.
- The line through the foci is the **focal axis**.
- The line through the center perpendicular to the focal axis is the **conjugate axis**.

A circle is an ellipse whose foci coincide. If $F_1 = F_2$, Eq. (1) reduces to $PF_1 = K/2$, which defines a circle of radius $K/2$ and center F_1.

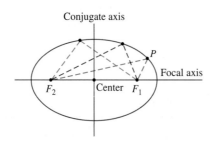

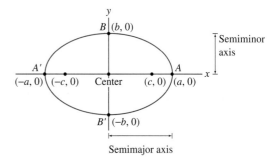

(A) The ellipse consists of all points P such that $PF_1 + PF_2 = $ constant.

(B) An ellipse in standard position with equation
$$\left(\frac{x}{a}\right)^2 + \left(\frac{y}{b}\right)^2 = 1.$$

FIGURE 2

The ellipse is said to be in **standard position** if the focal and conjugate axes are the x- and y-axes, as shown in Figure 2(B). In this case, the foci have coordinates $F_1 = (c, 0)$ and $F_2 = (-c, 0)$ for some $c > 0$. Let us prove that the equation of this ellipse has the particularly simple form

$$\left(\frac{x}{a}\right)^2 + \left(\frac{x}{b}\right)^2 = 1 \qquad \boxed{2}$$

where $a = K/2$ and $b = \sqrt{a^2 - c^2}$. By definition, a point $P = (x, y)$ lies on the ellipse if $PF_1 + PF_2 = 2a$ and therefore, by the distance formula,

$$\sqrt{(x + c)^2 + y^2} + \sqrt{(x - c)^2 + y^2} = 2a \qquad \boxed{3}$$

Move the second term on the left over to the right and square both sides:

$$(x + c)^2 + y^2 = 4a^2 - 4a\sqrt{(x - c)^2 + y^2} + (x - c)^2 + y^2$$

$$4a\sqrt{(x - c)^2 + y^2} = 4a^2 + (x - c)^2 - (x + c)^2 = 4a^2 - 4cx$$

Now divide by 4, square, and simplify:

Strictly speaking, we must also show that every point (x, y) satisfying (4) also satisfies (3). When we begin with (4) and reverse the algebraic steps, the process of taking square roots leads to the relation

$$\sqrt{(x - c)^2 + y^2} \pm \sqrt{(x + c)^2 + y^2} = \pm 2a$$

However, since $a > c$, this equation cannot hold unless both signs are positive.

$$a^2(x^2 - 2cx + c^2 + y^2) = a^4 - 2a^2cx + c^2x^2$$

$$(a^2 - c^2)x^2 + a^2y^2 = a^4 - a^2c^2 = a^2(a^2 - c^2)$$

$$\frac{x^2}{a^2} + \frac{y^2}{a^2 - c^2} = 1 \qquad \boxed{4}$$

This is Eq. (2) with $b^2 = a^2 - c^2$ as claimed.

The axes intersect the ellipse in four points called **vertices** [labeled A, A', B, and B' in Figure 2(B)]. The segments $\overline{AA'}$ and $\overline{BB'}$ are called the major and minor axes of the ellipse. Following common usage, the numbers a and b are referred to as the **semimajor axis** and **semiminor axis** (even though they are numbers rather than axes).

THEOREM 1 Equation of an Ellipse in Standard Position   Let $a, b > 0$ be constants with $a > b > 0$, and set $c = \sqrt{a^2 - b^2}$. Then

$$\left(\frac{x}{a}\right)^2 + \left(\frac{y}{b}\right)^2 = 1 \qquad \boxed{5}$$

is an equation of the ellipse

$$PF_1 + PF_2 = 2a$$

with foci $F_1 = (c, 0)$ and $F_2 = (-c, 0)$. Furthermore, the ellipse has:

- Semimajor axis a, semiminor axis b.
- Vertices $(\pm a, 0)$, $(0, \pm b)$.

If $b > a > 0$, then (5) defines an ellipse with foci $(0, \pm c)$, where $c = \sqrt{b^2 - a^2}$.

■ **EXAMPLE 1** Find the equation of the ellipse with foci $(\pm\sqrt{11}, 0)$ and semimajor axis $a = 6$. Then find the semiminor axis and sketch the graph.

Solution The foci are $(\pm c, 0)$ with $c = \sqrt{11}$ and the semimajor axis is $a = 6$, so we may use the relation $c = \sqrt{a^2 - b^2}$ to find b:

$$b^2 = a^2 - c^2 = 6^2 - (\sqrt{11})^2 = 25$$

Thus, the semiminor axis is $b = 5$ and the ellipse has equation $\left(\frac{x}{6}\right)^2 + \left(\frac{y}{5}\right)^2 = 1$. To sketch an ellipse, plot the vertices $(\pm 6, 0)$ and $(0, \pm 5)$ and connect them as in Figure 3. ■

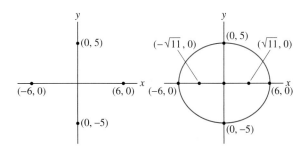

FIGURE 3 To sketch an ellipse, first draw the vertices.

We can easily write down the equation of an ellipse with axes parallel to the x- and y-axes and center translated to the point $C = (x_0, y_0)$ (Figure 4). The equation of the

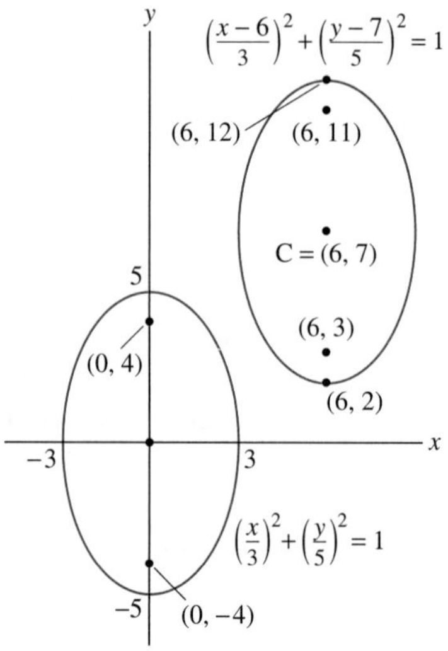

$$\left(\frac{x-6}{3}\right)^2 + \left(\frac{y-7}{5}\right)^2 = 1$$

(6, 12)   (6, 11)

$C = (6, 7)$

(6, 3)

(6, 2)

(0, 4)

$$\left(\frac{x}{3}\right)^2 + \left(\frac{y}{5}\right)^2 = 1$$

(0, −4)

FIGURE 4 An ellipse with vertical major axis and its translate with center $C = (6, 7)$.

translated ellipse is

$$\left(\frac{x - x_0}{a}\right)^2 + \left(\frac{y - y_0}{b}\right)^2 = 1$$

■ **EXAMPLE 2**   **Translating the Center of an Ellipse**   Find the equation of the ellipse with vertical focal axis, center at $C = (6, 7)$, semimajor axis 5, and semiminor axis 3. Where are the foci located?

Solution   First, we find the equation of the ellipse centered at the origin. Since the focal axis is vertical, we use Eq. (5) with $a < b$. We set $a = 3$ and $b = 5$ to obtain $\left(\frac{x}{3}\right)^2 + \left(\frac{y}{5}\right)^2 = 1$. Furthermore, $c = \sqrt{b^2 - a^2} = \sqrt{5^2 - 3^2} = 4$, so the foci are $F_1 = (0, 4)$ and $F_2 = (0, -4)$ on the y-axis. When we translate the ellipse so that its center is $(6, 7)$, the equation becomes

$$\left(\frac{x - 6}{3}\right)^2 + \left(\frac{y - 7}{5}\right)^2 = 1$$

The foci are still ± 4 vertical units from the center, so the foci of the translated ellipse are $F_1 = (6, 11)$ and $F_2 = (6, 3)$ (Figure 4). ■

A hyperbola is the set of all points P such that the difference of the distances to two fixed points F_1 and F_2 is $\pm K$, where $K > 0$ is a constant:

$$PF_1 - PF_2 = \pm K \qquad \boxed{6}$$

The points F_1 and F_2 are called the foci of the hyperbola. We assume that K is less than the distance $F_1 F_2$ between the foci. Note that a hyperbola consists of two branches corresponding to the choices of sign $\pm$ (Figure 5).

As before, the midpoint of $\overline{F_1 F_2}$ is the center of the hyperbola, the line through F_1 and F_2 is the focal axis, and the line through the center perpendicular to the focal axis is the conjugate axis. The vertices are the points where the focal axis intersects the hyperbola, labeled A and A' in Figure 5. The hyperbola is said to be in standard position

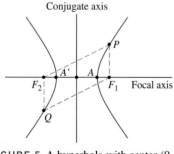

Conjugate axis

P

F_2 A' A F_1 Focal axis

Q

FIGURE 5 A hyperbola with center $(0, 0)$.

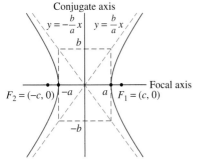

FIGURE 6 Hyperbola in standard position.

when the focal and conjugate axes are the x- and y-axes as in Figure 6. In this case, the foci are $(\pm c, 0)$ for some $c > 0$ and we can verify that the equation of the hyperbola has the simple form (7) below, with $a = K/2$ and $b = \sqrt{c^2 - a^2}$. The vertices are $A = (a, 0)$ and $A' = (-a, 0)$ (see Exercise 64).

THEOREM 2 Equation of a Hyperbola in Standard Position Let a, b be positive constants and set $c = \sqrt{a^2 + b^2}$. Then

$$\left(\frac{x}{a}\right)^2 - \left(\frac{y}{b}\right)^2 = 1 \qquad \boxed{7}$$

is the equation of the hyperbola

$$PF_1 - PF_2 = \pm 2a$$

with foci $F_1 = (c, 0)$ and $F_2 = (-c, 0)$.

A hyperbola has two **asymptotes** that may be determined by drawing the rectangle whose sides pass through $(\pm a, 0)$ and $(0, \pm b)$ as in Figure 6. Let us prove that the asymptotes are the diagonals of this rectangle, that is, the lines with equations $y = \pm \frac{b}{a} x$. Consider a point (x, y) on the hyperbola in the first quadrant. Thus we assume that $x > 0$ and $y > 0$. Equation (7) yields

$$y = \sqrt{\frac{b^2}{a^2} x^2 - b^2} = \frac{b}{a} \sqrt{x^2 - a^2}$$

To show that the point (x, y) on the hyperbola approaches the line $y = \frac{b}{a} x$ as $x \to \infty$, we verify that the following limit is zero:

$$\lim_{x \to \infty} \left(y - \frac{b}{a} x\right) = \frac{b}{a} \lim_{x \to \infty} \left(\sqrt{x^2 - a^2} - x\right)$$

$$= \frac{b}{a} \lim_{x \to \infty} \left(\sqrt{x^2 - a^2} - x\right) \left(\frac{\sqrt{x^2 - a^2} + x}{\sqrt{x^2 - a^2} + x}\right)$$

$$= \frac{b}{a} \lim_{x \to \infty} \left(\frac{-a^2}{\sqrt{x^2 - a^2} + x}\right) = 0$$

The asymptotic behavior in the remaining quadrants may be verified by a similar computation.

■ **EXAMPLE 3** Find the foci of the hyperbola $9x^2 - 4y^2 = 36$. Sketch its graph and asymptotes.

Solution First divide by 36 to write the equation in standard form:

$$\frac{x^2}{4} - \frac{y^2}{9} = 1 \qquad \text{or} \qquad \left(\frac{x}{2}\right)^2 - \left(\frac{y}{3}\right)^2 = 1$$

Thus, $a = 2$, $b = 3$, and $c = \sqrt{a^2 + b^2} = \sqrt{4 + 9} = \sqrt{13}$. The foci are $F_1 = (\sqrt{13}, 0)$ and $F_2 = (-\sqrt{13}, 0)$. To sketch the graph, we draw the rectangle through the points

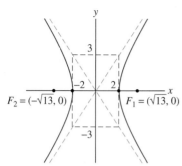

FIGURE 7 The hyperbola $9x^2 - 4y^2 = 36$.

$(\pm 2, 0)$ and $(0, \pm 3)$ as in Figure 7. The diagonals of the rectangle are the asymptotes $y = \pm \frac{3}{2}x$. We sketch the hyperbola so that it passes through the vertices $(\pm 2, 0)$ and approaches the asymptotes. ■

Finally, we consider the parabola. Unlike the ellipse and hyperbola, which are defined in terms of two foci, a parabola is the set of points P equidistant from a focus F and a line $\mathcal{D}$ called the **directrix**:

$$PF = P\mathcal{D} \qquad \boxed{8}$$

Here, when we speak of the *distance* from a point P to a line $\mathcal{D}$, we mean the distance from P to the point Q on $\mathcal{D}$ closest to P, obtained by dropping a perpendicular from P to $\mathcal{D}$ (Figure 8). We denote this distance by $P\mathcal{D}$.

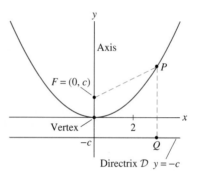

FIGURE 8 Parabola with focus $(0, c)$ and directrix $y = -c$.

The line through the focus F perpendicular to $\mathcal{D}$ is an axis of symmetry and is called the axis of the parabola. It intersects the parabola at a unique point called the vertex. We say that the parabola is in standard position if, for some c, the focus is $F = (0, c)$ and the directrix is $y = -c$, as shown in Figure 8. As we verify in Exercise 71, the vertex is then located at the origin and the equation of the parabola is $y = \dfrac{x^2}{4c}$. If $c < 0$, then the parabola opens downward.

> **THEOREM 3 Equation of a Parabola in Standard Position** Let $c \neq 0$. The parabola with focus $F = (0, c)$ and directrix $y = -c$ has equation
>
> $$y = \frac{1}{4c}x^2 \qquad \boxed{9}$$
>
> The vertex is located at the origin. If $c < 0$, then the parabola opens downward.

■ **EXAMPLE 4** Find the equation of the standard parabola with directrix $y = -2$. Where are its focus and vertex located? What is the equation if this parabola is translated so that its vertex is located at $(2, 4)$, and what are the directrix and focus in this case?

Solution We apply Eq. (9) with $c = 2$. The standard parabola with directrix $y = -2$ has equation $y = \dfrac{x^2}{8}$ (Figure 9). The focus is $(0, c) = (0, 2)$ and the vertex is the origin $(0, 0)$. Note that the focus is two units above the vertex.

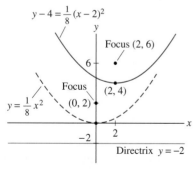

FIGURE 9 A parabola and its translate.

When the vertex is translated to (2, 4), the equation becomes

$$y - 4 = \frac{(x - 2)^2}{8} \qquad \text{or} \qquad y = \frac{1}{8}x^2 - \frac{1}{2}x + \frac{9}{2}$$

The focus F is still two units above the vertex, so now $F = (2, 6)$ (see Figure 9). The directrix is the horizontal line located two units below the vertex, namely $y = 2$. ∎

Eccentricity

Some ellipses are flatter than others, just as some hyperbolas are steeper. The "shape" of a conic section is measured by a quantity e called the **eccentricity**. For an ellipse or hyperbola,

$$e = \frac{\text{distance between foci}}{\text{distance between vertices on focal axis}}$$

A parabola is defined to have eccentricity $e = 1$.

The eccentricity of an ellipse or hyperbola in standard position is given by the formula $e = c/a$ in the notation used above. To check this, recall that for a standard ellipse with equation $\left(\frac{x}{a}\right)^2 + \left(\frac{y}{b}\right)^2 = 1$, where $a > b > 0$, we set $c = \sqrt{a^2 - b^2}$. The foci are located at $(\pm c, 0)$ and the vertices on the focal axis at $(\pm a, 0)$. Therefore,

$$e = \frac{\text{distance between foci}}{\text{distance between vertices on focal axis}} = \frac{2c}{2a} = \frac{c}{a}$$

Similarly, the standard hyperbola $\left(\frac{x}{a}\right)^2 - \left(\frac{y}{b}\right)^2 = 1$ has foci at $(\pm c, 0)$, where $c = \sqrt{a^2 + b^2}$, and the vertices on the focal axis at $(\pm a, 0)$. Again, we obtain $e = c/a$. In summary:

$$e = \frac{c}{a} \qquad \text{where} \quad c = \begin{cases} \sqrt{a^2 - b^2} & \text{for an ellipse} \\ \sqrt{a^2 + b^2} & \text{for a hyperbola} \end{cases} \qquad \boxed{10}$$

The eccentricity of an ellipse satisfies $0 \le e < 1$ since $c = \sqrt{a^2 - b^2} < a$. If $e = 0$, the ellipse is a circle because in this case, $a = b$. Circles are the "roundest" possible ellipses and, as we see in Figure 10(A), the larger the eccentricity, the flatter the ellipse. In fact, it is straightforward to check that, for an ellipse,

$$\frac{b}{a} = \sqrt{1 - e^2}$$

FIGURE 10 Effect of changing eccentricity.

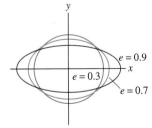

(A) Ellipse flattens as $e \to 1$.

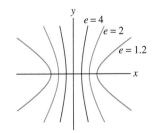

(B) Asymptotes of the hyperbola become more vertical as $e \to \infty$.

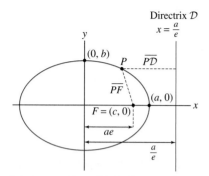

FIGURE 11 The ellipse consists of points P such that $PF = ePD$.

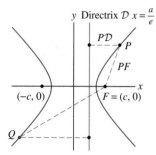

FIGURE 12 The hyperbola consists of points P such that $PF = ePD$.

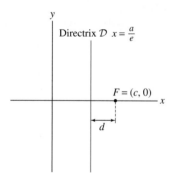

FIGURE 13

As $e \to 1$, the ratio b/a tends to zero. But b/a is the ratio of the semiminor axis to the semimajor axis, and thus a smaller value of b/a results in a flatter ellipse.

A hyperbola has eccentricity $e > 1$ since $c = \sqrt{a^2 + b^2} < a$. As e increases, the asymptotes of the hyperbola become steeper. We check this by observing that $b/a = \sqrt{1 + e^2}$. The asymptotes of the hyperbola have slopes $\pm b/a$ and thus the slopes of the asymptotes approach infinity as $e \to \infty$.

The notion of eccentricity may be used to give a unified focus-directrix definition of all three types of conics. Given a point F (the focus), a line $\mathcal{D}$ (the directrix), and a number $e > 0$, we consider the set of all points P such that

$$\boxed{PF = eP\mathcal{D}} \qquad \boxed{11}$$

For $e = 1$, this is precisely our definition of a parabola. According to the next theorem, (11) defines a conic section of eccentricity e for all $e > 0$ (Figures 11 and 12). Note, however, that there is no focus-directrix definition for circles ($e = 0$).

THEOREM 4 Focus-Directrix Definition Let F be a point, $\mathcal{D}$ a line, and let $e > 0$. Then the set of points satisfying (11) is a conic section of eccentricity e. Furthermore,

- **Ellipse:** Let $a > b > 0$ and $c = \sqrt{a^2 - b^2}$. The ellipse

$$\left(\frac{x}{a}\right)^2 + \left(\frac{y}{b}\right)^2 = 1$$

satisfies the focus-directrix definition (11) with focus $F = (c, 0)$, eccentricity $e = \dfrac{c}{a}$, and vertical directrix $x = \dfrac{a}{e}$.

- **Hyperbola:** Let $a, b > 0$ and $c = \sqrt{a^2 + b^2}$. The hyperbola

$$\left(\frac{x}{a}\right)^2 - \left(\frac{y}{b}\right)^2 = 1$$

satisfies the focus-directrix definition (11) with focus $F = (c, 0)$, eccentricity $e = \dfrac{c}{a}$, and vertical directrix $x = \dfrac{a}{e}$.

Proof We assume that $e > 1$ and prove that $PF = eP\mathcal{D}$ defines a hyperbola (the case $e < 1$ is similar, see Exercise 65). We may choose our coordinate axes so that the focus F lies on the x-axis and the directrix is vertical, lying to the left of F as in Figure 13. Anticipating the final result, we let d be the distance from the focus F to the directrix $\mathcal{D}$ and set

$$c = \frac{d}{1 - e^{-2}}, \quad a = \frac{c}{e}, \quad b = \sqrt{c^2 - a^2}$$

Since we are free to shift the y-axis, let us choose the y-axis so that the focus has coordinates $F = (c, 0)$. Then the directrix is the line

$$x = c - d = c - c(1 - e^{-2}) = c\,e^{-2} = \frac{a}{e}$$

Now, the equation $PF = eP\mathcal{D}$ for a point $P = (x, y)$ may be written

$$\underbrace{\sqrt{(x - c)^2 + y^2}}_{PF} = e\underbrace{\sqrt{(x - (a/e))^2}}_{P\mathcal{D}}$$

Algebraic manipulation yields

$$(x - c)^2 + y^2 = e^2(x - (a/e))^2 \qquad \text{(square)}$$

$$x^2 - 2cx + c^2 + y^2 = e^2x^2 - 2aex + a^2 \qquad \text{(note: } c = ae)$$

$$(e^2 - 1)x^2 - y^2 = c^2 - a^2 = a^2(e^2 - 1) \qquad \text{(rearrange)}$$

$$\frac{x^2}{a^2} - \frac{y^2}{a^2(e^2 - 1)} = 1 \qquad \text{(divide)}$$

Because $a^2(e^2 - 1) = c^2 - a^2 = b^2$, we obtain the equation of the hyperbola

$$\left(\frac{x}{a}\right)^2 - \left(\frac{y}{b}\right)^2 = 1$$

as claimed. ■

■ **EXAMPLE 5** Find the equation of the standard ellipse with eccentricity $e = 0.8$ and vertices $(\pm 10, 0)$. What are the foci and directrix?

Solution The vertices are $(\pm a, 0)$ with $a = 10$, so the ellipse has equation

$$\left(\frac{x}{10}\right)^2 + \left(\frac{y}{b}\right)^2 = 1$$

(Figure 14). By Theorem 4, $c = ae = 10(0.8) = 8$, and since $c = \sqrt{a^2 - b^2}$, we obtain $8 = \sqrt{10^2 - b^2}$ or $b = 6$. Thus, our ellipse has equation

$$\left(\frac{x}{10}\right)^2 + \left(\frac{y}{6}\right)^2 = 1$$

The foci are $(\pm c, 0) = (\pm 8, 0)$ and the directrix is $x = \dfrac{a}{e} = \dfrac{10}{0.8} = 12.5$. ■

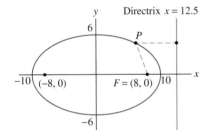

FIGURE 14 Ellipse of eccentricity $e = 0.8$ with focus at $(8, 0)$.

CONCEPTUAL INSIGHT We can prove that if two conic sections C_1 and C_2 have the same eccentricity e, then they have the same shape in the following precise sense: It is possible to scale C_1 so that C_1 and C_2 become congruent. By scaling, we mean changing the units along the x- and y-axes by a common positive factor. A curve scaled by a factor of 10 would have the same shape but would be ten times as large. This would correspond, for example, to changing units from centimeters to millimeters (smaller units make for a larger figure). On the other hand, by "congruent" we mean that after scaling, it is possible to move C_1 by a rigid motion (that is, without stretching or bending) so that it lies directly on top of C_2.

All circles ($e = 0$) have the same shape because scaling by a factor $r > 0$ transforms a circle of radius R into a circle of radius rR. Similarly, any two parabolas ($e = 1$) become congruent after suitable scaling. However, an ellipse of eccentricity $e = 0.5$ cannot be made congruent to an ellipse of eccentricity $e = 0.8$ by scaling (see Exercise 72).

In Section 13.6, we discuss the famous law of Johannes Kepler stating that the orbit of a planet around the sun is an ellipse with one focus at the sun. In this discussion, we will need to write the equation of an ellipse in polar coordinates. To derive the polar equations of the conic sections, it is convenient to use the focus-directrix definition with focus F at the origin O, and vertical line $x = d$ as directrix $\mathcal{D}$. By Eq. (11), the equation

FIGURE 15 Focus-directrix definition of the ellipse in polar coordinates.

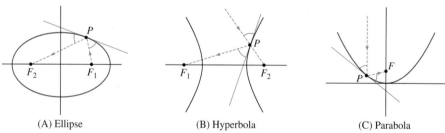

FIGURE 16 The paraboloid shape of this radio telescope directs the incoming signal to the focus.

of the ellipse is $PF = ePD$. Referring to Figure 15, we see that if $P = (r, \theta)$, then

$$PF = r, \qquad PD = d - r\cos\theta$$

Thus, the equation of the ellipse $PF = ePD$ becomes $r = e(d - r\cos\theta)$ or $r(1 + e\cos\theta) = ed$. This proves the following result, which is also valid for the hyperbola and parabola (see Exercise 66).

> **Polar Equation of a Conic Section** The conic section of eccentricity $e > 0$ with focus at the origin and directrix at $x = d$ has polar equation
>
> $$r = \frac{ed}{1 + e\cos\theta}$$
>
> **12**

Reflective Properties of Conic Sections

The conic sections have numerous geometric properties that are of importance in applications, especially in optics and communications (for example, in antenna and telescope design; Figure 16). Here, we limit ourselves to mentioning (without proof) the *reflective properties* of each family of conics. See Exercises 67–69 for a proof of the reflective property of an ellipse.

- **Ellipse:** The segments F_1P and F_2P make equal angles with the tangent line at a point P on the ellipse. Therefore, a beam of light originating at focus F_1 is reflected off the ellipse toward the second focus F_2 [Figure 17(A)].

(A) Ellipse   (B) Hyperbola   (C) Parabola

FIGURE 17

- **Hyperbola:** The tangent line at a point P on the hyperbola bisects the angle formed by the segments F_1P and F_2P. Therefore, a beam of light directed toward F_2 is reflected off the hyperbola toward the second focus F_1 [Figure 17(B)].
- **Parabola:** The segment FP and the line through P perpendicular to the directrix make equal angles with the tangent line at a point P on the parabola. Therefore, a beam of light approaching P from above, in the direction perpendicular to the directrix, is reflected off the parabola toward the focus F [Figure 17(C)].

General Equations of Degree Two

We now consider briefly the general equation of degree two in x and y:

$$Ax^2 + Bxy + Cy^2 + Dx + Ey + F = 0$$

13

where A, B, C, D, E, F are constants with A, B, C not all zero. The equations of the conic sections in standard position are special cases of this general quadratic equation. For example, the equation of a standard ellipse may be written in the form

$$Ax^2 + Cy^2 - 1 = 0$$

It turns out that the general quadratic equation does not give rise to any new types of curves. Apart from certain "degenerate cases," Eq. (13) defines a conic section with arbitrary center whose focal and conjugate axes may be rotated relative to the coordinate axes. For example, the equation

$$6x^2 - 8xy + 8y^2 - 12x - 24y + 38 = 0$$

defines an ellipse with center at $(3, 3)$ whose axes are rotated (Figure 18).

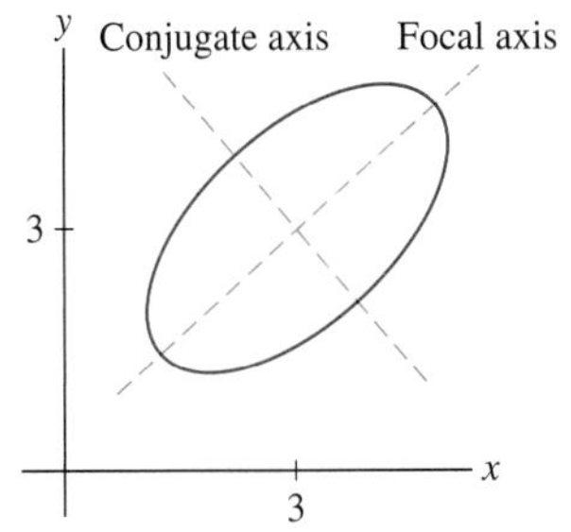

We say that Eq. (13) is degenerate if the set of solutions is a pair of intersecting lines, a pair of parallel lines, a single line, a point, or the empty set. For example:

- $x^2 - y^2 = 0$ defines a pair of intersecting lines $y = x$ and $y = -x$.
- $x^2 - x = 0$ defines a pair of parallel lines $x = 0$ and $x = 1$.
- $x^2 = 0$ defines a single line (the y-axis).
- $x^2 + y^2 = 0$ has just one solution $(0, 0)$.
- $x^2 + y^2 = -1$ has no solutions.

Now assume that Eq. (13) is nondegenerate. The term Bxy is called the *cross term*. When the cross term is zero (i.e., $B = 0$), we can "complete the square" to show that Eq. (13) defines a translate of conic in standard position. In other words, the axes of the conic are parallel to the coordinate axes. This is illustrated in the next example.

■ **EXAMPLE 6** Completing the Square Show that $4x^2 + 9y^2 + 24x - 72y + 144 = 0$ defines a translate of a conic section in standard position (Figure 19).

Solution Since there is no cross term, we may complete the square of the terms involving x and y terms separately:

$$4x^2 + 9y^2 + 24x - 72y + 144 = 4(x^2 + 6x) + 9(y^2 - 8y) + 144 = 0$$

$$4(x + 3)^2 - 36 + 9(y - 4)^2 - 144 + 144 = 0$$

$$4(x + 3)^2 + 9(y - 4)^2 = 36$$

Therefore, this quadratic equation can be rewritten in the form

$$\left(\frac{x + 3}{3}\right)^2 + \left(\frac{y - 4}{2}\right)^2 = 1$$ ■

When the cross-term Bxy is nonzero, Eq. (13) defines a conic whose axes are rotated relative to the coordinate axes. The marginal note describes how this may be verified in general. We illustrate with the following example.

■ **EXAMPLE 7** Show that $2xy = 1$ defines a conic section whose focal and conjugate axes are rotated relative to the coordinate axes.

Solution Figure 21(A) shows axes labeled x' and y' that are rotated by $45°$ relative to the standard coordinate axes. A point P with coordinates (x, y) may also be described by coordinates (x', y') relative to these rotated axes. Applying (14) and (15) with $\theta = \frac{\pi}{4}$, we find that (x, y) and (x', y') are related by the formulas

$$x = \frac{x' - y'}{\sqrt{2}}, \qquad y = \frac{x' + y'}{\sqrt{2}}$$

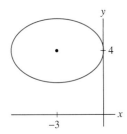

FIGURE 18 The ellipse with equation $6x^2 - 8xy + 8y^2 - 12x - 24y + 38 = 0$.

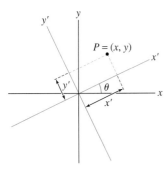

FIGURE 19 The ellipse with equation $4x^2 + 9y^2 + 24x - 72y + 144 = 0$.

FIGURE 20

More generally, if (x', y') are coordinates relative to axes rotated by an angle θ as in Figure 20, then

$$x = x' \cos\theta - y' \sin\theta \qquad \boxed{14}$$

$$y = x' \sin\theta + y' \cos\theta \qquad \boxed{15}$$

In Exercise 73, we show that the cross term disappears when Eq. (13) is rewritten in terms of x' and y' for the angle

$$\theta = \frac{1}{2} \cot^{-1} \frac{A - C}{B} \qquad \boxed{16}$$

Therefore, if $P = (x, y)$ lies on the hyperbola, that is, if $2xy = 1$, then

$$2xy = 2\left(\frac{x' - y'}{\sqrt{2}}\right)\left(\frac{x' + y'}{\sqrt{2}}\right) = x'^2 - y'^2 = 1$$

Thus, the coordinates (x', y') satisfy the equation of the standard hyperbola $x'^2 - y'^2 = 1$ whose focal and conjugate axes are the x'- and y'-axes, respectively. ∎

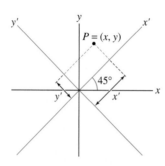

FIGURE 21 The x'- and y'-axes are rotated at a 45° angle relative to the x- and y-axes.

(A) The point $P = (x, y)$ may also be described by coordinates (x', y') relative to the rotated axis.

(B) The hyperbola $2xy = 1$ is in standard position relative to the x', y' axes.

We conclude our discussion of conics by stating the Discriminant Test. Suppose that the equation

$$Ax^2 + Bxy + Cy^2 + Dx + Ey + F = 0$$

is nondegenerate and thus defines a conic section. The type of conic is determined by the **discriminant** D (not to be confused with the coefficient D in the equation), defined by

$$D = B^2 - 4AC$$

According to the Discriminant Test (which we state without proof), the curve is

- an ellipse if $D < 0$.
- a hyperbola if $D > 0$.
- a parabola if $D = 0$.

For example, the discriminant of the equation $2xy = 1$ is

$$D = B^2 - 4AC = 2^2 - 0 = 4 > 0$$

According to the Discriminant Test, $2xy = 1$ defines a hyperbola. This agrees with our conclusion in Example 7.

11.5 SUMMARY

- An *ellipse* with foci F_1 and F_2 is the set of points P such that $PF_1 + PF_2 = K$, where $K > 0$ is a constant. The equation of an ellipse in standard position is

$$\left(\frac{x}{a}\right)^2 + \left(\frac{y}{b}\right)^2 = 1$$

	Focal Axis	Foci	Vertices
$a > b$	x-axis	$(\pm c, 0)$ with $c = \sqrt{a^2 - b^2}$	$(\pm a, 0), (0, \pm b)$
$a < b$	y-axis	$(0, \pm c)$ with $c = \sqrt{b^2 - a^2}$	$(\pm a, 0), (0, \pm b)$

- A *hyperbola* with foci F_1 and F_2 is the set of points P such that

$$PF_1 - PF_2 = \pm K$$

where $K > 0$ is a constant. The equation of a hyperbola in standard position is

$$\left(\frac{x}{a}\right)^2 - \left(\frac{y}{b}\right)^2 = 1$$

The focal axis is the x-axis and the foci are $(\pm c, 0)$ with $c = \sqrt{a^2 + b^2}$. The vertices are $(\pm a, 0)$ and the asymptotes are the lines $y = \pm \frac{b}{a}x$.

- A *parabola* with focus F and directrix $\mathcal{D}$ is the set of points P such that $PF = PD$. A parabola in standard position, with focus $F = (0, c)$ and directrix $y = -c$ with $c \neq 0$, has equation $y = \frac{1}{4c}x^2$. The vertex of a parabola in standard position is the origin $(0, 0)$.

- To translate a conic so that its center is located at (x_0, y_0) (while keeping its axes parallel to the x- and y-axes), replace x and y by $x - x_0$ and $y - y_0$, respectively, in the equation of the conic.

- The *eccentricity* e of an ellipse or a hyperbola is the quantity

$$e = \frac{\text{distance betweeen foci}}{\text{distance between vertices on focal axis}}$$

In standard position, $e = \frac{c}{a}$. For an ellipse, $0 \le e < 1$ (with $e = 0$ a circle); for a hyperbola, $e > 1$. By definition, a parabola has eccentricity $e = 1$.

- *Focus-directrix definition of conic sections:* Given a point P (a focus), a line $\mathcal{D}$ (a directrix), and a number $e > 0$, the set of points P such that $PF = ePD$ is a conic section of eccentricity e.

- The conic section of eccentricity $e > 0$ with focus at the origin and directrix $x = d$ has polar equation

$$r = \frac{ed}{1 + e \cos \theta}$$

11.5 EXERCISES

Preliminary Questions

1. Which of the following equations defines an ellipse? Which does not define a conic section?

(a) $4x^2 - 9y^2 = 12$
(b) $-4x + 9y^2 = 0$
(c) $4y^2 + 9x^2 = 12$
(d) $4x^3 + 9y^3 = 12$

2. For which conic sections do the vertices lie between the foci?

3. What are the foci of $\left(\frac{x}{a}\right)^2 + \left(\frac{y}{b}\right)^2 = 1$ if $a < b$?

4. For a hyperbola in standard position, the set of points equidistant from the foci is the y-axis. Use the definition $PF_1 - PF_2 = \pm K$ to explain why the hyperbola does not intersect the y-axis.

5. What is the geometric interpretation of the quantity $\frac{b}{a}$ in the equation of a hyperbola in standard position?

Exercises

In Exercises 1–8, find the vertices and foci of the conic section.

1. $\left(\frac{x}{9}\right)^2 + \left(\frac{y}{4}\right)^2 = 1$

2. $\left(\frac{x}{4}\right)^2 + \left(\frac{y}{9}\right)^2 = 1$

3. $\frac{x^2}{9} + \frac{y^2}{4} = 1$

4. $\left(\frac{x}{9}\right)^2 - \left(\frac{y}{4}\right)^2 = 1$

5. $\left(\frac{x}{4}\right)^2 - \left(\frac{y}{9}\right)^2 = 1$

6. $\frac{x^2}{9} - \frac{y^2}{4} = 1$

7. $\left(\frac{x-3}{7}\right)^2 - \left(\frac{y+1}{4}\right)^2 = 1$

8. $\left(\frac{x-3}{4}\right)^2 + \left(\frac{y+1}{7}\right)^2 = 1$

In Exercises 9–12, consider the ellipse

$$\left(\frac{x-12}{5}\right)^2 + \left(\frac{y-9}{7}\right)^2 = 1$$

Find the equation of the translated ellipse.

9. Translated so that its center is at the origin

10. Translated four units to the right

11. Translated three units down

12. Translated so its center is $(-4, -9)$

In Exercises 13–16, find the equation of the ellipse with the given properties.

13. Vertices at $(\pm 9, 0)$ and $(0, \pm 16)$

14. Foci $(\pm 6, 0)$ and two vertices at $(\pm 10, 0)$

15. Foci $(0, \pm 6)$ and two vertices at $(\pm 4, 0)$

16. Foci $(0, \pm 3)$ and eccentricity $\frac{3}{4}$

In Exercises 17–22, find the equation of the hyperbola with the given properties.

17. Vertices $(\pm 3, 0)$ and foci at $(\pm 5, 0)$

18. Vertices $(0, \pm 5)$ and foci $(0, \pm 8)$

19. Vertices $(\pm 4, 0)$ and asymptotes $y = \pm 3x$

20. Vertices $(\pm 3, 0)$ and asymptotes $y = \pm\frac{1}{2}x$

21. Vertices $(0, -5)$, $(0, 4)$ and foci $(0, -8)$, $(0, 7)$

22. Foci $(0, \pm 3)$ and eccentricity $\sqrt{2}$

In Exercises 23–30, find the equation of the parabola with the given properties.

23. Vertex $(0, 0)$, focus $(2, 0)$

24. Vertex $(0, 0)$, focus $(0, \frac{1}{2})$

25. Vertex $(0, 0)$, directrix $y = -5$

26. Vertex $(0, 0)$, directrix $y = -\frac{1}{8}$

27. Focus $(0, 4)$, directrix $y = -4$

28. Focus $(0, -4)$, directrix $y = 4$

29. Focus $(2, 0)$, directrix $x = -2$

30. Focus $(-2, 0)$, directrix $x = 2$

In Exercises 31–40, find the vertices, foci, axes, center (if an ellipse or a hyperbola) and asymptotes (if a hyperbola) of the conic section.

31. $x^2 + 4y^2 = 16$

32. $\left(\frac{x-3}{16}\right)^2 - \left(\frac{y+5}{49}\right)^2 = 1$

33. $4x^2 + y^2 = 16$

34. $3x^2 - 27y^2 = 12$

35. $4x^2 - 3y^2 + 8x + 30y = 215$

36. $y = 4x^2$

37. $y = 4(x-4)^2$

38. $8y^2 + 6x^2 - 36x - 64y + 134 = 0$

39. $4x^2 + 25y^2 - 8x - 10y = 20$

40. $y^2 - 2x^2 + 8x - 9y - 12 = 0$

In Exercises 41–44, use the Discriminant Test to determine the type of the conic section defined by the equation. You may assume that the equation is nondegenerate. Plot the curve if you have a computer algebra system.

41. $4x^2 + 5xy + 7y^2 = 24$

42. $2x^2 - 8xy + 3y^2 - 4 = 0$

43. $2x^2 - 8xy - 3y^2 - 4 = 0$

44. $x^2 - 2xy + y^2 + 24x - 8 = 0$

45. Show that $\dfrac{b}{a} = \sqrt{1 - e^2}$ for a standard ellipse of eccentricity e.

46. Show that the eccentricity of a hyperbola in standard position is $e = \sqrt{1 + m^2}$, where $\pm m$ are the slopes of the asymptotes.

47. Explain why the dots in Figure 22 lie on a parabola. Where are the focus and directrix located?

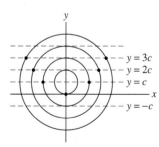

FIGURE 22

48. Show that the equation of the tangent line to the hyperbola $\left(\dfrac{x}{a}\right)^2 - \left(\dfrac{y}{b}\right)^2 = 1$ at a point (x_0, y_0) is

$$Ax - By = 1$$

where $A = \dfrac{x_0}{a^2}$ and $B = \dfrac{y_0}{b^2}$.

49. Kepler's First Law states that the orbits of the planets around the sun are ellipses with the sun at one focus. The orbit of Pluto has an eccentricity of approximately $e = 0.25$ and the **perihelion** (closest distance to the sun) of Pluto's orbit is approximately 2.7 billion miles. Find the **aphelion** (farthest distance from the sun).

50. Kepler's Third Law states that the ratio $T/a^{3/2}$ is equal to a constant C for all planetary orbits around the sun, where T is the period (time for a complete orbit) and a is the semimajor axis.

(a) Compute C in units of days and kilometers, given that the semimajor axis of the earth's orbit is 150×10^6 km.

(b) Compute the period of Saturn's orbit, given that its semimajor axis is approximately 1.43×10^9 km.

(c) Saturn's orbit has eccentricity $e = 0.056$. Find the perihelion and aphelion of Saturn.

In Exercises 51–54, find the polar equation of the conic with given eccentricity and directrix.

51. $e = \frac{1}{2}$, $\quad x = 3$

52. $e = \frac{1}{2}$, $\quad x = -3$

53. $e = 1$, $\quad x = 4$

54. $e = \frac{3}{2}$, $\quad x = -4$

In Exercises 55–58, identify the type of conic, the eccentricity, and the equation of the directrix.

55. $r = \dfrac{8}{1 + 4\cos\theta}$

56. $r = \dfrac{8}{4 + \cos\theta}$

57. $r = \dfrac{8}{4 + 3\cos\theta}$

58. $r = \dfrac{12}{4 + 3\cos\theta}$

59. Show that $r = f_1(\theta)$ and $r = f_2(\theta)$ define the same curves in polar coordinates if $f_1(\theta) = -f_2(\theta + \pi)$, and use this to show that the following define the same conic section:

$$r = \frac{de}{1 - e\cos\theta}, \qquad r = \frac{-de}{1 + e\cos\theta}$$

60. Find a polar equation for the hyperbola with focus at the origin, directrix $x = -2$, and eccentricity $e = 1.2$.

61. Find the equation of the ellipse $r = \dfrac{4}{2 + \cos\theta}$ in rectangular coordinates.

62. Let $\mathcal{C}$ be the ellipse $r = \dfrac{de}{1 + e\cos\theta}$, where $e < 1$. Express the x-coordinates of the vertices A, A', the center C, and the second focus F_2 in terms of d and e (Figure 23).

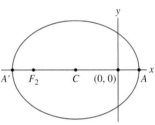

FIGURE 23

63. Let $e > 1$. Show that the vertices of the hyperbola $r = \dfrac{de}{1 + e\cos\theta}$ have x-coordinates $\dfrac{ed}{e + 1}$ and $\dfrac{ed}{e - 1}$.

Further Insights and Challenges

64. Verify Theorem 2.

65. Verify Theorem 4 in the case $0 < e < 1$. *Hint:* Repeat the proof of Theorem 4, but set $c = d/(e^{-2} - 1)$.

66. Verify that if $e > 1$, then Eq. (12) defines a hyperbola of eccentricity e, with its focus at the origin and directrix at $x = d$.

Reflective Property of the Ellipse   In Exercises 67–69, we prove that the focal radii at a point on an ellipse make equal angles with the tangent line. Let $P = (x_0, y_0)$ be a point on the ellipse $\left(\dfrac{x}{a}\right)^2 + \left(\dfrac{y}{b}\right)^2 = 1$ $(a > b)$ with foci $F_1 = (-c, 0)$ and $F_2 = (c, 0)$, and eccentricity e (Figure 24).

67. Show that $PF_1 = a + x_0e$ and $PF_2 = a - x_0e$. Hints:

(a) Show that $PF_1^2 - PF_2^2 = 4x_0c$.

(b) Divide the previous relation by $PF_1 + PF_2 = 2a$, and conclude that $PF_1 - PF_2 = 2x_0e$.

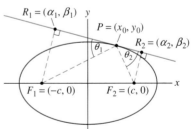

FIGURE 24 The ellipse $\left(\dfrac{x}{a}\right)^2 + \left(\dfrac{y}{b}\right)^2 = 1$.

68. Show that the equation of the tangent line at P is $Ax + By = 1$, where $A = \dfrac{x_0}{a^2}$ and $B = \dfrac{y_0}{b^2}$.

69. Define R_1 and R_2 as in the figure, so that $\overline{F_1 R_1}$ and $\overline{F_2 R_2}$ are perpendicular to the tangent line.

(a) Show that $\dfrac{\alpha_1 + c}{\beta_1} = \dfrac{\alpha_2 - c}{\beta_2} = \dfrac{A}{B}$.

(b) Use (a) and the distance formula to show that

$$\frac{F_1 R_1}{F_2 R_2} = \frac{\beta_1}{\beta_2}$$

(c) Solve for β_1 and β_2:

$$\beta_1 = \frac{B(1 + Ac)}{A^2 + B^2}, \qquad \beta_2 = \frac{B(1 - Ac)}{A^2 + B^2}$$

(d) Show that $\dfrac{F_1 R_1}{F_2 R_2} = \dfrac{P F_1}{P F_2}$. Conclude that $\theta_1 = \theta_2$.

70. Show that the length QR in Figure 25 is independent of the point P.

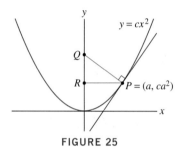

FIGURE 25

71. Show that $y = \dfrac{x^2}{4c}$ is the equation of a parabola with directrix $y = -c$, focus $(0, c)$, and the vertex at the origin, as stated in Theorem 3.

72. Consider two ellipses in standard position:

$$E_1 : \quad \left(\frac{x}{a_1}\right)^2 + \left(\frac{y}{b_1}\right)^2 = 1$$

$$E_2 : \quad \left(\frac{x}{a_2}\right)^2 + \left(\frac{y}{b_2}\right)^2 = 1$$

We say that E_1 is similar to E_2 under scaling if there exists a factor $r > 0$ such that for all (x, y) on E_1, the point (rx, ry) lies on E_2. Show that E_1 and E_2 are similar under scaling if and only if they have the same eccentricity. Show that any two circles are similar under scaling.

73. If we rewrite the general equation of degree two (13) in terms of variables x' and y' that are related to x and y by equations (14) and (15), we obtain a new equation of degree two in x' and y' of the same form but with different coefficients:

$$A'x^2 + B'xy + C'y^2 + D'x + E'y + F' = 0$$

(a) Show that $B' = B \cos 2\theta + (C - A) \sin 2\theta$.

(b) Show that if $B \neq 0$, then we obtain $B' = 0$ for

$$\theta = \frac{1}{2} \cot^{-1} \frac{A - C}{B}$$

This proves that it is always possible to eliminate the cross term Bxy by rotating the axes through a suitable angle.

CHAPTER REVIEW EXERCISES

1. Which of the following curves pass through the point $(1, 4)$?

(a) $c(t) = (t^2, t + 3)$ (b) $c(t) = (t^2, t - 3)$

(c) $c(t) = (t^2, 3 - t)$ (d) $c(t) = (t - 3, t^2)$

2. Find parametric equations for the line through $P = (2, 5)$ perpendicular to the line $y = 4x - 3$.

3. Find parametric equations for the circle of radius 2 with center $(1, 1)$. Use the equations to find the points of intersection of the circle with the x- and the y-axes.

4. Find a parametrization $c(t)$ of the line $y = 5 - 2x$ such that $c(0) = (2, 1)$.

5. Find a parametrization $c(\theta)$ of the unit circle such that $c(0) = (-1, 0)$.

6. Find a path $c(t)$ that traces the parabolic arc $y = x^2$ from $(0, 0)$ to $(3, 9)$ for $0 \leq t \leq 1$.

7. Find a path $c(t)$ that traces the line $y = 2x + 1$ from $(1, 3)$ to $(3, 7)$ for $0 \leq t \leq 1$.

8. Sketch the graph $c(t) = (1 + \cos t, \sin 2t)$ for $0 \leq t \leq 2\pi$ and draw arrows specifying the direction of motion.

In Exercises 9–12, express the parametric curve in the form $y = f(x)$.

9. $c(t) = (4t - 3, 10 - t)$ **10.** $c(t) = (t^3 + 1, t^2 - 4)$

11. $c(t) = \left(3 - \dfrac{2}{t}, t^3 + \dfrac{1}{t}\right)$ **12.** $x = \tan t, \quad y = \sec t$

13. Find all points visited twice by the path $c(t) = (t^2, \sin t)$. Plot $c(t)$ with a graphing utility.

In Exercises 14–17, calculate $\dfrac{dy}{dx}$ at the point indicated.

14. $c(t) = (t^3 + t, t^2 - 1), \quad t = 3$

15. $c(\theta) = (\tan^2 \theta, \cos \theta), \quad \theta = \frac{\pi}{4}$

16. $c(t) = (e^t - 1, \sin t), \quad t = 20$

17. $c(t) = (\ln t, 3t^2 - t), \quad P = (0, 2)$

18. $\boxed{CAS}$ Find the point on the cycloid $c(t) = (t - \sin t, 1 - \cos t)$ where the tangent line has slope $\frac{1}{2}$.

19. Find the points on $(t + \sin t, t - 2 \sin t)$ where the tangent is vertical or horizontal.

20. Find the equation of the Bézier curve with control points

$$P_0 = (-1, -1), \quad P_1 = (-1, 1), \quad P_2 = (1, 1), \quad P_3(1, -1)$$

21. Find the speed at $t = \frac{\pi}{4}$ of a particle whose position at time t seconds is $c(t) = (\sin 4t, \cos 3t)$.

22. Find the speed (as a function of t) of a particle whose position at time t seconds is $c(t) = (\sin t + t, \cos t + t)$. What is the particle's maximal speed?

23. Find the length of $(3e^t - 3, 4e^t + 7)$ for $0 \le t \le 1$.

In Exercises 24–25, let $c(t) = (e^{-t} \cos t, e^{-t} \sin t)$.

24. Show that $c(t)$ for $0 \le t < \infty$ has finite length and calculate its value.

25. Find the first positive value of t_0 such that the tangent line to $c(t_0)$ is vertical and calculate the speed at $t = t_0$.

26. $\boxed{CAS}$ Plot $c(t) = (\sin 2t, 2 \cos t)$ for $0 \le t \le \pi$. Express the length of the curve as a definite integral and approximate it using a computer algebra system.

27. Convert the points $(x, y) = (1, -3), (3, -1)$ from rectangular to polar coordinates.

28. Convert the points $(r, \theta) = \left(1, \frac{\pi}{6}\right), \left(3, \frac{5\pi}{4}\right)$ from polar to rectangular coordinates.

29. Write $(x + y)^2 = xy + 6$ as an equation in polar coordinates.

30. Write $r = \dfrac{2 \cos \theta}{\cos \theta - \sin \theta}$ as an equation in rectangular coordinates.

31. Show that $r = \dfrac{4}{7 \cos \theta - \sin \theta}$ is the polar equation of a line.

32. $\boxed{GU}$ Convert the equation

$$9(x^2 + y^2) = (x^2 + y^2 - 2y)^2$$

to polar coordinates and plot with a graphing utility.

33. Calculate the area of the circle $r = 3 \sin \theta$ bounded by the rays $\theta = \frac{\pi}{3}$ and $\theta = \frac{2\pi}{3}$.

34. $\boxed{GU}$ Plot the graph of $r = \sin 4\theta$ and calculate the area of one petal.

35. The equation $r = \sin(n\theta)$, where $n \ge 2$ is even, is a "rose" of $2n$ petals (Figure 1). Compute the total area of the flower and show that it does not depend on n.

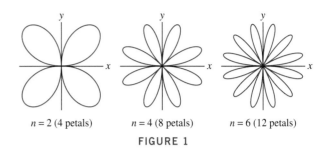

$n = 2$ (4 petals) $n = 4$ (8 petals) $n = 6$ (12 petals)

FIGURE 1

36. Calculate the total area enclosed by the curve $r^2 = \cos \theta e^{\sin \theta}$ (Figure 2).

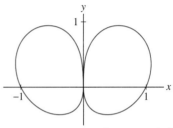

FIGURE 2 Graph of $r^2 = \cos \theta e^{\sin \theta}$.

37. Find the shaded area in Figure 3.

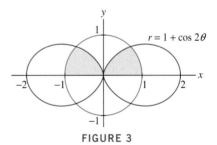

FIGURE 3

38. Calculate the length of the curve with polar equation $r = \theta$ in Figure 4.

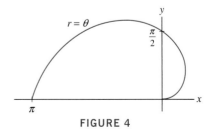

FIGURE 4

39. $\boxed{\textit{CAS}}$ Figure 5 shows the graph of $r = e^{0.5\theta} \sin\theta$ for $0 \le \theta \le 2\pi$. Use a CAS to approximate the difference in length between the outer and inner loops.

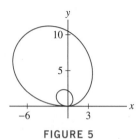

FIGURE 5

In Exercises 40–43, identify the conic section. Find the vertices and foci.

40. $\left(\dfrac{x}{3}\right)^2 + \left(\dfrac{y}{2}\right)^2 = 1$

41. $x^2 - 2y^2 = 4$

42. $(2x + \frac{1}{2}y)^2 = 4 - (x - y)^2$

43. $(y - 3)^2 = 2x^2 - 1$

44. Find the equation of a standard ellipse with two vertices at $(\pm 8, 0)$ and foci $(\pm\sqrt{3}, 0)$.

45. Find the equation of a standard hyperbola with vertices at $(\pm 8, 0)$ and asymptotes $y = \pm\dfrac{3}{4}x$.

46. Find the equation of a standard parabola with focus $(8, 0)$ and directrix $x = -8$.

47. Find the equation of a standard ellipse with foci at $(\pm 8, 0)$ and eccentricity $\frac{1}{8}$.

48. Find the asymptotes of the hyperbola $3x^2 + 6x - y^2 - 10y = 1$.

49. Show that the "conic section" with equation $x^2 - 4x + y^2 + 5 = 0$ has no points.

50. Show that the relation $\dfrac{dy}{dx} = (e^2 - 1)\dfrac{x}{y}$ holds on a standard ellipse or hyperbola of eccentricity e.

51. The orbit of Jupiter is an ellipse with the sun at a focus. Find the eccentricity of the orbit if the perihelion (closest distance to the sun) equals 740×10^6 km and the aphelion (farthest distance to the sun) equals 816×10^6 km.

52. Refer to Figure 24 in Section 11.5. Prove that the product of the perpendicular distances F_1R_1 and F_2R_2 from the foci to a tangent line of an ellipse is equal to the square b^2 of the semiminor axes.

A | THE LANGUAGE OF MATHEMATICS

One of the challenges in learning calculus is growing accustomed to its precise language and terminology, especially in the statements of theorems. In this section, we analyze a few details of logic that are helpful, and indeed essential, in understanding and applying theorems properly.

Many theorems in mathematics involve an **implication**. If A and B are statements, then the implication $A \implies B$ is the assertion that A implies B:

$$A \implies B: \qquad \text{If } A \text{ is true, then } B \text{ is true.}$$

Statement A is called the **hypothesis** (or premise) and statement B the **conclusion** of the implication. Here is an example: *If m and n are even integers, then $m + n$ is an even integer*. This statement may be divided into a hypothesis and conclusion:

$$\underbrace{m \text{ and } n \text{ are even integers}}_{A} \implies \underbrace{m + n \text{ is an even integer}}_{B}$$

In everyday speech, implications are often used in a less precise way. An example is: *If you work hard, then you will succeed.* Furthermore, some statements that do not initially have the form $A \implies B$ may be restated as implications. For example, the statement "Cats are mammals" can be rephrased:

$$\text{Let } X \text{ be an animal.} \quad \underbrace{X \text{ is a cat}}_{A} \implies \underbrace{X \text{ is a mammal}}_{B}$$

When we say that an implication $A \implies B$ is true, we do not claim that A or B is necessarily true. Rather, we are making the conditional statement that *if* A happens to be true, *then* B is also true. In the above, if X does not happen to be a cat, the implication tells us nothing.

The **negation** of a statement A is the assertion that A is false and is denoted $\neg A$.

Statement A	Negation $\neg A$
X lives in California.	X does not live in California.
$\triangle ABC$ is a right triangle.	$\triangle ABC$ is not a right triangle.

The negation of the negation is the original statement: $\neg(\neg A) = A$. To say that X does *not not live in California* is the same as saying that X *lives in California*.

■ **EXAMPLE 1** State the negation of:

(a) The door is open and the dog is barking.

(b) The door is open or the dog is barking (or both).

Solution

(a) The first statement is true if two conditions are satisfied (door open and dog barking), and it is false if at least one of these conditions is not satisfied. So the negation is

Either the door is not open *OR* the dog is not barking *(or both)*.

(b) The second statement is true if at least one of the conditions (door open or dog barking) is satisfied, and it is false if neither condition is satisfied. So the negation is

The door is not open AND the dog is not barking. ■

Contrapositive and Converse

Keep in mind that when we form the contrapositive, we reverse the order of A and B. The contrapositive of $A \implies B$ is NOT $\neg A \implies \neg B$.

Two important operations are the formation of the contrapositive and converse of a statement. The **contrapositive** of $A \implies B$ is the statement "If B is false, then A is false":

The contrapositive of $A \implies B$ is $\neg B \implies \neg A$.

Here are some examples:

Statement	Contrapositive
If X is a cat, then X is a mammal.	If X is not a mammal, then X is not a cat.
If you work hard, then you will succeed.	If you did not succeed, then you did not work hard.
If m and n are both even, then $m + n$ is even.	If $m + n$ is not even, then m and n are not both even.

A key observation is this:

The contrapositive and the original implication are equivalent.

The fact that $A \implies B$ is equivalent to its contrapositive is a general rule of logic that does not depend on what A and B happen to mean. This rule belongs to the subject of "formal logic," which deals with logical relations between statements without concern for the actual content of these statements.

In other words, if an implication is true, then its contrapositive is automatically true and vice versa. In essence, an implication and its contrapositive are two ways of saying the same thing. For example, the contrapositive "If X is not a mammal, then X is not a cat" is a roundabout way of saying that cats are mammals.

The **converse** of $A \implies B$ is the *reverse* implication $B \implies A$:

Implication: $A \implies B$	Converse $B \implies A$
If A is true, then B is true.	If B is true, then A is true.

The converse plays a very different role than the contrapositive because *the converse is NOT equivalent to the original implication.* The converse may be true or false, even if the original implication is true. Here are some examples:

True Statement	Converse	Converse True or False?
If X is a cat, then X is a mammal.	If X is a mammal, then X is a cat.	False
If m is even, then m^2 is even.	If m^2 is even, then m is even.	True

■ **EXAMPLE 2**   An Example Where the Converse Is False   Show that the converse of "If m and n are even, then $m + n$ is even" is false.

A counterexample is an example that satisfies the hypothesis but not the conclusion of a statement. If a single counterexample exists, then the statement is false. However, we cannot prove that a statement is true merely by giving an example.

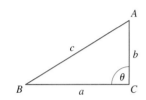

FIGURE 1

Solution The converse is "If $m + n$ is even, then m and n are even." To show that the converse is false, we display a counterexample. Take $m = 1$ and $n = 3$ (or any other pair of odd numbers). The sum is even (since $1 + 3 = 4$) but neither 1 nor 3 is even. Therefore, the converse is false. ■

■ **EXAMPLE 3** An Example Where the Converse Is True State the contrapositive and converse of the Pythagorean Theorem. Are either or both of these true?

Solution Consider a triangle with sides a, b, and c, and let θ be the angle opposite the side of length c as in Figure 1. The Pythagorean Theorem states that if $\theta = 90°$, then $a^2 + b^2 = c^2$. Here are the contrapositive and converse:

Pythagorean Theorem	$\theta = 90° \Longrightarrow a^2 + b^2 = c^2$	True
Contrapositive	$a^2 + b^2 \neq c^2 \Longrightarrow \theta \neq 90°$	Automatically true
Converse	$a^2 + b^2 = c^2 \Longrightarrow \theta = 90°$	True (but not automatic)

The contrapositive is automatically true because it is just another way of stating the original theorem. The converse is not automatically true since there could conceivably exist a nonright triangle that satisfies $a^2 + b^2 = c^2$. However, the converse of the Pythagorean Theorem is, in fact, true. This follows from the Law of Cosines (see Exercise 38). ■

When both a statement $A \Longrightarrow B$ and its converse $B \Longrightarrow A$ are true, we write $A \Longleftrightarrow B$. In this case, A and B are **equivalent**. We often express this with the phrase

$$A \Longleftrightarrow B \qquad A \text{ is true } \textit{if and only if } B \text{ is true.}$$

For example,

$$a^2 + b^2 = c^2 \qquad \text{if and only if} \qquad \theta = 90°$$

$$\text{It is morning} \qquad \text{if and only if} \qquad \text{the sun is rising.}$$

We mention the following variations of terminology involving implications that you may come across:

Statement	Is Another Way of Saying
A is true <u>if</u> B is true.	$B \Longrightarrow A$
A is true <u>only if</u> B is true.	$A \Longrightarrow B$ (A cannot be true unless B is also true.)
For A to be true, <u>it is necessary</u> that B be true.	$A \Longrightarrow B$ (A cannot be true unless B is also true.)
For A to be true, <u>it is sufficient</u> that B be true.	$B \Longrightarrow A$
For A to be true, it is <u>necessary and sufficient</u> that B be true.	$B \Longleftrightarrow A$

Analyzing a Theorem

To see how these rules of logic arise in the study of calculus, consider the following result from Section 4.2.

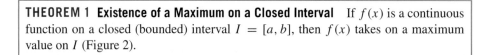

THEOREM 1 Existence of a Maximum on a Closed Interval   If $f(x)$ is a continuous function on a closed (bounded) interval $I = [a, b]$, then $f(x)$ takes on a maximum value on I (Figure 2).

To analyze this theorem, let's write out the hypotheses and conclusion separately:

Hypotheses A: $f(x)$ is continuous and I is closed.

Conclusion B: $f(x)$ takes on a maximum value on I.

A first question to ask is: "Are the hypotheses necessary?" Is the conclusion still true if we drop one or both assumptions? To show that both hypotheses are necessary, we provide the counterexamples:

- **The continuity of $f(x)$ is a necessary hypothesis.** Figure 3(A) shows the graph of a function on a closed interval $[a, b]$ that is not continuous. This function has no maximum value on $[a, b]$, which shows that the conclusion may fail if the continuity hypothesis is not satisfied.
- **The hypothesis that I is closed is necessary.** Figure 3(B) shows the graph of a continuous function on an *open interval* (a, b). This function has no maximum value, which shows that the conclusion may fail if the interval is not closed.

We see that both hypotheses in Theorem 1 are necessary. In stating this, we do not claim that the conclusion *always* fails when one or both of the hypotheses are not satisfied. We claim only that the conclusion *may* fail when the hypotheses are not satisfied. Next, let's analyze the contrapositive and converse:

- **Contrapositive $\neg B \Rightarrow \neg A$ (automatically true):** If $f(x)$ does not have a maximum value on I, then either $f(x)$ is not continuous or I is not closed (or both).
- **Converse $B \Rightarrow A$ (in this case, false):** If $f(x)$ has a maximum value on I, then $f(x)$ is continuous and I is closed.

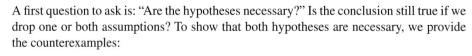

FIGURE 2 is shown at the left of the page:

FIGURE 2 A continuous function on a closed interval $I = [a, b]$ has a maximum value.

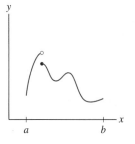

(A) The interval is closed but the function is not continuous. The function has no maximum value.

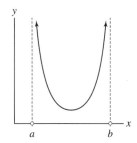

(B) The function is continuous but the interval is open. The function has no maximum value.

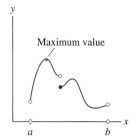

(C) This function is not continuous and the interval is not closed, but the function does have a maximum value.

FIGURE 3

The technique of proof by contradiction *is also known by its Latin name* reductio ad absurdum *or "reduction to the absurd." The ancient Greek mathematicians used proof by contradiction as early as the fifth century* BC, *and Euclid (325–265* BC*) employed it in his classic treatise on geometry entitled* The Elements. *A famous example is the proof that* $\sqrt{2}$ *is irrational in Example 4. The philosopher Plato (427–347* BC*) wrote: "He is unworthy of the name of man who is ignorant of the fact that the diagonal of a square is incommensurable with its side."*

FIGURE 4 The diagonal of the unit square has length $\sqrt{2}$.

One of the most famous problems in mathematics is known as "Fermat's Last Theorem." It states that the equation

$$x^n + y^n = z^n$$

has no solutions in positive integers if $n \geq 3$. In a marginal note written around 1630, Fermat claimed to have a proof, and over the centuries, that assertion was verified for many values of the exponent n. However, only in 1994 did the British-American mathematician Andrew Wiles, working at Princeton University, find a complete proof.

As we know, the contrapositive is merely a way of restating the theorem, so it is automatically true. The converse is not automatically true, and in fact, in this case it is false. The function in Figure 3(C) provides a counterexample to the converse: $f(x)$ has a maximum value on $I = (a, b)$, but $f(x)$ is not continuous and I is not closed.

Mathematicians have devised various general strategies and methods for proving theorems. The method of proof by induction is discussed in Appendix C. Another important method is **proof by contradiction**, also called **indirect proof**. Suppose our goal is to prove that $A \implies B$. In a proof by contradiction, we start by assuming that A is true but B is false, and we then show that this leads to a contradiction. Therefore, B must be true (to avoid the contradiction).

■ **EXAMPLE 4** Proof by Contradiction   The number $\sqrt{2}$ is irrational (Figure 4).

Solution   Assume that the theorem is false, namely that $\sqrt{2} = p/q$, where p and q are whole numbers. We may assume that p/q is in lowest terms and, therefore, at most one of p and q is even.

The relation $\sqrt{2} = p/q$ implies that $2 = p^2/q^2$ or $p^2 = 2q^2$. This shows that p must be even. But if p is even, say, $p = 2m$, then $4m^2 = 2q^2$ or $q^2 = 2m^2$. This shows that q is also even. This contradicts our assumption that p/q is in lowest terms. We conclude that our original assumption, that $\sqrt{2} = p/q$, must be false. Therefore, $\sqrt{2}$ is irrational.

■

CONCEPTUAL INSIGHT   The hallmark of mathematics is precision and rigor. A theorem is established, not through observation or experimentation, but by a proof that consists of a chain of reasoning with no gaps.

This approach to mathematics comes down to us from the ancient Greek mathematicians, especially Euclid, and it remains the standard in contemporary research. In recent decades, the computer has become a powerful tool for mathematical experimentation and data analysis. Researchers may use experimental data to discover potential new mathematical facts, but the title "theorem" is not bestowed until someone writes down a proof.

This insistence on theorems and proofs distinguishes mathematics from the other sciences. In the natural sciences, facts are established through experiment and are subject to change or modification as more knowledge is acquired. In mathematics, theories are also developed and expanded, but previous results are not invalidated. The Pythagorean Theorem was discovered in antiquity and is a cornerstone of plane geometry. In the nineteenth century, mathematicians began to study more general types of geometry (of the type that eventually led to Einstein's four-dimensional space-time geometry in the Theory of Relativity). The Pythagorean Theorem does not hold in these more general geometries, but its status in plane geometry is unchanged.

A. SUMMARY

- The implication $A \implies B$ is the assertion "If A is true, then B is true."
- The *contrapositive* of $A \implies B$ is the implication $\neg B \implies \neg A$, which says "If B is false, then A is false." An implication and its contrapositive are equivalent (one is true if and only if the other is true).
- The *converse* of $A \implies B$ is $B \implies A$. An implication and its converse are not necessarily equivalent. One may be true and the other false.
- A and B are *equivalent* if $A \implies B$ and $B \implies A$ are both true.

• In a proof by contradiction (in which the goal is to prove statement A), we start by assuming that A is false and show that this assumption leads to a contradiction.

A. EXERCISES

Preliminary Questions

1. Which is the contrapositive of $A \implies B$?

(a) $B \implies A$ **(b)** $\neg B \implies A$

(c) $\neg B \implies \neg A$ **(d)** $\neg A \implies \neg B$

2. Which of the choices in Question 1 is the converse of $A \implies B$?

3. Suppose that $A \implies B$ is true. Which is then automatically true, the converse or the contrapositive?

4. Restate as an implication: "A triangle is a polygon."

Exercises

1. Which is the negation of the statement "The car and the shirt are both blue"?

(a) Neither the car nor the shirt is blue.

(b) The car is not blue and/or the shirt is not blue.

2. Which is the contrapositive of the implication "If the car has gas, then it will run"?

(a) If the car has no gas, then it will not run.

(b) If the car will not run, then it has no gas.

In Exercises 3–8, state the negation.

3. The time is 4 o'clock.

4. $\triangle ABC$ is an isosceles triangle.

5. m and n are odd integers.

6. Either m is odd or n is odd.

7. x is a real number and y is an integer.

8. $f(x)$ is a linear function.

In Exercises 9–14, state the contrapositive and converse.

9. If m and n are odd integers, then mn is odd.

10. If today is Tuesday, then we are in Belgium.

11. If today is Tuesday, then we are not in Belgium.

12. If $x > 4$, then $x^2 > 16$.

13. If m^2 is divisible by 3, then m is divisible by 3.

14. If $x^2 = 2$, then x is irrational.

In Exercise 15–18, give a counterexample to show that the converse of the statement is false.

15. If m is odd, then $2m + 1$ is also odd.

16. If $\triangle ABC$ is equilateral, then it is an isosceles triangle.

17. If m is divisible by 9 and 4, then m is divisible by 12.

18. If m is odd, then $m^3 - m$ is divisible by 3.

In Exercise 19–22, determine whether the converse of the statement is false.

19. If $x > 4$ and $y > 4$, then $x + y > 8$.

20. If $x > 4$, then $x^2 > 16$.

21. If $|x| > 4$, then $x^2 > 16$.

22. If m and n are even, then mn is even.

In Exercises 23–24, state the contrapositive and converse (it is not necessary to know what these statements mean).

23. If $f(x)$ and $g(x)$ are differentiable, then $f(x)g(x)$ is differentiable.

24. If the force field is radial and decreases as the inverse square of the distance, then all closed orbits are ellipses.

*In Exercises 25–28, the **inverse** of $A \implies B$ is the implication $\neg A \implies \neg B$.*

25. Which of the following is the inverse of the implication "If she jumped in the lake, then she got wet"?

(a) If she did not get wet, then she did not jump in the lake.

(b) If she did not jump in the lake, then she did not get wet.

Is the inverse true?

26. State the inverses of these implications:

(a) If X is a mouse, then X is a rodent.

(b) If you sleep late, you will miss class.

(c) If a star revolves around the sun, then it's a planet.

27. Explain why the inverse is equivalent to the converse.

28. State the inverse of the Pythagorean Theorem. Is it true?

29. Theorem 1 in Section 2.4 states the following: "If $f(x)$ and $g(x)$ are continuous functions, then $f(x) + g(x)$ is continuous." Does it follow logically that if $f(x)$ and $g(x)$ are not continuous, then $f(x) + g(x)$ is not continuous?

30. Write out a proof by contradiction for this fact: There is no smallest positive rational number. Base your proof on the fact that if $r > 0$, then $0 < r/2 < r$.

31. Use proof by contradiction to prove that if $x + y > 2$, then $x > 1$ or $y > 1$ (or both).

In Exercises 32–35, use proof by contradiction to show that the number is irrational.

32. $\sqrt{\dfrac{1}{2}}$ **33.** $\sqrt{3}$ **34.** $\sqrt[3]{2}$ **35.** $\sqrt[4]{11}$

36. An isosceles triangle is a triangle with two equal sides. The following theorem holds: If $\triangle$ is a triangle with two equal angles, then $\triangle$ is an isosceles triangle.

(a) What is the hypothesis?

(b) Show by providing a counterexample that the hypothesis is necessary.

(c) What is the contrapositive?

(d) What is the converse? Is it true?

37. Consider the following theorem: Let $f(x)$ be a quadratic polynomial with a positive leading coefficient. Then $f(x)$ has a minimum value.

(a) What are the hypotheses?

(b) What is the contrapositive?

(c) What is the converse? Is it true?

Further Insights and Challenges

38. Let a, b, and c be the sides of a triangle and let θ be the angle opposite c. Use the Law of Cosines (Theorem 1 in Section 1.4) to prove the converse of the Pythagorean Theorem.

39. Carry out the details of the following proof by contradiction, due to R. Palais, that $\sqrt{2}$ is irrational. If $\sqrt{2}$ is rational, then $n\sqrt{2}$ is a whole number for some whole number n. Let n be the smallest such whole number and let $m = n\sqrt{2} - n$.

(a) Prove that $m < n$.

(b) Prove that $m\sqrt{2}$ is a whole number.

Explain why (a) and (b) imply that $\sqrt{2}$ is irrational.

40. Generalize the argument of Exercise 39 to prove that $\sqrt{A}$ is irrational if A is a whole number but not a perfect square. *Hint:* Choose n

as before and let $m = n\sqrt{A} - n[\sqrt{A}]$, where $[x]$ is the greatest integer function.

41. Generalize further and show that for any whole number r, the rth root $\sqrt[r]{A}$ is irrational unless A is an rth power. *Hint:* Let $x = \sqrt[r]{A}$. Show that if x is rational, then we may choose a smallest whole number n such that nx^j is a whole number for $j = 1, \ldots, r - 1$. Then consider $m = nx - n[x]$ as before.

42. Given a finite list of prime numbers $p_1, \ldots, p_N$, let $M = p_1 \cdot p_2 \cdots p_N + 1$. Show that M is not divisible by any of the primes $p_1, \ldots, p_N$. Use this and the fact that every number has a prime factorization to prove that there exist infinitely many prime numbers. This argument was advanced by Euclid in *The Elements*.

B | PROPERTIES OF REAL NUMBERS

"The ingenious method of expressing every possible number using a set of ten symbols (each symbol having a place value and an absolute value) emerged in India. The idea seems so simple nowadays that its significance and profound importance is no longer appreciated. Its simplicity lies in the way it facilitated calculation and placed arithmetic foremost amongst useful inventions. The importance of this invention is more readily appreciated when one considers that it was beyond the two greatest men of Antiquity, Archimedes and Apollonius."

—Pierre-Simon Laplace,
one of the great French mathematicians
of the eighteenth century

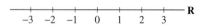

FIGURE 1 The real number line.

In this appendix, we discuss the basic properties of real numbers. First, let us recall that a real number is a number that may be represented by a finite or infinite decimal expansion. The set of all real numbers is denoted **R** and is often visualized as the "number line" (Figure 1). The two algebraic operations of addition and multiplication are defined on **R** and satisfy the Commutative, Associative, and Distributive Laws (Table 1).

TABLE 1   Algebraic Laws

Commutative Laws:	$a + b = b + a, \quad ab = ba$
Associative Laws:	$(a + b) + c = a + (b + c), \quad (ab)c = a(bc)$
Distributive Law:	$a(b + c) = ab + ac$

Every real number x has an additive inverse $-x$ such that $x + (-x) = 0$ and every nonzero real number x has a multiplicative inverse x^{-1} such that $x(x^{-1}) = 1$. We do not regard subtraction and division as separate algebraic operations because they are defined in terms of inverses. By definition, the difference $x - y$ is equal to $x + (-y)$ and the quotient x/y is equal to $x(y^{-1})$ for $y \neq 0$.

In addition to the algebraic operations, there is an **order relation** on **R**: For any two real numbers a and b, precisely one of the following is true:

$$\text{Either} \quad a = b, \quad \text{or} \quad a < b, \quad \text{or} \quad a > b$$

To distinguish between the conditions $a \leq b$ and $a < b$, we often refer to $a < b$ as a **strict inequality**. Similar conventions hold for $>$ and $\geq$. The rules given in Table 2 allow us to manipulate inequalities.

TABLE 2   Order Properties

If $a < b$ and $b < c$,	then $a < c$.
If $a < b$ and $c < d$,	then $a + c < b + d$.
If $a < b$ and $c > 0$,	then $ac < bc$.
If $a < b$ and $c < 0$,	then $ac > bc$.

The last order property says that an inequality reverses direction when multiplied by a negative number c. For example,

$$-2 < 5 \quad \text{but} \quad (-3)(-2) > (-3)5$$

The algebraic and order properties of real numbers are certainly familiar. We now discuss the less familiar **Least Upper Bound (LUB) Property** of the real numbers. This property is one way of expressing the so-called **completeness** of the real numbers. There are other ways of formulating completeness (such as the so-called nested interval property discussed in any book on analysis) that are equivalent to the LUB Property and serve the same purpose. Completeness is used in calculus to construct rigorous proofs of basic theorems about continuous functions such as the Intermediate Value Theorem (IVT) or the existence of extreme values on a closed interval. The underlying idea is that the real number line "has no holes." We elaborate on this idea below. First, we introduce the necessary definitions.

Suppose that S is a nonempty set of real numbers. A number M is called an **upper bound** for S if

$$x \leq M \qquad \text{for all } x \in S$$

If S has an upper bound, we say that S is **bounded above**. A **least upper bound** L is an upper bound for S such that every other upper bound M satisfies $M \geq L$. For example (Figure 2),

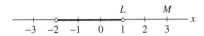

FIGURE 2 $M = 3$ is an upper bound for the set $S = (-2, 1)$. The LUB is $L = 1$.

- $M = 3$ is an upper bound for the open interval $S = (-2, 1)$.
- $L = 1$ is the LUB for $S = (-2, 1)$.

We now state the LUB Property of the real numbers.

THEOREM 1 Existence of a Least Upper Bound Let S be a nonempty set of real numbers that is bounded above. Then S has an LUB.

In a similar fashion, we say that a number B is a **lower bound** for S if $x \geq B$ for all $x \in S$. We say that S is **bounded below** if S has a lower bound. A **greatest lower bound** (GLB) is a lower bound M such that every other lower bound B satisfies $B \leq M$. The set of real numbers also has the GLB Property: If S is a nonempty set of real numbers that is bounded below, then S has a GLB. This may be deduced immediately from Theorem 1. For any nonempty set of real numbers S, let $-S$ be the set of numbers of the form $-x$ for $x \in S$. Then $-S$ has an upper bound if S has a lower bound. Consequently, $-S$ has an LUB L by Theorem 1, and $-L$ is a GLB for S.

CONCEPTUAL INSIGHT Theorem 1 may appear quite reasonable, but perhaps it is not clear why it is useful. We suggested above that the LUB Property expresses the idea that **R** is "complete" or "has no holes." To illustrate this idea, let's compare **R** to the set of rational numbers, denoted **Q**. Intuitively, **Q** is not complete because the irrational numbers are missing. For example, **Q** has a "hole" where the irrational number $\sqrt{2}$ should be located (Figure 3). This hole divides **Q** into two disconnected halves (the half to the left and the half to the right of $\sqrt{2}$). Because of this, **Q** is not a truly "continuous" object. In the following example, we will see that the existence of $\sqrt{2}$ as a real number is directly related to the LUB Property. Another consequence of completeness is the IVT, whose proof below relies on the LUB Property.

FIGURE 3 The rational numbers have a "hole" at the location $\sqrt{2}$.

■ **EXAMPLE 1** Show that 2 has a square root by applying the LUB Property to the set

$$S = \{x : x^2 < 2\}$$

Solution First, we note that S is bounded with the upper bound $M = 2$. Indeed, if $x > 2$, then x satisfies $x^2 > 4$, and hence, x does not belong to S. By the LUB Property, S has a least upper bound. Call it L. We claim that $L = \sqrt{2}$ or, equivalently, that $L^2 = 2$ (Figure 4). We prove this by showing that $L^2 \geq 2$ and $L^2 \leq 2$.

If $L^2 < 2$, let $b = L + h$, where $h > 0$. Then

$$b^2 = L^2 + 2Lh + h^2 = L^2 + h(2L + h) \qquad \boxed{1}$$

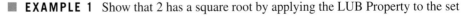

FIGURE 4 The set $S = \{x : x^2 < 2\}$ is bounded with the LUB $L = \sqrt{2}$.

We can make the quantity $h(2L + h)$ as small as desired by choosing $h > 0$ small enough. In particular, we may choose a positive h so that $h(2L + h) < 2 - L^2$. For this choice, $b^2 < L^2 - (2 - L)^2 = 2$ by Eq. (1). Therefore, $b \in S$. But $b > L$ since $h > 0$, and thus L is not an upper bound for S, in contradiction to our hypothesis on L. We conclude that $L^2 \geq 2$.

If $L^2 > 2$, let $b = L - h$, where $h > 0$. Then

$$b^2 = L^2 - 2Lh + h^2 = L^2 - h(2L + h)$$

Now choose h positive but small enough so that $h(2L + h) < L^2 - 2$. Then $b < L$ and $b^2 > L^2 - (L^2 - 2) = 2$. But in this case, b is a smaller lower bound for S. Indeed, if $x \geq b$, then $x^2 \geq b^2 > 2$, and x does not belong to S. This contradicts our hypothesis that L is the LUB. We conclude that $L^2 \leq 2$, and since we have already shown that $L^2 \geq 2$, we have $L^2 = 2$ as claimed. ∎

We now prove three important theorems, the third of which is used in the proof of the LUB Property below.

THEOREM 2 Bolzano–Weierstrass Theorem   Let S be a bounded, infinite set of real numbers. Then there exists a sequence of distinct elements $\{a_n\}$ in S such that the limit $L = \lim\limits_{n \to \infty} a_n$ exists.

Proof For simplicity of notation, we assume that S is contained in the unit interval $[0, 1]$ (a similar proof works in general). If $k_1, k_2, \ldots, k_n$ is a sequence of n digits (i.e., each k_j is a whole number and $0 \leq k_j \leq 9$), let

$$S(k_1, k_2, \ldots, k_n)$$

be the set of $x \in S$ whose decimal expansion begins $0.k_1 k_2 \ldots k_n$. The set S is the union of the subsets $S(0), S(1), \ldots, S(9)$, and since S is infinite, at least one of these subsets must be infinite. Therefore, we may choose k_1 so that $S(k_1)$ is infinite. In a similar fashion, at least one of the set $S(k_1, 0), S(k_2, 1), \ldots, S(k_1, 9)$ must be infinite, so we may choose k_2 so that $S(k_1, k_2)$ is infinite. Continuing in this way, we obtain an infinite sequence $\{k_n\}$ such that $S(k_1, k_2, \ldots, k_n)$ is infinite for all n. We may choose a sequence of elements $a_n \in S(k_1, k_2, \ldots, k_n)$ with the property that a_n differs from $a_1, \ldots, a_{n-1}$ for all n. Let L be the infinite decimal $0.k_1 k_2 k_3 \ldots$. Then $\lim\limits_{n \to \infty} a_n = L$ since $|L - a_n| < 10^{-n}$ for all n. ∎

We use the Bolzano–Weierstrass Theorem to prove two important results about sequences $\{a_n\}$. Recall that an upper bound for $\{a_n\}$ is a number M such that $a_j \leq M$ for all j. If an upper bound exists, $\{a_n\}$ is said to be bounded from above. Lower bounds are defined similarly and $\{a_n\}$ is said to be bounded from below if a lower bound exists. A sequence is bounded if it is bounded from above and below. A **subsequence** of $\{a_n\}$ is a sequence of elements $a_{n_1}, a_{n_2}, a_{n_3}, \ldots$, where $n_1 < n_2 < n_3 < \cdots$.

Now consider a bounded sequence $\{a_n\}$. If infinitely many of the a_n are distinct, the Bolzano–Weierstrass Theorem implies that there exists a subsequence $\{a_{n_1}, a_{n_2}, \ldots\}$ such that $\lim\limits_{n \to \infty} a_{n_k}$ exists. Otherwise, infinitely many of the a_n must coincide and these terms form a convergent subsequence. This proves the next result.

THEOREM 3   Every bounded sequence has a convergent subsequence.

THEOREM 4 **Bounded Monotonic Sequences Converge**

- If $\{a_n\}$ is increasing and $a_n \leq M$ for all n, then $\{a_n\}$ converges and $\lim\limits_{n\to\infty} a_n \leq M$.
- If $\{a_n\}$ is decreasing and $a_n \geq M$ for all n, then $\{a_n\}$ converges and $\lim\limits_{n\to\infty} a_n \geq M$.

Proof Suppose that $\{a_n\}$ is increasing and bounded above by M. Then $\{a_n\}$ is automatically bounded below by $m = a_1$ since $a_1 \leq a_2 \leq a_3 \cdots$. Hence, $\{a_n\}$ is bounded, and by Theorem 3, we may choose a convergent subsequence $a_{n_1}, a_{n_2}, \ldots$. Let

$$L = \lim_{k\to\infty} a_{n_k}$$

Observe that $a_n \leq L$ for all n. For if not, then $a_n > L$ for some n and then $a_{n_k} \geq a_n > L$ for all k such that $n_k \geq n$. But this contradicts that $a_{n_k} \to L$. Now, by definition, for any $\epsilon > 0$, there exists $N_\epsilon > 0$ such that

$$|a_{n_k} - L| < \epsilon \qquad \text{if } n_k > N_\epsilon$$

Choose m such that $n_m > N_\epsilon$. If $n \geq n_m$, then $a_{n_m} \leq a_n \leq L$, and therefore,

$$|a_n - L| \leq |a_{n_m} - L| < \epsilon \qquad \text{for all } n \geq n_m$$

This proves that $\lim\limits_{n\to\infty} a_n = L$ as desired. It remains to prove that $L \leq M$. If $L > M$, let $\epsilon = (L - M)/2$ and choose N so that

$$|a_n - L| < \epsilon \qquad \text{if } k > N$$

Then $a_n > L - \epsilon = M + \epsilon$. This contradicts our assumption that M is an upper bound for $\{a_n\}$. Therefore, $L \leq M$ as claimed. ∎

Proof of Theorem 1 We now use Theorem 4 to prove the LUB Property (Theorem 1). If x is a real number, we write $x(d)$ for the real number obtained by truncating the decimal expansion after the dth digit to the right of the decimal point. We call $x(d)$ the truncation of x of length d. Thus,

$$\text{If } x = 1.41569, \text{ then } x(3) = 1.415.$$

We say that x is a *decimal of length d* if $x = x(d)$. Any two distinct decimals of length d differ by at least 10^{-d}. It follows that for any two real numbers $A < B$, there are at most finitely many decimals of length d between A and B.

Now let S be a nonempty set of real numbers with an upper bound M. We shall prove that S has an LUB. Let $S(d)$ be the set of truncations of length d:

$$S(d) = \{x(d) : x \in S\}$$

We claim that $S(d)$ has a maximum element. To verify this, choose any $a \in S$. If $x \in S$ and $x(d) > a(d)$, then

$$a(d) \leq x(d) \leq M$$

Thus, by the remark of the previous paragraph, there are at most finitely many values of $x(d)$ in $S(d)$ larger than $a(d)$. The largest of these is the maximum element in $S(d)$.

For $d = 1, 2, \ldots$, choose an element x_d such that $x_d(d)$ is the maximum element in $S(d)$. By construction, $\{x_d(d)\}$ is an increasing sequence (since the largest dth truncation cannot get smaller as d increases). Furthermore, $x_d(d) \leq M$ for all d. We now apply

Theorem 4 to conclude that $\{x_d(d)\}$ converges to a limit L. We claim that L is the LUB of S. Observe first that L is an upper bound for S. Indeed, if $x \in S$, then $x(d) \leq L$ for all d and thus $x \leq L$. To show that L is the LUB, suppose that M is an upper bound such that $M < L$. Then $x_d \leq M$ for all d and hence $x_d(d) \leq M$ for all d. But then

$$L = \lim_{d \to \infty} x_d(d) \leq M$$

This is a contradiction since $M < L$. Therefore L is the LUB of S.   ■

As mentioned above, the LUB Property is used in calculus to establish certain basic theorems about continuous functions. As an example, we prove the IVT. Another example is the theorem on the existence of extrema on a closed interval (see Appendix D).

THEOREM 5 Intermediate Value Theorem   If $f(x)$ is continuous on a closed interval $[a, b]$ and $f(a) \neq f(b)$, then for every value M between $f(a)$ and $f(b)$, there exists at least one value $c \in (a, b)$ such that $f(c) = M$.

Proof We may replace $f(x)$ by $f(x) - M$ to reduce to the case $M = 0$. Similarly, replacing $f(x)$ by $-f(x)$ if necessary, we may assume that $f(a) < 0$ and $f(b) > 0$. Now let

$$S = \{x \in [a, b] : f(x) < 0\}$$

Then $a \in S$ since $f(a) < 0$ and thus S is nonempty. Clearly, b is an upper bound for S. Therefore, by the LUB Property, S has an LUB L. We claim that $f(L) = 0$. If not, set $r = f(L)$. Assume first that $r > 0$.

Since $f(x)$ is continuous, there exists a number $\delta > 0$ such that

$$|f(x) - f(L)| = |f(x) - r| < \frac{1}{2}r \qquad \text{if} \qquad |x - L| < \delta$$

Equivalently,

$$\frac{1}{2}r < f(x) < \frac{3}{2}r \qquad \text{if} \qquad |x - L| < \delta$$

The number $\frac{1}{2}r$ is positive so we conclude that

$$f(x) > 0 \qquad \text{if} \qquad L - \delta < x < L + \delta$$

By definition of L, $f(x) \geq 0$ for all $x \in [a, b]$ such that $x > L$, and thus $f(x) \geq 0$ for all $x \in [a, b]$ such that $x > L - \delta$. Thus, $L - \delta$ is an upper bound for S. This is a contradiction since L is the LUB of S and it follows that $r = f(L)$ cannot satisfy $r > 0$. Similarly, r cannot satisfy $r < 0$. We conclude that $f(L) = 0$ as claimed.   ■

INDUCTION AND
C | THE BINOMIAL
THEOREM

The Principle of Induction is a method of proof that is widely used to prove that a given statement $P(n)$ is valid for all natural numbers $n = 1, 2, 3, \dots$. Here are two statements of this kind:

- $P(n)$: The sum of the first n odd numbers is equal to n^2.
- $P(n)$: $\dfrac{d}{dx} x^n = n x^{n-1}$.

The first statement claims that for all natural numbers n,

$$\underbrace{1 + 3 + \cdots + (2n - 1)}_{\text{Sum of first } n \text{ odd numbers}} = n^2 \qquad \boxed{1}$$

We can check directly that $P(n)$ is true for the first few values of n:

$$P(1) \text{ is the equality:} \qquad 1 = 1^2 \quad \text{(true)}$$
$$P(2) \text{ is the equality:} \qquad 1 + 3 = 2^2 \quad \text{(true)}$$
$$P(3) \text{ is the equality:} \qquad 1 + 3 + 5 = 3^2 \quad \text{(true)}$$

The Principle of Induction may be used to establish $P(n)$ for all n.

The Principle of Induction applies if $P(n)$ is an assertion defined for $n \geq n_0$, where n_0 is a fixed integer. Assume that

(i) *Initial step:* $P(n_0)$ is true.
(ii) *Induction step:* If $P(n)$ is true for $n = k$, then $P(n)$ is also true for $n = k + 1$.

Then $P(n)$ is true for all $n \geq n_0$.

THEOREM 1 Principle of Induction   Let $P(n)$ be an assertion that depends on a natural number n. Assume that:

(i) **Initial step:** $P(1)$ is true.
(ii) **Induction step:** If $P(n)$ is true for $n = k$, then $P(n)$ is also true for $n = k + 1$.

Then $P(n)$ is true for all natural numbers $n = 1, 2, 3, \dots$.

■ **EXAMPLE 1** Prove that $1 + 3 + \cdots + (2n - 1) = n^2$ for all natural numbers n.

Solution As above, we let $P(n)$ denote the equality

$$P(n): \qquad 1 + 3 + \cdots + (2n - 1) = n^2$$

Step 1. **Initial step: Show that $P(1)$ is true.**
We checked this above. $P(1)$ is the equality $1 = 1^2$.

Step 2. **Induction step: Show that if $P(n)$ is true for $n = k$, then $P(n)$ is also true for $n = k + 1$.**
Assume that $P(k)$ is true. Then

$$1 + 3 + \cdots + (2k - 1) = k^2$$

Add $2k + 1$ to both sides:

$$\left[1 + 3 + \cdots + (2k - 1)\right] + (2k + 1) = k^2 + 2k + 1 = (k + 1)^2$$

$$1 + 3 + \cdots + (2k + 1) = (k + 1)^2$$

This is precisely the statement $P(k + 1)$. Thus, $P(k + 1)$ is true whenever $P(k)$ is true. By the Principle of Induction, $P(k)$ is true for all k. ▮

The intuition behind the Principle of Induction is the following. If $P(n)$ were not true for all n, then there would exist a smallest natural number k such that $P(k)$ is false. Furthermore, $k > 1$ since $P(1)$ is true. Thus $P(k - 1)$ is true [otherwise, $P(k)$ would not be the smallest "counterexample"]. On the other hand, if $P(k - 1)$ is true, then $P(k)$ is also true by the induction step. This is a contradiction. So $P(k)$ must be true for all k.

▮ **EXAMPLE 2**   Use Induction and the Product Rule to prove that for all whole numbers n,

$$\frac{d}{dx}x^n = nx^{n-1}$$

Solution   Let $P(n)$ be the formula $\dfrac{d}{dx}x^n = nx^{n-1}$.

Step 1. **Initial step: Show that $P(1)$ is true.**
We use the limit definition to verify $P(1)$:

$$\frac{d}{dx}x = \lim_{h \to 0} \frac{(x + h) - x}{h} = \lim_{h \to 0} \frac{h}{h} = \lim_{h \to 0} 1 = 1$$

Step 2. **Induction step: Show that if $P(n)$ is true for $n = k$, then $P(n)$ is also true for $n = k + 1$.**
To carry out the induction step, assume that $\dfrac{d}{dx}x^k = kx^{k-1}$, where $k \geq 1$. Then, by the Product Rule,

$$\frac{d}{dx}x^{k+1} = \frac{d}{dx}(x \cdot x^k) = x\frac{d}{dx}x^k + x^k\frac{d}{dx}x = x(kx^{k-1}) + x^k$$

$$= kx^k + x^k = (k + 1)x^k$$

This shows that $P(k + 1)$ is true.

By the Principle of Induction, $P(n)$ is true for all $n \geq 1$. ▮

As another application of induction, we prove the Binomial Theorem, which describes the expansion of the binomial $(a + b)^n$. The first few expansions are familiar:

$$(a + b)^1 = a + b$$

$$(a + b)^2 = a^2 + 2ab + b^2$$

$$(a + b)^3 = a^3 + 3a^2b + 3ab^2 + b^3$$

In general, we have an expansion

$$(a + b)^n = a^n + \binom{n}{1}a^{n-1}b + \binom{n}{2}a^{n-2}b^2 + \binom{n}{3}a^{n-3}b^3 + \cdots + \binom{n}{n-1}ab^{n-1} + b^n$$

In Pascal's Triangle, the nth row displays the coefficients in the expansion of $(a + b)^n$:

n													
0						1							
1					1		1						
2				1		2		1					
3			1		3		3		1				
4		1		4		6		4		1			
5	1		5		10		10		5		1		
6	1		6	15		20		15		6		1	

The triangle is constructed as follows: Each entry is the sum of the two entries above it in the previous line. For example, the entry 15 in line $n = 6$ is the sum $10 + 5$ of the entries above it in line $n = 5$. The recursion relation guarantees that the entries in the triangle are the binomial coefficients.

$\boxed{2}$

where the coefficient of $x^{n-k}x^k$, denoted $\binom{n}{k}$, is called the **binomial coefficient**. Note that the first term in (2) corresponds to $k = 0$ and the last term to $k = n$; thus, $\binom{n}{0} = \binom{n}{n} = 1$. In summation notation,

$$(a + b)^n = \sum_{k=0}^{n} \binom{n}{k} a^k b^{n-k}$$

Pascal's Triangle (described in the marginal note) can be used to compute binomial coefficients if n and k are not too large. The Binomial Theorem provides the following general formula:

$$\binom{n}{k} = \frac{n!}{k!\,(n-k)!} = \frac{n(n-1)(n-2)\cdots(n-k+1)}{k(k-1)(k-2)\cdots 2 \cdot 1}$$

$\boxed{3}$

Before proving this formula, we prove a recursion relation for binomial coefficients. Note, however, that (3) is certainly correct for $k = 0$ and $k = n$ (recall that by convention, $0! = 1$):

$$\binom{n}{0} = \frac{n!}{(n-0)!\,0!} = \frac{n!}{n!} = 1, \qquad \binom{n}{n} = \frac{n!}{(n-n)!\,n!} = \frac{n!}{n!} = 1$$

THEOREM 2 Recursion Relation for Binomial Coefficients

$$\binom{n}{k} = \binom{n-1}{k} + \binom{n-1}{k-1} \qquad \text{for } 1 \le k \le n-1$$

Proof We write $(a + b)^n$ as $(a + b)(a + b)^{n-1}$ and expand in terms of binomial coefficients:

$$(a + b)^n = (a + b)(a + b)^{n-1}$$

$$\sum_{k=0}^{n} \binom{n}{k} a^{n-k} b^k = (a + b) \sum_{k=0}^{n-1} \binom{n-1}{k} a^{n-1-k} b^k$$

$$= a \sum_{k=0}^{n-1} \binom{n-1}{k} a^{n-1-k} b^k + b \sum_{k=0}^{n-1} \binom{n-1}{k} a^{n-1-k} b^k$$

$$= \sum_{k=0}^{n-1} \binom{n-1}{k} a^{n-k} b^k + \sum_{k=0}^{n-1} \binom{n-1}{k} a^{n-(k+1)} b^{k+1}$$

Replacing k by $k - 1$ in the second sum, we obtain

$$\sum_{k=0}^{n} \binom{n}{k} a^{n-k} b^k = \sum_{k=0}^{n-1} \binom{n-1}{k} a^{n-k} b^k + \sum_{k=1}^{n} \binom{n-1}{k-1} a^{n-k} b^k$$

On the right-hand side, the first term in the first sum is a^n and the last term in the second sum is b^n. Thus, we have

$$\sum_{k=0}^{n} \binom{n}{k} a^{n-k} b^k = a^n + \left(\sum_{k=1}^{n-1} \left(\binom{n-1}{k} + \binom{n-1}{k-1} \right) a^{n-k} b^k \right) + b^n$$

The recursion relation follows because the coefficients of $a^{n-k}b^k$ on the two sides of the equation must be equal. ∎

We now use induction to prove Eq. (3). Let $P(n)$ be the claim:

$$\binom{n}{k} = \frac{n!}{k!\,(n-k)!} \qquad \text{for } 0 \le k \le n$$

We have $\binom{1}{0} = \binom{1}{1} = 1$ since $(a+b)^1 = a+b$, so $P(1)$ is true. Furthermore,

$$\binom{n}{n} = \binom{n}{0} = 1$$ as observed above, since a^n and b^n have coefficient 1 in the expansion of $(a+b)^n$. For the inductive step, assume that $P(n)$ is true. By the recursion relation, for $1 \le k \le n$, we have

$$\binom{n+1}{k} = \binom{n}{k} + \binom{n}{k-1} = \frac{n!}{k!\,(n-k)!} + \frac{n!}{(k-1)!\,(n-k+1)!}$$

$$= n!\left(\frac{n+1-k}{k!\,(n+1-k)!} + \frac{k}{k!\,(n+1-k)!} \right) = n!\left(\frac{n+1}{k!\,(n+1-k)!} \right)$$

$$= \frac{(n+1)!}{k!\,(n+1-k)!}$$

Thus, $P(n+1)$ is also true and the Binomial Theorem follows by induction.

■ **EXAMPLE 3** Use the Binomial Theorem to expand $(x+y)^5$ and $(x+2)^3$.

Solution The fifth row in Pascal's Triangle yields

$$(x+y)^5 = x^5 + 5x^4 y + 10x^3 y^2 + 10x^2 y^3 + 5xy^4 + y^5$$

The third row in Pascal's Triangle yields

$$(x+2)^3 = x^3 + 3x^2(2) + 3x(2)^2 + 2^3 = x^3 + 6x^2 + 12x + 8$$ ■

C. EXERCISES

In Exercises 1–4, use the Principle of Induction to prove the formula for all natural numbers n.

1. $1 + 2 + 3 + \cdots + n = \dfrac{n(n+1)}{2}$

2. $1^3 + 2^3 + 3^3 + \cdots + n^3 = \dfrac{n^2(n+1)^2}{4}$

3. $\dfrac{1}{1 \cdot 2} + \dfrac{1}{2 \cdot 3} + \cdots + \dfrac{1}{n(n+1)} = \dfrac{n}{n+1}$

4. $1 + x + x^2 + \cdots + x^n = \dfrac{1 - x^{n+1}}{1 - x}$ for any $x \ne 1$

5. Let $P(n)$ be the statement $2^n > n$.
(a) Show that $P(1)$ is true.
(b) Observe that if $2^n > n$, then $2^n + 2^n > 2n$. Use this to show that if $P(n)$ is true for $n = k$, then $P(n)$ is true for $n = k+1$. Conclude that $P(n)$ is true for all n.

6. Use induction to prove that $n! > 2^n$ for $n \ge 4$.

Let $\{F_n\}$ be the Fibonacci sequence, defined by the recursion formula

$$F_n = F_{n-1} + F_{n-2}, \qquad F_1 = F_2 = 1$$

The first few terms are 1, 1, 2, 3, 5, 8, 13, ... In Exercises 7–10, use induction to prove the identity.

7. $F_1 + F_2 + \cdots + F_n = F_{n+2} - 1$

8. $F_1^2 + F_2^2 + \cdots + F_n^2 = F_{n+1} F_n$

9. $F_n = \dfrac{R_+^n - R_-^n}{\sqrt{5}}$, where $R_\pm = \dfrac{1 \pm \sqrt{5}}{2}$

10. $F_{n+1}F_{n-1} = F_n^2 + (-1)^n$. *Hint:* For the induction step, show that

$$F_{n+2}F_n = F_{n+1}F_n + F_n^2$$

$$F_{n+1}^2 = F_{n+1}F_n + F_{n+1}F_{n-1}$$

11. Use induction to prove that $f(n) = 8^n - 1$ is divisible by 7 for all natural numbers n. *Hint:* For the induction step, show that

$$8^{k+1} - 1 = 7 \cdot 8^k + (8^k - 1)$$

12. Use induction to prove that $n^3 - n$ is divisible by 3 for all natural numbers n.

13. Use induction to prove that $5^{2n} - 4^n$ is divisible by 7 for all natural numbers n.

14. Use Pascal's Triangle to write out the expansions of $(a + b)^6$ and $(a - b)^4$.

15. Expand $(x + x^{-1})^4$.

16. What is the coefficient of x^9 in $(x^3 + x)^5$?

17. Let $S(n) = \sum_{k=0}^{n} \binom{n}{k}$.

(a) Use Pascal's Triangle to compute $S(n)$ for $n = 1, 2, 3, 4$.

(b) Prove that $S(n) = 2^n$ for all $n \geq 1$. *Hint:* Expand $(a + b)^n$ and evaluate at $a = b = 1$.

18. Let $T(n) = \sum_{k=0}^{n} (-1)^k \binom{n}{k}$.

(a) Use Pascal's Triangle to compute $T(n)$ for $n = 1, 2, 3, 4$.

(b) Prove that $T(n) = 0$ for all $n \geq 1$. *Hint:* Expand $(a + b)^n$ and evaluate at $a = 1, b = -1$.

D | ADDITIONAL PROOFS

In this appendix, we provide proofs of several theorems that were stated or used in the text.

| Section 2.3

> **THEOREM 1 Basic Limit Laws** Assume that $\lim\limits_{x \to c} f(x)$ and $\lim\limits_{x \to c} g(x)$ exist. Then:
>
> **(i)** $\lim\limits_{x \to c} \big(f(x) + g(x) \big) = \lim\limits_{x \to c} f(x) + \lim\limits_{x \to c} g(x)$
>
> **(ii)** For any number k, $\lim\limits_{x \to c} k f(x) = k \lim\limits_{x \to c} f(x)$.
>
> **(iii)** $\lim\limits_{x \to c} f(x)g(x) = \Big(\lim\limits_{x \to c} f(x) \Big) \Big(\lim\limits_{x \to c} g(x) \Big)$
>
> **(iv)** If $\lim\limits_{x \to c} g(x) \neq 0$, then
>
> $$\lim\limits_{x \to c} \frac{f(x)}{g(x)} = \frac{\lim\limits_{x \to c} f(x)}{\lim\limits_{x \to c} g(x)}$$

Proof Let $L = \lim\limits_{x \to c} f(x)$ and $M = \lim\limits_{x \to c} g(x)$. The Sum Law (i) was proved in Section 2.6. Observe that (ii) is a special case of (iii), where $g(x) = k$ is a constant function. Thus, it will suffice to prove the Product Law (iii). We write

$$f(x)g(x) - LM = f(x)(g(x) - M) + M(f(x) - L)$$

and apply the Triangle Inequality to obtain

$$|f(x)g(x) - LM| \leq |f(x)(g(x) - M)| + |M(f(x) - L)| \qquad \boxed{1}$$

By the limit definition, we may choose $\delta > 0$ so that

$$|f(x) - L| < 1 \qquad \text{if } 0 < |x - c| < \delta$$

If follows that $|f(x)| < |L| + 1$ for $0 < |x - c| < \delta$. Now choose any number $\epsilon > 0$. Applying the limit definition again, we see that by choosing a smaller δ if necessary, we may also ensure that if $0 < |x - c| < \delta$, then

$$|f(x) - L| \leq \frac{\epsilon}{2(|M| + 1)} \quad \text{and} \quad |g(x) - M| \leq \frac{\epsilon}{2(|L| + 1)}$$

Using (1), we see that if $0 < |x - c| < \delta$, then

$$|f(x)g(x) - LM| \leq |f(x)|\,|g(x) - M| + |M|\,|f(x) - L|$$

$$\leq (|L| + 1)\frac{\epsilon}{2(|L| + 1)} + |M|\frac{\epsilon}{2(|M| + 1)}$$

$$\leq \frac{\epsilon}{2} + \frac{\epsilon}{2} = \epsilon$$

Since ϵ is arbitrary, this proves that $\lim_{x \to c} f(x)g(x) = LM$. To prove the Quotient Law (iv), it suffices to verify that if $M \neq 0$, then

$$\lim_{x \to c} \frac{1}{g(x)} = \frac{1}{M} \qquad \boxed{2}$$

For if (2) holds, then we may apply the Product Law to $f(x)$ and $g(x)^{-1}$ to obtain the Quotient Law:

$$\lim_{x \to c} \frac{f(x)}{g(x)} = \lim_{x \to c} f(x) \frac{1}{g(x)} = \left(\lim_{x \to c} f(x) \right) \left(\lim_{x \to c} \frac{1}{g(x)} \right)$$

$$= L \left(\frac{1}{M} \right) = \frac{L}{M}$$

We now verify (2). Since $g(x)$ approaches M and $M \neq 0$, we may choose $\delta > 0$ so that $|g(x)| \geq |M|/2$ if $0 < |x - c| < \delta$. Now choose any number $\epsilon > 0$. By choosing a smaller δ if necessary, we may also ensure that

$$|M - g(x)| < \epsilon |M| \left(\frac{|M|}{2} \right) \qquad \text{for } 0 < |x - c| < \delta$$

Then

$$\left| \frac{1}{g(x)} - \frac{1}{M} \right| = \left| \frac{M - g(x)}{Mg(x)} \right| \leq \left| \frac{M - g(x)}{M(M/2)} \right| \leq \frac{\epsilon |M|(|M|/2)}{|M|(|M|/2)} = \epsilon$$

Since ϵ is arbitrary, the limit (2) is proved. ◼

The following result was used in the text.

THEOREM 2 Limits Preserve Inequalities Let (a, b) be an open interval and let $c \in (a, b)$. Suppose that $f(x)$ and $g(x)$ are defined on (a, b), except possibly at c. Assume that

$$f(x) \leq g(x) \qquad \text{for } x \in (a, b), \quad x \neq c$$

and that the limits $\lim_{x \to c} f(x)$ and $\lim_{x \to c} g(x)$ exist. Then

$$\lim_{x \to c} f(x) \leq \lim_{x \to c} g(x)$$

Proof Let $L = \lim_{x \to c} f(x)$ and $M = \lim_{x \to c} g(x)$. To show that $L \leq M$, we use proof by contradiction. If $L > M$, let $\epsilon = \frac{1}{2}(L - M)$. By the formal definition of limits, we may choose $\delta > 0$ so that the following two conditions are satisfied:

$$|M - g(x)| < \epsilon \qquad \text{if } |x - c| < \delta$$

$$|L - f(x)| < \epsilon \qquad \text{if } |x - c| < \delta$$

But then

$$f(x) > L - \epsilon = M + \epsilon > g(x)$$

This is a contradiction since $f(x) \leq g(x)$. We conclude that $L \leq M$. ◼

THEOREM 3 Limit of a Composite Function   Assume that the following limits exist:

$$L = \lim_{x \to c} g(x) \qquad \text{and} \qquad M = \lim_{x \to L} f(x)$$

Then $\lim_{x \to c} f(g(x)) = M$.

Proof Let $\epsilon > 0$ be given. By the limit definition, there exists $\delta_1 > 0$ such that

$$|f(x) - M| < \epsilon \qquad \text{if } 0 < |x - L| < \delta_1 \qquad \boxed{3}$$

Similarly, there exists $\delta > 0$ such that

$$|g(x) - L| < \delta_1 \qquad \text{if } 0 < |x - c| < \delta \qquad \boxed{4}$$

We replace x by $g(x)$ in (3) and apply (4) to obtain

$$|f(g(x)) - M| < \epsilon \qquad \text{if } 0 < |x - c| < \delta$$

Since ϵ is arbitrary, this proves that $\lim_{x \to c} f(g(x)) = M$. ∎

▎ *Section 2.4*

THEOREM 4 Continuity of Composite Functions   Let $F(x) = f(g(x))$ be a composite function. If g is continuous at $x = c$ and f is continuous at $x = g(c)$, then $F(x)$ is continuous at $x = c$.

Proof By definition of continuity,

$$\lim_{x \to c} g(x) = g(c) \qquad \text{and} \qquad \lim_{x \to g(c)} f(x) = f(g(c))$$

Therefore, we may apply Theorem 3 to obtain

$$\lim_{x \to c} f(g(x)) = f(g(c))$$

This proves that $f(g(x))$ is continuous at $x = c$. ∎

▎ *Section 2.6*

THEOREM 5 Squeeze Theorem   Assume that for $x \neq c$ (in some open interval containing c),

$$l(x) \leq f(x) \leq u(x) \qquad \text{and} \qquad \lim_{x \to c} l(x) = \lim_{x \to c} u(x) = L$$

Then $\lim_{x \to c} f(x)$ exists and

$$\lim_{x \to c} f(x) = L$$

Proof Let $\epsilon > 0$ be given. We may choose $\delta > 0$ such that

$$|l(x) - L| < \epsilon \quad \text{and} \quad |u(x) - L| < \epsilon \qquad \text{if } 0 < |x - c| < \delta$$

In principle, a different δ may be required to obtain the two inequalities for $l(x)$ and $u(x)$, but we may choose the smaller of the two deltas. Thus, if $0 < |x - c| < \delta$, we have

$$L - \epsilon < l(x) < L + \epsilon$$

and

$$L - \epsilon < u(x) < L + \epsilon$$

Since $f(x)$ lies between $l(x)$ and $u(x)$, it follows that

$$L - \epsilon < l(x) \le f(x) \le u(x) < L + \epsilon$$

and therefore $|f(x) - L| < \epsilon$ if $0 < |x - c| < \delta$. Since ϵ is arbitrary, this proves that $\lim_{x \to c} u(x) = L$ as desired.   ∎

I *Section 3.9*

> **THEOREM 6 Derivative of the Inverse**   Assume that $f(x)$ is differentiable and one-to-one with inverse $g(x)$. If b belongs to the domain of $g(x)$ and $f'(g(b)) \ne 0$, then $g'(b)$ exists and
>
> $$g'(b) = \frac{1}{f'(g(b))}$$

Proof The function $f(x)$ is one-to-one and continuous (since it is differentiable). It follows that $f(x)$ is monotonic increasing or decreasing. For if not, then $f(x)$ would have a local minimum or maximum at some point $x = x_0$. But then $f(x)$ would not be one-to-one in a small interval around x_0 by the IVT.

Suppose that $f(x)$ is increasing (the decreasing case is similar). We shall prove that $g(x)$ is continuous at $x = b$. Suppose that $f(a) = b$. Fix a small number $\epsilon > 0$. Since $f(x)$ is an increasing function, it maps the open interval $(a - \epsilon, a + \epsilon)$ to the open interval $(f(a - \epsilon), f(a + \epsilon))$ containing $f(a) = b$. We may choose a number $\delta > 0$ so that $(b - \delta, b + \delta)$ is contained in $(f(a - \epsilon), f(a + \epsilon))$. Then $g(x)$ maps $(b - \delta, b + \delta)$ back into $(a - \epsilon, a + \epsilon)$. It follows that

$$|g(y) - g(b)| < \epsilon \qquad \text{if } 0 < |y - b| < \delta$$

This proves that g is continuous at $x = b$.

To complete the proof, we must show that the following limit exists and is equal to $1/f'(g(b))$:

$$g'(a) = \lim_{y \to b} \frac{g(y) - g(b)}{y - b}$$

Let $a = g(b)$. By the inverse relationship, if $y = f(x)$, then $g(y) = x$, and since $g(y)$ is continuous, x approaches a as y approaches b. Thus, since $f(x)$ is differentiable and $f'(a) \ne 0$,

$$\lim_{y \to b} \frac{g(y) - g(b)}{y - b} = \lim_{x \to a} \frac{x - a}{f(x) - f(a)} = \frac{1}{f'(a)} = \frac{1}{f'(g(b))} \qquad ∎$$

I *Section 4.2*

> **THEOREM 7 Existence of Extrema on a Closed Interval**   If $f(x)$ is a continuous function on a closed (bounded) interval $I = [a, b]$, then $f(x)$ takes on a minimum and a maximum value on I.

Proof We prove that $f(x)$ takes on a maximum value in two steps (the case of a minimum is similar).

Step 1. **Prove that $f(x)$ is bounded from above.**

We use proof by contradiction. If $f(x)$ is not bounded from above, then there exist points $a_n \in [a, b]$ such that $f(a_n) \geq n$ for $n = 1, 2, \ldots$. By Theorem 3 in Appendix B, we may choose a subsequence of elements $a_{n_1}, a_{n_2}, \ldots$ that converges to a limit in $[a, b]$, say, $\lim_{k \to \infty} a_{n_k} = L$. Since $f(x)$ is continuous, there exists $\delta > 0$ such that

$$|f(x) - f(L)| < 1 \qquad \text{if} \quad x \in [a, b] \quad \text{and} \quad |x - L| < \delta$$

Therefore,

$$f(x) < f(L) + 1 \qquad \text{if} \quad x \in [a, b] \quad \text{and} \quad x \in (L - \delta, L + \delta) \qquad \boxed{5}$$

For k sufficiently large, a_{n_k} lies in $(L - \delta, L + \delta)$ because $\lim_{k \to \infty} a_{n_k} = L$. By (5), $f(a_{n_k})$ is bounded by $f(L) + 1$. However, $f(a_{n_k}) = n_k$ tends to infinity as $k \to \infty$. This is a contradiction. Hence, our assumption that $f(x)$ is not bounded from above is false.

Step 2. **Prove that $f(x)$ takes on a maximum value.**

The range of $f(x)$ on $I = [a, b]$ is the set

$$S = \{f(x) : x \in [a, b]\}$$

By the previous step, S is bounded from above and therefore has a least upper bound M by the LUB Property. To complete the proof, we show that $f(c) = M$ for some $c \in [a, b]$.

By definition, $M - 1/n$ is not an upper bound for $n \geq 1$, and therefore, we may choose a point b_n in $[a, b]$ such that

$$M - \frac{1}{n} \leq f(b_n) \leq M$$

Again by Theorem 3 in Appendix B, there exists a subsequence of elements $\{b_{n_1}, b_{n_2}, \ldots\}$ in $\{b_1, b_2, \ldots\}$ that converges to a limit, say,

$$\lim_{k \to \infty} b_{n_k} = c$$

To show that $f(c) = M$, let $\epsilon > 0$. Since $f(x)$ is continuous, we may choose k so large that the following two conditions are satisfied: $|f(c) - f(b_{n_k})| < \epsilon/2$ and $n_k > 2/\epsilon$. Then

$$|f(c) - M| \leq |f(c) - f(b_{n_k})| + |f(b_{n_k}) - M| \leq \frac{\epsilon}{2} + \frac{1}{n_k} \leq \frac{\epsilon}{2} + \frac{\epsilon}{2} = \epsilon$$

Thus, $|f(c) - M|$ is smaller than ϵ for all positive numbers ϵ. But this is not possible unless $|f(c) - M| = 0$. Thus $f(c) = M$ as desired. ∎

| Section 5.2

THEOREM 8 Continuous Functions Are Integrable   If $f(x)$ is continuous on $[a, b]$, then $f(x)$ is integrable over $[a, b]$.

Proof We shall make the simplifying assumption that $f(x)$ is differentiable and that its derivative $f'(x)$ is bounded. In other words, we assume that $|f'(x)| \leq K$ for some constant K. This assumption is used to show that $f(x)$ cannot vary too much in a small interval. More precisely, let us prove that if $[a_0, b_0]$ is any closed interval contained in

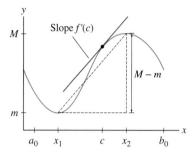

FIGURE 1 Since
$M - m = f'(c)(x_2 - x_1)$, we conclude
that $M - m \leq K(b_0 - a_0)$.

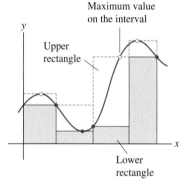

FIGURE 2 Lower and upper rectangles
for a partition of length $N = 4$.

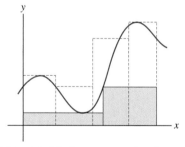

FIGURE 3 The lower rectangles always
lie below the upper rectangles, even when
the partitions are different.

$[a, b]$ and if m and M are the minimum and maximum values of $f(x)$ on $[a_0, b_0]$, then

$$|M - m| \leq K|b_0 - a_0| \qquad \boxed{6}$$

Figure 1 illustrates the idea behind this inequality. Suppose that $f(x_1) = m$ and $f(x_2) = M$, where x_1 and x_2 lie in $[a_0, b_0]$. If $x_1 \neq x_2$, then by the Mean Value Theorem (MVT), there is a point c between x_1 and x_2 such that

$$\frac{M - m}{x_2 - x_1} = \frac{f(x_2) - f(x_1)}{x_2 - x_1} = f'(c)$$

Since x_1, x_2 lie in $[a_0, b_0]$, we have $|x_2 - x_1| \leq |b_0 - a_0|$, and thus,

$$|M - m| = |f'(c)|\,|x_2 - x_1| \leq K|b_0 - a_0|$$

This proves (6).

We divide the rest of the proof into two steps. Consider a partition P:

$$P: \qquad x_0 = a < x_1 < \quad \cdots \quad < x_{N-1} < x_N = b$$

Let m_i be the minimum value of $f(x)$ on $[x_{i-1}, x_i]$ and M_i the maximum on $[x_{i-1}, x_i]$. We define the *lower* and *upper* Riemann sums

$$L(f, P) = \sum_{i=1}^{N} m_i \,\Delta x_i, \qquad U(f, P) = \sum_{i=1}^{N} M_i \,\Delta x_i,$$

These are the particular Riemann sums in which the intermediate point in $[x_{i-1}, x_i]$ is the point where $f(x)$ takes on its minimum or maximum on $[x_{i-1}, x_i]$. Figure 2 illustrates the case $N = 4$.

Step 1. **Prove that the lower and upper sums approach a limit.**
We observe that

$$L(f, P_1) \leq U(f, P_2) \quad \text{for any two partitions } P_1 \text{ and } P_2 \qquad \boxed{7}$$

Indeed, if a subinterval I_1 of P_1 overlaps with a subinterval I_2 of P_2, then the minimum of f on I_1 is less than or equal to the maximum of f on I_2 (Figure 3). In particular, the lower sums are bounded above by $U(f, P)$ for all partitions P. Let L be the least upper bound of the lower sums. Then for all partitions P,

$$L(f, P) \leq L \leq U(f, P) \qquad \boxed{8}$$

According to Eq. (6), $|M_i - m_i| \leq K\,\Delta x_i$ for all i. Since $\|P\|$ is the largest of the widths Δx_i, we see that $|M_i - m_i| \leq K\|P\|$ and

$$|U(f, P) - L(f, P)| \leq \sum_{i=1}^{N} |M_i - m_i|\,\Delta x_i$$

$$\leq K\|P\| \sum_{i=1}^{N} \Delta x_i = K\|P\|\,|b - a| \qquad \boxed{9}$$

Let $c = K|b - a|$. Using (8), we obtain

$$|L - U(f, P)| \leq |U(f, P) - L(f, P)| \leq c\|P\|$$

and we conclude that $\displaystyle\lim_{\|P\|\to 0}|L - U(f, P)| = 0$. Similarly,

$$|L - L(f, P)| \leq c\|P\|$$

and

$$\lim_{\|P\|\to 0}|L - L(f, P)| = 0$$

Thus, we have

$$\lim_{\|P\|\to 0} U(f, P) = \lim_{\|P\|\to 0} L(f, P) = L$$

Step 2. Prove that $\displaystyle\int_a^b f(x)\,dx$ exists and has value L.

Recall that for any choice C of intermediate points $c_i \in [x_{i-1}, x_i]$, we define the Riemann sum by

$$R(f, P, C) = \sum_{i=1}^{N} f(x_i)\Delta x_i$$

We have

$$L(f, P) \leq R(f, P, C) \leq U(f, P)$$

Indeed, since $c_i \in [x_{i-1}, x_i]$, we have $m_i \leq f(c_i) \leq M_i$, for all i and

$$\sum_{i=1}^{N} m_i\,\Delta x_i \leq \sum_{i=1}^{N} f(c_i)\,\Delta x_i \leq \sum_{i=1}^{N} M_i\,\Delta x_i$$

It follows that

$$|L - R(f, P, C)| \leq |U(f, P) - L(f, P)| \leq c\|P\|$$

This shows that $R(f, P, C)$ converges to L as $\|P\| \to 0$.   ■

▐ Section 10.1

THEOREM 9 If $f(x)$ is continuous and $\{a_n\}$ is a sequence such that the limit $\displaystyle\lim_{n\to\infty} a_n = L$ exists, then

$$\lim_{n\to\infty} f(a_n) = f(L)$$

Proof Choose any $\epsilon > 0$. Since $f(x)$ is continuous, there exists $\delta > 0$ such that

$$|f(x) - f(L)| < \epsilon \qquad \text{if } 0 < |x - L| < \delta$$

Since $\displaystyle\lim_{n\to\infty} a_n = L$, there exists $N > 0$ such that $|a_n - L| < \delta$ for $n > N$. Thus,

$$|f(a_n) - f(L)| < \epsilon \qquad \text{for } n > N$$

It follows that $\displaystyle\lim_{n\to\infty} f(a_n) = f(L)$.   ■

ANSWERS TO ODD-NUMBERED EXERCISES

Chapter 1

Section 1.1 Preliminary

1. $a = -3, b = 1$

2. The numbers $a \geq 0$ satisfy $|a| = a$ and $|-a| = a$. The numbers $a \leq 0$ satisfy $|a| = -a$.

3. $a = -3, b = 1$ **4.** $(9, -4)$

5. **(a)** First quadrant **(b)** Second quadrant **(c)** Fourth quadrant **(d)** Third quadrant

6. The radius is 3. **7.** **(b)**

8. Symmetry with respect to the origin.

Section 1.1 Exercises

1. $r = 9.8696$ **3.** $|x| \leq 2$ **5.** $|x - 2| < 2$ **7.** $|x - 3| \leq 2$

9. $-8 < x < 8$ **11.** $-3 < x < 2$ **13.** $(-4, 4)$ **15.** $(2, 6)$

17. $\left[-\frac{7}{4}, \frac{9}{4}\right]$ **19.** $(-\infty, 2) \cup (6, \infty)$

21. $(-\infty, -\sqrt{3}) \cup (\sqrt{3}, \infty)$

23. **(a)** **(i)** **(b)** **(iii)** **(c)** **(v)** **(d)** **(vi)** **(e)** **(ii)** **(f)** **(iv)**

27. $|a + b - 13| = |(a - 5) + (b - 8)| \leq |a - 5| + |b - 8|$

$$< \frac{1}{2} + \frac{1}{2} = 1$$

29. **(a)** 9 **(b)** $|x^2 - 16| = |x - 4| \cdot |x + 4| \leq 1 \cdot 9 = 9$

31. $r_1 = \frac{3}{11}, r_2 = \frac{4}{15}$

33. Let $a = 1$ and $b = .\overline{9}$. The decimal expansions of a and b do not agree, but $1 - .\overline{9} = 0 < 10^{-k}$ for any k.

35. $(x - 2)^2 + (y - 4)^2 = 3^2 = 9$

37. Located a distance 5 from the origin:

$(5, 0)$	$(-5, 0)$	$(0, 5)$	$(0, -5)$
$(3, 4)$	$(-3, 4)$	$(3, -4)$	$(-3, -4)$
$(4, 3)$	$(-4, 3)$	$(4, -3)$	$(-4, -3)$

Located a distance 5 from the point $(2, 3)$:

$(7, 3)$	$(-3, 3)$	$(2, 8)$	$(2, -2)$
$(5, 7)$	$(-1, 7)$	$(5, -1)$	$(-1, -1)$
$(6, 6)$	$(-2, 6)$	$(6, 0)$	$(-2, 0)$

39. Example: $f : \{a, b, c\} \to \{1, 2\}$ where $f(a) = 1, f(b) = 1, f(c) = 2$. There is no function whose domain has two elements and range has three elements.

41. D : All reals; $R : \{y : y \geq 0\}$

43. $D : \{t : t \leq 2\}; R : \{y : y \geq 0\}$

45. $D : \{s : s \neq 0\}; R : \{y : y \neq 0\}$

47. D : All reals; $R : \{y : y \geq \sqrt{2}\}$ **49.** $(-1, \infty)$ **51.** $(0, \infty)$

53. Zeros: ± 2; Inc: $x > 0$; Dec: $x < 0$; Sym: $f(-x) = f(x)$ (even function); y-axis symmetry.

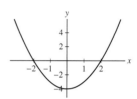

55. Zeros: $0, \pm 2$; Sym: $f(-x) = -f(x)$ (odd function); origin symmetry.

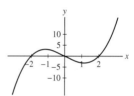

57. Zero: $x = \sqrt[3]{2}$; this is an x-axis reflection of x^3 moved up 2 units.

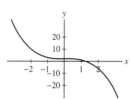

59. B **61.** $D : [0, 4]; R : [0, 4]$

63.

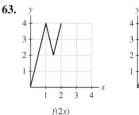 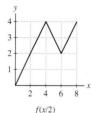 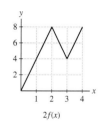

$f(2x)$ $f(x/2)$ $2f(x)$

65.

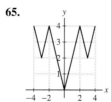

67. (a) $D : [4, 8]$, $R : [5, 9]$ **(b)** $D : [1, 5]$, $R : [2, 6]$
(c) $D : \left[\frac{4}{3}, \frac{8}{3}\right]$, $R : [2, 6]$ **(d)** $D : [4, 8]$, $R : [6, 18]$
69. (a) $h(x) = \sin(2x - 10)$ **(b)** $h(x) = \sin(2x - 5)$
71.

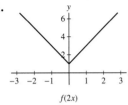
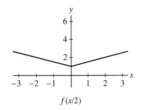

73.

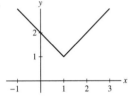

$D : \mathcal{R}$, $R : \{y | y \geq 1\}$
$f(x) = |x - 1| + 1$

75. Even:

$$(f + g)(-x) = f(-x) + g(-x) \overset{\text{even}}{=} f(x) + g(x)$$

$$= (f + g)(x)$$

Odd:

$$(f + g)(-x) = f(-x) + g(-x) \overset{\text{odd}}{=} -f(x) + -g(x)$$

$$= -(f + g)(x)$$

77. A circle of radius 1 with its center at the origin.
81. (a) An example:

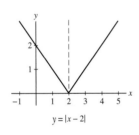

$y = |x - 2|$

(b) Let $g(x) = f(x + a)$. Then
$g(-x) = f(-x + a) = f(a + x) = g(x)$. Thus, $g(x)$ is even.

Section 1.2 Preliminary

1. -4 **2.** No
3. Parallel to the y-axis when $b = 0$ and parallel to the x-axis when $a = 0$.
4. $\Delta y = 9$ **5.** -4 **6.** $(x - 0)^2 + 1$

Section 1.2 Exercises

1. $m = 3$, $y = 12$, $x = -4$ **3.** $m = -\frac{4}{9}$, $y = \frac{1}{3}$, $x = \frac{3}{4}$
5. $m = 3$ **7.** $m = -\frac{3}{4}$ **9.** $y = 3x + 8$ **11.** $y = 3x - 12$
13. $y = -2$ **15.** $y = 3x - 2$ **17.** $y = \frac{5}{3}x - \frac{1}{3}$ **19.** $y = 4$
21. $y - 3 = -2(x - 3)$ **23.** $3x + 4y = 12$

25. (a) $c = -\frac{1}{4}$ **(b)** $c = -2$ **(c)** Such a constant does not exist.
(d) $c = 0$
27. (a) $L = 40.0248$ cm **(b)** $L = 64.9597$ in
(c) $L = 65(1 + \alpha(T - 100))$
29. $b = 4$
31. No, because the slopes between consecutive data points are not equal.
33. (a) $x = 1$ or $-\frac{1}{4}$ **(b)** $x = 1 \pm \sqrt{2}$
35. Minimum value is 0. **37.** Minimum value is -7.
39. Maximum value is $\frac{137}{16}$. **41.** Maximum value is $\frac{1}{3}$.
43.

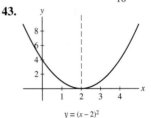

$y = (x - 2)^2$

45. A double root occurs when $c = \pm 2$. There are no real roots when $-2 < c < 2$.
47. For all $x > 0$, $0 \leq (x^{1/2} - x^{-1/2})^2 = x - 2 + \frac{1}{x}$.
51. $4 + 2\sqrt{2}$ and $4 - 2\sqrt{2}$
55. For x^2, $\frac{\Delta y}{\Delta x} = \frac{x_2^2 - x_1^2}{x_2 - x_1} = x_2 + x_1$.
59. $(x - \alpha)(x - \beta) = x^2 - \alpha x - \beta x + \alpha\beta$
$$= x^2 + (-\alpha - \beta)x + \alpha\beta$$

Section 1.3 Preliminary

1. An example: $\frac{3x^2 - 2}{7x^3 + x - 1}$
2. $|x|$ is not a polynomial. $|x^2 + 1|$ is a polynomial.
3. The domain of $f(g(x))$ is the empty set.
4. Decreasing **5.** An example: $f(x) = e^x - \sin x$

Section 1.3 Exercises

1. $x \geq 0$ **3.** All real numbers **5.** $t \neq -2$ **7.** $u \neq \pm 2$
9. $x \neq 0, 1$ **11.** $y > 0$ **13.** Polynomial **15.** Algebraic
17. Transcendental **19.** Rational **21.** Transcendental
23. Rational **25.** Yes
27. $f(g(x)) = \sqrt{x + 1}$; $D: x \geq -1$; $g(f(x)) = \sqrt{x} + 1$; $D: x \geq 0$
29. $f(g(x)) = 2^{x^2}$; $D: \mathcal{R}$; $g(f(x)) = (2^x)^2 = 2^{2x}$; $D: \mathcal{R}$
31. $f(g(x)) = \cos(x^3 + x^2)$; $D: \mathcal{R}$
$g(f(\theta)) = \cos^3 \theta + \cos^2 \theta$; $D: \mathcal{R}$
33. $f(g(t)) = \frac{1}{\sqrt{-t^2}}$; D: Not valid for any t

$g(f(t)) = -\left(\frac{1}{\sqrt{t}}\right)^2 = -\frac{1}{t}$; $D: t > 0$
35. $P(t + 10) = 30 \cdot 2^{0.1(t+10)} = 30 \cdot 2^{0.1t+1}$
$$= 2(30 \cdot 2^{0.1t}) = 2P(t)$$

$g\left(t + \frac{1}{k}\right) = a2^{k(t+1/k)} = a2^{kt+1}$

$$= 2a2^{kt} = 2g(t)$$

37. $f(x) = x^2$: $\delta f(x) = f(x+1) - f(x)$
$$= (x+1)^2 - x^2 = 2x + 1$$
$f(x) = x$: $\delta f(x) = x + 1 - x = 1$
$f(x) = x^3$: $\delta f(x) = (x+1)^3 - x^3 = 3x^2 + 3x + 1$

39. $\delta(f + g) = (f(x+1) + g(x+1)) - (f(x) - g(x))$
$$= (f(x+1) - f(x)) + (g(x+1) - g(x))$$
$$= \delta f(x) + \delta g(x)$$
$$\delta(cf) = cf(x+1) - cf(x) = c(f(x+1) - f(x))$$
$$= c\delta f(x).$$

Section 1.4 Preliminary

1. Two rotations that differ by a whole number of full revolutions will have the same ending radius.

2. $\frac{9\pi}{4}$ and $\frac{41\pi}{4}$ **3.** $-\frac{5\pi}{3}$ **4.** (a)

5. Let O denote the center of the unit circle, and let P be a point on the unit circle such that the radius $\overline{OP}$ makes an angle θ with the positive x-axis. Then, $\sin \theta$ is the y-coordinate of the point P.

Section 1.4 Exercises

1. $5\pi/4$

3. (a) $\frac{180°}{\pi} \approx 57.1°$ (b) $60°$ (c) $\frac{75°}{\pi} \approx 23.87°$ (d) $-135°$

5. $s = r\theta = 3.6$; $s = r\phi = 8$

7.

θ	$(\cos\theta, \sin\theta)$	θ	$(\cos\theta, \sin\theta)$
$\frac{\pi}{2}$	$(0, 1)$	$\frac{5\pi}{4}$	$\left(\frac{-\sqrt{2}}{2}, \frac{-\sqrt{2}}{2}\right)$
$\frac{2\pi}{3}$	$\left(\frac{-1}{2}, \frac{\sqrt{3}}{2}\right)$	$\frac{4\pi}{3}$	$\left(\frac{-1}{2}, \frac{-\sqrt{3}}{2}\right)$
$\frac{3\pi}{4}$	$\left(\frac{-\sqrt{2}}{2}, \frac{\sqrt{2}}{2}\right)$	$\frac{3\pi}{2}$	$(0, -1)$
$\frac{5\pi}{6}$	$\left(\frac{-\sqrt{3}}{2}, \frac{1}{2}\right)$	$\frac{5\pi}{3}$	$\left(\frac{1}{2}, \frac{-\sqrt{3}}{2}\right)$
π	$(-1, 0)$	$\frac{7\pi}{4}$	$\left(\frac{\sqrt{2}}{2}, \frac{-\sqrt{2}}{2}\right)$
$\frac{7\pi}{6}$	$\left(\frac{-\sqrt{3}}{2}, \frac{-1}{2}\right)$	$\frac{11\pi}{6}$	$\left(\frac{\sqrt{3}}{2}, \frac{-1}{2}\right)$

9. $\theta = \frac{\pi}{3}, \frac{5\pi}{3}$ **11.** $\theta = \frac{3\pi}{4}, \frac{7\pi}{4}$ **13.** $x = \frac{\pi}{3}, \frac{2\pi}{3}$

15.

θ	$\frac{\pi}{6}$	$\frac{\pi}{4}$	$\frac{\pi}{3}$	$\frac{\pi}{2}$	$\frac{2\pi}{3}$	$\frac{3\pi}{4}$	$\frac{5\pi}{6}$
$\tan\theta$	$\frac{1}{\sqrt{3}}$	1	$\sqrt{3}$	und	$-\sqrt{3}$	-1	$-\frac{1}{\sqrt{3}}$
$\sec\theta$	$\frac{2}{\sqrt{3}}$	$\sqrt{2}$	2	und	-2	$-\sqrt{2}$	$-\frac{2}{\sqrt{3}}$

17. $\cos\theta = \frac{1}{\sec\theta} = \frac{1}{\sqrt{1+\tan^2\theta}} = \frac{1}{\sqrt{1+c^2}}$

19. $\sin\theta = \frac{12}{13}$ and $\tan\theta = \frac{12}{5}$

21. $\sin\theta = \frac{2}{\sqrt{53}}$ and $\sec\theta = \frac{\sqrt{53}}{7}$

23. $\cos 2\theta = 23/25$

25. $\cos\theta = -\frac{\sqrt{21}}{5}$ and $\tan\theta = -\frac{2\sqrt{21}}{21}$

27. Figure 23(A):
Point in first quadrant: $\sin\theta = 0.918$, $\cos\theta = 0.3965$, and $\tan\theta = \frac{0.918}{0.3965} = 2.3153$.
Point in second quadrant: $\sin\theta = 0.3965$, $\cos\theta = -0.918$, and $\tan\theta = \frac{0.3965}{-0.918} = -0.4319$.
Point in third quadrant: $\sin\theta = -0.918$, $\cos\theta = -0.3965$, and $\tan\theta = \frac{-0.918}{-0.3965} = 2.3153$.
Point in fourth quadrant: $\sin\theta = -0.3965$, $\cos\theta = 0.918$, and $\tan\theta = \frac{-0.3965}{0.918} = -0.4319$.

Figure 23(B):
Point in first quadrant: $\sin\theta = 0.918$, $\cos\theta = 0.3965$, and $\tan\theta = \frac{0.918}{0.3965} = 2.3153$.
Point in second quadrant: $\sin\theta = 0.918$, $\cos\theta = -0.3965$, and $\tan\theta = \frac{0.918}{0.3965} = -2.3153$.
Point in third quadrant: $\sin\theta = -0.918$, $\cos\theta = -0.3965$, and $\tan\theta = \frac{-0.918}{-0.3965} = 2.3153$.
Point in fourth quadrant: $\sin\theta = -0.918$, $\cos\theta = 0.3965$, and $\tan\theta = \frac{-0.918}{0.3965} = -2.3153$.

29. $\cos\psi = 0.3$, $\sin\psi = \sqrt{0.91}$, $\cot\psi = \frac{0.3}{\sqrt{0.91}}$, and $\csc\psi = \frac{1}{\sqrt{0.91}}$

31. 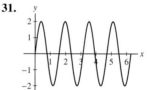 **33.**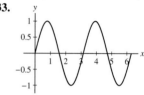

35. If $|c| > 1$, the horizontal line and the graph never intersect. If $|c| = 1$, then they intersect once. If $|c| < 1$, they intersect twice.

37. $\theta = 0, \frac{2\pi}{5}, \frac{4\pi}{5}, \pi, \frac{6\pi}{5}, \frac{8\pi}{5}$ **39.** $\theta = \frac{\pi}{6}, \frac{\pi}{2}, \frac{5\pi}{6}, \frac{7\pi}{6}, \frac{3\pi}{2}, \frac{11\pi}{6}$

41. Starting from the double angle formula for cosine, $\cos^2\theta = \frac{1}{2}(1 + \cos 2\theta)$, solve for $\cos 2\theta$.

43. Substitute $x = \theta/2$ into the double angle formula for sine, $\sin^2 x = \frac{1}{2}(1 - \cos 2x)$, and then take the square root of both sides.

45. $\cos(\theta + \pi) = \cos\theta\cos\pi - \sin\theta\sin\pi = \cos\theta(-1) = -\cos\theta$

47. $\tan(\pi - \theta) = \frac{\sin(\pi - \theta)}{\cos(\pi - \theta)} = \frac{-\sin(-\theta)}{-\cos(-\theta)} = \frac{\sin\theta}{-\cos\theta} = -\tan\theta$

49. $\frac{\sin 2x}{1 + \cos 2x} = \frac{2\sin x\cos x}{1 + 2\cos^2 x - 1} = \frac{2\sin x\cos x}{2\cos^2 x} = \frac{\sin x}{\cos x} = \tan x$

53. 16.928

Section 1.5 Preliminary

1. (a) Yes (b) Yes (c) No (d) No (e) No (f) Yes

2. No

3. The function is not one-to-one because many teenagers have the same last name.

4. Yes; $f^{-1}(6:27) =$ Hamilton Township.

5. The graph of the inverse is the reflection of the graph of $y = f(x)$ through the line $y = x$.

6. (a) $-\pi/6$ (b) Undefined (c) Undefined (d) $\pi/6$

7. Any angle $\theta < 0$ or $\theta > \pi$ will work. No, this does not contradict the definition of inverse function.

Section 1.5 Exercises

1. $f^{-1}(x) = \frac{1}{7}(x+4)$ **3.** $[-\pi/2, \pi/2]$

5. $f(g(x)) = \left((x-3)^{1/3}\right)^3 + 3 = x - 3 + 3 = x$
$g(f(x)) = (x^3 + 3 - 3)^{1/3} = (x^3)^{1/3} = x$

7. $v^{-1}(R) = \frac{2GM}{R^2}$

9. One-to-one for all real numbers; $f^{-1}(x) = 4 - x$

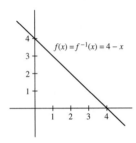

11. One-to-one on its entire domain, $\{x : x \neq \frac{3}{7}\}$; $f^{-1}(x) = \frac{1}{7x} + \frac{3}{7}$

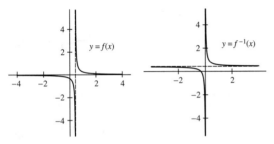

13. If we restrict the domain to $\{x : x \geq 0\}$, then $f^{-1}(x) = \frac{\sqrt{1-x^2}}{x}$;
If we restrict the domain to $\{x : x \leq 0\}$, then $f^{-1}(x) = -\frac{\sqrt{1-x^2}}{x}$

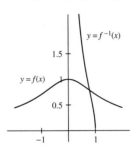

15. One-to-one on its entire domain, $\{x : x \geq -9^{1/3}\}$;
$f^{-1}(x) = (x^2 - 9)^{1/3}$

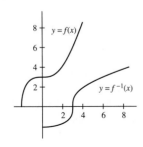

17. Figures (B) and (C) would not change when reflected around the line $y = x$. Therefore, these two satisfy $f^{-1} = f$.

19. **(a)** $f'(x) = 7x^6 + 1 > 0$ for all x. Thus, $f(x)$ is a strictly increasing function and by Example 3, it is one-to-one. Because f is one-to-one, by Theorem 1, f^{-1} exists.

(b) The domain of $f^{-1}(x)$ is the range of $f(x) : (-\infty, \infty)$.

(c) Note that $f(1) = 1^7 + 1 + 1 = 3$; therefore, $f^{-1}(3) = 1$.

21. If the domain of f is restricted to $x \leq 1$, then $f^{-1}(x) = 1 - \sqrt{x-1}$; if the domain of f is restricted to $x \geq 1$, then $f^{-1}(x) = 1 + \sqrt{x-1}$.

23. 0   **25.** $\frac{\pi}{4}$ **27.** $\frac{\pi}{3}$ **29.** $\frac{\pi}{3}$ **31.** $\frac{\pi}{2}$ **33.** $-\frac{\pi}{4}$ **35.** π

37. Undefined   **39.** $\frac{\sqrt{1-x^2}}{x}$ **41.** $\frac{1}{\sqrt{x^2-1}}$ **43.** $\frac{\sqrt{5}}{3}$ **45.** $\frac{4}{3}$

47. $\sqrt{3}$ **49.** $\frac{1}{20}$

Section 1.6 Preliminary

1. **(a)** Correct   **(b)** Correct   **(c)** Incorrect   **(d)** Correct

2. 2   **3.** $0 < x < 1$ **4.** Not defined

5. The phrase "The logarithm converts multiplication into addition" is a verbal translation of the property $\log(ab) = \log a + \log b$

6. Domain: $x > 0$; range: all real numbers

7. $\cosh x$ and $\operatorname{sech} x$ **8.** $\sinh x$ and $\tanh x$

9. Parity, identities, and derivative formulas

Section 1.6 Exercises

1. **(a)** 1   **(b)** 29   **(c)** 1   **(d)** 81   **(e)** 16   **(f)** 0

3. $x = 1$ **5.** $x = -1/2$ **7.** $x = -1/3$ **9.** $k = 9$

11. 3   **13.** $\frac{5}{3}$ **15.** $\frac{1}{3}$ **17.** $\frac{5}{6}$ **19.** 1   **21.** 7

23. **(a)** $\ln 1,600$ **(b)** $\ln 9x^{7/2}$

25. $t = \frac{1}{5} \ln\left(\frac{100}{7}\right)$ **27.** $x = -1, x = 3$ **29.** $x = e$

31.

	-3	0	5
$\sinh x$	-10.0179	0	74.203
$\cosh x$	10.0677	1	74.210

33. $\sinh x$ is increasing for all x; $\cosh x$ is decreasing for $x < 0$ and increasing for $x > 0$.

35. $t = \frac{\ln 2}{0.06} \approx 11.55$ yrs

37. **(a)** $E = 10^{4.8+1.5M}$ **(c)** $\frac{dE}{dM} = 1.5 \ln 10 \cdot 10^{4.8+1.5M}$

39. $\cosh x = \frac{\sqrt{41}}{5}$, $\tanh x = \frac{4\sqrt{41}}{41}$

41. $\sinh(2x) = \sinh x \cosh x + \cosh x \sinh x = 2 \sinh x \cosh x$
$\cosh(2x) = \cosh x \cosh x + \sinh x \sinh x = \cosh^2 x + \sinh^2 x$

43. $\log_a b \cdot \log_b a = \frac{\ln b}{\ln a} \cdot \frac{\ln a}{\ln b} = 1$

Section 1.7 Preliminary

1. No

2. (a) The screen will show nothing.

(b) The screen will show the portion of the parabola between the points $(0, 3)$ and $(1, 4)$.

3. No

4. Experiment with the viewing window to zoom in on the lowest point on the graph of the function. The y-coordinate of the lowest point on the graph is the minimum value of the function.

Section 1.7 Exercises

1.

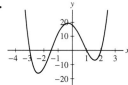

$x = -3, x = -1.5,$ $x = 1,$ and $x = 2$

3. Two positive solutions   **5.** There are no solutions.

7. The display will show nothing. An appropriate viewing rectangle: $[-50, 150]$ by $[1,000, 2,000]$

9.

Asymptotes are at: $x = 4, y = -1$.

11.

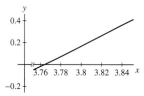

13. The following table and graphs suggest that as n gets large, $n^{1/n}$ approaches 1.

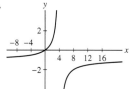

n	$n^{1/n}$
10	1.258925412
10^2	1.047128548
10^3	1.006931669
10^4	1.000921458
10^5	1.000115136
10^6	1.000013816

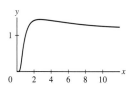

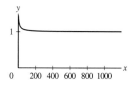

15. The following table and graphs suggest that as n gets large, $f(n)$ tends toward ∞.

n	$\left(1 + \frac{1}{n}\right)^{n^2}$
10	13780.61234
10^2	$1.635828711 \times 10^{43}$
10^3	$1.195306603 \times 10^{434}$
10^4	$5.341783312 \times 10^{4342}$
10^5	$1.702333054 \times 10^{43429}$
10^6	$1.839738749 \times 10^{434294}$

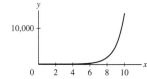

17. The following table and graphs suggest that as x gets large, $f(x)$ approaches 1.

x	$\left(x \tan \frac{1}{x}\right)^x$
10	1.033975759
10^2	1.003338973
10^3	1.000333389
10^4	1.000033334
10^5	1.000003333
10^6	1.000000333

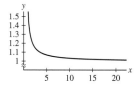

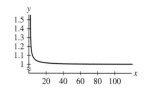

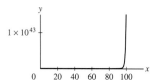

19.

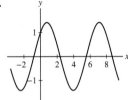

(A, B) = (1, 1)

(A, B) = (1, 2)

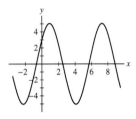

(A, B) = (3, 4)

21. $x \in (-2, 0) \cup (3, \infty)$

23. $f_3(x) = \frac{x^2+6x+1}{4(x+1)}$, $f_4(x) = \frac{x^4+28x^3+70x^2+28x+1}{8(1+x)(1+6x+x^2)}$, and $f_5(x) = \frac{1+120x+1{,}820x^2+8{,}008x^3+12{,}870x^4+8{,}008x^5+1{,}820x^6+120x^7+x^8}{16(1+x)(1+6x+x^2)(1+28x+70x^2+28x^3+x^4)}$.
It seems as if the f_n are asymptotic to $\sqrt{x}$.

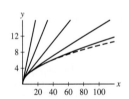

Chapter 1 Review

1. $\{x : |x - 7| < 3\}$ **3.** $\{x : 2 \le |x - 1| \le 6\} = [-5, -1] \cup [3, 7]$
5. $(x, 0)$ with $x \ge 0$. $(0, y)$ with $y < 0$
7.

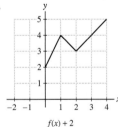

$f(x) + 2$ $\qquad$ $f(x + 2)$

9.

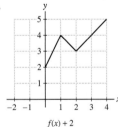

11. $D : \{x : x \ge -1\}$, $R : \{y : y \ge 0\}$
13. $D : \{x : x \ne 3\}$, $R : \{y : y \ne 0\}$
15. **(a)** Decreasing **(b)** Neither **(c)** Neither **(d)** Increasing
17. $2x - 3y = -14$ **19.** $6x - y = 53$ **21.** $x + y = 5$ **23.** Yes

25. Roots: $x = -2$, $x = 0$, and $x = 2$. Function is decreasing when $x < -1.4$ or $0 < x < 1.4$.

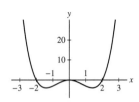

27. $f(x) = 10x^2 + 2x + 5$. Its minimum value is $\frac{49}{10}$.

29.

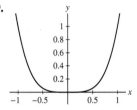

31.

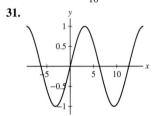

33.

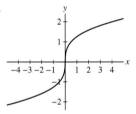

35. Let $g(x) = f\left(\frac{1}{3}x\right)$. Then

$$g(x - 3b) = f\left(\frac{1}{3}(x - 3b)\right) = f\left(\frac{1}{3}x - b\right).$$

The graph of $y = |\frac{1}{3}x - 4|$:

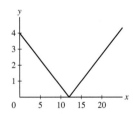

37. $f(t) = t^4$, $g(t) = 12t + 9$ **39.** 4π
41. **(a)** $a = b = \pi/2$ **(b)** $a = \pi$
43. $x = \pi/2$, $x = 7\pi/6$, $x = 3\pi/2$, and $x = 11\pi/6$
45. There are no solutions.
47. **(a)** No match **(b)** No match **(c)** (i): $(2^a)^b = 2^{ab}$ **(d)** (iii): $2^{a-b}3^{b-a} = 2^{a-b}\left(\frac{1}{3}\right)^{a-b} = \left(\frac{2}{3}\right)^{a-b}$
49. $f^{-1}(x) = \sqrt[3]{x + 8}$; domain and range are all real numbers
51. If we restrict the domain to $\{t : t \ge 3\}$, then $h^{-1}(t) = 3 + \sqrt{t}$; if we restrict the domain to $\{t : t \le 3\}$, then $h^{-1}(t) = 3 - \sqrt{t}$
53. **(a)** (iii) **(b)** (iv) **(c)** (ii) **(d)** (i)

Chapter 2

Section 2.1 Preliminary

1. The ratio of distance traveled to time elapsed.

2. On the graph of position as a function of time.

3. No, it is defined as the limit of average velocity as time elapsed shrinks to zero.

4. The slope of the line tangent to the graph of position as a function of time at $t = t_0$.

5. The slope of the secant line over the interval $[x_0, x_1]$ approaches the slope of the tangent line at $x = x_0$.

6. The graph of atmospheric temperature as a function of altitude. Possible units for this rate of change are °F/ft or °C/m.

Section 2.1 Exercises

1. **(a)** $\Delta s = 36$ ft   **(b)** $\frac{\Delta s}{\Delta t} = 72$ ft/s

(c)	Time interval	[2, 2.01]	[2, 2.005]	[2, 2.001]	[2, 2.00001]
	Average velocity	64.16	64.08	64.016	64.00016

The instantaneous velocity is 64 ft/s.

3. The instantaneous rate of change is approximately 0.57735 m/(s · K).

5. $\frac{\Delta h}{\Delta t} = 0.3$

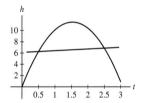

7. **(a)** Dollars/year

(b)	Time interval	[0, 0.5]	[0, 1]
	Average rate of change	7.8461	8

(c) The rate of change at $t = 0.5$ is approximately \$8/yr.

9. 16   **11.** −0.062   **13.** 1   **15.** 0.3333

17. The ROC is a constant −.00356 degrees per foot since T is a linear function of h with slope −0.00356.

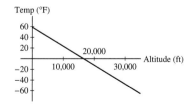

19. **(a)** ROC = 3,000 cells/hour

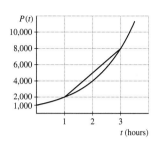

(b) $m = \frac{4,000}{3}$ cells/hour, which represents the instantaneous ROC at $t = 1$ hour.

21. **(a)** Seconds per meter; measures the amount by which the period changes when the length is changed

(b) The slope of the line B represents the average rate of change in T from $L = 1$ m to $L = 3$ m. The slope of the line A represents the instantaneous rate of change of T at $L = 3$ m.

(c) 0.4330 s/m

23. **(a)** D   **(b)** B   **(c)** A   **(d)** C

25. **(a)** The slope of line A is the average rate of change over the interval [4, 6], whereas the slope of the line B is the instantaneous rate of change at $t = 6$. Thus, the slope of the line $A \approx (0.28 - 0.19)/2 = 0.045$, whereas the slope of the line $B \approx (0.28 - 0.15)/6 = 0.0217$.

(b) $t = 3$ **(c)** $t = 4$

27.

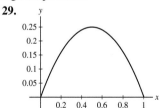

As the graph progresses to the right, the graph bends progressively downward, meaning that the ROC of v with respect to T is lower at high temperatures.

29.

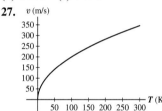

(a) 0   **(b)** 0   **(c)** $0 < x < 0.5$

31. **(a)** $h(1) = 48$ ft;
$h(t) - h(1) = -16t^2 + 64t - 48 = -16(t - 1)(t - 3)$
(b) $\frac{h(t)-h(1)}{t-1} = \frac{-16(t-1)(t-3)}{t-1} = -16(t - 3)$ **(c)** 32 ft/s

33. $\frac{f(x)-f(1)}{x-1} = \frac{x^3-1}{x-1} = \frac{(x-1)(x^2+x+1)}{x-1} = x^2 + x + 1$

At $x = 1$, the instantaneous ROC is $1^2 + 1 + 1 = 3$.

Section 2.2 Preliminary

1. $\lim\limits_{x \to \pi} 1 = 1$ **2.** $\lim\limits_{t \to \pi} t = \pi$ **3.** Yes, $\lim\limits_{x \to 1} \frac{x^2-1}{x-1}$

4. $\lim_{x \to 10} 20 = 20$ **5.** $\lim_{x \to 1-} f(x) = \infty$ and $\lim_{x \to 1+} f(x) = 3$

6. No, to determine whether $\lim_{x \to 5} f(x)$ exists, we must examine values of $f(x)$ on both sides of $x = 5$.

7. Yes **8.** The information in (a), (b), or (c) would be sufficient.

Section 2.2 Exercises

1.

x	0.998	0.999	0.9995	0.99999
$f(x)$	1.498501	1.499250	1.499625	1.499993

x	1.00001	1.0005	1.001	1.002
$f(x)$	1.500008	1.500375	1.500750	1.501500

$\lim_{x \to 1} f(x) = \frac{3}{2}$

3.

y	1.998	1.999	1.9999	2.0001	2.001	2.02
$f(y)$	0.59984	0.59992	0.599992	0.600008	0.60008	0.601594

$\lim_{y \to 2} f(y) = \frac{3}{5}$

5.

x	-0.5	-0.1	-0.05	-0.01
$f(x)$	0.426123	0.483742	0.491770	0.498338

x	0.01	0.05	0.1	0.5
$f(x)$	0.501671	0.508439	0.517092	0.594885

$\lim_{x \to 1} f(x) = \frac{1}{2}$

7. $\lim_{x \to 0.5} f(x) = 1.5$ **9.** $\lim_{x \to 21} f(x) = 21$ **21.** $\lim_{x \to 1} f(x) = \frac{1}{2}$

23. $\lim_{x \to 2} f(x) = \frac{5}{3}$ **25.** $\lim_{x \to 0} f(x) = 2$

27. The limit does not exist. **29.** The limit does not exist.

31. $\lim_{h \to 0} f(h) = 0.693$. (The exact answer is ln 2.)

33. $\lim_{x \to 1+} f(x) = 1.414$. (The exact answer is $\sqrt{2}$.)

35. The limit does not exist.

37. $\lim_{x \to 2-} f(x) = 2$, $\lim_{x \to 2+} f(x) = 1$

39. **(a)** The one-sided limits exist for all real values of c.
(b) $\lim_{x \to c} [x]$ exists when c is a noninteger.

41. $\lim_{x \to 0-} f(x) = -1$, $\lim_{x \to 0+} f(x) = 1$

43. $\lim_{x \to 0-} f(x) = \infty$, $\lim_{x \to 0+} f(x) = \frac{1}{6}$

45. $\lim_{x \to 2-} f(x) = \infty$, $\lim_{x \to 2+} f(x) = \infty$, $\lim_{x \to 4-} f(x) = -\infty$, and $\lim_{x \to 4+} f(x) = 10$

47.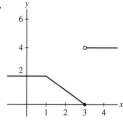

49.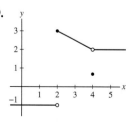

51. 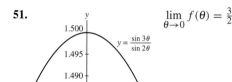 $\lim_{\theta \to 0} f(\theta) = \frac{3}{2}$

53. 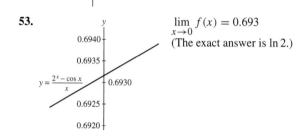 $\lim_{x \to 0} f(x) = 0.693$ (The exact answer is ln 2.)

55. 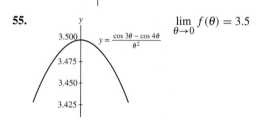 $\lim_{\theta \to 0} f(\theta) = 3.5$

57. If $R = 2$,

θ	-0.01	-0.005	0.005	0.01
$I(\theta)$	0.998667 I_m	0.9999667 I_m	0.9999667 I_m	0.9998667 I_m

If $R = 3$, the table becomes:

θ	-0.01	-0.005	0.005	0.01
$I(\theta)$	0.999700 I_m	0.999925 I_m	0.999925 I_m	0.999700 I_m

59.

x	-0.1	-0.01	-0.001	0.001	0.01	0.1
$\frac{5^x - 1}{x}$	1.486601	1.596556	1.608144	1.610734	1.622459	1.746189

Note $\ln 5 \approx 1.6094$.

x	-0.1	-0.01	-0.001	0.001	0.01	0.1
$\frac{3^x - 1}{x}$	1.040415	1.092600	1.098009	1.099216	1.104669	1.161232

Note $\ln 3 \approx 1.0986$.

61. For $k = 1$, the limit is 0; for $k = 2$, the limit is 1; for odd $k > 2$, the limit does not exist; for even $k > 2$, the limit is ∞.

63. **(a)**

x	-0.3	-0.2	-0.1	0.1	0.2	0.3
$f(x)$	-0.980506	-0.998049	-0.999998	0.999998	0.998049	0.980506

(b) As $x \to 0-$, $f(x) \to -1$, whereas as $x \to 0+$, $f(x) \to 1$.

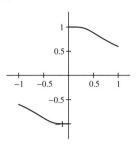

Section 2.3 Preliminary

1. Suppose $\lim\limits_{x \to c} f(x)$ and $\lim\limits_{x \to c} g(x)$ both exist. The Sum Law states that $\lim\limits_{x \to c} (f(x) + g(x)) = \lim\limits_{x \to c} f(x) + \lim\limits_{x \to c} g(x)$. Provided $\lim\limits_{x \to c} g(x) \neq 0$, the Quotient Law states that $\lim\limits_{x \to c} \dfrac{f(x)}{g(x)} = \dfrac{\lim\limits_{x \to c} f(x)}{\lim\limits_{x \to c} g(x)}$.

2. (b)   **3.** (a) and (d)   **4.** (a) and (c)

Section 2.3 Exercises

1. $\lim\limits_{x \to 9} x = 9$ **3.** $\lim\limits_{x \to 9} 14 = 14$ **5.** $\lim\limits_{x \to -3} (3x + 4) = -5$

7. $\lim\limits_{y \to -3} (y + 14) = 11$ **9.** $\lim\limits_{t \to 4} (3t - 14) = -2$

11. $\lim\limits_{x \to \frac{1}{2}} (4x + 1)(2x - 1) = 0$ **13.** $\lim\limits_{x \to 2} x(x + 1)(x + 2) = 24$

15. $\lim\limits_{t \to 9} \frac{t}{t+1} = \frac{9}{10}$ **17.** $\lim\limits_{x \to 3} \frac{1-x}{1+x} = -\frac{1}{2}$ **19.** $\lim\limits_{t \to 2} t^{-1} = \frac{1}{2}$

21. $\lim\limits_{x \to 3} (x^2 + 9x^{-3}) = \frac{28}{3}$

23. $\lim\limits_{x \to c} \left(\dfrac{1}{f(x)} \right) = \dfrac{\left(\lim\limits_{x \to c} 1 \right)}{\left(\lim\limits_{x \to c} f(x) \right)} = \dfrac{1}{\lim\limits_{x \to c} f(x)}$

25. $\lim\limits_{x \to -4} f(x)g(x) = 3$ **27.** $\lim\limits_{x \to -4} \frac{g(x)}{x^2} = \frac{1}{16}$

29. No, because $\lim\limits_{x \to 0} x = 0$.

31. $f(x) = 1/x$, $g(x) = -1/x$

35. $\lim\limits_{x \to c} f(x) = \lim\limits_{x \to c} \dfrac{f(x)g(x)}{g(x)} = \dfrac{\lim\limits_{x \to c} f(x)g(x)}{\lim\limits_{x \to c} g(x)} = \dfrac{L}{M}$

37. $\lim\limits_{t \to 3} h(t) = \lim\limits_{t \to 3} t \, \dfrac{h(t)}{t} = \left(\lim\limits_{t \to 3} t \right) \left(\lim\limits_{t \to 3} \dfrac{h(t)}{t} \right) = 3 \cdot 5 = 15$

41. $\lim\limits_{h \to 0} \dfrac{f(ah)}{h} = \lim\limits_{h \to 0} \left(a \cdot \dfrac{f(ah)}{ah} \right) = a \lim\limits_{h \to 0} \dfrac{f(ah)}{ah} = aL$, using the result from the previous exercise.

43. (b) is the correct law; $\lim\limits_{x \to 2} \sin(g(x)) = \lim\limits_{x \to \pi/6} \sin x = \frac{1}{2}$

Section 2.4 Preliminary

1. We can conclude that $\lim\limits_{x \to 2} x^3 = 8$ because x^3 is continuous at $x = 2$.

2. $f(3) = \frac{1}{2}$ **3.** f cannot be continuous at $x = 0$.   **4.** No

5. **(a)** False: "$f(x)$ is continuous at $x = a$ if the left- and right-hand limits of $f(x)$ as $x \to a$ exist and equal $f(a)$."

(b) True

(c) False: "If the left- and right-hand limits of $f(x)$ as $x \to a$ are equal but not equal to $f(a)$, then f has a removable discontinuity at $x = a$."

(d) True

(e) False: "If $f(x)$ and $g(x)$ are continuous at $x = a$ and $g(a) \neq 0$, then $f(x)/g(x)$ is continuous at $x = a$."

Section 2.4 Exercises

1. Left-continuous at $x = 1$; neither left-continuous nor right-continuous at $x = 3$; left-continuous at $x = 5$.

3. $c = 3$; $f(c) = f(3) = 4.5$ makes f continuous at $x = 3$.

5. **(a)** $x = 0$; $\lim\limits_{x \to 0-} f(x) = \infty$; $\lim\limits_{x \to 0+} f(x) = 2$

$x = 2$; $\lim\limits_{x \to 2-} f(x) = 6$; $\lim\limits_{x \to 2+} f(x) = 6$

(b) The discontinuity at $x = 2$ is removable; $f(2) = 6$.

7. x and $\sin x$ are continuous, so $x + \sin x$ is continuous by Continuity Law (i).

9. x and $\sin x$ are continuous, so $3x$ and $4 \sin x$ are continuous by Continuity Law (iii). Thus $3x + 4 \sin x$ is continuous by Continuity Law (i).

11. x is continuous, so x^2 is continuous by Continuity Law (ii). Constant functions are continuous, so $x^2 + 1$ is continuous. Finally, $\frac{1}{x^2+1}$ is continuous by Continuity Law (iv) because $x^2 + 1$ is never 0.

13. The functions 3^x, 1 and 4^x are each continuous. Therefore, $1 + 4^x$ is continuous by Continuity Law (i). Because $1 + 4^x$ is never zero, it follows that $\frac{3^x}{1+4^x}$ is continuous by Continuity Law (iv).

15. e^x and $\cos 3x$ are continuous, so $e^x \cos 3x$ is continuous by Continuity Law (ii).

17. Infinite discontinuity at $x = 0$.

19. Infinite discontinuity at $x = 1$.

21. Jump discontinuities (right-continuous and not left-continuous) at $x = n$ for every integer n.

23. Infinite discontinuities at $t = -1$ and $t = 1$.

25. Right-continuous at $x = 0$ (not defined for $x < 0$).

27. Infinite discontinuities at $z = -2$ and $z = 3$.

29. Jump discontinuity at $x = 2$.

31. Infinite discontinuities at $x = \pm\sqrt{n\pi}$ where n is a positive integer.

33. Continuous everywhere

35. Infinite discontinuity at $x = 0$

37. Continuous on $|x| \leq 3$. Not defined elsewhere.

39. Continuous on $x \geq 0$. Not defined elsewhere.

41. Continuous for all real numbers.

43. Continuous on $x \neq 0$. Not defined at $x = 0$.

45. Continuous on $x \neq \pm(2n - 1)\pi/2$ where n is a positive integer. Not defined elsewhere.

47. Continuous for all real numbers.

49. Continuous on $x \neq \pm 1$. Not defined elsewhere.

51. $f(0) = \lim\limits_{x \to 0-} f(x) = -4$; $f(0) = \lim\limits_{x \to 0+} f(x) = 2$

53. 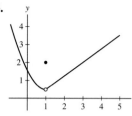 **55.**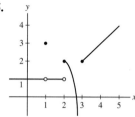

59. $\lim\limits_{x \to 5} x^2 = 25$ **61.** $\lim\limits_{x \to -1} (2x^3 - 4) = -6$

63. $\lim\limits_{x \to 0} \frac{x+9}{x-9} = -1$ **65.** $\lim\limits_{x \to \pi} \sin\left(\frac{x}{2} - \pi\right) = -1$

67. $\lim\limits_{x \to \frac{\pi}{4}} \tan(3x) = -1$ **69.** $\lim\limits_{x \to 4} x^{-5/2} = \frac{1}{32}$

71. $\lim\limits_{x \to -1} (1 - 8x^3)^{3/2} = 27$ **73.** $\lim\limits_{x \to 3} 10^{x^2 - 2x} = 1{,}000$

75. $\lim\limits_{x \to 1} e^{x^2 - x} = 1$ **77.** $\lim\limits_{x \to 4^-} \sin^{-1}\left(\frac{x}{4}\right) = \frac{\pi}{2}$

79.

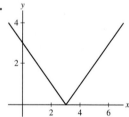

The function f is continuous everywhere.

81.

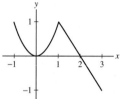

The function f is continuous everywhere.

83. $c = \frac{5}{3}$ **85.** $a - b = \frac{8 - 2\sqrt{2}}{\pi}$

87.

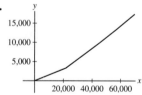

89. Answers may vary. $f(x) = C$ and $g(x)$ is defined for all x.

Section 2.5 Preliminary

1. $\frac{x^2 - 1}{\sqrt{x+3} - 2}$

2. (a) Let $f(x) = \frac{x^2 - 1}{x - 1}$. At $x = 1$, f is indeterminate of the form $\frac{0}{0}$ but

$$\lim_{x \to 1^-} \frac{x^2 - 1}{x - 1} = \lim_{x \to 1^-} (x + 1) = 2 = \lim_{x \to 1^+} (x + 1) = \lim_{x \to 1^+} \frac{x^2 - 1}{x - 1}.$$

(b) Again, let $f(x) = \frac{x^2 - 1}{x - 1}$. Then

$$\lim_{x \to 1} f(x) = \lim_{x \to 1} \frac{x^2 - 1}{x - 1} = \lim_{x \to 1} (x + 1) = 2$$

but $f(1)$ is indeterminate of the form $\frac{0}{0}$.

(c) Let $f(x) = \frac{1}{x}$. Then f is undefined at $x = 0$ but does not have an indeterminate form at $x = 0$.

3. The "simplify and plug-in" strategy is based on simplifying a function which is indeterminate to a continuous function. Once the simplification has been made, the limit of the remaining continuous function is obtained by evaluation.

Section 2.5 Exercises

1. $\lim\limits_{x \to 5} \frac{x^2 - 25}{x - 5} = \lim\limits_{x \to 5} (x + 5) = 10$ **3.** $\lim\limits_{t \to 7} \frac{2t - 14}{5t - 35} = \lim\limits_{t \to 7} \frac{2}{5} = \frac{2}{5}$

5. $\lim\limits_{x \to 8} \frac{x^2 - 64}{x - 8} = 16$ **7.** $\lim\limits_{x \to 2} \frac{x^2 - 3x + 2}{x - 2} = 1$ **9.** $\lim\limits_{x \to 2} \frac{x - 2}{x^3 - 4x} = \frac{1}{8}$

11. $\lim\limits_{h \to 0} \frac{(1+h)^3 - 1}{h} = 3$ **13.** $\lim\limits_{x \to 2} \frac{3x^2 - 4x - 4}{2x^2 - 8} = 1$

15. $\lim\limits_{y \to 2} \frac{(y-2)^3}{y^3 - 5y + 2} = 0$ **17.** $\lim\limits_{h \to 0} \frac{\frac{1}{3+h} - \frac{1}{3}}{h} = -\frac{1}{9}$

19. The limit does not exist.

21. $\lim\limits_{x \to 2} \frac{x - 2}{\sqrt{x} - \sqrt{4 - x}} = \sqrt{2}$ **23.** $\lim\limits_{x \to 2} \frac{\sqrt{x^2 - 1} - \sqrt{x + 1}}{x - 3} = 0$

25. $\lim\limits_{x \to 4} \left(\frac{1}{\sqrt{x} - 2} - \frac{4}{x - 4}\right) = \frac{1}{4}$ **27.** $\lim\limits_{x \to 0} \frac{\cot x}{\csc x} = 1$

29. $\lim\limits_{x \to \frac{\pi}{4}} \frac{\sin x - \cos x}{\tan x - 1} = \frac{\sqrt{2}}{2}$ **31.** $\lim\limits_{x \to 0} \frac{e^x - e^{2x}}{1 - e^x} = 1$

33. $\lim\limits_{x \to \frac{\pi}{3}} \frac{2\cos^2 x + 3\cos x - 2}{2\cos x - 1} = \frac{5}{2}$

35. $\lim\limits_{x \to 2} f(x) \approx 1.41 \approx \sqrt{2}$

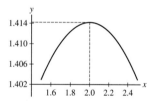

37. $\lim\limits_{x \to 1} \frac{x^3 - 1}{x - 1} = 3$ **39.** $\lim\limits_{x \to 1} \frac{x^2 - 3x + 2}{x^3 - 1} = -\frac{1}{3}$

41. $\lim\limits_{x \to 1} \frac{x^4 - 1}{x^3 - 1} = \frac{4}{3}$ **43.** $\lim\limits_{x \to 0} (2a + x) = 2a$

45. $\lim\limits_{t \to -1} (4t - 2at + 3a) = -4 + 5a$ **47.** $\lim\limits_{x \to 0} \frac{2(x+h)^2 - 2x^2}{h} = 2h$

49. $\lim\limits_{x \to a} \frac{\sqrt{x} - \sqrt{a}}{x - a} = \frac{1}{2\sqrt{a}}$ **51.** $\lim\limits_{x \to 0} \frac{(x+a)^3 - a^3}{x} = 3a^2$

53. $c = -1$, $c = 6$ **55.** +

Section 2.6 Preliminary

1. $\lim\limits_{x \to 0} f(x) = 0$; because $\lim\limits_{x \to 1/2} -x^4 = -\frac{1}{16} \neq \frac{1}{4} = \lim\limits_{x \to 1/2} x^2$, we do not have enough information to determine $\lim\limits_{x \to 1/2} f(x)$.

2. Assume that for $x \neq c$ (in some open interval containing c), $l(x) \leq f(x) \leq u(x)$ and that $\lim\limits_{x \to c} l(x) = \lim\limits_{x \to c} u(x) = L$. Then $\lim\limits_{x \to c} f(x)$ exists and $\lim\limits_{x \to c} f(x) = L$.

3. Yes   **4.** (a)

Section 2.6 Exercises

1. *Squeezed* at $x = 3$ and *trapped* at $x = 2$.

3. The Squeeze Theorem guarantees that $\lim\limits_{x \to 7} f(x) = 6$ provided $f(x)$ is squeezed in an open interval containing $x = 7$.

5. $\lim\limits_{x \to 0} x \cos \frac{1}{x} = 0$ **7.** $\lim\limits_{x \to 0+} \sqrt{x} e^{\cos(\pi/x)} = 0$

9. $\lim\limits_{x \to 0} \frac{\sin x \cos x}{x} = 1$ **11.** $\lim\limits_{t \to 0} \frac{\sin^2 t}{t} = 0$ **13.** $\lim\limits_{x \to 0} \frac{x^2}{\sin^2 x} = 1$

15. $\lim\limits_{t \to \frac{\pi}{4}} \frac{\sin t}{t} = \frac{2\sqrt{2}}{\pi}$

17. $\lim\limits_{x \to 0} \frac{\sin 10x}{x} = \lim\limits_{\theta \to 0} \frac{\sin \theta}{(\theta/10)} = \lim\limits_{\theta \to 0} \cdot 10 \frac{\sin \theta}{\theta}$
$= \lim\limits_{\theta \to 0} 10 \cdot \frac{\sin \theta}{\theta} = 10$

19. $\lim\limits_{h \to 0} \frac{\sin 6h}{h} = 6$ **21.** $\lim\limits_{h \to 0} \frac{\sin 6h}{6h} = 1$ **23.** $\lim\limits_{x \to 0} \frac{\sin 7x}{3x} = \frac{7}{3}$

25. $\lim\limits_{x \to 0} \frac{\tan 4x}{9x} = \frac{4}{9}$ **27.** $\lim\limits_{t \to 0} \frac{\tan 4t}{t \sec t} = 4$ **29.** $\lim\limits_{z \to 0} \frac{\sin(z/3)}{\sin z} = \frac{1}{3}$

31. $\lim\limits_{x \to 0} \frac{\tan 4x}{\tan 9x} = \frac{4}{9}$ **33.** $\lim\limits_{x \to 0} \frac{\sin 5x \sin 2x}{\sin 3x \sin 5x} = \frac{2}{3}$

35. $\lim\limits_{h \to 0} \frac{1 - \cos 2h}{h} = 0$ **37.** $\lim\limits_{t \to 0} \frac{1 - \cos t}{\sin t} = 0$

39. $\lim\limits_{h \to \frac{\pi}{2}} \frac{1 - \cos 3h}{h} = \frac{2}{\pi}$

41. (a) $\lim\limits_{x \to 0+} \frac{\sin x}{|x|} = 1$ **(b)** $\lim\limits_{x \to 0-} \frac{\sin x}{|x|} = -1$

43. Since $|\cos\left(\frac{1}{x}\right)| \leq 1$, it follows that $|\tan x \cos\left(\frac{1}{x}\right)| \leq |\tan x|$, which is equivalent to $-|\tan x| \leq \tan x \cos\left(\frac{1}{x}\right) \leq |\tan x|$;
$\lim\limits_{x \to 0} \tan x \cos\left(\frac{1}{x}\right) = 0$

45.

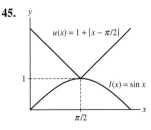

Any function $f(x)$ satisfying $l(x) \leq f(x) \leq u(x)$ for all x near $\pi/2$ will satisfy $\lim\limits_{x \to \pi/2} f(x) = 1$.

47.

h	-0.1	-0.01	0.01	0.1
$\frac{1 - \cos h}{h^2}$	0.499583	0.499996	0.499996	0.499583

The limit is $\frac{1}{2}$.

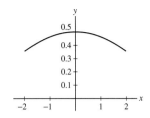

$$\lim\limits_{h \to 0} \frac{1 - \cos h}{h^2} = \lim\limits_{h \to 0} \frac{1 - \cos^2 h}{h^2 (1 + \cos h)}$$
$$= \lim\limits_{h \to 0} \left(\frac{\sin h}{h}\right)^2 \frac{1}{1 + \cos h} = \frac{1}{2}$$

49. $\lim\limits_{h \to 0} \frac{\cos 3h - 1}{\cos 2h - 1} = \frac{9}{4}$

53. (a) The area of each triangular piece of the n-gon is $\frac{1}{2} \sin \frac{2\pi}{n}$, so the area of the n-gon is $\frac{1}{2} n \sin \frac{2\pi}{n}$.

(b) As n increases, the difference between the n-gon and the unit circle shrinks to zero; hence, as n increases, $A(n)$ approaches the area of the unit circle.

(c) $\lim\limits_{n \to \infty} A(n) = \pi$

Section 2.7 Preliminary

1. $f(x) = x^2$ is continuous on $[0, 1]$ with $f(0) = 0$ and $f(1) = 1$. Because $f(0) < 0.5 < f(1)$, the Intermediate Value Theorem guarantees there is a $c \in [0, 1]$ such that $f(c) = 0.5$.

2. We must assume that temperature is a continuous function of time.

3. If f is continuous on $[a, b]$, then the horizontal line $y = k$ for every k between $f(a)$ and $f(b)$ intersects the graph of $y = f(x)$ at least once.

4.

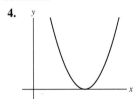

Section 2.7 Exercises

1. f is continuous everywhere with $f(1) = 2$ and $f(2) = 10$.

3. g is continuous for all $t \in [0, \frac{\pi}{4}]$ with $g(0) = 0$ and $g(\frac{\pi}{4}) = \frac{\pi^2}{16} > \frac{1}{2}$.

5. $f(x) = x - \cos x$ is continuous everywhere with $f(0) = -1$ and $f(1) = 1 - \cos 1 \approx 0.46$.

7. $f(x) = \sqrt{x} + \sqrt{x + 1} - 2$ is continuous on $\left[\frac{1}{4}, 1\right]$ with $f(\frac{1}{4}) = \sqrt{\frac{1}{4}} + \sqrt{\frac{5}{4}} - 2 \approx -0.38$ and $f(1) = \sqrt{2} - 1 \approx 0.41$.

11. $f(x) = x^k - \cos x$ is continuous on $\left[0, \frac{\pi}{2}\right]$ with $f(0) = -1$ and $f(\frac{\pi}{2}) = \left(\frac{\pi}{2}\right)^k > 0$.

15. $f(x) = e^x + \ln x$ is continuous on $[0.1, 1]$ with $f(0.1) < 0$ and $f(1) = e > 0$.

17. Note that $f(x)$ is continuous for all x.
(a) $f(1) = 1$, $f(1.5) = 2^{1.5} - (1.5)^3 < 3 - 3.375 < 0$. Hence, $f(x) = 0$ for some x between 0 and 1.5.
(b) $f(1.25) \approx 0.4253 > 0$ and $f(1.5) < 0$. Hence, $f(x) = 0$ for some x between 1.25 and 1.5.
(c) $f(1.375) \approx -0.0059$. Hence, $f(x) = 0$ for some x between 1.25 and 1.375.

19. $[0, 0.25]$

21. 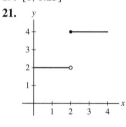 **23.**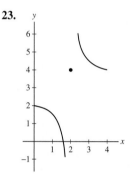

25. If $[a, b]$ contains $[-1, 1]$, $f(a) > 0$ and $f(b) > 0$. If we were using the corollary, we would guess that $f(x)$ has no roots in $[a, b]$.

Section 2.8 Preliminary

1. (c) **2.** (b) and (d)

Section 2.8 Exercises

1. (a) $|f(x) - 35| = |8x + 3 - 35| = |8x - 32| = |8(x - 4)|$
$= 8|x - 4|$.
(b) Given $\epsilon > 0$, let $\delta = \epsilon/8$ and suppose $|x - 4| < \delta$. By part **(a)**, $|f(x) - 35| = 8|x - 4| < 8\delta$. Substituting $\delta = \epsilon/8$, we see $|f(x) - 35| < 8\epsilon/8 = \epsilon$.

3. (a) If $0 < |x - 2| < \delta = 0.01$, then $|x| < 3$ and
$$|x^2 - 4| = |x - 2||x + 2| \le |x - 2|(|x| + 2) < 5|x - 2| < 0.05$$
(b) If $0 < |x - 2| < \delta = 0.0002$, then $|x| < 2.0002$ and
$$|x^2 - 4| \le |x - 2|(|x| + 2) < 4.0002|x - 2| < 0.00080004 < 0.0009$$
(c) $\delta = 10^{-5}$

5. $\delta = 6 \times 10^{-4}$

7. $\delta = 0.005$

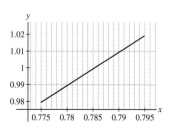

9. For $\epsilon = 0.5$, take $\delta = 0.18$; for $\epsilon = 0.2$, take $\delta = 0.075$; for $\epsilon = 0.1$, take $\delta = 0.035$

11. (b) $\delta = 0.0002$

13. $c = 3$, $L = 10.1$, $\epsilon = 0.3$, $\delta = 0.1$

15. Given $\epsilon > 0$, let $\delta = \epsilon/|a|$. Then, whenever $|x - c| < \delta$,
$$|f(x) - (ac + b)| = |a| \, |x - c| < |a| \cdot \epsilon/|a| = \epsilon.$$

Chapter 2 Review

1. Average velocity: $\frac{\sqrt{26} - \sqrt{5}}{3} \approx 0.954$ m/s; instantaneous velocity: 0.894 m/s.

3. The tangent line drawn in the figure appears to pass through the points $(15, 140)$ and $(10.5, 40)$; therefore, ROC $\approx \frac{200}{9}$.

5. $\lim_{x \to 0} \frac{1 - \cos^3(x)}{x^2} \approx 1.50$ **7.** $\lim_{x \to 2} \frac{x^x - 4}{x^2 - 4} \approx 1.69$

9. $\lim_{x \to 4} \left(3 + x^{\frac{1}{2}}\right) = 5$ **11.** $\lim_{x \to 2} \frac{4}{x^3} = -\frac{1}{2}$ **13.** $\lim_{x \to -1} \frac{x^3 - x}{x - 1} = 2$

15. $\lim_{t \to 9} \frac{\sqrt{t} - 3}{t - 9} = \frac{1}{6}$ **17.** $\lim_{x \to 3} \frac{\sqrt{x+1} - 2}{x - 3} = \frac{1}{4}$

19. $\lim_{h \to 0} \frac{2(a+h)^2 - 2a^2}{h} = 4a$ **21.** The two-sided limit does not exist.

23. $\lim_{a \to b} \frac{a^2 - 3ab + 2b^2}{a - b} = -b$ **25.** $\lim_{x \to 1} \frac{x^3 - ax^2 + ax - 1}{x - 1} = 3 - a$

27. The limit does not exist. **29.** $\lim_{x \to 4.3} \frac{1}{x - [x]} = \frac{10}{3}$

31. $\lim_{x \to 0+} \frac{[x]}{x} = 0$ **33.** The limit does not exist.

35. $\lim_{\theta \to 0} \frac{\sin 5\theta}{\theta} = 5$ **37.** The limit does not exist.

39. The limit does not exist. **41.** $\lim_{x \to 0} \frac{\sin 7x}{\sin 3x} = \frac{7}{3}$

43. $\lim_{\theta \to 0} \frac{\tan \theta - \sin \theta}{\sin^3 \theta} = \frac{1}{2}$

45.

47.

49. Because e^x and $\sin x$ are continuous for all real numbers, their composition, $e^{\sin x}$ is continuous for all real numbers. Moreover, x is continuous for all real numbers, so the product $xe^{\sin x}$ is continuous for all real numbers. Thus, $f(x) = xe^{\sin x}$ is continuous for all real numbers.

51. (a) $\lim_{x \to 3} (f(x) - 2g(x)) = -2$ (b) $\lim_{x \to 3} x^2 f(x) = 54$

(c) $\lim_{x \to 3} \frac{f(x)}{g(x)+x} = \frac{6}{7}$ (d) $\lim_{x \to 3} (2g(x)^3 - g(x)^2) = 112$

53. Let $f(x) = \frac{1}{(x-a)^3}$ and $g(x) = \frac{1}{(x-a)^5}$. Then, neither A nor B exists but $L = 0$.

57. Let $r_1 = \frac{20}{1+\sqrt{1-20a}}$ and $r_2 = \frac{20}{1-\sqrt{1-20a}}$.

59. $\delta = 0.55$

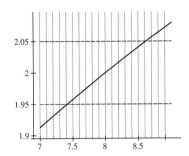

61. Given $\epsilon > 0$, let $\delta = \min\{1, \epsilon/6\}$. Because $\delta \leq 1$, $|x - 3| < \delta$ guarantees $|x + 2| < 6$. Then, whenever $|x - 3| < \delta$,

$$|f(x) - 6| = |x - 3|\,|x + 2| < 6|x - 3| < 6\delta \leq \epsilon.$$

63. $f(x)$ is continuous on $[0, 2]$ with $f(0) = -\frac{2}{3} < 0$ and $f(2) \approx 4.21 > 0$

65. 2.65

Chapter 3

Section 3.1 Preliminary

1. A and D **2.** $\frac{f(x)-f(a)}{x-a}$ or $\frac{f(a+h)-f(a)}{h}$ **3.** $x = 7$

4. (a) The difference in height between the points $(0.9, \sin 0.9)$ and $(1.3, \sin 1.3)$.

(b) The slope of the secant line between the points $(0.9, \sin 0.9)$ and $(1.3, \sin 1.3)$ on the graph.

(c) The slope of the tangent line to the graph at $x = 1.3$.

5. $a = 3$ and $h = 2$ **6.** $f(x) = \tan x$ at $x = \frac{\pi}{4}$

Section 3.1 Exercises

1. $f(2 + h) = 3(2 + h)^2 = 3(4 + 4h + h^2) = 12 + 12h + 3h^2$;

$\frac{f(2+h)-f(2)}{h} = \frac{12+12h+3h^2-12}{h} = \frac{12h+3h^2}{h} = 12 + 3h$;

$f'(2) = 12$

3. $f'(0) = 9$ **5.** $f'(-1) = -2$

7. $f(0) \approx 6$ and $f(2.5) \approx 2$, so the slope is -1.6.

9. For $h = 0.5$, the estimate is 2 [greater than $f'(2)$]; for $h = -0.5$, the estimate is 1 [smaller than $f'(2)$].

11. $f'(1) = f'(2) = 0$; $f'(4) = \frac{1}{2}$; $f'(7) = 0$

13. $f'(5.5)$ is larger than $f'(6.5)$. **15.** $f'(x) = 3$ for all x

17. $f'(t) = -1$ for all t

19. $f(-2 + h) = \frac{1}{-2+h}$. The difference quotient is $-\frac{1}{3}$.

21. $f'(9) = \frac{1}{6}$ **23.** $f'(2) = 14$; $y = 14x - 12$

25. $f'(2) = 12$; $y = 12x - 16$ **27.** $f'(0) = 1$; $y = x$

29. $f'(3) = -\frac{1}{9}$; $y = \left(\frac{1}{3}\right) - \left(\frac{1}{9}\right)(x - 3)$

31. $f'(-7) = 9$; $y = 9x - 4$ **33.** $f'(-2) = -1$; $y = -x - 1$

35. $f'(-1) = \frac{1}{2}$; $y = \frac{x}{2} + \frac{3}{2}$ **37.** $f'(1) = -3$; $y = 4 - 3t$

39. $f'(9) = -\frac{1}{54}$; $y = \left(-\frac{x}{54}\right) + \left(\frac{1}{2}\right)$ **41.** $y = 5 + 2(x - 3)$

43. The tangent at any point is $y = 2x + 8$.

45. $f(1) = \frac{1}{1+1^2} = \frac{1}{2}$; $L(1) = \frac{1}{2} + m(1 - 1) = \frac{1}{2}$;

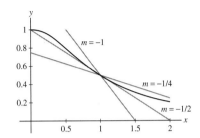

$f'(1) = -\frac{1}{2}$

47.

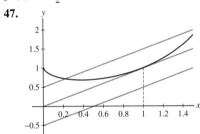

(a) The graph of $y = x$ appears to be tangent to the graph of $f(x)$ at $x = 1$. We therefore estimate that $f'(1) = 1$.

(b) With $x_0 = 1$, we generate the table

h	0.01	0.001	0.0001
$(f(x_0 + h) - f(x_0))/h$	1.010050	1.001001	1.000100

49. (a) $P'(350)$ is larger than $P'(300)$.

(b) $P'(303) \approx 0.00265$ atm/K;

$P'(313) \approx 0.004145$ atm/K;

$P'(323) \approx 0.006295$ atm/K;

$P'(333) \approx 0.00931$ atm/K;

$P'(343) \approx 0.013435$ atm/K.

51. $V'(120) \approx 9.75$ **53.** $f(x) = x^3$ and $a = 5$

55. $f(x) = \sin x$ and $a = \frac{\pi}{6}$ **57.** $f(x) = 5^x$ and $a = 2$

59. $f'\left(\frac{\pi}{2}\right) = 0$

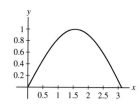

h	-0.01	-0.001	-0.0001	0.0001	0.001	0.01
$\dfrac{\sin\left(\frac{\pi}{2}+h\right)-1}{h}$	0.005	0.0005	0.00005	-0.00005	-0.0005	-0.005

61. (a) $f'(0) \approx -0.68$

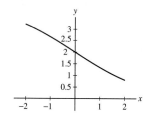 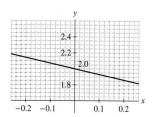

(b) $y = 2 - 0.68x$

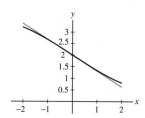

63. $f'\left(\frac{\pi}{2}\right) \approx -1$

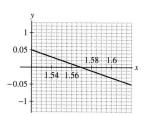

65. $f'\left(\frac{\pi}{4}\right) \approx 0.7071$

67. (b) $f'(4) \approx 20.0000$

(c) $y = 20x - 48$

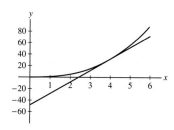

69. $W'(4) \approx 0.9$ kg/yr; horizontal tangent: $t = 10$ and $t = 11.6$; negative slope: $10 < t < 11.6$

71. $v'(10) \approx 7.9125;\quad i(10) = 0.495485$

77. Figure (A) satisfies the inequality.

Section 3.2 Preliminary

1. The slope is 8.

2. $(f - g)'(1) = -2;\quad (3f + 2g)'(1) = 19$; can't evaluate $(fg)'(1)$

3. (a) Yes **(b)** No **(c)** Yes **(d)** No **(e)** No **(f)** Yes

4. $x^n - a^n = (x - a)(x^{n-1} + x^{n-2}a + x^{n-3}a^2 + \cdots + xa^{n-2} + a^{n-1})$

5. Yes **6.** Vertical tangent

7. The line tangent to $f(x) = e^x$ at $x = 0$ has slope equal to 1.

8. 1; 7

Section 3.2 Exercises

1. $f'(x) = 4$ **3.** $f'(x) = -4x$ **5.** $f'(x) = -\dfrac{1}{x^2}$

7. $f'(x) = \dfrac{1}{2\sqrt{x}}$ **9.** $\dfrac{d}{dx}x^4\Big|_{x=-2} = -32$ **11.** $\dfrac{d}{dt}t^{2/3}\Big|_{t=8} = \dfrac{1}{3}$

13. $\dfrac{d}{dx}x^{0.35} = 0.35x^{-0.65}$ **15.** $\dfrac{d}{dt}t^{\sqrt{17}} = \sqrt{17}\,t^{\sqrt{17}-1}$

17. $f'(1) = 5;\quad y = 5(x - 1) + 1$

19. $f'(9) = \dfrac{17}{2};\quad y = \dfrac{17}{2}x + \dfrac{9}{2}$

21. (a) $\dfrac{d}{dx}9e^x = 9e^x$ **(b)** $\dfrac{d}{dt}(3t - 4e^t) = 3 - 4e^t$

(c) $\dfrac{d}{dt}e^{t+2} = e^{t+2}$

23. $\dfrac{d}{dx}(x^3 + x^2 - 12) = 3x^2 + 2x$

25. $\dfrac{d}{dx}(2x^3 - 10x^{-1}) = 6x^2 + 10x^{-2}$

27. $\dfrac{d}{dz}(7z^{-3} + z^2 + 5) = 2z - 21z^{-4}$

29. $f'(s) = \dfrac{1}{4}s^{-3/4} + \dfrac{1}{3}s^{-2/3}$

31. $\dfrac{d}{dx}((x + 1)^3) = 3x^2 + 6x + 3$

33. $\dfrac{d}{dz}((3z - 1)(2z + 1)) = 12z + 1$

35. $\dfrac{d}{dx}e^2 = 0$ **37.** $\dfrac{d}{dt}5e^{t-3} = 5e^{t-3}$

39. $f'(2) = -\dfrac{3}{8}$ **41.** $\dfrac{dT}{dC}\Big|_{C=8} = 1$

43. $\dfrac{ds}{dz}\Big|_{z=2} = -60$ **45.** $\dfrac{dp}{dh}\Big|_{h=4} = 7e^2$

47. Graph (A) matches the graph in (III).
Graph (B) matches the graph in (I).
Graph (C) matches the graph in (II).
Graph (D) matches the graph in (III).
The graph in (D) is just a vertical translation of the graph in (A) which means the two functions differ by a constant. The derivative of a constant is zero, so the two functions end up with the same derivative.

49.

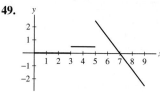

51. (a) $\frac{d}{dR}R = 1$ **(b)** $\frac{d}{dR}r = 0$ **(c)** $\frac{d}{dR}r^2R^3 = 3r^2R^2$

53. $x = \frac{1}{2}$ **55.** $x = 0$ or $x = \frac{3}{4}$ **57.** $a = 2$ and $b = -3$

59. The two parallel tangent lines with slope $m = 2$ are shown with the graph of $f(x)$ here:

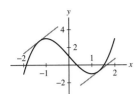

61. $f'(x) = -\frac{2}{x^3}$ **63.** $m_4 \approx 1.386$; $y'(0) \approx 1.386$; $y'(2) \approx 22.176$

65. $\frac{d}{dT}v_{\text{avg}}\Big|_{T=300°\text{K}} = 0.74234$

67. $\frac{dK}{dm}\Big|_{m=68} = 0.00315966$; $\frac{d}{dm}\left(\frac{K}{m}\right)\Big|_{m=68} = -8.19981 \times 10^{-6}$

69. The area is 2 regardless.

71. Graph (A) matches (III). Graph (B) matches (I). Graph (C) matches (II).

73. 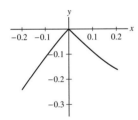 **75.** $g(x)$ is the derivative of $f(x)$.

77. $x = 1$ **79.** $x = 0$ **81.** $x = -1$; $x = 1$

83. f is not differentiable at $x = 0$. The tangent line does not exist.

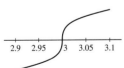

85. f is not differentiable at $x = 3$. The tangent line appears to be vertical.

87. f is not differentiable at $x = 0$. The tangent line does not exist at this point.

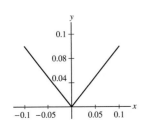

89. This statement is false. The function $|x|$ is not differentiable at $x = 0$, but it is continuous.

91. tangent line: $y = 2ax - a^2$; x-intercept: $\frac{a}{2}$

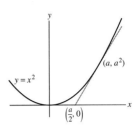

93.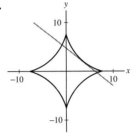

The tangent line to f at $x = a$ is

$$y = \left(4 - a^{2/3}\right)^{3/2} - \frac{\sqrt{4-a^{2/3}}(x-a)}{a^{1/3}}$$

The y-intercept of this line is the point $P = \left(0, 4\sqrt{4 - a^{2/3}}\right)$, its x-intercept is the point $Q = \left(4a^{1/3}, 0\right)$, and the distance between P and Q is 8.

95. Radius $\leq 1/2$ **97.** $\frac{d}{dx}(x^k) = kx^{k-1}$ for negative integers k.

101. (a) The equation has a unique solution when $\lambda < 0$ and when $\lambda = e$.

(b) The equation has at least one solution when $\lambda < 0$ and when $\lambda \geq e$.

Section 3.3 Preliminary

1. (a) False: The notation fg denotes the function whose value at x is $f(x)g(x)$.

(b) True **(c)** False: The derivative of the product fg is $f'(x)g(x) + f(x)g'(x)$.

2. (a) False: $\frac{d}{dx}(fg)\Big|_{x=4} = f(4)g'(4) + g(4)f'(4)$.

(b) False: $\frac{d}{dx}(f/g)\Big|_{x=4} = [g(4)f'(4) - f(4)g'(4)]/g(4)^2$.

(c) True

3. $\frac{d}{dx}(f/g)\Big|_{x=1} = -1$ **4.** $g(1) = 5$

Section 3.3 Exercises

1. $f'(x) = 3x^2 + 1$ **3.** $f'(x) = e^x(x^3 + 3x^2 - 1)$

5. $\left.\frac{dy}{dt}\right|_{t=3} = 82$ **7.** $f'(x) = \frac{-2}{(x-2)^2}$ **9.** $\left.\frac{dg}{dt}\right|_{t=-2} = \frac{8}{9}$

11. $g'(x) = -\frac{e^x}{(1+e^x)^2}$ **13.** $f'(t) = 6t^2 + 2t - 4$

15. $g'(x) = 2x + 2 - 6x^{-3}$

17. $f'(x) = 6x^5 + 5x^4 + 4x^3 - 8x - 4$ **19.** $\left.\frac{dy}{dx}\right|_{x=2} = -\frac{1}{36}$

21. $f'(x) = 1$ **23.** $\left.\frac{dy}{dx}\right|_{x=2} = -80$ **25.** $\left.\frac{dz}{dx}\right|_{x=1} = -\frac{3}{4}$

27. $h'(t) = -\frac{10t^{11} + 8t^9 + 3t^4 + t^2}{(t^{11} + t^9 + t^4 + t^2)^2}$ **29.** $f'(t) = 0$

31. $f'(x) = 3x^2 - 6x - 13$ **33.** $f'(x) = \frac{e^x(x - e^x)}{(e^x + 1)^2(x+1)^2}$

35. $g'(z) = 2z - 1$ **37.** $\frac{d}{dt}\left(\frac{xt-4}{t^2-x}\right) = \frac{-xt^2 + 8t - x^2}{(t^2-x)^2}$

39. $f'(x) > 0$ for $-1 < x < 1$, $f'(x) < 0$ for $|x| > 1$. $f'(x)$ is plotted on the left, $f(x)$ on the right.

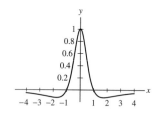 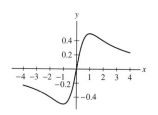

41. $\frac{dP}{dr} = -\frac{2V^2R}{(R+r)^3}$

43. (a) $\left.\frac{dI}{dR}\right|_{R=6} = -\frac{2}{3}$ **(b)** $\left.\frac{dV}{dR}\right|_{R=6} = 4$

45. At $x = -1$, the tangent line is $y = \frac{1}{2}x + 1$; at $x = 1$, the tangent line is $y = 1 - \frac{1}{2}x$.

47. $(f^2)' = (ff)' = ff' + ff' = 2ff'$

49. $(fg)'(4) = -17$; $\left(\frac{f}{g}\right)'(4) = -\frac{13}{25}$

51. $G'(4) = -58$ **53.** $F'(0) = -7$

55. $(fgh)' = f(gh' + hg') + ghf' = fgh' + fg'h + f'gh$

59. Let $g(x) = 1/f(x)$. Then by the quotient rule,

$$g'(x) = \frac{f(x)\cdot 0 - f'(x)}{f^2(x)} = -\frac{f'(x)}{f^2(x)}.$$

63. (a) $c = -1$ is a multiple root

(b) $c = -1$ is a root, but not a multiple root

Section 3.4 Preliminary

1. (a) atmospheres/meter **(b)** moles/(liter · hour)

2. $f(5) = 13$ **3.** 90 mph **4.** $f(26) \approx 43.75$

5. (a) Rate of change of the population of Freedonia in the year 1933.

(b) If $P'(1933) = 0.2$, then $P(1934) \approx 5.2$ million. If $P'(1933) = 0$, then $P(1934) \approx 5$ million.

Section 3.4 Exercises

1. When $s = 3$, the area changes at a rate of 6 square units per unit increase. When $s = 5$, the area changes at a rate of 10 square units per unit increase.

3. If $x = 1$, the ROC is -1. If $x = 10$, the ROC is $-\frac{1}{100}$.

5. $dV/dr = 3\pi r^2$ **7.** $dV/dr = 4\pi r^2$

9. (a) Average velocity over the interval [0.5, 1] is 50 km/h.

(b) Average velocity is greater over [2, 3]. **(c)** $t = 2.5$ h

11. The tangent line sketched in the figure below appears to pass through the points (10, 3) and (30, 4). Thus, the ROC of voltage at $t = 20$ seconds is approximately

$$\frac{4-3}{30-10} = 0.05 \text{ V/s}.$$

As we move to the right of the graph, the tangent lines to it grow shallower, indicating that the voltage changes more slowly as time goes on.

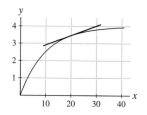

13. (a) 35.25 ft **(b)** -7 ft/s **(c)** ≈ 2.36 s

15. -15 degrees Fahrenheit per minute

17. $F'(6.77 \times 10^6) = -1.93 \times 10^{-4}$ N/m

19. $P'(3) = -0.1312$ W/Ω; $P'(5) = -0.0609$ W/Ω

21. (a) $\frac{dV}{dv} = -1$ **(b)** $\frac{dV}{dt} = -4$

23. (a) 1.0618 million; 10,600 people per year **(b)** Year 2010

27. (a)

t	Jan	Feb	Mar	Apr	May	Jun	Jul	Aug	Sep	Oct	Nov
$P'(t)$	146	155	175	191	210	227	252	242	233	219	185

(b) Here is a plot of these estimates (1 = Jan, 2 = Feb, etc.)

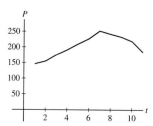

(c) "U.S. Growth Rate Declines After Midsummer Peak"

29. Denote $f(w) = \sqrt{hw}/60$ so that: $f'(w) = \frac{\sqrt{5}}{20\sqrt{w}}$;

$f'(70) = \frac{\sqrt{14}}{280} \approx 0.0133631 \frac{\text{m}^2}{\text{kg}}$; $f'(80) = \frac{1}{80}\frac{\text{m}^2}{\text{kg}}$

31. Falling at 76.68 m/s

33. $v_0 = 19.6$ m/s. Peak height is 19.6 m.

35. Either the 16th or 17th floor.

39. Approximation: $\sqrt{2} - \sqrt{1} \approx \frac{1}{2}$
Value up to six decimal places: $\sqrt{2} - \sqrt{1} = 0.414214$
Approximation: $\sqrt{101} - \sqrt{100} \approx 0.05$
Value up to six decimal places: $\sqrt{101} - \sqrt{100} = 0.0498756$

41. $F(65) = 198.25$ ft
Approximation: $F(66) - F(65) \approx F'(65) = 5$ ft
Actual value: $F(66) - F(65) = 5.03$ ft

43. $C(2000) = \$796$
Approximation: $C(2001) - C(2000) \approx C'(2000) = \$0.244$
Actual value: $C(2001) - C(2000) = \$0.244$

45. $D(40) = 22.5$ barrels per year. An increase in oil prices of a dollar leads to a decrease in demand of 0.5625 barrels a year, and a decrease of a dollar leads to an increase in demand of 0.5625 barrels a year.

49. **(a)** 7.09306 km/yr **(b)** 0.75992 kg

51. **(a)** Growing when $P = 250$; Shrinking when $P = 350$

(b) 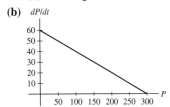 **(c)** Graph (A)

53. At point A, average cost is greater than marginal cost.
At point B, average cost is greater than marginal cost.
At point C, average cost and marginal cost are nearly the same.
At point D, average cost is less than marginal cost.

Section 3.5 Preliminary

1. The second derivative of GNP is negative.
The first derivative is positive.

2. The first derivative of stock prices is postive.
The second derivative is negative.

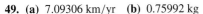

3. True **4.** Polynomials of the form $ax + b$. **5.** e^x

Section 3.5 Exercises

1. $y'' = 28$ and $y''' = 0$ **3.** $y'' = 12x^2 - 50$ and $y''' = 24x$

5. $y'' = 8\pi r$ and $y''' = 8\pi$

7. $y'' = -\frac{16}{5}t^{-6/5} + \frac{4}{3}t^{-4/3}$ and $y''' = \frac{96}{25}t^{-11/15} - \frac{16}{9}t^{-7/3}$

9. $y'' = -2z^{-3}$ and $y''' = 6z^{-4}$

11. $y'' = 20x^3 + 12x^2 + 2$ and $y''' = 60x^2 + 24x$

13. $y'' = e^x$ and $y''' = e^x$

15. $f^{(4)}(1) = 24$ **17.** $\left.\frac{d^2y}{dt^2}\right|_{t=1} = 54$ **19.** $h'''(9) = \frac{1}{648}$

21. $\left.\frac{d^4x}{dt^4}\right|_{t=16} = \frac{3,465}{134,217,728}$ **23.** $g''(1) = -\frac{1}{4}$ **25.** $h''(1) = \frac{1}{8}$

27. $f''(0) = -\frac{1}{2}$

29. $y^{(0)}(0) = d$, $y^{(1)}(0) = c$, $y^{(2)}(0) = 2b$, $y^{(3)}(0) = 6a$,
$y^{(4)}(0) = 24$, and $y^{(5)}(0) = 0$

31. $\frac{d^6}{dx^6}x^{-1} = 720x^{-7}$ **33.** $f^{(n)}(x) = (-1)^n n!(x+1)^{-(n+1)}$

35. $f^{(n)}(x) = (-1)^n \frac{(2n-1)\times(2n-3)\times\cdots\times1}{2^n} x^{-(2n+1)/2}$

37. **(a)** $a(5) = -90$ ft/min^2

(b) As the acceleration of the helicopter is negative, the velocity of the helicopter must be decreasing. Because the velocity is positive, the helicopter is slowing down.

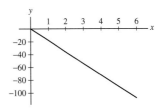

39. **(A)** f'' **(B)** f' **(C)** f

41. From time 15 to time 20 and from time 32 on.

43. $f(x) = x^2 - 2x$

45. **(a)** Does not apply. **(b)** Can apply. **(c)** Can apply.

47. $\frac{dS}{dQ} = -2882Q^{-2} - 0.052$; $\frac{d^2S}{dQ^2} = 5764Q^{-3}$

49. **(a)** Set $\frac{d^2s}{dt^2} = 0$ and solve for $v = \frac{ds}{dt}$.

(b) If $D = 0.003$ ft, then $v_{term} \approx 13.8564$ ft/s; if $D = 0.0003$ ft, then $v_{term} \approx 4.38178$ ft/s.

(c) Larger velocities correspond to lower acceleration.

51. $f'(x) = -\frac{3}{(x-1)^2} = (-1)^1 \frac{3\cdot1}{(x-1)^{1+1}}$;
$f''(x) = \frac{6}{(x-1)^3} = (-1)^2 \frac{3\cdot2\cdot1}{(x-1)^{2+1}}$;
$f'''(x) = -\frac{18}{(x-1)^4} = (-1)^3 \frac{3\cdot3!}{(x-1)^{3+1}}$; and
$f^{(4)}(x) = \frac{72}{(x-1)^5} = (-1)^4 \frac{3\cdot4!}{(x-1)^{4+1}}$.
We conjecture the general formula for $f^{(k)}(x)$ is
$f^{(k)}(x) = (-1)^k \frac{3\cdot k!}{(x-1)^{k+1}}$.

53. $p^{(99)}(x) = 99!$

55. $(fg)''' = f'''g + 3f''g' + 3f'g'' + fg'''$; We conjecture the general formula for $(fg)^{(n)}$ is $(fg)^{(n)} = \sum_{k=0}^{n} \binom{n}{k} f^{(n-k)}g^{(k)}$.

57. $f'(x) = (x^2 + 2x)e^x$;
$f''(x) = (x^2 + 4x + 2)e^x$;
$f'''(x) = (x^2 + 6x + 6)e^x$; and
$f^{(4)}(x) = (x^2 + 8x + 12)e^x$;
We conjecture the general formula for $f^{(n)}(x)$ is
$f^{(n)}(x) = (x^2 + 2nx + n(n-1))e^x$.

Section 3.6 Preliminary

1. (a) $\frac{d}{dx}(\sin x + \cos x) = -\sin x + \cos x$

(b) $\frac{d}{dx}\tan x = \sec^2 x$ (c) $\frac{d}{dx}\sec x = \sec x \tan x$

(d) $\frac{d}{dx}\cot x = -\csc^2 x$

2. (a) and (c) using the Product Rule.

3. $\frac{d}{dx}(\sin^2 x + \cos^2 x) = 0$

4. The difference quotient for the function $\sin x$ involves the expression $\sin(x + h)$. The addition formula for the sine function is used to expand this expression as
$\sin(x + h) = \sin x \cos h + \sin h \cos x$.

Section 3.6 Exercises

1. $y - \frac{\sqrt{2}}{2} = \frac{\sqrt{2}}{2}\left(x - \frac{\pi}{4}\right)$ 3. $y - 1 = 2\left(x - \frac{\pi}{4}\right)$

5. $f'(x) = -\sin^2 x + \cos^2 x$ 7. $f'(x) = 2\sin x \cos x$

9. $f'(x) = x^3 \cos x + 3x^2 \sin x$

11. $f'(\theta) = (\tan^2 \theta + \sec^2 \theta)\sec \theta$

13. $h'(\theta) = -e^{-\theta}\cos\theta(\cos\theta + 2\sin\theta)$

15. $f'(x) = (x - x^2)(-\csc^2 x) + \cot x(1 - 2x)$

17. $f'(x) = \frac{x \sec x \tan x - 2\sec x}{x^3}$

19. $g'(t) = \cos t - 2\sec t \tan t$

21. $f'(x) = \frac{2 + \sin x - x \cos x}{(2 + \sin x)^2}$ 23. $f'(x) = \frac{2\cos x}{(1 - \sin x)^2}$

25. $g'(x) = \frac{(x\tan x - 1)\sec x}{x^2}$ 27. $f''(x) = -3\sin x - 4\cos x$

29. $g''(\theta) = 2\cos\theta - \theta\sin\theta$ 31. $y = x$

33. $y = \left(1 - \frac{3\sqrt{3}}{2}\right)x + \sqrt{3} + \frac{3}{2} + \frac{\sqrt{3}}{2}\pi - \frac{\pi}{3}$

35. $y = (2 - \sqrt{2})x + \sqrt{2} - 1 + \frac{\pi}{4}(\sqrt{2} - 2)$

37. $y = x + 1$

43. $f'(x) = -\sin x$, $f''(x) = -\cos x$, $f'''(x) = \sin x$, $f^{(4)}(x) = \cos x$, and $f^{(5)}(x) = -\sin x$; $f^{(8)}(x) = \cos x$; $f^{(37)}(x) = -\sin x$.

45. $f''(x) = 2\sec^2 x \tan x$
$f'''(x) = 2\sec^4 x + 4\sec^2 x \tan^2 x$

47. (a) (b) $c_0 = 4.493409$

(c)

49. $v\left(\frac{\pi}{3}\right) = 20$ cm/s; $a\left(\frac{\pi}{3}\right) = -20\sqrt{3}$ cm/s^2

55. (a) $f'(x) = x \cos x + (\sin x) \cdot 1 = g(x) + \sin x$ and $g'(x) = (x)(-\sin x) + (\cos x) \cdot 1 = -f(x) + \cos x$

(b) $f''(x) = g'(x) + \cos x = -f(x) + 2\cos x$ and $g''(x) = -f'(x) - \sin x = -g(x) - 2\sin x$

Section 3.7 Preliminary

1. (a) Outer function is $\sqrt{x}$, inner function is $4x + 9x^2$.

(b) Outer function is $\tan x$, inner function is $x^2 + 1$.

(c) Outer function is x^5, inner function is $\sec x$.

(d) Outer function is x^4, inner function is $1 + e^x$.

2. The functions $\frac{x}{x+1}$, $\sqrt{x} \cdot \sec x$, and xe^x.

3. (b) $5f'(5x)$

4. (a) Once (b) Twice (c) Three times

(d) Twice

5. We do not have enough information—value of $f'(1)$ is missing.

Section 3.7 Exercises

1.
$f(g(x))$	$f'(u)$	$f'(g(x))$	$g'(x)$	$(f \circ g)'$
$(x^4 + 1)^{3/2}$	$\frac{3}{2}u^{1/2}$	$\frac{3}{2}(x^4 + 1)^{1/2}$	$4x^3$	$6x^3(x^4 + 1)^{1/2}$

3.
$f(g(x))$	$f'(u)$	$f'(g(x))$	$g'(x)$	$(f \circ g)'$
$\tan(x^4)$	$\sec^2 u$	$\sec^2(x^4)$	$4x^3$	$4x^3 \sec^2(x^4)$

5. $\frac{d}{dx}(x + \sin x)^4 = 4(x + \sin x)^3(1 + \cos x)$

7. (a) $\frac{d}{dx}\cos(9 - x^2) = 2x \sin(9 - x^2)$

(b) $\frac{d}{dx}\cos(x^{-1}) = \frac{\sin(x^{-1})}{x^2}$

(c) $\frac{d}{dx}\cos(\tan x) = -\sec^2 x \sin(\tan x)$

9. $\frac{d}{dx}(x^2 + 9)^4 = 4(x^2 + 9)^3(2x)$ 11. $\frac{d}{dx}\sqrt{11x + 4} = \frac{11}{2\sqrt{11x + 4}}$

13. $\frac{d}{dx}e^{10 - x^2} = -2xe^{10 - x^2}$ 15. $\frac{d}{dx}\sin(2x + 1) = 2\cos(2x + 1)$

17. $\frac{d}{dx}e^{x + x^{-1}} = \left(1 - x^{-2}\right)e^{x + x^{-1}}$

19. $\frac{d}{dx}f(g(x)) = -\sin(x^2 + 1)(2x)$; $\frac{d}{dx}g(f(x)) = -2\sin x \cos x$

21. $y' = 2x \cos(x^2)$ 23. $y' = -8t \csc^2(4t^2 + 9)$

25. $y' = -\frac{5(2t + 3)}{2(t^2 + 3t + 1)^{7/2}}$ 27. $y' = \frac{8(1 + x)^3}{(1 - x)^5}$

29. $y' = -12e^{4\theta}\cos^2(e^{4\theta})\sin(e^{4\theta})$ 31. $y' = \frac{x\sin(x^2)}{(1 + \cos(x^2))^{3/2}}$

33. $y' = -\frac{1}{x^2}e^{1/x}$ 35. $\frac{dy}{dx} = 5\sec^2(5x)$

37. $y' = 3x\sin(1 - 3x) + \cos(1 - 3x)$ 39. $\frac{dy}{dt} = 2(4t + 9)^{-1/2}$

41. $y' = 4(\sin x - 3x^2)(x^3 + \cos x)^{-5}$ 43. $y' = \frac{\sqrt{2}\cos 2x}{2\sqrt{\sin 2x}}$

45. $\frac{dy}{dz} = (z + 1)^3(2z - 1)^2(14z + 2)$

47. $y' = \frac{1}{2x^2\sqrt{x+1}}(3x^3 + 2x^2 - x - 2)$ 49. $y' = \frac{x\cos(x^2) - 3\sin 6x}{\sqrt{\cos 6x + \sin(x^2)}}$

51. $y' = 3(x^2\sec^2(x^3) + \sec^2 x \tan^2 x)$ 53. $\frac{dy}{dz} = \frac{-1}{\sqrt{z+1}(z-1)^{3/2}}$

55. $\frac{dy}{dx} = \frac{\sin(-1) - \sin(1 + x)}{(1 + \cos x)^2}$ 57. $\frac{dy}{dx} = -35x^4 \cot^6(x^5)\csc^2(x^5)$

59. $\frac{dy}{dx} = 30x(1 + (x^2 + 2)^5)^2(x^2 + 2)^4$

61. $\frac{dy}{dx} = -4e^{-x} - 14e^{-2x}$

63. $\frac{dy}{dx} = 24(2e^{3x} + 3e^{-2x})^3(e^{3x} - e^{-2x})$

65. $\frac{dy}{dt} = e^{-2t}(2t - 1)\sin(te^{-2t})$

67. $\frac{dy}{dx} = 4(x + 1)(x^2 + 2x + 3)e^{(x^2+2x+3)^2}$

69. $\frac{dy}{dx} = \dfrac{1}{8\sqrt{x}\sqrt{1+\sqrt{x}}\sqrt{1+\sqrt{1+\sqrt{x}}}}$

71. $y' = \frac{k}{2\sqrt{kx+b}}$ **73.** $\frac{df}{dx} = 12$

75. (a) $\frac{d}{d\theta}\underline{\sin\theta}\Big|_{\theta=60°} = \frac{\pi}{360}$

(b) $\frac{d}{d\theta}(\theta + \underline{\tan\theta})\Big|_{\theta=45°} = 1 + \frac{\pi}{90}$

77. $\frac{d}{dx}h(\sin x)\Big|_{x=\frac{\pi}{6}} = 5\sqrt{3}$ **79.** $\frac{d}{dx}f(g(x))\Big|_{x=6} = 12$

81. $\frac{d}{dx}g(\sqrt{x})\Big|_{x=16} = \frac{1}{16}$

83. $\frac{d^2}{dx^2}\sin(x^2) = 2\cos(x^2) - 4x^2\sin(x^2)$

85. $\frac{d^3}{dx^3}(3x + 9)^{11} = 26{,}730(3x + 9)^8$

87. (a) $f'(x) = -2x\csc^2(x^2)$
$f''(x) = 2\csc^2(x^2)(4x^2\cot(x^2) - 1)$
$f'''(x) = -8x\csc^2(x^2)(6x^2\cot^2(x^2) - 3\cot(x^2) + 2x^2)$

(b) $f'(x) = \dfrac{3x^2}{2\sqrt{x^3 + 1}}$

$f''(x) = \dfrac{3x(x^3 + 4)}{4(x^3 + 1)^{3/2}}$

$f'''(x) = -\dfrac{3(x^6 + 20x^3 - 8)}{8(x^3 + 1)^{5/2}}$

89. $\frac{d}{dx}\sin(g(x))\Big|_{x=2} = -11\sqrt{2}$ **91.** $\frac{d}{dt}(Ri^2)\Big|_{t=2} = 0$

93. $\frac{dF}{dt}\Big|_{t=10} = -1{,}600/(41^3) \approx -0.0232$ m/s

Section 3.8 Preliminary

1. The Chain Rule

2. (c) is incorrect; $\frac{d}{dx}\sin(y^2) = 2y\cos(y^2)\frac{dy}{dx}$.

3. He made two mistakes.
He did not use the product rule on the second term:
$\frac{d}{dx}(2xy) = 2x\frac{dy}{dx} + 2y$.
He did not use the general power rule on the third term:
$\frac{d}{dx}y^3 = 3y^2\frac{dy}{dx}$.

4. (b)

Section 3.8 Exercises

1. $y' = \frac{-2x}{6y^2}$. At $(2, 1)$, $y' = -\frac{2}{3}$

3. $\frac{d}{dx}(x^2y^3) = 3x^2y^2y' + 2xy^3$

5. $\frac{d}{dx}(x^2 + y^2)^{3/2} = 3(x + yy')\sqrt{x^2 + y^2}$

7. $\frac{d}{dx}(z + z^2) = z' + 2z(z')$ **9.** $y' = -\frac{2x}{9y^2}$

11. $y' = \frac{1-2xy-2y^2}{x^2+4xy-1}$ **13.** $y' = \frac{3-2xy}{x^2+4y^3}$ **15.** $z' = -x^3/z^3$

17. $y' = \sqrt{xy^5}$ **19.** $s' = -\frac{s^2(x^2+2\sqrt{x+s})}{x^2(s^2+2\sqrt{x+s})}$ **21.** $y' = \frac{y}{x-y^2}$

23. $y' = \frac{1-\cos(x+y)}{\cos(x+y)+\sin y}$ **25.** $y' = \frac{e^y-2y}{2x+3y^2-xe^y}$

27. $y' = \frac{1}{4}$ **29.** $y = 4 - x$ **31.** $y = 2 - x$ **33.** $y = \frac{8}{5} - \frac{3}{5}x$

35. (b) Setting $y' = 0$ we have $0 = 3x^2 - 3$, so $x = 1$ or $x = -1$.

(c) If we return to the equation $y^2 = x^3 - 3x + 1$ and substitute $x = 1$, we obtain the equation $y^2 = -1$, which has no real solutions.

(d) The tangent line is horizontal at the points $(-1, \sqrt{3})$ and $(-1, -\sqrt{3})$.

37. $y' = \frac{2x-3y}{3x-2y} = 0$ when $y = \frac{2}{3}x$. Substituting $y = \frac{2}{3}x$ into the equation of the implicit curve gives $-\frac{5}{9}x^2 = 1$, which has *no* real solutions.

39. $\frac{dx}{dy} = \frac{y(2y^2-1)}{x}$. The points where the tangent line is vertical are:
$(1, 0), (-1, 0), \left(\frac{\sqrt{3}}{2}, \frac{\sqrt{2}}{2}\right), \left(-\frac{\sqrt{3}}{2}, \frac{\sqrt{2}}{2}\right), \left(\frac{\sqrt{3}}{2}, -\frac{\sqrt{2}}{2}\right),$
$\left(-\frac{\sqrt{3}}{2}, -\frac{\sqrt{2}}{2}\right).$

41. $\frac{dy}{dt} = -\frac{x^2+y^2}{2xy}\frac{dx}{dt}$ **43.** $\frac{dy}{dt} = -\frac{x+y}{x+2y^3}\frac{dx}{dt}$ **45.** $y = \frac{4}{5}x + \frac{4}{5}$

47. (a)

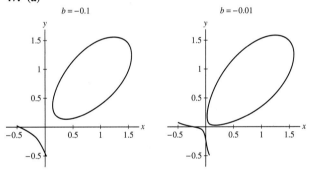

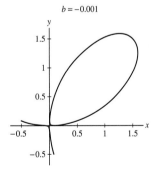

$b = 0.1$

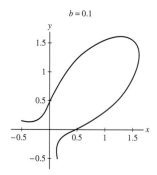

$b = 0.01$

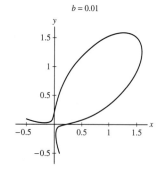

$b = 0.001$

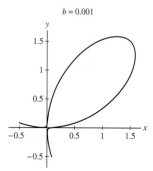

(b) At the point where $y = 0$, $y' = \sqrt[3]{b}$. Therefore, $y' \to \infty$ when $b \to \infty$.

$b = 01$

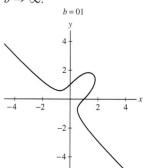

$b = 10$

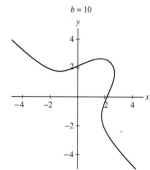

$b = 100$

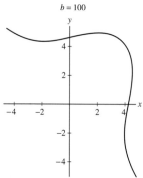

49. (b) $x = \frac{1}{2}, 1 \pm \sqrt{2}$

51. At $(1, 2)$, $y = \frac{1}{3}(x - 1) + 2$.

At $(1, -2)$, $y = -\frac{1}{3}(x - 1) - 2$.

At $\left(1, \frac{1}{2}\right)$, $y = \frac{11}{12}(x - 1) + \frac{1}{2}$.

At $\left(1, -\frac{1}{2}\right)$, $y = -\frac{11}{12}(x - 1) - \frac{1}{2}$.

53.

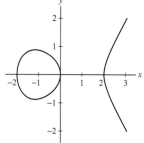

$\frac{dx}{dy} = \frac{2y}{3x^2 - 4} = 0$ when $y = 0$, so the tangent line to this curve is vertical at the points where the curve intersects the x-axis.

55. $y' = -\frac{x}{y}$; thus

$$y'' = -\frac{y \cdot 1 - xy'}{y^2} = -\frac{y - x\left(-\frac{x}{y}\right)}{y^2} = -\frac{y^2 + x^2}{y^3}$$
$$= -\frac{1}{y^3} = -y^{-3}$$

57. At the point $(\frac{2}{3}, \frac{4}{3})$, $y'' = -\frac{162}{125}$

59. $\left(-\frac{\sqrt{6}}{2}, -\frac{\sqrt{2}}{2}\right), \left(-\frac{\sqrt{6}}{2}, \frac{\sqrt{2}}{2}\right), \left(\frac{\sqrt{6}}{2}, -\frac{\sqrt{2}}{2}\right)$, and $\left(\frac{\sqrt{6}}{2}, \frac{\sqrt{2}}{2}\right)$

Section 3.9 Preliminary

1. 2 **2.** $\frac{1}{3}$ **3.** $g(x) = \tan^{-1} x$

4. Angles whose sine and cosine are x are complementary.

Section 3.9 Exercises

1. $g(x) = \sqrt{x^2 - 9}$, $g'(x) = \frac{x}{\sqrt{x^2 - 9}}$ **3.** $g'(x) = \frac{1}{7}$

5. $g'(x) = -\frac{1}{5}x^{-6/5}$ **7.** $g'(x) = (1 - x)^{-2}$ **9.** $g(7) = 1$, $g'(7) = \frac{1}{5}$ **11.** $g(1) = 0$, $g'(1) = 1$

13. $g(4) = 2$, $g'(4) = \frac{4}{5}$ **15.** $g(\frac{1}{4}) = 3$, $g'(\frac{1}{4}) = -16$

17. $g'(x) = \frac{1}{n}x^{1/n - 1}$; agrees with the Power Rule

19. $y'(\frac{1}{2}) = \frac{4}{5}$ **21.** $y'(\frac{1}{5}) = -\frac{20}{3}$ **23.** $y' = \frac{3}{x^2 + 9}$

25. $y' = \frac{1}{|t + 1|\sqrt{t^2 + 2t}}$ **27.** $y' = -\frac{\pi}{2\sqrt{1 - x^2}(\sin^{-1} x)^2}$

29. $y' = \frac{1}{\sqrt{1 - x^2}}$ **31.** $y' = \frac{1}{t^2 + 1}$ **33.** $y' = -\frac{e^{\cos^{-1} x}}{\sqrt{1 - x^2}}$

35. $y' = \frac{1}{\sqrt{1 - t^2} \sin^{-1} t}$ **37.** $y' = x^{\sin^{-1} x}\left(\frac{\sin^{-1} x}{x} + \frac{\ln x}{\sqrt{1 - x^2}}\right)$

Section 3.10 Preliminary

1. $\ln 4$ **2.** $b = e^2$ **3.** $\frac{1}{10}$ **4.** $b = e^3$

5. $y^{(100)}(x) = \cosh x$; $y^{(101)}(x) = \sinh x$

Section 3.10 Exercises

1. $\ln x + 1$ **3.** $\frac{2}{x}\ln x$ **5.** $\frac{3x^2 + 3}{x^3 + 3x + 1}$ **7.** $\frac{\cos t}{\sin t + 1}$ **9.** $\frac{1 - \ln x}{x^2}$

11. $\frac{1}{x \ln x}$ **13.** $\frac{3}{x \ln x}$ **15.** $\frac{1}{\sin x \cos x}$ **17.** $5^x \ln 5$

19. $f'(x) = \frac{1}{x \ln 2}$ **21.** $f'(3) = \frac{1}{3 \ln 5}$

23. $y = 64 \ln 4 (x - 3) + 64$

25. $y = 7 \ln 3 \cdot 3^{14} (t - 2) + 3^{14}$ **27.** $y = 5^8$

29. $y = 1 + \ln 4 - t$ **31.** $y = \frac{9}{5}(x - 2) + 4 \ln 20$

33. $y = \frac{1}{2 \ln 5}(x - 2) + \frac{\ln 2}{\ln 5}$ **35.** $y' = x^{2x}(2 + 2 \ln x)$

37. $y' = x e^x \left(\frac{e^x}{x} + e^x \ln x \right)$ **39.** $y' = x^{2^x} \left(\frac{2^x}{x} + 2^x \ln 2 \cdot \ln x \right)$

41. $y' = 2x + 6$ **43.** $y' = \frac{(x+1)^3}{(3x-1)^2} + \frac{3x(x+1)^2}{(3x-1)^2} - \frac{6x(x+1)^3}{(3x-1)^3}$

45. $y' = 4x^2(2x + 1)\sqrt{x - 9} \left(\frac{2}{2x+1} + \frac{2}{x} + \frac{1}{2(x-9)} \right)$

47. $y' = (x^2 + 1)(x^2 + 2)(x^2 + 3)^2 \left(\frac{2x}{x^2+1} + \frac{2x}{x^2+2} + \frac{4x}{x^2+3} \right)$

49. $(\ln x)' = \frac{1}{x}$ and $(\ln 2x)' = \frac{2}{2x} = \frac{1}{x}$; note $\ln 2x = \ln 2 + \ln x$

51. $y' = 2x \cosh(x^2)$ **53.** $y' = 2t \operatorname{sech}^2(t^2 + 1)$

55. $y' = \sinh x + \tanh x \operatorname{sech} x$ **57.** $y' = \tanh x$

59. $y' = \frac{\cosh(\ln x)}{x}$ **61.** $y' = e^x \operatorname{sech}^2 e^x$

63. $y' = \sinh x \cosh(\cosh x)$ **65.** $y' = -2x \operatorname{csch}^2 x^2$

67. $y' = -\operatorname{csch} x \coth x$ **69.** $y' = x^{\sinh x} \left(\cosh x \ln x + \frac{\sinh x}{x} \right)$

71. $y' = \frac{2x}{\sqrt{x^4 + 1}}$ **73.** $y' = \frac{e^{\cosh^{-1} x}}{\sqrt{x^2 - 1}}$ **75.** $y' = \frac{1}{(1 - x^2) \tanh^{-1} x}$

79. (a) $E = 10^{4.8 + 1.5M}$ **(c)** $\frac{dE}{dM} = (1.5 \ln 10) 10^{4.8 + 1.5M}$

Section 3.11 Preliminary

1. The restatement is "Determine $\frac{dV}{dt}$ if $\frac{ds}{dt} = 0.5$ cm/s."

2. $\frac{dV}{dt} = 4\pi r^2 \frac{dr}{dt}$.

3. The restatement is "Determine $\frac{dh}{dt}$ if $\frac{dV}{dt} = 2$ cm^3/min."

4. The restatement is "Determine $\frac{dV}{dt}$ if $\frac{dh}{dt} = 1$ cm/min."

Section 3.11 Exercises

1. 0.039 ft/min

3. (a) $100\pi \approx 314.16$ m^2/min   **(b)** $24\pi \approx 75.40$ m^2/min

5. $3,584\pi$ in^3/min   **7.** 896π in^2/min   **9.** $18\sqrt{10}$ mph

11. 1.68 m^3/min

13. (a) Issac traveled $\frac{27}{5}$ miles. Sonya traveled $\frac{88}{15}$ miles.

(b) $\frac{5,003}{\sqrt{14,305}} \approx 41.83$ mph

15. (a) 499.86 mph   **(b)** 0 mph

17. 0.61 mi/min   **19.** $\frac{-11}{\sqrt{15}}$ ft/s

21. $x = 4\sqrt{15}$ ft; $\frac{dx}{dt} = \frac{4}{\sqrt{15}} \approx 1.03$ ft/s

23. $\frac{dh}{dt} = -\frac{x}{h} \frac{dx}{dt}$; as $\frac{dx}{dt}$ is constant and positive and x approaches the length of the ladder as it falls, $-\frac{x}{h} \frac{dx}{dt}$ gets arbitrarily large as $h \to 0$.

25. 1,047.20 cm^3/s   **27.** $4,800\pi \approx 15,079.64$ mph

29. 495.25 mph   **31.** $\frac{dy}{dt} = -10.5$ ft/s   **33.** $\frac{dP}{dt} = -1.92$ kPa/min

35. $l'(t) = \frac{9x(2x^2 + 3)}{\sqrt{x^4 + 3x^2 + 1}}$ ft/s (where x is a function of t).

37. $-\frac{1}{1,440\pi} \approx -2.21 \times 10^{-4}$ cm/min

39. (a) $\frac{dx}{dt} = \frac{h}{x} \frac{dh}{dt}$ where the rate at which Henry pulls the rope is $-dh/dt$.

(b) $-\sqrt{13}/2 \approx -1.80$ ft/s

41. 300 mph   **43.** -8.94 ft/s   **45.** ≈ -4.68 m/s

Chapter 3 Review

1. ROC of $f(x)$ over $[0, 2]$ is 3.

3. The estimation is $\frac{8}{3}$ and it is larger than $f'(0.7)$.

5. $f'(1) = 1$; Tangent line: $y = x - 1$

7. $f'(4) = -\frac{1}{16}$; Tangent line: $y = \frac{1}{2} - \frac{1}{16}x$

9. $\frac{dy}{dx} = -2x$ **11.** $\frac{dy}{dx} = \frac{1}{(2-x)^2}$ **13.** $f'(1)$ where $f(x) = \sqrt{x}$

15. $f'(\pi)$ where $f(t) = \sin(t) \cos(t)$ **17.** $f'(4) = 3$; $f(4) = -2$

19. (C)

21. (a) 8.05 cm/yr   **(b)** Larger over the first half.

(c) $h'(3) \approx 7.8$ cm/yr; $h'(8) \approx 6.0$ cm/yr

23. $A'(t)$ is the rate of change in automobile production in the United States.
$A'(1971) \approx 0.25$ million automobiles/year.
$A'(1974)$ would be negative.

25. (b) $(\ln 2) 2^x$

27. $g'(8) = \frac{1}{12}$ **29.** $y' = -6x^{-5/2}$ **31.** $y' = 8x + 2x^{-3}$

33. $y' = -\frac{19}{(4t-9)^2}$ **35.** $y' = 6(6t - 60t^{-4})(3t^2 + 20t^{-3})^5$

37. $y' = (7x + 16)(x + 1)^2(x + 4)^3$ **39.** $y' = -\frac{3}{x^2}\left(1 + \frac{1}{x}\right)^2$

41. $y' = \frac{5 - 3x}{2(1-x)^2(2-x)^{3/2}}$ **43.** $y' = 12 \sin(2 - 3x)$

45. $y' = \frac{x}{\sqrt{x^2+1}} \cos(\sqrt{x^2 + 1})$ **47.** $y' = -\frac{4}{\theta^2} \cos\left(\frac{4}{\theta}\right)$

49. $y' = -9z \csc(9z + 1) \cot(9z + 1) + \csc(9z + 1)$

51. $y' = -\sin x \sec^2(\cos x)$

53. $y' = -\sin(\cos(\cos \theta)) \sin(\cos \theta) \sin \theta$

55. $f'(x) = \frac{8x}{4x^2 + 1}$ **57.** $f'(x) = \frac{1 + e^x}{x + e^x}$ **59.** $G'(s) = \frac{2}{s}$

61. $g'(t) = (2t - 1)e^{1/t}$ **63.** $f'(\theta) = \frac{\cos(\ln \theta)}{\theta}$

65. $f'(x) = 2 \sin x \cos x e^{\sin^2 x}$ **67.** $h'(y) = \frac{2e^y}{(1 - e^y)^2}$

69. $G'(s) = \frac{1}{2\sqrt{s}(1+s)}$ **71.** $f'(x) = \frac{e^{\sec^{-1} x}}{|x|\sqrt{x^2 - 1}}$

73. $h'(y) = 4y \operatorname{sech}^2(4y) + \tanh(4y)$

75. $g'(t) = \sqrt{\frac{t^2 - 1}{t^2 + 1}} + \frac{t \sinh^{-1} t}{\sqrt{t^2 - 1}}$ **79.** $S'(2) = -27$

81. $R'(2) = -\frac{57}{16}$ **83.** $F'(2) = -18$ **85.** $(-1, -1)$ and $(3, 7)$

87. $b = -76, 32$

89. (a) $a = \frac{1}{6}$ **(b)**

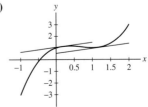

91. $y'' = 72x - 10$ **93.** $y'' = -(2x + 3)^{-3/2}$

95. $y'' = 8x^2 \sec^2(x^2)\tan(x^2) + 2\sec^2(x^2)$

97. For the plot on the left, the red, green and blue curves, respectively, are the graphs of f, f' and f''. For the plot on the right, the green, red and blue curves, respectively, are the graphs of f, f' and f''.

99. (b) $E(8) = -3$ **101.** $\frac{dy}{dx} = \frac{x^2}{y^2}$ **103.** $\frac{dy}{dx} = \frac{y^2+4x}{1-2xy}$

105. $\frac{dy}{dx} = \frac{y-2x}{y-x}$ **107.** $\frac{dy}{dx} = \frac{y-\sec^2(x+y)}{\sec^2(x+y)-x}$

109. $(1, 1)$, $\left(-\sqrt[3]{3}, \sqrt[3]{9}\right)$ **111.** $\alpha = 0$, $\alpha > 1$

113. $\frac{1}{15}$ m/min **115.** 0.75 ft/s

117. (a) -0.72 rad/s (b) -1.50 cm/s

119. $y' = \frac{(x+1)(x+2)^2}{(x+3)(x+4)}\left(\frac{1}{x+1} + \frac{2}{x+2} - \frac{1}{x+3} - \frac{1}{x+4}\right)$

121. $y' = \frac{e^x \sin^{-1}x}{\ln x}\left(1 + \frac{1}{\sqrt{1-x^2}\sin^{-1}x} - \frac{1}{x\ln x}\right)$

123. $y' = x^{\sqrt{x}}\left(x^{\ln x}\right)\left(\frac{\ln x}{2\sqrt{x}} + \frac{1}{\sqrt{x}} + \frac{2\ln x}{x}\right)$

Chapter 4

Section 4.1 Preliminary

1. $g(1.2) - g(1) \approx -0.8$ **2.** $f(2.1) \approx 1.3$ **3.** 55 ft
5. $f'(2) = 2$; $f(2) = 8$

Section 4.1 Exercises

1. $\Delta f \approx 0.12$ **3.** $\Delta f \approx -0.00222$

5. $\Delta f \approx 0.04e^6 \approx 16.137$ **7.** $\Delta f \approx \frac{1}{10e^2} \approx .013534$

9. $\Delta f \approx 0.166667$; error: 0.004389; percentage error: 2.70%
11. $\Delta f \approx 0.02$; error: 1.3×10^{-6}; percentage error: 0.0065%
13. $\Delta f \approx -0.021213$; error: 0.000315; percentage error: 1.46%
15. $\Delta f \approx 0.133333$; error: 0.007513; percentage error: 5.33%
17. $\Delta f \approx 0.015625$. The error is ≈ 0.00018.
19. $\Delta f \approx 0.001$. The error is $\approx 1.53 \times 10^{-5}$.
21. $\Delta f \approx 0.023$. The error is $\approx 2.03 \times 10^{-6}$.
23. $\Delta f \approx -0.03$. The error is $\approx 4.59 \times 10^{-4}$.
25. $\Delta f \approx 0.0074074$ **27.** $\Delta f \approx 0.008727$
29. (a) $\Delta P \approx -0.434906$ kilopascals
(b) $P(20.5) - P(20) = -0.418274$ kilopascals; percentage error in linear approximation: 3.98%
31. $\Delta A \approx 0.5$ ft^2
33. (a) $\Delta h \approx 0.78125$ ft (b) $\Delta h \approx 0.9375$ ft
35. $\Delta F \approx 4.88$ ft; $\Delta F \approx 7.04$ ft
37. Weight loss ≈ 0.066 pounds; pilot weighs 129.5 pounds at an altitude of approximately 7.6 miles.
39. (a) $P = 4.8$ atm, $\Delta P \approx 0.48$ atm (b) $|\Delta V| < 0.520833$ L
41. $L(x) = x$ **43.** $L(x) = -\frac{1}{2}x + 1$ **45.** $L(x) = 1$
47. $L(x) = \frac{2\sqrt{3}}{3}\left(x - \frac{1}{2}\right) + \frac{\pi}{6}$ **49.** $L(x) = e(x - 1)$

51. $\sqrt{16.2} \approx 4.025$

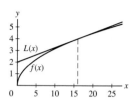

53. $\sqrt{17} \approx 4.125$. The percentage error is $\approx 0.046\%$.
55. $17^{1/4} \approx 2.03125$. The percentage error is $\approx 0.035\%$.
57. $27.001^{\frac{1}{3}} \approx 3.0000370370370$. The percentage error is 0.000000015%.
59. $\ln(1.07) \approx 0.07$. The percentage error is $\approx 3.46\%$.
61. $\cos^{-1}(0.52) \approx 1.024104$. The percentage error is $\approx 0.015\%$.
65. For $v = 25$ ft/s, $\Delta x \approx -0.118$ ft; for $v = 30$ ft/s, $\Delta x \approx -0.170$ ft. The shot is more sensitive to angle at larger velocities.

67.

h	$E = \left|\sqrt{9+h} - 3 - \frac{1}{6}h\right|$	$0.006h^2$
10^{-1}	4.604×10^{-5}	6.00×10^{-5}
10^{-2}	4.627×10^{-7}	6.00×10^{-7}
10^{-3}	4.629×10^{-9}	6.00×10^{-9}
10^{-4}	4.627×10^{-11}	6.00×10^{-11}

69. $(1.02)^{0.7} \approx 1.014$; $(1.02)^{-0.3} \approx 0.994$
73. (a) $dy/dx = -1$; $L(x) = 4 - x$
(b)

(c) $y \approx 0.9$; too large
(d) $y = 0.890604$; percentage error in approximation from part (c): 1.06%

Section 4.2 Preliminary

1. (c) **2.** (b) **4.** True **6.** (b)

Section 4.2 Exercises

1. (a) Three (b) 6 (c) 5 at $x = 5$
(d) One example is [4, 8]. (e) One example is [0, 2].
3. $x = 1$ **5.** $x = -3$ and $x = 6$ **7.** $x = \pm 1$ **9.** $x = 0$
11. $x = 1/e$ **13.** $x = \pm\frac{\sqrt{3}}{2}$
15. (a) $c = 2$; $f(c) = -3$ (b) $f(0) = f(4) = 1$
(c) Maximum 1; Minimum -3 (d) Maximum 1; Minimum -2
17. $\tan^{-1}\left(-\frac{1}{4}\right) \approx -0.244979$

19. Critical points at $t = 0$ and $t = \pm 1$; on $[0, 1]$, the maximum value is $h(1) = 0$ and the minimum is $h(0) = -1$; on $[0, 2]$, the maximum value is $h(2) = 3^{1/3} \approx 1.44225$ and the minimum is $h(0) = -1$

21. Critical points at approximately $x = 0.2$, $x = 1.4$ and $x = 4.8$; No minimum value as $\lim\limits_{x \to 0+} f(x) = -\infty$. The maximum value is $f(4.8) = 6.5$.

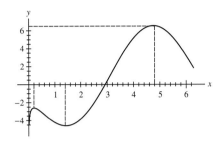

23. The minimum is $f(1) = 0$. The maximum is $f(3) = 8$.

25. The minimum is $f(2) = -9$. The maximum is $f(-2) = 15$.

27. The minimum is $f(-1) = -3$. The maximum is $f\left(\frac{3}{8}\right) = 4.5625$.

29. The minimum is $f(1) = -4$. The maximum is $f(-1) = 6$.

31. The minimum is $f(1) = -3$. The maximum is $f(-1) = 13$.

33. The minimum is $f(1) = -3$. The maximum is $f(4) = 78$.

35. The minimum is $f(-1) = -4$.
The maximum value is $f(5) = 3{,}050$.

37. The minimum is $f(6) = 18.5$. The maximum is $f(5) = 26$.

39. The minimum is $f(1) = -1$. The maximum is $f(0) = f(3) = 0$.

41. The minimum is $f(0) = 2\sqrt{6}$. The maximum is $f(2) = 4\sqrt{2}$.

43. The minimum is $f\left(\frac{\sqrt{3}}{2}\right) \approx -0.589980$.
The maximum is $f(4) \approx 0.472136$.

45. The minimum is $f(0) = f\left(\frac{\pi}{2}\right) = 0$. The maximum is $f\left(\frac{\pi}{4}\right) = \frac{1}{2}$.

47. The minimum is $f(0) = -1$.
The maximum is $f\left(\frac{\pi}{4}\right) = \sqrt{2}\left(\frac{\pi}{4} - 1\right)$.

49. The minimum is $g\left(\frac{\pi}{3}\right) = \frac{\pi}{3} - \sqrt{3}$.
The maximum is $g\left(\frac{5}{3}\pi\right) = \frac{5}{3}\pi + \sqrt{3}$.

51. The minimum is $f\left(\frac{\pi}{4}\right) = 1 - \frac{\pi}{2}$. The maximum is $f(0) = 0$.

53. The minimum is $f(1) = 0$.
The maximum is $f(e) = 1/e \approx 0.367879$.

55. The minimum is $f(5) = 5\tan^{-1} 5 - 5 \approx 1.867004$.
The maximum is $f(2) = 5\tan^{-1} 2 - 2 \approx 3.535744$.

57. (d) The critical points in the range $[0, 2\pi]$ are $\frac{\pi}{6}$, $\frac{\pi}{2}$, $\frac{5\pi}{6}$, $\frac{7\pi}{6}$, $\frac{3\pi}{2}$, and $\frac{11\pi}{6}$. On this interval, the maximum value is
$f\left(\frac{\pi}{6}\right) = f\left(\frac{7\pi}{6}\right) = \frac{3\sqrt{3}}{2}$ and the minimum value is
$f\left(\frac{5\pi}{6}\right) = f\left(\frac{11\pi}{6}\right) = -\frac{3\sqrt{3}}{2}$.

(e)

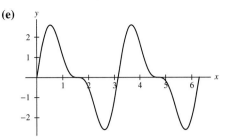

59. critical point: $x = 2$; minimum is $f(2) = 0$; maximum is $f(0) = 2$

61. critical point $x = 2$; minimum is $f(2) = 0$; maximum is $f(0) = 12$

63. $f(-2) = -6 + 8 = 2$ and $f(1) = 3 - 1 = 2$; therefore, there exists a $c \in (-2, 1)$ at which $f'(c) = 0$; take $c = -1$

65. $f\left(\frac{\pi}{4}\right) = f\left(\frac{3\pi}{4}\right) = \frac{\sqrt{2}}{2}$; take $c = \frac{\pi}{2}$

67. $f\left(\frac{\pi}{4}\right) = f\left(\frac{3\pi}{4}\right) = 0$; take $c = \frac{\pi}{2}$

73. (a) critical points: $v = 0$ and $v = 3v_r/2$

(b) Take $v_r = 10$. Then the positive critical point is at $v = 15$.

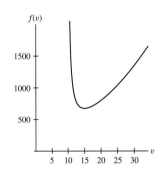

75. critical point: $x = 1$; minimum is $f(-5) = \frac{1}{7}$; maximum is $f(1) = 1$

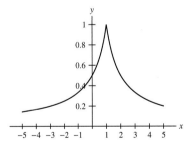

77. critical points: $x = \pm 1$ and $x = \pm 2$; minimum is $f(2) = -\frac{2}{3}$; maximum is $f(-2) = \frac{2}{3}$

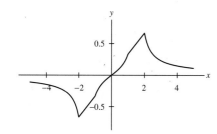

79. 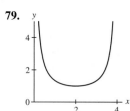 **81.**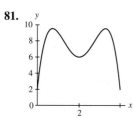

87. $b \geq \frac{1}{4}a^2$

Section 4.3 Preliminary

1. $m = 3$ **2.** (c)
3. Yes

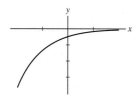

4. **(a)** $f(c)$ is a local maximum. **(b)** No

Section 4.3 Exercises

1. $c = 2$ **3.** $c = 2$ **5.** $c = 2\sqrt{7} - 1$ **7.** $c = \frac{4}{e}$
9. $c = 0$ **11.** $c \approx 0.62$
13. $f(x)$ increasing over $(0, 2)$ and $(4, 6)$, and decreasing over $(2, 4)$.
15. 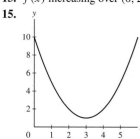 **17.**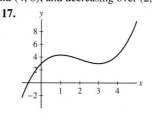

19. Maximum **21.** Minimum
23. $f(2)$ is a local maximum. $f(4)$ is a local minimum.
25. The critical point is $c = \frac{7}{2}$.

x	$\left(-\infty, \frac{7}{2}\right)$	$\frac{7}{2}$	$\left(\frac{7}{2}, \infty\right)$
f'	$+$	0	$-$
f	↗	M	↘

27. The critical points are $c = 0, 4$.

x	$(-\infty, 0)$	0	$(0, 4)$	4	$(4, \infty)$
f'	$+$	0	$-$	0	$+$
f	↗	M	↘	m	↗

29. The critical points are $c = -2, -1, 1$.

x	$(-\infty, -2)$	-2	$(-2, -1)$	-1	$(-1, 1)$	1	$(1, \infty)$
f'	$-$	0	$+$	0	$-$	0	$+$
f	↘	m	↗	M	↘	m	↗

31. The critical points are $c = -2, -1$.

x	$(-\infty, -2)$	-2	$(-2, -1)$	-1	$(-1, \infty)$
f'	$+$	0	$-$	0	$+$
f	↗	M	↘	m	↗

33. The critical points are $c = 0, -\frac{3}{4}$.

x	$\left(-\infty, -\frac{3}{4}\right)$	$-\frac{3}{4}$	$\left(-\frac{3}{4}, 0\right)$	0	$(0, \infty)$
f'	$-$	0	$+$	0	$+$
f	↘	m	↗	¬	↗

35. The critical point is $c = 0$.

x	$(-\infty, 0)$	0	$(0, \infty)$
f'	$+$	0	$-$
f	↗	M	↘

37. The critical point is $c = 1$.

x	$(0, 1)$	1	$(1, \infty)$
f'	$-$	0	$+$
f	↘	m	↗

39. The critical point is $x = \frac{16}{25}$.

x	$\left(0, \frac{16}{25}\right)$	$\frac{16}{25}$	$\left(\frac{16}{25}, \infty\right)$
f'	$-$	0	$+$
f	↘	m	↗

41. The critical points are $\theta = \frac{\pi}{4}$, $\theta = \frac{3\pi}{4}$, $\theta = \frac{5\pi}{4}$, and $\theta = \frac{7\pi}{4}$.

x	$\left(0, \frac{\pi}{4}\right)$	$\frac{\pi}{4}$	$\left(\frac{\pi}{4}, \frac{3\pi}{4}\right)$	$\frac{3\pi}{4}$	$\left(\frac{3\pi}{4}, \frac{5\pi}{4}\right)$	$\frac{5\pi}{4}$	$\left(\frac{5\pi}{4}, \frac{7\pi}{4}\right)$	$\frac{7\pi}{4}$	$\left(\frac{7\pi}{4}, 2\pi\right)$
f'	+	0	−	0	+	0	−	0	+
f	↗	M	↘	m	↗	M	↘	m	↗

43. The critical point is $x = \frac{\pi}{2}$.

x	$\left[0, \frac{\pi}{2}\right)$	$\frac{\pi}{2}$	$\left(\frac{\pi}{2}, 2\pi\right]$
f'	+	0	+
f	↗	¬	↗

45. The critical point is $x = 0$.

x	$(-\infty, 0)$	0	$(0, \infty)$
f'	−	0	+
f	↘	m	↗

47. The critical point is $x = -\frac{\pi}{4}$.

x	$\left[-\frac{\pi}{2}, -\frac{\pi}{4}\right)$	$-\frac{\pi}{4}$	$\left(-\frac{\pi}{4}, \frac{\pi}{2}\right]$
f'	+	0	−
f	↗	M	↘

49. The critical points are $x = \pm 1$.

x	$(-\infty, -1)$	-1	$(-1, 1)$	1	$(1, \infty)$
f'	−	0	+	0	−
f	↘	m	↗	M	↘

51. The critical point is $x = 1$.

x	$(0, 1)$	1	$(1, \infty)$
f'	−	0	+
f	↘	m	↗

55. $a \geq 0$

57. **(a)** MVT **(b)** IVT

63. The MVT, applied to the interval $[2, 4]$, guarantees there exists a $c \in (2, 4)$ such that $f'(c) = \frac{f(4) - f(2)}{4 - 2}$ or $f(4) - f(2) = 2f'(c)$. Because $f'(x) \geq 5$, it follows that $f(4) - f(2) \geq 10$, or $f(4) \geq f(2) + 10 = 8$.

65. Let $f(x) = x^3 + ax^2 + bx + c$. Then $f'(x) = 3x^2 + 2ax + b = 3\left(x + \frac{a}{3}\right)^2 - \frac{a^2}{3} + b > 0$ for all x if $b - \frac{a^2}{3} > 0$. Therefore, if $b > a^2/3$, then $f(x)$ is increasing on $(-\infty, \infty)$.

67. Let $f(x) = x$ and $g(x) = \sin x$. Then $f(0) = g(0) = 0$ and $f'(x) = 1 \geq \cos x = g'(x)$ for $x \geq 0$. Apply the result of the previous exercise to conclude that $x \geq \sin x$ or $\sin x \leq x$ for $x \geq 0$.

71. **(b)** No

Section 4.4 Preliminary

1. (a) **2.** (b) **3.** False **4.** True **5.** False **6.** True
7. True **8.** True **9.** **(c)** and **(d)**

Section 4.4 Exercises

1. **(a)** C **(b)** A **(c)** B **(d)** D

3. Increasing function (left); Decreasing function (right)

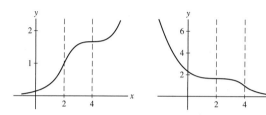

5. b, e; $b < x < e$

7. Concave up everywhere; no points of inflection.

9. $x = \frac{1}{2}(2n + 1)\pi$; concave up $\left(-\frac{\pi}{2} + 2n\pi, \frac{\pi}{2} + 2n\pi\right)$; concave down $\left(\frac{\pi}{2} + 2n\pi, \frac{3\pi}{2} + 2n\pi\right)$

11. Concave down for $0 < x < 9$; concave up for $x > 9$; inflection point at $x = 9$

13. Concave down for $|x| < 1$; concave up for $|x| > 1$; inflection point at $x = \pm 1$

15. Concave down for $x < 2$; concave up for $x > 2$; $x = 2$

17. Concave down for $-2 - \sqrt{5} < x < -2 + \sqrt{5}$; concave up for $x < -2 - \sqrt{5}$ and for $x > -2 + \sqrt{5}$; $x = -2 \pm \sqrt{5}$

19. f has no points of inflection.

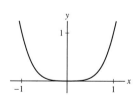

21. The growth rate at the point of inflection is 5.5 cm/day. First derivative (left); Second derivative (right)

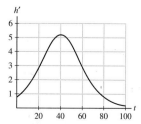
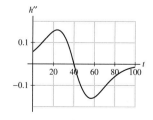

23. $f(3)$ is a local maximum, and $f(5)$ is a local minimum.

25. $f(0)$ is a local minimum. Second derivative test is inconclusive for $x = 1$.

27. $f\left(\sqrt{\frac{3}{5}}\right)$ is a local minimum. $f\left(-\sqrt{\frac{3}{5}}\right)$ is a local maximum.

29. $f(x)$ has local minima when x is an even multiple of π. $f(x)$ has local maxima when x is an odd multiple of π.

31. $f\left(\frac{1}{9}\right)$ is a local minimum. Second derivative test cannot be applied at $x = 0$.

33. $f\left(\frac{\sqrt{2}}{2}\right)$ is a local maximum, and $f\left(-\frac{\sqrt{2}}{2}\right)$ is a local minimum.

35. $f(e^{-1/3})$ is a local minimum.

37.

x	$\left(-\infty, \frac{1}{3}\right)$	$\frac{1}{3}$	$\left(\frac{1}{3}, 1\right)$	1	$(1, \infty)$
f'	+	0	−	0	+
f	↗	M	↘	m	↗

x	$\left(-\infty, \frac{2}{3}\right)$	$\frac{2}{3}$	$\left(\frac{2}{3}, \infty\right)$
f''	−	0	+
f	⌢	I	⌣

39.

t	$(-\infty, 0)$	0	$\left(0, \frac{2}{3}\right)$	$\frac{2}{3}$	$\left(\frac{2}{3}, \infty\right)$
f'	−	0	+	0	−
f	↘	m	↗	M	↘

t	$\left(-\infty, \frac{1}{3}\right)$	$\frac{1}{3}$	$\left(\frac{1}{3}, \infty\right)$
f''	+	0	−
f	⌣	I	⌢

41.

x	0	$(0, 0.396850)$	0.396850	$(0.396850, \infty)$
f'	U	−	0	+
f	M	↘	m	↗

Concave up everywhere; no points of inflection.

43.

t	$(-\infty, 0)$	0	$(0, \infty)$
f'	+	0	−
f	↗	M	↘

t	$\left(-\infty, -\sqrt{\frac{1}{3}}\right)$	$-\sqrt{\frac{1}{3}}$	$\left(-\sqrt{\frac{1}{3}}, \sqrt{\frac{1}{3}}\right)$	$\sqrt{\frac{1}{3}}$	$\left(\sqrt{\frac{1}{3}}, \infty\right)$
f''	+	0	−	0	+
f	⌣	I	⌢	I	⌣

45.

θ	$[0, \pi)$	π	$(\pi, 2\pi]$
f'	+	0	+
f	↗	¬	↗

θ	0	$(0, \pi)$	π	$(\pi, 2\pi)$	2π
f''	0	−	0	+	0
f	¬	⌢	I	⌣	¬

47.

x	$(0, 2\pi)$
f'	+
f	↗

x	$(0, \pi)$	π	$(\pi, 2\pi)$
f''	+	0	−
f	⌣	I	⌢

49.

x	$\left(-\frac{\pi}{2}, -\frac{\pi}{4}\right)$	$-\frac{\pi}{4}$	$\left(-\frac{\pi}{4}, \frac{3\pi}{4}\right)$	$\frac{3\pi}{4}$	$\left(\frac{3\pi}{4}, \frac{3\pi}{2}\right)$
f'	+	0	−	0	+
f	↗	M	↘	m	↗

x	$\left(-\frac{\pi}{2}, 0\right)$	0	$(0, \pi)$	π	$\left(\pi, \frac{3\pi}{2}\right)$
f''	−	0	+	0	−
f	⌢	I	⌣	I	⌢

51. **(a)** Near the beginning, R is concave up. Near the end, R is concave down.

(b) "Epidemic subsiding: number of new cases declining."

53. The point of inflection should occur when the water level is equal to the radius of the sphere.

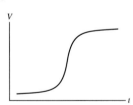

55. $\frac{d}{dx} \tan^{-1} x = \frac{1}{1+x^2} > 0$ for all x, so $\tan^{-1} x$ is always increasing. $\frac{d^2}{dx^2} \tan^{-1} x = \frac{-2x}{(1+x^2)^2}$, so $\tan^{-1} x$ is concave down for $x > 0$ and is concave up for $x < 0$.

57.

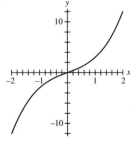

59. **(a)** False. If $f(x)$ is increasing, then $f^{-1}(x)$ is increasing.

(b) True.

(c) False. If $f(x)$ is concave up, the $f^{-1}(x)$ is concave down.

(d) True.

Section 4.5 Preliminary

1.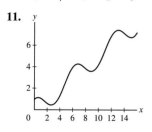

2. (c)

3. (a) $\lim\limits_{x \to \infty} x^3 = \infty$ **(b)** $\lim\limits_{x \to -\infty} x^3 = -\infty$

(c) $\lim\limits_{x \to -\infty} x^4 = \infty$

4. Negative **5.** Negative

6. The function f does not have a point of inflection at $x = 4$ because $x = 4$ is not in the domain of f.

Section 4.5 Exercises

1. A, $f' < 0$ and $f'' > 0$. B, $f' > 0$ and $f'' > 0$. C, $f' > 0$ and $f'' < 0$. D, $f' < 0$ and $f'' < 0$. E, $f' < 0$ and $f'' > 0$. F, $f' > 0$ and $f'' > 0$. G, $f' > 0$ and $f'' < 0$.

3.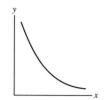

5.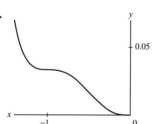

7.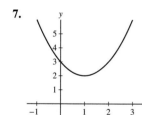

9.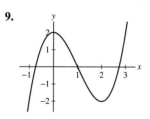

11.

13.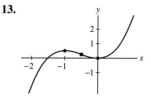

15.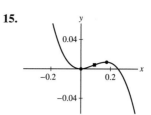

17.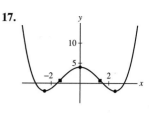

19.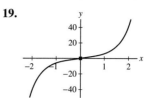

21.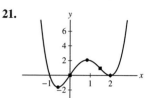

23.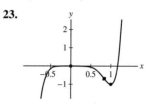

25.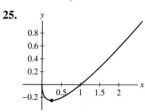

27.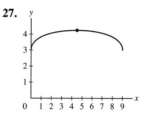

29.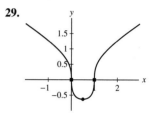

31.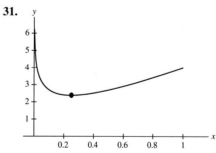

33.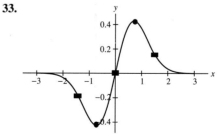

35.

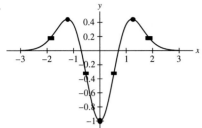

37.

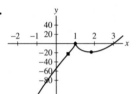

39.

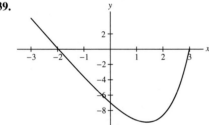

The graph has a local minimum at $2 \ln 2$.

41.

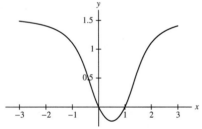

The graph has a local minimum at $\frac{1}{2}$ and inflection points at approximately -0.327199 and 1.327199.

43.

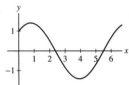

The graph has a local maximum at $\frac{\pi}{4}$, a local minimum at $\frac{5\pi}{4}$, and inflection points at $\frac{3\pi}{4}$ and $\frac{7\pi}{4}$.

45.

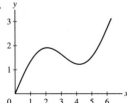

The graph has a local maximum at $\frac{2\pi}{3}$, a local minimum at $\frac{4\pi}{3}$, and an

inflection point at π.

47.

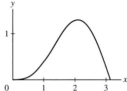

The graph has a local maximum at $x = \frac{2\pi}{3}$ and a point of inflection at $x = \cos^{-1}\frac{1}{4}$.

49. (a) 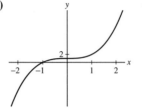   **(b)**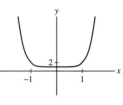

51. $\lim\limits_{x \to \infty} \frac{x}{x+9} = 1$ **53.** $\lim\limits_{x \to \infty} \frac{3x^2 + 20x}{2x^4 + 3x^3 - 29} = 0$

55. $\lim\limits_{x \to \infty} \frac{7x-9}{4x+3} = \frac{7}{4}$ **57.** $\lim\limits_{x \to -\infty} \frac{7x^2 - 9}{4x+3} = -\infty$

59. $\lim\limits_{x \to -\infty} \frac{x^2 - 1}{x+4} = -\infty$ **61.** $\lim\limits_{x \to \infty} \frac{\sqrt{x^2+1}}{x+1} = 1$

63. $\lim\limits_{x \to \infty} \frac{x+1}{\sqrt[3]{x^2+1}} = \infty$ **65.** $\lim\limits_{x \to -\infty} \frac{x}{(x^6+1)^{1/3}} = 0$

67. $\lim\limits_{x \to 0+} \frac{\ln x}{x^2} = -\infty$ **69.** $\lim\limits_{x \to \infty} \frac{e^{-x^2}}{x^2 - 5} = 0$ **71.** (A)

73. (a) D   **(b)** A   **(c)** B   **(d)** C

75. 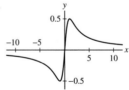   **77.**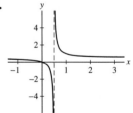

79.

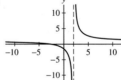

81.

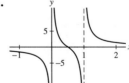

The graph has a point of inflection at $x = \frac{1}{2}$.

83.

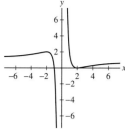

The graph has a local maximum at $x = -2$, a local minimum at $x = 2$ and points of inflection at $x = \pm 2\sqrt{2}$.

85.

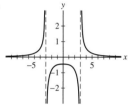

The graph has a local minimum at $x = 1$.

87.

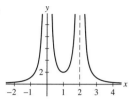

The graph has a local maximum at $x = 0$.

89.

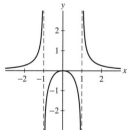

The graph has a local maximum at $x = 0$.

91.

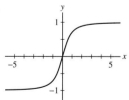

The graph has a point of inflection at $x = 0$.

93. **(a)** $y = x - 2$ **(b)** $y = x - 1$

95. $f(x)$ approaches the slant asymptote from above in both directions.

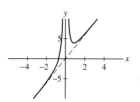

Section 4.6 Preliminary

1. $b + h + \sqrt{b^2 + h^2} = 10$ **2.** r, h **3.** No

4. If the function tends to infinity at the endpoints of the interval, then the function must take on a minimum value at a critical point.

Section 4.6 Exercises

1. **(a)** $y = 25 - x$ **(b)** $25x - x^2$ **(c)** Closed interval
(d) $x = y = \frac{25}{2}$ in

3. $x = 1$ **5.** $x = y = 4$

7. **(a)** $x = 10$ and $y = 10$ **(b)** No

9. width $= \frac{1,200}{4+\pi}$ and length $= \frac{600}{4+\pi}$

11. length of brick wall $= 10\sqrt{5}$ ft, length of adjacent side $= 20\sqrt{5}$ ft

13. $P \approx (0.59, 0.35)$

15. square base of side length 3.42 ft, height of 1.71 ft

17. $a = 4 + \frac{4\sqrt{3}}{3}, b = 4 - \frac{4\sqrt{3}}{3}$ **19.** $r = \frac{1}{3^{1/4}\sqrt{\pi}}, h = \frac{\sqrt{2}}{3^{1/4}\sqrt{\pi}}$

21. height $= \frac{2R}{\sqrt{3}}$, radius $= \sqrt{\frac{2}{3}}R$

23. $\theta = \cos^{-1}\left(\frac{1-\sqrt{3}}{2}\right)$ **25.** $A = \frac{18,000,000}{n+1}$

27. $A = 8 - 4\sqrt{3}$ **29.** $A = \frac{1}{2}$

31. **(b)** Net cash intake is $C(r) = 200r - \frac{1}{10}r^2 - 19,000$; maximized when $r = \$1,000$.

33. For $n = 2, x = \frac{1}{2}(x_1 + x_2)$; for $n = 3, x = \frac{1}{3}(x_1 + x_2 + x_3)$; for general $n, x = \frac{1}{n}\sum_{k=1}^{n} x_k$.

35. **(a)** $v = \frac{175\sqrt{6}}{3+1.5\sqrt{6}} \approx 64.2$ mph

(b) The cost as a function of speed is shown below for $L = 100$. The optimal speed is clearly around 64 mph.

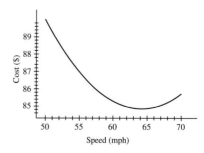

(c) We expect v to increase if P goes down to $2 per gallon.

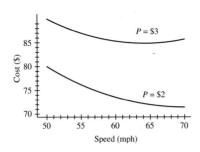

37. $r = \frac{1}{(4\pi)^{1/3}}, h = 4^{2/3}\pi^{-1/3}$

39. $h = 3$; dimensions: $9 \times 18 \times 3$ **41.** $A = B = 12$ in

45. $N = 9$ **47.** $A = \frac{3\sqrt{3}}{4}r^2$

51. (b) The underwater observer sees what appears to be a silvery mirror due to total internal reaction.

55. (a) If $b < \sqrt{3}a$, then $d = a - b/\sqrt{3} > 0$. If $b \geq \sqrt{3}a$, then $d = 0$.

(b)

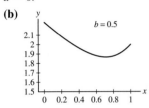

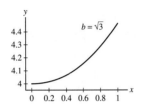

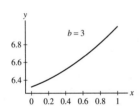

For $b = 3$ and $b = \sqrt{3}$, the minimum is 0. For $b = 0.5$, the minimum is 0.711.

59. $4\sqrt{1 + 2^{2/3}}(2^{2/3} + 1)$ **61.** $\theta = \pi/6$

67. (a) $d \approx 630.13$ ft   **(b)** That a different model is required.

69. (e) When $r = 1.5$ cm, the minimum is $S(0) = 8.4954$ cm^2; when $r = 2$ cm, the minimum is $S(x_2) = 12.852$ cm^2.

Section 4.7 Preliminary

1. $\lim\limits_{x \to 4} \frac{3x-12}{x^2-16}$

2. $\lim\limits_{x \to 0} \frac{x^2-2x}{3x-2}$ is not indeterminate, so L'Hôpital's Rule does not apply.

Section 4.7 Exercises

1. 5   **3.** $\frac{1}{15}$ **5.** $\frac{1}{114}$ **7.** 2   **9.** 0   **11.** $-\frac{1}{4}$ **13.** 0   **15.** 0

17. $\frac{4}{3}$ **19.** 1   **21.** 0   **23.** $-\frac{3}{5}$ **25.** $-\frac{7}{3}$ **27.** $\frac{3}{4}$ **29.** $\frac{1}{2}$

31. -1 **33.** 2   **35.** e **37.** $\frac{1}{2}$ **39.** 0   **41.** $\frac{1}{15}$ **43.** 1

45. 0   **47.** 1   **49.** 1   **51.** $\frac{1}{\pi}$

53. $\lim\limits_{x \to \pi/2} \frac{\cos mx}{\cos nx} = \begin{cases} (-1)^{(m-n)/2}, & m, n \text{ even} \\ \text{does not exist}, & m \text{ even}, n \text{ odd} \\ 0 & m \text{ odd}, n \text{ even} \\ (-1)^{(m-n)/2}\frac{m}{n}, & m, n \text{ odd} \end{cases}$

55. Since $\sin(1/x)$ oscillates as $x \to 0+$, L'Hôpital's Rule cannot be applied. Both numerical and graphical investigations suggest that the limit does not exist due to the oscillation.

x	1	0.1	0.01	0.001	0.0001	0.00001
$x^{\sin(1/x)}$	1	3.4996	10.2975	0.003316	16.6900	0.6626

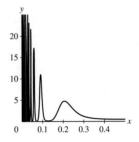

57. (a) $\lim\limits_{x \to 0+} f(x) = 0$; $\lim\limits_{x \to \infty} f(x) = 1$

(b) $f(x)$ is increasing for $0 < x < e$, is decreasing for $x > e$, and achieves a maximum at $x = e$. The maximum value is $f(e) = e^{1/e} \approx 1.444668$

(c) Because $(e, e^{1/e})$ is the only maximum, no solution exists for $c > e^{1/e}$ and only one solution exists for $c = e^{1/e}$. Moreover, because $f(x)$ increases from 0 to $e^{1/e}$ as x goes from 0 to e and then decreases from $e^{1/e}$ to 1 as x goes from e to $+\infty$, it follows that there are two solutions for $1 < c < e^{1/e}$, but only one solution for $0 < c \leq 1$.

(d) Observe that if we sketch the horizontal line $y = c$, this line will intersect the graph of $y = f(x)$ only once for $0 < c \leq 1$ and $c = e^{1/e}$ and will intersect the graph of $y = f(x)$ twice for $1 < c < e^{1/e}$. There are no points of intersection for $c > e^{1/e}$.

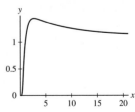

63. $\lim\limits_{x \to \infty} x^n e^{-x} = \lim\limits_{x \to \infty} \frac{x^n}{e^x} = \lim\limits_{x \to \infty} \frac{nx^{n-1}}{e^x}$
$= \lim\limits_{x \to \infty} \frac{n(n-1)x^{n-2}}{e^x}$
$= \cdots = \lim\limits_{x \to \infty} \frac{n!}{e^x} = 0.$

65. Because we are interested in the limit as $t \to +\infty$, we will restrict attention to $t > 1$. Then, for all k,

$$0 \leq t^k e^{-t^2} \leq t^k e^{-t}$$

As $\lim\limits_{t\to\infty} t^k e^{-t} = 0$, it follows from the Squeeze Theorem that

$$\lim_{t\to\infty} t^k e^{-t^2} = 0$$

71. (a) $\lim\limits_{\omega\to\lambda} y(t) = C\dfrac{\lambda t \cos(\lambda t) - \sin(\lambda t)}{-2\lambda}$

(c) The graphs below were produced with $C = 1$ and $\lambda = 1$. Moving from left to right and from top to bottom, $\omega = 0.5, 0.8, 0.9, 0.99, 0.999, 1$.

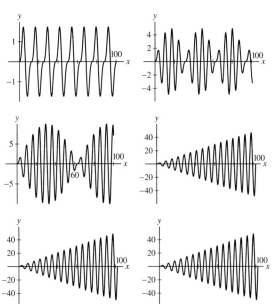

75. (a) $e^{-1/6}$ (b) $\frac{1}{3}$

Section 4.8 Preliminary

1. One   **2.** Sequence will remain x_0.   **3.** x_1 will not be defined.
4. Yes

Section 4.8 Exercises

1.

n	1	2	3
x_n	1.5	1.416666667	1.414215686

3.

n	1	2	3
x_n	1.717708333	1.710010702	1.709975947

5. $x_0 = -1.4$

n	1	2	3
x_n	-1.330964467	-1.328272820	-1.328268856

7. The roots are approximately 0.259 and 2.543.

9.

n	1	2	3
x_n	3.16667	3.162280702	3.16227766

3.16227766

11.

n	1	2	3	4
x_n	1.75	1.710884354	1.709976429	1.709975947

1.709975947

13. 2.093064358

15. $x_0 = 1$

n	1	2	3
x_n	1.25	1.189379699	1.184171279

17. $\theta_0 = 0.8$; The smallest positive solution is 0.787.

19. $x_1 \approx 3.0, x_2 \approx 2.2$ **21.** $x = 4.49341$

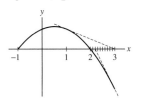 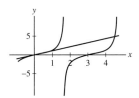

23. (a) $P \approx \$156.69$ (b) $b \approx 1.02121$; $r \approx 0.25452$
25. (c) $\theta = 1.76696$. Yes.
27. The sequence of iterates is $-2, 2, -2, 2, \dots$, which diverges by oscillation.
29. -3.58291867 for $x_0 = 1.85$ and $x_0 = 1.7$; 0.251322863 for $x_0 = 1.55$.
31. $\theta = 1.2757$; $h = L\dfrac{1-\cos\theta}{2\sin\theta} \approx 1.11181$
33. (a) $a \approx 46.95$ (b) $s \approx 29.23$
35. (a) $a \approx 28.46$
(b) For $\Delta L = 1$, $\Delta s \approx 0.61$; for $\Delta L = 5$, $\Delta s \approx 3.04$
(c) $s(161) - s(160) \approx 0.6068$; $s(165) - s(160) \approx 3.0155$

Section 4.9 Preliminary

1. $\ln(-x)$ **2.** Any constant function   **3.** No difference
4. No   **5.** $\sin x + C$; at $x = 0$, the value of the antiderivative is C.
6. (a) False   (b) True   (c) False
7. $y = x^2$ is *not* a solution.

Section 4.9 Exercises

1. $F(x) = 6x^2 + C$ **3.** $F(x) = \frac{1}{3}x^3 + \frac{3}{2}x^2 + 2x + C$
5. $F(x) = -\frac{8}{3}x^{-3} + C$ **7.** $F(x) = 5e^x + \frac{1}{3}x^3 + C$
9. (b)   **11.** (a)
13. $F(x) = \frac{1}{2}x^2 + x + C$ **15.** $F(t) = \frac{1}{6}t^6 + \frac{3}{2}t^2 + 2t + C$
17. $F(t) = -\frac{5}{4}t^{-4/5} + C$ **19.** $F(x) = 2x + C$
21. $F(t) = \frac{5}{2}t^2 - 9t + C$ **23.** $F(x) = -\frac{1}{x} + C$
25. $F(x) = -\frac{1}{x+3} + C$ **27.** $F(z) = -\frac{3}{4}z^{-4} + C$
29. $F(x) = \frac{2}{5}x^{5/2} - \frac{2}{3}x^{3/2} + C$ **31.** $F(t) = \frac{2}{3}t^{3/2} - 14t^{1/2} + C$
33. $F(x) = -4\cos x - 3\sin x + C$ **35.** $F(t) = \frac{1}{6}\sin(6t + 4) + C$
37. $F(t) = -\frac{1}{4}\sin(3 - 4t) + C$ **39.** $F(x) = \sin x - e^x + C$

41. $F(x) = 5e^{5x} + C$ **43.** (A)

45. (a) $F(x) = \frac{1}{3}\tan(3x) + C$ (b) $F(x) = \sec(x + 3) + C$

47. $y = \frac{1}{2}\sin 2x + 3$ **49.** $y = \frac{1}{2}x^2 + 5$

51. $y = 5t - \frac{2}{3}t^3 - \frac{7}{3}$ **53.** $y = 2t^2 + 9t + 1$

55. $y = 1 - \cos x$ **57.** $y = 3 + \frac{1}{5}\sin 5x$

59. $y = e^x + 3$ **61.** $y = \frac{1}{5}e^{5x} - \frac{16}{5}$

63. $f(x) = \frac{1}{6}x^3 + x$ **65.** $f(x) = \frac{1}{20}x^5 - \frac{1}{3}x^3 + \frac{1}{2}x^2 - \frac{1}{4}x + \frac{121}{30}$

67. $f(\theta) = -\cos\theta + 6$

71. $\frac{ds}{dt} = \frac{1}{2}t^2 - t$, $s(0) = 0$, $s(t) = \frac{1}{6}t^3 - \frac{1}{2}t^2$

73. (a) $s(t) = \frac{25}{2}t^2 - \frac{1}{3}t^3 + 5$ (b) $s(t) = \frac{25}{2}t^2 - \frac{1}{3}t^3 - \frac{127}{3}$

75. $t = 6$ s, $s = 252$ ft   **77.** $v(500) \approx 764$ m/s

79. $c_1 = c_2 = 1$

83. (a) $\frac{d}{dx}\left(\frac{1}{2}F(2x)\right) = \frac{1}{2}F'(2x) \cdot 2 = F'(2x) = f(2x)$

(b) $\frac{1}{k}F(kx) + C$

Chapter 4 Review

1. $8.1^{1/3} - 2 \approx 0.00833333$. The error is 3.445×10^{-5}.

3. $625^{1/4} - 624^{1/4} \approx 0.002$. The error is 1.201×10^{-6}.

5. $\frac{1}{1.02} \approx L(1.02) = 0.98$. The error is 3.922×10^{-4}.

7. $L(x) = 5 + \frac{1}{10}(x - 25)$ **9.** $L(r) = 36\pi(r - 2)$

11. $L(x) = \frac{1}{\sqrt{e}}(2 - x)$ **13.** $\Delta s \approx 0.632$

15. 60 MP3 players will be sold per week when the price is \$80; 104 MP3 players will be sold per week when the price is \$69.

17. $\sqrt{26} \approx 5.1$. The error is 9.80×10^{-4}.

23. $f\left(\frac{2}{3}\right)$ is a local maximum. $f(2)$ is a local minimum.

25. $f(-2)$ is neither a local maximum nor a local minimum, $f\left(-\frac{4}{5}\right)$ is a local maximum, and $f(0)$ is a local minimum.

27. The maximum value is 21. The minimum value is -11.

29. The minimum value is -1. The maximum value is $\frac{5}{4}$.

31. The minimum value is -1. The maximum value is 3.

33. The minimum value is $12 - 12\ln 12 \approx -17.818880$. The maximum value is $40 - 12\ln 40 \approx -4.266553$.

35. Critical points $x = 1$, $x = 3$. The minimum value is 2. The maximum value is 17.

37. $x = \frac{4}{3}$ **39.** $x = \pm\frac{2}{\sqrt{3}}$ **41.** $x = 1$ and $x = 4$

43. (a) ii   (b) i   (c) iii

45. $L = \infty$ **47.** $L = \infty$ **49.** $L = \frac{1}{4}$ **51.** $L = \infty$

53. $L = \frac{1}{2}$ **55.** $L = -\infty$ **57.** $L = -\infty$ **59.** $L = 0$

61.

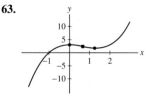

63.

65.

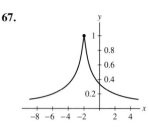

67.

69.

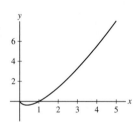

71. $V = \frac{16}{81}\pi R^2 H$

77. $\sqrt[3]{25} = 2.9240$ **79.** $F(x) = x^4 - \frac{2}{3}x^3 + C$

81. $F(\theta) = -\cos(\theta - 8) + C$ **83.** $F(t) = -2t^{-2} + 4t^{-3} + C$

85. $F(x) = \tan x + C$

87. $F(y) = \frac{1}{5}(y + 2)^5$

89. $F(x) = e^x - \frac{1}{2}x^2 + C$ **91.** $F(x) = 4\ln|x| + C$

93. $y(x) = x^4 + 3$ **95.** $y(x) = 2x^{1/2} - 1$ **97.** $y(x) = 4 - e^{-x}$

99. $f(t) = \frac{1}{2}t^2 - \frac{1}{3}t^3 - t + 2$ **101.** $f(0) = \frac{2}{e}$ is a local minimum

103. Local minimum at $x = e^{-1}$; no points of inflection; $\lim\limits_{x \to 0+} x\ln x = 0$; $\lim\limits_{x \to \infty} x\ln x = \infty$

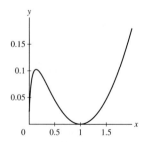

105. Local minimum at $x = 1$; local maximum at $x = 10^{-2/\ln 10} = e^{-2}$; point of inflection at $x = 10^{-1/\ln 10} = \frac{1}{e}$; $\lim\limits_{x \to 0+} x(\log x)^2 = 0$; $\lim\limits_{x \to \infty} x(\log x)^2 = \infty$

107. As $x \to \infty$, both $2x - \sin x$ and $3x + \cos 2x$ tend toward infinity, so L'Hôpital's Rule applies to $\lim\limits_{x \to \infty} \frac{2x - \sin x}{3x + \cos 2x}$; however, the resulting limit, $\lim\limits_{x \to \infty} \frac{2 - \cos x}{3 - 2\sin 2x}$, does not exist due to the oscillation of $\sin x$ and $\cos x$. $\lim\limits_{x \to \infty} \frac{2x - \sin x}{3x + \cos 2x} = \frac{2}{3}$

109. 4   **111.** 0   **113.** 3   **115.** $\ln 2$ **117.** $\frac{1}{6}$ **119.** 2

Chapter 5

Section 5.1 Preliminary

1. The right-endpoints are $\frac{5}{2}, 3, \frac{7}{2}, 4, \frac{9}{2}, 5$. The left-endpoints are 2, $\frac{5}{2}, 3, \frac{7}{2}, 4, \frac{9}{2}$.
2. L_2 **3.** (b)
5. (a) $1, \frac{15}{4}$ (b) $\frac{5}{4}, \frac{3}{2}$
6. (a) False   (b) True   (c) True

Section 5.1 Exercises

1. $10\frac{1}{2}$ mi   **3.** 11.4 in.   **5.** $R_6 = 54.5$ ft, $L_6 = 44.5$ ft, $M_3 = 51$ ft
7. (a) $L_6 = 16.5$, $R_6 = 19.5$ (b) exact area: $A = 18$; $L_6 - A = -1.5$; $R_6 - A = 1.5$
9. $L_6 = 2.925$, $R_6 = 2.4875$, $M_6 = 2.75$
11.

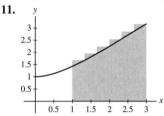

13. $R_8 = 5.75$ **15.** $M_4 = 0.328125$ **17.** $R_6 = 47.5$
19. $L_5 \approx 0.745635$ **21.** $L_4 \approx 0.361372$
23. $M_6 \approx 0.386871$ **25.** $0.518 \le A \le 0.768$
29. $L_{100} = 0.6614629$, $R_{100} = 0.6714629$; $L_{150} = 0.6632220$, $R_{150} = 0.6698887$
31. (a) 15   (b) 18   (c) 99   (d) -1 (e) $\frac{11}{6}$ (f) 40
33. 15,050   **35.** $\sum_{k=2}^{5} (k^2 + k)$ **37.** $\sum_{k=1}^{n} \sqrt{k + k^3}$
39. $\sum_{k=1}^{n} e^{\pi/k}$ **41.** 440   **43.** 379,507,500   **45.** $\frac{46,168}{3}$
47. 26,475   **49.** 3   **51.** 11   **53.** $\frac{1}{4}$ **55.** $-\frac{79}{4}$ **57.** 20   **59.** 0.5
61. $R_N = \frac{1}{4} + \frac{1}{2N} + \frac{1}{4N^2}$, $\lim\limits_{N\to\infty} R_N = \frac{1}{4}$
63. $R_N = 1 - \frac{1}{4} - \frac{1}{2N} - \frac{1}{4N^2}$, $\lim\limits_{N\to\infty} R_N = \frac{3}{4}$
65. $R_N = 128 + \frac{136}{N} + \frac{32}{N^2}$, $\lim\limits_{N\to\infty} R_N = 128$
67. $R_N = \frac{56}{3} + \frac{12}{N} + \frac{4}{3N^2}$, $\lim\limits_{N\to\infty} R_N = \frac{56}{3}$
69.
$$R_N = a^2(b - a) + a(b - a)^2 + \frac{a(b-a)^2}{N} + \frac{(b-a)^3}{3}$$
$$+ \frac{(b-a)^3}{2N} + \frac{(b-a)^3}{6N^2},$$
$$\lim\limits_{N\to\infty} R_N = \frac{1}{3}b^3 - \frac{1}{3}a^3$$
73. The area between the graph of $f(x) = x^4$ and the x-axis over the interval [2, 5].
75. The area between the graph of $f(x) = \sin x$ and the x-axis over the interval $\left[\frac{\pi}{3}, \frac{5\pi}{6}\right]$.

77. $\lim\limits_{N\to\infty} \frac{\pi}{N} \sum\limits_{k=1}^{N} \sin\left(\frac{k\pi}{N}\right)$
79. $\lim\limits_{N\to\infty} \frac{1}{2N} \sum\limits_{j=1}^{N} \tan\left(\frac{1}{2} + \frac{1}{2N}\left(j - \frac{1}{2}\right)\right)$
81. $\lim\limits_{N\to\infty} \frac{7\pi}{8N} \sum\limits_{j=0}^{N-1} \cos\left(\frac{\pi}{8} + j\frac{7\pi}{8N}\right)$ **85.** R_{501}, L_{500}, M_{10}
87. Left/right-endpoint approximation, $n = 2$ No.

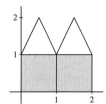

93. $N > 30,000$

Section 5.2 Preliminary

1. $b - a$
2. (a) False   (b) True   (c) True
4. No   **5.** 2

Section 5.2 Exercises

1. Area = 0

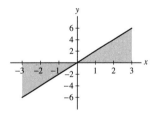

3. Area = 7.5

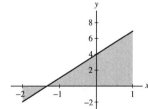

5. Area = 0

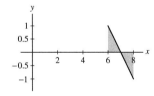

7. Area = $\frac{25\pi}{4}$

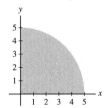

9. Area = 4

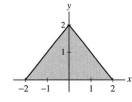

11. Area = 6

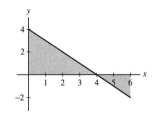

13. (a) $-\frac{\pi}{2}$ (b) $\frac{3\pi}{2}$ (c) $\frac{3}{4}\pi$ (d) $\frac{9}{4}\pi$
15. $a = 4$, $b = 1$, $c = 4$

17.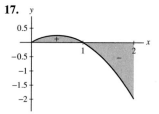

19.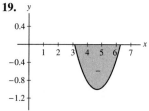

21.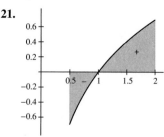

23. Positive **25.** Negative

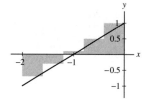

27. $R(f, P, C) = 1.59$ **29.** $R(f, P, C) = 0.24$

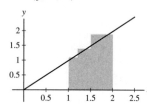

31. $\frac{64}{3}$ **33.** $\frac{51}{2}$ **35.** $-\frac{2}{3}$ **37.** $\frac{1}{3}a^3 - \frac{1}{2}a^2 + \frac{5}{6}$ **39.** $e^4 - 1$
43. $\frac{20}{3}$ **45.** -2 **47.** -24 **49.** $-\frac{9}{4}$ **51.** 53 **53.** -45
55. 8 **57.** -7 **59.** $\int_0^7 f(x)\,dx$ **61.** $\int_5^9 f(x)\,dx$ **63.** $\frac{2}{3}$
65. -3
69. Here is a graphical example of this phenomenon.

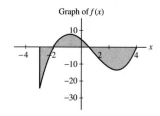

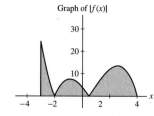

71. 9 **73.** 0.5
75. On the interval $[0, 1]$, $x^5 \leq x^4$, so $\int_0^1 x^5\,dx \leq \int_0^1 x^4\,dx$. On the other hand, $x^4 \leq x^5$ for $x \in [1, 2]$, so $\int_1^2 x^4\,dx \leq \int_1^2 x^5\,dx$.
81. The assertion is false. A counterexample: $a = 0, b = 1$, $f(x) = x, g(x) = 2$.

Section 5.3 Preliminary

1. 4
2. Signed area between the graph of $y = f(x)$ and the x-axis over the interval $[1, 4]$.
3. $\int_0^7 f(x)\,dx = 6$, $\int_2^7 f(x)\,dx = 2$
4. (a) False (b) False (c) False
5. 0

Section 5.3 Exercises

1. Area $= \frac{1}{3}$ **3.** Area $= 1$

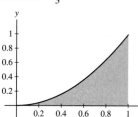

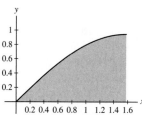

5. $\frac{27}{2}$ **7.** $\frac{35}{3}$ **9.** $e^5 - e^3$ **11.** $2e^{-2} - 8$ **13.** $\frac{34}{3}$ **15.** $\frac{133}{3}$
17. 0 **19.** $\frac{162\sqrt{3}}{5} - \frac{82}{45}$ **21.** $\frac{3}{4}$ **23.** 60 **25.** 4 **27.** $\frac{1}{3}\ln 50$
29. $-\frac{3}{8}$ **31.** $60\sqrt{3} - \frac{8}{3}$ **33.** 2 **35.** $\sqrt{2}$ **37.** 1 **39.** $2 - \frac{2}{3}\sqrt{3}$
41. $\int_{-2}^0 -x\,dx + \int_0^1 x\,dx = \frac{5}{2}$ **43.** $\int_{-2}^0 (-x^3)\,dx + \int_0^3 x^3\,dx = \frac{97}{4}$
45. $\int_0^{\pi/2} \cos x\,dx + \int_{\pi/2}^{\pi} (-\cos x)\,dx = 2$
47. $\frac{1}{4}(b^4 - 1)$ **49.** $\frac{1}{6}(b^6 - 1)$ **51.** $\ln 5$ **57.** $\frac{707}{12}$

Section 5.4 Preliminary

1. $A(-2) = 0$
2. (a) No (b) Yes
3. (c) **4.** False **5.** (b), (e), (f) are correct
6. $\frac{d}{dx}\left(\int_1^3 t^3\,dt\right)\Big|_{x=2} = 0$

Section 5.4 Exercises

1. $A(x) = \int_{-2}^x (2t + 4)\,dt = x^2 + 4x + 4$
3. (a) $G(1) = \int_1^1 (t^2 - 2)\,dt = 0$ (b) $G'(1) = -1$ and $G'(2) = 2$
(c) $G(x) = \frac{1}{3}x^3 - 2x + \frac{5}{3}$
5. $G(1) = 0, G'(0) = 0, G'(\frac{\pi}{4}) = 1$ **7.** $F(x) = \frac{1}{4}x^4 - 4$
9. $F(x) = \frac{1}{2}x^4 - \frac{1}{2}$ **11.** $F(x) = e^5 - e^x$
13. $F(x) = \tan x + 1$ **15.** $F(x) = \int_3^x \sqrt{t^4 + 1}\,dt$
17. $F(x) = \int_0^x \sec t\,dt$ **19.** $\frac{d}{dx}\int_0^x (t^3 - t)\,dt = x^3 - x$
21. $\frac{d}{dt}\int_{100}^t \cos 5x\,dx = \cos 5t$
23.

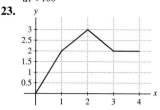

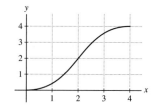

25.

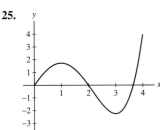

27. $G'(x) = 3x^2 \tan x^3$

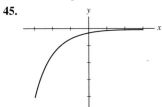

29. $\frac{d}{dx} \int_0^{x^2} \sin^2 t \, dt = 2x \sin^2(x^2)$

31. $\frac{d}{ds} \int_{-6}^{\cos s} (u^4 - 3u) \, du = -\sin s (\cos^4 s - 3 \cos s)$

33. $\frac{d}{dx} \int_{x^3}^0 \sin^2 t \, dt = -3x^2 \sin^2 x^3$

35. $\frac{d}{dx} \int_{\sqrt{x}}^{x^2} \tan t \, dt = 2x \tan x^2 - \frac{\tan(\sqrt{x})}{2\sqrt{x}}$

37. Minimum: $A(1.5) = -1.25$; maximum: $A(4.5) = 1.25$

39. (a) No   (b) R   (c) S   (d) True

43. (a) Increasing on $(0, 4)$ and $(8, 12)$. Decreasing on $(4, 8)$ and $(12, \infty)$.

(b) Local minimum at $x = 8$. Local maximum at $x = 4$ and at $x = 12$.

(c) $x = 2$, $x = 6$, and $x = 10$

(d) Concave up $(0, 2)$ and $(6, 10)$. Concave down $(2, 6)$ and $(10, \infty)$.

45.

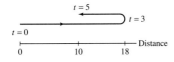

47. $x = \pi^{2/3}$, $F(x)$ changes from concave down to concave up.

49. $g(x) = 2x + 1$, $c = 2$, or $c = -3$

53. $a = -3$ and $b = 3$

Section 5.5 Preliminary

1. 350 mi   **2.** The total drop in temperature

3. (a) and (c) represented as derivatives. (b) and (d) represented as integrals.

4. Difference in distance

Section 5.5 Exercises

1. 15,062.5 gal   **3.** 682 insects   **5.** 186 bicycles   **7.** 12 ft

9. Displacement = 10 ft, distance = 26 ft

11. Displacement = 0 m, distance = 1 m

13. 132 liters   **15.** 205 ft/s   **17.** 9,200 cars

19. The total cost of reducing the amount of CO_2 by 3 tons.

21. (a) Net migration during 1988–1991.   (b) Net outflow of 66,500 people.   (c) 1989

23. (b) $\frac{1,300 - 100\sqrt{2}}{3} \approx 386.19$ J

25. 3,517 families

Section 5.6 Preliminary

1. (a), (b)

2. (a) $c = \frac{1}{2}$, $g(u) = u^4$, and $u(x) = x^2 + 9$

(b) $c = \frac{1}{3}$, $g(u) = \sin u$, and $u(x) = x^3$

(c) $c = -1$, $g(u) = u^2$, and $u(x) = \cos x$

3. (c)

Section 5.6 Exercises

1. $du = -2x \, dx$ **3.** $du = 3x^2 \, dx$ **5.** $du = -2x \sin(x^2) \, dx$

7. $\int u^3 \, du = \frac{1}{4}(x - 7)^4 + C$ **9.** $\int u^{-2} \, du = -\frac{1}{x+1} + C$

11. $\frac{1}{2} \int \sin u \, du = -\frac{1}{2} \cos(2x - 4) + C$

13. $\frac{1}{2} \int \frac{1}{u^3} \, du = \frac{-1}{4(x^2+2x)^2} + C$

15. $\frac{1}{4} \int u^{1/2} \, du = \frac{1}{6}(4x - 1)^{3/2} + C$

17. $\frac{1}{64} \int (u^{5/2} + 2u^{3/2} + u^{1/2}) \, du = $
$\frac{1}{224}(4x - 1)^{7/2} + \frac{1}{160}(4x - 1)^{5/2} + \frac{1}{96}(4x - 1)^{3/2} + C$

19. $\int u^2 \, du = \frac{1}{3} \sin^3 x + C$

21. $-\frac{1}{2} \int \frac{du}{u} = -\frac{1}{2} \ln |\cos 2x| + C$

23. $-\frac{1}{2} \int e^u \, du = -\frac{1}{2} e^{-x^2} + C$ **25.** $\int \frac{du}{(u+1)^2} = -\frac{1}{e^t+1} + C$

27. $\int u^{-2} \, du = -\frac{1}{\ln x} + C$ **29.** Take $u = x^4$ **31.** Take $u = x^{3/2}$

33. $\int (4x + 3)^4 \, dx = \frac{1}{20}(4x + 3)^5 + C$

35. $\int \frac{1}{\sqrt{x-7}} \, dx = 2\sqrt{x - 7} + C$

37. $\int x\sqrt{x^2 - 4} \, dx = \frac{1}{3}(x^2 - 4)^{3/2} + C$

39. $\int \frac{dx}{(x+9)^2} \, dx = -\frac{1}{x+9} + C$

41. $\int \frac{2x^2+x}{(4x^3+3x^2)^2} \, dx = -\frac{1}{6}(4x^3 + 3x^2)^{-1} + C$

43. $\int \frac{5x^4+2x}{(x^5+x^2)^3} \, dx = -\frac{1}{2} \frac{1}{(x^5+x^2)^2} + C$

45. $\int (3x + 9)^{10} \, dx = \frac{1}{33}(3x + 9)^{11} + C$

47. $\int x(x + 1)^{1/4} \, dx = \frac{4}{9}(x + 1)^{9/4} - \frac{4}{5}(x + 1)^{5/4} + C$

49. $\int x^3(x^2 - 1)^{3/2} \, dx = \frac{1}{7}(x^2 - 1)^{7/2} + \frac{1}{5}(x^2 - 1)^{5/2} + C$

51. $\int \sin^5 x \cos x \, dx = \frac{1}{6} \sin^6 x + C$

53. $\int \tan 3x \, dx = -\frac{1}{3} \ln |\cos 3x| + C$

55. $\int \sec^2(4x + 9) \, dx = \frac{1}{4} \tan(4x + 9) + C$

57. $\int \frac{\cos 2x}{(1+\sin 2x)^2} \, dx = -\frac{1}{2}(1 + \sin 2x)^{-1} + C$

59. $\int \cos x (3 \sin x - 1)\, dx = \frac{1}{6}(3 \sin x - 1)^2 + C$

61. $\int \sec^2 x (4 \tan^3 x - 3 \tan^2 x)\, dx = \tan^4 x - \tan^3 x + C$

63. $\int (x + 1)e^{x^2+2x}\, dx = \frac{1}{2}e^{x^2+2x} + C$

65. $\int \frac{e^x\, dx}{(e^x+1)^4} = -\frac{1}{3}(e^x + 1)^{-3} + C$

67. $\int \frac{(\ln x)^4\, dx}{x} = \frac{1}{5}(\ln x)^5 + C$

69. $\int \frac{dx}{x \ln x} = \ln|\ln x| + C$

71. $\int \sqrt{x^3 + 1}\, x^5\, dx = \frac{2}{15}(x^3 + 1)^{5/2} - \frac{2}{9}(x^3 + 1)^{3/2} + C$

77. (d)   **79.** $\frac{38}{3}$ **81.** $\frac{42}{5}$ **83.** $\frac{72}{289}$ **85.** 3   **87.** $\frac{1}{2}\ln(\sec 1)$

89. $-\frac{1}{3}$ **91.** $\frac{1}{3}$ **93.** $\frac{20}{3}\sqrt{5} - \frac{32}{5}\sqrt{3}$

95. $\int \frac{f'(x)}{f(x)^2}\, dx = -\frac{1}{f(x)} + C$ **97.** $\frac{1}{n+1}$ **99.** $\frac{\pi}{4}$

Section 5.7 Preliminary

1. (a) $\frac{2^x}{\ln 2} + C$ (b) $\ln|x| + C$ (c) $\sin^{-1} x + C$

2. (a) $b = 3$ (b) $b = e^3$

3. $b = \sqrt{3}$ **4.** (b): Use the substitution $u = 3x$ **5.** $x = 4u$

Section 5.7 Exercises

1. $\ln 2$ **3.** 1   **5.** -1 **7.** $\frac{\pi}{6}$ **9.** $\frac{\pi}{6}$ **13.** $\frac{\pi}{8}$

15. $\sin^{-1}\left(\frac{t}{4}\right) + C$ **17.** $\frac{1}{2}\sin^{-1}\left(\frac{2t}{5}\right) + C$ **19.** $\frac{1}{2}\sin^{-1} 2x + C$

21. $-\sqrt{1 - x^2} + \sin^{-1} x + C$ **23.** $\tan^{-1} e^x + C$

25. $\frac{1}{2}(\tan^{-1} x)^2 + C$ **27.** $\frac{2}{\ln 3}$ **29.** $\frac{1}{\ln 2}$ **31.** $-\frac{1}{\ln 9}\cos(9^x) + C$

33. $e^x + 2x + C$ **35.** $-\frac{7^{-x}}{\ln 7} + C$ **37.** $\frac{1}{4}e^{4x} + x + C$

39. $-\frac{1}{9}e^{-9t} + C$ **41.** $\frac{1}{4}\sin^{-1}(4x) + C$ **43.** $\frac{2}{3}(e^t + 1)^{3/2} + C$

45. $7x - \frac{1}{10}e^{10x} + C$ **47.** $\sec^{-1}(5x) + C$ **49.** $-\frac{1}{8}e^{-4x^2} + C$

51. $2\sqrt{e^x + 1} + C$ **53.** $\frac{1}{2}\ln|2x + 4| + C$ **55.** $\frac{1}{3}\ln|x^3 + 2| + C$

57. $-\frac{1}{4}\ln|\cos(4x + 1)| + C$ **59.** $\frac{1}{2}\ln(2\sin x + 3) + C$

61. $\frac{1}{8}(4\ln x + 5)^2 + C$ **63.** $\ln|\ln x| + C$ **65.** $\frac{1}{2}\big(\ln(\ln x)\big)^2 + C$

67. $\frac{3^x}{\ln 3} + C$ **69.** $\frac{3^{\sin x}}{\ln 3} + C$

75. Assume $f(x) = e^x$ is a polynomial function of degree n. Then $f^{(n+1)}(x) = 0$. But we know that any derivative of e^x is e^x and $e^x \neq 0$. Hence, e^x cannot be a polynomial function.

81.

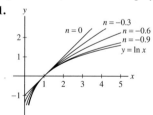

Section 5.8 Preliminary

1. The quantity with $k = 3.4$ doubles more rapidly.   **2.** No

3. They take the same amount of time.   **4.** $\frac{dS}{dn} = \frac{1}{2}S$

5. Too ancient   **6.** 12% compounded quarterly

7. Continuously: 1.09417; quarterly: 1.09308

8. (b) The amount you would have to invest today in order to receive N dollars at time T.

9. Decrease

10. If the interest rate goes up, the present value of $1,000 one year from today decreases. Therefore, Xavier will be sad if the interest rate has just increased from 6 to 7%.

Section 5.8 Exercises

1. (a) $P(0) = 2,000$ (b) $t = \frac{1}{1.3}\ln 5 \approx 1.24$ hr

3. $N'(t) = \frac{\ln 2}{3}N(t)$; after 10 minutes, there are 10 molecules.

5. $\frac{\ln 2}{0.13} \approx 5.33$ yr   **7.** $y(t) = Ce^{-5t}$; $y(t) = 3.4e^{-5t}$

9. $y(t) = 4e^{3(t-2)}$

11. (a) $\frac{\ln 2}{0.06} \approx 11.55$ yr   (b) $\frac{\ln 3}{0.06} \approx 18.31$ yr   (c) 23.10 yr

13. Data I   **17.** $k = \frac{\ln(3/10)}{-17} \approx 0.071$ yr^{-1}

19. $10^{-12}e^{-0.000121(32,400)} \approx 1.98 \times 10^{-14}$; $10^{-12}e^{-0.000121(29,700)} \approx 2.75 \times 10^{-14}$

21. (a) $k = -\frac{1}{10}\ln\left(\frac{2.13}{14.7}\right) \approx 0.193$ mi^{-1}

(b) $P(15) \approx 0.813$ lb/in^2

23. $t = \frac{1}{0.1155}\ln 1.5 \approx 3.5$ hr

25. (a) Yes; $k \approx \frac{\ln 18,666.67}{29} \approx 0.339$. Note: A better estimate can be found by calculating k for each time period and then averaging the k values.

(b)

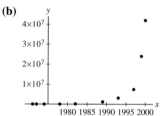

(c) $N(t) = 2,250e^{0.339t}$

(d) The doubling time is $\ln 2/0.339 \approx 2.04$ yr.

(e) $2,250e^{0.339(39)} \approx 1,241,623,327$

(f) No, you can't make a microchip smaller than an atom.

29. (a) Starting from $t_0 = 1$, P doubles when $t = \sqrt{2}$. Starting from $t_0 = 2$, P doubles when $t = 2\sqrt{2}$. Finally, starting from $t_0 = 3$, P doubles when $t = 3\sqrt{2}$.

(b) Starting from $t = t_0$, P doubles when $t = t_0\sqrt{2}$.

31. (a) $4,870.38   (b) $4,902.71   (c) $4,919.21

33. (a) 1.05   (b) 1.0508   (c) 1.0513

35. The account doubles when $P(t) = 2P_0 = P_0 e^{rt}$ so $2 = e^{rt}$ and $t = \frac{\ln 2}{r}$.

37. $26,629.59

39. $r = 0.08$: receive $1,000 now; $r = 0.03$: receive $1,300 four years from now

41. Yes, the investment is worthwhile.

43. (a) $22,252,915.21   (b) $2,747,084.79

45. $39,346.93   **47.** $3,296.80   **51.** $R = \$1{,}200$
53. $71,460.53   **55.** Potassium-49
57. **(d)** Let $g(c) = c \ln c - c$. Then,

c	0.01	0.001	0.0001	0.00001
$g(c)$	-0.056052	-0.007908	-0.001021	-0.000125

Thus, as $c \to 0+$, it appears that $g(c) \to 0$.
(e) $M = 1/k$ **(f)** $M = 5.52$ days

Chapter 5 Review

1. $L_4 = \frac{23}{4}$, $M_4 = 7$ **3.** [0, 2] for R_4, [2, 3] for L_4
5. $R_6 = \frac{463}{8}$ $M_6 = \frac{815}{16}$

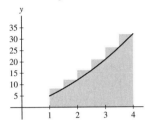
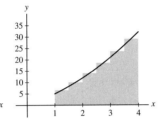

$L_6 = \frac{355}{8}$

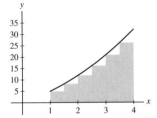

7. $L_N = \frac{32}{3} - \frac{12}{N} + \frac{4}{3N^2} + \frac{8(N-1)}{N}$, $\int_0^2 f(x)\,dx = \frac{32}{3}$
9. $R_6 \approx 0.742574$, $M_6 \approx 0.785977$, $L_6 \approx 0.825907$
11. R_5, $R_5 = L_5 = 90$
13. $\int (6x^3 - 9x^2 + 4x)\,dx = \frac{3}{2}x^4 - 3x^3 + 2x^2 + C$
15. $\int (2x^3 - 1)^2\,dx = \frac{4}{7}x^7 - x^4 + x + C$
17. $\int \frac{x^4+1}{x^2}\,dx = \frac{1}{3}x^3 - x^{-1} + C$ **19.** 30
21. $\int \csc^2 \theta\,d\theta = -\cot\theta + C$
23. $\int \sec^2(9t - 4)\,dt = \frac{1}{9}\tan(9t - 4) + C$
25. $\int (9t - 4)^{11}\,dt = \frac{1}{108}(9t - 4)^{12} + C$
27. $\int \sin^2(3\theta)\cos(3\theta)\,d\theta = \frac{1}{9}\sin^3(3\theta) + C$
29. $\int \frac{2x^3+3x}{(3x^4+9x^2)^5}\,dx = -\frac{1}{24}(3x^4 + 9x^2)^{-4} + C$
31. $\int \sin\theta\sqrt{4 - \cos\theta}\,d\theta = \frac{2}{3}(4 - \cos\theta)^{3/2} + C$
33. $\int y\sqrt{2y + 3}\,dy = \frac{1}{10}(2y + 3)^{5/2} - \frac{1}{2}(2y + 3)^{3/2} + C$
35. $\int_{-2}^{6} f(x)\,dx$ **37.** $x = \pm 1$

39. Daily consumption is 9.312 million gallons. 18:00–24:00 consumption is 1.68 million gallons.
41. $208,245   **45.** $\int_0^2 \sin x\,dx = 1 - \cos 2$
47. $\int_{\pi/2}^{3\pi/2} \sin x\,dx = 0$ **49.** $\int_0^1 x^k\,dx = \frac{1}{k+1}$
51. $\int_0^1 f(x)\,dx = \frac{13}{2}$ **53.** $\int x\cos x\,dx = x\sin x + \cos x + C$
55.

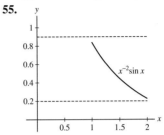

57. $A'(x) = \sin(x^3)$ **59.** $\frac{d}{dy}\int_{-2}^{y} 3^x\,dx = 3^y$ **61.** $G'(2) = 36$
65. $\frac{1}{3}(\ln x)^3 + C$ **67.** $-\sin^{-1}(e^{-x}) + C$ **69.** $\tan^{-1}(\ln t) + C$
71. $-\frac{1}{2}e^{9-2x} + C$ **73.** $\frac{1}{2}\cos(e^{-2x}) + C$ **75.** $\frac{1}{2}$
77. $\sin^{-1}\left(\frac{2}{3}\right) - \sin^{-1}\left(\frac{1}{3}\right)$ **79.** $\ln 2$ **81.** $\frac{1}{2}\sin 2$ **83.** $\frac{1}{2}\ln 2$
85. $\frac{1}{2}\sin^{-1} x^2 + C$ **87.** $\frac{1}{2}(\sin^{-1})^2 + C$ **89.** $-\frac{1}{4}\cos^4 x + C$
91. $\frac{\sqrt{2}}{2}\tan^{-1}(4\sqrt{2})$
95. Because $t > \ln t$ for $t > 2$,

$$F(x) = \int_2^x \frac{dt}{\ln t} > \int_2^x \frac{dt}{t} > \ln x$$

Thus, $F(x) \to \infty$ as $x \to \infty$. Moreover,

$$\lim_{x\to\infty} G(x) = \lim_{x\to\infty} \frac{1}{1/x} = \lim_{x\to\infty} x = \infty$$

Thus, $\lim_{x\to\infty} \frac{F(x)}{G(x)}$ is of the form ∞/∞ and L'Hôpital's Rule applies.
Finally,

$$L = \lim_{x\to\infty} \frac{F(x)}{G(x)} = \lim_{x\to\infty} \frac{\ln x}{\ln x - 1} = 1$$

97. approximately 6,066 years   **99.** $12,809.44
101. $\frac{dk}{dT} \approx 12.27$ hr^{-1}-K^{-1}; approximate change in k when T is raised from 500 to 510: 122.7 hr^{-1}.
103. If $r = 5\%$, this is a good investment. The largest interest rate that would make the investment worthwhile is $r \approx 10.13\%$.

Chapter 6

Section 6.1 Preliminary

1. Area of the region bounded between the graphs of $y = f(x)$ and $y = g(x)$, bounded on the left by the vertical line $x = a$ and on the right by the vertical line $x = b$.
2. Yes   **3.** $\int_0^3 \big(f(x) - g(x)\big)\,dx + \int_3^5 \big(g(x) - f(x)\big)\,dx$
4. Negative

Section 6.1 Exercises

1. 102

3. **(a)** $(-2, -2)$ and $(1, 1)$ **(b)** 4.5

5. $\sqrt{2} - 1$ **7.** $2\sqrt{2}$ **9.** $\frac{343}{3}$ **11.** $\frac{1}{2}e^2 - e + \frac{1}{2}$

13. $\frac{2\pi}{3}$

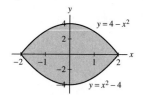

15. $\frac{160}{3}$ **17.** $\frac{5}{24}$ **19.** $\frac{81}{2}$ **21.** $2 - \frac{\pi}{2}$ **23.** $\frac{1,331}{6}$ **25.** $\frac{32}{3}$

27. $\frac{64}{3}$

29. $\frac{64}{3}$ **31.** $2 - \frac{\pi}{2}$

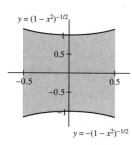

33. 2 **35.** $\frac{128\sqrt{2}}{15}$

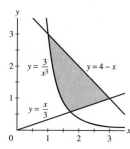

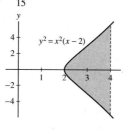

37. $\frac{44}{3}$ **39.** 12

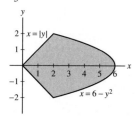

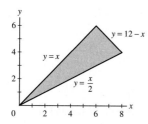

41. $\frac{32}{3}$ **43.** $\frac{16}{3}$

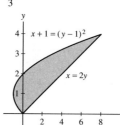

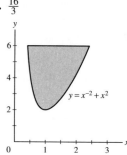

45. $2 - \sqrt{2}$

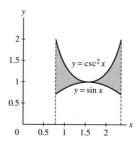

47. The region is $\frac{1}{4}$ of the unit circle; $\frac{\pi}{4}$

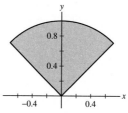

49. $\int_0^3 \left(f(x) - g(x)\right) dx + \int_3^5 f(x) \, dx - \int_5^9 f(x) \, dx$

51. $\int_{-1}^1 \left(\sqrt{2 - x^2} - \left(1 - \sqrt{1 - x^2}\right)\right) dx$

53. $x = 0.8278585215$. The area is 0.8244398727.

55. $x = 0.6662394325$. The area is 0.09393667698.

57. $m = 1 - \left(\frac{1}{2}\right)^{1/3} \approx 0.206299$

Section 6.2 Preliminary

1. 3 **2.** 15 **3.** Circles, circles

4. $2\pi \int_0^R r\rho(r) \, dr$, where $\rho(r)$ is the radial population density function.

5. Flow rate is the volume of fluid that passes through a cross-sectional area at a given point per unit time.

6. The fluid velocity depended only on the radial distance from the center of the tube.

Section 6.2 Exercises

1. **(a)** $\frac{4}{25}(20 - y)^2$ **(b)** $V = \frac{1,280}{3}$

3. $V = \frac{\pi r^2 h}{3}$ **5.** $V = \pi \left(Rh^2 - \frac{h^3}{3}\right)$ **7.** $V = \frac{1}{6}abc$ **9.** $V = \frac{8}{3}$

11. $V = 36$ **13.** $V = 18$ **15.** $V = \frac{\pi}{3}$ **17.** $V = \frac{s^3\sqrt{2}}{12}$

21. (a) $w = 2\sqrt{r^2 - y^2}$ **(b)** $4(r^2 - y^2)$ **(c)** $V = \frac{16}{3}r^3$

23. $V = 160\pi$ **25.** $M = 2$ kg **27.** $P \approx 4{,}423.59$ thousand

29. $P = 463$ **31.** $P \approx 61$ deer **33.** $Q = 128\pi$ cm^3/s

35. $Q = \frac{8\pi}{3}$ cm^3/s **37.** $\frac{1}{4}$ **39.** $\frac{2}{\pi}$ **41.** 0.1 **43.** 3 **45.** $\frac{\pi}{4}$

47. Mean Value Theorem for Integrals; $c = \frac{A}{\sqrt[3]{4}}$

49. Over $[0, 1]$, $f(x)$; over $[1, 2]$, $g(x)$.

51. Many solutions exist. One could be:

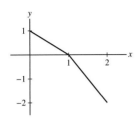

53. Over $[2, 6]$, the average temperature is 72.387324; over $[0, 24]$, the average temperature is 70.

55. $\frac{\pi}{3}$

57. Average acceleration $= \frac{1}{4}$ m/s^2; average velocity $= \frac{7}{60}$ m/s

59. $c = \frac{1{,}444}{225}$ **61.** $v_0/2$

Section 6.3 Preliminary

1. (a), (c) **2.** True

3. False. The cross sections will be washers. **4. (b)**

Section 6.3 Exercises

1. (a)

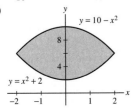

(b) Disk with radius $x + 1$ **(c)** $V = 21\pi$

3. (a)

(b) Disk with radius $\sqrt{x + 1}$ **(c)** $V = \frac{21\pi}{2}$

5. $V = \frac{81\pi}{10}$ **7.** $V = \frac{24{,}573\pi}{13}$ **9.** $V = \pi$ **11.** $\frac{\pi}{2}(e^2 - 1)$

13. (a)

(b) A washer with outer radius $R = 10 - x^2$ and inner radius $r = x^2 + 2$.

(c) $V = 256\pi$

15. (a)

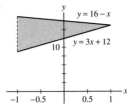

(b) A washer with outer radius $R = 16 - x$ and inner radius $r = 3x + 12$.

(c) $V = \frac{656\pi}{3}$

17. (a)

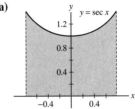

(b) A circular disk with radius $R = \sec x$. **(c)** $V = 2\pi$

19. $V = \frac{15\pi}{2}$ **21.** $V = \frac{3\pi}{10}$ **23.** $V = \frac{704\pi}{15}$ **25.** $V = \frac{128\pi}{5}$

27. $V = 40\pi$ **29.** $V = \frac{376\pi}{15}$ **31.** $V = \frac{824\pi}{15}$ **33.** $V = \frac{32\pi}{3}$

35. $V = \frac{1{,}872\pi}{5}$ **37.** $V = \frac{4\pi}{3}$ **39.** $V = \frac{1{,}472\pi}{15}$ **41.** $V = \frac{9\pi}{8}$

43. $V = \frac{32\pi}{35}$ **45.** $V = 7\pi(1 - \ln 2)$ **47.** $V = \frac{32\pi}{3}$

49. $V = \frac{32\pi}{105}$

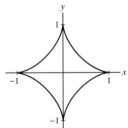

51. $V = 4\pi\sqrt{3}$ **53.** $V = \frac{4}{3}\pi a^2 b$ **55.** $V = \frac{\pi}{3}(1 - c^2)^{3/2}$

Section 6.4 Preliminary

1. (a) Radius h and height r. **(b)** Radius r and height h.

2. (a) With respect to x. **(b)** With respect to y.

Section 6.4 Exercises

1.

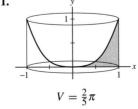

$V = \frac{2}{5}\pi$

3.

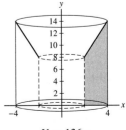

$V = 136\pi$

5.

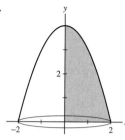

$$V = 8\pi$$

7.

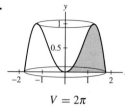

$$V = 2\pi$$

9.

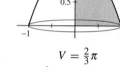

$$V = \frac{2}{3}\pi$$

11. $V = 16\pi$

13. $V = \frac{3}{10}\pi$

15. The point of intersection is $x = 1.376769504$; $V = 1.321975576$.

17.

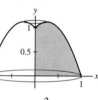

$$V = \frac{3\pi}{5}$$

19.

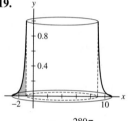

$$V = \frac{280\pi}{81}$$

21.

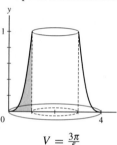

$$V = \frac{a^2(a+3b)}{3b^2}\pi$$

23. $V = \frac{\pi}{3}$

25. $V = \frac{38\pi}{5}$ **27.** $V = \frac{7\pi}{192}$

29. (a) $V = \frac{576\pi}{7}$ (b) $V = \frac{96\pi}{5}$

31. (a) The segment $\overline{AB}$ gives a disk with radius $R = h(y)$; the segment $\overline{CB}$ gives a shell with radius x and height $f(x)$.
(b) Shell method: $V = 2\pi \int_0^2 x f(x)\, dx$. Disk method: $V = \pi \int_0^{1.3} (h(y))^2\, dy$.

33. $V = 8\pi$ **35.** $V = \frac{40\pi}{3}$ **37.** $V = \frac{1,024\pi}{15}$ **39.** $V = 16\pi$

41. $V = \frac{32\pi}{3}$ **43.** $V = \frac{776\pi}{15}$ **45.** $V = \frac{4}{3}\pi r^3$ **47.** $V = 2\pi^2 ar^2$

49. (a) $V = \frac{563\pi}{30}$ (b) $V = \frac{1,748\pi}{35}$

Section 6.5 Preliminary

1. Because the required force is not constant through the stretching process.

2. The force involved in lifting a tank is the weight of the tank, which is constant.

3. $\frac{1}{2}kx^2$

Section 6.5 Exercises

1. $W = 627.2$ J   **3.** $W = 1.08$ J   **5.** $W = 1.5$ J   **7.** $W = 11.25$ J

9. 6 in.   **11.** $W = 1.6 \times 10^6$ ft-lb   **13.** $W = \frac{128,000\pi}{3}$ ft-lb

15. $W = 1.399 \times 10^{12}$ ft-lb   **17.** $W \approx 1.18 \times 10^8$ J

19. $W = 9,800\ell\pi r^3$ J   **21.** $W = 2.94 \times 10^6$ J

23. $W = \frac{1,684,375\pi}{2}$ J $\approx 2.65 \times 10^6$ J

25. (a) $F(y) = 9,800\pi(1,000 - y^3) - \frac{1,225\pi}{2}(10,000 - y^4)$ J

(b)

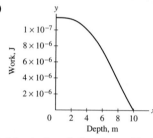

(c) Marginal work function   (d) 6.91 m

27. $W = 176.4$ J   **29.** $W = 600$ ft-lb   **31.** $W = 34,500$ ft-lb

33. $W \approx 1.99 \times 10^{10}$ J

35. (a) $k = 2.517382 \times 10^6$ lb-in^2 (b) $F = k(9\pi)^{-0.4}x^{-1.4}$

(c) $W = 74,677.8$ ft-lb

39. $v_0 = \sqrt{2GM_e\left(\frac{1}{r_e} - \frac{1}{r+r_e}\right)}$ m/s   **41.** $v_{esc} = \sqrt{\frac{2GM_e}{r_e}}$ m/s

Chapter 6 Review

1. $3\sqrt{2} - 1$ **3.** $\frac{9}{4}$ **5.** $e - \frac{3}{2}$ **7.** $\frac{1}{4}\pi - \frac{5}{12}$ **9.** 0.5

11. $x = 0.7145563847$, Area $= 0.08235024596$

13. $\frac{33}{2}$ lb   **15.** $\frac{9}{4}$ **17.** 2   **19.** $\frac{1}{2}\sinh 1$ **21.** 27   **23.** $V = \frac{2\pi}{15}$ m^5

25. (a) $\int_0^1 \left(\sqrt{1 - (x-1)^2} - \left(1 - \sqrt{1-x^2}\right)\right) dx$

(b) $\pi \int_0^1 \left((1 - (x-1)^2) - \left(1 - \sqrt{1-x^2}\right)^2\right) dx$

27. $V = \frac{2,048\pi}{3}$ **29.** $V = 8\pi$ **31.** $V = \frac{2\pi}{35}(12,032\sqrt{2} - 7,083)$

33. $V = \frac{56\pi}{15}$ **35.** $V = 2\pi\left(16\sqrt{3} - \frac{40}{3}\right)$

37. x-axis: $V = c\pi$; y-axis: $V = \frac{2\pi}{3}\left((1 + c^2)^{3/2} - c^3 - 1\right)$

39. $V = \frac{\pi}{6}$ **41.** $W = 74,900\pi$ ft-lb   **43.** $W = 83,040$ ft-lb

Chapter 7

Section 7.1 Preliminary

1. $T_1 = 6$; $T_2 = 7$

2. T_N overestimates the integral of $y = g(x)$; M_N overestimates the integral of $y = f(x)$.

3. 0; the Trapezoidal Rule integrates a linear function exactly.

4. $\frac{9}{32}$

5. The sum of the areas of the midpoint rectangles or the sum of the areas of the tangential trapezoids.

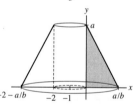

Section 7.1 Exercises

1. $T_4 = 2.75$; $M_4 = 2.625$ **3.** $T_6 = 64.6875$; $M_6 = 63.28125$

5. $T_6 \approx 1.40536$; $M_6 \approx 1.37693$

7. $T_6 \approx 0.883387$; $M_6 \approx 0.880369$

9. $T_5 \approx 1.12096$; $M_5 \approx 1.11716$

11. $T_8 \approx 2.96581$; $M_8 \approx 2.95302$ **13.** $S_4 \approx 5.25221$

15. $S_4 \approx 1.08055$ **17.** $S_6 \approx 0.746830$ **19.** $S_8 \approx 2.54499$

21. $S_{10} \approx 0.881377$ **23.** $V \approx 2.46740$ **25.** $V \approx 1.87691$

27. $\int_0^{\pi/2} \frac{\sin x}{x}\, dx \approx 1.37076$

29. $M_6 \approx 1.0028615$

(a) Because $f(x) = \cos x$ is concave down on $[0, \pi/2]$, M_6 is too large.

(b) $\text{Error}(M_6) \le \frac{K_2(b-a)^3}{24N^2} = \frac{\pi^3}{6{,}912} \approx 0.00448586$

(c) $\int_0^{\pi/2} \cos x\, dx = 1$; $\text{Error}(M_6) = |M_6 - 1| \approx 0.00286$

31. T_{20} overestimates the integral; $\text{Error}(T_{20}) \le 1.04167 \times 10^{-4}$

33. M_{20} overestimates the integral; $\text{Error}(M_{20}) \le 5.04659 \times 10^{-5}$

35. $N \ge 109.37$; $T_{110} \approx 0.4999990559$;
$\left| \int_0^{\pi/6} \cos x\, dx - T_{110} \right| \approx 9.441 \times 10^{-7}$.

37. $N \ge 1{,}500$; $T_{1{,}500} \approx 0.9502132468$;
$\left| \int_0^3 e^{-x}\, dx - T_{1{,}500} \right| \approx 3.15 \times 10^{-7}$.

39. (a) $S_6 \approx 0.432361$

(b) $\text{Error}(S_6) \le \frac{K_4(b-a)^5}{180N^4} = \frac{1}{14{,}580} \approx 6.85871 \times 10^{-5}$

(c) $\int_0^1 e^{-2x}\, dx = \frac{1-e^{-2}}{2} \approx 0.432332$; $\text{Error}(S_6) \approx 2.93 \times 10^{-5}$

41. $S_8 \approx 4.046655$; $\text{Error}(S_8) \le \frac{1}{120} \approx 0.0083333$; $N = 78$

43. With $K_4 = 15$, $\text{Error}(S_{40}) \le \frac{5}{49{,}152} \approx 1.0173 \times 10^{-4}$.

45. $N = 396$ **47.** $N = 186$

49. $\int_0^1 \frac{dx}{1+x^2} = \tan^{-1} x \big|_0^1 = \tan^{-1} 1 - \tan^{-1} 0 = \frac{\pi}{4}$

(a) Let $f(x) = (1+x^2)^{-1}$. Then $|f^{(4)}(x)| \le 24$ on $[0, 1]$.

(b) $N = 20$; $S_{20} \approx 0.785398163242$ and
$\left| 0.785398163242 - \frac{\pi}{4} \right| \approx 1.55 \times 10^{-10}$.

51.

Approximation	Error	Error Ratio
2.052344	0.052344	
2.012909	0.012909	0.246619
2.0032164	0.0032164	0.249160
2.00080342	0.00080342	0.249789
2.00020081	0.00020081	0.249944

53. Note $\text{Error}(S_{2N}) = \frac{K_4(b-a)^5}{180(2N)^4} = \frac{1}{16} \cdot \frac{K_4(b-a)^5}{180N^4} = \frac{1}{16}\, \text{Error}(S_N)$.

Approximation	Error	Error Ratio
2.004560	4.560×10^{-3}	
2.0002692	2.692×10^{-4}	0.059035
2.00001659	1.659×10^{-5}	0.061627
2.000001033	1.033×10^{-6}	0.062266
2.00000006453	6.453×10^{-8}	0.062469

55. $T_{\text{ave}} = 21.2°C$

61. Let $f(x) = ax^3 + bx^2 + cx + d$, with $a \ne 0$, be any cubic polynomial. Then, $f^{(4)}(x) = 0$, so we can take $K_4 = 0$. This yields $\text{Error}(S_N) \le \frac{0}{180N^4} = 0$. In other words, S_N is exact for all cubic polynomials for all N.

Section 7.2 Preliminary

1. Integration by parts is derived from the Product Rule.

2. (a) $\int x \cos(x^2)\, dx$; use the substitution $u = x^2$.

(b) $\int x \cos x\, dx$; use Integration by Parts.

(c) $\int x^2 e^x\, dx$; use Integration by Parts.

(d) $\int x e^{x^2}\, dx$; use the substitution $u = x^2$.

3. Transforming $v' = x$ into $v = \frac{1}{2}x^2$ increases the power of x and makes the new integral harder than the original.

Section 7.2 Exercises

1. $-x \cos x + \sin x + C$ **3.** $e^x(2x + 9) + C$

5. $\frac{x^4}{16}(4 \ln x - 1) + C$ **7.** $-e^{-x}(3x + 2) + C$

9. $e^x(x^2 - 2x + 2) + C$ **11.** $\frac{1}{2}x \sin 2x + \frac{1}{4} \cos 2x + C$

13. $-\frac{1}{2}e^{-x}(\sin x + \cos x) + C$ **15.** $\frac{x^2}{4}(2 \ln x - 1) + C$

17. $-\frac{1}{8x^8}\left(\ln x + \frac{1}{8}\right) + C$ **19.** $-x \sin(2 - x) + \cos(2 - x) + C$

21. $\frac{2^x}{\ln 2}\left(x - \frac{1}{\ln 2}\right) + C$ **23.** $x\big((\ln x)^2 - 2 \ln x + 2\big) + C$

25. $x \sin^{-1} x + \sqrt{1 - x^2} + C$ **27.** $\frac{5^x}{\ln 5}\left(x - \frac{1}{\ln 5}\right) + C$

29. $\frac{1}{2}x \sinh(2x) - \frac{1}{4}\cosh(2x) + C$ **31.** $x \sinh^{-1} x - \sqrt{1 + x^2} + C$

33. $2e^{\sqrt{x}}\left(\sqrt{x} - 1\right) + C$ **35.** $\frac{1}{4}x \sin 4x + \frac{1}{16}\cos 4x + C$

37. $\frac{2}{3}(x + 1)^{3/2} - 2(x + 1)^{1/2} + C$ **39.** $\sin x \ln(\sin x) - \sin x + C$

41. $2 \sin \sqrt{x} - 2\sqrt{x} \cos \sqrt{x} + C$ **43.** $\frac{1}{4}(\ln x)^2\big(2 \ln(\ln x) - 1\big) + C$

45. $\frac{1}{81}(17e^{18} + 1)$ **47.** $128/15$ **49.** $\frac{16}{3}\ln 4 - \frac{28}{9}$

51. $e^x(x^4 - 4x^3 + 12x^2 - 24x + 24) + C$

53. $\int x^n e^{-x}\, dx = -x^n e^{-x} + n \int x^{n-1} e^{-x}\, dx$

55. Integration by Parts with $u = \ln x$ and $v' = \sqrt{x}$.

57. Substitution with $u = 4 - x^2$, $du = -2x\, dx$ and $x^2 = 4 - u$. Follow this with algebraic manipulation.

59. Substitution with $u = x^2 + 3x + 6$ and $du = (2x + 3)\, dx$.

61. Integration by Parts with $u = x$ and $v' = \sin(3x + 4)$.

63. $x(\sin^{-1} x)^2 + 2\sqrt{1 - x^2} \sin^{-1} x - 2x + C$

65. $\frac{x^4}{4}\sin(x^4) + \frac{1}{4}\cos(x^4) + C$ **67.** $V = 2\pi(e^2 + 1)$

69. $PV = \frac{5{,}000r + 100 - e^{-10r}(6{,}000r + 100)}{r^2}$

71. $\int (\ln x)^2\, dx = x(\ln x)^2 - 2x \ln x + 2x + C$
$\int (\ln x)^3\, dx = x(\ln x)^3 - 3x(\ln x)^2 + 6x \ln x - 6x + C$

77. (d) $I(1, 1) = \frac{1}{6}$; $I(3, 2) = \frac{1}{60}$

79. (c) $I_3 = \frac{1}{2}x^2 \sin(x^2) + \frac{1}{2}\cos(x^2) + C$

Section 7.3 Preliminary

1. Rewrite $\sin^5 x = \sin x \sin^4 x = \sin x (1 - \cos^2 x)^2$ and then substitute $u = \cos x$.

2. Use a reduction formula.

3. No, a reduction formula is not needed because the sine function is raised to an odd power.

4. Write $\cos^2 x = 1 - \sin^2 x$ and then apply the reduction formula for powers of the sine function.

5. The first integral can be evaluated using a simple substitution; the second requires use of reduction formulas.

Section 7.3 Exercises

1. $\sin x - \frac{1}{3}\sin^3 x + C$ **3.** $-\frac{1}{3}\cos^3\theta + \frac{1}{5}\cos^5\theta + C$

5. $-\frac{1}{4}\cos^4 t + \frac{1}{6}\cos^6 t + C$ **7.** $A = 2$

9. $\frac{1}{4}\cos^3 y \sin y + \frac{3}{8}\cos y \sin y + \frac{3}{8}y + C$

11. $\frac{1}{6}\sin^5 x \cos x - \frac{1}{24}\sin^3 x \cos x - \frac{1}{16}\sin x \cos x + \frac{1}{16}x + C$

13. $\frac{1}{3}\tan^3 x + C$ **15.** $\frac{1}{5}\tan^5 x + C$ **17.** $\tan t - t + C$

19. $\frac{1}{2}\tan x \sec x + \frac{1}{2}\ln|\sec x + \tan x| + C$ **21.** $-\frac{1}{6}\cos^6 x + C$

23. $\frac{1}{12}\cos^3(3x)\sin(3x) + \frac{1}{8}\cos(3x)\sin(3x) + \frac{3}{8}x + C$

25. $\frac{1}{5\pi}\sin^5\pi\theta - \frac{1}{7\pi}\sin^7\pi\theta + C$

27. $-\frac{1}{12}\sin^3(3x)\cos(3x) - \frac{1}{8}\cos(3x)\sin(3x) + \frac{3}{8}x + C$

29. $\frac{1}{7}\ln|\sec 7t + \tan 7t| + C$ **31.** $\frac{1}{2}\tan^2 x + C$

33. $\frac{1}{8}\sec^8 x - \frac{1}{3}\sec^6 x + \frac{1}{4}\sec^4 x + C$

35. $\frac{1}{4}\tan x \sec^3 x - \frac{5}{8}\tan x \sec x + \frac{3}{8}\ln|\sec x + \tan x| + C$

37. $\frac{1}{6}\tan x \sec^5 x - \frac{7}{24}\tan x \sec^3 x + \frac{1}{16}\tan x \sec x$
$\quad + \frac{1}{16}\ln|\sec x + \tan x| + C$

39. $\frac{1}{4}\sin^2 2x + C$ **41.** $\frac{1}{4}\cos 2x - \frac{1}{12}\cos 6x + C$

43. $\frac{1}{2}\tan^2(\ln t) - \ln|\sec(\ln t)| + C$ **45.** π **47.** $\frac{5}{24}$

49. $\ln(\sqrt{2}+1) \approx 0.88137$ **51.** $\frac{1}{2}\ln 2 - \frac{1}{4} \approx 0.096574$ **53.** $-\frac{6}{7}$

57. $\csc x - \frac{1}{3}\csc^3 x + C$ **59.** $-\frac{1}{3}\cot^3 x + C$ **61.** $V = \frac{\pi^2}{2}$

63. $\frac{1}{8}x - \frac{1}{16}\sin 2x \cos 2x + C$

65. $\frac{1}{16}x - \frac{1}{48}\sin 2x - \frac{1}{32}\sin 2x \cos 2x + \frac{1}{48}\cos^2 2x \sin 2x + C$

69. $\int \csc x \, dx = \int \frac{\csc x(\csc x - \cot x)}{\csc x - \cot x} \, dx = \ln|\csc x - \cot x| + C$

73. $\pi/2$

Section 7.4 Preliminary

1. Substitute $u = 9 - x^2$; then, $du = -2x\,dx$.

2. (a) $x = 3\sin\theta$ **(b)** $x = 4\sec\theta$ **(c)** $x = 4\tan\theta$
(d) $x = \sqrt{5}\sec\theta$

3. Triangle (A)   **4.** $\frac{x}{\sqrt{9-x^2}}$ **5.** $\frac{3}{x}$

6. $\sin 2\theta = 2\sin\theta\cos\theta = 2x\sqrt{1-x^2}$

Section 7.4 Exercises

1. (b) $\sin^{-1}\left(\frac{x}{3}\right) + C$ **3. (c)** $\ln\left|x + \sqrt{x^2+9}\right| + C$

5. $2\sin^{-1}\left(\frac{x}{2}\right) + \frac{1}{2}x\sqrt{4-x^2} + C$ **7.** $\frac{1}{3}\sec^{-1}\left(\frac{x}{3}\right) + C$

9. $-\frac{x}{4\sqrt{x^2-4}} + C$

11. The substitution $u = x^2 - 4$ is not effective. Using the substitution $x = 2\sec\theta$,

$$\int \frac{x^2}{\sqrt{x^2-4}}\,dx = \frac{1}{2}x\sqrt{x^2-4} + 2\ln\left|x + \sqrt{x^2-4}\right| + C.$$

13. $\frac{9}{2}\sin^{-1}\left(\frac{x}{3}\right) - \frac{1}{2}x\sqrt{9-x^2} + C$

15. $x\sqrt{x^2+3} + 3\ln\left|\sqrt{x^2+3} + x\right| + C$

17. $\frac{t}{4\sqrt{4-t^2}} + C$ **19.** $-\frac{\sqrt{5-y^2}}{5y} + C$

21. $\frac{1}{16}\sec^{-1}\left(\frac{z}{2}\right) + \frac{\sqrt{z^2-4}}{8z^2} + C$

23. $\frac{1}{2}x\sqrt{x^2+1} - \frac{1}{2}\ln\left|\sqrt{x^2+1} + x\right| + C$

25. $\frac{1}{54}\tan^{-1}\left(\frac{x}{3}\right) + \frac{x}{18(x^2+9)} + C$

27. $-\frac{x}{8(x^2-4)} - \frac{1}{16}\ln\left|\frac{x-2}{\sqrt{x^2-4}}\right| + C$

29. $-3(9-x^2)^{3/2} + \frac{1}{5}(9-x^2)^{5/2} + C$

33. (b) $I = \ln\left|\sqrt{(x-2)^2+4} + x - 2\right| + C$

35. $\ln\left|\sqrt{x^2+4x+13} + x + 2\right| + C$

37. $\ln\left|2x + 1 + 2\sqrt{x^2+x}\right| + C$

39. $\frac{1}{2}(x-2)\sqrt{x^2-4x+3} - \frac{1}{2}\ln\left|x - 2 + \sqrt{x^2-4x+3}\right| + C$

41. $x\sec^{-1}x - \ln\left|x + \sqrt{x^2-1}\right| + C$

43. First, complete the square. Next, substitute $u = x + 3$, $du = dx$. Finally, use the trigonometric substitution $u = \sqrt{21}\sin\theta$.

45. Integration by Parts with $u = x$ and $v' = \sec^2 x$.

47. First, substitute $u = e^{2x}$, $du = 2e^{2x}\,dx$. Then, use the integration formula for the inverse tangent.

49. Use the integration formula for the cosecant function.

51. First, substitute $u = x + 2$, $du = dx$. Then, use the integration formula for power functions.

53. (a) Use the substitution $u = 1 - x^2$. Then $du = -2x\,dx$, $x^2 = 1 - u$ and $\int x^3\sqrt{1-x^2}\,dx = -\frac{1}{2}\int (1-u)u^{1/2}\,du$.

(b) Let $x = \sin\theta$. Then $dx = \cos\theta\,d\theta$, $1 - x^2 = \cos^2\theta$ and $\int x^2\sqrt{1-x^2}\,dx = \int \sin^2\theta\cos^2\theta\,d\theta$.

(c) Let $x = \sin\theta$. Then $dx = \cos\theta\,d\theta$, $1 - x^2 = \cos^2\theta$ and $\int \frac{x^4}{\sqrt{1-x^2}}\,dx = \int \sin^4\theta\,d\theta$.

(d) Use the substitution $u = 1 - x^2$. Then $du = -2x\,dx$, and $\int \frac{x}{\sqrt{1-x^2}} = -\frac{1}{2}\int u^{-1/2}\,du$.

55. $V = \pi^2/8$ **57.** $V = 8\pi\sqrt{3} - 6\pi\ln(2 + \sqrt{3}) \approx 18.707$

59. $x = 2.385$ in

Section 7.5 Preliminary

1. (a) $x = \sinh t$ **(b)** $x = 3 \sinh t$ **(c)** $3x = \sinh t$

2. $\sinh x$ and $\tanh x$ **3.** $\frac{1}{2} \ln \left| \frac{1+x}{1-x} \right|$

Section 7.5 Exercises

1. $\frac{1}{3} \sinh 3x + C$ **3.** $\frac{1}{2} \cosh(x^2 + 1) + C$

5. $-\frac{1}{2} \tanh(1 - 2x) + C$ **7.** $\frac{1}{2} \tanh^2 x + C$ **9.** $\ln \cosh x + C$

11. $\ln |\sinh x| + C$ **13.** $\frac{1}{16} \sinh(8x - 18) - \frac{1}{2}x + C$

15. $\frac{1}{32} \sinh(4x) - \frac{x}{8} + C$ **17.** $\cosh^{-1} 4 - \cosh^{-1} 2$

19. $\sinh^{-1}\left(\frac{x}{4}\right) + C$ **21.** $\frac{1}{2}x\sqrt{x^2 - 1} - \frac{1}{2}\cosh^{-1} x + C$

23. $2 \tanh^{-1}\left(\frac{1}{2}\right)$ **25.** $\sinh^{-1} 1$

27. $\frac{1}{4} \ln 3 - \frac{1}{4} \ln(4 + \sqrt{17})$ **29.** $\cosh^{-1} x - \frac{\sqrt{x^2-1}}{x} + C$

33. Using trigonometric substitution:
$\frac{1}{2}x\sqrt{x^2 + 16} + 8 \ln |x + \sqrt{x^2 + 16}| + C$

35. $\frac{1}{2} \ln \left| \frac{1+x}{1-x} \right| + C$

37. tangent line: $y = 1.048 - 0.976 * (x - 0.8)$

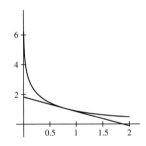

39. $\frac{1}{4} \cosh^3 x \sinh x + \frac{3}{8} \cosh x \sinh x + \frac{3}{8}x + C$

41. $x \sinh^{-1} x - \sqrt{x^2 + 1} + C$

43. $\frac{1}{2}x^2 \tanh^{-1} x - \frac{1}{2} \tanh^{-1} x + \frac{x}{2} + C$

45. $2 \tan^{-1}(\tanh(x/2)) + C$

Section 7.6 Preliminary

1. No, $f(x)$ cannot be a rational function because the integral of a rational function cannot contain a term with a non-integer exponent.

2. (a) No **(b)** Yes **(c)** Yes **(d)** No

3. (a) Square is already completed; irreducible

(b) Square is already completed; factors as $(x - \sqrt{5})(x + \sqrt{5})$

(c) $x^2 + 4x + 6 = (x + 2)^2 + 2$; irreducible

(d) $x^2 + 4x + 2 = (x + 2)^2 - 2$; factors as $(x + 2 - \sqrt{2})(x + 2 + \sqrt{2})$

4. (a) is a sum of logarithmic terms $A_i \ln(x - a_i)$ for some constants A_i.

Section 7.6 Exercises

1. (a) $\frac{x^2+4x+12}{(x+2)(x^2+4)} = \frac{1}{x+2} + \frac{4}{x^2+4}$

(b) $\frac{2x^2+8x+24}{(x+2)^2(x^2+4)} = \frac{1}{x+2} + \frac{2}{(x+2)^2} + \frac{-x+2}{x^2+4}$

(c) $\frac{x^2-4x+8}{(x-1)^2(x-2)^2} = \frac{-8}{x-2} + \frac{4}{(x-2)^2} + \frac{8}{x-1} + \frac{5}{(x-1)^2}$

(d) $\frac{x^4-4x+8}{(x+2)(x^2+4)} = x - 2 + \frac{4}{x+2} - \frac{4x-4}{x^2+4}$

3. $B = -2$

5. $\frac{x}{3x-9} = \frac{1}{3} + \frac{1}{x-3}$; so $\int \frac{x}{3x-9}\, dx = \frac{1}{3}x + \ln|x - 3| + C$

7. $\frac{x^3+x+1}{x-2} = x^2 + 2x + 5 + \frac{11}{x-2}$; so
$\int \frac{x^3+x+1}{x-2}\, dx = \frac{1}{3}x^3 + x^2 + 5x + 11 \ln|x - 2| + C$

9. $-\frac{1}{2} \ln|x - 2| + \frac{1}{2} \ln|x - 4| + C$

11. $\ln|x| - \ln|2x + 1| + C$

13. $-3 \ln|x - 2| + 5 \ln|x - 3| + C$

15. $2 \ln|x + 3| - \ln|x + 5| - \frac{2}{3} \ln|3x - 2| + C$

17. $3 \ln|x - 1| - 2 \ln|x + 1| - \frac{5}{x+1} + C$

19. $2 \ln|x - 1| - \frac{1}{x-1} - 2 \ln|x - 2| - \frac{1}{x-2} + C$

21. $3 \ln|x| - 3 \ln|x + 4| + \frac{12}{x+4} + C$

23. $-\frac{1}{9} \ln|x - 4| - \frac{1}{3(x-4)} + \frac{1}{9} \ln|x - 1| + C$

25. $\frac{11}{3} \ln|x| - \frac{2}{x} - \frac{9}{2} \ln|x - 1| + \frac{5}{6} \ln|x - 3| + C$

27. $\frac{3}{2}x^2 + 12x + 46 \ln|x - 4| + C$

29. $\ln|x| - \frac{1}{2} \ln(x^2 + 1) + C$ **31.** $x - \sqrt{3} \tan^{-1}\left(\frac{x}{\sqrt{3}}\right) + C$

33. $\frac{1}{2} \ln|x + 1| + \frac{1}{4} \ln(x^2 + 1) - \frac{1}{2} \tan^{-1} x + C$

35. $\frac{1}{27} \ln|3x + 7| + \frac{14}{27}(3x + 7)^{-1} - \frac{49}{54}(3x + 7)^{-2} + C$

37. $-\frac{1}{25x} - \frac{1}{125} \tan^{-1}\left(\frac{x}{5}\right) + C$

39. $\frac{1}{10} \ln|x + 1| - \frac{1}{20} \ln(x^2 + 9) + \frac{7}{135} \tan^{-1}\left(\frac{x}{3}\right) + \frac{x+9}{18(x^2+9)} + C$

41. $3 \ln|x - 3| - \frac{3}{2} \ln(x^2 + 1) - 4 \tan^{-1} x + \frac{5x+15}{x^2+1} + C$

43. $\ln|x + 1| - \frac{1}{2} \ln(x^2 - 2x + 6) + \frac{2}{\sqrt{5}} \tan^{-1}\left(\frac{x-1}{\sqrt{5}}\right) + C$

45. $\frac{3\sqrt{2}}{4} \tan^{-1}\left(\frac{x+1}{\sqrt{2}}\right) + \frac{x+3}{2(x^2+2x+3)} + C$

47. $\frac{1}{2} \ln|x - 1| - \frac{1}{2} \ln|x + 1| + C = \ln\left|\frac{x-1}{\sqrt{x^2-1}}\right| + C$

49. $2\sqrt{x} + \ln|\sqrt{x} - 1| - \ln|\sqrt{x} + 1| + C$

51. $-\frac{\sqrt{4-x^2}}{4x} + C$ **53.** $\ln|x| - \ln|x - 1| - \frac{1}{x-1} + C$

55. $\frac{1}{54} \tan^{-1}\left(\frac{x}{3}\right) + \frac{x}{18(x^2+9)} + C$

57. $\frac{1}{5} \sec^5 x - \frac{2}{3} \sec^3 x + \sec x + C$

59. $\ln\left|x + \sqrt{x^2 - 1}\right| - \frac{x}{\sqrt{x^2-1}} + C$

61. $-\ln|x| + \frac{1}{2} \ln|x - 1| + \frac{1}{2} \ln|x + 1| + C$

63. $\frac{2}{3} \tan^{-1}(x^{3/2}) + C$

65. $-\frac{1}{3} \ln|x^3| - \frac{1}{3x^3} + \frac{1}{3} \ln|x^3 + 1| + C$

67. $\frac{4}{5} \ln\left|2 \tan\left(\frac{\theta}{2}\right) + 1\right| - \frac{4}{5} \ln\left|\tan\left(\frac{\theta}{2}\right) - 2\right| + C$

71. (b) $\frac{3x-2}{x^2-4x-12} = \frac{2}{x-6} + \frac{1}{x+2}$

Section 7.7 Preliminary

1. (a) Converges **(b)** Diverges **(c)** Diverges **(d)** Converges

2. Yes, this is an improper integral because $\cot x$ is undefined at $x = 0$.

3. One choice is $b = 2$. **4.** $\int_0^\infty \frac{dx}{x+e^x} < \int_0^\infty e^{-x}\,dx$

5. $\frac{e^{-x}}{x} < \frac{1}{x}$ for $1 \le x < \infty$, but $\int_1^\infty \frac{dx}{x}$ diverges.

Section 7.7 Exercises

1. (a) Improper: The function $x^{-1/3}$ is infinite at $x = 0$.

(b) Improper: Infinite limit of integration.

(c) Improper: Infinite limit of integration.

(d) Proper: The function e^{-x} is continuous on the finite interval $[0, 1]$.

(e) Improper: The function $\sec x$ is infinite at $x = \pi/2$.

(f) Improper: Infinite limit of integration.

(g) Proper: The function $\sin x$ is continuous on the finite interval $[0, 1]$.

(h) Proper: The function $(3 - x^2)^{-1/2}$ is continuous on the finite interval $[0, 1]$.

(i) Improper: Infinite limit of integration.

(j) Improper: The function $\ln x$ is infinite at $x = 0$.

5. Integral does not converge. **7.** $10{,}000e^{0.0004}$

9. Integral does not converge. **11.** 4 **13.** $\frac{1}{8}$ **15.** 2

17. 1.25 **19.** $\frac{1}{3}e^{-12}$ **21.** $\frac{1}{3}$ **23.** $\frac{1}{375}$ **25.** $2\sqrt{2}$

27. Integral does not converge. **29.** $\frac{\pi}{4}$ **31.** $\frac{1}{2}$ **33.** $\frac{\pi}{2}$

35. Integral does not converge. **37.** Integral does not converge.

39. Integral does not converge. **41.** Integral does not converge.

43. $-\frac{1}{4}$ **45.** Integral does not converge.

49. Integral does not converge. **51.** π **53.** 0 **55.** $a < 0$

59. The integral $\int_1^\infty x^{-3}\,dx$ converges because $3 > 1$. Since $x^3 + 4 \ge x^3$, it follows that $\frac{1}{x^3+4} \le \frac{1}{x^3}$. Therefore, by the comparison test $\int_1^\infty \frac{dx}{x^3+4}$ converges.

63. Let $f(x) = \frac{1-\sin x}{x^2}$. Because $f(x) \le \frac{2}{x^2}$ and $\int_1^\infty 2x^{-2}\,dx = 2$, it follows by the comparison test that $\int_1^\infty f(x)\,dx$ converges.

65. Converges **67.** Diverges **69.** Converges **71.** Diverges

73. Converges **75.** \$3,571.43 **77.** \$2,000,000

79. (a) $V = \pi$

83. $A = \frac{1}{k}$; For Radon-222, $A = \frac{1}{k} = \frac{3.825}{\ln 2} \approx 5.518$ days

85. $J_4 = \frac{24}{\alpha^5}$

87. Let $\alpha = h/kT$. Then $E = \frac{8\pi h}{c^3} \int_0^\infty \frac{v^3}{e^{\alpha v}-1}\,dv$. Since $\alpha > 0$ and $8\pi h/c^3$ is a constant, it follows from the previous exercise that E is finite.

89. $C = k$; $\mu = \frac{1}{k}$ **93.** $\frac{s}{s^2+\alpha^2}$

Chapter 7 Review

1. $T_2 = -\frac{9}{8}$; $M_3 = -2$; $T_6 = -\frac{13}{8}$; $S_6 = -\frac{21}{12}$ **3.** 2.19 in

5. $T_3 \approx 25.976514$ **7.** $T_6 \approx 0.358016$ **9.** $S_8 \approx 0.608711$

11. With $K_2 = 1.5$, Error$(T_9) \le 9$.

13. With $K_2 = 30$, Error$(T_{24}) \le 0.016822$. **15.** $N = 4$

17. $\frac{1}{9}\sin^9\theta - \frac{1}{11}\sin^{11}\theta + C$

19. $\frac{1}{6}\tan\theta\sec^5\theta - \frac{7}{24}\tan\theta\sec^3\theta + \frac{1}{16}\tan\theta\sec\theta + \frac{1}{16}\ln|\sec\theta + \tan\theta| + C$

21. $-\frac{1}{\sqrt{x^2-1}} - \sec^{-1}x + C$ **23.** $\frac{1}{2}\ln(x^2 + 1) + C$

25. $\frac{1}{6}\ln|x - 1| - \frac{1}{6}\ln|x + 5| + C$

27. $-\frac{1}{2}x\sqrt{9 - x^2} + \frac{9}{2}\sin^{-1}\left(\frac{x}{3}\right) + C$ **29.** $\frac{1}{5}\tan^5\theta + C$

31. $\frac{1}{4}\ln|t + 1| - \frac{1}{4}\ln|t - 1| - \frac{1}{4(t-1)} - \frac{1}{4(t+1)} + C$

33. $\frac{1}{2}\sin^4\theta + C$

35. $(x + 1)\left(\ln(x + 1)\right)^2 - 2(x + 1)\ln(x + 1) + 2x + C$

37. $\frac{3}{8}x + \frac{1}{36}\sin(18x - 4) + \frac{1}{288}\sin(36x - 8) + C$

39. $2\ln|\sec x| + C$ **41.** $2\tan x + 2\sec x - x + C$

43. $-2\cot\left(\frac{\theta}{2}\right) - \theta + C$

45. $\frac{1}{49}\ln|t + 4| - \frac{1}{49}\ln|t - 3| - \frac{1}{7(t-3)} + C$

47. $\frac{1}{2}\sec^{-1}\left(\frac{x}{2}\right) + C$ **49.** $\frac{2}{\sqrt{a}}\tan^{-1}\left(\frac{\sqrt{x}}{\sqrt{a}}\right) + C$ (assuming $a > 0$)

51. $\ln|x + 2| + \frac{5}{x+2} - \frac{3}{(x+2)^2} + C$

53. $-\frac{2}{x-2} + \frac{1}{2}\ln(x^2 + 4) - \ln|x - 2| + C$

55. $\frac{1}{3}\tan^{-1}\left(\frac{x+4}{3}\right) + C$ **57.** $\ln|x + 2| + \frac{5}{x+2} - \frac{3}{(x+2)^2} + C$

59. $\frac{(x^2-2)\sqrt{x^2+4}}{24x^3} + C$ **61.** $-\frac{1}{9}(4 + 3x)e^{4-3x} + C$

63. $-\frac{\tan^{-1}x}{x} - \frac{1}{2}\ln(1 + x^2) + \ln|x| + C$

65. $\frac{1}{3}x^3(\ln x)^2 - \frac{2}{9}x^3\ln x + \frac{2}{27}x^3 + C$ **67.** $\frac{1}{2}(\tan^{-1}t)^2 + C$

69. $\frac{1}{2}\sin x\sinh x - \frac{1}{2}\cos x\cosh x + C$ **71.** $\frac{1}{5}\ln|\cosh 5x| + C$

73. $t + \frac{1}{4}\coth(1 - 4t) + C$ **75.** $\frac{\pi}{3}$ **77.** $\tan^{-1}(\tanh t) + C$

81. $a = 0$ **83.** Integral does not converge.

85. $\frac{5}{189}3^{1/5}$ **87.** $\frac{1}{4}e^{36}$ **89.** $\ln\left(\frac{6}{5}\right)$ **91.** $\frac{3}{2}3^{2/3}$

93. Converges **95.** Diverges **97.** Converges

99. (a) Infinite **(b)** Finite: π **(c)** Infinite

101. $\bar{E} = kT$ **103.** $\frac{2}{(s-\alpha)^3}$

Chapter 8

Section 8.1 Preliminary

1. $\int_0^\pi \sqrt{1 + \sin^2 x}\,dx$

2. The graph of $y = f(x) + C$ is a vertical translation of the graph of $y = f(x)$; hence, the two graphs should have the same arc length. We can explicitly establish this as follows:

$$\text{Length of } y = f(x) + C = \int_a^b \sqrt{1 + \left[\frac{d}{dx}(f(x) + C)\right]^2}\, dx$$

$$= \int_a^b \sqrt{1 + [f'(x)]^2}\, dx$$

$$= \text{length of } y = f(x).$$

Section 8.1 Exercises

1. $L = \int_2^6 \sqrt{1 + 16x^6}\, dx$ **3.** $\frac{13}{12}$ **5.** $3\sqrt{10}$

7. $\frac{1}{27}(22\sqrt{22} - 13\sqrt{13})$ **9.** $e^2 + \frac{\ln 2}{2} + \frac{1}{4}$

11. $\int_1^2 \sqrt{1 + x^6}\, dx \approx 3.957736$

13. $\int_1^2 \sqrt{1 + \frac{1}{x^4}}\, dx \approx 1.132123$ **15.** 6   **17.** $2\sqrt{3}$

19. Let h denote the length of the hypotenuse. Then, by the Pythagorean Theorem, $h^2 = (b - a)^2 + m^2(b - a)^2$ $= (b - a)^2(1 + m^2)$, or $h = (b - a)\sqrt{1 + m^2}$ since $b > a$. Moreover, $(f'(x))^2 = m^2$, so

$$L = \int_a^b \sqrt{1 + m^2}\, dx = (b - a)\sqrt{1 + m^2}.$$

25. $\sqrt{1 + e^{2a}} + \frac{1}{2}\ln\frac{\sqrt{1 + e^{2a}} - 1}{\sqrt{1 + e^{2a}} + 1} - \sqrt{2} + \frac{1}{2}\ln\frac{1 + \sqrt{2}}{\sqrt{2} - 1}$

27. $\ln(1 + \sqrt{2})$ **31.** 1.552248   **33.** $60\pi\sqrt{17}$ **35.** $\frac{384\pi}{5}$

37. $\pi\left(e\sqrt{1 + e^2} + \ln\left(e + \sqrt{1 + e^2}\right) - \sqrt{2} - \ln(1 + \sqrt{2})\right)$

39. $2\pi\left(\sqrt{2} + \ln(1 + \sqrt{2})\right)$

43. $\frac{\pi}{32}\left(1{,}032\sqrt{65} - \ln(8 + \sqrt{65})\right)$

45. $2\pi \int_0^{\pi/4} \tan x\sqrt{1 + \sec^4 x}\, dx \approx 3.839077$ **47.** $n = 13$

51. $2\pi b^2 + \frac{2\pi b a^2}{\sqrt{b^2 - a^2}}\ln\left|\frac{\sqrt{b^2 - a^2}}{a} + \frac{b}{a}\right|$

Section 8.2 Preliminary

1. Pressure is defined as force per unit area.

2. The factor of proportionality is the weight density of the fluid, $w = \rho g$.

3. Fluid force acts in the direction perpendicular to the side of the submerged object.

4. Pressure depends only on depth and does not change horizontally at a given depth.

5. When a plate is submerged vertically, the pressure is not constant along the plate, so the fluid force is not equal to the pressure times the area.

Section 8.2 Exercises

1. **(a)** Top: $F = 1{,}125$ lb; bottom: $F = 4{,}500$ lb

(b) $F \approx \sum_{j=1}^N 3wy_j\, \Delta y$ **(c)** $F = \int_2^8 3wy\, dy$ **(d)** $F = 5{,}625$ lb

3. **(a)** The width of the triangle varies linearly from 0 at a depth of $y = 3$ ft to 1 at a depth of $y = 5$ ft. Thus, $w = \frac{1}{2}(y - 3)$.

(b) The area of the strip at depth y is $\frac{1}{2}(y - 3)\Delta y$, and the pressure at depth y is wy, where $w = 62.5$ lb/ft^3. Thus, the fluid force acting on the strip at depth y is approximately equal to $\frac{1}{2}wy(y - 3)\Delta y$.

(c) $F \approx \sum_{j=1}^N \frac{1}{2}wy_j(y_j - 3)\, \Delta y \to \int_3^5 \frac{1}{2}wy(y - 3)\, dy$

(d) $F = \frac{1{,}625}{6}$ lb

5. **(b)** $F = \frac{19{,}600}{3}r^3$ N

7. $F = \frac{19{,}600}{3}r^3 + 4{,}900\pi m r^2$ N   **9.** $F = 441{,}000$ N

11. $F = \frac{64}{5}w$ **13.** $F = \left(1 - \frac{\pi}{8}\right)w$ **15.** $F = w$

17. $F = 595{,}350$ N   **19.** $F = \frac{200w\sqrt{3}}{3}$

21. $F = \frac{45}{2}hd\sqrt{(b - a)^2 + 4h^2}$ lb

23. Front and back: $F = \frac{62.5\sqrt{3}}{9}y^3$; slanted sides: $F = \frac{62.5\sqrt{3}}{3}\ell y^2$ where ℓ denotes the length of the trough.

Section 8.3 Preliminary

1. $M_x = M_y = 0$ **2.** $M_x = 21$ **3.** $M_x = 5$; $M_y = 10$

4. Because a rectangle is symmetric with respect to both the vertical line and the horizontal line through the center of the rectangle, the Symmetry Principle guarantees that the centroid of the rectangle must lie along both these lines. The only point in common to both lines of symmetry is the center of the rectangle, so the centroid of the rectangle must be the center of the rectangle.

Section 8.3 Exercises

1. **(a)** $\left(\frac{9}{4}, 1\right)$ **(b)** $\left(\frac{46}{17}, \frac{14}{17}\right)$

5. A sketch of the lamina is shown here.

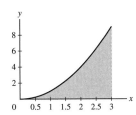

(a) $M_x = \frac{243}{10}$; $M_y = \frac{243}{4}$

(b) Area $= 27$; center of mass: $\left(\frac{9}{4}, \frac{9}{10}\right)$

7. $M_x = \frac{64\rho}{7}$; $M_y = \frac{32\rho}{5}$; center of mass : $\left(\frac{8}{5}, \frac{16}{7}\right)$

9. **(a)** $M_x = 24$ **(b)** $M = 12$, so $y_{cm} = 2$; center of mass: $(0, 2)$

11. $\left(\frac{633}{95}, \frac{195}{152}\right)$ **13.** $\left(\frac{9}{8}, \frac{18}{5}\right)$ **15.** $\left(\frac{1 - 5e^{-4}}{1 - e^{-4}}, \frac{1 - e^{-8}}{4(1 - e^{-4})}\right)$

17. $\left(\frac{\pi}{2}, \frac{\pi}{8}\right)$

19. A sketch of the region is shown here.

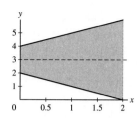

The region is clearly symmetric about the line $y = 3$, so we expect the centroid of the region to lie along this line. We find $M_x = 24$, $M_y = \frac{28}{3}$, centroid: $\left(\frac{7}{6}, 3\right)$.

21. $\left(\frac{9}{20}, \frac{9}{20}\right)$ **23.** $\left(\frac{1}{2(e-2)}, \frac{e^2-3}{4(e-2)}\right)$ **25.** $\left(\frac{\pi\sqrt{2}-4}{4(\sqrt{2}-1)}, \frac{1}{4(\sqrt{2}-1)}\right)$

27. A sketch of the region is shown here. Centroid: $\left(0, \frac{4}{7}\right)$

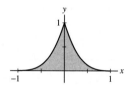

29. $\left(0, \frac{4b}{3\pi}\right)$ **31.** $\left(\frac{4}{3\pi}, \frac{4}{3\pi}\right)$

33. $\left(0, \dfrac{\frac{2}{3}(r^2-h^2)^{3/2}}{r^2\sin^{-1}\sqrt{1-h^2/r^2}-h\sqrt{r^2-h^2}}\right)$; with $r = 1$ and $h = \frac{1}{2}$:

$\left(0, \frac{3\sqrt{3}}{4\pi-3\sqrt{3}}\right) \approx (0, 0.71)$

35. $\left(0, \frac{16+3\pi\sqrt{3}}{\pi+2\sqrt{3}}\right)$

37. **(c)** $M_y^{\text{big}} = 0$, $M_y^{\text{small}} = -\pi r^3$, $M_y^S = \pi r^3$ **(d)** $\left(\frac{r}{3}, 0\right)$

41. Let A be the cross-sectional area of the rod. Place the rod with the m_1 at the origin and the rod lying along the positive x-axis. Then, the center of mass is located at

$$x = \frac{dm_2+\frac{1}{2}\rho A d^2}{m_1+m_2+\rho A d}$$

Section 8.4 Preliminary

1. $T_3(x) = 9 + 8(x - 3) + 2(x - 3)^2 + 2(x - 3)^3$
2. The polynomial graphed on the right is a Maclaurin polynomial.
3. A Maclaurin polynomial gives the value of $f(0)$ exactly.
4. The correct statement is **(b)**: $|T_3(2) - f(2)| \le \frac{2^3}{6}$.

Section 8.4 Exercises

1. $T_2(x) = x$; $T_3(x) = x - \frac{x^3}{6}$
3. $T_2(x) = 1 - x + x^2$; $T_3(x) = 1 - x + x^2 - x^3$
5. $T_2(x) = x$; $T_3(x) = x + \frac{x^3}{3}$
7. $T_2(x) = 1 - x^2$; $T_3(x) = 1 - x^2$
9. $T_2(x) = 1 + x + \frac{x^2}{2}$; $T_3(x) = 1 + x + \frac{x^2}{2} + \frac{x^3}{6}$
11. $T_2(x) = 2 - 3x + \frac{5x^2}{2}$; $T_3(x) = 2 - 3x + \frac{5x^2}{2} - \frac{3x^3}{2}$
13. $T_2(x) = x - \frac{x^2}{2}$; $T_3(x) = x - \frac{x^2}{2} + \frac{x^3}{3}$

15. $T_2(x) = 1 + x + \frac{x^2}{2}$; $|T_2(-0.5) - f(-0.5)| = 0.018531$
17. $T_2(x) = 1 - \frac{3(x-1)}{2} + \frac{15(x-1)^2}{8}$; $|T_2(0.3) - f(0.3)| = 3.11706$
19. Let $f(x) = e^x$. Then, for all n,

$$f^{(n)}(x) = e^x \quad \text{and} \quad f^{(n)}(0) = 1.$$

It follows that

$$T_n(x) = 1 + \frac{x}{1!} + \frac{x^2}{2!} + \cdots + \frac{x^n}{n!}.$$

21. $T_n(x) = 1 + x + x^2 + x^3 + \cdots + x^n$
23. $T_n(x) = e + e(x - 1) + \frac{e(x-1)^2}{2!} + \cdots + \frac{e(x-1)^n}{n!}$
25. $T_N(x) = 4\sqrt{2} + 5\sqrt{2}(x - 2) + \frac{15(x-2)^2}{4\sqrt{2}} + \frac{5(x-2)^3}{16\sqrt{2}} + \cdots$

$\quad + \frac{15}{8}(-1)^{n-3}\frac{1}{2^{n-3}}\left(\frac{(2n-7)!}{2^{n-3}(n-4)!n!}\right)2^{-(2n-7)/2}(x-2)^n$

27. The odd polynomials are concave down on the left and lie beneath the graph of $y = e^x$; the even polynomials are concave up on the left and lie above the graph of $y = e^x$.

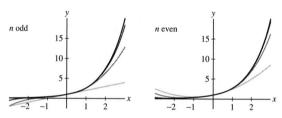

29. Error bound $= \frac{|0.3|^6}{6!} \approx 1.0125 \times 10^{-6}$;
$|\cos 0.3 - T_5(0.3)| = 1.010874 \times 10^{-6}$
31. $n = 3$
33. $T_3(x) = x - \frac{x^3}{3}$; $T_3\left(\frac{1}{2}\right) = \frac{11}{24}$. With $K = 5$,

$$\left|T_3\left(\frac{1}{2}\right) - \tan^{-1}\frac{1}{2}\right| \le \frac{5\left(\frac{1}{2}\right)^4}{4!} = \frac{5}{384}.$$

37. $n = 6$ **39.** $n = 9$
41. $1.5^{1.6} \approx T_2(0.5) = \frac{307}{288} \approx 1.065872$
43. $T_4(x) = x - \frac{2}{3}x^3$
45. $T_n(x) = 1 - x^2 + x^4 - x^6 + \cdots + (-x^2)^n$
47. At $a = 0$,

$$T_1(x) = -4 - x$$
$$T_2(x) = -4 - x + 2x^2$$
$$T_3(x) = -4 - x + 2x^2 + 3x^3 = f(x)$$
$$T_4(x) = T_3(x)$$
$$T_5(x) = T_3(x)$$

At $a = 1$

$$T_1(x) = 12(x - 1)$$

$$T_2(x) = 12(x - 1) + 11(x - 1)^2$$

$$T_3(x) = 12(x - 1) + 11(x - 1)^2 + 3(x - 1)^3$$

$$= -4 - x + 2x^2 + 3x^3 = f(x)$$

$$T_4(x) = T_3(x)$$

$$T_5(x) = T_3(x)$$

49. (b) By substitution

$$L = \lim_{x \to 0} \frac{x^2 + g_1(x)}{x^2/2 + g_2(x)} = \lim_{x \to 0} \frac{1 + g_1(x)/x^2}{1/2 + g_2(x)/x^2} = 2.$$

(c) Using L'Hôpital's Rule

$$L = \lim_{x \to 0} \frac{e^x + e^{-x} - 2}{1 - \cos x} = \lim_{x \to 0} \frac{e^x - e^{-x}}{\sin x} = \lim_{x \to 0} \frac{e^x + e^{-x}}{\cos x} = 2.$$

51. (c) $R_1 \approx \sqrt{\lambda L} = 0.04873$ cm and
$R_{100} \approx \sqrt{100\lambda L} = 0.4873$ cm

57. $\int_0^{1/2} \sin(x^2)\, dx \approx \int_0^{1/2} x^2\, dx = \frac{1}{24}$; on $[0, 1/2]$,

$$\left| \sin(x^2) - T_4(x) \right| \le \frac{(1/2)^6}{6},$$

so the error in the approximation is bounded by

$$\frac{(1/2)^7}{6} = 1.116 \times 10^{-3}$$

Chapter 8 Review

1. $\frac{2,918,074}{10,125}$ **3.** $4\sqrt{17}$ **7.** $24\pi\sqrt{2}$

9. $\frac{67\pi}{36}$ **11.** $12\pi + 4\pi^2$ **13.** 1,125 lb

15. $M_x = \frac{256}{3}$; $M_y = \frac{320}{3}$; center of mass: $\left(2, \frac{8}{5}\right)$

17. $\left(0, \frac{2}{\pi}\right)$ **19.** $\left(-\frac{44}{80+15\pi}, 0\right)$

21. $T_3(x) = 3(x + 2)^3 - 5(x + 2)$

23. $T_3(x) = 2 + \frac{1}{4}(x - 2) - \frac{1}{32}(x - 2)^2 + \frac{5}{768}(x - 2)^3$

25. $T_3(x) = -\frac{1}{2}x^2$

27. $\sqrt{e} \approx 1.6486979$; error $\approx 2.335 \times 10^{-5}$

29. $\left| f(17) - T_4(17) \right| \le \frac{105/32}{5!} 4^{-9} \approx 1.043 \times 10^{-7}$

31. $\left| f(1.2) - T_4(1.2) \right| \le \frac{6}{5!}(0.2)^5 = 1.6 \times 10^{-5}$

Chapter 9

Section 9.1 Preliminary

1. (a) First order   **(b)** First order   **(c)** Order 3

2. Yes   **3.** Example: $y' = y^2$ **4.** Example: $y' = y^2$

5. Example: $y' + xy = x^2$

Section 9.1 Exercises

1. (a) First order   **(b)** Not first order   **(c)** First order
(d) First order   **(e)** Not first order   **(f)** First order

3. Let $y = 4x^2$. Then $y' - 8x = 8x - 8x = 0$.

5. Let $y = \sqrt{12 - 4x^2}$. Then

$$yy' + 4x = \sqrt{12 - 4x^2} \frac{-4x}{\sqrt{12-4x^2}} + 4x$$

$$= -4x + 4x = 0$$

11. (b) $e^{-y} = -\sin x + C$ **(d)** $C = 1$

13. $y = \frac{-1}{\left(\frac{1}{2}\right)x^2 + C}$, where C is an arbitrary constant.

15. $y = Ae^{9x}$, where A is an arbitrary constant.

17. $y = Ae^{-3x} - \frac{2}{3}$, where A is an arbitrary constant.

19. $y = -\ln\left| C + \frac{1}{2}t^2 \right|$, where C is an arbitrary constant.

21. $y = \frac{1}{(1/3)x^3 - x + C}$, where C is an arbitrary constant.

23. $x = \tan(\tan^{-1} t + C)$ OR $x = \frac{t+C}{1-Ct}$ using the sum identities for tangents, where C is an arbitrary constant.

25. $y = \sin^{-1}\left(\frac{1}{2}x^2 + C\right)$, where C is an arbitrary constant.

27. $y = C \sec t$, where C is an arbitrary constant.

29. $y = 12e^{-2x}$ **31.** $y = -\sqrt{\ln(x^2 + e)}$ **33.** $y = e^{(1/2)x^2 - x} + 2$

35. $y = (e)e^{-e^{-t}} = e^{1 - e^{-t}}$ **37.** $y = \frac{et}{e^{1/t}} - 1$ **39.** $y = \frac{1}{1 - \sin^{-1} x}$

41. $y = \frac{1}{\cos x - (2/3)}$ **43.** $a = -4, 2$

45. (a) $t = (45/2)(\sqrt{10} - \sqrt{5}) \approx 20.840$ s
(b) $t = (45/2)\sqrt{10} \approx 71.15$ s

47. $y(t) = 20 - \left(\frac{9}{20}t + 10^{3/2}\right)^{2/3}$ and $t_e \approx 128.5$ s or 2.1 min

49. (a) $q = CV - Ae^{-t/(RC)}$
(b) As $t \to \infty$, $e^{-t/(RC)} \to 0$, so $q(t) \to CV$.
(c) $q(t) = CV - CVe^{-t/(RC)}$ and
$q(\tau) = CV(1 - e^{-RC/RC}) \approx CV(0.6321)$

51. $V = (kt/3 + C)^3$, V increases roughly with the cube of time.

53. $g(x) = Ae^{2x}$; $g(x) = \frac{A}{1-x}$ **57.** $y = \pm\sqrt{x^2 + C}$ and $y = A/x$

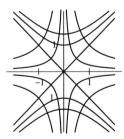

59. (a) $v(t) = -\alpha\left(\frac{1 - e^{-(2k\alpha/m)t}}{1 + e^{-(2k\alpha/m)t}}\right)$

61. (a) $(y')^2 + y^2 \ge 0$ and equals zero if and only if $y' = 0$ and $y = 0$
(b) $(y')^2 + y^2 + 1 \ge 1 > 0$ for all y' and y, so $(y')^2 + y^2 + 1 = 0$ has no solution

63. (c) $C = \frac{7\pi}{120B} R^{5/2}$

Section 9.2 Preliminary

1. $y(t) = b + ce^{-kt}$, where c is an arbitrary constant.

2. $y(t) = 5 - ce^{4t}$ for any positive constant c.

3. No **4.** True **5.** Material A

Section 9.2 Exercises

1. General solution: $y(t) = 10 + ce^{2t}$; solution satisfying
$y(0) = 25$: $y(t) = 10 + 15e^{2t}$; solution satisfying $y(0) = 5$:
$y(t) = 10 - 5e^{2t}$

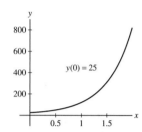
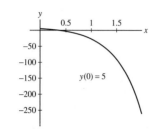

5. (a) $k = \frac{\ln 2}{40} \approx 0.017 \text{ s}^{-1}$ (b) $F'(t) = -0.017(F(t) - 60)$
(c) $F(t) = 60 + 56.2e^{-0.017t}$ (d) $116.2°\text{F}$

7. $50\frac{3^{4/5}-2}{3^{4/5}-1} \approx 14.49°\text{C}$

9. (a) $5.4°\text{C/min}$ (b) $0.54°\text{C}$ (c) $\frac{1}{0.09}\ln\left(\frac{70}{45}\right) \approx 4.91 \text{ min}$

11. $-58.8 \text{ m/s} \approx -131.5 \text{ mph}$ **13.** -25.0 ft/s

15. (a) $t = \frac{1}{0.05}\ln 4 \approx 28 \text{ yr}$ (b) $P_0 = \$20,000$

17. $\$10,000$ **19.** 10%

21. (b) $t = \frac{1}{0.09}\ln\left(\frac{13,333.33}{3,333.33}\right) \approx 15.4 \text{ yr}$ (c) No

23. (a) $b = \frac{V}{R}$ (b) $I(t) = \frac{V}{R}(1 - e^{-kt})$

Section 9.3 Preliminary

1. 7 **2.** $y = \pm\sqrt{t+1}$ **3.** False **4.** Yes **5.** $n = 20$

Section 9.3 Exercises

1.

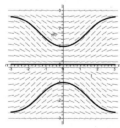

3.

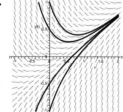

5.

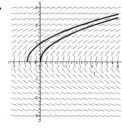

7. With $y(0) > 0$, as $t \to \infty$, y increases at an increasing rate, so
that $\lim_{t\to\infty} y = \infty$. On the other hand, if $y(0) < 0$, then y decreases at
an ever faster rate as $t \to \infty$, so $\lim_{t\to\infty} y = -\infty$.

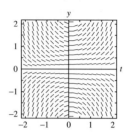

9. For $y' = t^2$, y' only depends on t. The isoclines of any slope c
will be the two vertical lines $t = \pm\sqrt{c}$. This indicates that the slope
field will be the one given in Figure 12(A). The solutions are sketched
in the following figure.

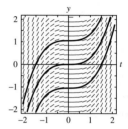

For $y' = y^2$, y' depends only on y. The isoclines of any slope c
will be the two *horizontal* lines $y = \pm\sqrt{c}$. The solutions are sketched
in the following figure.

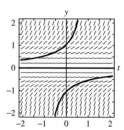

11. This differential equation cannot be solved by separation of
variables. See the solutions sketched in the following figure.

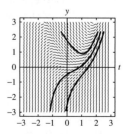

13. (a) Since the step width is $h = 0.1$, $y(1.0) \approx y_{10}$. We compute:

k	t_k	y_k
0	0	0
1	0.1	0
2	0.2	0.01
3	0.3	0.029801
4	0.4	0.0589202
5	0.5	0.0966314
6	0.6	0.142026
7	0.7	0.194082
8	0.8	0.251733
9	0.9	0.313929
10	1.0	0.379681

(b) $y = \ln\left|\frac{1}{2}t^2 + 1\right|$

(c) $\left|y(0.1) - y_1\right| = |0.00498754 - 0| = 0.00498754$

$\left|y(0.5) - y_5\right| = |0.117783 - 0.0966314| = 0.0211516$

$\left|y(1.0) - y_{10}\right| = |0.405465 - 0.379681| = 0.257844$

15. $y(1) \approx 0$ **17.** $y(1.5) \approx 2.478982$ **19.** $y(0.5) \approx 0.68098$

21. Euler's method: $y(1.5) \approx 4.177248$; Euler's midpoint method: $y(1.5) \approx 4.471304$; $y(t) = e^t$; Error from Euler's method: 0.304441; Error from Euler's midpoint method: 0.010385

23. $y(0.5) \approx 0$ **25.** $y(1) \approx 0.714081$

Section 9.4 Preliminary

1. (a) No **(b)** Yes **(c)** No **(d)** Yes

2. False **3.** True

4. (a) $\lim_{t \to \infty} y(t) = 3$ **(b)** $\lim_{t \to \infty} y(t) = 3$ **(c)** $\lim_{t \to \infty} y(t) = -\infty$

Section 9.4 Exercises

1. $y = \dfrac{5}{1 - e^{-3t}/C}$ and $y = \dfrac{5}{1 + (3/2)e^{-3t}}$ **3.** $\lim_{t \to \infty} y(t) = 2$

5. (a) $P(t) = \dfrac{2{,}000}{1 + 3e^{-0.6t}}$ **(b)** $t = \dfrac{1}{0.6}\ln 3 \approx 1.83$ yrs

7. $k = \ln\frac{81}{31} \approx 0.96$ yrs $- 1$; $t = \dfrac{\ln 9}{2\ln 9 - \ln 31} \approx 2.29$ yrs

9. After $t = 4\dfrac{\ln(217/2)}{\ln(23/2)} \approx 7.68$ hours, or at 3:41 PM

11. (a) $y_1(t) = \dfrac{10}{10 - 9e^{-t}}$ and $y_2(t) = \dfrac{1}{1 - 2e^{-t}}$

(b) $t = \ln\frac{9}{8}$ **(c)** $t = \ln 2$

13. (a) $u = \sqrt{M}\,\dfrac{Ce^{(k/\sqrt{M})t} - 1}{Ce^{(k/\sqrt{M})t} + 1}$

17. (d) $t = -\frac{1}{k}\left(\ln y_0 - \ln(y_0 - A)\right)$

Section 9.5 Preliminary

1. (a) Yes **(b)** No **(c)** Yes **(d)** No

2. The answer is (b).

Section 9.5 Exercises

1. (c) $y = \frac{x^4}{5} + \frac{C}{x}$ **(d)** $y = \frac{x^4}{5} - \frac{1}{5x}$

5. $y = \frac{1}{2}x + \frac{C}{x}$ **7.** $y = -\frac{1}{4}x^{-1} + Cx^{1/3}$

9. $y = \frac{1}{5}x^2 + \frac{1}{3} + Cx^{-3}$ **11.** $y = -x\ln|x| + Cx$

13. $y = \frac{1}{2}e^x + Ce^{-x}$ **15.** $y = x\cos x + C\cos x$

17. $y = x^x + Cx^x e^{-x}$ **19.** $y = \frac{1}{5}e^{2x} - \frac{6}{5}e^{-3x}$

21. $y = \dfrac{\ln|x|}{x+1} - \dfrac{1}{x(x+1)} + \dfrac{5}{x+1}$ **23.** $y = -\cos x + \sin x$

25. $y = \tanh x + 3\,\mathrm{sech}\,x$

27. For $m \neq -n$: $y = \dfrac{1}{m+n}e^{mx} + Ce^{-nx}$; for $m = -n$: $y = (x + C)e^{-nx}$

29. $y'(t) = 8 - \frac{y}{5}$, $y(0) = 10$; $y(t) = 40 - 30e^{-t/5}$; 0.4 lbs per gal

31. (a) $\dfrac{dV}{dt} = \dfrac{20}{1+t} - 5$ and $V(t) = 20\ln(1 + t) - 5t + 100$

(b) The maximum value is $V(3) = 20\ln 4 - 15 + 100 \approx 112.726$.

(c) Tank is empty after $t \approx 34.25$ min.

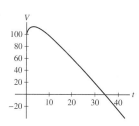

33. $I(t) = \dfrac{2 - 2e^{-20t}}{20} = \dfrac{1 - e^{-20t}}{10}$

35. (a) $I(t) = \dfrac{V}{R} - \dfrac{V}{R}e^{-(R/L)t}$

(c) Approximately 0.0184 s

37. (b) $y_2(t) = -4e^{-t/10} + 4e^{-t/20}$

$= 4e^{-t/20}(1 - e^{-t/20})$

(c) See accompanying figure.

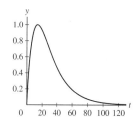

(d) $\frac{1}{200}$

Chapter 9 Review

1. (a) No, first order **(b)** Yes, first order

(c) No, order 3 **(d)** Yes, order 2

3. $y = \left(\frac{4}{3}t^3 + C\right)^{1/4}$, where C is an arbitrary constant.

5. $y = Cx - 1$, where C is an arbitrary constant.

7. $y = \frac{1}{2}\left(x + \frac{1}{2}\sin 2x\right) + \frac{\pi}{4}$ **9.** $y = \dfrac{2}{2 - x^2}$

11.

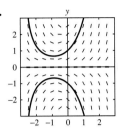

13. $y(t) = \tan t$

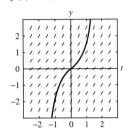

15. $y(0.1) \approx 1.1$; $y(0.2) \approx 1.209890$; $y(0.3) \approx 1.329919$

17. $y = x^2 + 2x$ **19.** $y = \frac{1}{2} + e^{-x} - \frac{11}{2}e^{-2x}$

21. $y = \cos x \sin x + C \cos x$, where C is an arbitrary constant

23. $\frac{1}{2}y^2 - y = \frac{1}{2}t^2 + 7$, or since we know that $y(1) < 0$, $y = 1 - \sqrt{15 + t^2}$ **25.** $w = \tan(k \ln x + \frac{\pi}{4})$

27. $y = -\cos x + \frac{\sin x}{x} + \frac{C}{x}$, where C is an arbitrary constant.

29. Solution satisfying $y(0) = 3$: $y(t) = 4 - e^{-2t}$; solution satisfying $y(0) = 4$: $y(t) = 4$

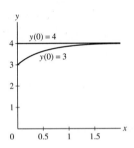

31. (a) 12

(b) ∞, if $y(0) > 12$; 12, if $y(0) = 12$; $-\infty$, if $y(0) < 12$

(c) -3

33. $40,000 - 35,000e^{0.1} \approx \$1,319.02$ **35.** \$40,000

39. $y(t) = \frac{4}{1 - (1/2)e^{-1.6t}}$ **41.** $k = \frac{\ln 2}{10} \approx 0.0693$

43. (a) $\frac{dP}{dt} = 0.12P - r$, $P(0) = 150$ **(b)** $r = 18$

45. (a) $\frac{dy}{dt} = 1.5 - \frac{4y}{20 - t}$, $y(0) = 0$

(b) $y(t) = 10 - \frac{1}{2}t - \frac{(20 - t)^4}{16,000}$

(c) Amount of toxin is maximal at $t = 20 - 10\sqrt[3]{2} \approx 7.4$ minutes.

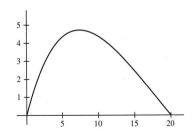

Chapter 10

Section 10.1 Preliminary

1. $a_4 = 12$ **2.** (c) **3.** $\lim\limits_{n \to \infty} a_n = \sqrt{2}$ **4.** (b)

5. (a) False. Counterexample: $a_n = \cos \pi n$. **(b)** True

(c) False. Counterexample: $a_n = (-1)^n$.

Section 10.1 Exercises

1. (a) (iv) **(b)** (i) **(c)** (iii) **(d)** (ii)

3. $c_1 = 2, c_2 = 2, c_3 = \frac{4}{3}, c_4 = \frac{2}{3}$

5. $a_1 = 3, a_2 = 10, a_3 = 101, a_4 = 10,202$

7. $c_1 = 1, c_2 = \frac{3}{2}, c_3 = \frac{11}{6}, c_4 = \frac{25}{12}$

9. $b_1 = 2, b_2 = 5, b_3 = 9, b_4 = 19$

11. (a) $a_n = \frac{(-1)^{n+1}}{n^3}$ **(b)** $a_n = \frac{n+1}{5+n}$

13. $\lim\limits_{n \to \infty} a_n = 4$ **15.** $\lim\limits_{n \to \infty} \left(5 - \frac{9}{n^2}\right) = 5$

17. $\lim\limits_{n \to \infty} (-2^{-n}) = 0$ **19.** $\lim\limits_{n \to \infty} \left(\frac{1}{3}\right)^n = 0$

21. The sequence diverges. **23.** $\lim\limits_{n \to \infty} \frac{n}{\sqrt{n^3 + 1}} = 0$

25. The sequence diverges.

27. (a) $M = 999$ **(b)** $M = 99,999$

31. $\lim\limits_{n \to \infty} 2^{1/n} = 1$ **33.** $\lim\limits_{n \to \infty} n^{1/n} = 1$ **35.** $\lim\limits_{n \to \infty} \left(1 + \frac{1}{n}\right)^n = e$

37. $\lim\limits_{n \to \infty} a_n = 0$ **39.** $\lim\limits_{n \to \infty} n \sin \frac{1}{n} = 1$ **41.** (b)

43. $\lim\limits_{n \to \infty} \frac{3n^2 + n + 2}{2n^2 - 3} = \frac{3}{2}$ **45.** $\lim\limits_{n \to \infty} 3 + \left(-\frac{1}{2}\right)^n = 3$

47. $\lim\limits_{n \to \infty} \tan^{-1}(1 - \frac{2}{n}) = \frac{\pi}{4}$ **49.** $\lim\limits_{n \to \infty} \ln\left(\frac{2n+1}{3n+4}\right) = \ln \frac{2}{3}$

51. $\lim\limits_{n \to \infty} \frac{e^n + 3^n}{5^n} = 0$ **53.** $\lim\limits_{n \to \infty} \frac{e^n}{2n^2} = 0$

55. $\lim\limits_{n \to \infty} \frac{n^3 + 2e^{-n}}{3n^3 + 4e^{-n}} = \frac{1}{3}$ **57.** The sequence diverges.

59. $\lim\limits_{n \to \infty} \frac{(-4,000)^n}{n!} = 0$ **61.** $\lim\limits_{n \to \infty} \cos \frac{\pi}{n} = 1$ **63.** $\lim\limits_{n \to \infty} \sqrt[n]{n} = 1$

65. Any number greater than or equal to 3 is an upper bound.

71. Example: $a_n = -n^2, b_n = n^2 + \frac{1}{n}$

75. $\lim\limits_{n \to \infty} a_n = \frac{1 + \sqrt{21}}{2}$ **77.** $\lim\limits_{n \to \infty} c_n = 1$

81. (a) $c_1 = \frac{3}{2}; c_2 = \frac{13}{12}; c_3 = \frac{19}{20}; c_4 = \frac{743}{840}$ **(c)** $\lim\limits_{n \to \infty} c_n = \ln 2$

Section 10.2 Preliminary

1. The sum of an infinite series is defined as the limit of the sequence of partial sums.

2. $S = \frac{1}{2}$ **3.** The result is negative hence the formula is invalid.

4. No **5.** No **6.** $N = 13$ **7.** No **8.** Example: $\sum\limits_{n=1}^{\infty} \frac{1}{n^{9/10}}$.

Section 10.2 Exercises

1. (a) $a_n = \frac{1}{3^n}$ (b) $a_n = \left(\frac{5}{2}\right)^{n-1}$

(c) $a_n = (-1)^{n+1} \frac{n^n}{n!}$ (d) $a_n = \frac{1 + \frac{(-1)^{n+1}+1}{2}}{n^2+1}$

3. $S_2 = \frac{5}{4}, S_4 = \frac{205}{144}, S_6 = \frac{5,369}{3,600}$ **5.** $S_2 = \frac{2}{3}, S_4 = \frac{4}{5}, S_6 = \frac{6}{7}$

7. $S_5 = 0.035672, S_{10} = 0.035352, S_{15} = 0.035413$. Yes.

9. $S_3 = \frac{3}{10}, S_4 = \frac{1}{3}, S_5 = \frac{5}{14}; \sum\limits_{n=1}^{\infty} \left(\frac{1}{n+1} - \frac{1}{n+2}\right) = \frac{1}{2}$

11. $\sum\limits_{n=3}^{\infty} \frac{1}{n(n-1)} = \sum\limits_{n=3}^{\infty} \left(\frac{1}{n-1} - \frac{1}{n}\right) = \frac{1}{2}$

13. $\lim\limits_{n\to\infty} (-1)^n n^2 \neq 0$

17. (a)   **19.** $S = \frac{1}{18}$ **21.** $S = \frac{27}{968}$ **23.** $S = \frac{7}{5}$ **25.** $S = \frac{e}{e^2-1}$

27. The series diverges.   **29.** $S = \frac{8}{5}$ **31.** $S = \frac{512}{49}$ **33.** (b), (c)

35. (a) $\sum\limits_{n=1}^{10} a_n = \frac{249}{50}, \sum\limits_{n=4}^{16} a_n = \frac{494}{2,304}$ (b) $a_3 = \frac{5}{18}$

(c) $a_n = \frac{2(2n-1)}{n^2(n-1)^2}$ (d) $\sum\limits_{n=1}^{\infty} a_n = 5$

39. 50 ft   **41.** $S = \frac{1}{2}$ **45.** $2 + \sqrt{2}$

Section 10.3 Preliminary

1. (b)   **2.** f must be positive, continuous and decreasing

3. Integral Test   **4.** Comparison Test

5. No; $\sum \frac{1}{n}$ diverges, but since $\frac{e^{-n}}{n} < \frac{1}{n}$, this tells us nothing about $\sum \frac{e^{-n}}{n}$.

Section 10.3 Exercises

1. $\int_1^{\infty} \frac{dx}{x^4}$ converges, so the series converges.

3. $\int_1^{\infty} x^{-1/3}\, dx = \infty$. The series diverges.

5. $\int_{25}^{\infty} \frac{x^2}{(x^3+9)^{5/2}}\, dx$ converges, so the series converges.

7. $\int_1^{\infty} \frac{dx}{x^2+1}$ converges, so the series converges.

9. $\int_1^{\infty} xe^{-x^2}\, dx$ converges, so the series converges.

11. $\int_1^{\infty} \frac{1}{2^{\ln x}}\, dx = \infty$. The series diverges.

13. $\int_1^{\infty} \frac{\ln x}{x^2}\, dx = \frac{1+\ln 2}{2}$. The series converges.

15. $\frac{1}{n^3+8n} \leq \frac{1}{n^3}$. The series converges.

19. $\frac{1}{n2^n} \leq \frac{1}{2^n}$. The series converges.

21. $\frac{k^{1/3}}{k^2+k} \leq \frac{k^{1/3}}{k^2} = \frac{1}{k^{5/3}}$. The series converges.

23. $\frac{1}{\sqrt{n^3+1}} \leq \frac{1}{\sqrt{n^3}} = \frac{1}{n^{3/2}}$. The series converges.

25. $0 \leq \frac{\sin^2 k}{k^2} \leq \frac{1}{k^2}$. The series converges.

27. $\frac{4}{m!+4^m} \leq \frac{4}{4^m} = 4 \cdot \left(\frac{1}{4}\right)^m$. The series converges.

29. $\frac{1}{2^{k^2}} \leq \frac{1}{2^k} = \left(\frac{1}{2}\right)^k$. The series converges.

31. $0 \leq \frac{\ln n}{n^3+3\ln n} \leq \frac{\ln n}{n^3}$. The series converges.

33. The general term of the series does not approach zero and the series diverges.

35. The series converges.   **37.** The series diverges.

39. The series converges.   **41.** The series converges.

47. The series converges for $a > 1$ and diverges for $a \leq 1$.

49. The series converges.   **51.** The series diverges.

53. The series converges.   **55.** The series converges.

57. The series diverges.   **59.** The series converges.

61. The series converges.   **63.** The series converges.

65. The series converges.   **67.** The series converges.

69. The series converges.   **71.** The series converges.

73. The series converges.   **81.** $\sum\limits_{n=1}^{\infty} \frac{1}{n^5} \approx 1.0369$

87. (c) $\sum\limits_{n=1}^{1,000} \frac{1}{n^2} = 1.643934568; 1 + \sum\limits_{n=1}^{100} \frac{1}{n^2(n+1)} = 1.644884890;$

the latter sum is a better approximation to $\pi^2/6$

Section 10.4 Preliminary

1. (c)   **2.** (b)   **3.** No   **4.** Example: $\sum \frac{(-1)^n}{\sqrt[3]{n}}$.

Section 10.4 Exercises

3. Converges conditionally   **5.** Converges absolutely

7. Converges conditionally   **9.** Converges absolutely

11. Converges absolutely

13. (a)

n	S_n
1	1.000000
2	0.875000
3	0.912037
4	0.896412
5	0.904412
6	0.899782
7	0.902698
8	0.900745
9	0.902116
10	0.901116

15. $S_5 = 0.947539$ **17.** $S_{44} = 0.0656746$

19. Converges (by Comparison Test)

21. Converges (by Leibniz Test)

23. Converges (by Linearity of Series)

25. Converges (by Leibniz Test)

27. Converges conditionally

Section 10.5 Preliminary

1. $\rho = \lim\limits_{n\to\infty} \left|\frac{a_{n+1}}{a_n}\right|$

2. The Ratio Test is conclusive for $\sum\limits_{n=1}^{\infty} \frac{1}{2^n}$ and inconclusive for $\sum\limits_{n=1}^{\infty} \frac{1}{n}$.

3. No

Section 10.5 Exercises

1. Converges (by the Ratio Test)

3. Converges absolutely (by the Ratio Test)

5. The Ratio Test is inconclusive.   7. Diverges (by the Ratio Test)

9. Converges (by the Ratio Test)

11. Converges (by the Ratio Test)

13. Diverges (by the Ratio Test)   15. The Ratio Test is inconclusive.

17. Converges (by the Ratio Test)   19. $\lim\limits_{n\to\infty}\left|\frac{a_{n+1}}{a_n}\right| = \frac{1}{3} < 1$

21. $\lim\limits_{n\to\infty}\left|\frac{a_{n+1}}{a_n}\right| = 2|x|$ 23. $\lim\limits_{n\to\infty}\left|\frac{a_{n+1}}{a_n}\right| = |r|$

27. Converges absolutely (by the Ratio Test)

29. The Ratio Test is inconclusive and the series may diverge/converge, depending on a_n.

31. Converges absolutely (by the Ratio Test)

33. The Ratio Test is inconclusive.

35. Converges (by the Root Test)   37. Converges (by the Root Test)

39. Converges (by the Root Test)

41. Converges (by the Ratio Test or by Linearity)

43. Converges (by the Ratio Test)

45. Converges (by the Limit Comparison Test)

47. Converges (by the Limit Comparison Test)

49. Converges (by the Limit Comparison Test)

51. Diverges (by the Divergence Test)

Section 10.6 Preliminary

1. Yes. The series converges for both $x = 4$ and $x = -3$.

2. Converges at $x = 8$; cannot determine convergence for $x = 0$, $x = 2$, $x = 12$.

3. $R = 4$ 4. $\sum\limits_{n=0}^{\infty}(n+1)^2 x^n$; $R = 1$

Section 10.6 Exercises

1. $R = 2$ 3. $R = 3$ for all series.

7. Converges in $|x| < 1$ and diverges elsewhere.

9. Converges in $-\frac{1}{2} \le x < \frac{1}{2}$ and diverges elsewhere.

11. Converges in $-1 \le x < 1$ and diverges elsewhere.

13. Converges for all x.

15. Converges in $-5 < x < -3$ and diverges elsewhere.

17. Converges in $-2 < x < 2$ and diverges elsewhere.

19. Converges in $3 \le x \le 5$ and diverges elsewhere.

21. Converges for $x = -5$ and diverges for $x \ne -5$.

23. Converges for $2 - \frac{1}{e} < x < 2 + \frac{1}{e}$ and diverges elsewhere.

25. Converges in $-1 \le x < 1$ and diverges elsewhere.

27. Expansion: $\sum\limits_{n=0}^{\infty} 3^n x^n$; the expansion is valid for $|x| < \frac{1}{3}$.

29. Expansion: $\sum\limits_{n=0}^{\infty} \frac{x^n}{3^{n+1}}$; the expansion is valid for $|x| < 3$.

31. Expansion: $\sum\limits_{n=0}^{\infty}(-1)^n x^{9n}$; the expansion is valid for $|x| < 1$.

33. Expansion: $\sum\limits_{n=0}^{\infty}(-1)^n 3^n x^{7n}$; the expansion is valid for $|x| < \frac{1}{3^{1/7}}$.

37. $\frac{1}{4-x} = \sum\limits_{n=0}^{\infty}(-1)^{n+1}(x-5)^n$ on the interval $(4, 6)$.

39. Example: $\sum\limits_{n=1}^{\infty}\frac{(x-4)^n}{n2^n}$ 41. N is at least 7; $S_7 = 0.405804$.

45. N is at least 7; $S_7 = 0.368056$.

47. $1 - \frac{1}{2}x^2 - \sum\limits_{n=2}^{\infty}\frac{1\cdot3\cdot5\cdots(2n-3)}{(2n)!}x^{2n}$; $R = \infty$

53. Converges for $-\frac{1}{3} < x < \frac{1}{3}$

Section 10.7 Preliminary

1. $f(0) = 3$ and $f'''(0) = 30$ 2. $f(-2) = 0$; $f^{(4)}(-2) = 48$

3. Substituting x^2 for the MacLaurin series for $\sin(x)$.

4. $f(x) = 4 + \sum\limits_{n=1}^{\infty}\frac{(x-3)^{n+1}}{n(n+1)}$ 5. (a) and (c)

Section 10.7 Exercises

1. $f(x) = 2 + 3x + 2x^2 + 2x^3 + \cdots$

3. $\frac{1}{1-2x} = \sum\limits_{n=0}^{\infty} 2^n x^n$ for $|x| < \frac{1}{2}$

5. $\cos(3x) = \sum\limits_{n=0}^{\infty}\frac{(-1)^n 9^n}{(2n)!}x^{2n}$ for all x

7. $\sin(x^2) = \sum\limits_{n=0}^{\infty}\frac{(-1)^n}{(2n+1)!}x^{4n+2}$ for all x

9. $\ln(1-x^2) = -\sum\limits_{n=1}^{\infty}\frac{x^{2n}}{n}$ for $|x| < 1$

11. $\tan^{-1}(x^2) = \sum\limits_{n=0}^{\infty}\frac{(-1)^n x^{4n+2}}{2n+1}$ for $|x| \le 1$

13. $e^{x-2} = \sum\limits_{n=0}^{\infty}\frac{1}{n!e^2}x^n$ for all x

15. $\ln(1-5x) = -\sum\limits_{n=1}^{\infty}\frac{5^n}{n}x^n$ for $|x| < \frac{1}{5}$ and $x = -\frac{1}{5}$

17. $\sinh(x) = \sum\limits_{n=0}^{\infty}\frac{1}{(2n+1)!}x^{2n+1}$ for all x

19. $\frac{1-\cos(x^2)}{x} = \sum\limits_{n=1}^{\infty}\frac{(-1)^{n+1}}{(2n)!}x^{4n-1}$ for $x \ne 0$

21. $e^x \sin(x) = x + x^2 + \frac{x^3}{3} - \frac{x^5}{30} + \cdots$

23. $e^x \ln(1-x) = -x - \frac{3x^2}{2} - \frac{4x^3}{3} - x^4 + \cdots$

25. $(1+x)^{-3/2} = 1 - \frac{3x}{2} + \frac{15x^2}{8} - \frac{35x^3}{16} + \frac{315x^4}{128} - \cdots$

27. $e^{e^x} = e + ex + ex^2 + \frac{5e}{6}x^3 + \cdots$

29. $\sin(x) = \sum\limits_{n=0}^{\infty}\frac{(-1)^n}{(2n)!}\left(x - \frac{\pi}{2}\right)^{2n}$ 31. $\frac{1}{x} = \sum\limits_{n=0}^{\infty}(-1)^n(x-1)^n$

33. $\frac{1}{1-x} = \sum_{n=0}^{\infty} \frac{(-1)^{n+1}(x-5)^n}{4^{n+1}}$

35. $x^4 + 3x - 1$

$$= 21 + 35(x-2) + 24(x-2)^2 + 8(x-2)^3 + (x-2)^4$$

37. $e^{3x} = \sum_{n=0}^{\infty} \frac{3^n e^{-3}}{n!}(x+1)^n$

39. $\frac{1}{1-x^2} = \sum_{n=0}^{\infty} \frac{(-1)^{n+1}(2^{n+1}-1)}{2^{2n+3}}(x-3)^n$, for $|x-3| < 2$

41. $f(x) = 1 + \sum_{n=0}^{\infty} \frac{1 \cdot 3 \cdot 5 \cdots (2n-1)}{2^n n!}(3x)^{2n}$ for $|x| < \frac{1}{3}$

43. $\sin^{-1} \frac{1}{2} \approx 0.52358519539$

47. (a) 5   (b) $\sum_{n=0}^{4} \frac{(-1)^n}{(2n+1)n!} = 0.747487$

49. $\int_0^1 \cos(x^2)\,dx = \sum_{n=0}^{\infty} \frac{(-1)^n}{(4n+1)(2n)!}$;

$\sum_{n=0}^{3} \frac{(-1)^n}{(4n+1)(2n)!} = 0.904522792$

51. $\int_0^2 e^{-x^3}\,dx = \sum_{n=0}^{\infty} \frac{(-1)^n 2^{3n+1}}{(3n+1)n!}$; $\sum_{n=0}^{24} \frac{(-1)^n 2^{3n+1}}{(3n+1)n!} = 0.892953509$

53. $\int_0^x \frac{1-\cos t}{t}\,dt = \sum_{n=1}^{\infty} \frac{(-1)^n}{2n(2n)!}x^{2n}$

55. $\int_0^x \ln(1+t^2)\,dt = \sum_{n=1}^{\infty} \frac{(-1)^{n-1}}{n(2n+1)}x^{2n+1}$

57. $f(x) = \frac{1}{1+2x}$ **59.** $x^2 - \frac{2}{3}x^6 + \frac{2}{15}x^{10} - \frac{4}{315}x^{14}$

61. $1 + x - \frac{1}{2}x^2 - \frac{1}{8}x^4$ **63.** $f(x) = e^{x^3}$

65. $f(x) = 1 - 5x + \sin(5x)$ **67.** $I(t) = \frac{V}{R} \sum_{n=1}^{\infty} \frac{(-1)^{n+1}}{n!}\left(\frac{Rt}{L}\right)^n$

69. $f(x) = \sum_{n=0}^{\infty} \frac{(-1)^n x^n}{(2n)!}$ and $f^{(5)}(0) = -\frac{1}{30{,}240}$

71. $f^{(8)}(0) = -1{,}575$

73. The expansion is not valid since the series only converges for $|x| < 1$.

77. 4; $\sum_{n=1}^{4} \frac{(-1)^{n-1}(0.2)^n}{n} = 0.182267$

79. $\frac{1}{(1-x)(1-2x)} = \sum_{n=0}^{\infty} (-1)^n \left(1 - \left(\frac{2}{3}\right)^{n+1}\right)(x-2)^n$, for $|x-2| < 1$

83. If a positive whole, $R = \infty$; if negative whole, $R = 1$.

Chapter 10 Review

1. (a) $a_1^2 = 4, a_2^2 = \frac{1}{4}, a_3^2 = 0$

(b) $b_1 = \frac{1}{24}, b_2 = \frac{1}{60}, b_3 = \frac{1}{240}$

(c) $a_1 b_1 = -\frac{1}{12}, a_2 b_2 = -\frac{1}{120}, a_3 b_3 = 0$

(d) $2a_2 - 3a_1 = 5, 2a_3 - 3a_2 = 1.5, 2a_4 - 3a_3 = \frac{1}{12}$

3. $\lim_{n\to\infty} (5a_n - 2a_n^2) = 2$ **5.** $\lim_{n\to\infty} e^{a_n} = e^2$

7. The limit does not exist.   **9.** $\lim_{n\to\infty} a_n = 0$

11. $\lim_{n\to\infty} a_n = 1$ **13.** The sequence diverges.

15. $\lim_{n\to\infty} b_n = \frac{\pi}{4}$ **17.** $\lim_{n\to\infty} b_n = \frac{1}{2}$ **19.** $\lim_{n\to\infty} a_n = 0$

21. $\lim_{n\to\infty} c_n = e^3$

25. (a) $\lim_{n\to\infty} a_n = \infty$ (b) $\lim_{n\to\infty} \frac{a_{n+1}}{a_n} = 3$

27. $S_4 = -\frac{11}{60} = -0.183333; S_7 = \frac{41}{630} = 0.0650794$

29. $\sum_{n=2}^{\infty} \left(\frac{2}{3}\right)^n = \frac{4}{3}$ **31.** $\sum_{n=0}^{\infty} \frac{2^{n+1}}{3^n} = 6$

33. Example: $a_n = \left(\frac{1}{2}\right)^n + 1, b_n = -1$

35. The sequence diverges.

37. $\int_1^{\infty} \frac{x^3}{e^{x^4}}\,dx = \frac{e^{-1}}{4}$. The sequence converges.

39. $\frac{1}{n^2} > \frac{1}{(n+1)^2}$. The sequence converges.

41. $\frac{n^2+1}{n^{3.5}-2} < \frac{4}{n^{1.5}}$. The sequence converges.

43. Since $\frac{\ln n}{1.5^n} < \frac{n}{1.5^n}$ for all $n > 1$. The sequence converges.

45. $\frac{1}{3^n - 2^n} < \frac{1}{2^n}$ for all $n \geq 2$. The sequence converges.

49. $0.3971162690 \leq S \leq 0.3971172688$. The maximum size of the error 10^{-6}.

51. Converges absolutely.   **53.** The series diverges.

55. 22; $K \approx \sum_{k=0}^{21} \frac{(-1)^k}{(2k+1)^2} = 0.91571$

57. (a) Converges   (b) Converges   (c) Diverges

(d) Converges

59. The Ratio Test is inconclusive.   **61.** The series converges.

63. The Ratio Test is inconclusive.   **65.** The series converges.

67. The series converges.   **69.** The series converges.

71. The series converges.   **73.** The series converges.

75. The series converges by Leibniz Test.

77. The series converges by Leibniz Test.

79. The series converges.   **81.** The series diverges.

83. The series converges.   **85.** The series converges for $x \in \mathbf{R}$.

87. The series converges for $x \in [2, 4]$.

89. The series converges for $x = 0$.

91. $\frac{2}{4-3x} = \frac{1}{2} \sum_{n=0}^{\infty} \left(\frac{3}{4}\right)^n x^n$. The series converges for $|x| < \frac{4}{3}$.

93. (c)

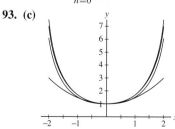

95. $\lim_{x\to 0} \frac{1 - \frac{x^2}{2} - \cos x}{x^4} = -\frac{1}{24}$ **97.** $f(x) = e^{-2} \sum_{n=0}^{\infty} \frac{2^n(x+1)^n}{n!}$

99. $f(x) = \sum\limits_{n=1}^{\infty} \frac{(-1)^{n+1}(x-2)^n}{n2^n}$

101. $f(x) = \sum\limits_{n=0}^{\infty} \frac{(-1)^n x^{n+1}}{2n+1}$ **103.** $f(x) = \sum\limits_{n=0}^{\infty} \frac{(x+1)^n}{n!e^2}$

105. $f^{(4)}(0) = -\frac{1}{2}$ **107.** $F(x) = \sum\limits_{n=1}^{\infty} \frac{x^n}{n \cdot n!}$

Chapter 11

Section 11.1 Preliminary

1. A circle of radius 3 centered at the origin.

2. The center is at $(4, 5)$. **3.** Maximum height: 4

4. Yes; no **5.** (a) ↔ (iii), (b) ↔ (ii), (c) ↔ (i)

Section 11.1 Exercises

1. $(t = 0) \ (1, 9); \ (t = 2) \ (9, -3); \ (t = 4) \ (65, -39)$

5. (a)   (b)

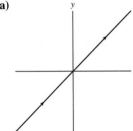

(c)   (d)

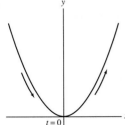 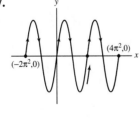

7. $y = 4x - 12$ **9.** $y = \tan^{-1}(x^3 + e^x)$

11. $y = \frac{6}{x^2}$ (where $x > 0$) **13.** $y = 2 - e^x$

15.   **17.**

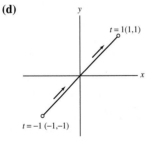

19. (a) ↔ (iv), (b) ↔ (ii), (c) ↔ (iii), (d) ↔ (i)

21. $\pi \le t \le 2\pi$ **23.** $c(t) = (t, 9 - 4t)$ **25.** $c(t) = \left(\frac{5+t^2}{4}, t\right)$

27. $c(t) = (-9 + 7\cos t, 4 + 7\sin t)$ **29.** $c(t) = \left(2 + t, 5 - \frac{1}{3}t\right)$

31. $c(t) = (4\cos t, 9\sin t)$ **33.** $c(t) = (t + 2, t^2 + 3)$

35. $y = 1 - x$ $(0 \le x \le 1)$ **37.** $c(t) = (t + 2, 3t + 2)$

39. $\frac{dy}{dx} = \coth t$; $x = a\cosh t$, $y = b\sinh t$ **41.** (I) $y(t)$; (II) $x(t)$

43.

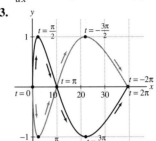

45. $\left.\frac{dy}{dx}\right|_{t=-4} = -\frac{1}{6}$ **47.** $\left.\frac{dy}{dx}\right|_{s=-1} = -\frac{3}{4}$

49. $y = -\frac{9}{2}x + \frac{11}{2}$; $\frac{dy}{dx} = -\frac{9}{2}$

51. $y = x^2 + x^{-1}$; $\frac{dy}{dx} = 2x - \frac{1}{x^2}$

53.

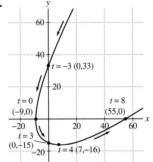

The graph is in: quadrant (i) for $t < -3$ or $t > 8$, quadrant (ii) for $-3 < t < 0$, quadrant (iii) for $0 < t < 3$, quadrant (iv) for $3 < t < 8$.

55. $(55, 0)$ **57.** $\left(\frac{x}{4}\right)^2 + \left(\frac{y}{7}\right)^2 = 1$; $m = -\frac{7}{4}$

59. $c(t) = (3 - 9t + 24t^2 - 16t^3, 2 + 6t^2 - 4t^3)$, $0 \le t \le 1$

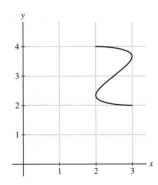

63.

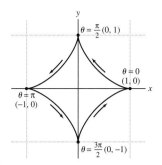

$y = -\sqrt{3}x + \frac{\sqrt{3}}{2}$

65. $((2k-1)\pi, 2), k = 0, \pm1, \pm2, \dots$

75. The coordinates of P, $(R\cos\theta, r\sin\theta)$ describe an ellipse for $0 \le \theta \le 2\pi$.

79. $\frac{d^2y}{dx^2}\Big|_{t=2} = -\frac{21}{512}$ **81.** $\frac{d^2y}{dx^2}\Big|_{t=-3} = 0$ **83.** Concave up: $t > 0$

87. Formula for the area under the graph of a positive function.

Section 11.2 Preliminary

1. $S = \int_a^b \sqrt{x'(t)^2 + y'(t)^2}\, dt$ **2.** The speed at time t

3. Displacement: 5; no **4.** $L = 180$ cm

Section 11.2 Exercises

1. $S = 3\pi$ **3.** $S = 10$ **5.** $S = 16\sqrt{13}$

7. $S = \frac{1}{2}(65^{3/2} - 5^{3/2}) \approx 256.43$ **9.** $S = 3\pi$ **11.** $S = 2$

15. $S = \pi\sqrt{1 + 4\pi^2} + \frac{1}{2}\ln(2\pi + \sqrt{1 + 4\pi^2}) \approx 21.256$

17. $\frac{ds}{dt}\Big|_{t=\frac{\pi}{4}} = 5\sqrt{9 + 55\sin^2(5 \cdot \frac{\pi}{4})} \approx 30.21$ m/s

19. $\frac{ds}{dt}\Big|_{t=1} = \sqrt{10} \approx 3.16$ m/s **21.** $\left(\frac{ds}{dt}\right)_{\min} \approx \sqrt{4.89} \approx 2.21$

23. $\frac{ds}{dt} = 8$

25.

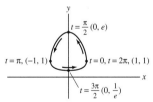

$M_{10} = 6.903734$, $M_{20} = 6.915035$, $M_{30} = 6.914949$, $M_{50} = 6.914951$

27. $M_{10} = 25.528309$,

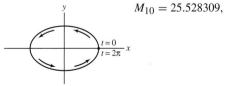

$M_{20} = 25.526999$, $M_{30} = 25.526999$, $M_{50} = 25.526999$

31. $S = \frac{6\pi}{5}$ **33.** $S = \frac{64\pi}{3}$

37. $R_m = 6{,}378\left(\frac{39{,}499.2}{84.5}\right)^{2/3} \approx 384{,}154$ km

Section 11.3 Preliminary

1. (b) **2.** Positive: $(r, \theta) = \left(1, \frac{\pi}{2}\right)$; Negative: $(r, \theta) = \left(-1, \frac{3\pi}{2}\right)$

3. Exactly one

4. (a) Equation of the circle of radius 2 centered at the origin.

(b) Equation of the circle of radius $\sqrt{2}$ centered at the origin.

(c) Equation of the vertical line through the point $(2, 0)$.

5. (a)

Section 11.3 Exercises

1. (a) $\left(3\sqrt{2}, \frac{3\pi}{4}\right)$ (b) $(3, \pi)$ (c) $(\sqrt{5}, \pi + 0.46) \approx (\sqrt{5}, 3.6)$

(d) $\left(\sqrt{2}, \frac{5\pi}{4}\right)$ (e) $\left(\sqrt{2}, \frac{\pi}{4}\right)$ (f) $\left(4, \frac{\pi}{6}\right)$ (g) $\left(4, \frac{11\pi}{6}\right)$

3. (a) $(1, 0)$ (b) $\left(\sqrt{12}, \frac{\pi}{6}\right)$ (c) $\left(\sqrt{8}, \frac{3\pi}{4}\right)$ (d) $\left(2, \frac{2\pi}{3}\right)$

5. (a) $\left(\frac{3\sqrt{3}}{2}, \frac{3}{2}\right)$ (b) $\left(-\frac{6}{\sqrt{2}}, \frac{6}{\sqrt{2}}\right)$ (c) $(0, -5)$

7. (b), (c) **9.** $\theta \approx 0.46$ **11.** $r = 2\csc\theta$

13. $x^2 + \left(y - \frac{1}{2}\right)^2 = \left(\frac{1}{2}\right)^2$ **15.** $y = 2$ **17.** $\frac{\left(x - \frac{1}{3}\right)^2}{\frac{4}{9}} + \frac{y^2}{\frac{1}{3}} = 1$

19. $r = 5\sec\theta$ **21.** $r^2 = 2\csc 2\theta$

23. A, $\theta = 0$; B, $\theta = \frac{\pi}{4}$; C, $\theta = \frac{\pi}{2}, \frac{3\pi}{2}$; D, $\theta = \frac{7\pi}{4}$

25. (a) $\leftrightarrow$ (iv), (b) $\leftrightarrow$ (iii), (c) $\leftrightarrow$ (i), (d) $\leftrightarrow$ (ii)

27. A, $\left(\frac{\pi}{2}, 1\right)$; B, $\left(\frac{3\pi}{4}, 0\right)$; C, $(0, 1)$; D, $\left(\frac{\pi}{4}, \sqrt{2}\right)$

29.

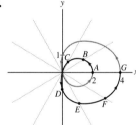

31. (a) A, $\theta = 0, r = 0$; B, $\theta = \frac{\pi}{4}, r = \sin\frac{2\pi}{4} = 1$; C, $\theta = \frac{\pi}{2}$, $r = 0$; D, $\theta = \frac{3\pi}{4}, r = \sin\frac{2\cdot3\pi}{4} = -1$; E, $\theta = \pi, r = 0$; F, $\theta = \frac{5\pi}{4}$, $r = 1$; G, $\theta = \frac{3\pi}{2}, r = 0$; H, $\theta = \frac{7\pi}{4}, r = -1$; I, $\theta = 2\pi, r = 0$

(b) $0 \le \theta \le \frac{\pi}{2}$ is in the first quadrant. $\frac{\pi}{2} \le \theta \le \pi$ is in the fourth quadrant. $\pi \le \theta \le \frac{3\pi}{2}$ is in the third quadrant. $\frac{3\pi}{2} \le \theta \le 2\pi$ is in the second quadrant.

33.

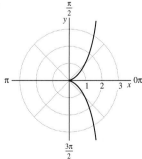

35. $\left(x - \frac{a}{2}\right)^2 + \left(y - \frac{b}{2}\right)^2 = \frac{a^2+b^2}{4}$, $r = \frac{\sqrt{a^2+b^2}}{2}$, centered at the point $\left(\frac{a}{2}, \frac{b}{2}\right)$.

37. $r^2 = \sec 2\theta$ **39.** $\left(x^2 + y^2\right)^2 = x^3 - 3y^2 x$

41. $r = 2\sec\left(\theta - \frac{\pi}{9}\right)$ **43.** $r = 2\sqrt{10}\sec(\theta - 4.39)$

45. $r = \frac{9}{4\cos\theta - \sin\theta}$ **53.** $y = \frac{x}{\sqrt{3}}$

Section 11.4 Preliminary

1. False **2.** (b) **3.** False **4.** (c)

Section 11.4 Exercises

1. $A = \frac{1}{2}\int_{\pi/2}^{\pi} r^2 \, d\theta = \frac{25\pi}{4}$

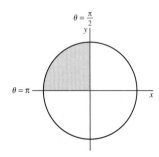

3. $A = \frac{1}{2}\int_0^{\pi} r^2 \, d\theta = 4\pi$ **5.** $A = \frac{3\pi}{2}$

7. $A = \frac{\pi}{8} \approx 0.39$ **9.** $A = 1$ **11.** $A = \frac{3\pi a^2}{2}$

13. $A = \frac{\sqrt{15}}{2} + 7\cos^{-1}\left(\frac{1}{4}\right) \approx 11.163$

15. $A = \pi - \frac{3\sqrt{3}}{2} \approx 0.54$ **17.** $A = \frac{\pi}{8} - \frac{1}{4} \approx 0.14$

19. $A = 4\pi$ **21.** $A = \frac{9\pi}{2} - 4\sqrt{2}$ **23.** $A = \frac{\pi^3}{48} - \frac{\pi}{8} \approx 0.25$

25. $L = \frac{A}{2}\sqrt{A^2 + 1} + \frac{1}{2}\ln\left|A + \sqrt{A^2 + 1}\right|$

27. $L = \tan A$

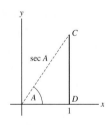

29. $L = 8$ **31.** $L = \sqrt{2}(e^{2\pi} - 1) \approx 755.9$

33. $L = \int_0^{\pi}\sqrt{1 + a^2}e^{a\theta}\, d\theta$

35. $L = \int_0^{2\pi}\sqrt{5 - 4\cos\theta}(2 - \cos\theta)^{-2}\, d\theta$

Section 11.5 Preliminary

1. (a) Hyperbola **(b)** Parabola **(c)** Ellipse
(d) Not a conic section

2. Hyperbola **3.** The points $(0, c)$ and $(0, -c)$

5. $\pm\frac{b}{a}$ are the slopes of the two asymptotes of the hyperbola.

Section 11.5 Exercises

1. $F_1 = \left(-\sqrt{65}, 0\right)$, $F_2 = \left(\sqrt{65}, 0\right)$; The vertices are $(9, 0)$, $(-9, 0)$, $(0, 4)$, $(0, -4)$.

3. $F_1 = \left(-\sqrt{5}, 0\right)$, $F_2 = \left(\sqrt{5}, 0\right)$; The vertices are $(3, 0)$, $(-3, 0)$, $(0, 2)$, $(0, -2)$.

5. $F_1 = \left(\sqrt{97}, 0\right)$, $F_2 = \left(-\sqrt{97}, 0\right)$; The vertices are $(4, 0)$ and $(-4, 0)$.

7. $F_1 = \left(\sqrt{65} + 3, -1\right)$, $F_2 = \left(-\sqrt{65} + 3, -1\right)$; The vertices are $(10, -1)$ and $(-4, -1)$.

9. $\frac{x^2}{5^2} + \frac{y^2}{7^2} = 1$ **11.** $\frac{(x-12)^2}{5^2} + \frac{(y-6)^2}{7^2} = 1$

13. $\frac{x^2}{9} + \frac{y^2}{16} = 1$ **15.** $\left(\frac{x}{4}\right)^2 + \left(\frac{y}{2\sqrt{13}}\right)^2 = 1$ **17.** $\frac{x^2}{9} + \frac{y^2}{16} = 1$

19. $\left(\frac{x}{4}\right)^2 - \left(\frac{y}{12}\right)^2 = 1$ **21.** $\left(\frac{y+(1/2)}{9/2}\right)^2 - \left(\frac{x}{6}\right)^2 = 1$

23. $x = \frac{y^2}{8}$ **25.** $y = \frac{1}{20}x^2$ **27.** $y = \frac{x^2}{16}$ **29.** $x = \frac{y^2}{8}$

31. Vertices $(\pm 4, 0)$, $(0, \pm 2)$. Foci $\left(-\sqrt{12}, 0\right)$, $\left(\sqrt{12}, 0\right)$. Focal axis is the x-axis. Conjugate axis is the y-axis. Centered at the origin.

33. Vertices $(\pm 2, 0)$, $(0, \pm 4)$. Foci $\left(0, \pm\sqrt{12}\right)$. Focal axis is the y-axis. Conjugate axis is the x-axis. Centered at the origin.

35. Vertices $(5, 5)$, $(-7, 5)$. Foci $\left(\sqrt{84} - 1, 5\right)$, $\left(-\sqrt{84} - 1, 5\right)$. Focal axis: $y = 5$. Conjugate axis: $x = -1$. Asymptotes $y = -1.15x + 3.85$, $y = 1.15x + 6.15$. Centered at $(-1, 5)$.

37. Vertex $(4, 0)$. Focus $\left(4, \frac{1}{16}\right)$. The axis is the vertical line $x = 4$.

39. Foci $\left(-\frac{\sqrt{21}}{2} + 1, \frac{1}{5}\right)$, $\left(\frac{\sqrt{21}}{2} + 1, \frac{1}{5}\right)$. Focal axis: $y = 0.2$. Conjugate axis: $x = 1$. Centered at $(1, 0.2)$.

41. $D = -87$; ellipse **43.** $D = 88$; hyperbola

47. Focus $(0, c)$. Directrix $y = -c$.

49. $\overline{A'F_0} = a + c = 3.6 + 0.9 = 4.5$ billion mi

51. $r = \frac{3}{2+\cos\theta}$ **53.** $r = \frac{4}{1+\cos\theta}$

55. Hyperbola, $e = 4$; directrix $x = 2$

57. Ellipse, $e = \frac{3}{4}$; directrix $x = \frac{8}{3}$

61. $\left(\frac{x+\frac{4}{3}}{\frac{8}{3}}\right)^2 + \left(\frac{y}{\frac{4}{\sqrt{3}}}\right)^2 = 1$

Chapter 11 Review

1. a, c

3. $c(t) = (1 + 2\cos t, 1 + 2\sin t)$. The intersection points with the y-axis are $\left(0, 1 \pm \sqrt{3}\right)$. The intersection points with the x-axis are $\left(1 \pm \sqrt{3}, 0\right)$.

5. $c(\theta) = (\cos(\theta + \pi), \sin(\theta + \pi))$ **7.** $c(t) = (1 + 2t, 3 + 4t)$

9. $y = -\frac{x}{4} + \frac{37}{4}$ **11.** $y = \frac{8}{(3-x)^3} + \frac{3-x}{2}$

13. $\left(\pi^2 k^2, 0\right)$, where $k \in \mathbb{Z}$

15. $\left.\dfrac{dy}{dx}\right|_{\theta=\frac{\pi}{4}} - \dfrac{1}{4\sqrt{2}}$ **17.** $\left.\dfrac{dy}{dx}\right|_{P} = 5$

19. Vertical: $t = \pi + 2\pi k$ where $k \in \mathbb{Z}$; horizontal: $t = \pm\frac{\pi}{3} + 2\pi k$ where $k \in \mathbb{Z}$

21. $\left.\dfrac{ds}{dt}\right|_{t=\frac{\pi}{4}} = \sqrt{20.5} \approx 4.53$ **23.** $L = 5(e - 1)$

25. $t_0 = \frac{3}{4}\pi$; $\left.\dfrac{ds}{dt}\right|_{t=\frac{3}{4}\pi} = e^{-3\pi/4}\sqrt{2}$

27. $\left(\sqrt{10}, 5.034\right); \left(\sqrt{10}, -0.321\right)$ **29.** $r^2 = \dfrac{12}{2+\sin 2\theta}$

31. $7x - y = 4$ **33.** $A = \frac{9}{4}\left(\frac{\pi}{3} + \frac{\sqrt{3}}{2}\right)$ **35.** $A = \frac{\pi}{2}$

37. $A = \frac{5\pi}{8} - 1$ **39.** $L_{\text{inner}} = 7.5087$; $L_{\text{outer}} = 36.121$

41. Foci $\left(\pm\sqrt{6}, 0\right)$. Vertices $(\pm 2, 0)$. Hyperbola centered at $(0, 0)$.

43. Hyperbola shifted 3 units on the y-axis. Foci $\left(\pm\sqrt{\frac{3}{2}}, 3\right)$.
Vertices $\left(\pm\frac{1}{\sqrt{2}}, 3\right)$.

45. $\left(\frac{x}{8}\right)^2 - \left(\frac{y}{6}\right)^2 = 1$ **47.** $\left(\frac{x}{64}\right)^2 + \left(\frac{y}{8\sqrt{63}}\right)^2 = 1$

51. $e = \frac{38}{778} \approx 0.049$

REFERENCES

The online source MacTutor History of Mathematics Archive (www-history.mcs.st-and.ac.uk) has been a valuable source of historical information.

Section 1.1

1. (EX 73) Adapted from *Calculus Problems for a New Century*, Robert Fraga, ed., Mathematical Association of America, Washington, DC, 1993, p. 9.

Section 1.2

1. (EX 25) Adapted from *Calculus Problems for a New Century*, Robert Fraga, ed., Mathematical Association of America, Washington, DC, 1993, p. 9.

Section 1.7

2. (EXMP 4) Adapted from B. Waits and F. Demana, "The Calculator and Computer Pre-Calculus Project," in *The Impact of Calculators on Mathematics Instruction*, University of Houston, 1994.

2. (EX 12) Adapted from B. Waits and F. Demana, "The Calculator and Computer Pre-Calculus Project," in *The Impact of Calculators on Mathematics Instruction*, University of Houston, 1994.

Section 2.1

3. (CO2) From Z. Toroczkai, G. Karolyi, Á. Péntek, T. Tè, C. Grebogi, and J. A. Yorke, "Wada Dye Boundaries in Open Hydrodynamical Flows," *Physica A* 239:235 (1997).

Section 2.2

4. (EX 61) Adapted from *Calculus Problems for a New Century*, Robert Fraga, ed., Mathematical Association of America, Washington, DC, 1993, Note 28.

Section 2.3

4. (EX 38) Adapted from *Calculus Problems for a New Century*, Robert Fraga, ed., Mathematical Association of America, Washington, DC, 1993, Note 28.

Chapter 2 Review

4. (EX 54) Adapted from *Calculus Problems for a New Century*, Robert Fraga, ed., Mathematical Association of America, Washington, DC, 1993, Note 28.

Section 3.1

5. (EX 42) Problem suggested by Dennis DeTurck, University of Pennsylvania.

5. (EX 75) Problem suggested by Dennis DeTurck, University of Pennsylvania.

Section 3.2

6. (EX 94) Problem suggested by Chris Bishop, SUNY Stony Brook.

6. (EX 95) Problem suggested by Chris Bishop, SUNY Stony Brook.

Section 3.4

7. (PQ 2) From *Calculus Problems for a New Century*, Robert Fraga, ed., Mathematical Association of America, Washington, DC, 1993, p. 25.

8. (EX 49) Karl J. Niklas and Brian J. Enquist, "Invariant Scaling Relationships for Interspecific Plant Biomass Production Rates and Body Size," *Proc. Natl. Acad. Sci.* 98, no. 5:2922–2927 (February 27, 2001).

Section 3.5

9. (EX 46) Adapted from a contribution by Jo Hoffacker, University of Georgia.

10. (EX 47–48) Adapted from a contribution by Thomas M. Smith, University of Illinois at Chicago, and Cindy S. Smith, Plainfield High School.

11. (EX 49) Adapted from Walter Meyer, *Falling Raindrops*, in *Applications of Calculus*, P. Straffin, ed., Mathematical Association of America, Washington, DC, 1993.

6. (EX 52, 56) Problem suggested by Chris Bishop, SUNY Stony Brook.

Section 3.6

12. (EX 56) Adapted from J. M. Gelfand and M. Saul, *Trigonometry*, Birkhäuser, Boston, 2001.

Section 3.11

13. (EX 34) Adapted from *Calculus Problems for a New Century*, Robert Fraga, ed., Mathematical Association of America, Washington, DC, 1993.

14. (EX 38) Problem suggested by Kay Dundas.

13. (EX 43, 45) Adapted from *Calculus Problems for a New Century*, Robert Fraga, ed., Mathematical Association of America, Washington, DC, 1993.

Chapter 3 Review

6. (EX 89–90, 111, 114) Problem suggested by Chris Bishop, SUNY Stony Brook.

Section 4.2

15. (MN, p. 225) From Pierre Fermat, *On Maxima and Minima and on Tangents*, translated by D. J. Struik (ed.), *A Source Book in Mathematics, 1200–1800*, Princeton University Press, Princeton, N.J., 1986.

Section 4.5

16. (EX 49, 98) Adapted from *Calculus Problems for a New Century*, Robert Fraga, ed., Mathematical Association of America, Washington, DC, 1993, p. 63.

Section 4.6

13. (EX 29–30) Adapted from *Calculus Problems for a New Century*, Robert Fraga, ed., Mathematical Association of America, Washington, DC, 1993.

17. (EX 32) Problem suggested by John Haverhals, Bradley University. *Source:* Illinois Agrinews.

13. (EX 35–36) Adapted from *Calculus Problems for a New Century*, Robert Fraga, ed., Mathematical Association of America, Washington, DC, 1993.

18. (EX 61) From Michael Helfgott, *Thomas Simpson and Maxima and Minima, Convergence Magazine*, published online by the Mathematical Association of America.

19. (EX 65–67) Adapted from B. Noble, *Applications of Undergraduate Mathematics in Engineering*, Macmillan, New York, 1967.

20. (EX 69) Adapted from Roger Johnson, "A Problem in Maxima and Minima," *American Mathematical Monthly*, 35:187–188 (1928).

Section 4.7

31. (EX 64) Adapted from Robert J. Bumcrot, "Some Subtleties in L'Hôpital's Rule," in *A Century of Calculus*, Part II, Mathematical Association of America, Washington, DC, 1992.

Section 4.8

21. (EX 19) Adapted from *Calculus Problems for a New Century*, Robert Fraga, ed., Mathematical Association of America, Washington, DC, 1993, p. 52.

22. (EX 30–31) Adapted from E. Packel and S. Wagon, *Animating Calculus*, Springer-Verlag, New York, 1997, p. 79.

Chapter 4 Review

13. (EX 76) Adapted from *Calculus Problems for a New Century*, Robert Fraga, ed., Mathematical Association of America, Washington, DC, 1993.

Section 5.1

23. (EX 3) Problem suggested by John Polhill, Bloomsburg University.

Section 5.2

5. (EX 79) Problem suggested by Dennis DeTurck, University of Pennsylvania.

6. (EX 84) Problem suggested by Chris Bishop, SUNY Stony Brook.

Section 5.4

24. (PQ 6) Adapted from *Calculus Problems for a New Century*, Robert Fraga, ed., Mathematical Association of America, Washington, DC, 1993, p. 103.

25. (EX 48–49) Adapted from *Calculus Problems for a New Century*, Robert Fraga, ed., Mathematical Association of America, Washington, DC, 1993, p. 102.

5. (EX 53) Problem suggested by Dennis DeTurck, University of Pennsylvania.

Section 5.5

26. (EX 24–25) M. Newman and G. Eble, "Decline in Extinction Rates and Scale Invariance in the Fossil Record," *Paleobiology* 25:434–439 (1999).

27. (EX 27) From H. Flanders, R. Korfhage, and J. Price, *Calculus*, Academic Press, New York, 1970.

Section 5.6

28. (EX 73) Adapted from *Calculus Problems for a New Century*, Robert Fraga, ed., Mathematical Association of America, Washington, DC, 1993, p. 121.

Section 6.1

29. (EX 49) Adapted from Tom Farmer and Fred Gass, "Miami University: An Alternative Calculus," in *Priming the Calculus Pump*, Thomas Tucker, ed., Mathematical Association of America, Washington, DC, 1990, Note 17.

13. (EX 59) Adapted from *Calculus Problems for a New Century*, Robert Fraga, ed., Mathematical Association of America, Washington, DC, 1993.

Section 6.3

30. (EX 52, 56) Adapted from G. Alexanderson and L. Klosinski, "Some Surprising Volumes of Revolution," *Two-Year College Mathematics Journal* 6, 3:13–15 (1975).

Section 7.1

33. See R. Courant and F. John, *Introduction to Calculus and Analysis*, Vol. 1, Springer-Verlag, New York, 1989.

Section 7.2

34. (EX 55–62) Problems suggested by Brian Bradie, Christopher Newport University.

13. (EX 66) Adapted from *Calculus Problems for a New Century*, Robert Fraga, ed., Mathematical Association of America, Washington, DC, 1993.

35. (EX 74) Adapted from J. L. Borman, "A Remark on Integration by Parts," *American Mathematical Monthly* 51:32–33 (1944).

Section 7.4

34. (EX 43–52) Problems suggested by Brian Bradie, Christopher Newport University.

36. (EX 59) Adapted from *Calculus Problems for a New Century*, Robert Fraga, ed., Mathematical Association of America, Washington, DC, 1993, p. 118.

Section 7.7

6. (EX 79) Problem suggested by Chris Bishop, SUNY Stony Brook.

Section 8.1

37. (EX 48) Adapted from G. Klambauer, *Aspects of Calculus*, Springer-Verlag, New York, 1986, Ch. 6.

Section 9.1

38. (EX 51) Adapted from E. Batschelet, *Introduction to Mathematics for Life Scientists*, Springer-Verlag, New York, 1979.

13. (EX 53) Adapted from *Calculus Problems for a New Century*, Robert Fraga, ed., Mathematical Association of America, Washington, DC, 1993.

39. (EX 54–55) Adapted from M. Tenenbaum and H. Pollard, *Ordinary Differential Equations*, Dover, New York, 1985.

Section 10.1

40. (EX 77) Adapted from G. Klambauer, *Aspects of Calculus*, Springer-Verlag, New York, 1986, p. 393.

Section 10.2

41. (EX 35) Adapted from *Calculus Problems for a New Century*, Robert Fraga, ed., Mathematical Association of America, Washington, DC, 1993, p. 137.

42. (EX 36) Adapted from *Calculus Problems for a New Century*, Robert Fraga, ed., Mathematical Association of America, Washington, DC, 1993, p. 138.

43. (EX 49) Adapted from George Andrews, "The Geometric Series in Calculus," *American Mathematical Monthly* 105, 1:36–40 (1998).

44. (EX 51) Adapted from Larry E. Knop, "Cantor's Disappearing Table," *The College Mathematics Journal* 16, 5:398–399 (1985).

Section 10.3

45. (EX 83) Adapted from *Calculus Problems for a New Century*, Robert Fraga, ed., Mathematical Association of America, Washington, DC, 1993, p. 141.

Section 10.4

46. (EX 27) Adapted from *Calculus Problems for a New Century*, Robert Fraga, ed., Mathematical Association of America, Washington, DC, 1993, p. 145.

Section 11.2

47. (EX 36) Adapted from Richard Courant and Fritz John, *Differential and Integral Calculus*, Wiley-Interscience, New York, 1965.

Section 11.3

13. (EX 56) Adapted from *Calculus Problems for a New Century*, Robert Fraga, ed., Mathematical Association of America, Washington, DC, 1993.

Appendix D

33. [Proof of Theorem 7] A proof without this simplifying assumption can be found in R. Courant and F. John, *Introduction to Calculus and Analysis*, Vol. 1, Springer-Verlag, New York, 1989.

PHOTO CREDITS

ELEMENTARY FUNCTIONS

Power Functions $f(x) = x^a$

$f(x) = x^n$, n a positive integer

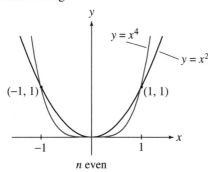

n even

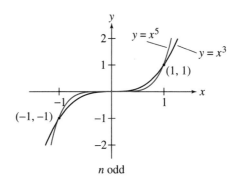

n odd

Asymptotic behavior of a polynomial function of even degree and positive leading coefficient

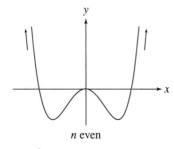

n even

Asymptotic behavior of a polynomial function of odd degree and positive leading coefficient

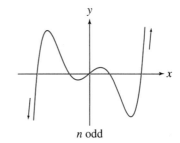

n odd

$f(x) = x^{-n} = \dfrac{1}{x^n}$

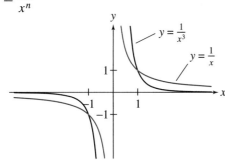

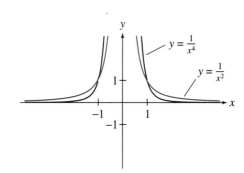

Inverse Trigonometric Functions

$\arcsin x = \sin^{-1} x = \theta$

$\Leftrightarrow \quad \sin\theta = x, \quad -\dfrac{\pi}{2} \le \theta \le \dfrac{\pi}{2}$

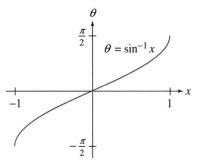

$\arccos x = \cos^{-1} x = \theta$

$\Leftrightarrow \quad \cos\theta = x, \quad 0 \le \theta \le \pi$

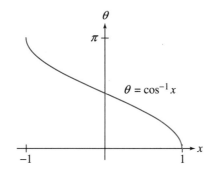

$\arctan x = \tan^{-1} x = \theta$

$\Leftrightarrow \quad \tan\theta = x, \quad -\dfrac{\pi}{2} < \theta < \dfrac{\pi}{2}$

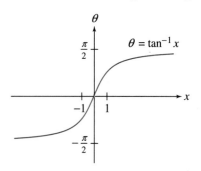

Exponential and Logarithmic Functions

$$\boxed{\log_a x = y \quad \Leftrightarrow \quad a^y = x}$$

$\log_a(a^x) = x \qquad a^{\log_a x} = x$

$\log_a 1 = 0 \qquad \log_a a = 1$

$$\boxed{\ln x = y \quad \Leftrightarrow \quad e^y = x}$$

$\ln(e^x) = x \qquad e^{\ln x} = x$

$\ln 1 = 0 \qquad \ln e = 1$

$\log_a(xy) = \log_a x + \log_a y$

$\log_a\left(\dfrac{x}{y}\right) = \log_a x - \log_a y$

$\log_a(x^r) = r \log_a x$

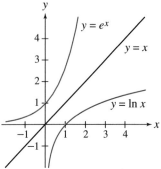

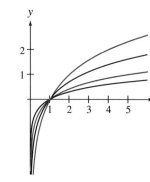

$\displaystyle\lim_{x\to\infty} a^x = \infty, \quad a > 1$

$\displaystyle\lim_{x\to\infty} a^x = 0, \quad 0 < a < 1$

$\displaystyle\lim_{x\to-\infty} a^x = 0, \quad a > 1$

$\displaystyle\lim_{x\to-\infty} a^x = \infty, \quad 0 < a < 1$

$\displaystyle\lim_{x\to 0^+} \log_a x = -\infty$

$\displaystyle\lim_{x\to\infty} \log_a x = \infty$

Hyperbolic Functions

$\sinh x = \dfrac{e^x - e^{-x}}{2} \qquad \operatorname{csch} x = \dfrac{1}{\sinh x}$

$\cosh x = \dfrac{e^x + e^{-x}}{2} \qquad \operatorname{sech} x = \dfrac{1}{\cosh x}$

$\tanh x = \dfrac{\sinh x}{\cosh x} \qquad \coth x = \dfrac{\cosh x}{\sinh x}$

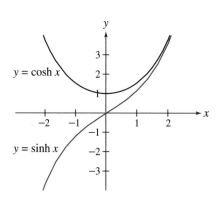

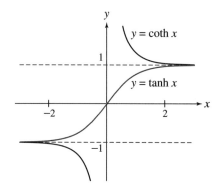

$\sinh(x + y) = \sinh x \cosh y + \cosh x \sinh y$

$\cosh(x + y) = \cosh x \cosh y + \sinh x \sinh y$

$\sinh 2x = 2 \sinh x \cosh x$

$\cosh 2x = \cosh^2 x + \sinh^2 x$

Inverse Hyperbolic Functions

$y = \sinh^{-1} x \quad \Leftrightarrow \quad \sinh y = x$

$y = \cosh^{-1} x \quad \Leftrightarrow \quad \cosh y = x \text{ and } y \geq 0$

$y = \tanh^{-1} x \quad \Leftrightarrow \quad \tanh y = x$

$\sinh^{-1} x = \ln\left(x + \sqrt{x^2 + 1}\right)$

$\cosh^{-1} x = \ln\left(x + \sqrt{x^2 - 1}\right) \quad x > 1$

$\tanh^{-1} x = \dfrac{1}{2} \ln\left(\dfrac{1 + x}{1 - x}\right) \quad -1 < x < 1$

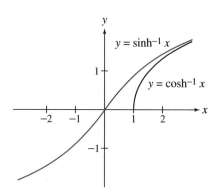

DIFFERENTIATION

Differentiation Rules

1. $\dfrac{d}{dx}(c) = 0$

2. $\dfrac{d}{dx}x = 1$

3. $\dfrac{d}{dx}(x^n) = nx^{n-1}$ (Power Rule)

4. $\dfrac{d}{dx}[cf(x)] = cf'(x)$

5. $\dfrac{d}{dx}[f(x) + g(x)] = f'(x) + g'(x)$

6. $\dfrac{d}{dx}[f(x)g(x)] = f(x)g'(x) + g(x)f'(x)$ (Product Rule)

7. $\dfrac{d}{dx}\left[\dfrac{f(x)}{g(x)}\right] = \dfrac{g(x)f'(x) - f(x)g'(x)}{[g(x)]^2}$ (Quotient Rule)

8. $\dfrac{d}{dx}f(g(x)) = f'(g(x))g'(x)$ (Chain Rule)

9. $\dfrac{d}{dx}f(x)^n = nf(x)^{n-1}f'(x)$ (General Power Rule)

10. $\dfrac{d}{dx}f(kx + b) = kf'(kx + b)$

11. $g'(x) = \dfrac{1}{f'(g(x))}$ where $g(x)$ is the inverse $f^{-1}(x)$

12. $\dfrac{d}{dx}\ln f(x) = \dfrac{f'(x)}{f(x)}$

Trigonometric Functions

13. $\dfrac{d}{dx}\sin x = \cos x$

14. $\dfrac{d}{dx}\cos x = -\sin x$

15. $\dfrac{d}{dx}\tan x = \sec^2 x$

16. $\dfrac{d}{dx}\csc x = -\csc x \cot x$

17. $\dfrac{d}{dx}\sec x = \sec x \tan x$

18. $\dfrac{d}{dx}\cot x = -\csc^2 x$

Inverse Trigonometric Functions

19. $\dfrac{d}{dx}(\sin^{-1} x) = \dfrac{1}{\sqrt{1 - x^2}}$

20. $\dfrac{d}{dx}(\cos^{-1} x) = -\dfrac{1}{\sqrt{1 - x^2}}$

21. $\dfrac{d}{dx}(\tan^{-1} x) = \dfrac{1}{1 + x^2}$

22. $\dfrac{d}{dx}(\csc^{-1} x) = -\dfrac{1}{x\sqrt{x^2 - 1}}$

23. $\dfrac{d}{dx}(\sec^{-1} x) = \dfrac{1}{x\sqrt{x^2 - 1}}$

24. $\dfrac{d}{dx}(\cot^{-1} x) = -\dfrac{1}{1 + x^2}$

Exponential and Logarithmic Functions

25. $\dfrac{d}{dx}(e^x) = e^x$

26. $\dfrac{d}{dx}(a^x) = (\ln a)a^x$

27. $\dfrac{d}{dx}\ln|x| = \dfrac{1}{x}$

28. $\dfrac{d}{dx}(\log_a x) = \dfrac{1}{(\ln a)x}$

Hyperbolic Functions

29. $\dfrac{d}{dx}(\sinh x) = \cosh x$

30. $\dfrac{d}{dx}(\cosh x) = \sinh x$

31. $\dfrac{d}{dx}(\tanh x) = \operatorname{sech}^2 x$

32. $\dfrac{d}{dx}(\operatorname{csch} x) = -\operatorname{csch} x \coth x$

33. $\dfrac{d}{dx}(\operatorname{sech} x) = -\operatorname{sech} x \tanh x$

34. $\dfrac{d}{dx}(\coth x) = -\operatorname{csch}^2 x$

Inverse Hyperbolic Functions

35. $\dfrac{d}{dx}(\sinh^{-1} x) = \dfrac{1}{\sqrt{1 + x^2}}$

36. $\dfrac{d}{dx}(\cosh^{-1} x) = \dfrac{1}{\sqrt{x^2 - 1}}$

37. $\dfrac{d}{dx}(\tanh^{-1} x) = \dfrac{1}{1 - x^2}$

38. $\dfrac{d}{dx}(\operatorname{csch}^{-1} x) = -\dfrac{1}{|x|\sqrt{x^2 + 1}}$

39. $\dfrac{d}{dx}(\operatorname{sech}^{-1} x) = -\dfrac{1}{x\sqrt{1 - x^2}}$

40. $\dfrac{d}{dx}(\coth^{-1} x) = \dfrac{1}{1 - x^2}$

INTEGRATION

Substitution

If an integrand has the form $f(u(x))u'(x)$, then rewrite the entire integral in terms of u and its differential $du = u'(x)\,dx$:

$$\int f(u(x))u'(x)\,dx = \int f(u)\,du$$

Integration by Parts Formula

$$\int u(x)v'(x)\,dx = u(x)v(x) - \int u'(x)v(x)\,dx$$

TABLE OF INTEGRALS

Basic Forms

1. $\int u^n\,du = \dfrac{u^{n+1}}{n+1} + C, \quad n \neq -1$

2. $\int \dfrac{du}{u} = \ln|u| + C$

3. $\int e^u\,du = e^u + C$

4. $\int a^u\,du = \dfrac{a^u}{\ln a} + C$

5. $\int \sin u\,du = -\cos u + C$

6. $\int \cos u\,du = \sin u + C$

7. $\int \sec^2 u\,du = \tan u + C$

8. $\int \csc^2 u\,du = -\cot u + C$

9. $\int \sec u \tan u\,du = \sec u + C$

10. $\int \csc u \cot u\,du = -\csc u + C$

11. $\int \tan u\,du = \ln|\sec u| + C$

12. $\int \cot u\,du = \ln|\sin u| + C$

13. $\int \sec u\,du = \ln|\sec u + \tan u| + C$

14. $\int \csc u\,du = \ln|\csc u - \cot u| + C$

15. $\int \dfrac{du}{\sqrt{a^2 - u^2}} = \sin^{-1}\dfrac{u}{a} + C$

16. $\int \dfrac{du}{a^2 + u^2} = \dfrac{1}{a}\tan^{-1}\dfrac{u}{a} + C$

Exponential and Logarithmic Forms

17. $\int ue^{au}\,du = \dfrac{1}{a^2}(au - 1)e^{au} + C$

18. $\int u^n e^{au}\,du = \dfrac{1}{a}u^n e^{au} - \dfrac{n}{a}\int u^{n-1}e^{au}\,du$

19. $\int e^{au}\sin bu\,du = \dfrac{e^{au}}{a^2 + b^2}(a\sin bu - b\cos bu) + C$

20. $\int e^{au}\cos bu\,du = \dfrac{e^{au}}{a^2 + b^2}(a\cos bu + b\sin bu) + C$

21. $\int \ln u\,du = u\ln u - u + C$

22. $\int u^n \ln u\,du = \dfrac{u^{n+1}}{(n+1)^2}[(n+1)\ln u - 1] + C$

23. $\int \dfrac{1}{u\ln u}\,du = \ln|\ln u| + C$

Hyperbolic Forms

24. $\int \sinh u\,du = \cosh u + C$

25. $\int \cosh u\,du = \sinh u + C$

26. $\int \tanh u\,du = \ln\cosh u + C$

27. $\int \coth u\,du = \ln|\sinh u| + C$

28. $\int \operatorname{sech} u\,du = \tan^{-1}|\sinh u| + C$

29. $\int \operatorname{csch} u\,du = \ln\left|\tanh\dfrac{1}{2}u\right| + C$

30. $\int \operatorname{sech}^2 u\,du = \tanh u + C$

31. $\int \operatorname{csch}^2 u\,du = -\coth u + C$

32. $\int \operatorname{sech} u \tanh u\,du = -\operatorname{sech} u + C$

33. $\int \operatorname{csch} u \coth u\,du = -\operatorname{csch} u + C$

Trigonometric Forms

34. $\int \sin^2 u\,du = \dfrac{1}{2}u - \dfrac{1}{4}\sin 2u + C$

35. $\int \cos^2 u\,du = \dfrac{1}{2}u + \dfrac{1}{4}\sin 2u + C$

36. $\int \tan^2 u\,du = \tan u - u + C$

37. $\int \cot^2 u\,du = -\cot u - u + C$

38. $\int \sin^3 u\,du = -\dfrac{1}{3}(2 + \sin^2 u)\cos u + C$

39. $\int \cos^3 u\,du = \dfrac{1}{3}(2 + \cos^2 u)\sin u + C$

40. $\int \tan^3 u\,du = \dfrac{1}{2}\tan^2 u + \ln|\cos u| + C$

41. $\int \cot^3 u\,du = -\dfrac{1}{2}\cot^2 u - \ln|\sin u| + C$

42. $\int \sec^3 u\,du = \dfrac{1}{2}\sec u \tan u + \dfrac{1}{2}\ln|\sec u + \tan u| + C$

43. $\int \csc^3 u \, du = -\frac{1}{2} \csc u \cot u + \frac{1}{2} \ln |\csc u - \cot u| + C$

44. $\int \sin^n u \, du = -\frac{1}{n} \sin^{n-1} u \cos u + \frac{n-1}{n} \int \sin^{n-2} u \, du$

45. $\int \cos^n u \, du = \frac{1}{n} \cos^{n-1} u \sin u + \frac{n-1}{n} \int \cos^{n-2} u \, du$

46. $\int \tan^n u \, du = \frac{1}{n-1} \tan^{n-1} u - \int \tan^{n-2} u \, du$

47. $\int \cot^n u \, du = \frac{-1}{n-1} \cot^{n-1} u - \int \cot^{n-2} u \, du$

48. $\int \sec^n u \, du = \frac{1}{n-1} \tan u \sec^{n-2} u + \frac{n-2}{n-1} \int \sec^{n-2} u \, du$

49. $\int \csc^n u \, du = \frac{-1}{n-1} \cot u \csc^{n-2} u + \frac{n-2}{n-1} \int \csc^{n-2} u \, du$

50. $\int \sin au \sin bu \, du = \frac{\sin(a-b)u}{2(a-b)} - \frac{\sin(a+b)u}{2(a+b)} + C$

51. $\int \cos au \cos bu \, du = \frac{\sin(a-b)u}{2(a-b)} + \frac{\sin(a+b)u}{2(a+b)} + C$

52. $\int \sin au \cos bu \, du = -\frac{\cos(a-b)u}{2(a-b)} - \frac{\cos(a+b)u}{2(a+b)} + C$

53. $\int u \sin u \, du = \sin u - u \cos u + C$

54. $\int u \cos u \, du = \cos u + u \sin u + C$

55. $\int u^n \sin u \, du = -u^n \cos u + n \int u^{n-1} \cos u \, du$

56. $\int u^n \cos u \, du = u^n \sin u - n \int u^{n-1} \sin u \, du$

57. $\int \sin^n u \cos^m u \, du$
$$= -\frac{\sin^{n-1} u \cos^{m+1} u}{n+m} + \frac{n-1}{n+m} \int \sin^{n-2} u \cos^m u \, du$$
$$= \frac{\sin^{n+1} u \cos^{m-1} u}{n+m} + \frac{m-1}{n+m} \int \sin^n u \cos^{m-2} u \, du$$

Inverse Trigonometric Forms

58. $\int \sin^{-1} u \, du = u \sin^{-1} u + \sqrt{1-u^2} + C$

59. $\int \cos^{-1} u \, du = u \cos^{-1} u - \sqrt{1-u^2} + C$

60. $\int \tan^{-1} u \, du = u \tan^{-1} u - \frac{1}{2} \ln(1+u^2) + C$

61. $\int u \sin^{-1} u \, du = \frac{2u^2-1}{4} \sin^{-1} u + \frac{u\sqrt{1-u^2}}{4} + C$

62. $\int u \cos^{-1} u \, du = \frac{2u^2-1}{4} \cos^{-1} u - \frac{u\sqrt{1-u^2}}{4} + C$

63. $\int u \tan^{-1} u \, du = \frac{u^2+1}{2} \tan^{-1} u - \frac{u}{2} + C$

64. $\int u^n \sin^{-1} u \, du = \frac{1}{n+1} \left[u^{n+1} \sin^{-1} u - \int \frac{u^{n+1} \, du}{\sqrt{1-u^2}} \right], \quad n \neq -1$

65. $\int u^n \cos^{-1} u \, du = \frac{1}{n+1} \left[u^{n+1} \cos^{-1} u + \int \frac{u^{n+1} \, du}{\sqrt{1-u^2}} \right], \quad n \neq -1$

66. $\int u^n \tan^{-1} u \, du = \frac{1}{n+1} \left[u^{n+1} \tan^{-1} u - \int \frac{u^{n+1} \, du}{1+u^2} \right], \quad n \neq -1$

Forms Involving $\sqrt{a^2 - u^2}, a > 0$

67. $\int \sqrt{a^2-u^2} \, du = \frac{u}{2} \sqrt{a^2-u^2} + \frac{a^2}{2} \sin^{-1} \frac{u}{a} + C$

68. $\int u^2 \sqrt{a^2-u^2} \, du = \frac{u}{8}(2u^2-a^2)\sqrt{a^2-u^2} + \frac{a^4}{8} \sin^{-1} \frac{u}{a} + C$

69. $\int \frac{\sqrt{a^2-u^2}}{u} \, du = \sqrt{a^2-u^2} - a \ln \left| \frac{a+\sqrt{a^2-u^2}}{u} \right| + C$

70. $\int \frac{\sqrt{a^2-u^2}}{u^2} \, du = -\frac{1}{u} \sqrt{a^2-u^2} - \sin^{-1} \frac{u}{a} + C$

71. $\int \frac{u^2 \, du}{\sqrt{a^2-u^2}} = -\frac{u}{2} \sqrt{a^2-u^2} + \frac{a^2}{2} \sin^{-1} \frac{u}{a} + C$

72. $\int \frac{du}{u\sqrt{a^2-u^2}} = -\frac{1}{a} \ln \left| \frac{a+\sqrt{a^2-u^2}}{u} \right| + C$

73. $\int \frac{du}{u^2\sqrt{a^2-u^2}} = -\frac{1}{a^2 u} \sqrt{a^2-u^2} + C$

74. $\int (a^2-u^2)^{3/2} \, du = -\frac{u}{8}(2u^2-5a^2)\sqrt{a^2-u^2} + \frac{3a^4}{8} \sin^{-1} \frac{u}{a} + C$

75. $\int \frac{du}{(a^2-u^2)^{3/2}} = \frac{u}{a^2\sqrt{a^2-u^2}} + C$

Forms Involving $\sqrt{u^2 - a^2}, a > 0$

76. $\int \sqrt{u^2-a^2} \, du = \frac{u}{2} \sqrt{u^2-a^2} - \frac{a^2}{2} \ln |u + \sqrt{u^2-a^2}| + C$

77. $\int u^2 \sqrt{u^2-a^2} \, du$
$$= \frac{u}{8}(2u^2-a^2)\sqrt{u^2-a^2} - \frac{a^4}{8} \ln |u + \sqrt{u^2-a^2}| + C$$

78. $\int \frac{\sqrt{u^2-a^2}}{u} \, du = \sqrt{u^2-a^2} - a \cos^{-1} \frac{a}{|u|} + C$

79. $\int \frac{\sqrt{u^2-a^2}}{u} \, du = -\frac{\sqrt{u^2-a^2}}{u} + \ln |u + \sqrt{u^2-a^2}| + C$

80. $\int \frac{du}{\sqrt{u^2-a^2}} = \ln |u + \sqrt{u^2-a^2}| + C$

81. $\int \frac{u^2 \, du}{\sqrt{u^2-a^2}} = \frac{u}{2} \sqrt{u^2-a^2} + \frac{a^2}{2} \ln |u + \sqrt{u^2-a^2}| + C$

82. $\int \frac{du}{u^2\sqrt{u^2-a^2}} = \frac{\sqrt{u^2-a^2}}{a^2 u} + C$

83. $\int \frac{du}{(u^2-a^2)^{3/2}} = -\frac{u}{a^2\sqrt{u^2-a^2}} + C$

Forms Involving $\sqrt{a^2 + u^2}, a > 0$

84. $\int \sqrt{a^2+u^2} \, du = \frac{u}{2} \sqrt{a^2+u^2} + \frac{a^2}{2} \ln(u + \sqrt{a^2+u^2}) + C$

85. $\int u^2 \sqrt{a^2+u^2} \, du$
$$= \frac{u}{8}(a^2+2u^2)\sqrt{a^2+u^2} - \frac{a^4}{8} \ln(u + \sqrt{a^2+u^2}) + C$$

86. $\int \frac{\sqrt{a^2+u^2}}{u} \, du = \sqrt{a^2+u^2} - a \ln \left| \frac{a+\sqrt{a^2+u^2}}{u} \right| + C$

87. $\int \frac{\sqrt{a^2+u^2}}{u^2} \, du = -\frac{\sqrt{a^2+u^2}}{u} + \ln(u + \sqrt{a^2+u^2}) + C$

88. $\displaystyle\int \frac{du}{\sqrt{a^2+u^2}} = \ln\left(u+\sqrt{a^2+u^2}\right) + C$

89. $\displaystyle\int \frac{u^2\,du}{\sqrt{a^2+u^2}} = \frac{u}{2}\sqrt{a^2+u^2} - \frac{a^2}{2}\ln\left(u+\sqrt{a^2+u^2}\right) + C$

90. $\displaystyle\int \frac{du}{u\sqrt{a^2+u^2}} = -\frac{1}{a}\ln\left|\frac{\sqrt{a^2+u^2}+a}{u}\right| + C$

91. $\displaystyle\int \frac{du}{u^2\sqrt{a^2+u^2}} = -\frac{\sqrt{a^2+u^2}}{a^2 u} + C$

92. $\displaystyle\int \frac{du}{(a^2+u^2)^{3/2}} = \frac{u}{a^2\sqrt{a^2+u^2}} + C$

Forms Involving $a + bu$

93. $\displaystyle\int \frac{u\,du}{a+bu} = \frac{1}{b^2}\left(a+bu-a\ln|a+bu|\right) + C$

94. $\displaystyle\int \frac{u^2\,du}{a+bu} = \frac{1}{2b^3}\left[(a+bu)^2 - 4a(a+bu) + 2a^2\ln|a+bu|\right] + C$

95. $\displaystyle\int \frac{du}{u(a+bu)} = \frac{1}{a}\ln\left|\frac{u}{a+bu}\right| + C$

96. $\displaystyle\int \frac{du}{u^2(a+bu)} = -\frac{1}{au} + \frac{b}{a^2}\ln\left|\frac{a+bu}{u}\right| + C$

97. $\displaystyle\int \frac{u\,du}{(a+bu)^2} = \frac{a}{b^2(a+bu)} + \frac{1}{b^2}\ln|a+bu| + C$

98. $\displaystyle\int \frac{du}{u(a+bu)^2} = \frac{1}{a(a+bu)} - \frac{1}{a^2}\ln\left|\frac{a+bu}{u}\right| + C$

99. $\displaystyle\int \frac{u^2\,du}{(a+bu)^2} = \frac{1}{b^3}\left(a+bu - \frac{a^2}{a+bu} - 2a\ln|a+bu|\right) + C$

100. $\displaystyle\int u\sqrt{a+bu}\,du = \frac{2}{15b^2}(3bu-2a)(a+bu)^{3/2} + C$

101. $\displaystyle\int u^n\sqrt{a+bu}\,du$
$$= \frac{2}{b(2n+3)}\left[u^n(a+bu)^{3/2} - na\int u^{n-1}\sqrt{a+bu}\,du\right]$$

102. $\displaystyle\int \frac{u\,du}{\sqrt{a+bu}} = \frac{2}{3b^2}(bu-2a)\sqrt{a+bu} + C$

103. $\displaystyle\int \frac{u^n\,du}{\sqrt{a+bu}} = \frac{2u^n\sqrt{a+bu}}{b(2n+1)} - \frac{2na}{b(2n+1)}\int \frac{u^{n-1}\,du}{\sqrt{a+bu}}$

104. $\displaystyle\int \frac{du}{u\sqrt{a+bu}} = \frac{1}{\sqrt{a}}\ln\left|\frac{\sqrt{a+bu}-\sqrt{a}}{\sqrt{a+bu}+\sqrt{a}}\right| + C,\quad \text{if } a>0$
$$= \frac{2}{\sqrt{-a}}\tan^{-1}\sqrt{\frac{a+bu}{-a}} + C,\quad \text{if } a<0$$

105. $\displaystyle\int \frac{du}{u^n\sqrt{a+bu}} = -\frac{\sqrt{a+bu}}{a(n-1)u^{n-1}} - \frac{b(2n-3)}{2a(n-1)}\int \frac{du}{u^{n-1}\sqrt{a+bu}}$

106. $\displaystyle\int \frac{\sqrt{a+bu}}{u}\,du = 2\sqrt{a+bu} + a\int \frac{du}{u\sqrt{a+bu}}$

107. $\displaystyle\int \frac{\sqrt{a+bu}}{u^2}\,du = -\frac{\sqrt{a+bu}}{u} + \frac{b}{2}\int \frac{du}{u\sqrt{a+bu}}$

Forms Involving $\sqrt{2au-u^2},\ a>0$

108. $\displaystyle\int \sqrt{2au-u^2}\,du = \frac{u-a}{2}\sqrt{2au-u^2} + \frac{a^2}{2}\cos^{-1}\left(\frac{a-u}{a}\right) + C$

109. $\displaystyle\int u\sqrt{2au-u^2}\,du$
$$= \frac{2u^2-au-3a^2}{6}\sqrt{2au-u^2} + \frac{a^3}{2}\cos^{-1}\left(\frac{a-u}{a}\right) + C$$

110. $\displaystyle\int \frac{du}{\sqrt{2au-u^2}} = \cos^{-1}\left(\frac{a-u}{a}\right) + C$

111. $\displaystyle\int \frac{du}{u\sqrt{2au-u^2}} = -\frac{\sqrt{2au-u^2}}{au} + C$

ESSENTIAL THEOREMS

Intermediate Value Theorem

If $f(x)$ is continuous on a closed interval $[a, b]$ and $f(a) \neq f(b)$, then for every value M between $f(a)$ and $f(b)$, there exists at least one value $c \in (a, b)$ such that $f(c) = M$.

Mean Value Theorem

If $f(x)$ is continuous on a closed interval $[a, b]$ and differentiable on (a, b), then there exists at least one value $c \in (a, b)$ such that
$$f'(c) = \frac{f(b)-f(a)}{b-a}$$

Extreme Values on a Closed Interval

If $f(x)$ is continuous on a closed interval $[a, b]$, then $f(x)$ attains both a minimum and a maximum value on $[a, b]$. Furthermore, if $c \in [a, b]$ and $f(c)$ is an extreme value (min or max), then c is either a critical point or one of the endpoints a or b.

The Fundamental Theorem of Calculus, Part I

Assume that $f(x)$ is continuous on $[a, b]$ and let $F(x)$ be an antiderivative of $f(x)$ on $[a, b]$. Then
$$\int_a^b f(x)\,dx = F(b) - F(a)$$

Fundamental Theorem of Calculus, Part II

Assume that $f(x)$ is a continuous function on $[a, b]$. Then the area function $A(x) = \displaystyle\int_a^x f(t)\,dt$ is an antiderivative of $f(x)$, that is,
$$A'(x) = f(x) \quad \text{or equivalently} \quad \frac{d}{dx}\int_a^x f(t)\,dt = f(x)$$

Furthermore, $A(x)$ satisfies the initial condition $A(a) = 0$.